Basic Technical Mathematics
with Calculus

Third Edition

Other books by the author

Essentials of Basic Mathematics (with Boyd & Plotkin)

Essentials of Algebra (with Boyd & Plotkin)

Introduction to Technical Mathematics, Second Edition

Basic Technical Mathematics, Third Edition

Basic Technical Mathematics with Calculus, Third Edition, Metric Version

Technical Calculus with Analytic Geometry

An Introduction to Calculus with Applications

Modules in Technical Mathematics

    Introductory Topics from Arithmetic
    Introduction to Algebra
    Introduction to Trigonometry
    Introduction to Geometry

Plane Trigonometry (with Edmond)

Basic Technical Mathematics
with Calculus
Third Edition

Allyn J. Washington

Dutchess Community College
Poughkeepsie, New York

The Benjamin/Cummings Publishing Company
Menlo Park, California • Reading, Massachusetts
London • Amsterdam • Don Mills, Ontario • Sydney

Sponsoring editor: Adrian Perenon
Production editor: Margaret Moore
Cover designer: Marjorie Spiegelman
Book designer: Peter Martin
Artist: Judith McCarty
Compositor: Typothetae

ISBN 0–8053–9521–0
abcdefghij–MU–798

The Benjamin / Cummings Publishing Company, Inc.
2727 Sand Hill Road
Menlo Park, California 94025

Preface

Scope of the Book

This book is intended primarily for students in technical and pre-engineering technology programs or where a coverage of basic mathematics is required.

Chapters 1 through 19 provide the necessary background in algebra and trigonometry for analytic geometry and calculus courses, and Chapters 1 through 20 provide the background necessary for calculus courses. There is an integrated treatment of mathematical topics, from algebra to calculus, which are necessary for a sound mathematical background for the technician. Numerous applications from many fields of technology are included primarily to indicate where and how mathematical techniques are used. However, it is not necessary that the student have a specific knowledge of the technical area from which any given problem is taken.

It is assumed that students using this text will have a background including algebra and geometry. However, the material is presented in sufficient detail for use by those whose background is possibly deficient to some extent in these areas. The material presented here is sufficient for three or four semesters.

One of the primary reasons for the arrangement of topics in this text is to present material in such an order that it is possible for a student to take courses in allied technical areas, such as physics and electricity, concurrently. These allied courses normally require a student to know certain mathematical topics by certain definite times, and yet, with the traditional mathematical order of topics, it is difficult to attain this coverage without loss of continuity. However, this material can be rearranged to fit any appropriate sequence of topics, if this is deemed necessary. Another feature of the material in this text is that certain topics which are traditionally included, primarily for mathematical completeness, have been omitted.

The approach used here is basically an intuitive one. It is not unduly rigorous mathematically, although all appropriate terms and concepts are introduced as needed and given an intuitive or algebraic foundation. The book's aim is to help the student develop a feeling for mathematical methods, not simply to have a collection of formulas when the work in the text has been completed. The text emphasizes that it is essential for the student to have a fluent background in algebra and trigonometry to understand and succeed in any subsequent work in mathematics.

New Features

This third edition of *Basic Technical Mathematics with Calculus* includes all the basic features of the first two editions. However, most sections have been rewritten to some degree to include additional explanatory material, examples, and exercises. Specifically, among the new features of the third edition are the following: (1) An appendix briefly discussing the use of a scientific calculator has replaced the appendix on the slide rule. Although no specific calculational device is required, it is assumed that most students will use a scientific calculator for most of their calculating work. Also, a number of problems have been designed for solution on a calculator. (2) A new table of trigonometric functions, using degrees and tenths of a degree, has been added. (3) New sections are Section 16–5, devoted to graphical solution of inequalities with two variables (and including an introduction to linear programming), and Section 18–4, devoted to the binomial theorem. (4) Certain sections of the second edition have been separated into two sections to provide more detailed coverage. Thus, Section 1–11 is a separate section covering literal equations and verbal problems, Section 3–5 is devoted completely to right triangle applications, Section 5–3 covers factoring trinomials, Section 8–1 introduces vectors and develops graphical methods in more detail, and Section 22–8 is devoted to differentiation of implicit functions. (5) Two sections have been moved to provide more timely coverage. Scientific notation is now in Chapter 1 in Section 1–5. Exponential and logarithmic functions are now included in Chapter 12. (6) Topics for which emphasis on coverage has been changed include logarithms and limits. Logarithms are presented more as a useful function and less as a calculational device. Continuity is briefly introduced along with limits. (7) There are more problems using metric units, and a discussion of the metric system (SI) is included in Appendix B. (8) Important formulas have been set off so that they are easily located and utilized. (9) There are now over 7500 exercises, an increase of about 25%. Each exercise group is designated by the number of the section which it follows. Exercises are generally grouped such that there is an even-numbered exercise equivalent to each odd-numbered exercise. (10) Many new examples are included, and more detail is included in examples which appeared in earlier editions. There are now approximately 1100 examples, an increase of about 15%.

Other features are: (1) The many examples included in this text are often used advantageously to introduce concepts, as well as to clarify and illustrate points made in the text. (2) There is extensive use of graphical methods. (3) Stated problems are included a few at a time to allow the student to develop techniques of solution. (4) Those topics which experience has shown to be more difficult for the student have been developed in more detail, with many examples. (5) The order of coverage can be changed in several places without loss of continuity. Also, certain sections may be omitted without loss of continuity. Any omissions or changes in order will, of course, depend on the type of course and the completeness required. (6) The chapter on statistics is included in order to introduce the student to statistical and empirical methods. (7) Review exercises are included after each chapter. These may be used either for additional problems or for review assignments. (8) The answers to all the odd-numbered exercises are given at the back of the book. Included are answers to graphical problems and other types which are not always included in textbooks.

Acknowledgments

The author wishes to acknowledge the help and suggestions given him by many of those who have used the first two editions of this text. In particular, I wish to thank Gail Brittain, Stephen Lange, Michael Mayer, John Davenport, Mario Triola of Dutchess Community College, and William K. Viertel, formerly of State University of New York Agricultural and Technical College at Canton. Also, Judith L. Gersting of Indiana University–Purdue University at Indianapolis, Richard C. Wheeler of Wentworth Institute, Edwin P. McCravy of Midlands Technical Education Center, Carl M. Schell of Sinclair Community College, and A. P. Paris of British Columbia Institute of Technology provided many valuable suggestions in their reviews of the text and manuscript. Many others contributed to this third edition through their response to my publisher's survey and to their field representatives. Their efforts are most appreciated, and there is space here only to thank them collectively.

In addition, I wish to thank Carolyn Edmond, formerly of Dutchess Community College, for her help in reading proof and checking answers for this third edition. The assistance, cooperation, and diligent efforts of the Benjamin/Cummings staff during the production of this text are also greatly appreciated. Finally, special mention is due my wife, who helped with her patience during the preparation of this text, and for checking many of the answers of the earlier editions.

A. J. W.

Contents

Chapter 13   Additional Types of Equations and Systems of Equations   324

Chapter 14   Equations of Higher Degree   341

123
14 - 20

Chapter 15   Determinants and Matrices   362

Chapter 16   Inequalities   403

Chapter 17   Variation   427

1

Fundamental Concepts
and Operations

1–1 Numbers and Literal Symbols

Mathematics has played a most important role in the development and understanding of the various fields of technology, and in the endless chain of technological and scientific advances of our time. With the mathematics we shall develop in this text, many kinds of applied problems can and will be solved. Of course, we cannot solve the more advanced types of problems which arise, but we can form a foundation for the more advanced mathematics which is used to solve such problems. Therefore, the development of a real understanding of the mathematics presented in this text will be of great value to you in your future work.

A thorough understanding of algebra is essential to the comprehension of any of the fields of elementary mathematics. It is important for the reader to learn and understand the basic concepts and operations presented here, or the development and the applications of later topics will be difficult to comprehend. Unless the algebraic operations are understood well, the result will be a weak foundation for further work in mathematics and in many of the technical areas where mathematics may be applied.

We shall begin our study of mathematics by reviewing some of the basic concepts and operations that deal with numbers and symbols. With these we shall be able to develop the topics in algebra which are necessary for further progress into other fields of mathematics, such as trigonometry and calculus.

The way we represent numbers today has been evolving for thousands of years. The first numbers used were those which stand for whole quantities, and these we call the **positive integers.** The positive integers are represented by the symbols 1, 2, 3, 4, and so forth.

Of course, it is necessary to have numbers to represent parts of certain quantities, and for this purpose fractional quantities are introduced. *The name* **positive rational number** *is given to any number that we can represent by the division of one positive integer by another. Numbers that cannot be designated by the division of one integer by another are termed* **irrational.**

Example A

The numbers 5, 18, and 1978 are positive integers. They are also rational numbers, since they may be written as $\frac{5}{1}$, $\frac{18}{1}$, and $\frac{1978}{1}$. Normally we do not write the 1's in the denominators.

The numbers $\frac{1}{2}$, $\frac{5}{8}$, $\frac{11}{3}$, and $\frac{106}{17}$ are positive rational numbers, since both the numerators and the denominators are integers.

The numbers $\sqrt{2}$, $\sqrt{3}$, and π are irrational. It is not possible to find any two integers which represent these numbers when one of the integers is divided by the other. For example, $\frac{22}{7}$ is not an *exact* representation of π; it is only an approximation.

In addition to the positive numbers, it is necessary to introduce **negative numbers,** not only because we need to have a numerical answer to problems such as $5 - 8$, but also because the negative sign is used to designate direction. *Thus,* -1, -2, -3, *and so on are the* **negative integers.** The number **zero** is an integer, but it is neither positive nor negative. *This means that the* **integers** *are the numbers* . . . , -3, -2, -1, 0, 1, 2, 3, . . . , *and so on.*

The integers, the rational numbers, and the irrational numbers, which include all such numbers which are zero, positive, or negative, constitute what we call the **real number system.** We shall use real numbers throughout this text, with one important exception. In the chapter on the *j*-operator, we shall be using **imaginary numbers,** *which is the name given to square roots of negative numbers.* The symbol j is used to designate $\sqrt{-1}$, which is not part of the real number system.

Example B

The number 7 is an integer. It is also a rational number since $7 = \frac{7}{1}$, and it is a real number since the real numbers include all of the rational numbers.

The number 3π is irrational, and it is real since the real numbers include all of the irrational numbers.

The numbers $\sqrt{-10}$ and $7j$ are imaginary.

The number $\frac{1}{8}$ is rational and real. The number $\sqrt{5}$ is irrational and real.

The number $\frac{-3}{7}$ is rational and real. The number $-\sqrt{7}$ is irrational and real.

The number $\sqrt{-7}$ is imaginary.

The number $\frac{\pi}{6}$ is irrational and real. The number $\frac{\sqrt{-3}}{2}$ is imaginary.

A **fraction** *may contain any number or symbol representing a number in its numerator or in its denominator. Thus, a fraction may be rational, irrational, or imaginary.*

Example C

The numbers $\frac{2}{7}$ and $\frac{-3}{2}$ are fractions, and they are also rational.

The numbers $\frac{\sqrt{2}}{9}$ and $\frac{6}{\pi}$ are fractions, but they are not rational numbers. It is not possible to express either as the ratio of one integer to another.

The number $\frac{\sqrt{-3}}{2}$ is a fraction, and it is also imaginary.

The real numbers may be represented as points on a line. We draw a horizontal line and designate some point on it by O, which we call the **origin** (see Fig. 1–1). The number zero, which is an integer, is located at this point. Then equal intervals are marked off from this point toward the right, and the positive integers are placed at these positions. The other rational numbers are located between the positions of the integers. It cannot be proved here, but the rational numbers do not take up all the positions on the line; the remaining points represent irrational numbers.

Now we can give the direction interpretation to negative numbers. By starting at the origin and proceeding to the left, in the **negative direction**, we locate all the negative numbers. As shown in Fig. 1–1, the positive numbers are to the right of the origin and the negative numbers to the left. Representing numbers in this way will be especially useful when we study graphical methods.

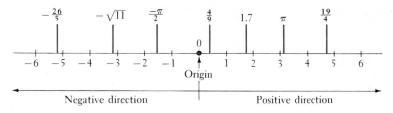

Figure 1—1

Another important mathematical concept we use in dealing with numbers is the **absolute value** of a number. By definition, *the absolute value of a positive number is the number itself, and the absolute value of a negative number is the corresponding positive number (obtained by changing its sign).* We may interpret the absolute value as being the number of units a given number is from the origin, regardless of direction. The absolute value is designated by $|\ |$ placed around the number.

Example D

The absolute value of 6 is 6, and the absolute value of −7 is 7. We designate this by $|6| = 6$ and $|-7| = 7$.

Other examples are $|-\pi| = \pi$, $|\frac{7}{5}| = \frac{7}{5}$, and $|-\sqrt{2}| = \sqrt{2}$.

On the number scale, if a first number is to the right of a second number, then the first number is said to be **greater than** the second. If the first number is to the left of the second, it is **less than** the second number. *"Greater than" is designated by* $>$, *and "less than" is designated by* $<$. These are called **signs of inequality**.

Example E

$$6 > 3, \quad 8 > -1, \quad 5 < 9, \quad 0 > -4, \quad -2 > -4, \quad -1 < 0$$

Every number, except zero, has a **reciprocal**. *The reciprocal of any number is 1 divided by that number.*

Example F

The reciprocal of 7 is $\frac{1}{7}$. The reciprocal of $\frac{2}{3}$ is

$$\frac{1}{\frac{2}{3}} = 1 \cdot \frac{3}{2} = \frac{3}{2}$$

The reciprocal of π is $1/\pi$. The reciprocal of -5 is $-\frac{1}{5}$. Notice that the negative sign is retained in the reciprocal.

In applications, numbers are often used to represent some type of measurement and therefore have certain units of measurement associated with them. Such numbers are referred to as **denominate numbers**. For a discussion of units of measurement, and the symbols which are used, see Appendix B. The following example illustrates the use of units and the symbols which represent them.

Example G

To indicate that an object is ten feet long, we represent the length as 10 ft.

To indicate that the speed of a projectile is 150 meters per second, we represent the speed as 150 m/s. (Note the use of s for second. We use s rather than sec, which is also common.)

To indicate that the volume of a container is 500 cubic centimeters, we represent the volume by 500 cm³. (Also common are cc and cu cm.)

To this point we have been dealing with numbers in their explicit form. However, it is normally more convenient to state definitions and operations on numbers in a generalized form. To do this we represent the numbers by letters, often referred to as **literal numbers**.

For example, we can say, "if a is to the right of b on the number scale, then a is greater than b, or $a > b$." This is more convenient than saying, "if a first number is to the right of a second number, then the first number is greater than the second number." The statement "the reciprocal of a number a is $1/a$" is another example of using letters to stand for numbers in general.

In an algebraic discussion, certain letters are sometimes allowed to take on any value, while other letters represent the same number throughout the discussion. *Those which may vary in a given problem are called* **variables,** *and those which are held fixed are called* **constants.**

Common usage today normally designates the letters near the end of the alphabet as variables, and letters near the beginning of the alphabet as constants. There are exceptions, but these are specifically noted. Letters in the middle of the alphabet are also used, but their meaning in any problem is specified.

Example H

The electric resistance R of a wire may be related to the temperature T by the equation $R = aT + b$. R and T may take on various values, and a and b are fixed for any particular wire. However, a and b may change if a different wire is considered. Here, R and T are the variables and a and b are constants.

Exercises 1–1

In Exercises 1 through 4 designate each of the given numbers as being an integer, rational, irrational, real, or imaginary. (More than one designation may be correct.)

1. $3, -\pi$ 2. $\dfrac{5}{4}, \sqrt{-4}$ 3. $6j, \dfrac{\sqrt{7}}{3}$ 4. $-\dfrac{7}{3}, \dfrac{\pi}{6}$

In Exercises 5 through 8 find the absolute value of each of the given numbers.

5. $3, \dfrac{7}{2}$ 6. $-4, \sqrt{2}$ 7. $-\dfrac{6}{7}, -\sqrt{3}$ 8. $-\dfrac{\pi}{2}, -\dfrac{19}{4}$

In Exercises 9 through 16 insert the correct sign of inequality ($>$ or $<$) between the given pairs of numbers.

9. $6 \quad 8$ 10. $7 \quad 5$ 11. $\pi \quad -1$ 12. $-4 \quad 3$

13. $-4 \quad -3$ 14. $-\sqrt{2} \quad -9$ 15. $-\dfrac{1}{3} \quad -\dfrac{1}{2}$ 16. $0.2 \quad 0.6$

In Exercises 17 through 20 find the reciprocals of the given numbers.

17. $3, -2$ 18. $\dfrac{1}{6}, -\dfrac{7}{4}$ 19. $-\dfrac{5}{\pi}, x$ 20. $-\dfrac{8}{3}, \dfrac{y}{b}$

In Exercises 21 through 24 locate the given numbers on a number line as in Fig. 1–1.

21. $2.5, -\dfrac{1}{2}$ 22. $\sqrt{3}, -\dfrac{12}{5}$ 23. $-\dfrac{\sqrt{2}}{2}, 2\pi$ 24. $\dfrac{123}{19}, -\dfrac{\pi}{6}$

In Exercises 25 through 34 answer the given questions.

25. List the following numbers in numerical order, starting with the smallest: $-1, 9, \pi, \sqrt{5}, |-8|, -|-3|, -18$.

26. List the following numbers in numerical order, starting with the smallest: $\frac{1}{5}, -\sqrt{10}, -|-6|, -4, 0.25, |-\pi|$.

27. The value of π expressed to four decimal places is 3.1416. Show by division that $\frac{22}{7}$ has a different value when expressed to four decimal places.

28. If a and b represent positive integers, what kind of number is represented by (a) $a + b$, (b) a/b, (c) $a \cdot b$?

29. Describe the location of a number x on the number line when (a) $x > 0$, (b) $x < -4$.

30. Describe the location of a number x on the number line when (a) $|x| < 1$, (b) $|x| > 2$.

31. The pressure P and volume V of a certain body of gas are related by the equation $P = c/V$ for certain conditions. Identify the symbols as variables or constants.

32. The velocity v and height h of an object are related by the equation $v = \sqrt{2ah}$. Identify the literal symbols as variables or constants.

33. In writing a laboratory report, a student wrote "$-20°C > -30°C$." Is this statement correct?

34. After 2 s, the current in a certain circuit is less than 3 A. Using t to represent time and i to represent current, this statement may be written, "for $t > 2$ s, $i < 3$ A." In this way write the statement, "less than three meters from the light source the illuminance is greater than eight lumens (lm) per square meter." (Let I represent illuminance and s represent distance.)

1—2 Fundamental Laws of Algebra

In performing the basic operations with numbers, we know that certain basic laws are valid. These basic statements are called the fundamental laws of algebra.

For example, we know that if two numbers are to be added, it does not matter in which order they are added. Thus $5 + 3 = 8$, as well as $3 + 5 = 8$. For this case we can say that $5 + 3 = 3 + 5$. This statement generalized and assumed correct for all possible combinations of numbers to be added, is called the **commutative law** for addition. The law states that *the sum of two numbers is the same, regardless of the order in which they are added.* We make no attempt to prove this in general, but accept its validity.

In the same way we have the **associative law** for addition, which states that *the sum of three or more numbers is the same, regardless of the manner in which they are grouped for addition.* For example,

$$3 + (5 + 6) = (3 + 5) + 6$$

The laws which we have just stated for addition are also true for multiplication. Therefore, *the product of two numbers is the same, regardless of the order in which they are multiplied,* and *the product of three or more numbers is the same, regardless of the manner in which they are grouped for multiplication.* For example, $2 \cdot 5 = 5 \cdot 2$ and $5 \cdot (4 \cdot 2) = (5 \cdot 4) \cdot 2$.

There is one more important law, called the **distributive law.** It states that *the product of one number and the sum of two or more other numbers is equal to the sum of the products of the first number and each of the other numbers of the sum.* For example,

$$4(3 + 5) = 4 \cdot 3 + 4 \cdot 5$$

In practice these laws are used intuitively. However, it is necessary to state them and to accept their validity, so that we may build our later results with them.

Not all operations are associative and commutative. For example, division is not commutative, since the indicated order of division of two numbers does matter. For example, $\frac{6}{5} \neq \frac{5}{6}$ ($\neq$ is read "does not equal").

Using literal symbols, the fundamental laws of algebra are as follows:

Commutative law of addition: $a + b = b + a$
Associative law of addition: $a + (b + c) = (a + b) + c$
Commutative law of multiplication: $ab = ba$
Associative law of multiplication: $a(bc) = (ab)c$
Distributive law: $a(b + c) = ab + ac$

Having identified the fundamental laws of algebra, we shall state the laws which govern the operations of addition, subtraction, multiplication, and division of signed numbers. These laws will be of primary and direct use in all of our work.

1. *To add two real numbers with like signs, add their absolute values and affix their common sign to the result.*

Example A

$$(+2) + (+6) = +(2 + 6) = +8$$
$$(-2) + (-6) = -(2 + 6) = -8$$

2. *To add two real numbers with unlike signs, subtract the smaller absolute value from the larger and affix the sign of the number with the larger absolute value to the result.*

Example B

$$(+2) + (-6) = -(6 - 2) = -4$$
$$(+6) + (-2) = +(6 - 2) = +4$$

3. *To subtract one real number from another, change the sign of the number to be subtracted, and then proceed as in addition.*

Example C

$$(+2) - (+6) = +2 + (-6) = -(6 - 2) = -4$$
$$(-a) - (-a) = -a + a = 0$$

The second part of Example C shows that subtracting a negative number from itself results in zero. *Subtracting the negative number is equivalent to adding a positive number of the same absolute value.* This reasoning is the basis of the rule which states, "the negative of a negative number is a positive number."

4. *The product (or quotient) of two real numbers of like signs is the product (or quotient) of their absolute values. The product (or quotient) of two real numbers of unlike signs is the negative of the product (or quotient) of their absolute values.*

Example D

$$\frac{+3}{+5} = +\left(\frac{3}{5}\right) = +\frac{3}{5}$$

$$(-3)(+5) = -(3 \cdot 5) = -15$$

$$\frac{-3}{-5} = +\left(\frac{3}{5}\right) = +\frac{3}{5}$$

When we have an expression in which there is a combination of the basic operations, we must be careful to perform them in the proper order. Generally it is clear by the grouping of numbers as to the proper order of performing these operations. However, *if the order of operations is not indicated by specific grouping, multiplications and divisions are performed first, and then the additions and subtractions are performed.*

Example E

The expression $20 \div (2 + 3)$ is evaluated by first adding $2 + 3$ and then dividing 20 by 5 to obtain the result 4. Here, the grouping of $2 + 3$ is clearly shown by the parentheses.

The expression $20 \div 2 + 3$ is evaluated by first dividing 20 by 2 and adding this quotient of 10 to 3 in order to obtain the result of 13. Here no specific grouping is shown, and therefore the division is performed before the addition.

Example F

$$(-6) - 2(-4) + \frac{25}{-5} = (-6) - (-8) + (-5) = -6 + 8 - 5 = -3$$

$$\frac{40}{(+7) + (-3)(+5)} = \frac{40}{+7 + (-15)} = \frac{40}{7 - 15} = \frac{40}{-8} = -5$$

$$\frac{(-8)(+3)}{2} - (-5)(+2)(+3) = \frac{-24}{2} - (-10)(+3)$$

$$= (-12) - (-30) = -12 + 30 = 18$$

In the first illustration, we see that the multiplication and division were performed first, and then the addition and subtraction were performed. Also, it can be seen that the addition and subtraction were changed to operations on unsigned (equivalent to positive) numbers. This is generally more convenient, especially when more than one addition or subtraction is involved. In the second illustration, the multiplication in the denominator was performed first, and then the addition was performed. It was necessary to evaluate the denominator before the division could be performed. In the third illustration, the left expression can be evaluated by performing either the multiplication or division first. Also, the order of multiplication in the right expression does not matter. However, these multiplications and divisions must be performed before the subtraction.

1–3 Operations with Zero

Since the basic operations with zero tend to cause some difficulty, we shall demonstrate them separately in this section.

If a represents any real number, the various operations with zero are defined as follows:

$$a \pm 0 = a \text{ (the symbol } \pm \text{ means ``plus or minus'')}$$

$$a \cdot 0 = 0$$

$$\frac{0}{a} = 0 \quad \text{if} \quad a \neq 0$$

Note that there is no answer defined for division by zero. To understand the reason for this, consider the problem of 4/0. If there were an answer to this expression, it would mean that the answer, which we shall call b, should give 4 when multiplied by 0. That is, $0 \cdot b = 4$. However, no such number b exists, since we already know that $0 \cdot b = 0$. Also, the expression 0/0 has no meaning, since $0 \cdot b = 0$ for any value of b which may be chosen. Thus **division by zero is not defined**. All other operations with zero are the same as for any other number.

Example A

$$5 + 0 = 5, \quad 7 - 0 = 7, \quad 0 - 4 = -4,$$

$$\frac{0}{6} = 0, \qquad \frac{0}{-3} = 0, \qquad \frac{5 \cdot 0}{7} = 0,$$

$$\frac{8}{0} \text{ is undefined}, \quad \frac{7 \cdot 0}{0 \cdot 6} \text{ is undefined}$$

There is no need for confusion in the operations with zero. They will not cause any difficulty if we remember that **division by zero is undefined** and that this is the only undefined operation.

Exercises 1–2, 1–3   In Exercises 1 through 36 evaluate each of the given expressions by performing the indicated operations.

1. $6 + 5$

2. $8 + (-4)$

3. $(-4) + (-7)$

4. $(-3) + (+9)$

5. $16 - 7$

6. $(+8) - (+11)$

7. $-9 - (-6)$

8. $8 - (-4)$

9. $(8)(-3)$

10. $(+9)(-3)$

11. $(-7)(-5)$

12. $(+5)(-8)$

13. $\dfrac{-9}{+3}$

14. $\dfrac{-18}{-6}$

15. $\dfrac{-60}{-3}$

16. $\dfrac{+28}{-7}$

17. $(-2)(+4)(-5)$

18. $(+3)(-4)(+6)$

19. $\dfrac{(+2)(-5)}{10}$

20. $\dfrac{-64}{(2)(-4)}$

21. $9 - 0$

22. $\dfrac{0}{-6}$

23. $\dfrac{+17}{0}$

24. $\dfrac{(+3)(0)}{0}$

25. $8 - 3(-4)$

26. $20 + 8 \div 4$

27. $3 - 2(6) + \dfrac{8}{2}$

28. $0 - (-6)(-8) + (-10)$

29. $\dfrac{(+3)(-6)(-2)}{0 - 4}$

30. $\dfrac{7 - (-5)}{(-1)(-2)}$

31. $\dfrac{24}{3 + (-5)} - 4(-9)$

32. $\dfrac{-18}{+3} - \dfrac{4 - 6}{-1}$

33. $(-7) - \dfrac{-14}{2} - 3(2)$

34. $-7(-3) + \dfrac{+6}{-3} - (-9)$

35. $\dfrac{(+3)(-9) - 2(-3)}{3 - 10}$

36. $\dfrac{(+2)(-7) - 4(-2)}{-9 - (-9)}$

In Exercises 37 through 44 determine which of the fundamental laws of algebra is demonstrated.

37. $(6)(7) = (7)(6)$

38. $6 + 8 = 8 + 6$

39. $6(3 + 1) = 6(3) + 6(1)$

40. $4(5 \cdot 7) = (4 \cdot 5)(7)$

41. $3 + (5 + 9) = (3 + 5) + 9$

42. $8(3 - 2) = 8(3) - 8(2)$

43. $(2 \times 3) \times 9 = 2 \times (3 \times 9)$

44. $(3 \cdot 6) \cdot 7 = 7 \cdot (3 \cdot 6)$

In Exercises 45 through 48 answer the given questions.

45. What is the sign of the product of an even number of negative numbers?

46. What is the sign of the product of an odd number of negative numbers?

47. Is subtraction commutative? Illustrate.

48. A boat travels at 6 mi/h in still water. A stream flows at 4 mi/h. If the boat travels downstream for 2 h, set up the expression which would be used to evaluate the distance traveled. What fundamental law is illustrated?

1–4 Exponents

We have introduced numbers and the fundamental laws which are used with them in the fundamental operations. Also, we have shown the use of literal numbers to represent numbers. In this section we shall introduce some basic terminology and notation which are important to the basic algebraic operations developed in the following sections.

In multiplication we often encounter a number which is to be multiplied by itself several times. Rather than writing this number over and over repeatedly, we use the notation a^n, where a is the number being considered and n is the number of times it appears in the product. *The number a is called the* **base**, *the number n is called the* **exponent**, *and, in words, the expression is read as the "nth* **power of a.**"

Example A

$$4 \cdot 4 \cdot 4 \cdot 4 \cdot 4 = 4^5 \quad \text{(the fifth power of 4)}$$
$$(-2)(-2)(-2)(-2) = (-2)^4 \quad \text{(the fourth power of } -2)$$
$$a \cdot a = a^2 \quad \text{(the second power of } a, \text{ called "}a \text{ squared")}$$
$$(\tfrac{1}{5})(\tfrac{1}{5})(\tfrac{1}{5}) = (\tfrac{1}{5})^3 \quad \text{(the third power of } \tfrac{1}{5}, \text{ called "}\tfrac{1}{5} \text{ cubed")}$$
$$8 \cdot 8 \cdot 8 \cdot 8 \cdot 8 \cdot 8 \cdot 8 \cdot 8 \cdot 8 = 8^9 \quad \text{(the ninth power of 8)}$$

The basic operations with exponents will now be stated symbolically. We first state them for positive integers as exponents, and then show how zero and negative integers are used as exponents. Therefore, if m and n are positive integers, we have the following important operations for exponents.

$$a^m \cdot a^n = a^{m+n} \tag{1–1}$$

$$\frac{a^m}{a^n} = a^{m-n} \quad (m > n, \ a \neq 0) \qquad \frac{a^m}{a^n} = \frac{1}{a^{n-m}} \quad (m < n, \ a \neq 0) \tag{1–2}$$

$$(a^m)^n = a^{mn} \tag{1–3}$$

$$(ab)^n = a^n b^n, \qquad \left(\frac{a}{b}\right)^n = \frac{a^n}{b^n} \quad (b \neq 0) \tag{1–4}$$

In applying Eqs. (1–1) and (1–2), the base a must be the same for the exponents to be added or subtracted. When a problem involves a product of different bases, *only exponents of the same base may be combined.* In the following three examples, Eqs. (1–1) to (1–4) are verified and illustrated.

Example B
Applying Eq. (1–1), we have

$a^3 \cdot a^5 = a^{3+5} = a^8$. We see that this result is correct since we can also write $a^3 \cdot a^5 = (a \cdot a \cdot a)(a \cdot a \cdot a \cdot a \cdot a) = a^8$.

Applying the first form of Eqs. (1–2), we have

$$\frac{a^5}{a^3} = a^{5-3} = a^2, \quad \frac{a^5}{a^3} = \frac{\cancel{a} \cdot \cancel{a} \cdot \cancel{a} \cdot a \cdot a}{\cancel{a} \cdot \cancel{a} \cdot \cancel{a}} = a^2$$

Applying the second form of Eqs. (1–2), we have

$$\frac{a^3}{a^5} = \frac{1}{a^{5-3}} = \frac{1}{a^2}, \quad \frac{a^3}{a^5} = \frac{\cancel{a} \cdot \cancel{a} \cdot \cancel{a}}{\cancel{a} \cdot \cancel{a} \cdot \cancel{a} \cdot a \cdot a} = \frac{1}{a^2}$$

Example C
Applying Eq. (1–3), we have

$$(a^5)^3 = a^{5(3)} = a^{15}, \quad (a^5)^3 = (a^5)(a^5)(a^5) = a^{5+5+5} = a^{15}$$

Applying the first form of Eqs. (1–4), we have

$$(ab)^3 = a^3 b^3, \quad (ab)^3 = (ab)(ab)(ab) = a^3 b^3$$

Applying the second form of Eqs. (1–4), we have

$$\left(\frac{a}{b}\right)^3 = \frac{a^3}{b^3}, \quad \left(\frac{a}{b}\right)^3 = \left(\frac{a}{b}\right)\left(\frac{a}{b}\right)\left(\frac{a}{b}\right) = \frac{a^3}{b^3}$$

Example D
Other illustrations of the use of Eqs. (1–1) to (1–4) are as follows:

$$\frac{(3 \cdot 2)^4}{(3 \cdot 5)^3} = \frac{3^4 2^4}{3^3 5^3} = \frac{3 \cdot 2^4}{5^3}$$

$$(-x^2)^3 = [(-1)x^2]^3 = (-1)^3(x^2)^3 = -x^6$$
$$ax^2(ax)^3 = ax^2(a^3 x^3) = a^4 x^5$$

$$\frac{(ry^3)^2}{r(y^2)^4} = \frac{r^2 y^6}{ry^8} = \frac{r}{y^2}$$

As we previously pointed out, Eqs. (1–1) to (1–4) were developed for use with positive integers as exponents. We shall now show how their use can be extended to include zero and negative integers as exponents.

In Eqs. (1–2), if $n = m$, we would have $\frac{a^m}{a^m} = a^{m-m} = a^0$. Also, $\frac{a^m}{a^m} = 1$, since any nonzero quantity divided by itself equals 1. Therefore, for Eqs. (1–2) to hold when $m = n$, we have

$$a^0 = 1, \quad (a \neq 0) \tag{1-5}$$

Equation (1–5) gives the definition of zero as an exponent. Since a has not been specified, this equation states that *any nonzero algebraic expression raised to the zero power is* 1. Also, the other laws of exponents are valid for this definition.

Example E

Equation (1–1) states that $a^m \cdot a^n = a^{m+n}$. If $n = 0$, we thus have $a^m \cdot a^0 = a^{m+0} = a^m$. Since $a^0 = 1$, this equation could be written as $a^m(1) = a^m$. This provides further verification for the validity of Eq. (1–5).

Example F

$$5^0 = 1, \qquad (2x)^0 = 1, \qquad (ax + b)^0 = 1$$
$$(a^2 x b^4)^0 = 1, \qquad (a^2 b^0 c)^2 = a^4 b^0 c^2 = a^4 c^2$$
$$2t^0 = 2(1) = 2$$

We note in the last illustration that only t is raised to the zero power. If the quantity $2t$ was raised to the zero power, it would be written as $(2t)^0$.

If we apply the first form of Eqs. (1–2) to the case where $n > m$, the resulting exponent is negative. This leads us to the definition of a negative exponent.

Example G

Applying the first form of Eqs. (1–2) to a^2/a^7, we have

$$\frac{a^2}{a^7} = a^{2-7} = a^{-5}$$

Applying the second form of Eqs. (1–2) to the same fraction leads to

$$\frac{a^2}{a^7} = \frac{1}{a^{7-2}} = \frac{1}{a^5}$$

In order that these results can be consistent, it must be true that

$$a^{-5} = \frac{1}{a^5}$$

Following the reasoning in Example G, if we define

$$a^{-n} = \frac{1}{a^n}, \qquad (a \neq 0) \tag{1–6}$$

then all of the laws of exponents will hold for negative integers.

Example H

$$3^{-1} = \frac{1}{3}, \qquad 4^{-2} = \frac{1}{16}, \qquad \frac{1}{a^{-3}} = a^3, \qquad a^4 = \frac{1}{a^{-4}}$$

Example I

$$(a^0b^2c)^{-2} = \frac{1}{(b^2c)^2} = \frac{1}{b^4c^2}$$

$$\left(\frac{a^3t}{b^2x}\right)^{-2} = \frac{(a^3t)^{-2}}{(b^2x)^{-2}} = \frac{(b^2x)^2}{(a^3t)^2} = \frac{b^4x^2}{a^6t^2}$$

When we discussed the operations with signed numbers in Section 1–2, we noted that multiplications and divisions are performed prior to additions and subtractions, unless grouping specifies another order. Since raising a number to a power is in essence a form of multiplication, this operation is also performed before additions and subtractions. Thus, *unless grouping specifies otherwise, the order of operations is powers, products, and quotients—and then additions and subtractions.*

Example J

$$8 - (-1)^2 - 2(-3)^2 = 8 - (+1) - 2(+9) = 8 - 1 - 18 = -11$$

Note that we squared -1 and -3 as the first operation. The next operation was finding the product in the last term. Finally the subtractions were performed. We did *not* change the sign of -1 before we squared it, for this would have been incorrect.

Exercises 1—4

In Exercises 1 through 44 simplify the given expressions. Express results with positive exponents only.

1. x^3x^4

2. y^2y^7

3. $2b^4b^2$

4. $3k(k^5)$

5. $\dfrac{m^5}{m^3}$

6. $\dfrac{x^6}{x}$

7. $\dfrac{n^5}{n^9}$

8. $\dfrac{s}{s^4}$

9. $(a^2)^4$

10. $(x^8)^3$

11. $(t^5)^4$

12. $(n^3)^7$

13. $(2n)^3$

14. $(ax)^5$

15. $(ax^4)^2$

16. $(3a^2)^3$

17. $\left(\dfrac{2}{b}\right)^3$

18. $\left(\dfrac{x}{y}\right)^7$

19. $\left(\dfrac{x^2}{2}\right)^4$

20. $\left(\dfrac{3}{n^3}\right)^3$

21. 7^0

22. $(8a)^0$

23. $3x^0$

24. $6v^0$

25. 6^{-1}

26. 10^{-3}

27. $\dfrac{1}{s^{-2}}$

28. $\dfrac{1}{t^{-5}}$

29. $(-t^2)^7$

30. $(-y^3)^5$

31. $(2x^2)^6$

32. $(-c^4)^4$

33. $(4xa^{-2})^0$

34. $3(ab^{-1})^0$

35. b^5b^{-3}

36. $2c^4c^{-7}$

37. $(5^0x^2a^{-1})^{-1}$

38. $(3m^{-2}n^4)^{-2}$

39. $\left(\dfrac{4a}{x}\right)^{-3}$

40. $\left(\dfrac{2b^2}{y^5}\right)^{-2}$

41. $(-8gs^3)^2$

42. $ax^2(-a^2x)^2$

43. $\dfrac{15a^2n^5}{3an^6}$

44. $\dfrac{(ab^2)^3}{a^2b^8}$

In Exercises 45 through 48 evaluate the given expressions.

45. $7(-4) - (-5)^2$ **46.** $\dfrac{12}{-3} - (-1)^3$

47. $6 + (-2)^5 - (-2)(8)$ **48.** $9 - 2(-3)^4 - (-7)$

In Exercises 49 and 50 solve the given problems.

49. In analyzing the deformation of a certain beam, it might be necessary to simplify the expression

$$\frac{wx(-2Lx^2)}{24EI}$$

Perform this simplification.

50. In a certain electric circuit, it might be necessary to simplify the expression

$$\frac{gM}{j\omega C(\omega^2 M^2)} \quad (\omega \text{ is the Greek omega})$$

Perform this simplification.

1—5 Scientific Notation

In technical and scientific work we often encounter numbers which are either very large or very small in magnitude. Illustrations of such numbers are given in the following example.

Example A

Television signals travel at about 30,000,000,000 cm/s. The mass of the earth is about 6,600,000,000,000,000,000,000 tons. A typical protective coating used on aluminum is about 0.0005 in. thick. The wavelength of some x-rays is about 0.000000095 cm.

Writing numbers such as these is inconvenient in ordinary notation, as shown in Example A, particularly when the numbers of zeros needed for the proper location of the decimal point are excessive. Therefore, a convenient notation, known as **scientific notation**, is normally used to represent such numbers.

A number written in scientific notation is expressed as the product of a number between 1 *and* 10 *and a power of ten.* Symbolically this can be written as

$$P \times 10^k$$

where $1 \le P < 10$, and k can take on any integral value. The following example illustrates how numbers are written in scientific notation.

Example B

$$340{,}000 = 3.4(100{,}000) = 3.4 \times 10^5$$

$$0.017 = \frac{1.7}{100} = \frac{1.7}{10^2} = 1.7 \times 10^{-2}$$

$$0.000503 = \frac{5.03}{10000} = \frac{5.03}{10^4} = 5.03 \times 10^{-4}$$

$$6.82 = 6.82(1) = 6.82 \times 10^0$$

From Example B we can establish a method for changing numbers from ordinary notation to scientific notation. The decimal point is moved so that only one nonzero digit is to its left. The number of places moved is the value of *k*. It is positive if the decimal point is moved to the left, and it is negative if it is moved to the right. Consider the illustrations in the following example.

Example C

$$340000 = 3.4 \times 10^5 \qquad 0.017 = 1.7 \times 10^{-2}$$
5 places 2 places

$$0.000503 = 5.03 \times 10^{-4} \qquad 6.82 = 6.82 \times 10^0$$
4 places 0 places

To change a number from scientific notation to ordinary notation, the procedure above is reversed. The following example illustrates the procedure.

Example D

To change 5.83×10^6 to ordinary notation, we must move the decimal point 6 places to the right. Therefore, additional zeros must be included for the proper location of the decimal point. This means we write $5.83 \times 10^6 = 5{,}830{,}000$.

To change 8.06×10^{-3} to ordinary notation, we must move the decimal point 3 places to the left. Again, additional zeros must be included. Thus, $8.06 \times 10^{-3} = 0.00806$.

An illustration of the importance of scientific notation is demonstrated by the metric system use of prefixes on units to denote certain powers of ten. These are illustrated in Appendix B.

As we see from the previous examples, scientific notation provides an important application of the use of exponents, positive and negative. Also, scientific notation provides a practical way to handle calculations involving numbers of very large or very small magnitudes. This can be seen by the fact that a feature on many calculators is that of scientific notation.

The calculation can be made by first expressing all numbers in scientific notation. Then the actual calculation can be performed on numbers between one and ten, with the laws of exponents used to find the proper power of ten for the result. It is proper to leave the result in scientific notation.

Example E

In determining the result of 95,600,000,000,000/0.0286, we may set up the calculation as

$$\frac{9.56 \times 10^{13}}{2.86 \times 10^{-2}} = \left(\frac{9.56}{2.86}\right) \times 10^{15} = 3.34 \times 10^{15}$$

The power of ten here is sufficiently large that we would normally leave the result in this form. Even on most calculators with the scientific notation feature, it would be necessary to at least express the numerator in scientific notation before performing the calculation.

Example F

In evaluating the product $(7.50 \times 10^9)(6.44 \times 10^{-3})$, we obtain

$$(7.50 \times 10^9)(6.44 \times 10^{-3}) = 48.3 \times 10^6$$

However, since a number in scientific notation is expressed as the product of a number between 1 and 10, and a power of ten, we should rewrite this result as

$$48.3 \times 10^6 = (4.83 \times 10)(10^6)$$
$$= 4.83 \times 10^7$$

Exercises 1–5

In Exercises 1 through 8 change the numbers from scientific notation to ordinary notation.

1. 4.5×10^4 2. 6.8×10^7 3. 2.01×10^{-3} 4. 9.61×10^{-5}

5. 3.23×10^0 6. 8.40×10^0 7. 1.86×10 8. 5.44×10^{-1}

In Exercises 9 through 16 change the given numbers from ordinary notation to scientific notation.

9. 40000   10. 5600000   11. 0.0087   12. 0.702

13. 6.89   14. 1.09   15. 0.063   16. 0.0000908

In Exercises 17 through 24 perform the indicated calculations by first expressing all numbers in scientific notation. (See Example E.)

17. (67000)(3040000000)   18. (56200)(0.00632)

19. (1280)(86500)(43.8)   20. (0.0000659)(0.00486)(3190000000)

21. $\dfrac{87400}{0.00895}$ 22. $\dfrac{0.00728}{670000}$

23. $\dfrac{(0.0732)(6700)}{0.00134}$ 24. $\dfrac{(2430)(97000)}{0.00452}$

In Exercises 25 through 36 change any numbers in ordinary notation to scientific notation or change any numbers in scientific notation to ordinary notation. (See Appendix B for an explanation of symbols used.)

25. The stress on a certain structure is 22,500 lb/in.^{2}.
26. The half-life of uranium 235 is 710,000,000 years.
27. The power of a radio signal received from a lunar probe is 1.6×10^{-12} W.
28. Some computers can perform an addition in 4.5×10^{-9} s.
29. A certain electrical resistor has a resistance of 4.5×10^3 Ω.
30. To attain an energy density of that in some laser beams, an object would have to be heated to about 10^{30} °C.
31. One foot of steel pipe will increase about 0.000011 ft for a 1°C rise in temperature.
32. The wave length of yellow light is about 0.00000059 m.
33. The mass of a proton is 1.67×10^{-24} g.
34. The average distance from the earth to the sun is 9.29×10^7 mi.
35. The diameter of the sun is 864,000 mi.
36. The total acreage of national forests in the United States is approximately 187,000,000 acres.

In Exercises 37 through 40 perform the indicated calculations by first expressing all numbers in scientific notation.

37. There are about 161,000 cm in one mile. What is the area in square centimeters of one square mile?
38. A particular virus is a sphere 0.0000175 cm in diameter. What volume does the virus occupy?
39. The resistance R, in ohms, of a wire of length l and cross-sectional area A is given by $R = \rho l/A$, where ρ (the Greek rho) is known as the resistivity. Find R for a wire for which $\rho = 0.0000000175$ $\Omega \cdot$ m, $l = 0.150$ m and $A = 0.000000435$ m^2.
40. For a certain gas, the product of the volume and pressure is 16.5 Pa·cm^3. If the pressure is 0.00108 Pa, what is the volume?

1—6 Roots and Radicals

A problem that is often encountered is, "what number multiplied by itself n times gives another specified number?" For example, we may ask, "what number squared is 9?" The answer to this question is the **square root** of 9, which is denoted by $\sqrt{9}$.

The general notation for the nth root of a is $\sqrt[n]{a}$. (When $n = 2$, it is common practice not to put the 2 where n appears.) The $\sqrt{}$ sign is called a **radical sign**.

Example A

$\sqrt{2}$ (the square root of two)

$\sqrt[3]{2}$ (the cube root of two)

$\sqrt[4]{2}$ (the fourth root of two)

$\sqrt[7]{6}$ (the seventh root of six)

$\sqrt[3]{8}$ (the cube root of 8, which also equals 2)

In considering the question "what number squared is 9?" we can easily see that either $+3$ or -3 gives a proper result. This would imply that both of these values equaled $\sqrt{9}$. To avoid this ambiguity, *we define the* **principal nth root** *of a to be positive if a is positive, and the principal nth root of a to be negative if a is negative and n is odd*. This means that $\sqrt{9} = 3$ and not -3, and that $-\sqrt{9} = -3$.

Example B

$$\sqrt{4} = 2 \quad (\sqrt{4} \neq -2), \qquad \sqrt{169} = 13 \quad (\sqrt{169} \neq -13),$$
$$-\sqrt{64} = -8, \qquad -\sqrt{81} = -9, \qquad -\sqrt[4]{256} = -4,$$
$$\sqrt[3]{27} = 3, \qquad \sqrt[3]{-27} = -3, \qquad -\sqrt[3]{27} = -(+3) = -3$$

Another important property of radicals is that *the square root of a product of positive numbers is the product of the square roots*. That is,

$$\sqrt{ab} = \sqrt{a} \cdot \sqrt{b} \tag{1-7}$$

This property is useful is simplifying radicals. It is most useful if either a or b is a perfect square. Consider the following example.

Example C

$$\sqrt{8} = \sqrt{(4)(2)} = \sqrt{4}\sqrt{2} = 2\sqrt{2}$$
$$\sqrt{75} = \sqrt{(25)(3)} = \sqrt{25}\sqrt{3} = 5\sqrt{3}$$
$$\sqrt{80} = \sqrt{(16)(5)} = \sqrt{16}\sqrt{5} = 4\sqrt{5}$$

Earlier we mentioned imaginary numbers. Until Chapter 11 it will be necessary only that we recognize imaginary numbers when they occur. This is done by recalling that the square root of a negative number is an imaginary number. Thus, if we have the square root of a negative number, we can write it as j times the square root of the corresponding positive number.

Example D

$$\sqrt{-4} = \sqrt{(4)(-1)} = \sqrt{4}\sqrt{-1} = 2j$$
$$\sqrt{-27} = \sqrt{(27)(-1)} = \sqrt{27}\sqrt{-1} = \sqrt{(9)(3)}\sqrt{-1}$$
$$= \sqrt{9}\sqrt{3}\sqrt{-1} = 3\sqrt{3}j$$

It should be emphasized that although the square root of a negative number gives an imaginary number, the cube root of a negative number gives a negative real number. More generally, *the even root of a negative number is imaginary, and the odd root of a negative number is real.* A more detailed discussion of exponents, radicals, and imaginary numbers is found in Chapters 10 and 11.

Example E

$$\sqrt{-64} = \sqrt{64}j = 8j, \qquad \sqrt[3]{-64} = -4$$

Exercises 1—6

In Exercises 1 through 24 simplify the given expressions.

1. $\sqrt{25}$	2. $\sqrt{81}$	3. $-\sqrt{121}$	4. $-\sqrt{36}$
5. $\sqrt[3]{125}$	6. $\sqrt[4]{16}$	7. $\sqrt[3]{-216}$	8. $\sqrt[5]{-32}$
9. $\sqrt{18}$	10. $\sqrt{32}$	11. $\sqrt{12}$	12. $\sqrt{50}$
13. $\sqrt{-9}$	14. $\sqrt{-25}$	15. $\sqrt{-12}$	16. $\sqrt{-28}$
17. $(\sqrt{5})^2$	18. $(\sqrt{19})^2$	19. $(\sqrt[3]{31})^3$	20. $(\sqrt[4]{53})^4$
21. $2\sqrt{48}$	22. $4\sqrt{108}$	23. $\dfrac{7^2\sqrt{81}}{3^2\sqrt{49}}$	24. $\dfrac{2^5\sqrt[5]{243}}{3\sqrt{144}}$

In Exercises 25 through 28 solve the given problems.

25. A square parcel of land has an area of 400 m². What is the length of a side of the parcel?
26. A cubical water tank holds 512 ft³. What is the length of an edge of the tank?
27. Is it always true that $\sqrt{a^2} = a$?
28. If $0 < x < 1$ (x between 0 and 1), is $x > \sqrt{x}$?

1—7 Addition and Subtraction of Algebraic Expressions

It is the basic characteristic of algebra that letters are used to represent numbers. Since we have used literal symbols to represent numbers, even if in a general sense, we may conclude that all operations valid for numbers are valid for these literal symbols. In this section we shall discuss the terminology and methods for combining literal symbols.

Addition, subtraction, multiplication, division, and taking of roots are known as **algebraic operations.** *Any combination of numbers and literal symbols which results from algebraic operations is known as an* **algebraic expression.**

When an algebraic expression consists of several parts connected by plus signs and minus signs, each part (along with its sign) is known as a **term** *of the expression. If a given expression is made up of the product of a number of quantities, each of these quantities, or any product of them, is called a* **factor** *of the expression.* **It is important to distinguish clearly between terms and factors since some operations that are valid for terms are not valid for factors, and conversely.**

Example A

$3xy + 6x^2 - 7x\sqrt{y}$ is an algebraic expression with three terms. They are $3xy$, $6x^2$, and $-7x\sqrt{y}$.

The first term $3xy$ has individual factors of 3, x, and y. Also, any product of these factors is also a factor of $3xy$. Thus, additional expressions which are factors of $3xy$ are $3x$, $3y$, xy, and $3xy$ itself.

Example B

$7x(y^2 + x) - \dfrac{x + y}{6x}$ is an algebraic expression with terms $7x(y^2 + x)$ and $\dfrac{-(x + y)}{6x}$.

The term $7x(y^2 + x)$ has individual factors of 7, x, and $(y^2 + x)$, as well as products of these factors. The factor $y^2 + x$ has two terms, y^2 and x.

The numerator of the term $-\dfrac{x + y}{6x}$ has two terms, and the denominator has individual factors of 2, 3, and x. The minus sign can be treated as a factor of -1.

An algebraic expression containing only one term is called a **monomial.** *An expression containing two terms is a* **binomial,** *and one containing three terms is a* **trinomial.** *Any expression containing two or more terms is called a* **multinomial.** Thus, any binomial or trinomial expression can also be considered as a multinomial.

In any given term, the numbers and literal symbols multiplying any given factor constitute the **coefficient** *of that factor. The product of all the numbers in explicit form is known as the* **numerical coefficient** *of the term. All terms which differ only in their numerical coefficients are known as* **similar** *or* **like** *terms.*

Example C

$7x^3\sqrt{y}$ is a monomial. It has a numerical coefficient of 7. The coefficient of $\sqrt{y}$ is $7x^3$, and the coefficient of x^3 is $7\sqrt{y}$.

Example D

$4 \cdot 2b + 81b - 6ab$ is a multinomial of three terms. The first term has a numerical coefficient of 8, the second has a numerical coefficient of 81, and the third has a numerical coefficient of -6 (the sign of the term is attached to the numerical coefficient). The first and second terms are similar, since they differ only in their numerical coefficient. The third term is not similar to either of the others, for it has the factor a.

In adding and subtracting algebraic expressions, we combine similar terms. In doing so, we are combining quantities which are alike. All of the similar terms may be combined into a single term, and the final simplified expression will be made up entirely of terms which are not similar.

Example E

$3x + 2x - 5y = 5x - 5y$. Since there are two similar terms in the original expression, they are added together, so the simplified result has two unlike terms.

$6a^2 - 7a + 8ax$ cannot be simplified, since none of the terms are like terms.

Similarly, $6a + 5c + 2a - c = 6a + 2a + 5c - c = 8a + 4c$. (Here we use the commutative and associative laws.)

In writing algebraic expressions, it is often necessary to group certain terms together. For this purpose we use **symbols of grouping.** In this text we shall use **parentheses** ( ), **brackets** [ ], and **braces** { }. The **bar,** which is used with radicals and fractions, also groups terms. The bar attached to the radical sign groups the terms under it, and the bar separating the numerator and denominator of a fraction groups the terms above and under it.

When adding and subtracting algebraic expressions, it is often necessary to remove symbols of grouping. To do so we must *change the sign of every term within the symbols if the grouping is preceded by a minus sign.* If the symbols of grouping are preceded by a plus sign, each term within the symbols retains its original sign. This is a result of the distributive law. Normally, *when several symbols of grouping are to be removed, it is more convenient to remove the innermost symbols first.* This is illustrated in Examples H and I.

Example F

(1) $2a - (3a + 2b) - b = 2a - 3a - 2b - b = -a - 3b$

(2) $-(2x - 3c) + (c - x) = -2x + 3c + c - x = 4c - 3x$

Example G

(1) $3 - 2(m^2 - 2) = 3 - 2m^2 + 4 = 7 - 2m^2$

(2) $4(t - 3 - 2t^2) - (6t + t^2 - 4) = 4t - 12 - 8t^2 - 6t - t^2 + 4$
$$= -9t^2 - 2t - 8$$

Example H

(1) $-[(4 - 5x) - (a - 2x - 7)] = -[4 - 5x - a + 2x + 7]$
$$= -[11 - 3x - a] = -11 + 3x + a$$

(2) $3ax - [ax - (5s - 2ax)] = 3ax - [ax - 5s + 2ax]$
$$= 3ax - ax + 5s - 2ax = 5s$$

Example I

$[a^2b - ab + (ab - 2a^2b)] - \{[(3a^2b + b) - (4ab - 2a^2b)] - b\}$
$$= [a^2b - ab + ab - 2a^2b] - \{[3a^2b + b - 4ab + 2a^2b] - b\}$$
$$= a^2b - ab + ab - 2a^2b - \{3a^2b + b - 4ab + 2a^2b - b\}$$
$$= a^2b - ab + ab - 2a^2b - 3a^2b - b + 4ab - 2a^2b + b$$
$$= -6a^2b + 4ab$$

One of the most common errors made by beginning students is changing the sign of only the first term when removing symbols of grouping preceded by a minus sign. *Remember, if the symbols are preceded by a minus sign, we must change the sign of* all *terms.*

Exercises 1—7

In the following exercises simplify the given algebraic expressions.

1. $5x + 7x - 4x$
2. $6t - 3t - 4t$
3. $2y - y + 4x$
4. $4c + d - 6c$
5. $2a - 2c - e + 3c - a$
6. $x - 2y + 3x - y + z$
7. $a^2b - a^2b^2 - 2a^2b$
8. $xy^2 - 3x^2y^2 + 2xy^2$
9. $v - (4 - 5x + 2v)$
10. $2a - (b - a)$
11. $2 - 3 - (4 - 5a)$
12. $\sqrt{x} + (y - 2\sqrt{x}) - 3\sqrt{x}$
13. $(a - 3) + (5 - 6a)$
14. $(4x - y) - (2x - 4y)$
15. $3(2r + s) - (5s - r)$
16. $3(a - b) - 2(a - 2b)$
17. $-7(6 - 3c) - 2(c + 4)$
18. $-(5t + a^2) - 2(3a^2 - 2st)$
19. $-[(6 - n) - (2n - 3)]$
20. $-[(a - b) - (b - a)]$
21. $2[4 - (t^2 - 5)]$
22. $3[3 - (a - 4)]$
23. $-2[-x - 2a - (a - x)]$
24. $-2[-3(x - 2y) + 4x]$
25. $a\sqrt{xy} - [3 - (a\sqrt{xy} + 4)]$
26. $9v - [6 - (v - 4) + 4v]$
27. $8c - \{5 - [2 - (3 + 4c)]\}$
28. $7y - \{y - [2y - (x - y)]\}$
29. $5p - (q - 2p) - [3q - (p - q)]$
30. $-(4 - x) - [(5x - 7) - (6x + 2)]$
31. $-2\{-(4 - x^2) - [3 + (4 - x^2)]\}$
32. $-\{-[-(x - 2a) - b] - a\}$

33. When discussing gear trains, the expression $-(-R - 1)$ is found. Simplify this expression.

34. In analyzing a certain electric circuit, the expression $I_1 + I_2 - (I_2 - I_3)$ is found. Simplify this expression. (The numbers below the I's are **subscripts.** Different subscripts denote different unknowns.)

35. Under certain conditions, the expression for finding the profit on given sales involves simplifying the expression $5(x + 1) - (400 + 2x)$. Simplify this expression.

36. In developing the theory for an elastic substance, we find the following expression:

$$[(B + \tfrac{4}{3}\alpha) + 2(B - \tfrac{2}{3}\alpha)] - [(B + \tfrac{4}{3}\alpha) - (B - \tfrac{2}{3}\alpha)]$$

Simplify this expression.

1—8 Multiplication of Algebraic Expressions

To find the product of two or more monomials, we use the laws of exponents as given in Section 1—4 and the laws for multiplying signed numbers as stated in Section 1—2. We first multiply the numerical coefficients to determine the numerical coefficient of the product. Then we multiply the literal factors, remembering that the exponents may be combined only if the base is the same. Consider the illustrations in the following example.

Example A

$$(1)\quad 3ac^3(4sa^2c) = 12a^3c^4s$$
$$(2)\quad (-2b^2y)(-9aby^5) = 18ab^3y^6$$
$$(3)\quad 2xy(-6cx^2)(3xcy^2) = -36c^2x^4y^3$$

We find the product of a monomial and a multinomial by using the distributive law, which states that we *multiply each term of the multinomial by the monomial.* We must be careful to assign the correct sign to each term of the result, using the rules for multiplication of signed numbers. Also, we must properly combine literal factors in each term of the result.

Example B

$$(1)\quad 2ax(3ax^2 - 4yz) = 2ax(3ax^2) - (2ax)(4yz) = 6a^2x^3 - 8axyz$$
$$(2)\quad 5cy^2(-7cx - ac) = (5cy^2)(-7cx) + (5cy^2)(-ac)$$
$$= -35c^2xy^2 - 5ac^2y^2$$

In practice, it is generally not necessary to write out the middle step as it appears in the example above. We can generally write the answer directly. For example, the first part of Example B would usually appear as

$$2ax(3ax^2 - 4yz) = 6a^2x^3 - 8axyz$$

We find the product of two or more multinomials by using the distributive law and the laws of exponents. The result is that we *multiply each term of one multinomial by each term of the other, and add the results.*

Example C

$$(1)\quad (x - 2)(x + 3) = x(x) + x(3) + (-2)(x) + (-2)(3)$$
$$= x^2 + 3x - 2x - 6 = x^2 + x - 6$$
$$(2)\quad (a^2 - 2ab)(xy^2 + x^2) = a^2(xy^2) + a^2(x^2) - 2ab(xy^2) - 2ab(x^2)$$
$$= a^2xy^2 + a^2x^2 - 2abxy^2 - 2abx^2$$
$$(3)\quad (x - 2y)(x^2 + 2xy + 4y^2) = x^3 + 2x^2y + 4xy^2 - 2x^2y - 4xy^2 - 8y^3$$
$$= x^3 - 8y^3$$

Finding the power of an algebraic expression is equivalent to multiplying the expression by itself the number of times indicated by the exponent. In practice, it is often convenient to write the power of an algebraic expression in this form before multiplying. Consider the following example.

Example D

$$(1) \quad (x + 5)^2 = (x + 5)(x + 5) = x^2 + 5x + 5x + 25$$
$$= x^2 + 10x + 25$$

$$(2) \quad (2a - b)^3 = (2a - b)(2a - b)(2a - b)$$
$$= (2a - b)(4a^2 - 4ab + b^2)$$
$$= 8a^3 - 8a^2b + 2ab^2 - 4a^2b + 4ab^2 - b^3$$
$$= 8a^3 - 12a^2b + 6ab^2 - b^3$$

Exercises 1–8

In the following exercises perform the indicated multiplications.

1. $(a^2)(ax)$
2. $(2xy)(x^2y^3)$
3. $-ac^2(acx^3)$
4. $-2s^2(-4cs)^2$
5. $(2ax^2)^2(-2ax)$
6. $6pq^3(3pq^2)^2$
7. $a(-a^2x)(-2a)$
8. $-2m^2(-3mn)(m^2n)^2$
9. $a^2(x + y)$
10. $2x(p - q)$
11. $-3s(s^2 - 5t)$
12. $-3b(2b^2 - b)$
13. $5m(m^2n + 3mn)$
14. $a^2bc(2ac - 3a^2b)$
15. $3x(x - y + 2)$
16. $b^2x^2(x^2 - 2x + 1)$
17. $ab^2c^4(ac - bc - ab)$
18. $-4c^2(9gc - 2c + g^2)$
19. $ax(cx^2)(x + y^3)$
20. $-2(-3st^3)(3s - 4t)$
21. $(x - 3)(x + 5)$
22. $(a + 7)(a + 1)$
23. $(x + 5)(2x - 1)$
24. $(4t + s)(2t - 3s)$
25. $(2a - b)(3a - 2b)$
26. $(4x - 3)(3x - 1)$
27. $(x^2 - 2x)(x + 4)$
28. $(2ab^2 - 5t)(-ab^2 - 6t)$
29. $(x + 1)(x^2 - 3x + 2)$
30. $(2x + 3)(x^2 - x + 5)$
31. $(4x - x^3)(2 + x - x^2)$
32. $(5a - 3c)(a^2 + ac - c^2)$
33. $2x(x - 1)(x + 4)$
34. $ax(x + 4)(7 - x^2)$
35. $(2x - 5)^2$
36. $(x - 3)^2$
37. $(x + 3a)^2$
38. $(2m + 1)^2$
39. $(xyz - 2)^2$
40. $(b - 2x^2)^2$
41. $(2 + x)(3 - x)(x - 1)$
42. $(3x - c^2)^3$
43. $3x(x + 2)^2(2x - 1)$
44. $[(x - 2)^2(x + 2)]^2$

45. Under given conditions, when determining the income from a business enterprise, the expression $(40 - x)(200 + 5x)$ is found. Perform the indicated multiplication.

46. In determining a certain chemical volume, the expression $a + b(1 - X) + c(1 - X)^2$ is found. Perform the indicated multiplications.

47. The analysis of the deflection of a certain concrete beam involves the expression $w(l^2 - x^2)^2$. Perform the indicated multiplication.

48. In finding the maximum power in a particular electric circuit, the expression $(R + r)^2 - 2r(R + r)$ is used. Multiply and simplify.

1–9 Division of Algebraic Expressions

To find the quotient of one monomial divided by another, we use the laws of exponents as given in Section 1–4 and the laws for dividing signed numbers as stated in Section 1–2. Again, the exponents may be combined only if the base is the same.

Example A

$$(1) \quad \frac{16x^3y^5}{4xy^2} = \frac{16}{4}(x^{3-1})(y^{5-2}) = 4x^2y^3$$

$$(2) \quad \frac{-6a^2xy^2}{2axy^4} = -\left(\frac{6}{2}\right)\frac{a^{2-1}x^{1-1}}{y^{4-2}} = -\frac{3a}{y^2}$$

As noted in the second illustration, we use only positive exponents in the final result, unless there are specific instructions otherwise.

The quotient of a multinomial divided by a monomial is found by dividing each term of the multinomial by the monomial and adding the results. This process is a result of the equivalent operation with arithmetic fractions, which can be shown as

$$\frac{a + b}{c} = \frac{a}{c} + \frac{b}{c}$$

(Be careful: although $\frac{a+b}{c} = \frac{a}{c} + \frac{b}{c}$, we must note that $\frac{c}{a+b}$ is not $\frac{c}{a} + \frac{c}{b}$.)

Example B

$$(1) \quad \frac{16r^3st^2 - 8r^2t^3}{4rt^2} = \frac{16r^3st^2}{4rt^2} - \frac{8r^2t^3}{4rt^2} = 4r^2s - 2rt$$

$$(2) \quad \frac{4x^3y - 8x^3y^2 + 6x^2y^4}{2x^2y} = \frac{4x^3y}{2x^2y} - \frac{8x^3y^2}{2x^2y} + \frac{6x^2y^4}{2x^2y}$$

$$= 2x - 4xy + 3y^3$$

$$(3) \quad \frac{a^3bc^4 - 6abc + 9a^2b^3c - 3}{3ab^2c^3} = \frac{a^2c}{3b} - \frac{2}{bc^2} + \frac{3ab}{c^2} - \frac{1}{ab^2c^3}$$

Usually in practice we would not write the middle step as shown in the first two illustrations of Example B. The divisions are done by inspection, and the expression would appear as shown in the third illustration. However, we must remember that each term in the numerator is divided by the monomial in the denominator.

If each term of an algebraic sum is a number or is of the form ax^n where n is a nonnegative integer, we call the expression a **polynomial** *in x.* The distinction between a multinomial and a polynomial is that a polynomial does not contain terms like $\sqrt{x}$ or $1/x^2$, whereas a multinomial may contain such terms. Also, a polynomial may consist of only one term, whereas a multinomial must have at least two terms. In a polynomial, *the greatest value of n which appears is the* **degree** *of the polynomial.*

Example C

> $3 + 2x^2 - x^3$ is a polynomial of degree 3,
> $x^4 - 3x^2 - \sqrt{x}$ is not a polynomial,
> $4x^5$ is a polynomial of degree 5.

The first two expressions are also multinomials since each contains more than one term. The second illustration is not a polynomial due to the presence of the $\sqrt{x}$ term. The third illustration is a polynomial since the exponent is a positive integer. It is also a monomial.

The problem often arises of dividing one polynomial by another. To solve this problem, we first arrange the dividend (the polynomial to be divided) and the divisor in descending powers of x. Then we divide the first term of the dividend by the first term of the divisor. The result gives the first term of the quotient. Next, we multiply the entire divisor by the first term of the quotient and subtract the product from the dividend. We divide the first term of this difference by the first term of the divisor. This gives the second term of the quotient. We multiply this term by each of the terms of the divisor and subtract this result from the first difference. We repeat this process until the remainder is either zero or a term which is of lower degree than the divisor. This process is similar to that of long division of numbers.

Example D
Divide $4x^3 + 6x^2 + 1$ by $2x - 1$. Since there is no x-term in the dividend, it is advisable to leave space for any x-terms which might arise.

$$
\begin{array}{r}
2x^2 + 4x \qquad\;\; + 2 \quad \text{(quotient)} \\
(\text{divisor}) \quad 2x - 1\overline{\smash{\big)}\,4x^3 + 6x^2 \qquad\;\; + 1} \quad \text{(dividend)} \\
\underline{4x^3 - 2x^2 \qquad\qquad\quad} \\
8x^2 \qquad\;\; + 1 \\
\underline{8x^2 - 4x \qquad} \\
4x + 1 \\
\underline{4x - 2} \\
3 \quad \text{(remainder)}
\end{array}
$$

Exercises 1-9

In Exercises 1 through 36 perform the indicated divisions.

1. $\dfrac{8x^3y^2}{-2xy}$

2. $\dfrac{-18b^7c^3}{bc^2}$

3. $\dfrac{-16r^3t^5}{-4r^5t}$

4. $\dfrac{51mn^5}{17m^2n^2}$

5. $\dfrac{(15x^2)(4bx)(2y)}{30bxy}$

6. $\dfrac{(5st)(8s^2t^3)}{10s^3t^2}$

7. $\dfrac{6(ax)^2}{-ax^2}$

8. $\dfrac{12a^2b}{(3ab^2)^2}$

9. $\dfrac{a^2x + 4xy}{x}$

10. $\dfrac{2m^2n - 6mn}{2m}$

11. $\dfrac{3rst - 6r^2st^2}{3rs}$

12. $\dfrac{-5a^2n - 10an^2}{5an}$

13. $\dfrac{4pq^3 + 8p^2q^2 - 16pq^5}{4pq^2}$

14. $\dfrac{a^2xy^2 + ax^3 - 4ax^2}{ax}$

15. $\dfrac{2\pi fL - \pi fR^2}{\pi fR}$

16. $\dfrac{2(ab)^4 - a^3b^4}{3(ab)^3}$

17. $\dfrac{3ab^2 - 6ab^3 + 9a^3b}{9a^2b^2}$

18. $\dfrac{4x^2y^3 + 8xy - 12x^2y^4}{2x^2y^2}$

19. $\dfrac{x^{n+2} + ax^n}{x^n}$

20. $\dfrac{3a(x + y)b^2 - (x + y)}{a(x + y)}$

21. $\dfrac{x^2 - 3x + 2}{x - 2}$

22. $\dfrac{2x^2 - 5x - 7}{x + 1}$

23. $\dfrac{x - 14x^2 + 8x^3}{2x - 3}$

24. $\dfrac{6x^2 + 6 + 7x}{2x + 1}$

25. $\dfrac{x^3 + 3x^2 - 4x - 12}{x + 2}$

26. $\dfrac{3x^3 + 19x^2 + 16x - 20}{3x - 2}$

27. $\dfrac{2x^4 + 4x^3 + 2}{x^2 - 1}$

28. $\dfrac{2x^3 - 3x^2 + 8x - 2}{x^2 - x + 2}$

29. $\dfrac{x^3 + 8}{x + 2}$

30. $\dfrac{x^3 - 1}{x - 1}$

31. $\dfrac{x^2 - 2xy + y^2}{x - y}$

32. $\dfrac{3a^2 - 5ab + 2b^2}{a - 3b}$

33. In determining the volume of a certain gas, the following expression is used:

$$\dfrac{RTV^2 - aV + ab}{RT}$$

Perform the indicated division.

34. In hydrodynamics the following expression is found:

$$\dfrac{2p + v^2d + 2ydg}{2dg}$$

Perform the indicated division.

35. The expression for the total resistance of three resistances in parallel in an electric circuit is

$$\frac{R_1 R_2 R_3}{R_2 R_3 + R_1 R_3 + R_1 R_2}$$

Find the reciprocal of this expression, and then perform the indicated division.

36. The following expression is found when analyzing the motion of a certain object:

$$\frac{s + 6}{s^2 + 8s + 25}$$

Find the reciprocal of this expression, and perform the indicated division.

1—10 Equations

The basic operations for algebraic expressions that we have developed are used in the important process of solving equations. In this section we show how the basic algebraic operations are used in solving equations, and in the following section we will demonstrate some of the important applications of equations.

An **equation** *is an algebraic statement that two algebraic expressions are equal.* It is possible that many values of the letter representing the **unknown** will **satisfy** the equation; that is, many values may produce equality when **substituted** in the equation. It is also possible that only one value for the unknown will satisfy the equation (and this will be true of the equations we solve in this section). Or possibly there may be no values which satisfy the equation, although the statement is still an equation.

Example A
The equation $x^2 - 4 = (x - 2)(x + 2)$ is true for all values of x. For example, if we substitute $x = 3$ we have $9 - 4 = (3 - 2)(3 + 2)$ or $5 = 5$. If we let $x = -1$, we have $-3 = -3$. *An equation that is true for all values of the unknown is termed an* **identity.**

The equation $x^2 - 2 = x$ is true if $x = 2$ or if $x = -1$, but it is not true for any other values of x. If $x = 2$ we obtain $2 = 2$, and if $x = -1$ we obtain $-1 = -1$. However, if we let $x = 4$, we obtain $14 = 4$, which of course is not correct. *An equation valid only for certain values of the unknown is termed a* **conditional equation.** These equations are those which are generally encountered.

Example B
The equation $3x - 5 = x + 1$ is true only for $x = 3$. When $x = 3$ we obtain $4 = 4$; if we let $x = 2$, we obtain $1 = 3$, which is not correct.

The equation $x + 5 = x + 1$ is not true for any value of x. For any value of x we try, we will find that the left side is 4 greater than the right side. However, it is still an equation.

To **solve** an equation we find the values of the unknown which satisfy it. There is one basic rule to follow when solving an equation: **Perform the same operation on both sides of the equation.** We do this to isolate the unknown and thus to find its values.

By performing the same operation on both sides of an equation, the two sides remain equal. Thus, *we may add the same number to both sides, subtract the same number from both sides, multiply both sides by the same number, or divide both sides by the same number (not zero).*

Example C
In solving the following equations, we note that we may isolate x, and thereby solve the equation, by performing the indicated operation.

$x - 3 = 12$	$x + 3 = 12$	$\dfrac{x}{3} = 12$	$3x = 12$
Add 3 to both sides.	Subtract 3 from both sides.	Multiply both sides by 3.	Divide both sides by 3.
$x - 3 + 3 = 12 + 3$	$x + 3 - 3 = 12 - 3$	$3\left(\dfrac{x}{3}\right) = 3(12)$	$\dfrac{3x}{3} = \dfrac{12}{3}$
$x = 15$	$x = 9$	$x = 36$	$x = 4$

Each can be checked by substitution in the original equation. (The term "transposing" is often used to denote the result of adding or subtracting a term from both sides of the equation. In transposing, a term is moved from one side of the equation to the other, and its sign is changed.)

The solution of an equation generally requires a combination of the basic operations. The following examples illustrate the solution of such equations.

Example D
Solve the equation $2t - 7 = 9$.

We are to perform basic operations to both sides of the equation to finally isolate t on one side. The steps to be followed are suggested by the form of the equation, and in this case are as follows.

$$2t - 7 = 9 \qquad \text{add 7 to both sides}$$
$$2t = 16 \qquad \text{divide both sides by 2}$$
$$t = 8$$

Therefore, we conclude that $t = 8$. Checking in the original equation, we see that we have $2(8) - 7 = 9$, $16 - 7 = 9$, or $9 = 9$. Therefore, the solution checks.

Example E
Solve the equation $3n + 4 = n - 6$.

$$2n + 4 = -6 \qquad n \text{ subtracted from both sides}$$
$$2n = -10 \qquad 4 \text{ subtracted from both sides}$$
$$n = -5 \qquad \text{both sides divided by 2}$$

Checking in the original equation, we have $-11 = -11$.

Example F
Solve the equation $x - 7 = 3x - (6x - 8)$.

$$x - 7 = 3x - 6x + 8 \qquad \text{parentheses removed}$$
$$x - 7 = -3x + 8 \qquad x\text{'s combined on right}$$
$$4x - 7 = 8 \qquad 3x \text{ added to both sides}$$
$$4x = 15 \qquad 7 \text{ added to both sides}$$
$$x = \tfrac{15}{4} \qquad \text{both sides divided by 4}$$

Checking in the original equation, we obtain (after simplifying) $-\tfrac{13}{4} = -\tfrac{13}{4}$.

Exercises 1–10

In Exercises 1 through 24 solve the given equations.

1. $x - 2 = 7$

2. $x - 4 = 1$

3. $x + 5 = 4$

4. $s + 6 = 3$

5. $\dfrac{t}{2} = 5$

6. $\dfrac{x}{4} = 2$

7. $4x = 20$

8. $2x = 12$

9. $3t + 5 = -4$

10. $5x - 2 = 13$

11. $5 - 2y = 3$

12. $8 - 5t = 18$

13. $3x + 7 = x$

14. $6 + 8y = 5 - y$

15. $2(s - 4) = s$

16. $3(n - 2) = -n$

17. $6 - (r - 4) = 2r$

18. $5 - (x + 2) = 5x$

19. $2(x - 3) - 5x = 7$

20. $4(x + 7) - x = 7$

21. $x - 5(x - 2) = 2$

22. $5x - 2(x - 5) = 4x$

23. $7 - 3(1 - 2p) = 4 + 2p$

24. $3 - 6(2 - 3t) = t - 5$

In Exercises 25 through 28 solve the given problems.

25. What conclusion can be made about the equation $2(x - 3) + 1 = 2x - 5$?

26. What conclusion can be made about the equation $1 - (3 - x) = x - 2$?

27. Show that the equation $3(x + 2) = 3x + 4$ is not valid for any values of x.

28. Show that the equation $7 - (2 - x) = x + 2$ is not valid for any values of x.

1—11 Applications of Equations

Equations and their solutions are of great importance in most fields of technology and science. They are used to attain, study, and confirm information of all kinds. One of the most important applications occurs in the use of formulas in mathematics, physics, engineering, and other fields. A formula is an algebraic statement that two expressions stand for the same number. For example, the formula for the area of a circle is $A = \pi r^2$. The symbol A stands for the area, as does the expression πr^2, but πr^2 expresses the area in terms of another quantity, the radius.

Often it is necessary to solve a formula for a particular letter or symbol which appears in it. We do this in the same manner as we solve any equation: we isolate the letter or symbol desired by use of the basic algebraic operations.

Example A

Solve $A = \pi r^2$ for π.

$$\frac{A}{r^2} = \pi \qquad \text{both sides divided by } r^2$$

$$\pi = \frac{A}{r^2} \qquad \text{since each side equals the other, it makes no difference which expression appears on the left}$$

Example B

A formula relating acceleration a, velocity v, initial velocity v_0, and time t, is $v = v_0 + at$. Solve for t.

$$v - v_0 = at \qquad v_0 \text{ subtracted from both sides}$$

$$t = \frac{v - v_0}{a} \qquad \text{both sides divided by } a \text{ and then sides are switched}$$

As we can see from Examples A and B, we can solve for the indicated literal number just as we solved for the unknown in the previous section. That is, we perform the basic algebraic operations on the various literal numbers which appear in the same way we perform them on explicit numbers. Another illustration appears in the following example.

Example C

The effect of temperature is important when accurate instrumentation is required. The volume V of a precision container at temperature T in terms of the volume V_0 at temperature T_0 is given by

$$V = V_0[1 + b(T - T_0)]$$

where b depends on the material of which the container is made. Solve for T.

Since we are to solve for T, we must isolate the term containing T. This can be done by first removing the grouping symbols, and then isolate the term with T.

$$V = V_0[1 + b(T - T_0)] \qquad \text{(original equation)}$$

$$V = V_0[1 + bT - bT_0] \qquad \text{(remove parentheses)}$$

$$V = V_0 + bTV_0 - bT_0V_0 \qquad \text{(remove brackets)}$$

$$V - V_0 + bT_0V_0 = bTV_0 \qquad \text{(subtract } V_0 \text{ and add } bT_0V_0 \text{ to both sides)}$$

$$T = \frac{V - V_0 + bT_0V_0}{bV_0} \qquad \text{(divide both sides by } bV_0 \text{ and switch sides)}$$

In practice it is often necessary to set up equations to be solved by using known formulas and given conditions. The most difficult part in solving such a stated problem is identifying the information which leads to the equation. Often this is due to the fact that some of the information is inferred, but not explicitly stated, in the problem.

Since a careful reading and analysis are important to the solution of stated problems, it is possible only to give a general guideline to follow. Thus, (1) *read the statement of the problem carefully;* (2) *clearly identify the unknown quantities, assign an appropriate letter to represent one of them, and specify the others in terms of this unknown;* (3) *analyze the statement clearly to establish the necessary equation;* and (4) *solve the equation, checking the solution in the original statement of the problem.* Carefully read the following examples.

Example D

Two machine parts together weigh 17 lb. If one weighs 3 lb more than the other, what is the weight of each?

Since the weight of each part is required, we write

let $w =$ the weight of the lighter part

as a way of establishing the unknown for the equation. Any appropriate letter could be used, and we could have let it represent the heavier part.

Also, since "one weighs 3 lb more than the other," we can write

let $w + 3 =$ the weight of the heavier part

Since the two parts together weigh 17 lb, we have the equation

$$w + (w + 3) = 17$$

This can now be solved.

$$2w + 3 = 17$$
$$2w = 14$$
$$w = 7$$

Thus, the lighter part weighs 7 lb and the heavier part weighs 10 lb. This checks with the original statement of the problem.

Example E

A man rowing x mi/h covers 8 mi in one hour when going downstream. By rowing twice as fast while going upstream, he is able to cover only 7 mi in one hour. Find the original rate of speed of his rowing, x, and the rate of flow of the stream.

In this problem we have let x equal the man's original rate of rowing. From the fact that the man was able to cover 8 mi in one hour while going downstream, we conclude that his rate plus the rate of the stream equals 8 mi/h. Thus, $8 - x$ is the rate of the stream. When he is going upstream, the stream retards his progress, which means that the rate at which he actually proceeds upstream is $2x - (8 - x)$. He goes at this rate for one hour, traveling 7 mi. Using the formula $d = rt$ (distance equals rate times time), we have

$$7 = [2x - (8 - x)](1)$$

Solving for x, we obtain

$$7 = 2x - (8 - x)$$
$$7 = 2x - 8 + x$$
$$15 = 3x$$
$$x = 5$$

Therefore, the original rate of rowing was 5 mi/h, and the stream flows at the rate of 3 mi/h. We see that this checks in that the distance covered in one hour going downstream is $(5 + 3)(1) = 8$ mi, and the distance covered in one hour going upstream is $[(2)(5) - 3] = 7$ mi.

Example F

A solution of alcohol and water contains 2 L of alcohol and 6 L of water. How much pure alcohol must be added to this solution so that the resulting solution will be $\frac{2}{5}$ alcohol?

First we let x equal the number of liters of alcohol to be added. The statement of the problem tells us that we want the volume of alcohol compared to the total volume of the final mixture to be $\frac{2}{5}$. The final total volume of alcohol will be $2 + x$, and the final total volume of the mixture of water and alcohol will be $8 + x$. This means that

$$\frac{2 + x}{8 + x} = \frac{2}{5}$$

Multiplying each side by $5(8 + x)$, we have

$$5(2 + x) = 2(8 + x)$$
$$10 + 5x = 16 + 2x$$
$$3x = 6$$
$$x = 2$$

Therefore, 2 L of alcohol are to be added to the solution. Note that this result checks, since there would be 4 L of alcohol of a total volume of 10 L when 2 L of pure alcohol are added to the original solution.

Exercises 1–11

In Exercises 1 through 8 solve for the indicated letter.

1. $ax = b$, for x
2. $cy + d = 0$, for y
3. $4n + 1 = m$, for n
4. $bt - 3 = a$, for t
5. $ax + 6 = 2ax - c$, for x
6. $s - 6n^2 = 3s + 4$, for s
7. $\frac{1}{2}t - (4 - a) = 2a$, for t
8. $7 - (p - \frac{1}{3}x) = 3p$, for x

In Exercises 9 through 24 each of the given formulas arises in the technical or scientific area of study listed. Solve for the indicated letter.

9. $E = IR$, for R (electricity)
10. $PV = RT$, for T (chemistry: gas law)
11. $A = \dfrac{M}{fjd}$, for d (mechanical design)
12. $V = \frac{4}{3}\pi r^3$, for π (geometry)
13. $v = v_0 - gt$, for g (physics: motion)
14. $A_1 = A(M + 1)$, for M (photography)
15. $a = V(k - PV)$, for k (biology: muscle contractions)
16. $s = s_0 + v_0 t - 16t^2$, for v_0 (physics: motion)
17. $l = a + (n - 1)d$, for n (mathematics: progressions)
18. $T_2 w = q(T_2 - T_1)$, for T_1 (chemistry: energy)
19. $L = \pi(r_1 + r_2) + 2d$, for r_1 (physics: pulleys)
20. $h = \dfrac{1}{r_e + r_c(1 - a)}$, for r_e (electricity: transistors)
21. $F = \dfrac{9C}{5} + 32$, for C (science: temperature)
22. $R = \dfrac{2(E - E_p)}{m_0 g}$, for E (modern physics)
23. $R = \dfrac{wL}{H(w + L)}$, for H (architecture: interior lighting)
24. $PV^2 = RT(1 - e)(V + b) - A$, for T (thermodynamics)

In Exercises 25 through 36 solve the given stated problems by first setting up an appropriate equation.

25. Together, two computers cost $7800 per month to rent. If one costs twice as much as the other, what is the monthly cost of each?
26. The combined capacity of two oil tanks is 1375 gallons. If one tank holds 275 gal more than the other tank, what is the capacity of each?
27. The sum of three electric currents is 12 A. If the smallest is 2 A less than the next, which in turn is 2 A less than the largest, what are the values of the three currents?
28. A vial contains 2000 mg which is to be used for two dosages. One patient is to be administered 660 mg more than the other. How much should be administered to each?

29. The length of a spring increases 0.25 ft for each pound it supports. If the spring is 8 ft long when 6 lb are hung from it, what was the original length of the spring?

30. A temperature measured in degrees Fahrenheit is 32 more than $\frac{9}{5}$ the corresponding reading in degrees Celsius. If the Celsius reading of a certain room is 25°C, what is the Fahrenheit reading?

31. An architect designs a rectangular window such that the width of the window is 18 in. less than the height. If the perimeter of the window is 180 in., what are its dimensions?

32. A square tract of land is enclosed with fencing and then divided in half by additional fencing parallel to two of the sides. If 75 m of fencing are used, what is the length of one side of the tract?

33. A car traveling at 30 mi/h leaves a certain point 2 h before a second car. If the second car travels 40 mi/h, when will it overtake the first car?

34. Two supersonic jets, originally 5400 mi apart, start at the same time and travel toward each other. Find the speed of each if one travels 400 mi/h faster than the other and they meet in 1.5 h.

35. A certain type of engine uses a fuel mixture of 15 parts of gasoline to one part of oil. How much gasoline must be mixed with a gasoline-oil mixture, which is 75% gasoline, to make 8 L of the required mixture for the engine?

36. An alloy weighing 20 lb is 30% copper. How many pounds of another alloy which is 80% copper must be added in order for the final alloy to be 50% copper?

1–12 Exercises for Chapter 1

In Exercises 1 through 12 simplify the given expressions.

1. $(-2) + (-5) - (+3)$

2. $(+6) - (+8) - (-4)$

3. $\dfrac{(-5)(+6)(-4)}{(-2)(+3)}$

4. $\dfrac{(-9)(-12)(-4)}{24}$

5. $-5 - 2(-6) + \dfrac{-15}{+3}$

6. $3 - 5(-2) - \dfrac{12}{-4}$

7. $\dfrac{18}{3-5} - (-4)^2$

8. $-(-3)^2 - \dfrac{-8}{(-2)-(-4)}$

9. $\sqrt{16} - \sqrt{64}$

10. $-\sqrt{144} + \sqrt{49}$

11. $(\sqrt{7})^2 - \sqrt[3]{8}$

12. $-\sqrt[4]{16} + (\sqrt{6})^2$

In Exercises 13 through 24 simplify the given expressions. Where appropriate, express results with positive exponents only.

13. $(-2rt^2)^2$

14. $(3x^4y)^3$

15. $\dfrac{18m^3n^4t}{3mn^5t^3}$

16. $\dfrac{15p^4q^2r}{5pq^5r}$

17. $(x^0y^{-1}z^3)^2$

18. $(3a^0b^{-2})^3$

19. $\dfrac{-16s^{-2}(st^2)}{-2st^{-1}}$

20. $\dfrac{-35x^{-1}y(x^2y)}{5xy^{-1}}$

21. $\sqrt{45}$

22. $\sqrt{48}$

23. $\sqrt{-20}$

24. $\sqrt{-18}$

In Exercises 25 through 52 perform the indicated operations.

25. $a - 3ab - 2a + ab$

26. $xy - y - 5y - 4xy$

27. $6xy - (xy - 3)$

28. $-(2x - b) - 3(x - 5b)$

29. $(2x - 1)(x + 5)$

30. $(x - 4y)(2x + y)$

31. $\dfrac{2h^3k^2 - 6h^4k^5}{2h^2k}$

32. $\dfrac{4a^2x^3 - 8ax^4}{2ax^2}$

33. $4a - [2b - (3a - 4b)]$

34. $3b - [2b + 3a - (2a - 3b)] + 4a$

35. $2xy - \{3z - [5xy - (7z - 6xy)]\}$

36. $x^2 + 3b + [(b - y) - 3(2b - y + z)]$

37. $(2x + 1)(x^2 - x - 3)$

38. $(x - 3)(2x^2 - 3x + 1)$

39. $-3y(x - 4y)^2$

40. $-s(4s - 3t)^2$

41. $3p[(q - p) - 2p(1 - 3q)]$

42. $3x[2y - r - 4(s - 2r)]$

43. $\dfrac{12p^3q^2 - 4p^4q + 6pq^5}{2p^4q}$

44. $\dfrac{27s^3t^2 - 18s^4t + 9s^2t}{9s^2t}$

45. $\dfrac{3x^3 - 7x^2 + 11x - 3}{3x - 1}$

46. $\dfrac{x^3 - 4x^2 + 7x - 12}{x - 3}$

47. $\dfrac{4x^4 + 10x^3 + 18x - 1}{x + 3}$

48. $\dfrac{8x^3 - 14x + 3}{2x + 3}$

49. $-3\{(r + s - t) - 2[(3r - 2s) - (t - 2s)]\}$

50. $(1 - 2x)(x - 3) - (x + 4)(4 - 3x)$

51. $\dfrac{2y^3 + 9y^2 - 7y + 5}{2y - 1}$

52. $\dfrac{6x^2 + 5xy - 4y^2}{2x - y}$

In Exercises 53 through 60 solve the given equations.

53. $3s + 8 = 5s$

54. $6n = 14 - n$

55. $3x + 1 = x - 9$

56. $4y - 3 = 5y + 7$

57. $6x - 5 = 3(x - 4)$

58. $-2(4 - y) = 3y$

59. $2s + 4(3 - s) = 6$

60. $-(4 - v) = 2(2v - 5)$

In Exercises 61 through 68 change any numbers in ordinary notation to scientific notation or change any numbers in scientific notation to ordinary notation. (See Appendix B for an explanation of symbols which are used.)

61. The escape velocity (the velocity required to leave the earth's gravitational field) of a rocket is in excess of 25,000 mi/h.

62. When the first pictures of the surface of Mars were transmitted to Earth, Mars was 213,000,000 mi from Earth.

63. The ratio of the charge to the mass of an electron is 1.76×10^{11} C/kg.

64. Atmospheric pressure is about 1.013×10^5 Pa.
65. An oil film on water is about 0.0000002 in. thick.
66. A typical capacitor has a capacitance of 0.00005 F.
67. The viscosity of air is about 1.8×10^{-5} N·s/m².
68. The electric field intensity in a certain electromagnetic wave is 2.5×10^{-3} V/m.

In Exercises 69 through 84 solve for the indicated letter. Where noted the given formula arises in the technical or scientific area of study listed.

69. $3s + 2 = 5a$, for s 70. $5 - 7t = 6b$, for t

71. $3(4 - x) = 8 + 2n$, for x 72. $6 - 3b = 5(7 - 2v)$, for v

73. $B = \dfrac{\phi}{A}$, for A (ϕ is Greek phi)   (electricity: magnetism)

74. $D = \dfrac{KI^2t}{A}$, for t (medicine: cell damage)

75. $I = P + Prt$, for t (business: interest)
76. $v^2 = v_0^2 + 2gh$, for h (physics: motion)
77. $I(r + nR) = nE$, for R (electricity)
78. $R(R_1 + R_2) = R_1R_2$, for R (electricity)

79. $D_p = \dfrac{ND_0}{N + 2}$, for D_0 (mechanics: gears)

80. $Y_{n+1} = \dfrac{R_D X_n + X_0}{R_D + 1}$, for X_n (chemistry: distillation)

81. $L = L_0[1 + \alpha(t_2 - t_1)]$, for α (the Greek alpha)   (physics: heat)

82. $P_a - P_b = \dfrac{32LV\mu}{g_c D^2}$, for μ (the Greek mu)   (fluid dynamics)

83. $S = \frac{1}{2}n[2a + (n - 1)d]$, for a (mathematics: progressions)
84. $M = Rx - P(x - a)$, for a (mechanics: beams)

In Exercises 85 through 88 perform the indicated operations.

85. When determining the center of mass of a certain object, the expression $[(8x - x^2) - (x^2 - 4x)]$ is used. Simplify this expression.

86. In determining the value of sales in a particular business enterprise, the expression $(700 + 100n)(12 - n)$ is found. Multiply and simplify this expression.

87. Simplify the following expression, which arises in determining the final temperature of a certain mixture of objects originally at different temperatures:

$$500(0.22)(T_f - 20) + 120(T_f - 20) - 22(75 - T_f)$$

88. In studying the dispersion of light, the expression $(k + 2)[(n - jK)^2 - 1]$ is found. Expand this expression.

In Exercises 89 through 94 solve the stated problems by first setting up an appropriate equation.

89. The combustion of carbon usually results in the production of both carbon monoxide and carbon dioxide. If 500 kg of oxygen are available for combustion and it is desired that 9 times as much oxygen be converted to carbon dioxide as is converted to carbon monoxide, how much oxygen would be converted to each of the compounds?

90. The electric current in one transistor is three times that in another transistor. If the sum of the currents is 0.012 A, what is the current in each?

91. The cost of producing a first type of pocket calculator is three times the cost of producing a second type. The total cost of producing two of the first type and three of the second type is $45. What is the cost of producing each type?

92. An air sample contains 4 ppm (parts per million) of two harmful pollutants. The concentration of one is four times the other. What is the concentration of each?

93. Two cars, 735 mi apart, start toward each other. One travels at the rate of 55 mi/h and the other at 43 mi/h. When will they meet?

94. Fifty pounds of a cement-sand mixture is 40% sand. How many pounds of sand must be added for the resulting mixture to be 60% sand?

2

Functions and Graphs

2—1 Functions

In Chapter 1 we established many of the basic algebraic operations. At the end of the chapter we discussed the solution of equations, with applications to formulas. In most of the formulas, one quantity was given in terms of one or more other quantities. It is obvious, then, that the various quantities are related by means of the formula. We see that in the study of natural phenomena, a relationship can be found to exist between certain quantities.

If we were to perform an experiment to determine whether or not a relationship exists between the distance an object drops and the time it falls, observation of the results would indicate (approximately, at least) that $s = 16t^2$, where s is the distance in feet and t is the time in seconds. We would therefore see that distance and time for a falling object are related.

A similar study of the pressure and the volume of a gas at constant temperature would show that as pressure increases, volume decreases according to the formula $PV = k$, where k is a constant. Electrical measurements of current and voltage with respect to a particular resistor would show that $V = kI$, where V is the voltage, I is the current, and k is a constant.

Considerations such as these lead us to the mathematical concept of a **function**, which is one of the most important and basic concepts in mathematics. Thus, *whenever a relationship exists between two variables x and y such that for every value of x there is only one corresponding value of y, we say that y is a function of x.* Here we call the variable x the **independent variable** (since x can be chosen arbitrarily so long as the value chosen produces a real number for y). Also, y is the **dependent variable** (since, once the value of x is chosen, the value of y is determined—that is, y depends on x). We must also realize that x and y are only representative—other literal symbols may be used as independent and dependent variables.

There are many ways to express functions. Formulas, such as those we have discussed, define functions. Other ways to express functions are by means of tables, charts, and graphs.

Example A

In the equation $y = 2x$, we see that y is a function of x, since for each value of x there is only one value of y. For example, if $x = 3$, $y = 6$ and no other value. The dependent variable is y and the independent variable is x.

Example B

The volume of a cube of edge e is given by $V = e^3$. Here V is a function of e, since for each value of e there is one value of V. The dependent variable is V and the independent variable is e.

If the equation relating the volume and edge of a cube was written as $e = \sqrt[3]{V}$, that is, if the edge of a cube was expressed in terms of its volume, we would say that e is a function of V. In this case e would be the dependent variable and V the independent variable.

Example C

The power P developed in a certain resistor by a current I is given by $P = 4I^2$. Here P is a function of I. The dependent variable is P, and the independent variable is I.

Example D

If the formula in Example C is written as $I = \frac{1}{2}\sqrt{P}$, then I is a function of P. The independent variable is P, and the dependent variable is I. Even though P is the independent variable, it is restricted to values which are positive or zero. (This is written as $P \geq 0$.) Otherwise, the values of I would be imaginary. Except when specified, we shall restrict ourselves to real numbers.

Example E

For the equation $y = 2x^2 - 6x$, we say that y is a function of x. The dependent variable is y, and the independent variable is x. Some of the values of y corresponding to chosen values of x are given in the following table.

x	-2	-1	0	$\frac{1}{2}$	1	2	3	π	10
y	20	8	0	$-\frac{5}{2}$	-4	-4	0	$2\pi^2 - 6\pi$	140

For convenience of notation, the phrase "function of x" is written as $f(x)$. This, in turn, is a simplification of the statement that "y is a function of x," so we may now write $y = f(x)$. [Notice that $f(x)$ does *not* mean f times x. The symbols must be written in this form, and not separated, to indicate a function.]

Example F

If $y = 6x^3 - 5x$ we may say that y is a function of x, where this function $f(x)$ is $6x^3 - 5x$. We may also write $f(x) = 6x^3 - 5x$. It is common to write functions in this form, rather than in the form $y = 6x^3 - 5x$. However, y and $f(x)$ represent the same expression.

One of the most important uses of this notation is to designate the value of a function for a particular value of the independent variable. That is, for the expression "the value of the function $f(x)$ when $x = a$" we may write $f(a)$.

Example G

For the function $f(x) = 3x^2 - 5$, the value of $f(x)$ for $x = 2$ may be represented as $f(2)$. Thus, substituting 2 for x, we have

$$f(2) = 3(2^2) - 5 = 7$$

In the same way, the value of $f(x)$ for $x = -1$ is

$$f(-1) = 3(-1)^2 - 5 = -2$$

In certain instances we need to define more than one function of x. Then we use different symbols to denote the functions. For example, $f(x)$ and $g(x)$ may represent different functions of x, such as $f(x) = 5x^2 - 3$ and $g(x) = 6x - 7$. Special functions are represented by particular symbols. For example, in trigonometry we shall come across the "sine of the angle ϕ," where the sine is a function of ϕ. This is designated by $\sin \phi$.

Example H

If $f(x) = \sqrt{3x} + x$ and $g(x) = ax^4 - 5x$, then

$$f(3) = \sqrt{3(3)} + 3 = 3 + 3 = 6$$

and

$$g(3) = a(3^4) - 5(3) = 81a - 15$$

There are occasions when we wish to evaluate a function in terms of a literal number rather than an explicit number. However, whatever number a represents in $f(a)$, we substitute a for x in $f(x)$.

Example I

If $g(t) = 4t^2 - 5t$, to find $g(a^3)$ we substitute a^3 for t in the function. Thus,

$$g(a^3) = 4(a^3)^2 - 5(a^3) = 4a^6 - 5a^3$$

For the same function,

$$g(b + 1) = 4(b + 1)^2 - 5(b + 1)$$
$$= 4(b^2 + 2b + 1) - 5(b + 1)$$
$$= 4b^2 + 8b + 4 - 5b - 5$$
$$= 4b^2 + 3b - 1$$

Example J

The resistance of a particular resistor as a function of temperature is $R = 10 + 0.1T + 0.001T^2$. If a given temperature T is increased by 10°C, what is the value of R for the increased temperature as a function of the temperature T?

We are to determine R for a temperature of $T + 10$. Since $f(T) = 10 + 0.1T + 0.001T^2$, we know that

$$f(T + 10) = 10 + 0.1(T + 10) + 0.001(T + 10)^2$$
$$= 10 + 0.1T + 1 + 0.001T^2 + 0.02T + 0.1$$
$$= 11.1 + 0.12T + 0.001T^2$$

A function may be looked upon as a set of instructions. These instructions tell us how to obtain the value of the dependent variable for a particular value of the independent variable, even if the set of instructions is expressed in literal symbols.

Example K

The function $f(x) = x^2 - 3x$ tell us to "square the value of the independent variable, multiply the value of the independent variable by 3, and subtract the second result from the first." An analogy would be a computer which was so programmed that whan a number was fed into the computer, it would square the number and then subtract 3 times the value of the number. This is represented in diagram form in Fig. 2–1.

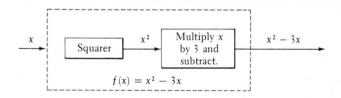

Figure 2–1

The functions $f(t) = t^2 - 3t$ and $f(n) = n^2 - 3n$ are the same as the function $f(x) = x^2 - 3x$, since the operations performed on the independent variable are the same. Although different literal symbols appear, this does not change the function.

As we mentioned earlier, we must be certain that the function is defined for any value of x that may be chosen. *Values of x which lead to division by zero or to imaginary values of y may not be chosen.* Example D and the following example illustrate this point.

Example L

The function $f(u) = \dfrac{u}{u - 4}$ is not defined if $u = 4$, since this would require division by zero. Therefore, the values of u are restricted to values other than 4.

The function $g(s) = \sqrt{3 - s}$ is not defined for real numbers if s is greater than 3, since such values would result in imaginary values for $g(s)$. Thus, the values of s are restricted to values equal to or less than 3.

Exercises 2–1

In Exercises 1 through 12 determine the appropriate functions.

1. Express the area A of a circle as a function of its radius r.
2. Express the area A of a circle as a function of its diameter d.
3. Express the circumference c of a circle as a function of its radius r.
4. Express the circumference c of a circle as a function of its diameter d.
5. Express the area A of a rectangle of width 5 as a function of its length l.
6. Express the volume V of a right circular cone of height 8 as a function of the radius r of the base.
7. Express the area A of a square as a function of its side s; express the side s of a square as a function of its area A.
8. Express the perimeter p of a square as a function of its side s; express the side s of a square as a function of its perimeter p.
9. A rocket weighs 3000 tons at liftoff. If the first-stage engines burn fuel at the rate of 10 tons per second, find the weight w of the rocket as a function of the time t while the first-stage engines operate.
10. A total of x ft is cut from a 24-ft board. Express the remaining length y as a function of x.
11. Express the simple interest I on \$200 at 4% per year as a function of the number of years t.
12. A taxi fare is 55¢ plus 10¢ for every $\frac{1}{5}$ mile traveled. Express the fare F as a function of the distance s traveled.

In Exercises 13 through 24 evaluate the given functions.

13. Given $f(x) = 2x + 1$, find $f(1)$ and $f(-1)$.
14. Given $f(x) = 5x - 9$, find $f(2)$ and $f(-2)$.
15. Given $f(x) = 5 - 3x$, find $f(-2)$ and $f(4)$.
16. Given $f(x) = 7 - 2x$, find $f(5)$ and $f(-4)$.
17. Given $f(n) = n^2 - 9n$, find $f(3)$ and $f(-5)$.
18. Given $f(v) = 2v^3 - 7v$, find $f(1)$ and $f(\frac{1}{2})$.
19. Given $\phi(x) = \dfrac{6 - x^2}{2x}$, find $\phi(1)$ and $\phi(-2)$.
20. Given $H(q) = \dfrac{8}{q} + 2\sqrt{q}$, find $H(4)$ and $H(16)$.

21. Given $g(t) = at^2 - a^2t$, find $g(-\frac{1}{2})$ and $g(a)$.

22. Given $s(y) = 6\sqrt{y} - 3$, find $s(9)$ and $s(a^2)$.

23. Given $K(s) = 3s^2 - s + 6$, find $K(-s)$ and $K(2s)$.

24. Given $T(t) = 5t + 7$, find $T(-2t)$ and $T(t + 1)$.

In Exercises 25 through 28 state the instructions of the function in words as in Example K.

25. $f(x) = x^2 + 2$ 26. $f(x) = 2x - 6$

27. $g(y) = 6y - y^3$ 28. $\phi(s) = 8 - 5s + s^5$

In Exercises 29 through 32 state any restrictions that might exist on the values of the independent variable.

29. $Y(y) = \dfrac{y + 1}{y - 1}$ 30. $G(z) = \dfrac{1}{(z - 4)(z + 2)}$

31. $F(y) = \sqrt{y - 1}$ 32. $X(x) = \dfrac{6}{\sqrt{1 - x}}$

In Exercises 33 through 40 solve the given problems.

33. The volume of a cylinder of height 6 cm as a function of the radius of the base is given by $V = 6\pi r^2$. What is the volume of the cylinder if the base has a radius of 3 cm?

34. The distance s that a freely falling body travels as a function of the time t is given by $s = 16t^2$, where s is measured in feet and t is measured in seconds. How far does an object fall in 2 s?

35. If the temperature within a certain refrigerator is maintained at 273 K (0°C), its *coefficient of performance* p as a function of the external temperature T (in kelvins) is

$$p = \frac{273}{T - 273}$$

What is its coefficient of performance at 308 K (35°C)?

36. The vertical distance of a point on a suspension cable from its lowest point as a function of the horizontal distance from the lowest point is given by

$$f(x) = \frac{x^4 + 600x^2}{2,000,000}$$

where x is measured in feet. How far above the lowest point is a point on the cable at a horizontal distance of 50 ft from the lowest point?

37. A piece of wire 60 in. long is cut into two pieces, one of which is bent into a circle and the other into a square. Express the total area of the two figures as a function of the perimeter of the square.

38. The net profit P made on selling 20 items, if each costs $15, as a function of the price p is $P = 20(p - 15)$. What is the profit if the price is $28? $12?

39. The voltage of a certain thermocouple as a function of the temperature is given by $E = 2.8T + 0.006T^2$. Find the voltage when the temperature is $T + h$.

40. The electrical resistance of a certain ammeter as a function of the resistance of the coil of the meter is

$$R = \frac{10R_c}{10 + R_c}$$

Find the function which represents the resistance of the meter if the resistance of the coil is doubled.

2–2 Rectangular Coordinates

One of the most valuable ways of representing functions is by graphical representation. By using graphs we are able to obtain a "picture" of the function, and by using this picture we can learn a great deal about the function.

To make a graphical representation, we recall that numbers can be represented by points on a line. For a function we have values of the independent variable and the corresponding values of the dependent variable. Therefore, it is necessary to use two lines to represent the values from these sets of numbers. We do this most conveniently by placing the lines perpendicular to each other.

We place one line horizontally and label it the **x-axis.** The numbers of the set for the independent variable are normally placed on this axis. The other line we place vertically, and label the **y-axis.** Normally the y-axis is used for values of the dependent variable. The point of intersection is called the **origin.** This is the **rectangular coordinate system.**

On the x-axis, positive values are to the right of the origin, and negative values are to the left of the origin. On the y-axis, positive values are above the origin, and negative values are below it. The four parts into which the plane is divided are called **quadrants,** which are numbered as in Fig. 2–2.

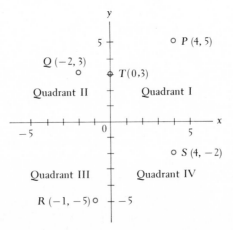

Figure 2–2

A point P in the plane is designated by the pair of numbers (x, y), where x is the value of the independent variable and y is the corresponding value of the dependent variable. *The x-value, called the* **abscissa,** *is the perpendicular distance of P from the y-axis. The y-value, called the* **ordinate,** *is the perpendicular distance of P from the x-axis.* The values x and y together, written as (x, y), are the **coordinates** of the point P.

Example A

The positions of points $P(4, 5)$, $Q(-2, 3)$, $R(-1, -5)$, $S(4, -2)$, and $T(0, 3)$ are shown in Fig. 2–2. Note that this representation allows for *one point for any pair of values* (x, y).

Example B

Three vertices of the rectangle in Fig. 2–3 are $A(-3, -2)$, $B(4, -2)$, and $C(4, 1)$. What is the fourth vertex?

We use the fact that opposite sides of a rectangle are equal and parallel to find the solution. Since both vertices of the base AB of the rectangle have a y-coordinate of -2, the base is parallel to the x-axis. Therefore, the top of the rectangle must also be parallel to the x-axis. Thus, the vertices of the top must both have a y-coordinate of 1, since one of them has a y-coordinate of 1. In the same way the x-coordinates of the left side must both be -3. Therefore, the fourth vertex is $D(-3, 1)$.

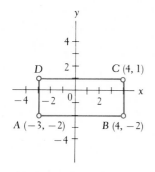

Figure 2–3

Example C

Where are all the points whose ordinates are 2?

All such points are 2 units above the x-axis; thus, the answer can be stated as "on a line 2 units above the x-axis."

Exercises 2−2

In Exercises 1 and 2 determine (at least approximately) the coordinates of the points specified in Fig. 2–4.

   1. A, B, C 2. D, E, F

In Exercises 3 and 4 plot (at least approximately) the given points.

   3. $A(2, 7)$, $B(-1, -2)$, $C(-4, 2)$ 4. $A(3, \frac{1}{2})$, $B(-6, 0)$, $C(-\frac{5}{2}, -5)$

In Exercises 5 and 6 plot the given points and then join these points, in the order given by straight-line segments. Name the geometric figure formed.

   5. $A(-1, 4)$, $B(3, 4)$, $C(1, -2)$ 6. $A(-5, -2)$, $B(4, -2)$, $C(6, 3)$, $D(-3, 3)$

In Exercises 7 and 8 find the indicated coordinates.

   7. Three vertices of a rectangle are $(5, 2)$, $(-1, 2)$, and $(-1, 4)$. What are the coordinates of the fourth vertex?

   8. Two vertices of an equilateral triangle are $(7, 1)$ and $(2, 1)$. What is the abscissa of the third vertex?

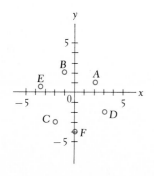

Figure 2–4

In Exercises 9 through 20 answer the given questions.

9. Where are all the points whose abscissas are 1?

10. Where are all the points whose ordinates are -3?

11. Where are all the points whose abscissas equal their ordinates?

12. Where are all the points whose abscissas equal the negative of their ordinates?

13. What is the abscissa of all points on the y-axis?

14. What is the ordinate of all points on the x-axis?

15. Where are all the points for which $x > 0$?

16. Where are all the points for which $y < 0$?

17. Where are all points (x, y) for which $x > 0$ and $y < 0$?

18. Where are all points (x, y) for which $x < 0$ and $y > 1$?

19. In which quadrants is the ratio y/x positive?

20. In which quadrants is the ratio y/x negative?

2–3 The Graph of a Function

Now that we have introduced the concepts of a function and the rectangular coordinate system, we are in a position to determine the graph of a function. In this way we shall obtain a visual representation of a function.

The graph of a function is the set of all points whose coordinates (x, y) satisfy the functional relationship $y = f(x)$. Since $y = f(x)$, we can write the coordinates of the points on the graph as $[x, f(x)]$. Writing the coordinates in this manner tells us exactly how to find them. We assume a certain value for x and then find the value of the function of x. These two numbers are the coordinates of the point.

Since there is no limit to the possible number of points which can be chosen, we normally select a few values of x, obtain the corresponding values of the function, and plot these points. These points are then connected by a *smooth* curve (not short straight lines from one point to the next), and are normally connected from left to right.

Example A

Graph the function $(3x - 5)$.

For purposes of graphing, we let y [or $f(x)$] $= 3x - 5$. We then let x take on various values and determine the corresponding values of y. Note that once we choose a given value of x, we have no choice about the corresponding y-value, as it is determined by evaluating the function. If $x = 0$ we find that $y = -5$. This means that the point $(0, -5)$ is on the graph of the function $3x - 5$. Choosing another value of x, for example, 1, we find that $y = -2$. This means that the point $(1, -2)$ is on the graph of the function $3x - 5$. Continuing to choose a few other values of x, we tabulate the results, as shown in Fig. 2–5. It is best to arrange the table so that the values of x increase; then there is no doubt how they are to be connected, for they are then connected in the order shown. Finally, we connect the points as shown in Fig. 2–5, and see that the graph of the function $3x - 5$ is a straight line.

x	y
-1	-8
0	-5
1	-2
2	1
3	4
4	7

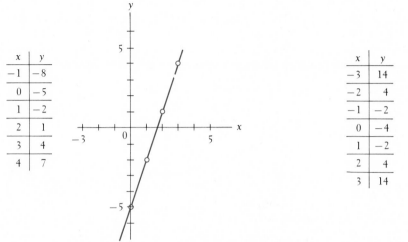

Figure 2–5

x	y
-3	14
-2	4
-1	-2
0	-4
1	-2
2	4
3	14

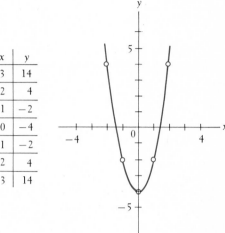

Figure 2–6

Example B

Graph the function $(2x^2 - 4)$.

First we let $y = 2x^2 - 4$ and tabulate the values as shown in Fig. 2–6. In determining the values in the table, we must take particular care to obtain the correct values of y for negative values of x. Mistakes are relatively common when dealing with negative numbers. We must carefully use the laws for signed numbers. For example, for $y = 2x^2 - 4$, if $x = -2$, we have $y = 2(-2)^2 - 4 = 2(4) - 4 = 8 - 4 = 4$. Once the values are obtained, we plot and connect the points with a smooth curve, as shown in Fig. 2–6.

There are some special points which should be noted. Since most common functions are smooth, any irregularities or sudden changes in the graph should be carefully checked. In these cases it usually helps to take values of x between those values where the question arises. Also, if the function is not defined for some value of x (remember, *division by zero is not defined, and only real values of the variables are permissible*), the function does not exist for that value of x. Finally, in applications, we must be careful to plot values that are meaningful; often negative values for quantities such as time are not physically meaningful. The following examples illustrate these points.

Example C
Graph the function $y = x - x^2$.

First we determine the values in the table as shown with Fig. 2–7. Again we must be careful with negative values of x. For $x = -1$, we have $y = (-1) - (-1)^2 = -1 - (+1) = -1 - 1 = -2$. Once all the values have been found and plotted, we note that $y = 0$ for both $x = 0$ and $x = 1$. The question arises—what happens between these values? Trying $x = \frac{1}{2}$, we find that $y = \frac{1}{4}$. Using this point completes the necessary information. Note that in plotting these graphs we do not stop the graph with the last point determined, but indicate that the curve continues.

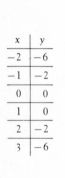

x	y
-2	-6
-1	-2
0	0
1	0
2	-2
3	-6

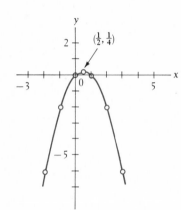

x	y
-4	$3/4$
-3	$2/3$
-2	$1/2$
-1	0
$-1/2$	-1
$-1/3$	-2
$1/3$	4
$1/2$	3
1	2
2	$3/2$
3	$4/3$
4	$5/4$

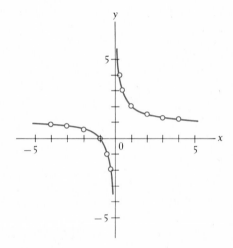

Figure 2–7   Figure 2–8

Example D

Graph the function $y = 1 + \dfrac{1}{x}$.

In finding the points on this graph as shown in Fig. 2–8, we note that y is not defined for $x = 0$. Thus we must be careful not to have any part of the curve cross the y-axis ($x = 0$). Although we cannot let $x = 0$, we can choose other values of x between -1 and 1 which are close to zero. In doing so, we find that as x gets closer to zero, the points get closer and closer to the y-axis, although they do not reach or touch it. In this case the y-axis is an **asymptote** of the curve.

Example E
Graph the function $y = \sqrt{x + 1}$.

When finding the points for the graph, we may not let x equal any negative value less than -1, for all such values of x lead to imaginary values for y. Also, since we have the positive square root indicated, all values of y are positive. See Fig. 2–9. Note that the graph stops at the point $(-1, 0)$.

x	y
−1	0
0	1
1	1.4
2	1.7
3	2
4	2.2
5	2.4
6	2.6
7	2.8
8	3

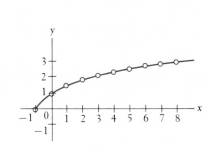

Figure 2–9

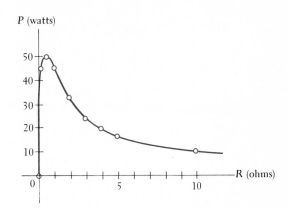

Figure 2–10

Example F

The power of a certain voltage source as a function of the load resistance is given by

$$P = \frac{100R}{(0.5 + R)^2}$$

where P is measured in watts and R in ohms. Plot the power as a function of the resistance.

Since a negative value for the resistance has no physical significance, we need to plot P for positive values of R only. The following table of values is obtained.

R(ohms)	0	0.25	0.50	1.0	2.0	3.0	4.0	5.0	10.0
P(watts)	0.0	44.4	50.0	44.4	32.0	24.5	19.8	16.5	9.1

The values 0.25 and 0.50 are used for R when it is found that P is less for $R = 2$ than for $R = 1$. In this way a smoother curve is obtained (see Fig. 2–10).

Empirical data may also be plotted in graphical form, although there may not be a formula connecting the values. When there is no possible formula which relates the sets of numbers, it is customary to connect the points with straight-line segments, merely to make them stand out better. The maximum temperature recorded weekly at a particular location would be an example of this. However, if it is reasonable that a functional relationship may exist, although it is unknown, the points may be connected by a smooth curve. Data from a physics experiment would be an example of this case. Illustrations of such cases are found in the exercises.

Functions of a particular type have graphs which are of a specific form, and many of them have been named. We noted that the graph of the function in Example A is a **straight line**. A **parabola** is illustrated in Example B and in Example C. The graph of the function in Example D is a **hyperbola**. Other types of graphs are found in the exercises and in later chapters. A more detailed analysis of several of these curves is found in Chapter 20. The use of graphs is extensive in mathematics and in nearly all areas of application. For example, in the section which follows we see how graphs can be used to solve equations. Many other uses of graphical methods appear in the later chapters.

Exercises 2−3

In Exercises 1 through 32 graph the given functions.

1. $y = 3x$
2. $y = -2x$
3. $y = 2x - 4$
4. $y = 3x + 5$
5. $y = 7 - 2x$
6. $y = 5 - 3x$
7. $y = \frac{1}{2}x - 2$
8. $y = 6 - \frac{1}{3}x$
9. $y = x^2$
10. $y = -2x^2$
11. $y = 3 - x^2$
12. $y = x^2 - 3$
13. $y = \frac{1}{2}x^2 + 2$
14. $y = 2x^2 + 1$
15. $y = x^2 + 2x$
16. $y = 2x - x^2$
17. $y = x^2 - 3x + 1$
18. $y = 2 + 3x + x^2$
19. $y = x^3$
20. $y = -2x^3$
21. $y = x^3 - x^2$
22. $y = 3x - x^3$
23. $y = x^4 - 4x^2$
24. $y = x^3 - x^4$

25. $y = \dfrac{1}{x}$
26. $y = \dfrac{1}{x + 2}$
27. $y = \dfrac{1}{x^2}$

28. $y = \dfrac{1}{x^2 + 1}$
29. $y = \sqrt{x}$
30. $y = \sqrt{4 - x}$

31. $y = \sqrt{16 - x^2}$
32. $y = \sqrt{x^2 - 16}$

In Exercises 33 through 40 graph the given functions. In each case plot the first mentioned variable as the ordinate and the second variable as the abscissa.

33. The velocity v (in feet per second) of an object under the influence of gravity, as a function of time t (in seconds), is given by $v = 100 - 32t$. If the object strikes the ground after 4 s, graph v as a function of t.

34. If $1000 is placed in an account earning 6% simple interest, the amount A in the account after t years is given by the function $A = 1000(1 + 0.06t)$. If the money is withdrawn from the account after 6 years, plot A as a function of t.

35. The surface area of a cube is given by $A = 6e^2$, where e is the side of the cube. Plot A as a function of e.

36. The energy in the electric field around an inductor is given by $E = \frac{1}{2}LI^2$, where I is the current in the inductor and L is the inductance. Plot E as a function of I (a) if $L = 1$ unit, (b) if $L = 0.1$ unit, and (c) if the E-axis is marked off in units of L.

37. The illuminance (in lumens per square meter) of a certain source of light as a function of the distance (in meters) from the source is given by $I = 400/r^2$. Plot I as a function of r.

38. The heat capacity (in joules per kilogram) of an organic liquid is related to the temperature (in degrees Celsius) by the equation $c_p = 2320 + 4.73T$ for the temperature range of $-40°C$ to $120°C$. Graphically show that this equation does not satisfy experimental data above $120°C$. Plot the graph of the equation and the following data points.

c_p	3030	3220	3350
T	140	160	200

39. If a rectangular tract of land has a perimeter of 600 m, its area as a function of its width is $A = 300w - w^2$. Plot A as a function of w.

40. The deflection y of a beam at a horizontal distance x from one end is given by $y = -k(x^4 - 30x^2 + 1000x)$, where k is a constant. Plot the deflection (in terms of k) as a function of x (in feet) if there are 10 ft between the end supports of the beam.

In Exercises 41 through 46 represent the data graphically.

41. The rainfall (in inches) in a certain city was recorded as follows.

Year	1970	1971	1972	1973	1974	1975	1976	1977	1978
Rainfall	35.4	36.7	40.4	40.2	38.2	32.8	33.4	41.2	40.4

42. The hourly temperatures (in degrees Fahrenheit) on a certain day were recorded as follows.

Hour	6 AM	7	8	9	10	11	12	1 PM	2	3	4
Temperature	16	18	20	25	32	36	39	41	39	42	36

43. The density (in kilograms per cubic meter) of water from 0° to 10°C is given in the following table.

Density	999.85	999.90	999.94	999.96	999.97	999.96
Temperature	0	1	2	3	4	5
Density	999.94	999.90	999.85	999.78	999.69	
Temperature	6	7	8	9	10	

44. The voltage (in volts) and current (in milliamperes) for a certain electrical experiment were measured as follows.

Voltage	10	20	30	40	50	60	70	80
Current	145	188	220	255	285	315	335	370

45. An experiment measuring the load (in kilograms) on a spring and the scale reading of the spring (in centimeters) produced the following results.

Load	0	1	2	3	4	5	6
Reading	7.0	7.6	8.2	8.8	9.4	10.1	12.8

46. The vapor pressure (in kilopascals) of ethane gas as a function of temperature (in degrees Celsius) is given by the following table.

Pressure	260	380	570	790	1060	1430	1840	2390	3270
Temperature	-70	-60	-50	-40	-30	-20	-10	0	15

2—4 Solving Equations Graphically

It is possible to solve equations by the use of graphs. This is particularly useful when algebraic methods cannot be applied conveniently to the equation. Before taking up graphical solutions, however, we shall briefly introduce the related concept of the zero of a function.

*Those values of the independent variable for which the function equals zero are known as the **zeros** of the function. To find the zeros of a given function, we must set the function equal to zero and solve the resulting equation.* Using functional notation, we may write this as $f(x) = 0$.

Example A

Find any zeros of the function $(3x - 9)$.

We write $f(x) = 3x - 9$ and then let $f(x) = 0$, which means that $3x - 9 = 0$. Thus we obtain the solution $x = 3$, which means that 3 is a zero of the function $(3x - 9)$.

Graphically, the zeros of a function may be found where the curve crosses the x-axis. These points are called the **x-intercepts** of the curve. The function is zero at these points since the x-axis represents all points for which y is zero, or $f(x) = 0$.

Example B

Graphically determine any zeros of the function $x^2 - 2x - 1$.

First we set $y = x^2 - 2x - 1$ and then find the points for the graph. After plotting the graph in Fig. 2–11, we see that the curve crosses the x-axis at approximately $x = -0.4$ and $x = 2.4$ (estimating to the nearest tenth). Thus the zeros of this function are approximately -0.4 and 2.4. Checking these values in the function, we have $f(-0.4) = -0.04$ and $f(2.4) = -0.04$, which shows that -0.4 and 2.4 are very close to the exact values.

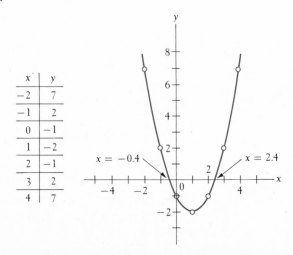

x	y
-2	7
-1	2
0	-1
1	-2
2	-1
3	2
4	7

Figure 2—11

Example C

Graphically determine any zeros of the function $x^2 + 1$.

Graphing the function $y = x^2 + 1$ in Fig. 2–12, we see that it does not cross the x-axis anywhere. Therefore, this function does not have any real zeros. We therefore can see that not all functions have real zeros.

We can now see how to solve equations graphically and how this is related to the zeros of a function. To solve an equation, we collect all terms on one side of the equals sign, giving the equation $f(x) = 0$. This equation is solved graphically by first setting $y = f(x)$ and graphing this function. We then find those values of x for which $y = 0$. This means that we are finding the zeros of this function. The following examples illustrate the method.

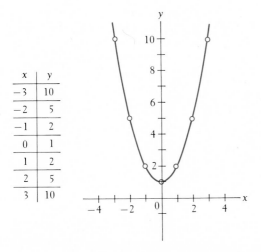

x	y
-3	10
-2	5
-1	2
0	1
1	2
2	5
3	10

Figure 2−12

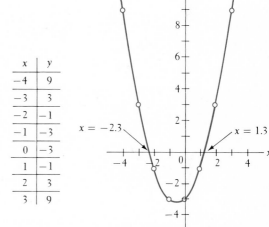

x	y
-4	9
-3	3
-2	-1
-1	-3
0	-3
1	-1
2	3
3	9

Figure 2−13

Example D

Solve the equation $3x = x(2 - x) + 3$ graphically.

We first collect algebraically all terms on the left side of the equals sign. This leads to the equation $x^2 + x - 3 = 0$. We then let $y = x^2 + x - 3$ and graph this function, as shown in Fig. 2–13. From the graph we see that $y = 0$ (which is equivalent to $x^2 + x - 3 = 0$) for approximately the values $x = -2.3$ and $x = 1.3$. Therefore, the solutions to the original equation are approximately $x = -2.3$ and $x = 1.3$. Checking these values in the original equation, we obtain $-6.9 = -6.89$ for $x = -2.3$ and $3.9 = 3.91$ for $x = 1.3$, which shows that the approximate solutions we obtained were very close to the exact solutions.

Example E

A box, whose volume is 30 in.^{3}, is made with a square base and a height which is 2 in. less than the length of a side of the base. To find the dimensions of the box we must solve the equation $x^2(x - 2) = 30$, where x is the length of a side of the base. (Verify the equation.) What are the dimensions of the box?

We are to solve the equation

$$x^2(x - 2) = 30$$

graphically. First we write the equation as $x^3 - 2x^2 - 30 = 0$. Now we set $y = x^3 - 2x^2 - 30$ and graph this equation as shown in Fig. 2–14. Only positive values of x are used since negative values of x have no meaning to the problem. From the graph we see that $x = 3.9$ is the approximate solution. Therefore, the approximate dimensions are 3.9 in., 3.9 in., and 1.9 in.

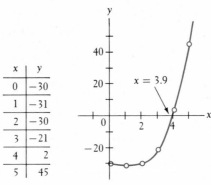

x	y
0	-30
1	-31
2	-30
3	-21
4	2
5	45

Figure 2–14

Exercises 2–4

In Exercises 1 through 4 find the zeros of the given functions algebraically as in Example A.

1. $5x - 10$ **2.** $7x - 4$ **3.** $4x + 9$ **4.** $2 - 3(x - 5)$

In Exercises 5 through 12 find the zeros of the given functions graphically. Check the solutions in Exercises 5 through 8 algebraically. Check the solutions in Exercises 9 through 12 by substituting in the function.

5. $2x - 7$ **6.** $3x - 2$ **7.** $5x - (3 - x)$ **8.** $3 - 2(2x - 7)$
9. $x^2 + x$ **10.** $2x^2 - x$ **11.** $x^2 - x + 3$ **12.** $x^2 + 3x - 5$

In Exercises 13 through 28 solve the given equations graphically.

13. $7x - 5 = 0$ **14.** $8x + 3 = 0$ **15.** $6x = 15$
16. $7x = -18$ **17.** $x^2 + x - 5 = 0$ **18.** $x^2 - 2x - 4 = 0$
19. $3x^2 + 2x = 2$ **20.** $2x^2 - x = 7$ **21.** $x(x - 4) = 9$
22. $x = 1 + (x + 2)^2$ **23.** $x^3 - 4x = 0$ **24.** $x^3 - 3x - 3 = 0$
25. $x^4 - 2x = 0$ **26.** $2x^5 - 5x = 0$

27. $y = \dfrac{1}{x^2 + 1}$ **28.** $x - 2 = \dfrac{1}{x}$

In Exercises 29 through 34 solve the given problems graphically.

29. Under certain conditions the velocity (in feet per second) of an object as a function of the time (in seconds) is given by the equation $v = 60 - 32t$. When is the velocity zero?

30. The perimeter of a field 50 m wide and l meters long is $p = 100 + 2l$. For what value of l is the perimeter 450 m?

31. Under given conditions a company finds that the profit p in producing x articles of a certain type is $p = 90x - x^2 - 1000$. For what values of x is the profit zero?

32. In the study of the velocities of nuclear fission fragments, it is found that under certain conditions the velocity would be zero if the expression $(1 + b)^2 - 3b$ were zero. For what values of b ($b > 0$), if any, is the velocity zero?

33. In order to find the distance x (in feet) from one end of a certain beam to the point where the deflection is zero, it is necessary to solve the equation $x^3 + x^2 - 5x = 0$. Determine where the deflection is zero.

34. If two electrical resistors, one of which is 1 Ω greater than the other, are placed in parallel, their combined resistance R_T as a function of the smaller resistance R is

$$R_T = \frac{R(R + 1)}{2R + 1}$$

What are the values of the resistors if $R_T = 10$ Ω?

2–5 Exercises for Chapter 2

In Exercises 1 through 4 determine the appropriate functions.

1. Find the volume of a right circular cylinder of height 8 ft as a function of the radius of the base.

2. Find the surface area A of a cube as a function of one of the edges e.

3. Find the temperature in degrees Fahrenheit as a function of degrees Celsius ($32°F = 0°C$ and $212°F = 100°C$).

4. Fencing 1000 m long is to be used to enclose three sides of a rectangular field. Express the area of the field as a function of its length (parallel to the fourth side).

In Exercises 5 through 12 evaluate the given functions.

5. Given $f(x) = 7x - 5$, find $f(3)$ and $f(-6)$.

6. Given $g(x) = 8 - 3x$, find $g(3)$ and $g(-4)$.

7. Given $F(u) = 3u - 2u^2$, find $F(-1)$ and $F(-3)$.

8. Given $h(y) = 2y^2 - y - 2$, find $h(5)$ and $h(-3)$.

9. Given $H(h) = \sqrt{1 - 2h}$, find $H(-4)$ and $H(2h)$.

10. Given $\phi(v) = \dfrac{3v - 2}{v + 1}$, find $\phi(-2)$ and $\phi(v + 1)$.

11. Given $f(x) = 3x^2 - 2x + 4$, find $f(x + h) - f(x)$.

12. Given $F(x) = x^3 + 2x^2 - 3x$, find $F(3 + h) - F(3)$.

In Exercises 13 through 24 graph the given functions.

13. $y = 4x + 2$　　　　　　　14. $y = 5x - 10$
15. $y = 4x - x^2$　　　　　　16. $y = x^2 - 8x - 5$
17. $y = 3 - x - 2x^2$　　　　18. $y = 6 + 4x + x^2$
19. $y = x^3 - 6x$　　　　　　20. $y = 3 - x^3$
21. $y = 2 - x^4$　　　　　　　22. $y = x^4 - 4x$

23. $y = \dfrac{x}{x + 1}$　　　　　　24. $y = \sqrt{25 - x^2}$

In Exercises 25 through 36 find any real zeros of the indicated functions by examining their graphs. Estimate the answer to the nearest tenth where necessary. Use the functions from Exercises 13 through 24 above as indicated.

25. Exercise 13　　　26. Exercise 14　　　27. Exercise 15
28. Exercise 16　　　29. Exercise 17　　　30. Exercise 18
31. Exercise 19　　　32. Exercise 20　　　33. Exercise 21
34. Exercise 22　　　35. Exercise 23　　　36. Exercise 24

In Exercises 37 through 44 solve the given equations graphically.

37. $7x - 3 = 0$　　　38. $2x + 11 = 0$　　　39. $x^2 + 1 = 6x$
40. $3x - 2 = x^2$　　41. $x^3 - x^2 = 2 - x$　　42. $5 - x^3 = 2x^2$

43. $\dfrac{1}{x} = 2x$　　　　44. $\sqrt{x} = 2x - 1$

In Exercises 45 through 48 answer the given questions.

45. Where are all the points (x, y) whose abscissas are 1 and for which $y > 0$?
46. Three vertices of a rectangle are $(6, 5)$, $(-4, 5)$, and $(-4, 2)$. What are the coordinates of the fourth vertex?
47..An equation which is found in electronics is

$$h = \frac{\alpha}{1 - \alpha}$$

Find h when $\alpha = 0.95$. That is, since $h = f(\alpha)$, find $f(0.95)$.

48. Under certain conditions, the distance s that an object is above the ground is given by $s = 96t - 16t^2$, where t is the time in seconds. When is the object 100 ft above the ground? Solve graphically.

In Exercises 49 through 58 plot the indicated functions.

49. A computer-leasing firm charges $150 plus $100 for every hour the computer is used during the month. What is the function relating the monthly charge C and the number of hours h that the computer is used? Plot a graph of the function for use up to 50 h.
50. There are 5000 L of oil in a tank which has a capacity of 100,000 L. It is filled at the rate of 7000 L/h. Determine the function relating the number of liters N and the time t while the tank is being filled. Plot N as a function of t.

51. A surveyor measuring the elevation of a point must consider the effect of the curvature of the earth. An approximate relation between the effect H (in feet) of this curvature, and the distance D (in miles) between the surveyor and the point of which he is measuring the elevation, is given by $H = 0.65D^2$. Plot H as a function of D.

52. The distance s (in feet) of an object above the ground as a function of the time t (in seconds) is $s = 100t - 16t^2$. Plot s as a function of t.

53. The resonant frequency f (in hertz) of a certain electric circuit is given by $f = 10,000/\sqrt{L}$, where L is the inductance (in henrys) in the circuit. Plot f as a function of L. (Take values of $L = 0.25, 0.49, 0.64, 1.00, 2.25$, and 4.00.)

54. The pneumatic resistance (in lb-s/ft^5) of a certain type of tubing is given by $R = 0.05/A^2$, where A is the cross-sectional area (in square feet). Plot R as a function of A. (Take values of A in the range of 0.001 to 0.01.)

55. The sales (in millions of dollars) of a certain corporation from 1970 through 1978 are shown in the following table.

Year	1970	1971	1972	1973	1974	1975	1976	1977	1978
Sales	12	14	17	18	20	24	32	35	39

56. The amplitude (in centimeters) of a certain pendulum, measured after each swing, as a function of time (in seconds), is given by the following table. Plot the graph of amplitude as a function of time.

Amplitude	5.0	2.8	1.6	0.9	0.5
Time	0.0	1.2	2.4	3.6	4.8

57. The length (in inches) of a brass rod is measured as a function of the temperature (in degrees Celsius). From the following table, plot the length as a function of temperature (make your scale meaningful).

Length	100.0	100.2	100.4	100.7	101.0	101.2
Temperature	0	100	200	300	400	500

58. The solubility (in kilograms of solute per cubic meter of water) of potassium nitrate as a function of temperature is given in the following table. Plot solubility as a function of temperature (in degrees Celsius).

Solubility	133	209	316	458	639	855	1100	1690
Temperature	0	10	20	30	40	50	60	80

3

The Trigonometric Functions

3–1 Introduction; Angles

The solution to a great many kinds of applied problems involves the use of triangles, especially right triangles. Problems which can be solved by the use of triangles include the determination of distances which cannot readily be determined directly, such as the widths of rivers and distances between various points in the universe. Also, problems involving forces and velocities lend themselves readily to triangle solution. Even certain types of electric circuits are analyzed by the use of triangles.

The basic properties of triangles will allow us to set up some very basic and useful functions involving their sides and angles. These functions are very important to the development of mathematics and to its applications in most fields of technology.

Thus we come to the study of **trigonometry,** the literal meaning of which is "triangle measurement." In this chapter we shall introduce the basic trigonometric functions and some of their elementary applications. Later chapters will consider additional topics in trigonometry. We shall begin our study by considering the concept of an angle.

An **angle** is defined as being *generated* by rotating a *half*-line about its endpoint from an initial position to a terminal position. A half-line is that portion of a line to one side of a fixed point on the line. We call the initial position of the half-line the **initial side,** and the terminal position of the half-line the **terminal side.** The fixed point is the **vertex.** The angle itself is the amount of rotation from the initial side to the terminal side.

If the rotation of the terminal side from the initial side is counterclockwise, the angle is said to be positive. If the rotation is clockwise the angle is negative. In Fig. 3–1, angle 1 is positive and angle 2 is negative.

There are many symbols used to designate angles. Probably the most widely used are certain Greek letters such as θ (theta), ϕ (phi), α (alpha), and β (beta).

Two measurements of angles are widely used. These are the **degree** and the **radian.** *A degree is defined as $\frac{1}{360}$ of a complete rotation.* The radian will be discussed in Chapter 7. The degree is divided into 60 equal parts called **minutes,** and each minute is divided into 60 equal parts called **seconds.** The symbols °, ′, and ″ are used to designate degrees, minutes, and seconds, respectively. Decimal parts of angles are also common, particularly with the use of electronic calculators.

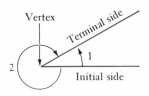

Vertex
Terminal side
1
2
Initial side

Figure 3–1

Example A

The angles $\theta = 30°$, $\phi = 140°$, $\alpha = 240°$, and $\beta = -120°$ are shown in Fig. 3–2.

In Fig. 3–2 we note that angles α and β have the same initial and terminal sides. Such angles are called **coterminal angles.** An understanding of coterminal angles is important in certain concepts in trigonometry.

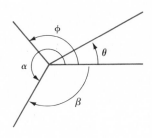

ϕ
θ
α
β

Figure 3–2

Example B

Determine the values of two angles which are coterminal with an angle of 145°32′.

Since there are 360° in a complete rotation, we can find one coterminal angle by considering the angle which is 360° larger than the given angle. This gives us an angle of 505°32′. Another method of finding a coterminal angle is to subtract 145°32′ from 360°, and then consider the resulting angle to be negative. This means that the original angle and the derived angle would make up one complete rotation, when put together. This method leads us to the angle of −214°28′ (see Fig. 3–3). These methods could be employed repeatedly to find other coterminal angles.

It is important to be able to change an angle expressed in degrees and minutes to an angle expressed in degrees and decimal parts of a degree, and conversely. One reason for this is that it has been very common to use degrees and minutes in tables, and most electronic calculators use degrees and decimal parts. The method of making these changes is illustrated in the following examples.

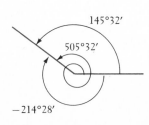

145°32′
505°32′
−214°28′

Figure 3–3

62   The Trigonometric Functions

Example C

To change the angle 43°24′ to decimal form, we use the fact that
$1' = (\frac{1}{60})°$. Therefore, $24' = (\frac{24}{60})° = 0.4°$. Therefore,

$$43°24' = 43.4°$$

Also, 17°53′ is changed to decimal form as follows.

$$53' = \left(\frac{53}{60}\right)° = 0.88°$$

Thus, 17°53′ = 17.88° (to the nearest hundredth), or 17°53′ = 17.9°
(to the nearest tenth). Here the results have been **rounded off**. (See
Appendix B.)

Example D

To change 154.36° to an angle measured to the nearest minute, we use
the fact that 1° = 60′. Therefore,

$$0.36° = 0.36(60') = 21.6'$$

To the nearest minute, we have 154.36° = 154°22′.

*If the initial side of the angle is the positive x-axis and the vertex is at
the origin, the angle is said to be in* **standard position.** The angle is then
determined by the position of the terminal side. If the terminal side is
in the first quadrant, the angle is called a "first-quadrant angle." Similar
terms are used when the terminal side is in the other quadrants. *If the
terminal side coincides with one of the axes, the angle is a* **quadrantal
angle.** When an angle is in standard position, the terminal side can be
determined if we know any point, other than the origin, on the terminal
side.

Example E

To draw a third-quadrant angle of 205°, we simply measure 205° from
the positive x-axis and draw the terminal side. See angle α in Fig. 3–4.

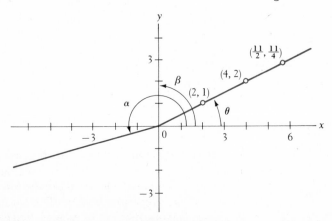

Figure 3–4

Also in Fig. 3–4, θ is in standard position and the terminal side is uniquely determined by knowing that it passes through the point (2, 1). The same terminal side passes through (4, 2) and $(\frac{11}{2}, \frac{11}{4})$, among other points. Knowing that the terminal side passes through any one of these points makes it possible to determine the terminal side.

Angle β in Fig. 3–4 is a quadrantal angle since its terminal side is the positive *y*-axis.

Exercises 3–1

In Exercises 1 through 4 draw the given angles.

1. 60°, 120°, −90°   2. 330°, −150°, 450°
3. 50°, −360°, −30°   4. 45°, 225°, −250°

In Exercises 5 through 12 determine one positive and one negative coterminal angle for each angle given.

5. 45°   6. 73°   7. 150°   8. 162°
9. 70°30′   10. 153°47′   11. 278.1°   12. 197.6°

In Exercises 13 through 20 change the given angles to equal angles expressed in decimal form. In Exercises 17 through 20 round off results to hundredths.

13. 15°12′   14. 246°48′   15. 86°3′   16. 157°39′
17. 301°16′   18. 4°47′   19. 96°8′   20. 38°28′

In Exercises 21 through 28 change the given angles to equal angles expressed to the nearest minute.

21. 47.5°   22. 315.8°   23. 19.75°   24. 84.55°
25. 5.62°   26. 238.21°   27. 24.92°   28. 142.87°

In Exercises 29 through 32 draw angles in standard position such that the terminal side passes through the given point.

29. (4, 2)   30. (−3, 8)   31. (−3, −5)   32. (6, −1)

3–2 Defining the Trigonometric Functions

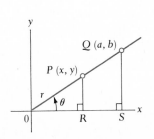

Figure 3–5

Let us place an angle θ in standard position and drop perpendicular lines from points on the terminal side to the *x*-axis as shown in Fig. 3–5. In doing this we set up similar triangles, each with one vertex at the origin. *Using the basic fact from geometry that corresponding sides of similar triangles are proportional, we may set up equal ratios of corresponding sides.*

There are three important distances for a given point on the terminal side of the angle involved in these ratios. *They are the* **abscissa** *(x-value), the* **ordinate** *(y-value), and the* **radius vector** *(the distance from the origin to the point).* The radius vector is denoted as *r*.

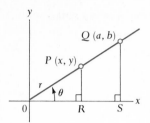

Figure 3–5

Example A

In Fig. 3–5 triangles *OPR* and *OQS* are similar. Therefore, the ratio of the abscissa (*x*-value) to the ordinate (*y*-value) of points *P* and *Q* is the same in each triangle. We can state this as

$$\frac{x}{y} = \frac{a}{b}$$

For any other point on the terminal side of θ that we might choose, the ratio of its abscissa to ordinate would still be the same as those already given.

For any angle θ in standard position, there are six different ratios which may be set up. Because of the property of similar triangles, these ratios are the same, regardless of the point on the terminal side which is chosen. For a different angle, with a different terminal side, the ratios have different values. Thus, the ratios depend on the position of the terminal side, which means that the ratios depend on the angle. In this way we see that *the ratios are functions of the angle. These functions are called the* **trigonometric functions,** *and are defined as follows:*

$$\text{sine } \theta = \frac{\text{ordinate of } P}{\text{radius vector of } P} = \frac{y}{r},$$

$$\text{cosine } \theta = \frac{\text{abscissa of } P}{\text{radius vector of } P} = \frac{x}{r},$$

$$\text{tangent } \theta = \frac{\text{ordinate of } P}{\text{abscissa of } P} = \frac{y}{x},$$

$$\text{cotangent } \theta = \frac{\text{abscissa of } P}{\text{ordinate of } P} = \frac{x}{y},$$

$$\text{secant } \theta = \frac{\text{radius vector of } P}{\text{abscissa of } P} = \frac{r}{x},$$

$$\text{cosecant } \theta = \frac{\text{radius vector of } P}{\text{ordinate of } P} = \frac{r}{y}$$

(3–1)

In this chapter we shall restrict our attention to the trigonometric functions of acute angles. However, it should be emphasized that the definitions in Eqs. (3–1) are general, and may be used for angles of any magnitude. Discussion of the trigonometric functions of angles in general, along with additional important properties, is found in Chapters 7 and 19.

For convenience, the names of the various functions are usually abbreviated to sin θ, cos θ, tan θ, cot θ, sec θ, and csc θ. Note that a given function is not defined if the denominator is zero. The denominator is zero in tan θ and sec θ for $x = 0$, and in cot θ and csc θ for $y = 0$. In all cases we will assume that $r > 0$. If $r = 0$ there would be no terminal side and therefore no angle.

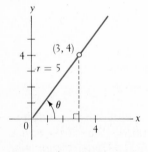

Figure 3–6

Example B

Determine the trigonometric functions of the angle with a terminal side passing through the point (3, 4).

By placing the angle in standard position, as shown in Fig. 3–6, and drawing the terminal side through (3, 4), we find that $r = 5$ (by use of the Pythagorean theorem). Using the values $x = 3$, $y = 4$, and $r = 5$, we find that

$$\sin \theta = \frac{4}{5}, \qquad \cos \theta = \frac{3}{5}$$

$$\tan \theta = \frac{4}{3}, \qquad \cot \theta = \frac{3}{4}$$

$$\sec \theta = \frac{5}{3}, \qquad \csc \theta = \frac{5}{4}$$

If one of the trigonometric functions is known, we can determine the other functions of the angle, using the Pythagorean theorem and the definitions of the functions. The following example illustrates the method.

Example C

If we know that $\sin \theta = \frac{3}{7}$ and that θ is a first quadrant angle, we know that the ratio of the ordinate to the radius vector (of y to r) is 3 to 7. Therefore, the point on the terminal side for which y is 3 can be determined by use of the Pythagorean theorem. The x-value for this point is

$$x = \sqrt{7^2 - 3^2} = \sqrt{49 - 9} = \sqrt{40} = 2\sqrt{10}$$

Therefore, the point $(2\sqrt{10},\ 3)$ is on the terminal side as shown in Fig. 3–7.

Therefore, using the values $x = 2\sqrt{10}$, $y = 3$, and $r = 7$ we have the other trigonometric functions of θ. They are

$$\cos \theta = \frac{2\sqrt{10}}{7}, \qquad \tan \theta = \frac{3}{2\sqrt{10}}, \qquad \cot \theta = \frac{2\sqrt{10}}{3},$$

$$\sec \theta = \frac{7}{2\sqrt{10}}, \qquad \csc \theta = \frac{7}{3}$$

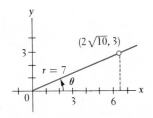

Figure 3−7

Exercises 3−2

In Exercises 1 through 8 determine the trigonometric functions of the angles whose terminal sides pass through the given points.

1. $(4, 3)$ 2. $(5, 12)$ 3. $(15, 8)$ 4. $(7, 24)$
5. $(1, \sqrt{15})$ 6. $(1, 1)$ 7. $(2, 5)$ 8. $(1, \frac{1}{2})$

In Exercises 9 through 16 use the given trigonometric functions to find the indicated trigonometric functions.

9. $\tan \theta = 1$, find $\sin \theta$ and $\sec \theta$ 10. $\sin \theta = \frac{1}{2}$, find $\cos \theta$ and $\csc \theta$
11. $\cos \theta = \frac{2}{3}$, find $\tan \theta$ and $\cot \theta$ 12. $\sec \theta = 3$, find $\tan \theta$ and $\sin \theta$
13. $\sin \theta = 0.7$, find $\cot \theta$ and $\csc \theta$ 14. $\cos \theta = \frac{5}{12}$, find $\sec \theta$ and $\tan \theta$
15. $\cot \theta = 0.25$, find $\cos \theta$ and $\csc \theta$ 16. $\csc \theta = 1.2$, find $\cos \theta$ and $\cot \theta$

In Exercises 17 through 20 each of the listed points is on the terminal side of an angle. Show that each of the indicated functions is the same for each of the points.

17. $(3, 4)$, $(6, 8)$, $(4.5, 6)$, $\sin \theta$ and $\tan \theta$
18. $(5, 12)$, $(15, 36)$, $(7.5, 18)$, $\cos \theta$ and $\cot \theta$
19. $(2, 1)$, $(4, 2)$, $(8, 4)$, $\tan \theta$ and $\sec \theta$
20. $(3, 2)$, $(6, 4)$, $(9, 6)$, $\csc \theta$ and $\cos \theta$

In Exercises 21 and 22 answer the given questions.

21. From the definitions of the trigonometric functions, it can be seen that $\csc \theta$ is the reciprocal of $\sin \theta$. What function is the reciprocal of $\cos \theta$? of $\cot \theta$?

22. Divide the expression for $\sin \theta$ by that for $\cos \theta$. Is the result the expression for any of the other functions?

3–3 Values of the Trigonometric Functions

We have been able to calculate the trigonometric functions if we knew one point on the terminal side of the angle. However, in practice it is more common to know the angle in degrees, for example, and to be required to find the functions of this angle. Therefore, we must be able to determine the trigonometric functions of angles in degrees.

One way to determine the functions of a given angle is to make a scale drawing. That is, we draw the angle in standard position using a protractor, and then measure the lengths of the values of x, y, and r for some point on the terminal side. By using the proper ratios we may determine the functions of this angle.

We may also use certain geometric facts to determine the functions of some particular angles. The following two examples illustrate this procedure.

Example A

From geometry we know that the side opposite a 30° angle in a right triangle is one-half the hypotenuse. By using this fact and letting $y = 1$ and $r = 2$ (see Fig. 3–8), we determine that $\sin 30° = \frac{1}{2}$. Also, by use of the Pythagorean theorem we determine that $x = \sqrt{3}$. Therefore, $\cos 30° = \frac{\sqrt{3}}{2}$ and $\tan 30° = \frac{1}{\sqrt{3}}$. In a similar way we may determine the values of the functions of 60° to be as follows: $\sin 60° = \frac{\sqrt{3}}{2}$, $\cos 60° = \frac{1}{2}$, and $\tan 60° = \sqrt{3}$.

Example B

Determine $\sin 45°$, $\cos 45°$, and $\tan 45°$.

From geometry we know that in an isosceles right triangle the angles are 45°, 45°, and 90°. We know that the sides are in proportion 1, 1,

$\sqrt{2}$, respectively. Putting the 45° angle in standard position, we find $x = 1$, $y = 1$, and $r = \sqrt{2}$ (see Fig. 3–9). From this we determine

$$\sin 45° = \frac{1}{\sqrt{2}}, \quad \cos 45° = \frac{1}{\sqrt{2}}, \quad \text{and} \quad \tan 45° = 1$$

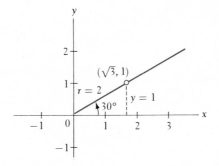

Figure 3–8

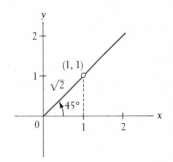

Figure 3–9

Summarizing the results for 30°, 45°, and 60°, we have the following table.

				(decimal approximations)		
θ	30°	45°	60°	30°	45°	60°
$\sin \theta$	$\dfrac{1}{2}$	$\dfrac{1}{\sqrt{2}}$	$\dfrac{\sqrt{3}}{2}$	0.500	0.707	0.866
$\cos \theta$	$\dfrac{\sqrt{3}}{2}$	$\dfrac{1}{\sqrt{2}}$	$\dfrac{1}{2}$	0.866	0.707	0.500
$\tan \theta$	$\dfrac{1}{\sqrt{3}}$	1	$\sqrt{3}$	0.577	1.000	1.732

The scale-drawing method is only approximate, and the geometric methods work only for certain angles. However, the values of the functions may be determined through more advanced methods (using calculus and what are known as power series).

We now refer to the tables of trigonometric functions presented in Appendix E. Since measuring angles in degrees and minutes or in degrees and decimal parts is common, a table of each type is given. Table 3 gives values of the trigonometric functions to each 10′, and Table 4 gives values to each 0.1°.

To obtain a value of a function from either table, we note that the angles from 0° to 45° are listed in the left-hand column and are read down. The angles from 45° to 90° are listed on the right-hand side and are read up. The functions for angles from 0° to 45° are listed along the top, and those for the angles from 45° to 90° are listed along the bottom.

Example C
Using Table 3, find sin 34°0′ and cos 72°0′.

Sin 34°0′ is found under sin θ (at top) and to the right of 34°00′ (at left). Thus, sin 34°0′ = 0.5592.

Cos 72°0′ is found over cos θ (at bottom) and to the left of 72°00′ (at right). Thus, cos 72°0′ = 0.3090.

Example D
Using Table 3, find tan 42°20′ and sin 64°40′.

Tan 42°20′ is found under tan θ (at top) and to the right of 20′ (which appears under 42°). Tan 42°20′ = 0.9110.

Sin 64°40′ is found over sin θ (at bottom) and to the left of 40′ (which appears *over* 64°). Sin 64°40′ = 0.9038.

Example E
Using Table 4, find cos 8.0° and cot 74.2°.

Cos 8.0° is found under cos θ (at top) and to the right of 8.0 (at left). Thus, cos 8.0° = 0.9903.

Cot 74.2° is found over cot θ (at bottom) and to the left of 0.2 (which appears above 74.0). Thus, cot 74.2° = 0.2830.

Not only are we able to find values of the functions if we know the angle, but also we can find the angle if we know the value of a function. This requires that we reverse the procedures mentioned above. In doing this, we are actually using another important type of mathematical function, an **inverse trigonometric function.** These are discussed in some detail in Chapter 19. For calculator purposes at this point, it is sufficient to recognize the notation which is used. The notation for "the angle whose sine is x" is Sin⁻¹ x or Arcsin x. Similar meanings are given to Cos⁻¹ x, Arccos x, Tan⁻¹ x, and Arctan x.

Example F
Given that sin θ = 0.2616, find θ to the nearest 10′.

We look for 0.2616, or the nearest number to it, in the columns for sin θ in Table 3. Since this number appears exactly under sin θ, we conclude that θ = 15°10′.

Example G
Given that tan θ = 2.375, find θ to the nearest 0.1°.

We look for 2.375, or the nearest number to it, in the columns for tan θ in Table 4. Since the nearest number which appears is 2.379, and this appears over tan θ, we conclude that θ = 67.2° to the nearest tenth of a degree.

There are occasions when it is necessary to use angles expressed to an accuracy greater than 10′ or 0.1°. This greater accuracy can be obtained by use of tables which give five or more digits for the values of the functions, or by scientific electronic calculators. It is also possible to obtain greater accuracy from the tables in Appendix E.

Using Table 3 we can obtain values of the trigonometric functions for angles expressed to the nearest minute. We can also find an angle to the nearest minute for a given value of a function. Since this table shows angles only to the nearest 10′, it is necessary to use **linear interpolation** for angles, expressed to the nearest minute, which are not listed directly.

Linear interpolation assumes that if a particular angle lies between two of those listed in the table, then the functions of that angle are at the same proportional distance between the functions listed. This assumption is not strictly correct, although it is a very good approximation.

Example H

To find sin 23°27′ we must interpolate between sin 23°20′ and 23°30′. Since 27′ is $\frac{7}{10}$ of the way between 20′ and 30′, we shall assume that sin 23°27′ is $\frac{7}{10}$ of the way between sin 23°20′ and sin 23°30′. These values are 0.3961 and 0.3987. The **tabular difference** between them is 26, and $\frac{7}{10}$ of this is 18. Adding 0.0018 to 0.3961, we obtain sin 23°27′ = 0.3979. Another method of indicating the interpolation is shown in Fig. 3–10. From the figure we see that

$$\frac{7}{10} = \frac{x}{26}$$

$$10x = 182$$
$$x = 18.2$$

Since we want to determine the value of the function to four decimal places (the limit of accuracy of the table), we must round off the value of x to the nearest integer. Therefore, $x = 18$, and

$$\sin 23°27′ = 0.3961 + 0.0018 = 0.3979$$

Example I

To find cos 76°14′, we first determine that we want the value $\frac{4}{10}$ of the way from cos 76°10′ to cos 76°20′. These values are 0.2391 and 0.2363, and their tabular difference is 28. Four-tenths of this is 11 (to the nearest unit). Thus, subtracting this (the value of cos 76°10′ is greater than cos 76°20′—the values of cos θ get *smaller* as θ gets larger) from 0.2391, we get cos 76°14′ = 0.2380. Again we can indicate the interpolation, as in Fig. 3–11. From the figure we see that $\frac{4}{10} = \frac{x}{28}$, or $x = 11$.

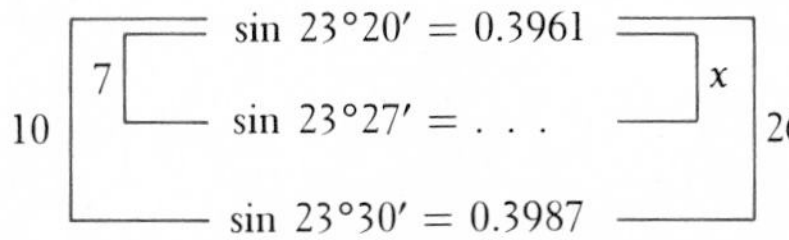

10 ⎰ 7 ⎱ sin 23°20′ = 0.3961 ⎰ x ⎱ 26
sin 23°27′ = . . .
sin 23°30′ = 0.3987

10 ⎰ 4 ⎱ cos 76°10′ = 0.2391 ⎰ x ⎱ 28
cos 76°14′ = . . .
cos 76°20′ = 0.2363

Figure 3—10 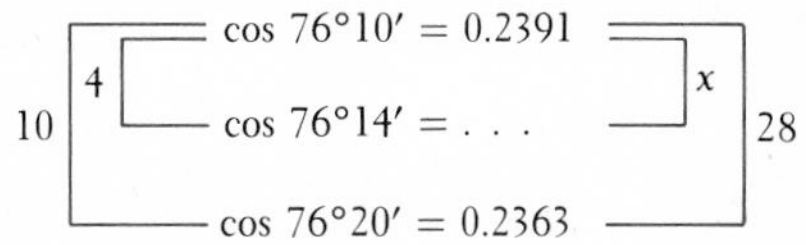   Figure 3—11

Example J

Given that cos θ = 0.8811, find θ to the nearest minute.

We find that this number lies between cos 28°10′ and cos 28°20′. These values are 0.8816 and 0.8802. The tabular difference between cos θ and cos 28°10′ is 5, and tabular difference between the two values given in Table 3 is 14. Thus, cos θ is assumed to be $\frac{5}{14}$ of the way from the first to the second. To the nearest tenth, this is $\frac{4}{10}$. Hence, θ = 28°14′ (to the nearest ′). The solution of this type of problem can also be indicated by means of a figure such as Fig. 3–12, from which we see that $\frac{5}{14} = \frac{x}{10}$, or x = 4.

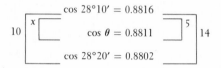

$$\begin{array}{c}
\text{cos } 28°10′ = 0.8816 \\
10 \quad x\ \boxed{} \quad \text{cos } \theta = 0.8811 \quad \boxed{}\ 5 \quad 14 \\
\text{cos } 28°20′ = 0.8802
\end{array}$$

Figure 3–12

If an electronic calculator is used, it may not be necessary to refer to the tables to obtain values of the trigonometric functions, or to use interpolation to obtain the additional accuracy. However, it should be noted that in this regard the calculator is another source from which these values are available, and does not eliminate the need to understand how these values are used. Also, although we have used interpolation only with Table 3 in the examples of this section, it is a method which can be used with a great many tables, including Table 4 and those for which values may not be available on a calculator. This is illustrated in the exercises which follow.

Exercises 3–3

In Exercises 1 through 4 use a protractor to draw the given angle. Measure off 10 units (centimeters are convenient) along the radius vector. Then measure the corresponding values of x and y. From these values determine the trigonometric functions of the angle.

1. 40°   2. 75°   3. 15°   4. 53°

In Exercises 5 through 12 find the value of each of the trigonometric functions from Table 3 in Appendix E.

5. sin 19°0′   6. cos 43°0′   7. tan 67°0′   8. cot 76°0′
9. cos 22°20′   10. tan 34°50′   11. sec 56°30′   12. csc 52°10′

In Exercises 13 through 20 find the value of each of the trigonometric functions from Table 4 in Appendix E.

13. cos 26.0°   14. cot 33.0°   15. sin 75.6°   16. cos 48.1°
17. tan 7.4°   18. sin 44.6°   19. cot 49.3°   20. tan 78.9°

In Exercises 21 through 28 use Table 3 to find θ to the nearest 10′ for each of the given trigonometric functions.

21. $\tan \theta = 0.8441$ **22.** $\sin \theta = 0.3854$ **23.** $\cos \theta = 0.4718$

24. $\sec \theta = 1.264$ **25.** $\csc \theta = 1.409$ **26.** $\tan \theta = 1.523$

27. $\sin \theta = 0.9175$ **28.** $\cos \theta = 0.1260$

In Exercises 29 through 36 use Table 4 to find θ to the nearest 0.1° for each of the given trigonometric functions.

29. $\sin \theta = 0.2385$ **30.** $\cot \theta = 1.819$ **31.** $\tan \theta = 4.870$

32. $\cos \theta = 0.1515$ **33.** $\cos \theta = 0.6800$ **34.** $\sin \theta = 0.9788$

35. $\cot \theta = 0.8433$ **36.** $\tan \theta = 1.926$

In Exercises 37 through 44 find the value of each of the trigonometric functions from Table 3, using interpolation.

37. $\tan 28°56′$ **38.** $\cos 48°44′$ **39.** $\sin 63°15′$ **40.** $\sec 71°47′$

41. $\cot 7°8′$ **42.** $\csc 12°14′$ **43.** $\cos 65°49′$ **44.** $\sin 57°52′$

In Exercises 45 through 52 use Table 3 and interpolation to find θ to the nearest 1′ for each of the given trigonometric functions.

45. $\cos \theta = 0.2960$ **46.** $\tan \theta = 0.1086$ **47.** $\sin \theta = 0.5755$

48. $\csc \theta = 1.168$ **49.** $\sec \theta = 1.289$ **50.** $\tan \theta = 0.6539$

51. $\cot \theta = 0.8070$ **52.** $\sin \theta = 0.9789$

In Exercises 53 through 56 additional problems involving interpolation are illustrated. In Exercises 53 and 54 use interpolation with Table 4 to evaluate the given trigonometric functions. In Exercises 55 and 56 use interpolation with the given table of values of the temperature of a cooling object as a function of time to obtain the required values.

53. $\tan 37.17°$ **54.** $\cos 58.72°$

Temperature (degrees Celsius)	150.0	145.6	141.6	137.9	134.6	131.6
Time (minutes)	0.0	1.0	2.0	3.0	4.0	5.0

55. Find T for $t = 1.3$ min.   **56.** Find T for $t = 3.8$ min.

In Exercises 57 through 60 solve the given problems.

57. When a projectile is fired into the air, its horizontal velocity v_x is related to the velocity v with which it is fired by the relation $v_x = v(\cos \theta)$, where θ is the angle between the horizontal and the direction in which it is fired. What is the horizontal velocity of a projectile fired with velocity 200 ft/s at an angle of 36°0′ with respect to the horizontal?

58. The coefficient of friction between an object moving down an inclined plane with constant speed and the plane equals the tangent of the angle that the plane makes with the horizontal. If the coefficient of friction between a metal block and a wooden plane is 0.150, at what angle (to the nearest tenth of a degree) is the plane inclined?

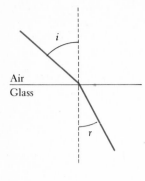

Air
Glass

Figure 3–13

59. The voltage at any instant in a coil of wire which is turning in a magnetic field is given by $E = E_{\max} (\cos \alpha)$, where $E_{\max}$ is the maximum voltage and α is the angle the coil makes with the field. If the maximum voltage produced by a certain coil is 120 V, what voltage is generated when the coil makes an angle of 55°35′ with the field?

60. When a light ray enters glass from the air it bends somewhat toward a line perpendicular to the surface. The *index of refraction* of the glass is defined as

$$n = \frac{\sin i}{\sin r}$$

where i is the angle between the perpendicular and the ray in the air and r is the angle between the perpendicular and the ray in the glass. A typical case for glass is $i = 59.0°$ and $r = 34.5°$. Find the index of refraction for this case. See Fig. 3–13.

3–4 The Right Triangle

From geometry we know that a triangle, by definition, consists of three sides and has three angles. If one side and any other two of these six parts of the triangle are known, it is possible to determine the other three parts. One of the three known parts must be a side, for if we know only the three angles, we can conclude only that an entire set of similar triangles has those particular angles.

Example A

Assume that one side and two angles are known. Then we may determine the third angle by the fact that the sum of the angles of a triangle is always 180°. Of all the possible similar triangles having these three angles, we have the one with the particular side which is known. Only one triangle with these parts is possible (in the sense that all triangles with the given parts are congruent and have equal corresponding angles and sides).

To **solve a triangle** *means that, when we are given three parts of a triangle (at least one a side), we are to find the other three parts.* In this section we are going to demonstrate the method of solving a right triangle. Since one angle of the triangle will be 90°, it is necessary to know one side and one other part. Also, we know that the sum of the three angles of a triangle is 180°, and this in turn tells us that the sum of the other two angles, both acute, is 90°. *Any two acute angles whose sum is 90° are said to be* **complementary.**

For consistency, when we are labeling the parts of the right triangle we shall use the letters A and B to denote the acute angles, and C to denote the right angle. The letters a, b, and c will denote the sides opposite these angles, respectively. Thus, side c is the hypotenuse of the right triangle.

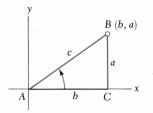

Figure 3—14

We shall find it convenient in solving right triangles to define the trigonometric functions of the acute angles in terms of the sides (see Fig. 3–14). Placing the vertex of angle A at the origin and the vertex of angle C on the positive x-axis, we obtain the following definitions:

$$\sin A = \frac{y}{r} = \frac{a}{c}, \qquad \cos A = \frac{x}{r} = \frac{b}{c}, \qquad \tan A = \frac{y}{x} = \frac{a}{b}$$

$$\cot A = \frac{x}{y} = \frac{b}{a}, \qquad \sec A = \frac{r}{x} = \frac{c}{b}, \qquad \csc A = \frac{r}{y} = \frac{c}{a}$$

(3–2)

If we should place the vertex of angle B at the origin, instead of the vertex of angle A, we would obtain the following definitions for the functions of angle B:

$$\sin B = \frac{b}{c}, \qquad \cos B = \frac{a}{c}, \qquad \tan B = \frac{b}{a}$$

$$\cot B = \frac{a}{b}, \qquad \sec B = \frac{c}{a}, \qquad \csc B = \frac{c}{b}$$

(3–3)

Inspecting these results, we may generalize our definitions of the trigonometric functions of acute angles of a right triangle to be as follows:

$$\sin \alpha = \frac{\text{opposite side}}{\text{hypotenuse}}, \qquad \cos \alpha = \frac{\text{adjacent side}}{\text{hypotenuse}}$$

$$\tan \alpha = \frac{\text{opposite side}}{\text{adjacent side}}, \qquad \cot \alpha = \frac{\text{adjacent side}}{\text{opposite side}}$$

$$\sec \alpha = \frac{\text{hypotenuse}}{\text{adjacent side}}, \qquad \csc \alpha = \frac{\text{hypotenuse}}{\text{opposite side}}$$

(3–4)

Using the definitions in this form, we can solve right triangles without placing the angle in standard position. The angle need only be a part of any right triangle.

We note from the above discussion that $\sin A = \cos B$, $\tan A = \cot B$, and $\sec A = \csc B$. From this we conclude that *cofunctions of acute complementary angles are equal.* The sine function and cosine functions are cofunctions, the tangent function and cotangent function are cofunctions, and the secant function and cosecant function are cofunctions. From this we can see how the tables of trigonometric functions are constructed. Since $\sin A = \cos (90° - A)$, the number representing either of these need appear only once in the tables.

Example B
Given $a = 4$, $b = 7$, and $c = \sqrt{65}$ ($C = 90°$), find sin A, cos A, and tan A. (See Fig. 3–15.)

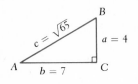

Figure 3–15

$$\sin A = \frac{\text{side opposite angle } A}{\text{hypotenuse}} = \frac{4}{\sqrt{65}} = 0.496$$

$$\cos A = \frac{\text{side adjacent angle } A}{\text{hypotenuse}} = \frac{7}{\sqrt{65}} = 0.868$$

$$\tan A = \frac{\text{side opposite angle } A}{\text{side adjacent angle } A} = \frac{4}{7} = 0.571$$

Example C
In Fig. 3–15, we have

$$\sin B = \frac{\text{side opposite angle } B}{\text{hypotenuse}} = \frac{7}{\sqrt{65}} = 0.868$$

$$\cos B = \frac{\text{side adjacent angle } B}{\text{hypotenuse}} = \frac{4}{\sqrt{65}} = 0.496$$

$$\tan B = \frac{\text{side opposite angle } B}{\text{side adjacent angle } B} = \frac{7}{4} = 1.75$$

We also note that sin A = cos B and cos A = sin B.

We are now ready to solve right triangles. We do this by expressing the unknown parts in terms of the known parts, as the following examples illustrate. For consistency, unless otherwise noted all results are rounded off to three significant digits, or to the nearest 10′ or 0.1° for angles. (Intermediate results in examples and exercises may contain additional significant digits. Discussions of rounding off and significant digits are given in Appendix B.)

Example D
Given $A = 50°0'$ and $b = 6.70$, solve the right triangle of which these are parts (see Fig. 3–16).
Since $\frac{a}{b} = \tan A$, we have $a = b \tan A$. Thus

$$a = 6.70 \ (\tan 50°0')$$
$$= 6.70 \ (1.192) = 7.99$$

Since $A = 50°0'$, $B = 40°0'$. Since $\frac{b}{c} = \cos A$, we have

$$c = \frac{b}{\cos A} = \frac{6.70}{0.6428} = 10.4$$

We have found that $a = 7.99$, $c = 10.4$, and $B = 40°0'$. It might be noted that we could have computed c by using an equation involving a. However, any error which may have been made in calculating a would also cause c to be in error. Thus, it is generally best to use given values in calculations.

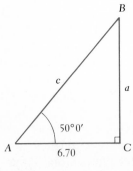

Figure 3–16

Example E

Given $b = 56.8$ and $c = 79.5$, solve the right triangle of which these are parts (see Fig. 3–17). Express angles in tenths.

Since $\cos A = \frac{b}{c}$, we have

$$\cos A = \frac{56.8}{79.5} = 0.7145$$

This means that

$$A = 44.4°$$

and

$$B = 90.0° - 44.4° = 45.6°$$

We solve for a by use of the Pythagorean theorem since in this way we can express a in terms of the given parts.

$$a^2 + b^2 = c^2 \text{ or } a = \sqrt{c^2 - b^2}$$

Thus,

$$a = \sqrt{79.5^2 - 56.8^2} = \sqrt{6320 - 3226} = \sqrt{3094}$$
$$= 55.6.$$

The above means we have determined that $a = 55.6$, $A = 44.4°$, and $B = 45.6°$.

Example F

Assuming that A and a are known, express the unknown parts of the right triangle in terms of a and A (see Fig. 3–18).

Since $\frac{a}{b} = \tan A$, we have $b = \dfrac{a}{\tan A}$. Since A is known, $B = 90° - A$.

Since $\frac{a}{c} = \sin A$, we have $c = \dfrac{a}{\sin A}$.

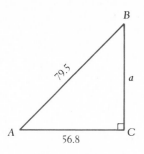

Figure 3—17

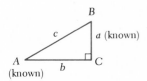

Figure 3—18

Exercises 3—4

In Exercises 1 through 4 draw appropriate figures and verify through observation that only one triangle may contain the given parts (that is, any others which may be drawn will be congruent—have equal corresponding sides and angles).

1. A 30° angle is included between sides of 3 in. and 6 in.
2. A side of 4 in. is included between angles of 40° and 70°.
3. A right triangle with a hypotenuse of 5 cm and a leg of 3 cm.
4. A right triangle with a 70° angle between the hypotenuse and a leg of 5 cm.

In Exercises 5 through 12 solve the right triangles which have the given parts. Express angles to the nearest 10′. Refer to Fig. 3–19.

5. $A = 32°0'$, $c = 56.8$
6. $B = 12°0'$, $c = 18.0$
7. $a = 56.7$, $b = 44.0$
8. $a = 9.98$, $c = 12.6$
9. $B = 37°40'$, $a = 0.886$
10. $A = 70°10'$, $a = 137$
11. $b = 86.7$, $c = 167$
12. $a = 6.85$, $b = 2.12$

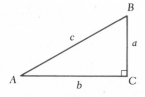

Figure 3—19

In Exercises 13 through 20 solve the right triangles which have the given parts. Express angles to the nearest 0.1°. Refer to Fig. 3–19.

13. $A = 77.8°$, $a = 6700$ 14. $A = 18.4°$, $c = 8.97$

15. $a = 150$, $c = 345$ 16. $a = 93.2$, $c = 124$

17. $B = 32.1°$, $c = 23.8$ 18. $B = 64.3°$, $b = 0.652$

19. $b = 82.1$, $c = 88.6$ 20. $a = 5920$, $b = 4110$

In Exercises 21 through 24 the parts listed refer to those in Fig. 3–19 and are assumed to be known. Express the other parts in terms of these known parts.

21. A, c 22. a, b 23. B, a 24. b, c

3–5 Applications of Right Triangles

Many applied problems can be solved by setting up the solutions in terms of right triangles. These applications are essentially the same as solving right triangles, although it is usually one specific part of the triangle that we wish to determine. The following examples illustrate some of the basic applications of right triangles.

Example A

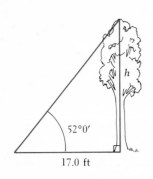

A tree has a shadow 17.0 ft long. From the point on the ground at the end of the shadow, the **angle of elevation** (the angle between the horizontal and the line of sight, when the object is above the horizontal) of the top of the tree is measured to be 52°0′. How tall is the tree?

By drawing an appropriate figure, as shown in Fig. 3–20, we note the information given and that which is required. Here we have let h be the height of the tree. Thus, we see that

$$\frac{h}{17.0} = \tan 52°0'$$

52°0′

17.0 ft

Figure 3–20

or

$$h = 17.0(\tan 52°0')$$
$$= 17.0(1.280)$$
$$= 21.8 \text{ ft}$$

Example B

From the roof of a building 46.0 ft high, the **angle of depression** (the angle between the horizontal and the line of sight, when the object is below the horizontal) of an object in the street is 74°0′. What is the distance of the observer from the object?

Again we draw an appropriate figure (Fig. 3–21). Here we let d represent the required distance. From the figure we see that

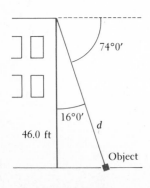

74°0′

16°0′ d

46.0 ft

Object

Figure 3–21

$$\frac{46.0}{d} = \cos 16°0'$$

$$d = \frac{46.0}{\cos 16°0'} = \frac{46.0}{0.9613}$$

$$= 47.9 \text{ ft}$$

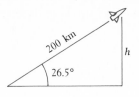

Figure 3–22

Example C

A missile is launched at an angle of 26.5° with respect to the horizontal. If it travels in a straight line over level terrain for 2 min, and its average speed is 6000 km/h, what is its altitude at this time?

In Fig. 3–22 we have let h represent the altitude of the missile after 2 min (altitude is measured on a perpendicular). Also, we determine that the missile has flown 200 km in a direct line from the launching site. This is found from the fact that it travels at 6000 km/h for $\frac{1}{30}$ h (2 min) and from the fact that $(6000 \text{ km/h})(\frac{1}{30} \text{ h}) = 200$ km. Therefore,

$$\frac{h}{200} = \sin 26.5°$$

$$h = 200(\sin 26.5°) = 200(0.4462)$$
$$= 89.2 \text{ km}$$

Example D

A shelf is supported by a straight 65.0-cm support attached to the wall at a point 53.5 cm below the bottom of the shelf. What angle does the support make with the wall? See Fig. 3–23.

In the figure we see that angle θ is to be determined. Therefore, we have

$$\cos \theta = \frac{53.5}{65.0} = 0.8231$$

or

$$\theta = 34.6°$$

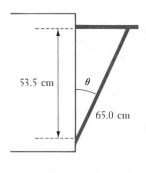

Figure 3–23

to the nearest tenth of a degree.

Example E

A television antenna is on the roof of a building. From a point on the ground 36.0 ft from the building, the angles of elevation of the top and the bottom of the antenna are 51°0′ and 42°0′, respectively. How tall is the antenna?

In Fig. 3–24 we let x represent the distance from the top of the building to the ground, and y represent the distance from the top of the antenna to the ground. Therefore,

$$\frac{x}{36.0} = \tan 42°0′$$

$$x = 36.0(\tan 42°0′) = 36.0(0.9004)$$
$$= 32.4 \text{ ft}$$

and

$$\frac{y}{36.0} = \tan 51°0′$$

$$y = 36.0(\tan 51°0′) = 36.0(1.235)$$
$$= 44.5 \text{ ft}$$

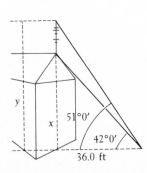

Figure 3–24

The length of the antenna is the difference of these distances, or 12.1 ft.

Exercises 3—5

In the following exercises solve the given problems. Draw an appropriate figure unless the figure is given.

1. The angle of elevation of the sun is 48°0′ at the time a television tower casts a shadow 346 ft long on level ground. Find the height of the tower.

2. A rope is stretched from the top of a vertical pole to a point 10.5 m from the foot of the pole. The rope makes an angle of 28°0′ with the pole. How tall is the pole?

3. The length of a kite string (assumed straight) is 560 ft. The angle of elevation of the kite is 64.0°. How high is the kite?

4. A 20.0-ft ladder leans against the side of a house. The angle between the ground and ladder is 70.0°. How far from the house is the foot of the ladder?

5. From the top of a cliff 110 m high the angle of depression to a boat is 23°20′. How far is the boat from the foot of the cliff?

6. A robot is on the surface of Mars. The angle of depression from a camera in the robot to a rock on the surface of Mars is 13.3°. The camera is 196 cm above the surface. How far is the camera from the rock?

7. On a blueprint, the walls of a rectangular room are 5.65 cm and 3.85 cm long, respectively. What is the angle (to the nearest 0.1°) between the longer wall and a diagonal across the room?

8. A roadway rises 120 ft for every 2200 ft along the road. Find the angle of inclination (to the nearest 10′) of the roadway.

9. A jet cruising at 590 mi/h climbs at an angle of 15°30′. What is its gain in altitude in 2 min?

10. Ten rivets are equally spaced on the circumference of a circular plate. If the center-to-center distance between two rivets is 6.25 cm, what is the radius of the circle?

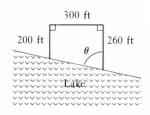

Figure 3—25

11. A loading platform is 3.25 ft above the ground. How long must a ramp be in order that it makes an angle of 20.0° with the ground?

12. In gauging screw threads a wire is placed in the thread. For the cross-section of wire and V-thread shown in Fig. 3—25, determine the distance OA if the diameter of the wire is 0.0550 in. and the angle at A is 72°40′.

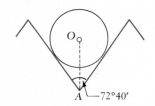

Figure 3—26

13. The level of a pond is 14.5 m above the level of a nearby stream. A straight 75.0-m drainage pipe extends from the pond surface to a point 3.0 m above the stream. What angle (to the nearest 0.1°) does the pipe make with the horizontal?

14. A lakefront property is shown in Fig. 3—26. What is the angle θ (to the nearest 10′) between the shoreline and the property line?

15. An astronaut circling the moon at an altitude of 100 mi notes that the angle of depression of the horizon is 23.8°. What is the radius of the moon?

16. A surveyor wishes to determine the width of a river. She sights a point B on the opposite side of the river from point C. She then measures off 400 ft from C to A such that C is a right angle. She then determines that $\angle A = 56°40′$. How wide is the river?

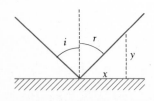

Figure 3—27

17. If a light ray strikes a reflecting surface (Fig. 3—27), the angle of reflection r equals the angle of incidence i. If a light ray has an angle of incidence of 42.0°, what is the distance y from the plane of the surface of a point on the reflected ray, if the horizontal distance x from the point of incidence is 7.42 cm?

18. A man considers building a dormer on his house. What are the dimensions x and y of the dormer in his plans (see Fig. 3—28)?

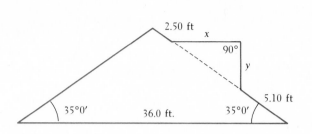

Figure 3—28   Figure 3—29

19. A machine part is indicated in Fig. 3—29. What is the angle θ (to the nearest 10′)?

20. A way of representing the impedance and resistance in an alternating-current circuit is equivalent to letting the impedance be the hypotenuse of a right triangle and the resistance be the side adjacent to the phase angle. If the resistance in a given circuit is 25.4 Ω and the phase angle is 24.5°, what is the impedance?

21. A surveyor sights two points directly ahead. Both are at an elevation 18.5 m lower than the point from which he observes them. How far apart are the points if the angles of depression are 13.5° and 21.3°, respectively?

22. A flagpole is atop a building. From a point on the ground 720 ft from the building, the angles of elevation of the top and bottom of the flagpole are 33°30′ and 31°10′, respectively. How tall is the flagpole?

3—6 Exercises for Chapter 3

In Exercises 1 through 4 find the smallest positive angle and the smallest negative angle (numerically) coterminal with, but not equal to the given angles.

1. 17°0′   2. 248°20′   3. −217.5°   4. −7.6°

In Exercises 5 through 8 change the given angles to equal angles expressed in decimal form.

5. 31°54′   6. 174°45′   7. 38°6′   8. 321°27′

In Exercises 9 through 12 change the given angles to equal angles expressed to the nearest minute.

9. 17.5°   10. 65.4°   11. 49.7°   12. 126.25°

In Exercises 13 through 16 determine the trigonometric functions of the angles whose terminal side passes through the given points.

13. (24, 7)   14. (5, 4)   15. (4, 4)   16. (1.2, 0.5)

In Exercises 17 through 20 using the given trigonometric functions, find the indicated trigonometric functions.

17. Given $\sin \theta = \frac{5}{13}$, find $\cos \theta$ and $\cot \theta$.
18. Given $\cos \theta = \frac{3}{8}$, find $\sin \theta$ and $\tan \theta$.
19. Given $\tan \theta = 2$, find $\cos \theta$ and $\csc \theta$.
20. Given $\cot \theta = 4$, find $\sin \theta$ and $\sec \theta$.

In Exercises 21 through 32 find the value of each of the given trigonometric functions.

21. $\sin 72°0'$ **22.** $\cos 40°10'$ **23.** $\tan 61°20'$ **24.** $\csc 19°30'$
25. $\tan 37.2°$ **26.** $\sin 49.9°$ **27.** $\cos 12.8°$ **28.** $\tan 20.6°$
29. $\tan 81°15'$ **30.** $\cot 37°17'$ **31.** $\cos 55°53'$ **32.** $\sec 58°54'$

In Exercises 33 through 36 find θ to the nearest $10'$ for each of the given trigonometric functions.

33. $\sin \theta = 0.5324$ **34.** $\tan \theta = 1.265$
35. $\cos \theta = 0.4669$ **36.** $\sec \theta = 2.107$

In Exercises 37 through 40 find θ to the nearest $0.1°$ for each of the given trigonometric functions.

37. $\cos \theta = 0.9500$ **38.** $\sin \theta = 0.6305$
39. $\tan \theta = 1.574$ **40.** $\cos \theta = 0.1345$

In Exercises 41 through 44 find θ to the nearest minute for each of the given trigonometric functions.

41. $\cot \theta = 1.132$ **42.** $\cos \theta = 0.7365$
43. $\sin \theta = 0.8666$ **44.** $\csc \theta = 1.533$

In Exercises 45 through 48 solve the right triangles which have the given parts. Express angles to the nearest $10'$. Refer to Fig. 3–30.

45. $A = 17°0', b = 6.00$ **46.** $B = 68°10', a = 1080$
47. $a = 81.0, b = 64.5$ **48.** $a = 1.06, c = 3.82$

In Exercises 49 through 52 solve the right triangles which have the given parts. Express angles to the nearest $0.1°$. Refer to Fig. 3–30.

49. $A = 37.5°, a = 12.0$ **50.** $B = 15.7°, c = 12.6$
51. $b = 6.50, c = 7.60$ **52.** $a = 72.1, b = 14.3$

In Exercises 53 through 56 solve the right triangles which have the given parts. Express angles to the nearest minute. Refer to Fig. 3–30.

53. $A = 49°43', c = 0.820$ **54.** $B = 4°26', b = 5.60$
55. $a = 10.0, c = 15.0$ **56.** $a = 724, b = 852$

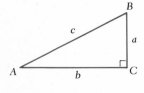

Figure 3–30

In Exercises 57 through 74 solve the given applied problems.

57. An approximate equation found in the diffraction of light through a narrow opening is

$$\sin \theta = \frac{\lambda}{d}$$

where λ (the Greek lambda) is the wavelength of the light and d is the width of the opening. If $\theta = 1°10'$ and $d = 30.0$ μm, what is the wavelength of the light?

58. An equation used for the instantaneous value of electric current in an alternating-current circuit is $i = I_m \cos \theta$. Calculate i for $I_m = 56.0$ mA and $\theta = 10.5°$.

59. In determining the height h of a building which is 220 m distant, a surveyor may use the equation $h = 220 \tan \theta$, where θ is the angle of elevation to the top of the building. What should θ be if $h = 130$ m?

60. A formula used with a certain type of gear is

$$D = \frac{N \sec \theta}{4}$$

where D is the pitch diameter of the gear, N is the number of teeth on the gear, and θ is called the spiral angle. If $D = 6.75$ in. and $N = 20$, find θ (to the nearest 10′).

61. A ship's captain, desiring to travel due south, discovers that, due to an improperly functioning instrument, he has gone 22.6 mi in a direction $4°0'$ east of south. How far from his course (to the east) is he?

62. A helicopter pilot notes that she is 150 m above a certain rooftop. If the angle of depression to the rooftop is $18°0'$, how far on a direct line from the rooftop is the helicopter?

63. An observer 3500 ft from the launch pad of a rocket measures the angle of elevation to the rocket soon after liftoff to be $54.0°$. How high is the rocket, assuming it has moved vertically?

64. A hemispherical bowl is standing level. Its inside radius is 6.50 in. and it is filled with water to a depth of 3.10 in. Through what angle may it be tilted before the water will spill?

65. A railroad embankment has a level top 22.0 ft wide, equal sloping sides of 14.5 ft, and a height of 7.20 ft. What is the width of the base of the embankment?

66. The horizontal distance between the extreme positions of a certain pendulum is 9.50 cm. If the length of the pendulum is 38.0 cm, through what angle does it swing?

67. The span of a roof is 32.0 ft. Its rise is 7.5 ft at the center of the span. What angle, to the nearest 10′, does the roof make with the horizontal?

68. If the impedance in a certain alternating-current circuit is 56.5 Ω and the resistance in the circuit is 17.0 Ω, what is the phase angle to the nearest $0.1°$? (See Exercise 20 of Section 3−5.)

69. A bridge is 12.5 m above the surrounding area. If the angle of elevation of the approach to the bridge is 4°40′, how long is the approach?

70. The windshield on an automobile is inclined 42.5° with respect to the horizontal. Assuming that the windshield is flat and rectangular, what is its area if it is 4.80 ft wide and the bottom is 1.50 ft in front of the top?

71. A person in a tall building observes an object drop from a window 20.0 ft away and directly opposite him. If the distance the object drops as a function of time is $s = 16t^2$, how far is the object from the observer (on a direct line) after 2.00 s?

72. The distance from ground level to the underside of a cloud is called the *ceiling*. One observer 1000 m from a searchlight aimed vertically notes that the angle of elevation of the spot of light on a cloud is 76°0′. What is the ceiling?

73. A laser beam is transmitted with a "width" of 0.002°. What is the diameter of a spot of the beam on an object 50,000 km distant? See Fig. 3–31. (Tan 0.001° = 1.745 × 10⁻⁵; see Exercise 29 of Section 7–4.)

74. What angle is subtended at the eye of an observer 5.00 mi from an airplane which is 150 ft long and is perpendicular to the line of sight?

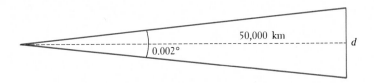

Figure 3–31

4

Systems of Linear Equations; Determinants

4—1 Linear Equations

In Chapter 1 we introduced the topic of solving equations and showed some of the types of technical problems which could be solved. Many of the equations we encountered at that time were examples of a very important type of equation, the **linear equation.** In general, an equation is termed linear in a given set of variables if each term contains only one variable, to the first power, or is a constant.

Example A
$5x - t + 6 = 0$ is linear in x and t, but $5x^2 - t + 6 = 0$ is not, due to the presence of x^2.

The equation $4x + y = 8$ is linear in x and y, but $4xy + y = 8$ is not, due to the presence of xy.

The equation $x - 6y + z - 4w = 7$ is linear in x, y, z, and w, but $x - \frac{6}{y} + z - 4w = 7$ is not, due to the presence of $\frac{6}{y}$ where y appears in the denominator.

An equation which can be written in the form

$$ax + b = 0 \tag{4-1}$$

is known as a **linear equation in one unknown.** We have already discussed the solution to this type of equation in Section 1–10. In general, *the* **solution,** *or* **root,** *of the equation is* $x = -b/a$. Also, it will be noted that finding the solution is equivalent to finding the zero of the **linear function** $ax + b$.

There also are a great number of applied problems which involve more than one unknown. When forces are analyzed in physics, equations with two unknowns (the forces) often result. In electricity, equations relating several unknown currents arise. Numerous kinds of stated problems from various technical areas involve equations with more than one unknown. The use of equations involving more than one unknown is well established in technical applications.

Example B
A very basic law of direct-current electricity, known as Kirchhoff's first law, may be stated as "The algebraic sum of the currents entering any junction in a circuit is zero." If three wires are joined at a junction, this law leads to the equation

$$i_1 + i_2 + i_3 = 0$$

where i_1, i_2 and i_3 are the currents in each of the wires.

When determining two forces F_1 and F_2 acting on a beam, we might encounter an equation such as

$$2F_1 + 4F_2 = 200$$

An equation which can be written in the form

$$ax + by = c \tag{4-2}$$

is known as a **linear equation in two unknowns.** In Chapter 2, when we were discussing functions we considered many functions of this type. We found that for each value of x there is a corresponding value of y. Each of these pairs of numbers is a **solution** to the equation, although we did not call it that at the time. A solution is any set of numbers, one for each variable, which satisfies the equation. When we represent the solutions in the form of a graph, we see that the graph of any linear equation in two unknowns is a straight line. Also, graphs of linear equations in one unknown, those for which $a = 0$ or $b = 0$, are also straight lines. Thus we see the significance of the name "linear."

Example C
The equation $4x + 8 = 0$ is a linear equation in one unknown, x. We find its solution by subtracting 8 from both sides and then dividing both sides by 4. This results in the root $x = -2$. Solving this equation is equivalent to finding the zero of the function $4x + 8$.

Example D
The equation $2x - y - 6 = 0$ is a linear equation in two unknowns, x and y. To sketch the graph of this equation, we write it in the more convenient form $y = 2x - 6$. The coordinates of each point on the

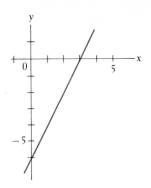

y

0

5

x

−5

Figure 4—1

graph, which is a straight line, are solutions of this equation. For example, the point $(1, -4)$ is a point on the line. This means that $x = 1$, $y = -4$ is a solution of the equation. In the same way, $x = 0$, $y = -6$ is a solution (see Fig. 4–1).

Two linear equations, each containing the same two unknowns,

$$a_1x + b_1y = c_1$$
$$a_2x + b_2y = c_2$$

(4-3)

are said to form a **system of simultaneous linear equations.** **A solution of the system** *is any pair of values (x, y) which satisfies both equations.* Methods of finding the solutions to such systems are the principal concern of this chapter.

Example E
The two linear equations

$$2x - y = 5$$
$$3x + 2y = 4$$

form a system of simultaneous linear equations. The solution of this system is $x = 2$, $y = -1$. These values satisfy both equations, since $2(2) - (-1) = 4 + 1 = 5$ and $3(2) + 2(-1) = 6 - 2 = 4$. This is the only pair of values which satisfies *both* equations. Methods for finding such solutions are taken up in the following sections.

Exercises 4—1

In Exercises 1 through 4 determine whether or not the given pairs of values are solutions of the given linear equations in two unknowns.

1. $2x + 3y = 9$; $(3, 1)$, $(5, \frac{1}{3})$

2. $5x + 2y = 1$; $(2, -4)$, $(1, -2)$

3. $-3x + 5y = 13$; $(-1, 2)$, $(4, 5)$

4. $x - 4y = 10$; $(2, -2)$, $(2,2)$

In Exercises 5 through 8, for each given value of x, determine the value of y which gives a solution to the given linear equations in two unknowns.

5. $5x - y = 6$; $x = 1$, $x = -2$

6. $2x + 7y = 8$; $x = -3$, $x = 2$

7. $x - 5y = 12$; $x = 3$, $x = -4$

8. $3x - 2y = 9$; $x = \frac{2}{3}$, $x = -3$

In Exercises 9 through 16 determine whether or not the given pair of values is a solution of the given system of simultaneous linear equations.

9. $x - y = 5$ $x = 4$, $y = -1$
   $2x + y = 7$

10. $2x + y = 8$ $x = -1$, $y = 10$
    $3x - y = -13$

11. $3x - 4y = -10$ $x = -2$, $y = 1$
    $x + 5y = -7$

12. $-3x + y = 1$ $x = \frac{1}{3}$, $y = 2$
    $6x - 3y = -4$

13. $2x - 5y = 0$ $x = \frac{1}{2}$, $y = -\frac{1}{5}$
    $4x + 10y = 4$

14. $3x - 4y = -1$ $x = 1$, $y = -1$
    $6x - y = 5$

15. $3x - 2y = 2.2$ $x = 0.6$, $y = -0.2$
    $5x + y = 2.8$

16. $x - 7y = -3.2$ $x = -1.1$, $y = 0.3$
    $2x + y = 2.5$

In Exercises 17 through 20 answer the given questions.

17. If a board 84 in. long is cut into two pieces such that one piece is 6 in. longer than the other, the lengths x and y of the two pieces are found by solving the system of equations

$$x + y = 84$$
$$x - y = 6$$

Are the resulting lengths 45 in. and 39 in.?

18. If two forces F_1 and F_2 support a 98-lb weight, the forces can be found by solving the system of equations

$$0.7F_1 - 0.6F_2 = 0, \ 0.7F_1 + 0.8F_2 = 98$$

Are the forces 60 lb and 70 lb?

19. Under certain conditions two electric currents i_1 and i_2 can be found by solving the system of equations

$$3i_1 + 4i_2 = 3, \ 3i_1 - 5i_2 = -6$$

Are the currents $-\frac{1}{3}$ A and 1 A?

20. Using the data that fuel consumption for transportation contributes a percent p_1 of pollution which is 16% less than the percent p_2 of all other sources combined, the equations

$$p_1 + p_2 = 100, \ p_2 - p_1 = 16$$

can be set up. Are the percents $p_1 = 58$ and $p_2 = 42$?

4—2 Solving Systems of Two Linear Equations in Two Unknowns Graphically

We shall now take up the problem of solving for the unknowns when we have a system of two simultaneous linear equations in two unknowns. In this section we shall show how the solution may be found graphically. The sections which follow will discuss other basic methods of solution.

Since a solution of a system of simultaneous linear equations in two unknowns is any pair of values (x, y) which satisfies both equations, graphically, *the solution would be the coordinates of the point of inter-section of the two lines.* This must be the case, for the coordinates of this point constitute the only pair of values to satisfy *both* equations. (In some special cases there may be no solution, in others there may be many solutions. See Examples D and E.)

Therefore, when we solve two simultaneous linear equations in two unknowns graphically, we must plot the graph of each line and determine the point of intersection. This may, of course, lead to approximate results if the lines cross at values between those chosen to determine the graph.

We may use the knowledge that each equation represents a straight line to advantage. By finding two points on the line, we can draw the line. Two points which are easily determined are those where the curve crosses the y-axis and the x-axis. These points are known as the **intercepts** of the line. These points are easily found because in each case one of the coordinates is zero. By setting $x = 0$ and $y = 0$, in turn, and determining the corresponding value of the other unknown, we obtain the coordinates of the intercepts. A third point should be found as a check. This method is sufficient unless the line passes through the origin. Then both intercepts are at the origin and one more point must be determined. Example A illustrates how a line is plotted by finding its intercepts.

Example A
Plot the graph of $2x - 3y = 6$ by finding its intercepts and one check point (see Fig. 4—2).

First let $x = 0$. This gives $-3y = 6$, or $y = -2$. Thus, the point $(0, -2)$ is on the graph. Next we let $y = 0$, and this gives $2x = 6$, or $x = 3$. Thus, the point $(3, 0)$ is on the graph. The point $(0, -2)$ is the y-intercept, and $(3, 0)$ is the x-intercept. These two points are sufficient to plot the line, but we should find another point as a check. Choosing $x = 1$, we find $y = -\frac{4}{3}$. Thus, the point $(1, -\frac{4}{3})$ should be on the line. From Fig. 4—2 we see that it is on the line.

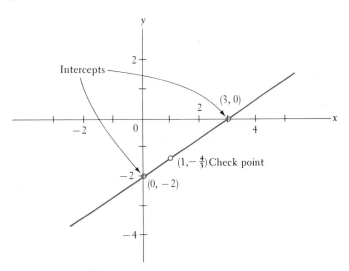

Figure 4—2

Now that we have discussed the meaning of a graphical solution of a system of simultaneous equations and the method of plotting a line, we are in a position to find graphical solutions of systems of linear equations. The following examples illustrate the method.

Example B

Solve the system of equations

$$2x + 5y = 10$$
$$3x - \ y = \ 6$$

We find that the intercepts of the first line are the points $(5, 0)$ and $(0, 2)$. A third point is $(-1, \frac{12}{5})$. The intercepts of the second line are $(2, 0)$ and $(0, -6)$. A third point is $(1, -3)$. Plotting these points and drawing the proper straight lines, we see that the lines cross at about $(2.3, 1.1)$. [The actual values are $(\frac{40}{17}, \frac{18}{17})$.] The solution of the system of equations is approximately $x = 2.3$, $y = 1.1$ (see Fig. 4–3).

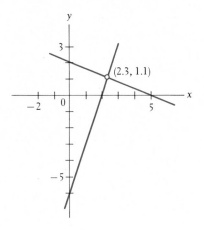

Figure 4–3   Figure 4–4

Example C

Solve the system of equations

$$x - 4y = 2$$
$$-2x + \ y = -4$$

The intercepts and a third point for the first line are $(2, 0)$, $(0, -0.5)$, and $(6, 1)$. For the second line they are $(2, 0)$, $(0, -4)$, and $(1, -2)$. Since they have one point in common, the point $(2, 0)$, we conclude that the exact solution to the system is $x = 2$, $y = 0$ (see Fig. 4–4).

Example D

Solve the system of equations

$$x - 2y = 6$$
$$3x - 6y = 6$$

The intercepts and a third point for the first line are $(6, 0)$, $(0, -3)$, and $(2, -2)$. For the second line they are $(2, 0)$, $(0, -1)$, and $(4, 1)$. The graphs of these two equations do not intersect (see Fig. 4–5). *Therefore, there are no solutions. Such a system is called* **inconsistent**.

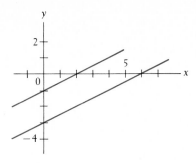

Figure 4-5   Figure 4-6

Example E

Solve the system of equations

$$x - 3y = 9$$
$$-2x + 6y = -18$$

The intercepts and a third point for the first line are $(9, 0)$, $(0, -3)$, and $(3, -2)$. In determining the intercepts for the second line, we find that they are $(9, 0)$ and $(0, -3)$, which are also the intercepts of the first line (see Fig. 4-6). As a check we note that $(3, -2)$ also satisfies the equation of the second line. This means the two lines are really the same line, and *the coordinates of any point on this common line constitute a solution of the system. Such a system is called* **dependent**.

Exercises 4-2

In the following exercises solve each system of equations graphically. Estimate the answer to the nearest tenth of a unit if necessary.

1. $x + y = 4$
   $x - y = 2$

2. $x - 2y = 2$
   $x + y = 8$

3. $2x - y = 6$
   $x + 3y = 3$

4. $-x + 2y = -8$
   $2x - y = -2$

5. $3x + 2y = 6$
   $x - 3y = 3$

6. $4x - 3y = -8$
   $6x + y = 6$

7. $2x - 5y = 10$
   $3x + 4y = -12$

8. $-5x + 3y = 15$
   $2x + 7y = 14$

9. $x - 4y = 8$
   $2x + 5y = 10$

10. $4x - y = 6$
    $2x - y = -4$

11. $y = -x + 3$
    $4x = 6 - 2y$

12. $x - 6 = 6y$
    $y = 3 - 3x$

13. $x - 4y = 6$
    $-x + 2y = 4$

14. $x + y = 3$
    $3x - 2y = 14$

15. $-2x + 2y = 7$
    $4x - 2y = 1$

16. $2x - 3y = -5$
    $3x + 2y = 12$

17. $x - 4y = 2$
    $-2x + 3y = 3$

18. $x - 2y = 4$
    $3x + 2y = 7$

19. $8x - 7y = 3$
    $7y + x = 7$

20. $5x - 2y = 7$
    $3x + 4y = 8$

21. $x - 5y = 10$
    $2x - 10y = 20$

22. $18x - 3y = 7$
    $2y = 1 + 12x$

23. $y = 3x$
    $x - 2y = 6$

24. $4x - y = 3$
    $2x + 3y = 0$

25. $5x = y + 3$
     $4x = 2y - 3$

26. $5x + 7y = 5$
     $2x - 3y = 4$

27. $3x = 8y + 12$
     $-6x + 16y = 6$

28. $y = 6x + 2$
     $12x - 2y = -4$

29. The perimeter of a rectangular area is 24 km, and the length is 6 km longer than the width. The dimensions l and w can be found by solving the equations

$$2l + 2w = 24$$
$$l - w = 6$$

(Notice that the first equation represents the perimeter of this rectangle and the second equation represents the relationship of the length to the width.)

30. To assemble a particular piece of machinery, 18 bolts are used. There are two kinds of bolts, and 4 more of one kind are used than the other. The numbers of each kind, a and b, can be found by solving the equations

$$a + b = 18$$
$$a - b = 4$$

31. In electricity, applying Ohm's law to a particular circuit gives the equations needed to find a specified voltage E and current I (in amperes) as

$$E - 4I = 0$$
$$E + 6I = 9$$

32. A computer requires 12 s to do two series of calculations. The first series of calculations requires twice as much time as the second series. The times t_1 and t_2 can be found by solving the equations

$$t_1 + t_2 = 12$$
$$t_1 = 2t_2$$

4–3 Solving Systems of Two Linear Equations in Two Unknowns Algebraically

We have just seen how a system of two linear equations can be solved graphically. This technique is good for obtaining a "picture" of the solution. Finding the solution of systems of equations by graphical methods has one difficulty: the results are usually approximate. If exact solutions are required, we must turn to other methods. In this section we shall present two algebraic methods of solution.

The first method involves elimination by **substitution.** To follow this method, we first solve one of the equations for one of the unknowns. This solution is then substituted into the other equation, resulting in one linear equation in one unknown. This equation can then be solved for the unknown it contains. By substituting this value into one of the original equations, we can find the corresponding value of the other unknown. The following two examples illustrate the method.

Example A

Use the method of elimination by substitution to solve the system of equations

$$x - 3y = 6$$
$$2x + 3y = 3$$

The first step is to solve one of the equations for one of the unknowns. The choice of which equation and which unknown depends on ease of manipulation. In this system, it is somewhat easier to solve the first equation for x. Therefore, performing this operation we have

$$x = 3y + 6$$

We then substitute this expression into the second equation in place of x, giving

$$2(3y + 6) + 3y = 3$$

Solving this equation for y we obtain

$$6y + 12 + 3y = \quad 3$$
$$9y = -9$$
$$y = -1$$

We now put the value $y = -1$ into the first of the original equations. Since we have already solved this equation for x in terms of y, we obtain

$$x = 3(-1) + 6 = 3$$

Therefore, the solution of the system is $x = 3$, $y = -1$. As a check, we substitute these values in the second equation. We obtain $2(3) + 3(-1) = 6 - 3 = 3$, which verifies the solution.

Example B

Use the method of elimination by substitution to solve the system of equations

$$-5x + 2y = -4$$
$$10x + 6y = \quad 3$$

It makes little difference which equation or which unknown is chosen. Therefore, choosing to solve the first equation for y, we obtain

$$2y = 5x - 4$$
$$y = \frac{5x - 4}{2}$$

Substituting this expression into the second equation, we have

$$10x + 6\left(\frac{5x - 4}{2}\right) = 3$$

We now proceed to solve this equation for x.

$$10x + 3(5x - 4) = 3$$
$$10x + 15x - 12 = 3$$
$$25x = 15$$
$$x = \frac{3}{5}$$

Substituting this value into the expression for y, we obtain

$$y = \frac{5(3/5) - 4}{2} = \frac{3 - 4}{2} = -\frac{1}{2}$$

Therefore, the solution of this system is $x = \frac{3}{5}$, $y = -\frac{1}{2}$. Substituting these values in the second equation shows that the solution checks.

The method of elimination by substitution is useful if one equation can easily be solved for one of the unknowns. However, the numerical coefficients often make this method somewhat cumbersome. So we use another method, that of elimination by **addition or subtraction**. To use this method we multiply each equation by a number chosen so that the coefficients for one of the unknowns will be numerically the same in both equations. If these numerically equal coefficients are opposite in sign, we *add* the two equations. If the numerically equal coefficients have the same signs, we subtract one equation from the other. That is, we subtract the left side of one equation from the left side of the other equation, and also do the same to the right sides. After adding or subtracting, we have a simple linear equation in one unknown, which we then solve for the unknown. We substitute this value into one of the original equations to obtain the value of the other unknown.

Example C
Use the method of elimination by addition or subtraction to solve the system of equations

$$x - 3y = 6$$
$$2x + 3y = 3$$

We look at the coefficients to determine the best way to eliminate one of the unknowns. In this case, since the coefficients of the y-terms are numerically the same and are opposite in sign, we may immediately add the two equations to eliminate y. Adding the left sides together and the right sides together, we obtain

$$x + 2x - 3y + 3y = 6 + 3$$
$$3x = 9$$
$$x = 3$$

Substituting this value into the first equation, we obtain

$$3 - 3y = 6$$
$$-3y = 3$$
$$y = -1$$

The solution $x = 3$, $y = -1$ agrees with the results obtained for the same problem illustrated in Example A of this section.

Example D
Use the method of elimination by addition or subtraction to solve the system of equations

$$3x - 2y = 4$$
$$x + 3y = 2$$

Probably the most convenient method is to multiply the second equation by 3. Then subtract the second equation from the first and solve for y. Substitute this value of y into the second equation and solve for x.

$$3x - 2y = \quad 4$$
$$\underline{3x + 9y = \quad 6}$$
$$-11y = -2$$
$$y = \frac{2}{11}$$
$$x + 3\left(\frac{2}{11}\right) = 2$$
$$11x + 6 = 22$$
$$x = \frac{16}{11}$$

We arrive at the solution $x = \frac{16}{11}$, $y = \frac{2}{11}$. Substituting these values into both of the original equations shows that the solution checks.

Example E
The system of Example D can also be solved by multiplying the first equation by 3 and the second by 2, thereby first eliminating y.

$$9x - 6y = 12$$
$$\underline{2x + 6y = \quad 4}$$
$$11x \qquad = 16 \qquad \text{Adding the equations}$$
$$x = \frac{16}{11}$$
$$3\left(\frac{16}{11}\right) - 2y = 4 \qquad \text{Substituting into the first of original equations}$$
$$48 - 22y = 44 \qquad \text{Multiplying each term by 11}$$
$$-22y = -4$$
$$y = \frac{2}{11}$$

Therefore, the solution is $x = \frac{16}{11}$, $y = \frac{2}{11}$, as we obtained in Example D.

Example F

Use the method of elimination by addition or subtraction to solve the system of equations

$$4x - 2y = 3$$
$$2x - y = 2$$

When we multiply the second equation by 2 and subtract, we obtain

$$4x - 2y = 3$$
$$\underline{4x - 2y = 4}$$
$$0 = -1$$

Since we know 0 does not equal -1, we conclude that there is no solution. When we obtain a result of $0 = a$ $(a \neq 0)$, the system of equations is inconsistent. If we obtain the result $0 = 0$, the system is dependent.

After solving systems of equations by these methods, it is always a good policy to check the results by substituting the values of the two unknowns into the other original equation to see that the values satisfy the equation. Remember, we are finding the one pair of values which satisfies both equations.

Linear equations in two unknowns are often useful in solving stated problems. In such problems we must read the statement carefully in order to identify the unknowns and the method of setting up the proper equations. Exercises 29, 30, and 32 of Section 4–2 give statements and the resulting equations, which the reader should be able to derive. The following example gives a complete illustration of the method.

Example G

By weight, one alloy is 70% copper and 30% zinc. Another alloy is 40% copper and 60% zinc. How many grams of each of these would be required to make 300 g of an alloy which is 60% copper and 40% zinc?

Let A = required number of grams of first alloy, and B = required number of grams of second alloy. We know that the total weight of the final alloy is 300 g, which leads us to the equation $A + B = 300$. We also know that the final alloy will contain 180 g of copper (60% of 300). The weight of copper from the first alloy is $0.70A$ and that from the second is $0.40B$. This leads to the equation $0.70A + 0.40B = 180$. These two equations can now be solved simultaneously.

$$A + B = 300$$
$$0.70A + 0.40B = 180$$

$4A + 4B = 1200$	Multiplying the first equation by 4
$\underline{7A + 4B = 1800}$	Multiplying the second equation by 10
$3A = 600$	Subtracting the first equation from the second
$A = 200 \text{ g}$	
$B = 100 \text{ g}$	By substituting into the first equation

Substitution shows that this solution checks with the given information.

Exercises 4−3

In Exercises 1 through 12 solve the given systems of equations by the method of elimination by substitution.

1. $x = y + 3$
   $x - 2y = 5$

2. $x = 2y + 1$
   $2x - 3y = 4$

3. $y = x - 4$
   $x + y = 10$

4. $y = 2x + 10$
   $2x + y = -2$

5. $x + y = -5$
   $2x - y = 2$

6. $3x + y = 1$
   $3x - 2y = 16$

7. $2x + 3y = 7$
   $6x - y = 1$

8. $2x + 2y = 1$
   $4x - 2y = 17$

9. $3x + 2y = 7$
   $-9x + 2y = 11$

10. $3x + 3y = -1$
    $-5x - 6y = 1$

11. $4x - 3y = 6$
    $2x + 4y = -5$

12. $5x + 4y = -7$
    $3x - 5y = -6$

In Exercises 13 through 24 solve the given systems of equations by the method of elimination by addition or subtraction.

13. $x + 2y = 5$
    $x - 2y = 1$

14. $x + 3y = 7$
    $2x + 3y = 5$

15. $2x - 3y = 4$
    $2x + y = -4$

16. $x - 4y = 17$
    $3x + 4y = 3$

17. $2x + 3y = 8$
    $x - 2y = -3$

18. $3x - y = 3$
    $4x - 3y = 14$

19. $x + 2y = 7$
    $2x + 4y = 9$

20. $3x - y = 5$
    $-9x + 3y = -15$

21. $2x - 3y = 4$
    $3x - 2y = -2$

22. $3x + 4y = -5$
    $5x - 3y = 2$

23. $3x - 7y = 4$
    $2x + 5y = 7$

24. $5x + 2y = -4$
    $3x - 5y = 6$

In Exercises 25 through 32 solve the given systems of equations by either method of this section.

25. $2x - y = 5$
    $6x + 2y = -5$

26. $3x + 2y = 4$
    $6x - 6y = 13$

27. $6x + 3y = -4$
    $9x + 5y = -6$

28. $5x - 6y = 1$
    $3x - 4y = 7$

29. $3x - 6y = 15$
    $4x - 8y = 20$

30. $2x + 6y = -3$
    $-6x - 18y = 5$

31. $7x + y = -9$
    $3x + 4y = -11$

32. $6x + 6y = -7$
    $3x - 12y = 13$

In Exercises 33 and 34 solve the given systems of equations by an appropriate algebraic method.

33. An electrical experiment results in the following equations for currents (in amperes) i_1 and i_2 of a certain circuit:

$$i_1 = 3i_2$$
$$4i_1 - 2i_2 = 5$$

Find i_1 and i_2.

34. In finding moments M_1 and M_2 (in foot-pounds) of certain forces acting on a given beam, the following equations are used:

$$10M_1 + 3M_2 = -140$$
$$4M_1 + 14M_2 = -564$$

Find M_1 and M_2.

In Exercises 35 through 42 set up appropriate systems of two linear equations in two unknowns, and solve the systems algebraically.

35. A piece of wire is 18 m long. Where must it be cut for one piece to be 3 m longer than the other piece?

36. The voltage across an electric resistor equals the current times the resistance. The sum of two resistances is 16 Ω. When a current of 4 A passes through the smaller resistance and 3 A through the larger resistance, the sum of the voltages is 52 V. What is the value of each resistance?

37. One electronic data-processing card sorter can sort a cards per minute, and a second card sorter can sort b cards per minute. If the first sorts for 3 min and the second for 2 min, 12,200 cards can be sorted. If the times are reversed, 11,300 cards can be sorted. Find the sorting rates a and b.

38. An airplane travels 1860 mi in 2 h with the wind and then returns in 2.5 h traveling against the wind. Find the speed of the airplane with respect to the air and the velocity of the wind.

39. A chemist has a 20% solution and an 8% solution of sulfuric acid. How many milliliters of each solution should he mix in order to obtain 180 mL of a 15% solution?

40. For proper dosage a drug must be a 10% solution. How many milliliters of a 5% solution and of a 25% solution should be mixed to obtain 1000 mL of the required solution?

41. The cost of booklets at a printing firm consists of a fixed charge and a charge for each booklet. The total cost of 1000 booklets is $550, and the total cost of 2000 booklets is $800. What is the fixed charge and the charge for each booklet?

42. Analyze the following statement from a student's laboratory report. "The sum of the currents was 10 A, and twice the first current was 3 A less twice the second current."

4–4 Solving Systems of Two Linear Equations in Two Unknowns by Determinants

Consider two linear equations in two unknowns,

$$a_1 x + b_1 y = c_1$$
$$a_2 x + b_2 y = c_2 \tag{4–3}$$

If we multiply the first of these equations by b_2 and the second by b_1, we obtain

$$a_1 b_2 x + b_1 b_2 y = c_1 b_2$$
$$a_2 b_1 x + b_2 b_1 y = c_2 b_1 \tag{4–4}$$

If we now subtract the second equation of (4–4) from the first, we obtain

$$a_1 b_2 x - a_2 b_1 x = c_1 b_2 - c_2 b_1$$

which by use of the distributive law may be written as

$$(a_1b_2 - a_2b_1)x = c_1b_2 - c_2b_1 \tag{4-5}$$

Solving Eq. (4–5) for x, we obtain

$$x = \frac{c_1b_2 - c_2b_1}{a_1b_2 - a_2b_1} \tag{4-6}$$

In the same manner, we may show that

$$y = \frac{a_1c_2 - a_2c_1}{a_1b_2 - a_2b_1} \tag{4-7}$$

If the denominator $a_1b_2 - a_2b_1 = 0$, there is no solution for Eqs. (4–6) and (4–7), since division by zero is not defined.

The expression $a_1b_2 - a_2b_1$, which appears in each of the denominators of Eqs. (4–6) and (4–7), is an example of a special kind of expression called a **determinant of the second order.** The determinant $a_1b_2 - a_2b_1$ is denoted by the symbol

$$\begin{vmatrix} a_1 & b_1 \\ a_2 & b_2 \end{vmatrix}$$

Thus, by definition, a determinant of the second order is given by

$$\begin{vmatrix} a_1 & b_1 \\ a_2 & b_2 \end{vmatrix} = a_1b_2 - a_2b_1 \tag{4-8}$$

The numbers a_1 and b_1 are called the **elements** of the first **row** of the determinant. The numbers a_1 and a_2 are the elements of the first **column** of the determinant. In the same manner, the numbers a_2 and b_2 are the elements of the second row, and the numbers b_1 and b_2 are the elements of the second column. The numbers a_1 and b_2 are the elements of the **principal diagonal,** and the numbers a_2 and b_1 are the elements of the **secondary diagonal.** Thus, one way of stating the definition indicated in Eq. (4–8) is that *the value of a determinant of the second order is found by taking the product of the elements of the principal diagonal and subtracting the product of the elements of the secondary diagonal.*

A diagram which is often helpful for remembering the expansion of a second-order determinant is shown in Fig. 4–7. The following examples illustrate how we carry out the evaluation of determinants.

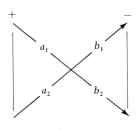

Figure 4—7

Example A

$$\begin{vmatrix} 1 & 4 \\ 3 & 2 \end{vmatrix} = 1(2) - 3(4) = 2 - 12 = -10$$

Example B

$$\begin{vmatrix} -5 & 8 \\ 3 & 7 \end{vmatrix} = (-5)(7) - 3(8) = -35 - 24 = -59$$

Example C

$$\begin{vmatrix} 4 & 6 \\ -3 & 17 \end{vmatrix} = 4(17) - (-3)(6) = 68 + 18 = 86$$

We note that the numerators of Eqs. (4–6) and (4–7) may also be written as determinants. The numerators of Eqs. (4–6) and (4–7) are

$$\begin{vmatrix} c_1 & b_1 \\ c_2 & b_2 \end{vmatrix} \quad \text{and} \quad \begin{vmatrix} a_1 & c_1 \\ a_2 & c_2 \end{vmatrix} \tag{4–9}$$

Therefore, the solutions for x and y of the system of equations (4–3) may be written directly in terms of determinants, without any algebraic operations, as

$$x = \frac{\begin{vmatrix} c_1 & b_1 \\ c_2 & b_2 \end{vmatrix}}{\begin{vmatrix} a_1 & b_1 \\ a_2 & b_2 \end{vmatrix}} \quad \text{and} \quad y = \frac{\begin{vmatrix} a_1 & c_1 \\ a_2 & c_2 \end{vmatrix}}{\begin{vmatrix} a_1 & b_1 \\ a_2 & b_2 \end{vmatrix}} \tag{4–10}$$

For this reason determinants provide a very quick and easy method of solution of systems of equations. Here again the denominator of each of Eqs. (4–10) is the same. *The determinant of the denominator is made up of the coefficients of x and y.* Also, we can see that *the determinant of the numerator of the solution for x may be obtained by replacing the column of a's by the column of c's. The numerator of the solution for y may be obtained from the determinant of the denominator by replacing the column of b's by the column of c's.* This result is often referred to as **Cramer's rule.**

The following examples illustrate the method of solving systems of equations by determinants.

Example D

Solve the following system of equations by determinants:

$$\begin{aligned} 2x + y &= 1 \\ 5x - 2y &= -11 \end{aligned}$$

First we set up the determinant for the denominator. This consists of the four coefficients in the system written as shown. Therefore, the determinant of the denominator is

$$\begin{vmatrix} 2 & 1 \\ 5 & -2 \end{vmatrix}$$

For finding x, the determinant in the numerator is obtained from this determinant by replacing the first column by the constants which appear on the right sides of the equations. Thus, the numerator for the solution for x is

$$\begin{vmatrix} 1 & 1 \\ -11 & -2 \end{vmatrix}$$

For finding y, the determinant in the numerator is obtained from the determinant of the denominator by replacing the second column by the constants which appear on the right sides of the equations. Thus, the determinant for the numerator for finding y is

$$\begin{vmatrix} 2 & 1 \\ 5 & -11 \end{vmatrix}$$

Now we set up the solutions for x and y using the determinants above.

$$x = \frac{\begin{vmatrix} 1 & 1 \\ -11 & -2 \end{vmatrix}}{\begin{vmatrix} 2 & 1 \\ 5 & -2 \end{vmatrix}} = \frac{1(-2) - (-11)(1)}{2(-2) - (5)(1)} = \frac{-2 + 11}{-4 - 5} = \frac{9}{-9} = -1$$

$$y = \frac{\begin{vmatrix} 2 & 1 \\ 5 & -11 \end{vmatrix}}{\begin{vmatrix} 2 & 1 \\ 5 & -2 \end{vmatrix}} = \frac{2(-11) - (5)(1)}{-9} = \frac{-22 - 5}{-9} = 3$$

Therefore, the solution to the system of equations is $x = -1$, $y = 3$.

Since the determinant in the denominators is the same, it needs to be evaluated only once. This means that three determinants are to be evaluated in order to solve the system.

Example E
Solve the following system of equations by determinants:

$$x - y = 4$$
$$2x + y = 11$$

$$x = \frac{\begin{vmatrix} 4 & -1 \\ 11 & 1 \end{vmatrix}}{\begin{vmatrix} 1 & -1 \\ 2 & 1 \end{vmatrix}} = \frac{4 - (-11)}{1 - (-2)} = 5 \qquad y = \frac{\begin{vmatrix} 1 & 4 \\ 2 & 11 \end{vmatrix}}{\begin{vmatrix} 1 & -1 \\ 2 & 1 \end{vmatrix}} = \frac{11 - 8}{3} = 1$$

Therefore, the solution is $x = 5$, $y = 1$.

Example F
Solve the following system of equations by determinants:

$$5x + 7y = 4$$
$$3x - 6y = 5$$

$$x = \frac{\begin{vmatrix} 4 & 7 \\ 5 & -6 \end{vmatrix}}{\begin{vmatrix} 5 & 7 \\ 3 & -6 \end{vmatrix}} = \frac{-24 - 35}{-30 - 21} = \frac{59}{51} \qquad y = \frac{\begin{vmatrix} 5 & 4 \\ 3 & 5 \end{vmatrix}}{\begin{vmatrix} 5 & 7 \\ 3 & -6 \end{vmatrix}} = \frac{25 - 12}{-51} = -\frac{13}{51}$$

Therefore, the solution is $x = \frac{59}{51}$, $y = -\frac{13}{51}$.

Example G
Two investments totaling $18,000 yield an annual income of $700. If the first investment has an interest rate of 5.5% and the second a rate of 3.0%, what is the value of each of the investments?

Let $x =$ the value of the first investment, and $y =$ the value of the second investment. We know that the total of the two investments is $18,000. This leads to the equation $x + y = 18,000$. The first investment yields $0.055x$ dollars annually, and the second yields $0.03y$ dollars annually. This leads to the equation $0.055x + 0.03y = 700$. These two equations are then solved simultaneously.

$$x + y = 18,000$$
$$0.055x + 0.030y = 700$$

$$x = \frac{\begin{vmatrix} 18,000 & 1 \\ 700 & 0.03 \end{vmatrix}}{\begin{vmatrix} 1 & 1 \\ 0.055 & 0.03 \end{vmatrix}} = \frac{540 - 700}{0.03 - 0.055} = \frac{160}{0.025} = 6400$$

The value of y can be found most easily by substituting this value of x into the first equation.

$$y = 18,000 - x = 18,000 - 6400 = 11,600$$

Therefore, the values invested are $6400 and $11,600, respectively.

Some other points should be made here. The equations must be in the form of Eqs. (4–3) before the determinants are set up. This is because the equations for the solutions in terms of determinants are based on that form of writing the system. Also, if either of the unknowns is missing from either equation, its coefficient is taken as zero, and zero is put in the appropriate position in that determinant. Finally, if the determinant of the denominator is zero, and that of the numerator is not zero, the system is inconsistent. If determinants of both numerator and denominator are zero, the system is dependent.

Exercises 4—4

In Exercises 1 through 12 evaluate the given determinants.

1. $\begin{vmatrix} 2 & 4 \\ 3 & 1 \end{vmatrix}$
2. $\begin{vmatrix} -1 & 3 \\ 2 & 6 \end{vmatrix}$
3. $\begin{vmatrix} 3 & -5 \\ 7 & -2 \end{vmatrix}$
4. $\begin{vmatrix} -4 & 7 \\ 1 & -3 \end{vmatrix}$

5. $\begin{vmatrix} 8 & -10 \\ 0 & 4 \end{vmatrix}$
6. $\begin{vmatrix} -4 & -3 \\ 9 & -2 \end{vmatrix}$
7. $\begin{vmatrix} -2 & 11 \\ -7 & -8 \end{vmatrix}$
8. $\begin{vmatrix} -6 & 12 \\ -15 & 3 \end{vmatrix}$

9. $\begin{vmatrix} 7 & -13 \\ 1 & 10 \end{vmatrix}$
10. $\begin{vmatrix} 20 & -5 \\ 28 & 9 \end{vmatrix}$
11. $\begin{vmatrix} 16 & -8 \\ 42 & -15 \end{vmatrix}$
12. $\begin{vmatrix} 43 & -7 \\ -81 & 16 \end{vmatrix}$

In Exercises 13 through 32 solve the given systems of equations by use of determinants. (These systems are the same as those for Exercises 13 through 32 of Section 4–3.)

13. $x + 2y = 5$
    $x - 2y = 1$

14. $x + 3y = 7$
    $2x + 3y = 5$

15. $2x - 3y = 4$
    $2x + y = -4$

16. $x - 4y = 17$
    $3x + 4y = 3$

17. $2x + 3y = 8$
    $x - 2y = -3$

18. $3x - y = 3$
    $4x - 3y = 14$

19. $x + 2y = 7$
    $2x + 4y = 9$

20. $3x - y = 5$
    $-9x + 3y = -15$

21. $2x - 3y = 4$
    $3x - 2y = -2$

22. $3x + 4y = -5$
    $5x - 3y = 2$

23. $3x - 7y = 4$
    $2x + 5y = 7$

24. $5x + 2y = -4$
    $3x - 5y = 6$

25. $2x - y = 5$
    $6x + 2y = -5$

26. $3x + 2y = 4$
    $6x - 6y = 13$

27. $6x + 3y = -4$
    $9x + 5y = -6$

28. $5x - 6y = 1$
    $3x - 4y = 7$

29. $3x - 6y = 15$
    $4x - 8y = 20$

30. $2x + 6y = -3$
    $-6x - 18y = 5$

31. $7x + y = -9$
    $3x + 4y = -11$

32. $6x + 6y = -7$
    $3x - 12y = 13$

In Exercises 33 and 34 solve the given systems of equations by use of determinants.

33. An object traveling at a constant velocity v (in feet per second) is 25 ft from a certain reference point after one second, and 35 ft from it after two seconds. Its initial distance s_0 from the reference point and its velocity can be found by solving the equations

$$s_0 + v = 25$$
$$s_0 + 2v = 35$$

Find s_0 and v.

34. When determining two forces F_1 and F_2 acting on a certain object, the following equations are obtained:

$$0.500F_1 + 0.600F_2 = 10$$
$$0.866F_1 - 0.800F_2 = 20$$

Find F_1 and F_2 (in newtons).

In Exercises 35 through 40 set up appropriate systems of two linear equations in two unknowns and then solve the system by use of determinants.

35. Two meshing gears together have 89 teeth. One of the gears has 4 less than twice the number of teeth of the other gear. How many teeth does each gear have?

36. A rocket is launched so that it averages 3000 mi/h. An hour later, another rocket is launched along the same path at an average speed of 4500 mi/h. Find the times of flight, t_1 and t_2, of the rockets when the second rocket overtakes the first.

37. A total of $8000 is invested, part at 3% and the remainder at 5%. Find the amount invested at each rate if the total annual income is $388.

38. A roof truss is in the shape of an isosceles triangle. The perimeter of the truss is 41 m, and the base is 5 m longer than a rafter (neglecting overhang). Find the length of the base and the length of a rafter.

39. The resistance of a certain wire as a function of temperature can be found from the following equation: $R = \alpha T + \beta$. If the resistance is 0.4 Ω at 20°C and 0.5 Ω at 80°C, find α and β, and then the resistance as a function of temperature.

40. The velocity of sound in steel is 15,900 ft/s faster than the velocity of sound in air. One end of a long steel bar is struck and an observer at the other end measures the time it takes for the sound to reach him. He finds that the sound through the bar takes 0.012 s to reach him and that the sound through the air takes 0.180 s. What are the velocities of sound in air and in steel?

4—5 Solving Systems of Three Linear Equations in Three Unknowns Algebraically

Many problems involve the solution of systems of linear equations which involve three, four, and occasionally even more unknowns. Solving such systems algebraically or by determinants is essentially the same as solving systems of two linear equations in two unknowns. Graphical solutions are not used, since a linear equation in three unknowns represents a plane in space. In this section we shall discuss the algebraic method of solving a system of three linear equations in three unknowns.

A system of three linear equations in three unknowns written in the form

$$
\begin{aligned}
a_1 x + b_1 y + c_1 z &= d_1 \\
a_2 x + b_2 y + c_2 z &= d_2 \\
a_3 x + b_3 y + c_3 z &= d_3
\end{aligned}
\qquad (4\text{--}11)
$$

has as its solution the set of values x, y, and z which satisfy all three equations simultaneously. The method of solution involves multiplying *two* of the equations by the proper numbers to eliminate *one* of the unknowns between these equations. We then repeat this process, using a *different pair* of the original equations, being sure that we eliminate the same unknown as we did between the first pair of equations. At this point we have two linear equations in two unknowns which can be solved by any of the methods previously discussed. The unknown originally eliminated may then be found by substitution into one of the original equations. It is wise to check these three values in one of the other original equations.

Example A

Solve the following system of equations:

$$
\begin{aligned}
(1) &\qquad x + 2y - z = -5 \\
(2) &\qquad 2x - y + 2z = 8 \\
(3) &\qquad 3x + 3y + 4z = 5
\end{aligned}
$$

$$
\begin{aligned}
(4) \quad & 2x + 4y - 2z = -10 \qquad &\text{(1) multiplied by 2} \\
& \underline{2x - y + 2z = 8} \qquad &\text{(2)} \\
(5) \quad & 4x + 3y = -2 \qquad &\text{Adding (4) and (2)}
\end{aligned}
$$

$$
\begin{aligned}
(6) \quad & 4x - 2y + 4z = 16 \qquad &\text{(2) multiplied by 2} \\
& \underline{3x + 3y + 4z = 5} \qquad &\text{(3)} \\
(7) \quad & x - 5y = 11 \qquad &\text{Subtracting}
\end{aligned}
$$

$$
\begin{aligned}
& 4x + 3y = -2 \qquad &\text{(5)} \\
(8) \quad & \underline{4x - 20y = 44} \qquad &\text{(7) multiplied by 4} \\
(9) \quad & 23y = -46 \qquad &\text{Subtracting} \\
(10) \quad & y = -2
\end{aligned}
$$

$$
\begin{aligned}
(11) \quad & x - 5(-2) = 11 \qquad &\text{Substituting (10) in (7)} \\
(12) \quad & x = 1
\end{aligned}
$$

$$
\begin{aligned}
(13) \quad & 1 + 2(-2) - z = -5 \qquad &\text{Substituting (12) and (10) in (1)} \\
(14) \quad & z = 2
\end{aligned}
$$

To check, we substitute the solution $x = 1$, $y = -2$, $z = 2$ in (2).

$$
\begin{aligned}
2(1) - (-2) + 2(2) &\overset{?}{=} 8 \\
8 &= 8 \quad \text{(It checks.)}
\end{aligned}
$$

It should be noted that Eqs. (1), (2), and (3) could be solved just as well by eliminating y between (1) and (2), and then again between (1) and (3). We would than have two equations to solve in x and z. Also, z could have been eliminated between Eqs. (1) and (3) to obtain the second equation in x and y.

Example B
Solve the following system of equations:

$$\begin{array}{llrrll}
(1) & 4x + & y + & 3z = & 1 \\
(2) & 2x - & 2y + & 6z = & 11 \\
(3) & -6x + & 3y + & 12z = & -4
\end{array}$$

(4)	$8x + 2y + 6z = 2$	(1) multiplied by 2
	$\underline{2x - 2y + 6z = 11}$	(2)
(5)	$10x \qquad + 12z = 13$	Adding

(6)	$12x + 3y + 9z = 3$	(1) multiplied by 3
	$\underline{-6x + 3y + 12z = -4}$	(3)
(7)	$18x \qquad - 3z = 7$	Subtracting

	$10x + 12z = 13$	(5)
(8)	$\underline{72x - 12z = 28}$	(7) multiplied by 4
(9)	$82x \qquad = 41$	Adding
(10)	$x = \dfrac{1}{2}$	

$$(11) \qquad 18\left(\frac{1}{2}\right) - 3z = 7 \qquad \text{Substituting (10) in (7)}$$

$$(12) \qquad -3z = -2$$

$$(13) \qquad z = \frac{2}{3}$$

$$(14) \quad 4\left(\frac{1}{2}\right) + y + 3\left(\frac{2}{3}\right) = 1 \qquad \text{Substituting (13) and (10) in (1)}$$

$$(15) \qquad 2 + y + 2 = 1$$
$$(16) \qquad y = -3$$

Therefore, the solution is $x = \frac{1}{2}$, $y = -3$, $z = \frac{2}{3}$. Checking the solution in (2) we have $2(\frac{1}{2}) - 2(-3) + 6(\frac{2}{3}) = 1 + 6 + 4 = 11$.

Example C
Three forces F_1, F_2, and F_3 are acting on a beam. Find the forces (in pounds). The forces are determined by solving the following equations:

$$\begin{array}{llrrll}
(1) & F_1 + & F_2 + & F_3 = & 25 \\
(2) & F_1 + & 2F_2 + & 3F_3 = & 59 \\
(3) & 2F_1 + & 2F_2 - & F_3 = & 5
\end{array}$$

(4)	$3F_1 + 3F_2 + 3F_3 = 75$	(1) multiplied by 3
	$\underline{F_1 + 2F_2 + 3F_3 = 59}$	(2)
(5)	$2F_1 + F_2 \qquad = 16$	Subtracting

(6) $3F_1 + 3F_2 \qquad = 30$ (1) added to (3)

$\quad\quad\; 2F_1 + \; F_2 \qquad = 16$ (5)

(7) $\dfrac{F_1 + \; F_2 \qquad\;\; = 10}{}$ (6) divided by 3

(8) $\quad F_1 \qquad\qquad = \;\; 6$ Subtracting

(9) $\qquad\qquad 6 + F_2 = 10$ (8) substituted in (7)

(10) $\qquad\qquad\quad\;\; F_2 = \;\; 4$

(11) $\qquad\; 6 + 4 + F_3 = 25$ Substituting (8) and (10) in (1)

(12) $\qquad\qquad\quad\;\; F_3 = 15$

Therefore, the three forces are 6 lb, 4 lb, and 15 lb, respectively. This solution can be checked by substituting in Eq. (2) or Eq. (3).

Example D

A triangle has a perimeter of 37 in. The longest side is 3 in. longer than the next longest, which in turn is 8 in. longer than the shortest side. Find the length of each side.

Let a = length of the longest side, b = length of the next-longest side, and c = length of the shortest side. Since the perimeter is 37 in., we have $a + b + c = 37$. The statement of the problem also leads to the equations $a = b + 3$ and $b = c + 8$. These equations are then put into standard form and solved simultaneously.

(1) $a + \; b + c = 37$

(2) $a - \; b \qquad\;\; = \;\; 3$ Rewriting the second equation

(3) $\dfrac{\qquad\; b - c = \;\; 8}{}$ Rewriting the third equation

(4) $a + 2b \qquad = 45$ Adding (1) and (3)

$\quad\quad \dfrac{a - \; b \qquad\;\; = \;\; 3}{}$ (2)

(5) $\qquad\; 3b \qquad = 42$ Subtracting

(6) $\qquad\qquad\;\; b = 14$

(7) $\qquad a - 14 = \;\; 3$ Substituting (6) in (2)

(8) $\qquad\qquad\;\; a = 17$

(9) $\qquad\; 14 - c = \;\; 8$ Substituting (6) in (3)

(10) $\qquad\qquad\;\; c = \;\; 6$

Therefore, the three sides of the triangle are 17 in., 14 in., and 6 in.

Checking the solution, the sum of the lengths of the three sides is 17 in. + 14 in. + 6 in. = 37 in., and the perimeter was given to be 37 in.

For systems of equations with more than three unknowns, the solution is found in a manner similar to that used with three unknowns. For example, with four unknowns one of the unknowns is eliminated between three different pairs of equations. The result is three equations in the remaining three unknowns. The solution then follows the procedure used with three unknowns.

Exercises 4—5

In Exercises 1 through 16 solve the given systems of equations.

1. $x + y + z = 2$
   $x - z = 1$
   $x + y = 1$

2. $x + y - z = -3$
   $x + z = 2$
   $2x - y + 2z = 3$

3. $2x + 3y + z = 2$
   $-x + 2y + 3z = -1$
   $-3x - 3y + z = 0$

4. $2x + y - z = 4$
   $4x - 3y - 2z = -2$
   $8x - 2y - 3z = 3$

5. $5x + 6y - 3z = 6$
   $4x - 7y - 2z = -3$
   $3x + y - 7z = 1$

6. $3r + s - t = 2$
   $r - 2s + t = 0$
   $4r - s + t = 3$

7. $2x - 2y + 3z = 5$
   $2x + y - 2z = -1$
   $4x - y - 3z = 0$

8. $2u + 2v + 3w = 0$
   $3u + v + 4w = 21$
   $-u - 3v + 7w = 15$

9. $3x - 7y + 3z = 6$
   $3x + 3y + 6z = 1$
   $5x - 5y + 2z = 5$

10. $8x + y + z = 1$
    $7x - 2y + 9z = -3$
    $4x - 6y + 8z = -5$

11. $x + 2y + 2z = 0$
    $2x + 6y - 3z = -1$
    $4x - 3y + 6z = -8$

12. $3x + 3y + z = 6$
    $2x + 2y - z = 9$
    $4x + 2y - 3z = 16$

13. $2x + 3y - 5z = 7$
    $4x - 3y - 2z = 1$
    $8x - y + 4z = 3$

14. $2x - 4y - 4z = 3$
    $3x + 8y + 2z = -11$
    $4x + 6y - z = -8$

15. $r - s - 3t - u = 1$
    $2r + 4s - 2u = 2$
    $3r + 4s - 2t = 0$
    $r + 2t - 3u = 3$

16. $3x + 2y - 4z + 2t = 3$
    $5x - 3y - 5z + 6t = 8$
    $2x - y + 3z - 2t = 1$
    $-2x + 3y + 2z - 3t = -2$

In Exercises 17 through 20 solve the given problems.

17. Show that the following system of equations has an unlimited number of solutions, and find one of them.

$$x - 2y - 3z = 2$$
$$x - 4y - 13z = 14$$
$$-3x + 5y + 4z = 0$$

18. Show that the following system of equations has no solution.

$$x - 2y - 3z = 2$$
$$x - 4y - 13z = 14$$
$$-3x + 5y + 4z = 2$$

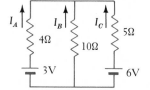

Figure 4−8

19. In applying Kirchhoff's laws (e.g., see Beiser, *Modern Technical Physics*, second ed., p. 445) to the electric circuit shown in Fig. 4−8, the following equations are found. Determine the indicated currents. (In Fig. 4−8, *I* signifies current, in amperes).

$$I_A + I_B + I_C = 0$$
$$4I_A - 10I_B = 3$$
$$-10I_B + 5I_C = 6$$

20. In a laboratory experiment to measure the acceleration of an object, the distances traveled by the object were recorded for three time intervals. These data led to the following equations:

$$s_0 + 2v_0 + 2a = 20$$
$$s_0 + 4v_0 + 8a = 54$$
$$s_0 + 6v_0 + 18a = 104$$

Here s_0 is the initial displacement (in feet), v_0 is the initial velocity (in feet per second), and a is the acceleration (in feet per second squared). Find s_0, v_0, and a.

In Exercises 21 through 24 set up systems of three linear equations and solve for the indicated quantities.

21. Three machines together produce 64 parts each hour. Three times the production of the first machine equals the production of the other two machines together. Five times the production of the second is 12 parts per hour more than twice the rate of the other two together. Find the production rates of the three machines.

22. Angle A of a triangle equals 20° less than the sum of angles B and C. (The triangle is not a right triangle.) Angle B is one-fifth the sum of angles A and C. Find the angles.

23. By volume, one alloy is 60% copper, 30% zinc, and 10% nickel. A second alloy has percentages 50, 30, and 20, respectively, of the three metals. A third alloy is 30% copper and 70% nickel. How much of each must be mixed so that 100 cm³ of the resulting alloy has percentages of 40, 15, and 45, respectively?

24. The budget of a certain corporation for three positions in a given department is $70,000. Position A pays as much as the other two positions together, and position A pays $5,000 more than twice position C. What do the three positions pay?

4—6 Solving Systems of Three Linear Equations in Three Unknowns by Determinants

Just as systems of two linear equations in two unknowns can be solved by the use of determinants, so can systems of three linear equations in three unknowns. The system as given in Eqs. (4–11) can be solved in general terms by the method of elimination by addition or subtraction. This leads to the following solutions for x, y, and z.

$$x = \frac{d_1 b_2 c_3 + d_3 b_1 c_2 + d_2 b_3 c_1 - d_3 b_2 c_1 - d_1 b_3 c_2 - d_2 b_1 c_3}{a_1 b_2 c_3 + a_3 b_1 c_2 + a_2 b_3 c_1 - a_3 b_2 c_1 - a_1 b_3 c_2 - a_2 b_1 c_3}$$

$$y = \frac{a_1 d_2 c_3 + a_3 d_1 c_2 + a_2 d_3 c_1 - a_3 d_2 c_1 - a_1 d_3 c_2 - a_2 d_1 c_3}{a_1 b_2 c_3 + a_3 b_1 c_2 + a_2 b_3 c_1 - a_3 b_2 c_1 - a_1 b_3 c_2 - a_2 b_1 c_3} \qquad (4\text{–}12)$$

$$z = \frac{a_1 b_2 d_3 + a_3 b_1 d_2 + a_2 b_3 d_1 - a_3 b_2 d_1 - a_1 b_3 d_2 - a_2 b_1 d_3}{a_1 b_2 c_3 + a_3 b_1 c_2 + a_2 b_3 c_1 - a_3 b_2 c_1 - a_1 b_3 c_2 - a_2 b_1 c_3}$$

The expression that appears in each of the denominators of Eqs. (4–12) is an example of a **determinant of the third order**. This determinant is denoted by the symbol

$$\begin{vmatrix} a_1 & b_1 & c_1 \\ a_2 & b_2 & c_2 \\ a_3 & b_3 & c_3 \end{vmatrix}$$

Therefore, a determinant of the third order is defined by the equation

$$\begin{vmatrix} a_1 & b_1 & c_1 \\ a_2 & b_2 & c_2 \\ a_3 & b_3 & c_3 \end{vmatrix} = a_1 b_2 c_3 + a_3 b_1 c_2 + a_2 b_3 c_1 - a_3 b_2 c_1 - a_1 b_3 c_2 - a_2 b_1 c_3 \qquad (4\text{–}13)$$

The elements, rows, columns, and diagonals of a third-order determinant are defined just as are those of a second-order determinant. For example, the principal diagonal is made up of the elements a_1, b_2, and c_3.

Probably the easiest way of remembering the method of determining the value of a third-order determinant is as follows (this method does *not* work for determinants of order higher than three): *Rewrite the first and second columns to the right of the determinant. The products of the elements of the principal diagonal and the two parallel diagonals to the right of it are then added. The products of the elements of the secondary diagonal and the two parallel diagonals to the right of it are subtracted from the first sum. The algebraic sum of these six products gives the value of the determinant* (see Fig. 4–9). Examples A through C illustrate the evaluation of third-order determinants.

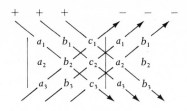

Figure 4—9

Example A

$$\begin{vmatrix} 1 & 5 & 4 \\ -2 & 3 & -1 \\ 2 & -1 & 5 \end{vmatrix} \begin{matrix} 1 & 5 \\ -2 & 3 \\ 2 & -1 \end{matrix} = 15 + (-10) + (+8) - (24) - (1) - (-50) = 38$$

Example B

$$\begin{vmatrix} -1 & 4 & -5 \\ 6 & 1 & 0 \\ 9 & -7 & 3 \end{vmatrix} \begin{matrix} -1 & 4 \\ 6 & 1 \\ 9 & -7 \end{matrix} = (-3) + 0 + 210 - (-45) - 0 - 72 = 180$$

Example C

$$\begin{vmatrix} 3 & -2 & 8 \\ -5 & 5 & -2 \\ 4 & 9 & -6 \end{vmatrix} \begin{matrix} 3 & -2 \\ -5 & 5 \\ 4 & 9 \end{matrix} = (-90) + 16 + (-360) - 160 - (-54) \\ - (-60) = -480$$

Inspection of Eqs. (4–12) reveals that the numerators of these solutions may also be written in terms of determinants. Thus, we may write the general solution to a system of three equations in three unknowns as

$$x = \frac{\begin{vmatrix} d_1 & b_1 & c_1 \\ d_2 & b_2 & c_2 \\ d_3 & b_3 & c_3 \end{vmatrix}}{\begin{vmatrix} a_1 & b_1 & c_1 \\ a_2 & b_2 & c_2 \\ a_3 & b_3 & c_3 \end{vmatrix}} \qquad y = \frac{\begin{vmatrix} a_1 & d_1 & c_1 \\ a_2 & d_2 & c_2 \\ a_3 & d_3 & c_3 \end{vmatrix}}{\begin{vmatrix} a_1 & b_1 & c_1 \\ a_2 & b_2 & c_2 \\ a_3 & b_3 & c_3 \end{vmatrix}} \qquad z = \frac{\begin{vmatrix} a_1 & b_1 & d_1 \\ a_2 & b_2 & d_2 \\ a_3 & b_3 & d_3 \end{vmatrix}}{\begin{vmatrix} a_1 & b_1 & c_1 \\ a_2 & b_2 & c_2 \\ a_3 & b_3 & c_3 \end{vmatrix}} \qquad (4-14)$$

If the determinant of the denominator is zero and the determinant of the numerator is not zero, the system is **inconsistent.** If the determinant of the denominator is not equal to zero, then there is a unique solution to the system of equations.

An analysis of Eqs. (4–14) shows that the situation is precisely the same as it was when we were using determinants to solve systems of two linear equations. That is, the determinants in the denominators in the expressions for x, y, and z are the same. They consist of elements which are the coefficients of the unknowns. The determinant of the numerator of the solution for x is the same as that of the denominator, except that the column of d's replaces the column of a's. The determinant in the numerator of the solution for y is the same as that of the denominator, except that the column of d's replaces the column of b's. The determinant of the numerator of the solution for z is the same as the determinant of the denominator, except that the column of d's replaces the column of c's. To summarize, we can state that *the determinants in the numerators are the same as those in the denominators, except that the column of d's replaces the column of coefficients of the unknown for which we are solving.* This again is Cramer's rule. Remember that the equations must be written in the standard form shown in Eqs. (4–11) before the determinants are formed.

Example D

Solve the following system of equations by determinants:

$$x + 2y + 2z = 1$$
$$2x - y + z = 3$$
$$4x + y + 2z = 0$$

$$x = \frac{\begin{vmatrix} 1 & 2 & 2 \\ 3 & -1 & 1 \\ 0 & 1 & 2 \end{vmatrix} \begin{matrix} 1 & 2 \\ 3 & -1 \\ 0 & 1 \end{matrix}}{\begin{vmatrix} 1 & 2 & 2 \\ 2 & -1 & 1 \\ 4 & 1 & 2 \end{vmatrix} \begin{matrix} 1 & 2 \\ 2 & -1 \\ 4 & 1 \end{matrix}} = \frac{-2 + 0 + 6 - 0 - 1 - 12}{-2 + 8 + 4 - (-8) - 1 - 8} = \frac{-9}{+9} = -1$$

Since the value of the denominator is already determined, there is no need to write the denominator in determinant form when solving for y and z.

$$y = \frac{\begin{vmatrix} 1 & 1 & 2 \\ 2 & 3 & 1 \\ 4 & 0 & 2 \end{vmatrix} \begin{matrix} 1 & 1 \\ 2 & 3 \\ 4 & 0 \end{matrix}}{9} = \frac{6 + 4 + 0 - 24 - 0 - 4}{9} = \frac{-18}{9} = -2$$

$$z = \frac{\begin{vmatrix} 1 & 2 & 1 \\ 2 & -1 & 3 \\ 4 & 1 & 0 \end{vmatrix} \begin{matrix} 1 & 2 \\ 2 & -1 \\ 4 & 1 \end{matrix}}{9} = \frac{0 + 24 + 2 - (-4) - 3 - 0}{9} = \frac{27}{9} = 3$$

As a check, we substitute these values into the first equation.

$$-1 + 2(-2) + 2(3) \overset{?}{=} 1, \quad 1 = 1. \text{ Thus, it checks.}$$

Example E
Solve the following system by determinants:

$$3x + 2y - 5z = -1$$
$$2x - 3y - z = 11$$
$$5x - 2y + 7z = 9$$

$$x = \frac{\begin{vmatrix} -1 & 2 & -5 \\ 11 & -3 & -1 \\ 9 & -2 & 7 \end{vmatrix}\begin{matrix} -1 & 2 \\ 11 & -3 \\ 9 & -2 \end{matrix}}{\begin{vmatrix} 3 & 2 & -5 \\ 2 & -3 & -1 \\ 5 & -2 & 7 \end{vmatrix}\begin{matrix} 3 & 2 \\ 2 & -3 \\ 5 & -2 \end{matrix}} = \frac{21 - 18 + 110 - 135 + 2 - 154}{-63 - 10 + 20 - 75 - 6 - 28}$$

$$= \frac{-174}{-162} = \frac{29}{27}$$

$$y = \frac{\begin{vmatrix} 3 & -1 & -5 \\ 2 & 11 & -1 \\ 5 & 9 & 7 \end{vmatrix}\begin{matrix} 3 & -1 \\ 2 & 11 \\ 5 & 9 \end{matrix}}{-162} = \frac{231 + 5 - 90 + 275 + 27 + 14}{-162} = \frac{462}{-162}$$

$$= -\frac{77}{27}$$

$$z = \frac{\begin{vmatrix} 3 & 2 & -1 \\ 2 & -3 & 11 \\ 5 & -2 & 9 \end{vmatrix}\begin{matrix} 3 & 2 \\ 2 & -3 \\ 5 & -2 \end{matrix}}{-162} = \frac{-81 + 110 + 4 - 15 + 66 - 36}{-162} = \frac{48}{-162}$$

$$= -\frac{8}{27}$$

Substituting into the second equation, we have

$$2\left(\frac{29}{27}\right) - 3\left(-\frac{77}{27}\right) - \left(-\frac{8}{27}\right) = \frac{58 + 231 + 8}{27} = \frac{297}{27}$$

$$= 11$$

which shows that the solution checks.

Example F

An 8% solution, a 10% solution, and a 20% solution of nitric acid are to be mixed in order to get 100 mL of a 12% solution. If the volume of acid from the 8% solution equals half the volume of acid from the other two solutions, how much of each is needed?

Let x = volume of 8% solution needed, y = volume of 10% solution needed, and z = volume of 20% solution needed.

We first use the fact that the sum of the volumes of the three solutions is 100 mL. This leads to the equation $x + y + z = 100$. Next we note that there are $0.08x$ mL of pure nitric acid from the first solution, $0.10y$ mL from the second, $0.20z$ mL from the third solution, and $0.12(100)$ mL in the final solution. This leads to the equation $0.08x + 0.10y + 0.20z = 12$. Finally, using the last stated condition, we have $0.08x = \frac{1}{2}(0.10y + 0.20z)$. These equations are then rewritten in standard form, simplified, and solved.

$$
\begin{aligned}
x + \quad y + \quad z &= 100 \\
0.08x + 0.10y + 0.20z &= \ \ 12 \\
0.08x \qquad\qquad\qquad &= 0.05y + 0.10z
\end{aligned}
$$

$$
\begin{aligned}
x + \quad y + \quad z &= 100 \\
4x + \quad 5y + \quad 10z &= 600 \\
8x - \quad 5y - \quad 10z &= \ \ \ 0
\end{aligned}
$$

$$
x = \cfrac{\begin{vmatrix} 100 & 1 & 1 \\ 600 & 5 & 10 \\ 0 & -5 & -10 \end{vmatrix} \begin{matrix} 100 & 1 \\ 600 & 5 \\ 0 & -5 \end{matrix}}{\begin{vmatrix} 1 & 1 & 1 \\ 4 & 5 & 10 \\ 8 & -5 & -10 \end{vmatrix} \begin{matrix} 1 & 1 \\ 4 & 5 \\ 8 & -5 \end{matrix}}
$$

$$
= \frac{-5000 + 0 - 3000 - 0 + 5000 + 6000}{-50 + 80 - 20 - 40 + 50 + 40} = \frac{3000}{60} = 50
$$

$$
y = \cfrac{\begin{vmatrix} 1 & 100 & 1 \\ 4 & 600 & 10 \\ 8 & 0 & -10 \end{vmatrix} \begin{matrix} 1 & 100 \\ 4 & 600 \\ 8 & 0 \end{matrix}}{60}
$$

$$
= \frac{-6000 + 8000 + 0 - 4800 - 0 + 4000}{60} = \frac{1200}{60} = 20
$$

$$z = \frac{\begin{vmatrix} 1 & 1 & 100 \\ 4 & 5 & 600 \\ 8 & -5 & 0 \end{vmatrix} \begin{matrix} 1 & 1 \\ 4 & 5 \\ 8 & -5 \end{matrix}}{60}$$

$$= \frac{0 + 4800 - 2000 - 4000 + 3000 - 0}{60} = \frac{1800}{60} = 30$$

Therefore, 50 mL of the 8% solution, 20 mL of the 10% solution, and 30 mL of the 20% solution are required to make the 12% solution. Substitution into the first equation shows that this answer is correct.

Additional techniques which are useful in solving systems of equations are taken up in Chapter 15. Methods of evaluating determinants which are particularly useful for higher order determinants are also discussed.

Exercises 4—6

In Exercises 1 through 12 evaluate the given third-order determinants.

1. $\begin{vmatrix} 5 & 4 & -1 \\ -2 & -6 & 8 \\ 7 & 1 & 1 \end{vmatrix}$

2. $\begin{vmatrix} -7 & 0 & 0 \\ 2 & 4 & 5 \\ 1 & 4 & 2 \end{vmatrix}$

3. $\begin{vmatrix} 8 & 9 & -6 \\ -3 & 7 & 2 \\ 4 & -2 & 5 \end{vmatrix}$

4. $\begin{vmatrix} -2 & 6 & -2 \\ 5 & -1 & 4 \\ 8 & -3 & -2 \end{vmatrix}$

5. $\begin{vmatrix} -3 & -4 & -8 \\ 5 & -1 & 0 \\ 2 & 10 & -1 \end{vmatrix}$

6. $\begin{vmatrix} 10 & 2 & -7 \\ -2 & -3 & 6 \\ 6 & 5 & -2 \end{vmatrix}$

7. $\begin{vmatrix} 4 & -3 & -11 \\ -9 & 2 & -2 \\ 0 & 1 & -5 \end{vmatrix}$

8. $\begin{vmatrix} 9 & -2 & 0 \\ -1 & 3 & -6 \\ -4 & -6 & -2 \end{vmatrix}$

9. $\begin{vmatrix} 5 & 4 & -5 \\ -3 & 2 & -1 \\ 7 & 1 & 3 \end{vmatrix}$

10. $\begin{vmatrix} 20 & 0 & -15 \\ -4 & 30 & 1 \\ 6 & -1 & 40 \end{vmatrix}$

11. $\begin{vmatrix} 0.1 & -0.2 & 0 \\ -0.5 & 1 & 0.4 \\ -2 & 0.8 & 2 \end{vmatrix}$

12. $\begin{vmatrix} 0.2 & -0.5 & -0.4 \\ 1.2 & 0.3 & 0.2 \\ -0.5 & 0.1 & -0.4 \end{vmatrix}$

In Exercises 13 through 28 solve the given systems of equations by use of determinants. (Exercises 15 through 28 are the same as Exercises 1 through 14 of Section 4–5.)

13. $2x + 3y + z = 4$
$3x - z = -3$
$x - 2y + 2z = -5$

14. $4x + y + z = 2$
$2x - y - z = 4$
$3y + z = 2$

15. $x + y + z = 2$
$x - z = 1$
$x + y = 1$

16. $x + y - z = -3$
$x + z = 2$
$2x - y + 2z = 3$

17. $2x + 3y + z = 2$
$-x + 2y + 3z = -1$
$-3x - 3y + z = 0$

18. $2x + y - z = 4$
$4x - 3y - 2z = -2$
$8x - 2y - 3z = 3$

19. $5x + 6y - 3z = 6$
$4x - 7y - 2z = -3$
$3x + y - 7z = 1$

20. $3r + s - t = 2$
$r - 2s + t = 0$
$4r - s + t = 3$

21. $2x - 2y + 3z = 5$
$2x + y - 2z = -1$
$4x - y - 3z = 0$

22. $2u + 2v + 3w = 0$
$3u + v + 4w = 21$
$-u - 3v + 7w = 15$

23. $3x - 7y + 3z = 6$
$3x + 3y + 6z = 1$
$5x - 5y + 2z = 5$

24. $8x + y + z = 1$
$7x - 2y + 9z = -3$
$4x - 6y + 8z = -5$

25. $x + 2y + 2z = 0$
$2x + 6y - 3z = -1$
$4x - 3y + 6z = -8$

26. $3x + 3y + z = 6$
$2x + 2y - z = 9$
$4x + 2y - 3z = 16$

27. $2x + 3y - 5z = 7$
$4x - 3y - 2z = 1$
$8x - y + 4z = 3$

28. $2x - 4y - 4z = 3$
$3x + 8y + 2z = -11$
$4x + 6y - z = -8$

In Exercises 29 through 32 solve the given problems by determinants. In Exercises 31 and 32 set up appropriate equations and then solve them.

29. In analyzing the forces on the bell-crank mechanism shown in Fig. 4–10, the following equations are obtained. Find the indicated forces.

$$A \quad - 0.6F = 80$$
$$B - 0.8F = 0$$
$$6A \quad - 10F = 0$$

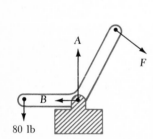
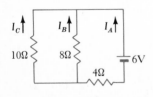

Figure 4–10

30. In applying Kirchhoff's laws (see Exercise 19 of Section 4–5) to the given electric circuit, these equations result.

$$I_A + I_B + I_C = 0$$
$$- 8I_B + 10I_C = 0$$
$$4I_A - 8I_B = 6$$

Determine the indicated currents, in amperes. (See Fig. 4–11).

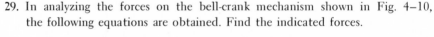

Figure 4–11

31. Twenty thousand dollars is invested part at 5.5%, part at 4.5%, and part at 4.0% (all of the amount is invested), yielding an annual interest of $1015. The income from the 5.5% investment yields $305 more annually than the other two investments combined. How much money is invested at each percentage?

32. A person traveled a total of 1870 mi from the time he left home until he reached his destination. He averaged 40 mi/h while driving to the airport. The plane averaged 600 mi/h, and he spent twice as long in the plane as in his car. The taxi averaged 30 mi/h between the airport and the destination. Assuming 40 min were used in making connections, what were the times he spent in his car, in the plane, and in the taxi, if the trip took 5.5 h?

4—7 Exercises for Chapter 4

In Exercises 1 through 4 evaluate the given determinants.

1. $\begin{vmatrix} -2 & 5 \\ 3 & 1 \end{vmatrix}$ 2. $\begin{vmatrix} 4 & 0 \\ -2 & -6 \end{vmatrix}$ 3. $\begin{vmatrix} -8 & -3 \\ -1 & 4 \end{vmatrix}$ 4. $\begin{vmatrix} 9 & -1 \\ 7 & -5 \end{vmatrix}$

In Exercises 5 through 12 solve the given systems of equations graphically.

5. $2x - y = 4$
   $3x + 2y = 6$

6. $3x + y = 3$
   $2x - y = 6$

7. $4x - y = 6$
   $3x + 2y = 12$

8. $2x - 5y = 10$
   $3x + y = 6$

9. $7x - 2y = 14$
   $4x + y = 4$

10. $5x + 3y = 15$
    $6x - y = 12$

11. $3x + 4y = 6$
    $2x - 3y = 2$

12. $5x + 2y = 5$
    $2x - 4y = 3$

In Exercises 13 through 24 solve the given systems of equations algebraically.

13. $x + 2y = 5$
    $x + 3y = 7$

14. $2x - y = 7$
    $x + y = 2$

15. $4x + 3y = -4$
    $2x - y = 3$

16. $x + 3y = -2$
    $-2x - 9y = 2$

17. $3x + 4y = 6$
    $9x + 8y = 11$

18. $3x - 6y = 5$
    $7x + 2y = 4$

19. $2x - 5y = 8$
    $5x - 3y = 7$

20. $3x + 4y = 8$
    $2x - 3y = 9$

21. $7x - 2y = -6$
    $4x + 7y = 12$

22. $5x + 3y = 8$
    $6x - 8y = 11$

23. $9x - 11y = 4$
    $6x - 3y = 5$

24. $3x - 4y = 9$
    $7x + 10y = -2$

In Exercises 25 through 36 solve the given systems of equations by use of determinants. (These systems are the same as for Exercises 13 through 24.)

25. $x + 2y = 5$
    $x + 3y = 7$

26. $2x - y = 7$
    $x + y = 2$

27. $4x + 3y = -4$
    $2x - y = 3$

28. $x + 3y = -2$
    $-2x - 9y = 2$

29. $3x + 4y = 6$
    $9x + 8y = 11$

30. $3x - 6y = 5$
    $7x + 2y = 4$

31. $2x - 5y = 8$
$5x - 3y = 7$

32. $3x + 4y = 8$
$2x - 3y = 9$

33. $7x - 2y = -6$
$4x + 7y = 12$

34. $5x + 3y = 8$
$6x - 8y = 11$

35. $9x - 11y = 4$
$6x - 3y = 5$

36. $3x - 4y = 9$
$7x + 10y = -2$

In Exercises 37 through 40 evaluate the given determinants.

37. $\begin{vmatrix} 4 & -1 & 8 \\ -1 & 6 & -2 \\ 2 & 1 & -1 \end{vmatrix}$

38. $\begin{vmatrix} -5 & 0 & -5 \\ 2 & 3 & -1 \\ -3 & 2 & 2 \end{vmatrix}$

39. $\begin{vmatrix} -2 & -4 & 7 \\ 1 & 6 & -3 \\ -7 & 2 & -1 \end{vmatrix}$

40. $\begin{vmatrix} 3 & 2 & -1 \\ 0 & -3 & 4 \\ 3 & -4 & -2 \end{vmatrix}$

In Exercises 41 through 48 solve the given systems of equations algebraically.

41. $2x + y + z = 4$
$x - 2y - z = 3$
$3x + 3y - 2z = 1$

42. $x + 2y + z = 2$
$3x - 6y + 2z = 2$
$2x - z = 8$

43. $3x + 2y + z = 1$
$9x - 4y + 2z = 8$
$12x - 18y = 17$

44. $2x + 2y - z = 2$
$3x + 4y + z = -4$
$5x - 2y - 3z = 5$

45. $2r + s + 2t = 8$
$3r - 2s - 4t = 5$
$-2r + 3s + 4t = -3$

46. $2u + 2v - w = -2$
$4u - 3v + 2w = -2$
$8u - 4v - 3w = 13$

47. $4x + 6y - z = -2$
$3x - 5y + 4z = 8$
$6x + 4y + 2z = 5$

48. $3t + 2u + 6v = 3$
$4t - 3u + 2v = 13$
$5t + u + v = 0$

In Exercises 49 through 56 solve the given systems of equations by use of determinants. (These systems are the same as for Exercises 41 through 48.)

49. $2x + y + z = 4$
$x - 2y - z = 3$
$3x + 3y - 2z = 1$

50. $x + 2y + z = 2$
$3x - 6y + 2z = 2$
$2x - z = 8$

51. $3x + 2y + z = 1$
$9x - 4y + 2z = 8$
$12x - 18y = 17$

52. $2x + 2y - z = 2$
$3x + 4y + z = -4$
$5x - 2y - 3z = 5$

53. $2r + s + 2t = 8$
$3r - 2s - 4t = 5$
$-2r + 3s + 4t = -3$

54. $2u + 2v - w = -2$
$4u - 3v + 2w = -2$
$8u - 4v - 3w = 13$

55. $4x + 6y - z = -2$
$3x - 5y + 4z = 8$
$6x + 4y + 2z = 5$

56. $3t + 2u + 6v = 3$
$4t - 3u + 2v = 13$
$5t + u + v = 0$

In Exercises 57 through 60, let $1/x = u$ and $1/y = v$. Solve for u and v, and then solve for x and y. In this way we will see how to solve systems of equations involving reciprocals.

57. $\dfrac{1}{x} - \dfrac{1}{y} = \dfrac{1}{2}$ 58. $\dfrac{1}{x} + \dfrac{1}{y} = 3$ 59. $\dfrac{2}{x} + \dfrac{3}{y} = 3$ 60. $\dfrac{3}{x} - \dfrac{2}{y} = 4$

$\dfrac{1}{x} + \dfrac{1}{y} = \dfrac{1}{4}$ $\dfrac{2}{x} + \dfrac{1}{y} = 1$ $\dfrac{5}{x} - \dfrac{6}{y} = 3$ $\dfrac{2}{x} + \dfrac{4}{y} = 1$

In Exercises 61 and 62 determine the value of a which makes the system dependent. In Exercises 63 and 64 determine the value of a which makes the system inconsistent.

61. $3x - ay = 6$ 62. $5x + 20y = 15$ 63. $ax - 2y = 5$ 64. $2x - 5y = 7$
$x + 2y = 2$ $2x + ay = 6$ $4x + 6y = 1$ $ax + 10y = 2$

Solve the systems of equations in Exercises 65 through 68 by any appropriate method.

65. In an experiment to determine the values of two electrical resistors, the following equations were determined:

$$2R_1 + 3R_2 = 16$$
$$3R_1 + 2R_2 = 19$$

Determine the resistances R_1 and R_2 (in ohms).

66. The production of nitric acid makes use of air and nitrogen compounds. In order to determine requirements as to size of equipment, a relationship between the air flow rate m (in moles per hour) and exhaust nitrogen rate n is often used. One particular operation produces the following equations:

$$1.58m + 41.5 = 38.0 + 2.00n$$
$$0.424m + 36.4 = 189 + 0.0728n$$

Solve for m and n.

67. In a certain machine there are three important types of parts. Considering the total number of parts used, their cost, and the time used in their manufacture, the number of each type used can be determined by solving the system of equations

$$a + b + c = 37$$
$$2a + 3b + 5c = 131$$
$$3a + 5b + 6c = 180$$

where a, b, and c are the numbers of each part used, respectively. Find a, b, and c.

68. In applying Kirchhoff's laws (see Exercise 19 of Section 4–5) to the given electric circuit, these equations result.

$$I_A + I_B + I_C = 0$$
$$6I_A \quad\quad - 10I_C = 8$$
$$6I_A - 2I_B \quad\quad = 5$$

Determine the indicated currents (in amperes). (See Fig. 4–12.)

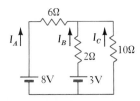

Figure 4–12

In Exercises 69 through 74 set up systems of equations and solve by any appropriate method.

69. A plane traveled 2000 mi, with the wind, in 4 h, and it made the return trip in 5 h. Determine the speed of the plane and that of the wind.

70. The relation between Fahrenheit temperature F and Celsius temperature C can be indicated by $F = aC + b$. If $0°C$ is equivalent to $32°F$ and $100°C$ is equivalent to $212°F$, find a and b.

71. If a lever is balanced by placing a single support (fulcrum) under a certain point, then the product of a weight on one side of the fulcrum and its distance from the fulcrum equals the product of a weight on the other side of the fulcrum times its distance from the fulcrum. A certain lever is balanced if a weight of 80 lb is put at one end and a weight of 30 lb at the other end. If the 80-lb weight is moved 1 ft closer to the fulcrum, it requires 20 lb at the other end to maintain the balance. How long is the lever? (Neglect the weight of the lever itself.)

72. The relative density of an object may be defined as its weight in air divided by the difference of its weight in air and its weight when submerged in water. If the sum of the weights in water and in air of an object of relative density equal to 10 is 30 lb, what is its weight in air?

73. An alloy important in the manufacture of electric transformers contains nickel, iron, and molybdenum. The percentage of nickel is 1% less than five times the percentage of iron. The percentage of iron is 1% more than three times the percentage of molybdenum. Find the percentage of each metal in the alloy.

74. A businessman is interested in a building site for a new factory. He determines that at least 30 acres of land are necessary and that over 40 acres would be too expensive. He learns of a 160-acre tract of land which has been subdivided according to terrain such that one portion is 16 acres less than the sum of the other two and that twice the area of the second portion is 8 acres more than the area of the third portion. Would any of the portions be of interest to him?

5

Factoring and Fractions

5—1 Special Products

In Chapter 1 we introduced certain fundamental algebraic operations. These have been sufficient for our purposes to this point. However, material we shall encounter later requires additional algebraic techniques. In this chapter we shall develop certain basic algebraic topics, which in turn will allow us to develop other topics having technical and scientific applications.

In working with algebraic expressions, we encounter certain types of products so frequently that we should become extremely familiar with them. These products are obtained by the methods of multiplication of algebraic expressions developed in Chapter 1 and are stated here in general form.

$$a(x + y) = ax + ay \tag{5–1}$$

$$(x + y)(x - y) = x^2 - y^2 \tag{5–2}$$

$$(x + y)^2 = x^2 + 2xy + y^2 \tag{5–3}$$

$$(x - y)^2 = x^2 - 2xy + y^2 \tag{5–4}$$

$$(x + a)(x + b) = x^2 + (a + b)x + ab \tag{5–5}$$

$$(ax + b)(cx + d) = acx^2 + (bc + ad)x + bd \tag{5–6}$$

These **special products** should be known thoroughly such that the multiplications represented are easily and clearly recognized. When this is the case, they allow us to perform many multiplications quickly and easily, often by inspection. We must realize that they are written in their most concise form. Any of the literal numbers appearing in these products may represent an expression which in turn represents a number.

Example A

Using Eq. (5–1) in the following product, we have

$$6(3r + 2s) = 6(3r) + 6(2s) = 18r + 12s$$

Using Eq. (5–2), we have

$$(3r + 2s)(3r - 2s) = (3r)^2 - (2s)^2 = 9r^2 - 4s^2$$

When we use Eq. (5–1) in the first illustration, we see that $a = 6$. In both illustrations $3r = x$ and $2s = y$.

Example B

Using Eqs. (5–3) and (5–4) in the following products, we have

$$(5a + 2)^2 = (5a)^2 + 2(5a)(2) + 2^2 = 25a^2 + 20a + 4$$
$$(5a - 2)^2 = (5a)^2 - 2(5a)(2) + 2^2 = 25a^2 - 20a + 4$$

In these illustrations, we have let $x = 5a$ and $y = 2$. It should also be emphasized that $(5a + 2)^2$ is not $(5a)^2 + 2^2$, or $25a^2 + 4$. We must be careful to properly follow the correct form of Eqs. (5–3) and (5–4) and include the middle term, $20a$.

Example C

Using Eqs. (5–5) and (5–6) in the following products, we have

$$(x + 5)(x - 3) = x^2 + [(5 + (-3)]x + (5)(-3) = x^2 + 2x - 15$$
$$(4x + 5)(2x - 3) = (4x)(2x) + [5(2) + 4(-3)]x + (5)(-3)$$
$$= 8x^2 - 2x - 15$$

Generally, when we use these special products, we do the middle step as shown in each of the examples above mentally and write down the result directly. This is indicated in the following example.

Example D

$$2(x - 6) = 2x - 12$$
$$(y - 5)(y + 5) = y^2 - 25$$
$$(3x - 2)^2 = 9x^2 - 12x + 4$$
$$(x - 4)(x + 7) = x^2 + 3x - 28$$

At times these special products may appear in combinations. When this happens it may be necessary to indicate an intermediate step.

Example E

In expanding $7(a + 2)(a - 2)$, we use Eqs. (5–2) and (5–1), preferably in that order. Performing this operation, we have

$$7(a + 2)(a - 2) = 7(a^2 - 4) = 7a^2 - 28$$

Example F

In determining the product $(x + y - 2)^2$, we may group the quantity $(x + y)$ in an intermediate step. This leads to

$$(x + y - 2)^2 = [(x + y) - 2]^2 = (x + y)^2 - 2(x + y)(2) + 2^2$$
$$= x^2 + 2xy + y^2 - 4x - 4y + 4$$

In this example we used Eqs. (5–3) and (5–4).

There are four other special products which occur less frequently. However, they are sufficiently important that they should be readily recognized. They are shown in Eqs. (5–7) through (5–10).

$$(x + y)^3 = x^3 + 3x^2y + 3xy^2 + y^3 \tag{5–7}$$
$$(x - y)^3 = x^3 - 3x^2y + 3xy^2 - y^3 \tag{5–8}$$
$$(x + y)(x^2 - xy + y^2) = x^3 + y^3 \tag{5–9}$$
$$(x - y)(x^2 + xy + y^2) = x^3 - y^3 \tag{5–10}$$

The following examples illustrate the use of Eqs. (5–7) through (5–10).

Example G

$$(x + 4)^3 = x^3 + 3(x^2)(4) + 3(x)(4^2) + 4^3$$
$$= x^3 + 12x^2 + 48x + 64$$
$$(2x - 5)^3 = (2x)^3 - 3(2x)^2(5) + 3(2x)(5^2) - 5^3$$
$$= 8x^3 - 60x^2 + 150x - 125$$

Example H

$$(x + 3)(x^2 - 3x + 9) = x^3 + 3^3 = x^3 + 27$$
$$(x - 2)(x^2 + 2x + 4) = x^3 - 2^3 = x^3 - 8$$

Exercises 5–1

In Exercises 1 through 28 find the indicated products directly *by inspection*. It should not be necessary to write down intermediate steps.

1. $40(x - y)$
2. $2x(a - 3)$
3. $2x^2(x - 4)$
4. $3a^2(2a + 7)$
5. $(y + 6)(y - 6)$
6. $(s + 2t)(s - 2t)$
7. $(3v - 2)(3v + 2)$
8. $(ab - c)(ab + c)$
9. $(5f + 4)^2$
10. $(i_1 + 3)^2$
11. $(2x + 7)^2$
12. $(5a + 2b)^2$
13. $(x - 2y)^2$
14. $(a - 5p)^2$
15. $(6s - t)^2$
16. $(3p - 4q)^2$
17. $(x + 1)(x + 5)$
18. $(y - 8)(y + 5)$

19. $(3 + c)(6 + c)$ 20. $(t - 1)(t - 7)$ 21. $(3x - 1)(2x + 5)$

22. $(2x - 7)(2x + 1)$ 23. $(4x - 5)(5x + 1)$ 24. $(2y - 1)(3y - 1)$

25. $(5v - 3)(4v + 5)$ 26. $(7s + 6)(2s + 5)$

27. $(3x + 7y)(2x - 9y)$ 28. $(8x - y)(3x + 4y)$

Use the special products of this section to determine the products of Exercises 29 through 48. You may need to write down one or two intermediate steps.

29. $2(x - 2)(x + 2)$ 30. $5(n - 5)(n + 5)$

31. $2a(2a - 1)(2a + 1)$ 32. $4c(2c - 3)(2c + 3)$

33. $6a(x + 2b)^2$ 34. $4y^2(y + 6)^2$

35. $4a(2a - 3)^2$ 36. $6t^2(5t - 3s)^2$

37. $(x + y + 1)^2$ 38. $(x + 2 + 3y)^2$

39. $(3 - x - y)^2$ 40. $2(x - y + 1)^2$

41. $(5 - t)^3$ 42. $(2s + 3)^3$

43. $(2x + 5t)^3$ 44. $(x - 5y)^3$

45. $(x + 2)(x^2 - 2x + 4)$ 46. $(a - 3)(a^2 + 3a + 9)$

47. $(4 - 3x)(16 + 12x + 9x^2)$ 48. $(2x + 3a)(4x^2 - 6ax + 9a^2)$

Use the special products of this section to determine the products in Exercises 49 through 54. Each comes from the technical area indicated.

49. $R(i_1 + i_2)^2$ (electricity)

50. $Fa(L - a)(L + a)$ (mechanics: force on a beam)

51. $16(4 + t)(3 - t)$ (physics: motion)

52. $V^2(V - b)^2$ (thermodynamics)

53. $P(1 + r)^3$ (business: compound interest)

54. $p^2(1 - p)^3$ (mathematics: probability)

5–2 Factoring: Common Factor and Difference of Squares

We often find that we want to determine what expressions can be multiplied together to equal a given algebraic expression. We know from Section 1–7 that when an algebraic expression is the product of two or more quantities, each of these quantities is called a **factor** of the expression. *Therefore, determining these factors, which is essentially reversing the process of finding a product, is called* **factoring.**

In our work on factoring we shall consider only the factoring of polynomials (see Section 1–9) which have integers as coefficients for all terms. Also, all factors will have integral coefficients. *A polynomial or a factor is called* **prime** *if it contains no factors other than* $+1$ *or* -1 *and plus or minus itself. We say that an expression is* **factored completely** *if it is expressed as a product of its prime factors.*

Example A
When we factor the expression $12x + 6x^2$ as

$$12x + 6x^2 = 2(6x + 3x^2)$$

we see that it has not been factored completely. The factor $6x + 3x^2$ is not prime, for it may be factored as

$$6x + 3x^2 = 3x(2 + x)$$

Therefore, the expression $12x + 6x^2$ is factored completely as

$$12x + 6x^2 = 6x(2 + x)$$

Here the factors x and $2 + x$ are prime. The numerical coefficient, 6, could be factored into $(2)(3)$, but it is normal practice not to factor numerical coefficients.

To factor expressions easily, we must be familiar with algebraic multiplication, particularly the special products of the preceding section. The solution of factoring problems is heavily dependent on the recognition of special products. The special products also provide methods of checking answers and deciding whether or not a given factor is prime.

Often an algebraic expression contains a monomial that is common to each term of the expression. Therefore, in accordance with Eq. (5–1), *the first step in factoring any expression should be to factor out any* **common monomial factor** that may exist.

Example B
In factoring the expression $6x - 2y$, we note that each term contains the factor 2. Therefore,

$$6x - 2y = 2(3x - y)$$

Here, 2 is the common monomial factor, and $2(3x - y)$ is the required factored form of $6x - 2y$.

Example C
Factor: $4ax^2 + 2ax$.

The numerical factor 2 and the literal factors a and x are common to each term. Therefore, the common monomial factor of $4ax^2 + 2ax$ is $2ax$. This means that

$$4ax^2 + 2ax = 2ax(2x + 1)$$

Note the presence of the 1 in the factored form. Since each of the factors of the second term is also a factor of the common monomial factor, we must include the factor of 1. Otherwise, if the factored form is multiplied out, we would not obtain the proper expression.

Example D

Factor: $6a^5x^2 - 9a^3x^3 + 3a^3x^2$.

After inspecting each term, we determine that each contains a factor of 3, a^3, and x^2. Thus, the common monomial factor is $3a^3x^2$. This means that

$$6a^5x^2 - 9a^3x^3 + 3a^3x^2 = 3a^3x^2(2a^2 - 3x + 1)$$

Another important form for factoring is based on the special product of Eq. (5–2). In Eq. (5–2) we see that the product of the sum and the difference of two numbers results in the difference between the squares of the numbers. Therefore, *factoring the difference between squares gives factors which are the sum and the difference of the numbers.*

Example E

In factoring $x^2 - 16$, we note that x^2 is the square of x and that 16 is the square of 4. Therefore,

$$x^2 - 16 = x^2 - 4^2 = (x + 4)(x - 4)$$

Usually in factoring an expression of this type we do not actually write out the middle step as shown.

Example F

Since $4x^2$ is the square of $2x$ and 9 is the square of 3, we may factor $4x^2 - 9$ as

$$4x^2 - 9 = (2x + 3)(2x - 3)$$

In the same way,

$$16x^4 - 25y^2 = (4x^2 + 5y)(4x^2 - 5y)$$

where we note that $16x^4 = (4x^2)^2$ and $25y^2 = (5y)^2$.

As indicated previously, *if it is possible to factor out a common monomial factor, this factoring should be done first.* We should then inspect the resulting factors to see if further factoring can be done. It is possible, for example, that the resulting factor is a difference of squares. Thus, complete factoring often requires more than one step. Be sure to include all prime factors in writing the result.

Example G

In factoring $20x^2 - 45$, we note that there is a common factor of 5 in each term. Therefore, $20x^2 - 45 = 5(4x^2 - 9)$. However, the factor $4x^2 - 9$ itself is the difference of squares. Therefore, $20x^2 - 45$ is completely factored as

$$20x^2 - 45 = 5(4x^2 - 9) = 5(2x + 3)(2x - 3)$$

In factoring $x^4 - y^4$, we note that we have the difference of squares. Therefore, $x^4 - y^4 = (x^2 + y^2)(x^2 - y^2)$. However, the factor $x^2 - y^2$ is also the difference of squares. This means that

$$x^4 - y^4 = (x^2 + y^2)(x^2 - y^2) = (x^2 + y^2)(x + y)(x - y)$$

The factor $x^2 + y^2$ is prime.

Exercises 5–2

In Exercises 1 through 32 factor the given expressions completely.

1. $6x + 6y$

2. $3a - 3b$

3. $5a - 5$

4. $2x^2 + 2$

5. $3x^2 - 9x$

6. $4s^2 + 20s$

7. $7b^2y - 28b$

8. $5a^2 - 20ax$

9. $2x + 4y - 8z$

10. $10a - 5b + 15c$

11. $3ab^2 - 6ab + 12ab^3$

12. $4pq - 14q^2 - 16pq^2$

13. $12pq^2 - 8pq - 28pq^3$

14. $27a^2b - 24ab - 9a$

15. $2a^2 - 2b^2 + 4c^2 - 6d^2$

16. $5a + 10ax - 5ay + 20az$

17. $x^2 - 4$

18. $r^2 - 25$

19. $100 - y^2$

20. $49 - z^2$

21. $81s^2 - 25t^2$

22. $36s^2 - 121t^2$

23. $144n^2 - 169p^4$

24. $36a^2b^2 - 169c^2$

25. $2x^2 - 8$

26. $5a^2 - 125$

27. $3x^2 - 27z^2$

28. $4x^2 - 100y^2$

29. $x^4 - 16$

30. $y^4 - 81$

31. $x^8 - 1$

32. $2x^4 - 8y^4$

In Exercises 33 through 36 the expressions are to be factored by a method known as **factoring by grouping**. An illustration of this method is

$$2x - 2y + ax - ay = (2x - 2y) + (ax - ay)$$
$$= 2(x - y) + a(x - y) = (2 + a)(x - y)$$

The terms are put into groups, a common factor is factored from each group, and then the factoring is continued.

33. $3x - 3y + bx - by$

34. $am + an + cn + cm$

35. $a^2 + ax - ab - bx$

36. $2y - y^2 - 6y^4 + 12y^3$

Factor the expressions given in Exercises 37 through 42. Each comes from the technical area indicated.

37. $iR_1 + iR_2 + ir$ (electricity)

38. $P + Prt$ (business: interest)

39. $mv_1^2 - mv_2^2$ (mechanics: energy)

40. $kD^2 - 4kr^2$ (hydrodynamics)

41. The difference in the expressions for the volume of a cube with edge e and a rectangular solid of edges 2, 2 and e is $e^3 - 4e$. Factor this expression.

42. The difference of the energy radiated by an electric light filament at temperature T_2 and that radiated by a filament at temperature T_1 is given by the formula

$$R = kT_2^4 - kT_1^4$$

Factor the right-hand side of this formula.

5–3 Factoring Trinomials

In the previous section we introduced the concept of factoring, and considered factoring based on special products of Eqs. (5–1) and (5–2). We now note that the special products formed from Eqs. (5–3) through (5–6) all result in trinomial (three term) polynomials. Thus, trinomials of the types formed from these products are important expressions to be factored, and this section is devoted to them.

Factoring expressions based on the special product of Eq. (5–5) will result in factors of the form $x + a$ and $x + b$. The numbers a and b are found by analyzing the constant and the coefficient of x in the expression to be factored.

Example A

In factoring $x^2 + 3x + 2$, the constant term, 2, suggests that the only possibilities for a and b are 2 and 1. The plus sign before the 2 indicates that the sign before the 1 and the 2 in the factors must be the same, either both plus or both minus. Since the coefficient of the middle term is the sum of a and b, the plus sign before the 3 tells us that the sign before the 2 and 1 must be plus. Therefore,

$$x^2 + 3x + 2 = (x + 2)(x + 1)$$

For an expression containing x^2 and 2 to be factored, the middle term must be $3x$. No other combination of a and b gives the proper middle term. Therefore, the expression

$$x^2 + 4x + 2$$

cannot be factored. The a and b would have to be 2 and 1, but the middle term would not be $4x$.

Example B

In order to factor $x^2 + 7x - 8$, we must find two integers whose product is -8 and whose sum is $+7$. The possible factors of -8 are -8 and $+1$, $+8$ and -1, -4 and $+2$, and $+4$ and -2. Inspecting these we see that only $+8$ and -1 have the sum of $+7$. Therefore,

$$x^2 + 7x - 8 = (x + 8)(x - 1)$$

In the same way, we have

$$x^2 - x - 12 = (x - 4)(x + 3)$$

and

$$x^2 - 5xy + 6y^2 = (x - 3y)(x - 2y)$$

In the last illustration we note that we were to find second terms of each factor such that their product was $6y^2$ and sum was $-5xy$. Thus, each term must contain a factor of y.

Example C

In order to factor $x^2 + 10x + 25$, we must find two integers whose sum is $+10$ and whose product is $+25$. Since $5^2 = 25$ we note that this expression may fit the form of Eq. (5–3). This could be the case only if the first and third terms were perfect squares. We see that the sum of $+5$ and $+5$ is $+10$, which means

$$x^2 + 10x + 25 = (x + 5)(x + 5)$$

or

$$x^2 + 10x + 25 = (x + 5)^2$$

Factoring expressions based upon the special product of Eq. (5–6) often requires some trial and error. The coefficient of x^2 gives the possibilities for the coefficients a and c in the factors. The constant gives the possibilities for the numbers b and d in the factors. It is then necessary to try possible combinations to determine which combination provides the middle term of the given expression.

Example D
When factoring the expression $2x^2 + 11x + 5$, the coefficient 2 indicates that the only possibilities for the x-terms in the factors are $2x$ and x. The 5 indicates that only 5 and 1 may be the constants. Therefore, possible combinations are $2x + 1$, and $x + 5$ or $2x + 5$ and $x + 1$. The combination which gives the coefficient of the middle term, 11, is

$$2x^2 + 11x + 5 = (2x + 1)(x + 5)$$

According to this analysis, the expression $2x^2 + 10x + 5$ is not factorable, but the following expression is:

$$2x^2 + 7x + 5 = (2x + 5)(x + 1)$$

Example E
Other examples of factoring based upon the special product of Eq. (5–6) are as follows:

$$4x^2 + 4x - 3 = (2x - 1)(2x + 3)$$
$$3x^2 - 13x + 12 = (3x - 4)(x - 3)$$
$$6s^2 + 19st - 20t^2 = (6s - 5t)(s + 4t)$$

In the first illustration, possible factors with x terms of $4x$ and x can be shown to be incorrect if tried. In the second illustration, other possible factorizations of 12, such as 6×2 and 12×1, can be shown to give improper middle terms. In the third illustration, there are numerous possibilities for the combination of 6 and 20. *We must remember to check carefully that the middle term of the expression is the proper result of the factors we have chosen.*

Example F
In factoring $9x^2 - 6x + 1$, we note that $9x^2$ is the square of $3x$ and 1 is the square of 1. Therefore, we recognize that this expression might fit the perfect square form of Eq. (5–4). This leads us to tentatively factor it as

$$9x^2 - 6x + 1 = (3x - 1)^2$$

However, before we can be certain that this factorization is correct, we must check to see if the middle term of the expansion of $(3x - 1)^2$ is $-6x$. Since $-6x$ properly fits the form of Eq. (5–4), the factorization is correct.
In the same way, we have

$$36x^2 + 84xy + 49y^2 = (6x + 7y)^2$$

As we pointed out in Section 5–2, we must be careful to see that we have factored an expression completely. We look for common monomial factors first, and then check each resulting factor. This check of each factor should be made every time we complete a step in factoring.

Example G

When factoring $2x^2 + 6x - 8$, we first note the common monomial factor of 2. This leads to

$$2x^2 + 6x - 8 = 2(x^2 + 3x - 4)$$

We now notice that $x^2 + 3x - 4$ is also factorable. Therefore,

$$2x^2 + 6x - 8 = 2(x + 4)(x - 1)$$

Now each factor is prime.

Exercises 5–3

In Exercises 1 through 32 factor the given expressions completely.

1. $x^2 + 5x + 4$ 2. $x^2 - 5x - 6$
3. $s^2 - s - 42$ 4. $a^2 + 14a - 32$
5. $x^2 + 2x + 1$ 6. $y^2 + 8y + 16$
7. $x^2 - 4x + 4$ 8. $b^2 - 12b + 36$
9. $3x^2 - 5x - 2$ 10. $2n^2 - 13n - 7$
11. $3y^2 - 8y - 3$ 12. $5x^2 + 9x - 2$
13. $3t^2 - 7tu + 4u^2$ 14. $3x^2 + xy - 14y^2$
15. $9x^2 + 7xy - 2y^2$ 16. $4r^2 + 11rs - 3s^2$
17. $4m^2 + 20m + 25$ 18. $16q^2 + 24q + 9$
19. $4x^2 - 12x + 9$ 20. $a^2c^2 - 2ac + 1$
21. $9t^2 - 15t + 4$ 22. $6x^2 + x - 12$
23. $8b^2 + 31b - 4$ 24. $12n^2 + 8n - 15$
25. $4p^2 - 25pq + 6q^2$ 26. $12x^2 + 4xy - 5y^2$
27. $12x^2 + 47xy - 4y^2$ 28. $8r^2 - 14rs - 9s^2$
29. $2x^2 - 14x + 12$ 30. $6y^2 - 33y - 18$
31. $4x^2 + 14x - 8$ 32. $12x^2 + 22x - 4$

In Exercises 33 through 36 factor the given expression by referring directly to the special products in Eqs. (5–7) through (5–10), respectively. These expressions are not trinomials, but their factorization depends on the proper recognition of their forms, as with Eqs. (5–3) and (5–4).

33. $x^3 + 3x^2 + 3x + 1$ 34. $x^3 - 6x^2 + 12x - 8$
35. $8x^3 + 1$ 36. $x^3 - 27$

Factor the expressions given in Exercises 37 through 40. Each comes from the technical area indicated.

37. $16t^2 - 32t - 128$ (physics: motion)
38. $2p^2 - 108p + 400$ (business)
39. $x^2 - 3Lx + 2L^2$ (mechanics: beams)
40. $V^2 - 2nBV + n^2B^2$ (chemistry)

5—4 Equivalent Fractions

When we deal with algebraic expressions, we must be able to work effectively with fractions. Since algebraic expressions are representations of numbers, the basic operations on fractions from arithmetic will form the basis of our algebraic operations. In this section we shall demonstrate a very important property of fractions, and in the following two sections we shall establish the basic algebraic operations with fractions.

This important property of fractions, often referred to as the **fundamental principle of fractions,** is that *the value of a fraction is unchanged if both numerator and denominator are multiplied or divided by the same number, provided this number is not zero.* Two fractions are said to be **equivalent** if one can be obtained from the other by use of the fundamental theorem.

Example A
If we multiply the numerator and denominator of the fraction $\frac{6}{8}$ by 2, we obtain the equivalent fraction $\frac{12}{16}$. If we divide the numerator and denominator of $\frac{6}{8}$ by 2, we obtain the equivalent fraction $\frac{3}{4}$. Therefore, the fractions $\frac{6}{8}$, $\frac{3}{4}$, and $\frac{12}{16}$ are equivalent.

Example B
We may write

$$\frac{ax}{2} = \frac{3a^2x}{6a}$$

since the right fraction is obtained from the left fraction by multiplying the numerator and the denominator by $3a$. Therefore, the fractions are equivalent.

One of the most important operations to be performed on a fraction is that of reducing it to its **simplest form,** or **lowest terms.** *A fraction is said to be in its simplest form if the numerator and the denominator have no common factors other than $+1$ or -1.* In reducing a fraction to its simplest form, we use the fundamental theorem by dividing both the numerator and the denominator by all factors which are common to each. (It will be assumed throughout this text that if any of the literal symbols were to be evaluated, numerical values would be restricted so that none of the denominators would be zero. Thereby, we avoid the undefined operation of division by zero.)

Example C
In order to reduce the fraction

$$\frac{16ab^3c^2}{24ab^2c^5}$$

to its lowest terms, we note that both the numerator and the denominator contain the factor $8ab^2c^2$. Therefore, we may write

$$\frac{16ab^3c^2}{24ab^2c^5} = \frac{2b(8ab^2c^2)}{3c^3(8ab^2c^2)} = \frac{2b}{3c^3}$$

In Example C we divided out the common factor. The resulting fraction is in lowest terms since there are no common factors in the numerator and the denominator other than $+1$ or -1.

We must note very carefully that in simplifying fractions, *we divide both the numerator and the denominator by the common factor.* This process is called **cancellation**. However, many students are tempted to try to remove any expression which appears in both the numerator and the denominator. If a *term* is removed in this way, it is an incorrect application of the cancellation process. The following example illustrates this common error in the simplification of fractions.

Example D
When simplifying the expression

$$\frac{x^2(x-2)}{x^2-4}$$

many students would "cancel" the x^2 from the numerator and the denominator. This is *incorrect*, since x^2 is only a term of the denominator.

In order to simplify the fraction above properly, we should factor the denominator. We obtain

$$\frac{x^2(x-2)}{(x-2)(x+2)} = \frac{x^2}{x+2}$$

Here, the common *factor* $x-2$ has been divided out.

The following examples illustrate the proper simplification of fractions.

Example E

$$\frac{4a}{2a^2x} = \frac{2}{ax}$$

We divide out the common factor of $2a$.

$$\frac{4a}{2a^2+x}$$

This cannot be reduced, since there are no common *factors* in the numerator and the denominator.

Example F

$$\frac{x^2-4x+4}{x^2-4} = \frac{(x-2)(x-2)}{(x+2)(x-2)} = \frac{x-2}{x+2}$$

In this simplification, the numerator and the denominator have each been factored first and then the common factor $x-2$ has been divided out. In the final form, neither the x's nor 2's may be cancelled, since they are not common *factors*.

Example G

$$\frac{4x^2 + 14x - 30}{8x - 12} = \frac{2(2x^2 + 7x - 15)}{4(2x - 3)} = \frac{2(2x - 3)(x + 5)}{4(2x - 3)} = \frac{x + 5}{2}$$

Here, the factors common to the numerator and the denominator are 2 and $(2x - 3)$.

In simplifying fractions we must be able to distinguish between factors which differ only in *sign*. Since $-(y - x) = -y + x = x - y$, we have

$$x - y = -(y - x) \tag{5–11}$$

Here we see that **factors $x - y$ and $y - x$ differ only in sign.** The following examples illustrate the simplification of fractions where a change of signs is necessary.

Example H

$$\frac{x^2 - 1}{1 - x} = \frac{(x - 1)(x + 1)}{-(x - 1)} = \frac{x + 1}{-1} = -(x + 1)$$

In the second fraction, we replaced $1 - x$ with the equal expression $-(x - 1)$. In the third fraction the common factor $x - 1$ was divided out. Finally, we expressed the result in the more convenient form by dividing $x + 1$ by -1, which makes the quantity $x + 1$ negative.

Example I

$$\frac{2x^3 - 32x}{20 + 7x - 3x^2} = \frac{2x(x^2 - 16)}{(4 - x)(5 + 3x)} = \frac{2x(x - 4)(x + 4)}{-(x - 4)(3x + 5)}$$

$$= -\frac{2x(x + 4)}{3x + 5}$$

Again, the factor $4 - x$ has been replaced by the equal expression $-(x - 4)$. This allows us to recognize the common factor of $x - 4$. Notice also that the order of the terms of the factor $5 + 3x$ has been changed to $3x + 5$. This is merely an application of the commutative law of addition.

Exercises 5–4

In Exercises 1 through 8 multiply the numerator and the denominator of each of the given fractions by the given factor and obtain an equivalent fraction.

1. $\dfrac{2}{3}$ (by 7)

2. $\dfrac{7}{5}$ (by 9)

3. $\dfrac{ax}{y}$ (by $2x$)

4. $\dfrac{2x^2y}{3n}$ (by $2xn^2$)

5. $\dfrac{2}{x + 3}$ (by $x - 2$)

6. $\dfrac{7}{a - 1}$ (by $a + 2$)

7. $\dfrac{a(x - y)}{x - 2y}$ (by $x + y$)

8. $\dfrac{x - 1}{x + 1}$ (by $x - 1$)

In Exercises 9 through 16 divide the numerator and the denominator of each of the given fractions by the given factor and obtain an equivalent fraction.

9. $\dfrac{28}{44}$ (by 4)

10. $\dfrac{25}{65}$ (by 5)

11. $\dfrac{4x^2y}{8xy^2}$ (by $2x$)

12. $\dfrac{6a^3b^2}{9a^5b^4}$ (by $3a^2b^2$)

13. $\dfrac{2(x-1)}{(x-1)(x+1)}$ (by $x-1$)

14. $\dfrac{(x+5)(x-3)}{3(x+5)}$ (by $x+5$)

15. $\dfrac{x^2-3x-10}{2x^2+3x-2}$ (by $x+2$)

16. $\dfrac{6x^2+13x-5}{6x^3-2x^2}$ (by $3x-1$)

In Exercises 17 through 40 reduce each fraction to its simplest form.

17. $\dfrac{2a}{8a}$

18. $\dfrac{6x}{15x}$

19. $\dfrac{18x^2y}{24xy}$

20. $\dfrac{2a^2xy}{6axyz^2}$

21. $\dfrac{a+b}{5a^2+5ab}$

22. $\dfrac{t-a}{t^2-a^2}$

23. $\dfrac{6a-4b}{4a-2b}$

24. $\dfrac{5r-20s}{10r-5s}$

25. $\dfrac{4x^2+1}{4x^2-1}$

26. $\dfrac{x^2-y^2}{x^2+y^2}$

27. $\dfrac{x^2-8x+16}{x^2-16}$

28. $\dfrac{4a^2+12ab+9b^2}{4a^2+6ab}$

29. $\dfrac{2x^2+5x-3}{x^2+11x+24}$

30. $\dfrac{4r^2-8rs-5s^2}{6r^2-17rs+5s^2}$

31. $\dfrac{x^4-16}{x+2}$

32. $\dfrac{2x^2-8}{4x+8}$

33. $\dfrac{x^2y^4-x^4y^2}{y^2-2xy+x^2}$

34. $\dfrac{8x^3+8x^2+2x}{4x+2}$

35. $\dfrac{(x-1)(3+x)}{(3-x)(1-x)}$

36. $\dfrac{(2x-1)(x+6)}{(x-3)(1-2x)}$

37. $\dfrac{y-x}{2x-2y}$

38. $\dfrac{x^2-y^2}{y-x}$

39. $\dfrac{(x+5)(x-2)(x+2)(3-x)}{(2-x)(5-x)(3+x)(2+x)}$

40. $\dfrac{(2x-3)(3-x)(x-7)(3x+1)}{(3x+2)(3-2x)(x-3)(7+x)}$

In Exercises 41 through 44 reduce each fraction to its simplest form. This will require the use of Eqs. (5–7) through (5–10).

41. $\dfrac{x^3-y^3}{x^2-y^2}$

42. $\dfrac{x^3-8}{x^2+2x+4}$

43. $\dfrac{x^3+3x^2+3x+1}{x^3+1}$

44. $\dfrac{a^3-6a^2+12a-8}{a^2-4a+4}$

In Exercises 45 through 48 determine which fractions are in simplest form.

45. (a) $\dfrac{x^2(x+2)}{x^2+4}$ (b) $\dfrac{x^4+4x^2}{x^4-16}$

46. (a) $\dfrac{2x+3}{2x+6}$ (b) $\dfrac{2(x+6)}{2x+6}$

47. (a) $\dfrac{x^2 - x - 2}{x^2 - x}$ (b) $\dfrac{x^2 - x - 2}{x^2 + x}$ 48. (a) $\dfrac{x^3 - x}{1 - x}$ (b) $\dfrac{2x^2 + 4x}{2x^2 + 4}$

5—5 Multiplication and Division of Fractions

From arithmetic we recall that *the product of two fractions is a fraction whose numerator is the product of the numerators and whose denominator is the product of the denominators of the given fractions.* Also, we recall that *we can find the quotient of two fractions by inverting the divisor and proceeding as in multiplication.* Symbolically, multiplication of fractions is indicated by

$$\frac{a}{b} \cdot \frac{c}{d} = \frac{ac}{bd}$$

and division is indicated by

$$\frac{\dfrac{a}{b}}{\dfrac{c}{d}} = \frac{a}{b} \cdot \frac{d}{c} = \frac{ad}{bc}$$

The rule for division may be verified by use of the fundamental principle of fractions. By multiplying the numerator and denominator of the fraction

$$\frac{\dfrac{a}{b}}{\dfrac{c}{d}} \quad \text{by} \quad \frac{d}{c} \quad \text{we obtain} \quad \frac{\dfrac{a}{b} \cdot \dfrac{d}{c}}{\dfrac{c}{d} \cdot \dfrac{d}{c}}$$

In the resulting denominator $\dfrac{c}{d} \cdot \dfrac{d}{c}$ becomes 1, and therefore the fraction is written as $\dfrac{ad}{bc}$.

Example A

$$\frac{3}{5} \cdot \frac{2}{7} = \frac{(3)(2)}{(5)(7)} = \frac{6}{35}$$

$$\frac{3a}{5b} \cdot \frac{15b^2}{a} = \frac{(3a)(15b^2)}{(5b)(a)} = \frac{45ab^2}{5ab} = \frac{9b}{1} = 9b$$

In the second illustration, we have divided out the common factor of $5ab$ to reduce the resulting fraction to its simplest form.

We shall usually want to express the result in its simplest form, which is generally its most useful form. Since all factors in the numerators and all factors in the denominators are to be multiplied, we should *first only indicate the multiplication and then factor the numerator and the denominator.* In this way we can easily identify any factors common to both. If we were to multiply out the numerator and the denominator before factoring, it is very possible that we would be unable to factor the result to simplify it. The following example illustrates this point.

Example B

In performing the multiplication

$$\frac{3(x-y)}{(x-y)^2} \cdot \frac{(x^2-y^2)}{6x+9y}$$

if we multiply out the numerators and the denominators before performing any factoring, we would have to simplify the fraction

$$\frac{3x^3 - 3x^2y - 3xy^2 + 3y^3}{6x^3 - 3x^2y - 12xy^2 + 9y^3}$$

It is possible to factor the numerator and the denominator, but finding any common factors this way is very difficult. If we first indicate the multiplications and then factor completely, we have

$$\frac{3(x-y)}{(x-y)^2} \cdot \frac{(x^2-y^2)}{6x+9y} = \frac{3(x-y)(x^2-y^2)}{(x-y)^2(6x+9y)} = \frac{3(x-y)(x+y)(x-y)}{(x-y)^2(3)(2x+3y)}$$

$$= \frac{3(x-y)^2(x+y)}{3(x-y)^2(2x+3y)}$$

$$= \frac{x+y}{2x+3y}$$

The common factor of $3(x-y)^2$ is readily recognized using this procedure.

Example C

$$\frac{2x-4}{4x+12} \cdot \frac{2x^2+x-15}{3x-1} = \frac{2(x-2)(2x-5)(x+3)}{4(x+3)(3x-1)}$$

$$= \frac{(x-2)(2x-5)}{2(3x-1)}$$

Here the common factor is $2(x+3)$. It is permissible to multiply out the final form of the numerator and the denominator, but it is often preferable to leave the numerator and denominator in factored form, as indicated.

The following examples illustrate the division of fractions.

Example D

$$\frac{6x}{7} \div \frac{5}{3} = \frac{6x}{7} \cdot \frac{3}{5} = \frac{18x}{35}$$

$$\frac{\dfrac{3a^2}{5c}}{\dfrac{2c^2}{a}} = \frac{3a^2}{5c} \cdot \frac{a}{2c^2} = \frac{3a^3}{10c^3}$$

Example E

$$\frac{x+y}{3} \div \frac{2x+2y}{6x+15y} = \frac{x+y}{3} \cdot \frac{6x+15y}{2x+2y} = \frac{(x+y)(3)(2x+5y)}{3(2)(x+y)}$$

$$= \frac{2x+5y}{2}$$

Example F

$$\frac{\dfrac{4-x^2}{x^2-3x+2}}{\dfrac{x+2}{x^2-9}} = \frac{4-x^2}{x^2-3x+2} \cdot \frac{x^2-9}{x+2}$$

$$= \frac{(2-x)(2+x)(x-3)(x+3)}{(x-2)(x-1)(x+2)}$$

$$= \frac{-(x-2)(x+2)(x-3)(x+3)}{(x-2)(x-1)(x+2)}$$

$$= -\frac{(x-3)(x+3)}{x-1} \quad \text{or} \quad \frac{(x-3)(x+3)}{1-x}$$

Note the use of Eq. (5–11) in the simplification and in expressing an alternate form of the result. The factor $2 - x$ was replaced by its equivalent $-(x - 2)$, and then $x - 1$ was replaced by $-(1 - x)$.

Exercises 5–5

In Exercises 1 through 28 perform the indicated operations and simplify.

1. $\dfrac{3}{8} \cdot \dfrac{2}{7}$

2. $\dfrac{11}{5} \cdot \dfrac{13}{33}$

3. $\dfrac{4x}{3y} \cdot \dfrac{9y^2}{2}$

4. $\dfrac{18sy^3}{ax^2} \cdot \dfrac{(ax)^2}{3s}$

5. $\dfrac{2}{9} \div \dfrac{4}{7}$

6. $\dfrac{5}{16} \div \dfrac{25}{13}$

7. $\dfrac{xy}{az} \div \dfrac{bz}{ay}$

8. $\dfrac{sr^2}{2t} \div \dfrac{st}{4}$

9. $\dfrac{4x+12}{5} \cdot \dfrac{15t}{3x+9}$

10. $\dfrac{y^2+2y}{6z} \cdot \dfrac{z^3}{y^2-4}$

11. $\dfrac{u^2-v^2}{u+2v} \cdot \dfrac{3u+6v}{u-v}$

12. $(x-y)\dfrac{x+2y}{x^2-y^2}$

13. $\dfrac{2a + 8}{15} \div \dfrac{a^2 + 8a + 16}{25}$

14. $\dfrac{a^2 - a}{3a + 9} \div \dfrac{a^2 - 2a + 1}{a^2 - 9}$

15. $\dfrac{x^2 - 9}{x} \div (x + 3)^2$

16. $\dfrac{9x^2 - 16}{x + 1} \div (4 - 3x)$

17. $\dfrac{3ax^2 - 9ax}{10x^2 + 5x} \cdot \dfrac{2x^2 + x}{a^2x - 3a^2}$

18. $\dfrac{2x^2 - 18}{x^3 - 25x} \cdot \dfrac{3x - 15}{2x^2 + 6x}$

19. $\dfrac{x^4 - 1}{8x + 16} \cdot \dfrac{2x^2 - 8x}{x^3 + x}$

20. $\dfrac{2x^2 - 4x - 6}{x^2 - 3x} \cdot \dfrac{x^3 - 4x^2}{4x^2 - 4x - 8}$

21. $\dfrac{ax + x^2}{2b - cx} \div \dfrac{a^2 + 2ax + x^2}{2bx - cx^2}$

22. $\dfrac{x^2 - 11x + 28}{x + 3} \div \dfrac{x - 4}{x + 3}$

23. $\dfrac{35a + 25}{12a + 33} \div \dfrac{28a + 20}{36a + 99}$

24. $\dfrac{2a^3 + a^2}{2b^3 + b^2} \div \dfrac{2ab + a}{2ab + b}$

25. $\dfrac{7x^2}{3a} \div \left(\dfrac{a}{x} \cdot \dfrac{a^2x}{x^2}\right)$

26. $\left(\dfrac{3u}{8v^2} \div \dfrac{9u^2}{2w^2}\right) \cdot \dfrac{2u^4}{15vw}$

27. $\left(\dfrac{4t^2 - 1}{t - 5} \div \dfrac{2t + 1}{2t}\right) \cdot \dfrac{2t^2 - 50}{4t^2 + 4t + 1}$

28. $\dfrac{2x^2 - 5x - 3}{x - 4} \div \left(\dfrac{x - 3}{x^2 - 16} \cdot \dfrac{1}{3 - x}\right)$

In Exercises 29 through 32 perform the indicated operations and simplify. Exercises 29 and 30 require the use of Eqs. (5–7) through (5–10), and Exercises 31 and 32 require the use of factoring by grouping.

29. $\dfrac{x^3 - y^3}{2x^2 - 2y^2} \cdot \dfrac{x^2 + 2xy + y^2}{x^2 + xy + y^2}$

30. $\dfrac{x^3 + 3x^2 + 3x + 1}{6x - 6} \div \dfrac{5x + 5}{x^2 - 1}$

31. $\dfrac{ax + bx + ay + by}{p - q} \cdot \dfrac{3p^2 + 4pq - 7q^2}{a + b}$

32. $\dfrac{x^4 + x^5 - 1 - x}{x - 1} \div \dfrac{x + 1}{x}$

In Exercises 33 and 34 solve the given problems.

33. A rectangular metal plate expands when heated. For small values of Celsius temperature, the length and width of a certain plate, as functions of the temperature, are

$$\dfrac{20{,}000 + 300T + T^2}{1600 - T^2} \quad \text{and} \quad \dfrac{16{,}000 + 360T - T^2}{400 + 2T}$$

respectively. Find the resulting expression for the area of the plate as a function of the temperature.

34. The current in a simple electric circuit is the voltage in the circuit divided by the resistance. Given that the voltage and resistance in a certain circuit are expressed as functions of time

$$V = \dfrac{5t + 10}{2t + 1} \quad \text{and} \quad R = \dfrac{t^2 + 4t + 4}{2t}$$

find the formula for the current as a function of time.

5—6 Addition and Subtraction of Fractions

From arithmetic we recall that *the sum of a set of fractions that all have the same denominator is the sum of the numerators divided by the common denominator.* Since algebraic expressions represent numbers, this fact is also true in algebra. Addition and subtraction of such fractions are illustrated in the following example.

Example A

$$\frac{5}{9} + \frac{2}{9} - \frac{4}{9} = \frac{5 + 2 - 4}{9} = \frac{3}{9} = \frac{1}{3}$$

$$\frac{b}{ax} - \frac{1}{ax} + \frac{2b - 1}{ax} = \frac{b - 1 + (2b - 1)}{ax} = \frac{b - 1 + 2b - 1}{ax}$$

$$= \frac{3b - 2}{ax}$$

If the fractions to be combined do not all have the same denominator, we must first change each to an equivalent fraction so that the resulting fractions do have the same denominator. Normally the denominator which is most convenient and useful is the **lowest common denominator.** This is the product of all of the prime factors which appear in the denominators, with each factor raised to the highest power to which it appears in any one of the denominators. Thus, *the lowest common denominator is the simplest algebraic expression into which all the given denominators will divide evenly.* The following two examples illustrate the method used in finding the lowest common denominator of a set of fractions.

Example B

Find the lowest common denominator of the fractions

$$\frac{3}{4a^2b}, \qquad \frac{5}{6ab^3}, \qquad \text{and} \qquad \frac{1}{4ab^2}$$

Expressing each of the denominators in terms of powers of the prime factors, we have

$$4a^2b = 2^2a^2b, \qquad 6ab^3 = 2 \cdot 3 \cdot ab^3, \qquad \text{and} \qquad 4ab^2 = 2^2ab^2$$

The prime factors to be considered are 2, 3, a, and b. The largest exponent of 2 which appears is 2. This means that 2^2 is a factor of the lowest common denominator. The largest exponent of 3 which appears is 1 (understood in the second denominator). Therefore, 3 is a factor of the lowest common denominator. The largest exponent of a which appears is 2, and the largest exponent of b which appears is 3. Thus, a^2 and b^3 are factors of the lowest common denominator. Therefore, the lowest common denominator of the fractions is $2^2 \cdot 3 \cdot a^2b^3 = 12a^2b^3$. This is the simplest expression into which *each* of the denominators above will divide evenly.

Example C
Find the lowest common denominator of the following fractions:

$$\frac{x-4}{x^2-2x+1}, \qquad \frac{1}{x^2-1}, \qquad \frac{x+3}{x^2-x}$$

Factoring each of the denominators, we find that the fractions are

$$\frac{x-4}{(x-1)^2}, \qquad \frac{1}{(x-1)(x+1)}, \qquad \text{and} \qquad \frac{x+3}{x(x-1)}$$

The factor $(x-1)$ appears in all of the denominators. It is squared in the denominator of the first fraction and appears to the first power only in the other two fractions. Thus, we must have $(x-1)^2$ as a factor of the common denominator. We do not need a higher power of $x-1$ since, as far as this factor is concerned, each denominator will divide into it evenly. Next, the second denominator contains a factor of $(x+1)$. Therefore, the common denominator must also contain a factor of $(x+1)$, otherwise the second denominator would not divide into it evenly. Finally, the third denominator indicates that a factor of x is also required in the common denominator. The lowest common denominator is, therefore, $x(x+1)(x-1)^2$. All three denominators will divide evenly into this expression, and there is no simpler expression for which this is true.

Once we have found the lowest common denominator for the fractions, we multiply the numerator and denominator of each fraction by the proper quantity to make the resulting denominator in each case the common denominator. After this step, it is necessary only to add the numerators, place this result over the common denominator, and simplify.

Example D

Combine $\dfrac{2}{3r^2} + \dfrac{4}{rs^3} - \dfrac{5}{3s}$.

By looking at the denominators, we see that the factors necessary in the lowest common denominator are 3, r, and s. The 3 appears only to the first power, the largest exponent of r is 2, and the largest exponent of s is 3. Therefore, the lowest common denominator is $3r^2s^3$. We now wish to write each fraction with this quantity as the denominator. Since the denominator of the first fraction already contains factors of 3 and r^2, it is necessary to introduce the factor of s^3. In other words, we must multiply the numerator and denominator of this fraction by s^3. For similar reasons, we must multiply the numerators and the denominators of the second and third fractions by $3r$ and r^2s^2, respectively. This leads to

$$\frac{2}{3r^2} + \frac{4}{rs^3} - \frac{5}{3s} = \frac{2(s^3)}{(3r^2)(s^3)} + \frac{4(3r)}{(rs^3)(3r)} - \frac{5(r^2s^2)}{(3s)(r^2s^2)}$$

$$= \frac{2s^3}{3r^2s^3} + \frac{12r}{3r^2s^3} - \frac{5r^2s^2}{3r^2s^3}$$

$$= \frac{2s^3 + 12r - 5r^2s^2}{3r^2s^3}$$

Example E

$$\frac{a}{x-1} + \frac{a}{x+1} = \frac{a(x+1)}{(x-1)(x+1)} + \frac{a(x-1)}{(x+1)(x-1)}$$

$$= \frac{ax + a + ax - a}{(x+1)(x-1)}$$

$$= \frac{2ax}{(x+1)(x-1)}$$

When we multiply each fraction by the quantity required to obtain the proper denominator, we do not actually have to write the common denominator under each numerator. Placing all the products which appear in the numerators over the common denominator is sufficient. Hence the illustration in this example would appear as

$$\frac{a}{x-1} + \frac{a}{x+1} = \frac{a(x+1) + a(x-1)}{(x-1)(x+1)} = \frac{ax + a + ax - a}{(x-1)(x+1)}$$

$$= \frac{2ax}{(x-1)(x+1)}$$

Example F

$$\frac{x-1}{x^2-25} - \frac{2}{x-5} = \frac{(x-1) - 2(x+5)}{(x-5)(x+5)} = \frac{x - 1 - 2x - 10}{(x-5)(x+5)}$$

$$= \frac{-(x+11)}{(x-5)(x+5)}$$

Example G

$$\frac{3x}{x^2-x-12} - \frac{x-1}{x^2-8x+16} - \frac{6-x}{2x-8}$$

$$= \frac{3x}{(x-4)(x+3)} - \frac{x-1}{(x-4)^2} - \frac{6-x}{2(x-4)}$$

$$= \frac{3x(2)(x-4) - (x-1)(2)(x+3) - (6-x)(x-4)(x+3)}{2(x-4)^2(x+3)}$$

$$= \frac{6x^2 - 24x - 2x^2 - 4x + 6 + x^3 - 7x^2 - 6x + 72}{2(x-4)^2(x+3)}$$

$$= \frac{x^3 - 3x^2 - 34x + 78}{2(x-4)^2(x+3)}$$

One note of caution must be sounded here. In doing this kind of problem, many errors may arise in the use of the minus sign. Remember, if a minus sign precedes a given expression, the signs of *all* terms in that expression must be changed before they can be combined with the other terms.

Example H
Simplify the fraction

$$\frac{1 + \dfrac{2}{x - 1}}{\dfrac{x^2 + x}{x^2 + x - 2}}$$

Before performing the indicated division, we first perform the indicated addition in the numerator. The numerator becomes

$$\frac{(x - 1) + 2}{x - 1} \quad \text{or} \quad \frac{x + 1}{x - 1}$$

This expression now replaces the numerator of the original fraction. Making this substitution and inverting the divisor, we then proceed with the simplification:

$$\frac{x + 1}{x - 1} \cdot \frac{x^2 + x - 2}{x^2 + x} = \frac{(x + 1)(x + 2)(x - 1)}{(x - 1)(x)(x + 1)} = \frac{x + 2}{x}$$

This is an example of what is known as a **complex fraction**. In a complex fraction the numerator, the denominator, or both numerator and denominator contain fractions.

Exercises 5—6

In Exercises 1 through 36 perform the indicated operations and simplify.

1. $\dfrac{3}{5} + \dfrac{6}{5}$

2. $\dfrac{2}{13} + \dfrac{6}{13}$

3. $\dfrac{1}{x} + \dfrac{7}{x}$

4. $\dfrac{2}{a} + \dfrac{3}{a}$

5. $\dfrac{1}{2} + \dfrac{3}{4}$

6. $\dfrac{5}{9} - \dfrac{1}{3}$

7. $\dfrac{3}{4x} + \dfrac{7a}{4}$

8. $\dfrac{t - 3}{a} - \dfrac{t}{2a}$

9. $\dfrac{a}{x} - \dfrac{b}{x^2}$

10. $\dfrac{2}{s^2} + \dfrac{3}{s}$

11. $\dfrac{6}{5x^3} + \dfrac{a}{25x}$

12. $\dfrac{a}{6y} - \dfrac{2b}{3y^4}$

13. $\dfrac{2}{5a} + \dfrac{1}{a} - \dfrac{a}{10}$

14. $\dfrac{2}{a} - \dfrac{6}{b} - \dfrac{9}{c}$

15. $\dfrac{x + 1}{x} - \dfrac{x - 3}{y} - \dfrac{2 - x}{xy}$

16. $5 + \dfrac{1 - x}{2} - \dfrac{3 + x}{4}$

17. $\dfrac{3}{2x-1} + \dfrac{1}{4x-2}$

18. $\dfrac{5}{6y+3} - \dfrac{a}{8y+4}$

19. $\dfrac{4}{x(x+1)} - \dfrac{3}{2x}$

20. $\dfrac{3}{ax+ay} - \dfrac{1}{a^2}$

21. $\dfrac{s}{2s-6} + \dfrac{1}{4} - \dfrac{3s}{4s-12}$

22. $\dfrac{2}{x+2} - \dfrac{3-x}{x^2+2x} + \dfrac{1}{x}$

23. $\dfrac{3x}{x^2-9} - \dfrac{2}{x+3}$

24. $\dfrac{2}{x^2+4x+4} - \dfrac{3}{x+2}$

25. $\dfrac{3}{x^2-8x+16} - \dfrac{2}{4-x}$

26. $\dfrac{1}{a^2-1} - \dfrac{2}{1-a}$

27. $\dfrac{3}{x^2-11x+30} - \dfrac{2}{x^2-25}$

28. $\dfrac{x-1}{2x^3-4x^2} + \dfrac{5}{x-2}$

29. $\dfrac{x-1}{3x^2-13x+4} - \dfrac{3x+1}{4-x}$

30. $\dfrac{x}{4x^2-12x+5} + \dfrac{2x-1}{4x^2-4x-15}$

31. $\dfrac{t}{t^2-t-6} - \dfrac{2t}{t^2+6t+9} + \dfrac{t}{t^2-9}$

32. $\dfrac{5}{2x^3-3x^2+x} - \dfrac{x}{x^4-x^2} + \dfrac{2-x}{2x^2+x-1}$

33. $\dfrac{1+\dfrac{1}{x}}{1-\dfrac{1}{x}}$

34. $\dfrac{x-\dfrac{1}{x}}{1-\dfrac{1}{x}}$

35. $\dfrac{x-\dfrac{1}{x}-\dfrac{2}{x+1}}{\dfrac{1}{x^2+2x+1}-1}$

36. $\dfrac{\dfrac{2}{a}-\dfrac{1}{4}-\dfrac{3}{4a-4b}}{\dfrac{1}{4a^2-4b^2}-\dfrac{2}{b}}$

The expression $f(x+h) - f(x)$ is frequently used in the study of calculus. In Exercises 37 through 40 determine and then simplify this expression for the given functions.

37. $f(x) = \dfrac{x}{x+1}$

38. $f(x) = \dfrac{3}{2x-1}$

39. $f(x) = \dfrac{1}{x^2}$

40. $f(x) = \dfrac{2}{x^2+4}$

In Exercises 41 through 48 simplify the given expressions.

41. Using the definitions of the trigonometric functions given in Section 3–2, find an expression equivalent to $(\tan \theta)(\cot \theta) + (\sin \theta)^2 - \cos \theta$, in terms of x, y, and r.

42. Using the definitions of the trigonometric functions given in Section 3–2, find an expression equivalent to $\sec \theta - (\cot \theta)^2 + \csc \theta$, in terms of x, y, and r.

43. If $f(x) = 2x - x^2$, find $f(\frac{1}{a})$.

44. If $f(x) = x^2 + x$, find $f(a + \frac{1}{a})$.

45. The analysis of the forces acting on a certain type of concrete slab gives the expression

$$1 - \frac{4c}{\pi l} + \frac{c^3}{3l^3}$$

Combine and simplify.

46. Experimentation to determine the velocity of light may use the expression

$$1 + \frac{v^2}{2c^2} + \frac{3v^4}{4c^4}$$

Combine and simplify.

47. In finding an expression to describe a magnetic field, the following expression is found.

$$\frac{b}{x^2 + y^2} - \frac{2bx^2}{(x^2 + y^2)^2}$$

Perform the indicated subtraction.

48. The expression for the volumetric expansion of liquids in terms of density ρ and temperature T is

$$\frac{\dfrac{1}{\rho_2} - \dfrac{1}{\rho_1}}{(T_2 - T_1)\left(\dfrac{1}{\rho_1} + \dfrac{1}{\rho_2}\right)}{2}$$

Simplify this expression.

5–7 Equations Involving Fractions

Many important equations in science and technology have fractions in them. Although the solution of these equations will still involve the use of the basic operations stated in Section 1–10, an additional procedure can be used to eliminate the fractions and thereby help lead to the solution. The method is to **multiply each term of the equation by the lowest common denominator.** The resulting equation will not involve fractions and can be solved by methods previously discussed. The following examples illustrate how to solve equations involving fractions.

Example A

Solve for x: $\dfrac{x}{12} - \dfrac{1}{8} = \dfrac{x + 2}{6}$.

We first note that the lowest common denominator of the terms of the equation is 24. Therefore, we multiply each term by 24. This gives

$$\frac{24(x)}{12} - \frac{24(1)}{8} = \frac{24(x+2)}{6}$$

We reduce each term to its lowest terms, and solve the resulting equation.

$$2x - 3 = 4(x + 2)$$
$$2x - 3 = 4x + 8$$
$$-2x = 11$$
$$x = -\frac{11}{2}$$

When we check this solution in the original equation, we obtain $-\frac{7}{12}$ on each side of the equals sign. Therefore, the solution is correct.

Example B

Solve for x: $\dfrac{x}{2} - \dfrac{1}{b^2} = \dfrac{x}{2b}$.

We first determine that the lowest common denominator of the terms of the equation is $2b^2$. We then multiply each term by $2b^2$ and continue with the solution.

$$\frac{2b^2(x)}{2} - \frac{2b^2(1)}{b^2} = \frac{2b^2(x)}{2b}$$

$$b^2x - 2 = bx$$

$$b^2x - bx = 2$$

To complete the solution for x, we must factor x from the terms on the left. Therefore, we have

$$x(b^2 - b) = 2$$

$$x = \frac{2}{b^2 - b}$$

Checking shows that each side of the original equation is $1/b^2(b-1)$.

Example C

When developing the equations which describe the motion of the planets, the equation

$$\frac{1}{2}v^2 - \frac{GM}{r} = -\frac{GM}{2a}$$

is found. Solve for M.

We first determine that the lowest common denominator of the terms of the equation is $2ar$. Multiplying each term by $2ar$ and proceeding with the solution, we have

$$\frac{2ar(v^2)}{2} - \frac{2ar(GM)}{r} = -\frac{2ar(GM)}{2a}$$

$$arv^2 - 2aGM = -rGM$$

$$rGM - 2aGM = -arv^2$$

$$M(rG - 2aG) = -arv^2$$

$$M = -\frac{arv^2}{rG - 2aG} \quad \text{or} \quad \frac{arv^2}{2aG - rG}$$

The second form of the result is obtained by using Eq. (5–11). Again, note the use of factoring to arrive at the final result.

Example D

Solve for x: $\dfrac{2}{x+1} - \dfrac{1}{x} = -\dfrac{2}{x^2+x}$.

Multiplying each term by the lowest common denominator $x(x + 1)$, and continuing with the solution, we have

$$\frac{2(x)(x+1)}{x+1} - \frac{x(x+1)}{x} = -\frac{2x(x+1)}{x(x+1)}$$

$$2x - (x+1) = -2$$

$$2x - x - 1 = -2$$

$$x = -1$$

Checking this solution *in the original equation*, we see that we have zero in the denominators of the first and third terms of the equation. Since division by zero is undefined (see Section 1–3), $x = -1$ cannot be a solution. *Thus there is no solution to this equation.* This example points out clearly why it is necessary to check solutions in the original equation. It also shows that whenever we multiply each term by a common denominator which *contains the unknown*, it is possible to obtain a value which is not a solution of the original equation. Such a value is termed an **extraneous solution.** Only certain equations will lead to extraneous solutions, but we must be careful to identify them when they occur.

Example E

One pipe can fill a certain oil storage tank in 4 h, while a second pipe can fill it in 6 h. How long will it take to fill the tank if both pipes operate together?

First, we let $x = $ the number of hours required to fill the tank with both pipes operating.

We know that it takes the first pipe 4 h to fill the tank. Therefore, it fills $\frac{1}{4}$ of the tank each hour it operates. This means that it fills $\frac{1}{4}x$ of the tank in x hours. In the same way, the second pipe fills $\frac{1}{6}x$ of the tank in x hours. When x hours have passed, the two pipes will have filled the whole tank (1 represents *one* tank).

$$\frac{x}{4} + \frac{x}{6} = 1$$

Multiplying each term by 12, we have

$$\frac{12x}{4} + \frac{12x}{6} = 12(1)$$

$$3x + 2x = 12$$

$$5x = 12$$

$$x = \frac{12}{5} = 2.4 \text{ h}$$

Therefore, it takes the two pipes 2.4 h to fill the tank when operating together.

Exercises 5−7

In Exercises 1 through 28 solve the given equations and check the results.

1. $\dfrac{x}{2} + 6 = 2x$

2. $\dfrac{x}{5} + 2 = \dfrac{15 + x}{10}$

3. $\dfrac{x}{6} - \dfrac{1}{2} = \dfrac{x}{3}$

4. $\dfrac{3x}{8} - \dfrac{3}{4} = \dfrac{x - 4}{2}$

5. $\dfrac{1}{2} - \dfrac{t - 5}{6} = \dfrac{3}{4}$

6. $\dfrac{2x - 7}{3} + 5 = \dfrac{1}{5}$

7. $\dfrac{3x}{7} - \dfrac{5}{21} = \dfrac{2 - x}{14}$

8. $\dfrac{x - 3}{12} - \dfrac{2}{3} = \dfrac{1 - 3x}{2}$

9. $\dfrac{3}{x} + 2 = \dfrac{5}{3}$

10. $\dfrac{1}{2y} - \dfrac{1}{2} = 4$

11. $3 - \dfrac{x - 2}{x} = \dfrac{1}{3}$

12. $\dfrac{1}{2x} - \dfrac{1}{3} = \dfrac{2}{x}$

13. $\dfrac{2y}{y - 1} = 5$

14. $\dfrac{x}{2x - 3} = 4$

15. $\dfrac{2}{s} = \dfrac{3}{s - 1}$

16. $\dfrac{5}{n + 2} = \dfrac{3}{2n}$

17. $\dfrac{5}{2x + 4} + \dfrac{3}{x + 2} = 2$

18. $\dfrac{3}{4x - 6} + \dfrac{1}{4} = \dfrac{5}{2x - 3}$

19. $\dfrac{4}{4 - x} + 2 - \dfrac{2}{12 - 3x} = \dfrac{1}{3}$

20. $\dfrac{2}{z - 5} - \dfrac{3}{10 - 2z} = 3$

21. $\dfrac{1}{x} + \dfrac{3}{2x} = \dfrac{2}{x + 1}$

22. $\dfrac{3}{t + 3} - \dfrac{1}{t} = \dfrac{5}{2t + 6}$

23. $\dfrac{1}{2x+3} = \dfrac{5}{2x} - \dfrac{4}{2x^2+3x}$

24. $\dfrac{7}{y} = \dfrac{3}{y-4} + \dfrac{7}{2y^2-8y}$

25. $\dfrac{1}{x^2-x} - \dfrac{1}{x} = \dfrac{1}{x-1}$

26. $\dfrac{2}{x^2-1} - \dfrac{2}{x+1} = \dfrac{1}{x-1}$

27. $\dfrac{2}{x^2-4} - \dfrac{1}{x-2} = \dfrac{1}{2x+4}$

28. $\dfrac{2}{2x^2+5x-3} - \dfrac{1}{4x-2} + \dfrac{3}{2x+6} = 0$

In Exercises 29 through 36 solve for the indicated letter.

29. $2 - \dfrac{1}{b} + \dfrac{3}{c} = 0$, for c

30. $\dfrac{2}{3} - \dfrac{h}{x} = \dfrac{1}{2x}$, for x

31. $\dfrac{t-3}{b} - \dfrac{t}{2b-1} = \dfrac{1}{2}$, for t

32. $\dfrac{1}{a^2+2a} - \dfrac{y}{2a} = \dfrac{2y}{a+2}$, for y

33. An equation used in nuclear physics is

$$E = V_0 + \dfrac{(m+M)V^2}{2} + \dfrac{p^2}{2I}$$

Solve for M.

34. An equation obtained in analyzing a certain electric circuit is

$$\dfrac{V-6}{5} + \dfrac{V-8}{15} + \dfrac{V}{10} = 0$$

Solve for V.

35. Under specified conditions, the combined resistance R of resistances R_1, R_2, and r is given by the equation

$$\dfrac{1}{R} = \dfrac{1}{R_1+r} + \dfrac{1}{R_2}$$

Solve for R_1.

36. An equation used in hydrodynamics is

$$F = PA + \dfrac{dQ^2}{gA_1} - \dfrac{dQ^2}{gA_2}$$

Solve for A.

In Exercises 37 through 40 set up appropriate equations and solve the given stated problems.

37. One data-processing card sorter can sort a certain number of cards in 6 min, and a second sorter can sort the same number in 9 min. How long would it take the two sorters together to sort this number of cards?

38. One steamshovel can excavate a certain site in 5 days, while it takes a second steamshovel 8 days. How long would it take the two working together?

39. The width of a particular rectangular land area is $\frac{2}{3}$ that of the length. If the perimeter is 192 m, find the dimensions.

40. The current in a certain stream flows at 3 mi/h. A motorboat can travel downstream 23 mi in the same time it can travel 11 mi upstream. What is the boat's rate in still water?

5—8 Exercises for Chapter 5

In Exercises 1 through 12 find the products *by inspection*. No intermediate steps should be necessary.

1. $3a(4x + 5a)$
2. $-7xy(4x^2 - 7y)$
3. $(2a + 7b)(2a - 7b)$
4. $(x - 4z)(x + 4z)$
5. $(2a + 1)^2$
6. $(4x - 3y)^2$
7. $(b - 4)(b + 7)$
8. $(y - 5)(y - 7)$
9. $(2x + 5)(x - 9)$
10. $(4ax - 3)(5ax + 7)$
11. $(2c + d)(8c - d)$
12. $(3s - 2t)(8s + 3t)$

In Exercises 13 through 44 factor the given expressions completely. Exercises 37 through 40 illustrate Eqs. (5–7) through (5–10), and Exercises 41 through 44 illustrate factoring by grouping.

13. $3s + 9t$
14. $7x - 28y$
15. $a^2x^2 + a^2$
16. $3ax - 6ax^4 - 9a$
17. $x^2 - 144$
18. $900 - n^2$
19. $400r^2 - t^4$
20. $25s^4 - 36t^2$
21. $9t^2 - 6t + 1$
22. $4x^2 - 12x + 9$
23. $25t^2 + 10t + 1$
24. $4x^2 + 36xy + 81y^2$
25. $x^2 + x - 56$
26. $x^2 - 4x - 45$
27. $t^2 - 5t - 36$
28. $n^2 - 11n + 10$
29. $2x^2 - x - 36$
30. $5x^2 + 2x - 3$
31. $4x^2 - 4x - 35$
32. $9x^2 + 7x - 16$
33. $10b^2 + 23b - 5$
34. $12x^2 - 7xy - 12y^2$
35. $4x^2 - 64$
36. $4a^2x^2 + 26a^2x + 36a^2$
37. $x^3 + 9x^2 + 27x + 27$
38. $x^3 - 3x^2 + 3x - 1$
39. $8x^3 + 27$
40. $x^3 - 125$
41. $ab^2 - 3b^2 + a - 3$
42. $axy - ay + ax - a$
43. $nx + 5n - x^2 + 25$
44. $ty - 4t + y^2 - 16$

In Exercises 45 through 64 perform the indicated operations and express results in simplest form.

45. $\dfrac{48ax^3y^6}{9a^3xy^6}$
46. $\dfrac{-39r^2s^4t^8}{52rs^5t}$

47. $\dfrac{6x^2 - 7x - 3}{4x^2 - 8x + 3}$
48. $\dfrac{x^2 - 3x - 4}{x^2 - x - 12}$

49. $\dfrac{4x + 4y}{35x^2} \cdot \dfrac{28x}{x^2 - y^2}$
50. $\dfrac{6x - 3}{x^2} \cdot \dfrac{4x^2 - 12x}{12x - 6}$

51. $\dfrac{18 - 6x}{x^2 - 6x + 9} \div \dfrac{x^2 - 2x - 15}{x^2 - 9}$
52. $\dfrac{6x^2 - xy - y^2}{2x^2 + xy - y^2} \div \dfrac{4x^2 - 16y^2}{x^2 + 3xy + 2y^2}$

53. $\dfrac{\dfrac{3x}{7x^2 + 13x - 2}}{\dfrac{6x^2}{x^2 + 4x + 4}}$
54. $\dfrac{\dfrac{3x - 3y}{2x^2 + 3xy - 2y^2}}{\dfrac{3x^2 - 3y^2}{x^2 + 4xy + 4y^2}}$

55. $\dfrac{x + \dfrac{1}{x} + 1}{x^2 - \dfrac{1}{x}}$

56. $\dfrac{\dfrac{4}{y} - 4y}{2 - \dfrac{2}{y}}$

57. $\dfrac{4}{9x} - \dfrac{5}{12x^2}$

58. $\dfrac{3}{10a^2} + \dfrac{1}{4a^3}$

59. $\dfrac{6}{x} - \dfrac{7}{2x} + \dfrac{3}{xy}$

60. $\dfrac{4}{a^2 b} - \dfrac{5}{2ab} + \dfrac{1}{2b}$

61. $\dfrac{a + 1}{a + 2} - \dfrac{a + 3}{a}$

62. $\dfrac{2x - 1}{4 - x} + \dfrac{x + 2}{5x - 20}$

63. $\dfrac{3x}{x^2 + 2x - 3} - \dfrac{2}{x^2 + 3x} + \dfrac{x}{x - 1}$

64. $\dfrac{3}{y^4 - 2y^3 - 8y^2} + \dfrac{y - 1}{y^2 + 2y} - \dfrac{y - 3}{y^2 - 4y}$

In Exercises 65 through 72 solve the given equations.

65. $\dfrac{x}{2} - 3 = \dfrac{x - 10}{4}$

66. $\dfrac{x}{6} - \dfrac{1}{2} = \dfrac{3 - x}{12}$

67. $\dfrac{2x}{c} - \dfrac{1}{2c} = \dfrac{3}{c} - x$, for x

68. $\dfrac{x}{a} - b + \dfrac{x}{c} = \dfrac{a}{b} - c$, for x

69. $\dfrac{2}{t} - \dfrac{1}{at} = 2 + \dfrac{a}{t}$, for t

70. $\dfrac{3}{a^2 y} - \dfrac{1}{ay} = \dfrac{9}{a}$, for y

71. $\dfrac{2x}{x^2 - 3x} - \dfrac{3}{x} = \dfrac{1}{2x - 6}$

72. $\dfrac{3}{x^2 + 3x} - \dfrac{1}{x} = \dfrac{1}{x + 3}$

In Exercises 73 through 88 perform the indicated operations.

73. Show that

$$xy = \frac{1}{4}[(x + y)^2 - (x - y)^2]$$

74. Show that

$$x^2 + y^2 = \frac{1}{2}[(x + y)^2 + (x - y)^2]$$

75. To find the side of a rectangle of a given area, it is necessary to factor the expression $x^2 - 3x - 70$. Factor this expression.

76. Under certain conditions, in order to find the total profit of an article selling for p dollars, one must factor the expression $2p^2 - 126p + 360$. Factor this expression.

77. If the edge of one cube is $x + 4$ and the edge of another cube is x, the difference in their volumes is $(x + 4)^3 - x^3$. Expand $(x + 4)^3$, simplify the resulting expression, and then factor.

78. An expression which occurs in the study of nuclear physics is

$$pa^2 + (1 - p)b^2 - [pa + (1 - p)b]^2$$

Expand the third term and then factor by grouping.

79. In finding the velocity of an object subject to specified conditions, it is necessary to simplify the expression

$$\frac{(t + 1)^2 - 2t(t + 1)}{(t + 1)^4}$$

Simplify this expression.

80. An expression found in solving a problem related to alternating-current power is

$$\frac{\dfrac{s + 10}{10}}{\left(\dfrac{s + 20}{20}\right)\left(\dfrac{s + 60}{60}\right)}$$

Simplify this expression.

81. An expression found in determining the tension in a certain cable is

$$1 + \frac{w^2x^2}{6T^2} - \frac{w^4x^4}{40T^4}$$

Combine and simplify.

82. An expression found in the analysis of the dynamics of missile firing is

$$\frac{1}{s} - \frac{1}{s + 4} + \frac{8}{(s + 4)^2}$$

Perform the indicated operations.

83. An expression found in the study of electronic amplifiers is

$$\frac{\left(\dfrac{\mu}{\mu + 1}\right)R}{\dfrac{r}{\mu + 1} + R}$$

Simplify this expression. (μ is the Greek letter mu.)

84. An expression which arises when finding the path between two points requiring the least time is

$$\frac{\dfrac{u^2}{2g} - x}{\dfrac{1}{2gc^2} - \dfrac{u^2}{2g} + x}$$

Simplify this expression.

85. The focal length f of a lens, in terms of its image distance q and object distance p, is given by

$$\frac{1}{f} = \frac{1}{p} + \frac{1}{q}$$

Solve for q.

86. The combined capacitance C of three capacitors connected in series is

$$\frac{1}{C} = \frac{1}{C_1} + \frac{1}{C_2} + \frac{1}{C_3}$$

Solve for C_1.

87. An equation used in studying the deflection of a beam is

$$\theta = \frac{wL^3}{24EI} - \frac{ML}{6EI}$$

Solve for M.

88. An equation determined during the study of the characteristics of a certain chemical solution is

$$X = \frac{H}{RT_1} - \frac{H}{RT}$$

Solve for T.

In Exercises 89 through 92 set up appropriate equations and solve the given stated problems.

89. If one riveter can do a certain job in 12 d, and a second riveter can do it in 16 d, how long will it take them to do it together?

90. Two crews are working on an oil pipeline. Crew A can lay pipe at the rate of 2000 m/d, and crew B can lay pipe at the rate of 2500 m/d. How long will it take the two crews together to lay 10,000 m of pipe?

91. One quality control inspector can properly inspect 50 parts per day, and a second inspector can inspect 30 parts per day. They are put on a project together to inspect 200 parts. After 1.5 d the second inspector becomes ill and leaves the project. How long does the project take?

92. A person travels from city A to city B on a train which averages 40 mi/h. He spends 4 h in city B and then returns to city A on a jet which averages 600 mi/h. If the total trip takes 20 h, how far is it from city A to city B?

6

Quadratic Equations

6–1 Quadratic Equations; Solution by Factoring

The solution of simple equations was first introduced in Chapter 1. Then, in Chapter 4, we extended the solution of equations to systems of linear equations. With the development of the algebraic operations in Chapter 5, we are now in a position to solve another important type of equation, the **quadratic equation**.

Given that a, b, and c are constants, the equation

$$ax^2 + bx + c = 0 \tag{6-1}$$

is called the **general quadratic equation in x.** From Eq. (6–1) we can see that the left side of the equation is a polynomial function of degree 2. *This function, $ax^2 + bx + c$, is known as the* **quadratic function.**

Quadratic equations and quadratic functions are found in applied problems of many technical fields of study. For example, in describing projectile motion, the equation $s_0 + v_0 t - 16t^2 = 0$ is found; in analyzing electric power, the function $EI - RI^2$ is found; and in determining the forces on beams, the function $ax^2 + bLx + cL^2$ is used.

Since it is the x^2 term that distinguishes the quadratic equation from other types of equations, the equation is not quadratic if $a = 0$. However, b or c or both may be zero, and the equation is quadratic. We should recognize a quadratic equation even when it does not initially appear in the form of Eq. (6–1). The following examples illustrate the recognition of quadratic equations.

Example A
The following are quadratic equations.

$x^2 - 4x - 5 = 0$	($a = 1$, $b = -4$, and $c = -5$)
$3x^2 - 6 = 0$	($a = 3$, $b = 0$, and $c = -6$)
$2x^2 + 7x = 0$	($a = 2$, $b = 7$, and $c = 0$)
$(a - 3)x^2 - ax + 7 = 0$	(The constants in Eq. (6–1) may include literal expressions. In this case, $a - 3$ takes the place of a, $-a$ takes the place of b, and $c = 7$.)
$4x^2 - 2x = x^2$	(After all the terms have been collected on the left side, the equation becomes $3x^2 - 2x = 0$.)
$(x + 1)^2 = 4$	(Expanding the left side, and collecting all terms on the left, we have $x^2 + 2x - 3 = 0$.)

Example B
The following are not quadratic equations.

$bx - 6 = 0$	(There is no x^2-term.)
$x^3 - x^2 - 5 = 0$	(There should be no term of degree higher than 2. Thus there can be no x^3-term in a quadratic equation.)
$x^2 + x - 7 = x^2$	(When terms are collected, there will be no x^2-term.)

From our previous work, we recall that the solution of an equation consists of all numbers which, when substituted in the equation, produce equality. Normally there are two such numbers for a quadratic equation, although occasionally there is only one number. In any case, there cannot be more than two roots of a quadratic equation. Also, due to the presence of the x^2-term, the roots may be imaginary numbers.

Example C
The quadratic equation

$$3x^2 - 7x + 2 = 0$$

has the roots $x = \frac{1}{3}$ and $x = 2$. This can be seen by substituting these values into the equation.

$$3\left(\frac{1}{3}\right)^2 - 7\left(\frac{1}{3}\right) + 2 = 3\left(\frac{1}{9}\right) - \frac{7}{3} + 2 = \frac{1}{3} - \frac{7}{3} + 2 = \frac{0}{3} = 0$$

$$3(2)^2 - 7(2) + 2 = 3(4) - 14 + 2 = 12 - 14 + 2 = 0$$

The quadratic equation

$$4x^2 - 4x + 1 = 0$$

has the *double root* (both roots the same) of $x = \frac{1}{2}$. This can be seen to be a solution by substitution:

$$4\left(\frac{1}{2}\right)^2 - 4\left(\frac{1}{2}\right) + 1 = 4\left(\frac{1}{4}\right) - 2 + 1 = 1 - 2 + 1 = 0$$

The quadratic equation

$$x^2 + 9 = 0$$

has roots of $x = 3j$ and $x = -3j$. Remembering that $j = \sqrt{-1}$, which means that $j^2 = -1$, we have

$$(3j)^2 + 9 = 9j^2 + 9 = 9(-1) + 9 = -9 + 9 = 0$$
$$(-3j)^2 + 9 = (-3)^2 j^2 + 9 = 9j^2 + 9 = 9(-1) + 9 = 0$$

In this section we shall deal only with those quadratic equations whose quadratic expression is factorable. Therefore, all roots will be rational. *To solve a quadratic equation by factoring, we collect all terms on the left so that the equation will be in the general form of Eq. (6–1). Then we factor the left side and set each factor, individually, equal to zero.* Here we are using the fact that a product is zero if any of its factors is zero. The solutions of the resulting linear equations constitute the solution of the quadratic equation.

Example D

$$x^2 - x - 12 = 0$$
$$(x - 4)(x + 3) = 0$$
$$x - 4 = 0 \qquad \text{or} \quad x = 4$$
$$x + 3 = 0 \qquad \text{or} \quad x = -3$$

The roots are $x = 4$ and $x = -3$. We can check them in the original equation by substitution. For the root $x = 4$, we have

$$(4)^2 - (4) - 12 \stackrel{?}{=} 0$$
$$0 = 0$$

For the root $x = -3$, we have

$$(-3)^2 - (-3) - 12 \stackrel{?}{=} 0$$
$$0 = 0$$

Both roots satisfy the original equation.

Example E

$$2x^2 + 7x - 4 = 0$$
$$(2x - 1)(x + 4) = 0$$

$$2x - 1 = 0 \quad \text{or} \quad x = \frac{1}{2}$$

$$x + 4 = 0 \quad \text{or} \quad x = -4$$

Therefore, the roots are $x = \frac{1}{2}$ and $x = -4$. These roots can be checked by the same procedure used in Example D.

Example F

$$x^2 + 4 = 4x$$
$$x^2 - 4x + 4 = 0$$
$$(x - 2)(x - 2) = 0$$
$$x - 2 = 0 \quad \text{or} \quad x = 2$$

Since both factors are the same, there is a double root of $x = 2$.

It is essential for the expression on the left to be equal to zero, because if a product equals a nonzero number, there is no assurance that either of the factors equals this number. Again, the first step must be to write the equation in the form of Eq. (6–1).

Example G

A car travels to and from a city 180 mi distant in 8.5 h. If the average speed on the return trip is 5 mi/h less than on the trip to the city, what was the average speed of the car when it was going toward the city?

Let x = average speed of car going to the city, and t = time to travel to the city. By our choice of unknowns we may state that $xt = 180$ (speed times time equals distance). Also, we know that the speed on the return trip was $x - 5$ and that the required time for the return trip was $8.5 - t$. Since the distance traveled returning was also 180 mi, we may state that $(x - 5)(8.5 - t) = 180$. Because we wish to find x, we can eliminate t between the equations by substitution.

$$(x - 5)\left(8.5 - \frac{180}{x}\right) = 180$$

$(x - 5)(17x - 360) = 360x$	Each side multiplied by $2x$.
$17x^2 - 360x - 85x + 1800 = 360x$	Remove parentheses.
$17x^2 - 805x + 1800 = 0$	Collect all terms to one side.
$(17x - 40)(x - 45) = 0$	Factor the quadratic expression.

$$17x - 40 = 0 \quad \text{or} \quad x = \frac{40}{17}$$

$$x - 45 = 0 \quad \text{or} \quad x = 45$$

The factors lead to two possible solutions, but only one of them has meaning for this problem. The solution $x = \frac{40}{17}$ cannot be the solution, since the return rate of 5 mi/h less would then be negative. Therefore, the solution is $x = 45$ mi/h. By substitution it is found that this solution satisfies the given conditions.

Exercises 6—1

In Exercises 1 through 8 determine whether or not the given equations are quadratic by performing algebraic operations which could put each in the form of Eq. (6–1). If the resulting form is quadratic, identify a, b, and c, with $a > 0$.

1. $x^2 + 5 = 8x$
2. $5x^2 = 9 - x$
3. $x(x - 2) = 4$
4. $(3x - 2)^2 = 2$
5. $x^2 = (x + 2)^2$
6. $x(2x + 5) = 7 + 2x^2$
7. $x(x^2 + x - 1) = x^3$
8. $(x - 7)^2 = (2x + 3)^2$

In Exercises 9 through 40 solve the given quadratic equations by factoring.

9. $x^2 - 4 = 0$
10. $x^2 - 400 = 0$
11. $x^2 - 8x - 9 = 0$
12. $s^2 + s - 6 = 0$
13. $x^2 - 7x + 12 = 0$
14. $x^2 - 11x + 30 = 0$
15. $x^2 = -2x$
16. $x^2 = 7x$
17. $4y^2 - 9 = 0$
18. $5p^2 - 80 = 0$
19. $3x^2 - 13x + 4 = 0$
20. $7x^2 + 3x - 4 = 0$
21. $x^2 + 8x + 16 = 0$
22. $4x^2 - 20x + 25 = 0$
23. $6x^2 = 13x - 6$
24. $6z^2 = 6 + 5z$
25. $4x^2 - 3 = -4x$
26. $10t^2 = 9 - 43t$
27. $x^2 - x - 1 = 1$
28. $2x^2 - 7x + 6 = 3$
29. $x^2 - 4b^2 = 0$
30. $a^2x^2 - 1 = 0$
31. $40x - 16x^2 = 0$
32. $18t^2 - 48t + 32 = 0$
33. $(x + 2)^3 = x^3 + 8$
34. $x(x^2 - 4) = x^2(x - 1)$
35. $(x + a)^2 - b^2 = 0$
36. $x^2(a^2 + 2ab + b^2) - x(a + b) = 0$

37. A projectile is fired vertically into the air. The distance (in feet) above the ground, as a function of the time (in seconds), is given by $s = 160t - 16t^2$. How long will it take the projectile to hit the ground?

38. In a certain electric circuit there is a resistance R of 2 Ω and a voltage E of 60 V. The relationship between current i (in amperes), E, and R is $i^2R + iE = 8000$. What current $i (i > 0)$ flows in the circuit?

39. Under certain conditions, the motion of an object suspended by a helical spring requires the solution of the equation $D^2 + 8D + 15 = 0$. Solve for D.

40. In determining the side x of a rectangle under specified conditions, the equation $2x^2 + 5x - 52 = 0$ is found. What is the side x (in centimeters) of the rectangle?

In Exercises 41 through 44 set up the appropriate quadratic equations and solve.

41. In electricity the equivalent resistance of two resistances connected in parallel is given by

$$\frac{1}{R} = \frac{1}{R_1} + \frac{1}{R_2}$$

Two resistances connected in series have an equivalent resistance given by $R = R_1 + R_2$. If two resistances connected in parallel have an equivalent resistance of 3 Ω and the same two resistances have an equivalent resistance of 16 Ω when connected in series, what are the resistances? (This equation is not quadratic. However, after the proper substitution is made, a fractional equation will exist. After fractions have been cleared, a quadratic equation will exist.)

42. The formula that relates the object distance p, image distance q, and focal length f of a lens is

$$\frac{1}{p} + \frac{1}{q} = \frac{1}{f}$$

Determine the positive value of p if $q = p + 3$ cm and $f = p - 1$ cm. (See Exercise 41.)

43. A certain rectangular machine part has a length which is 4 mm longer than its width. If the area of the part is 96 mm², what are its dimensions?

44. A jet, by increasing its speed by 200 mi/h, could decrease the time needed to cover 4000 mi by one hour. What is its speed?

6–2 Completing the Square

Many quadratic equations cannot be solved by factoring. This is true of most quadratic equations which arise in applied situations. In this section, therefore, we develop a method which can be used to solve any quadratic equation. The method is called **completing the square**. In the following section we shall use completing the square to develop a formula which also may be used to solve any quadratic equation.

In the first example which follows we show the solution of a type of quadratic equation which arises while using the method of completing the square. In the examples which follow it, the method itself is used and described.

Example A

In solving $x^2 = 16$, we note that either $+4$ or -4 satisfies the equation. These roots may be obtained by finding the square root of 16. Both the principal root and its negative give us roots, since each value satisfies the equation. Therefore, the roots of $x^2 = 16$ are $+4$ and -4.

We may solve $(x - 3)^2 = 16$ in a similar way by finding the square root of both sides of the equation, using the square root, and its negative, of 16, Thus,

$$x - 3 = 4 \quad \text{or} \quad x - 3 = -4$$

Solving these equations, we obtain the roots 7 and -1.

We may solve $(x - 3)^2 = 17$ in the same way. Thus,

$$x - 3 = \sqrt{17} \quad \text{or} \quad x - 3 = -\sqrt{17}$$

The roots are therefore $3 + \sqrt{17}$ and $3 - \sqrt{17}$. Decimal values of these roots are 7.123 and -1.123.

Example B

We wish to find the roots of the quadratic equation

$$x^2 - 6x - 8 = 0$$

First we note that this equation is not factorable. However, we do recognize that $x^2 - 6x$ is part of one of the special products. If 9 were added to this expression, we would have $x^2 - 6x + 9$, which is $(x - 3)^2$. We can solve this expression for x by taking a square root, which would leave us with $x - 3$. This expression can then be solved for x by any proper method of solving a linear equation. Thus, by creating an expression which is a perfect square and then taking the square root, we may solve the problem as an ordinary linear equation.

We may write the original equation as

$$x^2 - 6x = 8$$

and then add 9 to both sides of the equation. The result is

$$x^2 - 6x + 9 = 17$$

The left side of this equation may be rewritten, giving

$$(x - 3)^2 = 17$$

Taking square roots of both sides of the equation (which is the same as that solved in the third illustration of Example A), we arrive at

$$x - 3 = \pm \sqrt{17}$$

The $\pm$ sign is necessary, since by the definition of a square root, $(-\sqrt{17})^2 = 17$ and $(+\sqrt{17})^2 = 17$. Now, adding 3 to both sides, we obtain

$$x = 3 \pm \sqrt{17}$$

which means that $x = 3 + \sqrt{17}$ and $x = 3 - \sqrt{17}$ are the two roots of the equation.

How do we determine the number which must be added to complete the square? The answer to this question is based on the special products in Eqs. (5–3) and (5–4). We rewrite these in the form

$$(x + a)^2 = x^2 + 2ax + a^2 \tag{6–2}$$

and

$$(x - a)^2 = x^2 - 2ax + a^2 \tag{6–3}$$

The coefficient of x in each case is numerically $2a$, and the number added to complete the square is a^2. Thus *if we take half the coefficient of the x-term and square this result, we have the number which completes the square.* In our example, the numerical coefficient of the x-term was 6, and 9 was added to complete the square. We must be certain that the coefficient of the x^2 term is 1 before we start to complete the square. The following example outlines the steps necessary to complete the square.

Example C

Solve the following quadratic equation by the method of completing the square:

$$2x^2 + 16x - 9 = 0$$

First we divide each term by 2 so that the coefficient of the x^2 term becomes 1.

$$x^2 + 8x - \frac{9}{2} = 0$$

Now we put the constant term on the right-hand side by adding $\frac{9}{2}$ to both sides of the equation.

$$x^2 + 8x = \frac{9}{2}$$

Next we divide the coefficient of the x-term, 8, by 2, which gives us 4. We square 4 and obtain 16, which is the number to be added to both sides of the equation.

$$x^2 + 8x + 16 = \frac{9}{2} + 16 = \frac{41}{2}$$

We write the left side as the square of $(x + 4)$.

$$(x + 4)^2 = \frac{41}{2}$$

Taking the square root of both sides of the equation, we obtain

$$x + 4 = \pm\sqrt{\frac{41}{2}}$$

Solving for x, we have

$$x = -4 \pm \sqrt{\frac{41}{2}}$$

Since $\sqrt{\frac{41}{2}} = \sqrt{\frac{82}{4}} = \frac{1}{2}\sqrt{82}$, we may write the solution without a radical in the denominator as

$$x = \frac{-8 \pm \sqrt{82}}{2}$$

Therefore, the roots are $\frac{1}{2}(-8 + \sqrt{82})$ and $\frac{1}{2}(-8 - \sqrt{82})$. If we approximate $\sqrt{82}$ with 9.055, the approximate values of the roots are 0.528 and -8.528.

Example D

Solve $4x^2 - 12x + 5 = 0$ by completing the square.

$$4x^2 - 12x + 5 = 0$$

$$x^2 - 3x + \frac{5}{4} = 0$$

$$x^2 - 3x \quad = -\frac{5}{4}$$

$$x^2 - 3x + \frac{9}{4} = -\frac{5}{4} + \frac{9}{4}$$

$$\left(x - \frac{3}{2}\right)^2 = \frac{4}{4} = 1$$

$$x - \frac{3}{2} = \pm 1$$

$$x = \frac{3}{2} \pm 1$$

$$x = \frac{5}{2}, \; x = \frac{1}{2}$$

This equation could have been solved by factoring. However, at this point, we want to illustrate the method of completing the square.

Exercises 6-2

In Exercises 1 through 8 solve the given quadratic equations by finding the appropriate square roots as in Example A.

1. $x^2 = 25$ 2. $x^2 = 100$ 3. $x^2 = 7$
4. $x^2 = 15$ 5. $(x - 2)^2 = 25$ 6. $(x + 2)^2 = 100$
7. $(x + 3)^2 = 7$ 8. $(x - 4)^2 = 15$

In Exercises 9 through 24 solve the given quadratic equations by completing the square. Exercises 9 through 12 and 15 through 18 may be checked by factoring.

9. $x^2 + 2x - 8 = 0$ 10. $x^2 - x - 6 = 0$ 11. $x^2 + 3x + 2 = 0$
12. $t^2 + 5t - 6 = 0$ 13. $x^2 - 4x + 2 = 0$ 14. $x^2 + 10x - 4 = 0$
15. $v^2 + 2v - 15 = 0$ 16. $x^2 - 8x + 12 = 0$ 17. $2s^2 + 5s - 3 = 0$
18. $4x^2 + x - 3 = 0$ 19. $3y^2 - 3y - 2 = 0$ 20. $3x^2 + 4x - 3 = 0$
21. $2y^2 - y - 2 = 0$ 22. $9v^2 - 6v - 2 = 0$ 23. $x^2 + 2bx + c = 0$
24. $px^2 + qx + r = 0$

6–3 The Quadratic Formula

We shall now use the method of completing the square to derive a general formula which may be used for the solution of any quadratic equation.

Consider Eq. (6–1), the general quadratic equation

$$ax^2 + bx + c = 0$$

with $a > 0$. When we divide through by a, we obtain

$$x^2 + \frac{b}{a}x + \frac{c}{a} = 0$$

Subtracting c/a from each side, we have

$$x^2 + \frac{b}{a}x = -\frac{c}{a}$$

Half of b/a is $b/2a$, which squared is $b^2/4a^2$. Adding $b^2/4a^2$ to each side gives us

$$x^2 + \frac{b}{a}x + \frac{b^2}{4a^2} = -\frac{c}{a} + \frac{b^2}{4a^2}$$

Writing the left side as a perfect square, and combining fractions on the right side, we have

$$\left(x + \frac{b}{2a}\right)^2 = \frac{b^2 - 4ac}{4a^2}$$

Taking the square root of each side results in

$$x + \frac{b}{2a} = \frac{\pm\sqrt{b^2 - 4ac}}{2a}$$

When we subtract $b/2a$ from each side and simplify the resulting expression, we obtain the **quadratic formula**:

$$x = \frac{-b \pm \sqrt{b^2 - 4ac}}{2a} \tag{6–4}$$

To solve a quadratic equation by using the quadratic formula we need only to write the equation in standard form (see Eq. 6–1), identify a, b, and c, and substitute these numbers directly into the formula. We shall use the quadratic formula to solve the quadratic equations in the following examples.

Example A

$$x^2 - 5x + 6 = 0$$

In this equation $a = 1$, $b = -5$, and $c = 6$. Thus, we have

$$x = \frac{-(-5) \pm \sqrt{25 - 4(1)(6)}}{2} = \frac{5 \pm 1}{2} = 3, 2$$

The roots are $x = 3$ and $x = 2$. (This particular equation could have been solved by the method of factoring.)

Example B

$$2x^2 - 7x + 5 = 0$$

In this equation $a = 2$, $b = -7$, and $c = 5$. Hence,

$$x = \frac{7 \pm \sqrt{49 - 4(2)(5)}}{4} = \frac{7 \pm 3}{4} = \frac{5}{2}, 1$$

Thus, the roots are $x = \frac{5}{2}$ and $x = 1$.

Example C

$$9x^2 + 24x + 16 = 0$$

In this example, $a = 9$, $b = 24$, and $c = 16$. Thus,

$$x = \frac{-24 \pm \sqrt{576 - 4(9)(16)}}{18} = \frac{-24 \pm 0}{18} = -\frac{4}{3}$$

Here both roots are $-\frac{4}{3}$, and the answer should be written as $x = -\frac{4}{3}$ and $x = -\frac{4}{3}$.

Example D

$$3x^2 - 5x + 4 = 0$$

In this example, $a = 3$, $b = -5$, and $c = 4$. Therefore,

$$x = \frac{5 \pm \sqrt{25 - 4(3)(4)}}{6} = \frac{5 \pm \sqrt{-23}}{6}$$

Here we note that the roots contain imaginary numbers. This happens if $b^2 < 4ac$.

Example E

$$2x^2 = 4x + 3$$

First we must put the equation in the proper form. This is

$$2x^2 - 4x - 3 = 0.$$

Now we identify $a = 2$, $b = -4$, $c = -3$, which leads to the solution

$$x = \frac{-(-4) \pm \sqrt{(-4)^2 - 4(2)(-3)}}{2(2)} = \frac{4 \pm \sqrt{16 + 24}}{4}$$

$$= \frac{4 \pm \sqrt{40}}{4} = \frac{4 \pm 2\sqrt{10}}{4} = \frac{2(2 \pm \sqrt{10})}{4}$$

$$= \frac{2 \pm \sqrt{10}}{2}$$

Here we used the method of simplifying radicals as introduced in Section 1-6.

Example F

$$dx^2 - (3 + d)x + 4 = 0$$

In this example, $a = d$, $b = -(3 + d)$, and $c = 4$. We can use the quadratic formula to solve quadratic equations which have literal coefficients. Thus,

$$x = \frac{3 + d \pm \sqrt{[-(3 + d)]^2 - 4(d)(4)}}{2d} = \frac{3 + d \pm \sqrt{9 - 10d + d^2}}{2d}$$

Example G

A square field has a diagonal which is 10 m longer than one of the sides. What is the length of a side?

Let $x =$ the length of a side of the field, and $y =$ the length of the diagonal. Using the Pythagorean theorem, we know that $y^2 = x^2 + x^2$. From the given information we know that $y = x + 10$. Thus, we have

$$(x + 10)^2 = x^2 + x^2$$

We can now simplify and solve this equation as follows.

$$x^2 + 20x + 100 = 2x^2$$
$$x^2 - 20x - 100 = 0$$

$$x = \frac{20 \pm \sqrt{400 + 400}}{2} = 10 \pm 10\sqrt{2}$$

The negative solution has no meaning in this problem. This means that the solution is $10 + 10\sqrt{2} = 24.1$ m.

The quadratic formula provides a quick general method for solving quadratic equations. Proper recognition and substitution of the coefficients a, b, and c is all that is required to complete the solution, regardless of the nature of the roots.

Exercises 6–3

In Exercises 1 through 32 solve the given quadratic equations using the quadratic formula. Exercises 1 through 14 are the same as Exercises 9 through 22 of Section 6–2.

1. $x^2 + 2x - 8 = 0$ 2. $x^2 - x - 6 = 0$
3. $x^2 + 3x + 2 = 0$ 4. $t^2 + 5t - 6 = 0$
5. $x^2 - 4x + 2 = 0$ 6. $x^2 + 10x - 4 = 0$
7. $v^2 + 2v - 15 = 0$ 8. $x^2 - 8x + 12 = 0$
9. $2s^2 + 5s - 3 = 0$ 10. $4x^2 + x - 3 = 0$
11. $3y^2 - 3y - 2 = 0$ 12. $3x^2 + 4x - 3 = 0$
13. $2y^2 - y - 2 = 0$ 14. $9v^2 - 6v - 2 = 0$
15. $2x^2 - 7x + 4 = 0$ 16. $3x^2 - 5x - 4 = 0$
17. $2t^2 + 10t = -15$ 18. $2d^2 + 7 = 4d$

19. $3s^2 = s + 9$

20. $6r^2 - 6r - 1 = 0$

21. $4x^2 - 9 = 0$

22. $x^2 - 6x = 0$

23. $15 + 4z - 32z^2 = 0$

24. $4x^2 - 12x = 7$

25. $x^2 + 2cx - 1 = 0$

26. $x^2 - 7x + (6 + a) = 0$

27. $b^2x^2 - (b + 1)x + (1 - a) = 0$

28. $c^2x^2 - x - 1 = x^2$

29. Under certain conditions, the partial pressure P of a certain gas (in pascals) is found by solving the equation $P^2 - 3P + 1 = 0$. Solve for P such that $P < 1$ Pa.

30. A projectile is fired vertically into the air. The distance (in feet) above the ground as a function of time (in seconds) is given by the formula $s = 300 - 100t - 16t^2$. When will the projectile hit the ground?

31. The total surface area of a right circular cylinder is found by the formula $A = 2\pi r^2 + 2\pi rh$. If the height of the cylinder is 4 in., how much is the radius if the area is 9π in.²?

32. In calculating the current in an electric circuit with an inductance L (in henrys), a resistance R (in ohms) and capacitance C (in farads), it is necessary to solve the equation $Lx^2 + Rx + \frac{1}{C} = 0$. Find x in terms of L, R, and C.

In Exercises 33 through 36 set up appropriate equations and solve the given stated problems.

33. A metal cube expands when heated. If the volume changes by 6.00 mm³ and each edge is 0.20 mm longer after being heated, what was the original length of an edge of the cube?

34. After a laboratory experiment, a student reported that two particular resistances had a combined resistance of 4 Ω when connected in parallel, and a combined resistance of 7 Ω when connected in series. What values would she obtain for the resistances? (See Exercise 41 of Section 6–1.)

35. To cover a given floor with square tiles of a certain size, it is found that 648 tiles are needed. If the tiles were 1 in. larger in both dimensions, only 512 tiles would be required. What is the length of a side of one of the smaller tiles?

36. A jet pilot flies 2400 mi at a given speed. If the speed were increased by 300 mi/h, the trip would take one hour less. What would be the speed of the jet?

6–4 Exercises for Chapter 6

In Exercises 1 through 12 solve the given quadratic equations by factoring.

1. $x^2 + 3x - 4 = 0$

2. $x^2 + 3x - 10 = 0$

3. $x^2 - 10x + 16 = 0$

4. $x^2 - 6x - 27 = 0$

5. $3x^2 + 11x = 4$

6. $6y^2 = 11y - 3$

7. $6t^2 = 13t - 5$

8. $3x^2 + 5x + 2 = 0$

9. $6s^2 = 25s$

10. $6n^2 - 23n - 35 = 0$

11. $4x^2 - 8x = 21$

12. $6x^2 = 8 - 47x$

In Exercises 13 through 24 solve the given quadratic equations by using the quadratic formula.

13. $x^2 - x - 110 = 0$

14. $x^2 + 3x - 18 = 0$

15. $x^2 + 2x - 5 = 0$

16. $x^2 - 7x - 1 = 0$

17. $2x^2 - x - 36 = 0$

18. $3x^2 + x - 14 = 0$

19. $4x^2 - 3x - 2 = 0$

20. $5x^2 + 7x - 2 = 0$

21. $2x^2 + 2x + 5 = 0$

22. $3x^2 - 4x - 1 = 0$

23. $6x^2 = 9 - 4x$

24. $8x^2 = 5x + 2$

In Exercises 25 through 36 solve the given quadratic equations by any appropriate method.

25. $x^2 + 4x - 4 = 0$

26. $x^2 + 3x + 1 = 0$

27. $3x^2 + 8x + 2 = 0$

28. $3p^2 = 28 - 5p$

29. $4v^2 = v + 5$

30. $6x^2 - x + 2 = 0$

31. $2x^2 + 3x + 7 = 0$

32. $4y^2 - 5y = 8$

33. $a^2x^2 + 2ax + 2 = 0$

34. $16r^2 - 8r + 1 = 0$

35. $ax^2 = a^2 - 3x$

36. $2bx = x^2 - 3b$

In Exercises 37 through 40 solve the given quadratic equations by completing the square.

37. $x^2 - x - 30 = 0$

38. $x^2 - 2x - 5 = 0$

39. $2x^2 - x - 4 = 0$

40. $4x^2 - 8x - 3 = 0$

In Exercises 41 through 44 solve the equations involving fractions. After multiplying through by the lowest common denominator, quadratic equations should result.

41. $\dfrac{x-4}{x-1} = \dfrac{2}{x}$

42. $\dfrac{x-1}{3} = \dfrac{5}{x} + 1$

43. $\dfrac{x^2 - 3x}{x - 3} = \dfrac{x^2}{x + 2}$

44. $\dfrac{x-2}{x-5} = \dfrac{15}{x^2 - 5x}$

In Exercises 45 through 52 solve the given quadratic equations by any appropriate method.

45. To determine the electric current in a certain alternating-current circuit, it is necessary to solve the equation $m^2 + 10m + 2000 = 0$. Solve for m.

46. Under specified conditions, the deflection of a beam requires the solution of the equation $40x - x^2 - 400 = 0$, where x is the distance (in feet) from one end of the beam. Solve for x.

47. Under specified conditions, the power developed in an element of an electric circuit is $P = EI - RI^2$, where P is the power, E is the specified voltage, and R is a specified resistance. Assuming that P, E, and R are constants, solve for I.

48. For laminar flow of fluids, the coefficient K_e, used to calculate energy loss due to sudden enlargements, is given by

$$K_e = 1.00 - 2.67\frac{S_a}{S_b} + \left(\frac{S_a}{S_b}\right)^2$$

where S_a/S_b is the ratio of cross-sectional areas. If $K_e = -0.500$, what is the value of S_a/S_b ?

49. A company determines that the cost C (in dollars) of manufacturing x units of a certain product is given by $C = 0.1x^2 + 0.8x + 7$. How many units can be made for $25?

50. In determining the width w of a parcel of land, the equation $w^2 + 60w = 5000$ is used. What is the width (in meters) of the parcel?

51. In the theory to study the motion of biological cells and viruses, the equation $n = 2.5p - 12.6p^2$ is used. Solve for p in terms of n.

52. A general formula for the distance s traveled by an object, given an initial velocity v and acceleration a in time t, is $s = vt + \frac{1}{2}at^2$. Solve for t.

In Exercises 53 through 58 set up appropriate equations and solve the given stated problems.

53. The sum of two electric voltages is 20 V, and their product is 96 V^2. What are the voltages?

54. The length of one field is 400 m more than the side of a square field. The width is 100 m more than the side of the square field. If the rectangular field has twice the area of the square field, what are the dimensions of each field?

55. The manufacturer of a disk-shaped machine part of radius 1.00 in. discovered that he could prevent taking a loss in its production if the amount of material used in each was reduced by 20%. If the thickness remains the same, by how much must the radius be reduced in order to prevent a loss?

56. A roof truss is in the shape of a right triangle with the hypotenuse along the base. If one rafter (neglect overhang) is 4 ft longer than the other, and the base is 36 ft, what are the lengths of the rafters?

57. An electric utility company is placing utility poles along a road. It is determined that five less poles per kilometer would be necessary if the distance between poles were increased by 10 m. How many poles are being placed each kilometer?

58. A rectangular duct in a building's ventilating system is made of sheet metal 7 ft wide and has a cross-sectional area of 3 ft^2. What are the cross-sectional dimensions of the duct?

7

Trigonometric Functions
of Any Angle

7–1 Signs of the Trigonometric Functions

When we were dealing with trigonometric functions in Chapter 3, we restricted ourselves primarily to the functions of acute angles measured in degrees. Since we did define the functions in general, we can use these same definitions for finding the functions of any possible angle. We shall not only find the trigonometric functions of angles measured in degrees, but we shall introduce radian measure as well. From there we shall show how the trigonometric functions can be defined for numbers. In this section we shall determine the signs of the trigonometric functions in each of the four quadrants.

We recall the definitions of the trigonometric functions which were given in Section 3–2: Here the point (x, y) is a point on the terminal side of angle θ, and r is the radius vector.

$$\sin \theta = \frac{y}{r}, \qquad \cos \theta = \frac{x}{r}, \qquad \tan \theta = \frac{y}{x}$$

$$\cot \theta = \frac{x}{y}, \qquad \sec \theta = \frac{r}{x}, \qquad \csc \theta = \frac{r}{y}$$

(7–1)

We see that the functions are defined so long as we know the abscissa, ordinate, and radius vector of the terminal side of θ. Remembering that r is always considered positive, we can see that the various functions will vary in sign, depending on the signs of x and y.

If the terminal side of the angle is in the first or second quadrant, the value of $\sin \theta$ will be positive, but if the terminal side is in the third or fourth quadrant, $\sin \theta$ is negative. This is because y is positive if the point defining the terminal side is above the x-axis, and y is negative if this point is below the x-axis.

Example A

The value of $\sin 20°$ is positive, since the terminal side of $20°$ is in the first quadrant. The value of $\sin 160°$ is positive, since the terminal side of $160°$ is in the second quadrant. The values of $\sin 200°$ and $\sin 340°$ are negative, since the terminal sides of these angles are in the third and fourth quadrants, respectively.

The sign of $\tan \theta$ depends upon the ratio of y to x. In the first quadrant both x and y are positive, and therefore the ratio y/x is positive. In the third quadrant both x and y are negative, and therefore the ratio y/x is positive. In the second and fourth quadrants either x or y is positive and the other is negative, and so the ratio of y/x is negative.

Example B

The values of $\tan 20°$ and $\tan 200°$ are positive, since the terminal sides of these angles are in the first and third quadrants, respectively. The values of $\tan 160°$ and $\tan 340°$ are negative, since the terminal sides of these angles are in the second and fourth quadrants, respectively.

The sign of $\cos \theta$ depends upon the sign of x. Since x is positive in the first and fourth quadrants, $\cos \theta$ is positive in these quadrants. In the same way, $\cos \theta$ is negative in the second and third quadrants.

Example C

The values of $\cos 20°$ and $\cos 340°$ are positive, since these angles are first and fourth quadrant angles, respectively. The values of $\cos 160°$ and $\cos 200°$ are negative, since these angles are second- and third-quadrant angles, respectively.

Since $\csc \theta$ is defined in terms of r and y, as is $\sin \theta$, the sign of $\csc \theta$ is the same as that of $\sin \theta$. For the same reason, $\cot \theta$ has the same sign as $\tan \theta$, and $\sec \theta$ has the same sign as $\cos \theta$. A method for remembering the signs of the functions in the four quadrants is as follows:

All functions of first-quadrant angles are positive. Sin θ and csc θ are positive for second-quadrant angles. Tan θ and cot θ are positive for third-quadrant angles. Cos θ and sec θ are positive for fourth-quadrant angles. All others are negative.

This discussion does not include the quadrantal angles, those angles with terminal sides on one of the axes. They will be discussed in the following section.

Example D

sin 50°, sin 150°, sin (−200°), cos 8°, cos 300°, cos (−40°), tan 220°, tan (−100°), cot 260°, cot (−310°), sec 280°, sec (−37°), csc 140°, and csc (−190°) are all positive.

Example E

sin 190°, sin 325°, cos 100°, cos (−95°), tan 172°, tan 295°, cot 105°, cot (−6°), sec 135°, sec (−135°), csc 240°, and csc 355° are all negative.

Example F

Determine the trigonometric functions of θ if the terminal side of θ passes through $(-1, \sqrt{3})$.

We know that $x = -1$, $y = +\sqrt{3}$, and from the Pythagorean theorem we find that $r = 2$. Therefore, the trigonometric functions of θ are:

$$\sin \theta = +\frac{\sqrt{3}}{2}, \qquad \cos \theta = -\frac{1}{2}, \qquad \tan \theta = -\sqrt{3}$$

$$\cot \theta = -\frac{1}{\sqrt{3}}, \qquad \sec \theta = -2, \qquad \csc \theta = +\frac{2}{\sqrt{3}}$$

We note that the point $(-1, \sqrt{3})$ is on the terminal side of a second-quadrant angle, and that the signs of the functions of θ are those of a second-quadrant angle.

Exercises 7—1

In Exercises 1 through 8 determine the algebraic sign of the given trigonometric functions.

1. sin 60°, cos 120°, tan 320°
2. tan 185°, sec 115°, sin (−36°)
3. cos 300°, csc 97°, cot (−35°)
4. sin 100°, sec (−15°), cos 188°
5. cot 186°, sec 280°, sin 470°
6. tan (−91°), csc 87°, cot 103°
7. cos 700°, tan (−560°), csc 530°
8. sin 256°, tan 321°, cos (−370°)

In Exercises 9 through 16 find the trigonometric functions of θ, where the terminal side of θ passes through the given point.

9. (2, 1)
10. (−1, 1)
11. (−2, −3)
12. (4, −3)
13. (−5, 12)
14. (−3, −4)
15. (5, −2)
16. (3, 5)

In Exercises 17 through 24 determine the quadrant in which the terminal side of θ lies, subject to the given conditions.

17. sin θ positive, cos θ negative
18. tan θ positive, cos θ negative
19. sec θ negative, cot θ negative
20. cos θ positive, csc θ negative
21. csc θ negative, tan θ negative
22. sec θ positive, csc θ positive
23. sin θ negative, tan θ positive
24. cot θ negative, sin θ negative

7–2 Trigonometric Functions of Any Angle

The trigonometric functions of acute angles were discussed in Section 3–3, and in the last section we determined the signs of the trigonometric functions in each of the four quadrants. In this section we shall show how we can find the trigonometric functions of an angle of any magnitude. This information will be very important in Chapter 8 when we discuss oblique triangles and in Chapter 9 when we graph the trigonometric functions.

Any angle in standard position is coterminal with some positive angle less than 360°. Since the terminal sides of coterminal angles are the same, the trigonometric functions of coterminal angles are the same. Therefore, we need consider only the problem of finding the values of the trigonometric functions of positive angles less than 360°.

Example A
The following pairs of angles are coterminal.

390° and 30°,   −60° and 300°   900° and 180°,   −150° and 210°

From this we conclude that the trigonometric functions of both angles in these pairs are equal. That is, for example, $\sin 390° = \sin 30°$ and $\tan(−150°) = \tan 210°$.

Considering the definitions of the functions, we see that the values of the functions depend only on the values of x, y, and r. The values of the functions of second-quadrant angles are numerically equal to the functions of corresponding first-quadrant angles. For example, considering the angles shown in Fig. 7–1, for angle θ_2 with terminal side passing through $(−3, 4)$, $\tan \theta_2 = −\frac{4}{3}$, and for angle θ_1 with terminal side passing through $(3, 4)$, $\tan \theta_1 = \frac{4}{3}$. In Fig. 7–1, we see that the triangles containing angles θ_1 and α are congruent, which means that θ_1 and α are equal. We know that the trigonometric functions of θ_1 and θ_2 are numerically equal. This means that

$$|F(\theta_2)| = |F(\theta_1)| = |F(\alpha)| \qquad (7\text{–}2)$$

where F represents any of the trigonometric functions.

The angle labeled α is called the **reference angle**. *The reference angle of a given angle is the acute angle formed by the terminal side of the angle and the x-axis.*

Using Eq. (7–2) and the fact that $\alpha = 180° − \theta_2$, we may conclude that the value of any trigonometric function of any second-quadrant angle is found from

$$F(\theta_2) = \pm F(180° − \theta_2) = \pm F(\alpha) \qquad (7\text{–}3)$$

The sign to be used depends on whether the *function* is positive or negative in the second quadrant.

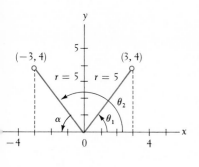

Figure 7–1

Example B

In Fig. 7–1, the trigonometric functions of θ_2 are as follows:

$$\sin \theta_2 = +\sin (180° - \theta_2) = +\sin \alpha = +\sin \theta_1 = \frac{4}{5}$$

$$\cos \theta_2 = -\cos \theta_1 = -\frac{3}{5}$$

$$\tan \theta_2 = -\frac{4}{3}, \qquad \cot \theta_2 = -\frac{3}{4}$$

$$\sec \theta_2 = -\frac{5}{3}, \qquad \csc \theta_2 = +\frac{5}{4}$$

In the same way we may derive the formulas for finding the trigonometric functions of any third- or fourth-quadrant angle. Considering the angles shown in Fig. 7–2, we see that the reference angle α is found by subtracting 180° from θ_3 and that functions of α and θ_1 are numerically equal. Considering the angles shown in Fig. 7–3, we see that the reference angle α is found by subtracting θ_4 from 360°. Therefore, we have

and
$$F(\theta_3) = \pm F(\theta_3 - 180°) \qquad (7\text{–}4)$$
$$F(\theta_4) = \pm F(360° - \theta_4) \qquad (7\text{–}5)$$

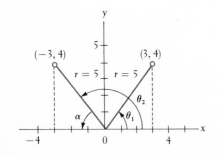

Figure 7–1

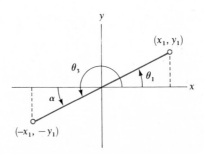

Figure 7–2

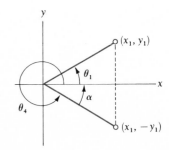

Figure 7–3

Example C

In Fig. 7–2, if $\theta_3 = 210°$, the trigonometric functions of θ_3 are found by using Eq. (7–4) as follows.

$$\sin 210° = -\sin (210° - 180°) = -\sin 30° = -\frac{1}{2} = -0.5000$$

$$\cos 210° = -\cos 30° = -0.8660$$
$$\tan 210° = +0.5774 \qquad \cot 210° = +1.732$$
$$\sec 210° = -1.155 \qquad \csc 210° = -2.000$$

Example D

In Fig. 7–3, if $\theta_4 = 315°$, the trigonometric functions of θ_4 are found by using Eq. (7–5) as follows.

$$\sin 315° = -\sin (360° - 315°) = -\sin 45° = -0.7071$$
$$\cos 315° = +\cos 45° = +0.7071$$
$$\tan 315° = -1.000 \qquad \cot 315° = -1.000$$
$$\sec 315° = +1.414 \qquad \csc 315° = -1.414$$

Example E

Other illustrations of the use of Eqs. (7–3), (7–4), and (7–5) are as follows.

$$\sin 160° = +\sin (180° - 160°) = \sin 20° = 0.3420$$
$$\tan 110° = -\tan (180° - 110°) = -\tan 70° = -2.747$$
$$\cos 225° = -\cos (225° - 180°) = -\cos 45° = -0.7071$$
$$\cot 260° = +\cot (260° - 180°) = \cot 80° = 0.1763$$
$$\sec 304° = +\sec (360° - 304°) = \sec 56° = 1.788$$
$$\sin 357° = -\sin (360° - 357°) = -\sin 3° = -0.0523$$

The following examples illustrate how Eqs. (7–3) through (7–5) are used to determine θ when a function of θ is given.

Example F

Given that $\sin \theta = 0.2250$, find θ for $0° < \theta < 360°$.

Here we are asked to find any angles between 0° and 360° for which $\sin \theta = 0.2250$. Since $\sin \theta$ is positive for first- and second-quadrant angles, there will be two such angles.

From the tables we find $\theta = 13°0'$. We also know that $\theta = 180°0' - 13°0' = 167°0'$. These are the two required answers.

Example G

Given that $\tan \theta = 2.050$ and $\cos \theta < 0$, find θ when $0° < \theta < 360°$.

Since $\tan \theta$ is positive and $\cos \theta$ is negative, θ must be in the third quadrant. We note from the tables that $2.050 = \tan 64°0'$. Therefore, $\theta = 180°0' + 64°0' = 244°0'$. Since the required angle is to be between 0° and 360°, this is the only possible answer.

If angles are expressed to the nearest 10' or 0.1°, we find the nearest value in the appropriate table as we did in Chapter 3. Also, in expressing angles to the nearest 1', interpolation is used if the value is obtained from Table 3. It is also possible to find these values on a scientific calculator.

Example H

To find $\sin 251.4°$, we first determine that the angle is a third-quadrant angle. Thus,

$$\sin 251.4° = -\sin (251.4° - 180.0°) = -\sin 71.4°$$
$$= -0.9478$$

To find tan 103°37′, we first determine that the angle is a second-quadrant angle. Thus,

$$\tan 103°37′ = -\tan(180°0′ - 103°37′) = -\tan 76°23′$$
$$= -4.129$$

The value is obtained by use of interpolation in Table 3.

Given that $\cos \theta = 0.1354$ for $0° < \theta < 360°$, $\theta = 82°10′$ to the nearest 10′ from Table 3. Also, since the cosine is positive in the fourth quadrant, θ may also equal $360°0′ - 82°10′ = 277°50′$. Thus, the two answers are 82°10′ and 277°50′. If we expressed θ to the nearest 0.1° from Table 4, we would obtain 82.2° and 277.8° for θ.

With the use of Eqs. (7–3) through (7–5) we may find the value of any function, as long as the terminal side of the angle lies *in* one of the quadrants. This problem reduces to finding the function of an acute angle. We are left with the case of the terminal side being along one of the axes, a **quadrantal angle**. Using the definitions of the functions, and remembering that $r > 0$, we arrive at the values in the following table.

θ	$\sin \theta$	$\cos \theta$	$\tan \theta$	$\cot \theta$	$\sec \theta$	$\csc \theta$
0°	0.000	1.000	0.000	Undef.	1.000	Undef.
90°	1.000	0.000	Undef.	0.000	Undef.	1.000
180°	0.000	−1.000	0.000	Undef.	−1.000	Undef.
270°	−1.000	0.000	Undef.	0.000	Undef.	−1.000
360°	Same as the functions of 0° (same terminal side)					

The values in the table may be verified by referring to the figures in Fig. 7–4.

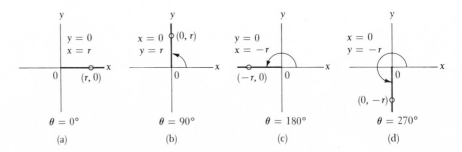

Figure 7–4

Example 1

Since $\sin \theta = y/r$, from Fig. 7–4 (a) we see that $\sin 0° = 0/r = 0$.

Since $\tan \theta = y/x$, from Fig. 7–4 (b) we see that $\tan 90° = r/0$, which is undefined due to the division by zero.

Since $\cos \theta = x/r$, from Fig. 7–4 (c) we see that $\cos 180° = -r/r = -1$.

Since $\cot \theta = x/y$, from Fig. 7–4 (d) we see that $\cot 270° = 0/-r = 0$.

Exercises 7–2

In Exercises 1 through 8 express the given trigonometric functions in terms of the same function of a positive acute angle.

1. sin 160°,   cos 220°
2. tan 91°,   sec 345°
3. tan 105°,   csc 302°
4. cos 190°,   cot 290°
5. sin (−123°),   cot 174°
6. sin 98°,   sec (−315°)
7. cos 400°,   tan (−400°)
8. tan 920°,   csc (−550°)

In Exercises 9 through 28 determine the values of the given trigonometric functions by use of tables.

9. sin 195°0′
10. tan 311°0′
11. cos 106°0′
12. sin 254°0′
13. cot 136°0′
14. cos 297°0′
15. sec (−115°0′)
16. csc (−193°0′)
17. tan 193°10′
18. sin 311°50′
19. cos 206°40′
20. sec 328°20′
21. sin 103.3°
22. tan 219.1°
23. cot 330.5°
24. cos 198.8°
25. sin 322°52′
26. cot 254°17′
27. tan 118°33′
28. cos (−67°5′)

In Exercises 29 through 36 find θ to the nearest 10′ for $0 < \theta < 360°$.

29. $\tan \theta = 0.5317$
30. $\cos \theta = 0.6428$
31. $\sin \theta = -0.3638$
32. $\cot \theta = -1.319$
33. $\sin \theta = 0.8708$
34. $\csc \theta = 2.311$
35. $\cos \theta = -0.1207$
36. $\tan \theta = -2.368$

In Exercises 37 through 44 find θ to the nearest 0.1° for $0 < \theta < 360°$.

37. $\sin \theta = -0.8480$
38. $\cot \theta = -0.2126$
39. $\cos \theta = 0.4003$
40. $\tan \theta = -1.830$
41. $\cot \theta = 0.5265$
42. $\sin \theta = 0.6374$
43. $\tan \theta = 0.2833$
44. $\cos \theta = -0.9287$

In Exercises 45 through 48 find θ to the nearest minute for $0 < \theta < 360°$.

45. $\sin \theta = -0.9527$
46. $\cos \theta = 0.8727$
47. $\cot \theta = -0.7144$
48. $\tan \theta = -2.664$

In Exercises 49 through 52 determine the function which satisfies the given conditions.

49. Find $\tan \theta$ when $\sin \theta = -0.5736$ and $\cos \theta > 0$.
50. Find $\sin \theta$ when $\cos \theta = 0.4226$ and $\tan \theta < 0$.
51. Find $\cos \theta$ when $\tan \theta = -0.8098$ and $\csc \theta > 0$.
52. Find $\cot \theta$ when $\sec \theta = 1.122$ and $\sin \theta < 0$.

In Exercises 53 through 56 insert the proper sign, $>$ or $<$ or $=$, between the given expressions.

53. sin 90°   2 sin 45°
54. cos 360°   2 cos 180°
55. tan 180°   tan 0°
56. sin 270°   3 sin 90°

In Exercises 57 through 60 evaluate the given expressions.

57. Under specified conditions, a force F (in pounds) is determined by solving the following equation for F:

$$\frac{F}{\sin 115.0°} = \frac{46.0}{\sin 35.0°}$$

Find the magnitude of the force.

58. A certain ac voltage can be found from the equation $V = 100 \cos 565.0°$. Find the voltage V.

59. A formula for finding the area of a triangle, knowing sides a and b, and angle C is $A = \frac{1}{2}ab \sin C$. Find the area of a triangle for which $a = {}^{\prime}37.2$, $b = 57.2$, and $C = 157.0°$.

60. In calculating the area of a triangular tract of land, a surveyor used the formula in Exercise 59. He used the values $a = 273$ m, $b = 156$ m, and $C = 112.5°$. Find the required area.

In Exercises 61 through 64 the trigonometric functions of negative angles are considered. In Exercises 62, 63, and 64 use the equations derived in Exercise 61.

61. From Fig. 7–5 we see that $\sin \theta = y/r$ and $\sin (-\theta) = -y/r$. From this we conclude that $\sin (-\theta) = -\sin \theta$. In the same way, verify the remaining Eqs. (7–6):

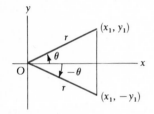

Figure 7–5

$$\begin{aligned}
\sin (-\theta) &= -\sin \theta, & \cos (-\theta) &= \cos \theta \\
\tan (-\theta) &= -\tan \theta, & \cot (-\theta) &= -\cot \theta \\
\sec (-\theta) &= \sec \theta, & \csc (-\theta) &= -\csc \theta
\end{aligned} \qquad (7\text{–}6)$$

62. Find (a) $\sin (-60°)$ and (b) $\cos (-176°)$.
63. Find (a) $\tan (-100°)$ and (b) $\cot (-215°)$.
64. Find (a) $\sec (-310°)$ and (b) $\csc (-35°)$.

7–3 Radians

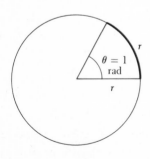

Figure 7–6

For many problems in which trigonometric functions are used, particularly those involving the solution of triangles, the degree measurement of angles is quite sufficient. However, in numerous other types of applications and in more theoretical discussions, another way of expressing the measure of angle is more meaningful and convenient. This unit measurement is the **radian**. *A radian is the measure of an angle with its vertex at the center of a circle and with an intercepted arc on the circle equal in length to the radius of the circle.* See Fig. 7–6.

Since the circumference of any circle in terms of its radius is given by $c = 2\pi r$, the ratio of the circumference to the radius is 2π. This means that the radius may be laid off 2π (about 6.28) times along the circumference, regardless of the length of the radius. Therefore, we see that

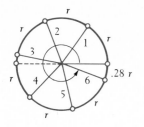

Figure 7-7

radian measure is independent of the radius of the circle. In Fig. 7–7 the numbers on each of the radii indicate the number of radians in the angle measured in standard position. The circular arrow shows an angle of 6 radians.

Since the radius may be laid off 2π times along the circumference, it follows that there are 2π radians in one complete rotation. Also, there are 360° in one complete rotation. Therefore, 360° is *equivalent* to 2π radians. It then follows that the relation between degrees and radians is 2π rad = 360°, or

$$\pi \text{ rad} = 180° \qquad\qquad (7-7)$$

From this relation we find that

$$1° = \frac{\pi}{180} \text{ rad} = 0.01745 \text{ rad} \qquad\qquad (7-8)$$

and that

$$1 \text{ rad} = \frac{180°}{\pi} = 57.3° \qquad\qquad (7-9)$$

We see from Eqs. (7–7) through (7–9) that (1) *to convert an angle measured in degrees to the same angle measured in radians, we multiply the number of degrees by $\pi/180$*, and (2) *to convert an angle measured in radians to the same angle measured in degrees, we multiply the number of radians by $180/\pi$.*

Example A

$$18.0° = \left(\frac{\pi}{180}\right)(18.0) = \frac{\pi}{10.0} = \frac{3.14}{10.0} = 0.314 \text{ rad}$$

$$120° = \left(\frac{\pi}{180}\right)(120) = \frac{6.28}{3.00} = 2.09 \text{ rad}$$

$$0.400 \text{ rad} = \left(\frac{180°}{\pi}\right)(0.400) = \frac{72.0°}{3.14} = 22.9°$$

$$2.00 \text{ rad} = \left(\frac{180°}{\pi}\right)(2.00) = \frac{360°}{3.14} = 114.6°$$

Due to the nature of the definition of the radian, it is very common to express radians in terms of π. Expressing angles in terms of π is illustrated in the following example.

Example B

$$30° = \left(\frac{\pi}{180}\right)(30) = \frac{\pi}{6} \text{ rad}$$

$$45° = \left(\frac{\pi}{180}\right)(45) = \frac{\pi}{4} \text{ rad}$$

$$\frac{\pi}{2} \text{ rad} = \left(\frac{180°}{\pi}\right)\left(\frac{\pi}{2}\right) = 90°$$

$$\frac{3\pi}{4} \text{ rad} = \left(\frac{180°}{\pi}\right)\left(\frac{3\pi}{4}\right) = 135°$$

We wish now to make a very important point. Since π is a special way of writing the number (slightly greater than 3) that is the ratio of the circumference of a circle to its diameter, it is the ratio of one distance to another. Thus radians really have no units and *radian measure amounts to measuring angles in terms of numbers.* It is this property of radians that makes them useful in many situations. Therefore, when radians are being used, it is customary that no units are indicated for the angle. *When no units are indicated, the radian is understood to be the unit of angle measurement.*

Example C

$$60.0° = \left(\frac{\pi}{180}\right)(60.0) = \frac{\pi}{3.00} = 1.05$$

$$2.50 = \left(\frac{180°}{\pi}\right)(2.50) = \frac{450°}{3.14} = 143°$$

Since no units are indicated for 1.05 or 2.50 in this example, they are known to be radian measure.

We can also use Table 3 directly to find the function of an acute angle given in radians. The following examples illustrate this use of the table.

Example D

In determining the value of sin 0.4538, we locate 0.4538 in the column labeled radians, and opposite it we note 0.4384 in the sine column. Therefore, sin 0.4538 = 0.4384.

In the same manner we find, using the nearest radian value shown in Table 3, that

$$\tan 0.9977 = 1.550, \quad \cos 0.6813 = 0.7771, \quad \text{and} \quad \sec 1.1368 = 2.381$$

If we wish to find the value of a function of an angle greater than $\frac{\pi}{2}$, we must first determine which quadrant the angle is in, and then find the reference angle. In this determination we should note the radian measure equivalents of 90°, 180°, 270°, and 360°. For 90° we have

$\frac{\pi}{2} = 1.571$; for 180° we have $\pi = 3.142$; for 270° we have $\frac{3}{2}\pi = 4.712$; and for 360° we have $2\pi = 6.283$.

Example E
(a) Find sin 3.402.
   Since 3.402 is greater than 3.142, but less than 4.712, we know that this angle is in the third quadrant and that it has a reference angle of $3.402 - 3.142 = 0.260$. Thus,

$$\sin 3.402 = -\sin 0.260 = -0.2560$$

using the nearest value shown in the table.
   (b) Find cos 5.210.
   Since 5.210 is between 4.712 and 6.283, we know that this angle is in the fourth quadrant and that its reference angle is $6.283 - 5.210 = 1.073$. Thus,

$$\cos 5.210 = \cos 1.073 = 0.4772$$

using the nearest value shown in the table.

   Often when one first encounters radian measure, expressions such as sin 1 and sin $\theta = 1$ are confused. The first is equivalent to sin 57.3°, since 57.3° = 1 (radian). The second means that θ is the angle for which the sine is 1. Since sin 90° = 1, we can say that $\theta = 90°$ or that $\theta = \pi/2$. The following examples give additional illustrations of evaluating expressions involving radians.

Example F

$$\sin \frac{\pi}{3} = \frac{\sqrt{3}}{2} \quad \text{since } \frac{\pi}{3} = 60°$$

$$\sin 0.605 = 0.5688 \quad \text{(We note that } 0.605 = 34°40'.)$$

$$\tan \theta = 1.709 \text{ means that } \theta = 59°40' \text{ (smallest positive } \theta)$$

Since 59°40' = 1.04, we can state that tan 1.04 = 1.709. This can also be determined directly from the radian column of Table 3.

Example G
Express θ in radians, such that $\cos \theta = 0.8829$ and $0 < \theta < 2\pi$.
   We are to find θ in radians for the given value of the cos θ. Also, since θ is restricted to values between 0 and 2π, we must find a first-quadrant angle and a fourth-quadrant angle (cos θ is positive in the first and the fourth quadrants). From the table we see that

$$\cos 0.4887 = 0.8829$$

Therefore, for the fourth-quadrant angle,

$$\cos (2\pi - 0.4887) = \cos (6.283 - 0.4887) = \cos (5.794)$$

This means

$$\theta = 0.4887 \quad \text{or} \quad \theta = 5.794$$

Exercises 7—3

In Exercises 1 through 8 express the given angle measurements in terms of π.

1. 15°, 150°   2. 12°, 225°   3. 75°, 330°   4. 36°, 315°
5. 210°, 270°   6. 240°, 300°   7. 160°, 260°   8. 66°, 350°

In Exercises 9 through 16 the given numbers express angle measure. Express the measure of each angle in terms of degrees.

9. $\dfrac{2\pi}{5}, \dfrac{3\pi}{2}$ 10. $\dfrac{3\pi}{10}, \dfrac{5\pi}{6}$ 11. $\dfrac{\pi}{18}, \dfrac{7\pi}{4}$ 12. $\dfrac{7\pi}{15}, \dfrac{4\pi}{3}$

13. $\dfrac{17\pi}{18}, \dfrac{5\pi}{3}$ 14. $\dfrac{11\pi}{36}, \dfrac{5\pi}{4}$ 15. $\dfrac{\pi}{12}, \dfrac{3\pi}{20}$ 16. $\dfrac{7\pi}{30}, \dfrac{4\pi}{15}$

In Exercises 17 through 24 express the given angles in radian measure. (Use 3.14 as an *approximation* for π.)

17. 23°0′   18. 54°0′   19. 252.0°   20. 104.0°
21. 333°30′   22. 168°40′   23. 178.5°   24. 86.1°

In Exercises 25 through 32 the given numbers express angle measure. Express the measure of each angle in terms of degrees to the nearest 0.1°.

25. 0.750   26. 0.240   27. 3.00   28. 1.70
29. 2.45   30. 34.4   31. 16.4   32. 100

In Exercises 33 through 40 evaluate the given trigonometric functions by first changing the radian measure to degree measure to the nearest 0.1°. When using Table 4, choose the nearest value shown.

33. $\sin \dfrac{\pi}{4}$ 34. $\cos \dfrac{\pi}{6}$ 35. $\tan \dfrac{5\pi}{12}$ 36. $\sin \dfrac{7\pi}{18}$

37. $\cot \dfrac{5\pi}{6}$ 38. $\tan \dfrac{4\pi}{3}$ 39. $\cos 4.59$ 40. $\cot 3.27$

In Exercises 41 through 48 evaluate the given trigonometric functions directly, without first changing the radian measure to degree measure. When using Table 3, choose the nearest value shown.

41. $\tan 0.7359$ 42. $\sec 0.9308$ 43. $\cot 4.24$ 44. $\tan 3.47$
45. $\cos 2.07$ 46. $\cot 2.34$ 47. $\sin 4.86$ 48. $\csc 6.19$

In Exercises 49 through 56 find θ for $0 < \theta < 2\pi$. In Table 3 use the nearest value shown.

49. $\sin \theta = 0.3090$ 50. $\cos \theta = -0.9135$ 51. $\tan \theta = -0.2126$
52. $\sin \theta = -0.0436$ 53. $\cos \theta = 0.6742$ 54. $\tan \theta = 1.860$
55. $\sec \theta = -1.307$ 56. $\csc \theta = 3.940$

In Exercises 57 through 60 evaluate the given expressions.

57. In optics, when determining the positions of maximum light intensity under specified conditions, the equation $\tan \alpha = \alpha$ is found. Show that a solution to this equation is $\alpha = 1.43\pi$.

58. The instantaneous voltage in a 120-V, 60-Hz power line is given approximately by the equation $V = 170 \sin 377t$, where t is the time (in seconds) after the generator started. Calculate the instantaneous voltage (a) after 0.001 s and (b) after 0.010 s.

59. The velocity v of an object undergoing simple harmonic motion at the end of a spring is given by

$$v = A\sqrt{\frac{k}{m}} \cos\sqrt{\frac{k}{m}}t$$

Here m is the mass of the object (in grams), k is a constant depending on the spring, A is the maximum distance the object moves, and t is the time (in seconds). Find the velocity (in centimeters per second) after 0.100 s of a 36.0-g object at the end of a spring for which $k = 400$ g/s², if $A = 5.00$ cm.

60. At a point x ft from the base of a building, the angle of elevation of the top of the building is θ. An excellent approximation to the error e in measuring the height of the building due to a small error $(\theta_1 - \theta)$ in measuring θ is given by $e = x(\theta_1 - \theta) \sec^2\theta$. Here it is necessary for $(\theta_1 - \theta)$ to be measured in radians. Calculate the error e (in feet), if $x = 180$ ft, $\theta_1 = 30.5°$, and $\theta = 30.0°$.

7—4 Applications of the Use of Radian Measure

Radian measure has numerous applications in mathematics and technology, some of which were indicated in the last four exercises of the previous section. In this section we shall illustrate the usefulness of radian measure in certain specific geometric and physical applications.

From geometry we know that the length of an arc on a circle is proportional to the central angle, and that the length of the arc of a complete circle is the circumference. Letting s stand for the length of arc, we may state that $s = 2\pi r$ for a complete circle. Since 2π is the central angle (in radians) of the complete circle, we have

$$s = \theta r \tag{7-10}$$

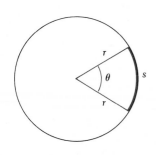

Figure 7—8

for any circular arc with central angle θ. If we know the central angle in radians and the radius of a circle, we can find the length of a circular arc directly by using Eq. (7–10). (See Fig. 7–8.)

Example A

If $\theta = \pi/6$ and $r = 3.00$ in.,

$$s = \left(\frac{\pi}{6}\right)(3.00) = \frac{\pi}{2.00} = 1.57 \text{ in.}$$

If the arc length is 7.20 cm for a central angle of 150° on a certain circle, we may find the radius of the circle by

$$7.20 = (150)\left(\frac{\pi}{180}\right)r = \frac{5.00\pi}{6.00}r \quad \text{or} \quad r = \frac{(6.00)(7.20)}{(5.00)(3.14)} = 2.75 \text{ cm}$$

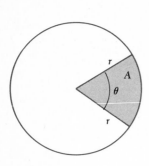

Figure 7–9

Another geometric application of radians is in finding the area of a sector of a circle. (See Fig. 7–9.) We recall from geometry that areas of sectors of circles are proportional to their central angles. The area of a circle is given by $A = \pi r^2$. This can be written as $A = \frac{1}{2}(2\pi)r^2$. We now note that the angle for a complete circle is 2π, and therefore the area of any sector of a circle in terms of the radius and the central angle is

$$A = \frac{1}{2}\theta r^2 \qquad\qquad (7\text{--}11)$$

Example B

The area of a sector of a circle with central angle 18.0° and a radius of 5.00 in. is

$$A = \frac{1}{2}(18.0)\left(\frac{\pi}{180}\right)(5.00)^2 = \frac{1}{2}\left(\frac{\pi}{10.0}\right)(25.0) = 3.93 \text{ in.}^2$$

Given that the area of a sector is 75.5 ft² and the radius is 12.2 ft, we can find the central angle by

$$75.5 = \frac{1}{2}\theta(12.2)^2, \qquad \theta = \frac{2(75.5)}{(12.2)^2} = \frac{151}{149} = 1.01$$

This means that the central angle is 1.01 rad, or 57.9°.

The next illustration deals with velocity. We know that average velocity is defined by the equation $v = s/t$, where v is the average velocity, s is the distance traveled, and t is the elapsed time. If an object is moving around a circular path with constant speed, the actual distance traveled is the length of arc traversed. Therefore, if we divide both sides of Eq. (7–10) by t, we obtain

$$\frac{s}{t} = \frac{\theta}{t}r$$

or

$$v = \omega r \qquad\qquad (7\text{--}12)$$

Equation (7—12) expresses the relationship between the **linear velocity** v and the **angular velocity** ω of an object moving around a circle of radius r. The most convenient units for ω are radians per unit of time. In this way the formula can be used directly. However, in practice, ω is often given in revolutions per minute, or in some similar unit. In cases like these, it is necessary to convert the units of ω to radians per unit of time before substituting in Eq. (7—12).

Example C
An object is moving about a circle of radius 6.00 m with an angular velocity of 4.00 rad/s. The linear velocity is

$$v = (6.00)(4.00) = 24.0 \text{ m/s}$$

(Remember that radians are numbers and are not included in the final set of units.) This means that the object is moving along the circumference of the circle at 24.0 m/s.

Example D
A flywheel rotates with an angular velocity of 20.0 r/min. If its radius is 18.0 in., find the linear velocity of a point on the rim.
   Since there are 2π radians in each revolution,

$$20.0 \text{ r/min} = 40.0\pi \text{ rad/min}$$

Therefore,

$$v = (40.0)(3.14)(18.0) = 2260 \text{ in./min}$$

This means that the linear velocity is 2260 in./min, which is equivalent to 188 ft/min or 3.13 ft/s.

Example E
A pulley belt 10.0 ft long takes 2.00 s to make one complete revolution. The radius of the pulley is 6.00 in. What is the angular velocity (in revolutions per minute) of a point on the rim of the pulley?
   Since the linear velocity of a point on the rim of the pulley is the same as the velocity of the belt, $v = 10.0/2.00 = 5.00$ ft/s. The radius of the pulley is $r = 6.00$ in. $= 0.500$ ft, and we can find ω by substituting into Eq. (7—12). This gives us

$$5.00 = \omega(0.500)$$

or

$$\omega = 10.0 \text{ rad/s}$$
$$= 600 \text{ rad/min}$$
$$= 95.5 \text{ r/min}$$

As it is shown in Appendix B, the change of units can be handled algebraically as

$$10.0\frac{\text{rad}}{\text{s}} \times 60\frac{\text{s}}{\text{min}} = 600\frac{\text{rad}}{\text{min}}, \quad \frac{600 \text{ rad/min}}{2\pi \text{ rad/r}} = 600\frac{\text{rad}}{\text{min}} \times \frac{1}{2\pi}\frac{\text{r}}{\text{rad}}$$

Exercises 7—4

In Exercises 1 through 28 solve the given problems.

1. In a circle of radius 10.0 in., find the length of arc intercepted on the circumference by a central angle of 60°0′.

2. In a circle of diameter 4.50 ft, find the length of arc intercepted on the circumference by a central angle of 42°0′.

3. Find the area of the circular sector indicated in Exercise 1.

4. Find the area of a sector of a circle, given that the central angle is 120.0° and the diameter is 56.0 cm.

5. Find the radian measure of an angle at the center of a circle of radius 5.00 in. which intercepts an arc length of 60.0 in.

6. Find the central angle of a circle which intercepts an arc length of 780 mm when the radius of the circle is 520 mm.

7. Two concentric (same center) circles have radii of 5.00 and 6.00 in. Find the portion of the area of the sector of the larger circle which is outside the smaller circle when the central angle is 30.0°.

8. In a circle of radius 6.00 m, the length of arc of a sector is 10.0 m. What is the area of the sector?

9. A pendulum 3.00 ft long oscillates through an angle of 5.0°. Find the distance through which the end of the pendulum swings in going from one extreme position to the other.

10. The radius of the earth is about 3960 mi. What is the length, in miles, of an arc of the earth's equator for a central angle of 1.0°?

11. In turning, an airplane traveling at 540 km/h moves through a circular arc for 2.00 min. What is the radius of the circle, given that the central angle is 8°0′?

12. An ammeter needle is deflected 52°0′ by a current of 0.200 A. The needle is 3.00 in. long and a circular scale is used. How long is the scale for a maximum current of 1.00 A?

13. A flywheel rotates at 300 r/min. If the radius is 6.00 cm, through what total distance does a point on the rim travel in 30.0 s?

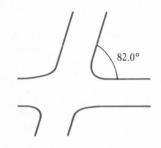
82.0°

14. For the flywheel in Exercise 13, how far does a point halfway out along a radius, move in one second?

15. Two streets meet at an angle of 82.0°. What is the length of the piece of curved curbing at the intersection if it is constructed along the arc of a circle 15.0 ft in radius? See Fig. 7–10.

Figure 7—10

16. In traveling one-third of the way along a traffic circle a car travels 0.125 km. What is the radius of the traffic circle?

17. An automobile is traveling at 60.0 mi/h (88.0 ft/s). The tires are 28.0 in. in diameter. What is the angular velocity of the tires in rad/s?

18. Find the velocity, due to the rotation of the earth, of a point on the surface of the earth at the equator (see Exercise 10).

19. An astronaut in a spacecraft circles the moon once each 1.95 h. If his altitude is constant at 70.0 mi, what is his velocity? The radius of the moon is 1080 mi.

20. What is the linear velocity of the point in Exercise 13?

21. The armature of a dynamo is 1.38 ft in diameter and is rotating at 1200 r/min. What is the linear velocity of a point on the rim of the armature?

22. A pulley belt 38.5 cm long takes 2.50 s to make one complete revolution. The diameter of the pulley is 11.0 cm. What is the angular velocity, in r/min, of the pulley?

23. The moon is about 240,000 mi from the earth. It takes the moon about 28 days to make one revolution. What is its angular velocity about the earth, in rad/s?

24. A phonograph record 6.90 in. in diameter rotates 45.0 times per minute. What is the linear velocity of a point on the rim in ft/s?

25. A 1500-kW wind turbine (windmill) rotates at 40.0 r/min. What is the linear velocity of a point on the end of a blade, if the blade is 100 ft long (from the center of rotation)?

26. The propeller of an airplane is 2.44 m in diameter and rotates at 2200 r/min. What is the linear velocity of a point on the tip of the propeller?

27. A circular sector whose central angle is 210° is cut from a circular piece of sheet metal of diameter 12.0 cm. A cone is then formed by bringing the two radii of the sector together. What is the lateral surface area of the cone?

28. A conical tent is made from a circular piece of canvas 15.0 ft in diameter, with a sector of central angle 120° removed. What is the surface area of the tent?

In Exercises 29 through 32 another use of radians is illustrated.

29. It can be shown through advanced mathematics that an excellent approximate method of evaluating $\sin \theta$ or $\tan \theta$ is given by

$$\sin \theta = \tan \theta = \theta \qquad (7\text{–}13)$$

for small values of θ (the equivalent of a few degrees or less), if θ is expressed in radians. (Note the values for θ, $\sin \theta$, and $\tan \theta$ in Table 3 for small values of θ.) Equation (7–13) is particularly useful for very small values of θ—even some scientific calculators cannot adequately handle angles of 1″ or 0.001° or less. Using Eq. (7–13), evaluate $\sin 1''$.

30. Using Eq. (7–13), evaluate $\tan 0.001°$.

31. An astronomer observes that a star 12.5 light years away moves through an angle of 0.2″ in one year. Assuming it moved in a straight line perpendicular to the initial line of observation, how many miles did the star move? (One light year $= 5.88 \times 10^{12}$ mi.)

32. In calculating a back line of a lot a surveyor discovers an error of 0.05° in an angle measurement. If the lot is 136.0 m deep, by how much is the back line calculation in error? See Fig. 7–11.

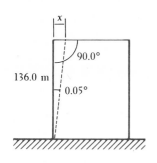

Figure 7–11

7–5 Exercises for Chapter 7

In Exercises 1 through 4 find the trigonometric functions of θ given that the terminal side of θ passes through the given point.

1. $(6, 8)$ 2. $(-12, 5)$ 3. $(7, -2)$ 4. $(-2, -3)$

In Exercises 5 through 8 express the given trigonometric functions in terms of the same function of a positive acute angle.

5. $\cos 132°$, $\quad \tan 194°$ **6.** $\sin 243°$, $\quad \cot 318°$

7. $\sin 289°$, $\quad \sec(-15°)$ **8.** $\cos 103°$, $\quad \csc(-100°)$

In Exercises 9 through 12 express the given angle measurements in terms of π.

9. $40°$, $\quad 153°$ **10.** $22.5°$, $\quad 324°$ **11.** $48°$, $\quad 202.5°$ **12.** $27°$, $\quad 162°$

In Exercises 13 through 20 the given numbers represent angle measure. Express the measure of each angle in terms of degrees.

13. $\dfrac{7\pi}{5}$, $\dfrac{13\pi}{18}$ **14.** $\dfrac{3\pi}{8}$, $\dfrac{7\pi}{20}$ **15.** $\dfrac{\pi}{15}$, $\dfrac{11\pi}{6}$ **16.** $\dfrac{17\pi}{10}$, $\dfrac{5\pi}{4}$

17. 0.560 **18.** 1.35 **19.** 3.60 **20.** 14.5

In Exercises 21 through 28 express the given angles in radians. (Do not answer in terms of π.)

21. $100°$ **22.** $305°$ **23.** $20°30'$ **24.** $148°30'$

25. $262°25'$ **26.** $18°47'$ **27.** $136.2°$ **28.** $385.4°$

In Exercises 29 through 48 determine the values of the given trigonometric functions.

29. $\cos 245°0'$ **30.** $\sin 141°0'$ **31.** $\cot 295.0°$ **32.** $\tan 184.0°$

33. $\csc 247°30'$ **34.** $\sec 96°20'$ **35.** $\sin 205.2°$ **36.** $\cos 326.7°$

37. $\tan 301.4°$ **38.** $\cot 103.9°$ **39.** $\tan 256°42'$ **40.** $\cos 162°32'$

41. $\sin \dfrac{9\pi}{5}$ **42.** $\sec \dfrac{5\pi}{8}$ **43.** $\cos \dfrac{7\pi}{6}$ **44.** $\tan \dfrac{23\pi}{12}$

45. $\sin 0.590$ **46.** $\tan 0.800$ **47.** $\csc 2.15$ **48.** $\cot 5.19$

In Exercises 49 through 52 find θ to the nearest $0.1°$ for $0° < \theta < 360°$.

49. $\tan \theta = 0.1817$ **50.** $\sin \theta = -0.9323$

51. $\cos \theta = -0.4730$ **52.** $\cot \theta = 1.196$

In Exercises 53 through 56 find θ to the nearest $10'$ for $0° < \theta < 360°$.

53. $\sin \theta = 0.2924$ **54.** $\cot \theta = -2.560$

55. $\cos \theta = 0.3297$ **56.** $\tan \theta = -0.7730$

In Exercises 57 through 60 find θ to the nearest minute for $0° < \theta < 360°$.

57. $\cos \theta = -0.7222$ **58.** $\tan \theta = -1.683$

59. $\cot \theta = 0.4291$ **60.** $\sin \theta = 0.2626$

In Exercises 61 through 64 find θ for $0 < \theta < 2\pi$.

61. $\cos \theta = 0.8387$ **62.** $\sin \theta = 0.1045$

63. $\sin \theta = -0.8650$ **64.** $\tan \theta = 2.840$

In Exercises 65 through 80 solve the given problems.

65. The voltage in a certain alternating-current circuit is given by the equation $v = V \cos 25t$, where V is the maximum possible voltage and t is the time in seconds. Find v for $t = 0.1$ s and $V = 150$ V.

66. The displacement (distance from equilibrium position) of a particle moving with simple harmonic motion is given by $d = A \sin 5t$, where A is the maximum displacement and t is the time. Find d, given that $A = 16.0$ cm and $t = 0.460$ s.

67. A pendulum 5.00 ft long swings through an angle of 4.50°. Through what distance does the bob swing in going from one extreme position to the other?

68. Two pulleys have radii of 10.0 in. and 6.00 in., and their centers are 40.0 in. apart. If the pulley belt is uncrossed, what must be the length of the belt?

69. A piece of circular filter paper 15.0 cm in diameter is folded such that its effective filtering area is the same as that of a sector with central angle of 220°. What is the filtering area?

70. A funnel is made from a circular piece of sheet metal 10.0 in. in radius, from which two pieces were removed. The first piece removed was a circle of radius 1.00 in. at the center, and the second piece removed was a sector of central angle 200°. What is the surface area of the funnel?

71. If the apparatus shown in Fig. 7—12 is rotating at 2.00 r/s, what is the linear velocity of the ball?

72. A lathe is to cut material at the rate of 350 ft/min. Calculate the radius of a cylindrical piece that is turned at the rate of 120 r/min.

73. A thermometer needle passes through 55.0° for a temperature change of 40°C. If the needle is 5.00 cm long and the scale is circular, how long must the scale be for a maximum temperature change of 150°C?

74. Under certain conditions an electron will travel in a circular path when in a magnetic field. If an electron is moving with a linear velocity of 20,000 m/s in a circular path of radius 0.500 m, how far does it travel in 0.100 s?

75. A cam is constructed so that part of it is a circular arc with a central angle of 72.0° and a radius of 5.30 mm. What is the length of arc along this part of the cam?

76. A horizontal water pipe has a radius of 6.00 ft. If the depth of water in the pipe is 3.00 ft, what percentage of the volume of the pipe is filled?

77. An electric fan blade 15.0 cm in radius rotates at 900 r/min. What is the linear velocity of a point at the end of the blade?

78. A circular saw blade 8.20 in. in diameter rotates at 1500 r/min. What is the linear velocity of a point at the end of one of the teeth?

79. A laser beam is transmitted with a "width" of 0.0008°, and makes a circular spot of radius 2.50 km on a distant object. How far is the object from the source of the laser beam? (See Exercise 29 of Section 7—4.)

80. The plant Venus subtends an angle of 15″ to an observer on Earth. If the distance between Venus and Earth is 1.04×10^8 mi, what is the diameter of Venus? (See Exercise 29 of Section 7—4.)

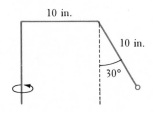

10 in.

10 in.

30°

Figure 7—12

8

Vectors and Oblique Triangles

8–1 Introduction to Vectors

A great many quantities with which we deal may be described by specifying their magnitudes. Generally, one can describe lengths of objects, areas, time intervals, monetary amounts, temperatures, and numerous other quantities by specifying a number: the magnitude of the quantity. Such quantities are know as **scalar** quantities.

Many other quantities are fully described only when both their magnitude and direction are specified. Such quantities are known as **vectors.** Examples of vectors are velocity, force, and momentum. Vectors are of utmost importance in many fields of science and technology. The following example illustrates a vector quantity and the distinction between scalars and vectors.

Example A

A jet flies over a certain point traveling at 600 mi/h. From this statement alone we know only the *speed* of the jet. Speed is a scalar quantity, and it designates only the *magnitude* of the rate.

If we were to add the phrase "in a direction 10° south of west" to the sentence above about the jet, we would be specifying the direction of travel as well as the speed. We then know the *velocity* of the jet; that is, we know the *direction* of travel as well as the *magnitude* of the rate at which the jet is traveling. Velocity is a vector quantity.

Let us analyze an example of the action of two vectors: Consider a boat moving in a stream. For purposes of this example, we shall assume that the boat is driven by a motor which can move it at 4 mi/h in still water. We shall assume the current is moving downstream at 3 mi/h. We immediately see that the motion of the boat depends on the direction in which it is headed. If the boat heads downstream, it can travel at 7 mi/h, for the current is going 3 mi/h and the boat moves at 4 mi/h with respect to the water. If the boat heads upstream, however, it progresses at the rate of only 1 mi/h, since the action of the boat and that of the stream are counter to each other. If the boat heads across the stream, the point which it reaches on the other side will not be directly opposite the point from which it started. We can see that this is so because we know that as the boat heads across the stream, the stream is moving the boat downstream *at the same time*.

This last case should be investigated further. Let us assume that the stream is $\frac{1}{2}$ mi wide where the boat is crossing. It will then take the boat $\frac{1}{8}$ h to cross. In $\frac{1}{8}$ h the stream will carry the boat $\frac{3}{8}$ mi downstream. Therefore, when the boat reaches the other side it will be $\frac{3}{8}$ mi downstream. From the Pythagorean theorem, we find that the boat traveled $\frac{5}{8}$ mi from its starting point to its finishing point.

$$d^2 = \left(\frac{4}{8}\right)^2 + \left(\frac{3}{8}\right)^2 = \frac{16+9}{64} = \frac{25}{64}; \qquad d = \frac{5}{8}\,\text{mi}$$

Since this $\frac{5}{8}$ mi was traveled in $\frac{1}{8}$ h, the magnitude of the velocity of the boat was actually

$$v = \frac{d}{t} = \frac{5/8}{1/8} = \frac{5}{8}\cdot\frac{8}{1} = 5 \ \text{mi/h}$$

Also, we see that the direction of this velocity can be represented along a line making an angle θ with the line directed across the stream (see Fig. 8–1).

We have just seen two velocity vectors being *added*. Note that these vectors are not added the way numbers are added. We have to take into account their direction as well as their magnitude. Reasoning along these lines, let us now define the sum of two vectors.

We will represent a vector quantity by a letter printed in boldface type. The same letter in italic (lightface) type represents the magnitude only. Thus, **A** is a vector of magnitude A. In handwriting, one usually places an arrow over the letter to represent a vector, such as $\vec{A}$.

Let **A** and **B** represent vectors directed from O to P and P to Q, respectively (see Fig. 8–2). *The vector sum* **A** + **B** *is the vector* **R**, *from the* **initial point** O *to the* **terminal point** Q. *Here, vector* **R** *is called the* **resultant**. *In general, a resultant is a single vector which can replace any number of other vectors and still produce the same physical effect.*

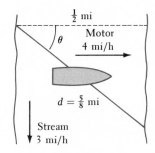

Figure 8—1

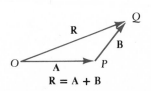

Figure 8—2

There are two common methods of adding vectors by means of a diagram. The first is illustrated in Fig. 8–3. To add **B** to **A**, we shift **B** parallel to itself until its tail touches the head of **A**. In doing so we must be careful to keep the magnitude and direction of **B** unchanged. The vector sum **A** + **B** is the vector **R**, which is drawn from the tail of **A** to the head of **B**.

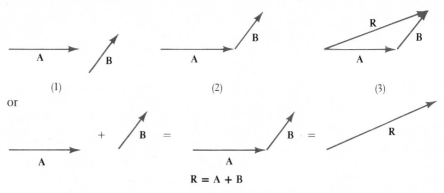

$$R = A + B$$

Figure 8–3

Three or more vectors may also be added in the same general manner. We place the initial point of the second vector at the terminal point of the first vector, the initial point of the third vector at the terminal point of the second vector, and so forth. The resultant is the vector from the initial point of the first vector to the terminal point of the last vector. The order in which they are placed together does not matter, as shown in the following example.

Example B
The addition of vectors **A**, **B**, and **C** is shown in Fig. 8–4.

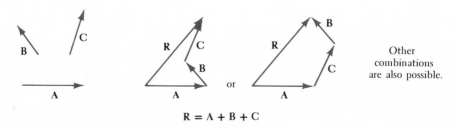

$$R = A + B + C$$

Figure 8–4

Another method which is convenient when two vectors are being added is to let the two vectors being added be the sides of a parallelogram. The resultant is then the diagonal of the parallelogram. The initial point of the resultant is the common initial point of the vectors being added. In using this method the vectors are first placed tail to tail. See the following example.

Example C
Using the parallelogram method, add vectors **A** and **B** of Fig. 8–3. See Fig. 8–5.

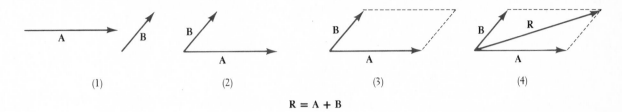

(1)   (2)   (3)   (4)

$$R = A + B$$

Figure 8–5

If vector **A** has the same direction as vector **B**, and **A** also has a magnitude n times that of **B**, we may state that **A** $= n$**B**. Thus, 2**A** represents a vector twice as long as **A**, but in the same direction.

Example D
For the vectors **A** and **B** in Fig. 8–3, find 3**A** + 2**B**. See Fig. 8–6.

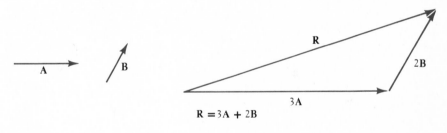

$$R = 3A + 2B$$

Figure 8–6

Vector **B** may be subtracted from vector **A** by reversing the direction of **B** and proceeding as in vector addition. Thus, **A** − **B** = **A** + (−**B**), where the minus sign indicates that vector −**B** has the opposite direction of vector **B**. Vector subtraction is shown in the following example.

Example E
For vectors **A** and **B**, find 2**A** − **B**. See Fig. 8–7.

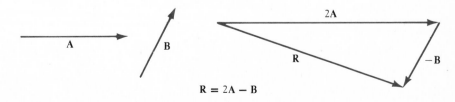

$$R = 2A - B$$

Figure 8–7

In addition to being able to add and subtract vectors, we often need to consider a given vector as the sum of two other vectors. *Two vectors which when added together give the original vector are called the* **components** *of the original vector.* In the example of the boat, the velocities of 4 mi/h cross-stream and 3 mi/h downstream are components of the 5 mi/h vector directed at the angle θ.

In practice, there are certain components of a vector which are of particular importance. If a vector is so placed that its initial point is at the origin of a rectangular coordinate system, and its direction is indicated by an angle in standard position, we may find its *x*- and *y*-components. These components are vectors directed along the coordinate axes which, when added together, result in the given vector. *Finding these component vectors is called* **resolving** *the vector into its components.*

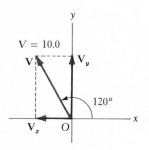

Figure 8—8

Example F

Resolve a vector 10.0 units long and directed at an angle of 120°0′ into its *x*- and *y*-components (see Fig. 8–8).

Placing the initial point of the vector at the origin, and putting the angle in standard position, we see that the vector directed along the *x*-axis, $\mathbf{V}_x$, is related to the vector $\mathbf{V}$, of magnitude V by

$$V_x = V \cos 120°0′ = -V \cos 60°0′$$

(The minus sign indicates that the *x*-component is directed in the negative direction, that is, to the left.) Since the vector directed along the *y*-axis, $\mathbf{V}_y$, could also be placed along the dashed line, it is related to the vector $\mathbf{V}$ by

$$V_y = V \sin 120°0′ = V \sin 60°0′$$

Thus, the vectors $\mathbf{V}_x$ and $\mathbf{V}_y$ have the magnitudes

$$V_x = -10.0(0.5000) = -5.00, \qquad V_y = 10.0(0.8660) = 8.66$$

Therefore, we have resolved the given vector into two components, one directed along the negative *x*-axis of magnitude 5.00, and the other directed along the positive *y*-axis of magnitude 8.66.

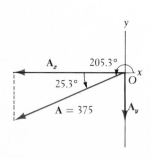

Figure 8—9

Example G

Resolve vector $\mathbf{A}$ of magnitude 375 and direction $\theta \doteq 205.3°$ into its *x*- and *y*-components. See Fig. 8–9.

By placing $\mathbf{A}$ such that θ is in standard position, we see that

$$A_x = A \cos 205.3° = 375(-0.9041) = -339$$

and

$$A_y = A \sin 205.3° = 375(-0.4274) = -160$$

Here, values for the trigonometric functions were found directly by use of a calculator.

We can also calculate these as follows:

$$A_x = A \cos 205.3° = 375(-\cos 25.3°) = 375(-0.9041) = -339$$

and

$$A_y = A \sin 205.3° = 375(-\sin 25.3°) = 375(-0.4274) = -160$$

Thus, **A** has two components, one directed along the negative x-axis of magnitude 339, and the other directed along the negative y-axis of magnitude 160.

Exercises 8—1

In Exercises 1 through 4 add the given vectors by drawing the appropriate resultant. Use the parallelogram method in Exercises 3 and 4.

1.　　　　　2.　　　　　3.　　　　　4.

In Exercises 5 through 24 find the indicated vector sums and differences with the given vectors by means of diagrams.

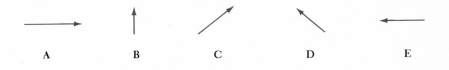

A　　　　B　　　　C　　　　D　　　　E

5. A + B	6. B + C
7. C + D	8. D + E
9. A + C + E	10. B + D + A
11. A + D + E	12. B + E + C
13. B + 3E	14. A + 2C
15. 3C + E	16. 2C + D
17. A − B	18. C − D
19. E − B	20. D − A
21. 3B − 2D	22. 2A − 3E
23. B + 2C − E	24. A + 4E − 3B

In Exercises 25 through 32 find the *x*- and *y*-components of the given vectors by use of the trigonometric functions.

25. Magnitude 8.60, $\theta = 68°0'$

26. Magnitude 9750, $\theta = 243°0'$

27. Magnitude 76.8, $\theta = 145.0°$

28. Magnitude 0.0998, $\theta = 296.0°$

29. Magnitude 9.04, $\theta = 283°30'$

30. Magnitude 16,000, $\theta = 156.5°$

31. Magnitude 2.65, $\theta = 197.3°$

32. Magnitude 67.8, $\theta = 22.5°$

8–2 Vector Addition by Components

Adding vectors by diagrams gives only approximate results. By use of the trigonometric functions and the Pythagorean theorem, it is possible to obtain accurate numerical results for the sum of vectors. In the following example we shall illustrate how this is done in the case when the two given vectors are at right angles.

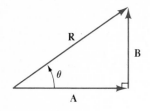

Figure 8–10

Example A

Add vectors **A** and **B**, with $A = 14.5$ and $B = 9.10$. The vectors are at right angles as shown in Fig. 8–10.

We can find the magnitude R of the resultant vector **R** by use of the Pythagorean theorem. This leads to

$$R = \sqrt{A^2 + B^2} = \sqrt{(14.5)^2 + (9.10)^2}$$
$$= \sqrt{210 + 82.8} = \sqrt{293} = 17.1$$

We shall now determine the direction of **R** by specifying its direction as the angle θ, that is, the angle **R** makes with **A**. Therefore,

$$\tan \theta = \frac{B}{A} = \frac{9.10}{14.5} = 0.6276$$

To the nearest 10′, $\theta = 32°10'$, or to the nearest 0.1°, $\theta = 32.1°$. Thus, **R** is a vector with magnitude $R = 17.1$ and in a direction 32°10′ from vector **A**.

If vectors are to be added and they are not at right angles, we first place each with tail at the origin. Next, we resolve each vector into its *x*- and *y*-components. We then add all of the *x*-components and add the *y*-components to determine the *x*- and *y*-components of the resultant. Then by use of the Pythagorean theorem and the tangent, as in Example A, we find the magnitude and direction of the resultant. *Remember, a vector is not completely specified unless both its magnitude and its direction are given.*

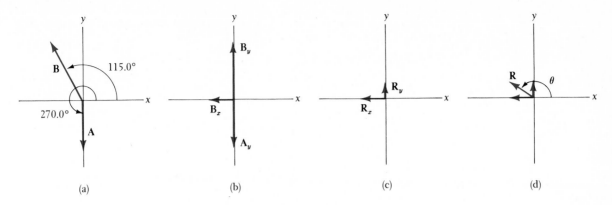

Figure 8—11

Example B
Find the resultant of two vectors **A** and **B** such that $A = 1200$, $\theta_A = 270.0°$, $B = 1750$, and $\theta_B = 115.0°$.

We first place the vectors on a coordinate system with the tail of each at the origin as shown in Fig. 8—11(a). We then resolve each into its x- and y-components, as shown in Fig. 8—11(b) and as calculated below. (Note that **A** is vertical and has no horizontal component.) Next, the components are combined as in Fig. 8—11(c) and as calculated. Finally, the magnitude and direction of the resultant, as shown in Fig. 8—11(d), are calculated.

$$A_x = A \cos 270.0° = 1200(0) = 0$$
$$A_y = A \sin 270.0° = 1200(-1.000) = -1200$$
$$B_x = B \cos 115.0° = 1750 \cos 115.0° = -1750 \cos 65.0°$$
$$= -1750(0.4226) = -740$$
$$B_y = B \sin 115.0° = 1750 \sin 115.0° = 1750 \sin 65.0°$$
$$= 1750(0.9063) = 1590$$
$$R_x = A_x + B_x = 0 - 740 = -740$$
$$R_y = A_y + B_y = -1200 + 1590 = 390$$
$$R = \sqrt{R_x^2 + R_y^2} = \sqrt{(-740)^2 + (390)^2} = \sqrt{547600 + 152100}$$
$$= \sqrt{699700} = 836$$

$$\tan \theta = \frac{R_y}{R_x} = \frac{390}{-740} = -0.5270, \quad \theta = 152.2°$$

Thus, the resultant has a magnitude of 836 and is directed at an angle of 152.2°. In determining θ, the reference angle was 27.8°, and we know that θ is a second quadrant angle since R_x is negative and R_y is positive.

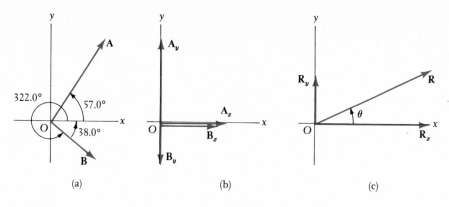

Figure 8-12

Example C

Find the resultant **R** of the two vectors given in Fig. 8–12(a), **A** of magnitude 8.00 and direction 57.0° and **B** of magnitude 5.00 and direction 322.0°.

$$A_x = (8.00)(\cos 57.0°) = (8.00)(0.5446) = 4.36$$
$$A_y = (8.00)(\sin 57.0°) = (8.00)(0.8387) = 6.71$$
$$B_x = (5.00)(\cos 38.0°) = (5.00)(0.7880) = 3.94$$
$$B_y = -(5.00)(\sin 38.0°) = -(5.00)(0.6157) = -3.08$$
$$R_x = A_x + B_x = 4.36 + 3.94 = 8.30$$
$$R_y = A_y + B_y = 6.71 - 3.08 = 3.63$$
$$R = \sqrt{(8.30)^2 + (3.63)^2} = \sqrt{68.9 + 13.2} = \sqrt{82.1} = 9.06$$

$$\tan \theta = \frac{R_y}{R_x} = \frac{3.63}{8.30} = 0.4373, \qquad \theta = 23.6°$$

The resultant vector is 9.06 units long and is directed at an angle of 23.6°, as shown in Fig. 8–12(c).

Some general formulas can be derived from the previous examples. For a given vector **A**, directed at an angle θ, and of magnitude A, with components A_x and A_y, we have the following relations:

$$A_x = A \cos \theta, \qquad A_y = A \sin \theta \tag{8-1}$$

$$A = \sqrt{A_x^2 + A_y^2} \tag{8-2}$$

$$\tan \theta = \frac{A_y}{A_x} \tag{8-3}$$

Example D

Find the resultant of the three given vectors with $A = 422$, $B = 405$, and $C = 210$, as shown in Fig. 8–13.

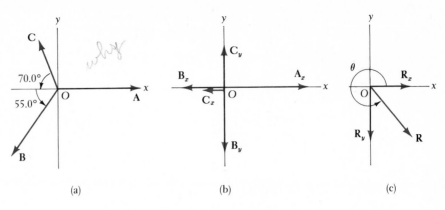

Figure 8-13

We can find the x- and y-components of the vectors by using Eq. (8-1). The following table is helpful for determining the necessary values.

Vector	x-component	y-component
A	$= +422$	$= 0$
B	$-405 \cos 55.0° = -232$	$-405 \sin 55.0° = -332$
C	$-210 \cos 70.0° = \underline{-72}$	$+210 \sin 70.0° = \underline{+197}$
R	$+118$	-135

From this table it is possible to compute R and θ:

$$R = \sqrt{(118)^2 + (-135)^2} = 179, \qquad \tan \theta = \frac{-135}{118} = -1.144$$

$$\theta = 311.2°$$

Exercises 8-2

In Exercises 1 through 4, vectors **A** and **B** are at right angles. Find the magnitude and direction of the resultant.

1. $A = 14.7$
   $B = 19.2$

2. $A = 592$
   $B = 195$

3. $A = 3.08$
   $B = 7.14$

4. $A = 1730$
   $B = 3290$

In Exercises 5 through 12 with the given sets of components, find R and θ.

5. $R_x = 5.18, R_y = 8.56$

6. $R_x = 89.6, R_y = -52.0$

7. $R_x = -0.982, R_y = 2.56$

8. $R_x = -729, R_y = -209$

9. $R_x = -646, R_y = 2030$

10. $R_x = -31.2, R_y = -41.2$

11. $R_x = 0.694, R_y = -1.24$

12. $R_x = 7.62, R_y = -6.35$

In Exercises 13 through 24 add the given vectors by using the trigonometric functions and the Pythagorean theorem.

13. $A = 18.0, \theta_A = 0°0'$
    $B = 12.0, \theta_B = 27°0'$

14. $A = 150, \theta_A = 90°0'$
    $B = 128, \theta_B = 43°0'$

15. $A = 56.0, \theta_A = 76.0°$
    $B = 24.0, \theta_B = 200.0°$

16. $A = 6.89, \theta_A = 123.0°$
    $B = 29.0, \theta_B = 260.0°$

17. $A = 9.82$, $\theta_A = 34°0'$
    $B = 17.4$, $\theta_B = 752°30'$

18. $A = 1.65$, $\theta_A = 36°30'$
    $B = 0.980$, $\theta_B = 253°0'$

19. $A = 12.6$, $\theta_A = 98.4°$
    $B = 15.1$, $\theta_B = 332.2°$

20. $A = 121$, $\theta_A = 292.0°$
    $B = 112$, $\theta_B = 198.7°$

21. $A = 21.9$, $\theta_A = 236.2°$
    $B = 96.7$, $\theta_B = 11.5°$
    $C = 62.9$, $\theta_C = 143.4°$

22. $A = 6300$, $\theta_A = 189.0°$
    $B = 1760$, $\theta_B = 320.0°$
    $C = 3240$, $\theta_C = 75.0°$

23. The vectors shown in Fig. 8–14

24. The vectors shown in Fig. 8–15

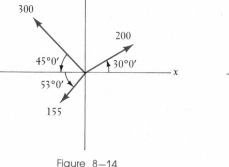

Figure 8–14

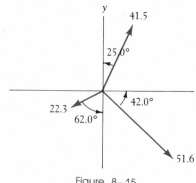

Figure 8–15

8–3 Application of Vectors

As we mentioned at the beginning of the chapter, vectors are important in science and technology. In this section a number of these applications are illustrated in the examples and exercises.

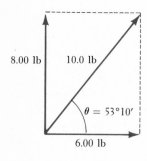

Figure 8–16

Example A

An object on a horizontal table is acted on by two horizontal forces. The two forces have magnitudes of 6.00 and 8.00 lb, and the angle between their lines of action is 90°0'. What is the resultant of these forces?

By means of an appropriate diagram (Fig. 8–16) we may better visualize the actual situation. We note that a good choice of axes (unless specified, it is often convenient to choose the x- and y-axes to fit the problem) is to have the x-axis in the direction of the 6.00-lb force and the y-axis in the direction of the 8.00-lb force. (This is possible since the angle between them is 90°.) With this choice we note that the two given forces will be the x- and y-components of the resultant. Therefore, we arrive at the following results:

$$F_x = 6.00 \text{ lb}, \qquad F_y = 8.00 \text{ lb}, \qquad F = \sqrt{36.0 + 64.0} = 10.0 \text{ lb}$$

$$\tan \theta = \frac{F_y}{F_x} = \frac{8.00}{6.00} = 1.333, \qquad \theta = 53°10'$$

We would state that the resultant has a magnitude of 10.0 lb and acts at an angle of 53°10' from the 6.00-lb force.

Example B
A ship sails 32.0 mi due east and then turns 40.0°N of E. After sailing an additional 16.0 mi, where is it with reference to the starting point?

The distance an object moves and the direction in which it moves give the **displacement** of an object. Therefore, in this problem we are to determine the resultant displacement of the ship from the two given displacements. The problem is diagrammed in Fig. 8–17, where the first displacement has been labeled vector **A** and the second as vector **B**.

Since east corresponds to the positive *x*-direction, we see that the *x*-component of the resultant is $A + B_x$, and the *y*-component of the resultant is B_y. Therefore, we have the following results:

$$R_x = A + B_x = 32.0 + 16.0 \cos 40.0°$$
$$= 32.0 + 16.0(0.7660) = 32.0 + 12.3$$
$$= 44.3 \text{ mi}$$
$$R_y = 16.0 \sin 40.0° = 16.0(0.6428) = 10.3 \text{ mi}$$
$$R = \sqrt{(44.3)^2 + (10.3)^2} = \sqrt{2069} = 45.5 \text{ mi}$$

$$\tan \theta = \frac{10.3}{44.3} = 0.2325$$

$$\theta = 13.1°$$

Therefore, the ship is 45.5 mi from the starting point, in a direction 13.1° N of E.

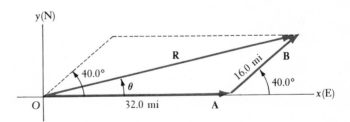

Figure 8–17

Example C
An airplane headed due east is in a wind which is blowing from the southeast. What is the resultant velocity of the plane with respect to the surface of the earth, if the plane's velocity with respect to the air is 600 km/h, and that of the wind is 100 km/h (see Fig. 8–18)?

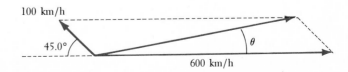

Figure 8–18

Let $\mathbf{v}_{px}$ be the velocity of the plane in the x-direction (east), $\mathbf{v}_{py}$ the velocity of the plane in the y-direction, $\mathbf{v}_{wx}$ the x-component of the velocity of the wind, $\mathbf{v}_{wy}$ the y-component of the velocity of the wind, and $\mathbf{v}_{pa}$ the velocity of the plane with respect to the air. Therefore,

$$v_{px} = v_{pa} - v_{wx} = 600 - 100\ (\cos 45.0°) = 600 - 70.7 = 529 \text{ km/h}$$

$$v_{py} = v_{wy} = 100\ (\sin 45.0°) = 70.7 \text{ km/h}$$

$$v = \sqrt{(529)^2 + (70.7)^2} = \sqrt{280{,}000 + 5000} = 534 \text{ km/h}$$

$$\tan \theta = \frac{v_{py}}{v_{px}} = \frac{70.7}{529} = 0.1336, \qquad \theta = 7.6°$$

We have determined that the plane is traveling 534 km/h and is flying in a direction 7.6° north of east. From this we observe that a plane does not necessarily head in the direction of its desired destination.

Example D

A block is resting on an inclined plane which makes an angle of 30.0° with the horizontal. If the block weighs 100 lb, what is the force of friction between the block and the plane?

The weight of the block is the force exerted on the block due to gravity. Therefore, the weight is directed vertically downward. The frictional force tends to oppose the motion of the block and is directed upward along the plane. The frictional force must be sufficient to counterbalance that component of the weight of the block which is directed down the plane for the block to be at rest. The plane itself "holds up" that component of the weight which is perpendicular to the plane. A convenient set of coordinates (Fig. 8–19) would be one with the x-axis directed up the plane and the y-axis perpendicular to the plane. The magnitude of the frictional force $\mathbf{F}_f$ is given by

$$F_f = 100 \sin 30.0° = 100(0.5000) = 50.0 \text{ lb}$$

(Since the component of the weight down the plane is 100 sin 30.0° and is equal to the frictional force, this relation is true.)

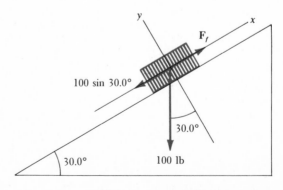

All forces are assumed to act
at the center of the block.

Figure 8–19

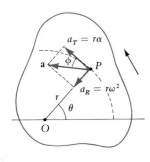

Figure 8–20

Example E

A 60.0-lb object hangs from a hook on a wall. If a horizontal force of 40.0 lb pulls the object away from the wall so that the object is in equilibrium (no resultant force in any direction), what is the tension T in the rope attached to the wall?

For the object to be in equilibrium, the tension in the rope must be equal and opposite to the resultant of the 40.0-lb force and the 60.0-lb force which is the weight of the object (see Fig. 8–20). Thus, the magnitude of the x-component of the tension is 40.0 lb and the magnitude of the y-component is 60.0 lb.

$$T = \sqrt{(40.0)^2 + (60.0)^2} = \sqrt{5200} = 72.1 \text{ lb}$$

$$\tan \theta = \frac{60.0}{40.0} = 1.500, \qquad \theta = 56.3°$$

Figure 8–21

Example F

If an object rotates about a point O, the tangential component of the acceleration $\mathbf{a}_T$ and the centripetal component of the acceleration $\mathbf{a}_R$ of a point P are given by the expressions shown in Fig. 8–21. The radius of the circle through which P is moving is r, the angular velocity at any instant is ω, and α is the rate at which the angular velocity ω is changing. Given that $r = 2.35$ m, $\omega = 5.60$ rad/s and $\alpha = 3.75$ rad/s², calculate the magnitude of the resultant acceleration and the angle it makes with the tangential component.

$$(\mathbf{a}_R \perp \mathbf{a}_T), \qquad a_R = r\omega^2 = (2.35)(5.60)^2 = 73.7 \text{ m/s}^2$$
$$a_T = r\alpha = (2.35)(3.75) = 8.81 \text{ m/s}^2$$
$$a = \sqrt{a_R^2 + a_T^2} = \sqrt{(73.7)^2 + (8.81)^2} = 74.2 \text{ m/s}^2$$

The angle ϕ between $\mathbf{a}$ and $\mathbf{a}_T$ is found from the relation $\cos \phi = a_T/a$. Hence,

$$\cos \phi = \frac{8.81}{73.7} = 0.1195 \quad \text{or} \quad \phi = 83.1°$$

Exercises 8–3

In Exercises 1 through 24 solve the given problems.

1. Two forces, one of 45.0 lb and the other of 68.0 lb, act on the same object and at right angles to each other. Find the resultant of these forces.

2. Two forces, one of 150 lb and the other of 220 lb, pull on an object. The angle between these forces is 90°0′. What is the resultant of these forces?

3. Two forces, one of 350 N and the other of 520 N, pull on an object. The angle between these forces is 25.0°. What is the resultant of these forces?

4. The angle between two forces acting on an object is 63°0′. If the two forces are 86.0 N and 103 N, respectively, what is their resultant?

5. A jet travels 450 mi due west from a city. It then turns and travels 240 mi south. What is its displacement from the city?

6. A ship sails 78.3 km due north after leaving its pier. It then turns and sails 51.2 km east. What is the displacement of the ship from its pier?

7. Town B is 52.0 mi southeast of town A. Town C is 45.0 mi due west of town B. What is the displacement of town C from town A?

8. A surveyor locates a tree 36.5 m northeast of her. She knows that it is 20.0 m north of a utility pole. What is the surveyor's displacement from the utility pole?

9. What are the horizontal and vertical components of the velocity of a stone thrown into the air with a velocity of 120 ft/s at an angle of 48°0′ with respect to the horizontal?

10. A rocket is traveling at an angle of 74.0° with respect to the horizontal at a speed of 3500 km/h. What are the horizontal and vertical components of the velocity?

11. A child weighing 60.0 lb sits in a swing and is pulled sideways by a horizontal force of 20.0 lb. What is the tension in each of the supporting ropes? What is the angle between the horizontal and the ropes?

12. A rope 10.0 ft long is fastened to supports which are 8.00 ft apart and at the same level. A 100-lb weight is then hung from its center. How much is the tension in the supporting rope?

13. A stone is thrown horizontally from a plane traveling at 200 m/s. If the stone is thrown at 50.0 m/s in a direction perpendicular to the direction of the plane, what is the velocity of the stone just after it is released?

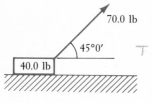

Figure 8–22

14. A plane is headed due north at a velocity of 500 km/h with respect to the air. If the wind is from the southwest at 80.0 km/h, what is the resultant speed of the plane, and in what direction is it traveling?

15. A 70.0-lb force is applied to a 40.0-lb box by a rigid metal rod at an angle of 45°0′ above the horizontal (see Fig. 8–22). Will the box be lifted from the ground?

16. Assume that the plane in Fig. 8–19 is frictionless. If the acceleration due to gravity is 9.80 m/s², what is the component of the acceleration of the object down the plane? (This is the acceleration the object will have, since it is restricted to moving down the plane.)

17. In Fig. 8–19, if the plane is inclined at 20.0° what is the force exerted on the 100-lb object by the plane itself?

18. In Fig. 8–19, if a horizontal 10.0-lb force is exerted to the right on the 100-lb object, what would the force of friction have to be so that the object did not move down the plane?

19. An object is dropped from a plane moving at 120 m/s. If the vertical velocity, as a function of time, is given by $v_y = 9.80t$, what is the velocity of the object after 4.50 s? In what direction is it moving?

20. The magnitude of the horizontal and vertical components of displacement of a certain projectile are given by $d_H = 120t$ and $d_V = 160t - 16t^2$, where t is the time in seconds. Find the displacement (in feet) of the object after 3.00 s.

21. In Fig. 8–21, given that $a = 56.4$ ft/s², $a_R = 37.9$ ft/s², and $r = 6.00$ ft, find α, the rate of change of angular velocity.

22. A boat travels across a river, reaching the opposite bank at a point directly opposite that from which it left. If the boat travels 6.00 km/h in still water, and the current of river flows at 3.00 km/h, what was the velocity of the boat in the water?

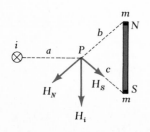

Figure 8–23

23. In Fig. 8–23, a long straight conductor perpendicular to the plane of the paper carries an electric current i. A bar magnet having poles of strength

m lies in the plane of the paper. The vectors $\mathbf{H}_i$, $\mathbf{H}_N$, and $\mathbf{H}_S$ represent the components of the magnetic intensity $\mathbf{H}$ due, respectively, to the current and to the N and S poles of the magnet. The magnitude of the components of $\mathbf{H}$ are given by

$$H_i = \frac{1}{2\pi}\frac{i}{a}, \qquad H_N = \frac{1}{4\pi}\frac{m}{b^2}, \qquad \text{and} \qquad H_S = \frac{1}{4\pi}\frac{m}{c^2}$$

Given that $a = 0.300$ m, $b = 0.400$ m, $c = 0.300$ m, the length of the magnet is 0.500 m, $i = 4.00$ A, and $m = 2.00$ A·m, calculate the resultant magnetic intensity $\mathbf{H}$. The component $\mathbf{H}_i$ is parallel to the magnet.

24. Solve the problem of Exercise 23 if $\mathbf{H}_i$ is directed away from the magnet, making an angle of $10°0'$ with the direction of the magnet.

8—4 Oblique Triangles, the Law of Sines

To this point we have limited our study of triangle solution to right triangles. However, many triangles which require solution do not contain a right angle. Such a triangle is termed an **oblique triangle.** Let us now discuss the solutions of oblique triangles.

In Section 3—4 we stated that we need three parts, at least one of them a side, in order to solve any triangle. With this in mind we may determine that there are four possible combinations of parts from which we may solve a triangle. These combinations are:

Case 1. Two angles and one side
Case 2. Two sides and the angle opposite one of them
Case 3. Two sides and the included angle
Case 4. Three sides

There are several ways in which oblique triangles may be solved, but we shall restrict our attention to the two most useful methods, the **Law of Sines** and the **Law of Cosines.** In this section we shall discuss the Law of Sines, and show that it may be used to solve Case 1 and Case 2.

Let ABC be an oblique triangle with sides a, b, and c opposite angles A, B, and C, respectively. By drawing a perpendicular h from B to side b, or its extension, we see from Fig. 8—24(a) that

$$h = c \sin A \qquad \text{or} \qquad h = a \sin C \qquad\qquad (8\text{–}4)$$

and from Fig. 8—24(b)

$$h = c \sin A \qquad \text{or} \qquad h = a \sin (180° - C) = a \sin C \qquad (8\text{–}5)$$

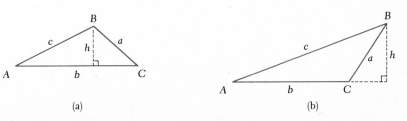

(a)   (b)

Figure 8—24

We note that the results are precisely the same in Eqs. (8–4) and (8–5). Setting the results for h equal to each other, we have

$$c \sin A = a \sin C \qquad \text{or} \qquad \frac{a}{\sin A} = \frac{c}{\sin C} \tag{8–6}$$

By dropping a perpendicular from A to a we also derive the result

$$c \sin B = b \sin C \qquad \text{or} \qquad \frac{b}{\sin B} = \frac{c}{\sin C} \tag{8–7}$$

Combining Eqs. (8–6) and (8–7) we have the **Law of Sines:**

$$\frac{a}{\sin A} = \frac{b}{\sin B} = \frac{c}{\sin C} \tag{8–8}$$

The Law of Sines is a statement of proportionality between the sides of a triangle and the sines of the angles opposite them.

Now we may see how the Law of Sines is applied to the solution of a triangle in which two angles and one side are known (Case 1). If two angles are known, the third may be found from the fact that the sum of the angles in a triangle is 180°. At this point we must be able to determine the ratio between the given side and the sine of the angle opposite it. Then, by use of the Law of Sines, we may find the other sides.

Example A
Given that $c = 6.00$, $A = 60°0'$, and $B = 40°0'$, find a, b, and C.
First we can see that

$$C = 180°0' - (60°0' + 40°0') = 80°0'$$

Thus,

$$\frac{a}{\sin 60°0'} = \frac{6.00}{\sin 80°0'} \qquad \text{or} \qquad a = \frac{(6.00)(0.8660)}{0.9848} = 5.28$$

$$\frac{b}{\sin 40°0'} = \frac{6.00}{\sin 80°0'} \qquad \text{or} \qquad b = \frac{(6.00)(0.6428)}{0.9848} = 3.92$$

Thus, $a = 5.28$, $b = 3.92$, and $C = 80°0'$.

Example B
Solve the triangle with the following given parts: $a = 63.7$, $A = 56°0'$, and $B = 97°0'$.
We may immediately determine that $C = 27°0'$. Thus,

$$\frac{b}{\sin 97°0'} = \frac{63.7}{\sin 56°0'} \qquad \text{or} \qquad b = \frac{63.7(0.9925)}{0.8290} = 76.3$$

and

$$\frac{c}{\sin 27°0'} = \frac{63.7}{\sin 56°0'} \qquad \text{or} \qquad c = \frac{63.7(0.4540)}{0.8290} = 34.9$$

Thus, $b = 76.3$, $c = 34.9$, and $C = 27°0'$.

Example C

Solve the triangle with the following given parts: $b = 5.06$, $A = 42.0°$, and $C = 28.5°$.

We determine that $B = 109.5°$. Thus,

$$\frac{a}{\sin 42.0°} = \frac{5.06}{\sin 109.5°} \quad \text{or} \quad a = \frac{5.06(0.6691)}{0.9426} = 3.59$$

and

$$\frac{c}{\sin 28.5°} = \frac{5.06}{\sin 109.5°} \quad \text{or} \quad c = \frac{5.06(0.4772)}{0.9426} = 2.56$$

Thus, $a = 3.59$, $c = 2.56$, and $B = 109.5°$.

If the given information is appropriate, the Law of Sines may be used to solve applied problems. The following example illustrates the use of the Law of Sines in such a problem.

Example D

A plane traveling at 650 mi/h with respect to the air is headed 30°0′ east of north. The wind is blowing from the south, which causes the actual course to be 27°0′ east of north. Find the velocity of the wind and the velocity of the plane with respect to the ground.

From the given information the angles are determined, as shown in Fig. 8—25. Then applying the Law of Sines, we have the relations

$$\frac{v_w}{\sin 3°0'} = \frac{v_{pg}}{\sin 150°0'} = \frac{650}{\sin 27°0'}$$

where v_w is the magnitude of the velocity of the wind and v_{pg} is the magnitude of the velocity of the plane with respect to the ground. Thus,

$$v_w = \frac{650(0.0523)}{0.4540} = 74.9 \text{ mi/h}$$

and

$$v_{pg} = \frac{650(0.5000)}{0.4540} = 716 \text{ mi/h}$$

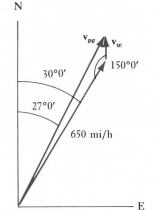

Figure 8—25

If we have information equivalent to Case 2 (two sides and the angle opposite one of them), we may find that there are *two* triangles which satisfy the given information. The following example illustrates this point.

Example E

Solve the triangle with the following given parts: $a = 60.0$, $b = 40.0$, and $B = 30.0°$.

By making a good scale drawing (Fig. 8—26), we note that the angle opposite a may be either at position A or A'. Both positions of this angle satisfy the given parts. Therefore, there are two triangles which result.

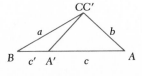

Figure 8—26

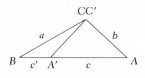

Figure 8–26

Using the Law of Sines, we solve the case in which A, opposite side a, is an acute angle.

$$\frac{60.0}{\sin A} = \frac{40.0}{\sin 30.0°} \quad \text{or} \quad \sin A = \frac{60.0(0.5000)}{40.0} = 0.7500$$

Therefore, $A = 48.6°$ and $C = 101.4°$. Using the Law of Sines again to find c, we have

$$\frac{c}{\sin 101.4°} = \frac{40.0}{\sin 30.0°} \quad \text{or} \quad c = \frac{40.0(0.9803)}{0.5000} = 78.4$$

The other solution is the case in which A', opposite side a, is an obtuse angle. Here we have

$$\frac{60.0}{\sin A'} = \frac{40.0}{\sin 30°0'}$$

which leads to $\sin A' = 0.7500$. Thus, $A' = 131.4°$. For this case we have C' (the angle opposite c when $A' = 131.4°$) as $18.6°$.

Using the Law of Sines to find c', we have

$$\frac{c'}{\sin 18.6°} = \frac{40.0}{\sin 30.0°} \quad \text{or} \quad c' = \frac{40.0(0.3190)}{0.5000} = 25.5$$

This means that the second solution is $A' = 131.4°$, $C' = 18.6°$, and $c' = 25.5$.

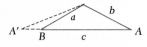

Figure 8–27

Example F

In Example E, if $b > 60.0$, only one solution would result. In this case, side b would intercept side c at A. It also intercepts the extension of side c, but this would require that angle B not be included in the triangle (see Fig. 8–27). Thus only one solution may result if $b > a$.

In Example E, there would be no solution if side b were not at least 30.0. For if this were the case, side b would not be sufficiently long to even touch side c. It can be seen that b must at least equal $a \sin B$. If it is just equal to $a \sin B$, there is one solution, a right triangle (Fig. 8–28).

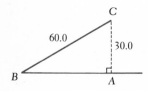

Figure 8–28

Summarizing the results for Case 2 as illustrated in Examples E and F, we make the following conclusions. Given sides a and b, and angle A (assuming here that a and A ($A < 90°$) are the given corresponding parts), we have:

(1) *no solution if $a < b \sin A$,*
(2) *a right triangle solution if $a = b \sin A$,*
(3) *two solutions if $b \sin A < a < b$,*
(4) *one solution if $a > b$.*

For the reason that two solutions may result from it, Case 2 is often referred to as the **ambiguous case.**

If we attempt to use the Law of Sines for the solution of Case 3 or Case 4, we find that we do not have sufficient information to complete one of the ratios. These cases can, however, be solved by the Law of Cosines, which we shall consider in the next section.

Example G
Given the 3 sides, $a = 5$, $b = 6$, $c = 7$, we would set up the ratios

$$\frac{5}{\sin A} = \frac{6}{\sin B} = \frac{7}{\sin C}$$

However, since there is no way to determine a complete ratio from these equations, we cannot find the solution of the triangle in this manner.

Exercises 8–4

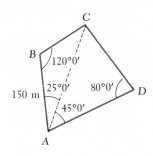

Figure 8–29

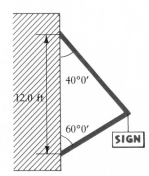

Figure 8–30

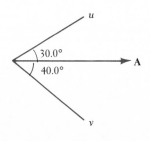

Figure 8–31

In Exercises 1 through 20 solve the triangles with the given parts.

1. $a = 45.7$, $A = 65°0'$, $B = 49°0'$
2. $b = 3.07$, $A = 26°0'$, $C = 120°0'$
3. $c = 4380$, $A = 37.0°$, $B = 34.0°$
4. $a = 93.2$, $B = 17.9°$, $C = 82.6°$
5. $a = 4.60$, $b = 3.10$, $A = 18.0°$
6. $b = 3.62$, $c = 2.94$, $B = 69.3°$
7. $b = 77.5$, $c = 36.4$, $B = 20.7°$
8. $a = 150$, $c = 250$, $C = 76.4°$
9. $b = 0.0742$, $B = 51°0'$, $C = 3°30'$
10. $c = 729$, $B = 121°0'$, $C = 44°10'$
11. $a = 63.8$, $B = 58.4°$, $C = 22.2°$
12. $a = 13.0$, $A = 55.2°$, $B = 67.5°$
13. $b = 438$, $B = 47.4°$, $C = 64.5°$
14. $b = 283$, $B = 13.7°$, $C = 76.3°$
15. $a = 5.24$, $b = 4.44$, $B = 48.1°$
16. $a = 89.4$, $c = 37.3$, $C = 15.6°$
17. $b = 2880$, $c = 3650$, $B = 31.4°$
18. $a = 0.841$, $b = 0.965$, $A = 57.1°$
19. $a = 45.0$, $b = 126$, $A = 64°0'$
20. $a = 10.0$, $c = 5.00$, $C = 30°0'$

In Exercises 21 through 28 use the Law of Sines to solve the given problems.

21. Find the lengths of the two steel supports of the sign shown in Fig. 8–29.
22. Find the unknown sides of the four-sided piece of land shown in Fig. 8–30.
23. The angles of elevation of an airplane, measured from points A and B, 7540 ft apart on the same side of the airplane (the airplane and points A and B are in the same vertical plane), are 32.0° and 44.0°. How far is point A from the airplane?
24. Resolve vector **A** ($A = 160$) into two components in the directions u and v, as shown in Fig. 8–31.
25. A ship leaves a port and travels due west. At a certain point it turns 30.0° N of W and travels an additional 42.0 mi to a point 63.0 mi from the port. How far from the port is the point where the ship turned?
26. City B is 40°0' south of east of city A. A pilot wishes to know what direction he should head the plane in flying from A to B if the wind is from the west at 40.0 km/h and his velocity with respect to the air is 300 km/h. What should his heading be?
27. A communications satellite is directly above the extension of a line between receiving towers A and B. It is determined from radio signals that the angle of elevation of the satellite from tower A is 89.2°, and the angle of elevation from tower B is 86.5°. If A and B are 1290 km apart, how far is the satellite from A? (Neglect the curvature of the earth.)
28. A person measures a triangular piece of land and reports the following information: "One side is 58.4 ft long and another side is 21.1 ft long. The angle opposite the shorter side is 24°0'." Could this information be correct?

8–5 The Law of Cosines

As we noted at the end of the preceding section, the Law of Sines cannot be used if the only information given is that of Case 3 or Case 4. There-fore, it is necessary to develop a method of finding at least one more part of the triangle. Here we can use the Law of Cosines. After obtaining another part by the Law of Cosines, we can then use the Law of Sines to complete the solution. We do this because the Law of Sines generally provides a simpler method of solution than the Law of Cosines.

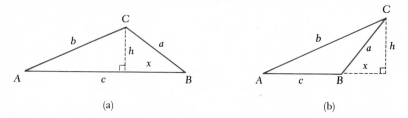

Figure 8–32

Consider any oblique triangle, for example either of the ones in Fig. 8–32. For each we obtain $h = b \sin A$. By using the Pythagorean theorem we obtain $a^2 = h^2 + x^2$ for each. Thus,

$$a^2 = b^2 \sin^2 A + x^2 \tag{8–9}$$

In Fig. 8–32(a), we have $c - x = b \cos A$, or $x = c - b \cos A$. In Fig. 8–32(b), we have $c + x = b \cos A$, or $x = b \cos A - c$. Substituting these relations into Eq. (8–9), we obtain

$$a^2 = b^2 \sin^2 A + (c - b \cos A)^2$$
and
$$a^2 = b^2 \sin^2 A + (b \cos A - c)^2 \tag{8–10}$$

respectively. When expanded, these give

$$a^2 = b^2 \sin^2 A + b^2 \cos^2 A + c^2 - 2bc \cos A$$
and
$$a^2 = b^2 (\sin^2 A + \cos^2 A) + c^2 - 2bc \cos A \tag{8–11}$$

Recalling the definitions of the trigonometric functions, we know that $\sin \theta = y/r$ and $\cos \theta = x/r$. Thus, $\sin^2\theta + \cos^2\theta = (y^2 + x^2)/r^2$. However, $x^2 + y^2 = r^2$, which means that

$$\sin^2 \theta + \cos^2 \theta = 1 \tag{8–12}$$

This equation holds for any angle θ, since we made no assumptions as to the properties of θ. By substituting Eq. (8–12) into Eq. (8–11), we arrive at the **Law of Cosines:**

$$a^2 = b^2 + c^2 - 2bc \cos A \tag{8–13}$$

Using the method above, we may also show that

$$b^2 = a^2 + c^2 - 2ac \cos B$$

and

$$c^2 = a^2 + b^2 - 2ab \cos C$$

Therefore, if we know two sides and the included angle (Case 3) we may directly solve for the side opposite the given angle. Then, by using the Law of Sines, we may complete the solution. If we are given all three sides (Case 4), we may solve for the angle opposite one of these sides by use of the Law of Cosines. Again we use the Law of Sines to complete the solution.

Example A

Solve the triangle with $a = 45.0$, $b = 67.0$, and $C = 35°0'$. Using the Law of Cosines, we have

$$c^2 = (45.0)^2 + (67.0)^2 - 2(45.0)(67.0)(0.8192)$$
$$= 2025 + 4489 - 4940 = 1574$$
$$c = 39.7$$

From the Law of Sines, we now have

$$\frac{45.0}{\sin A} = \frac{67.0}{\sin B} = \frac{39.7}{0.5736}$$

which leads to

$$\sin A = 0.6502 \quad \text{or} \quad A = 40°30'$$

We could solve for B from the above relation, or by use of the fact that the sum of all three angles is $180°$. Thus,

$$B = 104°30'$$

Example B

If, in Example A, $C = 145°0'$, we have

$$c^2 = (45.0)^2 + (67.0)^2 - 2(45.0)(67.0)(-0.8192)$$
$$= 2025 + 4489 + 4940 = 11450$$
$$c = 107$$

From the Law of Sines we then find $A = 14°0'$ and $B = 21°0'$.

Example C

Solve the triangle for which $a = 49.3$, $b = 21.6$, and $c = 42.6$.

$$\cos A = \frac{b^2 + c^2 - a^2}{2bc} = \frac{(21.6)^2 + (42.6)^2 - (49.3)^2}{2(21.6)(42.6)}$$

$$= \frac{467 + 1815 - 2430}{1840} = -0.0804$$

Since the value of cos A is negative, we know that A is between 90° and 180°. Thus,

$$A = 180.0° - 85.4° = 94.6°$$

We then find that $B = 25.9°$ and $C = 59.5°$, from the Law of Sines.

Example D

Find the resultant of two vectors having magnitudes of 78.0 and 45.0, and directed toward the east and 15.0° east of north, respectively (see Fig. 8–33).

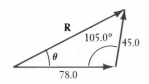

Figure 8–33

The magnitude of the resultant is given by

$$R = \sqrt{(78.0)^2 + (45.0)^2 - 2(78.0)(45.0)(\cos 105.0°)}$$
$$= \sqrt{6084 + 2025 + 1817} = 99.6$$

Also, $\theta = 25.9°$ is found from the Law of Sines.

Example E

A vertical radio antenna is to be built on a hill which makes an angle of 6.0° with the horizontal. If guy wires are to be attached at a point that is 150 ft up on the antenna and at points 100 ft from the base of the antenna, what will be the lengths of guy wires which are positioned directly up and directly down the hill?

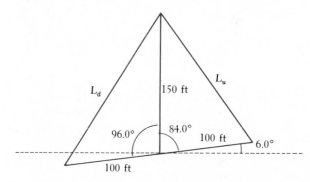

Figure 8–34

Making an appropriate figure such as Fig. 8–34, we are able to establish the equations necessary for the solution:

$$L_u{}^2 = (100)^2 + (150)^2 - 2(100)(150)\cos 84.0°$$
$$L_d{}^2 = (100)^2 + (150)^2 - 2(100)(150)\cos 96.0°$$
$$L_u{}^2 = 10,000 + 22,500 - 30,000(0.1045)$$
$$= 32,500 - 3135 = 29,365$$
$$L_u = 171 \text{ ft};$$
$$L_d{}^2 = 32,500 + 3135 = 35,635$$
$$L_d = 189 \text{ ft}$$

Exercises 8—5

In Exercises 1 through 20 solve the triangles with the given parts.

1. $a = 6.00, b = 7.56, C = 54.0°$
2. $b = 87.3, c = 34.0, A = 130.0°$
3. $a = 4530, b = 924, C = 98.0°$
4. $a = 0.0845, c = 0.116, B = 85.0°$
5. $a = 39.5, b = 45.2, c = 67.1$
6. $a = 23.3, b = 27.2, c = 29.1$
7. $a = 385, b = 467, c = 800$
8. $a = 0.243, b = 0.263, c = 0.153$
9. $a = 320, b = 847, C = 158.0°$
10. $b = 18.3, c = 27.1, A = 58.7°$
11. $a = 21.4, c = 4.28, B = 86.3°$
12. $a = 11.3, b = 5.10, C = 77.6°$
13. $a = 103, c = 159, C = 104.6°$
14. $a = 49.3, b = 54.5, B = 114.0°$
15. $a = 0.493, b = 0.595, c = 0.639$
16. $a = 69.7, b = 49.3, c = 56.2$
17. $a = 723, b = 598, c = 158$
18. $a = 1.78, b = 6.04, c = 4.80$
19. $a = 15.6, A = 15.1°, B = 150.5°$
20. $a = 17.5, b = 24.5, c = 37.0$

In Exercises 21 through 28 use the Law of Cosines to solve the given problems.

21. To measure the distance AC, a man walks 500 ft from A to B, then turns 30°0′ to face C, and walks 680 ft to C. What is the distance AC?

22. An airplane traveling at 700 km/h leaves the airport at noon going due east. At 2 PM the pilot turns 10.0° north of east. How far is the plane from the airport at 3 PM?

23. Two forces, one of 56.0 lb and the other of 67.0 lb, are applied to the same object. The resultant force is 82.0 lb. What is the angle between the two forces?

24. A triangular metal frame has sides of 8.00 ft, 12.0 ft, and 16.0 ft. What is the largest angle between parts of the frame?

25. A boat, which can travel 6.00 km/h in still water, heads downstream at an angle of 20°0′ with the bank. If the stream is flowing at the rate of 3.00 km/h, how fast is the boat traveling and in what direction?

26. An airplane's velocity with respect to the air is 520 mi/h, and it is headed 24.0° north of west. The wind is from due southwest and has a velocity of 55.0 mi/h. What is the true direction of the plane and what is its velocity with respect to the ground?

27. One end of a 13.1-m pole is 15.7 m from an observer's eyes and the other end is 19.3 m from his eyes. Through what angle does the observer see the pole?

28. A triangular piece of land is bounded by 134 ft of stone wall, 205 ft of road frontage, and 147 ft of fencing. What angle does the fence make with the road?

8—6 Exercises for Chapter 8

In Exercises 1 through 4 find the x- and y-components of the given vectors by use of the trigonometric functions.

1. $A = 65.0, \theta_A = 28.0°$
2. $A = 8.05, \theta_A = 149.0°$
3. $A = 0.920, \theta_A = 215°0′$
4. $A = 657, \theta_A = 343°0′$

In Exercises 5 through 8 vectors **A** and **B** are at right angles. Find the magnitude and direction of the resultant.

5. $A = 327$ 6. $A = 684$ 7. $A = 4960$ 8. $A = 26.5$
   $B = 505$ $B = 295$ $B = 3290$ $B = 89.8$

In Exercises 9 through 16 add the given vectors by use of the trigonometric functions and the Pythagorean theorem.

9. $A = 780, \theta_A = 28°0'$
   $B = 346, \theta_B = 320°0'$

10. $A = 0.0120, \theta_A = 10°30'$
    $B = 0.0078, \theta_B = 260°0'$

11. $A = 22.5, \theta_A = 130°10'$
    $B = 7.60, \theta_B = 200°0'$

12. $A = 18,700, \theta_A = 110°40'$
    $B = 4830, \theta_B = 350°20'$

13. $A = 51.3, \theta_A = 12.2°$
    $B = 42.6, \theta_B = 291.7°$

14. $A = 70.3, \theta_A = 122.5°$
    $B = 30.2, \theta_B = 214.8°$

15. $A = 75.0, \theta_A = 15.0°$
    $B = 26.5, \theta_B = 192.0°$
    $C = 54.8, \theta_C = 344.0°$

16. $A = 8100, \theta_A = 141.0°$
    $B = 1540, \theta_B = 165.0°$
    $C = 3470, \theta_C = 296.0°$

In Exercises 17 through 32 solve the triangles with the given parts.

17. $A = 48°0', B = 68°0', a = 14.5$
18. $A = 132°0', b = 7.50, C = 32°0'$
19. $a = 22.8, B = 33.5°, C = 125.3°$
20. $A = 71.0°, B = 48.5°, c = 8.42$
21. $b = 76.0, c = 40.5, B = 110°0'$
22. $A = 77°0', a = 12.0, c = 5.00$
23. $b = 14.5, c = 13.0, C = 56.6°$
24. $B = 40.0°, b = 7.00, c = 18.0$
25. $a = 186, B = 130.0°, c = 106$
26. $b = 750, c = 1100, A = 56.0°$
27. $a = 7.86, b = 2.45, C = 22.0°$
28. $a = 0.208, c = 0.697, B = 105.0°$
29. $a = 17.0, b = 12.0, c = 25.0$
30. $a = 900, b = 995, c = 1100$
31. $a = 5.30, b = 8.75, c = 12.5$
32. $a = 47.4, b = 40.0, c = 45.5$

In Exercises 33 through 50 solve the given problems.

33. A jet climbs at an angle $35.0°$ while traveling 600 km/h. How long will it take to climb to an altitude of 10,000 m?

34. A bullet is fired into the air at 2000 mi/h at an angle of $25°0'$ with the horizontal. What is the vertical component of the velocity?

35. A balloon is rising at the rate of 15.0 ft/s and at the same time is being blown horizontally by the wind at the rate of 22.5 ft/s. Find the resultant velocity.

36. A motorboat which travels at 8.00 km/h in still water heads directly across a stream which flows at 3.00 km/h. What is the resultant velocity of the boat?

37. A velocity vector is the resultant of two other vectors. If the given velocity is 450 mi/h and makes angles of $34°0'$ and $76°0'$ with the two components, find the magnitudes of the components.

38. A person on a hill in the middle of a plain relates that the angles of depression of two objects on the plain below (on directly opposite sides of the hill) are $30.0°$ and $40.0°$. She knows that the objects are 15,800 m apart. How far is she from the closest object?

39. In order to find the distance between points A and B on opposite sides of a river, a distance AC is measured off as 1000 ft, where point C is on the

same side of the river as *A*. Angle *BAC* is measured to be 102°0′ and angle *ACB* is 33°0′. What is the distance between *A* and *B*?

40. A 22.0-ft ladder leans against a wall, making an angle of 29°0′ with the wall. If the foot of the ladder is 10.7 ft from the foot of the wall, find the angle of inclination of the wall to the ground.

41. Two points on opposite sides of an obstruction are respectively 117 m and 88 m from a third point. The lines joining the first two points and the third point intersect at an angle of 115.0° at the third point. How far apart are the original two points?

42. A crate is being held aloft by two ropes which are tied at the same point on the crate. They are 14.5 m and 10.5 m long, respectively, and the angle between them is 104.0°. If the ropes are supported at the same level, how far apart are they?

43. Two cars are at the intersection of two straight roads. One travels 5.20 mi on one road, and the other car travels 3.75 mi on the other road. The drivers contact each other on CB radio and find they are at points known to be 4.50 mi apart. What angle do the roads make at the intersection?

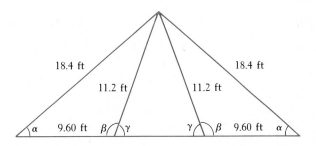

Figure 8—35

44. Determine the angles of the structure indicated in Fig. 8—35.

45. Atlanta is 290 mi and 51.0° south of east from Nashville. The pilot of an airplane due north of Atlanta radios Nashville and finds her plane is on a line 10.5° south of east from Nashville. How far is the plane from Nashville?

46. The edges of a saw tooth are 2.10 mm and 3.25 mm. The base of the tooth is 2.25 mm. At what angle do the edges of the tooth meet?

47. Determine the weight *w* being supported by the ropes shown in Fig. 8—36. The tension in the right support rope is 87.5 lb.

48. Boston is 650 km and 21.0° south of west from Halifax, Nova Scotia. Radio signals locate a ship 10.5° east of south from Halifax and 5.6° north of east from Boston. How far is the ship from each city?

49. Find the resultant of the forces indicated in Fig. 8—37.

50. A jet plane is traveling horizontally at 1200 ft/s. A missile is fired horizontally from it 30°0′ from the direction in which the plane is traveling. If the missile leaves the plane at 2000 ft/s, what is its velocity 10.0 s later if the vertical component is given by $v_V = -32t$ (in feet per second)?

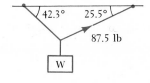

Figure 8—36

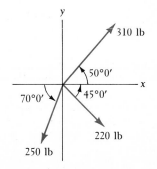

Figure 8—37

9

Graphs of the Trigonometric Functions

9–1 Graphs of $y = a \sin x$ and $y = a \cos x$

One of the clearest ways to demonstrate the variation of the trigonometric functions is by means of their graphs. The graphs are useful for analyzing properties of the trigonometric functions, and in themselves are valuable in many applications. Several of these applications are indicated in the exercises, particularly in the last two sections of this chapter.

The graphs are constructed on the rectangular coordinate system. In plotting the trigonometric functions, it is normal to express the angle in radians. In this way x and the function of x are expressed as *numbers*, and these numbers may have any desired unit of measurement. Therefore, in order to determine the graphs, it is necessary to be able to readily use angles expressed in radians. If necessary, Section 7–3 should be reviewed for this purpose.

In this section the graphs of the sine and cosine functions are demonstrated. We begin by constructing a table of values of x and y, for the function $y = \sin x$:

x	0	$\dfrac{\pi}{6}$	$\dfrac{\pi}{3}$	$\dfrac{\pi}{2}$	$\dfrac{2\pi}{3}$	$\dfrac{5\pi}{6}$	π	$\dfrac{7\pi}{6}$	$\dfrac{4\pi}{3}$	$\dfrac{3\pi}{2}$	$\dfrac{5\pi}{3}$	$\dfrac{11\pi}{6}$	2π
y	0	0.5	0.87	1	0.87	0.5	0	-0.5	-0.87	-1	-0.87	-0.5	0

Plotting these values, we obtain the graph shown in Fig. 9–1.

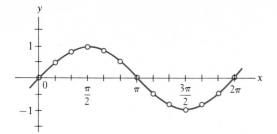

Figure 9–1

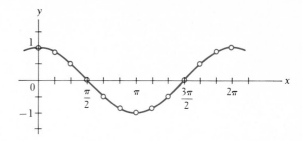

Figure 9–2

The graph of $y = \cos x$ may be constructed in the same manner. The table below gives the proper values for the graph of $y = \cos x$. The graph is shown in Fig. 9–2.

x	0	$\dfrac{\pi}{6}$	$\dfrac{\pi}{3}$	$\dfrac{\pi}{2}$	$\dfrac{2\pi}{3}$	$\dfrac{5\pi}{6}$	π	$\dfrac{7\pi}{6}$	$\dfrac{4\pi}{3}$	$\dfrac{3\pi}{2}$	$\dfrac{5\pi}{3}$	$\dfrac{11\pi}{6}$	2π
y	1	0.87	0.5	0	-0.5	-0.87	-1	-0.87	-0.5	0	0.5	0.87	1

The graphs are continued beyond the values shown in the table to indicate that they continue on indefinitely in each direction. From the values and the graphs, *it can be seen that the two graphs are of exactly the same shape, with the cosine curve displaced $\pi/2$ units to the left of the sine curve.* The shape of these curves should be recognized readily, with special note as to the points at which they cross the axes. This information will be especially valuable in "sketching" similar curves, since the basic shape always remains the same. We shall find it unnecessary to plot numerous points every time we wish to sketch such a curve.

To obtain the graph of $y = a \sin x$, we note that all of the y-values obtained for the graph of $y = \sin x$ are to be multiplied by the number a. In this case the greatest value of the sine function is $|a|$ instead of 1. *The number $|a|$ is called the* **amplitude** *of the curve and represents the greatest y-value of the curve.* This is also true for $y = a \cos x$.

Example A
Plot the curve of $y = 2 \sin x$.
The table of values to be used is as follows; Fig. 9–3 shows the graph.

x	0	$\dfrac{\pi}{6}$	$\dfrac{\pi}{3}$	$\dfrac{\pi}{2}$	$\dfrac{2\pi}{3}$	$\dfrac{5\pi}{6}$	π	$\dfrac{7\pi}{6}$	$\dfrac{4\pi}{3}$	$\dfrac{3\pi}{2}$	$\dfrac{5\pi}{3}$	$\dfrac{11\pi}{6}$
y	0	1	1.73	2	1.73	1	0	-1	-1.73	-2	-1.73	-1

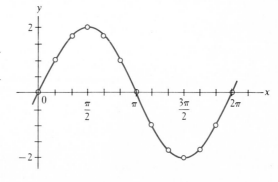

Figure 9–3

Example B

Plot the curve of $y = -3 \cos x$.

The table of values to be used is as follows; Fig. 9–4 shows the graph.

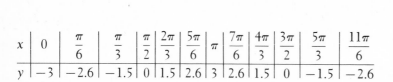

x	0	$\dfrac{\pi}{6}$	$\dfrac{\pi}{3}$	$\dfrac{\pi}{2}$	$\dfrac{2\pi}{3}$	$\dfrac{5\pi}{6}$	π	$\dfrac{7\pi}{6}$	$\dfrac{4\pi}{3}$	$\dfrac{3\pi}{2}$	$\dfrac{5\pi}{3}$	$\dfrac{11\pi}{6}$
y	-3	-2.6	-1.5	0	1.5	2.6	3	2.6	1.5	0	-1.5	-2.6

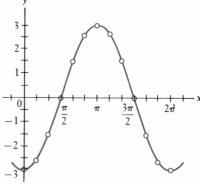

Figure 9–4

Note from Example B that *the effect of the minus sign before the number a is to invert the curve.* The effect of the number a can also be seen readily from these examples.

By knowing the general shape of the sine curve, where it crosses the axes, and the amplitude, we can rapidly *sketch* curves of form $y = a \sin x$ and $y = a \cos x$. There is generally no need to plot any more points than those corresponding to the values of the amplitude and those where the curve crosses the axes.

Example C

Sketch the graph of $y = 4 \cos x$.

First we set up a table of values for the points where the curve crosses the x-axis and for the highest and lowest points on the curve.

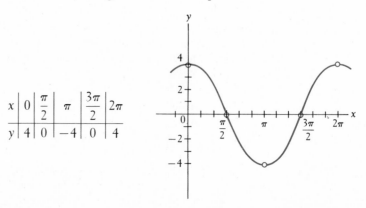

x	0	$\dfrac{\pi}{2}$	π	$\dfrac{3\pi}{2}$	2π
y	4	0	-4	0	4

Figure 9–5

Now we plot the above points and join them, knowing the basic shape of the curve. See Fig. 9–5.

Example D
Sketch the curve of $y = -2 \sin x$.
We list here the important values associated with this curve.

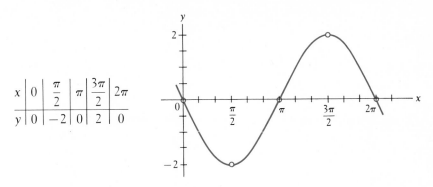

x	0	$\dfrac{\pi}{2}$	π	$\dfrac{3\pi}{2}$	2π
y	0	-2	0	2	0

Figure 9–6

Since we know the general shape of the sine curve, we can now sketch the graph as shown in Fig. 9–6. Note the inversion of the curve due to the minus sign.

Exercises 9–1

In Exercises 1 through 4 complete the following table for the given functions, and then plot the resulting graph.

x	$-\pi$	$-\dfrac{3\pi}{4}$	$-\dfrac{\pi}{2}$	$-\dfrac{\pi}{4}$	0	$\dfrac{\pi}{4}$	$\dfrac{\pi}{2}$	$\dfrac{3\pi}{4}$	π	$\dfrac{5\pi}{4}$	$\dfrac{3\pi}{2}$	$\dfrac{7\pi}{4}$	2π	$\dfrac{9\pi}{4}$	$\dfrac{5\pi}{2}$	$\dfrac{11\pi}{4}$	3π
y																	

1. $y = \sin x$ 2. $y = \cos x$ 3. $y = 3 \cos x$ 4. $y = -4 \sin x$

In Exercises 5 through 20 sketch the curves of the indicated functions.

5. $y = 3 \sin x$ 6. $y = 5 \sin x$ 7. $y = \frac{5}{2} \sin x$
8. $y = 0.5 \sin x$ 9. $y = 2 \cos x$ 10. $y = 3 \cos x$
11. $y = 0.8 \cos x$ 12. $y = \frac{3}{2} \cos x$ 13. $y = -\sin x$
14. $y = -3 \sin x$ 15. $y = -1.5 \sin x$ 16. $y = -0.2 \sin x$
17. $y = -\cos x$ 18. $y = -8 \cos x$ 19. $y = -2.5 \cos x$
20. $y = -0.4 \cos x$

Although units of π are often convenient, we must remember that π is really only a number. Numbers which are not multiples of π may be used as well. In Exercises 21 through 24 plot the indicated graphs by finding the values of y corresponding to the values of 0, 1, 2, 3, 4, 5, 6, and 7 for x. (Remember, the numbers 0, 1, 2, and so forth represent radian measure.) Values from Table 3 p A-42 may be used.

21. $y = \sin x$ 22. $y = 3 \sin x$ 23. $y = \cos x$ 24. $y = 2 \cos x$

9–2 Graphs of $y = a \sin bx$ and $y = a \cos bx$

In graphing the curve of $y = \sin x$ we note that the values of y start repeating every 2π units of x. This is because $\sin x = \sin (x + 2\pi) = \sin (x + 4\pi)$, and so forth. For any trigonometric function F, we say that it has a **period** P, if $F(x) = F(x + P)$. For functions which are periodic, such as the sine and cosine, *the period refers to the x-distance between any point and the next corresponding point for which the values of y start repeating.*

Let us now plot the curve $y = \sin 2x$. This means that we choose a value for x, multiply this value by two, and find the sine of the result. This leads to the following table of values for this function.

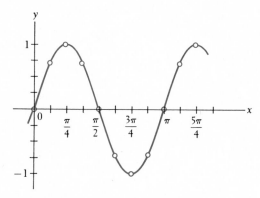

x	0	$\dfrac{\pi}{8}$	$\dfrac{\pi}{4}$	$\dfrac{3\pi}{8}$	$\dfrac{\pi}{2}$	$\dfrac{5\pi}{8}$	$\dfrac{3\pi}{4}$	$\dfrac{7\pi}{8}$	π	$\dfrac{9\pi}{8}$	$\dfrac{5\pi}{4}$
$2x$	0	$\dfrac{\pi}{4}$	$\dfrac{\pi}{2}$	$\dfrac{3\pi}{4}$	π	$\dfrac{5\pi}{4}$	$\dfrac{3\pi}{2}$	$\dfrac{7\pi}{4}$	2π	$\dfrac{9\pi}{4}$	$\dfrac{5\pi}{2}$
y	0	0.7	1	0.7	0	-0.7	-1	-0.7	0	0.7	1

Figure 9–7

Plotting these values, we have the curve shown in Fig. 9–7.

From the table and Fig. 9–7, we note that the function $y = \sin 2x$ starts repeating after π units of x. The effect of the 2 before the x has been to make the period of this curve half the period of the curve of $\sin x$. This leads us to the following conclusion: If the period of the trigonometric function $F(x)$ is P, then the period of $F(bx)$ is P/b. Since each of the functions $\sin x$ and $\cos x$ has a period of 2π, *each of the functions $\sin bx$ and $\cos bx$ has a period of $2\pi/b$.*

Example A
The period of $\sin 3x$ is $2\pi/3$, which means that the curve of the function $y = \sin 3x$ will repeat every $2\pi/3$ (approximately 2.09) units of x.

The period of $\cos 4x$ is $2\pi/4 = \pi/2$, and that of $\sin \frac{1}{2}x$ is $2\pi/\frac{1}{2} = 4\pi$.

Example B
The period of $\sin \pi x$ is $2\pi/\pi = 2$. That is, the curve of the function $\sin \pi x$ will repeat every 2 units. It is then noted that the periods of $\sin 3x$ and $\sin \pi x$ are nearly equal. This is to be expected since π is only slightly greater than 3.

The period of $\cos 3\pi x$ is $2\pi/3\pi = 2/3$.

Combining the result for the period with the results of Section 9–1, we conclude that *each of the functions $y = a \sin bx$ and $y = a \cos bx$ has an amplitude of $|a|$ and a period of $2\pi/b$*. These properties are very useful in sketching these functions, as it is shown in the following examples.

Example C

Sketch the graph of $y = 3 \sin 4x$ for $0 \le x \le \pi$.

We immediately conclude that the amplitude is 3 and the period is $2\pi/4 = \pi/2$. Therefore, we know that $y = 0$ when $x = 0$ and $y = 0$ when $x = \pi/2$. Also we recall that the sine function is zero halfway between these values, which means that $y = 0$ when $x = \pi/4$. The function reaches its amplitude values halfway between the zeros. Therefore, $y = 3$ for $x = \pi/8$ and $y = -3$ for $x = 3\pi/8$. A table for these important values of the function $y = 3 \sin 4x$ is as follows.

x	0	$\dfrac{\pi}{8}$	$\dfrac{\pi}{4}$	$\dfrac{3\pi}{8}$	$\dfrac{\pi}{2}$	$\dfrac{5\pi}{8}$	$\dfrac{3\pi}{4}$	$\dfrac{7\pi}{8}$	π
y	0	3	0	-3	0	3	0	-3	0

Using this table and the knowledge of the form of the sine curve, we sketch the function (Fig. 9–8).

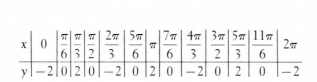

Figure 9–8

Example D

Sketch the graph of $y = -2 \cos 3x$ for $0 \le x \le 2\pi$.

We note that the amplitude is 2 and that the period is $2\pi/3$. Since the cosine curve is at its amplitude value for $x = 0$, we have $y = -2$ for $x = 0$ (the negative value is due to the minus sign before the function) and $y = -2$ for $x = 2\pi/3$. The cosine function also reaches its amplitude value halfway between these values, or $y = 2$ for $x = \pi/3$. The cosine function is zero halfway between the x-values listed for the amplitude; that is, $y = 0$ for $x = \pi/6$ and for $x = \pi/2$. A table of the important values follows.

x	0	$\dfrac{\pi}{6}$	$\dfrac{\pi}{3}$	$\dfrac{\pi}{2}$	$\dfrac{2\pi}{3}$	$\dfrac{5\pi}{6}$	π	$\dfrac{7\pi}{6}$	$\dfrac{4\pi}{3}$	$\dfrac{3\pi}{2}$	$\dfrac{5\pi}{3}$	$\dfrac{11\pi}{6}$	2π
y	-2	0	2	0	-2	0	2	0	-2	0	2	0	-2

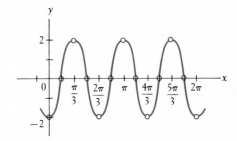

Figure 9–9

Using this table and the knowledge of the form of the cosine curve, we sketch the function as shown in Fig. 9–9.

Example E

Sketch the function $y = \cos \pi x$ for $0 \leq x \leq \pi$.

For this function the amplitude is 1; the period is $2\pi/\pi = 2$. Since the value of the period is not in terms of π, it is more convenient to use regular decimal units for x when sketching than to use units in terms of π as in the previous graphs. Therefore, we have the following table.

x	0	0.5	1	1.5	2	2.5	3
y	1	0	−1	0	1	0	−1

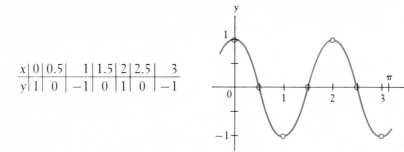

Figure 9—10

The graph of this function is shown in Fig. 9—10.

Exercises 9—2

In Exercises 1 through 20 find the period of each of the given functions.

1. $y = 2 \sin 6x$ 2. $y = 4 \sin 2x$ 3. $y = 3 \cos 8x$

4. $y = \cos 10x$ 5. $y = -2 \sin 12x$ 6. $y = -\sin 5x$

7. $y = -\cos 16x$ 8. $y = -4 \cos 2x$ 9. $y = 5 \sin 2\pi x$

10. $y = 2 \sin 3\pi x$ 11. $y = 3 \cos 4\pi x$ 12. $y = 4 \cos 10\pi x$

13. $y = 3 \sin \frac{1}{3}x$ 14. $y = -2 \sin \frac{2}{3}x$ 15. $y = -\frac{1}{2} \cos \frac{2}{3}x$

16. $y = \frac{1}{3} \cos \frac{1}{4}x$ 17. $y = 0.4 \sin \dfrac{2\pi x}{3}$ 18. $y = 1.5 \cos \dfrac{\pi x}{10}$

19. $y = 3.3 \cos \pi^2 x$ 20. $y = 2.5 \sin \dfrac{2x}{\pi}$

In Exercises 21 through 40 sketch the graphs of the given functions. For this, use the functions given for Exercises 1 through 20.

In Exercises 41 through 44 sketch the indicated graphs.

41. The electric current in a certain 60 Hz alternating-current circuit is given by $i = 10 \sin 120\pi t$, where i is the current (in amperes) and t is the time (in seconds). Sketch the graph of i vs. t for $0 \leq t \leq 0.1$ s.

42. A generator produces a voltage given by $V = 200 \cos 50\pi t$, where t is the time (in seconds). Sketch the graph of V vs. t for $0 \leq t \leq 0.1$ s.

43. The vertical displacement x of a certain object oscillating at the end of a spring is given by $x = 6 \cos 4\pi t$, where x is measured in inches and t in seconds. Sketch the graph of x vs. t for $0 \leq t \leq 1$ s.

44. The velocity of a piston in an engine is given by $v = 1200 \sin 1200\pi t$, where v is the velocity (in centimeters per second) and t is the time (in seconds). Sketch the graph of v vs. t for $0 \leq t \leq 0.01$ s.

9–3 Graphs of $y = a \sin (bx + c)$ and $y = a \cos (bx + c)$

There is one more important quantity to be discussed in relation to graphing the sine and cosine functions. This quantity is the **phase angle** of the function. In the function $y = a \sin (bx + c)$, c represents this phase angle. Its meaning is illustrated in the following example.

Example A

Sketch the graph of $y = \sin (2x + \frac{\pi}{4})$.

Note that $c = \frac{\pi}{4}$. This means that in order to obtain the values for the table we must assume a value of x, multiply it by two, add $\frac{\pi}{4}$ to this value, and then find the sine of this result. In this manner we arrive at the following table.

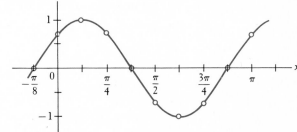

x	$-\dfrac{\pi}{8}$	0	$\dfrac{\pi}{8}$	$\dfrac{\pi}{4}$	$\dfrac{3\pi}{8}$	$\dfrac{\pi}{2}$	$\dfrac{5\pi}{8}$	$\dfrac{3\pi}{4}$	$\dfrac{7\pi}{8}$	π
y	0	0.7	1	0.7	0	-0.7	-1	-0.7	0	0.7

Figure 9–11

We use the value of $x = -\frac{\pi}{8}$ in the table, for we note that it corresponds to finding $\sin 0$. Now using the values listed in the table, we plot the graph of $y = \sin (2x + \frac{\pi}{4})$. See Fig. 9–11.

We can see from the table and from the graph that the curve of

$$y = \sin\left(2x + \frac{\pi}{4}\right)$$

is precisely the same as that of $y = \sin 2x$, except that it is shifted $\frac{\pi}{8}$ units to the left. The effect of c in the equation of $y = a \sin (bx + c)$ is to shift the curve of $y = a \sin bx$ to the left if $c > 0$, and to shift the curve to the right if $c < 0$. The amount of this shift is given by c/b. Due to its importance in sketching curves, *the quantity c/b is called the* **displacement.**

Therefore, the results above combined with the results of Section 9–2 may be used to sketch curves of the functions $y = a \sin (bx + c)$ and $y = a \cos (bx + c)$. These are the important quantities to determine:

(1) the amplitude (equal to $|a|$)

(2) the period $\left(\text{equal to } \dfrac{2\pi}{b}\right)$

(3) the displacement $\left(\text{equal to } \dfrac{c}{b}\right)$

By use of these quantities, the curves of these sine and cosine functions can be readily sketched.

Example B

Sketch the graph of $y = 2 \sin (3x - \pi)$ for $0 \le x \le \pi$.

First we note that $a = 2$, $b = 3$, and $c = -\pi$. Therefore, the amplitude is 2, the period is $\frac{2\pi}{3}$, and the displacement is $\frac{\pi}{3}$ to the right.

With this information we can tell that the curve "starts" at $x = \frac{\pi}{3}$ and "starts repeating" $\frac{2\pi}{3}$ units to the right of this point. (Be sure to grasp this point well. The period tells how many units there are along the x-axis *between* such corresponding points.) The value of y is zero when $x = \frac{\pi}{3}$ and when $x = \pi$. It is also zero halfway between these values of x, or when $x = \frac{2\pi}{3}$. The curve reaches the amplitude value of 2 midway between $x = \frac{\pi}{3}$ and $x = \frac{2\pi}{3}$, or when $x = \frac{\pi}{2}$. Extending the curve to the left to $x = 0$, we note that, since the period is $\frac{2\pi}{3}$, the curve passes through $(0, 0)$. Therefore, we have the following table of important values.

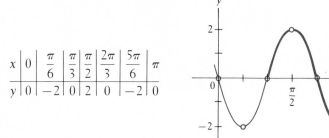

x	0	$\dfrac{\pi}{6}$	$\dfrac{\pi}{3}$	$\dfrac{\pi}{2}$	$\dfrac{2\pi}{3}$	$\dfrac{5\pi}{6}$	π
y	0	-2	0	2	0	-2	0

Figure 9–12

From these values we sketch the graph shown in Fig. 9–12.

Example C

Sketch the graph of the function $y = -\cos (2x + \frac{\pi}{6})$.

First we determine that the amplitude is 1, the period is $\frac{2\pi}{2} = \pi$, and that the displacement is $(\frac{\pi}{6}) \div 2 = \frac{\pi}{12}$ to the left ($c > 0$). From these values we construct the following table, remembering that the curve starts repeating π units to the right of $-\frac{\pi}{12}$.

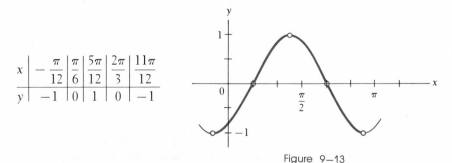

x	$-\dfrac{\pi}{12}$	$\dfrac{\pi}{6}$	$\dfrac{5\pi}{12}$	$\dfrac{2\pi}{3}$	$\dfrac{11\pi}{12}$
y	-1	0	1	0	-1

Figure 9–13

From this table we sketch the graph as shown in Fig. 9–13.

Example D

Sketch the graph of the function $y = 2 \cos \left(\frac{x}{2} - \frac{\pi}{6}\right)$.

From the values $a = 2$, $b = \frac{1}{2}$, and $c = -\frac{\pi}{6}$, we determine that the amplitude is 2, the period is $2\pi \div \frac{1}{2} = 4\pi$, and the displacement is $\frac{\pi}{6} \div \frac{1}{2} = \frac{\pi}{3}$ to the right. From these values we construct this table of values.

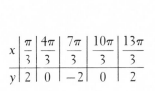

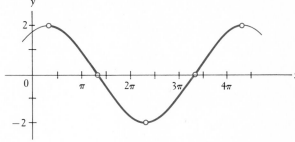

Figure 9–14

The graph is shown in Fig. 9–14. We note that when the coefficient of x is less than 1, the period is greater than 2π.

Example E

Sketch the graph of the function $y = 0.7 \sin (\pi x + \frac{\pi}{4})$.

From the values $a = 0.7$, $b = \pi$, and $c = \frac{\pi}{4}$, we can determine that the amplitude is 0.7, the period is $\frac{2\pi}{\pi} = 2$, and the displacement is $\left(\frac{\pi}{4}\right) \div \pi = \frac{1}{4}$ to the left. From these values we construct the following table of values.

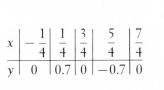

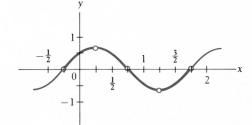

Figure 9–15

Since π is not used in the values of x, it is more convenient to use decimal number units for the graph (Fig. 9–15).

Each of the heavy portions of the graphs in Fig. 9–12, 9–13, 9–14, and 9–15 is called a **cycle** of the curve. *A cycle is the shortest section of the graph which includes one period.*

Exercises 9–3

In Exercises 1 through 24 determine the amplitude, period, and displacement for each of the functions. Then sketch the graphs of the functions.

1. $y = \sin\left(x - \dfrac{\pi}{6}\right)$

2. $y = 3 \sin\left(x + \dfrac{\pi}{4}\right)$

3. $y = \cos\left(x + \dfrac{\pi}{6}\right)$

4. $y = 2\,\cos\left(x - \dfrac{\pi}{8}\right)$

5. $y = 2\,\sin\left(2x + \dfrac{\pi}{2}\right)$

6. $y = -\sin\left(3x - \dfrac{\pi}{2}\right)$

7. $y = -\cos\,(2x - \pi)$

8. $y = 4\,\cos\left(3x + \dfrac{\pi}{3}\right)$

9. $y = \dfrac{1}{2}\,\sin\left(\dfrac{1}{2}x - \dfrac{\pi}{4}\right)$

10. $y = 2\,\sin\left(\dfrac{1}{4}x + \dfrac{\pi}{2}\right)$

11. $y = 3\,\cos\left(\dfrac{1}{3}x + \dfrac{\pi}{3}\right)$

12. $y = \dfrac{1}{3}\,\cos\left(\dfrac{1}{2}x - \dfrac{\pi}{8}\right)$

13. $y = \sin\left(\pi x + \dfrac{\pi}{8}\right)$

14. $y = -2\,\sin\,(2\pi x - \pi)$

15. $y = \dfrac{3}{4}\,\cos\left(4\pi x - \dfrac{\pi}{5}\right)$

16. $y = 6\,\cos\left(3\pi x + \dfrac{\pi}{2}\right)$

17. $y = -0.6\,\sin\,(2\pi x - 1)$

18. $y = 1.8\,\sin\left(\pi x + \dfrac{1}{3}\right)$

19. $y = 4\,\cos\,(3\pi x + 2)$

20. $y = 3\,\cos\,(6\pi x - 1)$

21. $y = \sin\,(\pi^2 x - \pi)$

22. $y = -\dfrac{1}{2}\sin\left(2x - \dfrac{1}{\pi}\right)$

23. $y = -\dfrac{3}{2}\cos\left(\pi x + \dfrac{\pi^2}{6}\right)$

24. $y = \pi\,\cos\left(\dfrac{1}{\pi}x + \dfrac{1}{3}\right)$

In Exercises 25 through 28 sketch the indicated curves.

25. A wave traveling in a string may be represented by the equation

$$y = A\,\sin\left(\dfrac{t}{T} - \dfrac{x}{\lambda}\right)$$

Here A is the amplitude, t is the time the wave has traveled, x is the distance from the origin, T is the time required for the wave to travel one *wavelength* λ (the Greek lambda). Sketch three cycles of the wave for which $A = 2.00$ cm, $T = 0.100$ s, $\lambda = 20.0$ cm, and $x = 5.00$ cm.

26. The cross-section of a particular water wave is

$$y = 0.5\,\sin\left(\dfrac{\pi}{2}x + \dfrac{\pi}{4}\right)$$

where x and y are measured in feet. Sketch two cycles of y vs. x.

27. A particular electromagnetic wave is described by the equation

$$y = a\,\cos\left(8\pi \times 10^{14}t + \dfrac{\pi}{6}\right)$$

Sketch two cycles of the graph of y (in centimeters) vs. t (in seconds).

28. The voltage in a certain alternating-current circuit is given by

$$y = 120 \cos\left(120\pi t + \frac{\pi}{6}\right)$$

where t represents the time (in seconds). Sketch three cycles of the curve.

9–4 Graphs of $y = \tan x$, $y = \cot x$, $y = \sec x$, $y = \csc x$

In this section we shall briefly consider the graphs of the other trigono-metric functions. We shall establish the basic form of each curve, and from these we shall be able to sketch other curves for these functions.

Considering the values and signs of the trigonometric functions as established in Chapter 7, we set up the following table for the function $y = \tan x$. The graph is shown in Fig. 9–16.

x	0	$\dfrac{\pi}{6}$	$\dfrac{\pi}{3}$	$\dfrac{\pi}{2}$	$\dfrac{2\pi}{3}$	$\dfrac{5\pi}{6}$	π	$\dfrac{7\pi}{6}$	$\dfrac{4\pi}{3}$	$\dfrac{3\pi}{2}$	$\dfrac{5\pi}{3}$	$\dfrac{11\pi}{6}$	2π
y	0	0.6	1.7	*	-1.7	-0.6	0	0.6	1.7	*	-1.7	-0.6	0

*Undefined.

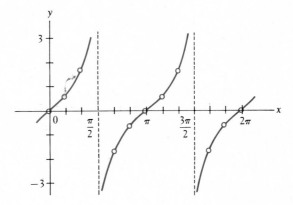

Figure 9–16

Since the curve is not defined for $x = \frac{\pi}{2}$, $x = \frac{3\pi}{2}$, and so forth, we look at the table and note that the value of $\tan x$ becomes very large when x approaches the value $\frac{\pi}{2}$. We must keep in mind, however, that there is no point on the curve corresponding to $x = \frac{\pi}{2}$. We note that the period of the tangent curve is π. This differs from the period of the sine and cosine functions.

By following the same procedure, we can set up tables for the graphs of the other functions. We present in Figures 9–17 through 9–20 the graphs of $y = \tan x$, $y = \cot x$, $y = \sec x$, and $y = \csc x$ (the graph of $y = \tan x$ is shown again to illustrate it more completely). The dashed lines in these figures are **asymptotes** (see Sections 2–3 and 20–6).

To sketch functions such as $y = a \sec x$, we first sketch $y = \sec x$, and then multiply each y-value by a. Here a is not an amplitude, since these functions are not limited in the values they take on, as are the sine and cosine functions.

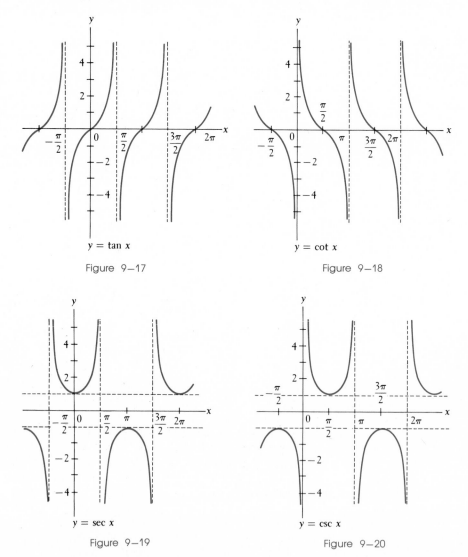

$y = \tan x$

Figure 9–17

$y = \cot x$

Figure 9–18

$y = \sec x$

Figure 9–19

$y = \csc x$

Figure 9–20

Example A

Sketch the graph of $y = 2 \sec x$.

First we sketch in $y = \sec x$, shown as the light curve in Fig. 9–21. Now we multiply the y-values of the secant function by 2 (approximately, of course). In this way we obtain the desired curve, shown as the heavy curve in Fig. 9–21.

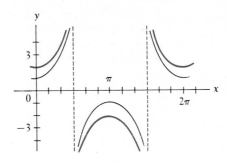

Figure 9–21

Example B
Sketch the graph of $y = -\frac{1}{2} \cot x$.

We sketch in $y = \cot x$, shown as the light curve in Fig. 9–22. Now we multiply each *y*-value by $-\frac{1}{2}$. The effect of the negative sign is to invert the curve. The resulting curve is shown as the heavy curve in Fig. 9–22.

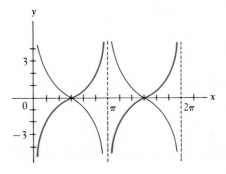

Figure 9–22

By knowing the graphs of the sine, cosine, and tangent functions, it is possible to graph the other three functions. Remembering the definitions of the trigonometric functions (Eqs. 7–1), we see that $\sin x$ and $\csc x$ are reciprocals, $\cos x$ and $\sec x$ are reciprocals, and $\tan x$ and $\cot x$ are reciprocals. That is,

$$\csc x = \frac{1}{\sin x}, \qquad \sec x = \frac{1}{\cos x}, \qquad \cot x = \frac{1}{\tan x} \qquad (9\text{–}1)$$

Thus, to sketch $y = \cot x$, $y = \sec x$, or $y = \csc x$, we sketch the corresponding reciprocal function, and from this graph determine the necessary values.

Example C

Sketch the graph of $y = \csc x$.

We first sketch in the graph of $y = \sin x$ (light curve). Where $\sin x$ is 1, $\csc x$ will also be 1, since $1/1 = 1$. Where $\sin x$ is 0, $\csc x$ is unde-fined, since $1/0$ is undefined. Where $\sin x$ is 0.5, $\csc x$ is 2, since $1/0.5 = 2$. Thus, as $\sin x$ becomes larger, $\csc x$ becomes smaller, and as $\sin x$ becomes smaller, $\csc x$ becomes larger. The two functions always have the same sign. We sketch the graph of $y = \csc x$ from this information, as shown by the heavy curve in Fig. 9–23.

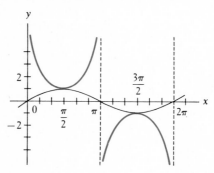

Figure 9–23

Exercises 9 – 4

In Exercises 1 through 4 fill in the following table for each function and then plot the curve from these points.

x	$-\frac{\pi}{2}$	$-\frac{\pi}{3}$	$-\frac{\pi}{4}$	$-\frac{\pi}{6}$	0	$\frac{\pi}{6}$	$\frac{\pi}{4}$	$\frac{\pi}{3}$	$\frac{\pi}{2}$	$\frac{2\pi}{3}$	$\frac{3\pi}{4}$	$\frac{5\pi}{6}$	π
y													

1. $y = \tan x$ 2. $y = \cot x$ 3. $y = \sec x$ 4. $y = \csc x$

In Exercises 5 through 12 sketch the curves of the given functions by use of the basic curve forms (Figs. 9–17, 9–18, 9–19, 9–20). See Examples A and B.

5. $y = 2 \tan x$ 6. $y = 3 \cot x$ 7. $y = \frac{1}{2}\sec x$

8. $y = \frac{3}{2}\csc x$ 9. $y = -2 \cot x$ 10. $y = -\tan x$

11. $y = -3 \csc x$ 12. $y = -\frac{1}{2} \sec x$

In Exercises 13 through 20 plot the graphs by first making an appropriate table for $0 \le x \le \pi$.

13. $y = \tan 2x$ 14. $y = 2 \cot 3x$ 15. $y = \frac{1}{2}\sec 3x$

16. $y = \csc 2x$ 17. $y = 2 \cot\left(2x + \frac{\pi}{6}\right)$ 18. $y = \tan\left(3x - \frac{\pi}{2}\right)$

19. $y = \csc\left(3x - \frac{\pi}{3}\right)$ 20. $y = 3 \sec\left(2x + \frac{\pi}{4}\right)$

In Exercises 21 through 24 sketch the given curves by first sketching the appropriate reciprocal function. See Example C.

21. $y = \sec x$ 22. $y = \cot x$ 23. $y = \csc 2x$ 24. $y = \sec \pi x$

In Exercises 25 through 28 construct the appropriate graphs.

25. For an object sliding down an inclined plane at constant speed, the coefficient of friction μ (the Greek mu) between the object and the plane is given by $\mu = \tan \theta$, where θ is the angle between the plane and the horizontal. Sketch a graph of the coefficient of friction vs. the angle of inclination of the plane for $0 \le \theta \le 60°$.

26. The tension T at any point in a cable supporting a distributed load is given by $T = T_0 \sec \theta$, where T_0 is the tension where the cable is horizontal, and θ is the angle between the cable and the horizontal at any point. Sketch a graph of T vs. θ for a cable for which $T_0 = 200$ lb.

27. From a point x meters from the base of a building 200 m high, the angle of elevation θ of the top of the building can be found from the equation $x = 200 \cot \theta$. Plot x as a function of θ.

28. An expression relating the initial velocity v_0 of a projectile, the time t of its flight, and the angle θ above the horizontal at which it is fired is $v_0 = gt \csc \theta$, where g is the acceleration due to gravity. Plot v_0 (in feet per second) for $g = 32.0$ ft/s² and $t = 10.0$ s.

9—5 Applications of the Trigonometric Graphs

There are a great many applications of the trigonometric functions and their graphs, a few of which have been indicated in the exercises of the previous sections. In this section we shall introduce an important physical concept and indicate some of the technical applications.

In Section 7—4, we discussed the velocity of an object moving in a circular path. The movement of the **projection** on a diameter of a particle revolving about a circle with constant velocity is known as **simple harmonic motion**. For example, this could be the motion of the shadow of an object which is moving around a circle. Another example would be the vertical (or horizontal) position of the end of a spoke of a wheel in motion.

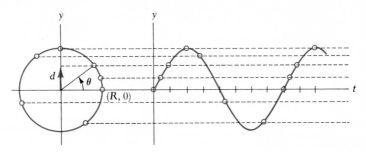

Figure 9—24

Example A
When we consider Fig. 9—24, let us assume that motion starts with the end of the radius at $(R, 0)$ and that it is moving with constant angular velocity ω. This means that the length of the projection of the radius

along the y-axis is given by $d = R \sin \theta$. The length of this projection is shown for a few different positions of the end of the radius. Since $\theta/t = \omega$, or $\theta = \omega t$, we have

$$d = R \sin \omega t \qquad\qquad (9-2)$$

as the equation for the length of the projection, with time as the independent variable. Normally, the position as a function of time is the important relationship.

For the case where $R = 10.0$ in. and $\omega = 4.00$ rad/s, we have

$$d = 10.0 \sin 4.00t$$

By sketching the graph of this function, we can readily determine the length of the projection d for a given time t. The graph is shown in Fig. 9–25.

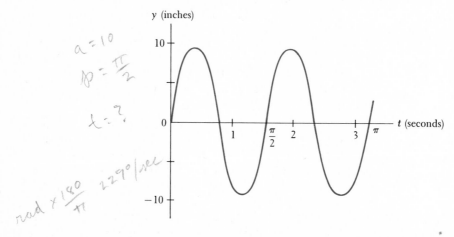

Figure 9–25

Example B

If the end of the radius is at $(\frac{R}{\sqrt{2}}, \frac{R}{\sqrt{2}})$, where $\theta = \frac{\pi}{4}$, when $t = 0$, we can express the projection d as a function of the time as

$$d = R \sin\left(\omega t + \frac{\pi}{4}\right)$$

If the end of the radius is at $(0, R)$, where $\theta = \frac{\pi}{2}$, when $t = 0$, we can express the projection d as a function of the time as

$$d = R \sin\left(\omega t + \frac{\pi}{2}\right) \quad \text{or} \quad d = R \cos \omega t$$

This can be seen from Fig. 9–24. If the motion started at the first maximum of the indicated curve, the resulting curve would be that of the cosine function.

Other examples of simple harmonic motion are (1) the movement of a pendulum bob through its arc (a very close approximation to simple harmonic motion), (2) the motion of an object on the end of a spring, (3) the motion of an object "bobbing" in the water, and (4) the movement of the end of a vibrating rod (which we hear as sound). Other phenomena which give rise to equations just like those for simple harmonic motion are found in the fields of optics, sound, and electricity. Such phenomena have the same mathematical form because they result from vibratory motion or motion in a circle.

Example C
A very important use of the trigonometric curves arises in the study of alternating current, which is caused by the motion of a wire passing through a magnetic field. If this wire is moving in a circular path, with angular velocity ω, the current i in the wire at time t is given by an equation of the form

$$i = I_m \sin (\omega t + \alpha) \qquad\qquad (9\text{-}3)$$

where I_m is the maximum current attainable and α is the phase angle. The current may be represented by a sine wave, as in the following example.

Example D
In Example C, given that $I_m = 6.00$ A, $\omega = 120\pi$ rad/s (60 Hz), and $\alpha = \pi/6$, we have the equation

$$i = 6.00 \sin\left(120\pi t + \frac{\pi}{6}\right)$$

From this equation we see that the amplitude is 6.00, the period is $\frac{1}{60}$ s, and the displacement is $\frac{1}{720}$ s to the left. From these values we draw the graph as shown in Fig. 9–26. Since the current takes on both positive and negative values, we conclude that it moves in opposite directions.

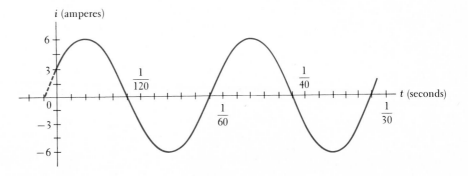

Figure 9–26

Exercises 9-5

In Exercises 1 and 2 draw two cycles of the curve of the projection of Example A as a function of time for the given values.

1. $R = 4.00$ in., $\omega = 1.00$ rad/s   2. $R = 8.00$ cm, $\omega = 0.500$ Hz

In Exercises 3 and 4, for the projection described in Example A, assume that the end of the radius starts at $(0, R)$. Draw two cycles of the projection as a function of time for the given values. (See Example B.)

3. $R = 2.00$ ft, $\omega = 1.00$ Hz   4. $R = 2.50$ in., $\omega = 0.300$ rad/s

In Exercises 5 and 6, for the projection described in Example A, assume that the end of the radius starts at the indicated point. Draw two cycles of the projection as a function of time for the given values. (See Example B.)

5. $R = 6.00$ cm, $\omega = 2.00$ Hz, starting point $(\frac{\sqrt{3}R}{2}, \frac{R}{2})$
6. $R = 3.20$ ft., $\omega = 0.200$ rad/s, starting point $(\frac{R}{2}, \frac{\sqrt{3}R}{2})$

In Exercises 7 and 8, for the alternating-current discussed in Example C, draw two cycles of the current as a function of time for the given values.

7. $I_m = 2.00$ A, $\omega = 60.0$ Hz, $\alpha = 0$
8. $I_m = 0.600$ A, $\omega = 100$ rad/s, $\alpha = \pi/4$

In Exercises 9 and 10, for an alternating-current circuit in which the voltage is given by

$$e = E \cos(\omega t + \alpha)$$

draw two cycles of the voltage as a function of time for the given values.

9. $E = 170$ V, $\omega = 50.0$ rad/s, $\alpha = 0$
10. $E = 110$ V, $\omega = 60.0$ Hz, $\alpha = -\pi/3$

In Exercises 11 through 16 draw the required curves.

11. Angular displacement θ of a pendulum bob is given in terms of its initial $(t = 0)$ displacement θ_0 by the equation $\theta = \theta_0 \cos \omega t$. If $\omega = 2.00$ rad/s and $\theta_0 = \pi/30$ rad, draw two cycles for the resulting equation.

12. Displacement of the end of a vibrating rod is given by $y = 1.50 \cos 200\pi t$. Sketch two cycles of y (in centimeters) vs. t (in seconds).

13. An object of weight w and cross-sectional area A is depressed a distance x_0 from its equilibrium position when in a liquid of density d and then released; its displacement as a function of time is given by

$$x = x_0 \cos\sqrt{\frac{dgA}{w}}t$$

where g (= 32.0 ft/s²) is the acceleration due to gravity. If a 4.00-lb object with a cross-sectional area of 2.00 ft² is depressed 3.00 ft in water (let $d = 62.4$ lb/ft³), find the equation which expresses the displacement as a function of time. Draw two cycles of the curve.

14. A wave is traveling in a string. The displacement, as a function of time, from its equilibrium position, is given by $y = A \cos(2\pi/T)t$. T is the period (measured in seconds) of the motion. If $A = 0.200$ in. and $T = 0.100$ s, draw two cycles of the displacement as a function of time.

15. The displacement, as a function of time, from the position of equilibrium, of an object on the end of a spring is given by $x = A \cos(\omega t + \alpha)$. Draw two cycles of the curve for displacement as a function of time for $A = 2.00$ in., $\omega = 0.500$ Hz, and $\alpha = \pi/6$.

16. The angular displacement θ of a pendulum bob (see Exercise 11) is given by $\theta = \theta_0 \sin(\omega t + \frac{\pi}{6})$. If $\theta_0 = 0.100$ rad and $\omega = \frac{\pi}{2}$ rad/s, sketch two cycles of the graph of θ vs. t.

9–6 Composite Trigonometric Curves

Many applications of trigonometric functions involve the combination of two or more functions. In this section we shall discuss two important methods in which trigonometric curves can be combined.

If we wish to find the curve of a function which itself is the sum of two other functions, we may find the resulting graph by first sketching the two individual functions, and then adding the y-values graphically. This method is called **addition of ordinates**, and is illustrated in the following examples.

Example A
Sketch the graph of $y = 2 \cos x + \sin 2x$.

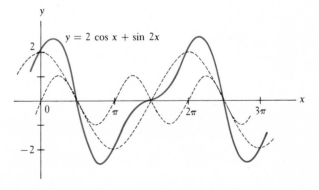

Figure 9–27

On the same set of coordinate axes we sketch the curves $y = 2 \cos x$ and $y = \sin 2x$. These are shown as dashed curves in Fig. 9–27. For various values of x, we determine the distance above or below the x-axis of each curve and add these distances, noting that those above the axis are positive and those below the axis are negative. We thereby graphically *add* the y-values of these two curves for these values of x to obtain the points on the resulting curve shown as a heavy curve in Fig. 9–27. We add the y-values for a sufficient number of x-values to obtain the proper representation. Some points are easily found. Where one curve crosses the x-axis, its y-value is zero, and therefore the resulting curve has its point on the other curve for this value of x. In this example, $\sin 2x$ is zero at $x = 0$, $\frac{\pi}{2}$, π, and so forth. We see that points on the resulting curve lie on the curve of $2 \cos x$. We should also add the values where each curve is at its amplitude values. In this case, $\sin 2x$ equals 1 at $\frac{\pi}{4}$, and the two y-values should be added together here to get a point on the resulting curve. At $x = \frac{5\pi}{4}$, we must take care in adding the values, since $\sin 2x$ is positive and $2 \cos x$ is negative. Reasonable care and accuracy are necessary to obtain a proper resulting curve.

Example B

Sketch the graph of $y = \dfrac{x}{2} - \cos x$.

The method of addition of ordinates is applicable regardless of the kinds of functions being added. Here we note that $y = \frac{x}{2}$ is a straight line, and that it is to be combined with a trigonometric curve.

We could graph the functions $y = \frac{x}{2}$ and $y = \cos x$ and then subtract the ordinates of $y = \cos x$ from the ordinates of $y = \frac{x}{2}$. But it is easier, and far less confusing, to add values, so we shall sketch $y = \frac{x}{2}$ and $y = -\cos x$ and add the ordinates to obtain points on the resulting curve. These graphs are shown as dashed curves in Fig. 9–28. The important points on the resulting curve are obtained by using the values of x corresponding to the zeros and amplitude values of $y = -\cos x$.

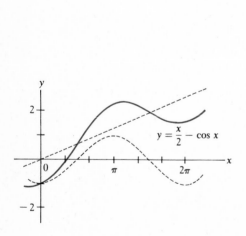

Figure 9–28

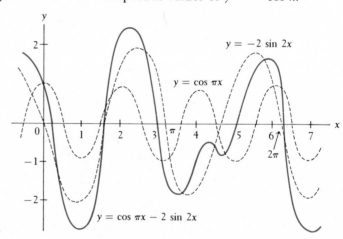

Figure 9–29

Example C

Sketch the graph of $y = \cos \pi x - 2 \sin 2x$.

The curves of $y = \cos \pi x$ and $y = -2 \sin 2x$ are shown as dashed curves in Fig. 9–29. Points for the resulting curve are found primarily at the x-values where each of the curves has its zero or amplitude values. Again, special care should be taken where one of the curves is negative and the other is positive.

Another important application of trigonometric curves is made when they are added at *right angles*. This can be accomplished in practice by applying different voltages to an oscilloscope. Let us consider the following examples.

Example D

Plot the graph for which the values of x and y are given by the equations $y = \sin 2\pi t$ and $x = 2 \cos \pi t$. (Equations given in this form, x and y in terms of a third variable, are called **parametric equations**.)

Since both x and y are given in terms of t, by assuming values for t we may find corresponding values of x and y, and use these values to plot the resulting points (see Fig. 9–30).

t	0	$\frac{1}{4}$	$\frac{1}{2}$	$\frac{3}{4}$	1	$\frac{5}{4}$	$\frac{3}{2}$	$\frac{7}{4}$	2	$\frac{9}{4}$
x	2	1.4	0	-1.4	-2	-1.4	0	1.4	2	1.4
y	0	1	0	-1	0	1	0	-1	0	1
Point number	1	2	3	4	5	6	7	8	9	10

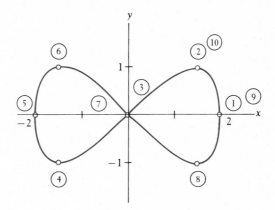

Figure 9–30

Since x and y are trigonometric functions of a third variable t, and since the x- and y-axes are at right angles, values of x and y obtained in this manner result in a combination of two trigonometric curves at right angles. Figures obtained in this manner are called **Lissajous figures**.

Example E

If we place a circle on the x-axis and another on the y-axis, we may represent the coordinates (x, y) for the curve of Example D by the lengths of the projections (see Example A of Section 9–5) of a point moving around each circle. A careful study of Fig. 9–31 will clarify this. We note that the radius of the circle giving the x-values is 2, whereas the radius of the other is 1. This is due to the manner in which x and y are defined. Also due to the definitions, the point revolves around the y-circle twice as fast as the corresponding point revolves around the x-circle.

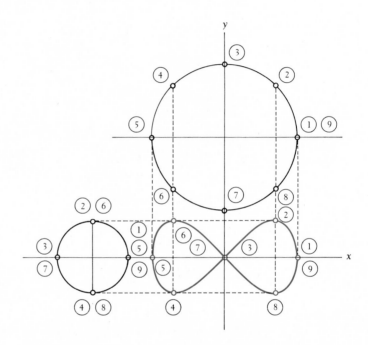

Figure 9–31

Example F

Plot the Lissajous figure for which the x- and y-values are given by the equation $x = 2 \sin 3t$ and $y = 3 \sin (t + \frac{\pi}{3})$.

Since values of t which are multiples of π give convenient values of x and y, the table is constructed with these values of t. Figure 9–32 shows the graph.

t	x	y	Point number
0	0	2.6	1
$\frac{\pi}{6}$	2	3	2
$\frac{\pi}{3}$	0	2.6	3
$\frac{\pi}{2}$	-2	1.5	4
$\frac{2\pi}{3}$	0	0	5
$\frac{5\pi}{6}$	2	-1.5	6
π	0	-2.6	7
$\frac{7\pi}{6}$	-2	-3	8
$\frac{4\pi}{3}$	0	-2.6	9
$\frac{3\pi}{2}$	2	-1.5	10
$\frac{5\pi}{3}$	0	0	11
$\frac{11\pi}{6}$	-2	1.5	12
2π	0	2.6	13

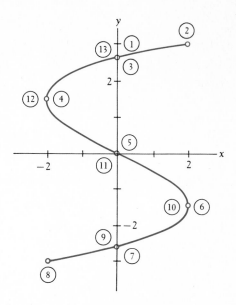

Figure 9-32

Exercises 9-6

In Exercises 1 through 20 use the method of addition of ordinates to sketch the given curves.

1. $y = x + \sin x$

2. $y = x + 2 \cos x$

3. $y = \frac{1}{3}x - \cos x$

4. $y = \frac{1}{4}x - \sin x$

5. $y = \frac{1}{10}x^2 + \sin 2x$

6. $y = \frac{1}{3}x^2 + \cos 3x$

7. $y = \dfrac{1}{x^2 + 1} - \sin \pi x$

8. $y = \frac{1}{5}x^3 - \cos \pi x$

9. $y = \sin x + \cos x$

10. $y = \sin x + \sin 2x$

11. $y = \sin x - \sin 2x$

12. $y = \cos 3x - \sin x$

13. $y = 2 \cos 2x + 3 \sin x$

14. $y = \frac{1}{2}\sin 4x + \cos 2x$

15. $y = 2 \sin x - \cos x$

16. $y = \sin \dfrac{x}{2} - \sin x$

17. $y = 2 \cos 4x - \cos\left(x - \dfrac{\pi}{4}\right)$

18. $y = \sin \pi x - \cos 2x$

19. $y = 2 \sin\left(2x - \dfrac{\pi}{6}\right) + \cos\left(2x + \dfrac{\pi}{3}\right)$

20. $y = 3 \cos 2\pi x + \sin \dfrac{\pi}{2}x$

In Exercises 21 through 28 plot the Lissajous figures.

21. $y = \sin t, \ x = \sin t$

22. $y = 2 \cos t, \ x = \cos t$

23. $y = \sin \pi t, \ x = \cos \pi t$

24. $y = \sin 2t, \ x = \cos\left(t + \dfrac{\pi}{4}\right)$

25. $y = 2 \sin \pi t, x = \cos \pi \left(t + \dfrac{1}{6} \right)$ 26. $y = \sin^2 \pi t, x = \cos 2\pi t$

27. $y = \cos 2t, x = 2 \cos 3t$ 28. $y = 3 \sin 3\pi t, x = 2 \sin \pi t$

In Exercises 29 through 34 sketch the appropriate figures.

29. The current in a certain electric circuit is given by the formula $i = 4 \sin 60\pi t + 2 \cos 120\pi t$. Sketch the curve representing the current (in amperes) as a function of time (in seconds).

30. In optics, two waves are said to interfere destructively if, when they pass through the same medium, the amplitude of the resulting wave is zero. Sketch the curve of $y = \sin x + \cos \left(x + \frac{\pi}{2} \right)$, and determine whether ðr not it would represent destructive interference of two waves.

31. An object oscillating on a spring, under specific conditions, has a displacement given by $y = 0.4 \sin 4t + 0.3 \cos 4t$. Plot y (in feet) versus t (in seconds).

32. The resultant voltage in a certain electric circuit is given by the formula $e = 50 \sin 50\pi t + 80 \sin 60\pi t$. Sketch the curve representing voltage as a function of time.

33. Two signals are being sent to an oscilloscope, and are seen on the oscilloscope as being at right angles. The equations governing the displacement of these signals are $x = 2 \cos 120\pi t$ and $y = 3 \cos 120\pi t$, respectively. Sketch the figure which would appear on the oscilloscope.

34. In the study of optics, light is said to be elliptically polarized if certain optic vibrations are out of phase. These may be represented by Lissajous figures. Determine the Lissajous figure for two waves of light given by $w_1 = \sin \omega t$, $w_2 = \sin \left(\omega t + \frac{\pi}{4} \right)$.

9–7 Exercises for Chapter 9

In Exercises 1 through 24 sketch the curves of the given trigonometric functions.

1. $y = \frac{2}{3} \sin x$ 2. $y = -4 \sin x$ 3. $y = -2 \cos x$

4. $y = 2.3 \cos x$ 5. $y = 2 \sin 3x$ 6. $y = 3 \sin \frac{1}{2} x$

7. $y = 2 \cos 2x$ 8. $y = 4 \cos 6x$ 9. $y = \sin \pi x$

10. $y = 3 \sin 4\pi x$ 11. $y = 5 \cos 2\pi x$ 12. $y = -\cos 3\pi x$

13. $y = 2 \sin \left(3x - \dfrac{\pi}{2} \right)$ 14. $y = 3 \sin \left(\dfrac{x}{2} + \dfrac{\pi}{2} \right)$

15. $y = -2 \cos (4x + \pi)$ 16. $y = 0.8 \cos \left(\dfrac{x}{6} - \dfrac{\pi}{2} \right)$

17. $y = -\sin \left(\pi x + \dfrac{\pi}{6} \right)$ 18. $y = 2 \sin (3\pi x - \pi)$

19. $y = 8 \cos \left(4\pi x - \dfrac{\pi}{2} \right)$ 20. $y = 3 \cos (2\pi x + \pi)$

21. $y = 3 \tan x$

22. $y = \frac{1}{4} \sec x$

23. $y = -\frac{1}{3} \csc x$

24. $y = -5 \cot x$

In Exercises 25 through 32 sketch the given curves by the method of addition of ordinates.

25. $y = \frac{1}{2} \sin 2x - x$

26. $y = \frac{1}{2}x - \cos \frac{1}{3}x$

27. $y = \sin 2x + 3 \cos x$

28. $y = \sin 3x + 2 \cos 2x$

29. $y = 2 \sin x - \cos 2x$

30. $y = \sin 3x - 2 \cos x$

31. $y = \cos\left(x + \frac{\pi}{4}\right) - 2 \sin 2x$

32. $y = 2 \cos \pi x + \cos (2\pi x - \pi)$

In Exercises 33 through 36 plot the Lissajous figures.

33. $y = 2 \sin \pi t, \; x = -\cos 2\pi t$

34. $y = \sin t, \; x = \sin\left(t + \frac{\pi}{6}\right)$

35. $y = \cos \pi t, \; x = \cos\left(2\pi t + \frac{\pi}{4}\right)$

36. $y = \cos\left(2t + \frac{\pi}{3}\right), \; x = \cos\left(t - \frac{\pi}{6}\right)$

In Exercises 37 through 48 sketch the appropriate figures.

37. A simple pendulum is started by giving it a velocity from its equilibrium position. The angle θ between the vertical and the pendulum is given by $\theta = a \sin (\sqrt{\frac{g}{l}}t)$, where a is the amplitude (in radians), g ($= 32.0$ ft/s²) is the acceleration due to gravity, l is the length of the pendulum (in feet), and t is the length of time of the motion. Sketch two cycles of θ as a function of t for the pendulum whose length is 2.00 ft and $a = 0.100$ rad.

38. The electric current in a certain circuit is given by $i = i_0 \sin (t/\sqrt{LC})$, where i_0 is the initial current in the circuit, L is an inductance, and C is a capacitance. Sketch two cycles of i as a function of t (in seconds) for the case where $i_0 = 0.500$ A, $L = 1.00$ H, and $C = 100 \; \mu$F.

39. A certain object is oscillating at the end of a spring. The displacement as a function of time is given by the relation $y = 0.200 \cos (8t + \frac{\pi}{3})$, where y is measured in meters and t in seconds. Plot the graph of y vs. t.

40. A circular disk suspended by a thin wire attached to the center at one of its flat faces is twisted through an angle θ. Torsion in the wire tends to turn the disk back in the opposite direction (thus the name "torsion pendulum" is given to this device). The angular displacement as a function of time is given by $\theta = \theta_0 \cos (\omega t + \alpha)$, where θ_0 is the maximum angular displacement, ω is a constant which depends on the properties of the disk and wire, and α is the phase angle. Plot the graph of θ vs. t if $\theta_0 = 0.100$ rad, $\omega = 2.50$ rad/s and $\alpha = \pi/4$.

41. The charge q on a certain capacitor as a function of time is given by $q = 0.001(1 - \cos 100t)$. Sketch two cycles of q as a function of t. Charge is measured in coulombs, and time is measured in seconds.

42. If the upper end of a spring is not fixed and is being moved with a sinusoidal motion, the motion of the bob at the end of the spring is affected. Plot the curve if the motion of the upper end of a spring is being moved by an external force and the bob moves according to the equation $y = 4 \sin 2t - 2 \cos 2t$.

43. Under certain conditions, the path of a certain moving particle is given by $x = 2 \cos 3\pi t$ and $y = \sin 3\pi t$. Plot the path of the object.

44. Two signals are applied to an oscilloscope. The equations governing the displacement of these signals are $x = 2 \cos 40\pi t$ and $y = \sin 120\pi t$. Sketch the figure which appears on the oscilloscope.

45. The height of a certain rocket ascending vertically is given by the formula $h = 800 \tan \theta$, where θ is the angle of elevation from an observer 800 m from the launch pad. Plot h (in meters) vs. θ.

46. An equation relating a force F and its x-component F_x is $F = F_x \sec \theta$, where θ is the angle between F and F_x. Plot F vs. θ for $F_x = 500$ lb.

47. The area of a rectangle as a function of a diagonal is $A = d^2 \sin \theta \cos \theta$, where θ is the angle between the diagonal and one of the sides. Plot A vs. θ for $d = 10.0$ m.

48. The instantaneous power in an electric circuit is defined as the product of the instantaneous voltage e and the instantaneous current i. If we have $e = 100 \cos 200t$ and $i = 2 \cos (200t + \frac{\pi}{4})$, plot the graph of the voltage and the graph of the current (in amperes), on the same coordinate system, vs. the time (in seconds). Then sketch the power (in watts) vs. time by multiplying appropriate values of e and i.

10

Exponents and Radicals

10−1 Integral Exponents

In Chapter 1 we introduced exponents and radicals. To this point only a basic understanding of the meaning and elementary operations with them has been necessary. However, in our future work a more detailed understanding of exponents and radicals, and operations on them, will be required. Therefore, in this chapter we shall develop the necessary operations.

The laws of exponents were given in Section 1–4. We now write them again for reference.

$$a^m \cdot a^n = a^{m+n} \tag{10–1}$$

$$\frac{a^m}{a^n} = a^{m-n} \quad \text{or} \quad \frac{a^m}{a^n} = \frac{1}{a^{n-m}}, \qquad a \neq 0 \tag{10–2}$$

$$(a^m)^n = a^{mn} \tag{10–3}$$

$$(ab)^n = a^n b^n, \qquad \left(\frac{a}{b}\right)^n = \frac{a^n}{b^n}, \qquad b \neq 0 \tag{10–4}$$

$$a^0 = 1, \qquad a \neq 0 \tag{10–5}$$

$$a^{-n} = \frac{1}{a^n}, \qquad a \neq 0 \tag{10–6}$$

Although Eqs. (10–1) through (10–4) were originally defined for positive integers as exponents, we showed in Section 1–4 that with the definitions given in Eqs. (10–5) and (10–6) they are valid for all integral exponents. Later in this chapter we shall show how fractions may be used as exponents. Since the equations above are very important to the development of the topics in this chapter, they should again be reviewed—and learned thoroughly.

In this chapter we review the use of exponents in using Eqs. (10–1) through (10–6). Then we show how they are used and handled in somewhat more involved expressions.

Example A

Applying Eq. (10–1), we have

$$a^5 \cdot a^{-3} = a^{5+(-3)} = a^{5-3} = a^2$$

Applying Eq. (10–1) and then Eq. (10–6), we have

$$a^3 \cdot a^{-5} = a^{3-5} = a^{-2} = \frac{1}{a^2}$$

We note that a final result is usually expressed with positive exponents, unless specified otherwise.

Example B

Applying Eqs. (10–2) and (10–5), we have

$$\frac{a^2 b^3 c^0}{a b^7} = \frac{a^{2-1}(1)}{b^{7-3}} = \frac{a}{b^4}$$

Applying Eqs. (10–4) and (10–3), we have

$$(x^{-2}y)^3 = (x^{-2})^3 (y^3) = x^{-6}y^3 = \frac{y^3}{x^6}$$

Here, the simplification was completed by the use of Eq. (10–6).

There are many occasions in using Eqs. (10–1) through (10–6) when more than one sequence of steps may be used in the simplification process. Consider the illustration in the following example.

Example C

$$\left(\frac{4}{a^2}\right)^{-3} = \frac{1}{\left(\dfrac{4}{a^2}\right)^3} = \frac{1}{\dfrac{4^3}{a^6}} = \frac{a^6}{4^3}$$

or

$$\left(\frac{4}{a^2}\right)^{-3} = \frac{4^{-3}}{a^{-6}} = \frac{a^6}{4^3}$$

In the first method we used Eq. (10–6) first, then Eq. (10–4), and finally, we inverted the divisor. In the second method we first used Eq. (10–4) and then Eq. (10–6).

In simplifying expressions, care must be taken to apply the laws of exponents properly. Certain relatively common problems are pointed out in the following two examples.

Example D

The expression $(-5x)^0$ equals 1, whereas the expression $-5x^0$ equals -5. For $(-5x)^0$ the parentheses indicate that the expression $-5x$ is raised to the zero power, whereas for $-5x^0$ only x is raised to the zero power and we have

$$-5x^0 = -5(1) = -5$$

Also, $(-5)^0 = 1$, whereas $-5^0 = -1$. Again note the use of parentheses. For $(-5)^0$ it is -5 which is raised to the zero power, whereas for -5^0 only 5 is raised to the zero power.

Example E

Simplify $(2a + b^{-1})^{-2}$.

In simplifying an expression we should give the result with positive exponents. For this expression we may use various orders of operations. One is as follows:

$$(2a + b^{-1})^{-2} = \frac{1}{(2a + b^{-1})^2} = \frac{1}{\left(2a + \dfrac{1}{b}\right)^2} = \frac{1}{\left(\dfrac{2ab + 1}{b}\right)^2}$$

$$= \frac{1}{\dfrac{(2ab + 1)^2}{b^2}} = \frac{b^2}{(2ab + 1)^2}$$

We may leave the result in this form, or multiply out the denominator and obtain

$$(2a + b^{-1})^{-2} = \frac{b^2}{(2ab + 1)^2} = \frac{b^2}{4a^2b^2 + 4ab + 1}$$

However, a common type of error is sometimes made with this type of expression. An incorrect step which is made is to express

$$(2a + b^{-1})^{-2} \quad \text{as} \quad (2a)^{-2} + (b^{-1})^{-2} \quad \text{or} \quad \frac{1}{4a^2} + b^2$$

Remember: As noted in Section 5–1, when raising a binomial (or any multinomial) to a power, we cannot simply raise each term to the power to obtain the result.

From the above examples, we see that *when a factor is moved from the denominator to the numerator of a fraction, or conversely, the sign of the exponent is changed.* We should heed carefully the word "factor"; this rule does not apply to moving terms in the numerator or denominator.

Example F

$$\frac{1}{x^{-1}}\left(\frac{x^{-1}-y^{-1}}{x^2-y^2}\right) = \frac{x}{1}\left(\frac{\dfrac{1}{x}-\dfrac{1}{y}}{x^2-y^2}\right) = x\left(\frac{\dfrac{y-x}{xy}}{x^2-y^2}\right)$$

$$= \frac{\dfrac{x(y-x)}{xy}}{(x-y)(x+y)} = \frac{x(y-x)}{xy} \cdot \frac{1}{(x-y)(x+y)}$$

$$= \frac{x(y-x)}{xy(x-y)(x+y)} = \frac{-(x-y)}{y(x-y)(x+y)}$$

$$= -\frac{1}{y(x+y)}$$

Note that in this example the x^{-1} and y^{-1} in the numerator could not be moved directly to the denominator with positive exponents because they are only terms of the original numerator.

Example G

$$3(x+4)^2(x-3)^{-2} - 2(x-3)^{-3}(x+4)^3$$

$$= \frac{3(x+4)^2}{(x-3)^2} - \frac{2(x+4)^3}{(x-3)^3} = \frac{3(x-3)(x+4)^2 - 2(x+4)^3}{(x-3)^3}$$

$$= \frac{(x+4)^2[3(x-3) - 2(x+4)]}{(x-3)^3} = \frac{(x+4)^2(x-17)}{(x-3)^3}$$

Expressions such as the one in this example are commonly found in problems in calculus.

Exercises 10−1

In Exercises 1 through 40 express each of the given expressions in the simplest form which contains only positive exponents.

1. $x^7 \cdot x^{-4}$
2. $y^9 \cdot y^{-2}$
3. $a^2 \cdot a^{-6}$
4. $s \cdot s^{-5}$

5. $(2ax^{-1})^2$
6. $(3xy^{-2})^3$
7. $(5an^{-2})^{-1}$
8. $(6s^2 t^{-1})^{-2}$

9. $(-4)^0$
10. -4^0
11. $-7x^0$
12. $(-7x)^0$

13. $\left(\dfrac{2}{n^3}\right)^{-1}$
14. $\left(\dfrac{3}{x^3}\right)^{-2}$
15. $\left(\dfrac{a}{b^{-2}}\right)^{-3}$
16. $\left(\dfrac{2n^{-2}}{m^{-1}}\right)^{-2}$

17. $(a+b)^{-1}$
18. $a^{-1}+b^{-1}$
19. $3x^{-2}+2y^{-2}$
20. $(3x+2y)^{-2}$

21. $\left(\dfrac{3a^2}{4b}\right)^{-3}\left(\dfrac{4}{a}\right)^{-5}$
22. $\left(\dfrac{2n}{p^2}\right)^{-2}\left(\dfrac{p}{4}\right)^{-1}$

23. $\left(\dfrac{y^{-1}}{2t}\right)^{-2}\left(\dfrac{t^2}{y^{-2}}\right)^{-3}$
24. $\left(\dfrac{a^{-2}}{b^2}\right)^{-3}\left(\dfrac{a^{-3}}{b^5}\right)^2$

25. $(x^2 y^{-1})^2 - x^{-4}$
26. $(st^{-2})^{-1} - s^{-2}$

27. $4a^{-2} + (3a^2)^{-2}$

28. $3(a^{-1}z^2)^{-3} + c^{-2}z^{-1}$

29. $(a^{-1} + b^{-1})^{-1}$

30. $(2a - b^{-2})^{-1}$

31. $(n^{-2} - 2n^{-1})^2$

32. $(2^{-3} - 4^{-1})^2$

33. $\dfrac{x - y^{-1}}{x^{-1} - y}$

34. $\dfrac{x^{-2} - y^{-2}}{x^{-1} - y^{-1}}$

35. $\dfrac{ax^{-2} + a^{-2}x}{a^{-1} + x^{-1}}$

36. $\dfrac{2x^{-2} - 2y^{-2}}{(xy)^{-3}}$

37. $2t^{-2} + t^{-1}(t + 1)$

38. $3x^{-1} + x^{-3}(y + 2)$

39. $(x - 1)^{-1} + (x + 1)^{-1}$

40. $4(2x - 1)(x + 2)^{-1} - (2x - 1)^2(x + 2)^{-2}$

In Exercises 41 through 44 perform the indicated operations.

41. When discussing electronic amplifiers, the expression $(1/r + 1/R)^{-1}$ is found. Simplify this expression.

42. Physical units associated with numbers are often expressed in terms of negative exponents (see Appendix B). If the units of a certain quantity are $(\text{m} \cdot \text{s}^{-1})^2$, express these units without the use of negative exponents.

43. An expression which is used for the focal length of a certain lens is $[(\mu - 1)(r_1^{-1} - r_2^{-1})]^{-1}$. Rewrite this expression without the use of negative exponents.

44. An expression encountered in the mathematics of finance is

$$\frac{p(1 + i)^{-1}[(1 + i)^{-n} - 1]}{(1 + i)^{-1} - 1}$$

where n is an integer. Simplify this expression.

10-2 Fractional Exponents

In Section 10-1 we reviewed the use of integral exponents, including exponents which are negative integers and zero. We now show how rational numbers may be used as exponents. With the appropriate definitions all of the laws of exponents are valid for all rational numbers as exponents.

Equation (10-3) states that $(a^m)^n = a^{mn}$. If we were to let $m = \frac{1}{2}$ and $n = 2$, we would have $(a^{1/2})^2 = a^1$. However, we already have a way of writing a quantity which when squared equals a. This is written as $\sqrt{a}$. To be consistent with previous definitions and to allow the laws of exponents to hold, we define

$$a^{1/n} = \sqrt[n]{a} \tag{10-7}$$

In order that Eqs. (10–3) and (10–7) may hold at the same time, we define

$$a^{m/n} = \sqrt[n]{a^m} \qquad (10\text{–}8)$$

It can be shown that these definitions are valid for all the laws of exponents.

Example A

We shall verify here that Eq. (10–1) holds for the above definitions:

$$a^{1/4}a^{1/4}a^{1/4}a^{1/4} = a^{(1/4)+(1/4)+(1/4)+(1/4)} = a^1$$

Now $a^{1/4} = \sqrt[4]{a}$, by definition. Also, by definition $\sqrt[4]{a}\,\sqrt[4]{a}\,\sqrt[4]{a}\,\sqrt[4]{a} = a$. Equation (10–1) is thereby verified for $n = 4$. Equation (10–3) is verified by the following:

$$a^{1/4}a^{1/4}a^{1/4}a^{1/4} = a^{4(1/4)} = a = \sqrt[4]{a^4}$$

We may interpret Eq. (10–8) as "the mth power of the nth root of a," as well as the way in which it is written, which is "the nth root of the mth power of a." This is illustrated in the following example.

Example B

$$(\sqrt[3]{a})^2 = \sqrt[3]{a^2} = a^{2/3}$$
$$8^{2/3} = (\sqrt[3]{8})^2 = (2)^2 = 4 \qquad \text{or} \qquad 8^{2/3} = \sqrt[3]{8^2} = \sqrt[3]{64} = 4$$

Although both interpretations of Eq. (10–8) are possible as indicated in Example B, in evaluating numerical expressions involving fractional exponents, it is almost always best to find the root first, as indicated by the denominator of the fractional exponent. This will allow us to find the root of the smaller number, which is normally easier to find.

Example C

To evaluate $(64)^{5/2}$, we should proceed as follows:

$$(64)^{5/2} = [(64)^{1/2}]^5 = 8^5 = 32{,}768$$

If we raised 64 to the fifth power first we would have

$$(64)^{5/2} = (64^5)^{1/2} = (1{,}073{,}741{,}824)^{1/2}$$

We would now have to evaluate the indicated square root. This demonstrates why it is preferable to find the indicated root first.

Example D

$$(16)^{3/4} = (16^{1/4})^3 = 2^3 = 8$$
$$4^{-1/2} = \frac{1}{4^{1/2}} = \frac{1}{2}, \qquad 9^{3/2} = (9^{1/2})^3 = 3^3 = 27$$

We note in the illustration of $4^{-1/2}$ that Eq. (10–6) must also hold for negative rational exponents. The only change in the exponent which is made in writing it as $1/4^{1/2}$ is that the sign of the exponent is changed.

The question may arise as to why we use fractional exponents, since we have already defined expressions which are equivalent to their meanings. The answer is that fractional exponents are often easier to handle in more complex expressions involving roots, and therefore any expression involving radicals can be solved by use of fractional exponents. Also, they can be used to find roots of numbers on a calculator.

Example E

$$(8a^2b^4)^{1/3} = [(8^{1/3})(a^2)^{1/3}(b^4)^{1/3}] = 2a^{2/3}b^{4/3}$$

$$a^{3/4}a^{4/5} = a^{3/4+4/5} = a^{31/20}$$

$$(25a^{-2}c^4)^{3/2} = [(25a^{-2}c^4)^{1/2}]^3 = \left(\frac{(25)^{1/2}(c^4)^{1/2}}{(a^2)^{1/2}}\right)^3 = \left(\frac{5c^2}{a}\right)^3$$

$$= \frac{125c^6}{a^3}$$

Example F

$$\left(\frac{4^{-3/2}x^{2/3}y^{-7/4}}{2^{3/2}x^{-1/3}y^{3/4}}\right)^{2/3} = \left(\frac{x^{2/3+1/3}}{2^{3/2}4^{3/2}y^{3/4+7/4}}\right)^{2/3}$$

$$= \frac{x^{(1)(2/3)}}{2^{(3/2)(2/3)}4^{(3/2)(2/3)}y^{(10/4)(2/3)}} = \frac{x^{2/3}}{8y^{5/3}}$$

Example G

$$(4x^4)^{-1/2} - 3x^{-3} = \frac{1}{(4x^4)^{1/2}} - \frac{3}{x^3}$$

$$= \frac{1}{2x^2} - \frac{3}{x^3}$$

$$= \frac{x - 6}{2x^3}$$

Example H

$$(2x + 1)^{1/2} + (x + 3)(2x + 1)^{-1/2}$$

$$= (2x + 1)^{1/2} + \frac{x + 3}{(2x + 1)^{1/2}}$$

$$= \frac{(2x + 1)^{1/2}(2x + 1)^{1/2} + (x + 3)}{(2x + 1)^{1/2}}$$

$$= \frac{(2x + 1) + (x + 3)}{(2x + 1)^{1/2}} = \frac{3x + 4}{(2x + 1)^{1/2}}$$

Exercises 10–2

In Exercises 1 through 28 evaluate the given expressions.

1. $(25)^{1/2}$ 2. $(49)^{1/2}$ 3. $(27)^{1/3}$

4. $(81)^{1/4}$ 5. $8^{4/3}$ 6. $(125)^{2/3}$

7. $(100)^{25/2}$ 8. $(16)^{5/4}$ 9. $8^{-1/3}$

10. $16^{-1/4}$ 11. $(64)^{-2/3}$ 12. $(32)^{-4/5}$

13. $5^{1/2}5^{3/2}$ 14. $8^{1/3}4^{1/2}$ 15. $(4^4)^{3/2}$

16. $(3^6)^{2/3}$ 17. $\dfrac{121^{-1/2}}{100^{1/2}}$ 18. $\dfrac{1000^{1/3}}{400^{-1/2}}$

19. $\dfrac{7^{-1/2}}{6^{-1}7^{1/2}}$ 20. $\dfrac{15^{2/3}}{5^2 15^{-1/3}}$ 21. $\dfrac{(-27)^{1/3}}{6}$

22. $\dfrac{(-8)^{2/3}}{-2}$ 23. $\dfrac{-8}{(-27)^{-1/3}}$ 24. $\dfrac{-4}{(-64)^{-2/3}}$

25. $(125)^{-2/3} - (100)^{-3/2}$ 26. $36^{-1/2} + 27^{-2/3}$

27. $\dfrac{25^{-1/2}}{5} + \dfrac{20^{-1/2}}{20^{1/2}}$ 28. $\dfrac{4^{-1}}{(36)^{-1/2}} - \dfrac{5^{-1/2}}{5^{1/2}}$

In Exercises 29 through 52 use the laws of exponents to simplify the given expressions. Express all answers with positive exponents.

29. $a^{2/3}a^{1/2}$ 30. $x^{5/6}x^{-1/3}$ 31. $\dfrac{y^{-1/2}}{y^{2/5}}$

32. $\dfrac{2r^{4/5}}{r^{-1}}$ 33. $\dfrac{s^{1/4}s^{2/3}}{s^{-1}}$ 34. $\dfrac{x^{3/10}}{x^{-1/5}x^2}$

35. $\dfrac{y^{-1}}{y^{1/3}y^{-1/4}}$ 36. $\dfrac{a^{-2/5}a^2}{a^{-3/10}}$ 37. $(8a^3b^6)^{1/3}$

38. $(8b^{-4}c^2)^{2/3}$ 39. $(16a^4b^3)^{-3/4}$ 40. $(32x^5y^4)^{-2/5}$

41. $\left(\dfrac{9t^{-2}}{16}\right)^{3/2}$ 42. $\left(\dfrac{a^{5/7}}{a^{2/3}}\right)^{7/4}$ 43. $\left(\dfrac{4a^{5/6}b^{-1/5}}{a^{2/3}b^2}\right)^{-1/2}$

44. $\left(\dfrac{a^0 b^8 c^{-1/8}}{ab^{63/64}}\right)^{32/3}$ 45. $\dfrac{6x^{-1/2}y^{2/3}}{18x^{-1}} \cdot \dfrac{2y^{1/4}}{x^{1/3}}$ 46. $\dfrac{3^{-1}a^{1/2}}{4^{-1/2}b} \div \dfrac{9^{1/2}a^{-1/3}}{2b^{-1/4}}$

47. $(x^{-1} + 2x^{-2})^{-1/2}$ 48. $(a^{-2} - a^{-4})^{-1/4}$ 49. $(a^3)^{-4/3} + a^{-2}$

50. $(4x^6)^{-1/2} - 2x^{-1}$ 51. $[(a^{1/2} - a^{-1/2})^2 + 4]^{1/2}$

52. $(3x - 1)^{-2/3}(1 - x) - (3x - 1)^{1/3}$

In Exercises 53 through 56 perform the indicated operations.

53. In determining the number of electrons involved in a certain calculation with semiconductors, the expression $9.60 \times 10^{18}T^{3/2}$ is used. Evaluate this expression for $T = 289K$.

54. In studying the properties of biological fluids, the expression $0.036M^{3/4}$ is used. Evaluate this expression for $M = 1.6 \times 10^5$.

55. An approximate expression for the efficiency of an engine is given by $E = 100(1 - R^{-2/5})$, where R is the compression ratio. What is the efficiency (in percent) of an engine for which $R = 243/32$?

56. An estimate of gas diffusivity may be made by the equation

$$D_m = 0.01 \frac{T^{1/2}}{(v_a^{1/3} + v_b^{1/3})^2} \left(\frac{1}{m_a} + \frac{1}{m_b} \right)^{1/2}$$

where T is the temperature in degrees Fahrenheit and the other symbols are constants which depend on the gases under consideration. Calculate the diffusivity of a gas in air at 484°F if $v_a = 27$, $v_b = 125$, $m_a = 25$, and $m_b = 144$. (The units of diffusivity are in lb·mol/ft·h.)

10—3 Simplest Radical Form

Radicals were first introduced in Section 1–6, and we used them again in developing the concept of a fractional exponent. As we mentioned in the preceding section, it is possible to use fractional exponents for any operation required with radicals. For operations involving multiplication and division, this method has certain advantages. But for adding and subtracting radicals, there is normally little advantage in changing form.

We shall now define the operations with radicals so that these definitions are consistent with the laws of exponents. This will enable us from now on to use either fractional exponents or radicals, whichever is more convenient.

$$\sqrt[n]{a^n} = (\sqrt[n]{a})^n = a \tag{10–9}$$

$$\sqrt[n]{a}\ \sqrt[n]{b} = \sqrt[n]{ab} \tag{10–10}$$

$$\sqrt[m]{\sqrt[n]{a}} = \sqrt[mn]{a} \tag{10–11}$$

$$\frac{\sqrt[n]{a}}{\sqrt[n]{b}} = \sqrt[n]{\frac{a}{b}}, \quad b \neq 0 \tag{10–12}$$

The number under the radical is called the **radicand,** *and the number indicating the root being taken is called the* **order** *of the radical.* To avoid difficulties with imaginary numbers (which are considered in the next chapter), we shall assume that all letters represent positive numbers.

Example A

$$\sqrt[3]{2}\,\sqrt[3]{3} = \sqrt[3]{6}, \qquad \sqrt[3]{\sqrt{5}} = \sqrt[6]{5}, \qquad \frac{\sqrt{7}}{\sqrt{3}} = \sqrt{\frac{7}{3}}$$

There are certain operations which should be performed on radicals to put them in their simplest form. The following examples will illustrate these operations.

Example B
To simplify $\sqrt{75}$, we recall that $75 = (25)(3)$ and that $\sqrt{25} = 5$. Using Eq. (10–10), we write $\sqrt{75} = \sqrt{25}\,\sqrt{3} = 5\sqrt{3}$. This illustrates one step which should always be carried out in simplifying radicals. *Always remove all perfect nth-power factors from the radicand of a radical of order n.*

Example C

$$\sqrt[3]{40} = \sqrt[3]{8}\,\sqrt[3]{5} = 2\sqrt[3]{5}$$
$$\sqrt{a^3 b^2} = \sqrt{a^2}\,\sqrt{a}\,\sqrt{b^2} = ab\sqrt{a}$$
$$\sqrt{72} = \sqrt{(36)(2)} = \sqrt{36}\,\sqrt{2} = 6\sqrt{2}$$
$$\sqrt[5]{64x^8 y^{12}} = \sqrt[5]{(32)(2)(x^5)(x^3)(y^{10})(y^2)} = \sqrt[5]{(32)(x^5)(y^{10})}\,\sqrt[5]{2x^3 y^2}$$
$$= 2xy^2\,\sqrt[5]{2x^3 y^2}$$

When working with fractions, an expression is not considered to be in simplest form if the denominator contains a radical. This includes the case in which the denominator of a fraction is included under the radical sign. We therefore rewrite it in an equivalent form, in which the denominator contains no radical expression. This procedure is called **rationalizing the denominator.** The resulting form is more convenient for purposes of calculation, although today this is of much less importance than it was before the common use of calculators. However, the rationalized form is often a more useful form.

Example D
To simplify $\sqrt{\frac{2}{5}}$ we write it in an equivalent form in which the denominator is not included under the radical sign. In order to do this we create a perfect square in the denominator by multiplying the numerator and the denominator under the radical by 5. This gives us $\sqrt{\frac{10}{25}}$ which may be written as $\frac{1}{5}\sqrt{10}$ or $\frac{\sqrt{10}}{5}$. These steps are written as follows:

$$\sqrt{\frac{2}{5}} = \sqrt{\frac{2\cdot 5}{5\cdot 5}} = \sqrt{\frac{10}{25}} = \frac{\sqrt{10}}{\sqrt{25}} = \frac{\sqrt{10}}{5}$$

Example E

$$\sqrt{\frac{5}{7}} = \sqrt{\frac{5\cdot 7}{7\cdot 7}} = \frac{\sqrt{35}}{\sqrt{49}} = \frac{\sqrt{35}}{7}, \qquad \frac{3}{\sqrt{8}} = \frac{3\sqrt{2}}{\sqrt{8\cdot 2}} = \frac{3\sqrt{2}}{\sqrt{16}} = \frac{3\sqrt{2}}{4}$$

$$\sqrt[3]{\frac{2}{3}} = \sqrt[3]{\frac{2\cdot 9}{3\cdot 9}} = \sqrt[3]{\frac{18}{27}} = \frac{\sqrt[3]{18}}{\sqrt[3]{27}} = \frac{\sqrt[3]{18}}{3}$$

In the second illustration a perfect square was made by multiplying by $\sqrt{2}$. We should try to choose the smallest possible number which can be used. In the third illustration we wanted a perfect cube in the denominator, since a cube root is being found.

The following examples illustrate another procedure which can be used to simplify certain radicals. The procedure is to *reduce the order of the radical*, when possible.

Example F

$$\sqrt[6]{8} = \sqrt[6]{2^3} = 2^{3/6} = 2^{1/2} = \sqrt{2}$$

In this example we started with a sixth root and ended with a square root. Thus the order of the radical was reduced. Fractional exponents are often helpful when we perform this operation.

Example G

$$\sqrt[8]{16} = \sqrt[8]{2^4} = 2^{4/8} = 2^{1/2} = \sqrt{2}$$

$$\frac{\sqrt[4]{9}}{\sqrt{3}} = \frac{\sqrt[4]{3^2}}{\sqrt{3}} = \frac{3^{2/4}}{3^{1/2}} = 1$$

$$\frac{\sqrt[6]{8}}{\sqrt{7}} = \frac{\sqrt[6]{2^3}}{\sqrt{7}} = \frac{2^{1/2}}{7^{1/2}} = \sqrt{\frac{2}{7}} = \frac{\sqrt{14}}{7}$$

$$\sqrt[9]{27x^6y^{12}} = \sqrt[9]{3^3x^6y^9y^3} = 3^{3/9}x^{6/9}y^{9/9}y^{3/9} = 3^{1/3}x^{2/3}y\,y^{1/3}$$
$$= y\sqrt[3]{3x^2y}$$

A radical is said to be *simplified* if the above steps are completed. That is,

(1) *all perfect nth-power factors are removed from a radical of order n,*
(2) *all denominators are rationalized, and*
(3) *if possible, the order of the radical is reduced.*

Example H

Simplify the radical $\sqrt{\dfrac{3a}{4b} - 2 + \dfrac{4b}{3a}}$, for $3a \geq 4b$.

$$\sqrt{\frac{3a}{4b} - 2 + \frac{4b}{3a}} = \sqrt{\frac{(3a)(3a) - 2(3a)(4b) + (4b)(4b)}{(3a)(4b)}}$$

$$= \sqrt{\frac{(3a - 4b)^2}{4(3ab)}} = \frac{3a - 4b}{2}\sqrt{\frac{1}{3ab}}$$

$$= \frac{3a - 4b}{6ab}\sqrt{3ab}$$

Exercises 10-3

In Exercises 1 through 56 write each expression in simplest radical form.

1. $\sqrt{24}$ 2. $\sqrt{150}$ 3. $\sqrt{45}$

4. $\sqrt{98}$ 5. $\sqrt{x^2y^5}$ 6. $\sqrt{s^3t^6}$

7. $\sqrt{pq^2r^7}$ 8. $\sqrt{x^2y^4z^3}$ 9. $\sqrt{5x^2}$

10. $\sqrt{12ab^2}$ 11. $\sqrt{18a^3bc^4}$ 12. $\sqrt{54m^5n^3}$

13. $\sqrt[3]{16}$ 14. $\sqrt[4]{48}$ 15. $\sqrt[6]{96}$

16. $\sqrt[3]{-16}$ 17. $\sqrt[3]{8a^2}$ 18. $\sqrt[3]{5a^4b^2}$

19. $\sqrt[4]{64r^3s^4t^5}$ 20. $\sqrt[5]{16x^5y^3z^{11}}$ 21. $\sqrt[3]{8}\sqrt[5]{4}$

22. $\sqrt[7]{4}\sqrt[7]{64}$ 23. $\sqrt[3]{ab^4}\sqrt[3]{a^2b}$ 24. $\sqrt[6]{3m^4n^5}\sqrt[6]{9m^2n^8}$

25. $\sqrt{\dfrac{3}{2}}$ 26. $\sqrt{\dfrac{6}{5}}$ 27. $\sqrt{\dfrac{a}{b}}$

28. $\sqrt{\dfrac{a}{b^3}}$ 29. $\sqrt[3]{\dfrac{3}{4}}$ 30. $\sqrt[4]{\dfrac{2}{5}}$

31. $\sqrt[5]{\dfrac{1}{9}}$ 32. $\sqrt[6]{\dfrac{5}{4}}$ 33. $\sqrt[4]{400}$

34. $\sqrt[8]{81}$ 35. $\sqrt[6]{64}$ 36. $\sqrt[9]{27}$

37. $\sqrt{4 \times 10^4}$ 38. $\sqrt{4 \times 10^5}$ 39. $\sqrt{4 \times 10^6}$

40. $\sqrt{16 \times 10^5}$ 41. $\sqrt[4]{4a^2}$ 42. $\sqrt[6]{b^2c^4}$

43. $\sqrt[4]{\dfrac{1}{4}}$ 44. $\dfrac{\sqrt[4]{80}}{\sqrt[4]{5}}$ 45. $\sqrt[4]{\sqrt[3]{16}}$

46. $\sqrt[5]{\sqrt[4]{9}}$ 47. $\sqrt{\sqrt{\sqrt{2}}}$ 48. $\sqrt{b^4\sqrt{a}}$

49. $\sqrt{\dfrac{1}{2} - \dfrac{1}{3}}$ 50. $\sqrt{\dfrac{5}{4} - \dfrac{1}{8}}$ 51. $\sqrt{\dfrac{1}{a^2} + \dfrac{1}{b}}$

52. $\sqrt{\dfrac{x}{y} + \dfrac{y}{x}}$ 53. $\sqrt{a^2 + 2ab + b^2}$ 54. $\sqrt{a^2 + b^2}$

55. $\sqrt{x^2 + \dfrac{1}{4}}$ 56. $\sqrt{\dfrac{1}{2} + 2r + 2r^2}$

In Exercises 57 through 60 perform the required operation.

57. The period (in seconds) for one cycle of a simple pendulum is given by $T = 2\pi\sqrt{L/g}$, where L is the length of the pendulum (in feet) and g is the acceleration due to gravity ($g = 32$ ft/s^2). If L is 3 ft, what is the period of the pendulum?

58. Under certain circumstances, the frequency in an electric circuit containing an inductance L and capacitance C is given by $f = 1/2\pi\sqrt{LC}$. If $L = 0.1$ H and $C = 250 \times 10^{-6}$ F, find f.

59. The distance between ion layers in a crystalline solid such as table salt is given by the expression $\sqrt[3]{M/2N\rho}$, where M is the molecular weight, N is called Avogadro's number, and ρ is the density. Express this in simplest form.

60. When dealing with the voltage of an electronic device, the expression $V(x/d)^{4/3}$ arises. Write this expression in rationalized radical form.

10—4 Addition and Subtraction of Radicals

When we first introduced the concept of adding algebraic expressions, we found that it was possible to combine similar terms, that is, those which differed only in numerical coefficients. The same is true of adding radicals. We must have similar radicals in order to perform the addition, rather than simply to be able to indicate addition. By similar radicals we mean radicals which differ only in their numerical coefficients, and which must therefore be of the same order and have the same radicand.

In order to add radicals, we first express each radical in its simplest form, and then add those which are similar. For those which are not similar, we can only indicate the addition.

Example A

$$2\sqrt{7} - 5\sqrt{7} + \sqrt{7} = -2\sqrt{7}$$
$$\sqrt[5]{6} + 4\sqrt[5]{6} - 2\sqrt[5]{6} = 3\sqrt[5]{6}$$
$$\sqrt{5} + 2\sqrt{3} - 5\sqrt{5} = 2\sqrt{3} - 4\sqrt{5}$$

We note that in the last illustration we are able only to indicate the final subtraction.

Example B

$$\sqrt{2} + \sqrt{8} = \sqrt{2} + 2\sqrt{2} = 3\sqrt{2}$$
$$\sqrt[3]{24} + \sqrt[3]{81} = 2\sqrt[3]{3} + 3\sqrt[3]{3} = 5\sqrt[3]{3}$$

Notice that $\sqrt{8}$, $\sqrt[3]{24}$, and $\sqrt[3]{81}$ were simplified before performing the addition.

We note in the illustrations of Example B that the radicals do not initially appear to be similar. However, after each is simplified we are able to recognize the similar radicals.

Example C

$$6\sqrt{7} - \sqrt{28} + 3\sqrt{63} = 6\sqrt{7} - 2\sqrt{7} + 3(3\sqrt{7})$$
$$= 6\sqrt{7} - 2\sqrt{7} + 9\sqrt{7} = 13\sqrt{7}$$
$$3\sqrt{125} - \sqrt{20} + \sqrt{27} = 3(5\sqrt{5}) - 2\sqrt{5} + 3\sqrt{3}$$
$$= 13\sqrt{5} + 3\sqrt{3}$$

Example D

$$\sqrt{24} + \sqrt{\frac{3}{2}} = 2\sqrt{6} + \frac{\sqrt{6}}{2} = \frac{4\sqrt{6} + \sqrt{6}}{2} = \frac{5}{2}\sqrt{6}$$

One radical was simplified by removing the perfect square factor and the other by rationalizing the denominator.

Example E

$$\sqrt{\frac{2}{3a}} - 2\sqrt{\frac{3}{2a}} = \frac{1}{3a}\sqrt{6a} - \frac{2}{2a}\sqrt{6a} = \frac{1}{3a}\sqrt{6a} - \frac{1}{a}\sqrt{6a}$$

$$= \frac{\sqrt{6a} - 3\sqrt{6a}}{3a} = \frac{-2\sqrt{6a}}{3a} = -\frac{2}{3a}\sqrt{6a}$$

Example F

$$\sqrt{\frac{4}{a} - 4 + a} + \sqrt{\frac{1}{a}} - \sqrt{16a^3} = \sqrt{\frac{4 - 4a + a^2}{a}} + \sqrt{\frac{1}{a}} - 4a\sqrt{a}$$

$$= \sqrt{\frac{(2-a)^2 \cdot a}{a \cdot a}} + \sqrt{\frac{1 \cdot a}{a \cdot a}} - 4a\sqrt{a}$$

$$= \frac{2-a}{a}\sqrt{a} + \frac{1}{a}\sqrt{a} - 4a\sqrt{a}$$

$$= \sqrt{a}\left(\frac{2-a}{a} + \frac{1}{a} - 4a\right) = \sqrt{a}\left(\frac{2 - a + 1 - 4a^2}{a}\right)$$

$$= \frac{(3 - a - 4a^2)\sqrt{a}}{a}$$

This simplification is valid for $a < 2$, since we let $\sqrt{(2-a)^2} = 2 - a$.

Exercises 10—4

In Exercises 1 through 36 perform the indicated operations and express the answers in simplest form.

1. $2\sqrt{3} + 5\sqrt{3}$

2. $8\sqrt{11} - 3\sqrt{11}$

3. $2\sqrt{7} + \sqrt{5} - 3\sqrt{7}$

4. $8\sqrt{6} - 2\sqrt{3} - 5\sqrt{6}$

5. $\sqrt{5} + \sqrt{20}$

6. $\sqrt{7} + \sqrt{63}$

7. $2\sqrt{3} - 3\sqrt{12}$

8. $4\sqrt{2} - \sqrt{50}$

9. $\sqrt{8} - \sqrt{32}$

10. $\sqrt{27} + 2\sqrt{18}$

11. $2\sqrt{28} + 3\sqrt{175}$

12. $5\sqrt{300} - 7\sqrt{48}$

13. $2\sqrt{20} - \sqrt{125} - \sqrt{45}$

14. $2\sqrt{44} - \sqrt{99} + \sqrt{176}$

15. $3\sqrt{75} + 2\sqrt{48} - 2\sqrt{18}$

16. $2\sqrt{28} - \sqrt{108} - 2\sqrt{175}$

17. $\sqrt{60} + \sqrt{\frac{5}{3}}$

18. $\sqrt{84} - \sqrt{\frac{3}{7}}$

19. $\sqrt{\frac{1}{2}} + \sqrt{\frac{25}{2}} - \sqrt{18}$

20. $\sqrt{6} - \sqrt{\frac{2}{3}} - \sqrt{18}$

21. $\sqrt[3]{81} + \sqrt[3]{3000}$

22. $\sqrt[3]{-16} + \sqrt[3]{54}$

23. $\sqrt[4]{32} - \sqrt[8]{4}$

24. $\sqrt[6]{\sqrt{2}} - \sqrt[12]{2^{13}}$

25. $\sqrt{a^3b} - \sqrt{4ab^3}$

26. $\sqrt{2x^2y} + \sqrt{8y^3}$

27. $\sqrt{6}\sqrt{5}\sqrt{3} - \sqrt{40a^2}$

28. $\sqrt{60n} + 2\sqrt{15b^2n} - b\sqrt{135n}$

29. $\sqrt[3]{24a^2b^4} - \sqrt[3]{3a^5b}$

30. $\sqrt[5]{32a^6b^4} + 3a\sqrt[5]{243ab^9}$

31. $\sqrt{\frac{a}{c^5}} - \sqrt{\frac{c}{a^3}}$

32. $\sqrt{\frac{2x}{3y}} + \sqrt{\frac{27y}{8x}}$

33. $\sqrt[3]{\dfrac{a}{b}} - \sqrt[3]{\dfrac{8b^2}{a^2}}$

34. $\sqrt[4]{\dfrac{c}{b}} - \sqrt[4]{bc}$

35. $\sqrt{\dfrac{a-b}{a+b}} - \sqrt{\dfrac{a+b}{a-b}}$

36. $\sqrt{\dfrac{16}{x} + 8 + x} - \sqrt{1 - \dfrac{1}{x}}$

In Exercises 37 and 38 solve the given problems.

37. Find the sum of the two roots of the quadratic equation $ax^2 + bx + c = 0$.

38. In the study of the kinetic theory of gases, the expression

$$a\sqrt{\dfrac{h^3 m^5}{\pi}} + b\sqrt{\dfrac{h^3 m^3}{\pi}}$$

is found. Perform the indicated addition.

10–5 Multiplication of Radicals

When multiplying expressions containing radicals, we use Eq. (10–10) along with the normal procedures of algebraic multiplication. Note that the orders of the radicals being multiplied in Eq. (10–10) are the same. The following examples illustrate the method.

Example A

$$\sqrt{5}\,\sqrt{2} = \sqrt{10},$$
$$\sqrt{33}\,\sqrt{3} = \sqrt{99} = \sqrt{9(11)} = \sqrt{9}\,\sqrt{11} = 3\sqrt{11}$$

We note that we must be careful to express the resulting radical in simplest form.

Example B

$$\sqrt[3]{6}\,\sqrt[3]{4} = \sqrt[3]{24} = \sqrt[3]{8}\,\sqrt[3]{3} = 2\sqrt[3]{3}$$
$$\sqrt[5]{8a^3 b^4}\,\sqrt[5]{8a^2 b^3} = \sqrt[5]{64a^5 b^7} = \sqrt[5]{32a^5 b^5}\,\sqrt[5]{2b^2} = 2ab\sqrt[5]{2b^2}$$

Example C

$$\sqrt{2}(3\sqrt{5} - 4\sqrt{2}) = 3\sqrt{2}\sqrt{5} - 4\sqrt{2}\sqrt{2} = 3\sqrt{10} - 4\sqrt{4}$$
$$= 3\sqrt{10} - 4(2) = 3\sqrt{10} - 8$$

When raising a single term radical expression to a power we use the basic meaning of the power. When raising a binomial to a power, we proceed as with any binomial and use Eq. (10–10).

Example D

$$(2\sqrt{7})^2 = 2^2(\sqrt{7})^2 = 4(7) = 28$$
$$(\sqrt{a} - \sqrt{b})^2 = (\sqrt{a})^2 - 2\sqrt{a}\sqrt{b} + (\sqrt{b})^2 = a + b - 2\sqrt{ab}$$

Example E

$$(5\sqrt{7} - 2\sqrt{3})(4\sqrt{7} + 3\sqrt{3})$$
$$= 20\sqrt{7}\,\sqrt{7} + 15\sqrt{7}\,\sqrt{3} - 8\sqrt{3}\,\sqrt{7} - 6\sqrt{3}\,\sqrt{3}$$
$$= 20(7) + 15\sqrt{21} - 8\sqrt{21} - 6(3)$$
$$= 140 + 7\sqrt{21} - 18$$
$$= 122 + 7\sqrt{21}$$

Example F

$$(\sqrt{6} - \sqrt{2} - \sqrt{3})(\sqrt{6} + \sqrt{2})$$
$$= (\sqrt{6} - \sqrt{2})(\sqrt{6} + \sqrt{2}) - \sqrt{3}(\sqrt{6} + \sqrt{2})$$
$$= (6 - 2) - \sqrt{18} - \sqrt{6} = 4 - 3\sqrt{2} - \sqrt{6}$$

Example G

$$\left(3\sqrt{\frac{a}{b}} - \sqrt{ab}\right)\left(2\sqrt{\frac{a}{b}} - \sqrt{ab}\right) = 6\frac{a}{b} - 5\sqrt{\frac{a^2b}{b}} + ab$$

$$= \frac{6a}{b} - 5a + ab = \frac{6a - 5ab + ab^2}{b}$$

$$= \frac{a(6 - 5b + b^2)}{b}$$

Again, we note that *to multiply radicals and combine them under one radical sign, it is necessary that the order of the radicals be the same.* If necessary we can make the order of each radical the same by appropriate operations on each radical separately. Fractional exponents are frequently useful for this purpose.

Example H

$$\sqrt[3]{2}\,\sqrt{5} = 2^{1/3}5^{1/2} = 2^{2/6}5^{3/6} = (2^2 5^3)^{1/6} = \sqrt[6]{500}$$
$$\sqrt[3]{4a^2b}\,\sqrt[4]{8a^3b^2} = (2^2a^2b)^{1/3}(2^3a^3b^2)^{1/4} = (2^2a^2b)^{4/12}(2^3a^3b^2)^{3/12}$$

$$= (2^8a^8b^4)^{1/12}(2^9a^9b^6)^{1/12} = (2^{17}a^{17}b^{10})^{1/12}$$

$$= 2a(2^5a^5b^{10})^{1/12}$$

$$= 2a\sqrt[12]{32a^5b^{10}}$$

Exercises 10—5

In Exercises 1 through 48 perform the indicated multiplications, expressing answers in simplest form.

1. $\sqrt{3}\,\sqrt{10}$ 2. $\sqrt{2}\,\sqrt{51}$ 3. $\sqrt{6}\,\sqrt{2}$ 4. $\sqrt{7}\,\sqrt{14}$

5. $\sqrt[3]{4}\,\sqrt[3]{2}$ 6. $\sqrt[3]{3}\,\sqrt[3]{27}$ 7. $\sqrt[5]{4}\,\sqrt[5]{16}$ 8. $\sqrt[3]{25}\,\sqrt[3]{50}$

9. $(5\sqrt{2})^2$ 10. $(3\sqrt{3})^2$ 11. $(2\sqrt[3]{2})^3$ 12. $(3\sqrt[3]{5})^3$

13. $\sqrt{\frac{2}{3}}\sqrt{5}$ 14. $\sqrt[3]{8}\sqrt{\frac{5}{2}}$ 15. $\sqrt{\frac{5}{6}}\sqrt{\frac{2}{11}}$ 16. $\sqrt{\frac{6}{7}}\sqrt{\frac{2}{3}}$

17. $\sqrt{3}(\sqrt{2} - \sqrt{5})$ 18. $\sqrt{5}(\sqrt{7} + \sqrt{2})$

19. $2\sqrt{2}(\sqrt{8} - 3\sqrt{6})$

20. $3\sqrt{5}(\sqrt{15} - 2\sqrt{5})$

21. $(2 - \sqrt{5})(2 + \sqrt{5})$

22. $(6 - \sqrt{3})(6 + \sqrt{3})$

23. $(6 - \sqrt{3})^2$

24. $(2 - \sqrt{5})^2$

25. $(3\sqrt{5} - 2\sqrt{3})(6\sqrt{5} + 7\sqrt{3})$

26. $(3\sqrt{7} - \sqrt{8})(\sqrt{7} + \sqrt{2})$

27. $(3\sqrt{11} - \sqrt{6})(2\sqrt{11} + 5\sqrt{6})$

28. $(2\sqrt{10} + 3\sqrt{15})(\sqrt{10} - 7\sqrt{15})$

29. $\sqrt{a}(\sqrt{ab} + \sqrt{c})$

30. $\sqrt{3x}(\sqrt{3x} - \sqrt{xy})$

31. $\sqrt{5n}(\sqrt{15n} + \sqrt{20m})$

32. $\sqrt{2x}(\sqrt{8xy} - 3\sqrt{y})$

33. $(\sqrt{2a} - \sqrt{b})(\sqrt{2a} + 3\sqrt{b})$

34. $(2\sqrt{mn} - 3\sqrt{n})(3\sqrt{mn} + 2\sqrt{n})$

35. $(\sqrt{2} + \sqrt{3} + \sqrt{5})(\sqrt{3} - \sqrt{5})$

36. $(2\sqrt{7} - \sqrt{5})(\sqrt{14} - 2\sqrt{5} + \sqrt{7})$

37. $\sqrt{2}\sqrt[3]{3}$

38. $\sqrt[5]{16}\sqrt[5]{8}$

39. $\sqrt[4]{ab}\sqrt[3]{bc}$

40. $\sqrt{2x}\sqrt[5]{16x}$

41. $(\sqrt[5]{\sqrt{6}} - \sqrt{5})(\sqrt[5]{\sqrt{6}} + \sqrt{5})$

42. $\sqrt[3]{5} - \sqrt{17}\sqrt[3]{5} + \sqrt{17}$

43. $(\sqrt{2x^2} - \sqrt[3]{y})(\sqrt{4x} + \sqrt[3]{y^2})$

44. $(\sqrt{a} - \sqrt[3]{b})(2\sqrt{a} - \sqrt[3]{b})$

45. $\left(\sqrt{\dfrac{2}{a}} + \sqrt{\dfrac{a}{2}}\right)\left(\sqrt{\dfrac{2}{a}} - 2\sqrt{\dfrac{a}{2}}\right)$

46. $\left(\sqrt{\dfrac{x}{y}} - \sqrt{xy}\right)\left(\sqrt{\dfrac{y}{x}} + \sqrt{xy} - 1\right)$

47. $(2x - \sqrt{x - 2y})^2$

48. $(3 + \sqrt{6 - 2a})(2 - \sqrt{6 - 2a})$

In Exercises 49 and 50 perform the indicated operations.

49. Determine the product of the two roots of the quadratic equation $ax^2 + bx + c = 0$.

50. Relationships involving mass transfer of liquid involve the expression

$$\sqrt{\dfrac{dG}{u}}\sqrt[3]{\dfrac{MD}{u}}$$

Express this in simplest radical form.

10—6 Division of Radicals

We have already dealt with some cases of division of radicals in the previous sections. When we have the indicated division of one radical by another, the result is considered to be in simplest form when the denominators contain no radicals. Therefore, the rationalization of denominators, as we did in Section 10—3, is the principal step to be carried out. This means that the process of division is one in which we change the fraction to an equivalent form in which the denominator is free of radicals. In doing so, multiplication of the numerator and the denominator by the appropriate quantity is a primary step.

Example A

$$\frac{\sqrt{3}}{\sqrt{5}} = \frac{\sqrt{3}\sqrt{5}}{\sqrt{5}\sqrt{5}} = \frac{\sqrt{15}}{5}, \quad \frac{\sqrt{a}}{\sqrt[3]{b}} = \frac{\sqrt{a}}{\sqrt[3]{b}} \cdot \frac{\sqrt[3]{b^2}}{\sqrt[3]{b^2}} = \frac{\sqrt{a}\sqrt[3]{b^2}}{b} = \frac{a^{3/6}b^{4/6}}{b} = \frac{\sqrt[6]{a^3 b^4}}{b}$$

Notice that the denominator was rationalized and that the factors of the numerator were written in terms of fractional exponents so they could be combined under one radical.

If the denominator is the sum (or difference) of two terms, at least one of which is a radical, the fraction can be rationalized by multiplying both the numerator and the denominator by the difference (or sum) of the same two terms, if the radicals are square roots.

Example B

The fraction $1/(\sqrt{3} - \sqrt{2})$ can be rationalized by multiplying the numerator and the denominator by $\sqrt{3} + \sqrt{2}$. In this way the radicals will be removed from the denominator.

$$\frac{1}{\sqrt{3} - \sqrt{2}} \cdot \frac{\sqrt{3} + \sqrt{2}}{\sqrt{3} + \sqrt{2}} = \frac{\sqrt{3} + \sqrt{2}}{(\sqrt{3})^2 - (\sqrt{2})^2} = \frac{\sqrt{3} + \sqrt{2}}{3 - 2} = \sqrt{3} + \sqrt{2}$$

The reason this technique works is that an expression of the form $a^2 - b^2$ is created in the denominator, where a or b (or both) is a radical. We see that the result is a denominator free of radicals.

Example C

$$\frac{\sqrt{2}}{2\sqrt{5} + \sqrt{3}} = \frac{\sqrt{2}}{2\sqrt{5} + \sqrt{3}} \cdot \frac{2\sqrt{5} - \sqrt{3}}{2\sqrt{5} - \sqrt{3}} = \frac{2\sqrt{2}\sqrt{5} - \sqrt{2}\sqrt{3}}{(2\sqrt{5})^2 - (\sqrt{3})^2}$$

$$= \frac{2\sqrt{10} - \sqrt{6}}{2^2(\sqrt{5})^2 - (\sqrt{3})^2} = \frac{2\sqrt{10} - \sqrt{6}}{20 - 3} = \frac{2\sqrt{10} - \sqrt{6}}{17}$$

Example D

$$\frac{\sqrt{x - y}}{1 - \sqrt{x - y}} = \frac{\sqrt{x - y}(1 + \sqrt{x - y})}{(1 - \sqrt{x - y})(1 + \sqrt{x - y})}$$

$$= \frac{\sqrt{x - y} + x - y}{1 - x + y} \qquad (x > y)$$

Example E

$$\frac{1 + \dfrac{\sqrt{3}}{2}}{1 - \dfrac{\sqrt{3}}{2}} = \frac{\dfrac{2 + \sqrt{3}}{2}}{\dfrac{2 - \sqrt{3}}{2}} = \frac{2 + \sqrt{3}}{2} \cdot \frac{2}{2 - \sqrt{3}} = \frac{2 + \sqrt{3}}{2 - \sqrt{3}}$$

$$= \frac{(2 + \sqrt{3})(2 + \sqrt{3})}{(2 - \sqrt{3})(2 + \sqrt{3})} = \frac{4 + 4\sqrt{3} + 3}{4 - 3} = 7 + 4\sqrt{3}$$

Exercises 10—6

In Exercises 1 through 32 perform the indicated operations and express the answers in simplest form.

1. $\dfrac{\sqrt{21}}{\sqrt{3}}$

2. $\dfrac{\sqrt{105}}{\sqrt{5}}$

3. $\sqrt{7} \div \sqrt{2}$

4. $3\sqrt{2} \div 2\sqrt{3}$

5. $\dfrac{\sqrt[3]{x^2}}{\sqrt[3]{24}}$

6. $\dfrac{\sqrt{6}}{\sqrt[3]{2}}$

7. $\dfrac{\sqrt{a}}{\sqrt[3]{4}}$

8. $\dfrac{\sqrt[4]{32}}{\sqrt[5]{b^3}}$

9. $\dfrac{\sqrt{2a}-b}{\sqrt{a}}$

10. $\dfrac{\sqrt{8x}+\sqrt{2}}{\sqrt{2}}$

11. $\dfrac{\sqrt{3a}-\sqrt{b}}{\sqrt{3}}$

12. $\dfrac{\sqrt{7x}-\sqrt{14}}{\sqrt{7}}$

13. $\dfrac{1}{\sqrt{7}+\sqrt{3}}$

14. $\dfrac{4}{\sqrt{6}+\sqrt{2}}$

15. $\dfrac{\sqrt{7}}{\sqrt{5}-\sqrt{2}}$

16. $\dfrac{\sqrt{8}}{2\sqrt{3}-\sqrt{5}}$

17. $\dfrac{3}{2\sqrt{5}-6}$

18. $\dfrac{\sqrt{7}}{4-2\sqrt{7}}$

19. $\dfrac{2\sqrt{3}}{3\sqrt{3}-1}$

20. $\dfrac{6\sqrt{5}}{5-2\sqrt{5}}$

21. $\dfrac{\sqrt{2}-1}{\sqrt{7}-3\sqrt{2}}$

22. $\dfrac{3-\sqrt{5}}{2\sqrt{2}+\sqrt{5}}$

23. $\dfrac{2-\sqrt{3}}{5-2\sqrt{3}}$

24. $\dfrac{2\sqrt{15}-3}{\sqrt{15}+4}$

25. $\dfrac{2\sqrt{3}-5\sqrt{5}}{\sqrt{3}+2\sqrt{5}}$

26. $\dfrac{2\sqrt{6}+\sqrt{11}}{\sqrt{6}-3\sqrt{11}}$

27. $\dfrac{\sqrt{7}-\sqrt{14}}{2\sqrt{7}-3\sqrt{14}}$

28. $\dfrac{\sqrt{15}-3\sqrt{5}}{2\sqrt{15}-\sqrt{5}}$

29. $\dfrac{8}{3\sqrt{a}-2\sqrt{b}}$

30. $\dfrac{6}{1+2\sqrt{x}}$

31. $\dfrac{\sqrt{x+y}}{\sqrt{x-y}-\sqrt{x}}$

32. $\dfrac{\sqrt{1+a}}{a-\sqrt{1-a}}$

In Exercises 33 and 34 perform the indicated operations.

33. In the theory of semiconductors, the expression $km^{3/2}(E - E_1)^{1/2}$ is found. Write this expression in simplified radical form.

34. In the theory of waves in wires, the following expression is found:

$$\frac{\sqrt{d_1}-\sqrt{d_2}}{\sqrt{d_1}+\sqrt{d_2}}$$

Evaluate this expression if $d_1 = 10$ and $d_2 = 3$.

10–7 Exercises for Chapter 10

In Exercises 1 through 28 express each of the given expressions in the simplest form which contains only positive exponents.

1. $2a^{-2}b^0$

2. $(2c)^{-1}z^{-2}$

3. $\dfrac{2c^{-1}}{d^{-3}}$

4. $\dfrac{-5x^0}{3y^{-1}}$

5. $3(25)^{3/2}$

6. $32^{2/5}$

7. $400^{-3/2}$

8. $1000^{-2/3}$

9. $\left(\dfrac{3}{t^2}\right)^{-2}$

10. $\left(\dfrac{2x^3}{3}\right)^{-3}$

11. $\dfrac{-8^{2/3}}{49^{-1/2}}$

12. $\dfrac{81^{-3/4}}{7^{-1}}$

13. $(2a^{1/3}b^{5/6})^6$

14. $(ax^{-1/2}y^{1/4})^8$

15. $(-32m^{15}n^{10})^{3/5}$

16. $(27x^{-6}y^9)^{2/3}$

17. $2x^{-2}-y^{-1}$

18. $a^{-1}+b^{-2}$

19. $\dfrac{2x^{-1}}{x^{-1} + y^{-1}}$

20. $\dfrac{3a}{(2a)^{-1} - a}$

21. $(a - 3b^{-1})^{-1}$

22. $(2s^{-2} + t)^{-2}$

23. $(x^3 - y^{-3})^{1/3}$

24. $(x^2 + 2xy + y^2)^{-1/2}$

25. $(8a^3)^{2/3}(4a^{-2} + 1)^{1/2}$

26. $\left[\dfrac{(9a)^0 (4x^2)^{1/3} (3b^{1/2})}{(2b^0)^2} \right]^{-6}$

27. $2x(x - 1)^{-2} - 2(x^2 + 1)(x - 1)^{-3}$

28. $4(1 - x^2)^{1/2} - (1 - x^2)^{-1/2}$

In Exercises 29 through 68 perform the indicated operations and express the answer in simplest radical form.

29. $\sqrt{68}$

30. $\sqrt{96}$

31. $\sqrt{ab^5 c^2}$

32. $\sqrt{x^3 y^4 z^6}$

33. $\sqrt{9a^3 b^4}$

34. $\sqrt{8x^5 y^2}$

35. $\sqrt{84st^3 u^2}$

36. $\sqrt{52x^2 y^5}$

37. $\dfrac{5}{\sqrt{2s}}$

38. $\dfrac{3a}{\sqrt{5x}}$

39. $\sqrt{\dfrac{11}{27}}$

40. $\sqrt{\dfrac{7}{8}}$

41. $\sqrt[4]{8m^6 n^9}$

42. $\sqrt[3]{9a^7 b^{-3}}$

43. $\sqrt[4]{\sqrt[3]{64}}$

44. $\sqrt{a^{-3} \sqrt[5]{b^{12}}}$

45. $\sqrt{200} + \sqrt{32}$

46. $2\sqrt{68} - \sqrt{153}$

47. $\sqrt{63} - 2\sqrt{112} - \sqrt{28}$

48. $2\sqrt{20} - \sqrt{80} - 2\sqrt{125}$

49. $a\sqrt{2x^3} + \sqrt{8a^2 x^3}$

50. $2\sqrt{m^2 n^3} - \sqrt{n^5}$

51. $\sqrt[3]{8a^4} + b\sqrt[3]{a}$

52. $\sqrt[4]{2xy^5} - \sqrt[4]{32xy}$

53. $\sqrt{5}(2\sqrt{5} - \sqrt{11})$

54. $2\sqrt{8}(5\sqrt{2} - \sqrt{6})$

55. $2\sqrt{2}(\sqrt{6} - \sqrt{10})$

56. $3\sqrt{5}(\sqrt{15} + 2\sqrt{35})$

57. $(2 - 3\sqrt{17})(3 + \sqrt{17})$

58. $(5\sqrt{6} - 4)(3\sqrt{6} + 5)$

59. $(2\sqrt{7} - 3\sqrt{3})(3\sqrt{7} + \sqrt{3})$

60. $(3\sqrt{2} - \sqrt{13})(5\sqrt{2} + 3\sqrt{13})$

61. $\dfrac{\sqrt{2}}{\sqrt{3} - 4\sqrt{2}}$

62. $\dfrac{4}{3 - 2\sqrt{7}}$

63. $\dfrac{\sqrt{7} - \sqrt{5}}{\sqrt{5} + 3\sqrt{7}}$

64. $\dfrac{4 - 2\sqrt{6}}{3 + 2\sqrt{6}}$

65. $\sqrt{a^{-1} + b^2}$

66. $\sqrt{a^{-2} + \dfrac{1}{b^2}}$

67. $\left(\dfrac{2 - \sqrt{15}}{2} \right)^2 - \left(\dfrac{2 - \sqrt{15}}{2} \right)$

68. $\sqrt{2 + \dfrac{b}{a} + \dfrac{a}{b}} + \sqrt{a^4 b^2 + 2a^3 b^2 + a^2 b^2}$

In Exercises 69 through 76 perform the indicated operations.

69. In the study of electricity, the expression $e^{-i(\omega t - \alpha t)}$ is found. Rewrite this expression so that it contains no minus signs in the exponent.

70. An expression found when convection of heat is discussed is $k^{-1}x + h^{-1}$. Write this expression without the use of negative exponents.

71. In the study of fluid flow in pipes, the expression $0.220N^{-1/6}$ is found. Evaluate this expression for $N = 64 \times 10^6$.

72. In the study of biological effects of sound, the expression $(2n/\omega r)^{-1/2}$ is found. Express this in simplest radical form.

73. The root-mean-square velocity of a gas molecule is given by $v = \sqrt{3RT/M}$, where R is the gas constant, T is the thermodynamic temperature, and M is the molecular weight. Express the velocity (in meters per second) of an oxygen molecule in simplest radical form if $T = 300$ K, $R = 8.31$ J/mol $\cdot$ K, and $M = 0.032$ kg/mol. Then calculate the value of the expression.

74. A surveyor measuring distances with a steel tape must be careful to correct for the tension which is applied to the tape and for the sag in the tape. If she applies what is known as "normal tension," these two effects will cancel each other. An expression involving the normal tension T_n which is found for a certain tape is

$$\frac{0.2\,W \sqrt{2.7 \times 10^5}}{\sqrt{T_n - 20}}$$

Express this in simplest radical form.

75. The frequency of a certain electric circuit is given by

$$\frac{1}{2\pi \sqrt{\dfrac{LC_1C_2}{C_1 + C_2}}}$$

Express this in simplest radical form.

76. In determining the deflection of a certain type of beam, the expression

$$l \sqrt{\frac{a}{2l + a}}$$

is used. Express this in simplest radical form.

11

The *j*-Operator

In Chapter 1, when we were introducing the topic of numbers, imaginary numbers were mentioned. Again, when we considered quadratic equations and their solutions in Chapter 6, we briefly came across this type of number. However, until now we have purposely avoided any extended discussion of imaginary numbers. In this chapter we shall discuss the properties of these numbers and show some of the ways in which they may be applied.

When we defined radicals we were able to define square roots of positive numbers easily, since any positive or negative number squared equals a positive number. For this reason we can see that it is impossible to square any real number and have the product equal a negative number. We must define a new number system if we wish to include square roots of negative numbers. With the proper definitions, we shall find that these numbers can be used to great advantage in certain applications.

If the radicand in a square root is negative, we can express the indicated root as the product of $\sqrt{-1}$ and the square root of a positive number. *The symbol $\sqrt{-1}$ is defined as the* **imaginary unit,** *and is denoted by the symbol j.* (The symbol *i* is also often used for this purpose, but in electrical work *i* usually represents current. Therefore, we shall use *j* for the imaginary unit to avoid confusion.) In keeping with the definition of *j*, we have

$$j^2 = -1 \tag{11-1}$$

Example A

$$\sqrt{-9} = \sqrt{(9)(-1)} = \sqrt{9}\sqrt{-1} = 3j$$
$$\sqrt{-16} = \sqrt{16}\sqrt{-1} = 4j$$

Example B

$$(\sqrt{-4})^2 = (\sqrt{4}j)^2 = 4j^2 = -4$$

We note that the simplification of this expression does not follow Eq. (10–10), which states that $\sqrt{ab} = \sqrt{a}\sqrt{b}$ for square roots. This is the reason it was noted as being valid only if *a* and *b* are positive. In fact, Eq. (10–10) does not necessarily hold in general for negative values for *a* or *b*. If $(\sqrt{-4})^2$ did follow Eq. (10–10) we would have

$$(\sqrt{-4})^2 = \sqrt{(-4)(-4)} = \sqrt{16} = 4$$

We note that we obtain 4, and do not obtain the correct result of -4.

Example C
To further illustrate the method of handling square roots of negative numbers, consider the difference between $\sqrt{-3}\sqrt{-12}$ and $\sqrt{(-3)(-12)}$. For these expressions we have

$$\sqrt{-3}\sqrt{-12} = (\sqrt{3}j)(\sqrt{12}j) = \sqrt{3}\sqrt{12}j^2 = \sqrt{36}j^2$$
$$= 6(-1) = -6$$

and

$$\sqrt{(-3)(-12)} = \sqrt{36} = 6$$

For $\sqrt{-3}\sqrt{-12}$ we have the product of square roots of negative numbers, whereas for $\sqrt{(-3)(-12)}$ we have the product of negative numbers under the radical. We must be careful to note the difference.

From Examples B and C we see that *when we are dealing with the square roots of negative numbers,* **each should be expressed in terms of *j* before proceeding.** To do this, for any positive real number *a* we write

$$\sqrt{-a} = \sqrt{a}j, \quad (a > 0) \tag{11-2}$$

Example D

$$\sqrt{-6} = \sqrt{(6)(-1)} = \sqrt{6}\sqrt{-1} = \sqrt{6}j$$
$$\sqrt{-18} = \sqrt{(18)(-1)} = \sqrt{(9)(2)}\sqrt{-1} = 3\sqrt{2}j$$

In working with imaginary numbers, we often need to be able to raise these numbers to some power. Therefore, using the definitions of exponents and of *j*, we have the following results:

$$j = j,$$
$$j^2 = -1,$$
$$j^3 = j^2j = -j,$$
$$j^4 = j^2j^2 = (-1)(-1) = 1$$
$$j^5 = j^4j = j$$
$$j^6 = j^4j^2 = (1)(-1) = -1$$

The powers of *j* go through the cycle of $j, -1, -j, 1, j, -1, -j, 1$, and so forth. Noting this and the fact that *j* raised to a power which is a multiple of 4 equals one allows us to raise *j* to any integral power almost on sight.

Example E

$$j^{10} = j^8j^2 = (1)(-1) = -1$$
$$j^{45} = j^{44}j = (1)(j) = j$$
$$j^{531} = j^{528}j^3 = (1)(-j) = -j$$

Using real numbers and the imaginary unit *j*, we define a new kind of number. A **complex number** *is any number which can be written in the form a + bj, where a and b are real numbers. If a* = 0, *we have a number of the form bj, which is a* **pure imaginary number.** *If b* = 0, *then a + bj is a real number. The form a + bj is known as the* **rectangular form** *of a complex number, where a is known as the* **real part** *and bj is known as the* **imaginary part.** We can see that complex numbers include all the real numbers and all of the pure imaginary numbers.

A comment here about the words "imaginary" and "complex" is in order. The choice of the names of these numbers is historical in nature, and unfortunately it leads to some misconceptions about the numbers. The use of "imaginary" does not infer that the numbers do not exist. Imaginary numbers do in fact exist, as they are defined above. In the same way, the use of "complex" does not infer that the numbers are complicated and therefore difficult to understand. With the appropriate definitions and operations, we can work with complex numbers, just as with any type of number.

For complex numbers written in terms of *j* to follow all the operations defined in algebra, we define equality of two complex numbers in a special way. Complex numbers are not positive or negative in the ordinary sense of these terms, but the real and imaginary parts of complex numbers *are* positive or negative. *We define two complex numbers to be equal if the real parts are equal and the imaginary parts are equal.* That is, two complex numbers, *a + bj* and *x + yj*, are equal if *a* = *x* and *b* = *y*.

Example F

$$a + bj = 3 + 4j \text{ if } a = 3 \text{ and } b = 4$$
$$x + yj = 5 - 3j \text{ if } x = 5 \text{ and } y = -3$$

Example G

What values of x and y satisfy the equation $4 - 6j - x = j + jy$?

One way to solve this is to rearrange the terms so that all the known terms are on the right and all the terms containing the unknowns x and y are on the left. This leads to $-x - jy = -4 + 7j$. From the definition of equality of complex numbers, $-x = -4$ and $-y = 7$, or $x = 4$ and $y = -7$.

Example H

What values of x and y satisfy the equation

$$x + 3(xj + y) = 5 - j - jy$$

Rearranging the terms so that the known terms are on the right and the terms containing x and y are on the left, we have

$$x + 3y + 3jx + jy = 5 - j$$

Next, factoring j from the two terms on the left will put the expression on the left into proper form. This leads to

$$(x + 3y) + j(3x + y) = 5 - j$$

Using the definition of equality, we have

$$x + 3y = 5 \quad \text{and} \quad 3x + y = -1$$

We now solve this system of equations. The solution is $x = -1$ and $y = 2$. Actually, the solution can be obtained at any point by writing each side of the equation in the form $a + bj$ and then equating first the real parts and then the imaginary parts.

The **conjugate** of the complex number $a + bj$ is the complex number $a - bj$. We see that the sign of the imaginary part of a complex number is changed to obtain its conjugate.

Example I

$3 - 2j$ is the conjugate of $3 + 2j$. We may also say that $3 + 2j$ is the conjugate of $3 - 2j$. Thus each is the conjugate of the other.

Exercises 11–1

In Exercises 1 through 8 express each number in terms of j.

1. $\sqrt{-81}$ 2. $\sqrt{-121}$ 3. $-\sqrt{-4}$ 4. $-\sqrt{-0.01}$

5. $\sqrt{-8}$ 6. $\sqrt{-48}$ 7. $\sqrt{-\frac{7}{4}}$ 8. $\sqrt{-\frac{5}{3}}$

In Exercises 9 through 12 simplify each of the given expressions.

9. $(\sqrt{-7})^2; \sqrt{(-7)^2}$ 10. $\sqrt{(-15)^2}; (\sqrt{-15})^2$

11. $\sqrt{(-2)(-8)}; \sqrt{-2}\sqrt{-8}$ 12. $\sqrt{-9}\sqrt{-16}; \sqrt{(-9)(-16)}$

In Exercises 13 through 20 simplify the given expressions.

13. j^7 **14.** j^{49} **15.** $-j^{22}$ **16.** j^{408}

17. $j^2 - j^6$ **18.** $2j^5 - j^7$ **19.** $j^{15} - j^{13}$ **20.** $3j^{48} + j^{200}$

In Exercises 21 through 28 perform the indicated operations and simplify each complex number to its rectangular form $a + bj$.

21. $2 + \sqrt{-9}$ **22.** $-6 + \sqrt{-64}$ **23.** $2j^2 + 3j$ **24.** $j^3 - 6$

25. $\sqrt{18} - \sqrt{-8}$ **26.** $\sqrt{-27} + \sqrt{12}$

27. $(\sqrt{-2})^2 + j^4$ **28.** $(2\sqrt{2})^2 - (\sqrt{-1})^2$

In Exercises 29 through 32 find the conjugate of each complex number.

29. $6 - 7j$ **30.** $-3 + 2j$ **31.** $2j$ **32.** 6

In Exercises 33 through 40 find the values of x and y which satisfy the given equations.

33. $7x - 2yj = 14 + 4j$ **34.** $2x + 3jy = -6 + 12j$

35. $6j - 7 = 3 - x - yj$ **36.** $9 - j = xj + 1 - y$

37. $x - y = 1 - xj - yj - j$ **38.** $2x - 2j = 4 - 2xj - yj$

39. $x + 2 + 7j = yj - 2xj$ **40.** $2x + 6xj + 3 = yj - y + 7j$

In Exercises 41 and 42 answer the given questions.

41. What condition must be satisfied if a complex number and its conjugate are to be equal?

42. What type of number is a complex number if it is equal to the negative of its conjugate?

11–2 Basic Operations with Complex Numbers

The basic operations of addition, subtraction, multiplication, and division are defined in the same way for complex numbers in rectangular form as they are for real numbers. These operations are performed without regard for the fact that j has a special meaning. However, we must be careful to **express all complex numbers in terms of j before performing these operations.** Once this is done, we may proceed as with real numbers. We have the following definitions for these operations on complex numbers.

Addition (and subtraction):

$$(a + bj) + (c + dj) = (a + c) + (b + d)j \tag{11-3}$$

Multiplication:

$$(a + bj)(c + dj) = (ac - bd) + (ad + bc)j \tag{11-4}$$

Division:

$$\frac{a + bj}{c + dj} = \frac{(a + bj)(c - dj)}{(c + dj)(c - dj)} = \frac{(ac + bd) + (bc - ad)j}{c^2 + d^2} \tag{11-5}$$

Recalling Examples B and C of Section 11–1, we see the reason for expressing all complex numbers in terms of j before proceeding with any indicated operations.

We note from Eq. (11–3) that the addition or subtraction of complex numbers is accomplished by combining the real parts and combining the imaginary parts. Consider the following examples.

Example A

$$(3 - 2j) + (-5 + 7j) = (3 - 5) + (-2 + 7)j = -2 + 5j$$

Example B

$$(7 + 9j) - (6 - 4j) = 7 + 9j - 6 + 4j = 1 + 13j$$

Example C

$$(3\sqrt{-4} - 4) - (6 - 2\sqrt{-25}) - \sqrt{-81}$$
$$= [3(2j) - 4] - [6 - 2(5j)] - 9j$$
$$= [6j - 4] - [6 - 10j] - 9j$$
$$= 6j - 4 - 6 + 10j - 9j$$
$$= -10 + 7j$$

Here we note that our first step was to express the numbers in terms of j.

When complex numbers are multiplied, Eq. (11–4) indicates that we proceed as in any algebraic multiplication, properly expressing numbers in terms of j and evaluating powers of j. This is illustrated in Examples C and D.

Example D

$$(6 - \sqrt{-4})(\sqrt{-9}) = (6 - 2j)(3j) = 18j - 6j^2$$
$$= 18j - 6(-1) = 6 + 18j$$

Example E

$$(-9 - 6j)(2 + j) = -18 - 9j - 12j - 6j^2$$
$$= -18 - 21j - 6(-1) = -12 - 21j$$

We note that our procedure in dividing two complex numbers is the same procedure that we used for rationalizing the denominator of a fraction with a radical in the denominator. We use this procedure so that we can express any answer in the form of a complex number. We need merely to multiply numerator and denominator by the conjugate of the denominator in order to perform this operation.

Example F

$$\frac{7-2j}{3+4j} = \frac{7-2j}{3+4j} \cdot \frac{3-4j}{3-4j} = \frac{21-28j-6j+8j^2}{9-16j^2}$$

$$= \frac{21-34j+8(-1)}{9-16(-1)} = \frac{13-34j}{25}$$

This could be written in the form $a + bj$ as $\frac{13}{25} - \frac{34}{25}j$, but is generally left as a single fraction.

Example G

$$\frac{6+j}{2j} = \frac{6+j}{2j} \cdot \frac{-2j}{-2j} = \frac{-12j-2j^2}{4} = \frac{2-12j}{4} = \frac{1-6j}{2}$$

Example H

$$\frac{j^3+2j}{1-j^5} = \frac{-j+2j}{1-j} = \frac{j}{1-j} \cdot \frac{1+j}{1+j} = \frac{-1+j}{2}$$

Exercises 11—2

In Exercises 1 through 48 perform the indicated operations, expressing all answers in the form $a + bj$.

1. $(3 - 7j) + (2 - j)$
2. $(-4 - j) + (-7 - 4j)$
3. $(7j - 6) - (3 + j)$
4. $(2 - 3j) - (2 + 3j)$
5. $(4 + \sqrt{-16}) + (3 - \sqrt{-81})$
6. $(-1 + 3\sqrt{-4}) + (8 - 4\sqrt{-49})$
7. $(5 - \sqrt{-9}) - (\sqrt{-4} + 5)$
8. $(\sqrt{-25} - 1) - \sqrt{-9}$
9. $j - (j - 7) - 8$
10. $(7 - j) - (4 - 4j) + (6 - j)$
11. $(2\sqrt{-25} - 3) - (5 - 3\sqrt{-36}) - (\sqrt{-49})$
12. $(6 - 2\sqrt{-64}) - \sqrt{-100} - (\sqrt{-81} - 5)$
13. $(7 - j)(7j)$
14. $(-2j)(j - 5)$
15. $\sqrt{-16}(2\sqrt{-1} - 5)$
16. $(\sqrt{-4} - 1)(\sqrt{-9})$
17. $(4 - j)(5 + 2j)$
18. $(3 - 5j)(6 + 7j)$
19. $(2\sqrt{-9} - 3)(3\sqrt{-4} + 2)$
20. $(5\sqrt{-64} - 5)(7 + \sqrt{-16})$
21. $\sqrt{-18}\sqrt{-4}\sqrt{-9}$
22. $\sqrt{-6}\sqrt{-12}\sqrt{3}$
23. $(\sqrt{-5})^5$
24. $(\sqrt{-36})^4$
25. $\sqrt{-108} - \sqrt{-27}$
26. $2\sqrt{-54} + \sqrt{-24}$
27. $3\sqrt{-28} - 2\sqrt{12}$
28. $5\sqrt{24} - 3\sqrt{-45}$
29. $7j^3 - 7\sqrt{-9}$
30. $6j - 5j^2\sqrt{-63}$
31. $j\sqrt{-7} - j^6\sqrt{112} + 3j$
32. $j^2\sqrt{-7} - \sqrt{-28} + 8$
33. $(3 - 7j)^2$
34. $(4j + 5)^2$
35. $(1 - j)^3$
36. $(1 + j)(1 - j)^2$
37. $\dfrac{6j}{2 - 5j}$
38. $\dfrac{4}{3 + 7j}$
39. $\dfrac{2}{6 - \sqrt{-1}}$
40. $\dfrac{\sqrt{-4}}{2 + \sqrt{-9}}$

41. $\dfrac{1-j}{1+j}$ 42. $\dfrac{9-8j}{j-1}$ 43. $\dfrac{\sqrt{-2}-5}{\sqrt{-2}+3}$ 44. $\dfrac{2+3\sqrt{-3}}{5-\sqrt{-3}}$

45. $\dfrac{\sqrt{-16}-\sqrt{2}}{\sqrt{2}+j}$ 46. $\dfrac{1-\sqrt{-4}}{2+9j}$ 47. $\dfrac{j^2-j}{2j-j^8}$ 48. $\dfrac{j^5-j^3}{3+j}$

In Exercises 49 through 52 demonstrate the indicated properties.

49. Show that the sum of a complex number and its conjugate is a real number.

50. Show that the product of a complex number and its conjugate is a real number.

51. Show that the difference between a complex number and its conjugate is an imaginary number.

52. Show that the reciprocal of the imaginary unit is the negative of the imaginary unit.

11–3 Graphical Representation of Complex Numbers

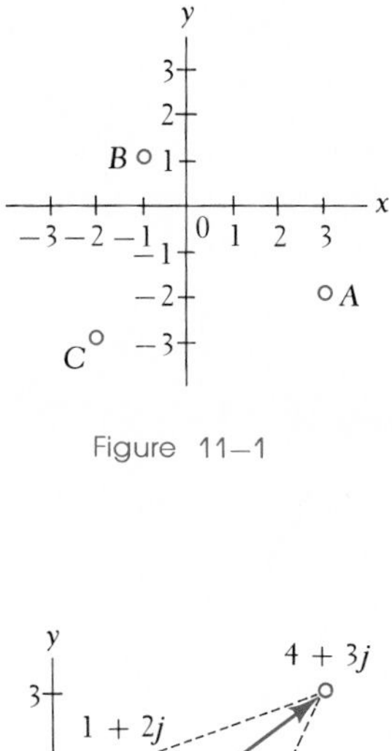

We showed in Section 1–1 how we could represent real numbers as points on a line. Because complex numbers include all real numbers as well as imaginary numbers, it is necessary to represent them graphically in a different way. Since there are two numbers associated with each complex number (the real part and the imaginary part), we find that we can represent complex numbers by representing the real parts by the x-values of the rectangular coordinate system, and the imaginary parts by the y-values. In this way *each complex number is represented as a point in the plane*, the point being designated as $a + bj$. When the rectangular coordinate system is used in this manner it is called the **complex plane**.

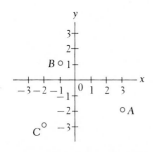

Figure 11–1

Example A

In Fig. 11–1, the point A represents the complex number $3 - 2j$; point B represents $-1 + j$; point C represents $-2 - 3j$. We note that these are equivalent to the points $(3, -2)$, $(-1, 1)$, and $(-2, -3)$. However, we must keep in mind that the meaning is different. Complex numbers were not included when we first learned to graph functions.

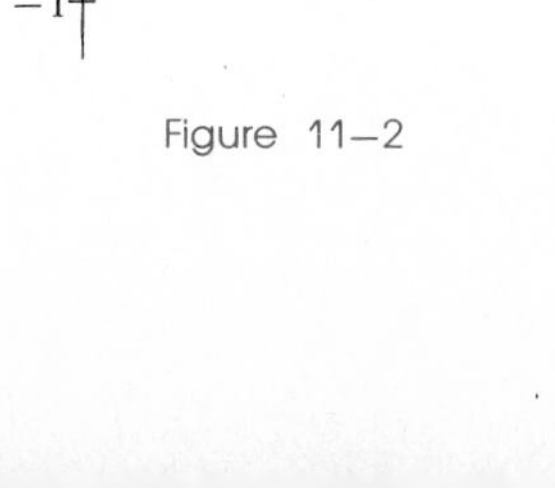

Let us represent two complex numbers and their sum in the complex plane. Consider, for example, the two complex numbers $1 + 2j$ and $3 + j$. By algebraic addition the sum is $4 + 3j$. When we draw lines from the origin to these points (see Fig. 11–2), we note that if we think of the complex numbers as being vectors, their sum is the vector sum. Because complex numbers can be used to represent vectors, these numbers are particularly important. Any complex number can be thought of as representing a vector from the origin to its point in the complex plane. To add two complex numbers graphically, we find the point corresponding to one of them and draw a line from the origin to this point. We

Figure 11–2

repeat this process for the second point. Next we complete a parallelo-gram with the lines drawn as adjacent sides. The resulting fourth vertex is the point representing the sum of the two complex numbers. Note that this is equivalent to adding vectors by graphical means.

Example B
Add the complex numbers $5 - 2j$ and $-2 - j$ graphically.

The solution is indicated in Fig. 11–3. We can see that the fourth vertex of the parallelogram is very near $3 - 3j$, which is, of course, the algebraic sum.

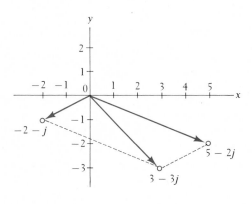

Figure 11–3   Figure 11–4

Example C
Subtract $4 - 2j$ from $2 - 3j$ graphically.

Subtracting $4 - 2j$ is equivalent to adding $-4 + 2j$. Thus, we com-plete the solution by adding $-4 + 2j$ and $2 - 3j$ (see Fig. 11–4). The result is $-2 - j$.

Example D
Show graphically that the sum of a complex number and its conjugate is a real number.

If we choose the complex number $a + bj$, we know that its conjugate is $a - bj$. The *y*-coordinate for the conjugate is as far below the *x*-axis as the *y*-coordinate of $a + bj$ is above it. Therefore, the sum of the imaginary parts must be zero and the sum of the two numbers must therefore lie on the *x*-axis, as shown in Fig. 11–5. Since any point on the *x*-axis is real, we have shown that the sum of $a + bj$ and $a - bj$ is real.

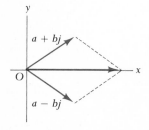

Figure 11–5

Exercises 11–3

In Exercises 1 through 4 locate the given complex numbers in the complex plane.

1. $2 + 6j$ 2. $-5 + j$ 3. $-4 - 3j$ 4. $3 - 4j$

In Exercises 5 through 20 perform the indicated operations graphically; check them algebraically.

5. $(5 - j) + (3 + 2j)$ 6. $(3 - 2j) + (-1 - j)$

7. $(2 - 4j) + (-2 + j)$ 8. $(-1 - 2j) + (6 - j)$
9. $(3 - 2j) - (4 - 6j)$ 10. $(2 - j) - j$
11. $(1 + 4j) - (3 + j)$ 12. $(-j - 2) - (-1 - 3j)$
13. $(4 - j) + (3 + 2j)$ 14. $(5 + 2j) - (-4 - 2j)$
15. $(3 - 6j) - (-1 + 5j)$ 16. $(-6 - 3j) + (2 - 7j)$
17. $(2j + 1) - 3j - (j + 1)$ 18. $(6 - j) - 9 - (2j - 3)$
19. $(j - 6) - j + (j - 7)$ 20. $j - (1 - j) + (3 + 2j)$

In Exercises 21 through 24 on the same coordinate system plot the given number, its negative, and its conjugate.

21. $3 + 2j$ 22. $-2 + 4j$ 23. $-3 - 5j$ 24. $5 - j$

11—4 Polar Form of a Complex Number

We have just seen the relationship between complex numbers and vectors. Since one can be used to represent the other, we shall use this fact to write complex numbers in another way. The new form has certain advantages when basic operations are performed on complex numbers.

By drawing a vector from the origin to the point in the complex plane which represents the number $x + yj$, we see the relation between vectors and complex numbers. Further observation indicates an angle in standard position has been formed. Also, the point $x + yj$ is r units from the origin. In fact, we can find any point in the complex plane by knowing this angle θ and the value of r. We have already developed the relations between x, y, r, and θ, in Eqs. (8–1) to (8–3). Let us rewrite these equations in a slightly different form. By referring to Eqs. (8–1) through (8–3) and to Fig. 11–6, we see that

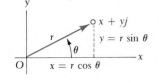

Figure 11—6

$$x = r \cos \theta, \qquad y = r \sin \theta \tag{11-6}$$

$$r^2 = x^2 + y^2, \qquad \tan \theta = \frac{y}{x} \tag{11-7}$$

Substituting Eqs. (11–6) into the rectangular form $x + yj$ of a complex number, we have

$$x + yj = r \cos \theta + j(r \sin \theta)$$

or

$$x + yj = r(\cos \theta + j \sin \theta) \tag{11-8}$$

The right side of Eq. (11–8) is called the **polar form** of a complex number. Sometimes it is referred to as the **trigonometric form**. Other notations which are used to represent the polar form are $r \angle \theta$ and r cis θ. The length r is called the **absolute value** or the **modulus**, and the angle θ is called the **argument** of the complex number. Therefore, Eq. (11–8), along with Eqs. (11–7), define the polar form of a complex number.

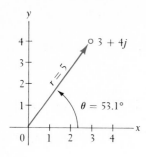

Figure 11–7

Example A

Represent the complex number $3 + 4j$ graphically, and give its polar form.

From the rectangular form $3 + 4j$ we see that $x = 3$ and $y = 4$. Using Eqs. (11–7), we have

$$r = \sqrt{3^2 + 4^2} = 5, \qquad \tan \theta = \frac{4}{3} = 1.333, \qquad \theta = 53.1°$$

Thus, the polar form is

$$5(\cos 53.1° + j \sin 53.1°)$$

The graphical representation is shown in Fig. 11–7.

A note on significant digits is in order here. In writing a complex number as $3 + 4j$ in Example A, no approximate values are intended. However, in expressing the polar form as $5(\cos 53.1° + j \sin 53.1°)$ we rounded off the angle to the nearest $0.1°$, as it is not possible to express the result exactly in degrees. Thus, in dealing with nonexact numbers we shall express trigonometric functions to four significant digits and angles to the nearest $0.1°$, for this is the accuracy in Table 4 in Appendix E. Other results, when approximate, will be expressed to three significant digits. Of course, in applied situations most numbers used are derived through measurement and are therefore approximate.

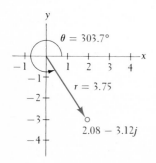

Figure 11–8

Example B

Represent the complex number $2.08 - 3.12j$ graphically, and give its polar form.

From Eqs. (11–7), we have

$$r = \sqrt{(2.08)^2 + (3.12)^2} = \sqrt{14.06} = 3.75$$

$$\tan \theta = \frac{-3.12}{2.08} = -1.500$$

Since we know that θ is a fourth-quadrant angle, we have $\theta = 303.7°$. Therefore, the polar form is

$$3.75(\cos 303.7° + j \sin 303.7°)$$

See Fig. 11–8.

Example C

Express the complex number $3.00(\cos 120.0° + j \sin 120.0°)$ in rectangular form.

From the given polar form, we know that $r = 3.00$ and $\theta = 120.0°$. Using Eqs. (11–6), we have

$$x = 3.00 \cos 120.0° = 3.00(-0.5000) = -1.50$$
$$y = 3.00 \sin 120.0° = 3.00(-0.8660) = 2.60$$

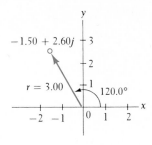

Figure 11-9

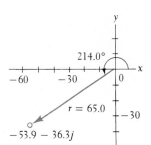

Figure 11-10

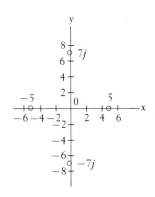

Figure 11-11

Therefore, the rectangular form is

$$-1.50 + 2.60j$$

See Fig. 11–9.

Example D

Express the complex number $65.0(\cos 214.0° + j \sin 214.0°)$ in rectangular form.

From the polar form, we have $r = 65.0$ and $\theta = 214.0°$. This means that

$$x = 65.0 \cos 214.0° = 65.0(-0.8290) = -53.9$$
$$y = 65.0 \sin 214.0° = 65.0(-0.5592) = -36.3$$

Therefore, the rectangular form is

$$-53.9 - 36.3j$$

See Fig. 11–10.

Example E

Represent the numbers 5, -5, $7j$, and $-7j$ in polar form.

Since any positive real number lies on the positive x-axis in the complex plane, real numbers are expressed in polar form by

$$a = a(\cos 0° + j \sin 0°)$$

Negative real numbers, being on the negative x-axis, are written as

$$a = |a|(\cos 180° + j \sin 180°)$$

Thus, $5 = 5(\cos 0° + j \sin 0°)$ and $-5 = 5(\cos 180° + j \sin 180°)$.

Positive pure imaginary numbers lie on the positive y-axis and are expressed in polar form by

$$bj = b(\cos 90° + j \sin 90°)$$

Similarly, negative pure imaginary numbers, being on the negative y-axis, are written as

$$-bj = |b|(\cos 270° + j \sin 270°)$$

Thus, $7j = 7(\cos 90° + j \sin 90°)$ and $-7j = 7(\cos 270° + j \sin 270°)$. The graphical representations of the *complex numbers* 5, -5, $7j$, and $-7j$ are shown in Fig. 11–11.

Exercises 11–4

In Exercises 1 through 16 represent each of the complex numbers graphically, and give the polar form of each number.

1. $8 + 6j$ 2. $-8 - 15j$ 3. $3 - 4j$

4. $-5 + 12j$ 5. $-2.00 + 3.00j$ 6. $7.00 - 5.00j$

7. $-5.50 - 2.40j$ 8. $4.60 - 4.60j$ 9. $1 + \sqrt{3}j$

10. $\sqrt{2} - \sqrt{2}j$ 11. $3.50 - 7.20j$ 12. $6.20 + 9.50j$

13. -3 14. 6 15. $9j$ 16. $-2j$

In Exercises 17 through 32 represent each of the complex numbers graphically, and give the rectangular form of each number.

17. $5.00(\cos 54.0° + j \sin 54.0°)$ 18. $3.00(\cos 232.0° + j \sin 232.0°)$

19. $1.60(\cos 150.0° + j \sin 150.0°)$ 20. $2.50(\cos 315.0° + j \sin 315.0°)$

21. $10.0(\cos 345.5° + j \sin 345.5°)$ 22. $220(\cos 155.2° + j \sin 155.2°)$

23. $6(\cos 180° + j \sin 180°)$ 24. $12(\cos 270° + j \sin 270°)$

25. $8(\cos 360° + j \sin 360°)$ 26. $15(\cos 0° + j \sin 0°)$

27. $4.75(\cos 172.8° + j \sin 172.8°)$ 28. $1.50(\cos 62.3° + j \sin 62.3°)$

29. $\cos 240.0° + j \sin 240.0°$ 30. $\cos 99.0° + j \sin 99.0°$

31. $28.0[\cos(-250.0°) + j \sin (-250.0°)]$ 32. $172[\cos(-105.0°) + j \sin(-105.0°)]$

In Exercises 33 and 34 solve the given problems.

33. Considering the relationship between complex numbers and vectors, what is the magnitude and direction of a displacement vector which is represented by $0.120 + 0.160j$ mm?

34. What is the magnitude and direction of a force which is represented by $3.72 - 5.12j$ lb? (See Exercise 33.)

11–5 Exponential Form of a Complex Number

Another important form of a complex number is known as the **exponential form**, which is written $re^{j\theta}$. In this expression r and θ have the same meaning as given in the last section and e represents a special irrational number equal to about 2.718. (In the calculus involved with exponential functions, the meaning of e is clarified.) We now define

$$re^{j\theta} = r(\cos \theta + j \sin \theta) \tag{11–9}$$

When θ is expressed in radians, the expression $j\theta$ is an actual exponent, and it can be shown to obey all the laws of exponents. For this reason and because it is more meaningful in applications, *we shall always express θ in radians when using the exponential form.* The following examples show how complex numbers can be changed to and from exponential form.

Example A

Express the number $3 + 4j$ in exponential form.

From Example A of Section 11–4, we know that this complex number may be written in polar form as $5(\cos 53.1° + j \sin 53.1°)$. Therefore, we know that $r = 5$. We now express $53.1°$ in terms of radians as

$$\frac{53.1\pi}{180} = 0.927 \text{ rad}$$

Thus, the exponential form is $5e^{0.927j}$. This means that

$$3 + 4j = 5(\cos 53.1° + j \sin 53.1°) = 5e^{0.927j}$$

Example B

Express the number 8.50(cos 136.3° + j sin 136.3°) in exponential form.

Since this complex number is in polar form we note that $r = 8.50$ and that we must express 136.3° in radians. Therefore, changing 136.3° to radians, we have

$$\frac{136.3\pi}{180} = 2.38 \text{ rad}$$

Therefore, the required exponential form is $8.50e^{2.38j}$. This means that

$$8.50(\cos 136.3° + j \sin 136.3°) = 8.50e^{2.38j}$$

Example C

Express the number $3.07 - 7.43j$ in exponential form.

From the rectangular form of the number, we have $x = 3.07$ and $y = -7.43$. Therefore,

$$r = \sqrt{(3.07)^2 + (-7.43)^2} = \sqrt{64.63} = 8.04$$

$$\tan \theta = \frac{-7.43}{3.07} = -2.420, \qquad \theta = 292.5°$$

Changing 292.5° to radians, we have 292.5° = 5.10 rad. Therefore, the exponential form is $8.04e^{5.10j}$. This means that

$$3.07 - 7.43j = 8.04e^{5.10j}$$

Example D

Express the complex number $2.00e^{4.80j}$ in polar and rectangular forms.

We first express 2.46 rad as 141.0°. From the exponential form we know that $r = 3.40$. Thus, the polar form is 3.40(cos 141.0° + j sin 141.0°). Next we find that cos 141.0° = -0.7771 and sin 141.0° = 0.6293. The rectangular form is $3.40(-0.7771 + 0.6293j) = -2.64 + 2.14j$. This means that

$$2.00e^{4.80j} = 2.00(\cos 275.0° + j \sin 275.0°) = 0.174 - 1.99j$$

Example E

Express the complex number $3.40e^{2.46j}$ in polar and rectangular forms.

We first express 2.46 rad as 141.0°. From the exponential form we know that $r = 3.40$. Thus, the polar form is 3.40(cos 141.0° + j sin 141.0°). Next we find that cos 141.0° = -0.7771 and sin 141.0° = 0.6293. The rectangular form is $3.40(-0.7771 + 0.6293j) = -2.64 + 2.14j$. This means that

$$3.40e^{2.46j} = 3.40(\cos 141.0° + j \sin 141.0°) = -2.64 + 2.14j$$

An important application of the use of complex numbers is in alternating current. When an alternating current flows through a given circuit, usually the current and voltage have different phases. That is, they do not reach their peak values at the same time. Therefore, one way of accounting for the magnitude as well as the phase of an electric current or voltage is to write it as a complex number. Here the modulus is the actual value of the current or voltage, and the argument is a measure of the phase.

Example F

A current of $2.00 - 4.00j$ amperes flows through a given circuit. Write this current in exponential form and determine the value of current in the circuit.

From the rectangular form, we have $x = 2.00$ and $y = -4.00$. Therefore, $r = \sqrt{(2.00)^2 + (-4.00)^2} = \sqrt{20.00} = 4.47$. Also, $\tan \theta = -\frac{4.00}{2.00} = -2.000$. Since $\tan 63.4° = 2.000$, $\theta = -63.4°$ (it is normal to express the phase in terms of negative angles). Changing $63.4°$ to radians, we have $63.4° = 1.11$ rad. Therefore, the exponential form of the current is $4.47e^{-1.11j}$. The modulus is 4.47, meaning the current is 4.47 A.

At this point we shall summarize the three important forms of a complex number.

Rectangular:	$x + yj$
Polar:	$r(\cos \theta + j \sin \theta)$
Exponential:	$re^{j\theta}$

It follows that

$$x + yj = r(\cos \theta + j \sin \theta) = re^{j\theta} \qquad (11\text{--}10)$$

where

$$r^2 = x^2 + y^2, \qquad \tan \theta = \frac{y}{x} \qquad (11\text{--}7)$$

Exercises 11—5

In Exercises 1 through 16 express the given complex numbers in exponential form.

1. $3.00(\cos 60.0° + j \sin 60.0°)$
2. $5.00(\cos 135.0° + j \sin 135.0°)$
3. $4.50(\cos 282.0° + j \sin 282.0°)$
4. $2.10(\cos 228.0° + j \sin 228.0°)$
5. $375(\cos 95.0° + j \sin 95.0°)$
6. $16.0(\cos 7.0° + j \sin 7.0°)$
7. $0.515(\cos 198.3° + j \sin 198.3°)$
8. $4650(\cos 326.5° + j \sin 326.5°)$
9. $3 - 4j$
10. $-1 - 5j$
11. $-3 + 2j$
12. $6 + j$
13. $5.90 + 2.40j$
14. $47.3 - 10.9j$
15. $-634 - 528j$
16. $-8570 + 5470j$

In Exercises 17 through 24 express the given complex numbers in polar and rectangular forms.

17. $3.00e^{0.500j}$
18. $2.00e^{1.00j}$
19. $4.64e^{1.85j}$
20. $2.50e^{3.84j}$
21. $3.20e^{5.41j}$
22. $0.800e^{3.00j}$
23. $0.172e^{2.39j}$
24. $820e^{3.49j}$

In Exercises 25 and 26 perform the indicated operations.

25. The electric current in a certain alternating-current circuit is given by $0.500 + 0.220j$ amperes. Write this current in exponential form and determine the magnitude of the current in the circuit.

26. The voltage in a certain alternating-current circuit is $125e^{1.31j}$. Determine the magnitude of the voltage in the circuit and the in-phase component (the real part) of the voltage.

11—6 Products, Quotients, Powers, and Roots of Complex Numbers

We have previously performed products and quotients using the rectangular forms of the given numbers. However, these operations can also be performed with complex numbers in polar and exponential forms. We find that these operations are convenient, and also useful for purposes of finding powers and roots of complex numbers.

We may find the product of two complex numbers by using the exponential form and the laws of exponents. Multiplying $r_1 e^{j\theta_1}$ by $r_2 e^{j\theta_2}$, we have

$$r_1 e^{j\theta_1} \cdot r_2 e^{j\theta_2} = r_1 r_2 e^{j\theta_1 + j\theta_2} = r_1 r_2 e^{j(\theta_1 + \theta_2)} \tag{11-11}$$

We use Eq. (11–11) to express the product of two complex numbers in polar form:

$$r_1 e^{j\theta_1} \cdot r_2 e^{j\theta_2} = r_1(\cos\theta_1 + j\sin\theta_1) \cdot r_2(\cos\theta_2 + j\sin\theta_2)$$

and

$$r_1 r_2 e^{j(\theta_1 + \theta_2)} = r_1 r_2[\cos(\theta_1 + \theta_2) + j\sin(\theta_1 + \theta_2)]$$

Therefore, the polar expressions are equal:

$$r_1(\cos\theta_1 + j\sin\theta_1)r_2(\cos\theta_2 + j\sin\theta_2) = r_1 r_2[\cos(\theta_1 + \theta_2) + j\sin(\theta_1 + \theta_2)] \tag{11-12}$$

Example A

Multiply the complex numbers $2 + 3j$ and $1 - j$ by using the polar form of each.

$$r_1 = \sqrt{4+9} = 3.61, \qquad \tan\theta_1 = 1.500, \qquad \theta_1 = 56.3°$$
$$r_2 = \sqrt{1+1} = 1.41, \qquad \tan\theta_2 = -1.000, \quad \theta_2 = 315.0°$$
$$(3.61)(\cos 56.3° + j\sin 56.3°)(1.41)(\cos 315.0° + j\sin 315.0°)$$
$$= (3.61)(1.41)[\cos(56.3° + 315.0°) + j\sin(56.3° + 315.0°)]$$
$$= 5.09(\cos 371.3° + j\sin 371.3°)$$
$$= 5.09(\cos 11.3° + j\sin 11.3°)$$

Example B
When we use the exponential form to multiply the two complex numbers in Example A, we have:

$$r_1 = 3.61, \qquad \theta_1 = 56.3° = 0.983 \text{ rad}$$
$$r_2 = 1.41, \qquad \theta_2 = 315° = 5.50 \text{ rad}$$
$$3.61\,e^{0.983j}1.41\,e^{5.50j} = 5.09e^{6.48j} = 5.09e^{0.20j}$$

If we wish to *divide* one complex number in exponential form by another, we arrive at the following result:

$$r_1 e^{j\theta_1} \div r_2 e^{j\theta_2} = \frac{r_1}{r_2} e^{j(\theta_1 - \theta_2)} \tag{11–13}$$

Therefore, the result of dividing one number in polar form by another is given by:

$$\frac{r_1}{r_2} \angle \theta_1 - \theta_2$$

$$\frac{r_1(\cos\theta_1 + j\sin\theta_1)}{r_2(\cos\theta_2 + j\sin\theta_2)} = \frac{r_1}{r_2}[\cos(\theta_1 - \theta_2) + j\sin(\theta_1 - \theta_2)] \tag{11–14}$$

Example C
Divide the first complex number of Example A by the second. Using polar form, we have the following:

$$\frac{3.61(\cos 56.3° + j\sin 56.3°)}{1.41(\cos 315.0° + j\sin 315.0°)} = \frac{3.61}{1.41}[\cos(56.3° - 315.0°)$$
$$+ j\sin(56.3° - 315.0°)]$$
$$= 2.56[\cos(-258.7°) + j\sin(-258.7°)]$$
$$= 2.56(\cos 101.3° + j\sin 101.3°)$$

Example D
Repeating Example C, using exponential forms, we obtain

$$\frac{3.61e^{0.983j}}{1.41e^{5.50j}} = 2.56e^{-4.52j} = 2.56e^{1.76j}$$

To raise a complex number to a power, we simply multiply one complex number by itself the required number of times. For example, in Eq. (11–11), if the two numbers being multiplied are equal, we have (letting $r_1 = r_2 = r$ and $\theta_1 = \theta_2 = \theta$)

$$(re^{j\theta})^2 = r^2 e^{j2\theta} \tag{11–15}$$

Multiplying the expression in Eq. (11–15) by $re^{j\theta}$ gives $r^3 e^{j3\theta}$. This leads to the general expression for raising a complex number to the *n*th power,

$$(re^{j\theta})^n = r^n e^{jn\theta} \tag{11–16}$$

Extending this to polar form, we have

$$[r(\cos \theta + j \sin \theta)]^n = r^n(\cos n\theta + j \sin n\theta) \qquad (11\text{–}17)$$

Equation (11–17) is known as **DeMoivre's theorem.** It is valid for all real values of n, and may also be used for finding the roots of complex numbers if n is a fractional exponent.

Example E

Using DeMoivre's theorem, find $(2 + 3j)^3$.

From Example A of this section, we know $r = 3.61$ and $\theta = 56.3°$. Thus, we have

$$[3.61 (\cos 56.3° + j \sin 56.3°)]^3$$
$$= (3.61)^3[\cos 3 \times 56.3° + j \sin 3 \times 56.3°]$$
$$= 47.0(\cos 168.9° + j \sin 168.9°)$$

From Example B we know that $\theta = 0.983$ rad. Thus, in exponential form,

$$(3.61e^{0.983j})^3 = (3.61)^3 e^{3 \times 0.983j} = 47.0e^{2.95j}$$

Therefore,

$$(2 + 3j)^3 = 47.0(\cos 168.9° + j \sin 168.9°) = 47.0e^{2.95j}$$

Example F

Find $\sqrt[3]{-1}$.

Since we know that -1 is a real number, we can find its cube root by means of the definition. That is, $(-1)^3 = -1$. We shall check this by DeMoivre's theorem. Writing -1 in polar form, we have

$$-1 = 1(\cos 180° + j \sin 180°)$$

Applying DeMoivre's theorem, with $n = \frac{1}{3}$, we obtain

$$(-1)^{1/3} = 1^{1/3}(\cos \frac{1}{3}180° + j \sin \frac{1}{3}180°) = \cos 60° + j \sin 60°$$

$$= 0.500 + 0.866j$$

We note that we did not obtain -1 as an answer. If we check the answer which was obtained, in the form $\frac{1}{2}(1 + \sqrt{3}j)$, by actually cubing it, we obtain -1! Thus it is a correct answer.

We should note that it is possible to take $\frac{1}{3}$ of any angle up to 1080° and still have an angle less than 360°. Since 180° and 540° have the same terminal side, let us try writing -1 as $1(\cos 540° + j \sin 540°)$. Using DeMoivre's theorem, we have

$$(-1)^{1/3} = 1^{1/3}(\cos \frac{1}{3}540° + j \sin \frac{1}{3}540°)$$

$$= \cos 180° + j \sin 180° = -1$$

We have found the answer we originally anticipated.

Angles of 180° and 900° also have the same terminal side, so we try

$$(-1)^{1/3} = 1^{1/3}(\cos \frac{1}{3}900° + j \sin \frac{1}{3}900°) = \cos 300° + j \sin 300°$$

$$= 0.500 - 0.866j$$

Checking this, we find that it is also a correct root. We may try 1260°, but $\frac{1}{3}(1260°) = 420°$, which has the same functional values as 60°, and would give us the answer $0.500 + 0.866j$ again.

We have found, therefore, three cube roots of -1. They are -1, $0.500 + 0.866j$, and $0.500 - 0.866j$. When this is generalized, it can be proved that there are n nth roots of any complex number. The method for finding the n roots is to use θ to find one root, and then add 360° to θ, $n - 1$ times, in order to find the other roots.

Example G

Find the square roots of j.

We must first properly write j in polar form so that we may use DeMoivre's theorem to find the roots. In polar form, j is

$$j = 1(\cos 90° + j \sin 90°)$$

To find the square roots, we apply DeMoivre's theorem with $n = \frac{1}{2}$.

$$j^{1/2} = 1^{1/2}\left(\cos \frac{90°}{2} + j \sin \frac{90°}{2}\right) = \cos 45° + j \sin 45° = 0.707 + 0.707j$$

To find the other square root using DeMoivre's theorem, we must write j in polar form as

$$j = 1[\cos(90° + 360°) + j \sin(90° + 360°)] = 1(\cos 450° + j \sin 450°).$$

Applying DeMoivre's theorem to j in this form, we have

$$j^{1/2} = 1^{1/2}\left(\cos \frac{450°}{2} + j \sin \frac{450°}{2}\right) = \cos 225° + j \sin 225°$$

$$= -0.707 - 0.707j$$

Thus, the two square roots of j are $0.707 + 0.707j$ and $-0.707 - 0.707j$.

Example H

Find the six 6th roots of 1.

Here we shall use directly the method for finding the roots of a number, as outlined at the end of Example F:

$$1 = 1(\cos 0° + j \sin 0°)$$

First root: $1^{1/6} = 1^{1/6}\left(\cos \frac{0°}{6} + j \sin \frac{0°}{6}\right) = \cos 0° + j \sin 0° = 1$

Second root: $1^{1/6} = 1^{1/6}\left(\cos \frac{0° + 360°}{6} + j \sin \frac{0° + 360°}{6}\right)$

$$= \cos 60° + j \sin 60° = \frac{1}{2} + j \frac{\sqrt{3}}{2}$$

Third root: $1^{1/6} = 1^{1/6}\left(\cos\dfrac{0° + 720°}{6} + j\sin\dfrac{0° + 720°}{6}\right)$

$$= \cos 120° + j\sin 120° = -\frac{1}{2} + j\frac{\sqrt{3}}{2}$$

Fourth root: $1^{1/6} = 1^{1/6}\left(\cos\dfrac{0° + 1080°}{6} + j\sin\dfrac{0° + 1080°}{6}\right)$

$$= \cos 180° + j\sin 180° = -1$$

Fifth root: $1^{1/6} = 1^{1/6}\left(\cos\dfrac{0° + 1440°}{6} + j\sin\dfrac{0° + 1440°}{6}\right)$

$$= \cos 240° + j\sin 240° = -\frac{1}{2} - j\frac{\sqrt{3}}{2}$$

Sixth root: $1^{1/6} = 1^{1/6}\left(\cos\dfrac{0° + 1800°}{6} + j\sin\dfrac{0° + 1800°}{6}\right)$

$$= \cos 300° + j\sin 300° = \frac{1}{2} - j\frac{\sqrt{3}}{2}$$

At this point we can see advantages for the various forms of writing complex numbers. Rectangular form lends itself best to addition and subtraction. Polar form is generally used for multiplying, dividing, raising to powers, and finding roots. Exponential form is used for theoretical purposes (e.g., deriving DeMoivre's theorem).

Exercises 11–6

In Exercises 1 through 12 perform the indicated operations. Leave the result in polar form.

1. $[4(\cos 60° + j\sin 60°)][2(\cos 20° + j\sin 20°)]$
2. $[3(\cos 120° + j\sin 120°)][5(\cos 45° + j\sin 45°)]$
3. $[0.5(\cos 140° + j\sin 140°)][6(\cos 110° + j\sin 110°)]$
4. $[0.4(\cos 320° + j\sin 320°)][5.5(\cos 150° + j\sin 150°)]$

5. $\dfrac{8(\cos 100° + j\sin 100°)}{4(\cos 65° + j\sin 65°)}$

6. $\dfrac{9(\cos 230° + j\sin 230°)}{3(\cos 80° + j\sin 80°)}$

7. $\dfrac{12(\cos 320° + j\sin 320°)}{5(\cos 210° + j\sin 210°)}$

8. $\dfrac{2(\cos 90° + j\sin 90°)}{4(\cos 75° + j\sin 75°)}$

9. $[2(\cos 35° + j\sin 35°)]^3$

10. $[3(\cos 120° + j\sin 120°)]^4$

11. $[2(\cos 135° + j\sin 135°)]^8$

12. $(\cos 142° + j\sin 142°)^{10}$

In Exercises 13 through 24 change each number to polar form and then perform the indicated operations. Express the final result in rectangular and polar forms. Check by performing the same operation in rectangular form.

13. $(3 + 4j)(5 - 12j)$

14. $(-2 + 5j)(-1 - j)$

15. $(7 - 3j)(8 + j)$

16. $(1 + 5j)(4 + 2j)$

17. $\dfrac{7}{1 - 3j}$

18. $\dfrac{8j}{7 + 2j}$

19. $\dfrac{3 + 4j}{5 - 12j}$ **20.** $\dfrac{-2 + 5j}{-1 - j}$ **21.** $(3 + 4j)^4$

22. $(-1 - j)^8$ **23.** $(2 + 3j)^5$ **24.** $(1 - 2j)^6$

In Exercises 25 through 32 use DeMoivre's theorem to find the indicated roots. Be sure to find all roots.

25. $\sqrt{4(\cos 60° + j \sin 60°)}$ **26.** $\sqrt[3]{27(\cos 120° + j \sin 120°)}$ **27.** $\sqrt[3]{3 - 4j}$

28. $\sqrt{-5 + 12j}$ **29.** $\sqrt[4]{1}$ **30.** $\sqrt[3]{8}$ **31.** $\sqrt[3]{-j}$ **32.** $\sqrt[4]{j}$

In Exercises 33 and 34 perform the indicated multiplications.

33. In Example F we showed that one cube root of -1 is $0.500 - 0.866j$. The exact form of this root is $\frac{1}{2}(1 - \sqrt{3}j)$. Cube this expression in rectangular form and show that the result is -1.

34. In Example G we showed that one of the square roots of j is equal to $0.707 + 0.707j$. The exact form of this root is $\frac{1}{2}\sqrt{2}(1 + j)$. Square this expression and show that the result is j.

11–7 An Application to Alternating-Current (AC) Circuits

omit.

We shall complete our study of the *j*-operator by showing its use in one aspect of alternating-current circuit theory. This application will be made to measuring voltage between any two points in a simple ac circuit, similar to the application mentioned in Section 11–5. We shall consider a circuit containing a resistance, a capacitance, and an inductance.

Briefly, a resistance is any part of a circuit which tends to obstruct the flow of electric current through the circuit. It is denoted by R (units in ohms, with symbol Ω) and in diagrams by —$\wedge\!\wedge$— . In essence, a capacitance is two nonconnected plates in a circuit; no current actually flows across the gap between them. In an ac circuit, an electric charge is continually going to and from each plate and, therefore, the current in the circuit is not effectively stopped. It is denoted by C (units in farads, with symbol F) and in diagrams by —$\dashv\vdash$— (Fig. 11–12). An inductance, basically, is a coil of wire in which current is induced because the current is continuously changing in the circuit. It is denoted by L (units in henrys, with symbol H) and in diagrams by —$\widehat{\text{mm}}$—. All these elements affect the voltage in an alternating-current circuit. We shall state here the relation each has to the voltage and current in the circuit.

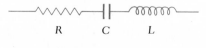

$$\quad R \qquad\quad C \qquad\quad L$$

Figure 11–12

In Chapter 9, when we were discussing the graphs of the trigonometric functions, we noted that the current and voltage in an ac circuit could be represented by a sine or cosine curve. Therefore, each reaches peak values periodically. If they reach their respective peak values at the same time, we say they are *in phase*. If the voltage reaches its peak before the current, we say that the voltage *leads* the current. If the voltage reaches its peak after the current, we say that the voltage *lags* the current. In the study of electricity it is shown that the voltage across a resistance is in phase with the current. The voltage across a capacitor lags the current by 90°, and the voltage across an inductance leads the current by 90°.

Each element in an ac circuit tends to offer a type of resistance to the flow of current. The effective resistance of any part of the circuit is called the **reactance,** and it is denoted by X. The voltage across any part of the circuit whose reactance is X is given $V = IX$, where I is the current (in amperes) and V is the voltage (in volts). Therefore, the voltage across a resistor, capacitor, and inductor, is, respectively

$$V_R = IX_R, \qquad V_C = IX_C, \qquad V_L = IX_L \qquad (11{-}18)$$

To determine the voltage across a combination of these elements of a circuit, we must account for the reactance as well as the phase of the voltage across the individual elements. Since the voltage across a resistor is in phase with the current, we shall represent X_R along the x-axis as a real number R (the actual value of the resistance). Since the voltage across an inductance *leads* the current by 90°, we shall represent this voltage as a positive, pure imaginary number. In the same way, by representing the voltage across a capacitor as a negative, pure imaginary number, we show that the voltage lags the current by 90°. These representations are meaningful since the positive y-axis (positive, pure imaginary numbers) is +90° from the positive x-axis, and the negative y-axis (negative, pure imaginary numbers) is −90° from the positive x-axis.

The total voltage across a combination of all three elements is given by $V_R + V_L + V_C$, which we shall represent by V_{RLC}. Therefore,

$$V_{RLC} = IR + IX_L j - IX_C j = I[R + j(X_L - X_C)]$$

This expression is also written as

$$V_{RLC} = IZ \qquad (11{-}19)$$

where the symbol Z is called the **impedance** of the circuit. It is the total effective resistance to the flow of current by a combination of the elements in the circuit, taking into account the phase of the voltage in each element. From its definition, we see that Z is a complex number,

$$Z = R + j(X_L - X_C) \qquad (11{-}20)$$

with a magnitude

$$|Z| = \sqrt{R^2 + (X_L - X_C)^2} \qquad (11{-}21)$$

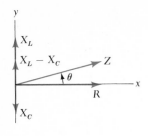

Figure 11–13

Also, as a complex number, it makes an angle θ with the x-axis given by

$$\tan \theta = \frac{X_L - X_C}{R} \qquad (11\text{--}22)$$

All these equations are based on phase relations of voltages with respect to the current. Therefore, the angle θ represents the phase angle between the current and the voltage (see Fig. 11–13).

In the examples and exercises of this section, the commonly used units and symbols for them are used. For a summary of these units and symbols, including prefixes, see Appendix B.

Example A

In the series circuit shown in Fig. 11–14(a), $R = 12.0\ \Omega$ and $X_L = 5.00\ \Omega$. A current of 2.00 A is in the circuit. Find the voltage across each element, the impedance, the voltage across the combination, and the phase angle between the current and voltage.

Since the voltage across any element is the product of the current and reactance, we have the voltage across the resistor (between points a and b) as $V_R = (2.00)(12.0) = 24.0$ V. The voltage across the inductor (between points b and c) is $V_L = (2.00)(5.00) = 10.0$ V. To find the voltage across the combination, between points a and c, we must first find the magnitude of the impedance. The voltage is *not* the arithmetic sum of V_R and V_L; we must account for the phase. By Eq. (11–20), the impedance is $Z = 12.0 + 5.00j$ with magnitude

$$|Z| = \sqrt{R^2 + X_L^2} = \sqrt{(12.0)^2 + (5.00)^2} = \sqrt{169} = 13.0\ \Omega$$

Therefore, the voltage across the combination is

$$V_{RL} = (2.00)(13.0) = 26.0\ \text{V}$$

The phase angle between the voltage and current is found by Eq. (11–22). This gives $\tan \theta = \frac{5.00}{12.0} = 0.4167$ which means that $\theta = 22.6°$. The voltage leads the current by 22.6°, as shown in Fig. 11–14(b).

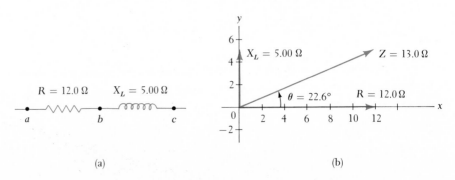

(a)   (b)

Figure 11–14

Example B

For a circuit in which $R = 8.00 \ \Omega$, $X_L = 7.00 \ \Omega$, and $X_C = 13.0 \ \Omega$, find the impedance and the phase angle between the current and the voltage.

By the definition of impedance, Eq. (11–20), we have

$$Z = 8.00 + (7.00 - 13.0)j = 8.00 - 6.00j$$

where the magnitude of the impedance is

$$|Z| = \sqrt{(8.00)^2 + (-6.00)^2} = \sqrt{64.0 + 36.0} = 10.0 \ \Omega$$

The phase angle is found by

$$\tan \theta = \frac{-6.00}{8.00} = -0.7500$$

Therefore, $\theta = -36.9°$ (negative angles are used, having a useful purpose in this type of problem). This means that the voltage lags the current by 36.9° (see Fig. 11–15).

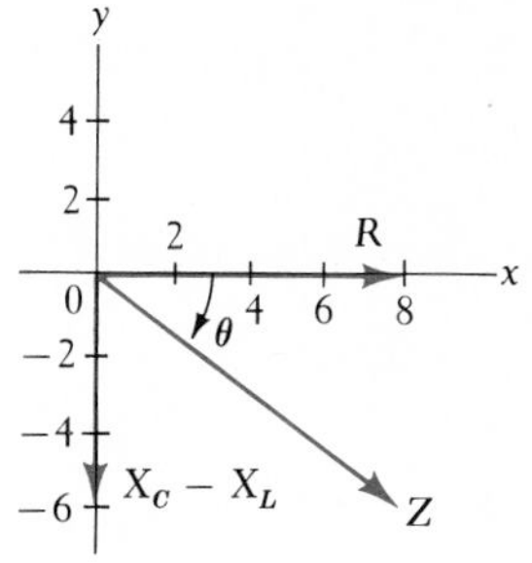

Example C

Let $R = 6.00 \ \Omega$, $X_L = 8.00 \ \Omega$, and $X_C = 4.00 \ \Omega$. Find the impedance and the phase angle between the current and voltage.

$$Z = 6.00 + (8.00 - 4.00)j = 6.00 + 4.00j$$

$$|Z| = \sqrt{(6.00)^2 + (4.00)^2} = \sqrt{36.0 + 16.0} = 7.21 \ \Omega$$

$$\tan \theta = \frac{4.00}{6.00} = 0.6667, \qquad \theta = 33.7°$$

The voltage leads the current by 33.7° (see Fig. 11–16).

Note that the resistance is represented in the same way as a vector along the positive x-axis. Actually resistance is not a vector quantity, but is represented in this manner in order to assign an angle as the phase of the current. The important concept in this analysis is that the phase *difference* between the current and voltage is constant, and therefore any direction may be chosen arbitrarily for one of them. Once this choice is made, other phase angles are measured with respect to this direction. A common choice, as above, is to make the phase angle of the current zero. If an arbitrary angle is chosen, it is necessary to treat the current, voltage, and impedance as complex numbers.

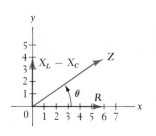

Figure 11—15

Figure 11—16

Example D

In a particular circuit, the current is $2.00 - 3.00j$ amperes and the impedance is $6.00 + 2.00j$ ohms. The voltage across this part of the circuit is

$$V = (2.00 - 3.00j)(6.00 + 2.00j) = 12.0 - 14.0j - 6.00j^2$$
$$= 12.0 - 14.0j + 6.00$$
$$= 18.0 - 14.0 \text{ volts}$$

The magnitude of the voltage is

$$|V| = \sqrt{(18.0)^2 + (-14.0)^2} = \sqrt{324 + 196} = \sqrt{520} = 22.8 \text{ V}$$

Since the voltage across a resistor is in phase with the current, this voltage can be represented as having a phase difference of zero with respect to the current. Therefore, the resistance is indicated as an arrow in the positive *x*-direction, denoting the fact that the current and voltage are in phase. Such a representation is called a **phasor**. The arrow denoted by *R*, as in Fig. 11–16, is actually the phasor representing the voltage across the resistor. Remember, the positive *x*-axis is arbitrarily chosen as the direction of the phase of the current.

To show properly that the voltage across an inductance leads the current by 90°, its reactance (effective resistance) is multiplied by *j*. We know that there is a positive 90° angle between a real number and a positive imaginary number. In the same way, by multiplying the capacitive reactance by −*j*, we show the 90° difference in phase between the voltage and current in a capacitor, with the current leading. Therefore, jX_L represents the phasor for the voltage across an inductor and $-jX_C$ is the phasor for the voltage across the capacitor. The phasor for the voltage across the combination of the resistance, inductance, and capacitance is *Z*, where the phase difference between the voltage and current for the combination is the angle θ.

This also points out well the significance of the word "operator" in the term "*j*-operator." Multiplying a phasor by *j* means to perform the operation of rotating it through 90°. We have seen that this is the same result obtained when a real number is multiplied by *j*.

An alternating current is produced by a coil of wire rotating through a magnetic field. If the angular velocity of this wire is ω, the capacitive and inductive reactances are given by the relations

$$X_C = \frac{1}{\omega C} \quad \text{and} \quad X_L = \omega L \tag{11–23}$$

Therefore, if ω, *C*, and *L* are known, the reactance of the circuit may be determined.

Example E

Given that $R = 12.0 \ \Omega$, $L = 0.300$ H, $C = 250 \ \mu$F, $\omega = 80.0$ rad/s, determine the impedance and the phase difference between the current and voltage.

$$X_C = \frac{1}{(80.0)(250 \times 10^{-6})} = 50.0 \ \Omega$$

$$X_L = (0.300)(80.0) = 24.0 \ \Omega$$

$$Z = 12.0 + (24.0 - 50.0)j = 12.0 - 26.0j$$

$$|Z| = \sqrt{(12.0)^2 + (-26.0)^2} = 28.6 \ \Omega$$

$$\tan \theta = \frac{-26.0}{12.0} = -2.167, \quad \theta = -65.2°$$

The voltage lags the current (see Fig. 11–17).

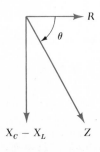

Figure 11–17

An important concept in the application of this theory is that of **resonance.** For resonance, the impedance of any circuit is a minimum, or the total impedance is R. Thus $X_L - X_C = 0$. Also, it can be seen that the current and voltage are in phase under these conditions. Resonance is required for the tuning of radio and television receivers.

Exercises 11—7

For Exercises 1 through 4 use the circuit shown in Fig. 11–18. A current of 3.00 A flows through the circuit. Determine the indicated quantities.

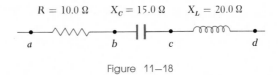

$$R = 10.0\ \Omega \qquad X_C = 15.0\ \Omega \qquad X_L = 20.0\ \Omega$$

Figure 11—18

1. Find the voltage across the resistor (between points a and b).
2. Find the voltage across the capacitor (between points b and c).
3. (a) Find the magnitude of the impedance across the resistor and the capacitor (between points a and c).
   (b) Find the phase angle between the current and voltage for this combination.
   (c) Find the voltage across this combination.
4. (a) Find the magnitude of the impedance across the resistor, capacitor, and inductor (between points a and d).
   (b) Find the phase angle between the current and voltage for this combination.
   (c) Find the voltage across this combination.

In Exercises 5 through 8 use the following information to find the required quantities. In a given circuit $R = 6.00\ \Omega$, $X_C = 10.0\ \Omega$, and $X_L = 7.00\ \Omega$. Find (a) the magnitude of the impedance and (b) the phase angle between the current and voltage, under the specified conditions.

5. With the resistor removed (an LC circuit)
6. With the capacitor removed (an RL circuit)
7. With the inductor removed (an RC circuit)
8. With all elements present (an RLC circuit)

In Exercises 9 through 18 find the required quantities.

9. Given that the current in a given circuit is $8.00 - 2.00j$ amperes and the impedance is $2.00 + 5.00j$ ohms, find the magnitude of the voltage.
10. Given that the voltage in a given circuit is $8.00 - 3.00j$ volts and the impedance is $2.00 - 1.00j$ ohms, find the magnitude of the current.
11. Given that $\omega = 1000$ rad/s, $C = 0.500\ \mu$F, and $L = 3.00$ H, find the capacitive and inductive reactances.
12. A coil of wire is rotating through a magnetic field at 60.0 Hz. Determine the capacitive reactance for a capacitor of 2.00 μF which is in the circuit.
13. If the capacitor and inductor in Exercise 11 are put in a series circuit with a resistance of 4000 Ω, what is the impedance of this combination? Find the phase angle.

14. Rework Exercise 13, assuming that the capacitor is removed.

15. If $\omega = 100$ rad/s and $L = 0.500$ H, what must be the value of C to produce resonance?

16. What is the frequency ω (in hertz) for resonance in a circuit for which $L = 2.00$ H and $C = 25.0$ μF?

17. The average power supplied to any combination of components in an ac circuit is given by the relation $P = VI \cos \theta$, where P is the power (in watts), V is the effective voltage, I is the effective current, and θ is the phase angle between the current and voltage. Assuming that the effective voltage across the resistor, capacitor, and inductor combination in Exercise 13 is 200 V, determine the power supplied to these elements.

18. Find the power supplied to a resistor of 120 Ω and a capacitor of 0.700 μF if $\omega = 1000$ rad/s and the effective voltage is 110 V.

11–8 Exercises for Chapter 11

In Exercises 1 through 16 perform the indicated operations, expressing all answers in the simplest rectangular form.

1. $(6 - 2j) + (4 + j)$

2. $(12 + 7j) + (-8 + 6j)$

3. $(18 - 3j) - (12 - 5j)$

4. $(-4 - 2j) - (-6 - 7j)$

5. $(2 + j)(4 - j)$

6. $(-5 + 3j)(8 - 4j)$

7. $(2j)(6 - 3j)(4 + 3j)$

8. $j(3 - 2j) - (j^3)(5 + j)$

9. $\dfrac{3}{7 - 6j}$

10. $\dfrac{4j}{2 + 9j}$

11. $\dfrac{6 - 4j}{7 - 2j}$

12. $\dfrac{3 - 2j}{4 + j}$

13. $\dfrac{5j - (3 - j)}{4 - 2j}$

14. $\dfrac{2 + (j - 6)}{5 - 2j}$

15. $\dfrac{j(7 - 3j)}{2 + j}$

16. $\dfrac{(2 - j)(3 + 2j)}{4 - 3j}$

In Exercises 17 through 20 find the values of x and y for which the equations are valid.

17. $3x - 2j = yj - 2$

18. $2xj - 2y = (y + 3)j - 3$

19. $2x - j + 4 = 6y + 2xj$

20. $3yj + xj = 6 + 3x + y$

In Exercises 21 through 24 perform the indicated operations graphically; check them algebraically.

21. $(-1 + 5j) + (4 + 6j)$

22. $(7 - 2j) + (5 + 4j)$

23. $(9 + 2j) - (5 - 6j)$

24. $(1 + 4j) - (-3 - 3j)$

In Exercises 25 through 32 give the polar and exponential forms of each of the complex numbers.

25. $1 - j$

26. $4 + 3j$

27. $-2 - 7j$

28. $6 - 2j$

29. $1.07 + 4.55j$

30. $-327 + 158j$

31. 10

32. $-4j$

In Exercises 33 through 40 give the rectangular form of each of the complex numbers.

33. $2(\cos 225° + j \sin 225°)$

34. $4(\cos 60° + j \sin 60°)$

35. $5.00(\cos 123.0° + j \sin 123.0°)$

36. $2.00(\cos 296.0° + j \sin 296.0°)$

37. $2.00e^{0.25j}$

38. $e^{3.62j}$

39. $5.37e^{1.90j}$

40. $4.47e^{6.04j}$

In Exercises 41 through 48 perform the indicated operations. Leave the result in polar form.

41. $[3(\cos 32° + j \sin 32°)][5(\cos 52° + j \sin 52°)]$

42. $[2.5(\cos 162° + j \sin 162°)][8(\cos 115° + j \sin 115°)]$

43. $\dfrac{24(\cos 165° + j \sin 165°)}{3(\cos 106° + j \sin 106°)}$

44. $\dfrac{18(\cos 403° + j \sin 403°)}{4(\cos 192° + j \sin 192°)}$

45. $[2(\cos 16° + j \sin 16°)]^{10}$

46. $[3(\cos 36° + j \sin 36°)]^{6}$

47. $[3(\cos 110.5° + j \sin 110.5°]^{3}$

48. $[5(\cos 220.3° + j \sin 220.3°)]^{4}$

In Exercises 49 through 52 change each number to polar form and then perform the indicated operations. Express the final result in rectangular and polar forms. Check by performing the same operation in rectangular form.

49. $(1 - j)^{10}$

50. $(\sqrt{3} + j)^{8}(1 + j)^{5}$

51. $\dfrac{(5 + 5j)^{4}}{(-1 - j)^{6}}$

52. $(\sqrt{3} - j)^{-8}$

In Exercises 53 through 56 use DeMoivre's theorem to find the indicated roots. Be sure to find all roots.

53. $\sqrt[3]{-8}$ 54. $\sqrt[3]{1}$ 55. $\sqrt[4]{-j}$ 56. $\sqrt[5]{-32}$

In Exercises 57 through 68 find the required quantities.

57. In a given circuit $R = 7.50\ \Omega$ and $X_C = 10.0\ \Omega$. Find the magnitude of the impedance and the phase angle between the current and voltage.

58. In a given circuit $R = 15.0\ \Omega$, $X_C = 27.0\ \Omega$, and $X_L = 35.0\ \Omega$. Find the magnitude of the impedance and the phase angle between the current and voltage.

59. A coil of wire is going around a circle at 60.0 r/s. If this coil generates a current in a circuit containing a resistance of $10.0\ \Omega$, an inductance of 0.010 H, and a capacitance of $500\ \mu$F, what is the magnitude of the impedance of the circuit? What is the angle between the current and voltage?

60. A coil of wire rotates at 120 r/s. If the coil generates a current in a circuit containing a resistance of $12.0\ \Omega$, an inductance of 0.040 H and an impedance of $22.0\ \Omega$, what must be the value of a capacitor (in farads) in the circuit?

61. In a given circuit the current is $5.00 - 2.00j$ amperes and the impedance is $6.00 + 3.00j$ ohms. Find the magnitude of the voltage.

62. In a given circuit $I = 6.00$ A and $V = 3.00 - 2.50j$ volts. Find the magnitude of the impedance.

63. What is the magnitude and direction of a force which is represented by $600 - 550j$ pounds?

64. What is the magnitude and direction of a velocity which is represented by $2500 + 1500j$ kilometers per hour?

65. In the study of shearing effects in the spinal column, the expression $\frac{1}{u + j\omega n}$ is found. Express this in rectangular form.

66. In the theory of light reflection on metals, the expression

$$\frac{\mu(1 - kj) - 1}{\mu(1 - kj) + 1}$$

is encountered. Simplify this expression.

67. Show that $e^{j\pi} = -1$.   68. Show that $(e^{j\pi})^{1/2} = j$.

12

Exponential and Logarithmic Functions

In Chapter 10, we dealt in some detail with exponents. There is a special use of exponents which is important in computational work and for theoretical purposes. Exponents used in this manner are given the name **logarithms.** Today, with the extensive use of computers and electronic calculators, logarithms are used much less for computation than in the past. However, their usefulness in advanced mathematics and in applications in technical fields remains of great importance.

To illustrate the importance of logarithms, many applications may be cited. The basic units used to measure the intensity of sound, and those used to measure the intensity of earthquakes, are defined in terms of logarithms. In chemistry, the distinction between a base and an acid is defined in terms of logarithms. In electrical transmission lines, power gains and losses are measured in terms of logarithmic units. In electronics and in mechanical systems, the use of exponential functions, which are closely related to logarithms, is extensive. Many of these applications are illustrated throughout the chapter.

Chapter 10 dealt with exponents in expressions of the form x^n, where we showed that n could be any rational number. Here we shall deal with expressions of the form b^x, where x is any real number. When we look at these expressions, we note the primary difference is that in the second expression *the exponent is variable.* We have not previously dealt with variable exponents. *Thus let us define the* **exponential function** *to be*

$$y = b^x \tag{12–1}$$

In Eq. (12–1), *x is called the logarithm of the number y to the base b.* In our work with logarithms we shall restrict all numbers to the real number system. This leads us to choose the base as a positive number other than 1. We know that 1 raised to any power will result in 1, which would make y a constant regardless of the value of x. Negative numbers for b would result in imaginary values for y if x were any fractional exponent with an even integer for its denominator.

Example A

$y = 2^x$ is an exponential function, where x is the logarithm of y to the base 2. This means that 2 raised to a given power gives us the corresponding value of y.

If $x = 2$, $y = 2^2 = 4$; this means that 2 is the logarithm of 4 to the base 2. If $x = 4$, $y = 2^4 = 16$; this means that 4 is the logarithm of 16 to the base 2. If $x = \frac{1}{2}$, $y = 2^{1/2} = 1.41$; this means that $\frac{1}{2}$ is the logarithm of 1.41 to the base 2.

Using the definition of a logarithm, Eq. (12–1) may be solved for x, and is written in the form

$$x = \log_b y \tag{12–2}$$

This equation is read in accordance with the definition of x in Eq. (12–1): **x equals the logarithm of y to the base b.** This means that x is the power to which the base b must be raised in order to equal the number y; that is, x is a logarithm, and a logarithm is an exponent. Note that Eqs. (12–1) and (12–2) state the same relationship, but in a different manner. Equation (12–1) is the **exponential form**, and Eq. (12–2) is the **logarithmic form.**

Example B

The equation $y = 2^x$ would be written as $x = \log_2 y$ if we put it in logarithmic form. When we choose values of y to find the corresponding values of x from this equation, we ask ourselves, "2 raised to what power gives y?" Hence if $y = 4$, we know that 2^2 is 4, and x would be 2. If $y = 8$, $2^3 = 8$, or $x = 3$.

Example C

$3^2 = 9$ in logarithmic form is $2 = \log_3 9$; $4^{-1} = \frac{1}{4}$ in logarithmic form is $-1 = \log_4 \left(\frac{1}{4}\right)$. Remember, the exponent may be negative. The base must be positive.

Example D

$(64)^{1/3} = 4$ in logarithmic form is $\frac{1}{3} = \log_{64} 4$,

$(32)^{3/5} = 8$ in logarithmic form is $\frac{3}{5} = \log_{32} 8$

Example E

$\log_2 32 = 5$ in exponential form is $32 = 2^5$,

$\log_6 \left(\frac{1}{36}\right) = -2$ in exponential form is $\frac{1}{36} = 6^{-2}$

Example F

Find b, given that $-4 = \log_b \left(\frac{1}{81}\right)$.

Writing this in exponential form, we have $\frac{1}{81} = b^{-4}$. Thus $b = 3$, since $3^4 = 81$.

Example G

Find y, given that $\log_4 y = \frac{1}{2}$.

In exponential form we have $y = 4^{1/2}$ or $y = 2$.

We see that exponential form is very useful for determining values written in logarithmic form. For this reason it is important that you learn to transform readily from one form to the other.

Exercises 12–1

In Exercises 1 through 12 express the given equations in logarithmic form.

1. $3^3 = 27$ 2. $5^2 = 25$ 3. $4^4 = 256$ 4. $8^2 = 64$

5. $4^{-2} = \frac{1}{16}$ 6. $3^{-2} = \frac{1}{9}$ 7. $2^{-6} = \frac{1}{64}$ 8. $(12)^0 = 1$

9. $8^{1/3} = 2$ 10. $(81)^{3/4} = 27$ 11. $\left(\frac{1}{4}\right)^2 = \frac{1}{16}$ 12. $\left(\frac{1}{2}\right)^{-2} = 4$

In Exercises 13 through 24 express the given equations in exponential form.

13. $\log_3 81 = 4$ 14. $\log_{11} 121 = 2$ 15. $\log_9 9 = 1$

16. $\log_{15} 1 = 0$ 17. $\log_{25} 5 = \frac{1}{2}$ 18. $\log_8 16 = \frac{4}{3}$

19. $\log_{243} 3 = \frac{1}{5}$ 20. $\log_{1/32} \left(\frac{1}{8}\right) = \frac{3}{5}$ 21. $\log_{10} 0.1 = -1$

22. $\log_7 \left(\frac{1}{49}\right) = -2$ 23. $\log_{0.5} 16 = -4$ 24. $\log_{1/3} 3 = -1$

In Exercises 25 through 40 determine the value of the unknown.

25. $\log_4 16 = x$ 26. $\log_5 125 = x$ 27. $\log_{10} 0.01 = x$

28. $\log_{16} \left(\frac{1}{4}\right) = x$ 29. $\log_7 y = 3$ 30. $\log_8 N = 3$

31. $\log_8 y = -\frac{2}{3}$ 32. $\log_7 y = -2$ 33. $\log_b 81 = 2$

34. $\log_b 625 = 4$ 35. $\log_b 4 = -\frac{1}{3}$ 36. $\log_b 4 = \frac{2}{3}$

37. $\log_{10} 10^{0.2} = x$ 38. $\log_5 5^{1.3} = x$ 39. $\log_3 27^{-1} = x$

40. $\log_b \left(\frac{1}{4}\right) = -\frac{1}{2}$

In Exercises 41 through 44 perform the indicated operations.

41. If there are initially 1000 bacteria in a culture, and the number of bacteria then doubles each hour, the number of bacteria as a function of time is $N = 1000(2^t)$. By writing this equation in logarithmic form, solve for t.

42. Under specified conditions, the instantaneous voltage E in a given circuit can be expressed as

$$E = E_m e^{-Rt/L}$$

By rewriting this equation in logarithmic form, solve for t.

43. An equation relating the number N of atoms of radium at any time t in terms of the number of atoms at $t = 0$, N_0, is $\log_e(N/N_0) = -kt$, where k is a constant. By expressing this equation in exponential form, solve for N.

44. In the theory dealing with the optical brightness of objects, the equation $D = \log_{10}(I_0/I)$ is found. By writing this equation in exponential form, solve for I.

12–2 Graphs of $y = b^x$ and $y = \log_b x$

When we are working with functions, we must keep in mind that a function is defined by the operation being performed on the independent variable, and not by the letter chosen to represent it. However, for consistency, it is normal practice to let y represent the dependent variable and x represent the independent variable. Therefore, the **logarithmic function** is

$$y = \log_b x \tag{12–3}$$

Equations (12–2) and (12–3) express the same *function*, the logarithmic function. They do not represent different functions, due to the difference in location of the variables, since they represent the same operation on the independent variable. Equation (12–3) simply expresses the function with the usual choice of variables.

Graphical representation of functions is often valuable when we wish to demonstrate their properties. We shall now show the graphs of the exponential function [Eq. (12–1)] and the logarithmic function [Eq. (12–3)].

Example A

Plot the graph of $y = 2^x$.

Assuming values for x and then finding the corresponding values for y, we obtain the following table.

x	-3	-2	-1	0	1	2	3	4
y	$\frac{1}{8}$	$\frac{1}{4}$	$\frac{1}{2}$	1	2	4	8	16

From these values we plot the curve, as shown in Fig. 12–1. We note that the x-axis is an asymptote of the curve.

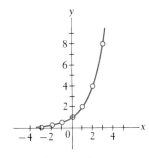

Figure 12–1

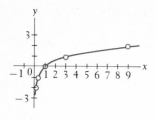

Figure 12–2

Example B

Plot the graph of $y = \log_3 x$.

We can find the points for this graph more easily if we first put the equation in exponential form: $x = 3^y$. By assuming values for y, we can find the corresponding values for x.

x	$\frac{1}{9}$	$\frac{1}{3}$	1	3	9	27
y	-2	-1	0	1	2	3

Using these values, we construct the graph seen in Fig. 12–2.

Any exponential or logarithmic curve, where $b > 1$, will be similar in shape to those of Examples A and B. From these curves we can draw certain conclusions:

(1) If $0 < x < 1$, $\log_b x < 0$; if $x = 1$, $\log_b 1 = 0$; if $x > 1$, $\log_b x > 0$.
(2) If $x > 1$, x increases more rapidly than $\log_b x$.
(3) For all values of x, $b^x > 0$.
(4) If $x > 1$, b^x increases more rapidly than x.

Although the bases important to applications are greater than 1, to understand how the curve of the exponential function differs somewhat if $b < 1$, let us consider the following example.

Example C

Plot the graph of $y = (\frac{1}{2})^x$.

The values are found for the following table; the graph is plotted in Fig. 12–3.

x	-3	-2	-1	0	1	2	3	4
y	8	4	2	1	$\frac{1}{2}$	$\frac{1}{4}$	$\frac{1}{8}$	$\frac{1}{16}$

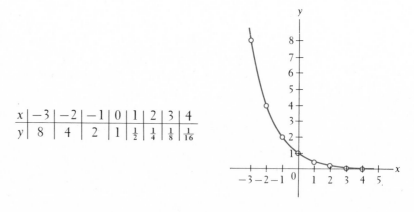

Figure 12–3

We note that as the values of x increase, the values of $(\frac{1}{2})^x$ decrease. This is different from the behavior of $y = b^x$, where $b > 1$.

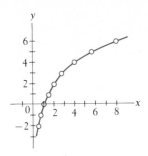

Figure 12–4

Example D

Plot the graph of $y = 2 \log_2 x$.

As in Example B, it is generally more convenient to work with the exponential form. In order to change this equation to exponential form, we first divide each side by 2. Thus, we have

$$\frac{y}{2} = \log_2 x, \qquad x = 2^{y/2}$$

Choosing values of y, we then calculate corresponding values of x and obtain the following table.

x	0.5	0.7	1	1.4	2	2.8	4	5.7	8
y	-2	-1	0	1	2	3	4	5	6

See Fig. 12–4.

Exercises 12–2

In Exercises 1 through 12 plot graphs of the given functions. Values of x from -3 or -2 through 2 or 3 are appropriate.

1. $y = 3^x$
2. $y = 4^x$
3. $y = 6^x$
4. $y = 10^x$
5. $y = (1.5)^x$
6. $y = (2.7)^x$
7. $y = (\frac{1}{3})^x$
8. $y = (\frac{1}{4})^x$
9. $y = 2(2^x)$
10. $y = 1.5(4^x)$
11. $y = 0.5(3^x)$
12. $y = 0.1(10^x)$

In Exercises 13 through 24 plot graphs of the given functions. Values of y from -3 or -2 through 2 or 3 are appropriate for Exercises 13 through 20. Care should be taken in selecting appropriate values used in Exercises 21 through 24.

13. $y = \log_2 x$
14. $y = \log_4 x$
15. $y = \log_6 x$
16. $y = \log_{10} x$
17. $y = \log_{1.5} x$
18. $y = \log_{2.7} x$
19. $y = \log_{32} x$
20. $y = \log_{1/2} x$
21. $y = 2 \log_3 x$
22. $y = 3 \log_2 x$
23. $y = 0.2 \log_4 x$
24. $y = 5 \log_{10} x$

In Exercises 25 through 28 plot the indicated graphs.

25. The electric current in a certain type of circuit is given by $i = I_0 e^{-Rt/L}$, where I_0 is the initial current, R is a resistance, and L is an inductance (see Section 11–7). Plot the graph for i vs. t for a circuit in which $I_0 = 5.0$ A, $R = 10 \ \Omega$, and $L = 5.0$ H. Use $e = 2.7$, and $0 \le t < 4$ s.

26. If $1000 is placed in a bank account in which interest is compounded daily with an effective 8% annual interest, the amount in the account after t years is $A = 1000(1.08)^t$. Plot A as a function of t for $0 \le t \le 5$ years.

27. Under certain conditions the temperature T (in degrees Celsius) of a cooling object as a function of time (in minutes) is $T = 50.0(10^{-0.1t})$. Plot T as a function of t.

28. In Exercise 20 the graph of $y = \log_{1/2} x$ is plotted. By inspecting the graph, and noting the properties of $\log_{1/2} x$, determine some of the differences of logarithms to a base less than 1 from those to a base greater than 1.

12–3 Properties of Logarithms

Since a logarithm is an exponent, it must follow the laws of exponents. Those laws which are of the greatest importance at this time are listed here for reference.

$$b^u \cdot b^v = b^{u+v} \tag{12-4}$$
$$b^u / b^v = b^{u-v} \tag{12-5}$$
$$(b^u)^n = b^{nu} \tag{12-6}$$

We shall now show how these laws for exponents give certain useful properties to logarithms.

If we let $u = \log_b x$ and $v = \log_b y$ and write these equations in exponential form, we have $x = b^u$ and $y = b^v$. Therefore, forming the product of x and y, we obtain

$$xy = b^u b^v = b^{u+v} \quad \text{or} \quad xy = b^{u+v}$$

Writing this last equation in logarithmic form yields

$$u + v = \log_b xy$$

or

$$\log_b x + \log_b y = \log_b xy \tag{12-7}$$

Equation (12–7) states the property that *the logarithm of the product of two numbers is equal to the sum of the logarithms of the numbers.*

Using the same definitions of u and v to form the quotient of x and y, we then have

$$\frac{x}{y} = \frac{b^u}{b^v} = b^{u-v} \quad \text{or} \quad \frac{x}{y} = b^{u-v}$$

Writing this last equation in logarithmic form, we have

$$u - v = \log_b\left(\frac{x}{y}\right)$$

or

$$\log_b x - \log_b y = \log_b\left(\frac{x}{y}\right) \tag{12-8}$$

Equation (12–8) states the property that *the logarithm of the quotient of two numbers is equal to the logarithm of the numerator minus the logarithm of the denominator.*

If we again let $u = \log_b x$ and write this in exponential form, we have $x = b^u$. To find the nth power of x, we write

$$x^n = (b^u)^n = b^{nu}$$

Expressing this equation in logarithmic form yields

$$nu = \log_b (x^n)$$

or

$$n \log_b x = \log_b (x^n) \tag{12–9}$$

This last equation states that *the logarithm of the nth power of a number is equal to n times the logarithm of the number.* The exponent n may be integral or fractional and, therefore, we may use Eq. (12–9) for finding powers and roots of numbers.

Example A
Using Eq. (12–7) we may express $\log_4 15$ as a sum of logarithms as follows:

$$\log_4 15 = \log_4 (3 \cdot 5) = \log_4 3 + \log_4 5$$

Using Eq. (12–8) we may express $\log_4 \left(\frac{5}{3}\right)$ as the difference of logarithms as follows:

$$\log_4 \left(\frac{5}{3}\right) = \log_4 5 - \log_4 3$$

Using Eq. (12–9) we may express $\log_4 9$ as twice $\log_4 3$ as follows:

$$\log_4 9 = \log_4 (3^2) = 2 \log_4 3$$

Example B
Using Eqs. (12–7) through (12–9) we may express a sum or difference of logarithms as the logarithm of a single quantity.

$$\log_4 3 + \log_4 x = \log_4 (3 \cdot x) = \log_4 3x$$

$$\log_4 3 - \log_4 x = \log_4 \left(\frac{3}{x}\right)$$

$$\log_4 3 + 2 \log_4 x = \log_4 3 + \log_4 (x^2) = \log_4 3x^2$$

$$\log_4 3 + 2 \log_4 x - \log_4 y = \log_4 \left(\frac{3x^2}{y}\right)$$

We may use Eq. (12–9) to find another important property of logarithms. Since $b = b^1$ in logarithmic form is $\log_b b = 1$, we have

$$\log_b (b^n) = n \log_b b = n(1) = n$$

or

$$\log_b (b^n) = n \tag{12–10}$$

Equation (12–10) may be used to evaluate certain logarithms when the exact values may be determined.

Example C

We may evaluate $\log_3 9$ in the following manner: Using Eq. (12–10), we have

$$\log_3 9 = \log_3(3^2) = 2$$

We can establish the exact value since the base of logarithms and the number being raised to the power are the same. Of course, this could have been evaluated directly from the definition of a logarithm.

However, we cannot establish an exact value of $\log_4 9$ in this way. We have

$$\log_4 9 = \log_4(3^2) = 2 \log_4 3$$

We must leave it in this form until we are able to evaluate $\log_4 3$. This can be done, and we shall consider this type of expression later in the chapter.

Example D

Calculate $\log_3(3^{0.4})$.

Using Eq. (12–10) we may calculate this logarithm. This gives us

$$\log_3(3^{0.4}) = 0.4$$

Although we did not calculate $3^{0.4}$ at this point, we were able to calculate $\log_3 3^{0.4}$.

Example E

We may also use Eq. (12–10) in combination with other properties of logarithms. The following illustration shows how it can be used alone or in combination with Eq. (12–8) to evaluate the indicated logarithm.

$$\log_5(\tfrac{1}{25}) = \log_5 1 - \log_5 25 = 0 - \log_5(5^2) = -2$$
$$\log_5(\tfrac{1}{25}) = \log_5(5^{-2}) = -2$$

Either method is appropriate.

In the following examples, the properties of logarithms of Eqs. (12–7) through (12–10) are further illustrated.

Example F

$$\log_2 6 = \log_2(2\cdot 3) = \log_2 2 + \log_2 3 = 1 + \log_2 3$$
$$\log_3(\tfrac{2}{9}) = \log_3 2 - \log_3 9 = \log_3 2 - \log_3(3^2) = \log_3 2 - 2$$
$$= -2 + \log_3 2$$

Example G

$$\log_{10}\sqrt{7} = \log_{10}(7^{1/2}) = \tfrac{1}{2}\log_{10}7$$

This demonstrates the property which is especially useful for finding roots of numbers. We see that we need merely to multiply the logarithm of the number by the fractional exponent representing the root to obtain the logarithm of the root.

Similarly,

$$\log_{10} \sqrt[3]{10x} = \log_{10}(10x)^{1/3} = \tfrac{1}{3}\log_{10}(10x)$$
$$= \tfrac{1}{3}(\log_{10}10 + \log_{10}x)$$
$$= \tfrac{1}{3}(1 + \log_{10}x)$$

Example H

Use the basic properties of logarithms to solve for y in terms of x:
$\log_b y = 2\log_b x + \log_b a$.

Using Eq. (12–9) and then Eq. (12–7), we have

$$\log_b y = \log_b(x^2) + \log_b a = \log_b(ax^2)$$

Now, since we have the logarithm to the base b of different expressions on each side of the resulting equation, the expressions must be equal. Therefore, $y = ax^2$.

Exercises 12–3

In Exercises 1 through 12 express each as a sum, difference, or multiple of logarithms. See Example A.

1. $\log_5 xy$
2. $\log_3 7y$
3. $\log_7\left(\tfrac{5}{a}\right)$
4. $\log_3\left(\tfrac{r}{8}\right)$
5. $\log_2(a^3)$
6. $\log_8(n^5)$
7. $\log_6 abc$
8. $\log_2\left(\tfrac{xy}{z}\right)$
9. $\log_5\sqrt[4]{y}$
10. $\log_4\sqrt[7]{x}$
11. $\log_2\left(\tfrac{\sqrt{x}}{a^2}\right)$
12. $\log_3\left(\tfrac{\sqrt{y}}{8}\right)$

In Exercises 13 through 20 express each as the logarithm of a single quantity. See Example B.

13. $\log_b a + \log_b c$
14. $\log_2 3 + \log_2 x$
15. $\log_5 9 - \log_5 3$
16. $\log_8 6 - \log_8 a$
17. $\log_b x^2 - \log_b\sqrt{x}$
18. $\log_4 3^3 + \log_4 9$
19. $2\log_e 2 + 3\log_e n$
20. $\tfrac{1}{2}\log_b a - 2\log_b 5$

In Exercises 21 through 28 determine the exact value of each of the given logarithms.

21. $\log_2\left(\tfrac{1}{32}\right)$
22. $\log_3\left(\tfrac{1}{81}\right)$
23. $\log_2(2^{2.5})$
24. $\log_5(5^{0.1})$
25. $\log_7\sqrt{7}$
26. $\log_6\sqrt[3]{6}$
27. $\log_3\sqrt[4]{27}$
28. $\log_5\sqrt[3]{25}$

In Exercises 29 through 40 express each as a sum, difference, or multiple of logarithms. In each case part of the logarithm may be determined exactly.

29. $\log_3 18$
30. $\log_5 75$
31. $\log_2\left(\tfrac{1}{6}\right)$
32. $\log_{10}(0.05)$
33. $\log_3\sqrt{6}$
34. $\log_2\sqrt[3]{24}$
35. $\log_2(4^2\cdot 3^3)$
36. $\log_7(7^4\cdot 3^5)$
37. $\log_{10}3000$
38. $\log_{10}(40^2)$
39. $\log_{10}\left(\tfrac{27}{100}\right)$
40. $\log_5\left(\tfrac{4}{125}\right)$

In Exercises 41 through 48 solve for y in terms of x.

41. $\log_b y = \log_b 2 + \log_b x$
42. $\log_b y = \log_b 6 + \log_b x$
43. $\log_4 y = \log_4 x - \log_4 5$
44. $\log_3 y = \log_3 7 - \log_3 x$
45. $\log_{10} y = 2\log_{10}7 - 3\log_{10}x$
46. $\log_b y = 3\log_b\sqrt{x} + 2\log_b 10$
47. $5\log_2 y - \log_2 x = 3\log_2 4 + \log_2 a$
48. $4\log_2 x - 3\log_2 y = \log_2 27$

In Exercises 49 through 52 using $\log_{10} 2 = 0.301$, evaluate each of the given expressions.

49. $\log_{10} 4$ 50. $\log_{10} 20$ 51. $\log_{10}(0.5)$ 52. $\log_{10} 8000$

In Exercises 53 through 56 plot the indicated graphs and perform the indicated operations.

53. Plot the graphs of $y = 2 \log_2 x$ and $y = \log_2 x^2$, and show that they are the same.

54. Plot the graphs of $y = \log_2 4x$ and $y = 2 + \log_2 x$, and show that they are the same.

55. An equation used in thermodynamics is $S = C \log_e T - nR \log_e P$. Express this equation with a single logarithm on the right side.

56. An equation used for a certain electric circuit is $\log_e i - \log_e I = -t/RC$. Solve for i.

12–4   Logarithms to the Base 10

In Section 12–1 we stated that a base of logarithms must be a positive number, not equal to one. In the examples and exercises of the previous sections we used a number of different bases. There are, however, only two bases which are used extensively. They are 10 and e, where e equals approximately 2.718. The number e was first introduced in Section 11–5.

Base 10 logarithms are used primarily for calculations and in certain types of applications. Base e logarithms are used extensively in technical and scientific work. In this section we discuss how to determine the base 10 logarithm of any given number. Base e logarithms are considered in Section 12–7.

In Section 1–5, we showed that any number may be expressed in scientific notation as the product of a number between 1 and 10 and a power of 10. Writing this as $N = P \times 10^k$, and taking logarithms of both sides of this equation, we have

$$\log_b N = \log_b(P \times 10^k) = \log_b P + \log_b 10^k = \log_b P + k \log_b 10$$

If we let $b = 10$, then $k \log_{10} 10 = k$, and this equation becomes

$$\log_{10} N = k + \log_{10} P \qquad (12\text{--}11)$$

Equation (12–11) shows us that if we have a method for finding logarithms to the base 10 of numbers between 1 and 10, then we can find the logarithm of *any* number to base 10. The value of k can be found by writing the number N in scientific notation, and P is a number between 1 and 10. Logarithms to the base 10 have been tabulated, and tables of these **common logarithms** may be found in Appendix E. Logarithms may be calculated for any base, but for purposes of computation, logarithms to the base 10 are the most convenient. From now on we shall not write the number 10 to indicate the base, and log N will be assumed to be to the base 10. Thus,

$$\log N = k + \log P \tag{12–12}$$

In Eq. (12–12), *k is called the* **characteristic,** *and log P is known as the* **mantissa.** Remember, k is the power of 10 of the number, when it is written in scientific notation, and the term $\log P$ is the logarithm of the number between 1 and 10.

Example A

For $N = 3600 = 3.6 \times 10^3$, we see that the characteristic $k = 3$, and the mantissa $\log P = \log 3.6$. Therefore, $\log 3600 = 3 + \log 3.6$.

For $N = 80.9 = 8.09 \times 10^1$, we see that $k = 1$ and $\log P = \log 8.09$. Therefore, $\log 80.9 = 1 + \log 8.09$.

Example B

For $N = 0.00543 = 5.43 \times 10^{-3}$, we see that $k = -3$ and $\log P = \log 5.43$. Therefore, $\log 0.00543 = -3 + \log 5.43$.

For $N = 0.741 = 7.41 \times 10^{-1}$, we see that $k = -1$ and $\log P = \log 7.41$. Therefore, $\log 0.741 = -1 + \log 7.41$.

To find $\log P$ we use Table 2 in Appendix E. The following examples illustrate how to use this table.

Example C

Find log 572.

We first write the number in scientific notation as 5.72×10^2. The characteristic is 2, and we must now find log 5.72. We look in the column headed N and find 57 (the first two significant digits). Then, to the right of this, we look under the column headed 2 (the third significant digit) and we find 7574. All numbers between 1 and 10 will have common logarithms between 0 and 1 (log 1 = 0 and log 10 = 1). Therefore, log 5.72 = 0.7574, and the logarithm of 572 = 2 + 0.7574. We then write this in the usual form of 2.7574, and write this result as

$$\log 572 = 2.7574$$

It should be emphasized that both the mantissa and characteristic of a logarithm must be found before the logarithm of a number is complete. The mantissa is found from Table 2, and the characteristic is found from the location of the decimal point. A rather common error is to assume the logarithm has been found when the mantissa is determined. However, the characteristic is just as much a proper part of the logarithm as the mantissa.

Example D
Find log 0.00485.

Writing this number in scientific notation gives us 4.85×10^{-3}, and we see that $k = -3$. From the tables we find that log $4.85 = 0.6857$. Thus, log $0.00485 = -3 + 0.6857$. We do *not* write this as -3.6857, for this would say that the mantissa was also negative, which it is not. To avoid this possible confusion, we shall write it in the form $7.6857 - 10$. We shall follow this policy whenever the characteristic is negative. That is, we shall write a negative characteristic as the appropriate positive number with 10 subtracted. For example, for a characteristic of -6, we write 4 before the mantissa and -10 after it. Thus,

$$\log 0.00485 = 7.6857 - 10$$

Example E
Other examples of logarithms are as follows:

$$89,000 = 8.9 \times 10^4: k = 4, \log 8.9 = 0.9494; \log 89000 = 4.9494$$
$$0.307 = 3.07 \times 10^{-1}: k = -1, \log 3.07 = 0.4871;$$
$$\log 0.307 = 9.4871 - 10$$
$$0.00629 = 6.29 \times 10^{-3}: k = -3, \log 6.29 = 0.7987;$$
$$\log 0.00629 = 7.7987 - 10$$

Table 2 is a four-place table, which means that we can obtain accuracy to four significant digits. However, only three digits may be read directly, and the fourth place is found by interpolation. We discussed this in Section 3–3, in reference to finding values from trigonometric tables. The method for finding values from logarithmic tables is the same. It is illustrated in the following examples.

Example F
Find log 686300.

In scientific notation, $686300 = 6.863 \times 10^5$. This means that the characteristic is 5. To find the mantissa from the table, we must interpolate, finding the value $\frac{3}{10}$ (since the fourth digit is 3) of the way between log 6.86 and log 6.87. These latter two values are 0.8363 and 0.8370. The tabular difference is 7, and $(\frac{3}{10})(7) = 2$ (to one significant digit). Adding this to the mantissa 0.8363 gives 0.8365. Hence, log $686300 = 5.8365$.

Example G
Find log 0.02178.

In scientific notation, $0.02178 = 2.178 \times 10^{-2}$. Therefore, the characteristic is -2. To find the mantissa, we must interpolate, finding the value $\frac{8}{10}$ of the way between log 2.17 and log 2.18. These two values are 0.3365 and 0.3385. The tabular difference is 20, and $(\frac{8}{10})(20) = 16$. Adding this to 0.3365, we find the mantissa to be 0.3381. Therefore, we have log $0.02178 = 8.3381 - 10$.

We may also use Table 2 to find N if we know log N. In this case we may refer to N as the **antilogarithm** of log N. The following examples illustrate the determination of antilogarithms.

Example H
Given log N = 1.5263, find N.
Direct observation of the given logarithm tells us that the characteristic is 1 and that $N = P \times 10^1$. In Table 2 we find 5263 opposite 33 and under 6. Thus, $P = 3.360$. This means that $N = 3.360 \times 10^1$, or in ordinary notation, $N = 33.60$.

Example I
Given log N = 8.2611 − 10, find N.
Using the method described in Example D, we determine that the characteristic is $8 - 10 = -2$. We look for 0.2611 in the tables, and find that it is between 2601 (log 1.82) and 2625 (log 1.83). These latter two values have a tabular difference of 24, and the difference between 2611 and 2601 is 10. Thus, the number we want is $\frac{10}{24}$, or 0.4 of the way between 1.82 and 1.83. Hence,

$$N = 1.824 \times 10^{-2} = 0.01824$$

The basic properties of logarithms allow us to find the logarithms of products, quotients, and roots. The following example illustrates the method.

Example J
Find log $\sqrt{0.846}$.
From the properties of logarithms we write log $\sqrt{0.846} = \frac{1}{2}$ log 0.846. From the tables we find that log 8.46 = 0.9274. Therefore, log 0.846 = 9.9274 − 10. To obtain the desired logarithm we must divide this by 2. This will result in a 5 to be subtracted. To assure our answer being in the usual form of a negative characteristic, we shall write log 0.846 as 19.9274 − 20. Thus, log $\sqrt{0.846}$ = 9.9637 − 10, when we divide through by 2. In this type of problem we choose that part of the characteristic which is to be subtracted so that 10 will result after division. This is done by *adding* the proper multiple of 10 to each part of the characteristic.

Exercises 12–4

In Exercises 1 through 16 find the common logarithm of each of the given numbers.

1. 567
2. 60.5
3. 0.0640
4. 0.000566
5. 9.24×10^6
6. 3.19×10^{15}
7. 1.17×10^{-4}
8. 8.04×10^{-8}
9. 1.053
10. 73.27
11. 0.2384
12. 0.004309
13. 7.331×10^8
14. 1.656×10^{-5}
15. $\sqrt{0.002006}$
16. $\sqrt[3]{38310000}$

In Exercises 17 through 32 find the antilogarithm N from the given logarithms.

17. 4.4378
18. 0.9294
19. 8.6955 − 10
20. 3.0212 − 10
21. 3.3010
22. 8.8241
23. 9.8597 − 10
24. 7.4409 − 10
25. 1.9495
26. 2.4367
27. 6.6090 − 10
28. 9.3755 − 10
29. 0.1543
30. 10.2750
31. 17.7625 − 20
32. 35.6641 − 40

In Exercises 33 through 40 find the logarithms of the given numbers.

33. A certain radar signal has a frequency of 1.15×10^9 Hz.
34. The earth travels about 595,000,000 mi in one year.
35. A certain bank charges 7.5% interest on loans that it makes.
36. The coefficient of thermal expansion of steel is about 1.2×10^{-5} per °C.
37. A typical x-ray tube operates with a voltage of 150,000 V.
38. The bending moment of a particular concrete column is 4.60×10^6 lb-in.
39. Electronic calculators which display numbers in scientific notation can calculate results which are as numerically small as 10^{-99}.
40. In an air sample taken in an urban area, $5/10^6$ of the air was carbon monoxide.

12–5 Computations Using Logarithms

Logarithms were developed in the seventeenth century for the purpose of making tedious and complicated calculations which arose in astronomy and navigation. Until recently (the mid 1970s), when calculators came into extensive use, logarithms were used commonly for calculational purposes. The slide rule was also used extensively, but its accuracy was generally limited to three significant digits. (The construction of a slide rule is based on logarithms.) Also, logarithms are used for some computations performed by calculators.

Logarithms can be used to make more complicated calculations by means of additions, subtractions, and basic multiplications and divisions. By means of logarithms and the basic arithmetic operations we can determine the answer to such problems as $\sqrt[5]{7.60}$ or $(89.1)^{0.3}$. In this section we show how logarithms are used to make such calculations to an accuracy up to four significant digits. (Greater accuracy is obtainable with more extensive tables.) Even if logarithms are not required for such calculations, performing them with logarithms provides an opportunity to better understand the meaning and properties of logarithms. The following examples illustrate the use of logarithms in computations.

Example A
By the use of logarithms, calculate the value of $(42.80)(215.0)$.

Equation (12–7) tells us that $\log xy = \log x + \log y$. If we find $\log 42.80$ and $\log 215.0$ and add them, we shall have the logarithm of the product. Using this result, we look up the antilogarithm, which is the desired product.

$$\log 42.80 = 1.6314$$
$$\log 215.0 = 2.3324$$
$$\log(42.80)(215.0) = 3.9638$$
$$\log 9200 = 3.9638$$

Thus, $(42.80)(215.0) = 9200$.

Example B

By the use of logarithms, calculate the value of $8.640 \div 45.55$.

From Eq. (12–8), we know that $\log (x/y) = \log x - \log y$. Therefore, by subtracting log 45.55 from log 8.640, we shall have the logarithm of the quotient. The antilogarithm gives the desired result.

$$\begin{aligned} \log 8.640 &= 10.9365 - 10 \\ \log 45.55 &= \underline{1.6585} \\ \log(8.640/45.55) &= 9.2780 - 10 \end{aligned}$$

We wrote the characteristic of log 8.640 as $10 - 10$, so that when we subtracted, the part of the result containing the mantissa would be positive although the characteristic was negative. The antilogarithm of $9.2780 - 10$ is 0.1897. Interpolation is required to determine the fourth significant digit. Therefore,

$$\frac{8.640}{45.55} = 0.1897$$

Example C

By the use of logarithms, calculate the value of $\sqrt[5]{0.03760}$.

From Eq. (12–9), we know that $\log x^n = n \log x$. Therefore, by writing $\sqrt[5]{0.03760} = (0.03760)^{1/5}$, we know that we can find the logarithm of the result by multiplying log 0.03760 by $\frac{1}{5}$. Now, we determine that $\log 0.03760 = 8.5752 - 10$. Since we wish to multiply this by $\frac{1}{5}$, we shall write this logarithm as

$$\log 0.03760 = 48.5752 - 50$$

by adding and subtracting 40. Multiplying by $\frac{1}{5}$, we have

$$\tfrac{1}{5} \log 0.03760 = \tfrac{1}{5}(48.5752 - 50) = 9.7150 - 10$$

The antilogarithm of $9.7150 - 10$ is 0.5188. Therefore,

$$\sqrt[5]{0.03760} = 0.5188$$

The following examples illustrate calculations which involve the use of a combination of the basic properties of logarithms.

Example D

Calculate the value of

$$N = \frac{6.875\sqrt{98.66}}{7.880}$$

$$\log N = \log 6.875 + \tfrac{1}{2} \log 98.66 - \log 7.880$$

$$\begin{aligned} \log 6.875 &= 0.8373 \qquad \log 98.66 = 1.9941 \\ \tfrac{1}{2} \log 98.66 &= \underline{0.9970} \\ &\ \ 1.8343 \\ \log 7.880 &= \underline{0.8965} \\ \log N &= 0.9378 \qquad N = 8.666 \end{aligned}$$

Example E
Calculate the value of

$$N = \left[\frac{(0.05325)\sqrt{0.8884}}{\sqrt[3]{895.3}} \right]^{0.3}$$

$\log N = 0.3(\log 0.05325 + \frac{1}{2}\log 0.8884 - \frac{1}{3}\log 895.3)$

$\log 0.05325 = 8.7263 - 10$ $\log 0.8884 = 19.9486 - 20$

$\frac{1}{2}\log 0.8884 = \underline{9.9743 - 10}$

$18.7006 - 20$

$\frac{1}{3}\log 895.3 = \underline{0.9840}$ $\log 895.3 = 2.9520$

$17.7166 - 20$

$0.3(97.7166 - 100) = 29.3150 - 30$ $N = 0.2065$

Example F
The velocity of an object moving with constant acceleration can be found from the equation $v = \sqrt{v_0^2 + 2as}$, where v_0 is the initial velocity, a is the acceleration, and s is the distance traveled. Determine the velocity of an object if $v_0 = 86.46$ ft/s, $a = 17.92$ ft/s^2, and $s = 136.7$ ft.

Before we can compute the square root, we must square v_0, determine the product $2as$, and add these results. Another calculation is then necessary to determine the square root:

$$\log v_0^2 = 2 \log 86.46 = 2(1.9368) = 3.8736 \qquad v_0^2 = 7475$$

$$\log 2as = \log 2 + \log 17.92 + \log 136.7$$

$\log 2 = 0.3010$
$\log 17.92 = 1.2534$
$\log 136.7 = \underline{2.1357}$
$\log 2as = 3.6901$ $\qquad\qquad 2as = \underline{4899}$
$\qquad\qquad\qquad\qquad\qquad v_0^2 + 2as = 12374$

$$\log v = \frac{1}{2}\log 12370 = \frac{1}{2}(4.0924) = 2.0462$$
$$v = 111.2 \text{ ft/s}$$

Note that 12374 was rounded off to 12370 for purposes of calculation, since only four significant digits can be used.

Exercises 12—5

In Exercises 1 through 32 use logarithms to perform the indicated calculations.

1. $(5.980)(14.30)$ 2. $(0.7640)(551.0)$ 3. $(0.8256)(0.04532)$

4. $(0.0008080)(2623)$ 5. $\dfrac{790.0}{8.020}$ 6. $\dfrac{31.60}{0.4540}$

7. $\dfrac{76.98}{43.82}$ 8. $\dfrac{0.008670}{0.6521}$ 9. $(6.750)^6$

10. $(0.9040)^5$ 11. $(89.00)^{0.3}$ 12. $(0.04030)^{0.6}$

13. $\sqrt[5]{7.600}$ 14. $\sqrt[3]{95.40}$ 15. $\sqrt{641.6}$

16. $\sqrt[8]{308.7}$

17. $\dfrac{(4510)(0.6120)}{738.0}$

18. $\dfrac{87.42}{(11.54)(0.9316)}$

19. $\dfrac{\sqrt{0.07530}}{86.02}$

20. $(\sqrt{5.270})(\sqrt[3]{42.19})$

21. $\dfrac{89.42\sqrt[3]{0.1142}}{0.04290}$

22. $\left(\dfrac{75.19}{900.5\sqrt{15.00}}\right)^{0.1}$

23. $(8.723)^{9.742}$ (Find log log 8.723.)

24. $(4.072)^{-10}$ (Be careful, especially if your method of solution leads to a negative "mantissa.")

25. What is the area of a rectangular field 325.5 m by 246.4 m?

26. In testing a new engine in order to determine its fuel economy, a car traveled 426.4 mi on 11.40 gallons of gasoline. What is the miles per gallon rating of the car's engine?

27. Plutonium is radioactive and disintegrates such that of 1000 mg originally present, $1000(0.5)^{0.0000410t}$ mg will remain after t years. Calculate the amount present after 10,000 years.

28. The molecular mass M of a gas may be calculated from the formula $PV = mRT/M$, where P is the pressure, V is the volume, m is the mass, R is a constant for all gases, and T is the thermodynamic temperature. Determine M (in kilograms) if you are given that $P = 1.081 \times 10^5$ Pa, $V = 2.485 \times 10^{-4}$ m^3, $R = 8.314$ J/mol·K, $T = 373.6$ K, and $m = 1.267 \times 10^{-3}$ kg.

29. The velocity of sound in air is given by $v = \sqrt{1.410p/d}$, where p is the pressure and d is the density. Given that $p = 1.013 \times 10^5$ Pa and $d = 1.293$ kg/m^3, find v (in meters per second).

30. In undergoing an adiabatic (no *heat* gained or lost) expansion, the relation between the initial and final temperatures and volumes is given by $T_f = T_i(V_i/V_f)^{0.4}$, where the temperatures are expressed in kelvins. Given that $V_i = 1.506$ cm^3, $V_f = 0.1290$ cm^3 and $T_i = 373.2$ K, find T_f.

31. Given the density of iron as 491.0 lb/ft^3, find the radius of a spherical iron ball which weighs 25.65 lb.

32. When a light ray is incident on glass, the percentage of light reflected is given by

$$I_r = 100\left(1 - \dfrac{4n_a n_g}{(n_a + n_g)^2}\right)$$

where n_a and n_g are the indices of refraction of air and glass, respectively. What percentage of light is reflected if $n_g = 1.532$ and $n_a = 1.000$?

12–6 Logarithms of Trigonometric Functions

omit

When we are working with trigonometric functions, there is often a great deal of calculational work to be done. Logarithms may be used to make these calculations. In such cases, if we wish to use logarithms, we could look up the function of the desired angle and then find the logarithm of this number to perform some operation on it.

306 Exponential and Logarithmic Functions

Example A

Find log sin 23°20′. (Use Tables 2 and 3 in Appendix E.)

$$\sin 23°20′ = 0.3961, \qquad \log 0.3961 = 9.5978 - 10$$

To facilitate work when using trigonometric functions, we can use tables of logarithms of the trigonometric functions, which allow us to find these logarithms in one step.

Example B

Find log sin 23°20′. (Use Table 5.)

By direct reading we find log sin 23°20′ = 9.5978 − 10.

Like other similar tables, these tables enable us to find by interpolation values not directly listed. We may also find an angle directly, if we know the logarithm of some function of that angle.

Example C

Find log tan 57°34′.

We find that log tan 57°30′ = 0.1958 and that log tan 57°40′ = 0.1986. The tabular difference is 28, and 0.4(28) = 11 (to 2 digits). Thus,

$$\log \tan 57°34′ = 0.1969$$

Example D

Given that log cos θ = 9.9049 − 10, find θ.

From Table 5 we find that

$$\log \cos 36°30′ = 9.9052 - 10$$

and

$$\log \cos 36°40′ = 9.9042 - 10$$

We find that the tabular difference between listed values is 10, and the tabular difference between 9.9049 and 9.9052 is 3. Therefore,

$$\theta = 36°33′$$

(Be careful—the values of the cosine and its logarithm *decrease* as θ increases.)

Example E

Solve the following oblique triangle, using logarithms for your calculations: a = 34.12, A = 31°20′, B = 52°43′.

We first determine that the solution may be completed by the law of sines. Next we find C = 95°57′, and then we can find sides b and c. From the law of sines we have

$$b = \frac{a \sin B}{\sin A} \qquad \text{and} \qquad c = \frac{a \sin C}{\sin A}$$

By using a and sin A in both calculations, we reduce the amount of information required from the tables.

$$\log b = \log 34.12 + \log \sin 52°43' - \log \sin 31°20'$$

$$
\begin{array}{rl}
\log 34.12 & = \quad 1.5330 \\
\log \sin 52°43' & = \quad \underline{9.9007 - 10} \\
& \quad\ \ 11.4337 - 10 \\
\log \sin 31°20' & = \quad \underline{9.7160 - 10} \\
& \quad\ \ \ 1.7177
\end{array}
$$

$$b = 52.20$$

$$\log c = \log 34.12 + \log \sin 95°57' - \log \sin 31°20'$$

$$
\begin{array}{rl}
\log 34.12 & = \quad 1.5330 \\
\log \sin 95°57' & = \quad \underline{9.9976 - 10} \ (\log \sin 84°3') \\
& \quad\ \ 11.5306 - 10 \\
\log \sin 31°20' & = \quad \underline{9.7160 - 10} \\
& \quad\ \ \ 1.8146
\end{array}
$$

$$c = 65.26$$

Example F

A ship passes a certain point at noon going north at 12.35 mi/h. A second ship passes the same point at 1 PM going east at 16.42 mi/h. How far apart are the ships at 3 PM?

At 3 PM the first ship is 37.05 mi from the point and the second ship is 32.84 mi from it (Fig. 12–5). Since the angle between their directions is 90°, the distance d between them can be found from the Pythagorean theorem. Also, by finding the angle α from tan α = 37.05/32.84, we can then solve for d by using d = 32.84/cos α. This second method has the advantage that once we determine α, we can find log cos α immediately from the table by shifting from the log tan column to the log cos column:

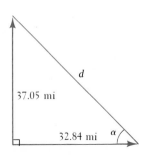

37.05 mi

32.84 mi α

Figure 12–5

$$\log \tan \alpha = \log 37.05 - \log 32.84$$

$$
\begin{array}{rl}
\log 37.05 & = 1.5688 \\
\log 32.84 & = \underline{1.5164} \\
\log \tan \alpha & = 0.0524 \\
\alpha & = 48°27'
\end{array}
$$

$$\log d = \log 32.84 - \log \cos \alpha$$

$$
\begin{array}{rl}
\log 32.84 & = 11.5164 - 10 \\
\log \cos \alpha & = \underline{9.8217 - 10} \\
\log d & = \quad 1.6947 \\
d & = 49.51 \text{ mi}
\end{array}
$$

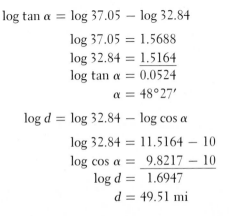

Exercises 12−6

In Exercises 1 through 12 use Table 5 to find the values of the indicated logarithms.

1. $\log \sin 22°10'$ 2. $\log \cos 31°40'$ 3. $\log \tan 52°0'$

4. $\log \cot 61°50'$ 5. $\log \sin 38°14'$ 6. $\log \cos 12°7'$

7. $\log \tan 56°45'$ 8. $\log \sin 75°42'$ 9. $\log \cos 322°17'$

10. $\log \cot 228'12'$ 11. $\log \cos 79°6'$ 12. $\log \tan 85°52'$

In Exercises 13 through 20 use Table 5 to find the smallest positive θ.

13. $\log \sin \theta = 9.6740 - 10$ 14. $\log \cot \theta = 9.9341 - 10$

15. $\log \cos \theta = 9.8056 - 10$ 16. $\log \tan \theta = 9.9140 - 10$

17. $\log \tan \theta = 0.0599$ 18. $\log \sin \theta = 8.9150 - 10$

19. $\log \cos \theta = 9.9998 - 10$ 20. $\log \cot \theta = 0.4767$

In Exercises 21 through 28 solve the given triangles by logarithms.

21. $A = 82°5'$, $C = 90°0'$, $c = 86.17$ 22. $B = 54°10'$, $C = 90°0'$, $b = 15.70$

23. $A = 65°40'$, $B = 72°10'$, $a = 9100$ 24. $A = 63°14'$, $C = 18°16'$, $c = 0.5320$

25. $A = 67°10'$, $B = 44°42'$, $b = 9.328$ 26. $A = 47°36'$, $a = 17.45$, $b = 10.29$

27. $a = 298.5$, $b = 382.6$, $C = 90°0'$ 28. $a = 7392$, $b = 4218$, $c = 4005$

In Exercises 29 through 36 solve the given problems by logarithms.

29. A 24.2-ft. vertical steel girder is supported by a cable connected at the top of the girder and at the level of the foot of the girder. If the angle between the girder and cable is 37.5°, how long is the cable?

30. A jet cruising at 730.0 km/h is descending at an angle of 9.5°. What is its loss in altitude in 3.000 min?

31. A 56.62-lb block is on an inclined plane which makes an angle of 22°42' with respect to the horizontal. What are the components of the weight parallel to and perpendicular to the plane?

32. A surveyor finds one side of a rectangular piece of land to be 137.8 ft, and the angle between this side and the diagonal to be 36°17'. What is the area of the piece of land?

33. Two spring balances support an object as shown in Fig. 12−6. What is the weight of the object? We can find T_2 by using the fact that there is no net force acting horizontally. This means that

$$T_1 \sin \alpha = T_2 \sin \beta$$

34. An airplane headed south has a speed with respect to the air of 418.5 mi/h. The speed with respect to the ground is 425.0 mi/h in a direction of 3°16' east of south. Find the direction and speed of the wind.

35. For a given electrical circuit, $R = 21.35\ \Omega$ and $X_C = 13.37\ \Omega$. Find the phase angle between the current and voltage (see Section 11–7).

36. The limiting distance d of resolution of a microscope is given by the formula $d = \lambda/2 \sin \theta$, where λ is the wavelength of the light being used and θ is an angle associated with the object and lens. Calculate d for $\lambda = 0.5893\ \mu\mathrm{m}$ and $\theta = 81°30'$.

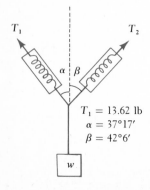

$T_1 = 13.62$ lb
$\alpha = 37°17'$
$\beta = 42°6'$

Figure 12−6

12–7 Logarithms to Bases Other Than 10; Natural Logarithms

As we mentioned earlier, another number which is important as a base of logarithms is the number e. *Logarithms to the base e are called natural logarithms.* Since e is an irrational number equal to approximately 2.718, it may appear to be a very unnatural choice as a base of logarithms. However, in the development of the calculus, the reason for its choice, and the fact that it is a very natural number for a base of logarithms, is shown.

Just as the notation $\log x$ refers to logarithms to the base 10, the notation $\ln x$ is used to denote logarithms to the base e. Due to the extensive use of natural logarithms, the notation $\ln x$ is more convenient than $\log_e x$, although they mean the same thing.

Since more than one base is important, there are times when it is useful to be able to change a logarithm in one base to another base. If $u = \log_b x$, then $b^u = x$. Taking logarithms of both sides of this last expression to the base a, we have

$$\log_a b^u = \log_a x$$
$$u \log_a b = \log_a x$$

Solving this last equation for u, we have

$$u = \frac{\log_a x}{\log_a b}$$

However, $u = \log_b x$, which means that

$$\log_b x = \frac{\log_a x}{\log_a b} \qquad (12\text{–}13)$$

Equation (12–13) allows us to change a logarithm in one base to a logarithm in another base. The following examples illustrate the method of performing this operation.

Example A

Change $\log 20 = 1.3010$ to a logarithm with base e; that is, find $\ln 20$.

In Eq. (12–13), if we let $b = e$ and $a = 10$, we have

$$\log_e x = \frac{\log_{10} x}{\log_{10} e}$$

or

$$\ln x = \frac{\log x}{\log e}$$

In this example, $x = 20$. Therefore,

$$\ln 20 = \frac{\log 20}{\log e} = \frac{\log 20}{\log 2.718} = \frac{1.3010}{0.4343} = 2.996$$

Here we divided 1.3010 by 0.4343 to obtain the final result. Even though 1.3010 and 0.4343 are logarithms, it is their quotient that we need. We can calculate the quotient of 1.301 and 0.4343 by any appropriate means, which, of course, includes logarithms. Using logarithms, we have

$$
\begin{array}{rl}
\log 1.301 &= 10.1143 - 10 \\
\log 0.4343 &= \underline{9.6378 - 10} \\
& \quad\; 0.4765
\end{array}
$$

The antilogarithm of 0.4765 is 2.996, which means that $\ln 20 = 2.996$.

Example B

Find $\log_5 560$.

In Eq. (12–13), if we let $b = 5$ and $a = 10$, we have

$$
\log_5 x = \frac{\log x}{\log 5}
$$

In this example, $x = 560$. Therefore, we have

$$
\log_5 560 = \frac{\log 560}{\log 5} = \frac{2.7482}{0.6990} = 3.932
$$

Therefore, we have found that $\log_5 560 = 3.932$.

Since natural logarithms are used extensively, it is often convenient to have Eq. (12–13) written specifically for use with logarithms to the base 10 and natural logarithms. Since $\log e = 0.4343$, and $\ln 10 = 2.3026$, we can write

$$
\boxed{\ln x = 2.3026 \log x} \tag{12–14}
$$

and

$$
\boxed{\log x = 0.4343 \ln x} \tag{12–15}
$$

Example C

Using Eq. (12–14), find $\ln 0.811$.

From Eq. (12–14), we have

$$
\ln 0.811 = 2.3026 \log 0.811
$$

In many applications of natural logarithms, it is preferable to write the logarithms in their explicit form, even when they are negative. Therefore, here we will express $\log 0.811$ as

$$
\log 0.811 = 9.9090 - 10 = -0.0910
$$

Thus,

$$
\begin{aligned}
\ln 0.811 &= 2.3026(-0.0910) \\
&= -0.2095
\end{aligned}
$$

Values of natural logarithms are also available from tables and scientific calculators. In tables of natural logarithms it is not possible to present a table of mantissas, since mantissas of numbers which differ only by a power of ten are not the same. This can be seen by taking natural logarithms of a number written in scientific notation. For $N = P \times 10^k$, we have

$$\ln N = \ln P + \ln 10^k$$
$$= \ln P + k \ln 10$$

or

$$\ln N = k \ln 10 + \ln P \tag{12–16}$$

Thus, a table of natural logarithms can be used directly only for the numbers which are tabulated. Its use can be extended by use of Eq. (12–16). In Appendix E, Table 6 is such a table of natural logarithms. The following example illustrates the use of this table and Eq. (12–16).

Example D
Using Table 6 and Eq. (12–16), determine the value of ln 820.
Since 820 does not appear in the column labeled n in Table 6, we cannot find its value directly from the table. However, we can write $820 = 8.2 \times 10^2$ and use Eq. (12–16). Here we have

$$\ln 820 = 2 \ln 10 + \ln 8.2$$

We determine ln 8.2 from the table, and note that the value of 2 ln 10 is also given. Thus,

$$\ln 820 = 4.6052 + 2.1041$$
$$= 6.7093$$

It is possible to use natural logarithms for calculations in the same way that we use common logarithms. By using Table 6 with interpolation, three-place accuracy is obtainable. Other applications of natural logarithms are found in many fields of technology. One such application is shown in the following example, and others are found in the exercises.

Example E
Under certain conditions, the electric current i in a circuit containing a resistance and an inductance (see Section 11–7) is given by $\ln \frac{i}{I} = -\frac{Rt}{L}$, where I is the current at $t = 0$, R is the resistance, t is the time, and L is the inductance. Calculate how long (in seconds) it takes i to reach 0.430 A, if $I = 0.750$ A, $R = 7.50$ Ω, and $L = 1.25$ H.
Solving for t, we have

$$t = -\frac{L \ln(i/I)}{R} = -\frac{L(\ln i - \ln I)}{R}$$

Thus, for the given values, we have

$$t = -\frac{1.25(\ln 0.430 - \ln 0.750)}{7.50} \text{ s}$$

$$\ln 0.430 = \frac{\log 0.430}{\log e} = \frac{-0.3665}{0.4343} = -0.8439$$

$$\ln 0.750 = \frac{\log 0.750}{\log e} = \frac{-0.1249}{0.4343} = -0.2876$$

$$t = -\frac{1.25(-0.8439 + 0.2876)}{7.50} = \frac{1.25(0.5563)}{7.50} = 0.0927 \text{ s}$$

Therefore, the current changes from 0.750 A to 0.430 A in 0.0927 s.

Exercises 12−7

In Exercises 1 through 8 use logarithms to the base 10 to find the natural logarithms of the given numbers.

use cal.

1. 26.0   2. 631   3. 1.56   4. 45.7
5. 0.501   6. 0.052   7. 0.00732   8. 0.000443

In Exercises 9 through 16 use logarithms to the base 10 to find the indicated logarithms.

9. $\log_7 42$ 10. $\log_2 86$ 11. $\log_5 245$ 12. $\log_3 706$
13. $\log_{12} 122$ 14. $\log_{20} 86$ 15. $\log_{40} 750$ 16. $\log_{100} 3720$

In Exercises 17 through 24 use Eq. (12−14) to find the natural logarithms of the indicated numbers.

17. 51.4   18. 293   19. 1.39   20. 65.6
21. 0.991   22. 0.0020   23. 0.0129   24. 0.0000608

In Exercises 25 through 28 use Eq. (12−15) and Table 6 to find the logarithms to the base 10 of the indicated numbers.

25. 2.7   26. 40   27. 87   28. 0.63

In Exercises 29 through 32 perform the indicated calculations by use of natural logarithms from Table 6.

29. 2.50×4700 30. $\dfrac{380}{0.900}$ 31. $\sqrt{75.0}$ 32. $(2.9)^{10}$

In Exercises 33 through 40 solve the given problems.

33. Solve for y in terms of x: $\ln y - \ln x = 1.0986$
34. Solve for y in terms of x: $\ln y + 2 \ln x = 1 + \ln 5$
35. If interest is compounded continuously (daily compounded interest closely approximates this), a bank account can double in t years according to the equation $i = \frac{\ln 2}{t}$, where i is the interest rate. What interest rate is required for an account to double in 8.5 years?

36. One approximate formula for world population growth is $T = 50.0 \ln 2$, where T is the number of years for the population to double. According to this formula, how long does it take for the population to double?

37. For the electric circuit of Example E, find how long it takes the current to reach 0.1 of the initial value of 0.750 A.

38. Under specific conditions, an equation relating the pressure P and volume V of a gas is $\ln P = C - \gamma \ln V$, where C and γ (the Greek gamma) are constants. Find P (in atmospheres) if $C = 3.000$, $\gamma = 1.50$, and $V = 2.20$ ft³.

39. If 100 mg of radium radioactively decays, an equation relating the amount Q which remains, and the time t is

$$\ln Q - \ln 100 = kt$$

where k is a constant. If $Q = 90.0$ mg, and $k = -0.000410$ per year, find t.

40. For a certain electric circuit, the voltage v is given by $v = e^{-0.1t}$. What is $\ln v$ after 2.00 s?

12-8 Exponential and Logarithmic Equations

In solving equations in which there is an unknown exponent, it is often advantageous to take logarithms of both sides of the equation and then proceed. In solving logarithmic equations, one should keep in mind the basic properties of logarithms, since these can often help transform the equation into a solvable form. There is, however, no general algebraic method for solving such equations, and here we shall solve only some special cases.

Example A
Solve the equation $2^x = 8$.
   By writing this in logarithmic form, we have

$$x = \log_2 8 = 3$$

We could also solve this equation by taking logarithms of both sides. This would yield

$$\log 2^x = \log 8$$
$$x \log 2 = \log 8$$

$$= \frac{\log 8}{\log 2} = \frac{0.9031}{0.3010} = 3.00$$

This last method is more generally applicable, since the first method is good only if we can directly evaluate the logarithm which results.

Example B
Solve the equation $3^{x-2} = 5$.

Taking logarithms of both sides, we have

$$\log 3^{x-2} = \log 5 \qquad \text{or} \qquad (x-2)\log 3 = \log 5$$

Solving this last equation for x, we have

$$x = 2 + \frac{\log 5}{\log 3} = 2 + \frac{0.6990}{0.4771} = 2 + 1.465 = 3.465$$

Thus, the solution to this equation is $x = 3.465$.

Example C
Solve the equation $2(4^{x-1}) = 17^x$.

By taking logarithms of both sides, we have the following:

$$\log 2 + (x-1)\log 4 = x \log 17$$
$$x \log 4 - x \log 17 = \log 4 - \log 2$$
$$x(\log 4 - \log 17) = \log 4 - \log 2$$
$$x = \frac{\log 4 - \log 2}{\log 4 - \log 17} = \frac{\log (4/2)}{\log 4 - \log 17}$$
$$= \frac{\log 2}{\log 4 - \log 17} = \frac{0.3010}{0.6021 - 1.2304}$$
$$= \frac{0.3010}{-0.6283} = -0.479$$

Some of the important measurements in scientific and technical work are defined in terms of logarithms. We shall consider here some of the important applications which are basic logarithmic formulas. In using them we shall solve some equations in which logarithms are involved. The following example illustrates one such area of application, and others are found in the exercises.

Example D
It has been found that the human ear responds to sound on a scale which is approximately proportional to the logarithm of the intensity of the sound. Thus, the loudness of sound, measured in decibels, is defined by the equation $b = 10 \log (I/I_0)$, where I is the intensity of the sound and I_0 is the minimum intensity detectable.

A busy city street has a loudness of 70 dB, and riveting has a loudness of 100 dB. How many times greater is the intensity of the sound of riveting I_r than the sound of the city street I_c?

First, we substitute the decibel readings into the above definition. This gives us

$$70 = 10 \log(I_c/I_o) \qquad \text{and} \qquad 100 = 10 \log(I_r/I_o)$$

To solve these equations for I_c and I_r, we divide each side by 10 and then use the exponential form. Thus, we have

$$7.0 = \log(I_c/I_o) \qquad \text{and} \qquad 10 = \log(I_r/I_o)$$

$$\frac{I_c}{I_o} = 10^{7.0} \qquad\qquad\qquad \frac{I_r}{I_o} = 10^{10}$$

$$I_c = I_o(10^{7.0}) \qquad\qquad\qquad I_r = I_o(10^{10})$$

Since we want the number of times I_r is greater than I_c, we divide I_r by I_c. This gives us

$$\frac{I_r}{I_c} = \frac{I_o(10^{10})}{I_o(10^{7.0})} = \frac{10^{10}}{10^{7.0}} = 10^{3.0}$$

or

$$I_r = 10^{3.0}I_c = 1000I_c$$

Thus, the sound of riveting is 1000 times as intense as the sound of the city street. This demonstrates that sound intensity levels are considerably greater than loudness levels. (See Exercise 28.)

The following examples illustrate the solution of other logarithmic equations.

Example E
Solve the equation $\log_2 7 - \log_2 14 = x$.
    Using the basic properties of logarithms, we arrive at the following result:

$$\log_2\left(\tfrac{7}{14}\right) = x$$
$$\log_2\left(\tfrac{1}{2}\right) = x \qquad \text{or} \qquad \tfrac{1}{2} = 2^x$$

Thus, $x = -1$.

Example F
Solve the equation $2 \log x - 1 = \log(1 - 2x)$.

$$\log x^2 - \log(1 - 2x) = 1$$

$$\log \frac{x^2}{1 - 2x} = 1$$

$$\frac{x^2}{1 - 2x} = 10^1$$

$$x^2 = 10 - 20x$$

$$x^2 + 20x - 10 = 0$$

$$x = \frac{-20 \pm \sqrt{400 + 40}}{2} = -10 \pm \sqrt{110}$$

Since logarithms of negative numbers are not defined, we have the result that $x = \sqrt{110} - 10$.

Exercises 12−8

In Exercises 1 through 24 solve the given equations.

1. $2^x = 16$ 2. $3^x = \frac{1}{81}$

3. $5^x = 4$ 4. $6^x = 15$

5. $6^{x+1} = 10$ 6. $5^{x-1} = 2$

7. $4(3^x) = 5$ 8. $14^x = 40$

9. $0.8^x = 0.4$ 10. $0.6^x = 100$

11. $(15.6)^{x+2} = 23^x$ 12. $5^{x+2} = 3^{2x}$

13. $2 \log_2 x = 4$ 14. $3 \log_8 x = 1$

15. $3 \log_8 x = 2$ 16. $5 \log_{32} x = -3$

17. $2 \log(3 - x) = 1$ 18. $3 \log(2x - 1) = 1$

19. $\frac{1}{2} \log(x + 2) + \log 5 = 1$ 20. $\frac{1}{2} \log(x - 1) - \log x = 0$

21. $\log_5 (x - 3) + \log_5 x = \log_5 4$ 22. $\log_7 x + \log_7 (2x - 5) = \log_7 3$

23. $\log(2x - 1) + \log(x + 4) = 1$ 24. $\log_2 x + \log_2 (x + 2) = 3$

In Exercises 25 through 32 determine the required quantities.

25. For a certain electric circuit, the current i is given by $i = 1.50e^{-200t}$. For what value of t (in seconds) is $i = 1.00$ A? (Use 2.718 to approximate e.)

26. The amount q of a certain radioactive substance remaining after t years is given by $q = 100(0.900)^t$. After how many years are there 50.0 mg of the substance remaining?

27. Referring to Example D, how many times I_0 is the intensity of sound of a jet plane which has a loudness of 110 dB?

28. Referring to Example D, show that if the difference in loudness of two sounds is d decibels, the louder sound is $10^{d/10}$ more intense than the quieter sound.

29. Measured on the Richter scale, the magnitude of an earthquake of intensity I is defined as $R = \log(I/I_0)$, where I_0 is a minimum level for comparison. How many times I_0 was the 1906 San Francisco earthquake whose magnitude was 8.25 on the Richter scale?

30. How many more times intense was the 1964 Alaska earthquake, $R = 7.5$, than the 1971 Los Angeles earthquake, $R = 6.7$? (See Exercise 29.)

31. In chemistry, the pH-value of a solution is a measure of its acidity. The pH-value is defined by the relation $pH = -\log (H^+)$, where H^+ is the hydrogen ion concentration. If the pH of a certain wine is 3.4065, find the hydrogen ion concentration. (If the pH-value is less than 7, the solution is acid. If the pH-value is above 7, the solution is basic.)

32. Referring to Exercise 31, find the hydrogen ion concentration for ammonia for which the pH is 10.8.

To solve more complicated problems, we may use graphical methods. For example, if we wish to solve the equation $2^x + 3^x = 50$, we can set up the function $y = 2^x + 3^x - 50$ and then determine its zeros graphically. Note that the given equation can be written as $2^x + 3^x - 50 = 0$, and therefore the zeros of the function which has been set up will give the desired solution. In Exercises 33 through 36 solve the given equations in this way.

33. $2^x + 3^x = 50$ 34. $3^{x+1} - 4^x = 1$

35. $3^x - 2x = 40$ 36. $4^x + x^2 = 25$

12—9 Graphs on Logarithmic and Semilogarithmic Paper

If, when we are graphing, the range of values of one variable is much greater than the corresponding range of values of the other variable, it is often convenient to use what is known as **semilogarithmic** paper. On this paper the y-axis (usually) is marked off in distances proportional to the logarithms of numbers. This means that the distances between numbers on this axis are not even, but this system does allow for a much greater range of values, and with much greater accuracy for many of the numbers. There is another advantage to this paper: Many equations which would exhibit more complex curves on ordinary graph paper will work out as straight lines on semilogarithmic paper. In many instances this makes the analysis of the curve easier.

If we wish to indicate a large range of values for each of the variables, we use what is known as **logarithmic** paper, or as **log-log** paper. Both axes are marked off with logarithmic scales. Again, the more complicated equations give simple curves or straight lines on this paper.

The following examples will illustrate the use of semilogarithmic and logarithmic paper.

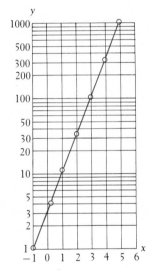

Figure 12—7

Example A

Construct the graph of $y = 4(3)^x$ on semilogarithmic paper.

First we construct a table of values.

x	-1	0	1	2	3	4	5
y	1.3	4	12	36	108	324	972

From the table we see that the range of y-values is large. If we plotted this curve on a regular coordinate system we would have large units for each interval along the y-axis. This would make the values of 1.3, 4, 12, and 36 appear at practically the same level. However, if we use semilogarithmic graph paper, we can label each axis such that all y-values are accurately plotted as well as the x-values.

The logarithmic scale is shown in cycles, and we must label the base line of the first cycle as 1 times a power of ten (0.01, 0.1, 1, 10, 100, etc.) with the following cycle labeled with the next power of ten. The lines between are labeled with 2, 3, 4, and so on, times the proper power of ten. See the vertical scale in Fig. 12—7. We now plot the points in the table on the graph. The resulting graph is a straight line, as we see in Fig. 12—7. Taking logarithms of both sides of the equation, we have

$$\log y = \log 4 + x \log 3$$

However, since $\log y$ was plotted automatically (because we used semilogarithmic paper), the graph really represents

$$u = \log 4 + x \log 3$$

where $u = \log y$; $\log 3$ and $\log 4$ are constants, and therefore this equation is of the form $u = ax + b$, which is a straight line (see Section 4—1). If we had sketched this graph on regular coordinate paper, the scale would be so reduced that the values of 0.5, 1.3, 4, 12, and 36 would appear at practically the same level.

Example B

Construct the graph of $x^4 y^2 = 1$ on logarithmic paper.

First we solve for y and then construct a table of values. Considering positive values of x and y, we have

$$y = \sqrt{\frac{1}{x^4}} = \frac{1}{x^2}$$

x	0.5	1	2	8	20
y	4	1	0.25	0.0156	0.0025

We now plot these values on log-log paper on which both scales are logarithmic, as shown in Fig. 12–8. We again note that we have a straight line. Taking logarithms of both sides of the equation, we have

$$4 \log x + 2 \log y = 0$$

If we let $u = \log y$ and $v = \log x$, we then have

$$4v + 2u = 0 \qquad \text{or} \qquad u = -2v$$

which is the equation of a straight line as shown in Fig. 12–8. It should be pointed out that not all graphs on logarithmic paper are straight lines.

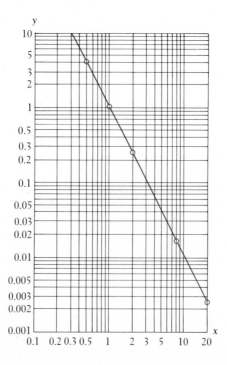

Figure 12–8

Example C
The deflection (in feet) of a certain cantilever beam as a function of the distance x from one end is

$$d = 0.0001(30x^2 - x^3)$$

If the beam is 20.0 ft long, plot a graph of d vs. x on log-log paper. Constructing a table of values we have

x (feet)	1.00	1.50	2.00	3.00	4.00	5.00	10.0	15.0	20.0
d (feet)	0.00290	0.00641	0.0112	0.0243	0.0416	0.0625	0.200	0.338	0.400

The graph is shown in Fig. 12—9.

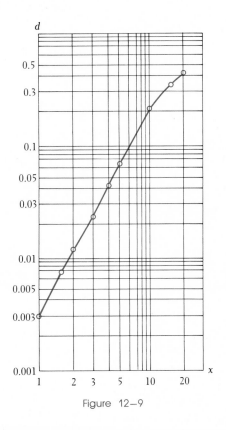

Figure 12—9

Logarithmic and semilogarithmic paper is often useful for plotting data derived from experimentation. Often the data cover too large a range of values to be plotted on ordinary graph paper. The following example illustrates how we use semilogarithmic paper to plot data.

Example D

The vapor pressure of water depends on the temperature. The following table gives the vapor pressure (in kilopascals) for corresponding values of temperature (in degrees Celsius).

Pressure	1.19	2.33	7.34	19.9	47.3	101	199	361	617
Temp.	10	20	40	60	80	100	120	140	160

These data are then plotted on semilogarithmic paper, as shown in Fig. 12–10. Intermediate values of temperature and pressure can then be read directly from the graph.

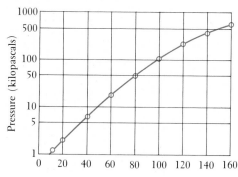

Figure 12–10

Exercises 12—9

In Exercises 1 through 12 plot the graphs of the given functions on semilogarithmic paper.

1. $y = 2^x$

2. $y = 5^x$

3. $y = 2(4^x)$

4. $y = 5(10^x)$

5. $y = 3^{-x}$

6. $y = 2^{-x}$

7. $y = x^3$

8. $y = x^5$

9. $y = 3x^2$

10. $y = 2x^4$

11. $y = 2x^3 + 4x$

12. $y = 4x^3 + 2x^2$

In Exercises 13 through 24 plot the graphs of the given functions on log-log paper.

13. $y = 0.01x^4$

14. $y = 0.02x^3$

15. $y = \sqrt{x}$

16. $y = x^{2/3}$

17. $y = x^2 + 2x$

18. $y = x + \sqrt{x}$

19. $xy = 4$

20. $xy^2 = 10$

21. $y^2x = 1$

22. $x^2y^3 = 1$

23. $x^2y^2 = 25$

24. $x^3y = 8$

In Exercises 25 through 30 plot the indicated graphs.

25. The atmospheric pressure p at a given height h is given by $p = p_0e^{-kh}$, where p_0 and k are constants. On semilogarithmic paper plot p (in atmospheres) vs. h (in feet) for $0 \le h \le 10^5$. Use $e = 2.7$, $p_0 = 1$ atm, and $k = 10^{-5}$ per foot.

26. Strontium 90 decays according to the equation $N = N_0e^{-0.028t}$, where N is the amount present after t years, and N_0 is the original amount. Plot N vs. t on semilogarithmic paper if $N_0 = 1000$ g.

27. A company estimates that the value of a piece of machinery is $V = 75000(2^{-0.15t})$, where V is the value (in dollars) t years after the purchase. Plot V vs. t on semilogarithmic paper.

28. At constant temperature, the relation between the volume V and pressure P of a gas is $PV = c$, where c is a constant. On logarithmic paper, plot the graph of P (in atmospheres) vs. V (in cubic feet) for $c = 4$ atm-ft^3. Use values of $0.1 \leq P \leq 10$.

29. One end of a very hot steel bar is sprayed with a stream of cool water. The rate of cooling (in degrees Fahrenheit per second) as a function of the distance (in inches) from the end of the bar is then measured. The following results are obtained.

Cooling rate	600	190	100	72	46	29	17	10	6
Distance	0.063	0.13	0.19	0.25	0.38	0.50	0.75	1.0	1.5

Plot the data on logarithmic paper. Such experiments are made to determine the hardenability of steel.

30. The magnetic intensity H (in amperes per meter) and flux density B (in teslas) of annealed iron are given in the following table.

H	10	50	100	150	200	500	1000	10000	100000
B	0.0042	0.043	0.67	1.01	1.18	1.44	1.58	1.72	2.26

Plot H versus B on logarithmic paper.

12–10 Exercises for Chapter 12

In Exercises 1 through 12 determine the value of x.

1. $\log_{10} x = 4$
2. $\log_9 x = 3$
3. $\log_5 x = -1$
4. $\log_4 x = -\frac{1}{2}$
5. $\log_2 64 = x$
6. $\log_{12} 144 = x$
7. $\log_8 32 = x$
8. $\log_9 27 = x$
9. $\log_x 36 = 2$
10. $\log_x 243 = 5$
11. $\log_x 10 = \frac{1}{2}$
12. $\log_x 8 = \frac{3}{4}$

In Exercises 13 through 24 express each as a sum, difference, or multiple of logarithms. Wherever possible, evaluate logarithms of the result.

13. $\log_7 2x$
14. $\log_5 \left(\frac{7}{a}\right)$
15. $\log_3 (t^2)$
16. $\log_6 \sqrt{5}$
17. $\log_2 28$
18. $\log_7 98$
19. $\log_3 \left(\frac{9}{x}\right)$
20. $\log_6 \left(\frac{5}{36}\right)$
21. $\log_4 \sqrt{48}$
22. $\log_6 \sqrt{72y}$
23. $\log_{10}(1000x^4)$
24. $\log_3 (9^2 \cdot 6^3)$

In Exercises 25 through 28, solve for y in terms of x.

25. $\log_6 y = \log_6 4 - \log_6 x$
26. $\log_3 y = \frac{1}{2}\log_3 7 + \frac{1}{2}\log_3 x$
27. $\log_2 y + \log_2 x = 3$
28. $6\log_4 y = 8\log_4 4 - 3\log_4 x$

In Exercises 29 through 32 graph the given functions.

29. $y = 0.5(5^x)$
30. $y = 3(2^x)$
31. $y = 0.5\log_4 x$
32. $y = 10\log_{16} x$

omit In Exercises 33 through 44 use logarithms to perform the indicated calculations.

33. $(13.60)(0.6930)$

34. $(0.07255)(4320)$

35. $\dfrac{9.826}{0.08004}$

36. $\dfrac{87.64}{108.2}$

37. $(5.670)^{20}$

38. $(0.9823)^{10}$

39. $\sqrt[4]{17.22}$

40. $(0.006247)^{0.2}$

41. $\dfrac{\sqrt{8645}}{19.49}$

42. $[(9.060)(13.45)]^{1/3}$

43. $\dfrac{\sqrt[5]{22.46}(14.98)}{\sqrt[3]{0.8664}}$

44. $(12.66)^{1.096}$

omit In Exercises 45 through 48 use logarithms to solve the given triangles.

45. $A = 36°20'$, $C = 90°0'$, $a = 15.60$ 46. $B = 14°50'$, $C = 90°0'$, $c = 4730$
47. $A = 45°0'$, $B = 67°10'$, $a = 76.50$ 48. $B = 123°0'$, $C = 15°43'$, $a = 0.9122$

Cal In Exercises 49 through 52, by using logarithms to the base 10, find the natural logarithms of the given numbers.

49. 8.86   50. 33.0   51. 2.07   52. 0.542

In Exercises 53 through 56 solve the given equations.

(12.8) 53. $3^{x+2} = 5^x$

54. $5^x = 10$

55. $\log_8(x + 2) + \log_8 2 = 2$

56. $\log(x + 2) + \log x = 0.4771$

In Exercises 57 and 58 plot the graphs of the given functions on semilogarithmic paper. In Exercises 59 and 60 plot the graphs of the given functions on log-log paper.

omit 57. $y = 6^x$ 58. $y = 5x^3$ 59. $y = \sqrt[3]{x}$ 60. $xy^4 = 16$

In Exercises 61 through 76 use logarithms to make any indicated calculations.

61. The vapor pressure P over a liquid may be related to temperature by the formula $\log P = a/T + b$, where a and b are constants. Solve for P.

62. The Beer–Lambert law of light absorption may be expressed as

$$\log \frac{I}{I_0} = -\alpha x$$

where I/I_0 is that fraction of the incident light beam which is transmitted, α is a constant, and x is the distance the light travels through the medium. Solve for I.

63. The edge of a cubical metal block expanded from 37.75 cm to 38.25 cm while being heated. What was the increase in volume?

64. The ratio of the rates of diffusion of two gases is given by the equation $r_1/r_2 = \sqrt{m_2}/\sqrt{m_1}$, where m_1 and m_2 are the masses of the molecules of the gases. Given that $m_1 = 31.44$ units of mass and $m_2 = 74.92$ units of mass, calculate the ratio of r_1 to r_2.

65. Under certain circumstances the efficiency of an internal combustion engine is given by

$$\text{eff (in percent)} = 100\left(1 - \frac{1}{(V_1/V_2)^{0.4}}\right)$$

where V_1 and V_2 are, respectively, the maximum and minimum volumes of air in a cylinder. The ratio V_1/V_2 is called the compression ratio. Compute the efficiency of an engine with a compression ratio of 6.550.

66. If P dollars are invested at an interest rate r which is compounded n times a year, the value A of the investment t years later is given by the formula $A = P(1 + r/n)^{nt}$. What is the value after 5 years of \$5636 invested at $5\frac{1}{2}\%$ and compounded quarterly?

67. Points A and B are on opposite sides of a lake. A third point C is found such that $AC = 402.5$ ft and $BC = 317.9$ ft. What is the distance AB if the angle BAC is $41°18'$?

68. The angle of depression of a fire noticed directly north of a 74.50-ft fire tower is $5°30'$. The angle of depression of a stream running east to west, and also north of the tower, is $13°0'$. How far is the fire from the stream?

69. Under certain conditions, the potential (in volts) due to a magnet is given by $V = -k \ln (1 + l/b)$, with l the length of the magnet and b the distance from the point where the potential is measured. Find V, if $k = 2$ units, $l = 5.00$ cm, and $b = 2.00$ cm.

70. The Nernst equation,

$$E = E_0 - \frac{0.05910}{n} \log Q$$

is used for oxidation-reduction reactions. In the equation, E and E_0 are voltages, n is the number of electrons involved in the reaction, and Q is a measure of the activity of reaction. Given that $E_0 = 1.1000$ V, $E = 1.1300$ V, and $n = 2$, what is the value of Q?

71. For first-order chemical reactions, concentration of a reacting chemical species is related to time by the expression

$$\log \frac{x_0}{x} = kt$$

where x_0 is the initial concentration and x is the concentration after time t. Determine the quantity of sucrose remaining after 3 h, if the initial concentration is 9.00 g-mol/L and $k = 0.00158$ per minute.

72. The power gain of an electronic device such as an amplifier is defined as $n = 10 \log (P_o/P_i)$, where n is measured in decibels, P_o is the power output, and P_i is the power input. If $P_o = 10.0$ W and $P_i = 0.125$ W, calculate the power gain. (See Example D of Section 12–8.)

73. In 1975 it was estimated that the total annual world demand for copper was $C = 9.0e^{0.08t}$, where C is in millions of tons of copper and t is the number of years after 1975. When will copper demand be 20 million tons?

74. The bacteria population in a certain culture is given by $N = 1000(1.5)^t$. How long does it take for the population to reach 10,000 if t is measured in hours?

75. Plot a semilogarithmic graph of N vs. t for the bacteria culture of Exercise 74.

76. The luminous efficiency (measured in lumens per watt) of a tungsten lamp as a function of its input power (in watts) is given by the following table. On semilogarithmic paper, plot efficiency versus power.

Efficiency	7.8	10.4	11.7	13.9	16.3	18.3	19.9	21.5
Power	10	25	40	60	100	200	500	1000

13

Additional Types of Equations and Systems of Equations

13—1 Graphical Solution of Systems of Equations

In Chapter 2 we learned how to graph a function as well as how to solve equations graphically. Since then we have dealt with methods for solving quadratic equations and systems of linear equations. Also, we have graphed the trigonometric, logarithmic, and exponential functions. In this section we shall introduce one more general type of equation: the general quadratic equation. We shall then discuss graphical solutions of systems of equations involving quadratic equations as well as other types of equations. Here again, as in solving systems of linear equations, we shall obtain the desired solution by finding the values of x and y which satisfy both equations in a system at the same time. From the standpoint of graphs, this means that we wish to find all points which the graphs of the given functions have in common.

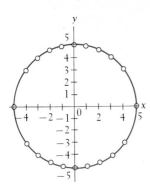

Figure 13–1

An equation of the form

$$ax^2 + bxy + cy^2 + dx + ey + f = 0 \qquad (13\text{–}1)$$

is called a **general quadratic equation in x and y.** We shall be interested primarily in some special cases of this equation. The graphs of the various possible forms of this equation result in curves known as **conic sections.** These curves are the circle, parabola, ellipse, and hyperbola. The following examples will illustrate these curves. A more complete discussion will be found in Chapter 20.

Example A
Plot the graph of the equation $x^2 + y^2 = 25$.
   We first solve this equation for y, obtaining $y = \pm \sqrt{25 - x^2}$. We now assume values for x and find the corresponding values for y.

x	0	± 1	± 2	± 3	± 4	± 5
y	± 5	± 4.9	± 4.6	± 4	± 3	0

If we try values greater than 5, we have imaginary numbers. These cannot be plotted, for we assume that both x and y are real. (The complex plane is only for *numbers* of the form $a + bj$ and does not represent pairs of numbers representing two variables.) When we give the value $x = \pm 4$ when $y = \pm 3$, this is simply a short way of representing 4 points. These points are $(4, 3)$, $(4, -3)$, $(-4, 3)$, $(-4, -3)$. We note in Fig. 13–1 that the resulting curve is a **circle.** A circle always results from an equation of the form $x^2 + y^2 = r^2$, and r is the radius of the circle.

Example B
Plot the graph of the equation $y = 3x^2 - 6x$.
   We plotted curves of this form in Section 2–3, and we follow the same method here.

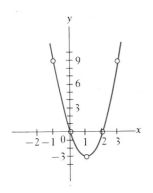

Figure 13–2

x	-1	0	1	2	3
y	9	0	-3	0	9

This curve (Fig. 13–2) is called a **parabola.** A parabola always results if the equation is of the form $y = ax^2 + dx + f$.

Example C
Plot the graph of the equation $2x^2 + 5y^2 = 10$.
   We first solve for y, then we construct the table of values.

$$y = \pm \sqrt{\frac{10 - 2x^2}{5}}$$

x	0	± 1	± 2	$\pm \sqrt{5} \; (= \pm 2.2)$
y	± 1.4	± 1.3	± 0.6	0

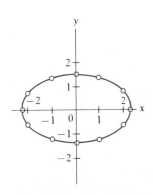

Figure 13–3

Values of x greater than $\sqrt{5}$ result in imaginary values of y. The curve (Fig. 13–3) is an **ellipse.** An ellipse results from an equation of the form $ax^2 + cy^2 = k$.

Example D

Plot the graph of the equation $xy = 4$.

   Solving for y, we obtain $y = 4/x$. Now, constructing the table of values, we have the following points.

x	-8	-4	-2	-1	$-\frac{1}{2}$	$\frac{1}{2}$	1	2	4	8
y	$-\frac{1}{2}$	-1	-2	-4	-8	8	4	2	1	$\frac{1}{2}$

Plotting these points, we obtain the curve in Fig. 13–4. This curve is called a **hyperbola.** A hyperbola always results if the equation is of the form $xy = k$. We also obtain a hyperbola if the equation is of the form $ax^2 - cy^2 = k$.

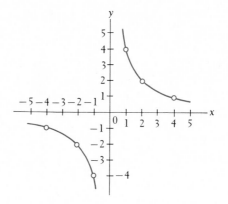

Figure 13–4

   To solve any system of equations graphically, we need only graph the equations and then find the points of intersection. If the curves do not intersect, the system has no real solution.

Example E

Graphically solve the system of equations

$$2x^2 - y^2 = 4$$
$$x - 3y = 6$$

   We should recognize the second equation as that of a straight line. Now constructing the tables, we solve $2x^2 - y^2 = 4$ for y and get $y = \pm\sqrt{2x^2 - 4}$. Therefore, we obtain the following table.

x	± 1.4	± 2	± 4	± 6
y	0	± 2	± 5.3	± 8.2

For the straight line we have the points

x	0	6	3
y	-2	0	-1

The solutions, as indicated on the graph in Fig. 13–5, are approximately $x = 1.8$, $y = -1.4$, and $x = -2.4$, $y = -2.8$.

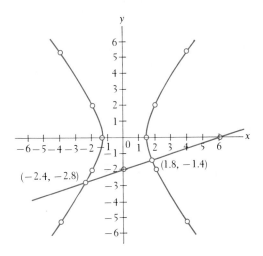

Figure 13–5

The following example illustrates the graphical solution of a system of equations in which one of the equations is not a general quadratic type.

Example F
Graphically solve the system of equations

$$9x^2 + 4y^2 = 36$$
$$y = 3^x$$

The first equation is of the form represented by an ellipse, as indicated in Example C. The second equation is an exponential function, as discussed in Chapter 12. Solving the first equation for y, we have $y = \pm\frac{1}{2}\sqrt{36 - 9x^2}$. Substituting values for x, we obtain the following table.

x	0	± 1	± 2
y	± 3	± 2.6	0

For the exponential function, we obtain the following table.

x	-3	-2	-1	0	1	2
y	$\frac{1}{27}$	$\frac{1}{9}$	$\frac{1}{3}$	1	3	9

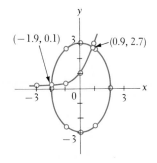

Figure 13–6

We plot these curves as shown in Fig. 13–6. The points of intersection are approximately $x = -1.9$, $y = 0.1$, and $x = 0.9$, $y = 2.7$.

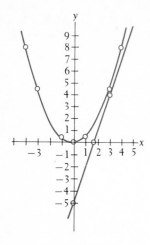

Figure 13–7

Example G

Graphically solve the system of equations

$$x^2 = 2y$$
$$3x - y = 5$$

We note that the two curves in this system are a parabola and a straight line. Solving the equation of the parabola for y, we obtain $y = \frac{1}{2}x^2$. We construct the following table.

x	0	± 1	± 2	± 3	± 4
y	0	$\frac{1}{2}$	2	$\frac{9}{2}$	8

For the straight line we have the following points.

x	0	$\frac{5}{3}$	3
y	-5	0	4

We plot these curves and see in Fig. 13–7 that they do not intersect, so we conclude that there are no real solutions to the system.

Exercises 13–1

In Exercises 1 through 20 solve the given systems of equations graphically.

1. $y = 2x$
   $x^2 + y^2 = 16$

2. $3x - y = 4$
   $y = 6 - 2x^2$

3. $x^2 + 2y^2 = 8$
   $x - 2y = 4$

4. $y = 3x - 6$
   $xy = 6$

5. $y = x^2 - 2$
   $4y = 12x - 17$

6. $x^2 + 4y^2 = 4$
   $2y = 12 - x$

7. $y = x^2$
   $xy = 4$

8. $y = -2x^2$
   $y = x^2 - 6$

9. $y = -x^2 + 4$
   $x^2 + y^2 = 9$

10. $y = 2x^2 - 1$
    $x^2 + 2y^2 = 16$

11. $x^2 - 4y^2 = 16$
    $x^2 + y^2 = 1$

12. $y = 2x^2 - 4x$
    $xy = -4$

13. $2x^2 + 3y^2 = 19$
    $x^2 + y^2 = 9$

14. $x^2 - y^2 = 4$
    $2x^2 + y^2 = 16$

15. $x^2 + y^2 = 1$
    $xy = \frac{1}{2}$

16. $x^2 + y^2 = 25$
    $x^2 - y^2 = 7$

17. $y = x^2$
    $y = \sin x$

18. $y = 2^x$
    $x^2 + y^2 = 4$

19. $x^2 - y^2 = 1$
    $y = \log_2 x$

20. $y = \cos x$
    $y = \log_3 x$

In Exercises 21 through 24 solve the indicated systems of equations graphically. In Exercises 23 and 24 the necessary systems of equations must be properly set up.

21. A rectangular field has a perimeter of 1140 m and an area of 75,600 m². Show that the two equations necessary to find the dimensions l and w are $l + w = 570$, $lw = 75600$. Graphically solve for l and w.

22. The diagonal of a rectangular metal plate is 10 in. and the area is 50 in.^{2}. Show that the two equations necessary to find the dimensions l and w are $l^2 + w^2 = 100$, $lw = 50$. Graphically solve for l and w.

23. The power developed in an electric resistor R is i^2R, where i is the current. If one current passes through a 2 Ω resistor, and the second current passes through a 3 Ω resistor, the total power produced is 12 W. The sum of the currents is 3 A. Find the currents.

24. A circular hole y cm in radius is cut from a square piece of wood x cm on a side, leaving an area of 10 cm^2. If $x - y = 4$ cm, find x and y graphically.

13–2 Algebraic Solution of Systems of Equations

Often the graphical method is the easiest way to solve a system of equations. However, this method does not usually give an exact answer. Using algebraic methods to find exact solutions for some systems of equations is either not possible or quite involved. There are some systems, however, which do lend themselves to relatively simple solutions by algebraic means. In this section we shall consider two useful methods, both of which we discussed before when we were studying systems of linear equations.

The first method is substitution. If we can solve one equation for one of its variables, we can substitute this solution into the other equation. We then have only one unknown in the resulting equation, and we can solve this equation by methods discussed in earlier chapters.

Example A
Solve by substitution the system of equations

$$2x - y = 4$$
$$x^2 - y^2 = 4$$

We solve the first equation for y, obtaining $y = 2x - 4$. We now substitute this into the second equation, getting $x^2 - (2x - 4)^2 = 4$. When simplified, this gives a quadratic equation.

$$x^2 - (4x^2 - 16x + 16) = 4$$
$$-3x^2 + 16x - 20 = 0$$

$$x = \frac{-16 \pm \sqrt{256 - 4(-3)(-20)}}{-6} = \frac{-16 \pm \sqrt{16}}{-6} = \frac{-16 \pm 4}{-6} = \frac{10}{3}, 2$$

We now find the corresponding values of y by substituting into $y = 2x - 4$. Thus, we have the solutions $x = \frac{10}{3}$, $y = \frac{8}{3}$, and $x = 2$, $y = 0$. By substitution, these values also satisfy the equation $x^2 - y^2 = 4$. (We do this as a check.)

Example B
Solve by substitution the system of equations

$$xy = -2$$
$$2x + y = 2$$

From the first equation we have $y = -2/x$. Substituting this into the second equation, we have

$$2x - \left(\frac{2}{x}\right) = 2$$
$$2x^2 - 2 = 2x$$
$$x^2 - x - 1 = 0$$
$$x = \frac{1 \pm \sqrt{1 + 4}}{2} = \frac{1 \pm \sqrt{5}}{2}$$

We find the corresponding values of y, and we have the solutions

$$x = \frac{1 + \sqrt{5}}{2}, \quad y = 1 - \sqrt{5} \quad \text{and} \quad x = \frac{1 - \sqrt{5}}{2}, \quad y = 1 + \sqrt{5}$$

Now let us use the other algebraic method of solution, that of addition or subtraction. This method can be used to great advantage if both equations have only squared terms and constants.

Example C
Solve, by addition or subtraction, the system of equations

$$2x^2 + y^2 = 9$$
$$x^2 - y^2 = 3$$

We note that if we add the two equations we get $3x^2 = 12$. Thus, $x = \pm 2$. For $x = 2$, we have two corresponding y-values, $y = \pm 1$. Also for $x = -2$, we have two corresponding y-values, $y = \pm 1$. Thus, we have four solutions: $x = 2$, $y = 1$; $x = 2$, $y = -1$; $x = -2$, $y = 1$; and $x = -2$, $y = -1$.

Example D
Solve, by addition or subtraction, the system of equations

$$3x^2 - 2y^2 = 5$$
$$x^2 + y^2 = 5$$

If we multiply the second equation by 2 and then add the two resulting equations, we get $5x^2 = 15$. Thus, $x = \pm \sqrt{3}$. The corresponding values of y for each value of x are $\pm \sqrt{2}$. Again we have four solutions: $x = \sqrt{3}$, $y = \sqrt{2}$; $x = \sqrt{3}$, $y = -\sqrt{2}$; $x = -\sqrt{3}$, $y = \sqrt{2}$; and $x = -\sqrt{3}$, $y = -\sqrt{2}$.

Example E

A certain number of machine parts cost $1000. If they cost $5 less per part, ten additional parts could be purchased for the same amount of money. What is the cost of each?

Since the cost of each part is required, we let $c =$ the cost per part. Also, we let $n =$ the number of parts. From the first statement of the problem, we see that $cn = 1000$. Also, from the second statement, we have $(c - 5)(n + 10) = 1000$. Therefore, we are to solve the system of equations

$$cn = 1000$$
$$(c - 5)(n + 10) = 1000$$

Solving the first equation for n, and multiplying out the second equation, we have

$$n = \frac{1000}{c}$$

$$cn + 10c - 5n - 50 = 1000$$

Now, substituting the expression for n into the second equation, we solve for c.

$$c\left(\frac{1000}{c}\right) + 10c - 5\left(\frac{1000}{c}\right) - 50 = 1000$$

$$1000 + 10c - \frac{5000}{c} - 50 = 1000$$

$$10c - \frac{5000}{c} - 50 = 0$$

$$10c^2 - 50c - 5000 = 0$$
$$c^2 - 5c - 500 = 0$$
$$(c + 20)(c - 25) = 0$$
$$c = -20, 25$$

Since a negative answer has no significance in this particular situation, we see that the solution is $c = 25 per part. Checking with the original statement of the problem, we see that this is correct.

Exercises 13–2

In Exercises 1 through 20 solve the given systems of equations algebraically.

1. $y = x + 1$
   $y = x^2 + 1$

2. $y = 2x - 1$
   $y = 2x^2 + 2x - 3$

3. $x + 2y = 3$
   $x^2 + y^2 = 26$

4. $y = x + 1$
   $x^2 + y^2 = 25$

5. $2x - y = 2$
   $2x^2 + 3y^2 = 4$

6. $6y - x = 6$
   $x^2 + 3y^2 = 36$

7. $xy = 3$
   $3x - 2y = -7$

8. $xy = -4$
   $2x + y = -2$

9. $y = x^2$
   $y = 3x^2 - 8$

10. $y = x^2 - 1$
    $2x^2 - y^2 = 2$

11. $x^2 - y = -1$
    $x^2 + y^2 = 5$

12. $x^2 + y = 5$
    $x^2 + y^2 = 25$

13. $x^2 - 1 = y$
    $x^2 - 2y^2 = 1$

14. $2y^2 - 4x = 7$
    $y^2 + 2x^2 = 3$

15. $x^2 + y^2 = 25$
    $x^2 - 2y^2 = 7$

16. $3x^2 - y^2 = 4$
    $x^2 + 4y^2 = 10$

17. $y^2 - 2x^2 = 6$
    $5x^2 + 3y^2 = 20$

18. $y^2 - 2x^2 = 17$
    $2y^2 + x^2 = 54$

19. $x^2 + 3y^2 = 37$
    $2x^2 - 9y^2 = 14$

20. $5x^2 - 4y^2 = 15$
    $3y^2 + 4x^2 = 12$

In Exercises 21 through 28 solve the indicated systems of equations algebraically. In Exercises 23 through 28 it is necessary to properly set up the systems of equations.

21. The vertical distance which a certain projectile travels from its starting point is given by $y = 60t - 16t^2$. When the horizontal distance it has traveled equals twice the vertical distance, $y = 40t$. Find the values of y (in feet) and t (in seconds) which satisfy these equations.

22. A 300-g block and a 200-g block collide. Using the physical laws of conservation of energy and conservation of momentum, along with certain given conditions, we can establish the following equations involving velocities of each block after collision:

$$150v_1^2 + 100v_2^2 = 1375000$$
$$300v_1 + 200v_2 = -5000$$

Find these velocities (in centimeters per second).

23. Find two positive numbers such that the sum of their squares is 233 and the difference between their squares is 105.

24. The length of a table is three times the width, and the area is 48 ft². Find the dimensions of the table.

25. Two ships leave a port, one traveling due south and the other due east. The ship going east travels twice as far as the other ship, at which time they are 10 km apart. How far does each travel?

26. To enclose a rectangular field of 11,200 ft² in area, 440 ft of fence are required. What are the dimensions of the field?

27. The radii of two spheres differ by 4 in., and the difference between the spherical surfaces is 320π in.². Find the radii. (The surface area of a sphere is $4\pi r^2$.)

28. Two cities are 2000 mi apart. If an airplane increases its usual speed between these two cities by 100 mi/h, the trip would take 1 h less. Find the normal speed of the plane and the normal time of the flight.

13—3 Equations in Quadratic Form

Often we encounter equations which can be solved by methods applicable to quadratic equations, even though these equations are not actually quadratic. They do have the property, however, that with a proper substitution they may be written in the form of a quadratic equation. All that is necessary is that the equation have terms including some quantity, its square, and perhaps a constant term. The following example illustrates these types of equations.

Example A

The equation $x - 2\sqrt{x} - 5 = 0$ is an equation in quadratic form, because if we let $y = \sqrt{x}$, we have the resulting equivalent equation $y^2 - 2y - 5 = 0$.

Other examples of equations in quadratic form are as follows:

$$t^{-4} - 5t^{-2} + 3 = 0;$$

by letting $y = t^{-2}$ we have $y^2 - 5y + 3 = 0$

$$t^3 - 3t^{3/2} - 7 = 0;$$

by letting $y = t^{3/2}$ we have $y^2 - 3y - 7 = 0$

$$(x + 1)^4 - (x + 1)^2 - 1 = 0;$$

by letting $y = (x + 1)^2$ we have $y^2 - y - 1 = 0$

$$x^{10} - 2x^5 + 1 = 0;$$

by letting $y = x^5$ we have $y^2 - 2y + 1 = 0$

$$(x - 3) + \sqrt{x - 3} - 6 = 0;$$

by letting $y = \sqrt{x - 3}$ we have $y^2 + y - 6 = 0$

The following examples illustrate the method of solving equations in quadratic form.

Example B

Solve the equation $x^4 - 5x^2 + 4 = 0$.

We first let $y = x^2$, and obtain the resulting equivalent equation $y^2 - 5y + 4 = 0$. This may be factored as $(y - 4)(y - 1) = 0$. Thus, we have the solutions $y = 4$ and $y = 1$. Therefore, $x^2 = 4$ and $x^2 = 1$, which means that we have $x = \pm 2$ and $x = \pm 1$. Substitution into the original equation verifies that each of these is a solution.

Example C

Solve the equation $x - \sqrt{x} - 2 = 0$.

By letting $y = \sqrt{x}$, we have the equivalent equation $y^2 - y - 2 = 0$. This is factorable into $(y - 2)(y + 1) = 0$. Therefore, we have $y = 2$ and $y = -1$. Since $y = \sqrt{x}$, we note that y cannot be negative, and this in turn tells us that $y = -1$ cannot lead to a solution. For $y = 2$ we have $x = 4$. Checking, we find that $x = 4$ satisfies the original equation. Thus, the only solution is $x = 4$.

Example C illustrates a very important point: *Whenever any operation involving the unknown is performed on an equation, this operation may introduce roots into a subsequent equation which are not roots of the original equation. Therefore, we must check all answers in the original equation.* Only operations involving constants—that is, adding, subtracting, multiplying by, or dividing by constants—are certain not to introduce these **extraneous roots**. Squaring both sides of an equation is a common way of introducing extraneous roots. We first encountered the concept of an extraneous root in Section 5–7, when we were discussing equations involving fractions.

Example D

Solve the equation $x^{-2} + 3x^{-1} + 1 = 0$.

By substituting $y = x^{-1}$, we have $y^2 + 3y + 1 = 0$. To solve this equation we may use the quadratic formula:

$$y = \frac{-3 \pm \sqrt{9 - 4}}{2} = \frac{-3 \pm \sqrt{5}}{2}$$

Thus, since $x = \frac{1}{y}$

$$x = \frac{2}{-3 + \sqrt{5}}, \quad \frac{2}{-3 - \sqrt{5}}$$

Rationalizing the denominators of the values of x, we have

$$x = \frac{-6 - 2\sqrt{5}}{4} = \frac{-3 - \sqrt{5}}{2} \quad \text{and} \quad x = \frac{-3 + \sqrt{5}}{2}$$

Checking these solutions, we have

$$\left(\frac{-3 - \sqrt{5}}{2}\right)^{-2} + 3\left(\frac{-3 - \sqrt{5}}{2}\right)^{-1} + 1 \overset{?}{=} 0 \quad \text{or} \quad 0 = 0$$

and

$$\left(\frac{-3 + \sqrt{5}}{2}\right)^{-2} + 3\left(\frac{-3 + \sqrt{5}}{2}\right)^{-1} + 1 \overset{?}{=} 0 \quad \text{or} \quad 0 = 0$$

Thus, these solutions check.

Example E

Solve the equation $(x^2 - x)^2 - 8(x^2 - x) + 12 = 0$.

By substituting $y = x^2 - x$, we have $y^2 - 8y + 12 = 0$. This is solved by factoring, which gives us the solutions $y = 2$ and $y = 6$. Thus, $x^2 - x = 2$ and $x^2 - x = 6$. Solving these, we find that $x = 2, -1, 3, -2$. Substituting these in the original equation, we find that all are solutions.

Example F illustrates a stated problem which leads to an equation in quadratic form.

Example F

A rectangular plate has an area of 60 cm². The diagonal of the plate is 13 cm. Find the length and width of the plate.

Since the required quantities are the length and width, let $l = $ the length of the plate and $w = $ the width of the plate. Now, since the area is 60 cm², $lw = 60$. Also, using the Pythagorean theorem and the fact that the diagonal is 13 cm, we have $l^2 + w^2 = 169$. Therefore, we are to solve the system of equations

$$lw = 60, \quad l^2 + w^2 = 169$$

Solving the first equation for l, we have $l = 60/w$. Substituting this expression into the second equation we have

$$\left(\frac{60}{w}\right)^2 + w^2 = 169$$

We now solve for w as follows:

$$\frac{3600}{w^2} + w^2 = 169$$

$$3600 + w^4 = 169w^2$$

$$w^4 - 169w^2 + 3600 = 0$$

Let $x = w^2$.

$$x^2 - 169x + 3600 = 0$$

$$(x - 144)(x - 25) = 0$$

$$x = 25, 144.$$

Therefore,

$$w^2 = 25, 144$$

Solving for w, we obtain $w = \pm 5$ or $w = \pm 12$. Only the positive values of w are meaningful in this problem. Therefore, if $w = 5$ cm, then $l = 12$ cm. Normally, we designate the length as the longer dimension. Checking in the original equation, we find that this solution is correct.

Exercises 13–3

In Exercises 1 through 16 solve the given equations.

1. $x^4 - 13x^2 + 36 = 0$

2. $x^4 - 20x^2 + 64 = 0$

3. $x^6 + 7x^3 - 8 = 0$

4. $x^6 - 19x^3 - 216 = 0$

5. $x^{-2} - 2x^{-1} - 8 = 0$

6. $10x^{-2} + 3x^{-1} - 1 = 0$

7. $x^{-4} + 2x^{-2} = 24$

8. $x^{-6} - 3x^{-3} - 10 = 0$

9. $x - 4\sqrt{x} + 3 = 0$

10. $2x + \sqrt{x} - 1 = 0$

11. $3\sqrt[3]{x} - 5\sqrt[6]{x} + 2 = 0$

12. $\sqrt{x} + 3\sqrt[4]{x} = 28$

13. $x^{2/3} - 2x^{1/3} - 15 = 0$

14. $x^3 + 2x^{3/2} - 80 = 0$

15. $(x - 1) - \sqrt{x - 1} - 2 = 0$

16. $(x + 1)^{-2/3} + 5(x + 1)^{-1/3} - 6 = 0$

17. $(x^2 - 2x)^2 - 11(x^2 - 2x) + 24 = 0$

18. $3(x^2 + 3x)^2 - 2(x^2 + 3x) - 5 = 0$

19. $x - 3\sqrt{x - 2} = 6$ (Let $y = \sqrt{x - 2}$.)

20. $(x^2 - 1)^2 + (x^2 - 1)^{-2} = 2$

In Exercises 21 through 24 solve the indicated equations. In Exercises 23 and 24 it is necessary to set up the required equation.

21. A manufacturer determines that the total profit P when producing x thousand television sets monthly is given by $P = 10,000(-x^4 + 8x^2 - 2)$, where P is measured in dollars. Determine the number of sets which can be produced for a profit of $\$140,000$.

22. In optics, in the theory which deals with interferometers, the equation $\sqrt{F} = 2\sqrt{p}/(1 - p)$ is found. Solve for p if $F = 16$.

23. Find the dimensions of a rectangular area having a diagonal of 40 ft and an area of 768 ft².

24. A metal plate is in the shape of an isosceles triangle. The length of the base equals the square root of one of the equal sides. Determine the lengths of the sides if the perimeter of the plate is 55 cm.

13−4 Equations with Radicals

Equations with radicals in them are normally solved by squaring both sides of the equation, or by a similar operation. However, when we do this, we often introduce extraneous roots. Thus, it is very important that all solutions be checked in the original equation.

Example A
Solve the equation $\sqrt{x - 4} = 2$.
   By squaring both sides of the equation, we have

$$(\sqrt{x - 4})^2 = 2^2 \quad \text{or} \quad x - 4 = 4 \quad \text{or} \quad x = 8$$

This solution checks when put into the original equation.

Example B
Solve the equation $\sqrt{x - 1} = x - 3$.
   Squaring both sides of the equation, we have

$$(\sqrt{x - 1})^2 = (x - 3)^2$$

(Remember, we are squaring each *side* of the equation, not just the terms separately on each side.) Hence,

$$x - 1 = x^2 - 6x + 9$$
$$x^2 - 7x + 10 = 0$$
$$(x - 5)(x - 2) = 0$$
$$x = 5 \quad \text{or} \quad x = 2$$

The solution $x = 5$ checks, but the solution $x = 2$ gives $1 = -1$. Thus, the solution is $x = 5$, and $x = 2$ is an extraneous root.

Example C
Solve the equation $\sqrt[3]{x - 8} = 2$.
   Cubing both sides of the equation, we have $x - 8 = 8$. Thus $x = 16$, which checks.

Example D
Solve the equation $\sqrt{x + 1} + \sqrt{x - 4} = 5$.

This is most easily solved by first placing one of the radicals on the right side and then squaring both sides of the equation:

$$\sqrt{x+1} = 5 - \sqrt{x-4}$$
$$(\sqrt{x+1})^2 = (5 - \sqrt{x-4})^2$$
$$x + 1 = 25 - 10\sqrt{x-4} + (x-4)$$

Now, isolating the radical on one side of the equation and squaring again, we have

$$10\sqrt{x-4} = 20$$
$$\sqrt{x-4} = 2$$
$$x - 4 = 4$$
$$x = 8$$

This solution checks.

Example E

Solve the equation $\sqrt{x} - \sqrt[4]{x} = 2$.

We can solve this most easily by handling it as an equation in quadratic form. By letting $y = \sqrt[4]{x}$, we have

$$y^2 - y - 2 = 0$$
$$(y-2)(y+1) = 0$$
$$y = 2 \qquad \text{or} \qquad y = -1$$
$$x = 16 \qquad \text{or} \qquad x = 1$$

The solution $x = 16$ checks, but the solution $x = 1$ does not. Thus, the only solution is $x = 16$.

Example F

The perimeter of a right triangle is 60 ft, and its area is 120 ft². Find the lengths of the three sides (see Fig. 13–8).

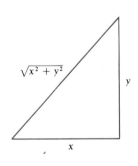

$\sqrt{x^2 + y^2}$

y

x

Figure 13–8

If we let the two legs of the triangle be x and y, the perimeter can be expressed as $p = x + y + \sqrt{x^2 + y^2}$. Also, the area is $A = \frac{1}{2}xy$. In this way we arrive at the equations $x + y + \sqrt{x^2 + y^2} = 60$ and $xy = 240$. Solving the first of these for the radical, we then have $\sqrt{x^2 + y^2} = 60 - x - y$. Squaring both sides and combining terms, we have

$$0 = 3600 - 120x - 120y + 2xy$$

Solving the second of the original equations for y, we have $y = 240/x$. Substituting, we have

$$0 = 3600 - 120x - 120\left(\frac{240}{x}\right) + 480$$

Multiplying each side by x, dividing through by 120, and rearranging the terms, we have $x^2 - 34x + 240 = 0$, which may be factored into $(x-10)(x-24) = 0$. Thus, x can be either 10 or 24. From the equation $xy = 240$, if $x = 10$, then $y = 24$. This means that the two legs are 10 ft and 24 ft, and the hypotenuse is 26 ft. The same solution is found by using the value of 24 ft for x.

Exercises 13–4

In Exercises 1 through 24 solve the given equations.

1. $\sqrt{x-8}=2$ 2. $\sqrt{x+4}=3$

3. $\sqrt{8-2x}=x$ 4. $\sqrt{3x+4}=x$

5. $\sqrt{x-2}=x-2$ 6. $\sqrt{5x-1}=x-3$

7. $\sqrt[3]{y-5}=3$ 8. $\sqrt[4]{5-x}=2$

9. $\sqrt{x}+12=x$ 10. $\sqrt{x}+3=4x$

11. $5\sqrt{x}+3=2x$ 12. $4\sqrt{x}=x+3$

13. $\sqrt{x+4}=x-8$ 14. $\sqrt{x+15}=x-5$

15. $2\sqrt{x+2}-\sqrt{3x+4}=1$ 16. $\sqrt{x-1}+\sqrt{x+2}=3$

17. $\sqrt{5x+1}-1=3\sqrt{x}$ 18. $\sqrt{2x+1}+\sqrt{3x}=11$

19. $\sqrt{2x-1}-\sqrt{x+11}=-1$ 20. $\sqrt{5x-4}-\sqrt{x}=2$

21. $\sqrt[3]{2x-1}=\sqrt[3]{x+5}$ 22. $\sqrt[4]{x+10}=\sqrt{x-2}$

23. $\sqrt{x-2}=\sqrt[4]{x-2}+12$ 24. $\sqrt{3x+\sqrt{3x+4}}=4$

In Exercises 25 through 28 solve the indicated equations for the indicated letter.

25. The velocity of an object falling under the influence of gravity in terms of its initial velocity v_0, the acceleration due to gravity g, and the height fallen, is given by $v=\sqrt{v_0^2-2gh}$. Solve this equation for h.

26. In measuring the velocity of water flowing through an opening under given conditions, the equation $v=\sqrt{2g(h_1-h_2)}$ arises. Solve for h_1.

27. An equation used in analyzing a certain type of concrete beam is $k=\sqrt{2np+(np)^2}-np$. Solve for p.

28. The theory of relativity states that the mass m of an object increases with velocity v according to the relation

$$m=\frac{m_0}{\sqrt{1-v^2/c^2}}$$

where m_0 is the "rest mass" and c is the velocity of light. Solve for v in terms of m.

In Exercises 29 through 32 set up the proper equations and solve them.

29. Find the dimensions of the rectangle for which the diagonal is 3 in. more than the longer side, which in turn is 3 in. longer than the shorter side.

30. The sides of a certain triangle are $\sqrt{x-1}$, $\sqrt{5x-1}$, and $x-1$. Find x when the perimeter of the triangle is 19.

31. An island is 3 mi offshore from the nearest point P on a straight beach. A person in a motorboat travels straight from the island to the beach x mi from P, and then travels 2 mi along the beach away from P. Find x if the person traveled a total of 8 mi.

32. The focal length f of a lens, in terms of its image distance q and object distance p, is given by

$$\frac{1}{f}=\frac{1}{p}+\frac{1}{q}$$

Find p and q if $f=4$ cm and $p=\sqrt{q}$.

13–5 Exercises for Chapter 13

In Exercises 1 through 8 solve the given systems of equations graphically.

1. $x + 2y = 6$
   $y = 4x^2$

2. $x + y = 3$
   $x^2 + y^2 = 25$

3. $3x + 2y = 6$
   $x^2 + 4y^2 = 4$

4. $x^2 - 2y = 0$
   $y = 3x - 5$

5. $y = x^2 + 1$
   $2x^2 + y^2 = 4$

6. $\dfrac{x^2}{4} + y^2 = 1$

   $x^2 - y^2 = 1$

7. $y = 4 - x^2$
   $y = 2x^2$

8. $xy = -2$
   $y = 1 - 2x^2$

In Exercises 9 through 16 solve the given systems of equations algebraically.

9. $y = 4x^2$
   $y = 8x$

10. $x + y = 2$
    $xy = 1$

11. $4x^2 + y = 3$
    $2x + 3y = 1$

12. $2x^2 + y^2 = 3$
    $x + 2y = 1$

13. $4x^2 - 7y^2 = 21$
    $x^2 + 2y^2 = 99$

14. $3x^2 + 2y^2 = 11$
    $2x^2 - y^2 = 30$

15. $4x^2 + 3xy = 4$
    $x + 3y = 4$

16. $\dfrac{6}{x} + \dfrac{3}{y} = 4$

    $\dfrac{36}{x^2} + \dfrac{36}{y^2} = 13$

In Exercises 17 through 32 solve the given equations.

17. $x^4 - 20x^2 + 64 = 0$

18. $x^6 - 26x^3 - 27 = 0$

19. $x^{3/2} - 9x^{3/4} + 8 = 0$

20. $x^{1/2} + 3x^{1/4} - 28 = 0$

21. $x^{-2} + 4x^{-1} - 21 = 0$

22. $x^{-4} - 5x^{-2} - 36 = 0$

23. $2x - 3\sqrt{x} - 5 = 0$

24. $(x^2 + 5x)^2 - 5(x^2 + 5x) = 6$

25. $\sqrt{x + 5} = 4$

26. $\sqrt[3]{x - 2} = 3$

27. $x - 1 = \sqrt{5x + 9}$

28. $x + 2 = \sqrt{11x - 2}$

29. $\sqrt{x + 1} + \sqrt{x} = 2$

30. $\sqrt{3 + x} + \sqrt{3x - 2} = 1$

31. $\sqrt{3x + 4} + \sqrt{x + 2} = 8$

32. $\sqrt{3x - 2} - \sqrt{x + 7} = 1$

In Exercises 33 through 36 solve for the indicated quantities.

33. The formula for the lateral surface area of a cone can be stated as $S = \pi r \sqrt{r^2 + h^2}$. Solve for r.

34. Under certain conditions, the frequency ω of an RLC circuit is given by

$$\omega = \frac{\sqrt{R^2 + 4(L/C)} + R}{2L}$$

Solve for C.

35. In an experiment, an object is allowed to fall, stopped, and then falls for twice the initial time. The total distance the object falls is 45 ft. The equations relating the times t_1 and t_2, in seconds, of fall are $16t_1^2 + 16t_2^2 = 45$, and $t_2 = 2t_1$. Find the times of fall.

36. If two objects collide and the kinetic energy remains constant, the collision is termed perfectly elastic. Under these conditions, if an object of mass m_1 and initial velocity u_1 strikes a second object (initially at rest) of mass m_2, such that the velocities after collision are v_1 and v_2, the following equations are found:

$$m_1 u_1 = m_1 v_1 + m_2 v_2$$
$$\tfrac{1}{2}m_1 u_1^2 = \tfrac{1}{2}m_1 v_1^2 + \tfrac{1}{2}m_2 v_2^2$$

Solve these equations for m_2 in terms of u_1, v_1, and m_1.

In Exercises 37 through 40 set up the appropriate equations and solve them.

37. Find the values of x and y in Fig. 13–9.

38. The perimeter of a rectangle is 36 cm. The length is x cm and the width is $\sqrt{x + 2}$ cm. Find the dimensions of the rectangle.

39. An object is dropped from the top of a building. Six seconds later it is heard to hit the street below. How high is the building? (The velocity of sound is 1100 ft/s and the distance the object falls as a function of time is $s = 16t^2$.)

40. Two trains are approaching the same crossing on tracks which are at right angles to each other. Each is traveling at 60 km/h. If one is 6 km from the crossing when the other is 3 km from it, how much later will they be 4 km apart?

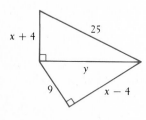

Figure 13–9

14

Equations of Higher Degree

14-1 The Remainder Theorem and the Factor Theorem

In previous chapters we have discussed methods of solving many kinds of equations. Except for special cases, however, we have not solved polynomial equations of degree higher than two (a polynomial equation of the first degree is a linear equation, and a polynomial equation of the second degree is a quadratic equation.) In this chapter we shall develop certain methods for solving polynomial equations, especially the higher-degree equations. Since we shall be discussing equations involving only polynomials, in this chapter $f(x)$ will be assumed to be a polynomial.

Any polynomial is a function of the form

$$f(x) = a_0 x^n + a_1 x^{n-1} + \cdots + a_n \qquad (14\text{--}1)$$

If we divide a polynomial by $x - r$, we find a result of the following form:

$$f(x) = (x - r)q(x) + R \qquad (14\text{--}2)$$

where $q(x)$ is the quotient and R is the remainder.

Example A

Divide $f(x) = 3x^2 + 5x - 8$ by $x - 2$.

$$
\begin{array}{r}
3x + 11 \\
x - 2 \overline{\smash{\big)}\ 3x^2 + 5x - 8} \\
\underline{3x^2 - 6x} \\
11x - 8 \\
\underline{11x - 22} \\
14
\end{array}
$$

Thus,

$$3x^2 + 5x - 8 = (x - 2)(3x + 11) + 14$$

If we now set $x = r$ in Eq. (14–2), we have

$$f(r) = q(r)(r - r) + R = R \qquad (14\text{–}3)$$

The equation above states that the remainder equals the function of *r*. This leads us to the **remainder theorem,** which states that *if a polynomial f(x) is divided by x − r until a constant remainder (R) is obtained, then f(r) = R.*

Example B

In Example A, $f(x) = 3x^2 + 5x - 8$, $R = 14$, $r = 2$.
We find that

$$f(2) = 3(4) + 5(2) - 8 = 14$$

Thus, $f(2) = 14$ verifies that $f(r) = R$.

Example C

Using the remainder theorem, determine the remainder when we divide $3x^3 - x^2 - 20x + 5$ by $x + 4$.

In using the remainder theorem, we determine the remainder when the function is divided by $x - r$ by evaluating the function for $x = r$. To have $x + 4$ in the proper form to identify *r*, we write it as $x - (-4)$. This means that $r = -4$, and we therefore are to evaluate the function $f(x) = 3x^3 - x^2 - 20x + 5$ for $x = -4$, or find $f(-4)$. Thus,

$$f(-4) = 3(-4)^3 - (-4)^2 - 20(-4) + 5 = -192 - 16 + 80 + 5$$
$$= -123$$

Thus, the remainder when $3x^3 - x^2 - 20x + 5$ is divided by $x + 4$ is -123.

The remainder theorem leads immediately to another important theorem known as the **factor theorem.** The factor theorem states that *if f(r) = R = 0, then x − r is a factor of f(x).* Inspection of Eq. (14–2) justifies this theorem. It is also true that if $x - r$ is a factor of $f(x)$, then *r* is a zero of $f(x)$. (The zero of a function was first introduced in Chapter 2.)

Example D

Is $x + 1$ a factor of $f(x) = x^3 + 2x^2 - 5x - 6$?
  Here $r = -1$, and thus

$$f(-1) = -1 + 2 + 5 - 6 = 0$$

Therefore, since $f(-1) = 0$, $x + 1$ is a factor of $f(x)$.

Example E

The expression $x + 2$ is not a factor of the function in Example D, since

$$f(-2) = -8 + 8 + 10 - 6 = 4$$

But $x - 2$ is a factor, since $f(2) = 8 + 8 - 10 - 6 = 0$.

  We now have one way of determining whether or not an expression of the form $x - r$ is a factor of a function. By finding $f(r)$, we can determine whether or not $x - r$ is a factor and whether or not r is a zero of the function.

Exercises 14–1

In Exercises 1 through 8 find the remainder R by long division and by the remainder theorem.

1. $(x^3 + 2x^2 - x - 2) \div (x - 1)$ 2. $(x^3 - 3x^2 - x + 2) \div (x - 2)$
3. $(x^3 + 2x + 3) \div (x + 1)$ 4. $(x^4 - 4x^3 - x^2 + x - 100) \div (x + 3)$
5. $(2x^5 - x^2 + 8x + 44) \div (x + 2)$ 6. $(x^3 + 4x^2 - 25x - 98) \div (x - 5)$
7. $(2x^4 - 3x^3 - 4x^2 + 2x - 5) \div (x - 3)$
8. $(2x^4 - 10x^2 + 30x - 60) \div (x + 4)$

In Exercises 9 through 16 find the remainder using the remainder theorem.

9. $(x^3 + 2x^2 - 3x + 4) \div (x + 1)$ 10. $(2x^3 - 4x^2 + x - 1) \div (x + 2)$
11. $(x^4 + x^3 - 2x^2 - 5x + 3) \div (x + 4)$ 12. $(2x^4 - x^2 + 5x - 7) \div (x - 3)$
13. $(2x^4 - 7x^3 - x^2 + 8) \div (x - 3)$
14. $(x^4 - 5x^3 + x^2 - 2x + 6) \div (x + 4)$
15. $(x^5 - 3x^3 + 5x^2 - 10x + 6) \div (x - 2)$
16. $(3x^4 - 12x^3 - 60x + 4) \div (x - 5)$

In Exercises 17 through 24 use the factor theorem to determine whether or not the second expression is a factor of the first.

17. $x^2 - 2x - 3$, $x - 3$ 18. $3x^3 + 2x^2 - 3x - 2$, $x + 2$
19. $4x^3 + x^2 - 16x - 4$, $x - 2$ 20. $3x^3 + 14x^2 + 7x - 4$, $x + 4$
21. $5x^3 - 3x^2 + 4$, $x - 2$ 22. $x^5 - 2x^4 + 3x^3 - 6x^2 - 4x + 8$, $x - 2$
23. $x^6 + 1$, $x + 1$ 24. $x^7 - 128$, $x + 2$

In Exercises 25 through 28 determine whether or not the given numbers are zeros of the given functions.

25. $f(x) = x^3 - 2x^2 - 9x + 18$;   2
26. $f(x) = 2x^3 + 3x^2 - 8x - 12$; $-\frac{3}{2}$
27. $f(x) = 4x^4 - 4x^3 + 23x^2 + x - 6$; $\frac{1}{2}$
28. $f(x) = 2x^4 + 3x^3 - 12x^2 - 7x + 6$; -3

In Exercises 29 and 30 answer the given questions.

29. By division, show that $2x - 1$ is a factor of $f(x) = 4x^3 + 8x^2 - x - 2$. May we therefore conclude that $f(1) = 0$?

30. By division, show that $x^2 + 2$ is a factor of $f(x) = 3x^3 - x^2 + 6x - 2$. May we therefore conclude that $f(-2) = 0$?

14–2 Synthetic Division

We shall now develop a method which greatly simplifies the procedure for dividing a polynomial by an expression of the form $x - r$. Using **synthetic division**, which is an abbreviated form of long division, we can determine the coefficients of the quotient as well as the remainder. Of course, for some values of r, we can easily calculate $f(r)$ directly. However, if the degree of the equation is high, this requires finding and combining high powers of r. Synthetic division therefore allows us to find $f(r)$ easily by finding the remainder. The method is developed in the following example.

Example A

Divide $x^4 + 4x^3 - x^2 - 16x - 14$ by $x - 2$.

We shall first perform this division in the usual manner.

$$
\begin{array}{r}
x^3 + 6x^2 + 11x + 6 \\
x - 2 \,\big)\, \overline{x^4 + 4x^3 - x^2 - 16x - 14} \\
\underline{x^4 - 2x^3} \\
6x^3 - x^2 \\
\underline{6x^3 - 12x^2} \\
11x^2 - 16x \\
\underline{11x^2 - 22x} \\
6x - 14 \\
\underline{6x - 12} \\
-2
\end{array}
$$

We now note that, when we performed this division, we repeated many terms. Also, the only quantities of importance in the function being divided are the coefficients. There is no real need to put in the powers of x all the time. Therefore, we shall now write the above example without any x's and also eliminate the writing of identical terms:

$$
\begin{array}{rrrrrr}
& 1 & 6 & 11 & 6 & \\
-2 \,\big)\, 1 & 4 & -1 & -16 & -14 \\
& \underline{-2} & & & & \\
& 6 & & & & \\
& & \underline{-12} & & & \\
& & 11 & & & \\
& & & \underline{-22} & & \\
& & & 6 & & \\
& & & & \underline{-12} & \\
& & & & -2 &
\end{array}
$$

All but the first of the numbers which represent coefficients of the quotient are repeated below. Also, all the numbers below the dividend may be written in two lines. Thus, we have the following form:

$$
\begin{array}{r|rrrrr}
-2 & 1 & 4 & -1 & -16 & -14 \\
& & -2 & -12 & -22 & -12 \\
\hline
& & 6 & 11 & 6 & -2
\end{array}
$$

All the coefficients of the actual quotient appear except the first, so we shall now repeat the 1 in the bottom line. Also, we shall change -2 to 2, which is the actual value of r. Then, to conform to the normal practice of writing r to the right, we have the following form:

$$
\begin{array}{rrrrr|l}
1 & 4 & -1 & -16 & -14 & 2 \\
& -2 & -12 & -22 & -12 & \\
\hline
1 & 6 & 11 & 6 & -2 &
\end{array}
$$

In this form the 1, 6, 11, and 6 represent the coefficients of the x^3, x^2, x, and constant terms of the quotient. The -2 is the remainder. Finally, we find it easier to use addition rather than subtraction in the process, so we change the signs of the numbers in the middle row. Remember that originally the bottom line was found by subtraction. Thus, we have

$$
\begin{array}{rrrrr|l}
1 & 4 & -1 & -16 & -14 & 2 \\
& 2 & 12 & 22 & 12 & \\
\hline
1 & 6 & 11 & 6 & -2 &
\end{array}
$$

When we inspect this form we find the following: The 1 multiplied by the 2 (r) gives 2, which is the first number of the middle row. The 4 and 2 (of the middle row) added is 6, which is the second number in the bottom row. The 6 multiplied by 2 (r) is 12, which is the second number in the middle row. This 12 and the -1 give 11. The 11 multiplied by 2 is 22. The 22 added to -16 is 6. This 6 multiplied by 2 gives 12. this 12 added to -14 is -2. When this process is followed in general, the method is called synthetic division.

Generalizing on this last example, we have the following procedure: We write down the coefficients of $f(x)$, being certain that the powers are in descending order and that zeros are placed in for missing powers. We write the value of r to the right. We carry down the left coefficient, multiply this number by r, and place this product under the second coefficient of the top line. We add the two numbers in this second column and place the result below; then we multiply this number by r and place the result under the third coefficient of the top line. We continue until the bottom row has as many numbers as the top row. The last number in the bottom row is the remainder, the other numbers being the respective coefficients of the quotient. The first term of the quotient is of degree one less than the dividend.

Example B

By synthetic division, divide $x^5 + 2x^4 - 4x^2 + 3x - 4$ by $x + 3$.

In writing down the coefficients of $f(x)$, we must be certain to include a zero for the missing x^3 term. Also, since the divisor is $x + 3$, we must recognize that $r = -3$. The setup and synthetic division follow.

1	2	0	-4	3	-4	$\lfloor -3$
	-3	3	-9	39	-126	
1	-1	3	-13	42	-130	

Thus, the quotient is $x^4 - x^3 + 3x^2 - 13x + 42$ and the remainder is -140. Notice the degree of the dividend is 5 and the degree of the quotient is 4.

Example C

By synthetic division, divide $3x^4 - 5x + 6$ by $x - 4$.

3	0	0	-5	6	$\lfloor 4$
	12	48	192	748	
3	12	48	187	754	

Thus, the quotient is $3x^3 + 12x^2 + 48x + 187$ and the remainder is 754.

Example D

By synthetic division, determine whether or not $x - 4$ is a factor of $x^4 + 2x^3 - 15x^2 - 32x - 16$.

1	2	-15	-32	-16	$\lfloor 4$
	4	24	36	16	
1	6	9	4	0	

Since the remainder is zero, $x - 4$ is a factor. We may also conclude that $f(x) = (x - 4)(x^3 + 6x^2 + 9x + 4)$, since the bottom line gives us the coefficients in the quotient.

Example E

By using synthetic division, determine whether $2x - 3$ is a factor of $2x^3 - 3x^2 + 8x - 12$.

We first note that the coefficient of x in the possible factor is not 1. Thus, we cannot use $r = 3$, since the factor is not of the form $x - r$. However, $2x - 3 = 2(x - \frac{3}{2})$, which means that if $2(x - \frac{3}{2})$ is a factor of the function, $2x - 3$ is a factor. If we use $r = \frac{3}{2}$, and find that the remainder is zero, then $x - \frac{3}{2}$ is a factor.

2	-3	8	-12	$\lfloor \frac{3}{2}$
	3	0	12	
2	0	8	0	

Since the remainder is zero, $x - \frac{3}{2}$ is a factor. Also, the quotient is $2x^2 + 8$, which may be factored into $2(x^2 + 4)$. Thus, 2 is also a factor of the function. This means that $2(x - \frac{3}{2})$ is a factor of the function, and this in turn means that $2x - 3$ is a factor.

Example F

By synthetic division, determine whether or not $\frac{1}{3}$ is a zero of the function $3x^3 + 2x^2 - 4x + 1$.

This problem is equivalent to dividing the function by $x - \frac{1}{3}$. If the remainder is zero, $\frac{1}{3}$ is a zero of the function.

$$
\begin{array}{rrrr|l}
3 & 2 & -4 & 1 & \frac{1}{3} \\
  & 1 & 1 & -1 & \\
\hline
3 & 3 & -3 & 0 &
\end{array}
$$

Since the remainder is zero, we conclude that $\frac{1}{3}$ is a zero of the function.

$$3x^3 + 2x^2 - 4x + 1 = (x - \tfrac{1}{3})(3x^2 + 3x - 3)$$
$$= 3(x - \tfrac{1}{3})(x^2 + x - 1).$$

Exercises 14–2

In Exercises 1 through 20 perform the required divisions by synthetic division. Exercises 1 through 16 are the same as those of Section 14–1.

1. $(x^3 + 2x^2 - x - 2) \div (x - 1)$ 2. $(x^3 - 3x^2 - x + 2) \div (x - 2)$
3. $(x^3 + 2x + 3) \div (x + 1)$ 4. $(x^4 - 4x^3 - x^2 + x - 100) \div (x + 3)$
5. $(2x^5 - x^2 + 8x + 44) \div (x + 2)$ 6. $(x^3 + 4x^2 - 25x - 98) \div (x - 5)$
7. $(2x^4 - 3x^3 - 4x^2 + 2x - 5) \div (x - 3)$
8. $(2x^4 - 10x^2 + 30x - 60) \div (x + 4)$
9. $(x^3 + 2x^2 - 3x + 4) \div (x + 1)$ 10. $(2x^3 - 4x^2 + x - 1) \div (x + 2)$
11. $(x^4 + x^3 - 2x^2 - 5x + 3) \div (x + 4)$
12. $(2x^4 - x^2 + 5x - 7) \div (x - 3)$ 13. $(2x^4 - 7x^3 - x^2 + 8) \div (x - 3)$
14. $(x^4 - 5x^3 + x^2 - 2x + 6) \div (x + 4)$
15. $(x^5 - 3x^3 + 5x^2 - 10x + 6) \div (x - 2)$
16. $(3x^4 - 12x^3 - 60x + 4) \div (x - 5)$
17. $(x^6 + 2x^2 - 6) \div (x - 2)$ 18. $(x^5 + 4x^4 - 8) \div (x + 1)$
19. $(x^7 - 128) \div (x - 2)$ 20. $(x^5 + 32) \div (x + 2)$

In Exercises 21 through 28 use the factor theorem and synthetic division to determine whether or not the second expression is a factor of the first.

21. $x^3 + x^2 - x + 2$; $x + 2$
22. $x^3 + 6x^2 + 10x + 6$; $x + 3$
23. $x^4 - 6x^2 - 3x - 2$; $x - 3$
24. $2x^4 - 5x^3 - 24x^2 + 5$; $x - 5$
25. $2x^4 - x^3 + 2x^2 - 3x + 1$; $2x - 1$
26. $6x^4 + 5x^3 - x^2 + 6x - 2$; $3x - 1$
27. $4x^4 + 2x^3 - 8x^2 + 3x + 12$; $2x + 3$
28. $3x^4 - 2x^3 + x^2 + 15x + 4$; $3x + 4$

In Exercises 29 through 32 use synthetic division to determine whether or not the given numbers are zeros of the given functions.

29. $x^4 - 5x^3 - 15x^2 + 5x + 14$;   7   30. $x^4 + 7x^3 + 12x^2 + x + 4$; -4
31. $9x^3 + 9x^2 - x + 2$; $-\frac{2}{3}$ 32. $2x^3 + 13x^2 + 10x - 4$; $\frac{1}{2}$

14–3 The Roots of an Equation

In this section we shall present certain theorems which are useful in determining the number of roots in the equation $f(x) = 0$, and the nature of some of these roots. In dealing with polynomial equations of higher degree, it is helpful to have as much of this kind of information as is readily obtainable before proceeding to solve for the roots.

The first of these theorems is so important that it is called **the fundamental theorem of algebra.** It states that *every polynomial equation has at least one (real or complex) root.* The proof of this theorem is of an advanced nature, and therefore we must accept its validity at this time. However, using the fundamental theorem, we can show the validity of other useful theorems.

Let us now assume that we have a polynomial equation $f(x) = 0$, and that we are looking for its roots. By the fundamental theorem, we know that it has at least one root. Assuming that we can find this root by some means (the factor theorem, for example), we shall call this root r_1. Thus,

$$f(x) = (x - r_1)f_1(x)$$

where $f_1(x)$ is the polynomial quotient found by dividing $f(x)$ by $(x - r_1)$. However, since the fundamental theorem states that any polynomial equation has at least one root, this must apply to $f_1(x) = 0$ as well. Let us assume that $f_1(x) = 0$ has the root r_2. Therefore, this means that $f(x) = (x - r_1)(x - r_2)f_2(x)$. Continuing this process until one of the quotients is a constant a, we have

$$f(x) = a(x - r_1)(x - r_2) \cdots (x - r_n)$$

Note that one linear factor appears each time a root is found, and that the degree of the quotient is one less each time. Thus there are n factors, if the degree of $f(x)$ is n. This leads us to two theorems. The first of these states that *each polynomial of the nth degree can be factored into n linear factors.* The second theorem states that *each polynomial equation of degree n has exactly n roots.*

Example A

Consider the equation $f(x) = 2x^4 - 3x^3 - 12x^2 + 7x + 6 = 0$.

$$2x^4 - 3x^3 - 12x^2 + 7x + 6 = (x - 3)(2x^3 + 3x^2 - 3x - 2)$$
$$2x^3 + 3x^2 - 3x - 2 = (x + 2)(2x^2 - x - 1)$$
$$2x^2 - x - 1 = (x - 1)(2x + 1)$$
$$2x + 1 = 2(x + \tfrac{1}{2})$$

Therefore,

$$2x^4 - 3x^3 - 12x^2 + 7x + 6 = 2(x - 3)(x + 2)(x - 1)(x + \tfrac{1}{2}) = 0$$

The degree of $f(x)$ is 4. There are 4 linear factors: $(x - 3)$, $(x + 2)$, $(x - 1)$, and $(x + \frac{1}{2})$. There are 4 roots of the equation: 3, -2, 1, and $-\frac{1}{2}$. Thus, we have verified each of the theorems above for this example.

It is not necessary for each root of an equation to be different from the others. For example, the equation $(x - 1)^2 = 0$ has two roots, both of which are 1. Such roots are referred to as multiple roots.

When we solve the equation $x^2 + 1 = 0$, we get two roots, j and $-j$. In fact, if we have any equation for which the roots are complex, for every root of the form $a + bj$ $(b \neq 0)$, there is also a root of the form $a - bj$. This is so because any quadratic equation can be solved by the quadratic formula. The solutions from the quadratic formula (for an equation of the form $ax^2 + bx + c = 0$) are

$$\frac{-b + \sqrt{b^2 - 4ac}}{2a} \quad \text{and} \quad \frac{-b - \sqrt{b^2 - 4ac}}{2a}$$

and the only difference between these roots is the sign before the radical. Thus we have the following theorem. *If a complex number $a + bj$ is the root of $f(x) = 0$, its conjugate $a - bj$ is also a root.*

Example B
Consider the equation $f(x) = (x - 1)^3(x^2 + x + 1) = 0$.

We observe directly (since three factors of $x - 1$ are already indicated) that there is a triple root of 1. To find the other two roots, we use the quadratic formula on the *factor* $(x^2 + x + 1)$. This is permissible, because what we are actually finding are those values of x which make $x^2 + x + 1 = 0$. For this we have

$$x = \frac{-1 \pm \sqrt{1 - 4}}{2}$$

Thus,

$$x = \frac{-1 + \sqrt{3}j}{2} \quad \text{and} \quad x = \frac{-1 - \sqrt{3}j}{2}$$

Therefore, the roots of $f(x)$ are

$$1, \ 1, \ 1, \ \frac{-1 + \sqrt{3}j}{2}, \ \frac{-1 - \sqrt{3}j}{2}$$

One further observation can be made from Example B. *Whenever enough roots are known so that the remaining factor is quadratic, it is always possible to find the remaining roots from the quadratic formula.* This is true for finding real or complex roots. If there are n roots and if we find $n - 2$ of these roots, the solution may be completed by using the quadratic formula.

Example C
Solve the equation $3x^3 + 10x^2 - 16x - 32 = 0$ given that $-\frac{4}{3}$ is a root.

Using synthetic division and the given root we have

$$
\begin{array}{rrrr|r}
3 & 10 & -16 & -32 & \underline{-\frac{4}{3}} \\
  & -4 & -8 & 32 & \\
\hline
3 & 6 & -24 & 0 & \\
\end{array}
$$

Thus, $3x^3 + 10x^2 - 16x - 32 = (x + \frac{4}{3})(3x^2 + 6x - 24)$. We know that $x + \frac{4}{3}$ is a factor from the given root, and that $3x^2 + 6x - 24$ is a factor from the synthetic division. This second factor has a common factor of 3, which means that $3x^2 + 6x - 24 = 3(x^2 + 2x - 8)$. We now see that $x^2 + 2x - 8 = (x + 4)(x - 2)$. Thus,

$$3x^3 + 10x^2 - 16x - 32 = 3(x + \tfrac{4}{3})(x + 4)(x - 2)$$

This means the roots are $-\frac{4}{3}$, -4, and 2.

Example D
Solve the equation $x^4 + 3x^3 - 4x^2 - 10x - 4 = 0$, given that -1 and 2 are roots.

Using synthetic division and the root -1, we have

$$
\begin{array}{rrrrr|r}
1 & 3 & -4 & -10 & -4 & \underline{-1} \\
  & -1 & -2 & 6 & 4 & \\
\hline
1 & 2 & -6 & -4 & 0 & \\
\end{array}
$$

Therefore, we now know that

$$x^4 + 3x^3 - 4x^2 - 10x - 4 = (x + 1)(x^3 + 2x^2 - 6x - 4)$$

We now know that $x - 2$ must be a factor of $x^3 + 2x^2 - 6x - 4$, since it is a factor of the original function. Again, using synthetic division and this time the root 2, we have the following:

$$
\begin{array}{rrrr|r}
1 & 2 & -6 & -4 & \underline{2} \\
  & 2 & 8 & 4 & \\
\hline
1 & 4 & 2 & 0 & \\
\end{array}
$$

Thus,

$$x^4 + 3x^3 - 4x^2 - 10x - 4 = (x + 1)(x - 2)(x^2 + 4x + 2)$$

The roots from this last factor are now found by the quadratic formula:

$$x = \frac{-4 \pm \sqrt{16 - 8}}{2} = \frac{-4 \pm 2\sqrt{2}}{2} = -2 \pm \sqrt{2}$$

Therefore, the roots are -1, 2, $-2 + \sqrt{2}$, $-2 - \sqrt{2}$.

Example E

Solve the equation $3x^4 - 26x^3 + 63x^2 - 36x - 20 = 0$, given that 2 is a double root.

Using synthetic division, we have

$$
\begin{array}{rrrrr|l}
3 & -26 & 63 & -36 & -20 & \underline{2} \\
  & 6 & -40 & 46 & 20 & \\
\hline
3 & -20 & 23 & 10 & 0 &
\end{array}
$$

Therefore, we know that

$$3x^4 - 26x^3 + 63x^2 - 36x - 20 = (x - 2)(3x^3 - 20x^2 + 23x + 10)$$

Also, since 2 is a double root, it must be a root of the quotient $3x^3 - 20x^2 + 23x + 10$. Using synthetic division again, we have

$$
\begin{array}{rrrr|l}
3 & -20 & 23 & 10 & \underline{2} \\
  & 6 & -28 & -10 & \\
\hline
3 & -14 & -5 & 0 &
\end{array}
$$

The quotient $3x^2 - 14x - 5$ factors into $(3x + 1)(x - 5)$. Therefore, the roots of the equation are 2, 2, $-\frac{1}{3}$, and 5.

Example F

Solve the equation $2x^4 - x^3 + 7x^2 - 4x - 4 = 0$, given that $2j$ is a root.

Since $2j$ is a root, we know that $-2j$ is also a root. Using synthetic division twice, we can then reduce the remaining factor to a quadratic function.

$$
\begin{array}{rrrrr|l}
2 & -1 & 7 & -4 & -4 & \underline{2j} \\
  & 4j & -8 - 2j & 4 - 2j & 4 & \\
\hline
2 & 4j - 1 & -1 - 2j & -2j & & \underline{-2j} \\
  & -4j & & 2j & 2j & \\
\hline
2 & -1 & -1 & & &
\end{array}
$$

The quadratic factor $2x^2 - x - 1$ factors into $(2x + 1)(x - 1)$. Therefore, the roots of the function are $2j$, $-2j$, 1, and $-\frac{1}{2}$.

Exercises 14–3

In Exercises 1 through 20 solve the given equations using synthetic division, given the roots indicated.

1. $x^3 + 2x^2 - x - 2 = 0$ $(r_1 = 1)$ 2. $x^3 + 2x^2 + x + 2 = 0$ $(r_1 = -2)$

3. $x^3 + x^2 - 8x - 12 = 0$ $(r_1 = -2)$ 4. $x^3 - 1 = 0$ $(r_1 = 1)$

5. $2x^3 + 11x^2 + 20x + 12 = 0$ $(r_1 = -\frac{3}{2})$

6. $2x^3 + 5x^2 - 11x + 4 = 0$ $(r_1 = \frac{1}{2})$

7. $3x^3 + 2x^2 + 3x + 2 = 0$ $(r_1 = j)$

8. $x^3 + 5x^2 + 9x + 5 = 0$ $(r_1 = -2 + j)$

9. $x^4 + x^3 - 2x^2 + 4x - 24 = 0$ $(r_1 = 2, r_2 = -3)$

10. $x^4 + 2x^3 - 4x^2 - 5x + 6 = 0$ $(r_1 = 1, r_2 = -2)$

11. $x^4 - 6x^2 - 8x - 3 = 0$ (-1 is a double root)
12. $4x^4 + 28x^3 + 61x^2 + 42x + 9 = 0$ (-3 is a double root)
13. $6x^4 + 5x^3 - 15x^2 + 4 = 0$ ($r_1 = -\frac{1}{2}, r_2 = \frac{2}{3}$)
14. $6x^4 - 5x^3 - 14x^2 + 14x - 3 = 0$ ($r_1 = \frac{1}{3}, r_2 = \frac{1}{2}$)
15. $2x^5 + 11x^4 + 16x^3 - 8x^2 - 32x - 16 = 0$ (-2 is a triple root)
16. $x^5 - 3x^4 + 4x^3 - 4x^2 + 3x - 1 = 0$ (1 is a triple root)
17. $2x^5 + x^4 - 15x^3 + 5x^2 + 13x - 6 = 0$ ($r_1 = 1, r_2 = -1, r_3 = \frac{1}{2}$)
18. $12x^5 - 7x^4 + 41x^3 - 26x^2 - 28x + 8 = 0$ ($r_1 = 1, r_2 = \frac{1}{4}, r_3 = -\frac{2}{3}$)
19. $x^6 + 2x^5 - 4x^4 - 10x^3 - 41x^2 - 72x - 36 = 0$ (-1 is a double root, $2j$ is a root)
20. $x^6 - x^5 - 2x^3 - 3x^2 - x - 2 = 0$ (j is a double root)

14–4 Rational Roots

If we form the product of the factors $(x + 2)(x - 4)(x + 3)$, we obtain $x^3 + x^2 - 14x - 24$. In forming this product, we find that the constant 24 which results is determined only by the numbers 2, 4, and 3. We note that these numbers represent the roots of the equation if the given function is set equal to zero. In fact, if we found all the integral roots of an equation, and represented the equation in the form

$$f(x) = (x - r_1)(x - r_2) \cdots (x - r_k)f_{k+1}(x) = 0$$

where all the roots indicated are integers, the constant term of $f(x)$ must have factors of $r_1, r_2, \ldots, r_k$. This leads us to the theorem which states that *if the coefficient of the highest power of x is 1, then any integral roots are factors of the constant term of the function.*

Example A
The equation $f(x) = x^5 - 4x^4 - 7x^3 + 14x^2 - 44x + 120 = 0$ can be written as

$$(x - 5)(x + 3)(x - 2)(x^2 + 4) = 0$$

We now note that $5(3)(2)(4) = 120$. Thus the roots 5, -3, and 2 are numerical factors of $|120|$. The theorem states nothing in regard to the signs involved.

If the coefficient of the highest-power term of $f(x)$ is not 1, then this coefficient can be factored from every term of $f(x)$. Thus any equation of the form $f(x) = a_0x^n + a_1x^{n-1} + \cdots + a_n = 0$ can be written in the form

$$f(x) = a_0\left(x^n + \frac{a_1}{a_0}x^{n-1} + \cdots + \frac{a_n}{a_0}\right) = 0 \tag{14–4}$$

This equation, along with the theorem above, now gives us another, more inclusive, theorem. *Any rational roots of a polynomial equation $f(x) = a_0x^n + a_1x^{n-1} + \cdots + a_n = 0$ must be integral factors of a_n divided* *by integral factors of a_0.* The same reasoning as previously stated, applied to the factor within parentheses of Eq. (14–4), leads us to this result.

Example B

If $f(x) = 4x^3 - 3x^2 - 25x - 6 = 0$, any rational roots, if they exist, must be integral factors of 6 divided by integral factors of 4. The integral factors of 6 are 1, 2, 3, and 6 and the integral factors of 4 are 1, 2, and 4. Forming all possible positive and negative quotients, any rational roots that exist will be found in the following list: ± 1, $\pm \frac{1}{2}$, $\pm \frac{1}{4}$, ± 2, ± 3, $\pm \frac{3}{2}$, $\pm \frac{3}{4}$, ± 6.

The roots of this equation are -2, 3, and $-\frac{1}{4}$.

There are 16 different possible rational roots in Example B. Since we have no way of telling which of these are the actual roots, we now present a rule which will help us to find these roots. This rule is known as **Descartes' rule of signs.** It states that *the number of positive roots of a polynomial equation $f(x) = 0$ cannot exceed the number of changes in sign in $f(x)$ in going from one term to the next in $f(x)$. The number of negative roots cannot exceed the number of sign changes in $f(-x)$.*

We can reason this way: If $f(x)$ has all positive terms, then any positive number substituted in $f(x)$ must give a positive value for the function. This indicates that the number substituted in the function is not a root. Thus, there must be at least one negative and one positive term in the function for any positive number to be a root. This is not a proof, but does indicate the type of reasoning which is used in developing the theorem.

Example C

By Descartes' rule of signs, determine the maximum number of positive and negative roots of $3x^3 - x^2 - x + 4 = 0$.

Here $f(x) = 3x^3 - x^2 - x + 4$. The first term is positive and the second negative, which indicates a change of sign. The third term is also negative; there is no change of sign from the second to the third term. The fourth term is positive, thus giving us a second change of sign, from the third to the fourth term. Hence there are two changes in sign, and therefore no more than two positive roots of $f(x) = 0$. Then we write

$$f(-x) = 3(-x)^3 - (-x)^2 - (-x) + 4 = -3x^3 - x^2 + x + 4$$

There is only one change of sign in $f(-x)$; therefore, there is one negative root. *When there is just one change of sign in $f(x)$ there is a positive root, and when there is just one change of sign in $f(-x)$ there is a negative root.*

Example D

For the equation $4x^5 - x^4 - 4x^3 + x^2 - 5x - 6 = 0$, we write

$$f(x) = 4x^5 - x^4 - 4x^3 + x^2 - 5x - 6$$

and

$$f(-x) = -4x^5 - x^4 + 4x^3 + x^2 + 5x - 6.$$

Thus, there are no more than three positive and two negative roots.

At this point let us summarize the information we can determine about the roots of a polynomial equation $f(x) = 0$ of degree n:

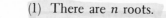

(1) There are n roots.
(2) Complex roots appear in conjugate pairs.
(3) Any rational roots must be factors of the constant term divided by factors of the coefficient of the highest-power term.
(4) The maximum number of positive roots is the number of sign changes in $f(x)$, and the maximum number of negative roots is the number of sign changes in $f(-x)$.
(5) Once we determine $n - 2$ of the roots, the remaining roots can be found by the quadratic formula.

Synthetic division is normally used to try possible roots. This is because synthetic division is relatively easy to perform, and when a root is found we have the quotient factor, which is of degree one less than the degree of the dividend. Each root we find makes the ensuing work simpler. The following examples indicate the complete method, as well as two other helpful rules.

Example E

Determine the roots of the equation $2x^3 + x^2 + 5x - 3 = 0$.

Since $n = 3$, there are three roots. If we can find one of these roots, we can use the quadratic formula to find the other two. We have $f(x) = 2x^3 + x^2 + 5x - 3$, and therefore there is one positive root. We also have $f(-x) = -2x^3 + x^2 - 5x - 3$, and therefore there are no more than two negative roots. The *possible* rational roots are ± 1, $\pm \frac{1}{2}$, $\pm \frac{3}{2}$, ± 3. Thus, using synthetic division, we shall try these. We first try the root 1 (always a possibility if there are positive roots).

$$
\begin{array}{rrrr|c}
2 & 1 & 5 & -3 & \underline{|1} \\
  & 2 & 3 & 8 & \\
\hline
2 & 3 & 8 & 5 &
\end{array}
$$

Thus we see that 1 is not a root, but we have gained some information, if we observe closely. If we try any positive number larger than 1, the results in the last row will be larger positive numbers than we now have. The products will be larger, and therefore the sums will also be larger. Thus there is no positive root larger than 1. This leads to the following rule: *When we are trying a root, if the bottom row contains all positive numbers, then there are no roots larger than the value tried.* This rule tells us that there is no reason to try $+\frac{3}{2}$ and $+3$ as roots. Therefore, let us now try $+\frac{1}{2}$.

$$
\begin{array}{rrrr|c}
2 & 1 & 5 & -3 & \underline{|\frac{1}{2}} \\
  & 1 & 1 & 3 & \\
\hline
2 & 2 & 6 & 0 &
\end{array}
$$

Hence $+\frac{1}{2}$ is a root, and the remaining factor is $2x^2 + 2x + 6$, which itself factors to $2(x^2 + x + 3)$. By the quadratic formula we find the remaining roots.

$$x = \frac{-1 \pm \sqrt{1 - 12}}{2} = \frac{-1 \pm \sqrt{11}j}{2}$$

The three roots are

$$\frac{1}{2}, \quad \frac{-1 + \sqrt{11}j}{2} \quad \text{and} \quad \frac{-1 - \sqrt{11}j}{2}$$

We note that there were actually no negative roots, because the non-positive roots are complex. Also, in proceeding in this way, we never found it necessary to try any negative roots. It must be admitted, however, that the solutions to all problems may not be so easily determined.

Example F

Determine the roots of the equation $x^4 - 7x^3 + 12x^2 + 4x - 16 = 0$.
We write

$$f(x) = x^4 - 7x^3 + 12x^2 + 4x - 16$$
$$f(-x) = x^4 + 7x^3 + 12x^2 - 4x - 16$$

We see that there are four roots; there are no more than three positive roots, and there is one negative root. The possible rational roots are ± 1, ± 2, ± 4, ± 8, ± 16. Since there is only one negative root, we shall look for this one first. Trying -2, we have

```
1    -7    +12    +4     -16      |-2
          -2    +18    -60    +112
------------------------------------
1    -9    +30    -56    +96
```

If we were to try any negative roots less than -2 (remember, -3 is less than -2), we would find that the numbers would still alternate from term to term in the quotient. Thus, we have this rule: *If the signs alternate in the bottom row, then there are no roots less than the value tried.* So we next try -1.

```
1    -7    +12    +4    -16      |-1
          -1     8   -20    16
------------------------------------
1    -8    20   -16     0
```

Thus, -1 is the negative root. Next we shall try $+1$.

```
1    -8    20   -16      |1
          1   -7    13
------------------------------
1    -7    13   -3
```

Since $+1$ is not a root, we next try $+2$.

```
1    -8    20   -16      |2
          2  -12    16
------------------------------
1    -6    8    0
```

Thus $+2$ is a root. It is not necessary to find any more roots by trial and error. We now may use the quadratic formula on the remaining factor $x^2 - 6x + 8$, and find that the roots are 2 and 4. Thus, the roots are -1, 2, 2, and 4. (Note that 2 is a double root.)

Exercises 14—4

In Exercises 1 through 20 solve the given equations.

1. $x^3 + 2x^2 - x - 2 = 0$

2. $x^3 + x^2 - 5x + 3 = 0$

3. $x^3 + 2x^2 - 5x - 6 = 0$

4. $x^3 + 1 = 0$

5. $2x^3 - 5x^2 - 28x + 15 = 0$

6. $2x^3 - x^2 - 3x - 1 = 0$

7. $3x^3 + 11x^2 + 5x - 3 = 0$

8. $4x^3 - 5x^2 - 23x + 6 = 0$

9. $x^4 - 11x^2 - 12x + 4 = 0$

10. $x^4 + x^3 - 2x^2 - 4x - 8 = 0$

11. $x^4 - 2x^3 - 13x^2 + 14x + 24 = 0$

12. $x^4 - x^3 + 2x^2 - 4x - 8 = 0$

13. $2x^4 - 5x^3 - 3x^2 + 4x + 2 = 0$

14. $2x^4 + 7x^3 + 9x^2 + 5x + 1 = 0$

15. $12x^4 + 44x^3 + 21x^2 - 11x - 6 = 0$

16. $9x^4 - 3x^3 + 34x^2 - 12x - 8 = 0$

17. $x^5 + x^4 - 9x^3 - 5x^2 + 16x + 12 = 0$

18. $x^6 - x^4 - 14x^2 + 24 = 0$

19. $2x^5 - 5x^4 + 6x^3 - 6x^2 + 4x - 1 = 0$

20. $2x^5 + 5x^4 - 4x^3 - 19x^2 - 16x - 4 = 0$

In Exercises 21 through 26 determine the required quantities. In Exercises 25 and 26 it is necessary to set up equations of higher degree.

21. Under certain conditions, the velocity of an object as a function of time is given by $v = 2t^3 - 11t^2 - 28t - 15$. For what values of t is $v = 0$?

22. The deflection y of a beam at a horizontal distance x from one end is given by $y = k(x^4 - 2Lx^3 + L^3x)$, where L is the length of the beam and k is a constant. For what values of x is the deflection zero?

23. In the theory of the motion of a sphere moving through a fluid, the expression $4r^3 - 3ar^2 - a^3$ is found. In terms of a, solve for r if this expression is zero.

24. Three electric resistors are connected in parallel. The second resistor is 1 Ω more than the first, and the third resistor is 4 Ω more than the first. The total resistance of the combination is 1 Ω. To find the first resistance R, we must solve the equation

$$\frac{1}{R} + \frac{1}{R + 1} + \frac{1}{R + 4} = 1$$

Find the values of the resistances.

25. A rectangular box is made from a piece of cardboard 8 in. by 12 in., by cutting a square from each corner and bending up the sides. How large is the side of the square cut out, if the volume of the box is 64 in.^{3}?

26. A slice 2 cm thick is cut off the side of a cube, leaving 75 cm^3 in the remaining volume. Find the length of the edge of the cube.

14—5 Irrational Roots by Linear Interpolation

When a polynomial equation has more than two irrational roots, we cannot find these roots by the methods just presented. Therefore, we must have some method for finding irrational roots for those equations in which we cannot reduce the given function to quadratic factors. Many methods have been developed for this purpose, but we shall discuss only one: **linear interpolation.** The basic assumption is the same as that involved in finding values which lie between listed values in tables;

namely, that if two points are sufficiently close to each other, a straight line joining the points will very nearly approximate the actual curve between the two points. This method is basically a graphical one, and is illustrated in the following examples.

Example A
Using the method of linear interpolation, find the irrational root of the equation $x^3 + 2x^2 + 8x - 2 = 0$ which lies between 0 and 1.

We know that wherever a curve crosses the x-axis, that value of x is a root. We are told that the root we want is between 0 and 1. Normally we would then expect to find the function to be either positive or negative when $x = 0$, and to have the opposite sign when $x = 1$. To check this, the remainder theorem may be used. We find that $f(0) = -2$ and $f(1) = 9$. We now *assume* the curve between these points can be approximated by a straight line between these points. If this were exactly correct, as the magnified scale drawing in Fig. 14–1 shows, the root would lie between 0.1 and 0.2, very close to 0.2. Hence we shall try these values, using synthetic division, and rounding off to two significant digits for now.

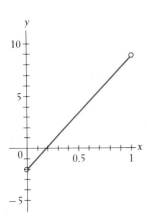

Figure 14—1

1	2	8	−2	|0.1
	0.1	0.2	0.8	
1	2.1	8.2	−1.2	

1	2	8	−2	|0.2
	0.2	0.4	1.7	
1	2.2	8.4	−0.3	

We note that the remainders for $x = 0.1$ and $x = 0.2$ are -1.2 and -0.3 respectively. Since the signs are the same, the root does not lie between 0.1 and 0.2. We do know, however, that it lies between 0.2 and 1. Thus we continue trying values of x nearer 1.

1	2	8	−2	|0.3
	0.3	0.7	2.6	
1	2.3	8.7	0.6	

Since the sign of the remainder for $x = 0.3$ is positive, we now know that the root lies between 0.2 and 0.3. Again we make a scale drawing (Fig. 14–2) to approximate the hundredths digit. The root is apparently between 0.23 and 0.24. We shall try these values (rounding off to three significant digits).

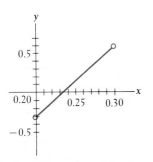

Figure 14—2

1	2	8	−2	|0.23
	0.23	0.51	1.96	
1	2.23	8.51	−0.04	

1	2	8	−2	|0.24
	0.24	0.54	2.05	
1	2.24	8.54	0.05	

Thus the root is between 0.23 and 0.24. Since the remainder is numerically smaller for $x = 0.23$, we may conclude from linear interpolation that this is the value of the root to two significant digits. The process may be continued to obtain greater accuracy if necessary.

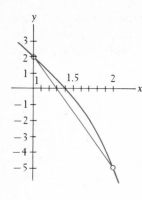

Figure 14–3

Example B

By linear interpolation, find the root between 1 and 2 of the equation $x^4 - 5x^3 + 6x^2 - 5x + 5 = 0$.

The method followed should be clear when we observe Fig. 14–3 and the steps shown.

$$
\begin{array}{rrrrr r}
1 & -5 & 6 & -5 & 5 & \underline{|1} \\
& 1 & -4 & 2 & -3 & \\
\hline
1 & -4 & 2 & -3 & +2 &
\end{array}
$$

$$
\begin{array}{rrrrr r}
1 & -5 & 6 & -5 & 5 & \underline{|2} \\
& 2 & -6 & 0 & -10 & \\
\hline
1 & -3 & 0 & -5 & -5 &
\end{array}
$$

$$
\begin{array}{rrrrr r}
1 & -5 & 6 & -5 & 5 & \underline{|1.2} \\
& 1.2 & -4.6 & 1.7 & -4.0 & \\
\hline
1 & -3.8 & 1.4 & -3.3 & +1.0 &
\end{array}
$$

$$
\begin{array}{rrrrr r}
1 & -5 & 6 & -5 & 5 & \underline{|1.3} \\
& 1.3 & -4.8 & 1.6 & -4.4 & \\
\hline
1 & -3.7 & 1.2 & -3.4 & +0.6 &
\end{array}
$$

$$
\begin{array}{rrrrr r}
1 & -5 & 6 & -5 & 5 & \underline{|1.4} \\
& 1.4 & -5.0 & 1.4 & -5.04 & \\
\hline
1 & -3.6 & +1.0 & -3.6 & -0.04 &
\end{array}
$$

$$
\begin{array}{rrrrr r}
1 & -5 & 6 & -5 & 5 & \underline{|1.39} \\
& 1.39 & -5.02 & 1.36 & -5.06 & \\
\hline
1 & -3.61 & 0.98 & -3.64 & -0.06 &
\end{array}
$$

$$
\begin{array}{rrrrr r}
1 & -5 & 6 & -5 & 5 & \underline{|1.38} \\
& 1.38 & -5.00 & 1.38 & -4.996 & \\
\hline
1 & -3.62 & 1.00 & -3.62 & +0.004 &
\end{array}
$$

Thus $r = 1.38$ (to three significant digits).

If we must find the roots of an equation without any specific information as to the location of the roots, we first make an approximate graph. In this way we can know the integers between which the roots lie. Then linear interpolation can be used to approximate them more accurately. The following example outlines the method.

Example C

Find the roots of the equation $2x^4 - 3x^3 - 6x^2 + 2x - 15 = 0$.

First we calculate values as shown in the following table so that we might graph the function

$$y = 2x^4 - 3x^3 - 6x^2 + 2x - 15$$

and thereby locate the roots approximately.

x	-2	-1	0	1	2	3
y	13	-18	-15	-20	-27	18

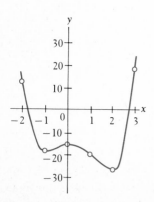

Figure 14–4

The graph is shown in Fig. 14–4. We note that the roots of the equation lie between $x = -2$ and $x = -1$, and between $x = 2$ and $x = 3$. By using linear interpolation these roots can be found more accurately. In this case, the roots are approximately $x = -1.79$ and $x = 2.79$.

Exercises 14—5

In Exercises 1 through 12 find by linear interpolation the irrational root (to two decimal places) between the indicated values of x.

1. $x^3 - 6x^2 + 10x - 4 = 0$ (0 and 1)   2. $x^3 - 3x^2 - 2x + 3 = 0$ (0 and 1)
3. $x^3 - 5x^2 + 7x - 2 = 0$ (0 and 1)   4. $x^3 + 5x^2 - 3x - 2 = 0$ (0 and 1)
5. $2x^3 + 2x^2 - 11x + 3 = 0$ (1 and 2)   6. $x^3 - 5x^2 + 3x + 4 = 0$ (1 and 2)
7. $3x^3 + 13x^2 + 3x - 4 = 0$ (-1 and 0)
8. $2x^4 - x^3 + 6x^2 - 4x - 8 = 0$ (-1 and 0)
9. $x^4 - x^3 - 3x^2 - x - 4 = 0$ (2 and 3)
10. $2x^4 - 2x^3 - 5x^2 - x - 3 = 0$ (2 and 3)
11. $x^4 - 2x^3 - 8x - 16 = 0$ (3 and 4)
12. $3x^4 - 3x^3 - 11x^2 - x - 4 = 0$ (-2 and -1)

In Exercises 13 through 16 find all of the real roots of the given equations to two decimal places.

13. $x^3 - 2x^2 - 5x + 4 = 0$ 14. $x^3 - 2x^2 - 2x - 7 = 0$
15. $x^4 - x^3 - 2x^2 - x - 3 = 0$ 16. $x^5 - 2x^4 + 4x^3 - 7x^2 + 4 = 0$

In Exercises 17 through 20 determine the required values. In Exercises 19 and 20 first set up the appropriate equation.

17. The ends of a 10-ft beam are supported at different levels. The deflection y of the beam is given by $y = kx^2(x^3 + 450x - 3500)$, where x is the horizontal distance from one end and k is a constant. Determine the values of x for which the deflection is zero.

18. In determining one of the dimensions d of the support columns of a building, the equation $3d^3 + 5d^2 - 400d - 3000 = 0$ is found. Determine, to one decimal place, this dimension (in inches).

19. A metal storage tank which holds 500 m^3 is spherical, with an outside diameter of 10.0 m. What is the thickness (to three decimal places) of the metal of which the tank is made?

20. The sides of a rectangular box are 3, 5, and 6 ft. If each side is increased by the same amount, the volume is doubled. By how much should each side be increased to accomplish this?

14—6 Exercises for Chapter 14

In Exercises 1 through 4 find the remainder of the indicated division by the remainder theorem.

1. $(2x^3 - 4x^2 - x + 4) \div (x - 1)$ 2. $(x^3 - 2x^2 + 9) \div (x + 2)$
3. $(4x^3 + x + 4) \div (x + 3)$ 4. $(x^4 - 5x^3 + 8x^2 + 15x - 2) \div (x - 3)$

In Exercises 5 through 8 use the factor theorem to determine whether or not the second expression is a factor of the first.

5. $x^4 + x^3 + x^2 - 2x - 3, x + 1$
6. $2x^3 - 2x^2 - 3x - 2, x - 2$
7. $x^4 + 4x^3 + 5x^2 + 5x - 6, x + 3$
8. $9x^3 + 6x^2 + 4x + 2, 3x + 1$

In Exercises 9 through 16 use synthetic division to perform the indicated divisions.

9. $(x^3 + 3x^2 + 6x + 1) \div (x - 1)$
10. $(3x^3 - 2x^2 + 7) \div (x - 3)$
11. $(2x^3 - 3x^2 - 4x + 3) \div (x + 2)$
12. $(3x^3 - 5x^2 + 7x - 6) \div (x + 4)$
13. $(x^4 - 2x^3 - 3x^2 - 4x - 8) \div (x + 1)$
14. $(x^4 - 6x^3 + x - 8) \div (x - 3)$
15. $(2x^5 - 46x^3 + x^2 - 9) \div (x - 5)$
16. $(x^6 + 63x^3 + 5x^2 - 9x - 8) \div (x + 4)$

In Exercises 17 through 20 use synthetic division to determine whether or not the given numbers are zeros of the given functions.

17. $x^3 + 8x^2 + 17x - 6; -3$
18. $2x^3 + x^2 - 4x + 4; -2$
19. $2x^4 - x^3 + 2x^2 + x - 1; \frac{1}{2}$
20. $6x^4 - 7x^3 + 2x^2 - 9x - 6; -\frac{2}{3}$

In Exercises 21 through 28 find all of the solutions of the given equations, with the aid of synthetic division and with the roots indicated.

21. $x^3 + 8x^2 + 17x + 6 = 0 \quad (r_1 = -3)$
22. $2x^3 + 7x^2 - 6x - 8 = 0 \quad (r_1 = -4)$
23. $3x^4 + 5x^3 + x^2 + x - 10 = 0 \quad (r_1 = 1, r_2 = -2)$
24. $x^4 - x^3 - 5x^2 - x - 6 = 0 \quad (r_1 = 3, r_2 = -2)$
25. $2x^4 + x^3 - 29x^2 - 34x + 24 = 0 \quad (r_1 = -2, r_2 = \frac{1}{2})$
26. $x^4 + x^3 - 11x^2 - 9x + 18 = 0 \quad (r_1 = -3, r_2 = 1)$
27. $x^5 + 3x^4 - x^3 - 11x^2 - 12x - 4 = 0 \quad (-1 \text{ is a triple root})$
28. $24x^5 + 10x^4 + 7x^2 - 6x + 1 = 0 \quad (r_1 = -1, r_2 = \frac{1}{4}, r_3 = \frac{1}{3})$

In Exercises 29 through 36 solve the given equations.

29. $x^3 + x^2 - 10x + 8 = 0$
30. $x^3 - 8x^2 + 20x - 16 = 0$
31. $2x^3 - x^2 - 8x - 5 = 0$
32. $2x^3 - 3x^2 - 11x + 6 = 0$
33. $6x^3 - x^2 - 12x - 5 = 0$
34. $6x^3 + 19x^2 + 2x - 3 = 0$
35. $2x^4 + x^3 + 3x^2 + 2x - 2 = 0$
36. $2x^4 + 5x^3 - 14x^2 - 23x + 30 = 0$

In Exercises 37 through 40, find the indicated irrational root by linear interpolation. Find the value to two decimal places.

37. $x^3 - 3x^2 - x + 2 = 0 \quad (\text{between 0 and 1})$
38. $x^3 + 3x^2 - 6x - 2 = 0 \quad (\text{between 1 and 2})$
39. $3x^3 - x^2 - 8x - 2 = 0 \quad (\text{between 1 and 2})$
40. $x^4 + 3x^3 + 6x + 4 = 0 \quad (\text{between } -1 \text{ and } 0)$

In Exercises 41 through 48 determine the required quantities. Where appropriate, set up the required equations.

41. The total profit P a manufacturer makes in producing x units of a commodity is given by $P = 2x^3 - 3x^2 - 5x - 150$. Less than how many units of production result in a loss?

42. Three electric capacitors are connected in series. The capacitance of the second is 1 μF (microfarad) more than the first, and the third is 2 μF more than the second. The capacitance of the combination is 1.33 μF. The equation used to determine C, the capacitance of the first capacitor, is

$$\frac{1}{C} + \frac{1}{C+1} + \frac{1}{C+3} = \frac{3}{4}$$

Find the values of the capacitances.

43. Where does the graph of function $f(x) = 6x^4 - 14x^3 + 5x^2 + 5x - 2$ cross the x-axis?

44. The hypotenuse of a right triangle is 2 m longer than one of the legs. The area of the triangle is 24 m². Find the sides of the triangle.

45. The width of a rectangular box equals the depth, and the length is 4 ft more than the width. The volume of the box is 539 ft³. What are the dimensions of the box?

46. A certain sphere floating in water sinks to a depth y which is given by the equation $y^3 - 6y^2 + 16 = 0$. Find, to one decimal place, the depth (in centimeters) to which the sphere sinks.

47. The area of a segment of a circle is given approximately by the equation

$$A = \frac{h^3}{2L} + \frac{2Lh}{3}$$

where L is the length of the chord and h is the altitude of the segment. Find h when $L = 3$ and $A = 4$. (Evaluate to two decimal places.)

48. If the edge of a cube is increased by 1 mm, its volume is doubled. Find the edge of the cube to two decimal places.

15

Determinants and Matrices

15–1 Determinants: Expansion by Minors

In Chapter 4 we first met the concept of a determinant and saw how it is used to solve systems of linear equations. However, at that time we limited our discussion to second- and third-order determinants. In the first two sections of this chapter we shall show methods of evaluating higher-order determinants. In the remainder of the chapter we shall develop another related concept and its use in solving systems of linear equations.

From Section 4–6, we recall that a third-order determinant is defined by the equation

$$\begin{vmatrix} a_1 & b_1 & c_1 \\ a_2 & b_2 & c_2 \\ a_3 & b_3 & c_3 \end{vmatrix} = a_1 b_2 c_3 + a_3 b_1 c_2 + a_2 b_3 c_1 - a_3 b_2 c_1 - a_1 b_3 c_2 - a_2 b_1 c_3 \qquad (15\text{--}1)$$

If we rearrange the terms on the right and factor a_1, a_2, and a_3 from the terms in which they are contained, we have

$$\begin{vmatrix} a_1 & b_1 & c_1 \\ a_2 & b_2 & c_2 \\ a_3 & b_3 & c_3 \end{vmatrix} = a_1 (b_2 c_3 - b_3 c_2) - a_2 (b_1 c_3 - b_3 c_1) + a_3 (b_1 c_2 - b_2 c_1) \qquad (15\text{--}2)$$

Recalling the definition of a second-order determinant we have

$$
\begin{vmatrix} a_1 & b_1 & c_1 \\ a_2 & b_2 & c_2 \\ a_3 & b_3 & c_3 \end{vmatrix} = a_1 \begin{vmatrix} b_2 & c_2 \\ b_3 & c_3 \end{vmatrix} - a_2 \begin{vmatrix} b_1 & c_1 \\ b_3 & c_3 \end{vmatrix} + a_3 \begin{vmatrix} b_1 & c_1 \\ b_2 & c_2 \end{vmatrix} \tag{15-3}
$$

In Eq. (15—3) we note that the third-order determinant is expanded with the terms of the expansion as products of the elements of the first column and specific second-order determinants. In each case the elements of the second-order determinant are those elements which are in neither the same row nor the same column as the element from the first column. These determinants are called **minors**.

In general, *the minor of a given element of a determinant is the determinant which results by deleting the row and the column in which the element lies.* Consider the following example.

Example A

Consider the determinant $\begin{vmatrix} 1 & 2 & 3 \\ 4 & 5 & 6 \\ 7 & 8 & 9 \end{vmatrix}$

We find the minor of the element 1 by deleting the elements in the first row and first column because the element 1 is located in the first row and in the first column. This minor is the determinant

$$
\begin{vmatrix} 5 & 6 \\ 8 & 9 \end{vmatrix}
$$

The minor for the element 2 is formed by deleting the elements in the first row and second column, for this is the location of the 2. The minor of 2 is the determinant

$$
\begin{vmatrix} 4 & 6 \\ 7 & 9 \end{vmatrix}
$$

The minor for the element 6 is the determinant

$$
\begin{vmatrix} 1 & 2 \\ 7 & 8 \end{vmatrix}
$$

The minor for the element 8 is the determinant

$$
\begin{vmatrix} 1 & 3 \\ 4 & 6 \end{vmatrix}
$$

We now see that Eq. (15—3) expresses the expansion of a third-order determinant as the sum of the products of the elements of the first column and their minors, with the second term assigned a minus sign.

Actually this is only one of several ways of expressing the expansion. However, it does lead to a general theorem regarding the expansion of a determinant of any order. The foregoing provides a basis for this theorem, although it cannot be considered as a proof. The theorem is as follows:

The value of a determinant of order n may be found by forming the n products of the elements of any column (or row) and their minors. A product is given a plus sign if the sum of the number of the column and the number of the row in which the element lies is even, and a minus sign if this sum is odd. The algebraic sum of the terms thus obtained is the value of the determinant.

The following examples illustrate the expansion of determinants by minors in accordance with the theorem above.

Example B

Evaluate $\begin{vmatrix} 1 & -3 & -2 \\ 4 & -1 & 0 \\ 4 & 3 & -5 \end{vmatrix}$ by expansion by minors.

Since we may expand by any column or row, let us select the first row. The expansion is as follows:

$$\begin{vmatrix} 1 & -3 & -2 \\ 4 & -1 & 0 \\ 4 & 3 & -5 \end{vmatrix} = +(1)\begin{vmatrix} -1 & 0 \\ 3 & -5 \end{vmatrix} - (-3)\begin{vmatrix} 4 & 0 \\ 4 & -5 \end{vmatrix} + (-2)\begin{vmatrix} 4 & -1 \\ 4 & 3 \end{vmatrix}$$

The first term of the expansion is assigned a plus sign since the element 1 is in column 1, row 1 and $1 + 1 = 2$ (even). The second term is assigned a minus sign since the element -3 is in column 2 and row 1, and $2 + 1 = 3$ (odd). The third term is assigned a plus sign since the element -2 is in column 3 and row 1, and $3 + 1 = 4$ (even). Actually, once the first sign has been properly determined, the others are known since the signs alternate from term to term. Using the definition of a second-order determinant, we complete the evaluation.

$$\begin{vmatrix} 1 & -3 & -2 \\ 4 & -1 & 0 \\ 4 & 3 & -5 \end{vmatrix} = +(1)(5 - 0) - (-3)(-20 - 0) + (-2)[12 - (-4)]$$

$$= 1(5) + 3(-20) - 2(16) = 5 - 60 - 32 = -87$$

Expansion of this same determinant by the third column is as follows:

$$\begin{vmatrix} 1 & -3 & -2 \\ 4 & -1 & 0 \\ 4 & 3 & -5 \end{vmatrix} = +(-2)\begin{vmatrix} 4 & -1 \\ 4 & 3 \end{vmatrix} - (0)\begin{vmatrix} 1 & -3 \\ 4 & 3 \end{vmatrix} + (-5)\begin{vmatrix} 1 & -3 \\ 4 & -1 \end{vmatrix}$$

$$= -2(12 + 4) + 0 - 5(-1 + 12)$$

$$= -2(16) - 5(11) = -32 - 55 = -87$$

This expansion has one advantage: since one of the elements of the third column is zero, its minor does not have to be evaluated because zero times whatever the value of the determinant will give the product of zero.

Example C
Evaluate

$$\begin{vmatrix} 3 & -2 & 0 & 2 \\ 1 & 0 & -1 & 4 \\ -3 & 1 & 2 & -2 \\ 2 & -1 & 0 & -1 \end{vmatrix}$$

Expanding by the third column, we have

$$\begin{vmatrix} 3 & -2 & 0 & 2 \\ 1 & 0 & -1 & 4 \\ -3 & 1 & 2 & -2 \\ 2 & -1 & 0 & -1 \end{vmatrix} = +(0)\begin{vmatrix} 1 & 0 & 4 \\ -3 & 1 & -2 \\ 2 & -1 & -1 \end{vmatrix} - (-1)\begin{vmatrix} 3 & -2 & 2 \\ -3 & 1 & -2 \\ 2 & -1 & -1 \end{vmatrix}$$

$$+(2)\begin{vmatrix} 3 & -2 & 2 \\ 1 & 0 & 4 \\ 2 & -1 & -1 \end{vmatrix} - (0)\begin{vmatrix} 3 & -2 & 2 \\ 1 & 0 & 4 \\ -3 & 1 & -2 \end{vmatrix}$$

It is not necessary to expand the minors in the first and fourth terms, since the element in each case is zero. The minors in the second and third terms can be expanded as third-order determinants or by minors. This illustrates well that expansion by minors effectively reduces the order of the determinant to be evaluated by one. Completing the evaluation by minors, we have

$$\begin{vmatrix} 3 & -2 & 0 & 2 \\ 1 & 0 & -1 & 4 \\ -3 & 1 & 2 & -2 \\ 2 & -1 & 0 & -1 \end{vmatrix} = \begin{vmatrix} 3 & -2 & 2 \\ -3 & 1 & -2 \\ 2 & -1 & -1 \end{vmatrix} + 2\begin{vmatrix} 3 & -2 & 2 \\ 1 & 0 & 4 \\ 2 & -1 & -1 \end{vmatrix}$$

$$= \left[3\begin{vmatrix} 1 & -2 \\ -1 & -1 \end{vmatrix} - (-3)\begin{vmatrix} -2 & 2 \\ -1 & -1 \end{vmatrix} + 2\begin{vmatrix} -2 & 2 \\ 1 & -2 \end{vmatrix} \right]$$

$$+ 2\left[-(-2)\begin{vmatrix} 1 & 4 \\ 2 & -1 \end{vmatrix} + 0\begin{vmatrix} 3 & 2 \\ 2 & -1 \end{vmatrix} - (-1)\begin{vmatrix} 3 & 2 \\ 1 & 4 \end{vmatrix} \right]$$

$$= [3(-1-2) + 3(2+2) + 2(4-2)] + 2[2(-1-8) + (12-2)]$$
$$= [-9 + 12 + 4] + 2[-18 + 10]$$
$$= +7 + 2(-8) = -9$$

We can use the expansion of determinants by minors to solve systems of linear equations. Cramer's rule for solving systems of equations, as stated in Section 4–6, is valid for any system of n equations in n unknowns. The following examples illustrate the solution of systems of equations.

Example D

The production of a particular computer component is done in three stages, taking a total of 7 h. The second stage is one hour less than the first, and the third stage is twice as long as the second stage. How long is each stage of production?

First, we let $a =$ the number of hours of the first production stage, $b =$ the number of hours of the second stage, and $c =$ the number of hours of the third stage.

The total production time of 7 h gives us $a + b + c = 7$. Since the second stage is one hour less than the first, we then have $b = a - 1$. The fact that the third stage is twice as long as the second gives us $c = 2b$. Stating these equations in standard form for solution, we have

$$a + b + c = 7$$
$$a - b \quad\;\; = 1$$
$$2b - c = 0$$

Using Cramer's rule, we have

$$a = \frac{\begin{vmatrix} 7 & 1 & 1 \\ 1 & -1 & 0 \\ 0 & 2 & -1 \end{vmatrix}}{\begin{vmatrix} 1 & 1 & 1 \\ 1 & -1 & 0 \\ 0 & 2 & -1 \end{vmatrix}} = \frac{-(1)\begin{vmatrix} 1 & 1 \\ 2 & -1 \end{vmatrix} + (-1)\begin{vmatrix} 7 & 1 \\ 0 & -1 \end{vmatrix} - 0\begin{vmatrix} 7 & 1 \\ 0 & 2 \end{vmatrix}}{-(1)\begin{vmatrix} 1 & 1 \\ 2 & -1 \end{vmatrix} + (-1)\begin{vmatrix} 1 & 1 \\ 0 & -1 \end{vmatrix} + 0\begin{vmatrix} 1 & 1 \\ 0 & 2 \end{vmatrix}}$$

$$= \frac{-(-1 - 2) - (-7)}{-(-1 - 2) - (-1)} = \frac{3 + 7}{3 + 1} = \frac{10}{4} = \frac{5}{2}$$

Here we expanded each determinant by minors of the second row. Now using the second equation we have $\frac{5}{2} - b = 1$, or $b = \frac{3}{2}$. Using the third equation we have $2(\frac{3}{2}) - c = 0$, or $c = 3$. Thus,

$$a = \frac{5}{2}\,\text{h}, \quad b = \frac{3}{2}\,\text{h}, \quad \text{and} \quad c = 3\,\text{h}$$

This means that the first stage takes 2.5 h, the second stage takes 1.5 h, and the third stage takes 3 h. We see that these times agree with the given information.

Example E

Solve the following system of equations.

$$
\begin{aligned}
x + 2y + z \qquad &= 5 \\
2x \qquad + z + 2t &= 1 \\
x - y + 3z + 4t &= -6 \\
4x - y \qquad - 2t &= 0
\end{aligned}
$$

$$
x = \frac{\begin{vmatrix} 5 & 2 & 1 & 0 \\ 1 & 0 & 1 & 2 \\ -6 & -1 & 3 & 4 \\ 0 & -1 & 0 & -2 \end{vmatrix}}{\begin{vmatrix} 1 & 2 & 1 & 0 \\ 2 & 0 & 1 & 2 \\ 1 & -1 & 3 & 4 \\ 4 & -1 & 0 & -2 \end{vmatrix}}
$$

$$
= \frac{-(0)\begin{vmatrix} 2 & 1 & 0 \\ 0 & 1 & 2 \\ -1 & 3 & 4 \end{vmatrix} + (-1)\begin{vmatrix} 5 & 1 & 0 \\ 1 & 1 & 2 \\ -6 & 3 & 4 \end{vmatrix} - (0)\begin{vmatrix} 5 & 2 & 0 \\ 1 & 0 & 2 \\ -6 & -1 & 4 \end{vmatrix} + (-2)\begin{vmatrix} 5 & 2 & 1 \\ 1 & 0 & 1 \\ -6 & -1 & 3 \end{vmatrix}}{(1)\begin{vmatrix} 0 & 1 & 2 \\ -1 & 3 & 4 \\ -1 & 0 & -2 \end{vmatrix} - 2\begin{vmatrix} 2 & 1 & 2 \\ 1 & 3 & 4 \\ 4 & 0 & -2 \end{vmatrix} + (1)\begin{vmatrix} 2 & 0 & 2 \\ 1 & -1 & 4 \\ 4 & -1 & -2 \end{vmatrix} - (0)\begin{vmatrix} 2 & 0 & 1 \\ 1 & -1 & 3 \\ 4 & -1 & 0 \end{vmatrix}}
$$

$$
= \frac{-(-26) - 2(-14)}{1(0) - 2(-18) + 1(18)} = \frac{26 + 28}{36 + 18} = \frac{54}{54} = 1
$$

In solving for x the determinant in the numerator was evaluated by expanding by the minors of the fourth row, since it contained two zeros. The determinant in the denominator was evaluated by expanding by the minors of the first row. Now we solve for y, and we again note two zeros in the fourth row of the determinant of the numerator.

$$
y = \frac{\begin{vmatrix} 1 & 5 & 1 & 0 \\ 2 & 1 & 1 & 2 \\ 1 & -6 & 3 & 4 \\ 4 & 0 & 0 & -2 \end{vmatrix}}{54} = \frac{-4\begin{vmatrix} 5 & 1 & 0 \\ 1 & 1 & 2 \\ -6 & 3 & 4 \end{vmatrix} + (-2)\begin{vmatrix} 1 & 5 & 1 \\ 2 & 1 & 1 \\ 1 & -6 & 3 \end{vmatrix}}{54}
$$

$$
= \frac{-4(-26) - 2(-29)}{54} = \frac{104 + 58}{54} = \frac{162}{54} = 3
$$

Substituting these values for x and y into the first equation, we can solve for z. This gives $z = -2$. Again, substituting the values for x and y into the fourth equation, we find $t = \frac{1}{2}$. Thus the required solution is $x = 1$, $y = 3$, $z = -2$, $t = \frac{1}{2}$. The solution can be checked by substituting these values into either the second or third equation.

Exercises 15−1

In Exercises 1 through 16 evaluate the given determinants by expansion by minors.

1. $\begin{vmatrix} 3 & 0 & 0 \\ -2 & 1 & 4 \\ 4 & -2 & 5 \end{vmatrix}$

2. $\begin{vmatrix} 10 & 0 & -3 \\ -2 & -4 & 1 \\ 3 & 0 & 2 \end{vmatrix}$

3. $\begin{vmatrix} -2 & -4 & 2 \\ 1 & 3 & 0 \\ -4 & 5 & 2 \end{vmatrix}$

4. $\begin{vmatrix} 5 & -1 & 2 \\ 8 & 3 & -4 \\ 0 & 2 & -6 \end{vmatrix}$

5. $\begin{vmatrix} -6 & -1 & 3 \\ 2 & -2 & -3 \\ 10 & 1 & -2 \end{vmatrix}$

6. $\begin{vmatrix} 9 & -3 & 1 \\ -1 & 2 & -1 \\ 2 & -1 & 3 \end{vmatrix}$

7. $\begin{vmatrix} -3 & -2 & 5 \\ 1 & 2 & -1 \\ 3 & -4 & 2 \end{vmatrix}$

8. $\begin{vmatrix} 4 & -3 & 3 \\ -3 & 5 & 6 \\ 2 & -1 & 2 \end{vmatrix}$

9. $\begin{vmatrix} 1 & 0 & 1 & 0 \\ 2 & 4 & -3 & 1 \\ 1 & 1 & 1 & 1 \\ 3 & 5 & 0 & 2 \end{vmatrix}$

10. $\begin{vmatrix} 2 & 0 & 3 & 1 \\ -1 & -1 & 4 & 0 \\ 1 & 2 & 1 & 2 \\ 3 & 3 & -2 & -1 \end{vmatrix}$

11. $\begin{vmatrix} 2 & -1 & 1 & -4 \\ 2 & 1 & 3 & -5 \\ 3 & -1 & -1 & 0 \\ 1 & 2 & 2 & 6 \end{vmatrix}$

12. $\begin{vmatrix} 3 & 6 & -2 & 4 \\ 2 & -5 & 2 & 6 \\ 5 & 3 & 4 & 0 \\ 1 & 2 & 0 & -1 \end{vmatrix}$

13. $\begin{vmatrix} 1 & 2 & -1 & -2 \\ 3 & 1 & 2 & 1 \\ -1 & 3 & -1 & 2 \\ 2 & 1 & 3 & -3 \end{vmatrix}$

14. $\begin{vmatrix} 3 & -1 & 2 & -5 \\ 1 & 4 & 2 & 5 \\ -1 & 1 & 1 & 3 \\ 1 & 2 & -1 & -2 \end{vmatrix}$

15. $\begin{vmatrix} 1 & 2 & 1 & 2 & 1 \\ 1 & 0 & 0 & 1 & 0 \\ 0 & 1 & 1 & 0 & 1 \\ 1 & 1 & 2 & 2 & 1 \\ 0 & 1 & 1 & 0 & 2 \end{vmatrix}$

16. $\begin{vmatrix} 3 & 1 & 1 & 1 & 2 \\ 1 & 1 & 0 & 0 & 1 \\ 1 & 1 & 2 & 2 & 3 \\ 0 & 2 & 1 & 0 & 3 \\ 1 & 1 & 0 & 1 & 0 \end{vmatrix}$

In Exercises 17 through 24 solve the given systems of equations by determinants. Evaluate the determinants by expansion by minors.

17. $2x + y + z = 6$
$x - 2y + 2z = 10$
$3x - y - z = 4$

18. $2x + y = -1$
$4x - 2y - z = 5$
$2x + 3y + 3z = -2$

19. $3x + 6y + 2z = -2$
    $x + 3y - 4z = 2$
    $2x - 3y - 2z = -2$

20. $x + 3y + z = 4$
    $2x - 6y - 3z = 10$
    $4x - 9y + 3z = 4$

21. $x + t = 0$
    $3x + y + z = -1$
    $2y - z + 3t = 1$
    $2z - 3t = 1$

22. $2x + y + z = 4$
    $2y - 2z - t = 3$
    $3y - 3z + 2t = 1$
    $6x - y + t = 0$

23. $x + 2y - z = 6$
    $y - 2z - 3t = -5$
    $3x - 2y + t = 2$
    $2x + y + z - t = 0$

24. $2x + 3y + z = 4$
    $x - 2y - 3z + 4t = -1$
    $3x + y + z - 5t = 3$
    $-x + 2y + z + 3t = 2$

In Exercises 25 through 28 solve the given problems by use of determinants, using methods of this section.

25. In applying Kirchhoff's laws (see Exercise 19 of Section 4–5) to the given electric circuit, the following equations are found. Determine the indicated currents in amperes (see Fig. 15–1).

$$I_A + I_B + I_C + I_D = 0$$
$$2I_A - I_B = -2$$
$$3I_C - 2I_D = 0$$
$$I_B - 3I_C = 6$$

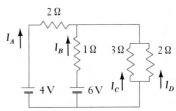

Figure 15–1

26. In analyzing the forces shown in Fig. 15–2, the following equations are derived:

$$F_1 = 20 + 0.8F_3$$
$$F_2 = 0.6F_3$$
$$3F_1 = 40 + 4F_3$$

Find forces F_1, F_2, and F_3.

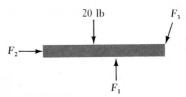

Figure 15–2

27. A 1% solution, a 5% solution, and a 10% solution of sulfuric acid are to be mixed in order to get 600 mL of a 6% solution. If the volume of the 5% solution equals the volume of the other two solutions together, how much of each is needed?

28. A company budgets $100,000 to buy a fleet of 20 cars, made up of four models costing $4000, $5000, $6000, and $9000 each, respectively. The budget calls for 15 of the $4000 and $5000 models. The total cost of the $5000 and $6000 models is $59,000. How many of each model are in the budget?

15–2 Some Properties of Determinants

Expansion of determinants by minors allows us to evaluate a determinant of any order. However, even for a fourth-order determinant, the amount of work necessary for the evaluation is usually excessive. There are a number of basic properties of determinants which allow us to perform the evaluation with considerably less work. We will present these properties here without proof, although each will be illustrated.

(1) *If each element below the principal diagonal of a determinant is zero, then the product of the elements of the principal diagonal is the value of the determinant.*

Example A

The value of the determinant

$$\begin{vmatrix} 2 & 1 & 5 & 8 \\ 0 & -5 & 7 & 9 \\ 0 & 0 & 4 & -6 \\ 0 & 0 & 0 & 3 \end{vmatrix}$$

equals the product $2(-5)(4)(3) = -120$. Since all of the elements below the principal diagonal are zero, there is no need to expand the determinant. It will be noted, however, that if the determinant is expanded by the first column, and successive determinants are expanded by their first columns, the same value is found. Performing the expansion we have

$$\begin{vmatrix} 2 & 1 & 5 & 8 \\ 0 & -5 & 7 & 9 \\ 0 & 0 & 4 & -6 \\ 0 & 0 & 0 & 3 \end{vmatrix} = 2 \begin{vmatrix} -5 & 7 & 9 \\ 0 & 4 & -6 \\ 0 & 0 & 3 \end{vmatrix}$$

$$= 2(-5) \begin{vmatrix} 4 & -6 \\ 0 & 3 \end{vmatrix} = 2(-5)(4)(3) = -120$$

(2) *If the corresponding rows and columns of a determinant are interchanged, the value of the determinant is unchanged.*

Example B

For the determinant

$$\begin{vmatrix} 1 & 3 & -1 \\ 2 & 0 & 4 \\ -2 & 5 & -6 \end{vmatrix}$$

if we interchange the first row and first column, the second row and second column, and the third row and third column, we obtain the determinant

$$\begin{vmatrix} 1 & 2 & -2 \\ 3 & 0 & 5 \\ -1 & 4 & -6 \end{vmatrix}$$

By expanding, we can show that the value of each is the same. We obtain very similar expansions if we expand the first by the first column and the second by the first row. These expansions are

$$\begin{vmatrix} 1 & 3 & -1 \\ 2 & 0 & 4 \\ -2 & 5 & -6 \end{vmatrix} = (1)\begin{vmatrix} 0 & 4 \\ 5 & -6 \end{vmatrix} - 2\begin{vmatrix} 3 & -1 \\ 5 & -6 \end{vmatrix} + (-2)\begin{vmatrix} 3 & -1 \\ 0 & 4 \end{vmatrix}$$

$$= (-20) - 2(-13) - 2(12) = -18$$

$$\begin{vmatrix} 1 & 2 & -2 \\ 3 & 0 & 5 \\ -1 & 4 & -6 \end{vmatrix} = (1)\begin{vmatrix} 0 & 5 \\ 4 & -6 \end{vmatrix} - 2\begin{vmatrix} 3 & 5 \\ -1 & -6 \end{vmatrix} + (-2)\begin{vmatrix} 3 & 0 \\ -1 & 4 \end{vmatrix}$$

$$= (-20) - 2(-13) - 2(12) = -18$$

Therefore, we see that $\begin{vmatrix} 1 & 3 & -1 \\ 2 & 0 & 4 \\ -2 & 5 & -6 \end{vmatrix} = \begin{vmatrix} 1 & 2 & -2 \\ 3 & 0 & 5 \\ -1 & 4 & -6 \end{vmatrix}$

(3) *If two columns (or rows) of a determinant are identical, the value of the determinant is zero.*

Example C
The value of the determinant

$$\begin{vmatrix} 3 & 5 & 2 \\ -4 & 6 & 9 \\ -4 & 6 & 9 \end{vmatrix}$$

is zero, since the second and third rows are identical. This is easily verified by expanding by the first row.

$$\begin{vmatrix} 3 & 5 & 2 \\ -4 & 6 & 9 \\ -4 & 6 & 9 \end{vmatrix} = 3\begin{vmatrix} 6 & 9 \\ 6 & 9 \end{vmatrix} - 5\begin{vmatrix} -4 & 9 \\ -4 & 9 \end{vmatrix} + 2\begin{vmatrix} -4 & 6 \\ -4 & 6 \end{vmatrix}$$

$$= 3(0) - 5(0) + 2(0) = 0$$

(4) *If two columns (or rows) of a determinant are interchanged, the value of the determinant is changed in sign.*

Example D
The values of the determinants

$$\begin{vmatrix} 3 & 0 & 2 \\ 1 & 1 & 5 \\ 2 & 1 & 3 \end{vmatrix} \quad \text{and} \quad \begin{vmatrix} 2 & 0 & 3 \\ 5 & 1 & 1 \\ 3 & 1 & 2 \end{vmatrix}$$

differ in sign, since the first and third columns are interchanged. We shall verify this by expanding each by the second column.

$$\begin{vmatrix} 3 & 0 & 2 \\ 1 & 1 & 5 \\ 2 & 1 & 3 \end{vmatrix} = -(0)\begin{vmatrix} 1 & 5 \\ 2 & 3 \end{vmatrix} + (1)\begin{vmatrix} 3 & 2 \\ 2 & 3 \end{vmatrix} - (1)\begin{vmatrix} 3 & 2 \\ 1 & 5 \end{vmatrix}$$

$$= 0 + (9 - 4) - (15 - 2) = 5 - 13 = -8$$

$$\begin{vmatrix} 2 & 0 & 3 \\ 5 & 1 & 1 \\ 3 & 1 & 2 \end{vmatrix} = -(0)\begin{vmatrix} 5 & 1 \\ 3 & 2 \end{vmatrix} + (1)\begin{vmatrix} 2 & 3 \\ 3 & 2 \end{vmatrix} - (1)\begin{vmatrix} 2 & 3 \\ 5 & 1 \end{vmatrix}$$

$$= 0 + (4 - 9) - (2 - 15) = -5 + 13 = 8$$

Therefore, $\begin{vmatrix} 3 & 0 & 2 \\ 1 & 1 & 5 \\ 2 & 1 & 3 \end{vmatrix} = -\begin{vmatrix} 2 & 0 & 3 \\ 5 & 1 & 1 \\ 3 & 1 & 2 \end{vmatrix}$

(5) *If all the elements of a column (or row) are multiplied by the same number k, the value of the determinant is multiplied by k.*

Example E
The value of the determinant

$$\begin{vmatrix} -1 & 0 & 6 \\ 2 & 1 & -2 \\ 0 & 5 & 3 \end{vmatrix}$$

is multiplied by 3 if the elements of the second row are multiplied by 3. That is,

$$\begin{vmatrix} -1 & 0 & 6 \\ 6 & 3 & -6 \\ 0 & 5 & 3 \end{vmatrix} = 3\begin{vmatrix} -1 & 0 & 6 \\ 2 & 1 & -2 \\ 0 & 5 & 3 \end{vmatrix}$$

By expansion, we can show that

$$\begin{vmatrix} -1 & 0 & 6 \\ 2 & 1 & -2 \\ 0 & 5 & 3 \end{vmatrix} = 47 \quad \text{and} \quad \begin{vmatrix} -1 & 0 & 6 \\ 6 & 3 & -6 \\ 0 & 5 & 3 \end{vmatrix} = 141$$

The validity of this property can be seen by expanding each determinant by the second row. In each case one element is three times the other corresponding element, but the minors are the same. Look at the element in the second row, first column and its minor for each determinant.

$$2\begin{vmatrix} 0 & 6 \\ 5 & 3 \end{vmatrix} \quad \text{and} \quad 6\begin{vmatrix} 0 & 6 \\ 5 & 3 \end{vmatrix}$$

The element, 6, in the second determinant is three times the corresponding element, 2, in the first determinant, and the minors are the same.

(6) *If all the elements of any column (or row) are multiplied by the same number k, and the resulting numbers are added to the corresponding elements of another column (or row), the value of the determinant is unchanged.*

Example F

The value of the determinant

$$\begin{vmatrix} 4 & -1 & 3 \\ 2 & 2 & 1 \\ 1 & 0 & -3 \end{vmatrix}$$

is unchanged if we multiply each element of the first row by 2, and add these numbers to the corresponding elements of the second row. This gives

$$\begin{vmatrix} 4 & -1 & 3 \\ 2+8 & 2+(-2) & 1+6 \\ 1 & 0 & -3 \end{vmatrix} = \begin{vmatrix} 4 & -1 & 3 \\ 10 & 0 & 7 \\ 1 & 0 & -3 \end{vmatrix}$$

or

$$\begin{vmatrix} 4 & -1 & 3 \\ 10 & 0 & 7 \\ 1 & 0 & -3 \end{vmatrix} = \begin{vmatrix} 4 & -1 & 3 \\ 2 & 2 & 1 \\ 1 & 0 & -3 \end{vmatrix}$$

When each determinant is expanded, the value -37 is obtained. The great value in property 6 is that by its use zeros can be purposely placed in the resulting determinant.

With the use of the properties above, determinants of higher order can be evaluated much more readily. The technique is to obtain zeros in a given column (or row) in all positions except one. We can than expand by this column (or row), thereby reducing the order of the determinant. The following example illustrates the method.

Example G
Evaluate

$$\begin{vmatrix} 3 & 2 & -1 & 1 \\ -1 & 1 & 2 & 3 \\ 2 & 2 & 1 & 4 \\ 0 & -1 & -2 & 2 \end{vmatrix}$$

The evaluation is as follows. The small circled numbers above the equals signs refer to the explanations given below the setup.

$$\begin{vmatrix} 3 & 2 & -1 & 1 \\ -1 & 1 & 2 & 3 \\ 2 & 2 & 1 & 4 \\ 0 & -1 & -2 & 2 \end{vmatrix} \overset{①}{=} \begin{vmatrix} 0 & 5 & 5 & 10 \\ -1 & 1 & 2 & 3 \\ 2 & 2 & 1 & 4 \\ 0 & -1 & -2 & 2 \end{vmatrix}$$

$$\overset{②}{=} \begin{vmatrix} 0 & 5 & 5 & 10 \\ -1 & 1 & 2 & 3 \\ 0 & 4 & 5 & 10 \\ 0 & -1 & -2 & 2 \end{vmatrix}$$

$$\overset{③}{=} -(-1)\begin{vmatrix} 5 & 5 & 10 \\ 4 & 5 & 10 \\ -1 & -2 & 2 \end{vmatrix} \overset{④}{=} 5\begin{vmatrix} 1 & 1 & 2 \\ 4 & 5 & 10 \\ -1 & -2 & 2 \end{vmatrix}$$

$$\overset{⑤}{=} 5(2)\begin{vmatrix} 1 & 1 & 1 \\ 4 & 5 & 5 \\ -1 & -2 & 1 \end{vmatrix} \overset{⑥}{=} 10\begin{vmatrix} 1 & 1 & 1 \\ 0 & 1 & 1 \\ -1 & -2 & 1 \end{vmatrix}$$

$$\overset{⑦}{=} 10\begin{vmatrix} 1 & 1 & 1 \\ 0 & 1 & 1 \\ 0 & -1 & 2 \end{vmatrix} \overset{⑧}{=} 10(1)\begin{vmatrix} 1 & 1 \\ -1 & 2 \end{vmatrix}$$

$$\overset{⑨}{=} 10(2+1) = 30$$

① Each element of the second row is multiplied by 3, and the resulting numbers are added to the corresponding elements of the first row. Here we have used property 6. In this way a zero has been placed in column 1, row 1.

② Each element of the second row is multiplied by 2, and the resulting numbers are added to the corresponding elements of the third row. Again, we have used property 6. Also, a zero has been placed in the first column, third row. We now have three zeros in the first column.

③ Expand the determinant by the first column. We have now reduced the determinant to a third-order determinant.

④ Factor 5 from each element of the first row. Here we are using property 5.

⑤ Factor 2 from each element of the third column. Again we are using property 5. Also, by doing this we have reduced the size of the numbers, and the resulting numbers are somewhat easier to work with.

⑥ Each element of the first row is multiplied by −4, and the resulting numbers are added to the corresponding elements of the second row. Here we are using property 6. We have placed a zero in the first column, second row.

⑦ Each element of the first row is added to the corresponding element of the third row. Again, we have used property 6. A zero has been placed in the first column, third row. We now have two zeros in the first column.

⑧ Expand the determinant by the first column.

⑨ Expand the second-order determinant.

A somewhat more systematic method is to place zeros below the principal diagonal and then use property 1. However, all such techniques are essentially equivalent.

Exercises 15—2

In Exercises 1 through 8 evaluate each of the determinants by inspection. Careful observation will allow evaluation by the use of one or more of the basic properties of this section.

1. $\begin{vmatrix} 4 & -5 & 9 \\ 0 & 3 & -8 \\ 0 & 0 & -5 \end{vmatrix}$

2. $\begin{vmatrix} 6 & 4 & 0 \\ 0 & -2 & 3 \\ 0 & 0 & -6 \end{vmatrix}$

3. $\begin{vmatrix} -2 & 0 & 0 \\ 15 & 4 & 0 \\ 2 & -7 & 7 \end{vmatrix}$

4. $\begin{vmatrix} 3 & 0 & 0 \\ 0 & 10 & 0 \\ -9 & -1 & -5 \end{vmatrix}$

5. $\begin{vmatrix} -2 & 0 & -1 \\ 5 & 0 & 3 \\ 3 & 0 & -4 \end{vmatrix}$

6. $\begin{vmatrix} -6 & -3 & 1 \\ 1 & 2 & -5 \\ 0 & 0 & 0 \end{vmatrix}$

7. $\begin{vmatrix} 3 & -2 & 4 \\ 5 & -1 & 2 \\ 3 & -2 & 4 \end{vmatrix}$

8. $\begin{vmatrix} -1 & -2 & -2 \\ 1 & 3 & 3 \\ -2 & 1 & 1 \end{vmatrix}$

In Exercises 9 through 20 evaluate the determinants using the properties given in this section. Do not evaluate directly more than one second-order determinant for each.

9. $\begin{vmatrix} 3 & 1 & 0 \\ -2 & 3 & -1 \\ 4 & 2 & 5 \end{vmatrix}$

10. $\begin{vmatrix} 6 & -1 & 3 \\ 0 & 2 & -2 \\ -1 & 4 & 3 \end{vmatrix}$

11. $\begin{vmatrix} 5 & -1 & -2 \\ 3 & -5 & -2 \\ 1 & 4 & 6 \end{vmatrix}$

12. $\begin{vmatrix} -4 & 3 & -2 \\ -2 & 2 & 4 \\ -1 & 5 & -3 \end{vmatrix}$ 13. $\begin{vmatrix} 4 & 3 & 6 & 0 \\ 3 & 0 & 0 & 4 \\ 5 & 0 & 1 & 2 \\ 2 & 1 & 1 & 7 \end{vmatrix}$ 14. $\begin{vmatrix} -2 & 1 & 3 & 0 \\ 1 & 3 & 0 & 0 \\ 0 & 2 & -3 & -1 \\ 4 & -1 & 2 & 1 \end{vmatrix}$

15. $\begin{vmatrix} 3 & 1 & 2 & -1 \\ 2 & -1 & 3 & -1 \\ 1 & 2 & 1 & 3 \\ 1 & -2 & -3 & 2 \end{vmatrix}$ 16. $\begin{vmatrix} 6 & -3 & -6 & 3 \\ -2 & 1 & 2 & -1 \\ 18 & 7 & -1 & 5 \\ 0 & -1 & 10 & 10 \end{vmatrix}$

17. $\begin{vmatrix} 1 & 3 & -3 & 5 \\ 4 & 2 & 1 & 2 \\ 3 & 2 & -2 & 2 \\ 0 & 1 & 2 & -1 \end{vmatrix}$ 18. $\begin{vmatrix} -2 & 2 & 1 & 3 \\ 1 & 4 & 3 & 1 \\ 4 & 3 & -2 & -2 \\ 3 & -2 & 1 & 5 \end{vmatrix}$

19. $\begin{vmatrix} 1 & 2 & 0 & 1 & 0 \\ 0 & 2 & 1 & 0 & 1 \\ 1 & 0 & -1 & 1 & -1 \\ -2 & 0 & -1 & 2 & 1 \\ 1 & 0 & 2 & -1 & -2 \end{vmatrix}$ 20. $\begin{vmatrix} -1 & 3 & 5 & 0 & -5 \\ 0 & 1 & 7 & 3 & -2 \\ 5 & -2 & -1 & 0 & 3 \\ -3 & 0 & 2 & -1 & 3 \\ 6 & 2 & 1 & -4 & 2 \end{vmatrix}$

In Exercises 21 through 28, solve the given systems of equations by determinants. Evaluate the determinants by the properties of determinants given in this section.

21. $2x - y + z = 5$
$x + 2y + 3z = 10$
$3x + 3y + 2z = 5$

22. $2x + y + z = 5$
$x + 3y - 3z = -13$
$3x + 2y - z = -1$

23. $3x + 2y + z = 1$
$9x + 2z = 5$
$6x - 4y - z = 3$

24. $3x + y + 2z = 4$
$x - y + 4z = 2$
$6x + 3y - 2z = 10$

25. $2x + y + z = 2$
$3y - z + 2t = 4$
$y + 2z + t = 0$
$3x + 2z = 4$

26. $2x + y + z = 0$
$x - y + 2t = 2$
$2y + z + 4t = 2$
$5x + 2z + 2t = 4$

27. $x + y + 2z = 1$
$2x - y + t = -2$
$x - y - z - 2t = 4$
$2x - y + 2z - t = 0$

28. $3x + y + t = 0$
$3z + 2t = 8$
$6x + 2y + 2z + t = 3$
$3x - y - z - t = 0$

In Exercises 29 through 32 solve the given problems by determinants. In Exercises 30 through 32 the necessary equations must be set up.

29. In applying Kirchhoff's laws (see Exercise 19 of Section 4-5) to the circuit shown in Fig. 15-3, the following equations are found. Determine the indicated currents, in amperes.

$$I_A + I_B + I_C + I_D + I_E = 0$$
$$-2I_A + 3I_B = 0$$
$$3I_B - 3I_C = 6$$
$$-3I_C + I_D = 0$$
$$-I_D + 2I_E = 0$$

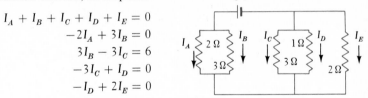

Figure 15-3

30. A land developer subdivides a tract of 100 acres into 120 building lots of 3 types. They have areas of $\frac{1}{2}$ acre, 1 acre, and 2 acres, to sell at $3000, $5000, and $8000, respectively. If all lots are sold, the gross income is $510,000. How many of each type are there in the development?

31. In testing for air pollution, a given air sample contained a total of 6 parts per million (ppm) of four pollutants, sulfur dioxide (SO_2), nitric oxide (NO), nitrogen dioxide (NO_2), and carbon monoxide (CO). The ppm of CO was ten times that of SO_2, which in turn equaled those of NO and NO_2. There was a total of 0.8 ppm of SO_2 and NO. How many ppm of each were present in the air sample?

32. A firm sells four types of appliances. Appliances A, B, C, and D respectively sell for $2, $3, $1, and $4 each. On a certain day it sold a total of 33 appliances, with receipts of $91. It sold twice as many of type B as type C, and twice as many of type D as type C. How many of each were sold on this day?

15-3 Matrices: Definitions and Basic Operations

Systems of linear equations occur in several areas of important technical and scientific applications. We indicated a few of these in Chapter 4 and in the first two sections of this chapter. Since a considerable amount of work is generally required to solve a system of equations, numerous methods have been developed for their solution.

Since the use of computers has been rapidly increasing in importance over the last several years, another mathematical concept which can be used to solve systems of equations is becoming used much more widely than in previous years. It is also used in numerous applications other than systems of equations, in such fields as business, economics, and psychology, as well as the scientific and technical areas. Since it is readily adaptable to use on a computer and is applicable to numerous areas, its importance will increase for some time to come. At this point, however, we shall only be able to introduce its definitions and basic operations.

A **matrix** *is an ordered rectangular array of numbers.* To distinguish such an array from a determinant, we shall enclose it within parentheses. As with a determinant, the individual numbers are called **elements** of the matrix.

Example A
Some examples of matrices are as follows:

$$\begin{pmatrix} 2 & 8 \\ 1 & 0 \end{pmatrix} \quad \begin{pmatrix} 2 & -4 & 6 \\ -1 & 0 & 5 \end{pmatrix} \quad \begin{pmatrix} 4 & 6 \\ 0 & -1 \\ -2 & 5 \\ 3 & 0 \end{pmatrix}$$

$$(-1 \quad 2 \quad 0 \quad 9) \quad \begin{pmatrix} -1 & 8 & 6 & 7 & 9 \\ 2 & 6 & 0 & 4 & 3 \\ 5 & -1 & 8 & 10 & 2 \end{pmatrix}$$

As we can see, it is not necessary for the number of columns and number of rows to be the same, although such is the case for a determinant. However, *if the number of rows does equal the number of columns, the matrix is called a* **square matrix.** We shall find that square matrices are of some special importance. *If all the elements of a matrix are zero, the matrix is called a* **zero matrix.** We shall find it convenient to designate a given matrix by a capital letter.

We must be careful to distinguish between a matrix and a determinant. *A matrix is simply any* **rectangular array** *of numbers, whereas a determinant is a specific value which is associated with a* **square** *matrix.*

Example B
Consider the following matrices:

$$A = \begin{pmatrix} 5 & 0 & -1 \\ 1 & 2 & 6 \\ 0 & -4 & -5 \end{pmatrix}, \quad B = \begin{pmatrix} 9 \\ 8 \\ 1 \\ 5 \end{pmatrix}, \quad C = (-1 \quad 6 \quad 8 \quad 9), \quad O = \begin{pmatrix} 0 & 0 \\ 0 & 0 \end{pmatrix}$$

Matrix A is an example of a square matrix, matrix B is an example of a matrix with one column, matrix C is an example of a matrix with one row, and matrix O is an example of a zero matrix.

To be able to refer to specific elements of a matrix, and to give a general representation, a double-subscript notation is usually employed. That is,

$$A = \begin{pmatrix} a_{11} & a_{12} & a_{13} \\ a_{21} & a_{22} & a_{23} \\ a_{31} & a_{32} & a_{33} \end{pmatrix}$$

We see that the first subscript refers to the row in which the element lies, and the second subscript refers to the column in which the element lies.

Two matrices are said to be equal if and only if they are identical. That is, they must have the same number of columns, the same number of rows, and the elements must respectively be equal. If these conditions are not satisfied, the matrices are not equal.

Example C

$$\begin{pmatrix} a_{11} & a_{12} & a_{13} \\ a_{21} & a_{22} & a_{23} \end{pmatrix} = \begin{pmatrix} 1 & -5 & 0 \\ 4 & 6 & -3 \end{pmatrix}$$

if and only if $a_{11} = 1$, $a_{12} = -5$, $a_{13} = 0$, $a_{21} = 4$, $a_{22} = 6$, and $a_{23} = -3$.

The matrices

$$\begin{pmatrix} 1 & 2 & 3 \\ -1 & -2 & -5 \end{pmatrix} \quad \text{and} \quad \begin{pmatrix} 1 & 2 & -5 \\ -1 & -2 & 3 \end{pmatrix}$$

are not equal, since the elements in the third column are reversed.

The matrices

$$\begin{pmatrix} 2 & 3 \\ -1 & 5 \end{pmatrix} \quad \text{and} \quad \begin{pmatrix} 2 & 3 \\ -1 & 5 \\ 0 & 0 \end{pmatrix}$$

are not equal, since the number of rows is different.

If two matrices have the same number of rows and the same number of columns, their **sum** *is defined as the matrix consisting of the sums of the corresponding elements.* If the number of rows or the number of columns of the two matrices is not equal, they cannot be added.

Example D

$$\begin{pmatrix} 8 & 1 & -5 & 9 \\ 0 & -2 & 3 & 7 \end{pmatrix} + \begin{pmatrix} -3 & 4 & 6 & 0 \\ 6 & -2 & 6 & 5 \end{pmatrix}$$

$$= \begin{pmatrix} 8 + (-3) & 1 + 4 & -5 + 6 & 9 + 0 \\ 0 + 6 & -2 + (-2) & 3 + 6 & 7 + 5 \end{pmatrix}$$

$$= \begin{pmatrix} 5 & 5 & 1 & 9 \\ 6 & -4 & 9 & 12 \end{pmatrix}$$

The matrices

$$\begin{pmatrix} 3 & -5 & 8 \\ 2 & 9 & 0 \\ 4 & -2 & 3 \end{pmatrix} \quad \text{and} \quad \begin{pmatrix} 3 & -5 & 8 & 0 \\ 2 & 9 & 0 & 0 \\ 4 & -2 & 3 & 0 \end{pmatrix}$$

cannot be added since the second matrix has one more column than the first matrix. The fact that the extra column contains only zeros does not matter.

The product of a number and a matrix is defined as the matrix whose elements are obtained by multiplying each element of the given matrix by the given number. That is, kA is the matrix obtained by multiplying the elements of matrix A by k. In this way A + A and 2A will result in the same matrix.

Example E
For the matrix

$$A = \begin{pmatrix} -5 & 7 \\ 3 & 0 \end{pmatrix}$$

we have

$$2A = \begin{pmatrix} 2(-5) & 2(7) \\ 2(3) & 2(0) \end{pmatrix} = \begin{pmatrix} -10 & 14 \\ 6 & 0 \end{pmatrix}$$

Also,

$$5A = \begin{pmatrix} -25 & 35 \\ 15 & 0 \end{pmatrix} \quad \text{and} \quad -A = \begin{pmatrix} 5 & -7 \\ -3 & 0 \end{pmatrix}$$

By combining the definitions for the addition of matrices and for the multiplication of a matrix by a number, we can define the difference of matrices. That is, $A - B = A + (-B)$. Therefore, we would change the sign of each element of matrix B, and proceed as in addition.

By the preceding definitions we can see that the operations of addition, subtraction, and multiplication by a number on matrices are like those for real numbers. For these operations, we say that the algebra of matrices is like the algebra of real numbers. Although it is not our primary purpose to develop the algebra of matrices, we can see that the following laws hold for matrices.

$A + B = B + A$	(commutative law)	(15–4)
$A + (B + C) = (A + B) + C$	(associative law)	(15–5)
$k(A + B) = kA + kB$		(15–6)
$A + O = A$		(15–7)

Here we have let O represent the zero matrix. We shall find in the next section that not all laws for the operations with matrices are similar to those for real numbers.

Exercises 15–3

In Exercises 1 through 4 determine the value of the literal symbols.

1. $\begin{pmatrix} a & b \\ c & d \end{pmatrix} = \begin{pmatrix} 1 & -3 \\ 4 & 7 \end{pmatrix}$

2. $\begin{pmatrix} x & y & z \\ r & -s & -t \end{pmatrix} = \begin{pmatrix} -2 & 7 & -9 \\ 4 & -4 & 5 \end{pmatrix}$

3. $\begin{pmatrix} x \\ x + y \end{pmatrix} = \begin{pmatrix} 2 \\ 5 \end{pmatrix}$

4. $(x \quad x + y \quad x + y + z) = (5 \quad 6 \quad 8)$

In Exercises 5 through 8 find the indicated sums of matrices.

5. $\begin{pmatrix} 2 & 3 \\ -5 & 4 \end{pmatrix} + \begin{pmatrix} -1 & 7 \\ 5 & -2 \end{pmatrix}$ 6. $\begin{pmatrix} 1 & 0 & 9 \\ 3 & -5 & -2 \end{pmatrix} + \begin{pmatrix} 4 & -1 & 7 \\ 2 & 0 & -3 \end{pmatrix}$

7. $\begin{pmatrix} 5 & -8 \\ -3 & 5 \\ -1 & 6 \end{pmatrix} + \begin{pmatrix} -5 & 8 \\ 4 & 1 \\ 2 & -6 \end{pmatrix}$ 8. $\begin{pmatrix} 4 & 2 & -9 \\ -6 & 4 & 7 \\ -1 & 0 & 5 \end{pmatrix} + \begin{pmatrix} -4 & -9 & -2 \\ 3 & 0 & 0 \\ 5 & 10 & -1 \end{pmatrix}$

In Exercises 9 through 16 use the following matrices to determine the indicated matrices.

$$A = \begin{pmatrix} -1 & 4 & -7 & 0 \\ 2 & -6 & -1 & 2 \end{pmatrix}, \quad B = \begin{pmatrix} 1 & 5 & -6 & 3 \\ 4 & -1 & 8 & -2 \end{pmatrix}, \quad C = \begin{pmatrix} 3 & -6 & 9 \\ -4 & 1 & 2 \end{pmatrix}$$

9. $A + B$ 10. $A - B$ 11. $A + C$ 12. $B + C$

13. $2A + B$ 14. $2B + A$ 15. $A - 2B$ 16. $3A - B$

In Exercises 17 through 20 use the given matrices to verify the indicated laws.

$$A = \begin{pmatrix} -1 & 2 & 3 & 7 \\ 0 & -3 & -1 & 4 \\ 9 & -1 & 0 & -2 \end{pmatrix}, \quad B = \begin{pmatrix} 4 & -1 & -3 & 0 \\ 5 & 0 & -1 & 1 \\ 1 & 11 & 8 & 2 \end{pmatrix}$$

17. $A + B = B + A$ 18. $A + 0 = A$

19. $-(A - B) = B - A$ 20. $3(A + B) = 3A + 3B$

In Exercises 21 and 22 perform the indicated matrix operations.

21. The contractor of a housing development constructs four different types of houses, with either a carport, one-car garage, or a two-car garage. The following matrix shows the number of houses of each type, and the type of garage.

	Type A	Type B	Type C	Type D
Carport	8	6	0	0
1-car garage	5	4	3	0
2-car garage	0	3	5	6

If the contractor builds two additional identical developments, find the matrix showing the total number of each house-garage type he built.

22. In taking inventory, a firm finds that it has in one warehouse 6 pieces of 20-ft brass pipe, 8 pieces of 30-ft brass pipe, 11 pieces of 40-ft brass pipe, 5 pieces of 20-ft steel pipe, 10 pieces of 30-ft steel pipe, and 15 pieces of 40-ft steel pipe. This inventory can be represented by the matrix

$$A = \begin{pmatrix} 6 & 8 & 11 \\ 5 & 10 & 15 \end{pmatrix}$$

In each of two other warehouses, the inventory of the same items is represented by the matrix

$$B = \begin{pmatrix} 8 & 3 & 4 \\ 6 & 10 & 5 \end{pmatrix}$$

By matrix addition and multiplication by a constant, find the matrix which represents the total number of each item in the three warehouses.

15–4 Multiplication of Matrices

The definition for the multiplication of matrices does not have an intuitive basis. However, through the solution of a system of linear equations we can, at least in part, show why multiplication is defined as it is. First, let us consider the following example.

Example A

If we solve the system of equations

$$2x + y = 1$$
$$7x + 3y = 5$$

we obtain the solution $x = 2$, $y = -3$. In checking this solution in each of the equations, we obtain

$$2(2) + 1(-3) = 1$$
$$7(2) + 3(-3) = 5$$

Let us represent the coefficients of the equations by the matrix $\begin{pmatrix} 2 & 1 \\ 7 & 3 \end{pmatrix}$

and the solutions by the matrix $\begin{pmatrix} 2 \\ -3 \end{pmatrix}$

If we now multiply these matrices as

$$\begin{pmatrix} 2 & 1 \\ 7 & 3 \end{pmatrix}\begin{pmatrix} 2 \\ -3 \end{pmatrix} = \begin{pmatrix} 2(2) + 1(-3) \\ 7(2) + 3(-3) \end{pmatrix} = \begin{pmatrix} 1 \\ 5 \end{pmatrix}$$

we note that we obtain a matrix which properly represents the right-side values of the equations. (Note carefully how the products and sums in the resulting matrix are formed.)

Following reasons along the lines indicated in Example A, we shall now define the **multiplication of matrices.** If the number of columns in a first matrix equals the number of rows in a second matrix, the product of these matrices is formed as follows: *The element in a specified row and a specified column of the product matrix is the sum of the products formed by multiplying each element in the specified row of the first matrix by the corresponding element in the specific column of the second matrix.* The product matrix will have the same number of rows as the first matrix and the same number of columns as the second matrix. Consider the following examples.

Example B

Find the product AB, where

$$A = \begin{pmatrix} 2 & 1 \\ -3 & 0 \\ 1 & 2 \end{pmatrix} \quad \text{and} \quad B = \begin{pmatrix} -1 & 6 & 5 & -2 \\ 3 & 0 & 1 & -4 \end{pmatrix}$$

To find the element in the first row and first column of the product, we find the sum of the products of corresponding elements of the first row of A and first column of B. To find the element in the first row and second column of the product, we find the sum of the products of corresponding elements in the first row of A and the second column of B. We continue this process until we have found the three rows (the number of rows in A) and the four columns (the number of columns in B) of the product. The product is formed as follows.

$$\begin{pmatrix} 2 & 1 \\ -3 & 0 \\ 1 & 2 \end{pmatrix} \begin{pmatrix} -1 & 6 & 5 & -2 \\ 3 & 0 & 1 & -4 \end{pmatrix}$$

$$= \begin{pmatrix} 2(-1)+1(3) & 2(6)+1(0) & 2(5)+1(1) & 2(-2)+1(-4) \\ -3(-1)+0(3) & -3(6)+0(0) & -3(5)+0(1) & -3(-2)+0(-4) \\ 1(-1)+2(3) & 1(6)+2(0) & 1(5)+2(1) & 1(-2)+2(-4) \end{pmatrix}$$

$$= \begin{pmatrix} 1 & 12 & 11 & -8 \\ 3 & -18 & -15 & 6 \\ 5 & 6 & 7 & -10 \end{pmatrix}$$

If we attempt to form the product BA, we find that B has four columns and A has 3 rows. Since the number of columns in B does not equal the number of rows in A, the product BA cannot be formed. In this way we see that $AB \neq BA$, which means that matrix multiplication is not commutative (except in special cases). Therefore, matrix multiplication differs from the multiplication of real numbers.

Example C
Find the product

$$\begin{pmatrix} -1 & 9 & 3 & -2 \\ 2 & 0 & -7 & 1 \end{pmatrix} \begin{pmatrix} 6 & -2 \\ 1 & 0 \\ 3 & -5 \\ 3 & 9 \end{pmatrix}$$

We can find the product because the first matrix has four columns and the second matrix has four rows. The product is found as follows.

$$\begin{pmatrix} -1 & 9 & 3 & -2 \\ 2 & 0 & -7 & 1 \end{pmatrix} \begin{pmatrix} 6 & -2 \\ 1 & 0 \\ 3 & -5 \\ 3 & 9 \end{pmatrix}$$

$$= \begin{pmatrix} -1(6)+9(1)+3(3)+(-2)(3) & -1(-2)+9(0)+3(-5)+(-2)(9) \\ 2(6)+0(1)+(-7)(3)+1(3) & 2(-2)+0(0)+(-7)(-5)+1(9) \end{pmatrix}$$

$$= \begin{pmatrix} -6+9+9-6 & 2+0-15-18 \\ 12+0-21+3 & -4+0+35+9 \end{pmatrix} = \begin{pmatrix} 6 & -31 \\ -6 & 40 \end{pmatrix}$$

There are two special matrices of particular importance in the multiplication of matrices. The first of these is the **identity matrix I,** *which is a square matrix with 1's for elements on the principal diagonal, with all other elements zero.* It has the property that if it is multiplied by another square matrix with the same number of rows and columns, then the second matrix equals the product matrix.

Example D

Show that $AI = IA = A$ for the matrix

$$A = \begin{pmatrix} 2 & -3 \\ 4 & 1 \end{pmatrix}$$

Since A has two rows and two columns, we choose I with two rows and two columns. Therefore, for this case

$$I = \begin{pmatrix} 1 & 0 \\ 0 & 1 \end{pmatrix}$$

Forming the indicated products, we have results as follows.

$$AI = \begin{pmatrix} 2 & -3 \\ 4 & 1 \end{pmatrix}\begin{pmatrix} 1 & 0 \\ 0 & 1 \end{pmatrix}$$

$$= \begin{pmatrix} 2(1) + (-3)(0) & 2(0) + (-3)(1) \\ 4(1) + 1(0) & 4(0) + 1(1) \end{pmatrix} = \begin{pmatrix} 2 & -3 \\ 4 & 1 \end{pmatrix}$$

$$IA = \begin{pmatrix} 1 & 0 \\ 0 & 1 \end{pmatrix}\begin{pmatrix} 2 & -3 \\ 4 & 1 \end{pmatrix}$$

$$= \begin{pmatrix} 1(2) + 0(4) & 1(-3) + 0(1) \\ 0(2) + 1(4) & 0(-3) + 1(1) \end{pmatrix} = \begin{pmatrix} 2 & -3 \\ 4 & 1 \end{pmatrix}$$

Therefore, we see that $AI = IA = A$.

For a given square matrix A, its **inverse A^{-1}** is the other important special matrix. The matrix A and its inverse have the property that

$$AA^{-1} = A^{-1}A = I \tag{15-8}$$

If the product of two matrices equals the identity matrix, the matrices are called inverses of each other. Under certain conditions the inverse of a given square matrix may not exist, although for most square matrices the inverse does exist. In the next section we shall develop the procedure for finding the inverse of a square matrix, and the section which follows shows how the inverse is used in the solution of systems of equations. At this point we shall simply show that the product of certain matrices equals the identity matrix, and that therefore these matrices are inverses of each other.

Example E

For the given matrices A and B, show that $AB = BA = I$, and therefore that $B = A^{-1}$.

$$A = \begin{pmatrix} 1 & -3 \\ -2 & 7 \end{pmatrix} \qquad B = \begin{pmatrix} 7 & 3 \\ 2 & 1 \end{pmatrix}$$

Forming the products AB and BA, we have the following:

$$AB = \begin{pmatrix} 1 & -3 \\ -2 & 7 \end{pmatrix}\begin{pmatrix} 7 & 3 \\ 2 & 1 \end{pmatrix} = \begin{pmatrix} 7-6 & 3-3 \\ -14+14 & -6+7 \end{pmatrix} = \begin{pmatrix} 1 & 0 \\ 0 & 1 \end{pmatrix}$$

$$BA = \begin{pmatrix} 7 & 3 \\ 2 & 1 \end{pmatrix}\begin{pmatrix} 1 & -3 \\ -2 & 7 \end{pmatrix} = \begin{pmatrix} 7-6 & -21+21 \\ 2-2 & -6+7 \end{pmatrix} = \begin{pmatrix} 1 & 0 \\ 0 & 1 \end{pmatrix}$$

Since $AB = I$, $B = A^{-1}$.

The following example illustrates one kind of application of the multiplication of matrices.

Example F

A particular firm produces three types of machines parts. On a given day it produces 40 of type X, 50 of type Y, and 80 of type Z. Each of type X requires 4 units of material and 1 man-hour to produce; each of type Y requires 5 units of material and 2 man-hours to produce; each of type Z requires 3 units of material and 2 man-hours to produce. By representing the number of each type produced as the matrix $A = (40 \quad 50 \quad 80)$ and the material and time requirements by the matrix

$$B = \begin{pmatrix} 4 & 1 \\ 5 & 2 \\ 3 & 2 \end{pmatrix}$$

the product AB gives the total number of units of material and the total number of man-hours needed for the day's production in a one-row, two-column matrix.

$$AB = (40 \quad 50 \quad 80)\begin{pmatrix} 4 & 1 \\ 5 & 2 \\ 3 & 2 \end{pmatrix}$$

$$= (160 + 250 + 240 \quad 40 + 100 + 160) = (650 \quad 300)$$

Therefore, 650 units of material and 300 man-hours are required.

We now have seen how multiplication is defined for matrices. We see that *matrix multiplication is not commutative;* that is, $AB \neq BA$ in general. This is a major difference from the multiplication of real numbers. Another difference is that it is possible that $AB = 0$, even though neither A nor B is 0 (see Exercise 8, below). There are some similarities, however, in that $AI = A$, where we make I and the number 1 equivalent for the two types of multiplication. Also, the distributive property $A(B + C) = AB + AC$ holds for matrix multiplication. This points out some more of the properties of the algebra of matrices.

Exercises 15—4

In Exercises 1 through 12 perform the indicated matrix multiplications.

1. $(4 \quad -2)\begin{pmatrix} -1 & 0 \\ 2 & 6 \end{pmatrix}$

2. $(-1 \quad 5 \quad -2)\begin{pmatrix} 6 & 3 \\ 2 & -1 \\ 0 & 2 \end{pmatrix}$

3. $\begin{pmatrix} 2 & -3 \\ 5 & -1 \end{pmatrix}\begin{pmatrix} 3 & 0 & -1 \\ 7 & -5 & 8 \end{pmatrix}$

4. $\begin{pmatrix} -7 & 8 \\ 5 & 0 \end{pmatrix}\begin{pmatrix} -9 & 10 \\ 1 & 4 \end{pmatrix}$

5. $\begin{pmatrix} 2 & -3 & 1 \\ 0 & 7 & -3 \end{pmatrix}\begin{pmatrix} 9 \\ -2 \\ 5 \end{pmatrix}$

6. $\begin{pmatrix} 0 & -1 & 2 \\ 4 & 11 & 2 \end{pmatrix}\begin{pmatrix} 3 & -1 \\ 1 & 2 \\ 6 & 1 \end{pmatrix}$

7. $\begin{pmatrix} -1 & -5 \\ 4 & 0 \\ 2 & 10 \end{pmatrix}\begin{pmatrix} 2 & 0 \\ 1 & 1 \end{pmatrix}$

8. $\begin{pmatrix} 12 & -4 \\ 3 & -1 \\ 6 & -2 \end{pmatrix}\begin{pmatrix} 2 & -1 & 3 \\ 6 & -3 & 9 \end{pmatrix}$

9. $\begin{pmatrix} -1 & 7 \\ 3 & 5 \\ 10 & -1 \\ -5 & 12 \end{pmatrix}\begin{pmatrix} 2 & 1 & 0 \\ 5 & -3 & 1 \end{pmatrix}$

10. $\begin{pmatrix} 3 & -1 & 8 \\ 0 & 2 & -4 \\ -1 & 6 & 7 \end{pmatrix}\begin{pmatrix} 7 & -1 \\ 0 & 3 \\ 1 & -2 \end{pmatrix}$

11. $\begin{pmatrix} -9 & -1 & 4 \\ 6 & 9 & -1 \end{pmatrix}\begin{pmatrix} 6 & -5 \\ 4 & 1 \\ -1 & 6 \end{pmatrix}$

12. $\begin{pmatrix} 1 & 2 & -6 & 6 & 1 \\ -2 & 4 & 0 & 1 & 2 \end{pmatrix}\begin{pmatrix} 1 \\ -1 \\ 0 \\ 5 \\ 2 \end{pmatrix}$

In Exercises 13 through 16 find, if possible, AB and BA.

13. $A = (1 \quad -3 \quad 8) \quad B = \begin{pmatrix} -1 \\ 5 \\ 7 \end{pmatrix}$

14.
$$A = \begin{pmatrix} -3 & 2 & 0 \\ 1 & -4 & 5 \end{pmatrix} \quad B = \begin{pmatrix} -2 & 0 \\ 4 & -6 \\ 5 & 1 \end{pmatrix}$$

15.
$$A = \begin{pmatrix} -1 & 2 & 3 \\ 5 & -1 & 0 \end{pmatrix} \quad B = \begin{pmatrix} 1 \\ -5 \\ 2 \end{pmatrix}$$

16.
$$A = \begin{pmatrix} -2 & 1 & 7 \\ 3 & -1 & 0 \\ 0 & 2 & -1 \end{pmatrix} \quad B = (4 \quad -1 \quad 5)$$

In Exercises 17 through 20 show that $AI = IA = A$.

17.
$$A = \begin{pmatrix} 1 & 8 \\ -2 & 2 \end{pmatrix}$$

18.
$$A = \begin{pmatrix} -3 & 4 \\ 1 & 2 \end{pmatrix}$$

19.
$$A = \begin{pmatrix} 1 & 3 & -5 \\ 2 & 0 & 1 \\ 1 & -2 & 4 \end{pmatrix}$$

20.
$$A = \begin{pmatrix} -1 & 2 & 0 \\ 4 & -3 & 1 \\ 2 & 1 & 3 \end{pmatrix}$$

In Exercises 21 through 24 determine whether or not $B = A^{-1}$.

21.
$$A = \begin{pmatrix} 5 & -2 \\ -2 & 1 \end{pmatrix} \quad B = \begin{pmatrix} 1 & 2 \\ 2 & 5 \end{pmatrix}$$

22.
$$A = \begin{pmatrix} 3 & -4 \\ 5 & -7 \end{pmatrix} \quad B = \begin{pmatrix} 7 & -4 \\ 5 & -2 \end{pmatrix}$$

23.
$$A = \begin{pmatrix} 1 & -2 & 3 \\ 2 & -5 & 7 \\ -1 & 3 & -5 \end{pmatrix} \quad B = \begin{pmatrix} 4 & -1 & 1 \\ 3 & -2 & -1 \\ 1 & -1 & -1 \end{pmatrix}$$

24.
$$A = \begin{pmatrix} 1 & -1 & 3 \\ 3 & -4 & 8 \\ -2 & 3 & -4 \end{pmatrix} \quad B = \begin{pmatrix} 8 & -5 & -4 \\ 4 & -2 & -1 \\ -1 & 1 & 1 \end{pmatrix}$$

In Exercises 25 through 28 determine by matrix multiplication whether or not A is the proper matrix of solution values.

25. $3x - 2y = -1$
$4x + y = 6$ $\quad A = \begin{pmatrix} 1 \\ 2 \end{pmatrix}$

26. $4x + y = -5$
$3x + 4y = 6$ $\quad A = \begin{pmatrix} -2 \\ 3 \end{pmatrix}$

27. $3x + y + 2z = 1$
$x - 3y + 4z = -3 \quad A = \begin{pmatrix} -1 \\ 2 \\ 1 \end{pmatrix}$
$2x + 2y + z = 1$

28. $2x - y + z = 7$
$x - 3y + 2z = 6 \quad A = \begin{pmatrix} 3 \\ -2 \\ -1 \end{pmatrix}$
$3x + y - z = 8$

In Exercises 29 and 30 perform the indicated matrix multiplications.

29. The firm referred to in Exercise 22 of Section 15–3 can determine the total number of feet of brass pipe and of steel pipe in each warehouse by multiplying the matrices of that exercise by the matrix

$$C = \begin{pmatrix} 20 \\ 30 \\ 40 \end{pmatrix}$$

Determine the total number of feet of each type of pipe (a) in the first warehouse and (b) in all three warehouses by matrix multiplication.

30. In the theory related to the reproduction of color photography, the equations

$$\begin{pmatrix} X \\ Y \\ Z \end{pmatrix} = \begin{pmatrix} 1.0 & 0.1 & 0 \\ 0.5 & 1.0 & 0.1 \\ 0.3 & 0.4 & 1.0 \end{pmatrix} \begin{pmatrix} x \\ y \\ z \end{pmatrix}$$

are found. The X, Y, and Z represent the red, green, and blue densities of the reproductions, respectively, and the x, y, and z represent the red, green, and blue densities, respectively, of the subject. Give the equations relating X, Y, and Z and x, y, and z.

15–5 Finding the Inverse of a Matrix

In the last section we introduced the concept of the inverse of a matrix. In this section we shall show how the inverse is found, and in the following section we shall show how this inverse is used in the solution of a system of linear equations.

We shall first show two methods of finding the inverse of a two-row, two-column (2×2) matrix. The first method is as follows:

(1) *Interchange the elements on the principal diagonal.*
(2) *Change the signs of the off-diagonal elements.*
(3) *Divide each resulting element by the determinant of the given matrix.*

This is illustrated in the following example.

Example A
Find the inverse of the matrix

$$A = \begin{pmatrix} 2 & -3 \\ 4 & -7 \end{pmatrix}$$

First we interchange the elements on the principal diagonal and change the signs of the off-diagonal elements. This gives us the matrix

$$\begin{pmatrix} -7 & 3 \\ -4 & 2 \end{pmatrix}$$

Now we find the determinant of the original matrix, which means we evaluate

$$\begin{vmatrix} 2 & -3 \\ 4 & -7 \end{vmatrix} = -2$$

(Note again that the matrix is the array of numbers, whereas the determinant of the matrix has a value associated with it.) We now divide each element of the second matrix by -2. This gives

$$\frac{1}{-2}\begin{pmatrix} -7 & 3 \\ -4 & 2 \end{pmatrix} = \begin{pmatrix} \dfrac{-7}{-2} & \dfrac{3}{-2} \\ \dfrac{-4}{-2} & \dfrac{2}{-2} \end{pmatrix} = \begin{pmatrix} \dfrac{7}{2} & -\dfrac{3}{2} \\ 2 & -1 \end{pmatrix}$$

This last matrix is the inverse of matrix A. Therefore,

$$A^{-1} = \begin{pmatrix} \frac{7}{2} & -\frac{3}{2} \\ 2 & -1 \end{pmatrix}$$

Check by multiplication gives

$$AA^{-1} = \begin{pmatrix} 2 & -3 \\ 4 & -7 \end{pmatrix}\begin{pmatrix} \frac{7}{2} & -\frac{3}{2} \\ 2 & -1 \end{pmatrix} = \begin{pmatrix} 7-6 & -3+3 \\ 14-14 & -6+7 \end{pmatrix} = \begin{pmatrix} 1 & 0 \\ 0 & 1 \end{pmatrix} = I$$

The second method involves transforming the given matrix into the identity matrix, while at the same time transforming the identity matrix into the inverse. There are two types of steps allowable in making these transformations.

(1) *Every element in any row may be multiplied by any given number other than zero.*

(2) *Any row may be replaced by a row whose elements are the sum of a nonzero multiple itself and a nonzero multiple of another row.*

Some reflection shows that these operations are those which are performed in solving a system of equations by addition or subtraction. The following example illustrates the method.

Example B

Find the inverse of the matrix

$$A = \begin{pmatrix} 2 & -3 \\ 4 & -7 \end{pmatrix}$$

First we set up the given matrix along with the identity matrix in the following manner.

$$\begin{pmatrix} 2 & -3 & | & 1 & 0 \\ 4 & -7 & | & 0 & 1 \end{pmatrix}$$

The vertical line simply shows the separation of the two matrices.

We wish to transform the left matrix into the identity matrix. Therefore, the first requirement is a 1 for element a_{11}. Therefore, we divide all elements of the first row by 2. This gives the following setup.

$$\begin{pmatrix} 1 & -\frac{3}{2} & | & \frac{1}{2} & 0 \\ 4 & -7 & | & 0 & 1 \end{pmatrix}$$

Next we want to have a zero for element a_{21}. Therefore, we shall subtract 4 times each element of row 1 from the corresponding element in row 2, replacing the elements of row 2. This gives us the following setup.

$$\begin{pmatrix} 1 & -\frac{3}{2} & | & \frac{1}{2} & 0 \\ 4-4(1) & -7-4(-\frac{3}{2}) & | & 0-4(\frac{1}{2}) & 1-4(0) \end{pmatrix}$$

or

$$\begin{pmatrix} 1 & -\frac{3}{2} & | & \frac{1}{2} & 0 \\ 0 & -1 & | & -2 & 1 \end{pmatrix}$$

Next, we want to have 1, not -1, for element a_{22}. Therefore, we multiply each element of row two by -1. This gives

$$\begin{pmatrix} 1 & -\frac{3}{2} & | & \frac{1}{2} & 0 \\ 0 & 1 & | & 2 & -1 \end{pmatrix}$$

Finally, we want zero for element a_{12}. Therefore, we add $\frac{3}{2}$ times each element of row two to the corresponding elements of row one, replacing row one. This gives

$$\begin{pmatrix} 1+\frac{3}{2}(0) & -\frac{3}{2}+\frac{3}{2}(1) & | & \frac{1}{2}+\frac{3}{2}(2) & 0+\frac{3}{2}(-1) \\ 0 & 1 & | & 2 & -1 \end{pmatrix}$$

or

$$\begin{pmatrix} 1 & 0 & \frac{7}{2} & -\frac{3}{2} \\ 0 & 1 & 2 & -1 \end{pmatrix}$$

At this point, we have transformed the given matrix into the identity matrix, and the identity matrix into the inverse. Therefore, the matrix to the right of the vertical bar in the last setup is the required inverse. Thus,

$$A^{-1} = \begin{pmatrix} \frac{7}{2} & -\frac{3}{2} \\ 2 & -1 \end{pmatrix}$$

This is the same matrix and inverse as illustrated in Example A.

The idea to be noted most carefully in Example B is the order in which the zeros and ones were placed in transforming the given matrix to the identity matrix. We shall now give another example of finding the inverse for a 2 × 2 matrix, and then we shall find the inverse for a 3 × 3 matrix with the same method. This method is applicable for a square matrix of any number of rows or columns.

Example C
Find the inverse of the matrix $\begin{pmatrix} -3 & 6 \\ 4 & 5 \end{pmatrix}$.

$$\begin{pmatrix} -3 & 6 & 1 & 0 \\ 4 & 5 & 0 & 1 \end{pmatrix} \qquad \text{original setup}$$

$$\begin{pmatrix} 1 & -2 & -\frac{1}{3} & 0 \\ 4 & 5 & 0 & 1 \end{pmatrix} \qquad \text{row 1 divided by } -3$$

$$\begin{pmatrix} 1 & -2 & -\frac{1}{3} & 0 \\ 0 & 13 & \frac{4}{3} & 1 \end{pmatrix} \qquad -4 \text{ times row 1 added to row 2}$$

$$\begin{pmatrix} 1 & -2 & -\frac{1}{3} & 0 \\ 0 & 1 & \frac{4}{39} & \frac{1}{13} \end{pmatrix} \qquad \text{row 2 divided by 13}$$

$$\begin{pmatrix} 1 & 0 & -\frac{5}{39} & \frac{2}{13} \\ 0 & 1 & \frac{4}{39} & \frac{1}{13} \end{pmatrix} \qquad 2 \text{ times row 2 added to row 1}$$

Therefore, $A^{-1} = \begin{pmatrix} -\frac{5}{39} & \frac{2}{13} \\ \frac{4}{39} & \frac{1}{13} \end{pmatrix}$, which can be checked by multiplication.

Example D
Find the inverse of the matrix $\begin{pmatrix} 1 & 2 & -1 \\ 3 & 5 & -1 \\ -2 & -1 & -2 \end{pmatrix}$

$$\left(\begin{array}{ccc|ccc} 1 & 2 & -1 & 1 & 0 & 0 \\ 3 & 5 & -1 & 0 & 1 & 0 \\ -2 & -1 & -2 & 0 & 0 & 1 \end{array} \right)$$ original setup

$$\left(\begin{array}{ccc|ccc} 1 & 2 & -1 & 1 & 0 & 0 \\ 0 & -1 & 2 & -3 & 1 & 0 \\ -2 & -1 & -2 & 0 & 0 & 1 \end{array} \right)$$ -3 times row 1 added to row 2

$$\left(\begin{array}{ccc|ccc} 1 & 2 & -1 & 1 & 0 & 0 \\ 0 & -1 & 2 & -3 & 1 & 0 \\ 0 & 3 & -4 & 2 & 0 & 1 \end{array} \right)$$ 2 times row 1 added to row 3

$$\left(\begin{array}{ccc|ccc} 1 & 2 & -1 & 1 & 0 & 0 \\ 0 & 1 & -2 & 3 & -1 & 0 \\ 0 & 3 & -4 & 2 & 0 & 1 \end{array} \right)$$ row 2 multiplied by -1

$$\left(\begin{array}{ccc|ccc} 1 & 0 & 3 & -5 & 2 & 0 \\ 0 & 1 & -2 & 3 & -1 & 0 \\ 0 & 3 & -4 & 2 & 0 & 1 \end{array} \right)$$ -2 times row 2 added to row 1

$$\left(\begin{array}{ccc|ccc} 1 & 0 & 3 & -5 & 2 & 0 \\ 0 & 1 & -2 & 3 & -1 & 0 \\ 0 & 0 & 2 & -7 & 3 & 1 \end{array} \right)$$ -3 times row 2 added to row 3

$$\left(\begin{array}{ccc|ccc} 1 & 0 & 3 & -5 & 2 & 0 \\ 0 & 1 & -2 & 3 & -1 & 0 \\ 0 & 0 & 1 & -\frac{7}{2} & \frac{3}{2} & \frac{1}{2} \end{array} \right)$$ row 3 divided by 2

$$\left(\begin{array}{ccc|ccc} 1 & 0 & 3 & -5 & 2 & 0 \\ 0 & 1 & 0 & -4 & 2 & 1 \\ 0 & 0 & 1 & -\frac{7}{2} & \frac{3}{2} & \frac{1}{2} \end{array} \right)$$ 2 times row 3 added to row 2

$$\left(\begin{array}{ccc|ccc} 1 & 0 & 0 & \frac{11}{2} & -\frac{5}{2} & -\frac{3}{2} \\ 0 & 1 & 0 & -4 & 2 & 1 \\ 0 & 0 & 1 & -\frac{7}{2} & \frac{3}{2} & \frac{1}{2} \end{array} \right)$$ -3 times row 3 added to row 1

Therefore, the required inverse matrix is

$$\begin{pmatrix} \frac{11}{2} & -\frac{5}{2} & -\frac{3}{2} \\ -4 & 2 & 1 \\ -\frac{7}{2} & \frac{3}{2} & \frac{1}{2} \end{pmatrix}$$

which may be checked by multiplication.

In transforming a matrix into the identity matrix, we work on one column at a time, transforming the columns in order from left to right. It is generally wisest to make the element on the principal diagonal for the column 1 first, and then to make all the other elements in the column 0. Looking back to Example D, we see that this procedure has been systematically followed, first on column one, then on column two, and finally on column three.

There are other methods of finding the inverse of a matrix. One of these other methods is shown in Exercises 25 through 28 which follow.

Exercises 15–5

In Exercises 1 through 8 find the inverse of each of the given matrices by the method of Example A of this section.

1. $\begin{pmatrix} 2 & -5 \\ -2 & 4 \end{pmatrix}$ 2. $\begin{pmatrix} -6 & 3 \\ 3 & -2 \end{pmatrix}$ 3. $\begin{pmatrix} -1 & 5 \\ 4 & 10 \end{pmatrix}$ 4. $\begin{pmatrix} 8 & -1 \\ -4 & -5 \end{pmatrix}$

5. $\begin{pmatrix} 0 & -4 \\ 2 & 6 \end{pmatrix}$ 6. $\begin{pmatrix} 7 & -2 \\ -6 & 2 \end{pmatrix}$ 7. $\begin{pmatrix} -5 & -4 \\ 2 & 8 \end{pmatrix}$ 8. $\begin{pmatrix} 7 & -3 \\ -1 & -5 \end{pmatrix}$

In Exercises 9 through 24 find the inverse of each of the given matrices by transforming the identity matrix, as in Examples B through D.

9. $\begin{pmatrix} 1 & 2 \\ 2 & 3 \end{pmatrix}$ 10. $\begin{pmatrix} 1 & 5 \\ -1 & -4 \end{pmatrix}$ 11. $\begin{pmatrix} 2 & 4 \\ -1 & -1 \end{pmatrix}$

12. $\begin{pmatrix} -2 & 6 \\ 3 & -4 \end{pmatrix}$ 13. $\begin{pmatrix} 2 & 5 \\ -1 & 2 \end{pmatrix}$ 14. $\begin{pmatrix} -2 & 3 \\ -3 & 5 \end{pmatrix}$

15. $\begin{pmatrix} 2 & -1 \\ 4 & 6 \end{pmatrix}$ 16. $\begin{pmatrix} 1 & -3 \\ 7 & -5 \end{pmatrix}$ 17. $\begin{pmatrix} 1 & -3 & -2 \\ -2 & 7 & 3 \\ 1 & -1 & -3 \end{pmatrix}$

18. $\begin{pmatrix} 1 & 2 & -1 \\ 3 & 7 & -5 \\ -1 & -2 & 0 \end{pmatrix}$ 19. $\begin{pmatrix} 1 & -1 & -3 \\ 0 & -1 & -2 \\ 2 & 1 & -1 \end{pmatrix}$ 20. $\begin{pmatrix} 1 & 4 & 1 \\ -3 & -13 & -1 \\ 0 & -2 & 5 \end{pmatrix}$

21. $\begin{pmatrix} 1 & 3 & 2 \\ -2 & -5 & -1 \\ 2 & 4 & 0 \end{pmatrix}$ 22. $\begin{pmatrix} 1 & 3 & 4 \\ -1 & -4 & -2 \\ 4 & 9 & 20 \end{pmatrix}$ 23. $\begin{pmatrix} 2 & 4 & 0 \\ 3 & 4 & -2 \\ -1 & 1 & 2 \end{pmatrix}$

24. $\begin{pmatrix} -2 & 6 & 1 \\ 0 & 3 & -3 \\ 4 & -7 & 3 \end{pmatrix}$

In Exercises 25 through 28 find the inverse of each of the given matrices (same as those for Exercises 21 through 24) by use of the following information. For matrix A, its inverse A^{-1} is found from

$$A = \begin{pmatrix} a_{11} & a_{12} & a_{13} \\ a_{21} & a_{22} & a_{23} \\ a_{31} & a_{32} & a_{33} \end{pmatrix}, \quad A^{-1} = \frac{1}{|A|} \begin{pmatrix} \begin{vmatrix} a_{22} & a_{23} \\ a_{32} & a_{33} \end{vmatrix} & -\begin{vmatrix} a_{12} & a_{13} \\ a_{32} & a_{33} \end{vmatrix} & \begin{vmatrix} a_{12} & a_{13} \\ a_{22} & a_{23} \end{vmatrix} \\ -\begin{vmatrix} a_{21} & a_{23} \\ a_{31} & a_{33} \end{vmatrix} & \begin{vmatrix} a_{11} & a_{13} \\ a_{31} & a_{33} \end{vmatrix} & -\begin{vmatrix} a_{11} & a_{13} \\ a_{21} & a_{23} \end{vmatrix} \\ \begin{vmatrix} a_{21} & a_{22} \\ a_{31} & a_{32} \end{vmatrix} & -\begin{vmatrix} a_{11} & a_{12} \\ a_{31} & a_{32} \end{vmatrix} & \begin{vmatrix} a_{11} & a_{12} \\ a_{21} & a_{22} \end{vmatrix} \end{pmatrix}$$

25. $\begin{pmatrix} 1 & 3 & 2 \\ -2 & -5 & -1 \\ 2 & 4 & 0 \end{pmatrix}$ 26. $\begin{pmatrix} 1 & 3 & 4 \\ -1 & -4 & -2 \\ 4 & 9 & 20 \end{pmatrix}$

27. $\begin{pmatrix} 2 & 4 & 0 \\ 3 & 4 & -2 \\ -1 & 1 & 2 \end{pmatrix}$ 28. $\begin{pmatrix} -2 & 6 & 1 \\ 0 & 3 & -3 \\ 4 & -7 & 3 \end{pmatrix}$

15–6 Matrices and Linear Equations

As we stated earlier, matrices can be used to solve systems of equations. In this section we shall show one of the methods of how this is done.

Let us consider the system of equations

$$a_1 x + b_1 y = c_1$$
$$a_2 x + b_2 y = c_2$$

Recalling the definition of equality of matrices, we can write this system directly in terms of matrices as

$$\begin{pmatrix} a_1 x + b_1 y \\ a_2 x + b_2 y \end{pmatrix} = \begin{pmatrix} c_1 \\ c_2 \end{pmatrix}$$

The left side of this equation can be written as the product of two matrices. If we let

$$A = \begin{pmatrix} a_1 & b_1 \\ a_2 & b_2 \end{pmatrix} \quad \text{and} \quad X = \begin{pmatrix} x \\ y \end{pmatrix} \tag{15–9}$$

then we have

$$AX = \begin{pmatrix} a_1 x + b_1 y \\ a_2 x + b_2 y \end{pmatrix} \tag{15–10}$$

Therefore, the system of equations in Eq. (15–10) can be written in terms of matrices as

$$AX = C \qquad\qquad (15\text{–}11)$$

where $C = \begin{pmatrix} c_1 \\ c_2 \end{pmatrix}$.

If we now multiply each side of this matrix equation by A^{-1}, we have

$$A^{-1}AX = A^{-1}C$$

Since $A^{-1}A = I$, we have

$$IX = A^{-1}C$$

However, $IX = X$. Therefore,

$$X = A^{-1}C \qquad\qquad (15\text{–}12)$$

Equation (15–12) states that *we can solve a system of linear equations by multiplying the one-column matrix of the constants on the right by the inverse of the matrix of the coefficients.* The result is a one-column matrix whose elements are the required values. The following examples illustrate the method.

Example A

Solve by matrices the system of equations

$$\begin{aligned} 2x - \;\; y &= 7 \\ 5x - 3y &= 18 \end{aligned}$$

We set up the matrix of the coefficients as

$$A = \begin{pmatrix} 2 & -1 \\ 5 & -3 \end{pmatrix}$$

By either of the methods of the previous section, we can determine the inverse of this matrix to be

$$A^{-1} = \begin{pmatrix} 3 & -1 \\ 5 & -2 \end{pmatrix}$$

We now form the matrix product $A^{-1}C$, where $C = \begin{pmatrix} 7 \\ 18 \end{pmatrix}$. This gives

$$A^{-1}C = \begin{pmatrix} 3 & -1 \\ 5 & -2 \end{pmatrix}\begin{pmatrix} 7 \\ 18 \end{pmatrix} = \begin{pmatrix} 21 - 18 \\ 35 - 36 \end{pmatrix} = \begin{pmatrix} 3 \\ -1 \end{pmatrix}$$

Since $X = A^{-1}C$, this means that

$$\begin{pmatrix} x \\ y \end{pmatrix} = \begin{pmatrix} 3 \\ -1 \end{pmatrix}$$

Therefore, the required solution is $x = 3$ and $y = -1$.

Example B

Solve by matrices the system of equations

$$2x - y = 3$$
$$6x + 4y = -5$$

Setting up matrices A and C, we have

$$A = \begin{pmatrix} 2 & -1 \\ 6 & 4 \end{pmatrix} \quad \text{and} \quad C = \begin{pmatrix} 3 \\ -5 \end{pmatrix}$$

We now find the inverse of A to be

$$A^{-1} = \begin{pmatrix} \frac{2}{7} & \frac{1}{14} \\ -\frac{3}{7} & \frac{1}{7} \end{pmatrix}$$

Therefore,

$$A^{-1}C = \begin{pmatrix} \frac{2}{7} & \frac{1}{14} \\ -\frac{3}{7} & \frac{1}{7} \end{pmatrix}\begin{pmatrix} 3 \\ -5 \end{pmatrix} = \begin{pmatrix} \frac{6}{7} - \frac{5}{14} \\ -\frac{9}{7} - \frac{5}{7} \end{pmatrix} = \begin{pmatrix} \frac{1}{2} \\ -2 \end{pmatrix}$$

Therefore, the required solution is $x = \frac{1}{2}$ and $y = -2$.

Example C

Solve by matrices the system of equations

$$x + 4y - z = 4$$
$$x + 3y + z = 8$$
$$2x + 6y + z = 13$$

Setting up matrices A and C, we have

$$A = \begin{pmatrix} 1 & 4 & -1 \\ 1 & 3 & 1 \\ 2 & 6 & 1 \end{pmatrix} \quad \text{and} \quad C = \begin{pmatrix} 4 \\ 8 \\ 13 \end{pmatrix}$$

To give another example of finding the inverse of a 3×3 matrix, we shall briefly show the steps for finding A^{-1}.

$$\left(\begin{array}{ccc|ccc} 1 & 4 & -1 & 1 & 0 & 0 \\ 1 & 3 & 1 & 0 & 1 & 0 \\ 2 & 6 & 1 & 0 & 0 & 1 \end{array}\right) \qquad \left(\begin{array}{ccc|ccc} 1 & 4 & -1 & 1 & 0 & 0 \\ 0 & -1 & 2 & -1 & 1 & 0 \\ 2 & 6 & 1 & 0 & 0 & 1 \end{array}\right)$$

$$\left(\begin{array}{ccc|ccc} 1 & 4 & -1 & 1 & 0 & 0 \\ 0 & -1 & 2 & -1 & 1 & 0 \\ 0 & -2 & 3 & -2 & 0 & 1 \end{array}\right) \qquad \left(\begin{array}{ccc|ccc} 1 & 4 & -1 & 1 & 0 & 0 \\ 0 & 1 & -2 & 1 & -1 & 0 \\ 0 & -2 & 3 & -2 & 0 & 1 \end{array}\right)$$

$$\left(\begin{array}{ccc|ccc} 1 & 0 & 7 & -3 & 4 & 0 \\ 0 & 1 & -2 & 1 & -1 & 0 \\ 0 & -2 & 3 & -2 & 0 & 1 \end{array}\right) \qquad \left(\begin{array}{ccc|ccc} 1 & 0 & 7 & -3 & 4 & 0 \\ 0 & 1 & -2 & 1 & -1 & 0 \\ 0 & 0 & -1 & 0 & -2 & 1 \end{array}\right)$$

$$\begin{pmatrix} 1 & 0 & 7 & | & -3 & 4 & 0 \\ 0 & 1 & -2 & | & 1 & -1 & 0 \\ 0 & 0 & 1 & | & 0 & 2 & -1 \end{pmatrix} \qquad \begin{pmatrix} 1 & 0 & 7 & | & -3 & 4 & 0 \\ 0 & 1 & 0 & | & 1 & 3 & -2 \\ 0 & 0 & 1 & | & 0 & 2 & -1 \end{pmatrix}$$

$$\begin{pmatrix} 1 & 0 & 0 & | & -3 & -10 & 7 \\ 0 & 1 & 0 & | & 1 & 3 & -2 \\ 0 & 0 & 1 & | & 0 & 2 & -1 \end{pmatrix} \qquad \text{Thus, } A^{-1} = \begin{pmatrix} -3 & -10 & 7 \\ 1 & 3 & -2 \\ 0 & 2 & -1 \end{pmatrix}$$

Therefore,

$$A^{-1}C = \begin{pmatrix} -3 & -10 & 7 \\ 1 & 3 & -2 \\ 0 & 2 & -1 \end{pmatrix} \begin{pmatrix} 4 \\ 8 \\ 13 \end{pmatrix} = \begin{pmatrix} -1 \\ 2 \\ 3 \end{pmatrix}$$

This means that $x = -1$, $y = 2$, $z = 3$.

Example D

Solve by matrices the system of equations

$$x + 2y - z = -4$$
$$3x + 5y - z = -5$$
$$-2x - y - 2z = -5$$

Setting up matrices A and C, we have

$$A = \begin{pmatrix} 1 & 2 & -1 \\ 3 & 5 & -1 \\ -2 & -1 & -2 \end{pmatrix} \qquad \text{and} \qquad C = \begin{pmatrix} -4 \\ -5 \\ -5 \end{pmatrix}$$

We now find the inverse of A to be

$$A^{-1} = \begin{pmatrix} \frac{11}{2} & -\frac{5}{2} & -\frac{3}{2} \\ -4 & 2 & 1 \\ -\frac{7}{2} & \frac{3}{2} & \frac{1}{2} \end{pmatrix}$$

(see Example D of Section 15–5). Therefore,

$$A^{-1}C = \begin{pmatrix} \frac{11}{2} & -\frac{5}{2} & -\frac{3}{2} \\ -4 & 2 & 1 \\ -\frac{7}{2} & \frac{3}{2} & \frac{1}{2} \end{pmatrix} \begin{pmatrix} -4 \\ -5 \\ -5 \end{pmatrix} = \begin{pmatrix} -2 \\ 1 \\ 4 \end{pmatrix}$$

This means that the solution is $x = -2$, $y = 1$, $z = 4$.

After having solved systems of equations in this manner, the reader may feel that the method is much longer and more tedious than previously developed techniques. The principal problem with this method is that a great deal of numerical computation is generally required. However, methods such as this one are easily programmed for use on a computer, which can do the arithmetic work very rapidly. Recalling that matrices are of particular importance in connection with computers, it is the method of solving the system of equations which is of primary importance.

Exercises 15—6

In Exercises 1 through 8 solve the given systems of equations by using the inverse of the coefficient matrix. The numbers in parentheses refer to exercises from Section 15–5 where the inverses may be checked.

1. $2x - 5y = -14$ (No. 1)
   $-2x + 4y = 11$

2. $-x + 5y = 4$ (No. 3)
   $4x + 10y = -4$

3. $2x + 4y = -9$ (No. 11)
   $-x - y = 2$

4. $2x + 5y = -6$ (No. 13)
   $-x + 2y = -6$

5. $x - 3y - 2z = -8$ (No. 17)
   $-2x + 7y + 3z = 19$
   $x - y - 3z = -3$

6. $x - y - 3z = -1$ (No. 19)
   $-y - 2z = -2$
   $2x + y - z = 2$

7. $x + 3y + 2z = 5$ (No. 21)
   $-2x - 5y - z = -1$
   $2x + 4y = -2$

8. $2x + 4y = -2$ (No. 23)
   $3x + 4y - 2z = -6$
   $-x + y + 2z = 5$

In Exercises 9 through 20 solve the given systems of equations by using the inverse of the coefficient matrix.

9. $2x + 7y = 16$
   $x + 4y = 9$

10. $4x - 3y = -13$
    $-3x + 2y = 9$

11. $2x - 3y = 3$
    $4x - 5y = 4$

12. $x + 2y = 3$
    $3x + 4y = 11$

13. $5x - 2y = -14$
    $3x + 4y = -11$

14. $4x - 3y = -1$
    $8x + 3y = 4$

15. $2x - y = 6$
    $4x + 3y = -10$

16. $4x - y = 3$
    $6x - 3y = 5$

17. $x + 2y + 2z = -4$
    $4x + 9y + 10z = -18$
    $-x + 3y + 7z = -7$

18. $x - 4y - 2z = -7$
    $-x + 5y + 5z = 18$
    $3x - 7y + 10z = 38$

19. $2x + 4y + z = 5$
    $-2x - 2y - z = -6$
    $-x + 2y + z = 0$

20. $4x + y = 2$
    $-2x - y + 3z = -18$
    $2x + y - z = 8$

In Exercises 21 through 24 solve the indicated systems of equations by using the inverse of the coefficient matrix. In Exercises 23 and 24 it is necessary to set up the appropriate equations.

21. Three forces F_1, F_2, and F_3 are acting on a certain beam. The forces (in pounds) can be found by solving the following equations.

$$F_1 + F_2 + F_3 = 30$$
$$4F_1 + F_2 - 4F_3 = 0$$
$$5F_2 - 3F_3 = 4$$

Determine these forces.

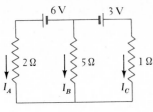

Figure 15–4

22. In applying Kirchhoff's laws (see Exercise 19 of Section 4–5) to the circuit shown in Fig. 15–4, the following equations are found. Determine the indicated currents, in amperes.

$$I_A + I_B + I_C = 0, \qquad 2I_A - 5I_B = 6, \qquad 5I_B - I_C = -3$$

23. Type A doors cost $10 each and type B doors cost $14 each. A builder was billed $220 for a shipment of these doors. He found that had he reversed the number of each on his order, he would have been billed $212. How many of each did he actually receive?

24. Fifty shares of stock A and 30 shares of stock B cost $2600. Thirty shares of stock A and 40 shares of stock B cost $2000. What is the price per share of each stock?

15–7 Exercises for Chapter 15

In Exercises 1 through 8 evaluate the given determinants by expansion by minors.

1.
$$\begin{vmatrix} 1 & 2 & -1 \\ 4 & 1 & -3 \\ -3 & -5 & 2 \end{vmatrix}$$

2.
$$\begin{vmatrix} 3 & -1 & 2 \\ 7 & -1 & 4 \\ 2 & 1 & -3 \end{vmatrix}$$

3.
$$\begin{vmatrix} -1 & 3 & -7 \\ 0 & 5 & 4 \\ 4 & -3 & -2 \end{vmatrix}$$

4.
$$\begin{vmatrix} 6 & -5 & -7 \\ -1 & 2 & 4 \\ 2 & -3 & 1 \end{vmatrix}$$

5.
$$\begin{vmatrix} 2 & 6 & 2 & 5 \\ 2 & 0 & 4 & -1 \\ 4 & -3 & 6 & 1 \\ 3 & -1 & 0 & -2 \end{vmatrix}$$

6.
$$\begin{vmatrix} 1 & -2 & 2 & 4 \\ 0 & 1 & 2 & 3 \\ 3 & 2 & 2 & 5 \\ 2 & 1 & -2 & 0 \end{vmatrix}$$

7.
$$\begin{vmatrix} 1 & 3 & -2 & 4 \\ 2 & 0 & 3 & -2 \\ 5 & -1 & 5 & -3 \\ -6 & 4 & -1 & 2 \end{vmatrix}$$

8.
$$\begin{vmatrix} 2 & 3 & -1 & -1 \\ -3 & -2 & 5 & -6 \\ 2 & 1 & -3 & 2 \\ 4 & 0 & -2 & 1 \end{vmatrix}$$

In Exercises 9 through 16 evaluate the determinants of Exercises 1 through 8 by using the basic properties of determinants.

In Exercises 17 through 20 evaluate the given determinants by using the basic properties of determinants.

17.
$$\begin{vmatrix} 1 & 0 & -3 & -2 \\ 1 & -1 & 2 & 0 \\ -1 & 1 & 1 & 1 \\ 5 & -1 & 2 & -1 \end{vmatrix}$$

18.
$$\begin{vmatrix} 2 & 6 & -2 & 4 \\ -2 & 2 & -3 & 3 \\ 3 & 2 & 2 & -2 \\ 2 & -6 & 4 & 1 \end{vmatrix}$$

19.
$$\begin{vmatrix} 1 & -1 & 3 & 0 & 2 \\ 4 & 0 & 4 & -2 & 2 \\ 0 & 4 & 0 & -1 & -1 \\ -2 & 2 & -1 & 4 & 0 \\ 1 & -1 & 2 & 0 & 1 \end{vmatrix}$$

20.
$$\begin{vmatrix} 1 & 4 & -3 & 3 & 0 \\ 3 & 1 & -1 & 2 & 2 \\ 1 & 2 & 1 & 1 & 1 \\ -3 & -5 & -5 & 0 & -6 \\ 2 & 2 & -2 & 3 & -2 \end{vmatrix}$$

In Exercises 21 through 28 use the given matrices and perform the indicated operations.

$$A = \begin{pmatrix} 2 & -3 \\ 4 & 1 \\ -5 & 0 \\ 2 & -3 \end{pmatrix}, \quad B = \begin{pmatrix} -1 & 0 \\ 4 & -6 \\ -3 & -2 \\ 1 & -7 \end{pmatrix}, \quad C = \begin{pmatrix} 5 & -6 \\ 2 & 8 \\ 0 & -2 \end{pmatrix}$$

21. $A + B$

22. $2C$

23. $-3B$

24. $B - A$

25. $A - C$

26. $2C - B$

27. $2A - 3B$

28. $2(A - B)$

In Exercises 29 through 32 perform the indicated matrix multiplications.

29. $\begin{pmatrix} 5 & -1 \\ 3 & 2 \end{pmatrix}\begin{pmatrix} 1 \\ -8 \end{pmatrix}$

30. $\begin{pmatrix} 6 & -4 & 1 & 0 \\ 2 & 0 & -4 & 3 \end{pmatrix}\begin{pmatrix} 7 & -1 & 6 \\ 4 & 0 & 1 \\ 3 & -2 & 5 \\ 9 & 1 & 0 \end{pmatrix}$

31. $\begin{pmatrix} -1 & 7 \\ 2 & 0 \\ 4 & -1 \end{pmatrix}\begin{pmatrix} 1 & -4 & 5 \\ 5 & 1 & 0 \end{pmatrix}$

32. $\begin{pmatrix} 0 & -1 & 6 \\ 8 & 1 & 4 \\ 7 & -2 & -1 \end{pmatrix}\begin{pmatrix} 5 & -1 & 7 & 1 & 5 \\ 0 & 1 & 0 & 4 & 1 \\ 1 & -2 & 3 & 0 & 1 \end{pmatrix}$

In Exercises 33 through 40 find the inverses of the given matrices.

33. $\begin{pmatrix} 2 & -5 \\ 2 & -4 \end{pmatrix}$

34. $\begin{pmatrix} -1 & -6 \\ 2 & 10 \end{pmatrix}$

35. $\begin{pmatrix} 7 & -1 \\ 4 & 8 \end{pmatrix}$

36. $\begin{pmatrix} 5 & -1 \\ 4 & -8 \end{pmatrix}$

37. $\begin{pmatrix} 1 & 1 & -2 \\ -1 & -2 & 1 \\ 0 & 3 & 4 \end{pmatrix}$

38. $\begin{pmatrix} -1 & -1 & 2 \\ 2 & 3 & 0 \\ 1 & 4 & 1 \end{pmatrix}$

39. $\begin{pmatrix} 2 & -4 & 3 \\ 4 & -6 & 5 \\ -2 & 1 & -1 \end{pmatrix}$

40. $\begin{pmatrix} 3 & 1 & -4 \\ -3 & 1 & -2 \\ -6 & 0 & 3 \end{pmatrix}$

In Exercises 41 through 48 solve the given systems of equations using the inverse of the coefficient matrix.

41. $2x - 3y = -9$
$4x - y = -13$

42. $5x - 7y = 62$
$6x + 5y = -6$

43. $3x + 5y = 29$
$4x - 7y = -57$

44. $4x - 2y = 1$
$8x + 2y = 5$

45. $2x - 3y + 2z = 7$
$3x + y - 3z = -6$
$x + 4y + z = -13$

46. $2x + 2y - z = 8$
$x + 4y + 2z = 5$
$3x - 2y + z = 17$

47. $x + 2y + 3z = 1$
    $3x - 4y - 3z = 2$
    $7x - 6y + 6z = 2$

48. $3x + 2y + z = 2$
    $2x + 3y - 6z = 3$
    $x + 3y + 3z = 1$

In Exercises 49 through 52 solve the given systems of equations by determinants. Use the basic properties of determinants.

49. $3x - 2y + z = 6$
    $2x + 3z = 3$
    $4x - y + 5z = 6$

50. $7x + y + 2z = 3$
    $4x - 2y + 4z = -2$
    $2x + 3y - 6z = 3$

51. $2x - 3y + z - t = -8$
    $4x + 3z + 2t = -3$
    $2y - 3z - t = 12$
    $x - y - z + t = 3$

52. $3x + 2y - 2z - 2t = 0$
    $5y + 3z + 4t = 3$
    $6y - 3z + 4t = 9$
    $6x - y + 2z - 2t = -3$

In Exercises 53 through 56 solve the given systems of equations by any appropriate method of this chapter.

53. To find the forces F_1 and F_2 shown in Fig. 15–5, it is necessary to solve the following equations.

$$F_1 + F_2 = 21$$
$$3F_1 - 4F_2 = 0$$

Find F_1 and F_2.

54. Two electric resistors, R_1 and R_2, are tested with currents and voltages such that the following equations are found.

$$2R_1 + 3R_2 = 26$$
$$3R_1 + 2R_2 = 24$$

Find the resistances R_1 and R_2 (in ohms).

55. A business executive, in pricing three different products, determines that a certain total income should be made on given combinations of sales of these products. Thus, if she sets prices p_1, p_2, and p_3, each respectively, she will arrive at the following equations.

$$p_1 + p_2 + p_3 = 200$$
$$p_1 + 2p_2 + 3p_3 = 430$$
$$4p_1 + 2p_2 + p_3 = 430$$

What are the prices necessary to meet her goals?

56. To find the electric currents (in amperes) indicated in Fig. 15–6, it is necessary to solve the following equations.

$$I_A + I_B + I_C = 0$$
$$5I_A - 2I_B = -4$$
$$2I_B - I_C = 0$$

Find I_A, I_B, and I_C.

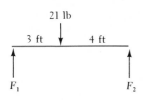

21 lb

3 ft   4 ft

F_1 F_2

Figure 15–5

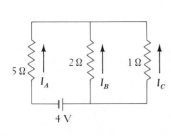

5 Ω   2 Ω   1 Ω

I_A I_B I_C

4 V

Figure 15–6

In Exercises 57 through 60 set up systems of linear equations and solve by any appropriate method illustrated in this chapter.

57. By mass three alloys have the following percentages of lead, zinc, and copper.

	Lead	Zinc	Copper
Alloy A	60%	30%	10%
Alloy B	40%	30%	30%
Alloy C	30%	70%	

How many grams of each of alloys A, B, and C must be mixed to get 100 g of an alloy which is 44% lead, 38% zinc, and 18% copper?

58. A sum of $9000 is invested, part at 6%, part at 5%, and part at 4%. The annual income from these investments is $460. The 5% investment yields $50 more than the 4% investment. How much is invested at each rate?

59. The angles A, B, C, and D of a quadrilateral are related in the following way. Angles B and C are supplementary; four times angle A equals 10° less than the sum of angles B, C, and D; six times angle A plus three times angle C equals twice the sum of all four angles. Find the angles.

60. In setting up salaries for personnel in a certain company, the following criteria were used. The total annual budget for salaries of the 3 managers, 15 research persons, 25 technicians, and 40 production workers was $1,230,000. Each manager receives $5000 more than each research person, $10,000 more than each technician, and $13,000 more than each production worker. What salary does each position pay?

In Exercises 61 and 62 perform the indicated operations on the required matrices.

61. An automobile maker has two assembly plants at which cars with 4 cylinders, 6 cylinders, or 8 cylinders, and with either standard transmission or automatic transmission are assembled. The annual production at the first plant of cars with number of cylinders-transmission type (standard-automatic) is as follows: 4—15,000, 10,000; 6—20,000, 18,000; 8—8,000, 30,000. At the second plant the production is 4—18,000, 12,000; 6—30,000, 22,000; 8—12,000, 40,000. Set up matrices for this information, and by matrix addition find the matrix for total production by the number of cylinders and type of transmission.

62. Set up a matrix representing the information given in Exercise 57. A given shipment contains 500 g of alloy A, 800 g of alloy B, and 700 g of alloy C. Set up a matrix for this information. By multiplying these matrices, obtain a matrix which gives the total weight of lead, zinc, and copper in the shipment.

16

Inequalities

16—1 Properties of Inequalities

Until now we have devoted a great deal of time to the solution of equations. Equation-solving does play an extremely important role in mathematics, but there are also times when we wish to solve inequalities. For example, we have often been faced with the problem of whether or not a given number is real or complex. We determine this from the quadratic formula, by observing the sign of the expression $b^2 - 4ac$. If this expression is positive, the resulting number is real, and if it is negative, the resulting number is complex. This is one of the important uses of inequalities. In this chapter we shall discuss some of the important properties of inequalities and methods of solving inequalities, and illustrate some of their applications in various fields of technology.

In Chapter 1 we first came across the signs of inequality. The expression $a < b$ is read as "a is less than b," and the expression $a > b$ is read as "a is greater than b." *These signs define what is known as the* **sense** *(indicated by the direction of the sign) of the inequality*. Two inequalities are said to have the same sense if the signs of inequality point in the same direction. They are said to have the opposite sense if the signs of inequality point in opposite directions. *The two sides of the inequality are called* **members** *of the inequality*.

Example A

The inequalities $x + 3 > 2$ and $x + 1 > 0$ have the same sense, as do the inequalities $3x - 1 < 4$ and $x^2 - 1 < 3$.

The inequalities $x - 4 < 0$ and $x > -4$ have the opposite sense, as do the inequalities $2x + 4 > 1$ and $3x^2 - 7 < 1$.

The **solution** *of an inequality consists of those values of the variable for which the inequality is satisfied.* Most inequalities with which we shall deal are known as **conditional inequalities.** That is, there are some values of the variable which satisfy the inequality, and also there are some values which do not satisfy it. Some inequalities are satisfied for all values of the variable. These are called **absolute inequalities.** Also, a solution of an inequality may consist of only real numbers, as the terms "greater than" and "less than" have not been defined for complex numbers.

Example B

The inequality $x + 1 > 0$ is satisfied by all values of x greater than -1. Thus, the values of x which satisfy this inequality are written as $x > -1$. This illustrates the difference between the solution of an equation and the solution of an inequality. The solution to an equation normally consists of a few specific numbers, whereas *the solution to an inequality normally consists of a range of values of the variable.* Any and all values within this range are termed the solution of the inequality.

Example C

The inequality $x^2 + 1 > 0$ is true for all values of x, since x^2 is never negative. This is an absolute inequality. The inequality shown in Example B is a conditional inequality.

There are occasions when it is convenient to combine an inequality with an equality. For such purposes, the symbols $\leq$, read "less than or equal to," and $\geq$, read "greater than or equal to," are used.

Example D

If we wish to state that x is positive, we would write $x > 0$. However, the value zero is not included in the solution. If we wished to state that x is not negative, that is, that zero is included as a part of the solution, we can write $x \geq 0$. In order to state that x is less than or equal to -5, we write $x \leq -5$.

We shall now present the basic operations performed on inequalities. These operations are the same as those performed on equations, but in certain cases the results take on a different form. The following are referred to as the **properties of inequalities:**

(1) *The sense of an inequality is not changed when the same number is added to—or subtracted from—both members of the inequality.* Symbolically this may be stated as "if $a > b$, then $a + c > b + c$, or $a - c > b - c$."

Example E

9 > 6; thus, 9 + 4 > 6 + 4, or 13 > 10. Also, 9 − 12 > 6 − 12 or −3 > −6.

(2) *The sense of an inequality is not changed if both members are multiplied or divided by the same positive number.* Symbolically this is stated as, "if $a > b$, then $ac > bc$, or $a/c > b/c$, provided that $c > 0$."

Example F

8 < 15; thus, 8(2) < 15(2) or 16 < 30. Also, $\frac{8}{2} < \frac{15}{2}$ or 4 < $\frac{15}{2}$.

(3) *The sense of an inequality is reversed if both members are multiplied or divided by the same negative number.* Symbolically this is stated as, "if $a > b$, then $ac < bc$, or $a/c < b/c$, provided that $c < 0$." Be very careful to note the *we obtain different results, depending on whether both members are multiplied by a positive or by a negative number.*

Example G

4 > −2; thus, 4(−3) < (−2)(−3), or −12 < 6. Also,

$$\frac{4}{-2} < \frac{-2}{-2} \qquad \text{or} \qquad -2 < 1$$

(4) *If both members of an inequality are positive numbers and n is a positive integer, then the inequality formed by taking the nth power of each member, or the nth root of each member, is in the same sense as the given inequality.* Symbolically this is stated as, "if $a > b$, then $a^n > b^n$, or $\sqrt[n]{a} > \sqrt[n]{b}$, provided that $n > 0$, $a > 0$, $b > 0$."

Example H

16 > 9; thus, $16^2 > 9^2$ or 256 > 81; also, $\sqrt{16} > \sqrt{9}$ or 4 > 3.

Many inequalities have more than two members. In fact, inequalities with three members are very common. All the operations stated above hold for inequalities with more than two members. Some care must be used, however, in stating inequalities with more than two members.

Example I

In order to state that 5 is less than 6, and also greater than 2, which says that 5 is between 2 and 6, we may write 2 < 5 < 6, or 6 > 5 > 2. However, generally the form with the *less than* inequality signs is preferred.

In order to state that a number x may be equal to or greater than 2, and also less than 6, we write $2 \leq x < 6$.

By writing $2 \leq x \leq 6$ we are stating that x is greater than or equal to 2, and at the same time less than or equal to 6.

By writing $x \leq -5$, $x > 7$ we are stating that x is less than or equal to −5, or greater than 7. This may not be stated as $7 < x \leq -5$, for this shows x as being less than −5, while at the same time greater than 7, and no such numbers exist.

Example J

The inequality $x^2 - 3x + 2 > 0$ is satisfied if x is either greater than 2 or less than 1. This would be written as $x > 2$ or $x < 1$, but it would be incorrect to state it as $1 > x > 2$. (If we wrote it this way, we would be saying that the same value of x is less than 1 and at the same time greater than 2. Of course, as we noted for this type of situation in Example I, no such number exists.) Any inequality must be valid for all values satisfying it. However, we could say that the inequality is not satisfied for $1 \leq x \leq 2$, which means those values of x between or equal to 1 and 2.

We shall now present two examples of other kinds of problems in which the basic properties of inequalities are used.

Example K

If $0 < x < 1$, prove that $x^2 < x$.

From the given inequality we see that x is a positive number less than 1. Thus, if we multiply the members of the given inequality by x, we have $0 < x^2 < x$, which gives the desired result if we consider the middle and right members. Note the meaning of this inequality. The square of any positive number less than 1 is less than the number itself.

Example L

State, by means of an inequality, the conditions that x must satisfy if a point in the xy-plane lies between the lines $x = 1$ and $x = 5$.

The x-coordinate of any point in this part of the plane is greater than 1, but at the same time less than 5. Thus we have $1 < x < 5$. See Fig. 16–1.

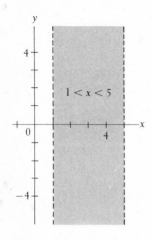

Figure 16–1

On occasion it is necessary to define a function in a different way for some values of the independent variable than for other values. Inequalities can then be used to denote the intervals over which the different definitions of the function are valid. The following example illustrates this use of inequalities, and includes a graphical representation.

Example M

The electrical intensity within a charged spherical conductor is zero. The intensity on the surface and outside of the sphere is equal to a constant divided by the square of the distance from the center of the sphere. State these relations by using inequalities, and make a graphical representation.

Let a = the radius of the sphere, r = the distance from the center of the sphere, and E = the electrical intensity.

The first statement may be written as $E = 0$ if $0 \leq r < a$, since this would be read as "the electric intensity is 0 if the distance from the center is less than the radius." Negative values of r are meaningless, which is the reason for saying that r is greater than or equal to zero. The second statement may be written as $E = k/r^2$ if $a \leq r$. Making a table of values for this equation, we have the following points:

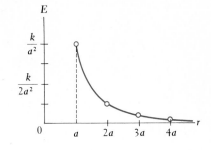

$$E \quad \left| \begin{array}{c|c|c|c|c} \dfrac{k}{a^2} & \dfrac{1}{4}\left(\dfrac{k}{a^2}\right) & \dfrac{1}{9}\left(\dfrac{k}{a^2}\right) & \dfrac{1}{16}\left(\dfrac{k}{a^2}\right) \\ \hline r & a & 2a & 3a & 4a \end{array} \right.$$

The graph of E versus r is shown in Fig. 16–2.

Figure 16–2

Exercises 16–1

In Exercises 1 through 8, for the inequality $4 < 9$, state the inequality resulting when the operations given are performed on both members.

1. Add 3.   2. Subtract 6.

3. Multiply by 5.   4. Multiply by -2.

5. Divide by -1.   6. Divide by 2.

7. Square both.   8. Take square roots.

In Exercises 9 through 16 give the inequalities which are equivalent to the following statements about a number x.

9. Greater than -2 10. Less than 7

11. Less than or equal to 4   12. Greater than or equal to -6

13. Greater than 1 and less than 7

14. Greater than or equal to -2 and less than 6

15. Less than -9, or greater than or equal to -4

16. Less than or equal to 8, or greater than or equal to 12

In Exercises 17 through 24 state the condition in terms of an inequality that x must satisfy to describe the location of the given point.

17. The point (x, y) lies to the right of the y-axis.

18. The point (x, y) lies to the right of the line $x = 1$.

19. The point (x, y) lies on or to the left of the y-axis.

20. The point (x, y) lies on or to the right of the line $x = -2$.

21. The point (x, y) lies outside of the region between the lines $x = -1$ and $x = 1$.

22. The point (x, y) lies in the region between the lines $x = -1$ and $x = 1$.

23. The point (x, y) lies on or to the right of the line $x = 2$ or to the left of the line $x = 6$.

24. The point (x, y) lies to the left of the line $x = -4$ or to the right of the line $x = 3$.

In Exercises 25 through 28 state the conditions in terms of inequalities that x, or y, or both, must satisfy to describe the location of the given point.

25. The point (x, y) lies in the first quadrant.
26. The point (x, y) lies in the region bounded by the lines $x = 1$, $x = 4$, $y = -3$, and $y = -1$.
27. The point (x, y) lies above the line $x = y$.
28. The point (x, y) lies within three units of the origin. [*Hint:* Use the Pythagorean theorem.]

In Exercises 29 through 32 prove the given inequalities.

29. If $x > 1$, prove that $x^2 > x$.   30. If $x > y > 0$, prove that $1/y > 1/x$.
31. If $0 < x < y$, prove that $\sqrt{xy} < y$.
32. If $x > x^2$, prove that $\sqrt{3x + 1} > x + 1$.

In Exercises 33 through 36 some applications of inequalities are shown.

33. A certain projectile is above 200 m from 3.5 s after it is launched until 15.3 s after it is launched. Express this statement in terms of inequalities in terms of the time t and height h.
34. An earth satellite put into orbit near the earth's surface will have an elliptic orbit if its velocity v is between 18,000 mi/h and 25,000 mi/h. State this as an inequality.
35. The electric potential V inside a charged spherical conductor equals a constant k divided by the radius a of the sphere. The potential on the surface of and outside the sphere equals the same constant k divided by the distance r from the center of the sphere. State these relations by the use of inequalities and make a graphical representation of V versus r.
36. A semiconductor diode, an electronic device, has the property that an electric current can flow through it in only one direction. Thus, if a diode is in a circuit with an alternating-current source, the current in the circuit exists only during the half-cycle when the direction is correct for the diode. If a source of current given by $i = 2 \sin 120\pi t$ milliamperes is connected in series with a diode, write the inequalities which are appropriate for the first four half-cycles and graph the resulting current versus the time. Assume that the diode allows a positive current to flow.

16—2 Graphical Solution of Inequalities

Equations can be solved by graphical and by algebraic means. This is also true of inequalities. In this section we shall take up graphical solutions, and in the following section we shall develop algebraic methods. The graphical method is shown in the following examples.

Example A

Graphically solve the inequality $3x - 2 > 4$.

This means that we want to locate all those values of x which make the left member of this inequality greater than the right member. By subtracting 4 from each member, we have the equivalent inequality $3x - 6 > 0$.

If we find the graph of $y = 3x - 6$, all those values of x for which y is positive would satisfy the inequality $3x - 6 > 0$. Thus, we graph the equation $y = 3x - 6$, and find those values of x for which y is positive. From the graph in Fig. 16–3, we see that values of $x > 2$ correspond to $y > 0$. Thus, the values of x which satisfy the inequality are given by $x > 2$.

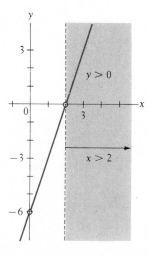

Figure 16–3

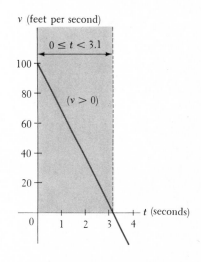

Figure 16–4

Example B

The velocity v (in feet per second) of a certain projectile in terms of the time t (in seconds) is given by $v = 100 - 32t$. For how long is the velocity positive? (This can be interpreted as "how long is the projectile moving upward?")

In terms of inequalities, we would like to know for what values of t is $v > 0$, or in other terms, solve the inequality $100 - 32t > 0$. Therefore, we graph the function $v = 100 - 32t$ as shown in Fig. 16–4. From the graph, we see that the values of t which correspond to $v > 0$ are $0 \leq t < 3.1$ s, which are therefore the required values of t. The last value is approximated from the graph.

Example C

Graphically solve the inequality $2x^2 < x + 3$.

Finding the equivalent inequality, with 0 for a right-hand member, we have $2x^2 - x - 3 < 0$. Thus, those values of x for which y is negative for the function $y = 2x^2 - x - 3$ will satisfy the inequality. So we graph the equation $y = 2x^2 - x - 3$, from the values given in the following table:

x	-3	-2	-1	0	1	2	3
y	18	7	0	-3	-2	3	12

From the graph in Fig. 16–5, we can see that the inequality is satisfied for the values $-1 < x < 1.5$.

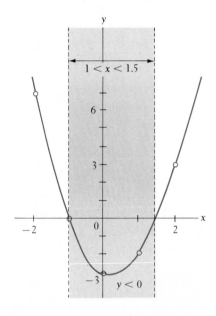

Figure 16–5

Summarizing the method for the graphical solution of an inequality, we see that we first write the inequality in an equivalent form with zero on the right. Next we set y equal to the left member and graph the resulting equation. Those values of x corresponding to the proper values of y (either above or below the x-axis) are those values which satisfy the inequality.

Example D

Graphically solve the inequality $x^3 > x^2 - 3$.

Finding the equivalent inequality with 0 on the right, we then have $x^3 - x^2 + 3 > 0$. By letting $y = x^3 - x^2 + 3$, we may solve the inequality by finding those values of x for which y is positive.

x	-2	-1	0	1	2
y	-9	1	3	3	7

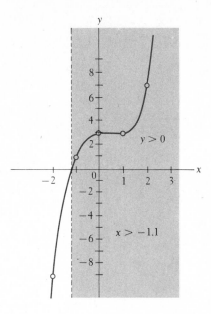

Figure 16–6

Approximating the root of the function to be -1.1, the values which satisfy the inequality are given by $x > -1.1$ (Fig. 16–6). If greater accuracy is required, the methods of Section 14–5 may be used to find the root. This example also points out the usefulness of graphical methods, since this inequality would prove to be beyond elementary methods if an algebraic solution were required.

Exercises 16–2

In Exercises 1 through 24 solve the given inequalities graphically.

1. $2x > 4$
2. $3x < -6$
3. $5x - 1 > 3x$
4. $2x - 3 < x$
5. $7x - 5 < 4x + 3$
6. $2x > 6x - 2$
7. $5x < 6x - 1$
8. $6 - x < x$
9. $x^2 > 2x$
10. $x^2 < x$
11. $2x + 3 < x^2$
12. $x - 1 < x^2$
13. $x^2 - 5x < -4$
14. $7x - 3 < -6x^2$
15. $3x^2 - 2x - 8 > 0$
16. $4x^2 - 2x > 5$
17. $x^3 > 1$
18. $x^3 > x + 4$
19. $x^4 < x^2 - 2x - 1$
20. $x^4 - 6x^3 + 7x^2 > 18 - 12x$
21. $2^x > 0$
22. $\log x > 1$
23. $\sin x < 0$ (limit the graph to the values $0 \le x \le 2\pi$)
24. $\cos 2x > 0$ (limit the graph as in Exercise 23)

In Exercises 25 through 28 answer the given questions by solving appropriate inequalities graphically.

25. A salesman receives \$300 monthly plus a 10% commission on sales. Therefore, his monthly income I in terms of his sales s is $I = 300 + 0.1s$. How much must his sales for the month be in order that his income is at least \$500?

26. The basic maintenance on a certain machine is $10 per day. It also costs $2 per hour while in operation. Therefore, the total cost C to operate this machine in terms of the number of hours t that it operates in a given day is $C = 10 + 2t$. How many hours can it operate without the cost exceeding $30 for the day?

27. The electrical resistance R of a certain material depends on the temperature, according to the relation $R = 40 + 0.1T + 0.01T^2$. For what values of the temperature $(T > 0°C)$ is the resistance over 42 Ω?

28. The height s of a certain object is given by $s = 140 + 60t - 16t^2$, where t is the time (in seconds). For what values of t is $s > 180$ ft?

16–3 Algebraic Solution of Inequalities

We learned that one important step in solving inequalities by graphical means was to set up an equivalent inequality with zero as the right member. This enabled us to find the values which satisfied the inequality by simply determining the *sign* of the function which was set up. A similar procedure is useful when we solve inequalities algebraically.

A linear function is either negative or positive for *all* values to the left of a particular value of x, and has the opposite sign for *all* values to the right of the same value of x. *This value of x which divides the positive and negative intervals is called the* **critical value**. *This critical value is found where the function is zero.* Thus, to determine where a linear function is positive and where it is negative, we need to find only this critical value, and then determine the *sign* of the function to the left and to the right of this value.

Example A
Solve the inequality $2x - 5 > 1$.

We first find the equivalent inequality with zero on the right. This is done by subtracting 1 from each member. Thus, we have $2x - 6 > 0$. We now set the left member *equal* to zero to find the critical value. Thus, the critical value is 3. We know that the function $2x - 6$ is of one sign for $x < 3$ and has the opposite sign for $x > 3$. Testing values in these intervals, we find that for $x > 3$, $2x - 6 > 0$. Thus, the values which satisfy the inequality are those for which $x > 3$.

It is possible, and appropriate, to solve linear inequalities such as this one just as we would solve an equation. That is, we could isolate x by adding 5 to each side, obtaining $2x > 6$, and then dividing both sides by 2, obtaining $x > 3$. However, the method used above is much more generally applicable. It can be used with higher-degree expressions and fractional expressions, as well as with linear expressions.

Example B
Solve the inequality $\frac{1}{2}x - 3 \le \frac{1}{3} - x$.

We note that this is an inequality combined with an equality. However, the solution proceeds essentially the same as with an inequality.

Multiplying by 6, we have $3x - 18 \leq 2 - 6x$. Subtracting $2 - 6x$ from each member, we have $9x - 20 \leq 0$. The critical value is $\frac{20}{9}$. If $x \leq \frac{20}{9}$, $9x - 20 \leq 0$. Thus, the values which satisfy the inequality are given by $x \leq \frac{20}{9}$.

The preceding analysis is especially useful for solving inequalities involving higher-degree functions or involving fractions with x in the denominator as well as in the numerator. When the equivalent inequality with zero on the right has been found, this inequality is then factored into linear factors (and those quadratic factors which lead to complex roots). Each linear factor can change sign only at its critical value, as all possible values of x are considered. Thus, the function on the left can change sign only where one of its factors changes sign. The function will have the same sign for all values of x less than the leftmost critical value. The sign of the function will also be the same within any given interval between two critical values. All values to the right of the rightmost critical value will also give the function the same sign. *Therefore, we must find all of the critical values and then determine the sign of the function to the left of the leftmost critical value, between the critical values, and to the right of the rightmost critical value. Those intervals in which we have the proper sign will satisfy the given inequality.*

Example C
Solve the inequality $x^2 - 3 > 2x$.

We first find the equivalent inequality with zero on the right. Thus, we have $x^2 - 2x - 3 > 0$. We then factor the left member, and have

$$(x - 3)(x + 1) > 0$$

We find the critical value for each of the factors, for these are the only values for which the function $x^2 - 2x - 3$ is zero. The left critical value is -1, and the right critical value is 3. All values of x to the left of -1 give the same sign for the function. All values of x between -1 and 3 give the function the same sign. All values of x to the right of 3 give the same sign to the function. Therefore, we must determine the sign for $x < -1$, $-1 < x < 3$, and $x > 3$. For the interval $x < -1$, we find that each of the factors is negative. However, the product of two negative numbers gives a positive number. Therefore, if $x < -1$, then $(x - 3)(x + 1) > 0$.

For the interval $-1 < x < 3$, we find that the left factor is negative, but the right factor is positive. The product of a negative and positive number gives a negative number. Thus, for the interval $-1 < x < 3$, $(x - 3)(x + 1) < 0$. For the interval $x > 3$, both factors are positive, making $(x - 3)(x + 1) > 0$. We tabulate the results.

$$\text{If} \qquad x < -1 \qquad (x - 3)(x + 1) > 0$$
$$\text{If} \quad -1 < x < 3 \qquad (x - 3)(x + 1) < 0$$
$$\text{If} \qquad x > 3 \qquad (x - 3)(x + 1) > 0$$

Thus, the inequality is satisfied for $x < -1$ or $x > 3$.

Example D

Solve the inequality $x^3 - 4x^2 + x + 6 < 0$.

By methods developed in Chapter 14, we factor the function of the left and obtain $(x + 1)(x - 2)(x - 3) < 0$. The critical values are -1, 2, 3. We wish to determine the sign of the left member for the intervals $x < -1$, $-1 < x < 2$, $2 < x < 3$, and $x > 3$. This is tabulated here, with the sign of the factors in each case indicated.

Interval	$(x + 1)(x - 2)(x - 3)$			Sign of $(x + 1)(x - 2)(x - 3)$
$x < -1$	$-$	$-$	$-$	$-$
$-1 < x < 2$	$+$	$-$	$-$	$+$
$2 < x < 3$	$+$	$+$	$-$	$-$
$x > 3$	$+$	$+$	$+$	$+$

Thus, the inequality is satisfied for $x < -1$ or $2 < x < 3$.

Example E

Solve the inequality $\dfrac{x - 3}{x + 4} \geq 0$.

The critical values are found from the factors, whether they are in the numerator or in the denominator. Thus, the critical values are -4 and 3. Considering now the *greater than* part of the $\geq$ sign, we set up the following table.

Interval	$\dfrac{x - 3}{x + 4}$	Sign of $\dfrac{x - 3}{x + 4}$
$x < -4$	$\dfrac{-}{-}$	$+$
$-4 < x < 3$	$\dfrac{-}{+}$	$-$
$x > 3$	$\dfrac{+}{+}$	$+$

Thus, the values which satisfy the greater than part of the problem are those for which $x < -4$ or for which $x > 3$. Now considering the equality part of the $\geq$ sign, we note that $x = 3$ is valid, for the fraction is zero. However, if $x = -4$, we have division by zero, and thus x may not equal -4. Therefore, the inequality is satisfied for $x < -4$ or $x \geq 3$.

Example F

Solve the inequality $x^3 - x^2 + x - 1 > 0$.

This leads to $(x^2 + 1)(x - 1) > 0$. There is only one linear factor with a critical value. The factor $x^2 + 1$ is never negative. The inequality is satisfied for $x > 1$.

Example G

Solve the inequality

$$\frac{(x - 2)^2(x + 3)}{4 - x} < 0$$

The critical values are -3, 2, and 4. Thus, we have the following table.

Interval	$\dfrac{(x - 2)^2(x + 3)}{4 - x}$	Sign of $\dfrac{(x - 2)^2(x + 3)}{4 - x}$
$x < -3$	$\dfrac{+ \quad -}{+}$	$-$
$-3 < x < 2$	$\dfrac{+ \quad +}{+}$	$+$
$2 < x < 4$	$\dfrac{+ \quad +}{+}$	$+$
$x > 4$	$\dfrac{+ \quad +}{-}$	$-$

The inequality is satisfied for $x < -3$ or $x > 4$.

Exercises 16–3

In Exercises 1 through 32 solve the given inequalities.

1. $x + 3 > 0$
2. $2x - 7 < 0$
3. $6x - 4 < 8 - x$

4. $2x - 6 < x + 4$
5. $3x - 7 \le x + 1$
6. $7 - x \ge x - 1$

7. $\dfrac{1}{3}(x - 6) > 4 - x$
8. $\dfrac{x}{4} - 7 > \dfrac{x}{2} + 3$

9. $x^2 - 1 < 0$
10. $x^2 - 4x - 5 > 0$

11. $3x^2 + 5x \ge 2$
12. $2x^2 - 12 \le -5x$

13. $6x^2 + 1 < 5x$
14. $9x^2 + 6x > -1$

15. $x^2 + 4 > 0$
16. $x^4 + 2 < 1$

17. $x^3 + x^2 - 2x > 0$
18. $x^3 - 2x^2 + x > 0$

19. $x^3 + 2x^2 - x - 2 > 0$
20. $x^4 - 2x^3 - 7x^2 + 8x + 12 < 0$

21. $\dfrac{x - 8}{3 - x} < 0$
22. $\dfrac{x + 5}{x - 1} > 0$
23. $\dfrac{2x - 3}{x + 6} \le 0$

24. $\dfrac{3x + 1}{x - 3} \ge 0$
25. $\dfrac{2}{x^2 - x - 2} < 0$
26. $\dfrac{-5}{2x^2 + 3x - 2} < 0$

27. $\dfrac{x^2 - 6x - 7}{x + 5} > 0$
28. $\dfrac{4 - x}{3 + 2x - x^2} > 0$

29. $\dfrac{6 - x}{3 - x - 4x^2} > 0$
30. $\dfrac{(x - 2)^2(5 - x)}{(4 - x)^3} < 0$

31. $\dfrac{x^4(9 - x)(x - 5)(2 - x)}{(4 - x)^5} > 0$
32. $\dfrac{x^3(1 - x)(x - 2)(3 - x)(4 - x)}{(5 - x)^2(x - 6)^3} < 0$

In Exercises 33 through 36 determine the values of x for which the given radicals represent real numbers.

33. $\sqrt{(x-1)(x+2)}$

34. $\sqrt{x^2 - 3x}$

35. $\sqrt{-x - x^2}$

36. $\sqrt{\dfrac{x^3 + 6x^2 + 8x}{3 - x}}$

In Exercises 37 through 40 answer the given questions by solving the appropriate inequalities.

37. The velocity (in feet per second) of a certain object in terms of the time t (in seconds) is given by $v = 120 - 32t$. For what values of t is the object ascending $(v > 0)$?

38. The relationship between Fahrenheit degrees and Celsius degrees is $F = \frac{9}{5}C + 32$. For what values of C is $F \geq 98.6$ (normal body temperature)?

39. Determine the values of T for which the resistor of Exercise 27 of Section 16–2 has a resistance between 41 and 42 Ω.

40. The deflection y of a certain beam 9 ft long is given by the equation $y = k(x^3 - 243x + 1458)$. For which values of x (the distance from one end of the beam) is the quantity y/k greater than 216 units?

16–4 Inequalities Involving Absolute Values

If we wish to write the inequality $|x| > 1$ without absolute-value signs, we must note that we are considering values of x which are *numerically* larger than 1. Thus we may write this inequality in the equivalent form $x < -1$ or $x > 1$. We now note that the original inequality, with an absolute value sign, can be written in terms of two equivalent inequalities, neither involving absolute values. If we are asked to write the inequality $|x| < 1$ without the absolute-value signs, we write $-1 < x < 1$ since we are considering values of x which are numerically less than 1.

Following reasoning similar to the above, whenever absolute values are involved in inequalities, the following two relations allow us to write equivalent inequalities without absolute values.

$$\text{If } |f(x)| > n, \text{ then } f(x) < -n \text{ or } f(x) > n. \qquad (16\text{--}1)$$
$$\text{If } |f(x)| < n, \text{ then } -n < f(x) < n. \qquad (16\text{--}2)$$

The use of these relations is indicated in the following examples.

Example A

Solve the inequality $|x - 3| < 2$.

Inspection of this inequality shows that we wish to find the values of x which are within 2 units of $x = 3$. Of course, such values are given by $1 < x < 5$. Let us now see how Eq. (16–2) gives us this result.

By using Eq. (16–2), we have

$$-2 < x - 3 < 2$$

By adding 3 to all three members of this inequality, we have

$$1 < x < 5$$

which is the proper interval.

Example B

Solve the inequality $|2x - 1| > 5$.

By using Eq. (16–1), we have

$$2x - 1 < -5 \quad \text{or} \quad 2x - 1 > 5$$

Completing the solution, we have

$$2x < -4 \quad \text{or} \quad 2x > 6$$
$$x < -2 \quad \text{or} \quad x > 3$$

This means that the given inequality is satisfied for $x < -2$ or for $x > 3$.

Example C

Solve the inequality $|3x + 2| \leq 4$.

Although there is a sign of equality involved, we may solve this inequality in the same way as indicated in Eq. (16–2). Since $f(x)$ is linear (and therefore no division by a factor containing x is involved), we may include the equals sign throughout the solution. Therefore, we have

$$-4 \leq 3x + 2 \leq 4$$
$$-6 \leq 3x \leq 2$$
$$-2 \leq x \leq \frac{2}{3}$$

We note that for $x = -2$ and for $x = \frac{2}{3}$, $|3x + 2| = 4$. Thus, the last inequality shown gives the values of x which satisfy the given inequality.

Example D

Solve the inequality $|x^2 + x - 4| > 2$.

By Eq. (16–1), we have $x^2 + x - 4 < -2$ or $x^2 + x - 4 > 2$, which means we want values of x which satisfy *either* of these inequalities. The first inequality becomes

$$x^2 + x - 2 < 0$$
$$(x + 2)(x - 1) < 0$$

which is satisfied for $-2 < x < 1$. The second inequality becomes

$$x^2 + x - 6 > 0$$
$$(x + 3)(x - 2) > 0$$

which is satisfied for $x < -3$ or for $x > 2$. Thus, the original inequality is satisfied for $x < -3$, or for $-2 < x < 1$, or for $x > 2$.

Example E
Solve the inequality $|x^2 + x - 4| < 2$.

By Eq. (16–2), we have $-2 < x^2 + x - 4 < 2$, which we can write as $x^2 + x - 4 > -2$ and $x^2 + x - 4 < 2$, as long as we remember that we want values of x which satisfy *both* of these at the same time. The first inequality can be written as

$$x^2 + x - 4 > -2$$
$$x^2 + x - 2 > 0$$
$$(x + 2)(x - 1) > 0$$

which is satisfied for $x < -2$ or for $x > 1$. The second inequality is

$$x^2 + x - 4 < 2$$
$$x^2 + x - 6 < 0$$

or

$$(x + 3)(x - 2) < 0$$

which is satisfied for $-3 < x < 2$. The values of x which satisfy both of the inequalities are those between -3 and -2 and those between 1 and 2. Thus, the original inequality is satisfied for $-3 < x < -2$ or for $1 < x < 2$.

Exercises 16—4

omit quad

In Exercises 1 through 16 solve the given inequalities.

1. $|x - 4| < 1$
2. $|x + 1| < 3$
3. $|3x - 5| > 2$
4. $|2x - 1| > 1$
5. $|6x - 5| \leq 4$
6. $|3 - x| \leq 2$
7. $|4x + 3| > 3$
8. $|3x + 1| > 2$
9. $|x + 4| < 6$
10. $|5x - 10| < 2$
11. $|2 - 3x| \geq 5$
12. $|3 - 2x| \geq 6$
13. $|x^2 + 3x - 1| < 3$
14. $|x^2 - 5x - 1| < 5$
15. $|x^2 + 3x - 1| > 3$
16. $|x^2 - 5x - 1| > 5$

In Exercises 17 through 20 use inequalities involving absolute values to solve the given problems.

17. The deflection y at a horizontal distance x from the left end of a beam is less than $\frac{1}{4}$ ft within 2 ft of a point 6 ft from the left end. State this with the use of an inequality involving absolute values.

18. A given projectile is at an altitude h between 40 and 60 m for the time between $t = 3$ s and $t = 5$ s. Write this using two inequalities involving absolute values.

19. An object is oscillating at the end of a spring which is suspended from a support. The distance x of the object from the support is given by the equation $|x - 8| < 3$. By solving this inequality, determine the distances (in inches) from the support of the extreme positions of the object.

20. The production p (in barrels) of an oil refinery for the coming month is estimated at $|p - 2,000,000| < 200,000$. By solving this inequality, determine the production which is anticipated.

16–5 Graphical Solution of Inequalities with Two Variables

To this point we have considered inequalities with one variable and certain methods of solving them. We may also graphically solve inequalities involving two variables, such as x and y. In this section we consider the solution of such inequalities, as well as one important type of application.

Let us consider the function $y = f(x)$. We know that the coordinates of points on the graph satisfy the equation $y = f(x)$. However, for points above the graph of the function, we have $y > f(x)$, and for points below the graph of the function we have $y < f(x)$. Consider the following example.

Example A

Consider the linear function $y = 2x - 1$. This equation is satisfied for points on the line. For example, the point $(2,3)$ is on the line and we have $3 = 2(2) - 1 = 3$. Therefore, for points on the line we have $y = 2x - 1$, or $y - 2x + 1 = 0$. The point $(2, 4)$ is above the line, since we have $4 > 2(2) - 1$, or $4 > 3$. Therefore, for points above the line we have $y > 2x - 1$, or $y - 2x + 1 > 0$. In the same way, for points below the line, $y < 2x - 1$ or $y - 2x + 1 < 0$. We note this is true for the point $(2,1)$, since $1 < 2(2) - 1$, or $1 < 3$. The line for which $y = 2x + 1$, and the regions for which $y > 2x - 1$, and for which $y < 2x - 1$ are shown in Fig. 16 – 7.

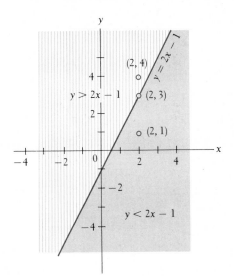

Figure 16–7

The illustration of Example A leads us to the graphical method of indicating the points which satisfy an inequality with two variables. First we solve the inequality for y and then determine the graph of the function $y = f(x)$. If we wish to solve the inequality $y > f(x)$, we indicate the appropriate points by shading in the region above the curve. For the inequality $y < f(x)$, we indicate the appropriate points by shading in the region below the curve. We note that the complete solution to the inequality consists of all points in an entire region of the plane.

Example B

Draw a sketch of the graph of the inequality $y < x + 3$.

First we graph the function $y = x + 3$, as shown in Fig. 16–8. Since we wish to find the points which satisfy the inequality $y < x + 3$, we show these points by shading in the region below the line. The line itself is shown as a dashed line since points on it do not satisfy the inequality.

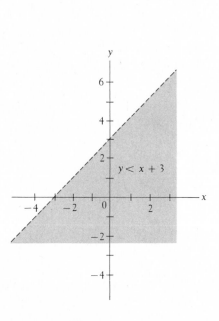

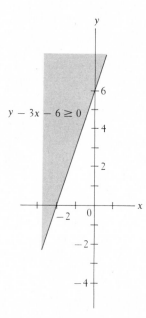

Figure 16–8   Figure 16–9

Example C

Draw a sketch of the graph of the inequality $y - 3x - 6 \geq 0$.

First we state the inequality as $y \geq 3x + 6$. Next the graph of the line $y = 3x + 6$ is drawn, as in Fig. 16–9. The points which satisfy the inequality consist of all points above the line and those points which are on the line. Therefore, we show the line as a solid line.

Example D

Draw a sketch of the graph of the inequality $y > x^2 - 4$.

Although the graph of $y = x^2 - 4$ is not a straight line, the method of solution is the same. We graph the function $y = x^2 - 4$ as shown in Fig. 16–10. We then shade in the region above the curve to indicate the points which satisfy the inequality.

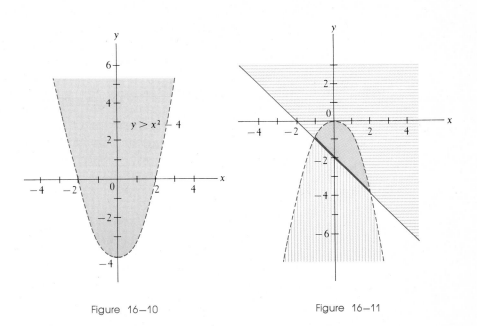

Figure 16–10   Figure 16–11

Example E

Draw a sketch of the region which is defined by the system of inequalities $y \geq -x - 2$ and $y + x^2 < 0$.

In this case we sketch the graph of both inequalities, and then determine the region common to both graphs. First we draw the graph of $y = -x - 2$ (the straight line), and shade in the region above the line. See Fig. 16–11. Next we have $y < -x^2$ and draw the graph of $y = -x^2$, shading the region below it. The sketch of the region which is defined by this system of inequalities is the darkly shaded region below the parabola which is above and on the line.

An important area in which graphs of inequalities with two or more variables are used is the branch of mathematics known as **linear programming** (in this context, "programming" has no relation to computer programming). This subject is widely applied in industry, business, economics, and technology. The analysis of many social problems can also be made by the use of linear programming.

omit

Linear programming is used to analyze problems such as those related to maximizing profit, minimizing costs, or the use of materials, with certain constraints of production. The following serves as an example of the use of linear programming.

Example F

A company makes two types of stereo speaker systems, their good quality system and their highest quality system. The production of the systems requires assembly of the speaker system itself and the production of the cabinets in which they are installed. The good quality system requires 3 worker-hours for speaker assembly and 2 worker-hours for cabinet production for each complete system. The highest quality system requires 4 worker-hours for speaker assembly and 6 worker-hours for cabinet production for each complete system. Available skilled labor allows for a maximum of 480 worker-hours per week for speaker assembly and a maximum of 540 worker-hours per week for cabinet production. It is anticipated that all systems will be sold and that the profit will be $10 for each good quality system and $25 for each highest quality system. How many of each should be produced to provide the greatest profit?

First, let x = the number of good quality systems and y = the number of highest quality systems made in one week. Thus, the profit p is given by

$$p = 10x + 25y$$

We know that negative numbers are not valid for either x or y, and therefore we have $x \geq 0$ and $y \geq 0$. Also, the number of available worker-hours per week for each part of the production restricts the number of systems which can be made. Both speaker assembly and cabinet production are required for all systems. The number of worker-hours needed to produce the x good quality systems is $3x$ in the speaker assembly shop. Also, $4y$ worker-hours are required in the speaker assembly shop for the highest quality systems. Thus,

$$3x + 4y \leq 480$$

since no more than 480 worker-hours are available in the speaker assembly shop. In the cabinet shop, we have

$$2x + 6y \leq 540$$

since no more than 540 worker-hours are available in the cabinet shop.

Therefore, we wish to maximize the profit p under the **constraints**

$$x \geq 0, \quad y \geq 0$$
$$3x + 4y \leq 480$$
$$2x + 6y \leq 540$$

In order to do this we sketch the region of points which satisfy this system of inequalities. From the previous examples, we see that the appropriate region is in the first quadrant (since $x \geq 0$ and $y \geq 0$) and under both lines. See Fig. 16–12.

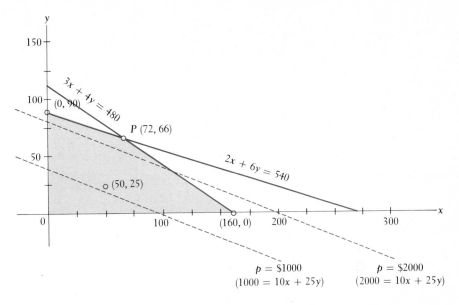

Figure 16–12

Any point in the shaded region which is defined by the preceding system of inequalities is known as a **feasible point**. In this case it means that it is possible to produce the number of systems of each type according to the coordinates of the point. For example, the point (50,25) is in the region, which means that it is possible to produce 50 good quality systems and 25 highest quality systems under the given constraints of available skilled labor. However, we wish to find the point which indicates the number of each kind of system which produces the greatest profit.

If we assume values for the profit, the resulting equations are straight lines. Thus, by finding the greatest value of p for which the line passes through a feasible point, we may solve the given problem. If $p = \$1000$, or if $p = \$2000$, we have the lines shown. Both are possible with various combinations of speaker systems being produced. However, we note the line for $p = \$2000$ passes through feasible points farther from the origin. It is also clear, since these lines are parallel, that the greatest profit attainable is given by the line passing through P, where $3x + 4y = 480$ and $2x + 6y = 540$ intersect. The coordinates of P are (72,66). Thus, the production should be 72 good quality systems and 66 highest quality systems to produce a weekly profit of $p = 10(72) + 25(66) = \$2370$.

For this type of problem, the solution will be given by one of the vertices of the region. However, it could be any one of them, which means it is possible that only one type of product should be produced. (See Exercise 27.) Thus, we can solve the problem by finding the appropriate region and then testing the coordinates of the vertex points. Here the vertex points are (160,0) which indicates a profit of \$1600, (0,90) which indicates a profit of \$2250, and (72,66) which indicates a profit of \$2370.

Exercises 16—5

In Exercises 1 through 20 draw a sketch of the graph of the given inequality.

1. $y > x - 1$

2. $y < 3x - 2$

3. $y \geq 2x + 5$

4. $y \leq 3 - x$

5. $2x + y < 5$

6. $4x - y > 1$

7. $3x + 2y + 6 > 0$

8. $x + 4y - 8 < 0$

9. $y < x^2$

10. $y > -2x^2$

11. $y \geq 1 - x^2$

12. $y \leq 2x^2 - 3$

13. $x^2 + 2x + y < 0$

14. $2x^2 - 4x - y > 0$

15. $4x^2 - x - 2y > 0$

16. $3x^2 + 6x + 2y < 0$

17. $y \leq x^3$

18. $y \geq 3x - x^3$

19. $y > x^4 - 8$

20. $y < 32x - x^4$

In Exercises 21 through 24 draw a sketch of the graph of the region in which the points satisfy the given system of inequalities.

21. $y > x$
$y > 1 - x$

22. $y \leq 2x$
$y \geq x - 1$

23. $y \leq 2x^2$
$y > x - 2$

24. $y > x^2$
$y < x + 4$

In Exercises 25 through 28 solve the given linear programming problems.

Constraints

$x \geq 0$

$y \geq 0$

$3x + 6y \leq 480$

$6x + 4y \leq 480$

$P = 8y + 10x$

25. A manufacturer makes two types of calculators, a business model and a scientific model. Each model is assembled in two sets of operations, where each operation is in production 8 h each day. The average time required for a business model in the first operation is 3 min, and 6 min is required in the second operation. The scientific model averages 6 min in the first operation and 4 min in the second operation. All calculators can be sold; the profit for a business model is $8, and the profit for a scientific model is $10. How many of each model should be made each day in order to maximize profit?

26. A company makes brands A and B of breakfast cereal, both of which are enriched with vitamins P and Q. The necessary information about these cereals is given in the following table.

	Cereal A	Cereal B	Min. daily requirement
Vitamin P (units/oz)	1	2	10
Vitamin Q (units/oz)	5	3	30
Cost per ounce	2¢	3¢	

Find the number of ounces of each cereal which together satisfies the minimum daily requirement of vitamins P and Q at the lowest cost. (Note: We wish to *minimize* cost; be careful in determining the feasible region.)

27. Using the information of Example F, with the single exception that the profit on each good quality system is $20, how many of each system should be made?

28. A company in competition with the company of Exercise 25 uses identical production methods. However, the profit it receives on its business calculators is only $4, although its profit on its scientific calculators is $10. How many of each type should this company make in order to maximize profit?

16—6 Exercises for Chapter 16

In Exercises 1 through 20 solve the given inequalities algebraically.

1. $2x - 12 > 0$

2. $5 - 3x < 0$

3. $3x + 5 \leq 0$

4. $\frac{1}{4}x - 2 \geq 3x$

5. $x^2 + 2x > 63$

6. $6x^2 - x > 35$

7. $x^3 + 4x^2 - x > 4$

8. $2x^3 + 4 \leq x^2 + 8x$

9. $\frac{x - 8}{2x + 1} \leq 0$

10. $\frac{3x + 2}{x - 3} > 0$

11. $\frac{(2x - 1)(3 - x)}{x + 4} > 0$

12. $\frac{(3 - x)^2}{2x + 7} \leq 0$

13. $x^4 + x^2 \leq 0$

14. $3x^3 + 7x^2 - 20x < 0$

15. $\frac{1}{x} < 2$

16. $\frac{1}{x - 2} < \frac{1}{4}$

17. $|x - 2| > 3$

18. $|2x - 1| > 5$

19. $|3x + 2| \leq 4$

20. $|4 - 3x| \leq 1$

In Exercises 21 through 28 solve the given inequalities graphically.

21. $6x - 3 < 0$

22. $5 - 8x > 0$

23. $3x - 2 > x$

24. $4 - 5x < 2$

25. $x^2 + 2x + 4 > 0$

26. $3x^2 + 5 > 16x$

27. $x^3 + x + 1 < 0$

28. $\frac{1}{x} > 2$

In Exercises 29 through 32 determine the values of x for which the given radicals represent real numbers.

29. $\sqrt{3 - x}$

30. $\sqrt{x + 5}$

31. $\sqrt{x^2 + 4x}$

32. $\sqrt{\frac{x - 1}{x + 2}}$

In Exercises 33 through 40 draw a sketch of the graph of the given inequality.

33. $y > 4 - x$

34. $y < \frac{1}{2}x + 2$

35. $2y - 3x - 4 \leq 0$

36. $3y - x + 6 \geq 0$

37. $y > x^2 + 1$

38. $y < 4x - x^2$

39. $y - x^3 + 1 < 0$

40. $2y + 2x^3 + 6x - 3 > 0$

In Exercises 41 through 44 prove the given inequalities.

41. If $x > 0$ and $y < 0$, prove that $\frac{1}{x} > \frac{1}{y}$.

42. If $x < -1$, prove that $\frac{1 - x}{x^2 + 1} > 0$.

43. If $y > 0$ and $x > y + 1$, prove that $\frac{x}{y} > \frac{y + 1}{x - 1}$.

44. If $x \neq y$, prove that $x^2 + y^2 > 2xy$.

In Exercises 45 through 52 solve the given problems using inequalities.

45. A piece of wire 100 ft long is to have a piece at least 20 ft cut from it. What lengths may the remaining piece be?

46. After conducting tests, it was determined that the stopping distance x (in feet) of a car traveling 60 mi/h was $|x - 290| \leq 35$. Express this inequality without absolute values, and determine the interval of stopping distances which were found in the tests.

47. One leg of a right triangle is 14 in. longer than the other leg. If the hypotenuse is to be greater than 34 in., what values of the other side are permissible?

48. The length of a rectangular lot is 20 m more than its width. If the area is to be at least 4800 m², what values may the width be?

49. Two resistors have a combined resistance of 8 Ω when connected in series. What are the permissible values if they are to have a combined resistance of at least $\frac{3}{2}$ Ω when connected in parallel? (See Exercise 41 of Section 6–1.)

50. City A is 300 mi from city B. One car starts from A for B one hour before a second car. The first car averages 45 mi/h and the second car averages 60 mi/h for the trip. For what times after the first car starts is the second car ahead of the first car?

51. For 1 g of ice to be melted, then heated to boiling water, and finally vaporized into steam, the relation between the temperature T (in degrees Celsius) and the number of joules absorbed Q is given by $T = 0$ if $Q < 335$, $T = Q - 335$ if $335 < Q < 750$, and $T = 100$ if $750 < Q < 3010$. Plot the graph of T (along the y-axis) versus Q (along the x-axis) if 3010 J of heat are absorbed by 1 g of ice originally at 0°C.

52. A company produces two types of cameras, the regular model and the deluxe model. For each regular model produced there is a profit of $8, and for each deluxe model the profit is $15. The same amount of materials is used to make each model, but the supply is sufficient only for 450 cameras per day. The deluxe model requires twice the time to produce as the regular model. If only regular models were made, there would be time enough to produce 600 per day. Assuming all models will be sold, how many of each model should be produced if the profit is to be a maximum?

17

Variation

17-1 Ratio and Proportion

In mathematics and its applications, we often come across the terms "ratio" and "proportion." In fact, in Chapter 3 when we first introduced the trigonometric functions of angles, we used ratios which had a specific meaning. In general, *a* **ratio** *of a number a to a number b* $(b \neq 0)$ *is the quotient a/b*. Thus, a fraction is a ratio.

Any measurement made is the ratio of the measured magnitude to an accepted unit of measurement. For example, when we say that an object is five feet long, we are saying that the length of that object is five times as long as an accepted unit of length, the foot. Other examples of ratios are density (weight/volume), relative density (density of object/density of water), and pressure (force/area). As these examples illustrate, ratios may compare quantities of the same kind, or they may express a division of magnitudes of different quantities.

Example A

The approximate airline distance from New York to San Francisco is 2500 mi, and the approximate airline distance from New York to Minneapolis is 1000 mi. The ratio of these distances is

$$\frac{2500 \text{ mi}}{1000 \text{ mi}} = \frac{5}{2}$$

Since the units in both are miles, the resulting ratio is a dimensionless number.

If a jet travels from New York to San Francisco in 4 h, its average speed is

$$\frac{2500 \text{ mi}}{4 \text{ h}} = 625 \text{ mi/h}$$

In this case we must attach the proper units to the resulting ratio.

As we noted in Example A, we must be careful to attach the proper units to the resulting ratio. Generally, the ratio of measurements of the same kind should be expressed as a dimensionless number. Consider the following example.

Example B

The length of a certain room is 24 ft, and the width of the room is 18 ft. Therefore, the ratio of the length to the width is $\frac{24}{18}$ or $\frac{4}{3}$.

If the width of the room were expressed as 6 yd, we should not express the ratio as 24 ft/6 yd = 4 ft/1 yd. It is much better and more meaningful to first change the units of one of the measurements. Changing the length from 6 yd to 18 ft, we express the ratio as $\frac{4}{3}$, as we saw above. From this ratio we can easily see that the length is $\frac{4}{3}$ as long as the width.

A statement of equality between two ratios is called a **proportion.** One way of denoting a proportion is $a:b = c:d$, which is read "*a* is to *b* as *c* is to *d*." (Another notation used for a proportion is $a:b::c:d$.) Of course, by the definition, $a/b = c/d$, which means that a proportion is an equation. Any operation applicable to an equation is also applicable to a proportion.

Example C

On a certain map 1 in. represents 10 mi. Thus on this map we have a ratio of 1 in./10 mi. To find the distance represented by 3.5 in., we can set up the proportion

$$\frac{3.5 \text{ in.}}{x \text{ mi}} = \frac{1 \text{ in.}}{10 \text{ mi}}$$

From this proportion we find the value of $x = 35$ mi.

Example D

The magnitude of an electric field E is defined as the ratio between the force F on a charge q and the magnitude of q. This can be written as $E = F/q$. If we know the force exerted on a particular charge at some point in the field, we can determine the force which would be exerted on another charge placed at the same point. For example, if we know that a force of 10 nN is exerted on a charge of 4 nC, we can then determine the force which would be exerted on a charge of 6 nC by the proportion

$$\frac{10 \times 10^{-9}}{4 \times 10^{-9}} = \frac{F}{6 \times 10^{-9}}$$

or

$$F = 1.5 \times 10^{-8}\,N = 15\ nN$$

Example E

A certain alloy is 5 parts tin and 3 parts lead. How many grams of each are there in 40 g of the alloy?

First, we let x = the number of grams of tin in the given amount of the alloy. Next we note that there are 8 total parts of alloy, of which 5 are tin. Thus, 5 is to 8 as x is to 40. This gives the equation

$$\frac{5}{8} = \frac{x}{40}$$

Multiplying each side by 40, we then find that $x = 25$ g. Therefore, there are 25 g of tin and 15 g of lead, and the ratio of 25 to 15 is the same as 5 to 3.

Exercises 17−1

In Exercises 1 through 8 express the ratios in the simplest form.

1. 18 V to 3 V   2. 27 ft to 18 ft   3. 48 in. to 3 ft
4. 120 s to 4 min   5. 20 qt to 25 gal   6. 4 lb to 8 oz
7. 14 kg to 350 g   8. 200 mm to 5 cm

In Exercises 9 through 16 find the required ratios.

9. A virus 3.0×10^{-5} cm long appears to be 1.2 cm long through a microscope. What is the *magnification* (ratio of image length to object length) of the microscope?

10. The efficiency of an engine is defined as the ratio of output to input. Find the efficiency, in percent, of an engine for which the input is 6000 W and output is 4500 W.

11. The ratio of the density of an object to the density of water is known as the *relative density* of the object. If the density of gold is 1200 lb/ft³ and the density of water is 62.4 lb/ft³, what is the relative density of gold?

12. The *atomic mass* of an atom is the ratio of the mass of the atom to the mass of an atom of carbon, which is 12 u. The ratio of the atomic mass of an atom of oxygen to that of an atom of carbon is $\frac{4}{3}$. What is the atomic mass of an atom of oxygen? (The symbol u represents the *unified atomic mass unit.*)

13. A certain body of water exerts a force of 18,000 lb on an area of 20 in.². Find the pressure (force per unit area) on this area.

14. The electric current in a given circuit is the ratio of the voltage to the resistance. What is the current (1 V/1 Ω = 1 A) for a circuit where the voltage is 24.0 V and the resistance is 10.0 Ω?

15. A student finds that 8 divisions can be done on a calculator in 20.0 s. What is the student's rate, in calculations per minute, in doing such division problems?

16. The *heat of vaporization* of a substance is the amount of heat required to change one unit amount of the substance from liquid to vapor. Experimentation shows that 7910 J are needed to change 3.50 g of water to steam. What is the heat of vaporization of water?

In Exercises 17 through 20 find the required quantities from the given proportions.

17. In an electric instrument called a "Wheatsone bridge" electric resistances are related by

$$\frac{R_1}{R_2} = \frac{R_3}{R_4}$$

Find R_2 if R_1 = 6.00 Ω, R_3 = 62.5 Ω, and R_4 = 15.0 Ω.

18. For two connected gears, the relation

$$\frac{d_1}{d_2} \cdot \frac{N_1}{N_2} \qquad \frac{d_1}{d_2} = \frac{N_1}{N_2}$$

holds, where d is the diameter of the gear and N is the number of teeth. Find N_1 if d_1 = 2.60 in., d_2 = 11.7 in., and N_2 = 45.

19. For two pulleys connected by a belt, the relation

$$\frac{d_1}{d_2} \cdot \frac{n_1}{n_2} \qquad \frac{d_1}{d_2} = \frac{n_2}{n_1}$$

holds, where d is the diameter of the pulley and n is the number of revolutions per unit time it makes. Find n_2 if d_1 = 4.60 in., d_2 = 8.30 in., and n_1 = 18.0 r/min.

20. In a transformer, an electric current in one coil of wire induces a current in a second coil. For a transformer

$$\frac{i_1}{i_2} = \frac{t_2}{t_1}$$

where i is the current and t is the number of windings in each coil, find i_2 for i_1 = 0.0350 A, t_1 = 560, and t_2 = 1500.

In Exercises 21 through 32 answer the given questions by setting up and solving the appropriate proportions.

21. If 1 lb = 454 g, what weight in grams is 20.0 lb?

22. If 9 ft² = 1 yd², what area in square yards is 45.0 ft²?

23. How many meters per second are equivalent to 45.0 km/h?

24. How many gallons per hour are equivalent to 500 quarts per minute?

25. The length of a picture is 48.0 in., and its width is 36.0 in. In a reproduction of the picture the length is 36.0 in. What is the width of the reproduction?

26. A car will travel 93.0 mi on 6.00 gal of gasoline. How far will it travel on 10.0 gal of gasoline?

27. In physics, power is defined as the time rate at which work is done. If a given motor does 6000 ft-lb of work in 20.0 s, what work will it do in 4.00 min?

28. It is known that 98.0 kg of sulfuric acid are required to neutralize 80.0 kg of sodium hydroxide. Given 37.0 kg of sulfuric acid, determine the amount of sodium hydroxide which will be neutralized.

29. A physician has 220 mg of medication, which is sufficient for just 2 dosages. The amount given to a patient should be proportional to the patient's weight. For persons of weights of 60 kg and 72 kg, what should be the dosages?

30. A person pays $4500 in state and federal income taxes. Find the amount paid for each if the state taxes were 30% of the taxes paid.

31. A board 10.0 ft long is cut into two pieces, the lengths of which are in the ratio 2:3. Find the lengths of the pieces.

32. A total of 322 bolts are in two containers. The ratio of the number in one container to the number in the other container is 5:9. How many are in each container?

17—2 Variation

Scientific laws are often stated in terms of ratios and proportions. For example, Charles' law can be stated as "for a perfect gas under constant pressure, the ratio of any two volumes this gas may occupy equals the ratio of the absolute temperatures." Symbolically this could be stated as $V_1/V_2 = T_1/T_2$. Thus, if the ratio of the volumes and one of the values of the temperature are known, we can easily find the other temperature.

By multiplying both sides of the proportion of Charles' law by V_2/T_1, we can change the form of the proportion to $V_1/T_1 = V_2/T_2$. This statement says that the ratio of the volume to the temperature (for constant pressure) is constant. Thus, if any pair of values of volume and temperature is known, this ratio of V_1/T_1 can be calculated. This ratio of V_1/T_1 can be called a constant k, which means that Charles' law can be written as $V/T = k$. We now have the statement that the ratio of the volume to temperature is always constant; or, as it is normally stated, "the volume is proportional to the temperature." Therefore, we write $V = kT$, the clearest and most informative statement of Charles' law.

Thus, for any two quantities always in the same proportion, we say that one *is proportional to* (or *varies directly as*) the second, and in general this is written as $y = kx$, where k is the **constant of proportionality**. This type of relationship is known as **direct variation**.

Example A

The circumference of a circle is proportional to (varies directly as) the radius. We write this symbolically as $C = kr$, where (in this case) $k = 2\pi$.

Example B

The fact that the electric resistance of a wire varies directly as (is proportional to) its length is stated as $R = kL$.

It is very common that, when two quantities are related, the product of the two quantities remains constant. In such a case $yx = k$, or $y = k/x$. This is stated as "*y varies inversely as x*," or "*y is inversely proportional to x*." This type of relationship is known as **inverse variation**.

Example C

Boyle's law states that "at a given temperature, the pressure of a gas varies inversely as the volume." This we write symbolically as $P = k/V$.

For many relationships, one quantity varies as a specified power of another. The terms "varies directly" and "varies inversely" are used in the following examples with the specified power of the relation.

Example D

The statement that the volume of a sphere varies directly as the cube of its radius is written as $V = kr^3$. In this case we know that $k = \frac{4}{3}\pi$.

Example E

The fact that the gravitational force of attraction between two bodies varies inversely as the square of the distance between them is written as $F = k/d^2$.

Finally, one quantity may vary as the product of two or more other quantities. Such variation is termed **joint variation**. Also, some quantities are related such that a combination of variations is involved.

Example F

The cost of sheet metal varies jointly as the area of the sheet and the cost per unit area of the metal. This we write as $C = kAc$.

Example G

Newton's law of gravitation states that "the force of gravitation between two objects varies jointly as the product of the masses of the objects, and inversely as the square of the distance between their centers." We write this symbolically as $F = km_1m_2/d^2$.

Once we know how to express the given statement in terms of the variables and the constant of proportionality, we may compute the value of k if one set of values of the variables is known. This value of k can then be used to compute values of one variable, when given the others.

Example H

If y varies inversely as x, and $x = 15$ when $y = 4$, find the value of y when $x = 12$.

First we write $y = k/x$ to denote that y varies inversely as x. Next we substitute $x = 15$ and $y = 4$ into the equation. This leads to

$$4 = \frac{k}{15}$$

or $k = 60$. Thus, for our present discussion the constant of proportionality is 60, and this may be substituted into $y = k/x$, giving

$$y = \frac{60}{x}$$

as the equation between y and x. Now, for any given value of x, we may find the value of y. For $x = 12$, we have

$$y = \frac{60}{12} = 5$$

Example I

The distance an object falls under the influence of gravity varies directly as the square of the time of fall. If an object falls 64.0 ft in 2.00 s, how far will it fall in 3.00 s?

We first write the relation $d = kt^2$. Then we use the given fact that $d = 64.0$ ft for $t = 2.00$ s. This gives $64.0 = 4.00k$. In this way we find that $k = 16.0$ ft/s². We now know a general relation between d and t, which is $d = 16.0t^2$. Now we put in the value 3.00 s for t, so that we can find d for this particular value for the time. This gives us $d = 16.0(9.00) = 144$ ft. We also note in this problem that k will normally have a certain set of units associated with it.

Example J

The kinetic energy of a moving object varies jointly as the mass of the object and the square of its velocity. If a 5.00-kg object, traveling at 10.0 m/s, has a kinetic energy of 250 J, find the kinetic energy of an 8.00-kg object traveling at 50.0 m/s.

We first write the relation $KE = kmv^2$. Then we use the known set of values to find k. We write $250 = k(5.00)(100)$ giving

$$k = 0.500 \frac{J}{kg(m/s)^2}$$

[actually 1 J $= 1$ kg (m/s)², which means that $k = 0.500$ with no physical units.] Next we find the desired value for KE by substituting the other given values, and the value of k, in the original relation:

$$KE = 0.500 \, mv^2 = 0.500(8.00)(2500) = 10,000 \text{ J}$$

Example K

The heat developed in a resistor varies jointly as the time and the square of the current in the resistor. If the heat developed in t_0 seconds with a current i_0 passing through the resistor is H_0, how much heat is developed if both the time and current are doubled?

First we set up the relation $H = kti^2$, where t is the time and i is the current. From the given information we can write $H_0 = kt_0 i_0^2$. Thus $k = H_0/t_0 i_0^2$, and the original equation becomes $H = H_0 ti^2/t_0 i_0^2$. We now let $t = 2t_0$ and $i = 2i_0$, so that we can find H when the time and current are doubled. With these values we obtain

$$H = \frac{H_0 (2t_0)(2i_0)^2}{t_0 i_0^2} = \frac{8H_0 t_0 i_0^2}{t_0 i_0^2} = 8H_0$$

This tells us that the heat developed is eight times as much as for the original values of i and t.

Exercises 17–2

In Exercises 1 through 4 express the given statements as equations.

1. y varies directly as z.

2. s varies inversely as the square of t.

3. w varies jointly as x and the cube of y.

4. q varies as the square of r and inversely as the fourth power of t.

In Exercises 5 through 8 give the equation relating the variables after evaluating the constant of proportionality for the given set of values.

5. r varies inversely as y, and $r = 2$ when $y = 8$.

6. y varies directly as the square root of x, and $y = 2$ when $x = 64$.

7. p is proportional to q and inversely proportional to the cube of r, and $p = 6$ when $q = 3$ and $r = 2$.

8. v is proportional to t and the square of s, and $v = 80$ when $s = 2$ and $t = 5$.

In Exercises 9 through 16 find the required value by setting up the general equation and then evaluating.

9. Find y when $x = 10$ if y varies directly as x and $y = 20$ when $x = 8$.

10. Find y when $x = 5$ if y varies directly as the square of x and $y = 6$ when $x = 8$.

11. Find s when $t = 10$ if s is inversely proportional to t and $s = 100$ when $t = 5$.

12. Find p for $q = 0.8$ if p is inversely proportional to the square of q and $p = 18$ when $q = 0.2$.

13. Find y for $x = 6$ and $z = 5$ if y varies directly as x and inversely as z and $y = 60$ when $x = 4$ and $z = 10$.

14. Find r when $n = 16$ if r varies directly as the square root of n and $r = 4$ when $n = 25$.

15. Find f when $p = 2$ and $c = 4$ if f varies jointly as p and the cube of c and $f = 8$ when $p = 4$ and $c = 0.1$.

16. Find v when $r = 2$, $s = 3$, and $t = 4$ if v varies jointly as r and s and inversely as the square of t and $v = 8$ when $r = 2$, $s = 6$, and $t = 6$.

In Exercises 17 through 36 solve the given applied problems.

17. Hooke's law states that the force needed to stretch a spring is proportional to the amount the spring is stretched. If 10.0 lb stretches a certain spring 4.00 in., how much will the spring be stretched by a force of 6.00 lb?

18. In modern physics we learn that the energy of a photon (a "particle" of light) is directly proportional to its frequency f. Given that the constant of proportionality is 6.63×10^{-34} J·s (this is known as *Planck's constant*), what is the energy of a photon whose frequency is 3.00×10^{15} Hz?

19. A particular type of automobile engine produces p cm³ of carbon monoxide proportional to the time t that it idles. Find the equation relating p and t if $p = 60,000$ cm³ for $t = 2.00$ min.

20. In biology it is found that under certain circumstances the rate of increase v of bacteria is proportional to the number N of bacteria present. If $v = 800$ bacteria/h when $N = 4000$ bacteria, what is v when $N = 7500$ bacteria?

21. The heat loss through rockwool insulation is inversely proportional to the thickness of the rockwool. If the loss through 6.00 in. of rockwool is 3200 BTU/h, find the loss through 2.50 in. of rockwool.

22. In economics it is often found that the demand D for a product varies inversely as the price P. If the demand is 500 units per week at a cost of $8.00, what would the demand be if the price were $10.00 each?

23. The illuminance of a light source varies inversely as the square of the distance from the source. Given that the illuminance is 25.0 units at a distance of 200 cm, find the general relation between the illuminance and distance for this light source.

24. The rate of emission of radiant energy from the surface of a body is proportional to the fourth power of the thermodynamic temperature. Given that a 25.0 W (the rate of emission) lamp has an operating temperature of 2500 K, what is the operating temperature of a similar 40.0 W lamp?

25. The electric resistance of a wire varies directly as its length and inversely as its cross-sectional area. Find the relation between resistance, length, and area for a wire which has a resistance of 0.200 Ω for a length of 200 ft and cross-sectional area of 0.0500 in.².

26. The general gas law states that the pressure of an ideal gas varies directly as the thermodynamic temperature and inversely as the volume. By first finding the relation between P, T, and V, find V for $P = 400$ kPa and $T = 400$ K, given that $P = 600$ kPa for $V = 10.0$ cm³ and $T = 300$ K.

27. The frequency of vibration of a wire varies directly as the square root of the tension on the wire. Express the relation between f and T. If $f = 400$ Hz when $T = 1.00$ N, find f when $T = 3.40$ N.

28. The period of a pendulum is directly proportional to the square root of its length. Given that a pendulum 2.00 ft long has a period of $\pi/2$ s, what is the period of a pendulum 4.00 ft long?

29. The distance s that an object falls due to gravity varies jointly as the acceleration due to gravity g and the square of the time t of fall. On Earth $g = 32.0$ ft/s² and on the moon $g = 5.50$ ft/s². On Earth an object falls 144 ft in 3.00 s. How far does an object fall in 7.00 s on the moon?

30. The power of an electric current varies jointly as the resistance and the square of the current. Given that the power is 10.0 W when the current is 0.500 A and the resistance is 40.0 Ω, find the power if the current is 2.00 A and the resistance is 20.0 Ω.

31. Under certain conditions the velocity of an object is proportional to the logarithm of the square root of the time elapsed. Given that $v = 18.0$ cm/s after 4.00 s, what is the velocity after 6.00 s?

32. The level of intensity of a sound wave is proportional to the logarithm of the ratio of the sound intensity to an arbitrary reference intensity, normally defined as 10^{-12} W/m². Assuming that the intensity level is 100 dB for $I = 10^{-2}$ W/m², what is the constant of proportionality?

33. When the volume of a gas changes very rapidly, an approximate relation is that the pressure varies inversely as the $\frac{3}{2}$ power of the volume. Express P as a function of V. Given the $P = 300$ kPa when $V = 100$ cm³, find P when $V = 25.0$ cm³.

34. The x-component of the velocity of an object moving around a circle with constant angular velocity ω varies jointly as ω and $\sin \omega t$. Given that ω is constant at $\pi/6$ rad/s, and the x-component of the velocity is -4π ft/s when $t = 1.00$ s, find the x-component of the velocity when $t = 9.00$ s.

35. The acceleration in the x-direction of the object referred to in Exercise 34 varies jointly as $\cos \omega t$ and the square of ω. Under the same conditions as stated in Exercise 34, calculate the x-component of the acceleration.

36. Under certain conditions, the natural logarithm of the ratio of an electric current at time t to the current at time $t = 0$ is proportional to the time. Given that the current is $1/e$ of its initial value after 0.100 s, what is its value after 0.200 s?

17–3 Exercises for Chapter 17

In Exercises 1 through 4 find the indicated ratios.

1. 1 km to 1 mm   2. 1 mL to 1 L

3. The area of the world needed for food production is about 5×10^{13} m², and the area needed for housing and industry is about 7×10^{12} m². Find the ratio of the food production area to that for housing and industry.

4. The capacitance C of a capacitor is defined as the ratio of its charge to the voltage. What is the capacitance of a capacitor (in farads) for which the charge is 5.00×10^{-6} C and the voltage is 200 V? (1 F = 1 C/1 V.)

In Exercises 5 through 12 answer the given questions by setting up and solving the appropriate proportions.

5. On a certain map, 1.00 in. represents 16.0 mi. What distance on the map represents 52.0 mi?

6. Given that 1.00 m equals 39.4 in., what length in inches is 2.45 m?

7. Given that 1.00 L equals 61.0 in.³, how many liters is 105 in.³?

8. Given that 1.00 BTU equals 1060 J, how many BTU equal 8190 J?

9. An electronic data-processing card sorter can sort 45,000 cards in 20.0 min. How many cards can it sort in 5.00 h?

10. A given machine can produce 80 bolts in 5.00 min. How many bolts can it produce in an hour?

11. Assuming that 195 g of zinc sulfide are used to produce 128 g of sulfur dioxide, how many grams of sulfur dioxide are produced by the use of 750 g of zinc sulfide?

12. Given that 32.0 lb of oxygen are required to burn 20.0 lb of a certain fuel gas, how much air (which can be assumed to be 21% oxygen) is required to burn 100 lb of the fuel gas?

In Exercises 13 through 16 give the equation relating the variables after evaluating the constant of proportionality for the given set of values.

13. y varies directly as the square of x, and $y = 27$ when $x = 3$.

14. f varies inversely as l, and $f = 5$ when $l = 8$.

15. v is directly proportional to x and inversely proportional to the cube of y, and $v = 10$ when $x = 5$ and $y = 4$.

16. r varies jointly as u, v, and the square of w, and $r = 8$ when $u = 2$, $v = 4$, and $w = 3$.

In Exercises 17 through 36 solve the given applied problems.

17. On a certain blueprint a measurement of 25.0 ft is represented by 2.00 in. What is the actual distance between two points if they are 5.75 in. apart on the blueprint?

18. A certain gasoline company sells regular gas and lead-free gas in the ratio of 7 to 2. If, in a month, the company sells a total of 18,000,000 gal, how many of each type were sold?

19. For a lever balanced at the fulcrum, the relation

$$\frac{F_1}{F_2} = \frac{L_2}{L_1}$$

holds, where F_1 and F_2 are forces on opposite sides of the fulcrum at distances L_1 and L_2, respectively. If $F_1 = 4.50$ lb, $F_2 = 6.75$ lb, and $L_1 = 17.5$ in., find L_2.

20. A company finds that the volume V of sales of a certain item and the price P of the item are related by

$$\frac{P_1}{P_2} = \frac{V_2}{V_1}$$

Find V_2 if $P_1 = \$8.00$, $P_2 = \$6.00$, and $V_1 = 3000$ per week.

21. The power of a gas engine is proportional to the area of the piston. If an engine with a piston area of 8.00 in.² can develop 30.0 hp, what power is developed by an engine with a piston area of 6.00 in.²?

22. Ohm's law states that the voltage across a given resistor is proportional to the current i in the resistor. Given that 18.0 V are across a certain resistor in which a current of 8.00 A is flowing, express the general relationship between voltage and current for this resistor.

23. An apartment owner charges rent R proportional to the floor area A of the apartment. Find the equation used relating R and A if an apartment of 900 ft² rents for $250 per month.

24. A manufacturer determines that the number r of aluminum cans that can be made by recycling n used cans is proportional to n. How many cans can be made from 50,000 used cans if $r = 115$ cans for $n = 125$ cans?

25. The surface area of a sphere varies directly as the square of its radius. The surface area of a certain sphere is 36π square units when the radius is 3.00 units. What is the surface area when the radius is 4.00 units?

26. The difference in pressure in a fluid between that at the surface and that at a point below varies jointly as the density of the fluid and the depth of the point. The density of water is 1000 kg/m³, and the density of alcohol is 800 kg/m³. This difference in pressure at a point 0.200 m below the surface of water is 1.96 kPa. What is the difference in pressure at a point 0.300 m below the surface of alcohol?

27. The velocity of a pulse traveling in a string varies directly as the square root of the tension of the string. Given that the velocity in a certain string is 450 ft/s when the tension is 20.0 lb, determine the velocity if the tension were 30.0 lb.

28. The velocity of a jet of fluid flowing from an opening in the side of a container is proportional to the square root of the depth of the opening. If the velocity of the jet from an opening at a depth of 4.00 ft is 16.0 ft/s, what is the velocity of a jet from an opening at a depth of 25.0 ft?

29. The crushing load of a pillar varies as the fourth power of its radius and inversely as the square of its length. Express L in terms of r and l for a pillar 20.0 ft tall and 1.00 ft in diameter which is crushed by a load of 20.0 tons.

30. The acoustical intensity of a sound wave is proportional to the square of the pressure amplitude and inversely proportional to the velocity of the wave. Given that the intensity is 0.474 W/m² for a pressure amplitude of 20.0 Pa and a velocity of 346 m/s, what is the intensity if the pressure amplitude is 15.0 Pa and the velocity is 320 m/s?

31. In any given electric circuit containing an inductance and capacitance, the resonant frequency is inversely proportional to the square root of the capacitance. If the resonant frequency in a circuit is 25.0 Hz and the capacitance is 100 μF, what is the resonant frequency of this circuit if the capacitance is 25.0 μF?

32. The safe uniformly distributed load on a horizontal beam, supported at both ends, varies jointly as the width and the square of the depth and inversely as the distance between supports. Given that one beam has double the dimensions of another, how many times heavier is the safe load it can support than the first can support?

33. Kepler's third law of planetary motion states that the square of the period of any planet is proportional to the cube of the mean radius (about the sun) of that planet, with the constant of proportionality being the same for all planets. Using the fact that the period of the earth is one year and its mean radius is 93.0 million mi, calculate the mean radius for Venus, given that its period is 7.38 months.

34. The amount of heat per unit of time passing through a wall t units thick is proportional to the temperature difference ΔT and the area A, and is inversely proportional to t. The constant of proportionality is called the coefficient of conductivity. Calculate the coefficient of conductivity of a 0.210-m-thick concrete wall if 0.419 J/s pass through an area of 0.0105 m² when the temperature difference is 10.5°C.

35. The percentage error in determining an electric current due to a small error in reading a galvanometer is proportional to $\tan \theta + \cot \theta$, where θ is the angular deflection of the galvanometer. Given that the percentage error is 2.00% when $\theta = 4.0°$, determine the percentage error as a function of θ.

36. Newton's law of gravitation is stated in Example G of Section 17–2. Here k is the same for any two objects. A spacecraft is traveling the 240,000 mi from the earth to the moon, whose mass is $\frac{1}{81}$ that of the earth. How far from the earth is the gravitational force of the earth on the spacecraft equal to the gravitational force of the moon on the spacecraft?

18

Progressions and
the Binomial Theorem

18—1 Arithmetic Progressions

In this chapter we are going to consider briefly the properties of certain sequences of numbers. In itself a sequence is some set of numbers arranged in some specified manner. The kinds of sequences we shall consider are those which form what are known as arithmetic progressions, those which form geometric progressions, and those which are used in the expansion of a binomial to a power. Applications of progressions can be found in many areas, including interest calculations and certain areas in physics. Also, progressions and binomial expansions are of importance in developing mathematical topics which in themselves have wide technical application.

An **arithmetic progression** (AP) *is a sequence of numbers in which each number after the first can be obtained from the preceding one by adding to it a fixed number called the* **common difference.**

Example A
The sequence 2, 5, 8, 11, 14, . . . is an AP with a common difference of 3. The sequence 7, 2, −3, −8, . . . is an AP with a common difference of −5.

If we know the first term of an AP, we can find any other term in the progression by successively adding the common difference enough times for the desired term to be obtained. This, however, is a very inefficient method, and we can learn more about the progression if we establish a general way of finding any particular term.

In general, if a is the first term and d the common difference, the second term is $a + d$, the third term is $a + 2d$, and so forth. If we are looking for the nth term, we note that we need only add d to the first term $n − 1$ times. Thus, the nth term l of an AP is given by

$$l = a + (n − 1)d \qquad (18–1)$$

Occasionally l is referred to as the last term of an AP, but this is somewhat misleading. In reality there is no actual limit to the possible number of terms in an AP, although we may be interested only in a particular number of them. For this reason it is clearer to call l the nth term, rather than the last term. Also, this is the reason for writing three dots after the last indicated term of an AP. These dots indicate that the progression may continue.

Example B
Find the tenth term of the progression 2, 5, 8, ...
By subtracting any given term from the following term, we find that the common difference is $d = 3$. From the terms given, we know that the first term is $a = 2$. From the statement of the problem, the desired term is the tenth, or $n = 10$. Thus, we may find the tenth term l by

$$l = 2 + (10 − 1)3 = 2 + 9·3 = 29$$

Example C
Find the number of terms in the progression for which $a = 5$, $l = −119$, and $d = −4$.
Substitution into Eq. (18–1) gives

$$−119 = 5 + (n − 1)(−4)$$

which leads to

$$−124 = −4n + 4$$
$$4n = 128$$
$$n = 32$$

Example D

How many numbers between 10 and 1000 are divisible by 6?

We must first find the smallest and the largest numbers in this range which are divisible by 6. These numbers are 12 and 996. Obviously the common difference between all multiples of 6 is 6. Thus, $a = 12$, $l = 996$, and $d = 6$. Therefore,

$$996 = 12 + (n - 1)6$$
$$6n = 990$$
$$n = 165$$

Thus there are 165 numbers between 10 and 1000 which are divisible by 6.

Another important quantity concerning an AP is the sum s of the first n terms. We can indicate this sum by either of the two equations,

$$s = a + (a + d) + (a + 2d) + \cdots + (l - d) + l$$

or

$$s = l + (l - d) + (l - 2d) + \cdots + (a + d) + a$$

If we now add these equations, we have

$$2s = (a + l) + (a + l) + (a + l) + \cdots + (a + l) + (a + l)$$

There is one factor $(a + l)$ for each term, and there are n terms. Thus,

$$s = \frac{n}{2}(a + l) \tag{18--2}$$

Example E

Find the sum of the first 1000 positive integers.

Here $a = 1$, $l = 1000$, and $n = 1000$. Thus,

$$s = \frac{1000}{2}(1 + 1000) = 500(1001) = 500{,}500$$

Example F

Find the sum of the AP for which $n = 10$, $a = 4$, and $d = -5$.

We first find the nth term. We write

$$l = 4 + (10 - 1)(-5) = 4 - 45 = -41$$

Now we can solve for s:

$$s = \frac{10}{2}(4 - 41) = 5(-37) = -185$$

Example G

For an AP, given that $a = 2$, $d = \frac{3}{2}$, and $s = 72$, find n and l.

First we substitute the given values into Eqs. (18–1) and (18–2) in order to identify what is known and how we may proceed. Substituting $a = 2$ and $d = \frac{3}{2}$ in Eq. (18–1), we obtain

$$l = 2 + (n - 1)\left(\frac{3}{2}\right)$$

Substituting $s = 72$ and $a = 2$ in Eq. (18–2), we obtain

$$72 = \frac{n}{2}(2 + l)$$

We note that n and l appear in both equations, which means that we must solve them simultaneously. Substituting the expression for l from the first equation into the second equation we proceed with the solution.

$$72 = \frac{n}{2}\left[2 + 2 + (n - 1)\left(\frac{3}{2}\right)\right]$$

$$72 = 2n + \frac{3n(n - 1)}{4}$$

$$288 = 8n + 3n^2 - 3n$$
$$3n^2 + 5n - 288 = 0$$

$$n = \frac{-5 \pm \sqrt{25 - 4(3)(-288)}}{6} = \frac{-5 \pm \sqrt{3481}}{6} = \frac{-5 \pm 59}{6}$$

Since n must be a positive integer we find that $n = \frac{-5 + 59}{6} = 9$. Using this value in the expression for l, we find

$$l = 2 + (9 - 1)\left(\frac{3}{2}\right) = 14$$

Therefore, $n = 9$ and $l = 14$.

Example H
Each swing of a pendulum is measured to be 3.00 in. shorter than the preceding swing. If the first swing is 10.0 ft, determine the total distance traveled by the pendulum bob in the first five swings.

In this problem each of the terms of the progression represents the distance traveled in each respective swing. This means that $a = 10.0$. Also, since 3.00 in. = 0.25 ft, we see that $d = -0.25$. Using these values, we find the distance traversed in the fifth swing to be

$$l = 10.0 + 4(-0.25) = 9.00$$

The total distance traversed is

$$s = \frac{5}{2}(10.0 + 9.00) = 47.5 \text{ ft}$$

Exercises 18−1   In Exercises 1 through 4 write five terms of the AP with the given values.

1. $a = 4$, $d = 2$ 2. $a = 6$, $d = -\frac{1}{2}$
3. Third term $= 5$, fifth term $= -3$
4. Second term $= -2$, fifth term $= 7$

In Exercises 5 through 12 find the nth term of the AP with the given values.

5. $1, 4, 7, \ldots n = 8$ 6. $-6, -4, -2, \ldots n = 10$
7. $18, 13, 8, \ldots n = 17$ 8. $2, \frac{1}{2}, -1, \ldots n = 25$
9. $a = -7$, $d = 4$, $n = 12$ 10. $a = \frac{2}{3}$, $d = \frac{1}{6}$, $n = 50$
11. $a = b$, $d = 2b$, $n = 25$ 12. $a = -c$, $d = 3c$, $n = 30$

In Exercises 13 through 16 find the indicated sum of the terms of the AP.

13. $n = 20$, $a = 4$, $l = 40$ 14. $n = 8$, $a = -12$, $l = -26$
15. $n = 10$, $a = -2$, $d = -\frac{1}{2}$ 16. $n = 40$, $a = 3$, $d = \frac{1}{3}$

In Exercises 17 through 28 find any of the values of a, d, l, n, or s that are missing.

17. $a = 5$, $d = 8$, $l = 45$ 18. $a = -2$, $n = 60$, $l = 28$
19. $a = \frac{5}{3}$, $n = 20$, $s = \frac{40}{3}$ 20. $a = 0.1$, $l = -5.9$, $s = -8.7$
21. $d = 3$, $n = 30$, $s = 1875$ 22. $d = 9$, $l = 86$, $s = 455$
23. $a = 74$, $d = -5$, $l = -231$ 24. $a = -\frac{9}{7}$, $n = 19$, $l = -\frac{36}{7}$
25. $a = -5$, $d = \frac{1}{2}$, $s = \frac{23}{2}$ 26. $d = -2$, $n = 50$, $s = 0$
27. $a = -c$, $l = \dfrac{b}{2}$, $s = 2b - 4c$ 28. $a = 3b$, $n = 7$, $d = \dfrac{b}{3}$

In Exercises 29 through 40 find the indicated quantities.

29. Sixth term $= 56$, tenth term $= 72$ (find a, d, s for $n = 10$).
30. Seventeenth term $= -91$, second term $= -73$ (find a, d, s for $n = 40$).
31. Fourth term $= 2$, tenth term $= 0$ (find a, d, s for $n = 10$)
32. Third term $= 1$, sixth term $= -8$ (find a, d, s for $n = 12$)
33. Find the sum of the first 100 integers.
34. Find the sum of the first 100 odd integers.
35. Find the sum of the first 200 multiples of 5.
36. Find the number of multiples of 8 between 99 and 999.
37. A man accepts a position which pays $8200 per year, and receives a raise of $450 each year. During what year of his association with the firm will his salary be $16,300?
38. A body falls 16.0 ft during the first second, 48.0 ft during the second second, 80.0 ft during the third second, etc. How far will it fall in the twentieth second?
39. For the object in Exercise 38, what is the total distance fallen in the first 20 s?
40. A well-driller charges $3 for drilling the first foot of a well, and for every foot thereafter he charges 1¢ more than the preceding foot. How much does he charge for drilling a 500-ft well?

18–2 Geometric Progressions

A second type of sequence is the **geometric progression.** *A geometric progression (GP) is a sequence of numbers in which each number after the first can be obtained from the preceding one by multiplying it by a fixed number called the* **common ratio.** One important application of geometric progressions is in computing interest on savings accounts. Other applications can be found in biology and physics.

Example A
The sequence 2, 4, 8, 16, . . . forms a GP with a common ratio of 2. The sequence 9, 3, 1, $\frac{1}{3}$, . . . forms a GP with a common ratio of $\frac{1}{3}$.

If we know the first term, we can then find any other desired term by multiplying by the common ratio a sufficient number of times. When we do this for a general GP, we can determine the nth term in terms of the first term a, the common ratio r, and n. Thus, the second term is ra, the third term is r^2a, and so forth. In general, the expression for the nth term is

$$l = ar^{n-1} \tag{18–3}$$

Example B
Find the eighth term of the GP 8, 4, 2, ...
Here $a = 8$, $r = \frac{1}{2}$, and $n = 8$. The eighth term is given by

$$l = 8\left(\frac{1}{2}\right)^{8-1} = \frac{8}{2^7} = \frac{1}{16}$$

Example C
Find the tenth term of a GP when the second term is 3, the fourth term is 9, and $r > 0$.
We can find r, if we let $a = 3$, $l = 9$, and $n = 3$ (we are at this time considering the progression made up of 3, the next number, and 9). Thus,

$$9 = 3r^2 \quad \text{or} \quad r = \sqrt{3}$$

We now can find a, by using just two terms (a and 3) for a progression:

$$3 = a(\sqrt{3})^{2-1} \quad \text{or} \quad a = \sqrt{3}$$

We now can find the tenth term directly:

$$l = \sqrt{3}(\sqrt{3})^{10-1} = \sqrt{3}(\sqrt{3})^9 = \sqrt{3}(3^4\sqrt{3}) = 3^5 = 243$$

We could have shortened this procedure one step by letting the second term be the first term of a new progression of 9 terms. If the first term is of no importance in itself, this is perfectly acceptable.

Example D

Under certain circumstances, 20% of a substance changes chemically each 10.0 min. If there are originally 100 g of a substance, how much will remain after an hour?

Let P = the portion of the substance remaining after each minute. From the statement of the problem, r = 0.8 (80% remains after each 10-min period), a = 100, and n represents the number of minutes elapsed. This means $P = 100(0.8)^{n/10}$. It is necessary to divide n by 10 because the ratio is given for a 10-min period. In order to find P when $n = 60$, we write

$$P = 100(0.8)^6$$
$$= 100(0.262)$$
$$= 26.2 \text{ g}$$

This means that 26.2 g are left after an hour.

A general expression for the sum of the first n terms of a geometric progression may be found by directly forming the sum and multiplying this equation by r. By doing this, we have

$$s = a + ar + ar^2 + \cdots + ar^{n-1}$$
$$rs = ar + ar^2 + ar^3 + \cdots + ar^n$$

If we now subtract the first of these equations from the second, we have $rs - s = ar^n - a$. All other terms cancel by subtraction. Solving this equation for s and writing both the numerator and the denominator in the final solution in the form generally used, we obtain

$$s = \frac{a(1 - r^n)}{1 - r} \qquad (r \neq 1) \tag{18-4}$$

Example E

Find the sum of the first 7 terms of the GP 2, 1, $\frac{1}{2}$, ...

Here $a = 2$, $r = \frac{1}{2}$, and $n = 7$. Hence

$$s = \frac{2(1 - (\frac{1}{2})^7)}{1 - \frac{1}{2}} = \frac{2(1 - \frac{1}{128})}{\frac{1}{2}} = 4\left(\frac{127}{128}\right) = \frac{127}{32}$$

Example F

If \$100 is invested each year at 5% interest compounded annually, what would be the total amount of the investment after 10 years (before the eleventh deposit is made)?

After one year the amount invested will have added to it the interest for the year. Thus, for the last \$100 invested, its value will become $\$100(1 + 0.05) = \$100(1.05) = \$105$. The next to last \$100 will have interest added twice. After one year its value becomes \$100(1.05), and after two years its value becomes $[\$100(1.05)](1.05) = \$100(1.05)^2$. In the same way, the value of the first \$100 becomes $\$100(1.05)^{10}$, since it will have interest added 10 times. This means we are asked to sum the progression

$$100(1.05) + 100(1.05)^2 + 100(1.05)^3 + \cdots + 100(1.05)^{10}$$

or

$$100[1.05 + (1.05)^2 + (1.05)^3 + \cdots + (1.05)^{10}]$$

For the progression in the brackets we have $a = 1.05$, $r = 1.05$, and $n = 10$. Thus,

$$s = \frac{1.05[1 - (1.05)^{10}]}{1 - 1.05} = \frac{1.05}{-0.05}(1 - 1.628895) = 13.2068$$

(This value is obtained by use of a calculator. If 4-place logarithms are used we would obtain 13.21.) Therefore, the total value of these $100 investments is $100(13.2068) = \$1320.68$. We see that $320.68 of interest has been earned.

Exercises 18–2

In Exercises 1 through 4 write down the first five terms of the GP with the given values.

1. $a = 45$, $r = \frac{1}{3}$ 2. $a = 9$, $r = -\frac{2}{3}$
3. $a = 2$, $r = 3$ 4. $a = -3$, $r = 2$

In Exercises 5 through 12 find the nth term of the GP which has the given values.

5. $\frac{1}{2}, 1, 2, \ldots$ $(n = 6)$ 6. $10, 1, 0.1, \ldots$ $(n = 8)$
7. $125, -25, 5, \ldots$ $(n = 7)$ 8. $0.1, 0.3, 0.9, \ldots$ $(n = 5)$
9. $a = -27$, $r = -\frac{1}{3}$, $n = 6$ 10. $a = 48$, $r = \frac{1}{2}$, $n = 9$
11. $a = 2$, $r = 10$, $n = 7$ 12. $a = -2$, $r = 2$, $n = 6$

In Exercises 13 through 16 find the sum of the n terms of the GP with the given values.

13. $a = 8$, $r = 2$, $n = 5$ 14. $a = 162$, $r = -\frac{1}{3}$, $n = 6$
15. $a = 192$, $l = 3$, $n = 4$ 16. $a = 9$, $l = -243$, $n = 4$

In Exercises 17 through 20 find any of the values of a, r, l, n, or s that are missing.

17. $l = 27$, $n = 4$, $s = 40$ 18. $a = 3$, $n = 5$, $l = 48$
19. $a = 75$, $r = \frac{1}{5}$, $l = \frac{3}{25}$ 20. $r = -2$, $n = 6$, $s = 42$

In Exercises 21 through 36 find the indicated quantities.

21. Find the tenth term of a GP if the fourth term is 8 and the seventh term 16.

22. Find the sum of the first 8 terms of the geometric progression for which the fifth term is 5, the seventh term is 10, and $r > 0$.

23. What is the value of an investment of $100 after 20 yr, if it draws interest of 4% annually?

24. What is the value of an investment of \$10,000 after 10 years if it earns 6% annual interest, compounded semiannually. (6% annual interest compounded semiannually means that 3% interest is added each six months.)

25. A person invests \$100 each year for 5 years. How much is the investment worth if the interest is 6% compounded annually?

26. A person invests \$1000 each year for 8 years. How much is the investment worth if the interest is 8% compounded quarterly? (See Exercise 24.)

27. If the population of a certain town increases 20% each year, how long will it take for the population to double?

28. If you decided to save money by putting away 1¢ on a given day, 2¢ one week later, 4¢ a week later, etc., how much would you have to put aside one year later?

29. A ball is dropped from a height of 8.00 ft, and on each rebound it rises $\frac{1}{2}$ of the height it last fell. What is the total distance the ball has traveled when it hits the ground for the fourth time?

30. How many direct ancestors (parents, grandparents, etc.) does a person have in the 10 generations which preceded him?

31. The American Wire Gauge standard of wire diameters is based on a geometric progression. The ratio of one diameter to the next is the 39th root of 92. If the diameter of No. 30 wire is 0.0100 in., what is the diameter of the wire which is 10 sizes larger (No. 20 wire)?

32. A certain object, after being heated, cools at such a rate that its temperature decreases 10% each minute. If the object is originally heated to 100°C, what is its temperature 10.0 min later?

33. The half-life of tungsten 176 is 80.0 min. This means that half of a given amount will disintegrate in 80.0 min. After 160 min three-fourths will have disintegrated. How much will disintegrate in 120 min?

34. The power on a satellite is supplied by a radioactive isotope. On a given satellite the power decreases by 0.2% each day. What percent of the initial power remains after one year?

35. Derive a formula for s in terms of a, r, and l.

36. Write down several terms of a general GP. Then verify the statement that, if the logarithm of each term is taken, the resulting sequence is an AP.

18–3 Geometric Progressions with Infinitely Many Terms

If we consider the sum of the first n terms of the GP with terms $1, \frac{1}{2}, \frac{1}{4}, \ldots$, we find that we get the values in the following table.

n	2	3	4	5	6	7	8	9	10
s	$\frac{3}{2}$	$\frac{7}{4}$	$\frac{15}{8}$	$\frac{31}{16}$	$\frac{63}{32}$	$\frac{127}{64}$	$\frac{255}{128}$	$\frac{511}{256}$	$\frac{1023}{512}$

We see that as n gets larger, the numerator of each fraction becomes more nearly twice the denominator. In fact, we would find that if we continued to compute s as n becomes larger, s can be found as close to the value 2 as desired, although it will never actually reach the value 2. For example, if $n = 100$, $s = 2 - 1.6 \times 10^{-30}$, which could be written as

$$1.99999999999999999999999999999984$$

to 32 significant figures. In the formula for the sum of n terms of a GP,

$$s = a\frac{1 - r^n}{1 - r}$$

the term r^n becomes exceedingly small, and if we consider n as being sufficiently large, we can see that this term is effectively zero. If this term were *exactly* zero, then the sum would be

$$s = 1\frac{1 - 0}{1 - \frac{1}{2}} = 2$$

The only problem is that we cannot find any number large enough for n to make $(\frac{1}{2})^n$ zero. There is, however, an accepted notation for this. This notation is

$$\lim_{n \to \infty} r^n = 0 \qquad (\text{if } |r| < 1)$$

and it is read as "the limit, as n *approaches* infinity, of r to the nth power is zero."

The symbol ∞ is read as **infinity**, but it must not be thought of as a number. It is simply a symbol which stands for a *process* of considering numbers which become large without bound. The number which is called the **limit** of the sums is simply the number which the sums get closer and closer to, as n is considered to approach infinity. This notation and terminology are of particular importance in the calculus.

If we consider values of r such that $|r| < 1$, and let the values of n become unbounded, we find that $\lim_{n \to \infty} r^n = 0$. The formula for the sum of a geometric progression with infinitely many terms then becomes

$$s = \frac{a}{1 - r} \qquad (18–5)$$

(If $r \geq 1$, s is unbounded in value.)

Example A

Find the sum of the geometric progression for which $a = 4$, $r = \frac{1}{8}$, and for which n increases without bound.

$$s = \frac{4}{1 - \frac{1}{8}} = \frac{4}{1} \cdot \frac{8}{7} = \frac{32}{7}$$

Example B
Find the fraction which has as its decimal form 0.44444444 . . . .
  This decimal form can be thought of as being

$$0.4 + 0.04 + 0.004 + 0.0004 + \cdots$$

which means that it can also be thought of as the sum of a GP with infinitely many terms, where $a = 0.4$ and $r = 0.1$. With these considerations, we have

$$s = \frac{0.4}{1 - 0.1} = \frac{0.4}{0.9} = \frac{4}{9}$$

Thus, the fraction $\frac{4}{9}$ and the decimal 0.4444 . . . represent the same number.

Example C
Find the fraction which has as its decimal form 0.121212 . . . .
  This decimal form can be considered as being

$$0.12 + 0.0012 + 0.000012 + \cdots$$

which means that we have a GP with infinitely many terms, and that $a = 0.12$ and $r = 0.01$. Thus

$$s = \frac{0.12}{1 - 0.01} = \frac{0.12}{0.99} = \frac{4}{33}$$

Therefore, the decimal 0.121212 . . . and the fraction $\frac{4}{33}$ represent the same number.

  The decimals in Examples B and C are called **repeating decimals,** because the numbers in the decimal form appear over and over again in a particular order. These two examples verify the theorem that any repeating decimal represents a rational number. However, all repeating decimals do not necessarily start repeating immediately. If numbers never do repeat, the decimal represents an irrational number. For example, there is no repeating decimal which represents π or $\sqrt{2}$.

Example D
Find the fraction which has as its decimal form the repeating decimal 0.50345345345 . . . .
  We first separate the decimal into the beginning, nonrepeating part, and the infinite repeating decimal which follows. Thus we have $0.50 + 0.00345345345$ . . . . This means that we are to add $\frac{50}{100}$ to the fraction which represents the sum of the terms of the GP $0.00345 + 0.00000345 + \cdots$. For this GP, $a = 0.00345$ and $r = 0.001$. We find the sum of this GP to be

$$s = \frac{0.00345}{1 - 0.001} = \frac{0.00345}{0.999} = \frac{115}{33300} = \frac{23}{6660}$$

Therefore,

$$0.50345345\ldots = \frac{5}{10} + \frac{23}{6660} = \frac{5(666) + 23}{6660} = \frac{3353}{6660}$$

Example E

Each swing of a certain pendulum bob is 95% as long as the preceding swing. How far does the bob travel in coming to rest if the first swing is 40.0 in. long?

We are to find the sum of a geometric progression with infinitely many terms, for which $a = 40.0$ and $r = 95\% = \frac{19}{20}$. Substituting these values into Eq. (18–5), we obtain

$$s = \frac{40.0}{1 - \frac{19}{20}} = \frac{40.0}{\frac{1}{20}} = (40.0)(20) = 800 \text{ in.}$$

Therefore, the pendulum bob travels 800 in. (about 67 ft) in coming to rest.

Exercises 18—3

In Exercises 1 through 12 find the sum of the given geometric progressions.

1. $4, 2, 1, \frac{1}{2}, \ldots$

2. $6, -2, \frac{2}{3}, \ldots$

3. $5, 1, 0.2, 0.04, \ldots$

4. $2, \sqrt{2}, 1, \ldots$

5. $20, -1, 0.05, \ldots$

6. $9, 8.1, 7.29, \ldots$

7. $1, \frac{7}{8}, \frac{49}{64}, \ldots$

8. $6, -4, \frac{8}{3}, \ldots$

9. $1, 0.0001, 0.00000001, \ldots$

10. $30, -9, 2.7, \ldots$

11. $2 + \sqrt{3}, 1, 2 - \sqrt{3}, \ldots$

12. $1 + \sqrt{2}, -1, \sqrt{2} - 1, \ldots$

In Exercises 13 through 24 find the fractions equal to the given decimals.

13. $0.33333\ldots$

14. $0.55555\ldots$

15. $0.404040\ldots$

16. $0.070707\ldots$

17. $0.181818\ldots$

18. $0.272727\ldots$

19. $0.273273273\ldots$

20. $0.792792792\ldots$

21. $0.366666\ldots$

22. $0.66424242\ldots$

23. $0.100841841841\ldots$

24. $0.184561845618456\ldots$

In Exercises 25 and 26 solve the given problems by use of the sum of a geometric progression with infinitely many terms.

25. If the ball in Exercise 29 of Section 18–2 is allowed to bounce indefinitely, what is the total distance it will travel?

26. An object suspended on a spring is oscillating up and down. If the first oscillation is 10.0 cm and each oscillation thereafter is nine-tenths of the preceding one, find the total distance the object travels.

18—4 The Binomial Theorem

If we wished to find the roots of the equation $(x + 2)^5 = 0$, we note that there are five factors of $x + 2$, which in turn tells us the only root is $x = -2$. However, if we wished to expand the expression $(x + 2)^5$, a number of repeated multiplications would be needed. This would be a relatively tedious operation. In this section we shall develop the **binomial theorem,** by which it is possible to expand binomials to any given power without direct multiplication. Such direct expansion can be helpful and labor-saving in developing certain mathematical topics. We may also expand certain expressions where direct multiplication is not actually possible. Also, the binomial theorem is used to develop the necessary expressions for use in certain technical applications.

By direct multiplication, we may obtain the following expansions of the binomial $a + b$.

$$(a + b)^0 = 1$$
$$(a + b)^1 = a + b$$
$$(a + b)^2 = a^2 + 2ab + b^2$$
$$(a + b)^3 = a^3 + 3a^2b + 3ab^2 + b^3$$
$$(a + b)^4 = a^4 + 4a^3b + 6a^2b^2 + 4ab^3 + b^4$$
$$(a + b)^5 = a^5 + 5a^4b + 10a^3b^2 + 10a^2b^3 + 5ab^4 + b^5$$

An inspection indicates certain properties which these expansions have, and which we shall assume are valid for the expansion of $(a + b)^n$, where n is any positive integer. These properties are as follows:

(1) There are $n + 1$ terms.
(2) The first term is a^n and the final term is b^n.
(3) Progressing from the first term to the last, the exponent of a decreases by 1 from term to term, the exponent of b increases by 1 from term to term, and the sum of the exponents of a and b in each term is n.
(4) The coefficients of terms equidistant from the ends are equal.
(5) If the coefficient of any term is multiplied by the exponent of a in that term, and this product is divided by the number of that term, we obtain the coefficient of the next term.

Example A
Using the above properties we shall develop the expansion for $(a + b)^5$.

From property (1) we know that there are 6 terms. From property (2) we know that the first term is a^5 and the final term is b^5. From property (3) we know that the factors of a and b in terms 2, 3, 4, and 5 are a^4b, a^3b^2, a^2b^3, and ab^4, respectively.

From property (5) we obtain the coefficients of terms 2, 3, 4, and 5. In the first term, a^5, the coefficient is 1. Multiplying by 5, the power of a, and dividing by 1, the number of the term, we obtain 5, which is the coefficient of the second term. Thus, the second term is $5a^4b$. Again using property (5) we obtain the coefficient of the third term. The coefficient of the second term is 5. Multiplying by 4, and dividing by 2, we obtain 10. This means that the third term is $10a^3b^2$.

From property (4) we know that the coefficient of the fifth term is the same as the second, and the coefficient of the fourth term is the same as the third. Thus,

$$(a + b)^5 = a^5 + 5a^4b + 10a^3b^2 + 10a^2b^3 + 5ab^4 + b^5$$

It is not necessary to use the above properties directly to expand a given binomial. If they are applied to $(a + b)^n$ we may obtain a general formula for the expansion of a binomial. Thus, the binomial theorem states that the following **binomial formula** is valid for all values of n (the binomial theorem is proven through advanced methods).

$$(a + b)^n = a^n + na^{n-1}b + \frac{n(n-1)}{2!}a^{n-2}b^2 + \frac{n(n-1)(n-2)}{3!}a^{n-3}b^3 + \cdots + b^n \qquad (18\text{–}6)$$

The notation $n!$ is read as "n **factorial**". It denotes the product of the first n integers. Thus, $2! = 1\cdot2$, $3! = 1\cdot2\cdot3$, $4! = 1\cdot2\cdot3\cdot4$, and so on. Evaluating these products we see that $2! = 2$, $3! = 6$, $4! = 24$, and so on.

Example B

By use of the binomial formula expand $(2x + 3)^6$.

In using the binomial formula for $(2x + 3)^6$ we use $2x$ for a, 3 for b, and 6 for n. Thus,

$$(2x + 3)^6 = (2x)^6 + 6(2x)^5(3) + \frac{(6)(5)}{2}(2x)^4(3^2)$$
$$+ \frac{(6)(5)(4)}{(2)(3)}(2x)^3(3^3) + \frac{(6)(5)(4)(3)}{(2)(3)(4)}(2x)^2(3^4)$$
$$+ \frac{(6)(5)(4)(3)(2)}{(2)(3)(4)(5)}(2x)(3^5) + 3^6$$
$$= 64x^6 + 576x^5 + 2160x^4 + 4320x^3 + 4860x^2 + 2916x + 729$$

For the first few integral powers of a binomial, the coefficients can be obtained by setting them up in the following pattern, known as **Pascal's triangle.**

$n = 0$						1						
$n = 1$					1		1					
$n = 2$				1		2		1				
$n = 3$			1		3		3		1			
$n = 4$		1		4		6		4		1		
$n = 5$	1		5		10		10		5		1	
$n = 6$	1	6		15		20		15		6		1

We note that the first and last coefficient shown in each row is 1, and the second and next-to-last coefficients are equal to n. Other coefficients are obtained by adding the two nearest coefficients in the row above. This pattern may be continued indefinitely, although the use of Pascal's triangle is cumbersome for high values of n.

Example C
By use of Pascal's triangle expand $(5s - 2t)^4$.
    Here we note that $n = 4$. Thus, the coefficients of the five terms are 1, 4, 6, 4, and 1, respectively. Also, here we use $5s$ for a and $-2t$ for b. We are expanding this as $[(5s) + (-2t)]^4$. Therefore,

$$(5s - 2t)^4 = (5s)^4 + 4(5s)^3(-2t) + 6(5s)^2(-2t)^2 + 4(5s)(-2t)^3 + (-2t)^4$$
$$= 625s^4 - 1000s^3t + 600s^2t^2 - 160st^3 + 16t^4$$

    In certain uses of a binomial expansion it is not necessary to obtain all terms. Only the first few terms are required. The following example illustrates finding the first four terms of an expansion.

Example D
Find the first four terms of the expansion of $(x + 7)^{12}$.
    Here we use x for a, 7 for b, and 12 for n. Thus, from the binomial formula we have

$$(x + 7)^{12} = x^{12} + 12x^{11}(7) + \frac{(12)(11)}{2}x^{10}(7^2) + \frac{(12)(11)(10)}{(2)(3)}x^9(7^3) + \cdots$$

$$= x^{12} + 84x^{11} + 3234x^{10} + 75460x^9 + \cdots$$

    If we let $a = 1$ and $b = x$ in the binomial formula, we obtain the **binomial series**

$$(1 + x)^n = 1 + nx + \frac{n(n - 1)}{2!}x^2 + \frac{n(n - 1)(n - 2)}{3!}x^3 + \cdots \qquad (18\text{--}7)$$

which through advanced methods can be shown to be valid for any real number n if $|x| < 1$. When n is either negative or a fraction, we obtain an infinite series. In such a case, we calculate as many terms as may be needed although such a series is not obtainable through direct multiplication. The binomial series may be used to find numerical approximations and to develop important expressions which are used in applications.

Example E

Approximate the value of $\sqrt[3]{1006}$ to five decimal places.

This approximation can be made by use of the first three terms of the binomial expansion if we write $\sqrt[3]{1006} = \sqrt[3]{1000(1.006)} = 10\sqrt[3]{1.006} = 10\sqrt[3]{1 + 0.006}$. By expanding $(1 + 0.006)^{1/3}$, we have

$$(1 + 0.006)^{1/3} = 1 + \frac{1}{3}(0.006) + \frac{\left(\frac{1}{3}\right)\left(-\frac{2}{3}\right)}{2}(0.006)^2 + \ldots$$

$$= 1 + 0.002 - 0.000004$$

$$= 1.001996$$

Therefore, $\sqrt[3]{1006} = 10(1.001996) = 10.01996$. The use of additional terms of the expansion will not affect the first five decimal places. The use of this type of approximation is helpful if a calculator is not available, or if accuracy beyond that available on the calculator is required.

Exercises 18–4

In Exercises 1 through 8 expand and simplify the given expressions by use of the binomial formula.

1. $(t + 1)^3$ 2. $(3x - 2)^3$ 3. $(2x - 1)^4$ 4. $(x^2 + 3)^4$

5. $(x + 2)^5$ 6. $(xy - z)^5$ 7. $(2a - b^2)^6$ 8. $\left(\frac{a}{x} + x\right)^6$

In Exercises 9 through 12 expand and simplify the given expressions by use of Pascal's triangle.

9. $(5x - 3)^4$ 10. $(b + 4)^5$ 11. $(2a + 1)^6$ 12. $(x - 3)^7$

In Exercises 13 through 16 find the first four terms of the indicated expansions.

13. $(x + 2)^{10}$ 14. $(x - 3)^8$ 15. $(1 - x)^{-2}$ 16. $(1 + x)^{-1/3}$

In Exercises 17 through 24 approximate the values of the given expressions to three decimal places by use of three terms of the appropriate binomial series.

17. $\sqrt{1.1}$ 18. $\sqrt{0.9}$ 19. $\sqrt[3]{994}$ (see Example E)

20. $\sqrt[3]{9}$ $\left(\sqrt[3]{9} = \sqrt[3]{8 + 1} = \sqrt[3]{8(1 + \frac{1}{8})} = 2\sqrt[3]{1 + \frac{1}{8}}\right)$

21. $\sqrt[4]{82}$ 22. $\sqrt[4]{15}$ 23. $(1.02)^{-4}$ 24. $(0.97)^{-2}$

In Exercises 25 through 28 find the indicated terms by use of the following information. The $r + 1$ term of the expansion of $(a + b)^n$ is given by

$$\frac{n(n - 1)(n - 2) \cdots (n - r + 1)}{r!} a^{n-r} b^r$$

25. The term involving b^5 in $(a + b)^8$.

26. The term involving y^6 in $(x + y)^{10}$.

27. The fifth term of $(2x - 3b)^{12}$.

28. The sixth term of $(a - b)^{14}$.

In Exercises 29 through 32 find the indicated expansions.

29. In determining the change of the rate of emission of energy from the surface of a body at temperature T, the expression $(T + h)^4$ is used. Expand this expression.

30. In determining the probability of a given number of heads or tails when 8 coins are tossed, the expression $(H + T)^8$ can be used. The various coefficients give the number of chances in 256 that a certain number of heads and tails will result. For example, the coefficient of the H^2T^6 term gives the chances in 256 that 2 heads and 6 tails will result. Expand this expression.

31. In the theory associated with the magnetic field due to an electric current, the expression $1 - \dfrac{x}{\sqrt{a^2 + x^2}}$ is found. By expanding $(a^2 + x^2)^{-1/2}$ find the first three nonzero terms which could be used to approximate the given expression.

32. Find the first four terms of the expansion of $(1 + x)^{-1}$ and then divide $1 + x$ into 1. Compare the results.

18–5 Exercises for Chapter 18

In Exercises 1 through 8 find the indicated term of each progression.

1. $1, 6, 11, \ldots$ (17th)
2. $1, -3, -7, \ldots$ (21st)
3. $\frac{1}{2}, 0.1, 0.02, \ldots$ (9th)
4. $0.025, 0.01, 0.004, \ldots$ (7th)
5. $8, \frac{7}{2}, -1, \ldots$ (16th)
6. $-1, -\frac{5}{3}, -\frac{7}{3}, \ldots$ (25th)
7. $\frac{3}{4}, \frac{1}{2}, \frac{1}{3}, \ldots$ (7th)
8. $\frac{2}{3}, 1, \frac{3}{2}, \ldots$ (7th)

In Exercises 9 through 12 find the sum of each progression with the indicated values.

9. $a = -4, n = 15, l = 17$ (AP)
10. $a = 3, d = -\frac{2}{3}, n = 10$
11. $a = 16, r = -\frac{1}{2}, n = 10$
12. $a = 64, l = 729, n = 7$ (GP, $r > 0$)

In Exercises 13 through 20 find the indicated quantities for the appropriate progressions.

13. $a = 17, d = -2, n = 9, s = ?$
14. $d = \frac{4}{3}, a = -3, l = 17, n = ?$
15. $a = 18, r = \frac{1}{2}, n = 6, l = ?$
16. $l = \frac{49}{8}, r = -\frac{2}{7}, s = \frac{17199}{288}, a = ?$
17. $a = -1, l = 32, n = 12, s = ?$ (AP)
18. $a = 1, l = 64, s = 325, n = ?$ (AP)
19. $a = 1, n = 7, l = 64, s = ?$
20. $a = \frac{1}{4}, n = 6, l = 8, s = ?$

In Exercises 21 through 28 find the fractions equal to the given decimals.

21. $0.77777\ldots$
22. $0.030303\ldots$
23. $0.757575\ldots$
24. $0.484848\ldots$
25. $0.123123123\ldots$
26. $0.0727272\ldots$
27. $0.166666\ldots$
28. $0.25399399399\ldots$

In Exercises 29 through 36 expand and simplify the given expressions. In Exercises 33 through 36 find the first four terms of the appropriate expansion.

29. $(x - 2)^4$ **30.** $(s + 2t)^4$ **31.** $(x^2 + 1)^5$ **32.** $(3n - a)^6$

33. $\sqrt{1 - a^2}$ **34.** $\sqrt{1 + b^4}$ **35.** $(1 - 2x)^{-3}$ **36.** $(1 + 4x)^{-1/4}$

In Exercises 37 through 40 approximate the values of the given expressions to three decimal places by use of three terms of the appropriate binomial series.

37. $\sqrt{908}$ **38.** $\sqrt[3]{61}$ **39.** $(8.04)^{-1}$ **40.** $(4.06)^{-1/2}$

In Exercises 41 through 52 solve the given problems by use of an appropriate progression or expansion.

41. Find the sum of the first 1000 positive even integers.

42. How many numbers divisible by 4 lie between 23 and 121?

43. Fifteen layers of logs are so piled that there are 20 logs in the bottom layer, and each layer contains one log less than the layer below it. How many logs are in the pile?

44. A contractor employed in the construction of a building was penalized for taking more time than the contract allowed. She forfeited $150 for the first day late, $225 for the second day, $300 for the third day, and so forth. If she forfeited a total of $6750, how many additional days did she require to complete the building?

45. What is the value after 20 years of an investment of $2500, if it draws interest at 5% compounded annually?

46. A person invests $1000 each year for 20 years. How much is the total investment worth if the annual interest is 6% and it is compounded quarterly?

47. A business estimates that the salvage value of a piece of machinery decreases by 20% each year. If the machinery is purchased for $80,000, what is its value after 10 years?

48. A tank contains 100 L of a given chemical. Thirty liters are drawn off and replaced with water. Then 30 L of the resulting solution are drawn off and replaced with water. If this operation is performed a total of 5 times, how much of the original chemical remains?

49. A ball, starting from rest, rolls down a uniform incline so that it covers 10 in. during the first second, 30 in. during the second second, 50 in. during the third second, and so on. How long will it take to cover $333\frac{1}{3}$ ft?

50. The successive distances traveled by a pendulum bob are 90 cm, 60 cm, 40 cm, ... Find the total distance the bob travels before it comes to rest.

51. In finding the partial pressure P_F of fluorine gas under certain conditions, the equation

$$P_F = \frac{(1 + 2 \times 10^{-10}) - \sqrt{1 + 4 \times 10^{-10}}}{2} \, \text{atm}$$

is found. By using three terms of the expansion for $\sqrt{1 + x}$ approximate the value of this expression.

52. The terms a, $a + 12$, $a + 24$ form an AP, and the terms a, $a + 24$, $a + 12$ form a GP. Find these progressions.

19

Additional Topics in Trigonometry

19—1 Fundamental Trigonometric Identities

The definitions of the trigonometric functions were first introduced in Section 3–2, and were again summarized in Section 7–1. If we take a close look at these definitions, we find that there are many relationships among the various functions. For example, the definition of the sine of an angle is $\sin \theta = y/r$, and the definition of the cosecant of an angle is $\csc \theta = r/y$. But we know that $1/(r/y) = y/r$, which means that $\sin \theta = 1/\csc \theta$. In writing this down we made no reference to any particular angle, and since the definitions hold for *any* angle, this relation between the $\sin \theta$ and $\csc \theta$ also holds for any angle. A *relation* such as this, *which holds for any value of the variable, is called an* **identity**. Of course, specific values where division by zero would be indicated are excluded.

Such trigonometric identities are important for a number of reasons. We have already actually made limited use of some of them in Section 9–4 when we graphed certain trigonometric functions. We also used an important identity in deriving the Law of Cosines in Chapter 8. When we consider equations with trigonometric functions later in this chapter, we will find that the solution of such equations depends on the proper use of identities. In the study of calculus, there are certain types of problems which require the use of trigonometric identities for solution (even problems in which trigonometric functions do not appear). Also, they are used in developing expressions and solving equations in certain technical areas.

Several important identities exist among the six trigonometric functions, and we shall develop these identities in this section. We shall also show how we can use the basic identities to verify other identities among the functions.

By the definitions, we have

$$\sin \theta \, \csc \theta = \frac{y}{r} \cdot \frac{r}{y} = 1 \quad \text{or} \quad \sin \theta = \frac{1}{\csc \theta} \quad \text{or} \quad \csc \theta = \frac{1}{\sin \theta}$$

$$\cos \theta \, \sec \theta = \frac{x}{r} \cdot \frac{r}{x} = 1 \quad \text{or} \quad \cos \theta = \frac{1}{\sec \theta} \quad \text{or} \quad \sec \theta = \frac{1}{\cos \theta}$$

$$\tan \theta \, \cot \theta = \frac{y}{x} \cdot \frac{x}{y} = 1 \quad \text{or} \quad \tan \theta = \frac{1}{\cot \theta} \quad \text{or} \quad \cot \theta = \frac{1}{\tan \theta}$$

$$\frac{\sin \theta}{\cos \theta} = \frac{y/r}{x/r} = \frac{y}{x} = \tan \theta; \qquad \frac{\cos \theta}{\sin \theta} = \frac{x/r}{y/r} = \frac{x}{y} = \cot \theta$$

Also, by the definitions and the Pythagorean theorem in the form of $x^2 + y^2 = r^2$, we arrive at the following identities:

By dividing the Pythagorean relation through by r^2, we have

$$\left(\frac{x}{r}\right)^2 + \left(\frac{y}{r}\right)^2 = 1 \quad \text{which leads us to } \cos^2\theta + \sin^2\theta = 1$$

By dividing the Pythagorean relation by x^2, we have

$$1 + \left(\frac{y}{x}\right)^2 = \left(\frac{r}{x}\right)^2 \quad \text{which leads us to } 1 + \tan^2\theta = \sec^2\theta$$

By dividing the Pythagorean relation by y^2, we have

$$\left(\frac{x}{y}\right)^2 + 1 = \left(\frac{r}{y}\right)^2 \quad \text{which leads us to } \cot^2\theta + 1 = \csc^2\theta$$

The term $\cos^2 \theta$ is the common way of writing $(\cos \theta)^2$, and thus it means to square the value of the cosine of the angle. Obviously the same holds true for the other functions.

Summarizing these results, we have the following important identities among the trigonometric functions:

(19–1)	$\sin \theta = \dfrac{1}{\csc \theta}$	$\cos \theta = \dfrac{1}{\sec \theta}$ (19–2)
(19–3)	$\tan \theta = \dfrac{1}{\cot \theta}$	$\tan \theta = \dfrac{\sin \theta}{\cos \theta}$ (19–4)
(19–5)	$\cot \theta = \dfrac{\cos \theta}{\sin \theta}$	$\sin^2\theta + \cos^2\theta = 1$ (19–6)
(19–7)	$1 + \tan^2\theta = \sec^2\theta$	$1 + \cot^2\theta = \csc^2\theta$ (19–8)

In using these identities, θ may stand for any angle or number or expression representing an angle or number.

Example A

$$\sin (x + 1) = \frac{1}{\csc (x + 1)}$$

$$\tan 157° = \frac{\sin 157°}{\cos 157°}, \quad 1 + \tan^2\left(\frac{\pi}{6}\right) = \sec^2\left(\frac{\pi}{6}\right)$$

Example B

We shall verify three of the identities for particular values of θ.

From Table 3, we find that $\cos 53° = 0.6018$ and $\sec 53° = 1.662$. Using Eq. (19–2), and dividing, we find that

$$\cos 53° = \frac{1}{\sec 53°} = \frac{1}{1.662} = 0.6018$$

and this value checks.

Using Table 3, we find that $\sin 157° = 0.3907$ and $\cos 157° = -0.9205$. Using Eq. (19–4), and dividing, we find that

$$\tan 157° = \frac{\sin 157°}{\cos 157°} = \frac{0.3907}{-0.9205} = -0.4245$$

Checking with Table 3, we see that this value checks.

From Section 3–3, we recall that

$$\sin 45° = \frac{1}{\sqrt{2}} = \frac{\sqrt{2}}{2} \quad \text{and} \quad \cos 45° = \frac{\sqrt{2}}{2}$$

Using Eq. (19–6), we have

$$\sin^2 45° + \cos^2 45° = \left(\frac{\sqrt{2}}{2}\right)^2 + \left(\frac{\sqrt{2}}{2}\right)^2 = \frac{1}{2} + \frac{1}{2} = 1$$

We see that this identity checks for these values.

A great many identities exist among the trigonometric functions. We are going to use the basic identities already developed in Eqs. (19—1) through (19—8), along with a few additional ones developed in later sections, to prove the validity of still other identities. **The ability to prove such identities depends to a large extent on being *very* familiar with the basic identities,** so that you can recognize them in somewhat different forms. If you do not learn these basic identities and learn them well, you will have difficulty in following the examples and doing the exercises. The more readily you recognize these forms, the more easily you will be able to prove such identities.

In proving identities, we should look for combinations which appear in, or are very similar to, those in the basic identities. Consider the following examples.

Example C

In proving the identity

$$\sin x = \frac{\cos x}{\cot x}$$

we know that $\cot x = \frac{\cos x}{\sin x}$. Since $\sin x$ appears on the left, substituting for $\cot x$ on the right will eliminate $\cot x$ and introduce $\sin x$. This should help us proceed in proving the identity. Thus,

$$\sin x = \frac{\cos x}{\cot x}$$

$$= \frac{\cos x}{\frac{\cos x}{\sin x}} = \frac{\cos x}{1} \cdot \frac{\sin x}{\cos x}$$

$$= \sin x$$

By showing that the right side may be changed exactly to $\sin x$, the expression on the left side, we have proved the identity.

Some important points should be made in relation to the proof of the identity of Example C. We must recognize what basic identities may be useful. The proof of an identity requires the use of basic algebraic operations, and these must be done carefully and correctly. Although in Example C we changed the right side to the form on the left, we could have changed the left to the form on the right. From this, and the fact that various substitutions are possible, we see that there is a variety of procedures which can be used to prove any given identity.

Example D

Prove that $\tan \theta \csc \theta = \sec \theta$.

In proving this identity we know that $\tan \theta = \dfrac{\sin \theta}{\cos \theta}$ and also that $\dfrac{1}{\cos \theta} = \sec \theta$. Thus, by substituting for $\tan \theta$ we introduce $\cos \theta$ in the denominator, which is equivalent to introducing $\sec \theta$ in the numerator. Therefore, changing only the left side, we have

$$\tan \theta \csc \theta = \sec \theta$$

$$\frac{\sin \theta}{\cos \theta} \csc \theta = \sec \theta$$

$$\frac{1}{\cos \theta} \sin \theta \csc \theta = \sec \theta$$

$$\sec \theta \sin \theta \frac{1}{\sin \theta} = \sec \theta$$

$$\sec \theta = \sec \theta$$

Many variations of the preceding steps are possible. Also, we could have changed only the right side to obtain the form on the left. For example,

$$\tan \theta \csc \theta = \sec \theta$$

$$= \frac{1}{\cos \theta} = \frac{\sin \theta}{\cos \theta \sin \theta} = \frac{\sin \theta}{\cos \theta} \frac{1}{\sin \theta}$$

$$= \tan \theta \csc \theta$$

In proving the identities of Examples C and D we have shown that the expression on one side of the equals sign can be changed into the expression on the other side. Although making the restriction that we change only one side is not entirely necessary, *we shall restrict the method of proof to changing only one side into the same form as the other side.* In this way, we know precisely what form we are to change to, and therefore by looking ahead we are better able to make the proper changes.

There is no set procedure which can be stated for working with identities. The most important factors are to be able to *recognize the proper forms,* to be able to *see what effect any change may have* before we actually perform it, and then *perform it correctly.* Normally *it is easier to change the more complicated side of an identity to the same form as the less complicated side.* If the two sides are of approximately the same complexity, a close look at each side usually suggests steps which will lead to the solution.

Example E

Prove the identity $\dfrac{\cos x \csc x}{\cot^2 x} = \tan x$.

First, we note that the lefthand side has several factors and the right-hand side has only one. Therefore, let us transform the lefthand side. Next, we note that we want tan x as the final result. We know that $\cot x = 1/\tan x$. Thus

$$\frac{\cos x \csc x}{\cot^2 x} = \frac{\cos x \csc x}{1/\tan^2 x} = \cos x \csc x \tan^2 x$$

At this point, we have two factors of tan x on the left. Since we want only one, let us factor out one. Therefore,

$$\cos x \csc x \tan^2 x = \tan x(\cos x \csc x \tan x)$$

Now, replacing tan x within the parentheses by sin $x/\cos x$, we have

$$\tan x(\cos x \csc x \tan x) = \frac{\tan x(\cos x \csc x \sin x)}{\cos x}$$

Now we may cancel cos x. Also, $\csc x \sin x = 1$ from Eq. (19–1). Finally,

$$\frac{\tan x(\cos x \csc x \sin x)}{\cos x} = \tan x\left(\frac{\cos x}{\cos x}\right)(\csc x \sin x)$$

$$= \tan x(1)(1) = \tan x$$

Since we have transformed the lefthand side into tan x, we have proven the identity. Of course, it is not necessary to rewrite expressions as we did in this example. This was done here only to include the explanations between steps.

Example F

Prove the identity $\sin^4 x - \cos^4 x = \sin^2 x - \cos^2 x$.

We note that either side of this identity may be factored, but that one of the factors of the left side is the expression which appears on the right. Therefore, by factoring, and the use of Eq. (19–6), we have (changing only the form of the left side)

$$\sin^4 x - \cos^4 x = \sin^2 x - \cos^2 x$$
$$(\sin^2 x - \cos^2 x)(\sin^2 x + \cos^2 x) = \sin^2 x - \cos^2 x$$
$$(\sin^2 x - \cos^2 x)(1) = \sin^2 x - \cos^2 x$$
$$\sin^2 x - \cos^2 x = \sin^2 x - \cos^2 x$$

Example G

Prove the identity $\dfrac{\sec^2 y}{\cot y} - \tan^3 y = \tan y$.

Here we shall simplify the left side. We note that we can remove cot y from the denominator since tan $y = 1/\cot y$. Also, the presence of $\sec^2 y$ suggests the use of Eq. (19–7). Therefore, we have

$$\frac{\sec^2 y}{\cot y} - \tan^3 y = \frac{\sec^2 y}{1/\tan y} - \tan^3 y = \sec^2 y \tan y - \tan^3 y$$

$$= \tan y(\sec^2 y - \tan^2 y) = \tan y(1) = \tan y$$

or

$$\frac{\sec^2 y}{\cot y} - \tan^3 y = \tan y$$

Here we have used Eq. (19–7) in the form $\sec^2 y - \tan^2 y = 1$.

Example H

Prove the identity $\dfrac{1 - \sin x}{\sin x \cot x} = \dfrac{\cos x}{1 + \sin x}$.

The combination $1 - \sin x$ also suggests $1 - \sin^2 x$, since multiplying $(1 - \sin x)$ by $(1 + \sin x)$ gives $1 - \sin^2 x$, which can then be replaced by $\cos^2 x$. Thus, changing only the left side we have

$$\frac{(1 - \sin x)}{\sin x \cot x} = \frac{\cos x}{1 + \sin x}$$

$$\frac{(1 - \sin x)(1 + \sin x)}{\sin x \cot x(1 + \sin x)} = \frac{\cos x}{1 + \sin x}$$

$$\frac{1 - \sin^2 x}{\sin x \left(\dfrac{\cos x}{\sin x}\right)(1 + \sin x)} = \frac{\cos x}{1 + \sin x}$$

$$\frac{\cos^2 x}{\cos x(1 + \sin x)} = \frac{\cos x}{1 + \sin x}$$

$$\frac{\cos x}{1 + \sin x} = \frac{\cos x}{1 + \sin x}$$

Example I

Prove the identity $\sec^2 x + \csc^2 x = \sec^2 x \csc^2 x$.

Here we note the presence of $\sec^2 x$ and $\csc^2 x$ on each side. This suggests the possible use of the square relationships. By replacing the $\sec^2 x$ on the righthand side by $1 + \tan^2 x$, we can create $\csc^2 x$ plus another term. The lefthand side is the $\csc^2x$ plus another term, so this procedure should help. Thus, changing only the right side,

$$\sec^2x + \csc^2x = \sec^2x \csc^2x$$
$$= (1 + \tan^2x)(\csc^2x)$$
$$= \csc^2x + \tan^2x \csc^2x$$

Now we note that $\tan x = \sin x/\cos x$ and $\csc x = 1/\sin x$. Thus,

$$\sec^2x + \csc^2x = \csc^2x + \left(\frac{\sin^2x}{\cos^2x}\right)\left(\frac{1}{\sin^2x}\right)$$

$$= \csc^2x + \frac{1}{\cos^2x}$$

$$= \csc^2x + \sec^2x$$

We could have used many other variations of this procedure, and they would have been perfectly valid.

Example J

Prove the identity $\dfrac{\csc x}{\tan x + \cot x} = \cos x$.

Here we shall simplify the lefthand side until we have the expression which appears on the righthand side.

$$\frac{\csc x}{\tan x + \cot x} = \frac{\csc x}{\tan x + \dfrac{1}{\tan x}} = \frac{\csc x}{\dfrac{\tan^2x + 1}{\tan x}}$$

$$= \frac{\csc x \tan x}{\tan^2x + 1} = \frac{\csc x \tan x}{\sec^2x}$$

$$= \frac{\dfrac{1}{\sin x} \cdot \dfrac{\sin x}{\cos x}}{\dfrac{1}{\cos^2x}} = \frac{1}{\sin x} \cdot \frac{\sin x}{\cos x} \cdot \frac{\cos^2x}{1}$$

$$= \cos x$$

Therefore, we have shown that $\dfrac{\csc x}{\tan x + \cot x} = \cos x$ which proves the identity.

Exercises 19–1

In Exercises 1 through 4 verify the indicated basic identities for the given angles.

1. Verify Eq. (19–3) for $\theta = 56°$ 2. Verify Eq. (19–5) for $\theta = 80°$

3. Verify Eq. (19–6) for $\theta = \dfrac{2\pi}{3}$ 4. Verify Eq. (19–8) for $\theta = \dfrac{7\pi}{6}$

In Exercises 5 through 48 prove the given identities.

5. $\dfrac{\cot \theta}{\cos \theta} = \csc \theta$ 6. $\dfrac{\tan y}{\sin y} = \sec y$ 7. $\dfrac{\sin x}{\tan x} = \cos x$ 8. $\dfrac{\csc \theta}{\sec \theta} = \cot \theta$

9. $\sin y \cot y = \cos y$ 10. $\cos x \tan x = \sin x$ 11. $\sin x \sec x = \tan x$

12. $\cot \theta \sec \theta = \csc \theta$ 13. $\csc^2 x (1 - \cos^2 x) = 1$ 14. $\cos^2 x (1 + \tan^2 x) = 1$

15. $\sin x (1 + \cot^2 x) = \csc x$ 16. $\sec \theta (1 - \sin^2\theta) = \cos \theta$

17. $\sin x (\csc x - \sin x) = \cos^2 x$ 18. $\cos y (\sec y - \cos y) = \sin^2 y$

19. $\tan y (\cot y + \tan y) = \sec^2 y$ 20. $\csc x (\csc x - \sin x) = \cot^2 x$

21. $\sin x \tan x + \cos x = \sec x$ 22. $\sec x \csc x - \cot x = \tan x$

23. $\cos \theta \cot \theta + \sin \theta = \csc \theta$ 24. $\csc x \sec x - \tan x = \cot x$

25. $\sec \theta \tan \theta \csc \theta = \tan^2\theta + 1$ 26. $\sin x \cos x \tan x = 1 - \cos^2 x$

27. $\cot \theta \sec^2\theta - \cot \theta = \tan \theta$ 28. $\sin y + \sin y \cot^2 y = \csc y$

29. $\tan x + \cot x = \sec x \csc x$ 30. $\tan x + \cot x = \tan x \csc^2 x$

31. $\cos^2 x - \sin^2 x = 1 - 2\sin^2 x$ 32. $\tan^2 y \sec^2 y - \tan^4 y = \tan^2 y$

33. $\dfrac{\sin x}{1 - \cos x} = \csc x + \cot x$ 34. $\dfrac{1 + \cos x}{\sin x} = \dfrac{\sin x}{1 - \cos x}$

35. $\dfrac{\sec x + \csc x}{1 + \tan x} = \csc x$ 36. $\dfrac{\cot x + 1}{\cot x} = 1 + \tan x$

37. $\tan^2 x \cos^2 x + \cot^2 x \sin^2 x = 1$ 38. $\cos^3 x \csc^3 x \tan^3 x = \csc^2 x - \cot^2 x$

39. $4 \sin x + \tan x = \dfrac{4 + \sec x}{\csc x}$ 40. $\dfrac{1 + \tan x}{\sin x} - \sec x = \csc x$

41. $\sec x + \tan x + \cot x = \dfrac{1 + \sin x}{\cos x \sin x}$

42. $\sec x (\sec x - \cos x) + \dfrac{\cos x - \sin x}{\cos x} + \tan x = \sec^2 x$

43. $2 \sin^4 x - 3 \sin^2 x + 1 = \cos^2 x (1 - 2 \sin^2 x)$ 44. $\dfrac{\sin^4 x - \cos^4 x}{1 - \cot^4 x} = \sin^4 x$

45. $\dfrac{\cot 2y}{\sec 2y - \tan 2y} - \dfrac{\cos 2y}{\sec 2y + \tan 2y} = \sin 2y + \csc 2y$

46. $\dfrac{1}{2} \sin 5y \left(\dfrac{\sin 5y}{1 - \cos 5y} + \dfrac{1 - \cos 5y}{\sin 5y} \right) = 1$

47. $1 + \sin^2 x + \sin^4 x + \cdots = \sec^2 x$ 48. $1 - \tan^2 x + \tan^4 x - \cdots = \cos^2 x$

In Exercises 49 through 52 prove the identities which arise in the given area of application.

49. For an object of weight w on an inclined plane the coefficient of friction μ, related to the frictional force between the plane and object, can be found from the equation $\mu w \cos \theta = w \sin \theta$, where θ is the angle between the plane and horizontal. Solve for μ and show that $\mu = \tan \theta$.

50. In finding the change in $\tan^2 x$ for a given change in x, the expression $2 \tan x \sec^2 x$ is used. Show that this expression is equal to $2 \sin x \sec^3 x$.

51. In determining the rate of radiation by an accelerated electric charge, it is necessary to show that $\sin^3\theta = \sin \theta - \cos^2\theta \sin \theta$. Show that this is valid, by transforming the lefthand side.

52. In determining the path of least time between two points under certain circumstances, it is necessary to show that

$$\sqrt{\frac{1 + \cos \theta}{1 - \cos \theta}} \sin \theta = (1 + \cos \theta)$$

Show this by transforming the lefthand side."

19–2 Sine and Cosine of the Sum and Difference of Two Angles

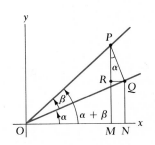

Figure 19–1

There are other important relations among the trigonometric functions. The most important and useful relations are those which involve twice an angle and half an angle. To obtain these relations, we shall first derive the expressions for the sine and cosine of the sum and difference of two angles. These expressions will lead directly to the desired relations of double and half angles.

In Fig. 19–1, the angle α is in standard position, and the angle β has as its initial side the terminal side of α. Thus the angle of interest, $\alpha + \beta$, is in standard position. From a point P, on the terminal side of $\alpha + \beta$, perpendiculars are dropped to the x-axis and to the terminal side of α, at the points M and Q respectively. Then perpendiculars are dropped from Q to the x-axis and to the line MP at points N and R respectively. By this construction, $\angle RPQ$ is equal to $\angle \alpha$. ($\angle RQO = \angle \alpha$ by alternate interior angles; $\angle RQO + \angle RQP = 90°$ by construction; $\angle RQP + \angle RPQ = 90°$ by the sum of the angles of a triangle being $180°$. Thus, $\angle \alpha = \angle RPQ$.) By definition,

$$\sin(\alpha + \beta) = \frac{MP}{OP} = \frac{MR + RP}{OP} = \frac{NQ}{OP} + \frac{RP}{OP}$$

These last two fractions do not define any function of either α or β, and therefore we multiply the first fraction (numerator and denominator) by OQ and the second fraction by QP. When we do this and rearrange the fractions, we have functions of α and β:

$$\sin(\alpha + \beta) = \frac{NQ}{OP} + \frac{RP}{OP} = \frac{NQ}{OQ} \cdot \frac{OQ}{OP} + \frac{RP}{QP} \cdot \frac{QP}{OP}$$

$$= \sin \alpha \cos \beta + \cos \alpha \sin \beta$$

Using the same figure, we can also obtain the expression for $\cos(\alpha + \beta)$. Thus we have the relations

$$\sin(\alpha + \beta) = \sin \alpha \cos \beta + \cos \alpha \sin \beta \qquad (19\text{–}9)$$

and

$$\cos(\alpha + \beta) = \cos \alpha \cos \beta - \sin \alpha \sin \beta \qquad (19\text{–}10)$$

Example A

Find sin 75° from sin 75° = sin(45° + 30°).

$$\sin 75° = \sin(45° + 30°) = \sin 45° \cos 30° + \cos 45° \sin 30°$$

$$= \frac{\sqrt{2}}{2} \cdot \frac{\sqrt{3}}{2} + \frac{\sqrt{2}}{2} \cdot \frac{1}{2} = \frac{\sqrt{6}}{4} + \frac{\sqrt{2}}{4} = \frac{\sqrt{6} + \sqrt{2}}{4}$$

$$= 0.9659$$

Example B

Verify that sin 90° = 1, by finding sin(60° + 30°).

$$\sin 90° = \sin(60° + 30°) = \sin 60° \cos 30° + \cos 60° \sin 30°$$

$$= \frac{\sqrt{3}}{2} \cdot \frac{\sqrt{3}}{2} + \frac{1}{2} \cdot \frac{1}{2} = \frac{3}{4} + \frac{1}{4} = 1$$

[It should be obvious from this example that $\sin(\alpha + \beta)$ is *not* equal to $\sin \alpha + \sin \beta$, something which many students assume before they are familiar with the formulas and ideas of this section. If we used such a formula, we would get $\sin 90° = \frac{1}{2}\sqrt{3} + \frac{1}{2} = 1.366$ for the combination (60° + 30°). This is not possible, since the values of the sine never exceed 1 in value. Also, if we used the combination (45° + 45°), we would get 1.414, a different value for the same number, sin 90°.]

From Eq. (19–9) and (19–10), we can easily find expressions for $\sin(\alpha - \beta)$ and $\cos(\alpha - \beta)$. This is done by finding $\sin(\alpha + (-\beta))$ and $\cos(\alpha + (-\beta))$. Thus, we have

$$\sin(\alpha - \beta) = \sin(\alpha + (-\beta)) = \sin \alpha \cos(-\beta) + \cos \alpha \sin(-\beta)$$

Since $\cos(-\beta) = \cos \beta$ and $\sin(-\beta) = -\sin \beta$ (see Exercise 61 of Section 7–2), we have

$$\sin(\alpha - \beta) = \sin \alpha \cos \beta - \cos \alpha \sin \beta \qquad (19\text{--}11)$$

In the same manner we find that

$$\cos(\alpha - \beta) = \cos \alpha \cos \beta + \sin \alpha \sin \beta \qquad (19\text{--}12)$$

Example C

Find cos 15° from cos(45° − 30°).

$$\cos 15° = \cos(45° - 30°) = \cos 45° \cos 30° + \sin 45° \sin 30°$$

$$= \frac{\sqrt{2}}{2} \cdot \frac{\sqrt{3}}{2} + \frac{\sqrt{2}}{2} \cdot \frac{1}{2} = \frac{\sqrt{6} + \sqrt{2}}{4} = 0.9659$$

We get the same result as in Example A, which should be the case since sin 75° = cos 15°. (See Section 3–4.)

By using Eqs. (19–9) and (19–10), we can determine expressions for $\tan(\alpha + \beta)$, $\cot(\alpha + \beta)$, $\sec(\alpha + \beta)$, and $\csc(\alpha + \beta)$. These expressions are less applicable than those for the sine and cosine, and therefore we shall not derive them here, although the expression for $\tan(\alpha + \beta)$' is found in the exercises at the end of this section. By using Eqs. (19–11) and (19–12), we can find similar expressions for the functions of $(\alpha - \beta)$.

Certain trigonometric identities can also be worked out by using the formulas derived in this section. The following examples illustrate the use of these formulas in identities.

Example D

Prove that $\sin(180° + x) = -\sin x$.

By using Eq. (19–9) we have

$$\sin(180° + x) = \sin 180° \cos x + \cos 180° \sin x$$

Since $\sin 180° = 0$ and $\cos 180° = -1$, we have

$$\sin 180° \cos x + \cos 180° \sin x = (0) \cos x + (-1) \sin x$$

or

$$\sin(180° + x) = -\sin x$$

Example E

Show that $\dfrac{\sin(\alpha - \beta)}{\sin \alpha \sin \beta} = \cot \beta - \cot \alpha$.

By using Eq. (19–11), we have

$$\frac{\sin(\alpha - \beta)}{\sin \alpha \sin \beta} = \frac{\sin \alpha \cos \beta - \cos \alpha \sin \beta}{\sin \alpha \sin \beta} = \frac{\sin \alpha \cos \beta}{\sin \alpha \sin \beta} - \frac{\cos \alpha \sin \beta}{\sin \alpha \sin \beta}$$

$$= \frac{\cos \beta}{\sin \beta} - \frac{\cos \alpha}{\sin \alpha} = \cot \beta - \cot \alpha$$

Example F

Show that

$$\sin\left(\frac{\pi}{4} + x\right)\cos\left(\frac{\pi}{4} + x\right) = \frac{1}{2}(\cos^2 x - \sin^2 x)$$

$$\sin\left(\frac{\pi}{4} + x\right)\cos\left(\frac{\pi}{4} + x\right)$$

$$= \left(\sin\frac{\pi}{4}\cos x + \cos\frac{\pi}{4}\sin x\right)\left(\cos\frac{\pi}{4}\cos x - \sin\frac{\pi}{4}\sin x\right)$$

$$= \sin\frac{\pi}{4}\cos\frac{\pi}{4}\cos^2 x - \sin^2\frac{\pi}{4}\sin x \cos x + \cos^2\frac{\pi}{4}\sin x \cos x - \sin^2 x \sin\frac{\pi}{4}\cos\frac{\pi}{4}$$

$$= \frac{\sqrt{2}}{2}\frac{\sqrt{2}}{2}\cos^2 x - \left(\frac{\sqrt{2}}{2}\right)^2 \sin x \cos x + \left(\frac{\sqrt{2}}{2}\right)^2 \sin x \cos x - \frac{\sqrt{2}}{2}\frac{\sqrt{2}}{2}\sin^2 x$$

$$= \frac{1}{2}\cos^2 x - \frac{1}{2}\sin^2 x = \frac{1}{2}(\cos^2 x - \sin^2 x)$$

Example G

Show that $\sin(x + y) \cos y - \cos(x + y) \sin y = \sin x$.

If we let $x + y = z$, we note that the lefthand side of the above expression becomes $\sin z \cos y - \cos z \sin y$, which is the proper form for $\sin(z - y)$. By replacing z with $x + y$, we obtain $\sin(x + y - y)$, which is $\sin x$. Therefore, the above expression has been shown to be true. We again see that proper recognition of a basic form leads to the solution.

Exercises 19–2

In Exercises 1 through 4 determine the values of the given functions as indicated.

1. Find $\sin 105°$ by using $105° = 60° + 45°$.
2. Find $\cos 75°$ by using $75° = 30° + 45°$.
3. Find $\cos 15°$ by using $15° = 60° - 45°$.
4. Find $\sin 15°$ by using $15° = 45° - 30°$.

In Exercises 5 through 8 evaluate the indicated functions with the following given information: $\sin \alpha = \frac{4}{5}$ (in first quadrant), and $\cos \beta = -\frac{12}{13}$ (in second quadrant).

5. $\sin(\alpha + \beta)$ 6. $\cos(\beta - \alpha)$ 7. $\cos(\alpha + \beta)$ 8. $\sin(\alpha - \beta)$

In Exercises 9 through 12 reduce each of the given expressions to a single term. Expansion of any term is not necessary; proper recognition of the form of the expression leads to the proper result.

9. $\sin x \cos 2x + \sin 2x \cos x$
10. $\sin 3x \cos x - \sin x \cos 3x$
11. $\cos(x + y)\cos y + \sin(x + y)\sin y$
12. $\cos(2x - y)\cos y - \sin(2x - y)\sin y$

In Exercises 13 through 24 prove the given identities.

13. $\sin(270° - x) = -\cos x$

14. $\sin(90° + x) = \cos x$

15. $\cos\left(\frac{\pi}{2} - x\right) = \sin x$

16. $\cos\left(\frac{3\pi}{2} + x\right) = \sin x$

17. $\cos(30° + x) = \dfrac{\sqrt{3} \cos x - \sin x}{2}$

18. $\sin(120° - x) = \dfrac{\sqrt{3} \cos x + \sin x}{2}$

19. $\sin\left(\frac{\pi}{4} + x\right) = \dfrac{\sin x + \cos x}{\sqrt{2}}$

20. $\cos\left(\frac{\pi}{3} + x\right) = \dfrac{\cos x - \sqrt{3} \sin x}{2}$

21. $\sin(x + y)\sin(x - y) = \sin^2 x - \sin^2 y$
22. $\cos(x + y)\cos(x - y) = \cos^2 x - \sin^2 y$
23. $\cos(\alpha + \beta) + \cos(\alpha - \beta) = 2 \cos \alpha \cos \beta$
24. $\cos(x - y) + \sin(x + y) = (\cos x + \sin x)(\cos y + \sin y)$

In Exercises 25 through 28 additional trigonometric identities are shown. Derive these in the indicated manner. Equations (19–13), (19–14), and (19–15) are known as the product formulas.

25. By dividing Eq. (19–9) by Eq. (19–10), show that

$$\tan(\alpha + \beta) = \frac{\tan \alpha + \tan \beta}{1 - \tan \alpha \tan \beta}$$

[*Hint*: Divide numerator and denominator by $\cos \alpha \cos \beta$.]

26. By adding Eqs. (19–9) and (19–11), derive the equation

$$\sin \alpha \cos \beta = \tfrac{1}{2}[\sin (\alpha + \beta) + \sin (\alpha - \beta)] \tag{19–13}$$

27. By adding Eqs. (19–10) and (19–12), derive the equation

$$\cos \alpha \cos \beta = \tfrac{1}{2}[\cos (\alpha + \beta) + \cos (\alpha - \beta)] \tag{19–14}$$

28. By subtracting Eq. (19–10) from Eq. (19–12), derive

$$\sin \alpha \sin \beta = \tfrac{1}{2}[\cos (\alpha - \beta) - \cos (\alpha + \beta)] \tag{19–15}$$

In Exercises 29 through 32 additional trigonometric identities are shown. Derive them by letting $\alpha + \beta = x$ and $\alpha - \beta = y$, which leads to $\alpha = \tfrac{1}{2}(x + y)$ and $\beta = \tfrac{1}{2}(x - y)$. The resulting equations are known as the factor formulas.

29. Use Eq. (19–13) and the substitutions above to derive the equation

$$\sin x + \sin y = 2 \sin \tfrac{1}{2}(x + y) \cos \tfrac{1}{2}(x - y) \tag{19–16}$$

30. Use Eqs. (19–9) and (19–11) and the substitutions above to derive the equation

$$\sin x - \sin y = 2 \sin \tfrac{1}{2}(x - y) \cos \tfrac{1}{2}(x + y) \tag{19–17}$$

31. Use Eq. (19–14) and the substitutions above to derive the equation

$$\cos x + \cos y = 2 \cos \tfrac{1}{2}(x + y) \cos \tfrac{1}{2}(x - y) \tag{19–18}$$

32. Use Eq. (19–15) and the substitutions above to derive the equation

$$\cos x - \cos y = -2 \sin \tfrac{1}{2}(x + y) \sin \tfrac{1}{2}(x - y) \tag{19–19}$$

In Exercises 33 through 36 use the equations of this section to solve the given problems.

33. In determining the motion of an object, the expression $\cos \alpha \sin (\omega t + \phi) - \sin \alpha \cos (\omega t + \phi)$ is found. Simplify this expression.

34. Under certain conditions, the current in an electric circuit is given by

$$i = I_m[\sin(\omega t + \alpha) \cos \phi + \cos(\omega t + \alpha) \sin \phi]$$

Simplify the expression on the right.

35. The displacements y_1 and y_2 of two waves traveling through the same medium are given by the equations $y_1 = A \sin 2\pi(t/T - x/\lambda)$ and $y_2 = A \sin 2\pi(t/T + x/\lambda)$. Find an expression for the displacement $y_1 + y_2$ of the combination of the waves.

36. In the analysis of the angles of incidence i and reflection r of a light ray subject to certain conditions, the following expression is found:

$$E_2\left(\frac{\tan r}{\tan i} + 1\right) = E_1\left(\frac{\tan r}{\tan i} - 1\right)$$

Show that an equivalent expression is

$$E_2 = E_1 \frac{\sin(r - i)}{\sin(r + i)}$$

19—3 Double-Angle Formulas

If we let $\beta = \alpha$ in Eqs. (19–9) and (19–10), we can derive the important double-angle formulas. Thus, by making this substitution in Eq. (19–9), we have

$$\sin(\alpha + \alpha) = \sin(2\alpha) = \sin \alpha \cos \alpha + \cos \alpha \sin \alpha = 2 \sin \alpha \cos \alpha$$

Using the same substitution in Eq. (19–10), we have

$$\cos(\alpha + \alpha) = \cos \alpha \cos \alpha - \sin \alpha \sin \alpha = \cos^2\alpha - \sin^2\alpha$$

By using the basic identity (19–6), other forms of this last equation may be derived. Thus, summarizing these formulas, we have

$$\sin 2\alpha = 2 \sin \alpha \cos \alpha \tag{19–20}$$
$$\cos 2\alpha = \cos^2\alpha - \sin^2\alpha \tag{19–21}$$
$$= 2 \cos^2\alpha - 1 \tag{19–22}$$
$$= 1 - 2 \sin^2\alpha \tag{19–23}$$

We should note carefully that these equations give expressions for the sine and cosine of twice an angle in terms of functions of the angle. They can be used any time we have expressed one angle as twice another. These double-angle formulas are widely used in applications of trigonometry, especially in the calculus. They should be known and recognized quickly in any of the various forms.

Example A

If $\alpha = 30°$, we have $\cos 2(30°) = \cos 60° = \cos^2 30° - \sin^2 30°$.
If $\alpha = 3x$, we have $\sin 2(3x) = \sin 6x = 2 \sin 3x \cos 3x$.
If $2\alpha = x$, we may write $\alpha = x/2$, which means that

$$\sin 2\left(\frac{x}{2}\right) = \sin x = 2 \sin \frac{x}{2} \cos \frac{x}{2}$$

Example B

Using the double-angle formulas, simplify the expression

$$\cos^2 2x - \sin^2 2x$$

By using Eq. (19–21) and letting $\alpha = 2x$, we have

$$\cos^2 2x - \sin^2 2x = \cos 2(2x) = \cos 4x$$

Example C
Verify the values of sin 90° and cos 90° by use of the functions of 45°.

$$\sin 90° = \sin 2(45°) = 2 \sin 45° \cos 45° = 2\left(\frac{\sqrt{2}}{2}\right)\left(\frac{\sqrt{2}}{2}\right) = 1$$

$$\cos 90° = \cos 2(45°) = \cos^2 45° - \sin^2 45° = \left(\frac{\sqrt{2}}{2}\right)^2 - \left(\frac{\sqrt{2}}{2}\right)^2 = 0$$

Example D
Given that $\cos \alpha = \frac{3}{5}$ (in the fourth quadrant), find $\sin 2\alpha$.

Knowing that $\cos \alpha = \frac{3}{5}$ for an angle in the fourth quadrant, we then determine that

$$\sin \alpha = -\frac{4}{5}$$

(see Fig. 19–2). Thus,

$$\sin 2\alpha = 2\left(-\frac{4}{5}\right)\left(\frac{3}{5}\right) = -\frac{24}{25}$$

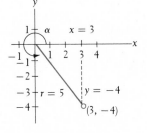

Figure 19–2

Example E

Prove the identity $\dfrac{2}{1 + \cos 2x} = \sec^2 x$.

$$\frac{2}{1 + \cos 2x} = \frac{2}{1 + (2 \cos^2 x - 1)} = \frac{2}{2 \cos^2 x} = \sec^2 x$$

Example F

Show that $\dfrac{\sin 3x}{\sin x} + \dfrac{\cos 3x}{\cos x} = 4 \cos 2x$.

The first step is to combine the two fractions on the left, so that we can see if any usable forms will emerge:

$$\frac{\sin 3x \cos x + \cos 3x \sin x}{\sin x \cos x} \overset{?}{=} 4 \cos 2x$$

We now note that the numerator is of the form $\sin (A + x)$, where $A = 3x$. Also, the denominator is $\frac{1}{2} \sin 2x$. Making these substitutions, we have

$$\frac{\sin(3x + x)}{\frac{1}{2} \sin 2x} = \frac{2 \sin 4x}{\sin 2x} \overset{?}{=} 4 \cos 2x$$

By expanding sin 4x into 2 sin 2x cos 2x, we obtain

$$\frac{2(2 \sin 2x \cos 2x)}{\sin 2x} = 4 \cos 2x$$

Therefore, the expression is shown to be valid.

Exercises 19–3

In Exercises 1 through 4 determine the values of the indicated functions in the given manner.

1. Find $\sin 60°$ by using the functions of $30°$.
2. Find $\sin 120°$ by using the functions of $60°$.
3. Find $\cos 120°$ by using the functions of $60°$.
4. Find $\cos 60°$ by using the functions of $30°$.

In Exercises 5 through 8 evaluate the indicated functions with the given information.

5. Find $\sin 2x$ if $\cos x = \frac{4}{5}$ (in first quadrant).
6. Find $\cos 2x$ if $\sin x = -\frac{12}{13}$ (in third quadrant).
7. Find $\cos 2x$ if $\tan x = \frac{1}{2}$ (in third quadrant).
8. Find $\sin 4x$ if $\sin x = \frac{3}{5}$ (in first quadrant) $[4x = 2(2x)]$.

In Exercises 9 through 12 reduce the given expressions to a single term. Expansion of any term is not necessary; proper recognition of the form of the expression leads to the proper result.

9. $4 \sin 4x \cos 4x$
10. $4 \sin^2 x \cos^2 x$
11. $1 - 2 \sin^2 4x$
12. $\sin^2 4x - \cos^2 4x$

In Exercises 13 through 20 prove the given identities.

13. $\cos^2 \alpha - \sin^2 \alpha = 2 \cos^2 \alpha - 1$
14. $\cos^2 \alpha - \sin^2 \alpha = 1 - 2 \sin^2 \alpha$
15. $\cos^4 x - \sin^4 x = \cos 2x$
16. $(\sin x + \cos x)^2 = 1 + \sin 2x$

17. $2 \csc 2x \tan x = \sec^2 x$
18. $2 \sin x + \sin 2x = \dfrac{2 \sin^3 x}{1 - \cos x}$

19. $\dfrac{\sin 3x}{\sin x} - \dfrac{\cos 3x}{\cos x} = 2$
20. $\dfrac{\sin 3x}{\sin x} + \dfrac{\cos 3x}{\cos x} = 4 \cos 2x$

In Exercises 21 and 22 prove the given identities by letting $3x = 2x + x$.

21. $\sin 3x = 3 \cos^2 x \sin x - \sin^3 x$
22. $\cos 3x = \cos^3 x - 3 \sin^2 x \cos x$

In Exercises 23 through 26 solve the given problems.

23. In Exercise 25 of Section 19–2, let $\beta = \alpha$, and show that

$$\tan 2\alpha = \frac{2 \tan \alpha}{1 - \tan^2 \alpha}$$

24. Given that $x = \cos 2\theta$ and $y = \sin \theta$, find the relation between x and y by eliminating θ.

25. The equation for the displacement of a certain object at the end of a spring is $y = A \sin 2t + B \cos 2t$. Show that this equation may be written as $y = C \sin(2t + \alpha)$ where $C = \sqrt{A^2 + B^2}$ and $\tan \alpha = B/A$. [*Hint:* Let $A/C = \cos \alpha$ and $B/C = \sin \alpha$.]

26. To find the horizontal range R of a projectile, the equation $R = vt \cos \alpha$ is used, where α is the angle between the line of fire and the horizontal, v is the initial velocity of the projectile, and t is the time of flight. It can be shown that $t = (2v \sin \alpha)/g$, where g is the acceleration due to gravity. Show that $R = (v^2 \sin 2\alpha)/g$.

19—4 Half-Angle Formulas

If we let $\theta = \alpha/2$ in the identity $\cos 2\theta = 1 - 2\sin^2\theta$ and then solve for $\sin(\alpha/2)$, we obtain

$$\sin \frac{\alpha}{2} = \pm \sqrt{\frac{1 - \cos \alpha}{2}} \qquad (19\text{–}24)$$

Also, with the same substitution in the identity $\cos 2\theta = 2\cos^2\theta - 1$, which is then solved for $\cos(\alpha/2)$, we have

$$\cos \frac{\alpha}{2} = \pm \sqrt{\frac{1 + \cos \alpha}{2}} \qquad (19\text{–}25)$$

In each of Eqs. (19–24) and (19–25), the sign chosen depends on the quadrant in which $\alpha/2$ lies.

We can use these half-angle formulas to find values of the functions of angles which are half of those for which the functions are known. Examples A through F which follow illustrate how these identities are used in evaluations and in identities.

Example A
We can find $\sin 15°$ by using the relation

$$\sin 15° = \sqrt{\frac{1 - \cos 30°}{2}} = \sqrt{\frac{1 - 0.8660}{2}} = 0.2588$$

Here the plus sign is used, since 15° is in the first quadrant.

Example B
We can find $\cos 165°$ by use of the relation

$$\cos 165° = -\sqrt{\frac{1 + \cos 330°}{2}} = -\sqrt{\frac{1 + 0.8660}{2}} = -0.9659$$

Here the minus sign is used, since 165° is in the second quadrant, and the cosine of a second-quadrant angle is negative.

Example C
Simplify the expression $\sqrt{18 - 18\cos 4x}$.

First we factor the 18 from each of the terms under the radical, and note that $18 = 9(2)$ and 9 is a perfect square. This leads to

$$\sqrt{18 - 18\cos 4x} = \sqrt{9(2)(1 - \cos 4x)} = 3\sqrt{2(1 - \cos 4x)}$$

This last expression is very similar to that for $\sin(\alpha/2)$, except that no 2 appears in the denominator. Therefore, multiplying the numerator and the denominator under the radical by 2 leads to the solution.

$$3\sqrt{2(1 - \cos 4x)} = 3\sqrt{\frac{4(1 - \cos 4x)}{2}} = 6\sqrt{\frac{1 - \cos 4x}{2}} = 6\sin\frac{4x}{2} = 6\sin 2x$$

Example D

Prove the identity $\sec \dfrac{\alpha}{2} + \csc \dfrac{\alpha}{2} = \dfrac{2\left(\sin \dfrac{\alpha}{2} + \cos \dfrac{\alpha}{2}\right)}{\sin \alpha}$

By expressing $\sin \alpha$ as $2 \sin \frac{\alpha}{2} \cos \frac{\alpha}{2}$, we have

$$\frac{2\left(\sin \dfrac{\alpha}{2} + \cos \dfrac{\alpha}{2}\right)}{2 \sin \dfrac{\alpha}{2} \cos \dfrac{\alpha}{2}} = \frac{1}{\cos \dfrac{\alpha}{2}} + \frac{1}{\sin \dfrac{\alpha}{2}} = \sec \dfrac{\alpha}{2} + \csc \dfrac{\alpha}{2}$$

Example E

We can find relations for the other functions of $\alpha/2$ by expressing these functions in terms of $\sin(\alpha/2)$ and $\cos(\alpha/2)$. For example:

$$\sec \frac{\alpha}{2} = \frac{1}{\cos \dfrac{\alpha}{2}} = \pm \frac{1}{\sqrt{\dfrac{1 + \cos \alpha}{2}}} = \pm \sqrt{\frac{2}{1 + \cos \alpha}}$$

Example F

Show that $2 \cos^2 \dfrac{x}{2} - \cos x = 1$.

The first step is to substitute for $\cos(x/2)$, which will result in each term containing x on the left being in terms of x, and no $x/2$ terms will exist. This might allow us to combine terms. So we perform this operation, and we have

$$2\left(\frac{1 + \cos x}{2}\right) - \cos x = 1$$

Combining terms, we can complete the proof:

$$1 + \cos x - \cos x = 1$$

Exercises 19–4

In Exercises 1 through 4 use the half-angle formulas to evaluate the given functions.

1. $\cos 15°$ **2.** $\sin 22.5°$ **3.** $\sin 75°$ **4.** $\cos 112.5°$

In Exercises 5 through 8 use the half-angle formulas to simplify the given expressions.

5. $\sqrt{\dfrac{1 - \cos 6\alpha}{2}}$ **6.** $\sqrt{\dfrac{4 + 4 \cos 8\beta}{2}}$

7. $\sqrt{8 + 8 \cos 4x}$ **8.** $\sqrt{2 - 2 \cos 16x}$

In Exercises 9 through 12 evaluate the indicated functions with the information given.

9. Find the value of $\sin(\alpha/2)$, if $\cos \alpha = \frac{12}{13}$ (in first quadrant).

10. Find the value of $\cos(\alpha/2)$, if $\sin \alpha = -\frac{4}{5}$ (in third quadrant).

11. Find the value of $\cos(\alpha/2)$, if $\tan \alpha = -\frac{7}{24}$ (in second quadrant).

12. Find the value of $\sin(\alpha/2)$, if $\cos \alpha = \frac{8}{17}$ (in fourth quadrant).

In Exercises 13 through 16 derive the required expressions.

13. Derive an expression for $\csc(\alpha/2)$ in terms of $\cos \alpha$. $\csc = \frac{1}{\sin}$

14. Derive an expression for $\sec(\alpha/2)$ in terms of $\sec \alpha$.

15. Derive an expression for $\tan(\alpha/2)$ in terms of $\sin \alpha$ and $\cos \alpha$.

16. Derive an expression for $\cot(\alpha/2)$ in terms of $\sin \alpha$ and $\cos \alpha$.

In Exercises 17 through 20 prove the given identities.

17. $\sin \dfrac{\alpha}{2} = \dfrac{1 - \cos \alpha}{2 \sin \dfrac{\alpha}{2}}$

18. $2 \cos \dfrac{x}{2} = (1 + \cos x) \sec \dfrac{x}{2}$

19. $2 \sin^2 \dfrac{\alpha}{2} - \cos^2 \dfrac{\alpha}{2} = \dfrac{1 - 3 \cos \alpha}{2}$

20. $\tan \dfrac{\alpha}{2} = \dfrac{\sin \alpha}{1 + \cos \alpha}$

In Exercises 21 and 22 use the half-angle formulas to solve the given problems.

21. In the kinetic theory of gases, the expression

$$\sqrt{(1 - \cos \alpha)^2 + \sin^2 \alpha \cos^2 \beta + \sin^2 \alpha \sin^2 \beta}$$

is found. Show that this expression equals $2 \sin \frac{\alpha}{2}$.

22. The index of refraction n, the angle of a prism A, and the minimum angle of refraction ϕ are related by

$$n = \dfrac{\sin \dfrac{A + \phi}{2}}{\sin \dfrac{A}{2}}$$

Show that an equivalent expression is

$$n = \sqrt{\dfrac{1 + \cos \phi}{2}} + \left(\cot \dfrac{A}{2}\right) \sqrt{\dfrac{1 - \cos \phi}{2}}$$

19–5 Trigonometric Equations

One of the most important uses of the trigonometric identities is in the solution of equations involving the trigonometric functions. When equations are written in terms of more than one function, the identities provide a way of transforming many of them to equations or factors involving only one function of the same angle. If we can accomplish this we can employ algebraic methods from then on to complete the solution. No general methods exist for the solution of such equations, but the following examples illustrate methods which prove to be useful.

Example A

Solve the equation $2 \cos \theta - 1 = 0$ for all values of θ such that $0 \leq \theta < 2\pi$.

Solving the equation for $\cos \theta$, we obtain $\cos \theta = \frac{1}{2}$. The problem asks for all values of θ between 0 and 2π that satisfy the equation. We know that the cosine of angles in the first and fourth quadrants is positive. Also, we know that $\cos (\pi/3) = \frac{1}{2}$. Therefore, $\theta = \pi/3$ and $\theta = 5\pi/3$.

Example B

Solve the equation $2 \cos^2 x - \sin x - 1 = 0$ $(0 \leq x < 2\pi)$.

By use of the identity $\sin^2 x + \cos^2 x = 1$, this equation may be put in terms of $\sin x$ only. Thus, we have

$$2(1 - \sin^2 x) - \sin x - 1 = 0$$
$$-2 \sin^2 x - \sin x + 1 = 0$$
$$2 \sin^2 x + \sin x - 1 = 0$$

or

$$(2 \sin x - 1)(\sin x + 1) = 0$$

Just as in solving algebraic equations, we can set each factor equal to zero to find valid solutions. Thus, $\sin x = \frac{1}{2}$ and $\sin x = -1$. For the range between 0 and 2π, the value $\sin x = \frac{1}{2}$ gives values of x as $\pi/6$ and $5\pi/6$, and $\sin x = -1$ gives the value $x = 3\pi/2$. Thus, the complete solution is $x = \pi/6$, $x = 5\pi/6$, and $x = 3\pi/2$.

Example C

Solve the equation $\sec^2 x + 2 \tan x - 6 = 0$ $(0 \leq x < 2\pi)$.

By use of the identity $1 + \tan^2 x = \sec^2 x$ we may express this equation in terms of $\tan x$ only. Therefore,

$$\tan^2 x + 1 + 2 \tan x - 6 = 0$$
$$\tan^2 x + 2 \tan x - 5 = 0$$

We note that this expression is not factorable. Therefore, using the quadratic formula, we obtain the following solution:

$$\tan x = \frac{-2 \pm \sqrt{4 + 20}}{2} = \frac{-2 \pm 4.899}{2}$$

Therefore, $\tan x = \frac{2.899}{2} = 1.450$ and $\tan x = \frac{-6.899}{2} = -3.450$. Expressing the results in radians, we find that $\tan x = 1.450$ for $x = 0.967$.

Since tan x is also positive in the third quadrant, we also have $x = 4.11$. In the same way, using tan $x = -3.450$, we obtain $x = 1.85$ and 4.99. Therefore, the correct solutions are $x = 0.967$, $x = 1.85$, $x = 4.11$, and $x = 4.99$.

Example D

Solve the equation sin $2x$ + sin $x = 0$ $(0 \leq x < 2\pi)$.

By using the double-angle formula for sin $2x$, we can write the equation in the form

$$2 \sin x \cos x + \sin x = 0 \qquad \text{or} \qquad \sin x(2 \cos x + 1) = 0$$

The first factor gives $x = 0$ or $x = \pi$. The second factor, for which $\cos x = -\frac{1}{2}$, gives $x = 2\pi/3$ and $x = 4\pi/3$. Thus, the complete solution is $x = 0$, $x = 2\pi/3$, $x = \pi$, and $x = 4\pi/3$.

Example E

Solve the equation $\cos \dfrac{x}{2} = 1 + \cos x$ $(0 \leq x < 2\pi)$.

By using the half-angle formula for cos $(x/2)$ and then squaring both sides of the resulting equation, this equation can be solved.

$$\pm\sqrt{\frac{1 + \cos x}{2}} = 1 + \cos x$$

$$\frac{1 + \cos x}{2} = 1 + 2 \cos x + \cos^2 x$$

Simplifying this last equation, we have

$$2 \cos^2 x + 3 \cos x + 1 = 0$$
$$(2 \cos x + 1)(\cos x + 1) = 0$$

The values of the cosine which come from these factors are $\cos x = -\frac{1}{2}$ and $\cos x = -1$. Thus, the values of x which satisfy the last equation are $x = 2\pi/3$, $x = 4\pi/3$, and $x = \pi$. However, when we solved this equation, we squared both sides of it. In doing this we may have introduced extraneous solutions (see Section 13–3). Thus, we must check each solution in the original equation to see if it is valid. Hence,

$$\cos \frac{\pi}{3} \stackrel{?}{=} 1 + \cos \frac{2\pi}{3} \qquad \text{or} \qquad \frac{1}{2} \stackrel{?}{=} 1 + \left(-\frac{1}{2}\right) \qquad \text{or} \qquad \frac{1}{2} = \frac{1}{2}$$

$$\cos \frac{2\pi}{3} \stackrel{?}{=} 1 + \cos \frac{4\pi}{3} \qquad \text{or} \qquad -\frac{1}{2} \stackrel{?}{=} 1 + \left(-\frac{1}{2}\right) \qquad \text{or} \qquad -\frac{1}{2} \neq \frac{1}{2}$$

$$\cos \frac{\pi}{2} \stackrel{?}{=} 1 + \cos \pi \qquad \text{or} \qquad 0 \stackrel{?}{=} 1 - 1 \qquad \text{or} \qquad 0 = 0$$

Thus, the apparent solution $x = 4\pi/3$ is not a solution of the original equation. The correct solutions are $x = 2\pi/3$ and $x = \pi$.

Example F

Solve the equation $\tan 3\theta - \cot 3\theta = 0$ $(0 \le \theta < 2\pi)$.

$$\tan 3\theta - \frac{1}{\tan 3\theta} = 0 \quad \text{or} \quad \tan^2 3\theta = 1 \quad \text{or} \quad \tan 3\theta = \pm 1$$

Thus,

$$3\theta = \frac{\pi}{4}, \frac{3\pi}{4}, \frac{5\pi}{4}, \frac{7\pi}{4}, \frac{9\pi}{4}, \frac{11\pi}{4}, \frac{13\pi}{4}, \frac{15\pi}{4}, \frac{17\pi}{4}, \frac{19\pi}{4}, \frac{21\pi}{4}, \frac{23\pi}{4}$$

Here we must include values of angles which when divided by 3 give angles between 0 and 2π. Thus, values of 3θ from 0 to 6π are necessary. The solutions are

$$\theta = \frac{\pi}{12}, \frac{\pi}{4}, \frac{5\pi}{12}, \frac{7\pi}{12}, \frac{3\pi}{4}, \frac{11\pi}{12}, \frac{13\pi}{12}, \frac{5\pi}{4}, \frac{17\pi}{12}, \frac{19\pi}{12}, \frac{7\pi}{4}, \frac{23\pi}{12}$$

It is noted that these values satisfy the original equation. Since we multiplied through by $\tan 3\theta$ in the solution, any value of θ which leads to $\tan 3\theta = 0$ would not be valid, since this would indicate division by zero in the original equation.

Example G

Solve the equation $\cos 3x \cos x + \sin 3x \sin x = 1$ $(0 \le x < 2\pi)$.

The left side of this equation is of the general form $\cos(A - x)$, where $A = 3x$. Therefore,

$$\cos 3x \cos x + \sin 3x \sin x = \cos(3x - x) = \cos 2x$$

The original equation becomes

$$\cos 2x = 1$$

This equation is satisfied if $2x = 0$ and $2x = 2\pi$. The solutions are $x = 0$ and $x = \pi$. Only through recognition of the proper trigonometric form can we readily solve this equation.

Exercises 19−5

In Exercises 1 through 28 solve the given trigonometric equations for values of x so that $0 \le x < 2\pi$.

1. $\sin x - 1 = 0$
2. $2 \sin x + 1 = 0$
3. $\tan x + 1 = 0$
4. $2 \cos x + 1 = 0$
5. $4 \cos^2 x - 1 = 0$
6. $\sin^2 x - 1 = 0$
7. $4 \sin^2 x - 3 = 0$
8. $3 \tan^2 x - 1 = 0$
9. $2 \sin^2 x - \sin x = 0$
10. $3 \cos x - 4 \cos^2 x = 0$
11. $\sin 4x - \cos 2x = 0$
12. $\sin 4x - \sin 2x = 0$
13. $\sin 2x \sin x + \cos x = 0$
14. $\cos 2x + \sin^2 x = 0$
15. $2 \sin x - \tan x = 0$
16. $\sin x - \sin \frac{x}{2} = 0$
17. $2 \cos^2 x - 2 \cos 2x - 1 = 0$
18. $2 \cos^2 2x + 1 = 3 \cos 2x$

19. $\sin^2 x - 2 \sin x - 1 = 0$
20. $\tan^2 x - 5 \tan x + 6 = 0$
21. $4 \tan x - \sec^2 x = 0$
22. $\tan^2 x - 2 \sec^2 x + 4 = 0$
23. $\sin 2x \cos x - \cos 2x \sin x = 0$
24. $\cos 3x \cos x - \sin 3x \sin x = 0$
25. $\sin 2x + \cos 2x = 0$
26. $2 \sin 4x + \csc 4x = 3$
27. $\tan x + 3 \cot x = 4$
28. $\sin x \sin \frac{1}{2} x = 1 - \cos x$

In Exercises 29 through 32 solve the indicated equations.

29. In finding the dimensions of the largest cylinder which can be inscribed in a sphere, the equation $\sin \theta - 3 \sin \theta \cos^2 \theta = 0$ must be solved for θ, for $0 < \theta < \pi/2$. Solve for the value of θ which satisfies the equation.

30. Vectors of magnitudes 3 and 2 are directed at an angle θ, such that the vertical component of the first vector equals the horizontal component of the second vector. Find the angle θ, such that $0 < \theta < \pi$.

31. The angular displacement θ of a certain pendulum in terms of the time t is given by $\theta = e^{-0.1t}(\cos 2t + 3 \sin 2t)$. What is the smallest value of t for which the displacement is zero?

32. The vertical displacement y of an object at the end of a spring, which itself is being moved up and down, is given by $y = 2 \cos 4t + \sin 2t$. Find the smallest value of t (in seconds) for which $y = 0$.

In Exercises 33 through 36 solve the given equations graphically.

33. $\sin 2x = x$
34. $\cos 2x = 4 - x^2$
35. In the study of light diffraction, the equation $\tan \theta = \theta$ is found. Solve this equation for $0 \leq \theta < 2\pi$.

36. An equation used in astronomy is $\theta - e \sin \theta = M$. Solve for θ for $e = 0.25$ and $M = 0.75$.

19–6 Introduction to the Inverse Trigonometric Functions

When we studied logarithms, we found that we often wished to change a given expression from exponential to logarithmic form, or from logarithmic to exponential form. Each of these forms has its advantages for particular purposes. We found that the exponential function $y = b^x$ can also be written in logarithmic form with x as a function of y, or $x = \log_b y$. We then represented both of these functions as y in terms of x, saying that the letter used for the dependent and independent variables did not matter, when we wished to express a functional relationship. Since y is normally the dependent variable, we wrote the logarithmic function as $y = \log_b x$.

These two functions, the exponential function $y = b^x$ and the logarithmic function $y = \log_b x$, are called **inverse functions**. This means that if we solve for the independent variable in terms of the dependent variable in one, we will arrive at the functional relationship expressed by the other. It also means that, for every value of x, there is only one corresponding value of y.

Just as we are able to solve $y = b^x$ for the exponent by writing it in logarithmic form, there are times when it is necessary to solve for the independent variable (the angle) in trigonometric functions. Therefore, we define the **inverse sine of x** by the relation

$$y = \arcsin x \quad \text{(the notation } y = \sin^{-1} x \text{ is also used)} \qquad (19\text{–}26)$$

Similar relations exist for the other inverse trigonometric relations. In Eq. (19–26), x is the value of the sine of the angle y, and therefore the most meaningful way of reading it is "y is the angle whose sine is x."

Example A

$y = \arccos x$ would be read as "y is the angle whose cosine is x."

The equation $y = \arctan 2x$ would be read as "y is the angle whose tangent is $2x$."

It is important to emphasize that $y = \arcsin x$ and $x = \sin y$ express the same relationship between x and y. The advantage of having both forms is that a trigonometric relation may be expressed in terms of a function of an angle or in terms of the angle itself.

If we consider closely the equation $y = \arcsin x$ and possible values of x, we note that there are an unlimited number of possible values of y for a given value of x. Consider the following example.

Example B

For $y = \arcsin x$, if $x = \frac{1}{2}$, we have $y = \arcsin \frac{1}{2}$. This means that we are to find an angle whose sine is $\frac{1}{2}$. We know that $\sin (\pi/6) = \frac{1}{2}$. Therefore, $y = \pi/6$.

However, we also know that $\sin (5\pi/6) = \frac{1}{2}$. Therefore, $y = 5\pi/6$ is also a proper value. If we consider negative angles, such as $-7\pi/6$, or angles generated by additional rotations, such as $13\pi/6$, we conclude that there are an unlimited number of possible values for y.

*To have a properly defined **function** in mathematics, there must be only one value of the dependent variable for a given value of the independent variable. A **relation**, on the other hand, may have more than one such value.* Therefore, we see that $y = \arcsin x$ is not really a function, although it is properly a relation. It is necessary to restrict the values of y in order to define the **inverse trigonometric functions,** and this is done in the following section. It is the purpose of this section to introduce the necessary notation and to develop an understanding of the basic concept. The following examples further illustrate the meaning of the notation.

Example C

If $y = \arccos 0$, y is the angle whose cosine is zero. The smallest positive angle for which this is true is $\pi/2$. Therefore, $y = \pi/2$ is an acceptable value.

If $y = \arctan 1$, an acceptable value for y is $\pi/4$. This is the same as saying $\tan \pi/4 = 1$.

Example D
Given that $y = \sec 2x$, solve for x.
   We first express the inverse relation as $2x = \text{arcsec } y$. Then we solve for x by dividing through by 2. Thus, we have $x = \frac{1}{2} \text{ arcsec } y$. Note that we first wrote the inverse relation by writing the expression for the angle, which in this case was $2x$. Just as sec $2x$ and 2 sec x are different relations, so are the arcsec $2x$ and 2 arcsec x.

Example E
Given that $4y = \text{arccot } 2x$, solve for x.
   Writing this as the cotangent of $4y$ (since the given expression means "$4y$ is the angle whose cotangent is $2x$"), we have

$$2x = \cot 4y \quad \text{or} \quad x = \frac{1}{2} \cot 4y$$

Example F
Given that $\pi - y = \text{arccsc } \frac{1}{3}x$, solve for x.

$$\frac{1}{3}x = \csc(\pi - y) \quad \text{or} \quad x = 3 \csc y$$

[since $\csc(\pi - y) = \csc y$].

Exercises 19—6

In Exercises 1 through 8 write down the meaning of each of the given equations. See Example A.

1. $y = \arctan x$ 2. $y = \text{arcsec } x$ 3. $y = \text{arccot } 3x$
4. $y = \text{arccsc } 4x$ 5. $y = 2 \arcsin x$ 6. $y = 3 \arctan x$
7. $y = 5 \arccos 2x$ 8. $y = 4 \arcsin 3x$

In Exercises 9 through 20 find the smallest positive angle (in terms of π) for each of the given expressions.

9. $\arccos \dfrac{1}{2}$ 10. $\arcsin 1$ 11. $\text{arcsec}(-\sqrt{2})$

12. $\arccos \dfrac{\sqrt{2}}{2}$ 13. $\arctan(-1)$ 14. $\text{arccsc}(-1)$

15. $\arctan \sqrt{3}$ 16. $\text{arcsec } 2$ 17. $\text{arccot}(-\sqrt{3})$

18. $\arcsin\left(-\dfrac{\sqrt{3}}{2}\right)$ 19. $\text{arccsc } \sqrt{2}$ 20. $\text{arccot } \dfrac{\sqrt{3}}{3}$

In Exercises 21 through 28 solve the given equations for x.

21. $y = \sin 3x$ 22. $y = \cos(x - \pi)$

23. $y = \arctan\left(\dfrac{x}{4}\right)$ 24. $y = 2 \arcsin\left(\dfrac{x}{6}\right)$

25. $y = 1 + \sec 3x$ 26. $4y = 5 - \csc 8x$
27. $1 - y = \arccos(1 - x)$ 28. $2y = \text{arccot } 3x - 5$

In Exercises 29 through 32 determine the required quadrants.

29. In which quadrants is arcsin x if $0 < x < 1$?
30. In which quadrants is arctan x if $0 < x < 1$?
31. In which quadrants is arccos x if $-1 < x < 0$?
32. In which quadrants is arcsin x if $-1 < x < 0$?

In Exercises 33 through 36 solve the given problems with the use of the inverse trigonometric relations.

33. A body is moving along a straight line in such a way that its acceleration is directed toward a fixed point, and is proportional to its distance from that point. Its position is given by $x = A \cos t \sqrt{k/m}$. Solve for k.
34. Under certain conditions the magnetic potential V at a distance r from a circuit with a current I is given by

$$V = \frac{kI \cos \theta}{r^2}$$

where k is a constant and θ is the angle between the radius vector r and the direction perpendicular to the plane of the circuit. Solve for θ.
35. The magnitude of a certain ray of polarized light is given by the equation $E = E_0 \cos \theta \sin \theta$. Solve for θ.
36. The equation for the displacement of a certain object oscillating at the end of a spring is given by $y = A \sin 2t + B \cos 2t$. Solve for t. [*Hint:* See Exercise 25 of Section 19–3.]

19—7 The Inverse Trigonometric Functions

We noted in the preceding section that we could find many values of y if we assumed some value for x in the relation $y = \text{arcsin } x$. As these relations were defined, any one of the various possibilities would be considered correct. This, however, does not meet a basic requirement for a function, and it also leads to ambiguity. In order to define the **inverse trigonometric functions** properly, so this ambiguity does not exist, there must be only a single value of y for any given value of x. Therefore, the following values are defined for the given functions.

$$-\frac{\pi}{2} \le \text{Arcsin } x \le \frac{\pi}{2}, \quad 0 \le \text{Arccos } x \le \pi, \quad -\frac{\pi}{2} < \text{Arctan } x < \frac{\pi}{2}$$

$$(19\text{–}27)$$

$$0 < \text{Arccot } x < \pi, \quad 0 \le \text{Arcsec } x \le \pi, \quad -\frac{\pi}{2} \le \text{Arccsc } x \le \frac{\pi}{2}$$

This means that when we are looking for a value of y to correspond to a given value for x, we must use a value of y as defined in Eqs. (19–27). The capital letter designates the use of the inverse trigonometric *function*.

Example A

$$\text{Arcsin}\left(\frac{1}{2}\right) = \frac{\pi}{6}$$

This is the only value of the function which lies within the defined range. The value $5\pi/6$ is not correct, since it lies outside the defined range of values.

Example B

$$\text{Arccos}\left(-\frac{1}{2}\right) = \frac{2\pi}{3}$$

Other values such as $4\pi/3$ and $-2\pi/3$ are not correct, since they are not within the defined range of values for the function Arccos x.

Example C

$$\text{Arctan}(-1) = -\frac{\pi}{4}$$

This is the only value within the defined range for the function Arctan x. We must remember that when x is negative for Arcsin x and Arctan x, the value of y is a fourth-quadrant angle, expressed as a *negative angle*. This is a direct result of the definition.

Example D

$$\text{Arcsin}\left(-\frac{\sqrt{3}}{2}\right) = -\frac{\pi}{3} \qquad \text{Arccos}(-1) = \pi$$

$$\text{Arctan}\ 0 = 0 \qquad\qquad \text{Arcsin}(-0.1564) = -\frac{\pi}{20}$$

$$\text{Arccos}(-0.8090) = \frac{4\pi}{5} \qquad \text{Arctan}(\sqrt{3}) = \frac{\pi}{3}$$

One might logically ask why these values are chosen when there are so many different possibilities. The values are so chosen that, if x is positive, the resulting answer gives an angle in the first quadrant. We must, however, account for the possibility that x might be negative. We could not choose second-quadrant angles for Arcsin x. Since the sine of a second-quadrant angle is also positive, this then would lead to ambiguity. The sine is negative for fourth-quadrant angles, and to have a continuous range of values we must express the fourth-quadrant angles in the form of negative angles. This range is also chosen for Arctan x, for similar reasons. However, Arccos x cannot be chosen in this way, since the cosine of fourth-quadrant angles is also positive. Thus, again to keep a continuous range of values for Arccos x, the second-quadrant angles are chosen for negative values of x.

As for the values for the other functions, we chose values such that if x is positive, the result is also an angle in the first quadrant. As for negative values of x, it rarely makes any difference, since either positive values of x arise, or we can use one of the other functions. Our definitions, however, are those which are generally used.

The graphs of the inverse trigonometric relations can be used to show the fact that many values of y correspond to a given value of x. We can also show the ranges used in defining the inverse trigonometric functions, and that these ranges are specific sections of the curves.

Since $y = \arcsin x$ and $x = \sin y$ are equivalent equations, we can obtain the graph of the inverse sine by sketching the sine curve *along the y-axis*. In Figs. (19–3), (19–4), and (19–5), the graphs of three inverse trigonometric relations are shown, with the heavier portions indicating the graphs of the inverse trigonometric functions. The graphs of the other inverse relations are found in the same way.

If we know the value of x for one of the inverse functions, we can find the trigonometric functions of the angle. If general relations are desired, a representative triangle is very useful. The following examples illustrate these methods.

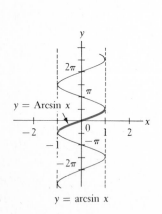

Figure 19–3

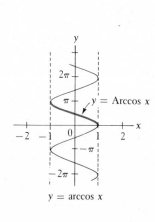

Figure 19–4

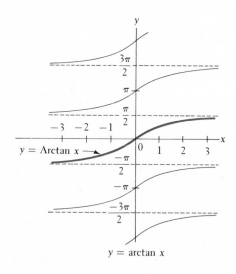

Figure 19–5

Example E

Find $\cos(\text{Arcsin } 0.5)$. [Remember again: the inverse functions yield *angles.*]

We know Arcsin 0.5 is a first-quadrant angle, since 0.5 is positive. Thus we find Arcsin $0.5 = \pi/6$. The problem now becomes one of finding $\cos(\pi/6)$. This is, of course, $\sqrt{3}/2$ or 0.8660.

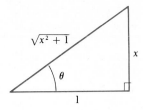

Figure 19–6

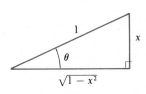

Figure 19–7

Example F

$$\sin(\text{Arccot } 1) = \sin\frac{\pi}{4} = \frac{\sqrt{2}}{2} = 0.7071$$

$$\tan[\text{Arccos}(-1)] = \tan \pi = 0$$

Example G

Find $\sin(\text{Arctan } x)$.

We know that Arctan x is another way of stating "the angle whose tangent is x." Thus, let us draw a right triangle (as in Fig. 19–6) and label one of the acute angles θ, the side opposite θ as x, and the side adjacent to θ as 1. In this way we see that, by definition, $\tan \theta = x/1$, or $\theta = \text{Arctan } x$, which means θ is the desired angle. By the Pythagorean theorem, the hypotenuse of this triangle is $\sqrt{x^2 + 1}$. Now we find that the sin θ, which is the same as sin (Arctan x), is $x/\sqrt{x^2 + 1}$, from the definition of the sine. Thus,

$$\sin(\text{Arctan } x) = \frac{x}{\sqrt{x^2 + 1}}$$

Example H

Find $\cos(2 \text{ Arcsin } x)$.

From Fig. 19–7, we see that $\theta = \text{Arcsin } x$. From the double-angle formulas, we have $\cos 2\theta = 1 - 2\sin^2\theta$. Thus, since $\sin \theta = x$, we have

$$\cos(2 \text{ Arcsin } x) = 1 - 2x^2$$

Exercises 19–7

In Exercises 1 through 24 evaluate the given expressions.

1. $\text{Arccos}(\frac{1}{2})$
2. $\text{Arcsin}(1)$
3. $\text{Arcsin } 0$
4. $\text{Arccos } 0$
5. $\text{Arctan}(-\sqrt{3})$
6. $\text{Arcsin}(-\frac{1}{2})$

7. $\text{Arcsec } 2$
8. $\text{Arccot } \sqrt{3}$
9. $\text{Arctan}\left(\frac{\sqrt{3}}{3}\right)$

10. $\text{Arctan } 1$
11. $\text{Arcsin}\left(-\frac{\sqrt{2}}{2}\right)$
12. $\text{Arccos}\left(-\frac{\sqrt{3}}{2}\right)$
13. $\text{Arccsc } \sqrt{2}$
14. $\text{Arccot } 1$
15. $\text{Arctan}(-3.732)$

16. $\text{Arccos}(-0.5878)$
17. $\sin(\text{Arctan } \sqrt{3})$
18. $\tan\left(\text{Arcsin } \frac{\sqrt{2}}{2}\right)$

19. $\cos[\text{Arctan}(-1)]$
20. $\sec[\text{Arccos}(-\frac{1}{2})]$
21. $\tan[\text{Arccos}(-0.6561)]$
22. $\cot[\text{Arcsin}(-0.3827)]$
23. $\cos(2 \text{ Arcsin } 1)$
24. $\sin(2 \text{ Arctan } 2)$

In Exercises 25 through 32 find an algebraic expression for each of the expressions given.

25. $\tan(\text{Arcsin } x)$
26. $\sin(\text{Arccos } x)$
27. $\cos(\text{Arcsec } x)$
28. $\cot(\text{Arccot } x)$
29. $\sec(\text{Arccsc } 3x)$
30. $\tan(\text{Arcsin } 2x)$
31. $\sin(2 \text{ Arcsin } x)$
32. $\cos(2 \text{ Arctan } x)$

In Exercises 33 and 34, prove that the given expressions are equal. This can be done by use of the relation for sin $(\alpha + \beta)$ and by showing that the sine of the sum of angles on the left equals the sine of the angle on the right.

33. Arcsin $\frac{3}{5}$ + Arcsin $\frac{5}{13}$ = Arcsin $\frac{56}{65}$ **34.** Arctan $\frac{1}{3}$ + Arctan $\frac{1}{2}$ = $\frac{\pi}{4}$

In Exercises 35 and 36 derive the given expressions.

35. For the triangle in Fig. 19–8, show that α = Arctan$(a + \tan \beta)$.

36. Show that the length of the pulley belt indicated in Fig. 19–9 is given by the expression

$$L = 24 + 11\pi + 10 \text{ Arcsin } (\tfrac{5}{13}).$$

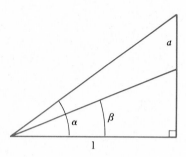

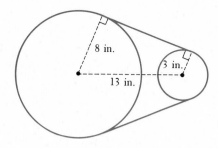

Figure 19–8   Figure 19–9

19–8 Exercises for Chapter 19

In Exercises 1 through 8 determine the values of the indicated functions in the given manner.

1. Find sin 120° by using 120° = 90° + 30°.

2. Find cos 30° by using 30° = 90° − 60°.

3. Find sin 135° by using 135° = 180° − 45°.

4. Find cos 225° by using 225° = 180° + 45°.

5. Find cos 180° by using 180° = 2(90°).

6. Find sin 180° by using 180° = 2(90°).

7. Find sin 45° by using 45° = $\frac{1}{2}$(90°).

8. Find cos 45° by using 45° = $\frac{1}{2}$(90°).

In Exercises 9 through 16 reduce each of the given expressions to a single term. Expansion of any term is not necessary; proper recognition of the form of the expression leads to the proper result.

9. $\sin 2x \cos 3x + \cos 2x \sin 3x$ **10.** $\cos 7x \cos 3x + \sin 7x \sin 3x$

11. $8 \sin 6x \cos 6x$ **12.** $10 \sin 5x \cos 5x$

13. $2 - 4 \sin^2 6x$ **14.** $\cos^2 2x - \sin^2 2x$

15. $\sqrt{2 + 2 \cos 2x}$ **16.** $\sqrt{32 - 32 \cos 4x}$

In Exercises 17 through 24 evaluate the given expressions.

17. $\text{Arcsin}(-1)$

18. $\text{Arcsec}\ \sqrt{2}$

19. $\text{Arccos}(0.9659)$

20. $\text{Arctan}(-0.6249)$

21. $\tan[\text{Arcsin}(-\tfrac{1}{2})]$

22. $\cos[\text{Arctan}(-\sqrt{3})]$

23. $\text{Arcsin}(\tan \pi)$

24. $\text{Arccos}\left[\tan\left(-\dfrac{\pi}{4}\right)\right]$

In Exercises 25 through 48 prove the given identities.

25. $\dfrac{\sec y}{\csc y} = \tan y$

26. $\cos \theta \csc \theta = \cot \theta$

27. $\sin x(\csc x - \sin x) = \cos^2 x$

28. $\cos y(\sec y - \cos y) = \sin^2 y$

29. $\dfrac{1}{\sin \theta} - \sin \theta = \cot \theta \cos \theta$

30. $\sin \theta \sec \theta \csc \theta \cos \theta = 1$

31. $\cos \theta \cot \theta + \sin \theta = \csc \theta$

32. $\dfrac{\sin x \cot x + \cos x}{\cot x} = 2 \sin x$

33. $\dfrac{\sec^4 x - 1}{\tan^2 x} = 2 + \tan^2 x$

34. $\cos^2 y - \sin^2 y = \dfrac{1 - \tan^2 y}{1 + \tan^2 y}$

35. $2 \csc 2x \cot x = 1 + \cot^2 x$

36. $\cos^8 x - \sin^8 x = (\cos^4 x + \sin^4 x)\cos 2x$

37. $\sin\dfrac{\theta}{2} \cos\dfrac{\theta}{2} = \dfrac{\sin \theta}{2}$

38. $\sin\dfrac{x}{2} = \dfrac{\sec x - 1}{2 \sec x \sin\left(\dfrac{x}{2}\right)}$

39. $\sec x + \tan x = \dfrac{\cos x}{1 - \sin x}$

40. $\dfrac{\cos \theta - \sin \theta}{\cos \theta + \sin \theta} = \dfrac{\cot \theta - 1}{\cot \theta + 1}$

41. $\cos(x - y)\cos y - \sin(x - y)\sin y = \cos x$

42. $\sin 3y \cos 2y - \cos 3y \sin 2y = \sin y$

43. $\sin 4x(\cos^2 2x - \sin^2 2x) = \dfrac{\sin 8x}{2}$

44. $\csc 2x + \cot 2x = \cot x$

45. $\dfrac{\sin x}{\csc x - \cot x} = 1 + \cos x$

46. $\cos x - \sin\dfrac{x}{2} = \left(1 - 2 \sin\dfrac{x}{2}\right)\left(1 + \sin\dfrac{x}{2}\right)$

47. $\dfrac{\sin(x + y) + \sin(x - y)}{\cos(x + y) + \cos(x - y)} = \tan x$

48. $\sec\dfrac{x}{2} + \csc\dfrac{x}{2} = \dfrac{2\left(\sin\dfrac{x}{2} + \cos\dfrac{x}{2}\right)}{\sin x}$

19-6

In Exercises 49 through 52 solve for x.

49. $y = 2 \cos 2x$

50. $y - 2 = 2 \tan\left(x - \dfrac{\pi}{2}\right)$

51. $y = \dfrac{\pi}{4} - 3 \arcsin 5x$

52. $2y = \operatorname{arcsec} 4x - 2$

19-5

In Exercises 53 through 60 solve the given equations for x such that $0 \le x < 2\pi$.

53. $\cos^2 2x - 1 = 0$

54. $2 \sin 2x + 1 = 0$

55. $4 \cos^2 x - 3 = 0$

56. $\cos 2x = \sin x$

57. $\sin^2 x - \cos^2 x + 1 = 0$

58. $\cos 3x \cos x + \sin 3x \sin x = 0$

59. $\sin^2\left(\dfrac{x}{2}\right) - \cos x + 1 = 0$

60. $\sin x + \cos x = 1$

In Exercises 61 through 64 find an algebraic expression for each of the expressions.

19-7

61. $\tan(\operatorname{Arccot} x)$

62. $\cos(\operatorname{Arccsc} x)$

63. $\sin(2 \operatorname{Arccos} x)$

64. $\cos(\pi - \operatorname{Arctan} x)$

In Exercises 65 through 76 use the formulas and methods of this chapter to solve the given problems.

65. In finding the area between a section of the curve of $y = \cos^3 x$ and the x-axis, it is necessary to transform $\cos^3 x$ to $\cos x - \sin^2 x \cos x$. Show that this is valid.

66. In the theory dealing with the motion of fluid in cylinders, the expression

$$4 \sin^2\alpha \cos^2\alpha + (\cos^2\alpha - \sin^2\alpha)^2$$

is found. Simplify this expression.

67. If the area bounded by $y = \sin x$ between 0 and π and the x-axis is rotated about the x-axis, a volume is generated. In finding this volume it is necessary to change $\sin^2 x$ into $\frac{1}{2}(1 - \cos 2x)$. Show that this change is valid.

68. In surveying, when determining an azimuth (a measure used for reference purposes), it might be necessary to simplify the expression

$$\frac{1}{2 \cos \alpha \cos \beta} - \tan \alpha \tan \beta$$

Perform this operation by expressing it in the simplest possible form when $\alpha = \beta$.

69. An object is under the influence of a central force (an example of one type of central force is the attraction of the sun for the earth). The y-coordinate of its path is given by $y = 20 \sin \frac{1}{3}t$. Solve for t.

70. In finding the pressure exerted by soil on a retaining wall, the expression $1 - \cos 2\varphi - \sin 2\varphi \tan \alpha$ is found. Show that this expression can also be written as $2 \sin \varphi(\sin \varphi - \cos \varphi \tan \alpha)$.

71. In the theory dealing with the reflection of light, the expression

$$\frac{\cos (\phi + \alpha)}{\cos (\phi - \alpha)}$$

is found. Express this in terms of functions of α if $\phi = 2\alpha$.

72. In developing an expression for the power in an alternating-current circuit, the expression $\sin \omega t \sin(\omega t + \phi)$ is found. Show that this expression can be written as $\frac{1}{2}[\cos \phi - \cos(2\omega t + \phi)]$.

73. In the theory of interference of light, the expression $1 - 2r^2\cos \beta + r^4$ is found. Show that this expression can be written as $(1 - r^2)^2 + 4r^2\sin^2(\beta/2)$.

74. In the theory of diffraction of light, an equation found is

$$y = R \sin 2\pi\left(\frac{t}{T} - \frac{a}{\lambda}\right)\cos \alpha - R \cos 2\pi\left(\frac{t}{T} - \frac{a}{\lambda}\right)\sin \alpha$$

Show that

$$y = R \sin 2\pi\left(\frac{t}{T} - \frac{a}{\lambda} - \frac{\alpha}{2\pi}\right)$$

75. If a plane surface inclined at angle θ moves horizontally, the angle for which the lifting force of the air is a maximum is found by solving the equation $2 \sin \theta \cos^2\theta - \sin^3\theta = 0$, where $0 < \theta < 90°$. Solve for θ.

76. The electric current as a function of the time for a particular circuit is given by $i = 8e^{-20t}(\sqrt{3} \cos 10t - \sin 10t)$. Find the time in seconds when the current is first zero.

20

Plane Analytic Geometry

20—1 Basic Definitions

We first introduced the graph of a function in Chapter 2, and since that time we have made extensive use of graphs for representing functions. We have used graphs to represent the trigonometric functions, the exponential and logarithmic functions, and the inverse trigonometric functions. We have also seen how graphs may be used in solving equations and systems of equations. In this chapter we shall consider certain basic principles relating to the graphs of functions. We shall also show how certain graphs can be constructed by recognizing the form of the equation.

It is necessary when studying many of the concepts of the calculus to have the ability to recognize certain curves and their basic characteristics. Also, curves have many technical applications, many of which are illustrated or indicated in the examples and exercises in this chapter.

The underlying principle of analytic geometry is the relationship of geometry to algebra. A great deal can be learned about a geometric figure if we can find the function which represents its graph. Also, by analyzing certain characteristics of a function, we can obtain useful information as to the nature of its graph. In this section we shall develop certain basic concepts which will be needed for future use in establishing the proper relationships between an equation and a curve.

The first of these concepts involves the distance between any two points in the coordinate plane. If these two points lie on a line parallel

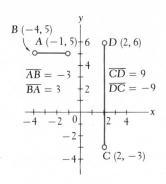

$\overline{AB} = -3$ $\overline{CD} = 9$
$\overline{BA} = 3$ $\overline{DC} = -9$

Figure 20–1

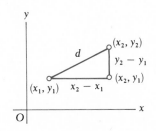

Figure 20–2

to the x-axis, the **directed distance** from the first point $A(x_1, y)$ to the second point $B(x_2, y)$ is denoted as $\overline{AB}$ and is defined as $x_2 - x_1$. We can see that $\overline{AB}$ *is positive if B is to the right of A, and it is negative if B is to the left of A.* Similarly, the directed distance between two points $C(x, y_1)$ and $D(x, y_2)$ on a line parallel to the y-axis is $y_2 - y_1$. We see that $\overline{CD}$ *is positive if D is above C, and it is negative if D is below C.*

Example A

The line segment joining $A(-1, 5)$ and $B(-4, 5)$ in Fig. 20–1 is parallel to the x-axis. Therefore, the directed distance $\overline{AB} = -4 - (-1) = -3$, and the directed distance $\overline{BA} = -1 - (-4) = 3$.

Also in Fig. 20–1, the line segment joining $C(2, -3)$ and $D(2, 6)$ is parallel to the y-axis. The directed distance $\overline{CD} = 6 - (-3) = 9$, and the directed distance $\overline{DC} = -3 - 6 = -9$.

We now wish to find the length of a line segment joining any two points in the plane. If these points are on a line which is not parallel to either of the axes (Fig. 20–2), we must use the Pythagorean theorem to find the distance between them. By making a right triangle with the line segment joining the two points as the hypotenuse, and line segments parallel to the axes as the legs, we have the formula which gives the distance between any two points in the plane. This formula, called the **distance formula,** is

$$d = \sqrt{(x_2 - x_1)^2 + (y_2 - y_1)^2} \tag{20–1}$$

Here we choose the positive square root since we are concerned only with the magnitude of the length of the line segment.

Example B

The distance between $(3, -1)$ and $(-2, -5)$ is given by

$$d = \sqrt{[(-2) - 3]^2 + [(-5) - (-1)]^2}$$
$$= \sqrt{(-5)^2 + (-4)^2} = \sqrt{25 + 16} = \sqrt{41}$$

It makes no difference which point is chosen as (x_1, y_1) and which is chosen as (x_2, y_2), since the difference in the x-coordinates (and y-co-ordinates) is squared. We also obtain $\sqrt{41}$ if we set up the distance as

$$d = \sqrt{[3 - (-2)]^2 + [(-1) - (-5)]^2}$$

Another important quantity which is defined for a line is its **slope.** *The slope gives a measure of the direction of a line, and is defined as the vertical directed distance from one point to another on the same straight line, divided by the horizontal directed distance from the first point to the second.* Thus, using the letter m to represent slope, we have

$$m = \frac{y_2 - y_1}{x_2 - x_1} \tag{20–2}$$

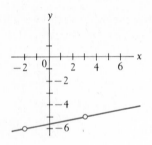

Figure 20-3

Example C

The slope of the line joining $(3, -5)$ and $(-2, -6)$ is

$$m = \frac{-6 - (-5)}{-2 - 3} = \frac{-6 + 5}{-5} = \frac{1}{5}$$

See Fig. 20–3. Again we may interpret either of the points as (x_1, y_1) and the other as (x_2, y_2). We can also obtain the slope of this same line from

$$m = \frac{-5 - (-6)}{3 - (-2)} = \frac{1}{5}$$

The larger the numerical value of the slope of a line, the more nearly vertical is the line. Also, a line rising to the right has a positive slope, and a line falling to the right has a negative slope.

Example D

The line through the two points in Example C has a positive slope, which is numerically small. From Fig. 20–3 it can be seen that the line rises slightly to the right.

The line joining $(3, 8)$ and $(4, -6)$ has a slope of -14. This line falls sharply to the right (see Fig. 20–4).

From the definition of slope, we may conclude that the slope of a line parallel to the y-axis cannot be defined (the x-coordinates of any two points on the line would be the same, which would then necessitate division by zero). This, however, does not prove to be of any trouble.

If a given line is extended indefinitely in either direction, it must cross the x-axis at some point unless it is parallel to the x-axis. *The angle measured from the x-axis in a positive direction to the line is called the* **inclination** *of the line* (see Fig. 20–5). The inclination of a line parallel to the x-axis is defined to be zero. An alternative definition of slope, in terms of the inclination, is

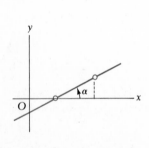

Figure 20–4

$$m = \tan \alpha, \qquad 0° \le \alpha < 180° \tag{20–3}$$

where α is the inclination. This can be seen from the fact that the slope can be defined in terms of any two points on the line. Thus, if we choose as one of these points that point where the line crosses the x-axis and any other point, we see from the definition of the tangent of an angle that Eq. (20–3) is in agreement with Eq. (20–2).

Figure 20–5

Example E

The slope of a line with an inclination of 45° is $m = \tan 45° = 1.000$. The slope of a line having an inclination of 120° is $m = \tan 120° = -\sqrt{3} = -1.732$. See Fig. 20–6.

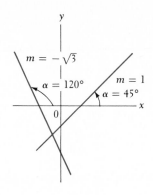

$m = -\sqrt{3}$

$\alpha = 120°$

$m = 1$

$\alpha = 45°$

Figure 20–6

Any two parallel lines crossing the x-axis will have the same inclination. Therefore, *the slopes of parallel lines are equal.* This can be stated as

$$m_1 = m_2 \quad \text{for } \| \text{ lines} \tag{20–4}$$

If two lines are perpendicular, this means that there must be 90° between their inclinations (Fig. 20–7). The relation between their inclinations is

$$\alpha_2 = \alpha_1 + 90°$$

which can be written as

$$90° - \alpha_2 = -\alpha_1$$

Taking the tangent of each of the angles in this last relation, we have

$$\tan(90° - \alpha_2) = \tan(-\alpha_1)$$

or

$$\cot \alpha_2 = -\tan \alpha_1$$

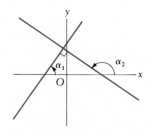

Figure 20–7

since a function of the complement of an angle is equal to the cofunction of that angle (see Section 3–4), and since $\tan(-\alpha) = -\tan \alpha$ (see Exercise 61 of Section 7–2). But $\cot \alpha = 1/\tan \alpha$, which means that $1/\tan \alpha_2 = -\tan \alpha_1$. Using the inclination definition of slope, we may write, as the relation between slopes of perpendicular lines,

$$\bullet \quad m_2 = -\frac{1}{m_1} \quad \text{or} \quad m_1 m_2 = -1 \quad \text{for } \perp \text{ lines} \tag{20–5}$$

Example F

The line through $(3, -5)$ and $(2, -7)$ has a slope of 2. The line through $(4, -6)$ and $(2, -5)$ has a slope of $-\frac{1}{2}$. These lines are perpendicular. See Fig. 20–8.

Using the formulas for distance and slope, we can show certain basic geometric relations. The following examples illustrate the use of the formulas, and thus show the use of algebra in solving problems which are basically geometric. This illustrates the methods of analytic geometry.

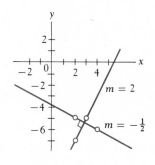

$m = 2$

$m = -\frac{1}{2}$

Figure 20–8

Example G

Show that line segments joining $A(-5, 3)$, $B(6, 0)$, and $C(5, 5)$ form a right triangle (see Fig. 20–9).

If these points are vertices of a right triangle, the slopes of two of the sides must be negative reciprocals. This would show perpendicularity. Thus, we find the slopes of the three lines to be

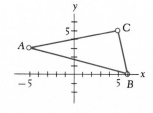

C

A

B

Figure 20–9

$$m_{AB} = \frac{3 - 0}{-5 - 6} = -\frac{3}{11}, \quad m_{AC} = \frac{3 - 5}{-5 - 5} = \frac{1}{5}, \quad m_{BC} = \frac{0 - 5}{6 - 5} = -5$$

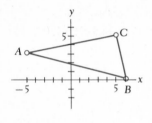

Figure 20-9

We see that the slopes of AC and BC are negative reciprocals, which means that $AC \perp BC$. From this we can conclude that the triangle is a right triangle.

Example H

Find the area of the triangle in Example G.

Since the right angle is at C, the legs of the triangle are AC and BC. The area is one-half the product of the lengths of the legs of a right triangle. The lengths of the legs are

$$d_{AC} = \sqrt{(-5 - 5)^2 + (3 - 5)^2} = \sqrt{104} = 2\sqrt{26}$$

and

$$d_{BC} = \sqrt{(6 - 5)^2 + (0 - 5)^2} = \sqrt{26}$$

Therefore, the area is

$$A = \frac{1}{2}(2\sqrt{26})(\sqrt{26}) = 26$$

Exercises 20-1

In Exercises 1 through 8 find the distances between the given pairs of points.

1. $(3, 8)$ and $(-1, -2)$ 2. $(-1, 3)$ and $(-8, -4)$
3. $(4, -5)$ and $(4, -8)$ 4. $(-3, 7)$ and $(2, 10)$
5. $(-1, 0)$ and $(5, -7)$ 6. $(15, -1)$ and $(-11, 1)$
7. $(-4, -3)$ and $(3, -3)$ 8. $(-2, 5)$ and $(-2, -2)$

In Exercises 9 through 16 find the slopes of the lines through the points in Exercises 1 through 8.

In Exercises 17 through 20 find the slopes of the lines with the given inclinations.

17. $30°$ 18. $60°$ 19. $150°$ 20. $135°$

In Exercises 21 through 24 find the inclinations of the lines with the given slopes.

21. 0.3640 22. 0.8243 23. -6.691 24. -1.428

In Exercises 25 through 28 determine whether the lines through the two pairs of points are parallel or perpendicular.

25. $(6, -1)$ and $(4, 3)$, and $(-5, 2)$ and $(-7, 6)$
26. $(-3, 9)$ and $(4, 4)$, and $(9, -1)$ and $(4, -8)$
27. $(-1, -4)$ and $(2, 3)$, and $(-5, 2)$ and $(-19, 8)$
28. $(-1, -2)$ and $(3, 6)$, and $(2, -6)$ and $(5, 0)$

In Exercises 29 through 32 determine the value of k.

29. The distance between $(-1, 3)$ and $(11, k)$ is 13.
30. The distance between $(k, 0)$ and $(0, 2k)$ is 10.

31. The points $(6, -1)$, $(3, k)$, and $(-3, -7)$ are all on the same line.
32. The points in Exercise 31 are the vertices of a right triangle, with the right angle at $(3, k)$.

In Exercises 33 through 36 show that the given points are vertices of the indicated geometric figures.

33. Show that the points $(2, 3)$, $(4, 9)$, and $(-2, 7)$ are the vertices of an isosceles triangle.
34. Show that $(-1, 3)$, $(3, 5)$, and $(5, 1)$ are the vertices of a right triangle.
35. Show that $(3, 2)$, $(7, 3)$, $(-1, -3)$, and $(3, -2)$ are the vertices of a parallelogram.
36. Show that $(-5, 6)$, $(0, 8)$, $(-3, 1)$, and $(2, 3)$ are the vertices of a square.

In Exercises 37 and 38 find the indicated areas.

37. Find the area of the triangle of Exercise 34.
38. Find the area of the square of Exercise 36.

20–2 The Straight Line

Using the definition of slope, we can derive the equation which always represents a straight line. This is another basic method of analytic geometry. That is, equations of a particular form can be shown to represent a particular type of curve. When we recognize the form of the equation, we know the kind of curve it represents. This can be of great assistance in sketching the graph.

A straight line can be defined as a "curve" with constant slope. By this we mean that for any two different points chosen on a given line, if the slope is calculated, the same value is always found. Thus, if we consider one point (x_1, y_1) on a line to be fixed (Fig. 20–10), and another point $P(x, y)$ which can *represent* any other point on the line, we have

$$m = \frac{y - y_1}{x - x_1}$$

which can be written as

$$y - y_1 = m(x - x_1) \tag{20–6}$$

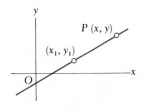

Figure 20–10

Equation (20–6) is known as the **point-slope form** of the equation of a straight line. It is useful when we know the slope of a line and some point through which the line passes. Direct substitution can then give us the equation of the line. Such information is often available about a given line.

Example A

Find the equation of the line which passes through $(-4, 1)$ with a slope of -2.

By using Eq. (20–6), we find that

$$y - 1 = (-2)(x + 4)$$

which can be simplified to

$$y + 2x + 7 = 0$$

This line is shown in Fig. 20–11.

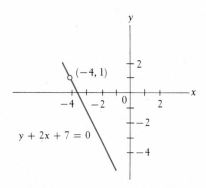

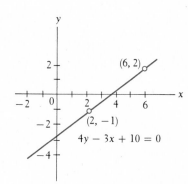

Figure 20–11   Figure 20–12

Example B

Find the equation of the line through $(2, -1)$ and $(6, 2)$.

We first find the slope of the line through these points:

$$m = \frac{2 + 1}{6 - 2} = \frac{3}{4}$$

Then, by using either of the two known points and Eq. (20–6), we can find the equation of the line:

$$y + 1 = \frac{3}{4}(x - 2)$$

or

$$4y + 4 = 3x - 6$$

or

$$4y - 3x + 10 = 0$$

This line is shown in Fig. 20–12.

Equation (20–6) can be used for any line except one parallel to the *y*-axis. Such a line has an undefined slope. However, it does have the property that all points have the same x-coordinate, regardless of the y-coordinate. We represent a line parallel to the y-axis as

$$\overline{x = a} \tag{20–7}$$

A line parallel to the x-axis has a slope of 0. From Eq. (20–6), we can find its equation to be $y = y_1$. To keep the same form as Eq. (20–7), we normally write this as

$$\overline{y = b} \tag{20–8}$$

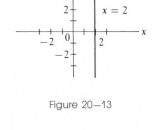

Figure 20–13

Example C
The line $x = 2$ is a line parallel to the y-axis and two units to the right of it. This line is shown in Fig. 20–13.

Example D
The line $y = -4$ is a line parallel to the x-axis and 4 units below it. This line is shown in Fig. 20–14.

If we choose the special point $(0, b)$, which is the y-intercept of the line, as the point to use in Eq. (20–6), we have

$$y - b = m(x - 0)$$

or

$$\overline{y = mx + b} \tag{20–9}$$

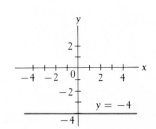

Figure 20–14

Equation (20–9) is known as the **slope-intercept form** of the equation of a straight line. Its usefulness lies in the fact that once we find the equation of a line and then write it in slope-intercept form, we know that the slope of the line is the coefficient of the x-term and that it crosses the y-axis with the coordinate indicated by the constant term.

Example E
Find the slope and the y-intercept of the straight line whose equation is $2y + 4x - 5 = 0$.
We write this equation in slope-intercept form:

$$2y = -4x + 5$$

$$y = -2x + \frac{5}{2}$$

Since the coefficient of x in this form is -2, the slope of the line is -2. The constant on the right is $\frac{5}{2}$, which means that the y-intercept is $\frac{5}{2}$.

From Eqs. (20–6) and (20–9), and from the examples of this section, we see that the equation of the straight line has certain characteristics: we have a term in y, a term in x, and a constant term if we simplify as much as possible. This form is represented by the equation

$$Ax + By + C = 0 \qquad\qquad (20\text{–}10)$$

which is known as the **general form** of the equation of the straight line. We have seen this form before in Chapter 4. Now we have shown why it represents a straight line.

Example F

What are the intercepts of the line $3x - 4y - 12 = 0$?

We know that the intercepts of a curve are those points where it crosses each of the axes. At the point where any curve crosses the x-axis, the y-coordinate must be 0. Thus, by letting $y = 0$ in the equation above, we solve for x and obtain $x = 4$. Hence one intercept is $(4, 0)$. By letting $x = 0$, we find the other intercept as $(0, -3)$.

In many physical situations a linear relationship exists between variables. A few examples of situations where such relationships exist are between (1) the distance traveled by an object and the elapsed time, when the velocity is constant, (2) the amount a spring stretches and the force applied, (3) the change in electric resistance and the change in temperature, (4) the force applied to an object and the resulting acceleration, and (5) the pressure at a certain point within a liquid and the depth of the point. The following example illustrates the use of a straight line in dealing with an applied problem.

Example G

Under the condition of constant acceleration, the velocity of an object varies linearly with the time. If after 1 s a certain object has a velocity of 40 ft/s, and 3 s later it has a velocity of 55 ft/s, find the equation relating the velocity and time, and graph this equation. From the graph determine the initial velocity (the velocity when $t = 0$) and the velocity after 6 s.

If we treat the velocity v as the dependent variable, and the time t as the independent variable, the slope of the straight line is

$$m = \frac{v_2 - v_1}{t_2 - t_1}$$

Using the information given in the problem, we have

$$m = \frac{55 - 40}{4 - 1} = 5$$

Then, using the point-slope form of the equation of a straight line, we have $v - 40 = 5(t - 1)$, or $v = 5t + 35$, which is the required equation (see Fig. 20–15). For purposes of graphing the line, the values given

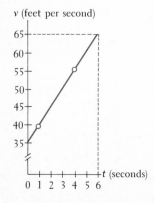

v (feet per second)

65
60
55
50
45
40
35

0 1 2 3 4 5 6 t (seconds)

Figure 20–15

are sufficient. Of course, there is no need to include negative values of t, since these have no physical meaning. From the graph we see that the line crosses the v-axis at 35. This means that the initial velocity is 35 ft/s. Also, when $t = 6$ we see that $v = 65$ ft/s.

Exercises 20–2

In Exercises 1 through 16 find the equation of each of the lines with the given properties.

1. Passes through $(-3, 8)$ with a slope of 4
2. Passes through $(-2, -1)$ with a slope of -2
3. Passes through $(-2, -5)$ and $(4, 2)$
4. Passes through $(-3, -5)$ and $(-2, 3)$
5. Passes through $(1, 3)$ and has an inclination of $45°$
6. Has a y-intercept of -2 and an inclination of $120°$
7. Passes through $(6, -3)$ and is parallel to the x-axis
8. Passes through $(-4, -2)$ and is perpendicular to the x-axis
9. Is parallel to the y-axis and is 3 units to the left of it
10. Is parallel to the x-axis and is 5 units below it
11. Has an x-intercept of 4 and a y-intercept of -6
12. Has an x-intercept of -3 and a slope of 2
13. Is perpendicular to a line with a slope of 3 and passes through $(1, -2)$
14. Is perpendicular to a line with a slope of -4 and has a y-intercept of 3
15. Is perpendicular to the line joining $(4, 2)$ and $(3, -5)$ and passes through $(4, 2)$
16. Is parallel to the line $2y - 6x - 5 = 0$ and passes through $(-4, -5)$

In Exercises 17 through 20 draw the lines with the given equations.

17. $4x - y = 8$
18. $2x - 3y - 6 = 0$
19. $3x + 5y - 10 = 0$
20. $4y = 6x - 9$

In Exercises 21 through 24 reduce the equations to slope-intercept form and determine the slope and y-intercept.

21. $3x - 2y - 1 = 0$
22. $4x + 2y - 5 = 0$
23. $5x - 2y + 5 = 0$
24. $6x - 3y - 4 = 0$

In Exercises 25 through 28 determine the value of k.

25. What is the value of k if the lines $4x - ky = 6$ and $6x + 3y + 2 = 0$ are to be parallel?
26. What must k equal in Exercise 25, if the given lines are to be perpendicular?
27. What must be the value of k if the lines $3x - y = 9$ and $kx + 3y = 5$ are to be perpendicular?
28. What must k equal in Exercise 27, if the given lines are to be parallel?

In Exercises 29 and 30 show that the given lines are parallel. In Exercises 31 and 32 show that the given lines are perpendicular.

29. $3x - 2y + 5 = 0$ and $4y = 6x - 1$
30. $3y - 2x = 4$ and $6x - 9y = 5$

31. $6x - 3y - 2 = 0$ and $x + 2y - 4 = 0$

32. $4x - y + 2 = 0$ and $2x + 8y - 1 = 0$

In Exercises 33 through 36 find the equations of the given lines.

33. The line which has an x-intercept of 4 and a y-intercept the same as the line $2y - 3x - 4 = 0$

34. The line which is perpendicular to the line $8x + 2y - 3 = 0$ and has the same x-intercept as this line

35. The line with a slope of -3 which also passes through the intersection of the lines $5x - y = 6$ and $x + y = 12$

36. The line which passes through the point of intersection of $2x + y - 3 = 0$ and $x - y - 3 = 0$ and through the point $(4, -3)$

In Exercises 37 through 50 some applications and methods involving straight lines are shown.

37. The average velocity of an object is defined as the change in displacement s divided by the corresponding change in time t. Find the equation relating the displacement s and time t for an object for which the average velocity is 50 m/s and $s = 10$ m when $t = 0$ s.

38. The acceleration of an object is defined as the change in velocity v divided by the corresponding change in time t. Find the equation relating the velocity v and time t for an object for which the acceleration is 20 ft/s² and $v = 5.0$ ft/s when $t = 0$ s.

39. Within certain limits, the amount which a spring stretches varies linearly with the amount of force applied. If a spring whose natural length is 15 in. stretches 2.0 in. when 3.0 lb of force are applied, find the equation relating the length of the spring and the applied force.

40. The electric resistance of a certain resistor increases by 0.005 Ω for every increase of 1°C. Given that its resistance is 2.000 Ω at 0°C, find the equation relating the resistance and temperature. From the equation find the resistance when the temperature is 50°C.

41. The amount of heat required to raise the temperature of water varies linearly with the increase in temperature. However, a certain quantity of heat is required to change ice (at 0°C) into water without a change in temperature. An experiment is performed with 10 g of ice at 0°C. It is found that ice requires 4.19 kJ to change it into water at 20°C. Another 1.26 kJ is required to warm the water to 50°C. How many kilojoules are required to melt the ice at 0°C into water at 0°C?

42. The pressure at a point below the surface of a body of water varies linearly with the depth of the water. If the pressure at a depth of 10.0 m is 199 kPa and the pressure at a depth of 30.0 m is 395 kPa, what is the pressure at the surface? (This is the atmospheric pressure of the air above the water.)

43. A survey of the traffic on a particular highway showed that the number of cars passing a particular point each minute varied linearly from 6:30 AM to 8:30 AM on workday mornings. The study showed that an average of 45 cars passed the point in one minute at 7 AM, and that 115 cars passed in one minute at 8 AM. If n is the number of cars passing the point in one minute and t is the number of minutes after 6:30 AM, find the equation relating n and t, and graph the equation. From the graph, determine n at 6:30 AM and at 8:30 AM.

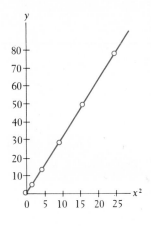

Figure 20–16

44. The total fixed cost for a company to operate a certain plant is $400 per day. It also costs $2 for each unit produced in the plant. Find the equation relating the total cost C of operating the plant and the number of units n produced each day. Graph the equation, assuming that $n \leq 300$.

45. Sometimes it is desirable to treat a nonlinear function as a linear function. For example, if $y = 2 + 3x^2$, this can be plotted as a straight line by plotting y vs. x^2. A table of values for this graph is

x	0	1	2	3	4	5
x^2	0	1	4	9	16	25
y	2	5	14	29	50	77

and the graph appears in Fig. 20–16. Using this method, sketch the function $y = 1 + \sqrt{x}$. That is, plot y vs. $\sqrt{x}$.

46. Using the method of Exercise 45, sketch the function $C = 10/E$ as a linear function. This is the equation between the voltage across a capacitor and the capacitance, assuming the charge on the capacitor is constant.

47. Another method of treating a nonlinear function as a linear one is to use logarithms. In Section 12–9 we noted that curves plotted on logarithmic and semilogarithmic paper often come out as straight lines. Since distances corresponding to log y and log x are plotted automatically if the graph is logarithmic in the respective directions, many nonlinear functions are linear when plotted on this paper. A function of the form $y = ax^n$ is straight when plotted on logarithmic paper, since $\log y = \log a + n \log x$ is in the form of a straight line. The variables are log y and log x; the slope can be found from $(\log y - \log a)/\log x = n$, and the intercept is a. (To get the slope from the graph, it is necessary to measure vertical and horizontal distances between two points. The log y intercept is found where log $x = 0$, and this occurs when $x = 1$.) Plot $y = 3x^4$ on logarithmic paper to verify this analysis.

48. A function of the form $y = a(b^x)$ is a straight line on semilogarithmic paper, since $\log y = \log a + x \log b$ is in the form of a straight line. The variables are log y and x, the slope is log b, and the intercept is a. [To get the slope from the graph, we must calculate $(\log y - \log a)/x$ for some set of values x and y. The intercept is read directly off the graph where $x = 0$.] Plot $y = 3(2^x)$ on semilogarithmic paper to verify this analysis.

49. If experimental data are plotted on logarithmic paper and the points lie on a straight line, it is possible to determine the function (see Exercise 47). The following data come from an experiment to determine the functional relationship between the tension in a string and the velocity of a wave in the string. From the graph on logarithmic paper, determine v as a function of T.

v (meters per second)	16.0	32.0	45.3	55.4	64.0
T (newtons)	0.100	0.400	0.800	1.20	1.60

50. If experimental data are plotted on semilogarithmic paper, and the points lie on a straight line, it is possible to determine the function (see Exercise 48). The following data come from an experiment designed to determine the relationship between the voltage across an inductor and the time, after the switch is opened. Determine v as a function of t.

v (volts)	40	15	5.6	2.2	0.8
t (milliseconds)	0.0	20	40	60	80

20–3 The Circle

We have found that we can obtain a general equation which represents a straight line by considering a fixed point on the line and then a general point $P(x, y)$ which can represent any other point on the same line. Mathematically we can state this as "the line is the **locus** of a point $P(x, y)$ which *moves* along the line." That is, the point $P(x, y)$ can be considered as a variable point which moves along the line.

In this way we can define a number of important curves. *The circle is defined as the locus of a point $P(x, y)$ which moves so that it is always equidistant from a fixed point. This fixed distance we call the* **radius,** *and the fixed point is called the* **center** *of the circle.* Thus, using this 'definition, calling the fixed point (h, k) and the radius r, we have

$$\sqrt{(x - h)^2 + (y - k)^2} = r$$

or, by squaring both sides, we have

$$(x - h)^2 + (y - k)^2 = r^2 \qquad (20\text{–}11)$$

Equation (20–11) is called the **standard equation** of a circle with center at (h, k) and radius r (Fig. 20–17).

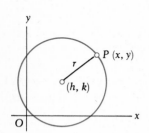

Figure 20–17

Example A

The equation $(x - 1)^2 + (y + 2)^2 = 16$ represents a circle with center at $(1, -2)$ and a radius of 4. We determine these values by considering the equation of this circle to be in the form of Eq. (20–11) as $(x - 1)^2 + (y - (-2))^2 = 4^2$. Note carefully the way in which we find the y-coordinate of the center. This circle is shown in Fig. 20–18.

If the center of the circle is at the origin, which means that the coordinates of the center are $(0, 0)$, the equation of the circle becomes

$$x^2 + y^2 = r^2 \qquad (20\text{–}12)$$

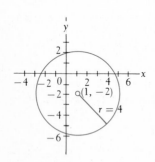

Figure 20–18

A circle of this type clearly exhibits an important property of the graphs of many equations. It is **symmetrical** to the x-axis and also to the y-axis. Symmetry to the x-axis can be thought of as meaning that the lower half of the curve is a reflection of the upper half, and conversely. It can be shown that *if $-y$ can replace y in an equation without changing the equation, the graph of the equation is symmetrical to the x-axis.* Symmetry to the y-axis has a similar meaning, so *if $-x$ can replace x in the equation without changing the equation, the graph is symmetrical to the y-axis.*

This type of circle is symmetrical to the origin as well as being symmetrical to both axes. The meaning of symmetry to the origin is that the origin is the midpoint of any two points (x, y) and $(-x, -y)$ which are on the curve. Thus, if $-x$ *can replace* x, *and* $-y$ *replace* y *at the same time, without changing the equation, the graph of the equation is symmetrical to the origin.*

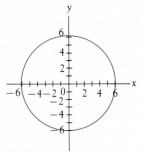

Figure 20–19

Example B

The equation of the circle with its center at the origin and with a radius of 6 is $x^2 + y^2 = 36$.

The symmetry of this circle can be shown analytically by the substitutions mentioned above. Replacing x by $-x$, we obtain $(-x)^2 + y^2 = 36$. Since $(-x)^2 = x^2$, this equation can be rewritten as $x^2 + y^2 = 36$. Since this substitution did not change the equation, the graph is symmetrical to the y-axis.

Replacing y by $-y$, we obtain $x^2 + (-y)^2 = 36$, which is the same as $x^2 + y^2 = 36$. This means that the curve is symmetrical to the x-axis.

Replacing x by $-x$, and simultaneously replacing y by $-y$, we obtain $(-x)^2 + (-y)^2 = 36$, which is the same as $x^2 + y^2 = 36$. This means that the curve is symmetrical to the origin. The circle is shown in Fig. 20–19.

Example C

Find the equation of the circle with center at (2, 1) and which passes through (4, 8).

In Eq. (20–11), we can determine the equation if we can find h, k, and r for this circle. From the given information, $h = 2$ and $k = 1$. To find r, we use the fact that *all points on the circle must satisfy the equation of the circle.* The point (4, 8) must satisfy Eq. (20–11), with $h = 2$ and $k = 1$. Thus, $(4 - 2)^2 + (8 - 1)^2 = r^2$. From this relation we find $r^2 = 53$. The equation of the circle is

$$(x - 2)^2 + (y - 1)^2 = 53$$

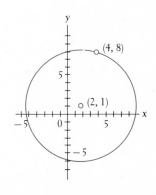

Figure 20–20

This circle is shown in Fig. 20–20.

If we multiply out each of the terms in Eq. (20–11), we may combine the resulting terms to obtain

$$x^2 - 2hx + h^2 + y^2 - 2ky + k^2 = r^2$$
$$x^2 + y^2 - 2hx - 2ky + (h^2 + k^2 - r^2) = 0 \qquad (20\text{–}13)$$

Since each of h, k, and r is constant for any given circle, the coefficients of x and y and the term within parentheses in Eq. (20–13) are constants. Equation (20–13) can then be written as

$$x^2 + y^2 + Dx + Ey + F = 0 \qquad (20\text{–}14)$$

Equation (20–14) is called the **general equation** of the circle. It tells us that any equation which can be written in that form will represent a circle.

Example D
Find the center and radius of the circle

$$x^2 + y^2 - 6x + 8y - 24 = 0$$

We can find this information if we write the given equation in standard form. To do so, we must complete the square in the *x*-terms and also in the *y*-terms. This is done by first writing the equation in the form

$$(x^2 - 6x \quad) + (y^2 + 8y \quad) = 24$$

To complete the square of the *x*-terms, we take half of 6, which is 3, square it, and add the result, 9, to each side of the equation. In the same way, we complete the square of the *y*-terms by adding 16 to each side of the equation, which gives

$$(x^2 - 6x + 9) + (y^2 + 8y + 16) = 24 + 9 + 16$$
$$(x - 3)^2 + (y + 4)^2 = 49$$

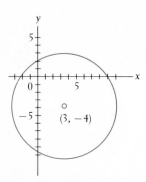

Figure 20−21

Thus, the center is $(3, -4)$, and the radius is 7 (see Fig. 20–21).

Example E
The equation $3x^2 + 3y^2 + 6x - 20 = 0$ can be seen to represent a circle by writing it in general form. This is done by dividing through by 3. In this way we have

$$x^2 + y^2 + 2x - \frac{20}{3} = 0$$

To determine the center and radius for this circle, we write the equation in standard form and then complete the necessary squares. This leads to

$$(x^2 + 2x + 1) + y^2 = \frac{20}{3} + 1$$

$$(x + 1)^2 + y^2 = \frac{23}{3}$$

Figure 20−22

Thus, the center is $(-1, 0)$; the radius is $\sqrt{23/3} = 2.77$ (see Fig. 20–22).

We can also see that this circle is symmetrical to the *x*-axis, but it is not symmetrical to the *y*-axis or to the origin. If we replace *y* by $-y$, the equation does not change, but if we replace *x* by $-x$, the 6x term in the original equation becomes negative, and the equation *does* change.

Exercises 20−3

In Exercises 1 through 4 determine the center and radius of each circle.

1. $(x - 2)^2 + (y - 1)^2 = 25$ 2. $(x - 3)^2 + (y + 4)^2 = 49$
3. $(x + 1)^2 + y^2 = 4$ 4. $x^2 + (y + 6)^2 = 64$

In Exercises 5 through 16 find the equation of each of the circles from the given information.

5. Center at $(0, 0)$, radius 3   6. Center at $(0, 0)$, radius 1
7. Center at $(2, 2)$, radius 4   8. Center at $(0, 2)$, radius 2

9. Center at $(-2, 5)$, radius $\sqrt{5}$ 10. Center at $(-3, -5)$, radius 8

11. Center at $(2, 1)$, passes through $(4, -1)$

12. Center at $(-1, 4)$, passes through $(-2, 3)$

13. Center $(-3, 5)$, tangent to the x-axis

14. Tangent to both axes, radius 4, and in the second quadrant

15. Center on the line $5x = 2y$, radius 5, tangent to the x-axis

16. The points $(3, 8)$ and $(-3, 0)$ are the ends of a diameter.

In Exercises 17 through 24 determine the center and radius of each circle. Sketch each circle.

17. $x^2 + y^2 - 25 = 0$ 18. $x^2 + y^2 - 9 = 0$

19. $x^2 + y^2 - 2x - 8 = 0$ $\;$ camb 19   20. $x^2 + y^2 - 4x - 6y - 12 = 0$

21. $x^2 + y^2 + 8x - 10y - 8 = 0$ 22. $x^2 + y^2 + 8x + 6y = 0$

23. $2x^2 + 2y^2 - 4x - 8y - 1 = 0$ 24. $3x^2 + 3y^2 - 12x + 4 = 0$

In Exercises 25 through 28 determine whether the circles with the given equations are symmetrical to either axis or the origin.

25. $x^2 + y^2 = 100$ 26. $x^2 + y^2 - 4x - 5 = 0$

27. $x^2 + y^2 + 8y - 9 = 0$ 28. $x^2 + y^2 - 2x + 4y - 3 = 0$

In Exercises 29 through 36 find the indicated quantities.

29. Determine where the circle $x^2 - 6x + y^2 - 7 = 0$ crosses the x-axis.

30. Find the points of intersection of the circle $x^2 + y^2 - x - 3y = 0$ and the line $y = x - 1$.

31. Find the locus of a point $P(x, y)$ which moves so that its distance from $(2, 4)$ is twice its distance from $(0, 0)$. Describe the locus.

32. Find the equation of the locus of a point $P(x, y)$ which moves so that the line joining it and $(2, 0)$ is always perpendicular to the line joining it and $(-2, 0)$. Describe the locus.

33. If an electrically charged particle enters a magnetic field of flux density B with a velocity v at right angles to the field, the path of the particle is a circle. The radius of the path is given by $R = mv/Bq$, where m is the mass of the particle and q is its charge. If a proton ($m = 1.67 \times 10^{-27}$ kg, $q = 1.60 \times 10^{-19}$ C) enters a magnetic field of flux density 1.50 T (tesla: $V \cdot s/m^2$) with a velocity of 2.40×10^8 m/s, find the equation of the path of the proton. (The fact that a charged particle travels in a circular path in a magnetic field is an important basis for the construction of a cyclotron.)

34. For a constant (magnitude) impedance and capacitive reactance, sketch the graph of resistance versus inductive reactance (see Section 11–7).

35. A cylindrical oil tank, 4.00 ft in diameter, is on its side. A circular hole, with center 18.0 in. below the center of the end of the tank, has been drilled in the end of the tank. If the radius of the hole is 3.00 in., what is the equation of the circle of the end of the tank and the equation of the circle of the hole? Use the center of the end of the tank as the origin.

36. A draftsman is drawing a friction drive in which two circular disks are in contact with each other. They are represented by circles in his drawing. The first has a radius of 10.0 cm and the second has a radius of 12.0 cm. What is the equation of each circle if the origin is at the center of the first circle and the x-axis passes through the center of the second circle?

20-4 The Parabola

Another important curve is the parabola. We have come across this curve several times in the past. In this section we shall find the form of the equation of the parabola and see that this is a familiar form.

The **parabola** is defined as the locus of a point $P(x, y)$ which moves so that it is always equidistant from a given line and a given point. The given line is called the **directrix**, and the given point is called the **focus**. The line through the focus which is perpendicular to the directrix is called the **axis** of the parabola. The point midway between the directrix and focus is the **vertex** of the parabola. Using the definition, we shall find the equation of the parabola for which the focus is the point $(p, 0)$ and the directrix is the line $x = -p$. By choosing the focus and directrix in this manner, we shall be able to find a general representation of the equation of a parabola with its vertex at the origin.

According to the definition of the parabola, the distance from a point $P(x, y)$ on the parabola to the focus $(p, 0)$ must equal the distance from $P(x, y)$ to the directrix $x = -p$. The distance from P to the focus can be found by use of the distance formula. The distance from P to the directrix is the perpendicular distance, and this can be found as the distance between two points on a line parallel to the x-axis. These distances are indicated in Fig. 20–23.

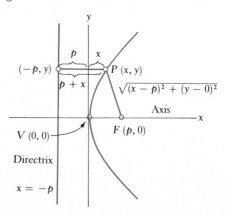

Figure 20–23

Thus, we have

$$\sqrt{(x - p)^2 + (y - 0)^2} = x + p$$

Squaring both sides of this equation, we have

$$(x - p)^2 + y^2 = (x + p)^2$$

or

$$x^2 - 2px + p^2 + y^2 = x^2 + 2px + p^2$$

Simplifying, we obtain

$$y^2 = 4px \tag{20-15}$$

Equation (20–15) is called the **standard form** of the equation of a parabola with its axis along the x-axis and the vertex at the origin. Its symmetry to the x-axis can be proven since $(-y)^2 = 4px$ is the same as
$y^2 = 4px$.

Example A
Find the coordinates of the focus, the equation of the directrix, and
sketch the graph of the parabola $y^2 = 12x$.

  From the form of the equation, we know that the vertex is at the
origin (Fig. 20–24). The coefficient 12 tells us that the focus is (3, 0),
since $p = \frac{12}{4}$ [note that the coefficient of x in Eq. (20–15) is $4p$]. Also,
this means that the directrix is the line $x = -3$.

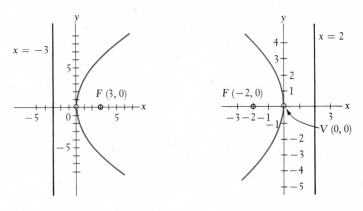

Figure 20−24   Figure 20−25

Example B
If the focus is to the left of the origin, with the directrix an equal distance to the right, the coefficient of the x-term will be negative. This
tells us that the parabola opens to the left, rather than to the right, as is
the case when the focus is to the right of the origin. For example, the
parabola $y^2 = -8x$ has its vertex at the origin, its focus at $(-2, 0)$, and
the line $x = 2$ as its directrix. This is consistent with Eq. (20–15), in
that $4p = -8$, or $p = -2$. The parabola opens to the left as shown in
Fig. 20–25.

  If we chose the focus as the point $(0, p)$ and the directrix as the line
$y = -p$, we would find that the resulting equation is

$$x^2 = 4py \tag{20-16}$$

This is the standard form of the equation of a parabola with the y-axis as its axis and the vertex at the origin. Its symmetry to the y-axis can be proven since $(-x)^2 = 4py$ is the same as $x^2 = 4py$. We note that the difference between this equation and Eq. (20–15) is that x is squared and y appears to the first power in Eq. (20–16), rather than the reverse, as in Eq. (20–15).

Example C
The parabola $x^2 = 4y$ has its vertex at the origin, focus at the point $(0, 1)$, and its directrix the line $y = -1$. The graph is shown in Fig. 20–26, and we see in this case that the parabola opens up.

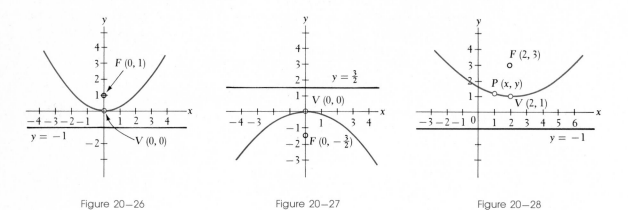

Figure 20–26   Figure 20–27   Figure 20–28

Example D
The parabola $x^2 = -6y$ has its vertex at the origin, its focus at the point $(0, -\frac{3}{2})$, and its directrix the line $y = \frac{3}{2}$. This parabola opens down, as Fig. 20–27 shows.

Equations (20–15) and (20–16) give us the general form of the equation of a parabola with its vertex at the origin and its focus on one of the coordinate axes. The following example illustrates the use of the definition of the parabola to find the equation of a parabola which has its vertex at a point other than at the origin.

Example E
Using the definition of the parabola, determine the equation of the parabola with its focus at $(2, 3)$ and its directrix the line $y = -1$ (see Fig. 20–28).

Choosing a general point $P(x, y)$ on the parabola, and equating the distances from this point to $(2, 3)$ and to the line $y = -1$, we have

$$\sqrt{(x - 2)^2 + (y - 3)^2} = y + 1$$

Squaring both sides of this equation and simplifying, we have

$$(x - 2)^2 + (y - 3)^2 = (y + 1)^2$$
$$x^2 - 4x + 4 + y^2 - 6y + 9 = y^2 + 2y + 1$$

or

$$8y = 12 - 4x + x^2$$

When we see it in this form, we note that this type of equation has appeared numerous times in earlier chapters. The x-term and the constant (12 in this case) are characteristic of a parabola which does not have its vertex at the origin if the directrix is parallel to the x-axis.

 Thus we can conclude that a parabola is characterized by the presence of the square of either (but not both) x or y, and a first-power term in the other. In this way we can recognize a parabola by inspecting the equation.

Exercises 20–4

In Exercises 1 through 12 determine the coordinates of the focus and the equation of the directrix of each of the given parabolas. Sketch each curve.

1. $y^2 = 4x$ 2. $y^2 = 16x$ 3. $y^2 = -4x$ 4. $y^2 = -16x$
5. $x^2 = 8y$ 6. $x^2 = 10y$ 7. $x^2 = -4y$ 8. $x^2 = -12y$
9. $y^2 = 2x$ 10. $x^2 = 14y$ 11. $y = x^2$ 12. $x = 3y^2$

In Exercises 13 through 20 find the equations of the parabolas satisfying the given conditions.

13. Focus $(3, 0)$, directrix $x = -3$ 14. Focus $(-2, 0)$, directrix $x = 2$
15. Focus $(0, 4)$, vertex $(0, 0)$ 16. Focus $(-3, 0)$, vertex $(0, 0)$
17. Vertex $(0, 0)$, directrix $y = -1$ 18. Vertex $(0, 0)$, directrix $y = \frac{1}{2}$
19. Vertex at $(0, 0)$, axis along the y-axis, passes through $(-1, 8)$
 20. Vertex at $(0, 0)$, axis along the x-axis, passes through $(2, -1)$

In Exercises 21 through 24 use the definition of the parabola to find the equations of the parabolas satisfying the given conditions. Sketch each curve.

21. Focus $(6, 1)$, directrix $x = 0$ 22. Focus $(1, -4)$, directrix $x = 2$
23. Focus $(1, 1)$, vertex $(1, 3)$ 24. Vertex $(-2, -4)$, directrix $x = 3$

In Exercises 25 through 32 find the indicated quantities.

25. In calculus it can be shown that a light ray emanating from the focus of a parabola will be reflected off parallel to the axis of the parabola. Suppose that a light ray from the focus of a parabolic reflector described by the equation $y^2 = 8x$ strikes the reflector at the point $(2, 4)$. Along what line does the incident ray move, and along what line does the reflected ray move?

26. The rate of development of heat H (measured in watts) in a resistor of resistance R (measured in ohms) of an electric circuit is given by $H = Ri^2$, where i is the current (measured in amperes) in the resistor. Sketch the graph of H versus i, if $R = 6.0 \ \Omega$.

27. Under certain conditions, a cable which hangs between two supports can be closely approximated as being parabolic. Assuming that this cable hangs in the shape of a parabola, find its equation if a point 10 ft horizontally from its lowest point is 1.0 ft above its lowest point. Choose the lowest point as the origin of the coordinate system.

28. A wire is fastened 36.0 ft up on each of two telephone poles which are 200 ft apart. Halfway between the poles the wire is 30.0 ft above the ground. Assuming the wire is parabolic, find the height of the wire 50.0 ft from either pole.

29. The period T (measured in seconds) of the oscillation for resonance in an electric circuit is given by $T = 2\pi\sqrt{LC}$, where L is the inductance (in henrys) and C is the capacitance (in farads). Sketch the graph of the period versus capacitance for a constant inductance of 1 H. Assume values of C from 1 μF to 250 μF for this is a common range of values.

30. The velocity v of a jet of water flowing from an opening in the side of a certain container is given by $v = 8\sqrt{h}$, where h is the depth of the opening. Sketch a graph of v (in feet per second) vs. h (in feet).

31. A small island is 4 km from a straight shoreline. A ship channel is equidistant between the island and the shoreline. Write an equation for the channel.

32. Under certain circumstances, the maximum power P in an electric circuit varies as the square of the voltage of the source E_0 and inversely as the internal resistance R_i of the source. If 10 W is the maximum power for a source of 2.0 V and internal resistance of 0.10 Ω, sketch the graph of P versus E_0, if R_i remains constant.

20–5 The Ellipse

The next important curve we shall discuss is the ellipse. *The ellipse is defined as the locus of a point $P(x, y)$ which moves so that the sum of its distances from two fixed points is constant.* These two fixed points are called the **foci** of the ellipse. Letting this fixed sum equal $2a$ and the foci be the points $(c, 0)$ and $(-c, 0)$, we have (see Fig. 20–29) from the definition of the ellipse

$$\sqrt{(x - c)^2 + y^2} + \sqrt{(x + c)^2 + y^2} = 2a$$

From the section on solving equations involving radicals (Section 13–4), we find that we should remove one of the radicals to the right side, and then square each side. Thus we have the following steps:

$$\sqrt{(x + c)^2 + y^2} = 2a - \sqrt{(x - c)^2 + y^2}$$
$$(x + c)^2 + y^2 = 4a^2 - 4a\sqrt{(x - c)^2 + y^2} + (\sqrt{(x - c)^2 + y^2})^2$$
$$x^2 + 2cx + c^2 + y^2 = 4a^2 - 4a\sqrt{(x - c)^2 + y^2} + x^2 - 2cx + c^2 + y^2$$
$$4a\sqrt{(x - c)^2 + y^2} = 4a^2 - 4cx$$
$$a\sqrt{(x - c)^2 + y^2} = a^2 - cx$$
$$a^2(x^2 - 2cx + c^2 + y^2) = a^4 - 2a^2cx + c^2x^2$$
$$(a^2 - c^2)x^2 + a^2y^2 = a^2(a^2 - c^2)$$

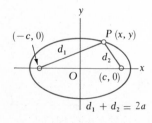

Figure 20–29

At this point we define (and the usefulness of this definition will be shown presently) $a^2 - c^2 = b^2$. This substitution gives us

$$b^2x^2 + a^2y^2 = a^2b^2$$

Dividing through by a^2b^2, we have

$$\frac{x^2}{a^2} + \frac{y^2}{b^2} = 1 \qquad (20\text{–}17)$$

If we let $y = 0$, we find that the x-intercepts are $(-a, 0)$ and $(a, 0)$. We see that the distance $2a$, originally chosen as the sum of the distances in the derivation of Eq. (20–17), is the distance between the x-intercepts. *These two points are known as the* **vertices** *of the ellipse, and the line between them is known as the* **major axis** *of the ellipse. Thus, a is called the semi-major axis.*

If we now let $x = 0$, we find the y-intercepts to be $(0, -b)$ and $(0, b)$. *The line joining these two intercepts is called the* **minor axis** *of the ellipse* (Fig. 20–30) *which means b is called the* **semi-minor axis.** The point $(0, b)$ is on the ellipse, and is also equidistant from $(-c, 0)$ and $(c, 0)$. Since the sum of the distances from these points to $(0, b)$ equals $2a$, the distance from $(c, 0)$ to $(0, b)$ must be a. Thus, we have a right triangle formed by line segments of lengths a, b, and c, with a as the hypotenuse. From this we have the relation

$$a^2 = b^2 + c^2 \qquad (20\text{–}18)$$

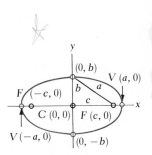

Figure 20–30

between the distances a, b, and c. This equation also shows us why b was defined as it was in the derivation.

Equation (20–17) is called the **standard equation** of the ellipse with its major axis along the x-axis and its center at the origin.

If we choose points on the y-axis as the foci, the standard equation of the ellipse, with its center at the origin and its major axis along the y-axis, is

$$\frac{y^2}{a^2} + \frac{x^2}{b^2} = 1 \qquad (20\text{–}19)$$

In this case the vertices are $(0, a)$ and $(0, -a)$, the foci are $(0, c)$ and $(0, -c)$, and the ends of the minor axis are $(b, 0)$ and $(-b, 0)$.

The symmetry of the ellipses given by Eqs. (20–17) and (20–19) to each of the axes and the origin can be proven since each of x and y can be replaced by its negative in these equations without changing the equations.

Example A
The ellipse

$$\frac{x^2}{25} + \frac{y^2}{9} = 1$$

has vertices at $(5, 0)$ and $(-5, 0)$. Its minor axis extends from $(0, 3)$ to $(0, -3)$, as we see in Fig. 20–31. This information is directly obtainable from this form of the equation, since the 25 tells us that $a^2 = 25$, or $a = 5$. In the same way we have $b = 3$. Since we know both a^2 and b^2, we can find c^2 from the relation $c^2 = a^2 - b^2$. Thus, $c^2 = 16$, or $c = 4$. This in turn tells us that the foci of this ellipse are at $(4, 0)$ and $(-4, 0)$.

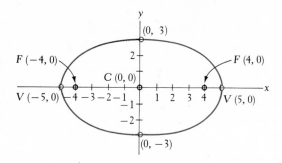

Figure 20–31

Example B
The ellipse

$$\frac{x^2}{4} + \frac{y^2}{9} = 1$$

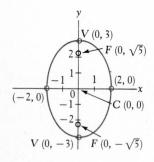

Figure 20–32

has vertices at $(0, 3)$ and $(0, -3)$. The minor axis extends from $(2, 0)$ to $(-2, 0)$. This we can find directly from the equation, since $a^2 = 9$ and $b^2 = 4$. One might ask how we chose $a^2 = 9$ in this example and $a^2 = 25$ in the preceding example. Since $a^2 = b^2 + c^2$, a is always larger than b. Thus, we can tell which axis (x or y) the major axis is along by seeing which number (in the denominator) is larger, when the equation is written in standard form. The larger one stands for a^2. In this example, the foci are $(0, \sqrt{5})$ and $(0, -\sqrt{5})$ (see Fig. 20–32).

Example C
Find the coordinates of the vertices, the ends of the minor axis, and the foci of the ellipse $4x^2 + 16y^2 = 64$.

This equation must be put in standard form first, which we do by dividing through by 64. When this is done, we obtain

$$\frac{x^2}{16} + \frac{y^2}{4} = 1$$

Thus, $a = 4$, $b = 2$, and $c = 2\sqrt{3}$. The vertices are at $(4, 0)$ and $(-4, 0)$. The ends of the minor axis are $(0, 2)$ and $(0, -2)$, and the foci are $(2\sqrt{3}, 0)$ and $(-2\sqrt{3}, 0)$ (see Fig. 20–33).

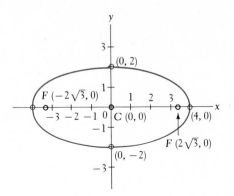

Figure 20–33

Example D
Find the equation of the ellipse for which the center is at the origin, the major axis is along the *y*-axis, the minor axis is 6 units long, and there are 8 units between foci.

Directly we know that $2b = 6$, or $b = 3$. Also $2c = 8$, or $c = 4$. Thus, we find that $a = 5$. Since the major axis is along the *y*-axis, we have the equation

$$\frac{y^2}{25} + \frac{x^2}{9} = 1$$

This ellipse is shown in Fig. 20–34.

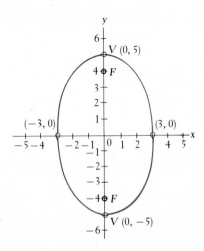

Figure 20–34

Equations (20–17) and (20–19) give us the standard form of the equation of an ellipse with its center at the origin and its foci on one of the coordinate axes. The following example illustrates the use of the definition of the ellipse to find the equation of an ellipse with its center at a point other than the origin.

Example E

Using the definition, find the equation of the ellipse with foci at (1, 3) and (9, 3), with a major axis of 10.

Using the same method as in the derivation of Eq. (20–17), we have the following steps. (Remember, the sum of distances in the definition equals the length of the major axis.)

$$\sqrt{(x-1)^2 + (y-3)^2} + \sqrt{(x-9)^2 + (y-3)^2} = 10$$
$$\sqrt{(x-1)^2 + (y-3)^2} = 10 - \sqrt{(x-9)^2 + (y-3)^2}$$
$$(x-1)^2 + (y-3)^2 = 100 - 20\sqrt{(x-9)^2 + (y-3)^2}$$
$$+ (\sqrt{(x-9)^2 + (y-3)^2})^2$$
$$x^2 - 2x + 1 + y^2 - 6y + 9 = 100 - 20\sqrt{(x-9)^2 + (y-3)^2}$$
$$+ x^2 - 18x + 81 + y^2 - 6y + 9$$
$$20\sqrt{(x-9)^2 + (y-3)^2} = 180 - 16x$$
$$5\sqrt{(x-9)^2 + (y-3)^2} = 45 - 4x$$
$$25(x^2 - 18x + 81 + y^2 - 6y + 9) = 2025 - 360x + 16x^2$$
$$25x^2 - 450x + 2250 + 25y^2 - 150y = 2025 - 360x + 16x^2$$
$$9x^2 - 90x + 25y^2 - 150y + 225 = 0$$

The additional x- and y-terms are characteristic of the equation of an ellipse whose center is not at the origin (see Fig. 20–35).

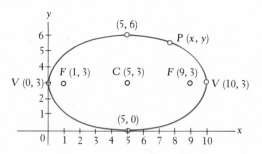

Figure 20–35

We can conclude that the equation of an ellipse is characterized by the presence of both an x^2- and a y^2-term, having different coefficients (in value but not in sign). The difference between the equation of an ellipse and that of a circle is that the coefficients of the squared terms in the equation of the circle are the same, whereas those of the ellipse differ.

Exercises 20-5

In Exercises 1 through 12 find the coordinates of the vertices and foci of the given ellipses. Sketch each curve.

1. $\dfrac{x^2}{4} + \dfrac{y^2}{1} = 1$

2. $\dfrac{x^2}{100} + \dfrac{y^2}{64} = 1$

3. $\dfrac{x^2}{25} + \dfrac{y^2}{36} = 1$

4. $\dfrac{x^2}{49} + \dfrac{y^2}{81} = 1$

5. $4x^2 + 9y^2 = 36$

6. $x^2 + 36y^2 = 144$

7. $49x^2 + 4y^2 = 196$

8. $25x^2 + y^2 = 25$

9. $8x^2 + y^2 = 16$

10. $2x^2 + 3y^2 = 6$

11. $4x^2 + 25y^2 = 25$

12. $9x^2 + 4y^2 = 9$

In Exercises 13 through 20 find the equations of the ellipses satisfying the given conditions. The center of each is at the origin.

13. Vertex $(15, 0)$, focus $(9, 0)$

14. Minor axis 8, vertex $(0, -5)$

15. Focus $(0, 2)$, major axis 6

16. Semi-minor axis 2, focus $(3, 0)$

17. Vertex $(8, 0)$, passes through $(2, 3)$

18. Focus $(0, 2)$, passes through $(-1, \sqrt{3})$

19. Passes through $(2, 2)$ and $(1, 4)$

20. Passes through $(-2, 2)$ and $(1, \sqrt{6})$

In Exercises 21 through 24 find the equations of the ellipses with the given properties by use of the definition of an ellipse.

21. Foci at $(-2, 1)$ and $(4, 1)$, a major axis of 10

22. Foci at $(-3, -2)$ and $(-3, 8)$, major axis 26

23. Vertices at $(1, 5)$ and $(1, -1)$, foci at $(1, 4)$ and $(1, 0)$

24. Vertices at $(-2, 1)$ and $(-2, 5)$, foci at $(-2, 2)$ and $(-2, 4)$

In Exercises 25 through 32 solve the given problems.

25. Show that the ellipse $2x^2 + 3y^2 - 8x - 4 = 0$ is symmetrical to the x-axis.

26. Show that the ellipse $5x^2 + y^2 - 3y - 7 = 0$ is symmetrical to the y-axis.

27. The arch of a bridge across a stream is in the form of half an ellipse above the water level. If the span of the arch at water level is 100 ft and the maximum height of the arch above water level is 30 ft, what is the equation of the arch? Choose the origin of the coordinate system at the most convenient point.

28. An elliptical gear (Fig. 20-36) which rotates about its center is kept continually in mesh with a circular gear which is free to move horizontally. If the equation of the ellipse of the gear (with the origin of the coordinate system at its center) in its present position is $3x^2 + 7y^2 = 20$, how far does the center of the circular gear move going from one extreme position to the other? (Assume the units are centimeters.)

29. An artificial satellite of the earth has a minimum altitude of 500 mi and a maximum altitude of 2000 mi. If the path of the satellite about the earth is an ellipse with the center of the earth at one focus, what is the equation of its path? (Assume the radius of the earth is 4000 mi.)

Figure 20-36

30. An electric current is caused to flow in a loop of wire rotating in a magnetic field. In a study of this phenomenon, a piece of wire is cut into two pieces, one of which is bent into a circle and the other into a square. If the sum of areas of the circle and square is always π units, find the relation between the radius r of the circle and the side x of the square. Sketch the graph of r versus x.

31. A vertical pipe 6.0 in. in diameter is to pass through a roof inclined at 45°. What are the dimensions of the elliptical hole which must be cut in the roof for the pipe?

32. The ends of a horizontal tank 20.0 ft long are ellipses, which can be described by the equation $9x^2 + 20y^2 = 180$, where x and y are measured in feet. The area of an ellipse is $A = \pi ab$. Find the volume of the tank.

20—6 The Hyperbola

The final curve we shall discuss in detail is the hyperbola. *The hyperbola is defined as the locus of a point $P(x, y)$ which moves so that the difference of the distances from two fixed points is a constant. These fixed points are the* foci *of the hyperbola.* Assuming that the foci of the hyperbola are the points $(c, 0)$ and $(-c, 0)$ (see Fig. 20–37) and the constant difference is $2a$, we have

$$\sqrt{(x + c)^2 + y^2} - \sqrt{(x - c)^2 + y^2} = 2a$$

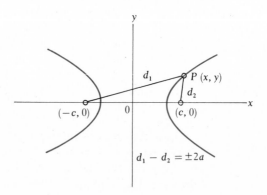

Figure 20—37

Following this same procedure as in the preceding section, we find the equation of the hyperbola to be

$$\frac{x^2}{a^2} - \frac{y^2}{b^2} = 1 \tag{20–20}$$

When we derive this equation, we have a definition of the relation between a, b, and c which is different from that for the ellipse. This relation is

$$c^2 = a^2 + b^2 \qquad\qquad (20\text{--}21)$$

An analysis of Eq. (20–20) will reveal the significance of a and b. First, by letting $y = 0$, we find that the x-intercepts are $(a, 0)$ and $(-a, 0)$, just as they are for the ellipse. These points are called the **vertices** of the hyperbola. By letting $x = 0$, we find that we have imaginary solutions for y, which means that there are no points on the curve which correspond to a value of $x = 0$.

To find the significance of b, we shall solve Eq. (20–20) for y in a particular form. First we have

$$\frac{y^2}{b^2} = \frac{x^2}{a^2} - 1$$

$$= \frac{x^2}{a^2} - \frac{a^2 x^2}{a^2 x^2}$$

$$= \frac{x^2}{a^2}\left(1 - \frac{a^2}{x^2}\right)$$

Multiplying through by b^2 and then taking the square root of each side, we have

$$y^2 = \frac{b^2 x^2}{a^2}\left(1 - \frac{a^2}{x^2}\right)$$

$$y = \pm\frac{bx}{a}\sqrt{1 - \frac{a^2}{x^2}} \qquad\qquad (20\text{--}22)$$

We note that, if large values of x are assumed in Eq. (20–22), the quantity under the radical becomes approximately 1. In fact, the larger x becomes, the nearer 1 this expression becomes, since the x^2 in the denominator of a^2/x^2 makes this term nearly zero. Thus, for large values of x, Eq. (20–22) is approximately

$$y = \pm\frac{bx}{a} \qquad\qquad (20\text{--}23)$$

Equation (20–23) can be seen to represent the equations for two straight lines, each of which passes through the origin. One has a slope of b/a and the other a slope of $-b/a$. *These lines are called the* **asymptotes** *of the hyperbola. An asymptote is a line which a curve approaches as one of the variables approaches some particular value.* The tangent curve also has asymptotes, as we can see in Fig. 9–17. We can designate this limiting procedure with notation introduced in Chapter 18, by saying that

$$y \to \frac{bx}{a} \text{ as } x \to \infty$$

Since straight lines are easily sketched, the easiest way to sketch a hyperbola is to draw its asymptotes and then to draw the hyperbola so that it comes closer and closer to these lines as x becomes larger numerically. To draw in the asymptotes, the usual procedure is to first draw a small rectangle, $2a$ by $2b$, with the origin in the center. Then straight lines are drawn through opposite vertices. These lines are the asymptotes (see Fig. 20–38). Thus we see that the significance of the value of b lies in the slope of the asymptotes of the hyperbola.

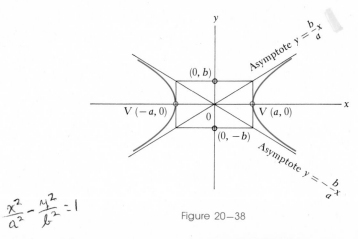

$$\frac{x^2}{a^2} - \frac{y^2}{b^2} = 1$$

Figure 20–38

Equation (20–20) is called the **standard equation** of the hyperbola with its center at the origin. *It has a* **transverse axis** *of length 2a along the x-axis and a* **conjugate axis** *of length 2b along the y-axis.* This means that a represents the length of the semi-transverse axis, and b represents the length of the semi-conjugate axis. The relation between a, b, and c is given in Eq. (20–21).

If the transverse axis is along the y-axis and the conjugate axis is along the x-axis, the equation of a hyperbola with its center at the origin is

$$\frac{y^2}{a^2} - \frac{x^2}{b^2} = 1 \qquad\qquad (20\text{–}24)$$

The symmetry of the hyperbolas given by Eqs. (20–20) and (20–24) to each of the axes and to the origin can be proven since each of x and y can be replaced by its negative in these equations without changing the equations.

Example A
The hyperbola

$$\frac{x^2}{16} - \frac{y^2}{9} = 1$$

has vertices at $(4, 0)$ and $(-4, 0)$. Its transverse axis extends from one vertex to the other. Its conjugate axis extends from $(0, 3)$ to $(0, -3)$. Since $c^2 = a^2 + b^2$, we find that $c = 5$, which means the foci are the points $(5, 0)$ and $(-5, 0)$. Drawing in the rectangle and then the asymptotes (Fig. 20–39), we draw the hyperbola from each vertex toward the asymptotes. Thus, the curve is sketched.

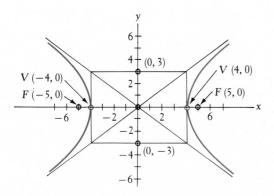

Figure 20–39

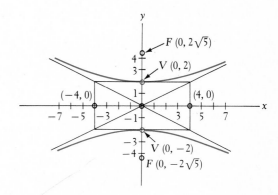

Figure 20–40

Example B
The hyperbola

$$\frac{y^2}{4} - \frac{x^2}{16} = 1$$

has vertices at $(0, 2)$ and $(0, -2)$. Its conjugate axis extends from $(4, 0)$ to $(-4, 0)$. The foci are $(0, 2\sqrt{5})$ and $(0, -2\sqrt{5})$. Since, with 1 on the right side of the equation, the y^2-term is the positive term, this hyperbola is in the form of Eq. (20–24). [Example A illustrates a hyperbola of the type of Eq. (20–20), since the x^2-term is the positive term.] Since $2a$ extends along the y-axis, we see that the equations of the asymptotes are $y = \pm(a/b)x$. This is not a contradiction of Eq. (20–23), but an extension of it for the case of a hyperbola with its transverse axis along the y-axis. The ratio a/b simply expresses the slope of the asymptote (see Fig. 20–40).

Example C

Determine the coordinates of the vertices and foci of the hyperbola $4x^2 - 9y^2 = 36$.

First, by dividing through by 36, we can put this equation in standard form. Thus, we have

$$\frac{x^2}{9} - \frac{y^2}{4} = 1$$

The transverse axis is along the x-axis with vertices at $(3, 0)$ and $(-3, 0)$, since $c^2 = a^2 + b^2$ and $c = \sqrt{13}$. This means that the foci are at $(\sqrt{13}, 0)$ and $(-\sqrt{13}, 0)$ (see Fig. 20–41).

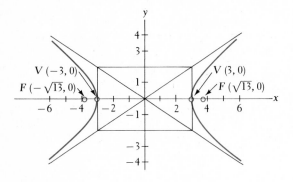

Figure 20–41

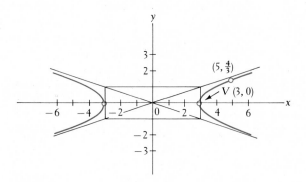

Figure 20–42

Example D

Find the equation of the hyperbola with its center at the origin, a vertex at $(3, 0)$, and which passes through $(5, \frac{4}{3})$.

When we know that the hyperbola has its center at the origin and a vertex at $(3, 0)$, we know that the standard form of its equation is given by Eq. (20–20). This also tells us that $a = 3$. By using the fact that the coordinates of the point $(5, \frac{4}{3})$ must satisfy the equation, we are able to find the value of b^2. Substituting, we have

$$\frac{25}{9} - \frac{(16/9)}{b^2} = 1$$

from which we find $b^2 = 1$. Therefore, the equation of the hyperbola is

$$\frac{x^2}{9} - \frac{y^2}{1} = 1$$

or

$$x^2 - 9y^2 = 9$$

This hyperbola is shown in Fig. 20–42.

x	y
-8	$-\frac{1}{2}$
-4	-1
-1	-4
$-\frac{1}{2}$	-8
$\frac{1}{2}$	8
1	4
4	1
8	$\frac{1}{2}$

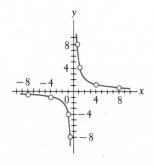

Figure 20—43

Equations (20–20) and (20–24) give us the standard form of the equation of the hyperbola with its center at the origin and its foci on one of the coordinate axes. There is one other important equation form which respresents a hyperbola, and that is

$$xy = c \qquad (20\text{--}25)$$

The asymptotes of this hyperbola are the coordinate axes, and the foci are on the line $y = x$, or on the line $y = -x$, if c is negative. This hyperbola is symmetrical to the origin, for if $-x$ replaces x, and $-y$ replaces y at the same time, we have $(-x)(-y) = c$, or $xy = c$. The equation remains unchanged. However, if either $-x$ replaces x, or if $-y$ replaces y, but not both, the sign on the left is changed. Therefore, it is not symmetrical to either axis. The c here represents a constant and is not related to the focus. The following example illustrates this type of hyperbola.

Example E

Plot the graph of the equation $xy = 4$.

We find the values of the table at the left, and then plot the appropriate points. Here it is permissible to use a limited number of points, since we know that the equation represents a hyperbola (Fig. 20–43). Thus, using $y = 4/x$, we obtain the values in the table.

We conclude that the equation of a hyperbola is characterized by the presence of both an x^2- and a y^2-term, having different signs, or by the presence of an xy-term with no squared terms.

Exercises 20—6

In Exercises 1 through 12 find the coordinates of the vertices and the foci of the given hyperbolas. Sketch each curve.

1. $\dfrac{x^2}{25} - \dfrac{y^2}{144} = 1$ 2. $\dfrac{x^2}{16} - \dfrac{y^2}{4} = 1$ 3. $\dfrac{y^2}{9} - \dfrac{x^2}{1} = 1$

4. $\dfrac{y^2}{2} - \dfrac{x^2}{2} = 1$ 5. $2x^2 - y^2 = 4$ 6. $3x^2 - y^2 = 9$

7. $2y^2 - 5x^2 = 10$ 8. $3y^2 - 2x^2 = 6$ 9. $4x^2 - y^2 + 4 = 0$
10. $9x^2 - y^2 - 9 = 0$ 11. $4x^2 - 9y^2 = 16$ 12. $y^2 - 9x^2 = 25$

In Exercises 13 through 20 find the equations of the hyperbolas satisfying the given conditions. The center of each is at the origin.

13. Vertex $(3, 0)$, focus $(5, 0)$ 14. Vertex $(0, 1)$, focus $(0, \sqrt{3})$
15. Conjugate axis $= 12$, vertex $(0, 10)$ 16. Focus $(8, 0)$, transverse axis $= 4$
17. Passes through $(2, 3)$, focus $(2, 0)$ 18. Passes through $(8, \sqrt{3})$, vertex $(4, 0)$
19. Passes through $(5, 4)$ and $(3, \frac{4}{5}\sqrt{5})$ 20. Passes through $(1, 2)$ and $(2, 2\sqrt{2})$

In Exercises 21 through 24 sketch the graphs of the hyperbolas given.

21. $xy = 2$ 22. $xy = 10$ 23. $xy = -2$ 24. $xy = -4$

45°   45°

In Exercises 25 through 28 find the equations of the hyperbolas with the given properties by use of the definition of the hyperbola.

25. Foci at (1, 2) and (11, 2), with a transverse axis of 8
26. Vertices (−2, 4) and (−2, −2), with conjugate axis of 4
27. Center at (2, 0), vertex at (3, 0), conjugate axis of 6
28. Center at (1, −1), focus at (1, 4), vertex at (1, 2)

In Exercises 29 through 34 solve the given problems.

29. Sketch the graph of impedance versus resistance if the reactance $X_L - X_C$ of a given electric circuit is constant at 60 Ω (see Section 11–7).
30. One statement of Boyle's law is that the product of the pressure and volume, for constant temperature, remains a constant for a perfect gas. If one set of values for a perfect gas under the condition of constant temperature is that the pressure is 300 kPa for a volume of 8.0 L, sketch a graph of pressure versus volume.
31. The relationship between the frequency f, wavelength λ, and the velocity v of a wave is given by $v = f\lambda$. The velocity of light is a constant, being 3.0×10^{10} cm/s. The visible spectrum ranges in wavelength from about 4.0×10^{-5} cm (violet) to about 7.0×10^{-5} cm (red). Sketch a graph of frequency f (in Hertz) as a function of wavelength for the visible spectrum.
32. Wavelengths of gamma rays vary from about 10^{-8} cm to about 10^{-13} cm, but the gamma rays have the same velocity as light. On logarithmic paper, plot the graph of frequency versus wavelength for gamma rays. What type of curve results when this type of hyperbola is plotted on logarithmic paper?
33. An electronic instrument located at point P records the sound of a rifle shot and the impact of the bullet striking the target at the same instant. Show that P lies on a branch of a hyperbola.
34. Two concentric hyperbolas are called conjugate hyperbolas if the transverse and conjugate axes of one are respectively the conjugate and transverse axes of the other. What is the equation of the hyperbola conjugate to the hyperbola of Exercise 14?

20–7 Translation of Axes

Until now, the equations considered for the parabola, the ellipse, and the hyperbola have been restricted to the particular cases in which the vertex of the parabola is at the origin and the center of the ellipse or hyperbola is at the origin. In this section we shall consider the equations of these curves for the cases in which the axis of the curve is parallel to one of the coordinate axes. This is done by **translation of axes.**

We choose a point (h, k) in the xy-coordinate plane and let this point be the origin of another coordinate system, the $x'y'$-coordinate system. The x'-axis is parallel to the x-axis and the y'-axis is parallel to the y-axis.

Every point in the plane now has two sets of coordinates associated with it, (x, y) and (x', y'). From Fig. 20–44 we see that

$$x = x' + h \quad \text{and} \quad y = y' + k \tag{20-26}$$

Equations (20–26) can also be written in the form

$$x' = x - h \quad \text{and} \quad y' = y - k \tag{20-27}$$

The following examples illustrate the use of Eqs. (20–27) in the analysis of equations of the parabola, ellipse, and hyperbola.

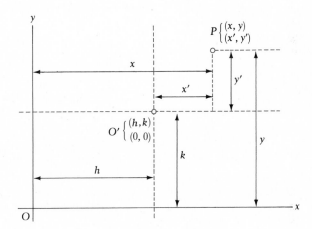

Figure 20—44

Example A

Describe the locus of the equation

$$\frac{(x - 3)^2}{25} + \frac{(y + 2)^2}{9} = 1$$

In this equation, given that $h = 3$ and $k = -2$, we have $x' = x - 3$ and $y' = y + 2$. In terms of x' and y', the equation is

$$\frac{(x')^2}{25} + \frac{(y')^2}{9} = 1$$

We recognize this equation as that of an ellipse (see Fig. 20–45) with a semi-major axis of 5 and a semi-minor axis of 3. The center of the ellipse is at $(3, -2)$ since this was the choice of h and k to make the equation fit a standard form.

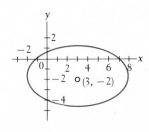

Figure 20—45

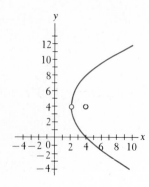

Figure 20—46

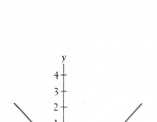

Figure 20—47

Example B

Find the equation of the parabola with vertex at (2, 4) and focus at (4, 4).

If we let the origin of the $x'y'$-coordinate system be the point (2, 4), the point (4, 4) would be the point (2, 0) in the $x'y'$-system. This means that $p = 2$ and $4p = 8$ (Fig. 20–46). In the $x'y'$-system, the equation is

$$(y')^2 = 8(x')$$

Since (2, 4) is the origin of the $x'y'$-system, this means that $h = 2$ and $k = 4$. Using Eq. (20–27), we have

$$(y - 4)^2 = 8(x - 2)$$

as the equation of the parabola in the xy-coordinate system. If this equation is multiplied out, and like terms are combined, we obtain

$$y^2 - 8x - 8y + 32 = 0$$

Example C

Find the center of the hyperbola $2x^2 - y^2 - 4x - 4y - 4 = 0$.

To analyze this curve, we first complete the square in the x-terms and in the y-terms. This will allow us to recognize properly the choice of h and k.

$$2x^2 - 4x - y^2 - 4y = 4$$
$$2(x^2 - 2x \quad) - (y^2 + 4y \quad) = 4$$
$$2(x^2 - 2x + 1) - (y^2 + 4y + 4) = 4 + 2 - 4$$
$$2(x - 1)^2 - (y + 2)^2 = 2$$

$$\frac{(x - 1)^2}{1} - \frac{(y + 2)^2}{2} = 1$$

Thus, if we let $h = 1$ and $k = -2$, the equation in the $x'y'$-system becomes

$$\frac{(x')^2}{1} - \frac{(y')^2}{2} = 1$$

This means that the center is at $(1, -2)$, since this point corresponds to the origin of the $x'y'$-coordinate system (see Fig. 20–47).

Example D

Find the vertex of the parabola $2x^2 - 12x - 3y + 15 = 0$.

First, we complete the square in the x terms, placing all other resulting terms on the right. By factoring the resulting expression on the right, we may determine the values of h and k.

$$2x^2 - 12x = 3y - 15$$
$$2(x^2 - 6x \quad) = 3y - 15$$
$$2(x^2 - 6x + 9) = 3y - 15 + 18$$
$$2(x - 3)^2 = 3(y + 1)$$

$$(x - 3)^2 = \frac{3}{2}(y + 1)$$

$$x'^2 = \frac{3}{2}y'$$

Therefore, the vertex is at $(3, -1)$, since this point corresponds to the origin of the $x'y'$-coordinate system. See Fig. 20–48.

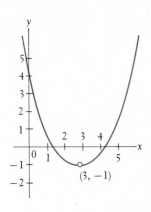

Figure 20–48

Example E

Glass beakers are to be made with a height of 3 in. Express the surface area of the beakers in terms of the radius of the base. Sketch the graph of area versus radius.

The total surface area of a beaker is the sum of the area of the base and the lateral surface area of the side. In general, this surface area S in terms of the radius r of the base and the height h of the side is

$$S = \pi r^2 + 2\pi rh$$

Since h is constant at 3 in., we have

$$S = \pi r^2 + 6\pi r$$

which is the desired relationship.

For the purposes of sketching the graph of S and r, we now complete the square of the r terms:

$$S = \pi(r^2 + 6r)$$
$$S + 9\pi = \pi(r + 6r + 9)$$
$$S + 9\pi = \pi(r + 3)^2$$
$$(r + 3)^2 = \frac{1}{\pi}(S + 9\pi)$$

We note that this equation represents a parabola with vertex at $(-3, -9\pi)$ for its coordinates (r, S). Since $4p = \frac{1}{\pi}$, $p = \frac{1}{4\pi}$ and the focus of this parabola is at $(-3, \frac{1}{4\pi} - 9\pi)$. Only positive values for S and r have meaning, and therefore the part of the graph for negative r is shown as a dashed curve (see Fig. 20–49).

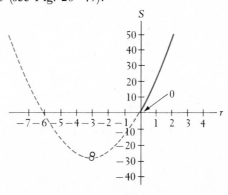

Figure 20–49

Exercises 20–7

In Exercises 1 through 8 describe the locus of each of the given equations. Identify the type of curve and its center (vertex if it is a parabola). Sketch each curve.

1. $(y - 2)^2 = 4(x + 1)$

2. $\dfrac{(x + 4)^2}{4} + \dfrac{(y - 1)^2}{1} = 1$

3. $\dfrac{(x - 1)^2}{4} - \dfrac{(y - 2)^2}{9} = 1$

4. $(y + 5)^2 = -8(x - 2)$

5. $\dfrac{(x + 1)^2}{1} + \dfrac{y^2}{9} = 1$

6. $\dfrac{(y - 4)^2}{16} - \dfrac{(x + 2)^2}{4} = 1$

7. $(x + 3)^2 = -12(y - 1)$

8. $\dfrac{x^2}{16} + \dfrac{(y + 1)^2}{1} = 1$

In Exercises 9 through 16 find the equation of each of the curves described by the given information.

9. Parabola: vertex $(-1, 3)$, $p = 4$, axis parallel to x-axis

10. Parabola: vertex $(2, -1)$, directrix $y = 3$

11. Parabola: vertex $(-3, 2)$, focus $(-3, 3)$

12. Parabola: focus $(2, 4)$, directrix $x = 6$

13. Ellipse: center $(-2, 2)$, focus $(-5, 2)$, vertex $(-7, 2)$

14. Ellipse: center $(0, 3)$, focus $(12, 3)$, major axis 26 units

15. Hyperbola: vertices $(2, 1)$ and $(-4, 1)$, focus $(-6, 1)$

16. Hyperbola: center $(1, -4,)$ focus $(1, 1)$, transverse axis 8 units

In Exercises 17 through 24 determine the center (or vertex if the curve is a parabola) of the given curves. Sketch each curve.

17. $x^2 + 2x - 4y - 3 = 0$

18. $y^2 - 2x - 2y - 9 = 0$

19. $4x^2 + 9y^2 + 24x = 0$

20. $2x^2 + 9y^2 + 8x - 72y + 134 = 0$

21. $9x^2 - y^2 + 8y - 7 = 0$

22. $5x^2 - 4y^2 + 20x + 8y = 4$

23. $2x^2 - 4x = 9y - 2$

24. $4x^2 + 16x = y - 20$

In Exercises 25 through 28 find the required equations.

25. Find the equation of the hyperbola with asymptotes $x - y = -1$ and $x + y = -3$, and vertex $(3, -1)$.

26. The circle $x^2 + y^2 + 4x - 5 = 0$ passes through the foci and the ends of the minor axis of an ellipse which has its major axis along the x-axis. Find the equation of the ellipse.

27. A first parabola has its vertex at the focus of a second parabola, and its focus at the vertex of the second parabola. If the equation of the second parabola is $y^2 = 4x$, find the equation of the first parabola.

28. Verify each of the following equations as being the standard form as indicated. Parabola, vertex at (h, k), axis parallel to the x-axis:

$$(y - k)^2 = 4p(x - h)$$

Parabola, vertex at (h, k), axis parallel to the y-axis:

$$(x - h)^2 = 4p(y - k)$$

Ellipse, center at (h, k), major axis parallel to the x-axis:

$$\frac{(x - h)^2}{a^2} + \frac{(y - k)^2}{b^2} = 1$$

Ellipse, center at (h, k), major axis parallel to the y-axis:

$$\frac{(y - k)^2}{a^2} + \frac{(x - h)^2}{b^2} = 1$$

Hyperbola, center at (h, k), transverse axis parallel to the x-axis:

$$\frac{(x - h)^2}{a^2} - \frac{(y - k)^2}{b^2} = 1$$

Hyperbola, center at (h, k), transverse axis parallel to the y-axis:

$$\frac{(y - k)^2}{a^2} - \frac{(x - h)^2}{b^2} = 1$$

In Exercises 29 through 32 solve the given problems.

29. The power supplied to a circuit by a battery with a voltage E and an internal resistance r is given by $P = EI - rI^2$, where P is the power (in watts) and I is the current (in amperes). Sketch the graph of P vs. I for a 6.0 V battery with an internal resistance of 0.30 Ω.

30. A calculator company determined that its total income I from the sale of a particular type of calculator is given by $I = 100x - x^2$, where x is the selling price of the calculator. Sketch a graph of I vs. x.

31. The planet Pluto moves about the sun in an elliptical orbit, with the sun at one focus. The closest that Pluto approaches the sun is 2.8 billion miles, and the farthest it gets from the sun is 4.6 billion miles. If the sun is at the origin of a coordinate system and the other focus is on the positive x-axis, what is the equation of the path of Pluto?

32. A rectangular tract of land is to have a perimeter of 800 m. Express the area in terms of its width and sketch the graph.

20–8 The Second-Degree Equation

The equations of the circle, parabola, ellipse, and hyperbola are all special cases of the same general equation. In this section we shall discuss this general equation and how to identify the particular form it takes when it represents a specific type of curve.

Each of these curves can be represented by a **second-degree equation** of the form

$$Ax^2 + Bxy + Cy^2 + Dx + Ey + F = 0 \tag{20–28}$$

[This equation is the same as Eq. (13–1).] The coefficients of the second-degree terms determine the type of curve which results. Recalling the discussions of the general forms of the equations of the circle, parabola, ellipse, and hyperbola from the previous sections of this chapter, we have the following results.

Equation (20–28) represents the indicated curve for the given conditions for A, B, and C.

[handwritten in margin: $ax^2 + Bxy + Cy^2 + Dx + Ey + F = 0$]

(1) If $A = C$, $B = 0$, a circle.
(2) If $A \neq C$ (but they have the same sign), $B = 0$, an ellipse.
(3) If A and C have different signs, $B = 0$, a hyperbola.
(4) If $A = 0$, $C = 0$, $B \neq 0$, a hyperbola.
(5) If either $A = 0$ or $C = 0$ (but not both), $B = 0$, a parabola.
(Special cases, such as a single point or no real locus, can also result.)

Another conclusion about Eq. (20–28) is that, if either $D \neq 0$ or $E \neq 0$ (or both), the center (or vertex of a parabola) of the curve is not at the origin. If $B \neq 0$, the axis of the curve has been rotated. We have considered only one such case (the hyperbola $xy = c$) in this chapter.

Example A
The equation $3x^2 = 6x - y^2 + 3$ represents an ellipse. Before we analyze the equation, we should put it in the form of Eq. (20–28). For the given equation, this form is

$$3x^2 + y^2 - 6x - 3 = 0$$

Here we see that $B = 0$ and $A \neq C$. Therefore, it is an ellipse. The $-6x$ term indicates that the center of the ellipse is not at the origin.

Example B
The equation $2(x + 3)^2 = y^2 + 2x^2$ represents a parabola. Putting it in the form of Eq. (20–28), we have

$$2(x^2 + 6x + 9) = y^2 + 2x^2$$
$$2x^2 + 12x + 18 = y^2 + 2x^2$$
$$y^2 - 12x - 18 = 0$$

We now note that $A = 0$, $B = 0$, and $C \neq 0$. This indicates that the equation represents a parabola. Here, the -18 term indicates that the vertex is not at the origin.

Example C
Identify the curve represented by the equation $2x^2 + 12x = y^2 - 14$. Determine the appropriate important quantities associated with the curve, and sketch the graph.

Writing this equation in the form of Eq. (20–28), we have

$$2x^2 - y^2 + 12x + 14 = 0$$

In this form we identify the equation as representing a hyperbola, since

A and *C* have different signs, and *B* = 0. We now write it in the standard form of a hyperbola.

$$2x^2 + 12x - y^2 = -14$$
$$2(x^2 + 6x \quad) - y^2 = -14$$
$$2(x^2 + 6x + 9) - y^2 = -14 + 18$$
$$2(x + 3)^2 - y^2 = 4$$

$$\frac{(x + 3)^2}{2} - \frac{y^2}{4} = 1$$

$$\frac{x'^2}{2} - \frac{y'^2}{4} = 1$$

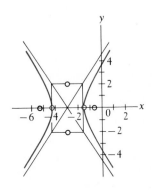

Thus, we see that the center (h, k) of the hyperbola is the point $(-3, 0)$. Also, $a = \sqrt{2}$ and $b = 2$. This means that the vertices are $(-3 + \sqrt{2}, 0)$ and $(-3 - \sqrt{2}, 0)$, and the conjugate axis extends from $(-3, 2)$ to $(-3, -2)$. Also, $c^2 = 2 + 4 = 6$, which means that $c = \sqrt{6}$. The foci are $(-3 + \sqrt{6}, 0)$ and $(-3 - \sqrt{6}, 0)$. The graph is shown in Fig. 20–50.

Figure 20-50

Example D

Identify the curve represented by the equation $4y^2 - 23 = 4(4x + 3y)$. Determine the appropriate important quantities associated with the curve, and sketch the graph.

Writing this equation in the form of Eq. (20–28), we have

$$4y^2 - 23 = 16x + 12y$$
$$4y^2 - 16x - 12y - 23 = 0$$

Therefore, we recognize the equation as representing a parabola, since $A = 0$ and $B = 0$. Now writing the equation in the standard form of a parabola, we have

$$4y^2 - 12y = 16x + 23$$
$$4(y^2 - 3y \quad) = 16x + 23$$

$$4\left(y^2 - 3y + \frac{9}{4}\right) = 16x + 23 + 9$$

$$4\left(y - \frac{3}{2}\right)^2 = 16(x + 2)$$

$$\left(y - \frac{3}{2}\right)^2 = 4(x + 2)$$

$$y'^2 = 4x'$$

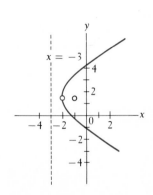

Figure 20-51

We now note that the vertex is the point $(-2, \frac{3}{2})$ and that $p = 1$. Also, it is symmetric to the x' axis. Therefore, the focus is $(-1, \frac{3}{2})$ and the directrix is $x = -3$. The graph is shown in Fig. 20–51.

In Chapter 13, when these curves were first introduced, they were referred to as **conic sections**. If a plane is passed through a cone, the intersection of the plane and the cone results in one of these curves, the curve formed depends on the angle of the plane with respect to the axis of the cone. This is indicated in Fig. 20–52.

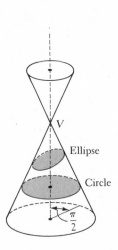
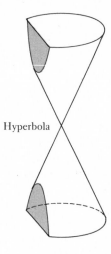
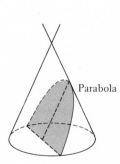

Figure 20–52

Exercises 20–8

In Exercises 1 through 16 identify each of the equations as representing either a circle, parabola, ellipse, or hyperbola.

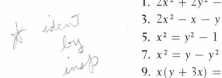

1. $2x^2 + 2y^2 - 3y - 1 = 0$ **2.** $x^2 - 2y^2 - 3x - 1 = 0$

3. $2x^2 - x - y = 1$ *e* **4.** $2x^2 + 4y^2 - y - 2x = 4$

5. $x^2 = y^2 - 1$ **6.** $3x^2 = 2y - 4y^2$

7. $x^2 = y - y^2$ **8.** $y = 3 - 6x^2$

9. $x(y + 3x) = x^2 + xy - y^2 + 1$ **10.** $x(2 - x) = y^2$

11. $2xy + x - 3y = 6$ **12.** $(y + 1)^2 = x^2 + y^2 - 1$

13. $2x(x - y) = y(3 - y - 2x)$ **14.** $2x^2 = x(x - 1) + 4y^2$

15. $y(3 - 2y) = 2(x^2 - y^2)$ **16.** $4x(x - 1) = 2x^2 - 2y^2 + 3$

In Exercises 17 through 24 identify the curve represented by each of the given equations. Determine the appropriate important quantities associated with the curve, and sketch the graph.

17. $x^2 = 8(y - x - 2)$ **18.** $x^2 = 6x - 4y^2 - 1$

19. $y^2 = 2(x^2 - 2x - 2y)$ **20.** $4x^2 + 4 = 9 - 8x - 4y^2$

21. $y^2 + 42 = 2x(10 - x)$ **22.** $x^2 - 4y = y^2 + 4(1 - x)$

23. $4(y^2 - 4x - 2) = 5(4y - 5)$ **24.** $2(2x^2 - y) = 8 - y^2$

In Exercises 25 through 28 set up the necessary equation and then determine the type of curve it represents.

25. For a given alternating-current circuit the resistance and capacitive reactance are constant. What type of curve is represented by the equation relating impedance and inductive reactance? (See Section 11–7.)

26. A room is 8.0 ft high, and the length is 6.5 ft longer than the width. Express the volume V of the (rectangular) room in terms of the width w. What type of curve is represented by the equation?

27. The sides of a rectangle are $2x$ and y, and the digaonal is $x + 5$. What type of curve is represented by the equation relating x and y?

28. The legs of a right triangle are x and y and the hypotenuse is $x + 2$. What type of curve is represented by the equation relating x and y?

20–9 Polar Coordinates

Thus far we have graphed all curves in one coordinate system. This system, the rectangular coordinate system, is probably the most useful and widely applicable system. However, for certain types of curves, other coordinate systems prove to be better adapted. These coordinate systems are widely used, especially when certain applications of higher mathematics are involved. We shall discuss one of these systems here.

Instead of designating a point by its x- and y-coordinates, we can specify its location by its radius vector and the angle which the radius vector makes with the x-axis. Thus, the r and θ that are used in the definitions of the trigonometric functions can also be used as the coordinates of points in the plane. The important aspect of choosing coordinates is that, for each set of values, there must be only one point which corresponds to this set. We can see that this condition is satisfied by the use of r and θ as coordinates. *In* **polar coordinates** *the origin is called the* **pole**, *and the positive x-axis is called the* **polar axis** (see Fig. 20–53).

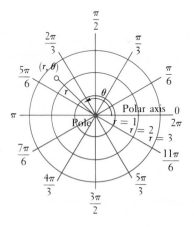

Figure 20–53

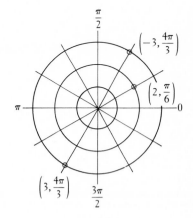

Figure 20–54

Example A

If $r = 2$ and $\theta = \pi/6$, we have the point as indicated in Fig. 20–54. The coordinates (r, θ) of this point are written as $(2, \pi/6)$ when polar coordinates are used. This point corresponds to $(\sqrt{3}, 1)$ in rectangular coordinates.

One difference between rectangular coordinates and polar coordinates is that, for each point in the plane, there are limitless possibilities for the polar coordinates of that point. For example, the point $(2, \pi/6)$ can also be represented by $(2, 13\pi/6)$ since the angles $\pi/6$ and $13\pi/6$ are coterminal. We also remove one restriction on r that we imposed in the definitions of the trigonometric functions. That is, r is allowed to take on positive and negative values. If r is considered negative, then the point is found on the opposite side of the pole from that on which it is positive.

Example B

The coordinates $(3, 4\pi/3)$ and $(3, -2\pi/3)$ represent the same point. However, the point $(-3, 4\pi/3)$ is on the opposite side of the pole, three units from the pole. Another possible set of coordinates for $(-3, 4\pi/3)$ is $(3, \pi/3)$ (see Fig. 20–54).

The relationships between the polar coordinates of a point and the rectangular coordinates of the same point come directly from the definitions of the trigonometric functions. Those most commonly used are

$$x = r \cos \theta, \qquad y = r \sin \theta \qquad\qquad (20\text{--}29)$$

and

$$\tan \theta = \frac{y}{x}, \qquad r = \sqrt{x^2 + y^2} \qquad\qquad (20\text{--}30)$$

The following examples show the use of Eqs. (20–29) and (20–30) in changing coordinates in one system to coordinates in the other system. Also, they are used to transform equations from one system to the other.

Example C

Using Eqs. (20–29), we can transform the polar coordinates of $(4, \pi/4)$ into the rectangular coordinates $(2\sqrt{2}, 2\sqrt{2})$, since

$$x = 4 \cos \frac{\pi}{4} = 4\left(\frac{\sqrt{2}}{2}\right) = 2\sqrt{2}$$

and

$$y = 4 \sin \frac{\pi}{4} = 4\left(\frac{\sqrt{2}}{2}\right) = 2\sqrt{2}$$

Example D

Using Eqs. (20–30), we can transform the rectangular coordinates $(3, -5)$ into polar coordinates.

$$\tan \theta = -\frac{5}{3} = -1.667, \qquad \theta = 5.25 \quad (\text{or } -1.03)$$

$$r = \sqrt{3^2 + (-5)^2} = \sqrt{34} = 5.83$$

We know that θ is a fourth-quadrant angle since x is positive and y is negative. Therefore, the point $(3, -5)$ in rectangular coordinates can be expressed as the point $(5.83, 5.25)$ in polar coordinates. Other polar coordinates for the point are also possible.

Example E

Find the polar equation of the circle $x^2 + y^2 = 2x$.

Here we are given the equation of a circle in rectangular coordinates x and y and are to change it into an equation expressed in polar coordinates r and θ. Since

$$r^2 = x^2 + y^2 \qquad \text{and} \qquad x = r \cos \theta$$

we have

$$r^2 = 2r \cos \theta$$

or

$$r = 2 \cos \theta$$

as the appropriate polar equation.

Example F

Find the rectangular equation of the "rose" $r = 4 \sin 2\theta$.

Using the relation $2 \sin \theta \cos \theta = \sin 2\theta$, we have $r = 8 \sin \theta \cos \theta$. Then, using Eqs. (20–29) and (20–30), we have

$$\sqrt{x^2 + y^2} = 8\left(\frac{y}{r}\right)\left(\frac{x}{r}\right) = \frac{8xy}{r^2} = \frac{8xy}{x^2 + y^2}$$

Squaring both sides, we obtain

$$x^2 + y^2 = \frac{64x^2y^2}{(x^2 + y^2)^2} \qquad \text{or} \qquad (x^2 + y^2)^3 = 64x^2y^2$$

From this example we can see that plotting the graph from the rectangular equation would be complicated.

Exercises 20–9

In Exercises 1 through 12 plot the given points on polar coordinate paper.

1. $\left(3, \dfrac{\pi}{6}\right)$ 2. $(2, \pi)$ 3. $\left(\dfrac{5}{2}, -\dfrac{2\pi}{5}\right)$ 4. $\left(5, -\dfrac{\pi}{3}\right)$

5. $\left(-2, \dfrac{7\pi}{6}\right)$ 6. $\left(-5, \dfrac{\pi}{4}\right)$ 7. $\left(-3, -\dfrac{5\pi}{4}\right)$ 8. $\left(-4, -\dfrac{5\pi}{3}\right)$

9. $\left(0.5, -\dfrac{8\pi}{3}\right)$ 10. $(2.2, -6\pi)$ 11. $(2, 2)$ 12. $(-1, -1)$

In Exercises 13 through 16 find a set of polar coordinates for each of the given points expressed in rectangular coordinates.

13. $(\sqrt{3}, 1)$ 14. $(-1, -1)$ 15. $\left(-\dfrac{\sqrt{3}}{2}, -\dfrac{1}{2}\right)$ 16. $(-5, 4)$

In Exercises 17 through 20 find the rectangular coordinates corresponding to the points for which the polar coordinates are given.

17. $\left(8, \frac{4\pi}{3}\right)$ 18. $(-4, -\pi)$ 19. $\left(3, -\frac{\pi}{8}\right)$ 20. $(-1, -1)$

In Exercises 21 through 28 find the polar equation of the given rectangular equations.

21. $x = 3$ 22. $y = 2$ 23. $x^2 + y^2 = a^2$
24. $x^2 + y^2 = 4y$ 25. $y^2 = 4x$ 26. $x^2 - y^2 = a^2$
27. $x^2 + 4y^2 = 4$ 28. $y = x^2$

In Exercises 29 through 36 find the rectangular equation of each of the given polar equations.

29. $r = \sin \theta$ 30. $r = 4 \cos \theta$ 31. $r \cos \theta = 4$
32. $r \sin \theta = -2$ 33. $r = 2(1 + \cos \theta)$ 34. $r = 1 - \sin \theta$
35. $r^2 = \sin 2\theta$ 36. $r^2 = 16 \cos 2\theta$

In Exercises 37 through 40 find the required equations.

37. If we refer back to Eqs. (7–10) and (7–11), we see that the length along the arc of a circle and the area within a circular sector vary with two variables. These variables are those which we refer to as the polar coordinates. Express the arc length s and the area A in terms of rectangular coordinates. (How must we express θ?)

38. Under certain conditions, the x- and y-components of a magnetic field B are given by the equations

$$B_x = \frac{-ky}{x^2 + y^2} \quad \text{and} \quad B_y = \frac{kx}{x^2 + y^2}$$

Write these equations in terms of polar coordinates.

39. Express the equation of the cable (see Exercise 27 of Section 20–4) in polar coordinates.

40. The polar equation of the path of an artificial satellite of the earth is

$$r = \frac{4800}{1 + 0.14 \cos \theta}$$

where r is measured in miles. Find the rectangular equation of the path of this satellite. The path is an ellipse, with the earth at one of the foci.

20–10 Curves in Polar Coordinates

The basic method for finding a curve in polar coordinates is the same as in rectangular coordinates. We assume values of θ and then find the corresponding values of r. These points are plotted and joined, thus forming the curve which represents the function. The following examples illustrate the method.

Example A

The graph of the polar equation $r = 3$ is a circle of radius 3, with center at the pole. This is the case since $r = 3$, regardless of the value of θ. It is not really necessary to find specific points for this circle. See Fig. 20–55.

The graph of $\theta = \pi/6$ is a straight line through the pole. It represents all points for which $\theta = \pi/6$, regardless of the value of r, positive or negative. See Fig. 20–55.

Example B

Plot the graph of $r = 1 + \cos \theta$.

We find the following table of values of r corresponding to the assumed values of θ.

θ	0	$\dfrac{\pi}{4}$	$\dfrac{\pi}{2}$	$\dfrac{3\pi}{4}$	π	$\dfrac{5\pi}{4}$	$\dfrac{3\pi}{2}$	$\dfrac{7\pi}{4}$	2π
r	2	1.7	1	0.3	0	0.3	1	1.7	2

We now see that the points on the curve start repeating, and it is unnecessary to find additional points. This curve is called a **cardioid** and is shown in Fig. 20–56.

Example C

Plot the graph of $r = 1 - 2 \sin \theta$.

θ	0	$\dfrac{\pi}{4}$	$\dfrac{\pi}{2}$	$\dfrac{3\pi}{4}$	π	$\dfrac{5\pi}{4}$	$\dfrac{3\pi}{2}$	$\dfrac{7\pi}{4}$	2π
r	1	-0.4	-1	-0.4	1	2.4	3	2.4	1

Particular care should be taken in plotting the points for which r is negative. This curve is known as a **limaçon** and is shown in Fig. 20–57.

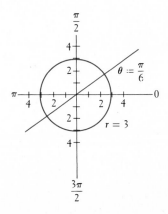

Figure 20–55

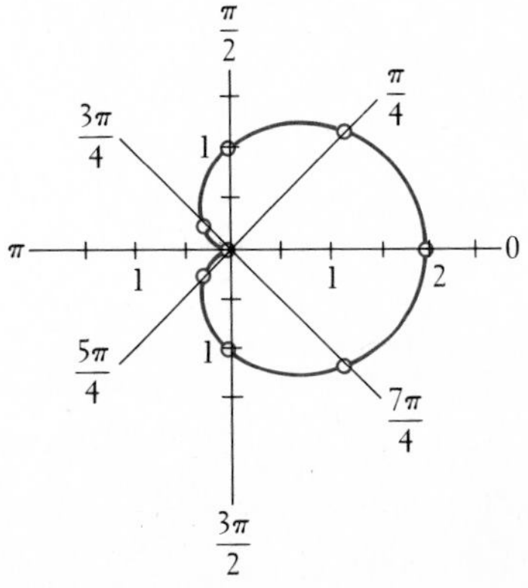

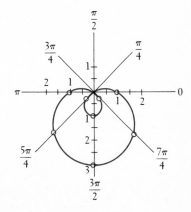

Figure 20–56

Figure 20–57

Example D

Plot the graph of $r = 2 \cos 2\theta$.

θ	0	$\dfrac{\pi}{12}$	$\dfrac{\pi}{6}$	$\dfrac{\pi}{4}$	$\dfrac{\pi}{3}$	$\dfrac{5\pi}{12}$	$\dfrac{\pi}{2}$	$\dfrac{7\pi}{12}$	$\dfrac{2\pi}{3}$	$\dfrac{3\pi}{4}$	$\dfrac{5\pi}{6}$	$\dfrac{11\pi}{12}$	π
r	2	1.7	1	0	-1	-1.7	-2	-1.7	-1	0	1	1.7	2

For the values of θ from π to 2π, the values of r repeat. We have a four-leaf **rose** (Fig. 20–58).

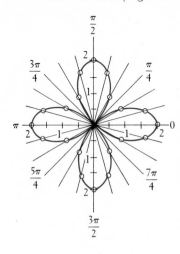

Figure 20–58

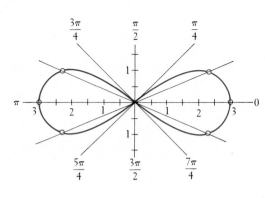

Figure 20–59

Example E

Plot the graph of $r^2 = 9 \cos 2\theta$.

θ	0	$\dfrac{\pi}{8}$	$\dfrac{\pi}{4}$	$\dots$	$\dfrac{3\pi}{4}$	$\dfrac{7\pi}{8}$	π
r	± 3	± 2.5	0		0	± 2.5	± 3

There are no values of r corresponding to values of θ in the range $\pi/4 < \theta < 3\pi/4$, since twice these angles are in the second and third quadrants, and the cosine is negative for such angles. The value of r^2 cannot be negative. Also, the values of r repeat for $\theta > \pi$. The figure is called a **lemniscate** (Fig. 20–59).

Exercises 20–10

In Exercises 1 through 24 plot the given curves in polar coordinates.

1. $r = 4$ 2. $r = 2$ 3. $\theta = \dfrac{3\pi}{4}$ 4. $\theta = \dfrac{5\pi}{3}$

5. $r = 4 \sec \theta$ 6. $r = 4 \csc \theta$ 7. $r = 2 \sin \theta$ 8. $r = 3 \cos \theta$

9. $r = 1 - \cos \theta$ (cardioid)   10. $r = \sin \theta - 1$ (cardioid)

11. $r = 2 - \cos \theta$ (limaçon)   12. $r = 2 + 3 \sin \theta$ (limaçon)

13. $r = 4 \sin 2\theta$ (rose)   14. $r = 2 \sin 3\theta$ (rose)

15. $r^2 = 4 \sin 2\theta$ (lemniscate)

16. $r^2 = 2 \sin \theta$

17. $r = 2^\theta$ (spiral)

18. $r = 10^\theta$ (spiral)

19. $r = -4 \sin 4\theta$ (rose)

20. $r = -\cos 3\theta$ (rose)

21. $r = \dfrac{1}{2 - \cos \theta}$ (ellipse)

22. $r = \dfrac{1}{1 - \cos \theta}$ (parabola)

23. $r = 3 - \sin 3\theta$

24. $r = 4 \tan \theta$

In Exercises 25 through 28 sketch the indicated graphs.

25. The charged particle (see Exercise 33 of Section 20–3) is originally traveling along the line $\theta = \pi$ toward the pole. When it reaches the pole it enters the magnetic field. Its path is then described by the polar equation $r = 2a \sin \theta$, where a is the radius of the circle. Sketch the graph of this path.

26. A cam is shaped such that the equation of the edge of the upper "half" is given by $r = 2 + \cos \theta$, and the equation of the edge of the lower "half" by $r = 3/(2 - \cos \theta)$. Plot the curve which represents the shape of the cam.

27. A satellite at a height proper to make one revolution per day around the earth will have for an excellent approximation of its projection on the earth of its path the curve

$$r^2 = R^2 \cos 2\left(\theta + \frac{\pi}{2}\right)$$

where R is the radius of the earth. Sketch the path of the projection.

28. Sketch the graph of the rectangular equation

$$4(x^6 + 3x^4y^2 + 3x^2y^4 + y^6 - x^4 - 2x^2y^2 - y^4) + y^2 = 0$$

[*Hint:* The equation can be written as $4(x^2 + y^2)^3 - 4(x^2 + y^2)^2 + y^2 = 0$. Transform to polar coordinates, and then sketch the curve.]

20—11 Exercises for Chapter 20

In Exercises 1 through 12 find the equation of the indicated curve subject to the given conditions. Sketch each curve.

1. Straight line: passes through $(1, -7)$ with a slope of 4

2. Straight line: passes through $(-1, 5)$ and $(-2, -3)$

3. Straight line: perpendicular to $3x - 2y + 8 = 0$ and has a y-intercept of -1

4. Straight line: parallel to $2x - 5y + 1 = 0$ and has an x-intercept of 2

5. Circle: center at $(1, -2)$, passes through $(4, -3)$

6. Circle: tangent to the line $x = 3$, center at $(5, 1)$

7. Parabola: focus $(3, 0)$, vertex $(0, 0)$

8. Parabola: directrix $y = -5$, vertex $(0, 0)$

9. Ellipse: vertex $(10, 0)$, focus $(8, 0)$, center $(0, 0)$

10. Ellipse: center $(0, 0)$, passes through $(0, 3)$ and $(2, 1)$

11. Hyperbola: vertex $(0, 13)$, center $(0, 0)$, conjugate axis of 24

12. Hyperbola: foci $(0, 10)$ and $(0, -10)$, vertex $(0, 8)$

In Exercises 13 through 24 find the indicated quantities for each of the given equations. Sketch each curve.

13. $x^2 + y^2 + 6x - 7 = 0$, center and radius

14. $x^2 + y^2 - 4x + 2y - 20 = 0$, center and radius

15. $x^2 = -20y$, focus and directrix

16. $y^2 = 24x$, focus and directrix

17. $16x^2 + y^2 = 16$, vertices and foci

18. $2y^2 - 9x^2 = 18$, vertices and foci

19. $2x^2 - 5y^2 = 8$, vertices and foci

20. $2x^2 + 25y^2 = 50$, vertices and foci

21. $x^2 - 8x - 4y - 16 = 0$, vertex and focus

22. $y^2 - 4x + 4y + 24 = 0$, vertex and directrix

23. $4x^2 + y^2 - 16x + 2y + 13 = 0$, center

24. $x^2 - 2y^2 + 4x + 4y + 6 = 0$, center

In Exercises 25 through 32 plot the given curves in polar coordinates.

25. $r = 4(1 + \sin\theta)$ 26. $r = 1 + 2\cos\theta$

27. $r = 4\cos 3\theta$ 28. $r = -3\sin\theta$

29. $r = \cot\theta$ 30. $r = \dfrac{1}{2(\sin\theta - 1)}$

31. $r = 2\sin\left(\dfrac{\theta}{2}\right)$ 32. $r = \theta$

In Exercises 33 through 36 find the polar equation of each of the given rectangular equations.

33. $y = 2x$ 34. $2xy = 1$ 35. $x^2 - y^2 = 16$ 36. $x^2 + y^2 = 7 - 6y$

In Exercises 37 through 40 find the rectangular equation of each of the given polar equations.

37. $r = 2\sin 2\theta$ 38. $r^2 = \sin\theta$

39. $r = \dfrac{4}{2 - \cos\theta}$ 40. $r = \dfrac{2}{1 - \sin\theta}$

In Exercises 41 through 68 solve the given problems.

41. Find the points of intersection of the ellipses $25x^2 + 4y^2 = 100$ and $4x^2 + 9y^2 = 36$.

42. Find the points of intersection of the hyperbola $y^2 - x^2 = 1$ and the ellipse $x^2 + 25y^2 = 25$.

43. In two ways show that the line segments joining $(-3, 11)$, $(2, -1)$, and $(14, 4)$ form a right triangle.

44. Show that the altitudes of the triangle with vertices $(2, -4)$, $(3, -1)$, and $(-2, 5)$ meet at a single point.

45. By means of the definition of a parabola, find the equation of the parabola with focus at $(3, 1)$ and directrix the line $y = -3$. Find the same equation by the method of translation of axes.

46. Repeat the instructions of Exercise 45 for the parabola with focus at $(0, 2)$ and directrix the line $y = 4$.

47. In an electric circuit the voltage V equals the product of the current I and the resistance R. Sketch the graph of V (in volts) vs. I (in amperes) for a circuit in which $R = 3\ \Omega$.

48. An airplane touches down when landing at 100 mi/h. Its velocity v while coming to a stop is given by $v = 100 - 20000t$, where t is the time in hours. Sketch the graph of v vs. t.

49. In a certain electric circuit, two resistors having resistances R_1 and R_2 respectively, act as a voltage divider. The relation between them is $\alpha = R_1/(R_1 + R_2)$, where α is a constant less than 1. Sketch the curve of R_1 vs. R_2 (both in ohms), if $\alpha = \frac{1}{2}$.

50. Let C and F denote corresponding Celsius and Fahrenheit temperature readings. If the equation relating the two is linear, determine this equation given that $C = 0$ when $F = 32$ and $C = 100$ when $F = 212$.

51. The arch of a small bridge across a stream is parabolic. If, at water level, the span of the arch is 80 ft and the maximum height of the span above water level is 20 ft, what is the equation of the arch? Choose the most convenient point for the origin of the coordinate system.

52. If a source of light is placed at the focus of a parabolic reflector, the reflected rays are parallel. Where is the focus of a parabolic reflector which is 8 cm across and 6 cm deep?

53. Sketch the curve of the total surface area of a right circular cylinder as a function of its radius, if the height is always 10 units.

54. At very low temperatures certain metals have an electric resistance of zero. This phenomenon is called superconductivity. A magnetic field also affects the superconductivity. A certain level of magnetic field, the threshold field, is related to the temperature T by

$$\frac{H_T}{H_0} = 1 - \left(\frac{T}{T_0}\right)^2$$

where H_0 and T_0 are specifically defined values of magnetic field and temperature. Sketch H_T/H_0 vs. T/T_0.

55. The vertical position of a projectile is given by $y = 120t - 16t^2$, and its hozizontal position is given by $x = 60t$. By eliminating the time t, determine the path of the projectile. Sketch this path for the length of time it would need to strike the ground, assuming level terrain.

56. The rate r in grams per second at which a substance is formed in a chemical reaction is given by $r = 10.0 - 0.010t^2$. Sketch the graph of r vs. t.

57. The inside of the top of an arch is a semi-ellipse with a major axis (width) of 26 ft and a minor axis of 10 ft. The arch is 7 ft thick at all points. Is the outside of the arch a portion of an ellipse? (*Hint:* Check the equation of the outer "ellipse" 1 ft to the right of center.)

58. The earth moves about the sun in an elliptical path. If the closest the earth gets to the sun is 90.5 million miles, and the farthest it gets from the sun is 93.5 million miles, find the equation of the path of the earth. The sun is at one focus of the ellipse. Assume the major axis to be along the x-axis and that the center of the ellipse is at the origin.

59. Soon after reaching the vicinity of the moon, Apollo 11 (the first spacecraft to land a man on the moon) went into an elliptical lunar orbit. The closest the craft was to the moon in this orbit was 70 mi, and the farthest it was

from the moon was 190 mi. What was the equation of the path if the moon was at one of the foci of the ellipse? Assume the major axis to be along the x-axis and that the center of the ellipse is at the origin. The radius of the moon is 1080 mi.

60. A machine-part designer wishes to make a model for an elliptical cam by placing two pins in her design board, putting a loop of string over the pins, and marking off the outline by keeping the string taut. (Note that she is using the definition of the ellipse.) If the cam is to measure 10 cm by 6 cm, how long should the loop of string be and how far apart should the pins be?

61. Under certain conditions the work W done on a wire by increasing the tension from T_1 to T_2 is given by $W = k(T_2^2 - T_1^2)$, where k is a constant depending on the properties of the wire. If the work is constant for various sets of initial and final tensions, sketch a graph of T_2 vs. T_1.

62. In order to study the relationship between velocity and pressure of a stream of water, a pipe is constructed such that its cross-section is a hyperbola. If the pipe is 10 ft long, 1 ft in diameter at the narrow point (middle) and 2 ft in diameter at the ends, what is the equation of the cross-section of the pipe? (In physics it is shown that where the velocity of a fluid is greatest, the pressure is the least.)

63. A hallway 16 ft wide has a ceiling whose cross-section is a semi-ellipse. The ceiling is 10 ft high at the walls and 14 ft high at the center. Find the height of the ceiling 4 ft from each wall.

64. A 60-ft rope passes over a pulley 10 ft above the ground and an object on the ground is attached at one end. A man holds the other end at a level of 4 ft above the ground. If the man walks away from the pulley, express the height of the object above the ground in terms of the distance the man is from directly below the object. Sketch the graph of distance vs. height.

65. Express the equation of the artificial satellite (see Exercise 29 of Section 20–5) in polar coordinates.

66. The x- and y-coordinates of the position of a certain moving object as functions of time are given by $x = 2t$ and $y = \sqrt{8t^2 + 1}$. Find the polar equation of the path of the object.

67. Under a force which varies inversely as the square of the distance from an attracting object (such as the force the sun exerts on the earth), it can be shown that the equation that an object follows is given in general by

$$\frac{1}{r} = a + b \cos \theta$$

where a and b are constant for a particular path. By transforming this equation to rectangular coordinates, show that this equation represents one of the conic sections, the particular section depending on the values of a and b. It is through this kind of analysis that we know the paths of the planets and comets are conic sections.

68. The sound produced by a jet engine was measured at a distance of 100 m in all directions. The loudness of the sound (in decibels) was found to be $d = 115 + 10 \cos \theta$, where the 0° line for the angle θ is directed in front of the engine. Sketch the graph of d vs. θ in polar coordinates (use d as r).

21

Introduction to Statistics
and Empirical Curve Fitting

<hr>

21—1 Probability

Decisions which we make about the course of action to be taken now or in the future are based on knowledge which involves at least some uncertainty as to the outcome. For example, a decision as to whether to play golf would normally be based on the probability of good weather. In the same way, knowledge of events in the past is usually incomplete, for not every fact about such events is obtainable. This can be illustrated in the statement that "from available information, the age of the Earth is estimated to be about four billion years." Also, events in recorded history often have contradictions and varied interpretations as to actual happenings.

The basic concepts of probability are widely used, at least in an intuitive way. In this section we deal with making determinations as to possible outcomes of events. We will consider certain basic types of situations in which the probability of outcomes can be determined by the nature of the event, or on what is known from past experience.

In the study of probability, a numerical value is given to the likelihood of some particular event actually happening. In determining such a value, we assume that all events are equally likely to occur, unless we have special knowledge to the contrary.

Example A

In considering the possible outcomes of the toss of a coin, we assume that the coin will land either heads or tails, and that either of these possibilities is equally likely.

When considering the drawing of a card from a deck of cards, we assume that the deck is thoroughly shuffled, and that any of the cards in the deck is as likely to be chosen as any other.

When a die is tossed, it is assumed that any of the six faces will come up on top with equal likelihood.

Example B

If a study were made of the percentages of usable and defective parts produced by a particular machine, it would normally be expected that the machine would not turn out as many defective as usable parts. Thus, by studying a random group of parts, we can count the number of defective parts produced. This number would then form a basis of the probable percentage of defective parts which the machine would produce.

Probability, as used in this chapter, is defined as follows: *If an event can turn out in any one of n equally likely ways, and s of these would be successful, then the probability of the event occurring successfully is*

$$P = \frac{s}{n} \tag{21-1}$$

If the number of equally likely ways an event may turn out cannot be determined from theoretical considerations and N trials are made, of which S are successful, the probability of success is given by

$$P = \frac{S}{N} \tag{21-2}$$

If the events are equally likely, as expressed in Eq. (21–1), the probability is called an *a priori* **probability**. This means that probabilities of possible outcomes are determined without any experimentation, and are based on a knowledge of the nature of the event. If the probability is based on past experience, as expressed in Eq. (21–2), the probability is termed **empirical probability**.

Example C

When a card is drawn from a bridge deck (the standard 52-card deck), the probability that this card is a diamond is $\frac{1}{4}$. Here we know that of the 52 cards, 13 of them are diamonds, and that the drawing of a diamond is a success. This in turn means that $n = 52$ and $s = 13$. Thus, for this case,

$$P = \frac{13}{52} = \frac{1}{4}$$

In the same way, the probability of drawing an ace is $\frac{1}{13}$, since there are 4 aces in the 52 cards. For this case, $n = 52$ and $s = 4$, which means

$$P = \frac{4}{52} = \frac{1}{13}$$

Example D

If a particular machine produced 20 defective parts from a lot of 1000, the empirical probability of a defective part being produced is $\frac{1}{50}$. In this case $N = 1000$ and $S = 20$. Thus,

$$P = \frac{20}{1000} = \frac{1}{50}$$

It should be pointed out here that the larger the number inspected and counted, the more accurate is the empirical probability. However, the number sampled must not be so large that it is impractical.

It should be noted that when we have calculated the probability of a certain event occurring, there is no assurance that it will, or will not occur as indicated by the value of the probability. For example, in Example D, we should not expect that exactly one of every fifty parts will be defective. However, we should expect that the more parts we consider, the more likely that the ratio of defective parts to total parts will be approximately $\frac{1}{50}$.

We can see from the definitions and examples that the value of a particular probability can extend from 0 (impossible) to 1 (the sure thing). A probability of $\frac{1}{2}$ expresses equal likelihood of success or failure.

There are cases in which it is more practical to compute the probability of the failure of an event, in order to calculate the probability of its success. This is based on the fact that the probability of failure is

$$F = 1 - \frac{s}{n}$$

Thus, the probability of success, in terms of the probability of failure, is

$$P = 1 - F \qquad\qquad (21-3)$$

Example E

A bag contains 3 red balls, 4 white balls, and 7 black balls. What is the probability of drawing a red or a black ball?

This can be computed directly, or it can be computed by first determining the probability of drawing a white ball, which would be the probability of failure.

Computing directly, we know that there are 10 balls which are either red or black. Thus, $s = 10$ and $n = 14$, which means that

$$P = \frac{10}{14} = \frac{5}{7}$$

Now, computing the probability of failure first, we know that there are 4 balls (the white ones) which would not be successful draws. Thus,

$$F = \frac{4}{14} = \frac{2}{7}$$

Using Eq. (21–3), we now have

$$P = 1 - \frac{2}{7} = \frac{5}{7}$$

We see that the results agree.

So far we have considered only the probability of success of a single event. The following examples illustrate how the probability of success of a combination of events is found.

Example F

What is the probability of a coin turning up heads in each of two successive tosses?

We can use the definition of probability to determine this result. On the first toss there are two possibilities. On the second toss there are also two possibilities, which means there are, in all, four possible ways in which the coin may fall in two successive tosses. These are HH, HT, TT, TH. Only one of these is successful (heads on two successive tosses). Thus, the probability is $\frac{1}{4}$. This is equivalent to multiplying the probability of success of the first toss by the probability of success of the second toss, or $(\frac{1}{2})(\frac{1}{2}) = \frac{1}{4}$. When we use this multiplication method it is not necessary to figure out all possibilities, a procedure which is often very lengthy, or even impossible from a practical point of view. This multiplication may be stated roughly as "there is a probability of $\frac{1}{2}$ (the second toss) of a probability of $\frac{1}{2}$ (the first toss) of success."

Based on the discussion and results of Example F, the probability of success of a compound event is given in terms of the probabilities of the separate events by

$$P_{1 \text{ and } 2} = P_1 P_2 \qquad\qquad (21\text{–}4)$$

Example G
Two cards are drawn from a deck of bridge cards. If the first card is not replaced before the second is drawn, what is the probability that both will be hearts?

The probability of drawing a heart on the first draw is $\frac{13}{52}$, or $\frac{1}{4}$. If this first card is a heart, there are only 12 hearts of 51 remaining cards for the second draw. Thus, the probability of success on the second draw is $\frac{12}{51}$. Multiplying these results, we have the probability of success for both, or

$$P = \left(\frac{1}{4}\right)\left(\frac{12}{51}\right) = \frac{1}{17}$$

This means there is a 1-in-17 chance of drawing two successive hearts in this manner.

Example H
In Example G determine the probability that both cards are hearts if the first card is replaced in the deck before the second card is drawn.

Again, the probability of drawing a heart on the first draw is $\frac{1}{4}$. However, since this card is replaced in the deck before the second draw is made, the probability that the second card is a heart is also $\frac{1}{4}$. Thus, the probability that both cards will be hearts is

$$P = \left(\frac{1}{4}\right)\left(\frac{1}{4}\right) = \frac{1}{16}$$

As we should expect, the probability of drawing two hearts in this way is slightly better than when the first card is not replaced.

Example I
In three tosses of a single die, what is the probability of tossing at least one 2?

Instead of calculating the various combinations, it is easier to calculate the probability of failure to get a 2, and subtract that from 1. This is due to the fact that the probability of failure is $\frac{5}{6}$ each time, and this must occur three times successively for failure of the compound event. Thus,

$$F = \left(\frac{5}{6}\right)\left(\frac{5}{6}\right)\left(\frac{5}{6}\right) = \frac{125}{216} \qquad \text{or} \qquad P = 1 - \frac{125}{216} = \frac{91}{216}$$

Example J

In eight tosses of a coin, what is the probability of tossing at least one head?

Again, it is easier to calculate the probability of failure to obtain the result. The probability of failure (tails) is $\frac{1}{2}$ for each toss. This must occur 8 successive times in order that a head does not appear. Thus,

$$F = \left(\frac{1}{2}\right)^8 = \frac{1}{256}$$

which means that

$$P = 1 - F = 1 - \frac{1}{256} = \frac{255}{256}$$

One misconception is often encountered when we are talking about probability: *In dealing with the probability that a single event will occur, we should remember that the occurrence of this event in the past does not alter the probability of the event occurring in the future.*

Example K

If a coin is tossed 7 times and it comes up tails each time, the probability that it will come up heads on the next toss is still $\frac{1}{2}$. On any given toss, the probability is $\frac{1}{2}$. In the previous example we showed that the probability of heads coming up at least once in 8 tosses is $\frac{255}{256}$. This, however, is a different problem from that of finding the probability of heads coming up on a particular toss of the coin.

The discussion of probability here has been restricted to certain basic cases. The probability of numerous other types of events can be determined. For example, it is possible to determine the probability of heads coming up exactly three times in five tosses of a coin. The general analysis of such cases, other than by specifying all possibilities, is beyond the scope of this discussion.

Exercises 21—1

In Exercises 1 through 8 consider a bag which contains 5 red balls, 6 white balls, and 9 black balls. What is the probability of drawing each of the following?

1. A red ball   2. A white ball

3. A white or black ball   4. A red or white ball

5. Two red balls on successive draws, if the first ball is replaced before the second draw is made

6. Two white balls on successive draws, if the first ball is replaced before the second draw is made

7. Two red balls on successive draws if the first ball is not replaced before the second draw is made

8. A red ball and then a white ball if the first ball is not replaced before the second draw is made

In Exercises 9 through 16 assume that we are tossing a single die. What is the probability of tossing the following?

9. A 4

10. A 2 or 4

11. Other than a 4

12. Other than a 2 or 4

13. Two successive 4's

14. Three successive 4's

15. At least one 4 in two successive tosses

16. At least one 4 in four successive tosses

In Exercises 17 through 24 assume that we are tossing two dice. An analysis of the possibilities shows that there are 36 different ways in which the dice may fall. What is the probability of tossing the following totals on the dice?

17. 2

18. 3

19. 7

20. 10

21. 7 or 11

22. 10, 11, or 12

23. 7 on two successive tosses

24. At least one 7 in two successive tosses

In Exercises 25 through 28 use the following information. An insurance company, in compiling mortality tables, found that of 10,000 10-year olds, 6,900 lived to be 60, and 4,700 lived to be 70.

25. What is the probability of a 10-year-old living to the age of 70?

26. What is the probability of a 60-year-old living to the age of 70?

27. What is the probability of two 10-year-old people living to the age of 60?

28. What is the probability of two 60-year-old people living to the age of 70?

In Exercises 29 through 32 use the following information:

The assembly of a certain product is done in three stages by machines A, B, and C. If any machine breaks down, the assembly cannot be done. The probabilities that the machines will not break down in a year are $\frac{4}{5}$, $\frac{9}{10}$, and $\frac{19}{20}$, respectively. Find the probabilities that production will cease at some point in a year due to breakdowns in the indicated machines.

29. A or B

30. B or C

31. A or C

32. Any of the machines

In Exercises 33 through 40 solve the given problems in probability.

33. What is the probability of drawing an ace on each of two successive draws from a standard bridge deck of cards, if the first card is not replaced before drawing the second card?

34. What is the probability of drawing at least one ace in two draws, if the first card is not replaced before drawing the second card?

35. For the machine in Example D, how many defective parts should we expect to find in 300 total parts?

36. A certain college, which accepts students only from the upper quarter of the high-school graduating class, finds that 12% of its students attain an average of 3.0 or better (based on a highest attainable score of 4.0). If all the graduates from a particular high school were to apply to this college, what is the probability of a particular one of them attaining an average of 3.0 or better in his classes at the college?

37. One of the first three trials in a series of complicated scientific experiments was unsuccessful due to a failure in a piece of equipment. Based on these three trials, what is the probability of success of the next two successive trials?

38. In Exercise 37, if the fourth trial is successful, what is the probability at this point of the fifth trial being successful?

39. In a random test of newly manufactured transistors, if there are two defective transistors in a particular group of 20, and two of the 20 are tested, what is the probability that at least one of those tested will be defective?

40. How many of the transistors of Exercise 39 would have to be tested in order for there to be a 50% chance of testing one of the defective ones?

21—2 Frequency Distributions and Measures of Central Tendency

When a mathematician wishes to state the relation between the area of a circle and its radius, he or she writes $A = \pi r^2$, and knows that this relation is true for all circles. When an engineer wishes to find the safe load which a particular cable is able to support, he or she cannot so simply write an equation for the relation, because the load which a cable may support depends on the diameter of the cable, the material of which it is made, the quality of material in the particular cable, any possible defects in its manufacture, and anything else which could make this cable different from all other cables. Thus, to determine the load which a cable can support, an engineer may test many cables of particular specifications. He or she will find that most of the cables can support approximately the same load, but that there is some variation, and occasionally perhaps a great variation for some reason or other. Any conclusions the person may draw will, by the nature of their source, yield probable knowledge of the cable.

The engineer, in testing a number of cables and thereby selecting a certain type of cable, is making use of the basic methods of **statistics.** That is, the engineer (1) collects data, (2) analyzes this data, and (3) interprets this data. In this section and the following section, we will discuss some of the methods of tabulating and analyzing this type of statistical information.

In statistics we are dealing with various sets of numbers. These could be measurements of length, test scores, weights of objects, or numerous other possibilities. We could deal with the entire set of numbers, but this is often too large to be practical. In such a case we deal with a selected (assumed to be representative) sample.

Example A

A company produces a machine part which is designed to be 3.80 cm long. If 10,000 are produced daily, it could be impractical to test each one to determine if it meets specifications. Therefore, it might be decided to test every tenth, or every hundredth, part.

If only ten such parts were produced daily, it might be possible to test each one for specifications.

One way of organizing data in order to develop some understanding of it is to tabulate the number of occurrences for each particular value within the set. This is called a **frequency distribution.**

Example B
A test station measured the loudness of the sound of jet aircraft taking off from a certain airport. The decibel readings of the first twenty jets were as follows: 110, 95, 100, 115, 105, 110, 120, 110, 115, 105, 90, 95, 105, 110, 100, 115, 105, 120, 95, 110.

We can see that it is difficult to determine any pattern to the readings. Therefore, we set up the following table to show the frequency distribution.

Decible reading	90	95	100	105	110	115	120
Frequency	1	3	2	4	5	3	2

We note that the distribution and pattern of readings is clearer in this form.

Just as graphs are useful in representing algebraic functions, so also are they a very convenient method of representing frequency distributions. There are several useful methods of graphing such distributions, among which the most important are the **histogram** and the **frequency polygon**. The following examples illustrate these graphical representations.

Example C
A histogram represents a particular set of data by use of rectangles. The width of the base of each rectangle is the same, and is labeled for one of the readings. The height of the rectangle represents the number of values of a given reading. In Fig. 21–1 a histogram representing the data of the frequency distribution of Example B is shown.

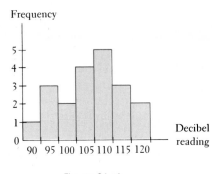

Figure 21–1

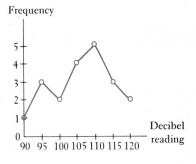

Figure 21–2

Example D
The frequency polygon is used to represent a set of data by plotting as abscissas (*x*-values) the values of the readings, and as ordinates (*y*-values) the number of occurrences of each reading (the frequency). The resulting points are joined by straight-line segments. Figure 21–2 shows a frequency polygon of the data of Example B.

Often the number of measurements is sufficiently large that it is not feasible to tabulate the number of occurrences for a particular value. In other cases such a table does not give a clear idea of the distribution. In these cases we may designate certain intervals, and tabulate the number of values within each interval. Frequency distributions, histograms, and frequency polygons can be used to represent data tabulated in this way.

Example E

A certain mathematics course had 80 students enrolled in it. After all the tests and exams were recorded, the instructors determined a numerical average (based on 100) for each student. The following is a list of the numerical grades, with the number of students receiving each.

$$22-1, 37-1, 40-2, 44-1, 47-1, 53-1, 55-3, 56-1, 60-3, 61-1,$$
$$63-4, 65-2, 66-1, 67-5, 68-2, 70-2, 71-4, 72-3, 74-5, 75-4,$$
$$77-7, 78-4, 79-1, 81-2, 82-4, 84-1, 85-1, 86-3, 87-2, 88-1,$$
$$90-2, 92-1, 93-2, 95-1, 97-1.$$

We can observe from this listing that it is difficult to see just how the grades were distributed. Therefore the instructors grouped the grades into intervals, which included five possible grades in each interval. This led to the following table.

Interval	20–24	25–29	30–34	35–39	40–44	45–49	50–54	55–59
Number in interval	1	0	0	1	3	1	1	4
Interval	60–64	65–69	70–74	75–79	80–84	85–89	90–94	95–99
Number in interval	8	10	14	16	7	7	5	2

We can see that the distribution of grades becomes much clearer in the above table. Finally, since the school graded students only with the letters A, B, C, D, and F, the instructors then grouped the grades in intervals below 60, from 60–69, from 70–79, from 80–89, and from 90–100, and assigned the appropriate letter grade to each numerical grade within each interval. This led to the following table.

Grade	A	B	C	D	F
Number receiving this grade	7	14	30	18	11

Thus, we can see the distribution according to letter grades. We can also note that, if the number of intervals were reduced much further, it would be difficult to draw any reasonable conclusions regarding the distribution of grades.

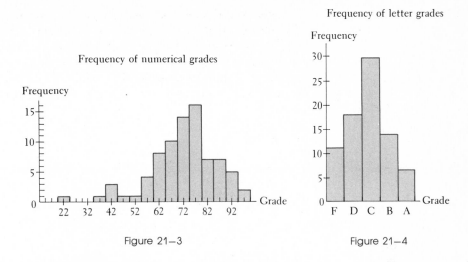

Figure 21–3   Figure 21–4

Histograms showing the frequency of numerical grades and letter grades are shown in Figs. 21–3 and 21–4. In Fig. 21–3, each interval is represented by one number, the middle number. In Fig. 21–5 a frequency polygon of numerical grades is shown, also with each interval represented by the middle number.

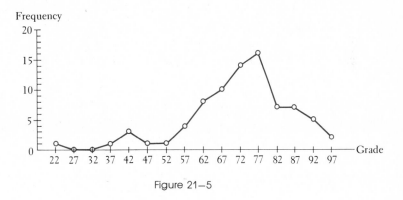

Figure 21–5

We can see from Example E that care must be taken in grouping data. If the number of intervals is too small, important characteristics of the data may not be apparent. If the number of intervals is too large, it may be difficult to analyze the data to determine the distribution and patterns.

Tables and graphical representations give a general description of data. However, it is often profitable and convenient to find representative values for the location of the center of distribution, and other numbers to give a measure of the deviation from this central value. In this way we can obtain an arithmetical description of the data. We shall now discuss the values commonly used to measure the location of the center of the distribution. These are referred to as "measures of central tendency."

The first of these measures of central tendency is the **median**. *The median is the middle number, that number for which there are as many above it as below it in the distribution.* If there is no middle number, the median is that number halfway between the two numbers nearest the middle of the distribution.

Example F

Given the numbers 5, 2, 6, 4, 7, 4, 7, 2, 8, 9, 4, 11, 9, 1, 3, we first arrange them in numerical order. This arrangement is

$$1, 2, 2, 3, 4, 4, 4, 5, 6, 7, 7, 8, 9, 9, 11.$$

Since there are 15 numbers, the middle number is the eighth. Since the eighth number is 5, the median is 5.

If the number 11 is not included in this set of numbers, and there are only 14 numbers in all, the median is that number halfway between the seventh and eighth numbers. Since the seventh is 4 and the eighth is 5, the median is 4.5.

Example G

In the distribution of grades given in Example E of this section, the median is 74. There are 80 grades in all, and if they are listed in numerical order we find that the 39th through the 43rd grade is 74. This means that the 40th and 41st grades are both 74. The number halfway between the 40th and 41st grades would be the median. The fact that both are 74 means that the median is also 74.

Another, and very widely applied, measure of central tendency is the **arithmetic mean**. *The mean is calculated by finding the sum of all the values and then dividing by the number of values.*

Example H

The arithmetic mean of the numbers given in Example F is found by finding the sum of all the numbers and dividing by 15. Thus, letting $\bar{x}$ (read as "x bar") represent the mean, we have

$$\bar{x} = \frac{5 + 2 + 6 + 4 + 7 + 4 + 7 + 2 + 8 + 9 + 4 + 11 + 9 + 1 + 3}{15}$$

$$= \frac{82}{15} = 5.5$$

Thus, the mean is 5.5.

If we wish to find the arithmetic mean of a large number of values, and if some of them appear more than once, the calculation can be somewhat simplified. By multiplying each number by its frequency, adding these results, and then dividing by the total number of values considered, the mean can be calculated. Letting $\bar{x}$ represent the mean of the values $x_1, x_2, \ldots, x_n$, which occur $f_1, f_2, \ldots, f_n$ times, respectively, we have

$$\bar{x} = \frac{x_1 f_1 + x_2 f_2 + \cdots + x_n f_n}{f_1 + f_2 + \cdots + f_n} \tag{21–5}$$

(This notation can be simplified by letting

$$x_1 f_1 + x_2 f_2 + \cdots + x_n f_n = \sum_{i=1}^{n} x_i f_i$$

and

$$f_1 + f_2 + \cdots + f_n = \sum_{i=1}^{n} f_i$$

This notation is often used to indicate a sum, where Σ is known as the **summation sign**.)

Example I
Using Eq. (21–5) to find the arithmetic mean of the numbers of Example F, we have

$$\bar{x} = \frac{1(1) + 2(2) + 3(1) + 4(3) + 5(1) + 6(1) + 7(2) + 8(1) + 9(2) + 11(1)}{15}$$

$$= \frac{82}{15} = 5.5$$

Example J
We find the arithmetic mean of the grades in Example E by

$$\bar{x} = \frac{22(1) + (37)(1) + (40)(2) + \cdots + (67)(5) + \cdots + (97)(1)}{80}$$

$$= \frac{5733}{80} = 71.7$$

Exercises 21–2   In Exercises 1 through 24 use the following sets of numbers.

        A: 3, 6, 4, 2, 5, 4, 7, 6, 3, 4, 6, 4, 5, 7, 3
        B: 25, 26, 23, 24, 25, 28, 26, 27, 23, 28, 25
        C: 48, 53, 49, 45, 55, 49, 47, 55, 48, 57, 51, 46
        D: 105, 108, 103, 108, 106, 104, 109, 104, 110, 108, 108, 104, 113, 106,
            107, 106, 107, 109, 105, 111, 109, 108

In Exercises 1 through 4 set up a frequency distribution table, indicating the frequency of each number given in the indicated set.

  1. Set A 2. Set B 3. Set C 4. Set D

In Exercises 5 through 8 set up a frequency distribution table, indicating the frequency of numbers for the given intervals of the given sets.

  5. Intervals 2–3, 4–5, and 6–7 for set A.
  6. Intervals 22–24, 25–27, and 28–30 for set B.
  7. Intervals 43–45, 46–48, 49–51, 52–54, and 55–57 for set C.
  8. Intervals 101–105, 106–110, and 111–115 for set D.

In Exercises 9 through 12 draw histograms for the data in the given exercise.

  9. Exercise 1   10. Exercise 4   11. Exercise 7   12. Exercise 8

In Exercises 13 through 16 draw frequency polygons for the data in the given exercise.

  13. Exercise 1   14. Exercise 4   15. Exercise 7   16. Exercise 8

In Exercises 17 through 20 determine the median of the numbers of the given set.

  17. Set A 18. Set B 19. Set C 20. Set D

In Exercises 21 through 24 determine the arithmetic mean of the numbers of the given set.

  21. Set A 22. Set B 23. Set C 24. Set D

In Exercises 25 through 36 find the indicated quantities.

  25. Form a histogram for the data given in the following table, which was compiled when a particular type of cable was being tested for its breaking load.

Load interval, pounds	830–839	840–849	850–859	860–869	870–879	880–889	890–899
Number of cables breaking	4	14	48	531	85	22	2

  26. Form a frequency polygon for the data given in Exercise 25.
  27. A researcher, testing an electric circuit, found the following values for the current (in milliamperes) in the circuit on successive trials: 3.44, 3.46, 3.39, 3.44, 3.48, 3.40, 3.29, 3.46, 3.41, 3.37, 3.45, 3.47, 3.43, 3.38, 3.50, 3.41, 3.42, 3.47. Form a histogram for the intervals 3.25–3.29, 3.30–3.34, etc.
  28. Form a histogram for the data given in Exercise 27 for the intervals 3.21–3.30, 3.31–3.40, etc.

29. Find the median of the data given in Exercise 27.

30. Find the mean of the data given in Exercise 27.

31. One hundred motorists were asked to maintain a miles per gallon record for their cars, all the same model. The records which were reported are shown in the following table.

Miles per gallon	19.0–19.4	19.5–19.9	20.0–20.4	20.5–20.9
Number reporting	4	9	14	24

Miles per gallon	21.0–21.4	21.5–21.9	22.0–22.4	22.5–22.9
Number reporting	28	16	4	1

Draw a histogram for this data.

32. Draw a frequency polygon for the data of Exercise 31.

33. Determine the arithmetic mean of the data of Exercise 31.

34. Determine the median of the data of Exercise 31.

35. Take two dice and toss them 100 times, recording at each toss the sum which appears. Draw a frequency polygon of the sum and the frequency with which it occurred. Compare this with the expected frequency as based on probability.

36. From the financial section of a newspaper, record (to the nearest dollar) the closing price of the first 50 stocks listed. Form a frequency table with intervals 0–9, 10–19, and so forth. Form a histogram of these data. Finally, find the median price.

In Exercises 37 through 40 use the following information. The **mode** of a frequency distribution, another measure of central tendency, is defined as the measure with the maximum frequency, if there is one. A set of numbers may have more than one mode. Using this definition, find the mode of the numbers in the indicated sets of numbers at the beginning of this exercise set.

37. Set A 38. Set B 39. Set C 40. Set D

21–3 Standard Deviation

In the preceding section we discussed measures of central tendency of sets of data. However, regardless of the measure which may be used, it does not tell us whether the data are grouped closely together or spread over a large range of values. Therefore, we also need some measure of the deviation, or dispersion, of the values from the median or mean. If the dispersion is small and the numbers are grouped closely together, the measure of central tendency is more reliable and descriptive of the data than the case in which the spread is greater.

There are several measures of dispersion, and we discuss in this section one which is very widely used. It is called the **standard deviation.**

The standard deviation of a set of numbers is given by the equation

$$s = \sqrt{\overline{(x - \bar{x})^2}}$$ (21–6)

The definition of s indicates that the following steps are to be taken in computing its value.

(1) Find the arithmetic mean $\bar{x}$ of the set of numbers.
(2) Subtract the mean from each of the numbers of the set.
(3) Square these differences.
(4) Find the arithmetic mean of these squares. (Note that a bar over any quantity signifies the mean of the set of values of that quantity.)
(5) Find the square root of this last arithmetic mean.

Defined in this way, s must be a positive number, and thus indicates a deviation from the mean, regardless of whether or not individual numbers are greater or less than the mean.

Example A

Find the standard deviation of the numbers 1, 5, 4, 2, 6, 2, 1, 1, 5, 3. The most efficient method of finding s is to make a table in which Steps (1) to (4) are indicated.

x	$x - \bar{x}$	$(x - \bar{x})^2$
1	-2	4
5	2	4
4	1	1
2	-1	1
6	3	9
2	-1	1
1	-2	4
1	-2	4
5	2	4
3	0	0
30		32

$$\bar{x} = \frac{30}{10} = 3$$

$$\overline{(x - \bar{x})^2} = \frac{32}{10} = 3.2$$

$$s = \sqrt{3.2} = 1.8$$

Example B

Find the standard deviation of the numbers given in Example F of Section 21–2.

Since several of the numbers appear more than once, it is helpful to use the frequency of each number in the table. Therefore, we have the following table and calculations.

x	f	fx	$x - \bar{x}$	$(x - \bar{x})^2$	$f(x - \bar{x})^2$
1	1	1	-4.5	20.25	20.25
2	2	4	-3.5	12.25	24.50
3	1	3	-2.5	6.25	6.25
4	3	12	-1.5	2.25	6.75
5	1	5	-0.5	0.25	0.25
6	1	6	0.5	0.25	0.25
7	2	14	1.5	2.25	4.50
8	1	8	2.5	6.25	6.25
9	2	18	3.5	12.25	24.50
11	1	11	5.5	30.25	30.25
	15	82			123.75

$$\bar{x} = \frac{82}{15} = 5.5$$

$$\overline{f(x - \bar{x})^2} = \frac{123.75}{15} = 8.25$$

$$s = \sqrt{8.25} = 2.9$$

We can see from Examples A and B that there is a great deal of calculational work required to find the standard deviation of a set of numbers. It is possible to reduce this work somewhat, and we now show how this may be done.

To find the mean of the sum of two sets of data, x_i and y_i, where the number of values in each set is the same, we can determine the mean of the x's and the mean of the y's and add the results. That is

$$\overline{x + y} = \overline{x} + \overline{y} \tag{21–7}$$

This is true by the meaning of the definition of the arithmetic mean.

$$\overline{x + y} = \frac{x_1 + x_2 + \cdots + x_n + y_1 + y_2 + \cdots + y_n}{n}$$

$$= \frac{x_1 + x_2 + \cdots + x_n}{n} + \frac{y_1 + y_2 + \cdots y_n}{n} = \overline{x} + \overline{y}$$

Also, when we find the arithmetic mean, if all the numbers contain a constant factor, this number can be factored before the mean is found. That is,

$$\overline{kx} = k\overline{x} \tag{21–8}$$

If we multiply out the quantity under the radical in Eq. (21–6), and then apply Eq. (21–7), we obtain

$$\overline{x^2 - 2x\overline{x} + \overline{x}^2} = \overline{x^2} - \overline{2x\overline{x}} + \overline{\overline{x}^2}$$

In this equation the number 2 and $\overline{x}$ are constants for any given problem. Thus, by using Eq. (21–8), we have

$$\overline{x^2} - \overline{2x\overline{x}} + \overline{\overline{x}^2} = \overline{x^2} - 2\overline{x}(\overline{x}) + \overline{x}^2(1) = \overline{x^2} - \overline{x}^2$$

Substituting this last result into Eq. (21–6), we have

$$s = \sqrt{\overline{x^2} - \overline{x}^2} \tag{21–9}$$

Equation (21–9) shows us that the standard deviation s may be found by finding the mean of the squares of the x's, subtracting the square of $\overline{x}$, and then finding the square root. This eliminates the step of subtracting $\overline{x}$ from each x, a step required by the original definition.

Example C

By using Eq. (21–9), find s for the numbers in Example A. Again, a table is the most convenient form.

x	x^2
1	1
5	25
4	16
2	4
6	36
2	4
1	1
1	1
5	25
3	9
30	122

$$\bar{x} = \frac{30}{10} = 3, \qquad \bar{x}^2 = 9$$

$$\overline{x^2} = \frac{122}{10} = 12.2$$

$$\overline{x^2} - \bar{x}^2 = 12.2 - 9 = 3.2$$

$$s = \sqrt{3.2} = 1.8$$

Example D

In an ammeter, two resistances are connected in parallel. Most of the current passing through the meter goes through the one called the shunt. In order to determine the accuracy of the resistance of shunts being made for ammeters, a manufacturer tested a sample of 100 shunts. The resistance of each, to the nearest hundredth of an ohm, is indicated in the following table. Calculate the standard deviation of the resistances of the shunts.

R (ohms)	f	fR	fR^2
0.200	1	0.200	0.0400
0.210	3	0.630	0.1323
0.220	5	1.100	0.2420
0.230	10	2.300	0.5290
0.240	17	4.080	0.9792
0.250	40	10.000	2.5000
0.260	13	3.380	0.8788
0.270	6	1.620	0.4374
0.280	3	0.840	0.2352
0.290	2	0.580	0.1682
	100	24.730	6.1421

$$\bar{R} = \frac{\Sigma\, fR}{\Sigma\, f} = \frac{24.73}{100} = 0.2473$$

$$\bar{R}^2 = 0.06116$$

$$\overline{R^2} = \frac{\Sigma\, fR^2}{\Sigma\, f} = \frac{6.142}{100} = 0.06142$$

$$\overline{R^2} - \bar{R}^2 = 0.00026$$

$$s = \sqrt{0.00026} = 0.016$$

The arithmetic mean of the resistances is 0.247 Ω, with a mean deviation of 0.016 Ω.

Example E

Find the standard deviation of the grades in Example E of Section 21–2. Use the frequency distribution as grouped in intervals 20–24, 25–29, and so forth, and then assume that each value in the interval is equal to the representative value (middle value) of the interval. (This method is not exact, but when a problem involves a large number of values, the method provides a very good approximation and eliminates a great deal of arithmetic work.)

x	f	fx	fx^2
22	1	22	484
27	0	0	0
32	0	0	0
37	1	37	1,369
42	3	126	5,292
47	1	47	2,209
52	1	52	2,704
57	4	228	12,996
62	8	496	30,752
67	10	670	44,890
72	14	1,008	72,576
77	16	1,232	94,864
82	7	574	47,068
87	7	609	52,983
92	5	460	42,320
97	2	194	18,818
	80	5,755	429,325

$$\bar{x} = \frac{\Sigma fx}{\Sigma f} = \frac{5,755}{80} = 71.9$$

$$\bar{x}^2 = 5,170$$

$$\overline{x^2} = \frac{\Sigma fx^2}{80} = \frac{429,325}{80} = 5,367$$

$$\overline{x^2} - \bar{x}^2 = 197$$

$$s = \sqrt{197} = 14.0$$

We have seen that the standard deviation is a measure of the dispersion of a set of data. Generally, if we make a great many measurements of a given type, we would expect to find a majority of them near the arithmetic mean. If this is the case, the distribution probably follows, at least to a reasonable extent, the **normal distribution curve**. See Fig. 21–6. This curve gives a theoretical distribution about the arithmetic mean, assuming that the number of values in the distribution becomes infinite, and its equation is

$$y = \frac{1}{\sqrt{2\pi}} e^{-x^2/2} \qquad\qquad (21\text{–}10)$$

An important feature of the standard deviation is that in a normal distribution, about 68% of the values are found within the interval $\bar{x} - s$ to $\bar{x} + s$.

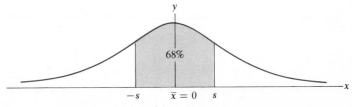

The normal distribution curve

Figure 21–6

Example F

In Example B, 9 of the 15 values, or 60%, are between 2.6 and 8.4. Here $\bar{x} = 5.5$ and $s = 2.9$. Thus, $\bar{x} - s = 2.6$ and $\bar{x} + s = 8.4$. Therefore, we see that the percentage in this interval is slightly less than in a normal distribution.

In Example E, using $\bar{x} = 72$ and $s = 14$, we have $\bar{x} - s = 58$ and $\bar{x} + s = 86$. We also note that 59 of the 80 values are in the interval from 58 to 86, which means that 74% of the values are in this interval.

Exercises 21—3

In Exercises 1 through 20 use the following sets of numbers. They are the same as those used in Exercises 21–2.

  A: 3, 6, 4, 2, 5, 4, 7, 6, 3, 4, 6, 4, 5, 7, 3
  B: 25, 26, 23, 24, 25, 28, 26, 27, 23, 28, 25
  C: 48, 53, 49, 45, 55, 49, 47, 55, 48, 57, 51, 46
  D: 105, 108, 103, 108, 106, 104, 109, 104, 110, 108, 108, 104, 113, 106, 107, 106, 107, 109, 105, 111, 109, 108

In Exercises 1 through 4 use Eq. (21–6) to find the standard deviation s for the indicated sets of numbers.

 1. Set A 2. Set B 3. Set C 4. Set D

In Exercises 5 through 8 use Eq. (21–9) to find the standard deviation s for the indicated sets of numbers.

 5. Set A 6. Set B 7. Set C 8. Set D

In Exercises 9 through 12 find the standard deviation s for the indicated sets of numbers.

 9. The weekly salaries (in dollars) for the workers in a small factory are as follows: 250, 350, 275, 225, 175, 300, 200, 350, 275, 400, 300, 225, 250, 300.
10. The measurements of electric current in Exercise 27 of Exercises 21–2
11. The measurements of cable breaking loads in Exercise 25 of Exercises 21–2
12. The measurements of miles per gallon in Exercise 31 of Exercises 21–2

In Exercises 13 through 20 find the percentage of values in the interval $\bar{x} - s$ to $\bar{x} + s$ for the indicated sets of numbers.

13. Set A 14. Set B 15. Set C 16. Set D
17. The salaries of Exercise 9
18. The measurements of electric current in Exercise 27 of Exercises 21–2
19. The measurements of cable breaking loads in Exercise 25 of Exercises 21–2
20. The measurements of miles per gallon in Exercises 31 of Exercises 21–2

21—4 Fitting a Straight Line to a Set of Points

We have considered statistical methods for dealing with one variable. We have discussed methods of tabulating, graphing, measuring the central tendency and the deviations from this value for one variable. We now shall discuss how to obtain a relationship between two variables for which a set of points is known.

In this section we shall show a method of "fitting" a straight line to a given set of points. In the following section fitting suitable nonlinear curves to given sets of points will be discussed. Some of the reasons for doing this are (1) to express a concise relationship between the variables, (2) to use the equation for the purpose of predicting certain fundamental results, (3) to determine the reliability of certain sets of data, and (4) to use the data for testing theoretical concepts.

We shall assume for the examples and exercises of the remainder of this chapter that there is some relationship between the variables. Often when we are analyzing statistics for variables between which we think a relationship might exist, the points are so scattered as to give no reasonable idea regarding the possible functional relationship. We shall assume here that such combinations of variables have been discarded in the analysis.

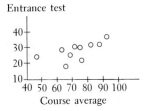

Entrance test

Course average

Figure 21—7

Example A

All the students enrolled in the mathematics course referred to in Example E of Section 21–2 took an entrance test in mathematics. To study the reliability of this test, an instructor tabulated the test scores of 10 students (selected at random), along with their course average, and made a graph of these figures (see table below, and Fig. 21–7).

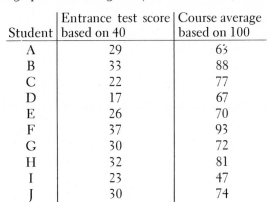

Student	Entrance test score based on 40	Course average based on 100
A	29	63
B	33	88
C	22	77
D	17	67
E	26	70
F	37	93
G	30	72
H	32	81
I	23	47
J	30	74

We now ask whether or not there is a functional relationship between the test scores and the course grades. Certainly no clear-cut relationship exists, but in general we see that the higher the test score, the higher the course grade. This leads to the possibility that there might be some straight line, from which none of the points would vary too significantly. If such a line could be found, then it could be the basis of predictions as to the possible success a student might have in the course, on the basis of his or her grade on the entrance test. Assuming that such a straight line exists, the problem is to find the equation of this line. Figure 21–8 shows two such possible lines.

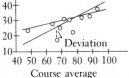

Entrance test

Course average

Figure 21—8

564   Introduction to Statistics and Empirical Curve Fitting

There are various methods of determining the line which best fits the given points. We shall employ the one which is most widely used: the **method of least squares.** *The basic principle of this method is that the sum of the squares of the deviation of all points from the best line (in accordance with this method) is the least it can be. By* **deviation** *we mean the difference between the y-value of the line and the y-value for the point (of original data) for a particular value of x.*

Example B

In Fig. 21–8 the deviation of one point is indicated. The point (67, 17) (student D of Example A) has a deviation of 8 from the indicated line of Fig. 21–8. Thus we square the value of this deviation to obtain 64. The method of least squares requires that the sum of all such squares be a minimum in order to determine the line which fits best.

Therefore, in applying this method, it is necessary to use the equation of a straight line and the coordinates of the points of the data. The deviation of these is indicated and squared. The problem then arises as to the method of determining the constants m and b in the equation of the line $y = mx + b$ for which these squares are a minimum. Since we are dealing with squared quantities, the problem is one of finding the minimum of a quadratic-type function. The following example indicates the method employed.

Example C

For what value of x is the quadratic function

$$y = x^2 - 8x + 19 \quad \text{a minimum?}$$

By inspection we see that this equation represents a parabola. From the properties of this type of parabola, we know that the minimum value of y will occur at the vertex. Thus, we are to find the x-coordinate of the vertex. To do this, we complete the square of the x-terms. And so we have

$$y = (x^2 - 8x + 16) + 3 \quad \text{or} \quad y = (x - 4)^2 + 3$$

We now see that if $x = 4$, $y = 3$. If x is anything other than 4, $y > 3$, since $(x - 4)$ is squared and is always positive for values other than $x = 4$. Therefore, the minimum value of this function is 3, and it occurs for $x = 4$.

The previous discussion indicates the type of reasoning used in developing the equations for the best straight line to fit a set of data. However, it is not a derivation, as a complete derivation requires more advanced mathematical methods.

By finding the deviations, squaring, and then applying the method above for finding the minimum of the sum of the squares, the equation of the **least-squares line**,

$$y = mx + b \qquad (21\text{–}11)$$

can be found by the values

$$m = \frac{\overline{xy} - \overline{x}\,\overline{y}}{s_x^2} \qquad (21\text{–}12)$$

and

$$b = \overline{y} - m\overline{x} \qquad (21\text{–}13)$$

In Eqs. (21–12) and (21–13), the x's and y's are those of the points in the data and s_x is the standard deviation of the x-values.

Example D

Find the least-squares line of the data of Example A.

Here the y-values will be the entrance-test scores, and the x-values the course averages. The results are best obtained by tabulating the necessary quantities, as we do in the following table.

x	y	xy	x^2
63	29	1,830	3,970
88	33	2,900	7,740
77	22	1,690	5,930
67	17	1,140	4,490
70	26	1,820	4,900
93	37	3,440	8,650
72	30	2,160	5,180
81	32	2,590	6,560
47	23	1,080	2,210
74	30	2,220	5,480
732	279	20,870	55,110

$$\overline{x} = \frac{732}{10} = 73.2, \quad \overline{x^2} = 5358$$

$$\overline{y} = \frac{279}{10} = 27.9, \quad \overline{x}\,\overline{y} = 2042$$

$$\overline{xy} = \frac{20,870}{10} = 2087, \quad \overline{x^2} = \frac{55,110}{10} = 5511$$

$$s_x^2 = \overline{x^2} - \overline{x}^2 = 153$$

$$m = \frac{\overline{xy} - \overline{x}\,\overline{y}}{s_x^2} = \frac{2087 - 2042}{153}$$

$$= \frac{45}{153} = 0.294$$

$$b = \overline{y} - m\overline{x} = 27.9 - (0.294)(73.2) = 27.9 - 21.5 = 6.4$$

$$y = 0.294x + 6.4$$

Thus, the equation of the line is $y = 0.294x + 6.4$. This is the line indicated in Fig. 21–9.

We note that when plotting the least-squares line, we know two points on it. These are $(0, b)$ and $(\bar{x}, \bar{y})$. We can see that $(\bar{x}, \bar{y})$ satisfies the equation from the solution for b [Eq. (21–13)].

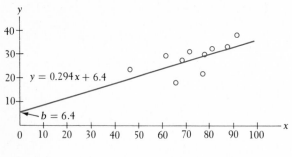

Figure 21–9

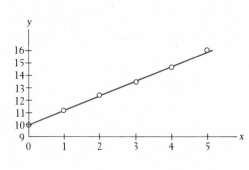

Figure 21–10

Example E

In an experiment to determine the relation between the load on a spring and the length of the spring, the following data were found.

Load (pounds)	0.0	1.0	2.0	3.0	4.0	5.0
Length (inches)	10.0	11.2	12.3	13.4	14.6	15.9

Find the least-squares line for this data which expresses the length as a function of the load.

x	y	xy	x^2
0.0	10.0	0.0	0.0
1.0	11.2	11.2	1.0
2.0	12.3	24.6	4.0
3.0	13.4	40.2	9.0
4.0	14.6	58.4	16.0
5.0	15.9	79.5	25.0
15.0	77.4	213.9	55.0

$$\bar{x} = \frac{15.0}{6} = 2.50, \quad \bar{x}^2 = 6.25$$

$$\bar{y} = \frac{77.4}{6} = 12.9, \quad \bar{x}\,\bar{y} = 32.25$$

$$\overline{xy} = \frac{213.9}{6} = 35.65, \quad \overline{x^2} = \frac{55.0}{6} = 9.17$$

$$s_x^2 = \overline{x^2} - \bar{x}^2 = 9.17 - 6.25 = 2.92$$

$$m = \frac{35.65 - 32.25}{2.92} = 1.16, \quad b = 12.9 - (1.16)(2.50) = 10.0$$

Therefore, the least-squares line in this case is

$$y = 1.16x + 10.0$$

where y is the length of the spring and x is the load. In Fig. 21–10, the line shown is the least-squares line, and the points are the experimental data points.

Exercises 21—4

In Exercises 1 through 8 find the equation of the least-squares line for the given data. In each case graph the line and the data points on the same graph.

1. The points in the following table:

x	4	6	8	10	12
y	1	4	5	8	9

2. The points in the following table:

x	1	2	3	4	5	6	7
y	10	17	28	37	49	56	72

3. The points in the following table:

x	20	26	30	38	48	60
y	160	145	135	120	100	90

4. The points in the following table:

x	1	3	6	5	8	10	4	7	3	8
y	15	12	10	8	9	2	11	9	11	7

5. In an electrical experiment, the following data were found for the values of current and voltage for a particular element of the circuit. Find the voltage as a function of the current.

Voltage (volts)	3.00	4.10	5.60	8.00	10.50
Current (milliamperes)	15.0	10.8	9.30	3.55	4.60

6. The velocity of a falling object was found each second by use of an electric device as shown. Find the velocity as a function of time.

Velocity (meters per second)	9.70	19.5	29.5	39.4	49.2	58.9	68.6
Time (seconds)	1.00	2.00	3.00	4.00	5.00	6.00	7.00

7. In an experiment on the photoelectric effect, the frequency of the light being used was measured as a function of the stopping potential (the voltage just sufficient to stop the photoelectric current) with the results given below. Find the least-squares line for V as a function of f. The frequency for $V = 0$ is known as the *threshold frequency*. From the graph, determine the threshold frequency.

V (volts)	0.350	0.600	0.850	1.10	1.45	1.80
f (petahertz)	0.550	0.605	0.660	0.735	0.805	0.880

8. If a gas is cooled under conditions of constant volume, it is noted that the pressure falls nearly proportionally as the temperature. If this were to happen until there was no pressure, the theoretical temperature for this case is referred to as *absolute zero*. In an elementary experiment, the following data were found for pressure and temperature for a gas under constant volume.

P (kilopascals)	133	143	153	162	172	183
T (degrees Celsius)	0.0	20	40	60	80	100

Find the least-squares line for P as a function of T, and, from the graph, determine the value of absolute zero found in this experiment.

The linear coefficient of correlation, a measure of the relatedness of two variables, is defined by $r = m(s_x/s_y)$. Due to its definition, the values of r lie in the range $-1 \leq r \leq 1$. If r is near 1, the correlation is considered good. For values of r between $-\frac{1}{2}$ and $+\frac{1}{2}$, the correlation is poor. If r is near -1, the variables are said to be negatively correlated; that is, one increases while the other decreases.

In Exercises 9 through 12 compute r for the given sets of data.

9. Exercise 1   10. Exercise 2   11. Exercise 4   12. Example A

21-5 Fitting Nonlinear Curves to Data

If the experimental points do not appear to be on a straight line, but we recognize them as being approximately on some other type of curve, the method of least squares can be extended to use on these other curves. For example, if the points are apparently on a parabola, we could use the function $y = a + bx^2$. To use the above method, we shall extend the least-squares line to

$$y = m[f(x)] + b \qquad (21-14)$$

Here, $f(x)$ must be calculated first, and then the problem can be treated as a least-squares line to find the values of m and b. Some of the functions $f(x)$ which may be considered for use are x^2, $1/x$, and 10^x.

Example A

Find the least-squares curve $y = mx^2 + b$ for the points in the following table.

y	1	5	12	24	53	76
x	0	1	2	3	4	5

The necessary quantities are tabulated for purposes of calculation.

x	y	$f(x) = x^2$	yx^2	$(x^2)^2$
0	1	0	0	0
1	5	1	5	1
2	12	4	48	16
3	24	9	216	81
4	53	16	848	256
5	76	25	1900	625
	171	55	3017	979

$$\overline{x^2} = \frac{55}{6} = 9.17, \quad \overline{x^2}^2 = 84.1$$

$$\overline{y} = \frac{171}{6} = 28.5, \quad \overline{x^2}\,\overline{y} = 261$$

$$\overline{yx^2} = \frac{3017}{6} = 503, \quad \overline{(x^2)^2} = \frac{979}{6} = 163.2$$

$$s_{(x^2)}^2 = 163.2 - 84.1 = 79.1$$

$$m = \frac{503 - 261}{79.1} = \frac{242}{79.1} = 3.06 \qquad b = 28.5 - (3.06)(9.17) = 0.40$$

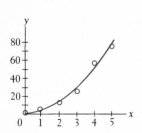

Figure 21-11

Therefore, the desired equation is $y = 3.06x^2 + 0.40$. The graph of this equation and the data points are shown in Fig. 21-11.

Example B

In a physics experiment designed to measure the pressure and volume of a gas at constant temperature, the following data were found. When the

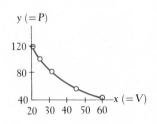

P (kilopascals)

Figure 21–12

points were plotted, they were seen to approximate the hyperbola $y = c/x$. Find the least-squares approximation to this hyperbola $[y = m(1/x) + b]$ (see Fig. 21–12).

P (kPa)	V (cm³)	$x (=V)$	$y (=P)$	$f(x) = \dfrac{1}{x}$	$y\left(\dfrac{1}{x}\right)$	$\left(\dfrac{1}{x}\right)^2$
120.0	21.0	21.0	120.0	0.04762	5.714	0.002268
99.2	25.0	25.0	99.2	0.04000	3.968	0.001600
81.3	31.8	31.8	81.3	0.03145	2.557	0.000989
60.6	41.1	41.1	60.6	0.02433	1.474	0.000592
42.7	60.1	60.1	42.7	0.01664	0.711	0.000277
			403.8	0.16004	14.424	0.005726

$$\overline{\frac{1}{x}} = 0.03201 \qquad \overline{\left(\frac{1}{x}\right)^2} = 0.001025 \qquad \overline{y} = 80.76 \qquad \overline{\frac{1}{x}}\,\overline{y} = 2.585$$

$$\overline{y\left(\frac{1}{x}\right)} = 2.885 \qquad \left(\overline{\frac{1}{x}}\right)^2 = 0.001145 \qquad s^2_{(1/x)} = 0.001145 - 0.001025 = 0.000120$$

$$m = \frac{2.885 - 2.585}{0.000120} = 2500 \qquad b = 80.76 - 2500(0.03201) = 0.74$$

Thus the equation of the hyperbola $y = m(1/x) + b$ is

$$y = \frac{2500}{x} + 0.74$$

The graph of this hyperbola and the points representing the data are shown in Fig. 21–13.

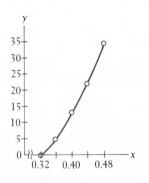

y (=P)

Figure 21–13

Example C

It has been found experimentally that the tensile strength of brass (a copper-zinc alloy) increases (within certain limits) with the percentage of zinc. The following table indicates the values which have been found (also see Fig. 21–14).

Tensile strength (10^5 lb/in.²)	0.32	0.36	0.40	0.44	0.48
Percentage of zinc	0	5	13	22	34

Fit a curve of the form $y = m(10^x) + b$ to the data. Let $x = $ tensile strength ($\times 10^5$) and $y = $ percentage of zinc.

x	y	$f(x) = 10^x$	$y(10^x)$	$(10^x)^2$
0.32	0	2.09	0.0	4.37
0.36	5	2.29	11.4	5.25
0.40	13	2.51	32.6	6.31
0.44	22	2.75	60.5	7.59
0.48	34	3.02	102.7	9.12
	74	12.66	207.2	32.64

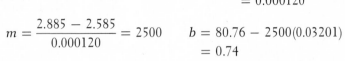

Figure 21–14

To find $f(x) = 10^x$, we may use either a calculator or logarithms. Using logarithms $x = 0.32$, $10^{0.32}$ is found by looking up the number for which 0.32 is the logarithm. This is due to the definition of a logarithm. Values of $(10^x)^2$ are found in the same manner. When $x = 0.32$, $(10^{0.32})^2 = (10)^{0.64}$, and the antilogarithm of 0.64 is required.

$$\overline{10^x} = 2.53 \qquad\qquad \overline{10^x y} = 37.4$$

$$\overline{10^{x^2}} = 6.40 \qquad\qquad \overline{y(10^x)} = 41.4$$

$$\bar{y} = 14.8 \qquad\qquad \overline{(10^x)^2} = 6.53$$

$$s^2_{(10^x)} = 6.53 - 6.40 = 0.13$$

$$m = \frac{41.4 - 37.4}{0.13} = 31$$

$$b = 14.8 - 31(2.53) = -64$$

The equation of the curve is $y = 31(10^x) - 64$. It must be remembered that for practical purposes y must be positive. The graph of the equation is shown in Fig. 21–15, with the solid portion denoting the meaningful part of the curve. The points of the data are also shown.

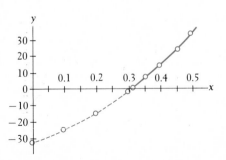

Figure 21–15

Exercises 21–5

In each of the following exercises find the indicated least-squares curve. Sketch the curve and plot the data points on the same graph.

1. For the points in the following table, find the least-squares curve $y = mx^2 + b$.

x	2	4	6	8	10
y	12	38	72	135	200

2. For the points in the following table, find the least-squares curve $y = m\sqrt{x} + b$.

x	0	4	8	12	16
y	1	9	11	14	15

3. For the points in the following table, find the least-squares curve $y = m(1/x) + b$.

x	1.10	2.45	4.04	5.86	6.90	8.54
y	9.85	4.50	2.90	1.75	1.48	1.30

4. For the points in the following table, find the least-squares curve $y = m(10^x) + b$.

x	0.00	0.200	0.500	0.950	1.325
y	6.00	6.60	8.20	14.0	26.0

5. The following data were found for the distance y that an object rolled down an inclined plane in time t. Determine the least-squares curve $y = mt^2 + b$.

y (centimeters)	6.0	23	55	98	148
t (seconds)	1.0	2.0	3.0	4.0	5.0

6. The increase in length of a certain metallic rod was measured in relation to particular increases in temperature. If y represents the increase in length for the corresponding increase in temperature x, find the least-squares curve $y = m(x^2) + b$ for these data.

x (degrees Celsius)	50.0	100	150	200	250
y (centimeters)	1.00	4.40	9.40	16.4	24.0

7. Use the data of Exercise 6 and determine the least-squares curve $y = m(10^z) + b$ where $z = x/1000$.

8. Measurements were made of the current in an electric circuit as a function of time. The circuit contained a resistance of 5 Ω and an inductance of 10 H. The following data were found.

i (amperes)	0.00	2.52	3.45	3.80	3.92
t (seconds)	0.00	2.00	4.00	6.00	8.00

Find the least-squares curve $i = m(e^{-0.5t}) + b$ for this data. Natural logarithms may be used. From the equation, determine the value the current approaches as t approaches infinity.

9. The resonant frequency of an electric circuit containing a $4\mu F$ capacitor was measured as a function of an inductance in the circuit. The following data were found.

f (hertz)	490	360	250	200	170
L (henrys)	1.0	2.0	4.0	6.0	9.0

Find the least-squares curve $f = m(1/\sqrt{L}) + b$.

10. The displacement of a pendulum bob from its equilibrium position as a function of time gave the following results.

Displacement (centimeters)	0.00	7.80	10.0	8.10	0.00
Time (seconds)	0.00	0.90	1.60	2.20	3.15

Find the least-squares curve of the form $y = m(\sin x) + b$, expressing the displacement as y and the time as x.

21—6 Exercises for Chapter 21

In Exercises 1 through 4 assume two dice are being tossed.

1. What is the probability of tossing a number greater than 4?
2. What is the probability of tossing a 6 or a 7?
3. What is the probability of tossing a 7 and then an 11 in successive tosses?
4. What is the probability of tossing a 2, a 3, and a 4 in successive tosses?

In Exercises 5 through 8 consider a person who has 6 pennies, 10 nickels, and 8 dimes in a drawer. Determine the probability of the person drawing at random from these coins:

5. Two pennies on successive draws, if the first coin is not replaced before the second is drawn
6. Two pennies on successive draws, if the first coin is replaced before the second is drawn
7. A nickel and then a dime, if the first coin is not replaced before the second coin is drawn
8. A penny, then a nickel, and then a dime, if the first and second coins are replaced before the following draw is made

In Exercises 9 through 12 use the following set of numbers:

$$2.3, \ 2.6, \ 4.2, \ 3.6, \ 3.5, \ 4.1, \ 4.8, \ 2.5, \ 3.0, \ 4.1, \ 3.8$$

9. Determine the median of the set of numbers.
10. Determine the arithmetic mean of the set of numbers.
11. Determine the standard deviation of the set of numbers.
12. Construct a frequency table with intervals 2.0–2.9, 3.0–3.9, and 4.0–4.9.

In Exercises 13 through 16 use the following data:

An important property of oil is its coefficient of viscosity, which gives a measure of how well it flows. In order to determine the viscosity of a certain motor oil, a refinery took samples from 12 different storage tanks and tested them at 50°C. The results (in pascal-seconds) were 0.24, 0.28, 0.29, 0.26, 0.27, 0.26, 0.25, 0.27, 0.28, 0.26, 0.26, 0.25.

13. Determine the arithmetic mean.   14. Determine the median.
15. Determine the standard deviation.   16. Make a histogram.

In Exercises 17 through 20 use the following data:

Two machine parts are considered satisfactorily assembled if their total thickness (to the nearest one-hundredth of an inch) is between or equal to 0.92 and 0.94 in. One hundred sample assemblies are tested, and the thicknesses, to the nearest one-hundredth of an inch, are given in the following table:

Total thickness	0.90	0.91	0.92	0.93	0.94	0.95	0.96
Number	3	9	31	38	12	5	2

17. Determine the arithmetic mean.   18. Determine the median.
19. Determine the standard deviation.   20. Make a frequency polygon.

In Exercises 21 through 24 use the following data:

A Geiger counter records the presence of high-energy nuclear particles. Even though no apparent radioactive source is present, a certain number of particles will be recorded. These are primarily cosmic rays, which are caused by very high-energy particles from outer space. In an experiment to measure the amount of cosmic radiation, the number of counts were recorded during 200 5-s intervals. The following table gives the number of counts, and the number of 5-s intervals having this number of counts. Draw a frequency curve for this data.

Counts	0	1	2	3	4	5	6	7	8	9	10
Intervals	3	10	25	45	29	39	26	11	7	2	3

21. Determine the median.
22. Determine the arithmetic mean.
23. Make a histogram.
24. Make a frequency polygon.

In Exercises 25 through 28 solve the given problems in probability.

25. An integer n, where $10 < n < 20$, is chosen at random. What is the probability that n is even?

26. What is the probability of drawing both red aces from a standard bridge deck of cards in two draws, if the first card is not replaced before drawing the second card?

27. A manufacturer finds that 3% of the parts produced by a certain machine are defective. A testing machine fails to operate properly in its determination of a defective part 0.1% of the time. What is the probability of a defective part not being detected by the testing machine?

28. A certain team won 12 of its first 20 games. Using the complete past record to determine the probability of winning a next game, what is the probability of this team winning the next two games?

In Exercises 29 through 34 find the indicated least-squares curves.

29. In a certain experiment, the resistance of a certain resistor was measured as a function of the temperature. The data found were as follows:

R (ohms)	25.0	26.8	28.9	31.2	32.8	34.7
T (degrees Celsius)	0.0	20.0	40.0	60.0	80.0	100

Find the least-squares line for this data, expressing R as a function of T. Sketch the line and data points on the same graph.

30. The solubility of sodium chloride (table salt) (in kilograms per cubic meter of water) as a function of temperature (degrees Celsius) is measured as follows:

Solubility	357	360	366	373	384
Temperature	0.0	20.0	40.0	60.0	80.0

Find the least-squares straight line for these data.

31. The coefficient of friction of an object sliding down an inclined plane was measured as a function of the angle of incline of the plane. The following results were obtained.

Coefficient of friction	0.16	0.34	0.55	0.85	1.24	1.82	2.80
Angle (degrees)	10.0	20.0	30.0	40.0	50.0	60.0	70.0

Find the least-squares curve of the form $y = m(\tan x) + b$, expressing the coefficient of friction as y and the angle as x.

32. An electric device recorded the total distance (in centimeters) which an object fell at 0.1 s intervals. Following are the data which were found.

Distance	4.90	19.5	44.0	78.1	122.0
Time	0.100	0.200	0.300	0.400	0.500

Find the least-squares curve of the form $y = mx^2 + b$, using y to represent distance and x to represent time.

33. The period of a pendulum as a function of its length was measured, giving the following results.

Period (seconds)	1.10	1.90	2.50	2.90	3.30
Length (feet)	1.00	3.00	5.00	7.00	9.00

Find the least-squares curve of the form $y = m\sqrt{x} + b$, which expresses the period as a function of the length.

34. In an elementary experiment which measured the wavelength of sound as a function of the frequency, the following results were obtained.

Wavelength (centimeters)	140	107	81.0	70.0	60.0
Frequency (hertz)	240	320	400	480	560

Find the least-squares curve of the form $y = m(1/x) + b$ for this data, expressing wavelength as y and frequency as x.

22

The Derivative

22–1 Limits

The problems which can be solved by the methods of algebra and trigonometry are numerous. There are, however, a great many problems which arise in the various fields of technology which require for their solution methods beyond those available from algebra and trigonometry. These traditional topics remain of definite importance, but it is necessary to develop additional methods of analyzing and solving problems.

One very important type of problem involves the rate of change of one quantity with respect to another. Examples of rates of change are velocity (the rate of change of distance with respect to time), the rate of change of the length of a metal rod with respect to temperature, the rate of change of light intensity with respect to the distance from the source, the rate of change of electric current with respect to time, and many other similar physical situations. The methods of **differential calculus,** which we shall start developing in this chapter, will enable us to solve problems involving these quantities.

Another principal type of problem which calculus allows us to solve is that of finding a function when its rate of change is known. This is **integral calculus.** One of its principal applications comes from electricity, where current is the time rate of change of electric charge. Integral calculus also leads to the solution of a great many apparently unrelated problems, including the determination of plane areas, volumes, and the physical concepts of work and pressure. The study of integral calculus starts in Chapter 24.

Before dealing directly with the rate of change of a function, we shall first discuss the concept of a **limit.** We encountered this concept in our discussion of geometric progressions with infinitely many terms and in our discussion of the asymptotes of a hyperbola. It is now necessary to develop this concept further.

In order to help understand and develop the concept of a limit, we consider briefly the **continuity** of a function. *For a function to be con-tinuous at a point,* *the function must exist at the point, and any small change in x produces only a small change in f(x).* In fact, the change in $f(x)$ can be made as small as we wish by restricting the change in x sufficiently, if the function is continuous. Also, *a function is said to be* **continuous over an interval** *if it is continuous at each point in the interval.*

Example A

The function $f(x) = 3x^2$ is continuous for all values of x. That is, $f(x)$ is defined for all values of x, and a small change in x for any given value produces only a small change in $f(x)$. If we choose $x = 2$ and then let x change by 0.1, 0.01, and so on, we obtain the values in the following table.

x	2	2.1	2.01	2.001
$f(x)$	12	13.23	12.1203	12.012003

We can also see that the change in $f(x)$ is made smaller by the smaller changes in x. This shows that $f(x)$ is continuous at $x = 2$. Since this type of result would be obtained for any other x we may choose, we see that $f(x)$ is continuous for all values, and therefore it is continuous over the interval of all values of x.

Example B

The function $f(x) = \frac{1}{x-2}$ is not continuous at $x = 2$. When we substitute 2 for x in the function, we have division by zero. This means that the function is not defined. Therefore, the condition that the function must exist is not satisfied.

Considering continuity from a graphical point of view, a function which is continuous over an interval will have no "breaks" in its graph over that interval. If a function is not continuous, either the graph of the function has a break in it or it does not exist for the values of x.

Example C

The graph of the function $f(x) = 3x^2$, which we determined to be continuous at all values of x in Example A, is shown in Fig. 22–1. We see that there are no breaks in the graph.

The graph of $f(x) = \frac{1}{x-2}$, which we determined to be not continuous at $x = 2$ in Example B, is shown in Fig. 22–2. We see that there is a break in the graph for $x = 2$.

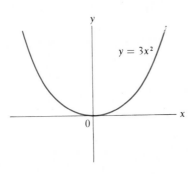

Figure 22–1

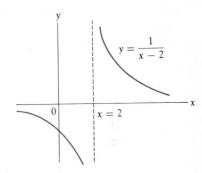

Figure 22–2

Example D

The function represented by the graph in Fig. 22–3 is continuous in the interval for which $x > 1$. The graph does not exist for values $x < 1$.

The function represented by the graph in Fig. 22–4 is not continuous at $x = 1$. The function is defined (by the solid circle point) for $x = 1$ (the open circle point indicates that point is not part of the graph). However, a small change from 1 may result in a change of at least 3 in $f(x)$, regardless of how small the change in x is made. Therefore, the small change condition is not satisfied.

The function represented by the graph in Fig. 22–5 is not continuous for $x = -2$. The open circle shows that point is not part of the graph, and therefore $f(x)$ is not defined for $x = -2$.

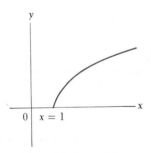

Figure 22–3

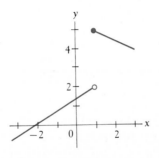

Figure 22–4

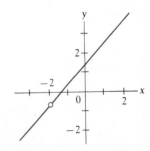

Figure 22–5

In our earlier discussions of infinite geometric series and the asymptotes of a hyperbola, we used the symbol → which means "approaches." *When we say that x → 2, we mean that x may take on any value as close to 2 as desired, but it also distinctly means that x cannot be set equal to 2.*

Example E
Let us consider the behavior of the function $f(x) = 2x + 1$ as $x \to 2$.

Since we are not to use $x = 2$, we set up tables to determine values of $f(x)$, as x gets close to 2.

x	1.000	1.500	1.900	1.990	1.999
$f(x)$	3.000	4.000	4.800	4.980	4.998

x	3.000	2.500	2.100	2.010	2.001
$f(x)$	7.000	6.000	5.200	5.020	5.002

We can see that $f(x)$ approaches 5, as x approaches 2, from above 2 and from below 2.

In Example E, since $f(x) \to 5$ as $x \to 2$, the number 5 is called the limit of $f(x)$ as $x \to 2$. This leads us to the meaning of the limit of a function. In general, *the* **limit of a function** $f(x)$ *is that value which the function approaches as x approaches a given value a.* This is written as

$$\lim_{x \to a} f(x) = L \tag{22-1}$$

where L is the value of the limit of the function. Remember, in approaching a, x may come as arbitrarily close as desired to a, but it may not equal a.

An important conclusion can be drawn from the limit in Example E. The function $f(x)$ is a continuous function, and $f(2)$ equals the value of the limit as $x \to 2$. In general, it is true that *the limit of $f(x)$ as $x \to a$ is the same as $f(a)$, if the function is continuous at $x = a$.*

Although we can evaluate the limit for a continuous function as $x \to a$ by evaluating $f(a)$, it is possible that a function is not continuous at $x = a$, and that the limit exists and can be determined. Thus, we must be able to determine the value of a limit without finding $f(a)$. The following example illustrates the evaluation of such a limit.

Example F
Find

$$\lim_{x \to 2} \frac{2x^2 - 3x - 2}{x - 2}$$

We note immediately that the function is not continuous at $x = 2$, for division by zero is indicated. Thus, we cannot evaluate the limit by substituting $x = 2$ into the function. By setting up tables, we determine the value which $f(x)$ approaches, as x approaches 2.

x	1.000	1.500	1.900	1.990	1.999
$f(x)$	3.000	4.000	4.800	4.980	4.998

x	3.000	2.500	2.100	2.010	2.001
$f(x)$	7.000	6.000	5.200	5.020	5.002

We see that the values obtained are identical to those in Example E. Since $f(x) \to 5$ as $x \to 2$, we have

$$\lim_{x \to 2} \frac{2x^2 - 3x - 2}{x - 2} = 5$$

Thus, the limit exists at $x = 2$, although the function does not exist at $x = 2$.

The reason that the functions in Examples E and F have the same limit is shown in the following example.

Example G
The function $\frac{2x^2 - 3x - 2}{x - 2}$ in Example F is the same as the function $2x + 1$ in Example E, except when $x = 2$. By factoring the numerator of the function of Example F, we have

$$\frac{2x^2 - 3x - 2}{x - 2} = \frac{(2x + 1)(x - 2)}{x - 2} = 2x + 1$$

The cancellation here is valid, as long as x does not equal 2, for we have division by zero at $x = 2$. Also, in finding the limit as $x \to 2$, we do not use the value $x = 2$. Therefore,

$$\lim_{x \to 2} \frac{2x^2 - 3x - 2}{x - 2} = \lim_{x \to 2} (2x + 1) = 5$$

The limits of the two functions are equal, since, again, in finding the limit, we do not let $x = 2$. The graphs of the two functions are shown in Fig. 22–6 (a) and (b). We can see from the graphs that the limits are the same, although one of the functions is not continuous.

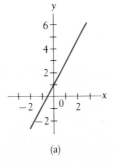

(a)

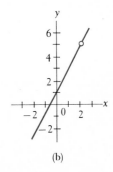

(b)

Figure 22—6

The limit of the function in Example F was determined by calculating values near $x = 2$ and by means of an algebraic change in the function. This illustrates that limits may be found through the meaning and definition, and through other procedures when the function is not continuous. The following examples further illustrate the evaluation of limits.

Example H
Find $\lim_{x \to 4} (x^2 - 7)$.

Since the function $x^2 - 7$ is continuous at $x = 4$, we may evaluate this limit by substitution. For $f(x) = x^2 - 7$, we have $f(4) = 9$. This means that

$$\lim_{x \to 4} (x^2 - 7) = 9$$

Example I
Find

$$\lim_{x \to 2} \left(\frac{x^2 - 4}{x - 2} \right)$$

Since

$$\frac{x^2 - 4}{x - 2} = \frac{(x - 2)(x + 2)}{x - 2} = x + 2$$

is valid as long as $x \neq 2$, we find that

$$\lim_{x \to 2} \left(\frac{x^2 - 4}{x - 2} \right) = \lim_{x \to 2} (x + 2) = 4$$

Again, we do not have to concern ourselves with the fact that the cancellation is not valid for $x = 2$. In finding the limit we do not consider the value of $f(x)$ at $x = 2$.

Example J
Find $\lim_{x \to 0} (x\sqrt{x - 3})$.

We see that $x\sqrt{x - 3} = 0$ if $x = 0$, but this function does not have real values for values of x less than 3 other than $x = 0$. *Since x cannot approach 0, f(x) does not approach 0 and the limit does not exist.* The point of this example is that even if $f(a)$ exists, we cannot evaluate the limit by finding $f(a)$, unless $f(x)$ is continuous at $x = a$. Here, $f(a)$ exists but the limit does not exist.

Limits as x approaches infinity are also of importance. When we write $x \to \infty$ we know that we are to consider values of x which are becoming large without bound. Consider the following examples.

Example K
Find

$$\lim_{x \to \infty} \left(3 + \frac{1}{x^2} \right)$$

As x becomes larger and larger, $1/x^2$ becomes smaller and smaller and approaches zero. This means that $f(x) \to 3$ as $x \to \infty$. Thus,

$$\lim_{x \to \infty} \left(3 + \frac{1}{x^2} \right) = 3$$

Example L
Find

$$\lim_{x \to \infty} \frac{x^2 + 1}{2x^2 + 3}$$

We note that as $x \to \infty$ both the numerator and denominator become large without bound. However, if we divide numerator and denominator by x^2, the function becomes

$$\frac{1 + \dfrac{1}{x^2}}{2 + \dfrac{3}{x^2}}$$

we see that $1/x^2$ and $3/x^2$ both approach zero as $x \to \infty$. This means that the numerator approaches 1 and the denominator approaches 2. Thus,

$$\lim_{x \to \infty} \frac{x^2 + 1}{2x^2 + 3} = \lim_{x \to \infty} \frac{1 + \dfrac{1}{x^2}}{2 + \dfrac{3}{x^2}} = \frac{1}{2}$$

The definitions and development of continuity and of a limit presented in this section are not mathematically completely rigorous. However, the development is consistent with a more rigorous development, and the concept of a limit is the principal concern.

Exercises 22–1

In Exercises 1 through 4 determine the values of x for which the function is continuous. If the function is not continuous, determine the reason.

1. $f(x) = 3x - 2$ 2. $f(x) = 9 - x^2$

3. $f(x) = \dfrac{1}{x + 3}$ 4. $f(x) = \dfrac{2}{x^2 - x}$

In Exercises 5 through 8 determine the values of x for which the function, as represented by the graph in Fig. 22–7, is continuous. If the function is not continuous, determine the reason.

5.

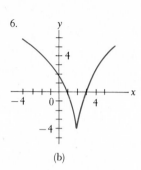

(a)

6.

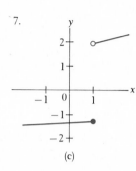

(b)

7.

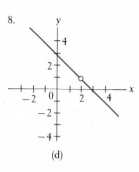

(c)

8.

(d)

Figure 22–7

In Exercises 9 through 12 evaluate the given functions for the given values of x. Do not change the form of the functions. Then find the indicated limits.

9. Evaluate $f(x) = 3x - 2$ for the following values of x: 4.000, 3.500, 3.100, 3.010, 3.001 and for 2.000, 2.500, 2.900, 2.990, 2.999. Find

$$\lim_{x \to 3} (3x - 2)$$

10. Evaluate $f(x) = 2x^2 - 4x$ for the following values of x: 0.4000, 0.4900, 0.4990, 0.4999 and for 0.6000, 0.5100, 0.5010, 0.5001. Find

$$\lim_{x \to 0.5} (2x^2 - 4x)$$

11. Evaluate $f(x) = \dfrac{x^3 - x}{x - 1}$ for the following values of x: 0.900, 0.990, 0.999 and for 1.100, 1.010, 1.001. Find

$$\lim_{x \to 1} \frac{x^3 - x}{x - 1}$$

12. Evaluate $f(x) = \dfrac{2 - \sqrt{x}}{4 - x}$ for the following values of x: 3.900, 3.990, 3.999 and for 4.100, 4.010, 4.001. Find

$$\lim_{x \to 4} \frac{2 - \sqrt{x}}{4 - x}$$

In Exercises 13 through 32 evaluate the given limits.

13. $\lim_{x \to 3} (x + 4)$

14. $\lim_{x \to -2} (2x - 1)$

15. $\lim_{x \to 2} \dfrac{x^2 - 1}{x + 1}$

16. $\lim_{x \to 5} \left(\dfrac{3}{x^2 + 2} \right)$

17. $\lim_{x \to 0} \dfrac{x^2 + x}{x}$

18. $\lim_{x \to 2} \dfrac{x^2 - 2x}{x - 2}$

19. $\lim_{x \to -1} \dfrac{x^2 - 1}{x + 1}$

20. $\lim_{x \to 5} \dfrac{x - 5}{x^2 - 25}$

21. $\lim_{x \to 1} \dfrac{x^3 - x}{x - 1}$

22. $\lim_{x \to 1/3} \dfrac{3x - 1}{3x^2 + 5x - 2}$

23. $\lim_{x \to 3} \dfrac{x^2 - 2x - 3}{3 - x}$

24. $\lim_{x \to 0} \dfrac{(2 + x)^2 - 4}{x}$

25. $\lim_{x \to -1} \sqrt{x}\,(x + 1)$

26. $\lim_{x \to 1} (x - 1)\,\sqrt{x^2 - 4}$

27. $\lim_{x \to \infty} \dfrac{\frac{2}{x}}{1 - 2x}$

28. $\lim_{x \to \infty} \dfrac{6}{1 + \dfrac{2}{x^2}}$

29. $\lim_{x \to \infty} \dfrac{3x^2 + 5}{x^2 - 2}$

30. $\lim_{x \to \infty} \dfrac{x - 1}{7x + 4}$

31. $\lim_{x \to \infty} \dfrac{2x - 6}{x^2 - 9}$

32. $\lim_{x \to \infty} \dfrac{2 - x^2}{3x^3 - 1}$

In Exercises 33 through 36 solve the given problems involving limits.

33. The magnetic field B at the center of a loop of wire due to a current i in the wire is given by $B = ki/r$, where r is the radius of the loop and k is a constant. If $k = 6.30 \times 10^{-7}$ Wb/A-m and $r = 0.100$ m, find $\lim_{i \to 0.3} B$ (in teslas) by evaluating the function for the following values of i: 0.200, 0.290, 0.299, 0.301, 0.310, 0.400.

34. Velocity can be determined by dividing the displacement of an object by the time elapsed in moving through the displacement. In a certain experiment the following values were determined for the displacements and elapsed times for the motion of an object. Determine the limit of the velocity.

displacement (centimeters)	0.480000	0.280000	0.029800	0.002998	0.00029998
time (seconds)	0.200000	0.100000	0.010000	0.001000	0.00010000

35. A certain object, after being heated, cools at such a rate that its temperature T (in degrees Celsius) decreases 10% each minute. If the object is originally heated to 100°C find $\lim_{t \to 10} T$ and $\lim_{t \to \infty} T$, where t is the time (in minutes).

36. Show that

$$\lim_{x \to 0^+} 2^{1/x} \neq \lim_{x \to 0^-} 2^{1/x}$$

where $\lim_{x \to 0^+}$ means to find the limit as x approaches zero through positive values only, and $\lim_{x \to 0^-}$ means to find the limit as x approaches zero through negative values only.

22–2 The Slope of a Tangent to a Curve

Having developed the basic operations with functions and the concept of a limit, we shall now turn our attention to a graphical interpretation of the rate of change of a function. This interpretation, basic to an understanding of the calculus, deals with the slope of a line tangent to the curve of a function.

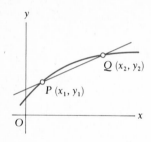

Figure 22–8

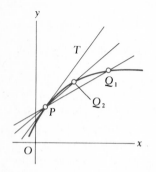

Figure 22–9

Consider the points $P(x_1, y_1)$ and $Q(x_2, y_2)$ in Fig. 22–8. From Chapter 20 we know that the slope of the line through these points is given by

$$m = \frac{y_2 - y_1}{x_2 - x_1}$$

This, however, represents the slope of the line through P and Q and no other line. If we now allow Q to be a point closer to P, the slope of PQ will more closely approximate the slope of a line drawn tangent to the curve at P (see Fig. 22–9). In fact, the closer Q is to P, the better this approximation becomes. It is not possible to allow Q to coincide with P, for then it would not be possible to define the slope of PQ in terms of two points. The slope of the tangent line, often referred to as the slope of the curve, is the limiting value of the slope of PQ as Q approaches P.

Example A

Find the slope of a line tangent to the curve $y = x^2 + 3x$ at the point $P(2, 10)$ by determining the limit of slopes of lines PQ as Q approaches P.

We shall let point Q have the x-values of 3.0, 2.5, 2.1, 2.01, and 2.001, and tabulate the necessary values. Since P is the point $(2, 10)$, $x_1 = 2$ and $y_1 = 10$. Thus, using the values of x_2 we tabulate the values of y_2, $y_2 - 10$, $x_2 - 2$, and thereby the values of the slope m.

Point	Q_1	Q_2	Q_3	Q_4	Q_5	P
x_2	3.0	2.5	2.1	2.01	2.001	2
y_2	18.0	13.75	10.71	10.0701	10.007001	10
$y_2 - 10$	8.0	3.75	0.71	0.0701	0.007001	
$x_2 - 2$	1.0	0.5	0.1	0.01	0.001	
$m = \dfrac{y_2 - 10}{x_2 - 2}$	8.0	7.5	7.1	7.01	7.001	

We can see from the above table that the slope of PQ approaches the value of 7 as Q approaches P. Thus, the slope of the tangent line at $(2,10)$ is 7. See Fig. 22–10.

With the proper notation, it is possible to express the coordinates of Q in terms of the coordinates of P. If we define the quantity Δx ("delta" x) by the equation

$$x_2 = x_1 + \Delta x \tag{22–2}$$

and Δy by the equation

$$y_2 = y_1 + \Delta y \tag{22–3}$$

the coordinates of Q become $(x_1 + \Delta x, y_1 + \Delta y)$. The quantities Δx and

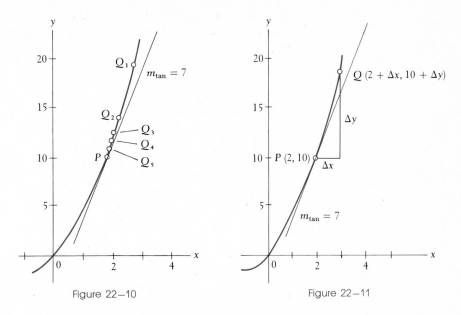

Figure 22–10   Figure 22–11

Δy represent the difference in the coordinates of P and Q. They are not to be throught of as "Δ times x," for the symbol Δ used here has no meaning by itself. *The name* **increment** *is given to the difference of the coordinates of two points, and therefore Δx and Δy are the increments in x and y, respectively.*

Using Eqs. (22–2) and (22–3) along with the definition of slope, we can express the slope of PQ as

$$m_{PQ} = \frac{(y_1 + \Delta y) - y_1}{(x_1 + \Delta x) - x_1} = \frac{\Delta y}{\Delta x} \tag{22–4}$$

By the previous discussion, as Q approaches P, the slope of the tangent line is more *nearly* approximated by $\Delta y/\Delta x$.

Example B
Find the slope of a line tangent to the curve $y = x^2 + 3x$ at the point $(2, 10)$ by the increment method indicated in Eq. (22–4). (This is the same slope as calculated in Example A.)

As in Example A, point P has the coordinates $(2, 10)$. Thus the coordinates of any other point Q can be expressed as $(2 + \Delta x, 10 + \Delta y)$. See Fig. 22–11. The slope of PQ then becomes

$$m_{PQ} = \frac{(10 + \Delta y) - 10}{(2 + \Delta x) - 2} = \frac{\Delta y}{\Delta x}$$

This expression itself does not enable us to find the slope, since values of Δx and Δy are not known. If, however, we can express Δy in terms of Δx, we might derive more information. Both P and Q are on the curve of the function, which means that the coordinates of each must satisfy the function. Using the coordinates of Q, we have

$$(10 + \Delta y) = (2 + \Delta x)^2 + 3(2 + \Delta x)$$
$$= 4 + 4\Delta x + (\Delta x)^2 + 6 + 3\Delta x$$

Subtracting 10 from each side, we have $\Delta y = 7\Delta x + (\Delta x)^2$. We then substitute this into the equation for m_{PQ}, which gives us

$$m_{PQ} = \frac{7\Delta x + (\Delta x)^2}{\Delta x} = 7 + \Delta x$$

As Q approaches P, Δx becomes smaller and smaller. The limiting value of Δx is zero. Using this fact, we can see that the slope of the tangent line is

$$m_{\text{tan}} = \lim_{\Delta x \to 0} m_{PQ} = 7$$

We see that this result agrees with that found in Example A.

Example C

Find the slope of a line tangent to the curve $y = 4x - x^2$ at the point (x_1, y_1).

The points P and Q are $P(x_1, y_1)$ and $Q(x_1 + \Delta x, y_1 + \Delta y)$. Using the coordinates of Q in the function, we obtain

$$(y_1 + \Delta y) = 4(x_1 + \Delta x) - (x_1 + \Delta x)^2$$
$$= 4x_1 + 4\Delta x - x_1^2 - 2x_1\Delta x - (\Delta x)^2$$

Using the coordinates of P, we obtain

$$y_1 = 4x_1 - x_1^2$$

Subtracting the second of these expressions from the first to solve for Δy, we obtain

$$(y_1 + \Delta y) - y_1 = (4x_1 + 4\Delta x - x_1^2 - 2x_1\Delta x - (\Delta x)^2) - (4x_1 - x_1^2)$$

or

$$\Delta y = 4\Delta x - 2x_1 \Delta x - (\Delta x)^2$$

Dividing through by Δx to obtain an expression for $\Delta y/\Delta x$, we obtain

$$\frac{\Delta y}{\Delta x} = 4 - 2x_1 - \Delta x$$

In this last equation, the desired expression is on the left, but all we can determine from $\Delta y/\Delta x$ itself is that the ratio will become one very small number divided by another very small number as $\Delta x \to 0$. The right side, however, approaches $4 - 2x_1$ as $\Delta x \to 0$. This indicates that the slope of a tangent at the point (x_1, y_1) is given by

$$m_{\text{tan}} = 4 - 2x_1$$

This method has an advantage over that of Example B—we now have a general expression for the slope of a tangent line for any value x_1. If

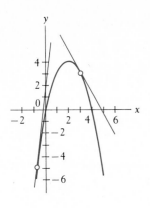

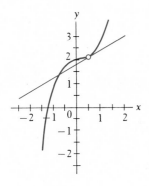

Figure 22–12

Figure 22–13

$x_1 = -1$, $m_{tan} = 6$, and if $x_1 = 3$, $m_{tan} = -2$. These tangent lines are indicated in Fig. 22–12.

Example D

Find the expression for the slope of a line tangent to the curve $y = x^3 + 2$ at the general point (x_1, y_1), and use this expression to determine the slope when $x = \frac{1}{2}$.

Using the coordinates of the point $Q(x_1 + \Delta x, y_1 + \Delta y)$ in the function, we obtain

$$y_1 + \Delta y = (x_1 + \Delta x)^3 + 2 = x_1^3 + 3x_1^2 \Delta x + 3x_1 (\Delta x)^2 + (\Delta x)^3 + 2$$

and using the coordinates of the point $P(x_1, y_1)$, we obtain $y_1 = x_1^3 + 2$. Subtracting this second expression from the first, we have

$$\Delta y = 3x_1^2 \Delta x + 3x_1 (\Delta x)^2 + (\Delta x)^3$$

Dividing by Δx, we have

$$\frac{\Delta y}{\Delta x} = 3x_1^2 + 3x_1 \Delta x + (\Delta x)^2$$

As $\Delta x \to 0$, the right side approaches the value $3x_1^2$. This means that

$$m_{tan} = 3x_1^2$$

When $x_1 = \frac{1}{2}$, we find that the slope of the tangent is $3(\frac{1}{4}) = \frac{3}{4}$. The curve and this tangent line are indicated in Fig. 22–13.

In interpreting this analysis as the rate of change of a function, we see that if $\Delta x = 1$ and the corresponding value of Δy is found, it can be said that y changes by the amount Δy as x changes one unit. If x changed by some lesser amount, we could still calculate a ratio for the amount of change in y for the given change in x. Therefore, as long as x changes at all, there will be a corresponding change in y. In this way the ratio $\Delta y/\Delta x$ is looked on as the **average rate of change** of y, which is a function of x, with respect to x. As $\Delta x \to 0$ the limit of the ratio $\Delta y/\Delta x$ gives us the **instantaneous rate of change** of y with respect to x.

Exercises 22–2

In Exercises 1 through 4 use the method of Example A to calculate the slope of a line tangent to the curve of each of the given functions. Let $Q_1, Q_2, Q_3,$ and Q_4 have the indicated x-values. Sketch the curves and tangent lines.

1. $y = x^2$; P is $(2, 4)$; let Q have x-values of 1.5, 1.9, 1.99, 1.999
2. $y = 1 - x^2$; P is $(2, -3)$; let Q have x-values of 1.5, 1.9, 1.99, 1.999
3. $y = x^2 + 2x$; P is $(-3, 3)$; let Q have x-values of $-2.5, -2.9, -2.99, -2.999$
4. $y = x^3 + 1$; P is $(-1, 0)$; let Q have x-values of $-0.5, -0.9, -0.99, -0.999$

In Exercises 5 through 8 use the method of Example B to calculate the slope of a line tangent to the curve of each of the given functions for the given points P. (These are the same functions and points P as for Exercises 1 through 4.)

5. $y = x^2$; P is $(2, 4)$
6. $y = 1 - x^2$; P is $(2, -3)$
7. $y = x^2 + 2x$; P is $(-3, 3)$
8. $y = x^3 + 1$; P is $(-1, 0)$

In Exercises 9 through 20 use the method of Example C to find a general expression for the slope of a tangent line to each of the indicated curves. Then find the slope for the given values of x. Sketch the curves and tangent lines.

9. $y = x^2$; $x = 2$, $x = -1$

10. $y = 1 - x^2$; $x = 2$, $x = -2$

11. $y = x^2 + 2x$; $x = -3$, $x = 1$

12. $y = 4 - 3x^2$; $x = 0$, $x = 2$

13. $y = x^2 + 4x + 5$; $x = -3$, $x = 2$

14. $y = 2x^2 - 4x$; $x = 1$, $x = 1.5$

15. $y = 6x - x^2$; $x = -2$, $x = 3$

16. $y = 2x - 3x^2$; $x = 0$, $x = 0.5$

17. $y = x^3 - 2x$; $x = -1$, $x = 0$, $x = 1$

18. $y = 3x - x^3$; $x = -2$, $x = 0$, $x = 2$

19. $y = x^4$; $x = 0$, $x = 0.5$, $x = 1$

20. $y = 1 - x^4$; $x = 0$, $x = 1$, $x = 2$

22-3 The Derivative

Using the terminology and techniques developed in the first section of this chapter, we are now ready to establish one of the fundamental definitions of calculus. Generalizing on the method of Example B of the preceding section, if we consider that the point $P(x, y)$ is held constant while the point $Q(x + \Delta x, y + \Delta y)$ approaches it, then both Δx and Δy approach zero. The points P and Q both lie on the curve of the function $f(x)$, which in turn means that the coordinates of each point must satisfy the function. For the point P this means that $y = f(x)$, and for the point Q this means that $y + \Delta y = f(x + \Delta x)$. Subtracting the first expression from the second, we have

$$(y + \Delta y) - y = f(x + \Delta x) - f(x)$$

$$\Delta y = f(x + \Delta x) - f(x) \tag{22-5}$$

From the discussion of the previous section, we saw that the slope of a line tangent to $f(x)$ at point $P(x, y)$ was found as the limiting value of the ratio $\Delta y / \Delta x$ as Δx approaches zero. Formally, *this limiting value of the ratio of $\Delta y / \Delta x$ is known as the* **derivative** *of the function.* Therefore, the derivative of a function $f(x)$ is defined by the relation

$$\lim_{\Delta x \to 0} \frac{\Delta y}{\Delta x} = \lim_{\Delta x \to 0} \frac{f(x + \Delta x) - f(x)}{\Delta x} \tag{22-6}$$

The process of finding the derivative of a function from its definition is called the Δ-process. *The general process of finding a derivative is called* **differentiation.**

Example A

Find the derivative of the function $y = 2x^2 + 3x$ by the delta-process.

To find Δy, we must first derive the quantity $f(x + \Delta x) - f(x)$. Thus, we find $y + \Delta y = f(x + \Delta x)$ first:

$$y + \Delta y = 2(x + \Delta x)^2 + 3(x + \Delta x)$$

Then we subtract the original function:

$$y + \Delta y - y = 2(x + \Delta x)^2 + 3(x + \Delta x) - 2x^2 - 3x$$

Simplifying, we find that the result is

$$\Delta y = 4x\Delta x + 2(\Delta x)^2 + 3\Delta x$$

Next, dividing through by Δx, we obtain

$$\frac{\Delta y}{\Delta x} = 4x + 2\Delta x + 3$$

When we find the limit as Δx approaches zero, we see that the $2\Delta x$-term on the right approaches zero. Thus,

$$\lim_{\Delta x \to 0} \frac{\Delta y}{\Delta x} = 4x + 3$$

Therefore, we find that the derivative of the function $y = 2x^2 + 3x$ is the function $4x + 3$. This means that we can calculate the slope of a tangent line for any point on the graph of $y = 2x^2 + 3x$ by substituting the x-coordinate into the expression $4x + 3$. For example, the slope of a tangent line is 7 if $x = 1$.

Example B

Find the derivative of $y = 6x - 2x^3$ by the Δ-process.
First we find $y + \Delta y = f(x + \Delta x)$.

$$
\begin{aligned}
y + \Delta y &= 6(x + \Delta x) - 2(x + \Delta x)^3 \\
&= 6x + 6(\Delta x) - 2[x^3 + 3x^2(\Delta x) + 3x(\Delta x)^2 + (\Delta x)^3] \\
&= 6x + 6(\Delta x) - 2x^3 - 6x^2(\Delta x) - 6x(\Delta x)^2 - 2(\Delta x)^3
\end{aligned}
$$

Next we subtract the original function.

$$
\begin{aligned}
y + \Delta y - y &= [6x + 6(\Delta x) - 2x^3 - 6x^2(\Delta x) - 6x(\Delta x)^2 - 2(\Delta x)^3] \\
&\quad - (6x - 2x^3) \\
\Delta y &= 6(\Delta x) - 6x^2(\Delta x) - 6x(\Delta x)^2 - 2(\Delta x)^3
\end{aligned}
$$

Dividing through by Δx is the next step.

$$
\begin{aligned}
\frac{\Delta y}{\Delta x} &= \frac{6(\Delta x) - 6x^2(\Delta x) - 6x(\Delta x)^2 - 2(\Delta x)^3}{\Delta x} \\
&= 6 - 6x^2 - 6x(\Delta x) - 2(\Delta x)^2
\end{aligned}
$$

Now we find the limit as Δx approaches zero.

$$
\lim_{\Delta x \to 0} \frac{\Delta y}{\Delta x} = \lim_{\Delta x \to 0} [6 - 6x^2 - 6x(\Delta x) - 2(\Delta x)^2]
$$

$$= 6 - 6x^2$$

Therefore, the derivative of $y = 6x - 2x^3$ is $6 - 6x^2$.

Example C

Find the derivative $y = \dfrac{1}{x}$ by the Δ-process.

We write

$$y + \Delta y = \frac{1}{x + \Delta x}$$

Then we find that

$$y + \Delta y - y = \frac{1}{x + \Delta x} - \frac{1}{x}$$

$$\Delta y = \frac{x - (x + \Delta x)}{x(x + \Delta x)} = \frac{-\Delta x}{x(x + \Delta x)}$$

$$\frac{\Delta y}{\Delta x} = \frac{-1}{x(x + \Delta x)}$$

$$\lim_{\Delta x \to 0} \frac{\Delta y}{\Delta x} = \lim_{\Delta x \to 0} \frac{-1}{x(x + \Delta x)}$$

$$\lim_{\Delta x \to 0} \frac{\Delta y}{\Delta x} = \frac{-1}{x^2}$$

Example D

Find the derivative of $y = x^2 + \dfrac{1}{x + 1}$ by the Δ-process.

$$y + \Delta y = (x + \Delta x)^2 + \frac{1}{x + \Delta x + 1}$$

$$y + \Delta y - y = x^2 + 2x\,\Delta x + (\Delta x)^2 + \frac{1}{x + \Delta x + 1} - x^2 - \frac{1}{x + 1}$$

The problem is handled algebraically most easily if the fractions are combined and the other terms are handled separately:

$$\Delta y = 2x\,\Delta x + (\Delta x)^2 + \frac{1}{x + \Delta x + 1} - \frac{1}{x + 1}$$

$$= 2x\,\Delta x + (\Delta x)^2 + \frac{(x + 1) - (x + \Delta x + 1)}{(x + \Delta x + 1)(x + 1)}$$

$$= 2x\,\Delta x + (\Delta x)^2 - \frac{\Delta x}{(x + \Delta x + 1)(x + 1)}$$

$$\frac{\Delta y}{\Delta x} = 2x + \Delta x - \frac{1}{(x + \Delta x + 1)(x + 1)}$$

$$\lim_{\Delta x \to 0} \frac{\Delta y}{\Delta x} = 2x - \frac{1}{(x + 1)^2}$$

Several shorter notations are used for the derivative. The notation of the definition is clumsy for general use, so the notations

$$y', \quad D_x y, \quad f'(x), \quad \text{and} \quad \frac{dy}{dx}$$

are commonly used. Thus, the answer to Example A could have been written as $y' = 4x + 3$, or as $dy/dx = 4x + 3$.

One might ask why, when we are finding a derivative, we take a limit as Δx approaches zero and do not simply let Δx equal zero. If we did this, the ratio $\Delta y/\Delta x$, would be exactly $0/0$, which would then require division by zero. As we know, this is an undefined operation in mathematics, and therefore Δx cannot *equal* zero. However, it can equal any value as near zero as necessary. This idea is inherent in the meaning of the word limit.

Example E

Find dy/dx of the function $y = \sqrt{x}$ by the Δ-process.

We first square both sides of the equation, thus obtaining $y^2 = x$. Substituting the coordinates of the point $(x + \Delta x, y + \Delta y)$ into this equation, we have

$$(y + \Delta y)^2 = x + \Delta x$$

Expanding the left side and then subtracting the equation $y^2 = x$, we have

$$y^2 + 2y\,\Delta y + (\Delta y)^2 - y^2 = x + \Delta x - x$$

or

$$2y\,\Delta y + (\Delta y)^2 = \Delta x$$

Next, dividing through by Δx, we obtain

$$2y\frac{\Delta y}{\Delta x} + \Delta y\frac{\Delta y}{\Delta x} = 1$$

Solving for $\Delta y/\Delta x$, we have

$$\frac{\Delta y}{\Delta x} = \frac{1}{2y + \Delta y}$$

Finally, taking the limit as $\Delta x \to 0$, we obtain

$$\lim_{\Delta x \to 0} \frac{\Delta y}{\Delta x} = \lim_{\Delta x \to 0} \frac{1}{2y + \Delta y} = \frac{1}{2y}$$

The Δy is omitted, since $\Delta y \to 0$ as $\Delta x \to 0$. Now, substituting in the original function, we obtain the final result:

$$\frac{dy}{dx} = \frac{1}{2\sqrt{x}}$$

Exercises 22−3

In Exercises 1 through 24 find the derivative of each of the functions by using the Δ-process.

1. $y = 3x - 1$ 2. $y = 6x + 3$ 3. $y = 1 - 2x$

4. $y = 2 - 5x$ 5. $y = x^2 - 1$ 6. $y = 4 - x^2$

7. $y = 5x^2$ 8. $y = -6x^2$ 9. $y = x^2 - 7x$

10. $y = x^2 + 4x$ 11. $y = 8x - 2x^2$ 12. $y = 3x - 5x^2$

13. $y = x^3 + 4x - 6$ 14. $y = 2x - 4x^3$ 15. $y = \dfrac{1}{x + 2}$

16. $y = \dfrac{1}{x + 1}$ 17. $y = x + \dfrac{1}{x}$ 18. $y = \dfrac{x}{x - 1}$

19. $y = \dfrac{2}{x^2}$ 20. $y = \dfrac{1}{x^2 + 1}$ 21. $y = x^4 + x^3 + x^2 + x$

22. $y = \frac{1}{3}x^3 + \frac{1}{2}x^2 + x$ 23. $y = x^4 - \dfrac{2}{x}$ 24. $y = \dfrac{1}{x} + \dfrac{1}{x^2}$

In Exercises 25 through 28 find dy/dx for the given functions by the method of Example E.

25. $y = \sqrt{x + 1}$ 26. $y = \sqrt{x - 2}$ 27. $y = \sqrt{1 - 3x}$ 28. $y = \sqrt{x^2 + 3}$

ex E p591

22−4 The Meaning of the Derivative

In Section 22−2, we saw that the slope of a curve was the limiting value of the slope of the line through points P and Q as Q approaches P. At the end of the section, we discussed the idea that $\Delta y/\Delta x$ indicated the rate of change of y with respect to x. In Section 22−3, we defined the limit of the ratio of $\Delta y/\Delta x$ as $\Delta x \to 0$ as the derivative. Thus, the derivative is a measure of the rate of change of y with respect to x at point P. However, P may represent any point, which means that the value of the derivative changes from one point on a curve to another point. **We therefore interpret the derivative as the *instantaneous rate of change of y with respect to x*.**

Example A
In Examples A and B of Section 22−2, y is changing at the rate of 7 units for every change of 1 unit in x, *when x is 2, and only when x is 2*. In Example C of Section 22−2, y is increasing 6 units for every increase of 1 unit of x *when* $x = -1$. When $x = 3$, y is decreasing 2 units for 1 unit of increase in x.

If a functional relationship exists between any two variables, then one can be thought of as varying with respect to the other. There are many applications of this principle, one of them being the velocity of an object. We shall consider here the case of rectilinear motion, that is, motion along a straight line.

As we have seen, the velocity of an object is found by dividing the change in displacement by the time required for this change. This, however, gives a value only for the **average velocity** for the specified time interval. If the time interval considered becomes smaller and smaller, then the average velocity which is calculated more nearly approximates the **instantaneous velocity** at some particular time. In the limit, the value of the average velocity gives the value of the instantaneous velocity. Using the symbols as defined for the derivative, the instantaneous velocity of an object moving in rectilinear motion at a particular time t is given by

$$v = \lim_{\Delta t \to 0} \frac{\Delta s}{\Delta t} \qquad\qquad (22\text{–}7)$$

where s is the displacement.

Example B

Find the instantaneous velocity when $t = 4$ s of an object for which the displacement s (in feet) is given by $s = 16t^2$ by calculating values of $\Delta s/\Delta t$ and determining the limit as Δt approaches zero.

   Here we shall let t take on values of 3.5, 3.9, 3.99, and 3.999 s. When $t = 4$ s, $s = 256$ ft. Therefore, we calculate Δt by subtracting values of t from 4, and Δs is calculated by subtracting values from 256. The values of velocity are then calculated by using $v = \Delta s/\Delta t$.

t (seconds)	3.5	3.9	3.99	3.999
s (feet)	196.0	243.36	254.7216	255.872016
Δs (feet)	60.0	12.64	1.2784	0.127984
Δt (seconds)	0.5	0.1	0.01	0.001
v (feet per second)	120.0	126.4	127.84	127.984

We can see that the limit is approaching 128 ft/s, which is therefore the instantaneous velocity when $t = 4$ s.

Example C

By use of the Δ-process, determine an expression for the instantaneous velocity of the object of Example B, for which the displacement s (in feet) is given by $s = 16t^2$ and t represents the time (in seconds). By use of the derived expression, determine the instantaneous velocity when $t = 2$ s and when $t = 4$ s.

   Applying the Δ-process to this function, we find the derivative of s with respect to t:

$$s + \Delta s = 16(t + \Delta t)^2 = 16t^2 + 32t\,\Delta t + 16(\Delta t)^2$$
$$\Delta s = 32t\,\Delta t + 16(\Delta t)^2$$

$$\frac{\Delta s}{\Delta t} = 32t + 16\,\Delta t$$

$$v = \lim_{\Delta t \to 0} \frac{\Delta s}{\Delta t} = \frac{ds}{dt} = 32t$$

If $t = 2$ s, the velocity is 64 ft/s. When $t = 4$ s, the velocity is 128 ft/s. We see that this second result agrees with that obtained in Example B.

By finding $\lim_{\Delta x \to 0} \Delta y/\Delta x$, we can find the instantaneous rate of change of y with respect to x. The expression $\lim_{\Delta t \to 0} \Delta s/\Delta t$ gives the velocity, or instantaneous rate of change of displacement with respect to time. Generalizing, we can say that *the derivative can be interpreted as giving the instantaneous rate of change of the dependent variable with respect to the independent variable*. This is true no matter what the variables represent, so long as a function is defined between the variables.

Example D
Find an expression for the rate of change of the volume of a sphere with respect to its radius.

$$V = \frac{4}{3}\pi r^3$$

$$V + \Delta V = \frac{4\pi}{3}(r + \Delta r)^3$$

$$= \frac{4\pi}{3}[r^3 + 3r^2\,\Delta r + 3r\,(\Delta r)^2 + (\Delta r)^3]$$

$$\Delta V = \frac{4\pi}{3}[3r^2\,\Delta r + 3r\,(\Delta r)^2 + (\Delta r)^3]$$

$$\frac{\Delta V}{\Delta r} = \frac{4\pi}{3}[3r^2 + 3r\,\Delta r + (\Delta r)^2]$$

$$\lim_{\Delta r \to 0}\frac{\Delta V}{\Delta r} = 4\pi r^2$$

We can see that, as the radius increases, the rate of change of the volume with respect to the radius also increases. This should be expected, since a sphere changing from a radius of 1 to a radius of 2 has an increase in volume from $4\pi/3$ to $32\pi/3$, whereas a change in the radius from 2 to 3 involves a volume increase from $32\pi/3$ to $108\pi/3$.

Example E
The power P produced by an electric current i in a resistor varies directly as the square of the current. Given that 1.2 W of power are produced by a current of 0.5 A in a particular resistor, find an expression for the rate of change of power with respect to current.

We must first find the functional relationship between power and current. This we can do by solving the indicated problem in variation:

$$P = ki^2, \qquad 1.2 = k(0.5)^2, \qquad k = 4.8 \text{ W/A}^2, \qquad P = 4.8i^2$$

Now, knowing the function, we may determine the expression for the rate of change of P with respect to i by use of the Δ-process:

$$P + \Delta P = 4.8(i + \Delta i)^2 = 4.8[i^2 + 2i(\Delta i) + (\Delta i)^2]$$

$$\Delta P = 4.8[2i(\Delta i) + (\Delta i)^2]$$

$$\frac{\Delta P}{\Delta i} = 4.8(2i + \Delta i)$$

$$\lim_{\Delta i \to 0} \frac{\Delta P}{\Delta i} = 9.6i$$

This last expression tells us that as the current increases, the power changes at a rate of $9.6i$ W/A. For example, if the current is 2 A, the power is changing at the rate of 19.2 W/A. The larger the current, the greater the increase in power. This should be expected, since the power varies as the square of the current.

Exercises 22—4

In Exercises 1 through 4 calculate the instantaneous velocity for the indicated value of the time (in seconds) of an object for which the displacement (in feet) is given by the indicated function. Use the method of Example A, and calculate values of $\Delta s / \Delta t$ for the given values of t, and determine the limit as Δt approaches zero.

1. $s = 4t + 10$; when $t = 3$; use values of t of 2.0, 2.5, 2.9, 2.99, 2.999
2. $s = 20 - 8t$; when $t = 2$; use values of t of 1.0, 1.5, 1.9, 1.99, 1.999
3. $s = 1 - 2t^2$; when $t = 4$; use values of t of 3.0, 3.5, 3.9, 3.99, 3.999
4. $s = 120t - 16t^2$; when $t = 0.5$; use values of t of 0.4, 0.45, 0.49, 0.499, 0.4999

In Exercises 5 through 8 use the Δ-process to find an expression for the instantaneous velocity of an object moving with rectilinear motion according to the given functions (the same as those of Exercises 1 through 4) relating s (in feet) and t (in seconds). Then calculate the instantaneous velocity for the given value of t.

5. $s = 4t + 10$; $t = 3$
6. $s = 20 - 8t$; $t = 2$
7. $s = 1 - 2t^2$; $t = 4$
8. $s = 120t - 16t^2$; $t = 0.5$

In Exercises 9 through 12 use the Δ-process to find an expression for the instantaneous velocity of an object moving with rectilinear motion according to the given functions relating s and t.

9. $s = 3t - \dfrac{2}{t}$
10. $s = \dfrac{1}{2t + 3}$
11. $s = t^3 - 6t + 2$
12. $s = s_0 + v_0 t - \frac{1}{2}at^2$ (s_0, v_0, and a are constants)

In Exercises 13 through 16 use the Δ-process to find an expression for the instantaneous acceleration a of an object moving with rectilinear motion according to the given functions. The instantaneous acceleration of an object is defined as the instantaneous rate of change of velocity with respect to time.

13. $v = 6t^2 - 4t + 2$ **14.** $v = v_0 - gt$ (v_0 and g are constants)

15. $s = 1 - 2t^2$ (find v, then find a)

16. $s = s_0 + v_0 t - \frac{1}{2}at^2$ (s_0, v_0, and a are constants) (find v, then find a)

In Exercises 17 through 28 find the indicated rates of change.

17. The perimeter p of a rectangular field is a function of its width w according to $p = 2(50 + w)$, where p and w are measured in meters. Find the expression for the instantaneous rate of change of p with respect to w.

18. The profit P in selling 100 items at price p, when the cost of each item is $10 is given by $P = 100(p - 10)$. Find the expression for the instantaneous rate of change of P with respect to p.

19. Find a general expression for the instantaneous rate of change of the area of a circle with respect to its radius. Then find the instantaneous rate of change of A with respect to r, when $r = 3$.

20. The centripetal acceleration a of an object moving in a circular path of radius r with a velocity v is given by $a = v^2/r$. Find the expression for the instantaneous rate of change of the acceleration with respect to the velocity for an object moving in a circular path of radius 2 ft. The acceleration is in feet per square second and the velocity is in feet per second.

21. The resistance of a certain resistor as a function of the temperature is given by the relation $R = 50(1 + 0.0053T + 0.00001T^2)$. Determine the instantaneous rate of change of resistance with respect to temperature (in ohms per degree Celsius) when the temperature is 20°C.

22. At very low temperatures, near absolute zero, the specific heat c of solids is given by the Debye equation $c = kT^3$, where T is the thermodynamic temperature and k is a constant dependent on the material. Determine the expression for the instantaneous rate of change for the specific heat with respect to temperature. The units of the specific heat are joules per kilogram kelvin and T is measured in kelvins.

23. A 5-Ω resistor and a variable resistor of resistance R are placed in parallel. The expression for the resulting resistance R_T is given by $R_T = 5R/(5 + R)$. Determine the instantaneous rate of change of R_T with respect to R, and then evaluate this expression for $R = 8\ \Omega$.

24. The value (in dollars) of a certain car is given by the function $V = \frac{6000}{t+1}$, where t is measured in years. Find a general expression for the rate of change of V with respect to t, and then evaluate this expression when $t = 3$ years.

25. The volume of a certain right circular cylinder is 100 in.^{3}. Find the expression for the instantaneous rate of change of the total surface area of the cylinder with respect to the radius of the base.

● 26. The total energy supplied to an inductor while the current increases from 0 to I is proportional to the square of the current I. If 8 J are supplied to a certain inductor as the current increases from 0 to 2 A, evaluate the expression for the instantaneous rate of change of energy with respect to I when $I = 1.5$ A.

27. The illuminance of a light source varies inversely as the square of the distance from the source. Given that the illuminance is 25 lx(lux) at a distance of 0.20 m, find the expression for the instantaneous rate of change of illuminance with respect to distance. Calculate this rate for distances of 0.01 m and 0.10 m.

28. The frequency of vibration of a wire varies directly as the square root of the tension on the wire. Find the expression for the instantaneous rate of change of frequency with respect to tension. Find the value of this rate of change for $T = 1$ N if the frequency is 400 Hz when $T = 1$ N.

22—5 Derivatives of Polynomials

The task of finding the derivative of a function can be considerably shortened from that involved in the direct use of the Δ-process. We can use the Δ-process to derive certain basic formulas for finding derivatives of particular types of functions. These formulas will then be used to find the derivatives. In this section, we shall derive the formulas for finding the derivatives of polynomial functions of the form $f(x) = a_0 x^n + a_1 x^{n-1} + \cdots + a_n$.

First we shall find the derivative of a constant. By letting $y = c$, and applying the Δ-process to this function, we can obtain the desired result:

$$y = c, \qquad y + \Delta y = c, \qquad \Delta y = 0, \qquad \frac{\Delta y}{\Delta x} = 0, \qquad \lim_{\Delta x \to 0} \frac{\Delta y}{\Delta x} = 0$$

From this we conclude that *the derivative of a constant is zero*. No assumption was made as to the value of the constant, and thus this result holds for all constants. Therefore, if $y = c$, $dy/dx = 0$. This is more commonly written as

$$\frac{dc}{dx} = 0 \qquad D_x\,C = 0 \tag{22—8}$$

Graphically this means that for any function of the type $y = c$ the slope is always zero. From our knowledge of straight lines, we know that $y = c$ represents a straight line parallel to the x-axis. From the definition of slope, we know that any line parallel to the x-axis has a slope of zero. We can see that the two results are consistent.

Next we shall find the derivative of any integral power of x. If $y = x^n$, where n is an integer, by use of the binomial theorem we have the following:

$$(y + \Delta y) = (x + \Delta x)^n$$

$$= x^n + nx^{n-1}\, \Delta x + \frac{n(n-1)}{2} x^{n-2}\, (\Delta x)^2 + \cdots + (\Delta x)^n$$

$$\Delta y = nx^{n-1}\, \Delta x + \frac{n(n-1)}{2} x^{n-2}\, (\Delta x)^2 + \cdots + (\Delta x)^n$$

$$\frac{\Delta y}{\Delta x} = nx^{n-1} + \frac{n(n-1)}{2} x^{n-2}\, \Delta x + \cdots + (\Delta x)^{n-1}$$

$$\lim_{\Delta x \to 0} \frac{\Delta y}{\Delta x} = nx^{n-1}$$

Thus, *the derivative of the nth power of x is found to be*

$$\frac{dx^n}{dx} = nx^{n-1} \qquad D_x\, X^n = n X^{n-1} \tag{22-9}$$

Example A

Find the derivative of the function $y = -5$.

Applying the result of Eq. (22–8), since -5 is a constant, we have

$$\frac{dy}{dx} = \frac{d(-5)}{dx} = 0$$

or

$$\frac{dy}{dx} = 0$$

Example B

Find the derivative of the function $y = x$.

Applying the result of Eq. (22–9) (here $n = 1$), we have

$$\frac{dy}{dx} = \frac{d(x)}{dx} = (1)x^{1-1} = 1$$

or simply

$$\frac{dy}{dx} = 1$$

Thus, the derivative of $y = x$ is 1, which means that the slope of the line $y = x$ is always 1. This is consistent with our previous discussion of the straight line.

Example C

Find the derivative of $y = x^3$.

We find that

$$\frac{dy}{dx} = \frac{d(x^3)}{dx}$$

$$= 3x^{3-1}$$

$$= 3x^2$$

We can see that this result is consistent with the results found in the examples and exercises of the previous sections in this chapter.

Example D

Find the derivative of $y = x^{10}$.

We find that

$$\frac{dy}{dx} = \frac{d(x^{10})}{dx}$$

$$= 10x^{10-1}$$

$$= 10x^9$$

Next we shall find the derivative of a constant times a function of x. If $y = cu$, where $u = f(x)$, we have the following result:

$$y + \Delta y = c(u + \Delta u)$$

(As x increases by Δx, u increases by Δu, since u is a function of x.) Then

$$\Delta y = c\,\Delta u, \qquad \frac{\Delta y}{\Delta x} = c\frac{\Delta u}{\Delta x}, \qquad \lim_{\Delta x \to 0}\frac{\Delta y}{\Delta x} = c\lim_{\Delta x \to 0}\frac{\Delta u}{\Delta x}$$

Therefore, *the derivative of the product of a constant and a function of x is the product of the constant and the derivative of the function of x.* This is written as

$$\frac{d(cu)}{dx} = c\frac{du}{dx} \qquad D_x\,cu = c\,D_x\,u \tag{22–10}$$

Example E

Find the derivative of $y = 3x^2$.

In this case $c = 3$ and $u = x^2$. Thus, $du/dx = 2x$. Therefore,

$$\frac{dy}{dx} = 3(2x) = 6x$$

Occasionally the derivative of a constant times a function of x is confused with the derivative of a constant (alone). The distinction between a constant multiplying a function and an isolated constant must be made.

Finally, if the types of functions for which we have found derivatives are added, the result is a polynomial function with more than one term. The derivative of such a function is found by letting $y = u + v$, where u and v are functions of x. Applying the Δ-process, since u and v are functions of x, each has an increment corresponding to an increment in x. Thus, we have the following result:

$$y + \Delta y = (u + \Delta u) + (v + \Delta v)$$

$$\Delta y = \Delta u + \Delta v$$

$$\frac{\Delta y}{\Delta x} = \frac{\Delta u}{\Delta x} + \frac{\Delta v}{\Delta x}$$

$$\lim_{\Delta x \to 0} \frac{\Delta y}{\Delta x} = \lim_{\Delta x \to 0} \frac{\Delta u}{\Delta x} + \lim_{\Delta x \to 0} \frac{\Delta v}{\Delta x}$$

This tells us that the derivative of *the sum of functions of x is the sum of the derivatives of the functions*. This is written as

$$\frac{d(u + v)}{dx} = \frac{du}{dx} + \frac{dv}{dx} \qquad D_x(u+v) = D_x U + D_x V \qquad (22\text{--}11)$$

Example F

Find the derivative of $y = 4x^2 + 5$.

Here $u = 4x^2$ and $v = 5$. Thus, $du/dx = 8x$ and $dv/dx = 0$. Hence we have

$$\frac{dy}{dx} = \frac{d(4x^2)}{dx} + \frac{d(5)}{dx}$$

$$= 8x + 0$$

$$= 8x$$

Example G

Find the derivative of $y = 5x^6 - 2x^3 - x + 7$.

$$\frac{dy}{dx} = \frac{d(5x^6)}{dx} - \frac{d(2x^3)}{dx} - \frac{dx}{dx} + \frac{d(7)}{dx}$$

$$= 30x^5 - 6x^2 - 1$$

Example H

Find the slope of a line which is tangent to the curve of the function $y = 4x^7 - x^4$ at the point $(1, 3)$.

We must find, and then evaluate, the derivative of the function for the value $x = 1$:

$$\frac{dy}{dx} = 28x^6 - 4x^3$$

For evaluating this, we write

$$\left.\frac{dy}{dx}\right|_{x=1} = 28(1) - 4(1) = 24$$

Thus, the slope of the tangent line is 24. Again, we note that the substitution $x = 1$ must be made after the differentiation has been performed.

Example 1

Find the instantaneous velocity when $t = 3$ s for an object moving with rectilinear motion according to the function $s = 8t - 2t^3$, where s is measured in centimeters.

In this example we must find the derivative of s with respect to t, ds/dt, and then evaluate it for $t = 3$. Thus,

$$s = 8t - 2t^3$$

$$\frac{ds}{dt} = 8 - 6t^2$$

$$\left.\frac{ds}{dt}\right|_{t=3} = 8 - 6(3^2) = 8 - 54 = -46 \text{ cm/s}$$

Therefore, the object is moving at -46 cm/s (it is moving in a negative direction) when $t = 3$ s.

Exercises 22-5

In Exercises 1 through 16 find the derivative of each of the given functions.

1. $y = x^5$ 2. $y = x^{12}$ 3. $y = 4x^9$ 4. $y = -7x^6$

5. $y = x^4 - 6$ 6. $y = 3x^5 - 1$ 7. $y = x^2 + 2x$ 8. $y = x^3 - 2x^2$

9. $y = 5x^3 - x - 1$ 10. $y = 6x^2 - 6x + 5$

11. $y = x^8 - 4x^7 - x$ 12. $y = 4x^4 - 2x + 9$

13. $y = 6x^7 - 5x^3 + 2$ 14. $y = 13x^4 - 6x^3 - x - 1$

15. $y = \frac{1}{3}x^3 + \frac{1}{2}x^2$ 16. $y = \frac{1}{4}x^8 - \frac{1}{2}x^4 - \sqrt{3}$

In Exercises 17 through 20 find the slope of a tangent line to the curve of each of the given functions for the given values of x.

17. $y = 2x^6 - 6x^2$ $(x = 2)$ 18. $y = 3x^3 - 9x$ $(x = 1)$

19. $y = 6x - 2x^4$ $(x = -1)$ 20. $y = x^4 - \frac{1}{2}x^2 + 2$ $(x = -2)$

In Exercises 21 through 24 find an expression for the instantaneous velocity of objects moving according to the functions given, if s represents displacement in terms of time t.

21. $s = 6t^5 - 5t + 2$.   22. $s = 4t^2 - 7t$

23. $s = 120t - 16t^2$ 24. $s = s_0 + v_0 t + \frac{1}{2}at^2$

In Exercises 25 through 32 solve the given problems by finding the appropriate derivative.

25. For what value(s) of x is the tangent to the curve of $y = 3x^2 - 6x$ parallel to the x-axis? (That is, where is the slope zero?)

26. For what value(s) of x is the tangent to the curve of $y = 4x^3 - 12x^2$ parallel to the x-axis? (See Exercise 25.)

27. If the radius of a right circular cylinder always equals its height, find an expression for the instantaneous rate of change of volume with respect to the radius.

28. If the length of a pillar is constant, the crushing load varies as the fourth power of its radius. If 18 tons will crush a certain pillar 5 in. in radius, find an expression which gives the instantaneous rate of change of crushing load with respect to a change in radius.

29. The voltage V produced by a certain thermocouple as a function of the temperature T is given by $V = 4.0T + 0.005T^2$. If the temperature changes slowly, find the instantaneous rate of change of voltage with respect to temperature when the temperature is 200°C.

30. The cost C (in dollars) to a firm producing x units is given by the equation $C = 0.02x^3 - 2.40x^2 + 100x$. Determine the instantaneous rate of change of C with respect to x for $x = 50$ units.

31. The altitude h (in meters) of an airplane as a function of the horizontal distance x (in kilometers) it has traveled is given by $h = 30x^2 - x^3$. Find the instantaneous rate of change of h with respect to x for $x = 10$ km.

32. The ends of a 10-ft beam are supported at different levels. The deflection y of the beam is given by $y = kx^2(x^3 + 450x - 3500)$, where x is the horizontal distance from one end and k is a constant. Determine the expression for the instantaneous rate of change of deflection with respect to x.

22—6 Derivatives of Products and Quotients of Functions

The formulas developed in the previous section are valid for polynomial functions. However, many functions are not polynomial in form. Some functions can best be expressed as the product of two or more simpler functions, others are the quotient of two simpler functions, and some are expressed as powers of a function. In this section, we shall develop the formula for the derivative of a product of functions and the formula for the derivative of the quotient of two functions.

Example A

The functions $f(x) = x^2 + 2$ and $g(x) = 3 - 2x$ can be combined to form new functions which represent the types mentioned above. For example, the function $p(x) = f(x)g(x) = (x^2 + 2)(3 - 2x)$ is an example of a function expressed as the product of two simpler functions. The function

$$q(x) = \frac{g(x)}{f(x)} = \frac{3 - 2x}{x^2 + 2}$$

is an example of a rational function which is the quotient of two other functions. The function $F(x) = (g(x))^3 = (3 - 2x)^3$ is an example of a power of a function.

If u and v both represent functions of x, the derivative of the product of u and v is found by applying the Δ-process. This leads to the following result:

$$y = u \cdot v$$
$$y + \Delta y = (u + \Delta u)(v + \Delta v)$$

(Since u and v are functions of x, each has an increment corresponding to an increment in x.) Then we have

$$\Delta y = u \, \Delta v + v \, \Delta u + \Delta u \, \Delta v$$

$$\frac{\Delta y}{\Delta x} = u \frac{\Delta v}{\Delta x} + v \frac{\Delta u}{\Delta x} + \Delta u \frac{\Delta v}{\Delta x}$$

$$\lim_{\Delta x \to 0} \frac{\Delta y}{\Delta x} = u \lim_{\Delta x \to 0} \frac{\Delta v}{\Delta x} + v \lim_{\Delta x \to 0} \frac{\Delta u}{\Delta x} + \lim_{\Delta x \to 0}\left(\Delta u \frac{\Delta v}{\Delta x}\right)$$

(The functions u and v are not affected by Δx approaching zero, but Δu and Δv both approach zero as Δx approaches 0.) Thus,

$$\frac{dy}{dx} = u \frac{dv}{dx} + v \frac{du}{dx} + 0 \frac{dv}{dx}$$

We conclude that *the derivative of the product of two functions equals the first function times the derivative of the second function plus the second function times the derivative of the first function.* This is written as

$$\frac{d(uv)}{dx} = u \frac{dv}{dx} + v \frac{du}{dx} \qquad (22\text{–}12)$$

$$D_x(uv) = u \, D_x v + v \, D_x u$$

Example B

For the product function in Example A, the derivative is found as follows:

$$y = (x^2 + 2)(3 - 2x), \qquad u = x^2 + 2, \qquad v = 3 - 2x$$

$$\frac{dy}{dx} = (x^2 + 2)(-2) + (3 - 2x)(2x) = -2x^2 - 4 + 6x - 4x^2$$

$$= -6x^2 + 6x - 4$$

Example C

Find the derivative of the function $y = (3 - x - 2x^2)(x^4 - x)$.

In this problem, $u = 3 - x - 2x^2$ and $v = x^4 - x$. Hence

$$\frac{dy}{dx} = (3 - x - 2x^2)(4x^3 - 1) + (x^4 - x)(-1 - 4x)$$

$$= 12x^3 - 3 - 4x^4 + x - 8x^5 + 2x^2 - x^4 - 4x^5 + x + 4x^2$$

$$= -12x^5 - 5x^4 + 12x^3 + 6x^2 + 2x - 3$$

In both these examples, we could have multiplied the functions first and then taken the derivative as a polynomial. However, we shall soon meet functions for which this latter method would not be applicable.

We shall now find the derivative of the quotient of two functions. We apply the Δ-process to the function $y = u/v$, and obtain the result as follows:

$$y + \Delta y = \frac{u + \Delta u}{v + \Delta v}$$

$$\Delta y = \frac{u + \Delta u}{v + \Delta v} - \frac{u}{v} = \frac{vu + v\,\Delta u - uv - u\,\Delta v}{v(v + \Delta v)}$$

$$\frac{\Delta y}{\Delta x} = \frac{v(\Delta u/\Delta x) - u(\Delta v/\Delta x)}{v(v + \Delta v)}$$

$$\lim_{\Delta x \to 0} \frac{\Delta y}{\Delta x} = \frac{v \lim_{\Delta x \to 0} (\Delta u/\Delta x) - u \lim_{\Delta x \to 0} (\Delta v/\Delta x)}{\lim_{\Delta x \to 0} v(v + \Delta v)}$$

Thus,

$$\frac{dy}{dx} = \frac{v\,(du/dx) - u\,(dv/dx)}{v^2}$$

This last result says that *the derivative of the quotient of two functions equals the denominator times the derivative of the numerator minus the numerator times the derivative of the denominator, all divided by the square of the denominator.* This is written as

$$\frac{d\dfrac{u}{v}}{dx} = \frac{v\dfrac{du}{dx} - u\dfrac{dv}{dx}}{v^2} \qquad\qquad (22\text{–}13)$$

$$D_x\left(\frac{u}{v}\right) = \frac{V\,D_x\,u - u\,D_x\,V}{V^2}$$

$$D_x\left(\frac{1}{v}\right) = -\frac{D_x\,V}{V^2}$$

Example D

Find the derivative of the quotient indicated in Example A.

$$y = \frac{3 - 2x}{x^2 + 2}, \qquad u = 3 - 2x, \qquad v = x^2 + 2$$

$$\frac{dy}{dx} = \frac{(x^2 + 2)(-2) - (3 - 2x)(2x)}{(x^2 + 2)^2} = \frac{-2x^2 - 4 - 6x + 4x^2}{(x^2 + 2)^2}$$

$$= \frac{2(x^2 - 3x - 2)}{(x^2 + 2)^2}$$

Example E

Find the derivative of

$$y = \frac{x^3 - x - 1}{1 - 2x^2}$$

$$\frac{dy}{dx} = \frac{(1 - 2x^2)(3x^2 - 1) - (x^3 - x - 1)(-4x)}{(1 - 2x^2)^2}$$

$$= \frac{3x^2 - 1 - 6x^4 + 2x^2 + 4x^4 - 4x^2 - 4x}{(1 - 2x^2)^2}$$

$$= \frac{-2x^4 + x^2 - 4x - 1}{(1 - 2x^2)^2}$$

Exercises 22—6

In Exercises 1 through 8 find the derivative of each function by Eq. (22–12). Then check by multiplying out before finding dy/dx, treating the function as a polynomial.

1. $y = x^2(3x + 2)$ 2. $y = 3x(x^3 + 1)$

3. $y = 6x(3x^2 - 5x)$ 4. $y = 2x^3(3x^4 + x)$

5. $y = (x + 2)(2x - 5)$ 6. $y = (3x - 1)(x^2 + 1)$

7. $y = (x^4 - 3x^2 + 3)(1 - 2x^3)$ 8. $y = (x^3 - 6x)(2 - 4x^3)$

In Exercises 9 through 16 find the derivative of each function by Eq. (22–13).

9. $y = \dfrac{x}{2x + 3}$ 10. $y = \dfrac{2x}{x + 1}$ 11. $y = \dfrac{1}{x^2 + 1}$ 12. $y = \dfrac{x + 2}{2x + 3}$

13. $y = \dfrac{x^2}{3 - 2x}$ 14. $y = \dfrac{x + 8}{x^2 + x + 2}$ 15. $y = \dfrac{3x}{4x^5 - 3x - 4}$ 16. $y = \dfrac{2}{3x^2 - 5x}$

In Exercises 17 through 28 solve the given problems by finding the appropriate derivatives.

17. Find the derivative of $y = \dfrac{x^2(1 - 2x)}{3x - 7}$, but do not multiply out the numerator before taking the derivative.

18. Find the derivative of $y = 4x^2 - \dfrac{1}{x - 1}$, but do not combine the terms before finding the derivative.

19. Find the slope of a tangent line to the curve $y = (4x + 1)(x^4 - 1)$ at the point $(-1, 0)$. Do not multiply the factors together before taking the derivative.

20. Find the slope of a line tangent to the curve of the function $y = (3x + 4)(1 - 4x)$ at the point $(2, -70)$. Do not multiply before finding the derivative.

21. For what value(s) of x is the slope of a tangent to the curve $y = \dfrac{x}{x^2 + 1}$ equal to zero?

22. Determine the sign of the derivative of the function $y = \dfrac{2x - 1}{1 - x^2}$ for the following values of x: $-2, -1, 0, 1, 2$. Is the slope of a tangent to this curve ever negative?

23. If the distance (in feet) traveled by a particle in a given time is found from the expression $s = 2/t^2$, find the velocity after 4 s. (Use the quotient rule.)

24. During a chemical change the number n of grams of a compound being formed is given by $n = \dfrac{6t}{t + 2}$, where t is measured in seconds. How many grams per second are being formed after 3 s?

25. The voltage across a resistor in an electric circuit is the product of the resistance and the current. If the current I (in amperes) varies with time (in seconds) according to the relation $I = 5.00 + 0.01t^2$, and the resistance varies with time according to the relation $R = 15.00 - 0.10t$, find the time rate of change of the voltage when $t = 5$ s.

26. In thermodynamics, an equation relating pressure p, volume V, and temperature T is

$$\left(p + \frac{a}{V^2}\right)(V - b) = RT$$

where a, b, and R are constants. This equation is known as the van der Waals equation. Solve this equation for p, and then find the expression for the rate of change of p with respect to V, assuming a constant temperature.

27. The electric power produced by a certain source is given by

$$P = \frac{E^2 r}{R^2 + 2Rr + r^2}$$

where E is the voltage of the source, R is the resistance of the source, and r is the resistance in the circuit. Find the expression for the rate of change of power with respect to the resistance in the circuit, assuming that the other quantities remain constant.

28. In the study of simple harmonic motion, under certain conditions the tangent of the angle by which the displacement lags the impress force is given by

$$\tan \varphi = \frac{30\omega}{600 - \omega^2}$$

where ω, (in hertz) is the frequency of vibration. Find the rate of change of $\tan \varphi$ with respect to ω when $\omega = 3$ Hz.

22–7 The Derivative of a Power of a Function

In Example A of Section 22–6 we illustrated $y = (3 - 2x)^3$ as the power of a function of x. Here, $3 - 2x$ is the function of x, and it is being raised to the third power. If we let $u = 3 - 2x$, we can write

$$y = u^3 \quad \text{where } u = 3 - 2x$$

Writing it this way, y is a function of u and u is a function of x. This means that y is a function of a function of x, often referred to as a **composite function**. However, y is still a function of x, since u is a function of x.

Since we will frequently need to find the derivative of the power of a function, we now develop the necessary formula. Considering $y = f(u)$ and $u = g(x)$, we can express the derivative dy/dx in terms of dy/du and du/dx. If Δx is the increment in x, then Δy and Δu are the corresponding increments in y and u, respectively. We may then write

$$\frac{\Delta y}{\Delta x} = \frac{\Delta y}{\Delta u} \cdot \frac{\Delta u}{\Delta x}$$

When Δx approaches zero, Δu and Δy both approach zero, for u and y are functions of x. Thus,

$$\lim_{\Delta x \to 0} \frac{\Delta y}{\Delta x} = \lim_{\Delta u \to 0} \frac{\Delta y}{\Delta u} \cdot \lim_{\Delta x \to 0} \frac{\Delta u}{\Delta x}$$

or

$$\frac{dy}{dx} = \frac{dy}{du} \cdot \frac{du}{dx} \tag{22–14}$$

(Here we have assumed $\Delta u \neq 0$, although it can be shown that this condition is not necessary). *Equation (22–14) is known as the* **chain rule** *for derivatives.*

Using Eq. (22–14) for $y = u^n$, where u is a function of x, we have

$$\frac{dy}{dx} = \frac{d(u^n)}{du} \cdot \frac{du}{dx}$$

or

$$D_x u^N = N u^{N-1} D_x u$$

$$\frac{du^n}{dx} = nu^{n-1}\left(\frac{du}{dx}\right) \tag{22–15}$$

By use of Eq. (22–15) we may find the derivative of a power of a function of x.

Example A

Find the derivative of $y = (3 - 2x)^3$.

For this function, we identify $n = 3$ and $u = 3 - 2x$. This means that $du/dx = -2$, and therefore

$$\frac{dy}{dx} = 3(3 - 2x)^2(-2)$$

$$= -6(3 - 2x)^2$$

A common type of error in finding this type of derivative is to omit the du/dx factor; in this case it is the -2. The derivative is incomplete, and therefore incorrect without this factor.

Example B

Find the derivative of $y = (1 - 3x^2)^4$.

In this example, $n = 4$ and $u = 1 - 3x^2$. Hence

$$\frac{dy}{dx} = 4(1 - 3x^2)^3(-6x) = -24x(1 - 3x^2)^3$$

(We must not forget the $-6x$.)

Example C

Find the derivative of

$$y = \frac{(3x - 1)^3}{1 - x}$$

Here we must use the quotient rule in combination with the power rule. We find the derivative of the numerator by using the power rule.

$$\frac{dy}{dx} = \frac{(1 - x)3(3x - 1)^2(3) - (3x - 1)^3(-1)}{(1 - x)^2}$$

$$= \frac{(3x - 1)^2(9 - 9x + 3x - 1)}{(1 - x)^2} = \frac{(3x - 1)^2(8 - 6x)}{(1 - x)^2}$$

$$= \frac{2(3x - 1)^2(4 - 3x)}{(1 - x)^2}$$

It is normally better to have the derivative in a factored, simplified form, since this is the only form from which useful information may readily be obtained. In this form we can determine where the derivative is undefined (denominator equal to zero) or where the slope is zero (numerator equal to zero); we can also make other required analyses of the derivative. Thus, all derivatives should be simplified.

So far, we have derived the formulas for the derivatives of powers of x and for functions of x for integral powers. We shall now establish that these formulas are also valid for any rational number used as an exponent. If $y = u^{p/q}$, and if each side is raised to the qth power, we have $y^q = u^p$. Applying the power rule to each side of this equation, we have

$$qy^{q-1}\left(\frac{dy}{dx}\right) = pu^{p-1}\left(\frac{du}{dx}\right)$$

Solving for dy/dx, we have

$$\frac{dy}{dx} = \frac{pu^{p-1}(du/dx)}{qy^{q-1}} = \frac{p}{q}\frac{u^{p-1}}{(u^{p/q})^{q-1}}\frac{du}{dx} = \frac{p}{q}\frac{u^{p-1}}{u^{p-p/q}}\frac{du}{dx}$$

$$= \frac{p}{q}u^{p-1-p+(p/q)}\frac{du}{dx}$$

Thus,

$$\frac{du^{p/q}}{dx} = \frac{p}{q}u^{(p/q)-1}\frac{du}{dx} \tag{22–16}$$

We see that in finding the derivative we multiply the function by the rational exponent and subtract 1 from it to find the exponent of the function in the derivative. This, of course, is the same rule derived for integral exponents. Since the only restriction on the values of p and q is that they are integers ($q \neq 0$), *the power rule can be used for all rational exponents, positive and negative.*

Example D
We can now find the derivative of $y = \sqrt{x^2 + 1}$.
   By use of Eq. (22–16), or Eq. (22–15), and writing the square root as the fractional exponent $\frac{1}{2}$, we can easily derive the result:

$$y = (x^2 + 1)^{1/2}$$

$$\frac{dy}{dx} = \frac{1}{2}(x^2 + 1)^{-1/2}(2x) = \frac{x}{(x^2 + 1)^{1/2}}$$

To avoid introducing apparently significant factors into the numerator, we do not usually rationalize such fractions.

   Having shown that we may use fractional exponents to find derivatives of roots of functions of x, we may also use them to find derivatives of roots of x itself. Consider the following example.

Example E
Find the derivative of $y = 6\sqrt[3]{x}$.
   We can write this function as $y = 6x^{1/3}$. In finding the derivative we may use Eq. (22–15), where we have $u = x$. This means that $du/dx = dx/dx = 1$. Therefore, using Eq. (22–15) is equivalent to using Eq. (22–9) for a power of x (not for a power of a function of x). Thus,

$$y = 6x^{1/3}$$

$$\frac{dy}{dx} = 6\left(\frac{1}{3}\right)x^{-2/3} = \frac{2}{x^{2/3}}$$

Example F
Find the derivative of

$$y = \frac{1}{(1 - 4x)^2}$$

   We can handle this more easily by direct use of the power rule, rather than the quotient rule. By writing the function as $y = (1 - 4x)^{-2}$, we can easily find the derivative:

$$\frac{dy}{dx} = (-2)(1 - 4x)^{-3}(-4) = \frac{8}{(1 - 4x)^3}$$

(Remember, subtracting 1 from -2 gives -3.)

We can now see the value of fractional exponents in calculus. They are useful in algebraic operations, but are almost essential in calculus. Without fractional exponents, it would be necessary to develop more formulas to find derivatives of radicals. To find the derivative of an algebraic function, we need only those equations which we have already developed. Often it is necessary to combine these, as we saw in Example B. Actually, most derivatives are combinations. The problem in finding the derivative is recognizing the form of the function with which you are dealing. When you have recognized the form, completing the problem is only a matter of mechanics and algebra. You should now see the importance of being able to handle algebraic operations with facility.

Example G

Find the derivative of

$$y = \frac{x}{\sqrt{1 - 4x}}$$

Here we have a quotient, and in order to find the derivative of this quotient, we must also use the power rule (and a derivative of a polynomial form). With sufficient practice in taking derivatives, we can recognize the rule to use almost automatically. Thus, we find the derivative:

$$\frac{dy}{dx} = \frac{(1 - 4x)^{1/2}(1) - x(\tfrac{1}{2})(1 - 4x)^{-1/2}(-4)}{1 - 4x}$$

$$= \frac{(1 - 4x)^{1/2} + \dfrac{2x}{(1 - 4x)^{1/2}}}{1 - 4x}$$

$$= \frac{\dfrac{(1 - 4x)^{1/2}(1 - 4x)^{1/2} + 2x}{(1 - 4x)^{1/2}}}{1 - 4x} = \frac{(1 - 4x) + 2x}{(1 - 4x)^{1/2}(1 - 4x)}$$

$$= \frac{1 - 2x}{(1 - 4x)^{3/2}}$$

Exercises 22–7

In Exercises 1 through 24 find the derivative of each of the given functions.

1. $y = \sqrt{x}$

2. $y = \sqrt[4]{x}$

3. $y = \dfrac{1}{x^2}$

4. $y = \dfrac{2}{x^4}$

5. $y = \dfrac{3}{\sqrt[3]{x}}$

6. $y = \dfrac{1}{\sqrt{x}}$

7. $y = x\sqrt{x} - \dfrac{1}{x}$

8. $y = 2x^{-3} - x^{-2}$

9. $y = (x^2 + 1)^5$

10. $y = (1 - 2x)^4$ **11.** $y = 2(7 - 4x^3)^8$ **12.** $y = 3(8x^2 - 1)^6$

13. $y = (2x^3 - 3)^{1/3}$ **14.** $y = (1 - 6x)^{3/2}$ **15.** $y = \dfrac{1}{(1 - x^2)^4}$

16. $y = \dfrac{1}{\sqrt{1 - x}}$ **17.** $y = 4(2x^4 - 5)^{3/4}$ **18.** $y = 5(3x^7 - 4)^{2/3}$

19. $y = \sqrt[4]{1 - 8x^2}$ **20.** $y = \sqrt[3]{4x^6 + 2}$ **21.** $y = x\sqrt{x - 1}$

22. $y = x^2(1 - 3x)^5$ **23.** $y = \dfrac{\sqrt{4x + 3}}{8x + 1}$ **24.** $y = \dfrac{x\sqrt{x - 1}}{2x + 1}$

In Exercises 25 through 36 solve the given problems by finding the appropriate derivative.

25. Find any values of x for which the derivative of $y = \dfrac{x^2}{\sqrt{x^2 + 1}}$ is zero.

26. Find any values of x for which the derivative of $y = \dfrac{x}{\sqrt{4x - 1}}$ is zero.

27. Find the slope of a line tangent to the parabola $y^2 = 4x$ at the point $(1, 2)$.

28. Find the slope of a tangent to the circle $x^2 + y^2 = 25$ at the point $(4, 3)$.

29. Find the velocity at $t = 4$, given that the distance in terms of the time for a particular object is $s = (t^3 - t)^{4/3}$.

30. The period of a pendulum is directly proportional to the square root of its length. Find the derivative of the period with respect to the length for a pendulum 2 ft long which has a period of $\pi/2$ s.

31. When the volume of a gas changes very rapidly, an approximate relation is that the pressure varies inversely as the $\frac{3}{2}$ power of the volume. If P is 300 kPa when $V = 100$ cm³, find the derivative of P with respect to V. Evaluate this derivative for $V = 100$ cm³.

32. Under certain conditions, the velocity of an object moving with simple harmonic motion is given by $v = k\sqrt{A^2 - x^2}$, where x is the x-coordinate of the position of the object, A is the amplitude, and k is a constant. Find the derivative of v with respect to x.

33. Under certain conditions, due to the presence of a charge q, the electric potential V along a line is given by

$$V = \dfrac{kq}{\sqrt{x^2 + b^2}}$$

where k is a constant and b is the minimum distance from the charge to the line. Find the expression for the rate of change of V with respect to x.

34. The current in a circuit containing a resistance R and an inductance L is found from the expression

$$I = \dfrac{V}{\sqrt{R^2 + (\omega L)^2}}$$

Find the expression for the rate of change of current with respect to L, assuming that the other quantities remain constant.

35. The magnetic field H due to a magnet of length l at a distance r is given by

$$H = \frac{k}{[r^2 + (l/2)^2]^{3/2}}$$

where k is a constant for a given magnet. Find the expression for the derivative of H with respect to r.

36. Under certain circumstances the velocity of a water wave is given by

$$v = \sqrt{\frac{a}{\lambda} + b\lambda}$$

where λ is the wavelength and a and b are constants. Find the derivative of v with respect to λ.

22—8 Differentiation of Implicit Functions

To this point the functions which we have differentiated have been of the form $y = f(x)$. There are, however, occasions when we need to find the derivative of a function determined by an equation which does not express the dependent variable explicitly in terms of the independent variable.

An equation in which y is not expressed explicitly in terms of x may determine one or more functions. *Any such function, where y is defined implicitly as a function of x, is called an* **implicit function.** Some equations defining implicit functions may be solved to determine the explicit functions, and for others it is not possible to solve for the explicit functions. Also, not all such equations define y as a function of x for real values of x.

Example A

The equation $3x + 4y = 5$ is an equation which defines a function, although it is not in explicit form. In solving for y as $y = -\frac{3}{4}x + \frac{5}{4}$, we have the explicit form of the function.

The equation $y^2 + x = 3$ is an equation which defines two functions, although we do not have the explicit forms. When we solve for y, we obtain the explicit functions $y = \sqrt{3 - x}$ and $y = -\sqrt{3 - x}$.

The equation $y^5 + xy^2 + 3x^2 = 5$ defines y as a function of x, although we cannot actually solve for the algebraic form of the explicit form of the function.

The equation $x^2 + y^2 + 4 = 0$ is not satisfied by any pair of real values of x and y.

Even when it is possible to determine the explicit form of a function given in implicit form, it is not always desirable to do so. In some cases the implicit form is more convenient than the explicit form.

The derivative of an implicit function may be found directly without having to solve for the explicit function. Thus, *to find dy/dx when y is defined as an implicit function of x, we differentiate each term of the equation with respect to x, regarding y as a function of x. We then solve for dy/dx, which will usually be in terms of x and y.*

Example B

Find dy/dx if $y^2 + 2x^2 = 5$.

Here we find the derivative of each term, and then solve for dy/dx. Thus,

$$\frac{d(y^2)}{dx} + \frac{d(2x^2)}{dx} = \frac{d(5)}{dx}$$

$$2y\frac{dy}{dx} + 4x = 0$$

$$\frac{dy}{dx} = -\frac{2x}{y}$$

The factor *dy/dx* arises from the derivative of the first term as a result of using the derivative of a power of a function of x (Eq. 22–15). The factor *dy/dx* corresponds to the *du/dx* of the formula. In the second term, no factor of *dy/dx* appears, since there are no y factors in the term.

Example C

Find dy/dx if $3y^4 + xy^2 + 2x^3 - 6 = 0$.

In finding the derivative, we note that the second term is a product, and we must use the product rule for derivatives on it. Thus, we have

$$\frac{d(3y^4)}{dx} + \frac{d(xy^2)}{dx} + \frac{d(2x^3)}{dx} - \frac{d(6)}{dx} = \frac{d(0)}{dx}$$

$$12y^3\frac{dy}{dx} + \left[x\left(2y\frac{dy}{dx}\right) + y^2(1)\right] + 6x^2 - 0 = 0$$

$$12y^3\frac{dy}{dx} + 2xy\frac{dy}{dx} + y^2 + 6x^2 = 0$$

$$(12y^3 + 2xy)\frac{dy}{dx} = -y^2 - 6x^2$$

$$\frac{dy}{dx} = \frac{-y^2 - 6x^2}{12y^3 + 2xy}$$

Example D

Find the slope of a line tangent to the curve of $2y^3 + xy + 1 = 0$ at the point $(-3, 1)$.

Here we must find dy/dx and evaluate it for $x = -3$ and $y = 1$.

$$\frac{d(2y^3)}{dx} + \frac{d(xy)}{dx} + \frac{d(1)}{dx} = \frac{d(0)}{dx}$$

$$6y^2\frac{dy}{dx} + x\frac{dy}{dx} + y + 0 = 0$$

$$\frac{dy}{dx} = \frac{-y}{6y^2 + x}$$

$$\frac{dy}{dx}\bigg|_{(-3,1)} = \frac{-1}{6(1^2) - 3} = \frac{-1}{6 - 3}$$

$$= -\frac{1}{3}$$

Thus, the slope is $-\frac{1}{3}$.

Exercises 22—8

In Exercises 1 through 16 find dy/dx by differentiating implicitly. When applicable, express the result in terms of x and y.

1. $3x + 2y = 5$ 2. $6x - 3y = 4$

3. $4y - 3x^2 = x$ 4. $x^5 - 5y = 6 - x$

5. $x^2 - y^2 - 9 = 0$ 6. $x^2 + 2y^2 - 11 = 0$

7. $y^5 = x^2 - 1$ 8. $y^4 = 3x^3 - x$

9. $y^2 + y = x^2 - 4$ 10. $2y^3 - y = 7 - x^4$

11. $y + 3xy - 4 = 0$ 12. $8y - xy - 7 = 0$

13. $xy^3 + 3y + x^2 = 9$ 14. $y^2x - x^2y + 3x = 4$

15. $2(x^2 + 1)^3 + (y^2 + 1)^2 = 17$ 16. $(2x + 1)(1 - 3y) + y^2 = 13$

In Exercises 17 through 20 solve the given problems by use of implicit differentiation.

17. Find the slope of a line tangent to the curve of $xy + y^2 + 2 = 0$ at the point $(-3, 1)$.

18. Find the velocity of an object moving such that its displacement s and the time t are related by the equation $s^3 - t^2 = 7$ for $s = 2$ and $t = 1$.

19. For a right triangle in which the hypotenuse is always a units in length, the legs x and y are related by the equation $x^2 + y^2 = a^2$. Find the expression for dy/dx.

20. The polar moment of inertia I of a rectangular area is given by $I = \frac{1}{12}(b^3h + bh^3)$, where b and h are the base and height, respectively, of the rectangle. If I is constant, find the expression for db/dh.

22—9 Exercises for Chapter 22

In Exercises 1 through 12 evaluate the given limits.

1. $\lim_{x \to 4} (8 - 3x)$ 2. $\lim_{x \to 3} (2x^2 - 10)$ 3. $\lim_{x \to -3} \frac{2x + 5}{x - 1}$

4. $\lim_{x \to 1} \frac{x^2 - 1}{x + 1}$ 5. $\lim_{x \to 2} \frac{4x - 8}{x^2 - 4}$ 6. $\lim_{x \to 5} \frac{x^2 - 25}{3x - 15}$

7. $\lim_{x \to 2} \frac{x^2 + 3x - 10}{x^2 - x - 2}$ 8. $\lim_{x \to 3} \frac{x^2 - x - 6}{x - 3}$ 9. $\lim_{x \to \infty} \frac{2 + \dfrac{1}{x + 4}}{3 - \dfrac{1}{x^2}}$

10. $\lim_{x \to \infty} \left(7 - \dfrac{1}{x + 1} \right)$
  11. $\lim_{x \to \infty} \dfrac{x - 2x^3}{1 + x^3}$
  12. $\lim_{x \to \infty} \dfrac{2x + 5}{3x^3 - 2x}$

In Exercises 13 through 20 use the Δ-process to find the derivative of each of the given functions.

13. $y = 7 + 5x$
  14. $y = 6x - 2$
  15. $y = 6 - 2x^2$

16. $y = 2x^2 - x^3$
  17. $y = \dfrac{2}{x^2}$
  18. $y = \dfrac{1}{1 - 4x}$

19. $y = \sqrt{x + 5}$
  20. $y = \dfrac{1}{\sqrt{x}}$

In Exercises 21 through 36 find the derivative of each of the given functions.

21. $y = 2x^7 - 3x^2 + 5$
  22. $y = 8x^5 - x - 1$

23. $y = 4\sqrt{x} - \dfrac{1}{x}$
  24. $y = \dfrac{3}{x^2} - 8\sqrt[4]{x}$

25. $y = \dfrac{x}{1 - x}$
  26. $y = \dfrac{2x - 1}{x^2 + 1}$

27. $y = (2 - 3x)^4$
  28. $y = (2x^2 - 3)^6$

29. $y = \dfrac{3}{(5 - 2x^2)^{3/4}}$
  30. $y = \dfrac{7}{(3x - 1)^3}$

31. $y = x^2\sqrt{1 - 6x}$
  32. $y = (x - 1)^3 (x^2 - 2)^2$

33. $y = \dfrac{\sqrt{4x + 3}}{2x}$
  34. $y = \dfrac{x}{\sqrt{x^2 + 1}}$

35. $(2x - 3y)^3 = x^2 - y$
  36. $x^2 y^2 = x^2 + y^2$

In Exercises 37 through 52 solve the given problems by finding the appropriate derivative or limit.

37. The combined capacitance for two capacitors connected in series is given by

$$C_T = \frac{C_1 C_2}{C_1 + C_2}$$

Find $\lim_{C_1 \to C_2} C_T$ and $\lim_{C_1 \to 0} C_T$.

38. The illuminance I is inversely proportional to the square of the distance x from the source. Find $\lim_{x \to 0} I$.

39. Find the slope of a line tangent to the curve of $y = 7x^4 - x^3$ at $(-1, 8)$.

40. Find the slope of a line tangent to the curve of $y = \sqrt[3]{3 - 8x}$ at $(-3, 3)$.

41. Find the velocity of an object after 3 s if its displacement s (in meters) is given by $s = \dfrac{t}{3t + 1}$.

42. The distance s (in feet) traveled by a subway train after the brakes are applied is given by $s = 40t - 5t^2$. How far does it travel, after the brakes are applied, in coming to a stop?

43. The voltage induced in an inductor L is given by

$$E = L\frac{dI}{dt}$$

where I is the current in the circuit and t is the time. Find the voltage induced in a 0.4-H inductor if the current I (in amperes) is related to the time (in seconds) by $I = 0.1t + 5.0$.

44. The specific heat c_p of a certain gas as a function of the temperature T is given by $c_p = 6.50 + 0.01T + 0.000002T^2$. Find the expression for the rate of change of the specific heat with respect to temperature.

45. The electric field E at a distance r from a point charge is $E = k/r^2$, where k is a constant. Find an expression for the rate of change of the electric field with respect to r.

46. The distance d between ion layers of a crystalline solid such as table salt is given by $d = \sqrt[3]{M/2N\rho}$ where M is the molecular weight, N is called Avogadro's number, and ρ is the density. Find the rate of change of d with respect to ρ.

47. The velocity of an object moving with constant acceleration can be found from the equation $v = \sqrt{v_0^2 + 2as}$, where v_0 is the initial velocity, a is the acceleration, and s is the distance traveled. Find dv/ds.

48. The time rate of change of angular velocity is angular acceleration. If a wire is moving through a magnetic field in a circular path such that its angular velocity is given by

$$\omega = \frac{2t^2}{1 + t}$$

where t is the time (in seconds), find the expression for the angular acceleration (in radians per second squared). What is the value of the angular acceleration when $t = 2$ s?

49. Under certain conditions, the efficiency of an internal combustion engine is given by

$$\text{eff (in percent)} = 100\left(1 - \frac{1}{(V_1/V_2)^{0.4}}\right)$$

where V_1 and V_2 are, respectively, the maximum and minimum volumes of air in a cylinder. Assuming that V_2 is kept constant, find the expression for the rate of change of efficiency with respect to V_1.

50. Water is being drained from a pond such that the volume of water in the pond (in cubic meters) after t hours is given by $V = 5000(60 - t)^2$. Find the rate at which the pool is being drained after 4 h.

51. A rectangle is inscribed in the first quadrant under the parabola $y = 4 - x^2$. Express the area of the rectangle in terms of x. Then find the expression for dA/dx.

52. An airplane flies over an observer with a velocity of 400 mi/h at an altitude of 2640 ft. If the plane flies horizontally in a straight line, find the rate at which the distance from the observer to the plane is changing 0.6 min after the plane passes over the observer.

23

Applications of the Derivative

23—1 Tangents and Normals

In Chapter 22 we developed the meaning of the derivative of a function and then went on to find several formulas by which we can differentiate functions. In establishing the concept of the derivative as an instantaneous rate of change, numerous applications of the derivative were indicated in the examples and exercises. It was not necessary to develop any of these applications in any detail, since finding the derivative was the primary concern. There are, however, certain types of problems in geometry and technology in which the derivative plays a key role in the solution, although other concepts also are involved. The first of these problems involves finding the equations of lines tangent and lines normal (perpendicular) to a given curve.

To find the equation of a line tangent to a curve at a given point, we first find the derivative of the function. The derivative is then evaluated at the point, and this gives us the slope of a line tangent to the curve at the point. Then, by using the point-slope form of the equation of a straight line, we find the equation of the tangent line. The following examples illustrate the method.

Point slope
$y - y_1 = m(x - x_1)$

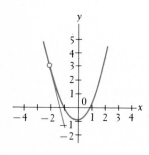

Figure 23–1

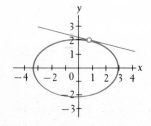

Figure 23–2

Example A

Find the equation of the line tangent to the parabola $y = x^2 - 1$ at the point $(-2, 3)$.

The derivative of this function is

$$\frac{dy}{dx} = 2x$$

The value of this derivative for the value $x = -2$ (the y-value of 3 is not used since the derivative does not directly contain y) is

$$\frac{dy}{dx} = -4$$

which means that the slope of the tangent line at $(-2, 3)$ is -4. Thus, by using the point-slope form of the equation of the straight line we obtain the desired equation. Thus, we have

$$y - 3 = -4(x + 2)$$
$$y - 3 = -4x - 8$$

or

$$y = -4x - 5$$

The parabola and the tangent line $y = -4x - 5$ are shown in Fig. 23–1.

Example B

Find the equation of the line tangent to the ellipse $4x^2 + 9y^2 = 40$ at the point $(1, 2)$.

The easiest method of finding the derivative of this equation is to treat it as an implicit function. In this way we have

$$8x + 18yy' = 0 \qquad \text{or} \qquad y' = -\frac{4x}{9y}$$

Evaluating this derivative at the point $(1, 2)$, we have

$$y' = -\frac{4}{18} = -\frac{2}{9}$$

Thus, the slope of the desired tangent line is $-\frac{2}{9}$. Using the point-slope form of the equation of the straight line we have

$$y - 2 = -\frac{2}{9}(x - 1)$$

$$9y - 18 = -2x + 2$$
$$2x + 9y - 20 = 0$$

The ellipse and the tangent line $2x + 9y - 20 = 0$ are shown in Fig. 23–2.

It might be noted that the derivative could also have been found by solving for y. In this way we would have to differentiate $y = \frac{2}{3}\sqrt{10 - x^2}$.

If we wish to obtain the equation of a line normal (perpendicular to a tangent) to a curve, we recall that the slopes of perpendicular lines are negative reciprocals. Thus, the derivative is found and evaluated at the specified point. Since this gives the slope of a tangent line, we take the negative reciprocal of this number to find the slope of the normal line. Then, by using the point-slope form of the equation of a straight line we find the equation of the normal. The following examples illustrate the method.

Example C
Find the equation of the line normal to the hyperbola $y = 2/x$ at the point $(2, 1)$.

The derivative of this function is $dy/dx = -2/x^2$, which evaluated at $x = 2$ gives $dy/dx = -\frac{1}{2}$. Therefore, the slope of a line normal to the curve at the point $(2, 1)$ is 2. The equation of the normal is then

$$y - 1 = 2(x - 2)$$

or

$$y = 2x - 3$$

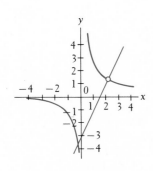

Figure 23–3

The hyperbola and the normal line are shown in Fig. 23–3.

Example D
Find the y-intercept of the line normal to the curve $y = 3x - x^3$ at $(3, -18)$.

The derivative of this function is $dy/dx = 3 - 3x^2$, which evaluated at $(3, -18)$ is

$$\left.\frac{dy}{dx}\right|_{x=3} = 3 - 3(9) = -24$$

Therefore, the slope of a normal line is $\frac{1}{24}$. The equation of the normal line is

$$y + 18 = \frac{1}{24}(x - 3)$$

or

$$24y - x + 435 = 0$$

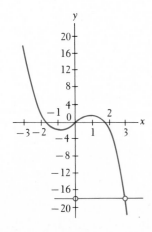

Figure 23–4

To find the y-intercept of this line we set $x = 0$. In this way we find the y-intercept to be $-\frac{145}{8}$. The curve, normal line, and intercept are indicated in Fig. 23–4.

Many of the applications of tangents and normals are geometric. However, some of the areas of technology where they may be applied are indicated in the exercises.

Exercises 23–1

In Exercises 1 through 4 find the equations of the lines tangent to the indicated curves at the given points. In each, sketch the curve and the tangent line.

1. $y = x^2 + 2$ at $(2, 6)$ 2. $y = x^3 - 2x$ at $(2, 4)$

3. $y = \dfrac{1}{x^2 + 1}$ at $(1, \frac{1}{2})$ 4. $x^2 + y^2 = 25$ at $(3, 4)$

In Exercises 5 through 8 find the equations of the lines normal to the indicated curves at the given points. In each, sketch the curve and the normal line.

5. $y = 6x - x^2$ at $(1, 5)$ 6. $y = 2x^3 + 10$ at $(-2, -6)$

7. $y = \dfrac{1}{(x^2 + 1)^2}$ at $(1, \frac{1}{4})$ 8. $x^2 - y^2 = 8$ at $(3, 1)$

In Exercises 9 through 12 find the equations of the tangent lines and the normal lines to the indicated curves.

9. $y = \dfrac{1}{\sqrt{x^2 + 1}}$ where $x = \sqrt{3}$ 10. $y = \dfrac{4}{(5 - 2x)^2}$ where $x = 2$

11. The parabola with vertex at $(0, 3)$ and focus at $(0, 0)$, where $x = -1$

12. The ellipse with focus at $(4, 0)$, vertex at $(5, 0)$, and center at $(0, 0)$ where $x = 2$

In Exercises 13 through 18 solve the given problems by finding the equation of the appropriate tangent or normal line.

13. Find the x-intercept of the line tangent to the parabola $y = 4x^2 - 8x$ at $(-1, 12)$.

14. Find the y-intercept of the line normal to the curve $y = x^{3/4}$ where $x = 16$.

15. A certain suspension cable with supports on the same level is closely approximated as being parabolic in shape. If the supports are 200 ft apart and the sag at the center is 30 ft, what is the equation of the line along which the tension acts (tangentially) at the right support. (Choose the origin of the coordinate system at the lowest point of the cable.)

16. In an electric field the lines of force are perpendicular to the curves of equal electric potential. A certain curve of equal electric potential is given by $y = 2x^{2/3}$. What is the equation of the line along which the force acts on an electron positioned on this curve where $x = 8$?

17. An object is moving in a horizontal circle of radius 2 ft at the rate of 2 rad/s. If the object is on the end of a string, and the string breaks after $\pi/3$ s, causing the object to travel along a line tangent to the circle, what is the equation of the path of the object after the string breaks? Sketch the path of the object. (Choose the origin of the coordinate system at the center of the circle and assume the object started on the positive x-axis moving counterclockwise.)

18. On a particular drawing, a pulley wheel can be described by the equation $x^2 + y^2 = 100$, where distances are measured in centimeters. The pulley belt is directed along the lines $y = -10$, and $4y - 3x - 50 = 0$ when first and last making contact with the wheel. What are the first and last points on the wheel where the belt makes contact?

23–2 Curvilinear Motion

A great many phenomena in technology involve the time rate of change of certain quantities. We encountered one of the most fundamental and most important of these when we discussed velocity, the time rate of change of displacement, in Section 22–4. In this section we further develop the concept of velocity and certain other concepts necessary for the discussion. Other time-rate-of-change problems are discussed in the next section.

When velocity was introduced in Section 22–4, the discussion was limited to rectilinear motion, or motion along a straight line. A more general discussion of velocity is necessary when we discuss the motion of an object in a plane. There are many important applications of motion in the plane, a principal one among these being the motion of a projectile.

An important concept in developing this topic is that of a vector. The necessary fundamentals related to vectors are taken up in Section 8–1. Although vectors can be used to represent many physical quantities, we shall restrict our attention to their use in describing the velocity and acceleration of an object moving in a plane along a specified path. Such motion is called **curvilinear motion.**

In describing an object undergoing curvilinear motion, it is very common to express the x and y coordinates of its position separately as functions of time. Equations given in this form, that is, *x and y both given in terms of a third variable (in this case t), are said to be in* **parametric form,** which we encountered in Section 9–6. The third variable, *t,* is called the **parameter.**

To find the velocity of an object whose coordinates are given in parametric form, we find its x-component of velocity v_x by determining dx/dt and its y-component of velocity v_y by determining dy/dt. These are then evaluated, and the resultant velocity is found from $v = \sqrt{v_x^2 + v_y^2}$. The direction in which the object is moving is found from $\tan \theta = v_y/v_x$. The following examples illustrate the method.

Example A

If the horizontal distance x that an object has moved is given by $x = 3t^2$, and the vertical distance y is given by $y = 1 - t^2$, find the resultant velocity when $t = 2$.

To find the resultant velocity, we must find v and θ, by first finding v_x and v_y as indicated above. Thus,

$$v_x = \frac{dx}{dt} = 6t \qquad v_x|_{t=2} = 12$$

$$v_y = \frac{dy}{dt} = -2t \qquad v_y|_{t=2} = -4$$

$$v = \sqrt{144 + 16} = \sqrt{160} = 12.6$$

$$\tan \theta = \frac{-4}{12} = -0.333 \qquad \theta = -18.4°$$

The path and velocity vectors are shown in Fig. 23–5.

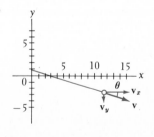

Figure 23–5

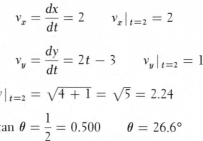

Figure 23–6

Example B

Find the velocity and direction of motion when $t = 2$ of an object moving such that its x and y coordinates of position are given by $x = 1 + 2t$ and $y = t^2 - 3t$.

$$v_x = \frac{dx}{dt} = 2 \qquad v_x|_{t=2} = 2$$

$$v_y = \frac{dy}{dt} = 2t - 3 \qquad v_y|_{t=2} = 1$$

$$v|_{t=2} = \sqrt{4 + 1} = \sqrt{5} = 2.24$$

$$\tan \theta = \frac{1}{2} = 0.500 \qquad \theta = 26.6°$$

These quantities are shown in Fig. 23–6.

Acceleration *is the time rate of change of velocity.* Therefore, if the velocity, or its components, is known as a function of time, the acceleration of an object can be found by taking the derivative of the velocity with respect to time. For this reason it may be necessary to find the derivative of a derivative. This is called a **second derivative.** The notation for the second derivative of x with respect to t, for example, is d^2x/dt^2.

Example C

Find the magnitude and direction of the acceleration when $t = 2$ for an object moving such that its x and y coordinates of position are given by $x = t^3$ and $y = 1 - t^2$:

$$v_x = \frac{dx}{dt} = 3t^2 \qquad a_x = \frac{dv_x}{dt} = \frac{d^2x}{dt^2} = 6t \qquad a_x|_{t=2} = 12$$

$$v_y = \frac{dy}{dt} = -2t \qquad a_y = \frac{dv_y}{dt} = \frac{d^2y}{dt^2} = -2 \qquad a_y|_{t=2} = -2$$

$$a|_{t=2} = \sqrt{144 + 4} = \sqrt{148} = 12.2$$

$$\tan \theta = \frac{a_y}{a_x} = -\frac{2}{12} = -0.167 \qquad \theta = -9.5°$$

The quadrant in which θ lies is determined from the fact that a_y is negative and a_x is positive. Thus, θ must be a fourth-quadrant angle (see Fig. 23–7).

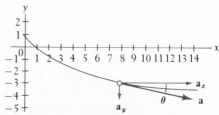

Figure 23–7

$$D_x\, u^n = n u^{n-1}\, D_x u$$

If the curvilinear path which an object follows is given as y as a function of x, the velocity (and acceleration) is found by taking derivatives of each term of the equation with respect to time. In finding the derivatives we must be very careful to use the power rule for finding derivatives, Eq. (22–15), so that the general factor du/dx is not neglected. The following example illustrates the method.

Example D

A particle moves along the parabola $y = \frac{1}{3}x^2$ so that its vertical velocity v_y is always 8. Find the magnitude of the resultant velocity when the particle is at the point $(2, \frac{4}{3})$.

Since both y and x change with time, both can be considered functions of time. Thus, we can take derivatives of $y = \frac{1}{3}x^2$ with respect to time. When we do this we obtain the following result:

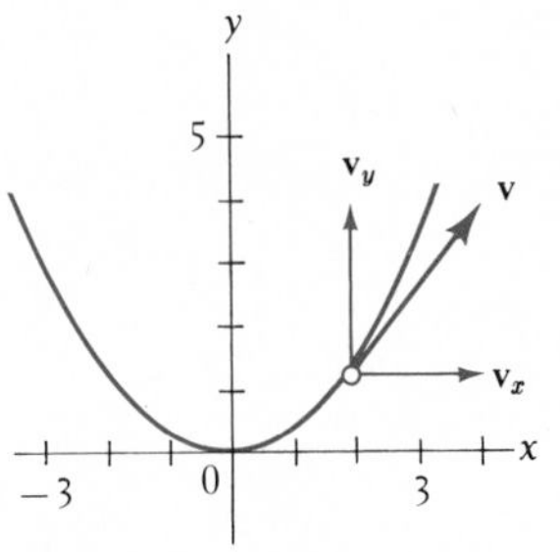

$$\frac{dy}{dt} = \frac{2}{3}x\,\frac{dx}{dt} \qquad v_y = \frac{2}{3}xv_x$$

At the point $(2, \frac{4}{3})$ we have $8 = \frac{2}{3}(2)v_x$. Thus, $v_x = 6$ at this point (see Fig. 23–8). Therefore, the resultant velocity is

$$v = \sqrt{64 + 36} = 10$$

Figure 23–8

Exercises 23–2

In Exercises 1 through 4, given that the x- and y-coordinates of a moving particle are given by the indicated parametric equations, find the magnitude and direction of the velocity for the specific value of t. Plot the curves and show the appropriate components of the velocity.

1. $x = 3t$, $y = 1 - t$, $t = 4$ 2. $x = 2t^3$, $y = 4t^2$, $t = 1$

3. $x = \dfrac{10}{2t + 3}$, $y = \dfrac{1}{t^2}$, $t = 2$ 4. $x = \sqrt{1 + 2t}$, $y = t - t^2$, $t = 4$

In Exercises 5 through 8 use the parametric equations and values of t of Exercises 1 through 4 to find the magnitude and direction of the acceleration in each case.

In Exercises 9 through 20 find the indicated velocities and accelerations.

9. A particle moves along the curve $y = 5x^3$ with a constant x component of velocity equal to 20. Find the magnitude and direction of the velocity at the point $(1, 5)$.

10. A particle moves along the curve $y = 2\sqrt{x}$. What is the ratio of the x-component of velocity to the y-component at $(9, 6)$?

11. A particle moves along the upper half of the parabola $y^2 = 2x + 1$ such that $v_x = \sqrt{2x + 3}$. Find v_y when $x = 4$.

12. A particle moves counterclockwise along the ellipse $x^2 + 9y^2 = 9$ such that $v_y = 2y$. Find v_x at the point $(2, \frac{1}{3}\sqrt{5})$.

13. A projectile moves according to the parametric equations $x = 120t$ and $y = 160t - 16t^2$, where distances are in feet and time is in seconds. Find the magnitude and direction of the velocity when $t = 3.0$ s and when $t = 6.0$ s.

14. A projectile moves according to the parametric equations $x = 100t$ and $y = -16t^2$, where distances are in feet and time is in seconds. Find the magnitude and direction of each of the velocity and acceleration when $t = 3.0$ s.

15. Find the acceleration at the specified times for the projectile in Exercise 13.

16. Find the magnitude and direction of the acceleration of the particle in Exercise 9.

17. A spacecraft moves such that it follows the equation $x = 10(\sqrt{1 + t^4} - 1)$, $y = 40t^{3/2}$ for the first hundred seconds after launch. Here, x and y are measured in meters and t is measured in seconds. Find the magnitude and direction of the velocity of the spacecraft 10 s and 100 s after launch.

18. An electron moves in an electric field according to the parametric equations

$$x = \frac{20}{\sqrt{1 + t^2}} \quad \text{and} \quad y = \frac{20t}{\sqrt{1 + t^2}}$$

where distances are in meters and time in seconds. Find the magnitude and direction of the velocity when $t = 1$ s.

19. A rocket follows a path given by $y = x - \frac{1}{90}x^3$ (distances in miles). If the horizontal velocity is given by $v_x = x$, find the magnitude and direction of the velocity when the rocket hits the ground (assume level terrain) if time is in minutes.

20. A meteor traveling toward the earth has a velocity inversely proportional to the square root of the distance from the earth's center. Show that its acceleration is inversely proportional to the square of the distance from the center of the earth.

23-3 Related Rates

Any two variables which vary with respect to time and between which a relation is known to exist, can have the time rate of change of one expressed in terms of the time rate of change of the other. We do this by taking the derivative with respect to time of the expression which relates the variables as we did in Example D of Section 23-2. Since the rates of change are related, this type of problem is referred to as a **related-rate** problem. The following examples illustrate the basic method of solution.

Example A
The voltage of a certain themocouple as a function of temperature is given by $E = 2.800T + 0.006T^2$. If the temperature is increasing at the rate $1.00°C/min$, how fast is the voltage increasing when $T = 100°C$?

Since we are asked to find the time rate of change of voltage, we first take derivatives with respect to time. This gives us

$$\frac{dE}{dt} = 2.800\frac{dT}{dt} + 0.012T\frac{dT}{dt}$$

again being careful to include the factor dT/dt. From the given information we know that $dT/dt = 1.00°C/min$ and that we wish to know

dE/dt when $T = 100°C$. Thus,

$$\left.\frac{dE}{dt}\right|_{T=100} = 2.800(1.00) + 0.012(100)(1.00) = 4.00 \text{ V/min}$$

The derivative must be taken before values are substituted. In this problem we are finding the time rate of change of the voltage for a specified value of T. For other values of T, dE/dt would have different values.

Example B

The distance q that an image is from a certain lens in terms of p, the distance the object is from the lens, is given by

$$q = \frac{10p}{p - 10}$$

If the object distance is increasing at the rate of 0.200 cm/s, how fast is the image distance changing when $p = 15.0$ cm?

Taking derivatives with respect to time, we have

$$\frac{dq}{dt} = \frac{(p - 10)\left(10\dfrac{dp}{dt}\right) - 10p\left(\dfrac{dp}{dt}\right)}{(p - 10)^2}$$

Now substituting $p = 15.0$ and $dp/dt = 0.200$, we have

$$\left.\frac{dq}{dt}\right|_{p=15} = \frac{5(10)(0.200) - 10(15.0)(0.2)}{25.0} = -0.800 \text{ cm/s}$$

Thus, the image distance is decreasing (the significance of the minus sign) at the rate of 0.800 cm/s when $p = 15.0$ cm.

In many related-rate problems the function is not given, but must be set up according to the statement of the problem. The following examples illustrate this type of problem.

Example C

A spherical balloon is being blown up at the constant rate of 2.00 ft³/min. Find the rate at which the radius is increasing when it is 3.00 ft.

We are asked to find the relation between the rate of change of the volume of a sphere with respect to time and the corresponding rate of change of the radius. Thus, we are to take derivatives of the expression for the volume of a sphere with respect to time.

$$V = \frac{4}{3}\pi r^3 \qquad \frac{dV}{dt} = 4\pi r^2\left(\frac{dr}{dt}\right)$$

From the information given, we know the value of dV/dt to be 2.00. We want to find dr/dt when $r = 3.00$. Substituting these values, we have

$$2.00 = 4\pi(3.00)^2\left(\frac{dr}{dt}\right) \qquad \text{or} \qquad \left.\frac{dr}{dt}\right|_{r=3} = \frac{1}{18.0\pi} = 0.0177 \text{ ft/min}$$

N

y   z

Port   x   —E

Figure 23–9

Example D
Two ships, *A* and *B*, leave a port at noon. Ship *A* travels north at 6.00 mi/h, and ship *B* travels east at 8.00 mi/h. How fast are they separating at 2 PM?

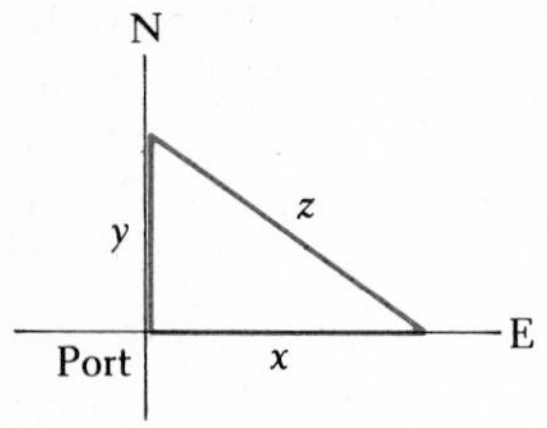

We are asked to find how fast the length of a line drawn directly from *A* to *B* is increasing at a particular time. First a relation must be found between the distances traveled. The Pythagorean theorem provides the necessary relation (see Fig. 23–9). If *y* is the distance traveled by *A* and *x* is the distance traveled by *B*, we can find the distance between them, *z*, from $z^2 = x^2 + y^2$. Taking derivatives of this expression, we find that

$$2z\frac{dz}{dt} = 2x\frac{dx}{dt} + 2y\frac{dy}{dt}$$

From the information given, we know that $dx/dt = 8.00$ mi/h, $dy/dt = 6.00$ mi/h, and for the specified time, $x = 16.0$ mi, $y = 12.0$ mi, and $z = 20.0$ mi. Thus we have

$$\left.\frac{dz}{dt}\right|_{z=20} = \frac{(16.0)(8.00) + (12.0)(6.00)}{20.0} = 10.0\text{ mi/h}$$

Exercises 23–3

In the following exercises solve the given problems in related rates.

1. The electric resistance of a certain resistor as a function of temperature is given by $R = 4.000 + 0.003\,T^2$, where *R* is measured in ohms and *T* in degrees Celsius. If the temperature is increasing at the rate of 0.100°C/s, find how fast the resistance changes when $T = 150°C$.

2. The kinetic energy of an object is given by $KE = \frac{1}{2}mv^2$, where *m* is the mass of the object and *v* is its velocity. If a 20.0-kg object is accelerating (remember that the time rate of change of velocity is acceleration) at 5.00 m/s², how fast is the kinetic energy (in joules) changing when the velocity is 30.0 m/s?

3. A firm found that its profit was *p* dollars for the production of *x* tons per week of a product according to the function $p = 30\sqrt{10x - x^2} - 50$. Determine the rate of change of profit if production is increasing such that $dx/dt = 0.200$ tons/week², when $x = 4.00$ tons/week.

4. A refinery is capable of producing two grades of oil. If *x* is the number of gallons of the lower-grade oil produced per day, and *y* is the number of gallons of the higher grade oil produced per day, where $y = \dfrac{100(20 - x)}{50 - x}$, find how fast *y* is changing if $dx/dt = 2.00$ gal/d², when $x = 8.00$ gal/d.

5. The length of a pendulum is slowly decreasing at the rate of 0.100 in./s. What is the rate of change of the period of the pendulum when the length is 16.0 in., if the relation between the period and length is $T = \pi\sqrt{L/96}$?

6. When air expands so that there is no change in heat (an adiabatic change) the relationship between the pressure and volume is $pv^{1.4} = k$, where *k* is a constant. At a certain instant, the pressure is 300 kPa and the volume is 10.0 cm³. The volume is increasing at the rate of 2.00 cm³/s. What is the time rate of change of pressure at this instant?

7. The relation between the voltage V which produces a current I in a wire of radius r is $V = 0.03I/r^2$. If the current increases at the rate of 0.020 A/s in a wire of 0.040 in. radius, find the rate at which the voltage is increasing.

8. Two resistances in parallel vary with time. At a certain instant R_1 is 8.00 Ω and is increasing at the rate of 0.500 Ω/s. At the same instant, R_2 is 6.00 Ω and is increasing at the rate of 0.600 Ω/s. What is the rate of change of the resistance R of the combination if

$$R = \frac{R_1 R_2}{R_1 + R_2}$$

9. The side of a square is increasing at the rate of 5.00 ft/min. How fast is the area changing when the side is 10.0 ft long?

10. A circular plate is contracting by being cooled. Find the rate of decrease in area if the radius decreases at the rate of 0.100 ft/h, when the radius is 4.00 ft.

11. The radius of a certain steam engine cylinder is 3.00 in. What is the speed of the piston if steam is entering the cylinder at 2.00 ft³/s? Assume no compression of the steam at this moment.

12. An ice cube is melting so that an edge decreases in length by 0.100 cm/min. How fast is the volume changing when an edge is 4.50 cm?

13. One statement of Boyle's law is that the pressure of a gas varies inversely as the volume for constant temperature. If a certain gas occupies 600 cm³ when the pressure is 200 kPa and the volume is increasing at the rate of 20.0 cm³/min, how fast is the pressure changing when the volume is 800 cm³?

14. The intensity of heat varies directly as the strength of the source and inversely as the square of the distance from the source. If an object approaches a heated object of strength of 8.00 units at the rate of 50.0 cm/s, how fast is the intensity changing when it is 100 cm from the source?

15. A spherical metal object is ejected from an Earth satellite and reenters the atmosphere. It heats up so that the radius increases at the rate of 5.00 mm/s. What is the rate of change of volume when the radius is 200 mm?

16. A pile of sand in the shape of a cone has a radius which always equals the altitude. If 100 ft³ of sand are poured onto the pile each minute, how fast is the radius increasing when the pile is 10.0 ft high?

17. Two cars leave the same point, traveling on straight roads which are at right angles. If the cars are traveling at 80.0 km/h and 100 km/h, how fast is the distance between them changing after 3.00 min (0.0500 h)?

18. A ladder is slipping down along a vertical wall. If the ladder is 10.0 ft long, and the top of it is slipping at the constant rate of 10.0 ft/s, how fast is the bottom of the ladder moving along the ground when the bottom is 6.00 ft from the wall?

19. A rope attached to a boat is being pulled in at a rate of 10.0 ft/s. If the water is 20.0 ft below the level at which the rope is being drawn in, how fast is the boat approaching the wharf when 36.0 ft of rope are yet to be pulled in?

20. A man 6.00 ft tall approaches a street light 15.0 ft above the ground at the rate of 5.00 ft/s. How fast is the end of the man's shadow moving when he is 10.0 ft from the base of the light?

23—4 Using Derivatives in Curve Sketching

Derivatives can be used effectively in the sketching of curves. An analysis of the first two derivatives can provide useful information as to the graph of a function. This information, possibly along with two or three key points on the curve, is often sufficient to obtain a good, although approximate, graph of the function. Graphs where this information must be supplemented with other analyses are the subject of the next section.

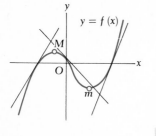

Figure 23—10

Considering the function $f(x)$ as shown in Fig. 23—10, we see that as x increases (from left to right) the y-values also increase until the point M is reached. From M to m the values of y decrease. To the right of m the values of y again increase. We also note that any tangent line to the left of M and to the right of m will have a positive slope. Any tangent line between M and m will have a negative slope. Since the derivative of a function determines the slope of a tangent line, we can conclude that, *as x increases, y increases if the derivative is positive, and decreases if the derivative is negative.* This can be stated as

$$f(x) \text{ increases if } f'(x) > 0$$

and

$$f(x) \text{ decreases if } f'(x) < 0$$

It is always assumed that x is increasing. Also, we assume in our present analysis that f(x) and its derivatives are continuous over the indicated interval.

Example A

Determine those values of x for which the function $f(x) = x^3 - 3x^2$ is increasing and those values for which it is decreasing.

We find the solution to this problem by determining the values of x for which the derivative is positive and those for which it is negative. Finding the derivative, we have

$$f'(x) = 3x^2 - 6x = 3x(x - 2)$$

This now becomes a problem of solving an inequality. To find the values of x for which $f(x)$ is increasing we must solve the inequality

$$3x(x - 2) > 0$$

We now recall that the solution of an inequality consists of *all* values of x which may satisfy it. Normally, this consists of certain intervals of values of x. These intervals are found by first setting the left side of the inequality equal to zero, thus obtaining the **critical values** of x. The function on the left will have the same *sign* for all values of x less than the leftmost critical value. The sign of the function will also be the same within any given interval between critical values and to the right of the rightmost critical value. Those intervals which give the proper sign will satisfy the inequality. In this case the critical values are $x = 0$ and $x = 2$. Therefore, we have the following analysis:

$$\text{If } x < 0, \quad 3x(x - 2) > 0 \quad \text{or} \quad f'(x) > 0$$

$$\text{If } 0 < x < 2, \quad 3x(x - 2) < 0 \quad \text{or} \quad f'(x) < 0$$

$$\text{If } x > 2, \quad 3x(x - 2) > 0 \quad \text{or} \quad f'(x) > 0$$

Therefore, the solution of the above inequality is $x < 0$ or $x > 2$, which means that for these values $f(x)$ is increasing. We can also see that for $0 < x < 2$, $f'(x) < 0$, which means that $f(x)$ is decreasing for these values of x. The solution is now complete.

The points M and m in Fig. 23–10 are called a **relative maximum point** and a **relative minimum point,** respectively. *This means that M has a greater y-value than any other point near it, and that m has the least y-value of any point near it.* This does not necessarily mean that M has the greatest y-value of any point on the curve, or that m has the least y-value of any point on the curve. However, the points M and m are the greatest or least values of y for that part of the curve (that is why we use the word "relative"). Observation of Fig. 23–10 verifies this point. *The characteristic of both M and m is that the derivative is zero at each point.* (We see that this is so since a tangent line would have a slope of zero at each.) *This is how relative maximum and relative minimum points are located. The derivative is found and then set equal to zero. The solutions of the resulting equation give the x-coordinates of the maximum and minimum points.*

It remains now to determine whether a given value of x, for which the derivative is zero, is the coordinate of a maximum or a minimum point (or neither, which is also possible). From the discussion of increasing and decreasing values for y, we can see that *the derivative changes sign from plus to minus when passing through a relative maximum point, and from minus to plus when passing through a relative minimum point.* Thus, we find maximum and minimum points by determining those values of x for which the derivative is zero and by properly analyzing the sign change of the derivative. If the sign of the derivative does not change, it is neither a maximum nor a minimum point. This is known as the **first-derivative test for maxima and minima.**

Example B

Find any maximum and minimum points of the function

$$y = 3x^5 - 5x^3$$

Finding the derivative and setting it equal to zero, we have

$$y' = 15x^4 - 15x^2 = 15x^2(x^2 - 1)$$
$$15x^2(x - 1)(x + 1) = 0 \quad \text{for} \quad x = 0, \quad x = 1 \quad \text{and} \quad x = -1$$

Thus, the sign of the derivative is the same for all points to the left of $x = -1$. For these values $y' > 0$ (thus y is increasing). For values of x between -1 and 0, $y' < 0$. For values of x between 0 and 1, $y' < 0$. For values of x greater than 1, $y' > 0$. Thus, the curve has a maximum at $(-1, 2)$ and a minimum at $(1, -2)$. The point $(0, 0)$ is neither a maximum nor a minimum, since the sign of the derivative did not change for this value of x.

Looking again at the slope of tangents drawn to a curve (see Fig. 23–11a), we see that the slope goes from large to small positive values and then to negative values for this curve. At the point I the values of the slope again start to increase so that values of the slope again become positive after the point m. *The conclusion is that the slope decreases as a point approaches I, and that the slope increases as a point moves to the right of I.* The curve in Fig. 23–11(b) is that of the derivative, and therefore indicates the values of the slope of $f(x)$. If the slope changes, this means that we are dealing with the rate of change of slope, or the rate of change of the derivative. This function is the second derivative.

The curve in Fig. 23–11(c) is that of the second derivative. This means that we wish to be able to find an expression for the derivative of a derivative. We encountered this type of function before, in the discussion of acceleration. *Where the second derivative of a function is negative, the slope is decreasing, or the curve is* **concave down** *(opens down). Where the second derivative is positive, the slope is increasing, or the curve is* **concave up** *(opens up).* This may be summarized as follows:

If $f''(x) > 0$, the curve is concave up.

If $f''(x) < 0$, the curve is concave down.

We can also now use this information in the determination of maximum and minimum points. By the nature of the definition of maximum and minimum points and of concavity, it is apparent that *a curve is concave down at a maximum point and concave up at a minimum point.* We can see these properties when we make a close analysis of the curves in Fig. 23–11. Therefore, at $x = a$,

if $f'(a) = 0$ and $f''(a) < 0$,

then $f(x)$ has a relative maximum at $x = a$, or

if $f'(a) = 0$ and $f''(a) > 0$,

then $f(x)$ has a relative minimum at $x = a$.

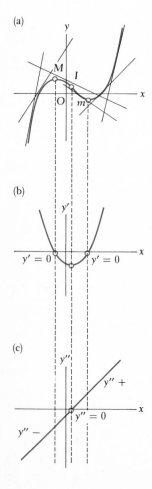

(a)

(b)

(c)

Figure 23–11

These statements comprise what is known as the **second-derivative test for maxima and minima.** This test is often easier to use than the first-derivative test. However, it can happen that $y'' = 0$ at a maximum or minimum point, and in such cases it is necessary that we use the first-derivative test.

The points at which the curve changes from concave up to concave down, or from concave down to concave up, are known as **points of inflection.** Thus point I in the figure is a point of inflection. Inflection points are found by determining those values of x for which the second derivative changes sign. This is analogous to finding maximum and minimum points by the first-derivative test.

Example C

Determine the concavity and find any points of inflection of the function $y = x^3 - 3x$.

This requires inspection and analysis of the second derivative. So we write

$$y' = 3x^2 - 3, \qquad y'' = 6x$$

The second derivative is positive where the function is concave up, and this occurs if $x > 0$. The curve is concave down for $x < 0$, since y'' is negative. Thus, $(0, 0)$ is a point of inflection, since the concavity changes there.

At this point we shall summarize the information regarding the derivatives of a function.

> $f(x)$ increases if $f'(x) > 0$; $f(x)$ decreases if $f'(x) < 0$.
> $f(x)$ increases if $f'(x) > 0$; $f(x)$ decreases if $f'(x) < 0$.
> $f''(x) < 0$.
> $f'(x) = 0$ at a minimum point if $f'(x)$ changes from $-$ to $+$, or if $f''(x) > 0$.
> $f''(x) > 0$ where $f(x)$ is concave up; $f''(x) < 0$ where $f(x)$ is concave down.
> $f''(x) = 0$ at a point of inflection if $f''(x)$ changes from $+$ to $-$ or $-$ to $+$.

The following examples illustrate how the above information is put together to obtain the graph of a function.

Example D

Sketch the graph of $y = 6x - x^2$.

Finding the first two derivatives, we have

$$y' = 6 - 2x = 2(3 - x)$$

and

$$y'' = -2$$

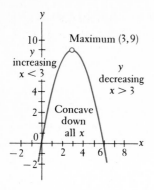

Figure 23–12

We now note that $y' = 0$ for $x = 3$. For $x < 3$ we see that $y' > 0$, which means that y is increasing over this interval. Also, for $x > 3$, we note that $y' < 0$, which means that y is decreasing over this interval.

Since y' changes from positive on the left of $x = 3$, to negative on the right of $x = 3$, the curve has a maximum point where $x = 3$. Since $y = 9$ for $x = 3$, this maximum point is $(3, 9)$.

Since $y'' = -2$, this means that its value remains constant for all values of x. Therefore, there are no points of inflection and the curve is concave down for all values of x. This also shows that the point $(3, 9)$ is a maximum point.

Summarizing, we know that y is increasing for $x < 3$, y is decreasing for $x > 3$, there is a maximum point at $(3, 9)$, and that the curve is always concave down. Using this information we sketch the curve shown in Fig. 23–12.

Example E
Sketch the graph of $y = 2x^3 + 3x^2 - 12x$.

Finding the first two derivatives, we have

$$y' = 6x^2 + 6x - 12 = 6(x + 2)(x - 1)$$

and

$$y'' = 12x + 6 = 6(2x + 1)$$

We note that $y' = 0$ when $x = -2$ and $x = 1$. Using these values in the second derivative we find that y'' is negative (-18) for $x = -2$ and y'' is positive $(+18)$ when $x = 1$. When $x = -2$, $y = 20$; and when $x = 1$, $y = -7$. Therefore, $(-2, 20)$ is a maximum point and $(1, -7)$ is a minimum point.

Next we see that $y' > 0$ if $x < -2$ or $x > 1$. Also, $y' < 0$ for the interval $-2 < x < 1$. Therefore, y is increasing if $x < -2$ or $x > 1$, and y is decreasing if $-2 < x < 1$.

Now we note that $y'' = 0$ when $x = -\frac{1}{2}$, and that $y'' < 0$ when $x < -\frac{1}{2}$, and $y'' > 0$ when $x > -\frac{1}{2}$. When $x = -\frac{1}{2}$, $y = \frac{13}{2}$. Therefore, there is a point of inflection at $(-\frac{1}{2}, \frac{13}{2})$, the curve is concave down if $x < -\frac{1}{2}$, and the curve is concave up if $x > -\frac{1}{2}$.

Finally, by locating the points $(-2, 20)$, $(-\frac{1}{2}, \frac{13}{2})$, and $(1, -7)$, we draw the curve *up* to $(-2, 20)$ and then *down* to $(-\frac{1}{2}, \frac{13}{2})$, with the curve *concave down*. Continuing *down*, but *concave up*, we draw the curve to $(1, -7)$ at which point we start *up* and continue up. This gives us an excellent graph indicating the shape of the curve. If more precision is required, additional points may be used (see Fig. 23–13).

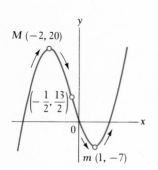

Figure 23–13

Example F
Sketch the graph of $y = x^5 - 5x^4$.

The first two derivatives are

$$y' = 5x^4 - 20x^3 = 5x^3(x - 4) \quad \text{and} \quad y'' = 20x^3 - 60x^2 = 20x^2(x - 3)$$

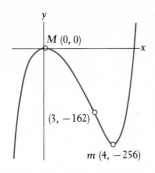

M (0, 0)

(3, −162)

m (4, −256)

Figure 23−14

We now see that $y' = 0$ when $x = 0$ and $x = 4$. When $x = 0$, $y'' = 0$ also, which means we cannot use the second derivative test for maximum and minimum points for this value of x. When $x = 4$, y'' is positive $(+320)$ which means $(4, -256)$ is a minimum point. Next we note that $y' > 0$ if $x < 0$ or $x > 4$ and $y' < 0$ if $0 < x < 4$. Thus, by the first derivative test, there is a maximum at $(0, 0)$. Also, we see that y is increasing if $x < 0$ or $x > 4$ and decreasing if $0 < x < 4$. The second derivative indicates that there is point of inflection at $(3, -162)$. It also indicates that the function is concave down if $x < 3$ $(x \neq 0)$ and concave up if $x > 3$. There is no point of inflection at $(0, 0)$ since the second derivative does not change sign when $x = 0$. From this information we sketch the curve in Fig. 23—14.

Exercises 23—4

In Exercises 1 through 4 determine those values of x for which the given functions are increasing and those values of x for which they are decreasing.

1. $y = x^2 - 2x$ 2. $y = 4 - x^2$
3. $y = 12x - x^3$ 4. $y = x^4 - 6x^2$

In Exercises 5 through 8 find any maximum or minimum points of the given functions. (These are the same functions as in Exercises 1 through 4.)

5. $y = x^2 - 2x$ 6. $y = 4 - x^2$
7. $y = 12x - x^3$ 8. $y = x^4 - 6x^2$

In Exercises 9 through 12 determine the values of x for which the curve of the given function is concave up, those values for which it is concave down, and any points of inflection. (These are the same functions as in Exercises 1 through 4.)

9. $y = x^2 - 2x$ 10. $y = 4 - x^2$
11. $y = 12x - x^3$ 12. $y = x^4 - 6x^2$

In Exercises 13 through 16 use the information from Exercises 1 through 12 to sketch the graphs of the given functions.

13. $y = x^2 - 2x$ 14. $y = 4 - x^2$
15. $y = 12x - x^3$ 16. $y = x^4 - 6x^2$

In Exercises 17 through 24 sketch the graphs of the given functions by determining the appropriate information and points from the first and second derivatives.

17. $y = 12x - 2x^2$ 18. $y = 3x^2 - 1$
19. $y = 2x^3 + 6x^2$ 20. $y = x^3 - 9x^2 + 15x + 1$
21. $y = x^5 - 5x$ 22. $y = x^4 + 8x + 2$
23. $y = 4x^3 - 3x^4$ 24. $y = x^5 - 20x^2$

In Exercises 25 through 28 sketch the indicated curves.

25. A certain projectile follows a path given by $y = x - 0.00025x^2$, where distances are measured in meters. By the methods of this section sketch the graph of the path of the projectile.

26. The deflection y of a beam at a horizontal distance x from one end is given by $y = k(x^3 - 60x^2)$. Sketch the curve representing the beam if it is 10 ft long and $k = 1/5000$.

27. A rectangular box is made from a piece of cardboard 8 in. by 12 in., by cutting a square from each corner and bending up the sides. Express the volume of the box as a function of the side of the square which is cut out, and then sketch the curve of the resulting equation.

28. A tool box whose end is square is to be made from 600 in.² of sheet metal. Express the volume of the box as a function of the side of the square of the end. Sketch the curve of the resulting function.

23–5 More on Curve Sketching

We are now in a position to combine the information from the derivative with information obtainable from the function itself to sketch the graph of a function. The information we can obtain from the function includes topics introduced previously. We can obtain intercepts, symmetry, the behavior of the curve as x becomes large, and the vertical asymptotes of the curve. Also, continuity is important in sketching certain functions. The examples which follow demonstrate how these concepts are used along with the information from the derivatives to sketch the graph of a function. We will find that some of these considerations are of much more value than others in graphing any particular curve.

Example A
Sketch the graph of

$$y = \frac{2}{x^2 + 1}$$

Intercepts: If $x = 0$, $y = 2$, which means that $(0, 2)$ is an intercept. We see that if $y = 0$, there is no corresponding value of x, since $2/(x^2 + 1)$ is a fraction greater than zero for all x. This also indicates that all points on the curve are above the x-axis.

Symmetry: The curve is symmetric to the y-axis since

$$y = \frac{2}{(-x)^2 + 1} \quad \text{is the same as} \quad y = \frac{2}{x^2 + 1}$$

The curve is not symmetric to the x-axis since

$$-y = \frac{2}{x^2 + 1} \quad \text{is not the same as} \quad y = \frac{2}{x^2 + 1}$$

The curve is not symmetric to the origin since

$$-y = \frac{2}{(-x)^2 + 1} \quad \text{is not the same as} \quad y = \frac{2}{x^2 + 1}$$

The value in knowing the symmetry is that we should find those portions of the curve on either side of the y-axis reflections of the other. It is possible to use this fact directly or to use it as a check.

Behavior as x becomes large: We note that as $x \rightarrow +\infty$, $y \rightarrow 0$ since $2/(x^2 + 1)$ is always a fraction greater than zero, but which becomes smaller as x becomes larger. Therefore, we see that $y = 0$ is an asymptote. Either from the symmetry or the function, we also see that $y \rightarrow 0$ as $x \rightarrow -\infty$.

Vertical asymptotes: From the discussion of the hyperbola we recall that an asymptote is a line which a curve approaches. We have already noted that $y = 0$ is an asymptote for this curve. This asymptote, the x-axis, is a horizontal line. *Vertical asymptotes, if any exist, are found by determining those values of x for which the denominator of any term is zero.* Such a value of x makes y undefined. Since $x^2 + 1$ cannot be zero, this curve has no vertical asymptotes. The next example illustrates a curve which has a vertical asymptote.

Derivatives: Since

$$y = \frac{2}{x^2 + 1} = 2(x^2 + 1)^{-1}$$

then

$$y' = -2(x^2 + 1)^{-2}(2x) = \frac{-4x}{(x^2 + 1)^2}$$

Since $(x^2 + 1)^2$ is positive for all values of x, the sign of y' is determined by the numerator. Thus, we note that $y' = 0$ for $x = 0$ and that $y' > 0$ for $x < 0$ and $y' < 0$ for $x > 0$. The curve, therefore, is increasing for $x < 0$, decreasing for $x > 0$, and has a maximum point at $(0, 2)$. Now,

$$y'' = \frac{(x^2 + 1)^2(-4) + 4x(2)(x^2 + 1)(2x)}{(x^2 + 1)^4} = \frac{-4(x^2 + 1) + 16x^2}{(x^2 + 1)^3}$$

$$= \frac{12x^2 - 4}{(x^2 + 1)^3} = \frac{4(3x^2 - 1)}{(x^2 + 1)^3}$$

We note that y'' is negative for $x = 0$, which confirms that $(0, 2)$ is a maximum point. Also, points of inflection are found for the values of x satisfying $3x^2 - 1 = 0$. Thus, $(-\frac{1}{3}\sqrt{3}, \frac{3}{2})$ and $(\frac{1}{3}\sqrt{3}, \frac{3}{2})$ are points of inflection. The curve is concave up if $x < -\frac{1}{3}\sqrt{3}$, or $x > \frac{1}{3}\sqrt{3}$, and concave down if $-\frac{1}{3}\sqrt{3} < x < \frac{1}{3}\sqrt{3}$.

Putting this information together we sketch the curve shown in Fig. 23–15. It might be noted that this curve could have been sketched primarily by use of the fact that $y \rightarrow 0$ as $x \rightarrow +\infty$ and as $x \rightarrow -\infty$, and the fact that a maximum point exists at $(0, 2)$. However, the other parts of the analysis, such as symmetry and concavity, serve as excellent checks and also make the curve more accurate.

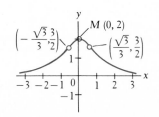

Figure 23–15

Example B

Sketch the graph of $y = x + \dfrac{4}{x}$.

Intercepts: If we set $x = 0$, y is undefined. This means that the curve is not continuous at $x = 0$ and there are no y-intercepts. If we set $y = 0$, $x + 4/x = (x^2 + 4)/x$ cannot be zero since $x^2 + 4$ cannot be zero. Therefore, there are no intercepts. This may seem to be of little value, but we must realize *this curve does not cross either axis.* This, in itself, will be of value when we sketch the graph in Fig. 23–16.

Symmetry: In testing for symmetry, we find that it is not symmetric to either axis. However, this curve does possess symmetry to the origin. This is determined by the fact that when $-x$ replaces x and at the same time $-y$ replaces y, the equation does not change.

Behavior as x becomes large: As $x \to +\infty$, and as $x \to -\infty$, $y \to x$ since $4/x \to 0$. Thus, $y = x$ is an asymptote of the curve.

Vertical asymptotes: As we noted in Example A, vertical asymptotes exist for values of x for which y is undefined. In this equation, $x = 0$ makes the second term on the right undefined and therefore y is undefined. In fact as $x \to 0$ from the positive side, $y \to +\infty$, and as $x \to 0$ from the negative side, $y \to -\infty$. This is derived from the sign of $4/x$ in each case.

Derivatives: Finding the first derivative we have

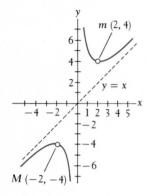

$$y' = 1 - \frac{4}{x^2} = \frac{x^2 - 4}{x^2}$$

The x^2 in the denominator indicates that the sign of the first derivative is the same as its numerator. The numerator is zero if $x = -2$ or $x = 2$. If $x < -2$ or $x > 2$, then $y' > 0$; and if $-2 < x < 2$, $x \neq 0$, $y' < 0$. Thus, y is increasing if $x < -2$ or $x > 2$, and also y is decreasing if $-2 < x < 2$, except at $x = 0$ (y is undefined). Also, $(-2, -4)$ is a maximum point and $(2, 4)$ is a minimum point. The second derivative is $y'' = 8/x^3$. This cannot be zero, but it is negative if $x < 0$ and positive if $x > 0$. Thus, the curve is concave down if $x < 0$ and concave up if $x > 0$. Using this information we have the graph shown in Fig. 23–16.

Figure 23–16

Example C

Sketch the graph of

$$y = \frac{1}{\sqrt{1 - x^2}}$$

Intercepts: If $x = 0$, $y = 1$. If $y = 0$, $1/\sqrt{1 - x^2}$ would have to be zero, but it cannot since it is a fraction with 1 as the numerator for all values of x. Thus, $(0, 1)$ is an intercept. We also note that if x is numerically greater than 1, we have a negative value under the radical. Thus, all points on the curve are between the lines $x = -1$ and $x = 1$.

Symmetry: The curve is symmetric to the y-axis.

Behavior as x becomes large: The values of x cannot be considered beyond 1 or -1, for any value of $x < -1$ or $x > 1$ gives imaginary values for y. Thus, the curve does not exist for values of $x < -1$ or $x > 1$.

Vertical asymptotes: If $x = 1$ or $x = -1$, y is undefined. In each case as $x \to 1$ and as $x \to -1$, $y \to +\infty$.

Derivatives:

$$y' = -\tfrac{1}{2}(1 - x^2)^{-3/2}(-2x) = \frac{x}{(1 - x^2)^{3/2}}$$

We see that $y' = 0$ if $x = 0$. If $-1 < x < 0$, $y' < 0$ and also if $0 < x < 1$, $y' > 0$. Thus the curve is decreasing if $-1 < x < 0$ and increasing if $0 < x < 1$. There is a minimum point at $(0, 1)$:

$$y'' = \frac{(1 - x^2)^{3/2} - x(\tfrac{3}{2})(1 - x^2)^{1/2}(-2x)}{(1 - x^2)^3} = \frac{(1 - x^2) + 3x^2}{(1 - x^2)^{5/2}}$$

$$= \frac{2x^2 + 1}{(1 - x^2)^{5/2}}$$

The second derivative cannot be zero since $2x^2 + 1$ is positive for all values of x. The second derivative is also positive for all permissible values of x which means the curve is concave up for these values.

Using this information we sketch the graph in Fig. 23–17.

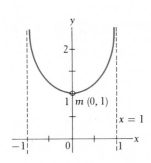

Figure 23–17

Exercises 23–5

In the following exercises use the method of the examples of this section to sketch the indicated curves.

1. $y = \dfrac{4}{x^2}$ 2. $y = \dfrac{x}{x - 2}$ 3. $y = x^2 + \dfrac{2}{x}$

4. $y = x + \dfrac{4}{x^2}$ 5. $y = \dfrac{x^2}{x + 1}$ 6. $y = \dfrac{9x}{x^2 + 9}$

7. $y = \dfrac{1}{x^2 - 1}$ 8. $y = x\sqrt{1 - x^2}$

9. Sketch R_T as a function of R for the function relating the combined resistance of the circuit given in Exercise 23 of Section 22–4.

10. A fence is to be constructed to enclose a rectangular area of 20,000 m². A previously constructed wall is to be used for one side. Sketch the length of fence to be built as a function of the side of the fence parallel to the wall.

11. A cylindrical oil drum is to be made such that it will contain 20 kL. Sketch the area of sheet metal required for construction as a function of the radius of the drum.

12. Sketch P as a function of r for the function relating power and resistance of the circuit given in Exercise 27 of Section 22–6. Assume values of $E = 6$ V and $R = 1$ Ω.

23–6 Applied Maximum and Minimum Problems

Problems from various applied situations frequently occur which require finding a maximum or minimum value of some function. If the function is known, the methods we have already discussed can be used directly. This is illustrated in the following example.

Example A

An automobile manufacturer, in testing a new engine on one of its models, found that the efficiency e of the engine as a function of the speed s of the car was given by $e = 0.768s - 0.00004s^3$. Here, e is given in percent and s is given in km/h. What is the maximum efficiency of the engine?

Since we wish to find a maximum value, we find the derivative of e with respect to s.

$$\frac{de}{ds} = 0.768 - 0.00012s^2$$

We then set the derivative equal to zero in order to find the value of s for which a maximum may occur.

$$0.768 - 0.00012s^2 = 0$$
$$0.00012s^2 = 0.768$$
$$s^2 = 6400$$
$$s = 80.0 \text{ km/h}$$

We know that s must be positive to have meaning in this problem. Therefore, the apparent solution of $s = -80$ is discarded. The second derivative is

$$\frac{d^2s}{de^2} = -0.00024s$$

which is negative for any positive value of s. Thus, a maximum occurs for $s = 80.0$. Substituting $s = 80.0$ in the function for e, we obtain

$$e = 0.768(80.0) - 0.00004(80.0^3)$$
$$= 61.44 - 20.48 = 40.96$$

Therefore, to the nearest tenth, the maximum efficiency is 41.0%, and this occurs for $s = 80.0$ km/h.

In many problems for which a maximum or minimum value is to be determined, the function is itself not given explicitly. It is necessary to find the function from the statement of the problem. The principal difficulty which arises in these problems is finding this function. Once we determine this function, we take a derivative with respect to a variable in which it is expressed, and then set the derivative equal to zero. Sufficient information must be given so that a unique solution is possible. Often this information is available, but we must analyze the wording of the problem carefully to find it. Also, many times it is more convenient to set up more than one function, based on the given information, and to find a derivative of each with respect to the same variable. We can then eliminate by substitution any undesired variables or derivatives which appear in each. The following examples illustrate these methods.

Example B

Find the number which exceeds its square by the greatest amount.

We must set up a function which expresses the difference between a number and its square. Let the number be x, and let the difference be D. Then $D = x - x^2$. Taking a derivative, we have

$$\frac{dD}{dx} = 1 - 2x$$

Setting the derivative equal to zero, we have $0 = 1 - 2x$. Solving for x, we obtain the result $x = \frac{1}{2}$. The second derivative gives $d^2D/dx^2 = -2$, which states that the second derivative is always negative. This means that whenever the first derivative is zero, it represents a maximum. In many problems it is not necessary to test for maximum or minimum, since the nature of the problem will indicate which must be the case. For example, in this problem we know that numbers greater than 1 do not exceed their squares at all. The same is true for all negative numbers. Thus, the answer must lie between 0 and 1.

Example C

Find the maximum possible area of a rectangle whose perimeter is 16 units.

When we reflect on this problem, we see that there are limitless possibilities for rectangles of a perimeter of 16 units and differing areas (see Fig. 23–18). For example, if the sides are 7 and 1, the area is 7. If the sides are 6 and 2, the area is 12. Therefore, we set up a function for the area of a rectangle in terms of its sides x and y:

$$A = xy$$

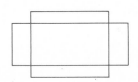

Figure 23–18

We know, however, one other important fact: that the perimeter is 16. Thus, $2x + 2y = 16$. Solving for y, we have $y = 8 - x$. Substituting this equation into the expression for the area, we have

$$A = x(8 - x) = 8x - x^2$$

Taking a derivative of this function and setting it equal to zero, we have

$$\frac{dA}{dx} = 8 - 2x$$

$$8 - 2x = 0$$

$$x = 4$$

By noting values of the derivative near 4 or by finding the second derivative, we can show that we have a maximum for $x = 4$. Thus, $x = 4$ and $y = 4$ give the maximum area of 16 square units.

Example D

Find the point on the parabola $y = x^2$ which is nearest the point $(6, 3)$.

In this example we must set up a function for this distance between a general point (x, y) on the parabola and the point $(6, 3)$. This relation is

$$D = \sqrt{(x - 6)^2 + (y - 3)^2}$$

However, to make it easier for us to take derivatives, we shall square both sides of this expression. If a function is a minimum, then so is its square. We shall also use the fact that the point (x, y) is on $y = x^2$ by replacing y by x^2. Thus, we have

$$D^2 = (x - 6)^2 + (x^2 - 3)^2 = x^2 - 12x + 36 + x^4 - 6x^2 + 9$$
$$= x^4 - 5x^2 - 12x + 45$$

$$\frac{dD^2}{dx} = 4x^3 - 10x - 12$$

$$0 = 2x^3 - 5x - 6$$

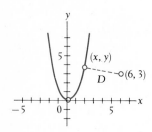

Figure 23–19

Using synthetic division or some similar method, we find that the solution to this equation is $x = 2$. Thus, the required point on the parabola is $(2, 4)$. (See Fig. 23–19.) We can show that we have a minimum by analyzing the first derivative, the second derivative, or by noting that points at much greater distances exist (therefore, it cannot be a maximum).

Example E

A company determines that it can sell 1000 units of a product per month if the price is $5 for each unit. It also estimates that for each 1-cent reduction in unit price, 10 more units can be sold. Under these conditions, what is the maximum possible income and what price per unit gives this income?

If we let $x =$ the number of units over 1000 sold, the total number of units sold is $1000 + x$. The price for each unit is $5 less 1 cent (0.01 dollars) for each block of ten units over 1000 which are sold. Thus, the price for each unit is

$$5 - 0.01\left(\frac{x}{10}\right) \quad \text{or} \quad 5 - 0.001x \text{ dollars}$$

The income I is the number of units sold times the price of each unit. Therefore, we have

$$I = (1000 + x)(5 - 0.001x)$$

Multiplying and finding the first derivative, we have

$$I = 5000 + 4x - 0.001x^2$$

$$\frac{dI}{dx} = 4 - 0.002x$$

Setting this derivative equal to zero, we have

$$0 = 4 - 0.002x$$
$$x = 2000$$

We note that if $x < 2000$, the derivative is positive, and if $x > 2000$, the derivative is negative. Therefore, if $x = 2000$, I is at a maximum. This means that the maximum income is derived if 2000 units over 1000 are sold, or 3000 units in all. This in turn means that the maximum income is $9000 and the price per unit is $3. These values are found by substituting $x = 2000$ into the expression for I and for the price.

Example F

Find the most economical dimensions for a cylindrical cup which holds a specified volume.

When we analyze the wording of the problem carefully, we see that we are to minimize the surface area of a right circular cylinder which has a bottom, but no top. Also the volume is to be considered as constant. Thus, we set up expressions for the surface area and volume:

$$A = \pi r^2 + 2\pi rh, \qquad V = \pi r^2 h$$

Since the volume is not specified numerically, it is easier to take the derivative of each of these formulas separately. The derivatives may be taken with respect to either r or h. Let us choose r, and write

$$\frac{dA}{dr} = 2\pi r + 2\pi r \frac{dh}{dr} + 2\pi h, \qquad \frac{dV}{dr} = 0 = 2\pi rh + \pi r^2 \left(\frac{dh}{dr}\right)$$

Setting dA/dr equal to zero, and at the same time eliminating the derivative dh/dr by substitution, we have

$$0 = 2\pi r + 2\pi r \left(-\frac{2h}{r}\right) + 2\pi h$$

or

$$0 = r - 2h + h$$

This last equation tells us that $h = r$, which is the relation between dimensions for the most economical cup (see Fig. 23—20).

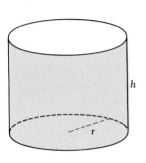

Figure 23—20

Example G

The illuminance of a light source at any point varies directly as the strength of the source and inversely as the square of the distance from the source. Given two sources, of strengths 8 units and 1 unit, respectively, which are 100 ft apart, determine at what point between them the illuminance is the least, assuming that the illuminance at any point is the sum of the illuminances of the two sources.

Let I = the sum of the illuminances and
x = the distance from the source of strength 8.

Then we find that

$$I = \frac{8}{x^2} + \frac{1}{(100 - x)^2}$$

is the function between the illuminance and the distance from the source of strength 8. We must now take a derivative of I with respect to x, set it equal to zero, and solve for x to find the point at which the illuminance is a minimum:

$$\frac{dI}{dx} = -\frac{16}{x^3} + \frac{2}{(100 - x)^3} = \frac{-16(100 - x)^3 + 2x^3}{x^3(100 - x)^3}$$

This function will be zero if the numerator is zero. Therefore, we have

$$2x^3 - 16(100 - x)^3 = 0 \quad \text{or} \quad x^3 = 8(100 - x)^3$$

Taking cube roots of each side, we have

$$x = 2(100 - x) \quad \text{or} \quad x = 66.7 \text{ ft}$$

The point where the illuminance is a minimum is 66.7 ft from the 8-unit source of illuminance.

Exercises 23–6

In the following exercises solve the given maximum and minimum problems.

1. The height s (in feet) of an object thrown vertically upward from the ground is given by $s = 112t - 16t^2$, where t is measured in seconds. What is the greatest height to which the object will go?

2. A company determines that the profit p (in dollars) from the sale of x units of a product, is given by $p = -2x^2 + 400x - 100$. What is the maximum profit which can be made from the sale of this product?

3. The cutting speed s (in feet per minute) of a saw in cutting a particular type of metal piece is given by $s = \sqrt{t - 4t^2}$ where t is the time in seconds. What is the maximum cutting speed in this operation?

4. The electric resistance (in ohms) of a particular type of resistor as a function of the temperature (in degrees Celsius) is given by

$$R = \sqrt{0.001 T^4 - 4T + 100}$$

What is the minimum resistance of this type of resistor?

5. The product of two positive numbers is 64. Find the numbers if their sum is a minimum.

6. The sum of two numbers is 40. Find the numbers such that their product is a maximum.

7. A rectangular field bounded on one side by a river is to be fenced in. If no fencing is required along the river, what is the maximum area which can be enclosed if 2000 ft of fencing are used?

8. An architect is designing a rectangular building in which the front wall costs twice as much per linear meter as the other three walls. The building is to cover 1350 m². What dimensions must it have such that the cost of the walls is a minimum?

9. Find the dimensions of the largest rectangle which can be inscribed in a quarter-circle of radius 4.

10. What is the area of the largest rectangle which can be inscribed in the first quadrant under the parabola $y = 4 - x^2$?

$10. \quad \dfrac{16\sqrt{3}}{9}$

11. If the hypotenuse of a right triangle is 10, what must the sides be so that the area is a maximum?

$12. \ 108 \, ft^3$

12. A box with a square base is open at the top. If 108 ft² of material are used, what is the maximum volume possible for the box?

13. What is the maximum slope of the curve $y = 6x^2 - x^3$?

14. What is the minimum slope of the curve $y = x^5 - 10x^2$?

15. A person is in a boat 4 km from the nearest point P on a straight shoreline. The person wishes to go to point A which is on the shore, 5 km from P. If the person can row at 3 km/h and walk at 5 km/h, how far along the shore from P towards A should the boat land in order that the person can reach A in the least time?

16. A poster is to have margins of 4 in. at the top and bottom and margins of 2 in. at each side. The printed matter of the poster is to cover 50 in.². Find the dimensions of the poster if the total area is to be a minimum.

17. An open box is to be made from a square piece of cardboard whose sides are 8 in. long by cutting equal squares from the corners and bending up the sides. Determine the side of the square which is to be cut out so that the volume of the box may be a maximum.

$18. \ r = \tfrac{1}{2}h$

18. Find the relation between the radius and the height of a right circular cylinder for which the total surface area is to be a minimum for a given volume.

19. A certain area is a rectangle with a semicircle at each end. Find the open side (not between the rectangle and the semicircle) of the rectangle if the perimeter is 8 units and the area of the rectangle is to be a maximum.

$20. \ 1250$

20. A company finds that there is a net profit of $10 for each of the first 1000 units produced each week. For each unit over 1000 produced, there is 2 cents less profit per unit. How many units should be produced each week to net the greatest profit?

21. The deflection y of a beam of length L at a horizontal distance x from one end is given by $y = k(2x^4 - 5Lx^3 + 3L^2x^2)$, where k is a constant. For what value of x does the maximum deflection occur?

$22. \ 0.600 \, \Omega$

22. The electric power (in watts) produced by a certain source is given by

$$P = \frac{144r}{(r + 0.6)^2}$$

where r is the resistance in the circuit. For what value of r is the power a maximum?

23. The strength of a rectangular beam is proportional to the product of its width and the square of its depth. Find the dimensions of the strongest beam that can be cut from a circular log 16 in. in diameter.

$24. \ r = 0.969 \, cm$
$\quad\ h = 10.6 \, cm$

24. A specially made cylindrical container is made of stainless steel sides and bottom and a silver top. If silver is ten times as expensive as stainless steel, what are the most economical dimensions of the container if it is to hold 10π cm³?

23–7 Exercises for Chapter 23

In Exercises 1 through 4 find the equations of the tangent and normal lines.

1. Find the equation of the line tangent to the parabola $y = 3x - x^2$ at the point $(-1, -4)$.

2. Find the equation of the line tangent to the curve $y = x^2 - \dfrac{2}{x}$ at the point $(2, 3)$.

3. Find the equation of the line normal to $y = \dfrac{x}{4x - 1}$ at the point $(1, \tfrac{1}{3})$.

4. Find the equation of the line normal to $y = \dfrac{1}{\sqrt{x - 2}}$ at the point $(6, \tfrac{1}{2})$.

In Exercises 5 through 8 determine the required velocities and accelerations.

5. Given that the x- and y-coordinates of a moving particle are given as a function of time by the parametric equations $x = t^4 - t$, $y = \dfrac{1}{t}$, find the magnitude and direction of the velocity when $t = 2$.

6. A particle moves along the curve of $y = \dfrac{1}{x + 2}$ with a constant velocity in the x-direction of 4 cm/s. Find v_y at $(2, \tfrac{1}{4})$.

7. Find the magnitude and direction of the acceleration for the particle in Exercise 5.

8. Find the magnitude and direction of the acceleration of the particle in Exercise 6.

In Exercises 9 through 16 sketch the graph of the given functions by information obtained from the function as well as information obtained from the derivatives.

9. $y = 4x^2 + 16x$ 10. $y = 2x^2 + x - 1$

11. $y = 27x - x^3$ 12. $y = x^3 + 2x^2 + x + 1$

13. $y = x^4 - 32x$ 14. $y = x^4 + 4x^3 - 16x$

15. $y = \dfrac{x}{x + 1}$ 16. $y = x^3 + \dfrac{3}{x}$

In Exercises 17 through 32 solve the given problems.

17. The parabolas $y = x^2 + 2$ and $y = 4x - x^2$ are tangent to each other. Find the equation of the line tangent to them at the point of tangency.

18. Find the equation of the line tangent to $y = x^4 - 8x$ and parallel to $y + 4x + 3 = 0$.

19. A projectile moves according to the equations $x = 1200t$ and $y = -490t^2$, where distances are measured in centimeters and time is measured in seconds. Find the magnitude and direction of velocity after 3.00 s.

20. An electron is moving along the path $y = 1/x$ at the constant velocity of 100 m/s. Find the velocity in the x-direction when the electron is at the point $(2, \tfrac{1}{2})$.

21. Sketch a continuous curve having the following characteristics.

$f(0) = 2$ $f'(x) < 0$ for $x < 0$

$f''(x) > 0$ for all x $f'(x) > 0$ for $x > 0$

22. Sketch a continuous curve having the following characteristics.

$f(0) = 1$ $f'(0) = 0$

$f''(x) < 0$ for $x < 0$ $f'(x) > 0$ for $|x| > 0$

$f''(x) > 0$ for $x > 0$

23. Side a of a triangle is increasing at the rate of 2.50 in./min, and side b is increasing at the rate of 3.70 in./min. If the angle between sides a and b is always 60°, find the rate of change of side c when $a = 5.60$ in. and $b = 8.00$ in.

24. The current I through a circuit with a resistance R and a battery whose voltage is E and whose internal resistance is r is given by $I = E/(R + r)$. If R changes at the rate of 0.250 Ω/min, how fast is the current changing when $R = 6.25$ Ω, if $E = 3.10$ V and $r = 0.230$ Ω?

25. A machine part is to be in the shape of a circular sector of radius r and central angle θ. Find r and θ if the area is one unit and the perimeter is a minimum.

26. A beam of rectangular cross section is to be cut from a log 2 ft in diameter. The stiffness varies as the width and the cube of the depth. What dimensions will give the beam maximum stiffness?

27. Sketch a graph of van der Waals equation (see Exercise 26 of Section 22–6), assuming the following values: $R = T = a = 1$ and $b = 0$. For many gases the value of a is much greater than that for b. Even though the values of $R = T = 1$ are not realistic, the *shape* of the curve will be correct for the assumed value of $b = 0$.

28. The weekly profits of a corporation for a particular year are given by $p = x^4 - 40x^3$, where x is the week of the year. Sketch the graph of p vs. x, assuming it to be a continuous function over the appropriate interval.

29. Two ships leave the same port at the same time, one traveling south at the rate of 12 km/h and the other traveling east at the rate of 5 km/h. At what rate is the distance between them changing two hours after they leave the port?

30. The base of a conical machine part is being milled such that the height is decreasing at the rate of 0.05 cm/min. If the part originally had a radius of 1 cm and height of 3 cm, how fast is the volume changing when the height is 2.8 cm?

31. An alpha-particle moves through a magnetic field along the parabolic path $y = x^2 - 4$. Determine the closest that the particle comes to the origin.

32. A Norman window has the form of a rectangle surmounted by a semicircle. Find the dimensions (radius of circular part and height of rectangular part) of the window that will admit the most light if the perimeter of the window is 12 ft.

24

Integration

24–1 Differentials

Physical investigations of phenomena often lead to information regarding the rate of change of a variable. Then to arrive at the functional relationship between variables, we have to reverse the process of differentiation. Other applications of this reverse process involve finding areas under curves and finding volumes of solids. As we shall see, these basic problems, although apparently rather distinct, have a very similar mathematical interpretation.

Although the primary concern of this chapter is the development of the inverse process of differentiation, it is necessary to first introduce certain concepts and notation in this section before proceeding to the inverse process. The material presented here is necessary to properly relate differentiation and the methods we shall develop.

We shall now define the **differential** of a function $y = f(x)$ as

$$dy = f'(x)\ dx \tag{24-1}$$

In Eq. (24–1), the quantity dy is the differential of y, and dx is the differential of x. The differential dx is defined as equal to Δx, the increment in x. We define it purposely in this way, so that $f'(x) = dy/dx$. In this way we can interpret the derivative as the ratio of the differential of y to the differential of x. That is, now the derivative can be considered as a fraction. Although we had previously used the notation dy/dx for the derivative, we had not interpreted it as a fraction.

Example A

Find the differential of $y = 3x^5 - x$.

In this example, $f(x) = 3x^5 - x$, which in turn means that $f'(x) = 15x^4 - 1$. Thus,

$$dy = (15x^4 - 1)\, dx$$

Example B

Find the differential of $y = (2x^3 - 1)^4$.

$$dy = 4(2x^3 - 1)^3 (6x^2)\, dx = 24x^2 (2x^3 - 1)^3\, dx$$

Example C

Find the differential of $y = \dfrac{4x}{x^2 + 4}$.

$$dy = \frac{(x^2 + 4)(4) - (4x)(2x)}{(x^2 + 4)^2}\, dx = \frac{4x^2 + 16 - 8x^2}{(x^2 + 4)^2}\, dx$$

$$= \frac{-4x^2 + 16}{(x^2 + 4)^2}\, dx = \frac{-4(x^2 - 4)}{(x^2 + 4)^2}\, dx$$

We shall find that the definition of the differential will clarify the connection between the various interpretations of the inverse process of differentiation. Although the principal purpose in introducing it here is to be able to use the notation, there are useful direct applications of the differential.

The applications of the differential are based upon the fact that the differential of y, dy, closely approximates the increment in y, Δy, if the differential of x, dx, is small. To understand this statement, let us look at Fig. 24–1. Recalling the meaning of Δx and Δy, we see that the points $P(x, y)$ and $Q(x + \Delta x, y + \Delta y)$ lie on the curve of $f(x)$. However, $f'(x) = dy/dx$ at P, which means if we draw a tangent line at P, its slope may be indicated by dy/dx. By choosing $\Delta x = dx$, we can see the difference between Δy and dy. It can be seen that as dx becomes smaller, Δy more nearly equals dy.

For given changes in x, it is necessary to use the Δ-process to find the exact change, Δy, in y. However, *for small values of Δx, dy can be used to approximate Δy closely.* Generally dy is much more easily determined than is Δy.

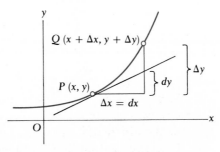

Figure 24–1

Example D

Calculate Δy and dy for $y = x^3 - 2x$ for $x = 3$ and $\Delta x = 0.1$.

By the Δ-process we find

$$\Delta y = 3x^2 \, \Delta x + 3x(\Delta x)^2 + (\Delta x)^3 - 2 \, \Delta x$$

Using the given values we find

$$\Delta y = 3(9)(0.1) + 3(3)(0.01) + (0.001) - 2(0.1) = 2.591$$

The differential of y is

$$dy = (3x^2 - 2) \, dx$$

Since $dx = \Delta x$, we have

$$dy = [3(9) - 2](0.1) = 2.5$$

Thus, $\Delta y = 2.591$ and $dy = 2.5$. We see that in this case dy is very nearly equal to Δy.

The fact that dy can be used to approximate Δy is useful in determining the error in a result, if the data are in error, or the equivalent problem of finding the change in a result if a change is made in the data. Even though such changes could be found by use of a calculator, the differential can be used to set up a general expression for the change in a particular function. The following example illustrates the method.

Example E

The edge of a cube was determined to be 9.00 in. From this value the volume was found. Later it was discovered that the value of the edge was 0.20 in. too small. Approximately by how much was the volume in error?

The volume V of a cube, in terms of an edge, e, is $V = e^3$. Since we wish to find the change in V for a given change in e, this means we want the value of dV for $de = 0.20$ in. Thus,

$$dV = 3e^2 \, de$$
$$dV = 3(9.00)^2(0.20) = 48.6 \text{ in.}^3$$

We see that in this case the volume was in error by about 49 in.^{3}. As long as de is small compared to e, we can calculate an error or change in the volume of a cube by calculating the value of $3e^2 \, de$.

Often when considering the error of a given value or result, the actual numerical value of the error, the **absolute error**, is not as important as its size is in relation to the size of the quantity itself. *The ratio of the absolute error to the size of the quantity itself is known as the* **relative error.** The relative error is commonly expressed as a percentage.

Example F

Referring to Example E, we see that the absolute error in the edge was 0.20 in. The relative error in the edge was $0.20/9.00 = 0.022 = 2.2\%$. The absolute error in the volume is 49 in.^{3}, whereas the relative error in the volume is $49/729 = 0.067 = 6.7\%$.

Exercises 24–1

In Exercises 1 through 8 find the differential of each of the given functions.

1. $y = x^5 + x$

2. $y = 3x^2 + 6$

3. $y = (x^2 - 1)^4$

4. $y = (4 + 3x)^{1/3}$

5. $y = x^2(1 - x)^3$

6. $y = x\sqrt{1 - 4x}$

7. $y = \dfrac{x}{x + 1}$

8. $y = \dfrac{1}{\sqrt{1 - x^3}}$

In Exercises 9 through 12 find the values of Δy and dy for the given values of x and Δx.

9. $y = 7x^2 + 4x$, $x = 4$, $\Delta x = 0.2$

10. $y = 2x^2 - 3x + 1$, $x = 5$, $\Delta x = 0.15$

11. $y = 2x^3 - 4x$, $x = 2.5$, $x = 0.05$

12. $y = x^4$, $x = 3.2$, $\Delta x = 0.08$

In Exercises 13 through 16 determine the value of dy for the given values of x and Δx. Compare with values of $f(x + \Delta x) - f(x)$ found by use of a calculator.

13. $y = (1 - 3x)^5$, $x = 1$, $\Delta x = 0.01$

14. $y = (x^2 + 2x)^3$, $x = 7$, $\Delta x = 0.02$

15. $y = x\sqrt{1 + 4x}$, $x = 12$, $\Delta x = 0.06$

16. $y = \dfrac{x}{\sqrt{6x - 1}}$, $x = 3.5$, $\Delta x = 0.025$

In Exercises 17 through 24 solve the given problems by finding the appropriate differential.

17. If the side of a square is measured to be 16.0 in., and an error of 0.5 in. is found in the measurement, approximately by how much is the original calculation of area in error?

18. If the radius of a circle is measured to be 8.00 cm, and the measurement has a maximum possible error of 0.03 cm, what is the maximum possible relative error in the area of the circle?

19. The voltage of a certain thermocouple as a function of temperature is given by $E = 6.2T + 0.0002T^3$. What is the approximate change in voltage if the temperature changes from 100°C to 101°C?

20. The velocity of an object rolling down a certain inclined plane is given by $v = \sqrt{100 + 16h}$, where h is the distance traveled along the plane by the object. What is the increase in velocity (in feet per second) of an object in moving from 20.0 to 20.5 ft along the plane? What is the relative change in the velocity?

21. What is the volume of metal used to make a right circular cylindrical container of radius 3.00 in. and height 4.00 in., if the thickness of sheet metal it was made from is 0.02 in.?

22. A precisely measured 10.0-Ω resistor (any error in its value will be considered negligible) is put in parallel with a variable resistor of resistance R. The combined resistance of the two resistors is $R_T = 10R/(10 + R)$. What is the relative error of the combined resistance, if R is measured at 40.0 Ω, with a possible error of 1.5 Ω?

23. Show that an error of 2% in the measurement of the side of a square results in an error of approximately 4% in the calculation of the area.

24. Show that the relative error in the calculation of the volume of a sphere is approximately three times the relative error in the measurement of the radius.

24–2 Antiderivatives

Since many kinds of problems in many areas, including science and technology, can be solved by reversing the process of finding a derivative or a differential, we shall introduce the basic technique of this procedure in this section. *This reverse process is known as* **antidifferentiation.** In the next section we shall formalize the process, but it is only the basic idea that is the topic of this section. Many of the applications are found in Chapter 25. The following example illustrates the method.

Example A
Find a function for which the derivative is $8x^3$. That is, find an antiderivative of $8x^3$.

We know that, when we set out to find the derivative of a polynomial, we reduce the power of x by 1. Also we multiply the coefficient by the power of x. Thus, the power of the function must have been 4, since the power in the derivative is 3. A factor of 4 of the derivative must also be divided out in the process of finding the function. If we write the derivative as $2(4x^3)$, we recognize $4x^3$ as the derivative of x^4. Therefore, the desired function is $2x^4$, which means that an antiderivative of $8x^3$ is $2x^4$. We can verify our result by finding the derivative.

Example B
Find an antiderivative of $x^2 + 2x$.

As for the x^2, we know that the power of x required in an antiderivative is 3. Also, to make the coefficient correct, we must multiply by $\frac{1}{3}$. The $2x$ should be recognized as the derivative of x^2. Therefore, we have the antiderivative as $\frac{1}{3}x^3 + x^2$.

In Examples A and B, we note that we could add any constant to the antiderivative given as the result, and still have a correct antiderivative. This is due to the fact that the derivative of a constant is zero. This is considered further in the following section. For the examples and exercises in this section, we will not include any constants in the results.

A great many functions to which we must apply the process of antidifferentiation are not polynomials. It is these functions which may cause more difficulty in the general process of antidifferentiation. The reader is advised to pay special attention to the following examples, for they illustrate a type of problem which *will* be found to be very important.

Example C
Find an antiderivative of $3(x^3 - 1)^2(3x^2)$.

Noting that we have a power of $x^3 - 1$ in the derivative, it is reasonable that the antiderivative may include a power of $x^3 - 1$. Since, in the derivative, $x^3 - 1$ is raised to the power 2, the antiderivative would then have $x^3 - 1$ raised to the power 3. Noting that the derivative of $(x^3 - 1)^3$ is $3(x^3 - 1)^2(3x^2)$, the desired antiderivative is $(x^3 - 1)^3$. We note that the factor of $3x^2$ does not appear in the antiderivative, for it was included from the process of finding a derivative. Therefore, it must be present for $(x^3 - 1)^3$ to be the proper antiderivative, but it must also be excluded in the process of antidifferentiation.

Example D
Find an antiderivative of $(2x + 1)^{1/2}$.

Here we note a power of $2x + 1$ in the derivative, which infers that the antiderivative has a power of $2x + 1$. Since in finding a derivative 1 is subtracted from the power of $2x + 1$, we should add 1 in finding the antiderivative. Thus, we should have $(2x + 1)^{3/2}$ as part of the antiderivative. Finding a derivative of $(2x + 1)^{3/2}$ we obtain $\frac{3}{2}(2x + 1)^{1/2}(2)$ $= 3(2x + 1)^{1/2}$. This differs from the given derivative by the factor of 3. Thus, if we write $(2x + 1)^{1/2} = \frac{1}{3}[3(2x + 1)^{1/2}]$, we have the required antiderivative as $\frac{1}{3}(2x + 1)^{3/2}$. Checking, the derivative of $\frac{1}{3}(2x + 1)^{3/2}$ is $\frac{1}{3}(\frac{3}{2})(2x + 1)^{1/2}(2) = (2x + 1)^{1/2}$.

Exercises 24–2

In the following exercises find antiderivatives of the given derivatives. That is, find functions for which the derivatives are given.

1. $3x^2$
2. $5x^4$
3. $6x^5$

4. $10x^9$
5. $6x^3 + 1$
6. $12x^5 + 2x$

7. $2x^2 - x$
8. $x^2 - 5$
9. $-\dfrac{1}{x^2}$

10. $-\dfrac{2}{x^3}$
11. $-\dfrac{6}{x^4}$
12. $-\dfrac{8}{x^5}$

13. $2x^4 + 1$
14. $3x^3 - 5x^2 + 3$
15. $6(2x + 1)^5(2)$

16. $3(x^2 + 1)^2(2x)$
17. $4(x^2 - 1)^3(2x)$
18. $5(2x^4 + 1)^4(8x^3)$

19. $x^3(2x^4 + 1)^4$
20. $x(1 - x^2)^7$
21. $\frac{3}{2}(6x + 1)^{1/2}(6)$

22. $\frac{5}{4}(1 - x)^{1/4}(-1)$
23. $(3x + 1)^{1/3}$
24. $(4x + 3)^{1/2}$

24–3 The Indefinite Integral

In the previous section, in developing the basic technique of finding an antiderivative, we noted that the results given are not unique. That is, we could have added any constant to the answers and the result would still have been correct. Again, this is the case since the derivative of a constant is zero.

Example A
The derivatives of x^3, $x^3 + 4$, $x^3 - 7$, and $x^3 + 4\pi$ are all $3x^2$. This means that any of the functions listed, as well as others, would be a proper answer to the problem of finding an antiderivative of $3x^2$.

From Section 24–1, we know that the differential of a function $F(x)$ can be written as $d[F(x)] = F'(x)dx$. Therefore, since finding a differential of a function is very closely related to finding the derivative, so is the antiderivative very closely related to the process of finding the function for which the differential is known.

The notation used for finding the general form of the antiderivative, the **indefinite integral,** *is written in terms of the differential.* Thus, the indefinite integral of a function $f(x)$, for which $dF(x)/dx = f(x)$, or $dF(x) = f(x)dx$, is defined as

$$\int f(x)\, dx = F(x) + C \qquad (24\text{--}2)$$

where C *is an arbitrary constant, called the* **constant of integration.** It represents any of the constants which may be attached to an antiderivative to have a proper result. We must have additional information beyond a knowledge of the differential to assign a specific value to C. *The symbol* $\int$ *is the* **integral sign,** *and indicates that the inverse of the differential is to be found. Determining the indefinite integral is called* **integration,** which we can see is essentially the same as finding an antiderivative.

Example B

$$\int 5x^4\, dx = x^5 + C.$$

We might think that the inclusion of this constant C would affect the derivative of the function x^5. However, the only effect of the C is to raise or lower the curve. The slope of $x^5 + 2$, $x^5 - 2$, or any function of the form $x^5 + C$ is the same for any given value of x. As Fig. 24–2 shows, tangents drawn to the curves are all parallel for the same value of x.

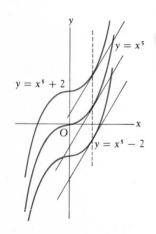

$y = x^5$

$y = x^5 + 2$

$y = x^5 - 2$

Figure 24–2

At this point we shall derive some basic formulas for integration. Since

$$\frac{d(cu)}{dx} = c\frac{du}{dx}$$

where u is a function of x, we can write

$$\int c\, du = c \int du = cu + C \qquad (24\text{--}3)$$

Also, since the derivative of a sum of functions equals the sum of the derivatives, we write

$$\int (du + dv) = u + v + C \qquad (24\text{--}4)$$

To find the derivative of a power of a function, we multiply by the power and subtract 1 from it. *To find the integral, we reverse this by adding 1 to the power of $f(x)$ in $f(x)$ dx, and dividing by this new power.* The power formula for integration is therefore

$$\int u^n \, du = \frac{u^{n+1}}{n+1} + C \qquad n \neq -1 \tag{24–5}$$

When we use the formulas above, we must be able to recognize not only the proper form but also the component parts of the given formula. Unless you can do this and unless you have a good knowledge of differentiation, you will have trouble in applying Eq. (24–5). Most of the difficulty, if it exists, arises from improper identification of du. The following examples illustrate the use of the above formulas.

Example C
Integrate: $\int 6x \, dx$.
   First we must identify u, n, du, and any multiplying constants. By noting that 6 is a multiplying constant, we identify x as u, which in turn means that dx must be du and $n = 1$. Hence,

$$\int 6x \, dx = 6 \int x \, dx = 6\left(\frac{x^2}{2}\right) + C = 3x^2 + C$$

We see that our result checks since the differential of $3x^2 + C$ is $6x \, dx$.

Example D
Integrate: $\int (5x^3 - 6x^2 + 1) \, dx$.
   In solving this problem, we have to use a combination of all three of the above formulas. Thus, we have

$$\int (5x^3 - 6x^2 + 1) \, dx = \int 5x^3 \, dx + \int (-6x^2) \, dx + \int dx$$

$$= 5 \int x^3 \, dx - 6 \int x^2 \, dx + \int dx$$

In the first of these integral, $u = x$, $n = 3$, and $du = dx$. In the second, $u = x$, $n = 2$, and $du = dx$. The third is a direct application of Eq. (24–3), with $c = 1$ and $du = dx$. Thus, we have

$$5 \int x^3 \, dx - 6 \int x^2 \, dx + \int dx = 5\left(\frac{x^4}{4}\right) - 6\left(\frac{x^3}{3}\right) + x + C$$

$$= \frac{5}{4}x^4 - 2x^3 + x + C$$

Example E

Integrate: $\int \left(\sqrt{x} - \dfrac{1}{x^3} \right) dx.$

We write

$$\int \left(\sqrt{x} - \frac{1}{x^3} \right) dx = \int x^{1/2} \, dx - \int x^{-3} \, dx$$

$$= \frac{1}{3/2} x^{3/2} - \frac{1}{-2} x^{-2} + C$$

$$= \frac{2}{3} x^{3/2} + \frac{1}{2} x^{-2} + C = \frac{2}{3} x^{3/2} + \frac{1}{2x^2} + C$$

Example F

Integrate: $\int (x^2 + 1)^3 (2x \, dx).$

We first note that $n = 3$, for this is the power involved in the function being integrated. If $n = 3$, then $x^2 + 1$ must be u. If $u = x^2 + 1$, then $du = 2x \, dx$. Thus the integral is in proper form for integration *as it stands*. Using the power formula, we have

$$\int (x^2 + 1)^3 (2x \, dx) = \frac{(x^2 + 1)^4}{4} + C$$

It is in this type of problem that recognition of the proper form of a differential is very important. Only a good knowledge of differential forms, along with sufficient experience, makes ready recognition of this type possible. *It cannot be overemphasized that the entire quantity* $(2x \, dx)$ *must be equated to du if we are to integrate properly.* Normally u and n are recognized first, and then the proper form of du is derived from u.

Example G

Integrate: $\int x^2 \sqrt{x^3 + 2} \, dx.$

We first note that $n = \frac{1}{2}$ and u is then $x^3 + 2$. Since $u = x^3 + 2$, $du = 3x^2 \, dx$. In order to integrate properly, we must group the quantity $3x^2 \, dx$ as du, with other factors isolated from this quantity. Since there is no 3 under the integral sign, we introduce one. In order not to change the numerical value, we also introduce a $\frac{1}{3}$, normally before the integral sign. In this way we take full advantage of the fact that a constant (and only a constant) factor may be moved across the integral sign. The next form we should write is

$$\int x^2 \sqrt{x^3 + 2} \, dx = \frac{1}{3} \int \sqrt{x^3 + 2} \, (3x^2 \, dx)$$

In this way we indicate the proper grouping to result in the correct form of Eq. (24–5). Thus,

$$\int x^2 \sqrt{x^3 + 2} \, dx = \frac{1}{3} \cdot \frac{2}{3} (x^3 + 2)^{3/2} + C = \frac{2}{9} (x^3 + 2)^{3/2} + C$$

The $1/\frac{3}{2}$ was written as $\frac{2}{3}$, since this form is generally more convenient when we are working with fractions.

We mentioned earlier that, in addition to knowing the differential, we need more information in order to find the constant of integration. Such information usually consists of a set of values which the function is known to satisfy. That is, a point through which the curve of the function passes would provide the necessary information.

Example H

Find y in terms of x, given that $dy/dx = 3x - 1$, and the curve passes through $(1, 4)$.

When we write the equation as $dy = (3x - 1)\ dx$, then indicate the integration $\int dy = \int (3x - 1)\ dx$, the result of this integration is $y = \frac{3}{2}x^2 - x + C$. We know that the required curve passes through $(1, 4)$, which means that the coordinates of this point must satisfy the equation. Thus, $4 = \frac{3}{2} - 1 + C$, or $C = \frac{7}{2}$. The complete solution is

$$y = \frac{3}{2}x^2 - x + \frac{7}{2}$$

or

$$2y = 3x^2 - 2x + 7$$

Example I

At each point of a certain curve, the slope is given by $x\sqrt{2 - x^2}$. Given that the curve passes through $(1, 2)$, find its equation.

We write

$$\frac{dy}{dx} = x\sqrt{2 - x^2}, \qquad y = \int x\sqrt{2 - x^2}\ dx$$

To integrate this last expression, we recognize that $n = \frac{1}{2}$, $u = 2 - x^2$, and that therefore $du = -2x\ dx$. Hence,

$$y = -\frac{1}{2}\int (2 - x^2)^{1/2}(-2x\ dx) = -\frac{1}{2} \cdot \frac{2}{3}(2 - x^2)^{3/2} + C$$

$$= -\frac{1}{3}(2 - x^2)^{3/2} + C$$

Using the coordinates of the point $(1, 2)$, we have

$$2 = -\frac{1}{3}(2 - 1)^{3/2} + C \qquad \text{or} \qquad C = \frac{7}{3}$$

and

$$3y = 7 - (2 - x^2)^{3/2}$$

In this section we have discussed the integration of certain basic types of functions. There are many other methods used to integrate other functions, and some of these methods are discussed in Chapter 27. Also, there are many functions which cannot be integrated.

Exercises 24—3

In Exercises 1 through 28 integrate each of the given expressions.

1. $\int 2x\,dx$ **2.** $\int 5x^4\,dx$ **3.** $\int x^7\,dx$

4. $\int x^5\,dx$ **5.** $\int x^{3/2}\,dx$ **6.** $\int \sqrt[3]{x}\,dx$

7. $\int x^{-4}\,dx$ **8.** $\int \frac{1}{\sqrt{x}}\,dx$ **9.** $\int (x^2 - x^5)\,dx$

10. $\int (1 - 3x)\,dx$ **11.** $\int (1 + 2x)^2\,dx$ **12.** $\int (1 - x)^2\,dx$

13. $\int (x\sqrt{x} - 5x^2)\,dx$ **14.** $\int \left(4x - \frac{2}{x^3}\right)dx$ **15.** $\int \sqrt{x}\,(x^2 - x)\,dx$

16. $\int (x^{1/3} + x^{1/5} + x^{-1/7})\,dx$ **17.** $\int (x^2 - 1)^5\,(2x\,dx)$

18. $\int (x^3 - 2)^6\,(3x^2\,dx)$ **19.** $\int (x^4 + 3)^4\,(4x^3\,dx)$ **20.** $\int (1 - 2x)^{1/3}\,(-2\,dx)$

21. $\int (x^5 + 4)^7 x^4\,dx$ **22.** $\int x^2(1 - x^3)^{4/3}\,dx$ **23.** $\int \sqrt{8x + 1}\,dx$

24. $\int \sqrt[3]{4 - 3x}\,dx$ **25.** $\int \frac{x\,dx}{\sqrt{6x^2 + 1}}$ **26.** $\int \frac{x^2\,dx}{\sqrt{2x^3 + 1}}$

27. $\int \frac{x - 1}{\sqrt{x^2 - 2x}}\,dx$ **28.** $\int (x^2 - x)(x^3 - \tfrac{3}{2}x^2)^8\,dx$

In Exercises 29 through 32 find y in terms of x.

29. $\frac{dy}{dx} = 6x^2$,   curve passes through $(0, 2)$

30. $\frac{dy}{dx} = x + 1$,   curve passes through $(-1, 4)$

31. $\frac{dy}{dx} = x^2(1 - x^3)^5$,   curve passes through $(1, 5)$

32. $\frac{dy}{dx} = 2x^3(x^4 - 6)^4$,   curve passes through $(2, 10)$

In Exercises 33 through 36 find the required equations.

33. Find the equation of the curve whose slope is $-x\sqrt{1 - 4x^2}$ and which passes through $(0, 7)$.

34. Find the equation of the curve whose slope is $\sqrt{6x - 3}$ and which passes through $(2, -1)$.

35. Find the equation of the curve for which the second derivative is 6. The curve passes through $(1, 2)$ with a slope of 8.

36. Find the equation of the curve for which the second derivative is $12x^2$. The curve passes through $(1, 6)$ with a slope of 4.

24–4 The Area Under a Curve

Another basic problem which can be solved by integration is that of finding the area under a curve. In geometry, there are methods and formulas for finding the area of regular figures. By means of the calculus we will find it possible to find the area between curves for which we know the equations. First let us look at an example which illustrates the basic idea behind the method.

Example A

Approximate the area in the first quadrant and to the left of the line $x = 4$ under the parabola $y = x^2 + 1$. First make this approximation by inscribing two rectangles of equal width in the area and finding the sum of the areas of these rectangles. Then improve the approximation by repeating the process using eight rectangles.

The area to be approximated is shown in Fig. 24–3(a). The area with two rectangles inscribed under the curve is shown in Fig. 24–3(b). The first approximation, admittedly small, of the area can be found by adding the areas of the two rectangles. Both rectangles have a width of 2. The left rectangle is 1 unit high and the right rectangle is 5 units high. Thus, the area of the two rectangles is

$$A = 2(1 + 5) = 12$$

A much better approximation is found by inscribing the eight rectangles as shown in Fig. 24–3(c). Each of these rectangles has a width of $\frac{1}{2}$. The leftmost rectangle has a height of 1. The next has a height of $\frac{5}{4}$, which is determined by finding y for $x = \frac{1}{2}$. The next rectangle has a height of 2, which is found by evaluating y for $x = 1$. Finding the

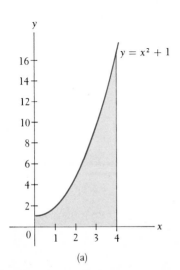

(a)

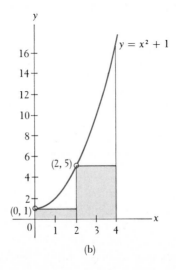

(b)

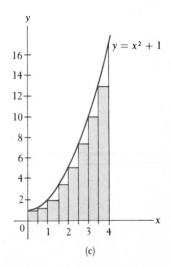

(c)

Figure 24–3

heights of all the rectangles and multiplying their sum by $\frac{1}{2}$ gives the area of the eight rectangles as

$$A = \frac{1}{2}\left(1 + \frac{5}{4} + 2 + \frac{13}{4} + 5 + \frac{29}{4} + 10 + \frac{53}{4}\right)$$

$$= \frac{43}{2} = 21.5$$

An even better approximation could be obtained by inscribing more rectangles under the curve. The greater the number of rectangles, the more nearly the sum of their areas equals the area under the curve. By a method involving integration developed later in this section we determine the *exact* area to be $\frac{76}{3} = 25\frac{1}{3}$.

We can now develop the basic method involved in finding an area under a curve. We consider the sum of the areas of rectangles inscribed under the curve as the number of rectangles is assumed to increase without bound. The reason for this last condition is that, as we saw in Example A, as the number of rectangles increases, the approximation of the area is better. Following is a specific example of this technique.

Example B

Find the area under the straight line $y = 2x$, above the x-axis, and to the left of the line $x = 4$.

We might mention at the beginning that this area can easily be found, since the figure is a triangle. However, the method we shall use here is the important concept at the moment. We first subdivide the interval from $x = 0$ to $x = 4$ into n inscribed rectangles of Δx in width. Then the extremities of the intervals are labeled $a, x_1, x_2, \ldots, b(=x_n)$, as shown in Fig. 24–4, where

$$x_1 = \Delta x$$
$$x_2 = 2 \Delta x$$
$$\vdots$$
$$x_{n-1} = (n-1) \Delta x$$
$$b = n \Delta x$$

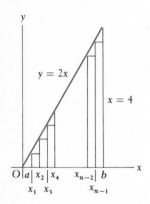

Figure 24–4

We then find the area for each of the rectangles. These areas are (for the respective rectangles):

First $f(a) \Delta x$, where $f(a) = f(0) = 2(0) = 0$ is the height
Second $f(x_1) \Delta x$, where $f(x_1) = 2(\Delta x) = 2 \Delta x$ is the height
Third $f(x_2) \Delta x$, where $f(x_2) = 2(2 \Delta x) = 4 \Delta x$ is the height
Fourth $f(x_3) \Delta x$, where $f(x_3) = 2(3 \Delta x) = 6 \Delta x$ is the height
$$\vdots$$
Last $f(x_{n-1}) \Delta x$, where $f[(n-1) \Delta x] = 2(n-1) \Delta x$ is the height

These areas are summed up as follows:

$$A_n = f(a)\, \Delta x + f(x_1)\, \Delta x + f(x_2)\, \Delta x + \cdots + f(x_{n-1})\, \Delta x$$
$$= 0 + 2\, \Delta x(\Delta x) + 4\, \Delta x(\Delta x) + \cdots + 2[(n-1)\, \Delta x]\, \Delta x$$
$$= \{2\, \Delta x[(1 + 2 + 3 + \cdots + (n-1)]\} \,\Delta x$$
$$= 2(\Delta x)^2[1 + 2 + 3 + \cdots + (n-1)]$$

Now $b = n\, \Delta x$, or $4 = n\, \Delta x$, or $\Delta x = 4/n$. Thus,

$$A_n = 2\left(\frac{4}{n}\right)^2 [1 + 2 + 3 + \cdots + (n-1)]$$

The sum $1 + 2 + 3 + \cdots + n - 1$ is the indicated sum of an arithmetic progression with the first term equal to 1 and the nth term equal to $n - 1$. This sum is

$$s = \frac{n-1}{2}(1 + n - 1) = \frac{n(n-1)}{2} = \frac{n^2 - n}{2}$$

Now the expression for the sum of the areas can be set forth as

$$A_n = \frac{32}{n^2}\left(\frac{n^2 - n}{2}\right) = 16\left(1 - \frac{1}{n}\right)$$

This expression is an approximation to the actual area under consideration. The larger n becomes, the better the approximation. If we let $n \to \infty$ (which is equivalent to letting $\Delta x \to 0$), the limit of this sum will equal the area in question. Thus,

$$A = \lim_{n \to \infty} 16\left(1 - \frac{1}{n}\right) = 16$$

(This checks with the geometric result.) The area under the curve can be considered as the limit of the sum of the inscribed rectangles as the number of rectangles approaches infinity.

The method indicated in Example B illustrates the interpretation of finding an area as a summation process, although it should not be considered as a proof. However, we shall find that integration proves to be a much more useful method for finding an area. Let us now see how integration can be used directly.

Let ΔA represent the area $BCEG$ under the curve, as indicated in Fig. 24–5. We see that the following inequality is true for the indicated areas:

$$A_{BCDG} < \Delta A < A_{BCEF}$$

If the point G is now designated as (x, y) and E as $(x + \Delta x, y + \Delta y)$, we have $y\, \Delta x < \Delta A < (y + \Delta y)\, \Delta x$. Dividing through by Δx, we have

$$y < \frac{\Delta A}{\Delta x} < y + \Delta y$$

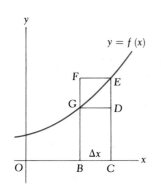

Figure 24–5

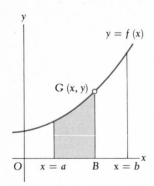

Figure 24–6

Now we take the limit as $\Delta x \rightarrow 0$ (Δy then also approaches 0). This results in the relation

$$\frac{dA}{dx} = y \qquad\qquad (24\text{–}6)$$

This is true since the left member of the inequality is y and the right member approaches y. Also remember that

$$\lim_{\Delta x \to 0} \frac{\Delta A}{\Delta x} = \frac{dA}{dx}$$

We shall now use Eq. (24–6) to show the method of finding the complete area under a curve. We now let $x = a$ be the left boundary of the desired area, and $x = b$ be the right boundary (Fig. 24–6). The area under the curve to the right of $x = a$ and bounded on the right by the line GB is now designated as A_{ax}. From Eq. (24–6), we have

$$dA_{ax} = [y \ dx]_a^x \qquad \text{or} \qquad A_{ax} = \left[\int y \ dx\right]_a^x$$

where $[\ ]_a^x$ is the notation used to indicate the boundaries of the area. Thus,

$$A_{ax} = \left[\int f(x) \ dx\right]_a^x = [F(x) + C]_a^x \qquad\qquad (24\text{–}7)$$

But we know that if $x = a$, then $A_{aa} = 0$. Thus, $0 = F(a) + C$, or $C = -F(a)$. Therefore,

$$A_{ax} = \left[\int f(x) \ dx\right]_a^x = F(x) - F(a) \qquad\qquad (24\text{–}8)$$

Now, to find the area under the curve that reaches from a to b, we write

$$A_{ab} = F(b) - F(a) \qquad\qquad (24\text{–}9)$$

Thus the area under the curve that reaches from a to b is given by

$$A_{ab} = \left[\int f(x) \ dx\right]_a^b = F(b) - F(a) \qquad\qquad (24\text{–}10)$$

This shows that *the area under the curve may be found by integrating the function $f(x)$ to find the function $F(x)$, which is then evaluated at each boundary value. The area is the difference between these values of $F(x)$.*

In Example B, we found an area under a curve by finding the limit of the sum of the inscribed rectangles as the number of rectangles approaches infinity. Equation (24–10) expresses the area under a curve in terms of integration. We can now see that we have obtained the area by summation and also expressed it in terms of integration. *Thus, we conclude that summations can be evaluated by integration.* Also, we have grasped the connection between the problem of finding the slope of a tangent to a curve (differentiation) and the problem of finding an area (integration). We would not normally suspect that these two problems would have solutions which lead to reverse processes. We have also seen that the definition of integration has much more application than originally anticipated.

Example C

Find the area under the curve $y = x^2 + 1$ between the y-axis and the line $x = 4$. This is the area of Example A, which is shown in Fig. 24–3(a) on page 657.

In Eq. (24–10), we note that $f(x) = x^2 + 1$. This means that

$$F(x) = \int (x^2 + 1)\ dx = \frac{1}{3}x^3 + x + C$$

Therefore, the area is given by

$$A_{0,4} = F(4) - F(0)$$

$$= \left[\frac{1}{3}(4^3) + 4 + C\right] - \left[\frac{1}{3}(0^3) + 0 + C\right]$$

$$= \frac{1}{3}(64) + 4 = \frac{76}{3}$$

We note that this is about 4 sq units greater than the value obtained by the approximation in Example A. This result means that the exact area is $25\frac{1}{3}$, as stated at the end of Example A.

Example D

Find the area under the curve $y = x^3$ between the lines $x = 1$ and $x = 2$.

In Eq. (24–10), $f(x) = x^3$. Therefore,

$$F(x) = \int x^3\ dx = \frac{1}{4}x^4 + C$$

$$A_{1,2} = F(2) - F(1) = \left[\frac{1}{4}(2^4) + C\right] - \left[\frac{1}{4}(1^4) + C\right]$$

$$= 4 - \frac{1}{4} = \frac{15}{4}$$

Therefore, the required area is $\frac{15}{4}$. Again, this is the exact area, not an approximation. See Fig. 24–7.

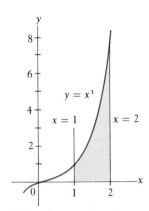

Figure 24–7

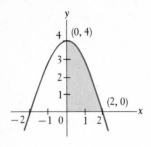

Figure 24—8

Example E

Find the area under the curve $y = 4 - x^2$ which lies in the first quadrant.

By solving the equation $4 - x^2 = 0$, we determine that the area to be found extends from $x = 0$ to $x = 2$ (see Fig. 24–8). Thus,

$$\int (4 - x^2)\,dx = 4x - \frac{x^3}{3} + C$$

$$A_{0,2} = \left(8 - \frac{8}{3} + C\right) - (0 - 0 + C) = \frac{16}{3}$$

We can see from these examples that we do not have to include the constant of integration when we are finding areas. It cancels out when the values are subtracted.

Exercises 24—4

In each of Exercises 1 through 8 find the approximate area under the given curves by dividing the indicated intervals into n subintervals, and add up the areas of the inscribed rectangles. There are two values of n for each, and therefore two approximations for each area. The height of each rectangle may be found by evaluating the function for the proper value of x. See Example A.

1. $y = 3x$, between $x = 0$ and $x = 3$, for (a) $n = 3$ ($\Delta x = 1$),
   (b) $n = 10$ ($\Delta x = 0.3$)

2. $y = 2x$, between $x = 0$ and $x = 2$, for (a) $n = 4$ ($\Delta x = 0.5$),
   (b) $n = 10$ ($\Delta x = 0.2$)

3. $y = x^2$, between $x = 0$ and $x = 2$, for (a) $n = 5$ ($\Delta x = 0.4$),
   (b) $n = 10$ ($\Delta x = 0.2$)

4. $y = x^2 + 2$, between $x = 0$ and $x = 3$, for (a) $n = 3$ ($\Delta x = 1$),
   (b) $n = 10$ ($\Delta x = 0.3$)

5. $y = 4x - x^2$, between $x = 1$ and $x = 4$, for (a) $n = 6$, (b) $n = 10$

6. $y = 1 - x^2$, between $x = 0.5$ and $x = 1$, for (a) $n = 5$, (b) $n = 10$

7. $y = \dfrac{1}{x^2}$, between $x = 1$ and $x = 5$, for (a) $n = 4$, (b) $n = 8$

8. $y = \sqrt{x}$, between $x = 1$ and $x = 4$, for (a) $n = 3$, (b) $n = 12$

In Exercises 9 through 16 find the exact area under the given curves between the indicated values of x. The functions are the same as those for which approximate areas were found in Exercises 1 through 8.

9. $y = 3x$, between $x = 0$ and $x = 3$

10. $y = 2x$, between $x = 0$ and $x = 2$

11. $y = x^2$, between $x = 0$ and $x = 2$

12. $y = x^2 + 2$, between $x = 0$ and $x = 3$

13. $y = 4x - x^2$, between $x = 1$ and $x = 4$

14. $y = 1 - x^2$, between $x = 0.5$ and $x = 1$

15. $y = \dfrac{1}{x^2}$, between $x = 1$ and $x = 5$

16. $y = \sqrt{x}$, between $x = 1$ and $x = 4$

24–5 The Definite Integral

Using reasoning similar to that in the preceding section, *we define the* **definite integral** *of a function* $f(x)$ *as*

$$\int_a^b f(x)\ dx = F(b) - F(a) \tag{24–11}$$

where $F'(x) = f(x)$. We call this a *definite integral* because the final result of integrating and evaluating is a number. (The *indefinite* integral had an arbitrary constant in the result.) *The numbers a and b are called the* **lower limit** *and the* **upper limit**, *respectively. We can see that the value of a definite integral is found by evaluting the function (found by integration) at the upper limit and subtracting the value of this function at the lower limit.*

We know, from the analysis in the preceding section, that this definite integral can be interpreted as a summation process, where the size of the subdivision approaches a limit of zero. This fact explains the choice of the $\int$ symbol for integration: It is an elongated S, representing the sum. It is this interpretation of integration which we shall apply to many kinds of problems.

Example A

Evaluate the integral $\int_0^2 x^4\ dx$.
  We write

$$\int_0^2 x^4\ dx = \frac{x^5}{5}\Big|_0^2 = \frac{2^5}{5} - 0 = \frac{32}{5}$$

Note that a vertical line—with the limits written at the top and the bottom—is the way the value is indicated after integration, but before evaluation.

Example B

Evaluate $\int_1^3 (x^{-2} - 1)\ dx$.
  We set this forth as

$$\int_1^3 (x^{-2} - 1)\ dx = -\frac{1}{x} - x\Big|_1^3 = \left(-\frac{1}{3} - 3\right) - (-1 - 1)$$

$$= 2 - \frac{10}{3} = -\frac{4}{3}$$

Example C

Evaluate $\int_0^1 5x(x^2 + 1)^5\ dx$.
  For purposes of integration, $n = 5$, $u = x^2 + 1$, and $du = 2x\ dx$. Hence

$$\int_0^1 5x(x^2 + 1)^5\ dx = \frac{5}{2}\int_0^1 (x^2 + 1)^5 (2x\ dx) = \frac{5}{2}\cdot\frac{1}{6}(x^2 + 1)^6\Big|_0^1$$

$$= \frac{5}{12}(2^6 - 1^6) = \frac{5(63)}{12} = \frac{105}{4}$$

Example D

Evaluate $\int_{-1}^{3} x(1 - 3x^2)^{1/3}\, dx$.

For purposes of integration, $n = \frac{1}{3}$, $u = 1 - 3x^2$, and $du = -6x\, dx$. Thus,

$$\int_{-1}^{3} x(1 - 3x^2)^{1/3}\, dx = -\frac{1}{6}\int_{-1}^{3} (1 - 3x^2)^{1/3}(-6x\, dx)$$

$$= -\frac{1}{6}\cdot\frac{3}{4}(1 - 3x^2)^{4/3}\Big|_{-1}^{3}$$

$$= -\frac{1}{8}(-26)^{4/3} + \frac{1}{8}(-2)^{4/3}$$

$$= \frac{1}{8}(2\sqrt[3]{2} - 26\sqrt[3]{26}) = \frac{1}{4}(\sqrt[3]{2} - 13\sqrt[3]{26})$$

Example E

Evaluate

$$\int_{0}^{4} \frac{x + 1}{(x^2 + 2x + 2)^3}\, dx$$

For purposes of integration,

$$n = -3,\; u = x^2 + 2x + 2 \qquad \text{and} \qquad du = (2x + 2)\, dx$$

Therefore,

$$\int_{0}^{4} (x^2 + 2x + 2)^{-3}(x + 1)\, dx = \frac{1}{2}\int_{0}^{4} (x^2 + 2x + 2)^{-3}[2(x + 1)\, dx]$$

$$= \frac{1}{2}\cdot\frac{1}{-2}(x^2 + 2x + 2)^{-2}\Big|_{0}^{4}$$

$$= -\frac{1}{4}(16 + 8 + 2)^{-2} + \frac{1}{4}(0 + 0 + 2)^{-2}$$

$$= \frac{1}{4}\left(-\frac{1}{26^2} + \frac{1}{2^2}\right) = \frac{1}{4}\left(\frac{1}{4} - \frac{1}{676}\right)$$

$$= \frac{1}{4}\left(\frac{168}{676}\right) = \frac{21}{338}$$

It should be pointed out that the definition of the definite integral is valid regardless of the source of $f(x)$. That is, we may apply the definite integral whenever we want to sum a function in a manner similar to that which we use to find an area.

Exercises 24−5   In the following exercises evaluate the given definite integrals.

1. $\int_0^1 2x \, dx$

2. $\int_0^2 3x^2 \, dx$

3. $\int_1^4 x^{5/2} \, dx$

4. $\int_4^9 (x^{3/2} - 1) \, dx$

5. $\int_{-1}^1 (1 - x)^{1/3} \, dx$

6. $\int_1^5 \sqrt{2x - 1} \, dx$

7. $\int_0^3 (x^4 - x^3 + x^2) \, dx$

8. $\int_1^2 (3x^5 - 2x^3) \, dx$

9. $\int_0^4 (1 - \sqrt{x})^2 \, dx$

10. $\int_1^4 \frac{x^2 + 1}{\sqrt{x}} \, dx$

11. $\int_1^2 2x(4 - x^2)^3 \, dx$

12. $\int_0^1 x(3x^2 - 1)^3 \, dx$

13. $\int_0^4 \frac{x \, dx}{\sqrt{x^2 + 9}}$

14. $\int_0^3 x^2(x^3 + 2)^{3/2} \, dx$

15. $\int_3^7 3\sqrt{4x - 3} \, dx$

16. $\int_{-5}^1 \sqrt{6 - 2x} \, dx$

17. $\int_0^2 2x(9 - 2x^2)^2 \, dx$

18. $\int_{-1}^0 x^3(1 - 2x^4)^3 \, dx$

19. $\int_0^1 (x^2 + 3)(x^3 + 9x + 6) \, dx$

20. $\int_2^3 \frac{x^2 + 1}{(x^3 + 3x)^2} \, dx$

24−6 Numerical Integration: The Trapezoidal Rule

For data and functions which cannot be directly integrated by available methods, it is possible to develop numerical methods of integration. These numerical methods are of greater importance today since they are readily adaptable for use on a calculator or computer. There are a great many such numerical techniques for approximating the value of an integral, but we shall develop only one of these, the trapezoidal rule. It is based upon the area under a curve.

We know from Sections 24−4 and 24−5 that we can interpret a definite integral as the area under a curve. We shall therefore show how we can approximate the value of the integral by approximating the appropriate area by a set of inscribed trapezoids. The basic idea here is very similar to that used when rectangles were inscribed under a curve. However, the use of trapezoids reduces the error and provides a better approximation.

The area to be found is subdivided into n intervals of equal width. Perpendicular lines are then dropped from the curve (or points, if only a given set of numbers is available). If the points on the curve are joined by straight-line segments, the area of successive parts under the curve is then approximated by finding the area of each of the trapezoids formed. However, if these points are not too far apart, the approximation will be very good (see Fig. 24–9). From geometry we recall that the area of a trapezoid equals one-half the product of the sum of the bases times the altitude. For these trapezoids the bases are the y-coordinates and the altitudes are Δx. When we thus indicate the sum of these trapezoidal areas, we have

$$A_T = \frac{1}{2}(y_0 + y_1)\,\Delta x + \frac{1}{2}(y_1 + y_2)\,\Delta x + \frac{1}{2}(y_2 + y_3)\,\Delta x + \cdots$$

$$+ \frac{1}{2}(y_{n-2} + y_{n-1})\,\Delta x + \frac{1}{2}(y_{n-1} + y_n)\,\Delta x$$

We note, when this addition is performed, that the result is

$$A_T = \left(\frac{1}{2}y_0 + y_1 + y_2 + \cdots + y_{n-1} + \frac{1}{2}y_n \right)\Delta x \qquad (24\text{–}12)$$

The y-values to be used are either derived from the function as $y_1 = f(x_1)$, or are the y-coordinates of a set of data.

Since A_T approximates the area under the curve, it also approximates the value of the definite integral, or

$$\int_a^b f(x)\,dx \approx \left(\frac{1}{2}y_0 + y_1 + y_2 + \cdots + y_{n-1} + \frac{1}{2}y_n \right)\Delta x \qquad (24\text{–}13)$$

Equation (24–13) is known as the **trapezoidal rule.**

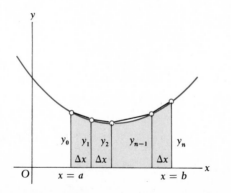

Figure 24–9

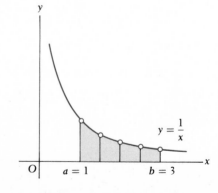

Figure 24–10

Example A

Approximate the value of $\int_1^3 \dfrac{1}{x}\,dx$ by the trapezoidal rule. Let $n = 4$.

We are to approximate the area under $y = 1/x$ from $x = 1$ to $x = 3$ by dividing the area into 4 trapezoids. This area is found by applying Eq. (24–12), which is the approximate value of the integral, as shown in Eq. (24–13). Figure 24–10 shows the graph. In this example, $f(x) = 1/x$, and

$$\Delta x = \frac{3-1}{4} = \frac{1}{2}, \qquad y_0 = f(a) = f(1) = 1$$

$$y_1 = f\left(\frac{3}{2}\right) = \frac{2}{3}, \qquad y_2 = f(2) = \frac{1}{2}$$

$$y_3 = f\left(\frac{5}{2}\right) = \frac{2}{5} \qquad \text{and} \qquad y_n = y_4 = f(b) = f(3) = \frac{1}{3}$$

And so we write

$$A_T = \left[\frac{1}{2}(1) + \frac{2}{3} + \frac{1}{2} + \frac{2}{5} + \frac{1}{2}\left(\frac{1}{3}\right)\right]\frac{1}{2}$$

$$= \left(\frac{15 + 20 + 15 + 12 + 5}{30}\right)\frac{1}{2} = \frac{67}{30}\left(\frac{1}{2}\right) = \frac{67}{60}$$

Therefore,

$$\int_1^3 \frac{1}{x}\,dx \approx \frac{67}{60}$$

Note that we cannot perform this integration directly by methods developed up to this point.

Example B

Approximate the value of $\int_0^1 \sqrt{x^2+1}\,dx$ by the trapezoidal rule. Let $n = 5$.

Figure 24–11 shows the graph. In this example,

$$\Delta x = \frac{1-0}{5} = 0.2$$

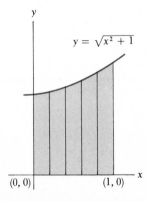

$y = \sqrt{x^2 + 1}$

Therefore,

$$y_0 = f(0) = 1 \qquad\qquad y_1 = f(0.2) = \sqrt{1.04} = 1.02$$
$$y_2 = f(0.4) = \sqrt{1.16} = 1.08 \qquad y_3 = f(0.6) = \sqrt{1.36} = 1.17$$
$$y_4 = f(0.8) = \sqrt{1.64} = 1.28 \qquad y_5 = f(1) = \sqrt{2.00} = 1.41$$

Hence we have

$$A_T = \left[\frac{1}{2}(1) + 1.02 + 1.08 + 1.17 + 1.28 + \frac{1}{2}(1.41)\right](0.2) = 1.15$$

This means that

$$\int_0^1 \sqrt{x^2+1}\,dx \approx 1.15$$

(0, 0)   (1, 0)

Figure 24–11

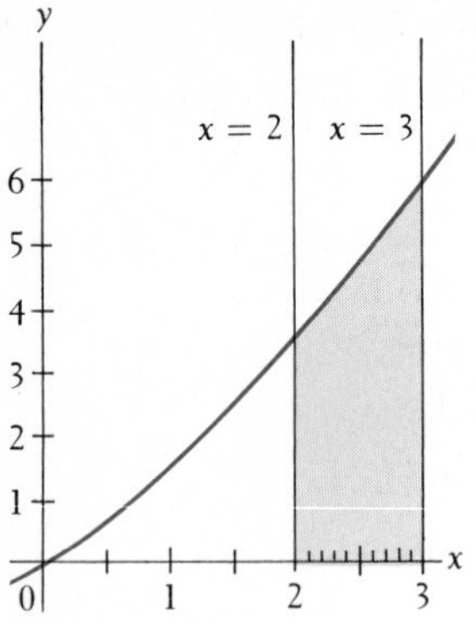

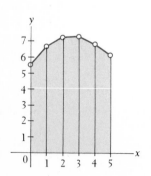

Figure 24–12

Example C

Approximate the value of $\int_2^3 x\sqrt{x+1}\ dx$ by use of the trapezoidal rule. Use $n = 10$.

In Fig. 24–12 the graph of the function and the area used in the trapezoidal rule are shown. From the given values we have $\Delta x = \frac{3-2}{10} = 0.1$. Therefore,

$$y_0 = f(2) = 2\sqrt{3} = 3.464 \qquad y_1 = f(2.1) = 2.1\sqrt{3.1} = 3.697$$
$$y_2 = f(2.2) = 2.2\sqrt{3.2} = 3.935 \qquad y_3 = f(2.3) = 2.3\sqrt{3.3} = 4.178$$
$$y_4 = f(2.4) = 2.4\sqrt{3.4} = 4.425 \qquad y_5 = f(2.5) = 2.5\sqrt{3.5} = 4.677$$
$$y_6 = f(2.6) = 2.6\sqrt{3.6} = 4.933 \qquad y_7 = f(2.7) = 2.7\sqrt{3.7} = 5.194$$
$$y_8 = f(2.8) = 2.8\sqrt{3.8} = 5.458 \qquad y_9 = f(2.9) = 2.9\sqrt{3.9} = 5.727$$
$$y_{10} = f(3) = 3\sqrt{4} = 6.000$$

$$A_T = \left[\frac{1}{2}(3.464) + 3.697 + 3.935 + \cdots + 5.727 + \frac{1}{2}(6.000)\right](0.1)$$

$$= 4.6956$$

Therefore, $\int_2^3 x\sqrt{x+1}\ dx \approx 4.6956$. (The actual value, to five significant digits, is 4.6954.)

Example D

The following points were found empirically.

x	0	1	2	3	4	5
y	5.68	6.75	7.32	7.35	6.88	6.24

Approximate the value of the integral of the function defined by these points between $x = 0$ and $x = 5$ by the trapezoidal rule.

In order to find A_T, we use the values of y_0, y_1, *etc.* directly from the table. We also note that $\Delta x = 1$. The graph is shown in Fig. 24–13. Therefore, we have

$$A_T = \left[\frac{1}{2}(5.68) + 6.75 + 7.32 + 7.35 + 6.88 + \frac{1}{2}(6.24)\right] = 34.26$$

Although we do not know the algebraic form of the function, we can state that

$$\int_0^5 f(x)\ dx \approx 34.26$$

Figure 24–13

Exercises 24–6

In Exercises 1 through 4, (a) approximate the value of each of the given integrals by use of the trapezoidal rule, using the given value of n, and (b) check by direct integration.

1. $\int_0^2 2x^2\ dx$, $n = 4$ 2. $\int_0^1 (1 - x^2)\ dx$, $n = 3$

3. $\int_1^4 (1 + \sqrt{x})\ dx$, $n = 6$ 4. $\int_3^8 \sqrt{1 + x}\ dx$, $n = 5$

In Exercises 5 through 12 approximate each of the given integrals by use of the trapezoidal rule, using the given value of n.

5. $\int_{2}^{3} \frac{1}{2x} \, dx, \; n = 2$

6. $\int_{2}^{4} \frac{dx}{x+3}, \; n = 4$

7. $\int_{0}^{2} \sqrt{4 - x^2} \, dx, \; n = 4$

8. $\int_{0}^{2} \sqrt{x^3 + 1} \, dx, \; n = 4$

9. $\int_{1}^{5} \frac{1}{x^2 + x} \, dx, \; n = 10$

10. $\int_{2}^{4} \frac{1}{x^2 + 1} \, dx, \; n = 10$

11. $\int_{0}^{4} 2^x \, dx, \; n = 12$

12. $\int_{0}^{1.5} 10^x \, dx, \; n = 15$

In Exercises 13 through 16 approximate the value of the integrals defined by the given sets of points.

13.
x	3	4	5	6	7
y	1.08	1.65	3.23	3.67	2.97

14.
x	2	4	6	8	10	12	14
y	0.67	2.34	4.56	3.67	3.56	4.78	6.87

15.
x	1.4	1.7	2.0	2.3	2.6	2.9	3.2
y	0.18	7.87	18.23	23.53	24.62	20.93	20.76

16.
x	0.4	0.8	1.2	1.6	2.0	2.4
y	1.24	3.54	5.52	4.40	3.63	3.88

24–7 Exercises for Chapter 24

In Exercises 1 through 20 evaluate the given integrals.

1. $\int (4x^3 - x) \, dx$

2. $\int (5 - 3x^2) \, dx$

3. $\int 2(x - x^{3/2}) \, dx$

4. $\int x(x - 3x^4) \, dx$

5. $\int_{1}^{4} \left(\sqrt{x} + \frac{1}{\sqrt{x}} \right) dx$

6. $\int_{1}^{2} \left(x + \frac{1}{x^2} \right) dx$

7. $\int_{0}^{2} x(4 - x) \, dx$

8. $\int_{0}^{1} x(x^3 + 1) \, dx$

9. $\int \frac{dx}{(2 - 5x)^2}$

10. $\int x(1 - 2x^2)^5 \, dx$

11. $\int (7 - 2x)^{3/4} \, dx$

12. $\int (3x - 1)^{1/3} \, dx$

13. $\int_{0}^{2} \frac{3x \, dx}{\sqrt[3]{1 + 2x^2}}$

14. $\int_{1}^{6} \frac{2 \, dx}{(3x - 2)^{3/4}}$

15. $\int x^2 (1 - 2x^3)^4 \, dx$

16. $\int 3x^3 (1 - 5x^4)^{1/3} \, dx$

17. $\int \frac{(2 - 3x^2) \, dx}{(2x - x^3)^2}$

18. $\int \frac{x^2 - 3}{\sqrt{6 + 9x - x^3}} \, dx$

19. $\int_{1}^{3} (x^2 + x + 2)(2x^3 + 3x^2 + 12x) \, dx$

20. $\int_{0}^{2} (4x + 18x^2)(x^2 + 3x^3)^2 \, dx$

In Exercises 21 through 24 find the differential of each of the given functions.

21. $y = \dfrac{1}{(x^2 - 1)^3}$

22. $y = \dfrac{1}{(2x - 1)^2}$

23. $y = x\sqrt[3]{1 - 3x}$

24. $y = \dfrac{3 + x}{4 - x^2}$

In Exercises 25 and 26 evaluate $\Delta y - dy$ for the given functions and values.

25. $y = x^3$, $x = 2$, $\Delta x = 0.1$ **26.** $y = 6x^2 - x$, $x = 3$, $\Delta x = 0.2$

In Exercises 27 and 28 find the required equations.

27. Find the equation of the curve which passes through $(-1, 3)$ for which the slope is given by $3 - x^2$.

28. Find the equation of the curve which passes through $(1, -2)$ for which the slope is $x(x^2 + 1)^2$.

In Exercises 29 and 30 use Eq. (24–10) to find the indicated areas.

29. The area under $y = 6x - 1$ between $x = 1$ and $x = 3$

30. The first quadrant area under $y = 8x - x^4$

In Exercises 31 and 32 solve the given problems by the trapezoidal rule.

31. Approximate $\displaystyle\int_1^3 \dfrac{dx}{2x - 1}$ with $n = 4$.

32. Approximate the value of the integral defined by the following set of points.

x	6.0	9.0	12	15	18	21
y	2.0	1.2	0.2	1.0	6.0	12

In Exercises 33 and 34 use the function $y = \dfrac{x}{x^2 + 2}$ and approximate the area under the curve in the first quadrant to the left of the line $x = 5$ by the indicated method.

33. Inscribe five rectangles and find the sum of the areas of the rectangles.

34. Use the trapezoidal rule with $n = 5$.

In Exercises 35 and 36 use the function $y = x\sqrt{x^3 + 1}$ and approximate the area under the curve between $x = 1$ and $x = 3$ by the indicated method.

35. Use the trapezoidal rule with $n = 10$.

36. Inscribe ten rectangles, and find the sum of the areas of the rectangles.

In Exercises 37 through 40 solve the given problems by finding the appropriate differentials.

37. A ball bearing 4.00 mm in radius is coated with a special metal of thickness 0.02 mm. Approximately what volume of the special metal is used?

38. The electric resistance of a certain resistor as a function of temperature is given by $R = 12.00 + 0.06T + 0.0001T^2$. Approximately by how much does the resistance (in ohms) change as T changes from $100°C$ to $102°C$?

39. The impedance Z of an electric circuit as a function of the resistance R and the reactance X is given by $Z = \sqrt{R^2 + X^2}$. Derive an expression for the relative error in impedance for an error in R and given value of X.

40. Show that the relative error of the nth root of a given measurement equals approximately $1/n$ of the relative error of the measurement.

25

Applications of Integration

25-1 Applications of the Indefinite Integral

The applications of integration in engineering and technology are numerous. In this section we shall present two basic applications of the indefinite integral, with other applications being indicated in the exercises. The sections which follow deal with many of the basic applications of the definite integral.

The first of these applications deals with velocity and acceleration. The concepts of velocity as a first derivative and acceleration as a second derivative were introduced in Chapters 22 and 23. Here we shall apply integration to the problem of finding the distance as a function of time, when we know the relationship between acceleration and time, as well as certain specific values of distance and velocity. These latter values are necessary for determining the values of the constants of integration which are introduced. Recalling now that the acceleration a of an object is given by $a = dv/dt$, we can find the expression for the velocity in terms of a, t, and the constant of integration. We write

$$dv = a\ dt \qquad \text{or}$$

$$v = \int a\ dt \qquad (25\text{-}1)$$

If the acceleration is constant, we have

$$v = at + C_1 \qquad (25\text{-}2)$$

Of course, Eq. (25–1) can be used in general to find the velocity as a function of time so long as we know the acceleration as a function of time. However, since the case of constant acceleration is often encountered, Eq. (25–2) is often encountered. If the velocity is known for some specified time, the constant C_1 may be evaluated.

Example A

Find the expression for the velocity if $a = 12t$, given that $v = 8$ when $t = 1$.

Using Eq. (25–1), we have

$$v = \int (12t) \; dt = 6t^2 + C_1$$

Substituting the known values, we obtain

$$8 = 6 + C_1 \quad \text{or} \quad C_1 = 2$$

Thus, $v = 6t^2 + 2$.

Example B

For an object falling under the influence of gravity, the acceleration due to gravity is essentially constant. Its value is -32 ft/s². (The minus sign is chosen so that all quantities directed up are positive, and all quantities directed down are negative.) Find the expression for the velocity of an object under the influence of gravity if $v = v_0$ when $t = 0$.

We write

$$v = \int (-32) \; dt = -32t + C_1 \qquad v_0 = 0 + C_1 \qquad v = v_0 - 32t$$

The velocity v_0 is called the initial velocity. If the object is given an initial upward velocity of 100 ft/s, $v_0 = 100$ ft/s. If the object is dropped, $v_0 = 0$. If the object is given an initial downward velocity of 100 ft/s, $v_0 = -100$ ft/s.

Once we obtain the expression for velocity, we can integrate to find the expression for displacement in terms of the time. Since $v = ds/dt$, we can write $ds = v \; dt$, or

$$s = \int v \; dt \tag{25–3}$$

Example C

Find the expression for displacement in terms of time, if $a = 6t^2$, $v = 0$ when $t = 2$, and $s = 4$ when $t = 0$.

$$v = \int 6t^2 \; dt = 2t^3 + C_1; \qquad 0 = 2(2^3) + C_1, \qquad C_1 = -16$$

$$v = 2t^3 - 16$$

$$s = \int (2t^3 - 16)\, dt = \frac{1}{2}t^4 - 16t + C_2; \qquad 4 = 0 - 0 + C_2, \qquad C_2 = 4$$

$$s = \frac{1}{2}t^4 - 16t + 4$$

Example D

Find the expression for the distance above the ground of an object, given a vertical velocity of v_0 from the ground.

From Example B, we know that $v = v_0 - 32t$. In this problem we know that $s = 0$ when $t = 0$ (given velocity v_0 *from the ground*) if distances are measured from ground level. Therefore,

$$s = \int (v_0 - 32t)\, dt = v_0 t - 16t^2 + C_2; \qquad 0 = 0 - 0 + C_2, \qquad C_2 = 0$$

$$s = v_0 t - 16t^2$$

Example E

An object is thrown vertically from the top of a building 200 ft high, and hits the ground 5 s later. What initial velocity was the object given?

Measuring vertical distances from the ground, we know that $s = 200$ ft when $t = 0$. Also, we know that $v = v_0 - 32t$. Thus,

$$s = \int (v_0 - 32t)\, dt = v_0 t - 16t^2 + C$$

$$200 = v_0 (0) - 16(0) + C, \qquad C = 200$$
$$s = v_0 t - 16t^2 + 200$$

We also know that $s = 0$ when $t = 5$ s. Thus,

$$0 = v_0 (5) - 16(5^2) + 200$$
$$5v_0 = 200$$
$$v_0 = 40 \text{ ft/s}$$

This means that the initial velocity was 40 ft/s upward.

The second basic application of the indefinite integral which we shall discuss comes from the field of electricity. By definition, *the current i in an electric circuit equals the time rate of change of the charge q (in coulombs) which passes a given point in the circuit*, or

$$i = \frac{dq}{dt} \tag{25–4}$$

Rewriting this expression in differential notation as $dq = i\, dt$ and integrating both sides of the equation we have

$$q = \int i\, dt \tag{25–5}$$

Now, the voltage V_C across a capacitor C is given by $V_C = q/C$. By combining equations, the voltage V_C is given by

$$V_C = \frac{1}{C} \int i \, dt \tag{25-6}$$

Here, V_C is measured in volts, C in farads, i in amperes, and t in seconds.

Example F
The current in a certain electric circuit as a function of time is given by $i = 6t^2 + 4$. Find an expression for the amount of charge which passes a point in the circuit as a function of time. Assuming that $q = 0$ when $t = 0$, determine the total charge which passes the point in two seconds.
   Since $q = \int i \, dt$, we have

$$q = \int (6t^2 + 4) \, dt$$

$$= 2t^3 + 4t + C$$

This last expression is the desired expression giving charge as a function of time. We note that when $t = 0$, then $q = C$, which means that the constant of integration represents the initial charge, or the change which passed a given point before we started timing. Using q_0 to represent this charge, we have

$$q = 2t^3 + 4t + q_0$$

Now, returning to the second part of the problem, we see that $q_0 = 0$. Therefore, evaluating q for $t = 2$ s, we have

$$q = 2(8) + 4(2) = 24 \text{ C}$$

(Here, the symbol C represents coulombs, and is not the C for capacitance of Eq. (25-6) or the constant of integration.) This is the charge which passes any specified point in the circuit in two seconds.

Example G
The voltage across a 5.0 μF capacitor is zero. What is the voltage after 20 ms if a current of 75 mA charges the capacitor?
   From Eq. (25-6) we have

$$V_C = \frac{1}{C} \int i \, dt$$

In substituting we must use the proper power of 10 which corresponds to each prefix which is used. Since 5.0 μF = 5.0×10^{-6} F and 75 mA = 7.5×10^{-2} A, we have

$$V_C = \frac{1}{5.0 \times 10^{-6}} \int 7.5 \times 10^{-2} dt$$

$$= 1.5 \times 10^4 \int dt$$

$$= 1.5 \times 10^4 t + C_1$$

From the given information we know that $V_C = 0$ when $t = 0$. Thus,

$$0 = 1.5 \times 10^4 (0) + C_1 \qquad \text{or} \qquad C_1 = 0$$

This means that

$$V_C = 1.5 \times 10^4 t$$

Evaluating this expression for $t = 20 \times 10^{-3}$ s, we have

$$V_C = 1.5 \times 10^4 (20 \times 10^{-3})$$
$$= 30 \times 10 = 300 \text{ V}$$

Example H

A certain capacitor is measured to have a voltage of 100 V across it. At this instant a current as a function of time given by $i = 0.06\sqrt{t}$ is sent through the circuit. After 0.25 s, the voltage across the capacitor is measured to be 140 V. What is the capacitance of the capacitor?

From Eq. (25–6) we have

$$V_C = \frac{1}{C} \int i \, dt$$

Substituting $i = 0.06\sqrt{t}$, we find that

$$V_C = \frac{1}{C} \int (0.06\sqrt{t} \, dt) = \frac{0.06}{C} \int t^{1/2} \, dt$$

$$= \frac{0.04}{C} t^{3/2} + C_1$$

From the given information we know that $V_C = 100$ V when $t = 0$. Thus,

$$100 = \frac{0.04}{C} (0) + C_1$$

or

$$C_1 = 100 \text{ V}$$

This means that

$$V_C = \frac{0.04}{C} t^{3/2} + 100$$

We also know that $V_C = 140$ V when $t = 0.25$ s. Therefore,

$$140 = \frac{0.04}{C} (0.25)^{3/2} + 100$$

$$40 = \frac{0.04}{C} (0.125)$$

or

$$C = 1.25 \times 10^{-4} \text{ F} = 125 \text{ } \mu\text{F}$$

Exercises 25-1

1. What is the velocity after 3 s of an object which is dropped and falls under the influence of gravity?

2. The acceleration of an object rolling down an inclined plane is 6.0 ft/s². If it is given an upward velocity along the plane of 20 ft/s, what is its velocity after 4 s?

3. The acceleration of an object is given by $6t$. What is the relation between the velocity and time for this object if $v = 8$ when $t = 1$?

4. An object starts from rest and has an acceleration given by $a = \sqrt{t + 1}$. Find the velocity (in meters per second) after 8 s.

5. An object moves in a straight line with a constant velocity of 80 cm/s. How far is it from its starting point after 7 s?

6. The velocity of an object is given by $v = 2t + 1$. Find the relation between s and t if $s = 0$ when $t = 0$.

7. If an object is given an initial upward velocity from the ground of 120 ft/s, what is its distance above the ground after 3 s?

8. An object is given an upward initial velocity of 60 ft/s from the ground. What is its distance above the ground after 2.5 s?

9. Standing on a cliff, a woman throws a ball with a vertical velocity of 64 ft/s upward. If the ball hits the ground at the base of the cliff 7 s after it is released, how far is it from the top of the cliff to the ground below?

10. A man wishes to throw an object to a height 100 ft above him. What initial vertical velocity must he give the object?

11. The acceleration of an object is given by $a = 24t$. Find the displacement in meters after 1.5 s if $s = 0$ and $v = 0$ when $t = 0$.

12. The acceleration of a certain particle is given by $a = t^4$. Find the displacement in centimeters of the particle after 2 s if $v = 6$ cm/s and $s = 0$ when $t = 0$.

13. The current in a certain electric circuit is 0.5 A. How many coulombs of charge pass a given point in two seconds?

14. The current in a certain circuit is a function of the time and is given by $i = 2t + 3$. How many coulombs pass a given point in the circuit in the first four seconds?

15. The current in a certain wire changes with time according to the relation $i = \sqrt[3]{1 + 3t}$. How many coulombs of charge pass a specified point in the first 3 seconds?

16. The current in a certain circuit as a function of time is given by $i = 8 - t$. If $q_0 = 0$, for what value of t, greater than zero, is $q = 0$? What interpretation can be given to this result?

17. The voltage across a 3.0 μF capacitor is zero. What is the voltage after 10 ms if a current of 0.20 A charges the capacitor?

18. The voltage across a 7.5 μF capacitor is zero. What is the voltage after 5.0 ms if a current of $i = 0.1t$ charges the capacitor?

19. The voltage across a 150 μF capacitor is 50 V. What is the voltage after 1 s if a current $i = 0.012t^{1/5}$ further charges the capacitor?

20. A current $i = t/\sqrt{t^2 + 1}$ is sent into a circuit containing a previously uncharged 4.0 μF capacitor. How long does it take for the capacitor voltage to be 100 V?

21. The angular velocity ω is the time rate of change of the angular displacement θ of a rotating object. If the angular velocity of an object is given by $\omega = 4t$, find an expression for the angular displacement θ if $\theta = 0$ when $t = 0$.

22. The angular acceleration α is the time rate of change of the angular velocity. If the angular acceleration of a rotating object is given by $\alpha = \sqrt{2t + 1}$, find the expression for the angular displacement if $\omega = 0$ and $\theta = 0$ when $t = 0$.

23. An inductor in an electric circuit is essentially a coil of wire in which the voltage is affected by a changing current. By definition, the voltage caused by the changing current is given by

$$V_L = L\frac{di}{dt}$$

where L is the inductance and is measured in henries. If $V_L = 12.0 - 0.2t$ for a 3-H inductor, find the current in the circuit after 20 s if the initial current was zero.

24. If the inner and outer walls of a container are at different temperatures, the rate of change of temperature with respect to the distance from one wall is a function of the distance from the wall. Symbolically this is stated as $dT/dx = f(x)$, where T is the temperature. If x is measured from the outer wall, at 20°C, and $f(x) = 72x^2$, find the temperature at the inner wall if the container walls are 0.5 cm thick.

25. Surrounding an electrically charged particle is an electric field. The rate of change of electric potential with respect to the distance from the particle creating the field equals the negative of the value of the electric field. That is,

$$\frac{dV}{dx} = -E$$

where E is the electric field. If $E = k/x^2$, where k is a constant, find the electric potential at a distance x_1 from the particle, if $V \to 0$ as $x \to \infty$.

26. The rate of change of the vertical deflection y with respect to the horizontal distance x from one end of a beam is a function of x. For a particular beam, this function is $k(x^5 + 1350x^3 - 7000x^2)$, where k is a constant. Find y as a function of x.

27. Fresh water is flowing into a brine solution, with an equal volume of mixed solution flowing out. The amount of salt in the solution decreases, but more slowly as time increases. Under certain conditions the time rate of change of mass of salt (in grams per minute) is given by $-1/\sqrt{t + 1}$. Find the mass of salt as a function of time if 1000 g were originally present. Under these conditions, how long would it take for all the salt to be removed?

28. The rate of change of resistance of a certain electric resistor with respect to the temperature is given by

$$\frac{0.002T}{\sqrt[3]{3T^2 + 1}}$$

Find the resistance as a function of temperature if $R = 0.5$ mΩ when $T = 0$°C.

25–2 Areas by Integration

In Section 24–4 we introduced the method of finding the area under a curve by means of integration. In the same section we also showed that the area under a curve can be found by a summation process performed on the rectangles inscribed under the curve. In this way it was shown that integration can be interpreted as a summation process. The basic applications of the definite integral use this summation interpretation of the integral. In this section we shall formulate a general procedure for finding the area for which the bounding curves are known. The method is based on the summing of the areas of inscribed rectangles and using integration for the summation.

The first step in finding any area is to make a sketch of the area to be found. Next a representative **element of area** dA (a typical rectangle) should be drawn. In Fig. 25–1 the width of the element is dx. The length of the element is determined by the y-coordinate (of the vertex of the element) of the point on the curve. Thus, we call the length y. The area of this element is $y\,dx$, which in turn means that $dA = y\,dx$, or

$$A = \int_a^b y\,dx = \int_a^b f(x)\,dx \tag{25–7}$$

This equation states that the elements are to be summed (this is the meaning of the integral sign) from a (the leftmost element) to b (the rightmost element).

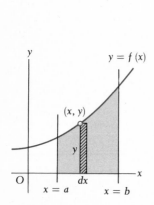

Figure 25–1

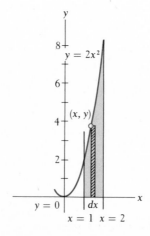

Figure 25–2

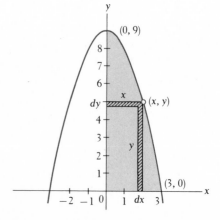

Figure 25–3

Example A
Find the area bounded by the curves $y = 2x^2$, $y = 0$, $x = 1$, and $x = 2$.
   The desired area is shown in Fig. 25–2. The rectangle shown is the representative element. Its area is $y \, dx$. The elements are to be summed from $x = 1$ to $x = 2$. Thus,

$$A = \int_1^2 y \, dx = \int_1^2 2x^2 \, dx = \frac{2}{3}x^3 \bigg|_1^2 = \frac{2}{3}(8) - \frac{2}{3}(1) = \frac{14}{3}$$

   In Figs. 25–1 and 25–2 the elements are vertical. It is also possible to use horizontal elements, and many problems are simplified by using them. The difference is that the length (longest dimension) is now measured in terms of the x-coordinate of the point on the curve, and the width becomes dy. In the following example, the area is found by both types of elements.

 Example B
Find the area in the first quadrant bounded by $y = 9 - x^2$.
   The area to be found is shown in Fig. 25–3. Using first the vertical element of length y and width dx, we have

$$A = \int_0^3 y \, dx = \int_0^3 (9 - x^2) \, dx = \left(9x - \frac{x^3}{3}\right)\bigg|_0^3 = (27 - 9) - 0 = 18$$

Now using the horizontal element of length x and width dy, we have

$$A = \int_0^9 x \, dy = \int_0^9 \sqrt{9 - y} \, dy = -\int_0^9 (9 - y)^{1/2}(-dy)$$

$$= -\frac{2}{3}(9 - y)^{3/2} \bigg|_0^9 = -\frac{2}{3}(9 - 9)^{3/2} + \frac{2}{3}(9 - 0)^{3/2} = \frac{2}{3}(27) = 18$$

   Note that the limits for the vertical elements were 0 and 3, while those for the horizontal elements were 0 and 9. These limits are determined by the direction in which the elements are summed. *By definition, vertical elements are summed from left to right, and horizontal elements are summed from bottom to top.* Doing it this way means that the summation will be in a positive direction.

   The choice of vertical or horizontal elements is determined by (1) which one leads to the simplest solution, or (2) the form of the resulting integral. In some problems it makes little difference which is chosen. However, our present methods of integration do not include many types of integrals.

   It is also possible to find the area between two curves if one is not an axis. In such a case, the length of the element becomes the difference in the y- or x-coordinates, depending on which element is used. The following examples show how we find this type of area.

Example C

Find the area bounded by $y = x^2$ and $y = x + 2$.

This area is indicated in Fig. 25–4. Here we choose vertical elements, since they are all bounded on the top by the line and on the bottom by the parabola. If we choose horizontal elements, the bounding curves are different above the point $(-1, 1)$ (points of intersection are found by solving equations simultaneously) than below this point. Choosing horizontal elements would thus require two separate integrals for solution. Therefore, using vertical elements, we have

$$A = \int_{-1}^{2} (y \text{ of line } - y \text{ of parabola}) \, dx$$

(The difference in y's is taken so that it is positive.) Continuing, we have

$$A = \int_{-1}^{2} (x + 2 - x^2) \, dx = \left(\frac{x^2}{2} + 2x - \frac{x^3}{3} \right) \Big|_{-1}^{2}$$

$$= \left(2 + 4 - \frac{8}{3} \right) - \left(\frac{1}{2} - 2 + \frac{1}{3} \right)$$

$$= \frac{10}{3} + \frac{7}{6} = \frac{27}{6} = \frac{9}{2}$$

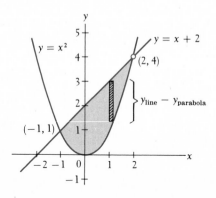

Figure 25–4

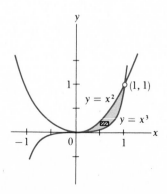

Figure 25–5

Example D

Find the area between $y = x^2$ and $y = x^3$.

Either horizontal or vertical elements may be used equally well (see Fig. 25–5). If we use horizontal elements, we have

$$A = \int_{0}^{1} (y^{1/3} - y^{1/2}) \, dy = \frac{3}{4} y^{4/3} - \frac{2}{3} y^{3/2} \Big|_{0}^{1} = \frac{3}{4} - \frac{2}{3} = \frac{1}{12}$$

It is important that the length of the element be positive. If the difference is taken incorrectly, the result will be negative. *Getting positive lengths can be assured for vertical elements if we subtract y of the lower* *curve from y of the upper curve. For horizontal elements we should sub-tract x of the left curve from x of the right curve.* This becomes of defi-nite importance if part of the area considered is above, and the remainder below, the x-axis. In such a case, the area is found by two integrals. The following example illustrates the necessity of this procedure.

Example E

Find the area between $y = x^3 - x$ and the x-axis.

We note from Fig. 25–6 that the area to the left of the origin is above the axis and the area to the right is below. If we find the area from

$$A = \int_{-1}^{1} (x^3 - x)\,dx = \frac{x^4}{4} - \frac{x^2}{2}\bigg|_{-1}^{1} = \left(\frac{1}{4} - \frac{1}{2}\right) - \left(\frac{1}{4} - \frac{1}{2}\right) = 0$$

we see that the apparent area is zero. From the figure we know this is not correct. Noting that the y-values (of the area) are negative to the right of the origin, we set up the integrals

$$A = \int_{-1}^{0} (x^3 - x)\,dx + \int_{0}^{1} -(x^3 - x)\,dx$$

$$= \left(\frac{x^4}{4} - \frac{x^2}{2}\right)\bigg|_{-1}^{0} - \left(\frac{x^4}{4} - \frac{x^2}{2}\right)\bigg|_{0}^{1}$$

$$= 0 - \left(\frac{1}{4} - \frac{1}{2}\right) - \left(\frac{1}{4} - \frac{1}{2}\right) + 0 = \frac{1}{2}$$

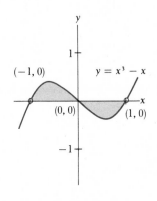

Figure 25–6

Exercises 25–2

In Exercises 1 through 28 find the areas bounded by the indicated curves.

1. $y = 4x$, $y = 0$, $x = 1$
2. $y = 2x$, $y = 0$, $x = 2$
3. $y = x^2$, $y = 0$, $x = 2$
4. $y = 3x^2$, $y = 0$, $x = 3$
5. $y = 6x$, $x = 0$, $y = 3$
6. $y = 8x$, $x = 0$, $y = 4$
7. $y = x^2$, $x = 0$, $y = 2$
8. $y = x^3$, $x = 0$, $y = 3$
9. $y = 2 - x$, $y = 0$, $x = 1$
10. $y = x^2 - 2x$, $y = 0$
11. $y = x^{-2}$, $y = 0$, $x = 2$, $x = 3$
12. $y = 16 - x^2$, $y = 0$, $x = 1$, $x = 2$
13. $y = \frac{1}{2}x$, $x = 0$, $y = 2$, $y = 4$
14. $y = \frac{1}{3}x$, $x = 0$, $y = 1$, $y = 5$
15. $y = \sqrt{x}$, $x = 0$, $y = 2$, $y = 3$
16. $x = y^2 - y$, $x = 0$
17. $y = 4 - 2x$, $x = 0$, $y = 0$, $y = 3$
18. $y = x$, $y = 2 - x$, $x = 0$
19. $y = x^2$, $y = 2 - x$, $x = 0$ (smaller)
20. $y = x^2$, $y = 2 - x$, $y = 2$ (smallest)
21. $y = x^4$, $y = 16$
22. $y = x^4 - 8x^2 + 16$, $y = 0$
23. $y = \sqrt{x - 1}$, $y = 0$, $x = 2$
24. $y = \sqrt{2x + 1}$, $y = 0$, $x = 4$
25. $y = x^2$, $y = \sqrt{x}$
26. $y = 8 - x^3$, $y = 7x$, $x = 0$
27. $y = x$, $x = -1$, $x = 1$, $y = 0$
28. $y = x^2 + 2x - 8$, $y = x + 4$

In Exercises 29 through 34 some applications of areas are shown.

29. Certain physical quantities are often represented as an area under a curve. By definition, power is defined as the time rate of change of performing work. Thus, $p = dw/dt$, or $dw = p\,dt$. Therefore, if $p = 12t - 4t^2$, find the work (in joules) performed in 3 s by finding the area under the curve of p versus t.

30. When considering the expansion of a gas, the expression for the work done in the expansion is $w = \int_{V1}^{V2} p\,dV$, where p is the pressure of the gas and V is its volume. For an adiabatic (no *heat* gained or lost) change, an approximate relation is $p = kV^{-1.4}$. Given that $k = 1$, find a numerical value for the work done if V changes from 1 to 4 units.

31. It is also possible to represent the change in displacement as an area. Since $s = \int v\,dt$, if $v = t^{2/3} + 1$, determine the distance (in feet) that an object moves from $t = 1$ s to $t = 8$ s, by finding the area under the curve of v versus t.

32. In designing a lawn area, an architect planned one such area of an estate to be between parabolas $y = 20 - 0.01x^2$ and $y = 0.005x^2 - 15$ and the lines $x = -27$ and $x = 27$ where dimensions are in meters. Find the area of this part of the lawn.

33. An experiment was performed by expanding a gas under the condition of constant temperature. The data found are tabulated below. How much work (in joules) was done in the expansion? Use the trapezoidal rule, with $n = 4$.

p (kilopascals)	102	77.0	60.0	47.0	43.0
V (cubic centimeters)	6.0	8.0	10.0	12.0	14.0

To obtain units of joules we must multiply the numerical value obtained by 10^{-3} to include the powers of 10 indicated by the prefix. 1 kPa $= 10^3$ Pa and 1 cm$^3 = (10^{-2}$ m$)^3 = 10^{-6}$ m^3.

34. A certain tract of land is bounded by a bend in a river and a straight fence. If the distances y from the fence to the river are given for various distances x along the fence from one end to the other as in the following table, find the approximate area of the piece of land by use of the trapezoidal rule, with $n = 8$.

y (feet)	0	15	50	70	95	80	35	20	0
x (feet)	0	10	20	30	40	50	60	70	80

25–3 Volumes by Integration

Consider an area and its representative element (see Fig. 25–7) to be rotated about the x-axis. When an area is rotated in this manner, it is said to generate a volume, which is also indicated in the figure. We shall now show methods of finding volumes which are generated in this manner.

As the area rotates about the x-axis, so does its representative element. The element generates a solid for which the volume is known. This is a thin right circular cylinder. We know that the volume of a right circular cylinder is π times the square of the radius times the height of the cylinder. We must now determine the radius and height of this cylinder.

Since the element is rotated about the x-axis, the y coordinate of the point on the curve which touches the element must represent the radius. Also, it can be seen that the height is dx (the cylinder is on its side). The representative **element of volume,** a circular **disk,** is $dV = \pi y^2 \, dx$. Summing these elements of volume from left to right, we have the total volume V as

$$V = \pi \int_a^b y^2 \, dx = \pi \int_a^b [f(x)]^2 \, dx \qquad (25\text{–}8)$$

Thus, by use of Eq. (25–8), we can find the volume generated by an area bounded by the x-axis, which is rotated about the x-axis.

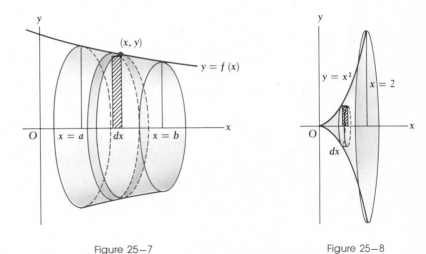

Figure 25–7   Figure 25–8

Example A
Find the volume generated by rotating the area bounded by $y = x^2$, $x = 2$, and $y = 0$ about the x-axis.
See Fig. 25–8. By Eq. (25–8), we have

$$V = \pi \int_0^2 (x^2)^2 \, dx = \pi \int_0^2 x^4 \, dx = \frac{\pi}{5}x^5 \Big|_0^2 = \frac{32\pi}{5}$$

If an area bounded by the y-axis is rotated about the y-axis, the volume generated is given by

$$V = \pi \int_c^d x^2 \, dy \qquad (25\text{–}9)$$

In this case the radius of the element of volume is the x-coordinate of the point on the curve, and the height of the disk is dy. One should always be careful to identify the radius and height properly.

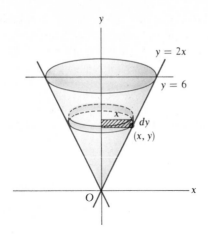

Figure 25–9   Figure 25–10

Example B

Find the volume generated by rotating the area bounded by $y = 2x$, $y = 6$, and $x = 0$ about the y-axis.

Figure 25–9 shows the volume to be found. We write

$$V = \pi \int_0^6 x^2 \, dy = \pi \int_0^6 \left(\frac{y}{2}\right)^2 dy = \frac{\pi}{4} \int_0^6 y^2 \, dy = \frac{\pi}{12} y^3 \Big|_0^6 = 18\pi$$

Since this volume is a right circular cone, it is possible to check the result:

$$V = \frac{1}{3}\pi r^2 h = \frac{1}{3}\pi (3^2)(6) = 18\pi$$

There is another method of finding a volume of a solid revolution. If the area in Fig. 25–8 is rotated about the y-axis, the element of area $y \, dx$ generates a different element of volume from that generated when it is rotated about the x-axis. We can see in Fig. 25–10 that this element of volume is a **cylindrical shell.** *The total volume is made up of an infinite number of concentric shells.* When the volumes of these shells are summed, we have the total volume generated. Thus we must now find the approximate volume dV of the representative shell. By finding the circumference of the base and multiplying this by the height, we can obtain an expression for the surface area of the shell. Then, by multiplying this by the thickness of the shell, we obtain its volume. The volume of the representative shell is

$$dV = 2\pi \, (\text{radius}) \cdot (\text{height}) \cdot (\text{thickness}) \tag{25–10}$$

Similarly, the volume of a disk is given by

$$dV = \pi \, (\text{radius})^2 \cdot (\text{height}) \tag{25–11}$$

The elements of volume should be remembered in the general forms given in Eqs. (25–10) and (25–11), and not in specific forms such as Eqs. (25–8) and (25–9). If we remember them in this manner, we can readily apply these methods to finding any such volume of a solid of revolution.

Example C

Use the method of cylindrical shells to find the volume generated by rotating the area bounded by $y = 4 - x^2$, $x = 0$, and $y = 0$ about the y-axis.

From Fig. 25–11, we identify the radius, height, and thickness: $r = x$, $h = y$, $t = dx$ (this determines the limits as $x = 0$ and $x = 2$). And so

$$V = 2\pi \int_0^2 xy\ dx = 2\pi \int_0^2 x(4 - x^2)\ dx = 2\pi \int_0^2 (4x - x^3)\ dx = 8\pi$$

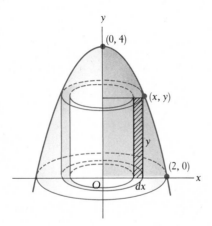

Figure 25–11   Figure 25–12

Example D

Use the disk method to find the indicated volume of Example C.

From Fig. 25–12, we identify the radius and height: $r = x$, $h = dy$ (which tells us that the limits are from $y = 0$ to $y = 4$). Thus,

$$V = \pi \int_0^4 x^2\ dy = \pi \int_0^4 (4 - y)\ dy = 8\pi$$

Example E

Use shells to find the volume if the area of Example C is rotated about the x-axis.

From Fig. 25–13, we see that $r = y$, $h = x$, $t = dy$, (thus, the limits are $y = 0$ to $y = 4$). Hence

$$V = 2\pi \int_0^4 xy\ dy = 2\pi \int_0^4 \sqrt{4 - y}(y\ dy) = \frac{256\pi}{15}$$

(The method of integrating this function has not yet been discussed. We present the answer here for the reader's information at this time.)

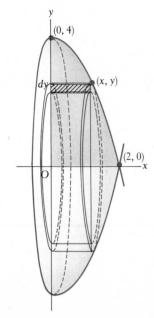

Figure 25–13

Example F

By the use of disks, find the volume indicated in Example E.

From Fig. 25–14, we see that $r = y$, $h = dx$ (the limits are $x = 0$ to $x = 2$). Therefore

$$V = \pi \int_0^2 y^2 \, dx = \pi \int_0^2 (4 - x^2)^2 \, dx$$

$$= \pi \int_0^2 (16 - 8x^2 + x^4) \, dx = \frac{256\pi}{15}$$

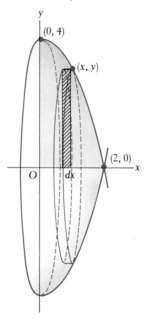

Figure 25–14

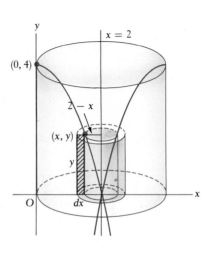

Figure 25–15

Example G

Find the volume generated if the area in Example C is rotated about the line $x = 2$.

Shells are convenient, since the volume of a shell can be expressed as a single integral. From Fig. 25–15, we see that

$$r = 2 - x \quad h = y \quad \text{and} \quad t = dx$$

Thus, we have

$$V = 2\pi \int_0^2 (2 - x) y \, dx = 2\pi \int_0^2 (2 - x)(4 - x^2) \, dx$$

$$= 2\pi \int_0^2 (8 - 2x^2 - 4x + x^3) \, dx = \frac{40\pi}{3}$$

(If the area had been rotated about the line $x = 3$, the only difference in the integral would have been that $r = 3 - x$. Everything else, including the limits, would have remained the same.)

Exercises 25–3

In Exercises 1 through 12 find the volume generated by the areas bounded by the given curves if they are rotated about the x-axis. Use the indicated method in each case.

1. $y = 1 - x$, $x = 0$, $y = 0$ (disks)
2. $y = x$, $y = 0$, $x = 2$ (disks)
3. Area of Exercise 1 (shells)
4. $y = x$, $x = 0$, $y = 2$ (shells)
5. $y = x^2 + 1$, $x = 0$, $x = 3$, $y = 0$ (disks)
6. $y = 2x - x^2$, $y = 0$ (disks)
7. $y = x^3$, $y = 8$, $x = 0$ (shells)
8. $y = x^2$, $y = x$ (shells)
9. $y = 3\sqrt{x}$, $y = 0$, $x = 4$ (disks)
10. $y = 6 - x - x^2$, $x = 0$, $y = 0$ (quad. I) (disks)
11. $x = 4y - y^2 - 3$, $x = 0$ (shells)
12. $y = x^4$, $x = 0$, $y = 1$, $y = 2$ (shells)

In Exercises 13 through 24 find the volume generated by the areas bounded by the given curves if they are rotated about the y-axis. Use the indicated method in each case.

13. Area of Exercise 1 (disks)
14. Area of Exercise 4 (disks)
15. Area of Exercise 1 (shells)
16. Area of Exercise 2 (shells)
17. $y = 2\sqrt{x}$, $x = 0$, $y = 2$ (disks)
18. $y^2 = x$, $y = 4$, $x = 0$ (disks)
19. Area of Exercise 6 (shells)
20. Area of Exercise 10 (shells)
21. Area of Exercise 11 (disks)
22. $x^2 + 4y^2 = 4$ (quad. I) (disks)
23. $y = \sqrt{4 - x^2}$ (quad. I) (shells)
24. $y = 8 - x^3$, $x = 0$, $y = 0$ (shells)

In Exercises 25 through 30 find the indicated volumes by integration.

25. Find the volume generated if the area of Exercise 6 is rotated about the line $x = 2$.
26. Find the volume generated if the area of Exercise 8 is rotated about the line $y = 4$.
27. Derive the expression for the volume of a right circular cone obtained by rotating the area bounded by $y = (r/h)x$, $y = 0$ and $x = h$ about the x-axis.
28. A conical funnel has an opening at the bottom of radius 0.5 cm. The diameter at the top of the funnel is 20 cm and the depth of the funnel is 19 cm. What is the maximum capacity of the funnel?
29. The water in a spherical tank 20 m in radius is 15 m deep at the deepest point. How much water is in the tank?
30. All horizontal cross-sections of a keg 4 ft tall are circular, and the sides of the keg are parabolic. The diameter at the top and the bottom is 2 ft, and the diameter in the middle is 3 ft. Find the volume that the keg holds.

25–4 Centroids

In the study of mechanics, a very important property of an object is its center of mass. In this section we shall show the meaning of center of mass and then show how integration is used to determine the center of mass for areas and solids of rotation.

If a mass m is at a distance d from a specified point O, the **moment** *of the mass about O is defined as* md. If several masses $m_1, m_2, \ldots, m_n$ are at distances $d_1, d_2, \ldots, d_n$, respectively, from point O, their moment (as a group) about O is defined as

$$m_1 d_1 + m_2 d_2 + \cdots + m_n d_n$$

If all the masses could be concentrated at one point $\overline{d}$ units from O, the moment would be $(m_1 + m_2 + \cdots + m_n)\overline{d}$: this is, however, what is meant by the above expression. Therefore, we may write

$$m_1 d_1 + m_2 d_2 + \cdots + m_n d_n = (m_1 + m_2 + \cdots + m_n)\overline{d} \qquad (25\text{--}12)$$

In Eq. (25–12), $\overline{d}$ *is the distance from O to the* **center of mass.** The moment of a mass is a measure of its tendency to rotate about a point. A weight far from the point of balance of a long rod is more likely to make the rod turn than if the same weight were placed near the point of balance. It is easier to open a door if you push near the door knob than if you push near the hinges. This is the type of physical property which the moment of mass measures.

Example A
On the x-axis a mass of 3 units is placed at (2, 0), another of 6 units at (5, 0) and a third, 7 units, at (6, 0). Find the center of mass of the three objects.

Taking the reference point as the origin, we find $d_1 = 2$, $d_2 = 5$, and $d_3 = 6$. Thus, $m_1 d_1 + m_2 d_2 + m_3 d_3 = (m_1 + m_2 + m_3)\overline{d}$ becomes

$$3(2) + 6(5) + 7(6) = (3 + 6 + 7)\overline{d} \qquad \text{or} \qquad \overline{d} = 4.88$$

This means that the center of mass of the three objects is at (4.88, 0). Therefore a mass of 16 units placed at this point has the same moment as the three masses as a unit (see Fig. 25–16).

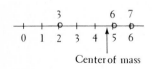

Center of mass

Figure 25–16

Example B
Find the center of mass of the area indicated in Fig. 25–17.

We first note that the center of mass is not *on* either axis. This can be seen from the fact that the major portion of the area is in the first quadrant. *We shall therefore measure the moments with respect to each axis to find the point which is the center of mass. This point is also called the* **centroid** *of the area.*

The easiest method of finding the centroid is to divide the area into rectangles, as indicated by the dashed line in Fig. 25–17, and assume that we may consider the mass of each rectangle to be concentrated at its center. In this way the left rectangle has its center $(-1, 1)$ and the right rectangle has its center at $(\frac{5}{2}, 2)$. The mass of each rectangle, assumed

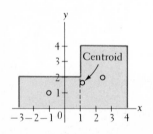

Figure 25–17

uniform, is porportional to the area. The area of the left rectangle is 8 units and that of the right rectangle is 12 units. Thus, taking moments with respect to the y-axis, we have

$$8(-1) + 12\left(\frac{5}{2}\right) = (8 + 12)\bar{x}$$

where $\bar{x}$ is the x-coordinate of the centroid. Solving for $\bar{x}$ we have $\bar{x} = \frac{11}{10}$.

Now taking moments with respect to the x-axis we have

$$8(1) + 12(2) = (8 + 12)\bar{y}$$

where $\bar{y}$ is the y-coordinate of the centroid. Thus, $\bar{y} = \frac{8}{5}$. This means that the coordinates of the centroid, the center of mass, are $(\frac{11}{10}, \frac{8}{5})$. This may be interpreted as meaning that an area of this shape would balance on a single support under this point. As an approximate check, we note from the figure that this point appears to be a reasonable balance point for the area.

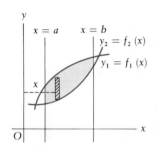

Figure 25–18

If an area is bounded on the curves of the functions $y_1 = f_1(x)$, $y_2 = f_2(x)$, $x = a$, and $x = b$, as shown in Fig. 25–18, the moment of the element of area about the y-axis is $(k\,dA)x$, where k is the mass per unit area. In this expression, $k\,dA$ is the mass of the element, and x is its distance (moment arm) from the y-axis. The element dA may be written as $(y_1 - y_2)\,dx$, which means that the moment may be written as $kx(y_1 - y_2)\,dx$. If we then sum up the moments of all the elements and express this as an integral (which, of course, means sum), we have $k \int_a^b x(y_1 - y_2)\,dx$. If we consider all the mass of the area to be concentrated at one point $\bar{x}$ units from the y-axis, the moment would be $(kA)\bar{x}$, where kA is the mass of the entire area, and $\bar{x}$ is the distance the center of mass is from the y-axis. By the previous discussion these two expressions should be equal. This means $k \int_a^b x(y_1 - y_2)\,dx = kA\bar{x}$. Since k appears on each side of the equation, we divide it out (we are assuming that the mass per unit area is constant). The area A is found by the integral $\int_a^b (y_1 - y_2)\,dx$. Therefore,

$$\bar{x} = \frac{\displaystyle\int_a^b x(y_1 - y_2)\,dx}{\displaystyle\int_a^b (y_1 - y_2)\,dx} \tag{25–13}$$

Equation (25–13) gives us the x-coordinate of the centroid of an area if vertical elements are used. It should be noted that the two integrals in Eq. (25–13) must be evaluated separately. We cannot cancel the apparent common factor $y_1 - y_2$, and we cannot combine quantities and perform one integration. The two integrals must be evaluated before cancellations are made.

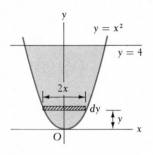

Figure 25–19

Following the same reasoning, if an area is bounded by $x_1 = g_1(y)$, $x_2 = g_2(y)$, $y = c$, and $y = d$ (Fig. 25–19), the y-coordinate of the centroid is

$$\overline{y} = \frac{\displaystyle\int_c^d y(x_2 - x_1)\, dy}{\displaystyle\int_c^d (x_2 - x_1)\, dy} \tag{25–14}$$

In this equation, horizontal elements are used.

In applying Eqs. (25–13) and (25–14), we should keep in mind that the denominators of the right-hand sides give the area, and once we have found this, we may use it for both $\overline{x}$ and $\overline{y}$. Also, we should utilize any symmetry a curve may have.

Example C

Find the coordinates of the centroid of the area bounded by $y = x^2$ and the line $y = 4$.

We sketch a graph indicating the area and an element of area (see Fig. 25–20). The curve is a parabola whose axis is the y-axis. Since the area is symmetrical to the y-axis, the centroid must be on this axis. This means that the x-coordinate of the centroid is zero, or $\overline{x} = 0$. To find the y-coordinate of the centroid, we use Eq. (25–14). For this area we have

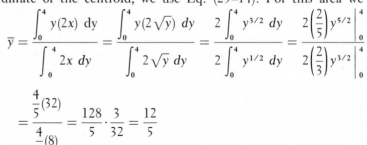

$$\overline{y} = \frac{\displaystyle\int_0^4 y(2x)\, dy}{\displaystyle\int_0^4 2x\, dy} = \frac{\displaystyle\int_0^4 y(2\sqrt{y})\, dy}{\displaystyle\int_0^4 2\sqrt{y}\, dy} = \frac{2\displaystyle\int_0^4 y^{3/2}\, dy}{2\displaystyle\int_0^4 y^{1/2}\, dy} = \frac{2\left(\dfrac{2}{5}\right)y^{5/2}\Big|_0^4}{2\left(\dfrac{2}{3}\right)y^{3/2}\Big|_0^4}$$

$$= \frac{\dfrac{4}{5}(32)}{\dfrac{4}{3}(8)} = \frac{128}{5} \cdot \frac{3}{32} = \frac{12}{5}$$

The coordinates of the centroid are $(0, \tfrac{12}{5})$. This area would balance if a single pointed support were to be put under this point.

Example D

Find the coordinates of the centroid of an isosceles right triangle with side a.

We must first set up this area in the xy-plane. One choice is to place the triangle with one vertex at the origin and the right angle on the x-axis (see Fig. 25–21). Since each side is a, the hypotenuse passes through the point (a, a). The equation of the hypotenuse is $y = x$. The x-coordinate of the centroid is

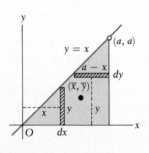

Figure 25–20

Figure 25–21

$$\overline{x} = \frac{\displaystyle\int_0^a xy\, dx}{\displaystyle\int_0^a y\, dx} = \frac{\displaystyle\int_0^a x(x)\, dx}{\displaystyle\int_0^a x\, dx} = \frac{\displaystyle\int_0^a x^2\, dx}{\dfrac{1}{2}x^2\Big|_0^a} = \frac{\dfrac{1}{3}x^3\Big|_0^a}{\dfrac{a^2}{2}} = \frac{\dfrac{a^3}{3}}{\dfrac{a^2}{2}} = \frac{2a}{3}$$

The y-coordinate of the centroid is

$$\bar{y} = \frac{\displaystyle\int_0^a y(a-x)\,dy}{\dfrac{a^2}{2}} = \frac{\displaystyle\int_0^a y(a-y)\,dy}{\dfrac{a^2}{2}} = \frac{\displaystyle\int_0^a (ay-y^2)\,dy}{\dfrac{a^2}{2}}$$

$$= \frac{\left.\dfrac{ay^2}{2} - \dfrac{y^3}{3}\right|_0^a}{\dfrac{a^2}{2}} = \frac{\dfrac{a^3}{6}}{\dfrac{a^2}{2}} = \frac{a}{3}$$

Thus, the coordinates of the centroid are $(\frac{2}{3}a, \frac{1}{3}a)$. The results indicate that the center of mass is $\frac{1}{3}a$ units from each of the equal sides.

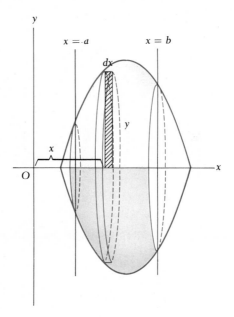

Figure 25–22

Another important figure for which we wish to find the centroid is the solid of revolution. If the density of the solid is constant, the coordinates of the centroid will be on the axis of rotation. The problem which remains is to find just where on the axis the centroid is located.

If a given area bounded by the x-axis, as shown in Fig. 25–22, is revolved about the x-axis, a vertical element of area generates a disk element of volume. The center of mass of the disk is at its center, and therefore, for purposes of finding moments, we may consider its mass concentrated there. The moment about the y-axis of a typical element is $x(k)(\pi y^2\,dx)$, where x is the moment arm, k is the density, and $\pi y^2\,dx$ is the volume. The sum of the moments of the elements can be expressed as an integral; it equals the volume times the density times the x-coordinate of the centroid of the volume. Since the density k and the factor of π would appear on each side of the equation, they need not be written. Therefore,

$$\bar{x} = \frac{\displaystyle\int_a^b xy^2 \, dx}{\displaystyle\int_a^b y^2 \, dx} \tag{25–15}$$

is the equation giving the x-coordinate of the centroid of a volume of a solid of revolution about the x-axis.

In the same manner we may find the y-coordinate of the centroid of a volume of a solid of revolution about the y-axis. It is

$$\bar{y} = \frac{\displaystyle\int_c^d yx^2 \, dy}{\displaystyle\int_c^d x^2 \, dy} \tag{25–16}$$

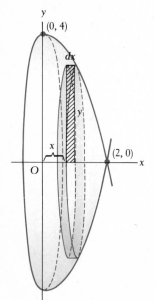

Figure 25–23

Example E

Find the coordinates of the centroid of the volume generated by rotating the first-quadrant area under the curve $y = 4 - x^2$ about the x-axis.

Since the curve (see Fig. 25–23) is rotated about the x-axis, the centroid is on the x-axis, which means that $\bar{y} = 0$. We find the x-coordinate by using Eq. (25–15):

$$\bar{x} = \frac{\displaystyle\int_0^2 xy^2 \, dx}{\displaystyle\int_0^2 y^2 \, dx} = \frac{\displaystyle\int_0^2 x(4 - x^2)^2 \, dx}{\displaystyle\int_0^2 (4 - x^2)^2 \, dx} = \frac{\displaystyle\int_0^2 (16x - 8x^3 + x^5) \, dx}{\displaystyle\int_0^2 (16 - 8x^2 + x^4) \, dx}$$

$$= \frac{8x^2 - 2x^4 + \dfrac{1}{6}x^6 \Big|_0^2}{16x - \dfrac{8}{3}x^3 + \dfrac{1}{5}x^5 \Big|_0^2} = \frac{32 - 32 + \dfrac{64}{6}}{32 - \dfrac{64}{3} + \dfrac{32}{5}} = \frac{5}{8}$$

The coordinates of the centroid are $(\frac{5}{8}, 0)$.

Example F

Find the coordinates of the centroid of the volume generated by rotating the area of Example E about the y-axis.

Since the curve is rotated about the y-axis, $\bar{x} = 0$ (see Fig. 25–24). The y-coordinate is found by Eq. (25–16):

$$\bar{y} = \frac{\displaystyle\int_0^4 yx^2 \, dy}{\displaystyle\int_0^4 x^2 \, dy} = \frac{\displaystyle\int_0^4 y(4-y) \, dy}{\displaystyle\int_0^4 (4-y) \, dy} = \frac{\displaystyle\int_0^4 (4y - y^2) \, dy}{\displaystyle\int_0^4 (4-y) \, dy} = \frac{2y^2 - \dfrac{1}{3}y^3 \Big|_0^4}{4y - \dfrac{1}{2}y^2 \Big|_0^4}$$

$$= \frac{32 - \dfrac{64}{3}}{16 - 8} = \frac{4}{3}$$

The coordinates of the centroid are $(0, \frac{4}{3})$.

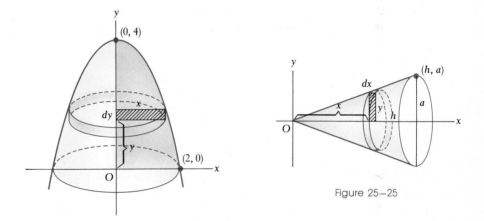

Figure 25—24

Figure 25—25

Example G

Find the centroid of a right circular cone of radius a and altitude h.

To generate a right circular cone, we may rotate a right triangle about one of its legs (Fig. 25–25). Placing the leg of length h along the x-axis, we rotate the right triangle whose hypotenuse is given by $y = (a/h)x$ about the x-axis. The x-coordinate of the centroid is

$$\bar{x} = \frac{\displaystyle\int_0^h xy^2 \, dx}{\displaystyle\int_0^h y^2 \, dx} = \frac{\displaystyle\int_0^h x\left[\left(\frac{a}{h}\right)x\right]^2 dx}{\displaystyle\int_0^h \left[\left(\frac{a}{h}\right)x\right]^2 dx} = \frac{\left(\dfrac{a^2}{h^2}\right)\left(\dfrac{1}{4}x^4\right)\Big|_0^h}{\left(\dfrac{a^2}{h^2}\right)\left(\dfrac{1}{3}x^3\right)\Big|_0^h} = \frac{3}{4}h$$

Therefore, the centroid is located along the altitude $\frac{3}{4}$ of the way from the vertex, or $\frac{1}{4}$ of the way from the base.

Exercises 25—4

In Exercises 1 through 4 find the center of mass of the particles of the given masses located at the given points.

1. 5 units at $(1, 0)$, 10 units at $(4, 0)$, 3 units at $(5, 0)$

2. 2 units at $(2, 0)$, 9 units at $(3, 0)$, 3 units at $(8, 0)$, 1 unit at $(12, 0)$

3. 4 units at $(-3, 0)$, 2 units at $(0, 0)$, 1 unit at $(2, 0)$, 8 units at $(3, 0)$

4. 2 units at $(-4, 0)$, 1 unit at $(-3, 0)$, 5 units at $(1, 0)$, 4 units at $(4, 0)$

In Exercises 5 through 8 find the coordinates of the centroid of the area shown.

5. Fig. 25—26(a)   6. Fig. 25—26(b)

7. Fig. 25—26(c)   8. Fig. 25—26(d)

(a)

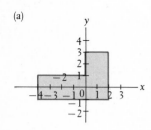

In Exercises 9 through 24 find the coordinates of the centroids of the indicated figures.

9. Find the coordinates of the centroid of the area bounded by $y = x^2$ and $y = 2$.

10. Find the coordinates of the centroid of a semicircular area.

11. Find the coordinates of the centroid of the area bounded by $y = 4 - x$ and the axes.

(b)

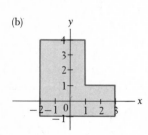

12. Find the coordinates of the centroid of the area bounded by $y = x^3$, $x = 2$, and the x-axis.

13. Find the coordinates of the centroid of the area bounded by $y = x^2$ and $y = x^3$.

14. Find the coordinates of the centroid of the area bounded by $y^2 = x$, $y = 2$, $x = 0$.

15. Find the coordinates of the centroid of the volume generated by rotating the area in the first quadrant bounded by $y^2 = 4x$, $y = 0$, and $x = 1$ about the y-axis.

(c)

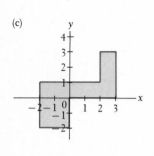

16. Find the coordinates of the centroid of the volume generated by rotating the area bounded by $y = x^2$, $x = 2$, and the x-axis about the x-axis.

17. Find the coordinates of the centroid of the volume generated by rotating the area bounded by $y^2 = 4x$ and $x = 1$ about the x-axis.

18. Find the coordinates of the centroid of the volume generated by rotating the area bounded by $x^2 - y^2 = 9$, $y = 4$, and the x-axis about the y-axis.

19. Find the coordinates of the centroid of the volume generated by rotating the area bounded by $y = 4/x^2$, $x = 1$, $x = 2$, and the x-axis about the x-axis.

(d)

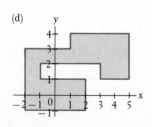

Figure 25—26

20. Find the coordinates of the centroid of the volume generated by rotating the first-quadrant area bounded by the ellipse $4x^2 + 9y^2 = 36$ around the y-axis.

21. Find the location of the centroid of a right triangle with legs a and b.

22. Find the coordinates of the centroid of an isosceles trapezoid with bases b_1 and b_2 and altitude a.

23. Find the location of the centroid of a hemisphere of radius a.

24. Find the location of the centroid of the frustum of a right circular cone if the radii are 2 and 6 in., and the altitude is 5 in.

25–5 Moments of Inertia

In applying the laws of physics, we encounter an important quantity when we are discussing the rotation of an object: its **moment of inertia.** The moment of inertia of an object rotating about an axis is analogous to the mass of an object moving through space in reference to some point. *In each case the quantity (moment of inertia or mass) is the measure of the tendency of an object to resist a change in motion.*

Suppose that a particle of mass m is rotating about some point; we define its moment of inertia as md^2, where d is the distance from the particle to the point. If a group of particles of masses $m_1, m_2, \ldots, m_n$ are rotating about some axis, the moment of inertia of the group with respect to that axis is

$$I = m_1 \, d_1^2 + m_2 \, d_2^2 + \cdots + m_n \, d_n^2$$

where the d's are the respective distances of the particles from the axis. If all of the masses were at the same distance R from the axis of rotation, so that the total moment of inertia were the same, we would have

$$m_1 \, d_1^2 + m_2 \, d_2^2 + \cdots + m_n \, d_n^2 = (m_1 + m_2 + \cdots + m_n)R^2 \qquad (25\text{–}17)$$

where R is called the **radius of gyration.**

Example A

Find the moment of inertia and radius of gyration of three masses, one of 3 units at $(-2, 0)$, the second of 5 units at $(1, 0)$, and the third of 4 units at $(4, 0)$ with respect to the origin (see Fig. 25–27).

The moment of inertia of the group is

$$I = 3(-2)^2 + 5(1)^2 + 4(4)^2 = 81$$

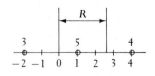

Figure 25–27

The radius of gyration is found from $I = (m_1 + m_2 + m_3)R^2$. Thus,

$$81 = (3 + 5 + 4)R^2 \qquad R^2 = \frac{81}{12} \qquad \text{or} \qquad R = 2.60$$

This means that a mass of 12 units placed at $(2.60, 0)$ [or at $(-2.60, 0)$] would have the same rotational inertia about the origin as the three objects as a unit.

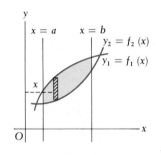

Figure 25–28

If an area is bounded by the curves of the functions $y_1 = f_1(x)$, $y_2 = f_2(x)$, and $x = a$ and $x = b$, as shown in Fig. 25–28, the moment of inertia of this area with respect to the y-axis, I_y, is given by the sum of the moments of inertia of the individual elements. The mass of each element is $k(y_1 - y_2)\,dx$ where k is the mass per unit area, and $(y_1 - y_2)\,dx$ is the area of the element. The distance of the element from the y-axis is x. Representing this sum as an integral, we have

$$I_y = k \int_a^b x^2 (y_1 - y_2) \, dx \qquad (25\text{–}18)$$

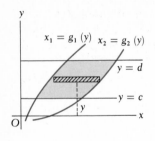

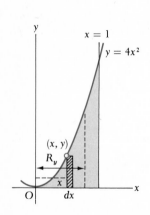

Figure 25–29

To *find the radius of gyration of the area with respect to the y-axis, R_y,
we would first find the moment of inertia, divide this by the mass of the
area, and take the square root of this result.*

In the same manner, the moment of inertia of an area, with respect to
the x-axis, bounded by $x_1 = g_1(y)$ and $x_2 = g_2(y)$ is given by

$$I_x = k \int_c^d y^2 (x_2 - x_1) \, dy \qquad (25\text{–}19)$$

We find the radius of gyration of the area with respect to the x-axis, R_x,
in the same manner as we find it with respect to the y-axis (see Fig.
25–29).

Example B
Find the moment of inertia and the radius of gyration of the area
bounded by $y = 4x^2$, $x = 1$, and the x-axis with respect to the y-axis.

We find the moment of inertia of the area (see Fig. 25–30) by using
Eq. (25–18):

$$I_y = k \int_0^1 x^2 y \, dx = k \int_0^1 x^2 (4x^2) \, dx = 4k \int_0^1 x^4 \, dx$$

$$= 4k \left(\frac{1}{5} x^5 \right) \Big|_0^1 = \frac{4k}{5}$$

To find the radius of gyration, we first determine the mass of the area:

$$m = k \int_0^1 y \, dx = k \int_0^1 (4x^2) \, dx = 4k \left(\frac{1}{3} x^3 \right) \Big|_0^1 = \frac{4k}{3}$$

$$R_y^2 = \frac{I_y}{m} = \frac{4k}{5} \cdot \frac{3}{4k} = \frac{3}{5} \qquad R_y = \sqrt{\frac{3}{5}} = \frac{\sqrt{15}}{5}$$

Figure 25–30

Example C
Find the moment of inertia of a right triangle with sides a and b with
respect to the side b. Assume that $k = 1$.

Placing the triangle as shown in Fig. 25–31, we see that the equation
of the hypotenuse is $y = (a/b)x$. The moment of inertia is

$$I_x = \int_0^a y^2 (b - x) \, dy = \int_0^a y^2 \left(b - \frac{b}{a} y \right) dy = b \int_0^a \left(y^2 - \frac{1}{a} y^3 \right) dy$$

$$= b \left(\frac{1}{3} y^3 - \frac{1}{4a} y^4 \right) \Big|_0^a = b \left(\frac{a^3}{3} - \frac{a^3}{4} \right) = \frac{ba^3}{12}$$

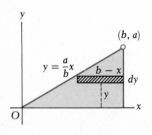

Figure 25–31

Among the most important moments of inertia are those of solids of
revolution. Since all parts of an element of mass should be at the same

distance from the axis, the most convenient element of volume to choose is the cylindrical shell (see Fig. 25–32). If the area bounded by the curves $y = f(x)$, $x = a$, $x = b$, and the x-axis is rotated about the y-axis, the moment of inertia of the element of volume is $k(2\pi xy\,dx)(x^2)$, where k is the density, $2\pi xy\,dx$ is the volume of the element, and x^2 is the square of its distance from the y-axis. Expressing the sum of the moments of the elements as an integral, the moment of inertia of the volume with respect to the y-axis, I_y, is

$$I_y = 2\pi k \int_a^b yx^3\,dx \tag{25–20}$$

The radius of gyration of the volume with respect to the y-axis R_y is found by determining (1) the moment of inertia, (2) the mass of the volume, and (3) the square root of the quotient of the moment of inertia divided by the mass.

The moment of inertia of the volume (see Fig. 25–33) generated by rotating the area bounded by $x = g(y)$, $y = c$, $y = d$, and the y-axis about the x-axis, I_x, is given by

$$I_x = 2\pi k \int_c^d xy^3\,dy \tag{25–21}$$

The radius of gyration of the volume with respect to the x-axis, R_x, is found in the same manner as R_y.

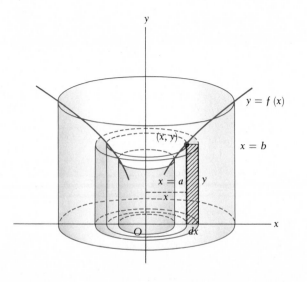

Figure 25–32

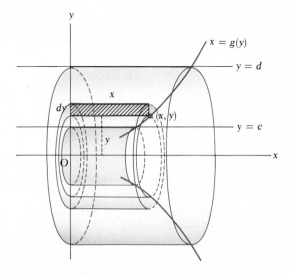

Figure 25–33

Example D

Find the moment of inertia and the radius of gyration of the solid (see Fig. 25–34) generated by rotating the area bounded by $y^3 = x$, $y = 2$, and the y-axis about the x-axis.

$$I_x = 2\pi k \int_0^2 xy^3 \, dy = 2\pi k \int_0^2 (y^3) y^3 \, dy = 2\pi k \left(\frac{1}{7} y^7 \right) \Bigg|_0^2 = \frac{256\pi k}{7}$$

$$m = 2\pi k \int_0^2 xy \, dy = 2\pi k \int_0^2 y^3 y \, dy = 2\pi k \left(\frac{1}{5} y^5 \right) \Bigg|_0^2 = \frac{64\pi k}{5}$$

$$R_x^2 = \frac{256\pi k}{7} \cdot \frac{5}{64\pi k} = \frac{20}{7} \qquad R_x = \sqrt{\frac{20}{7}} = \frac{2}{7} \sqrt{35}$$

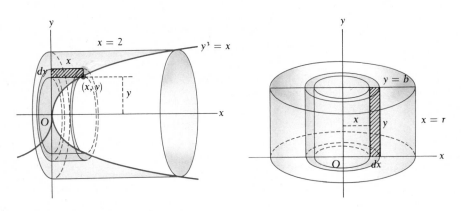

Figure 25–34   Figure 25–35

Example E

Find the moment of inertia of a disk of radius r with respect to its axis.

To generate a disk (see Fig. 25–35) we shall rotate the area bounded by the axes, $x = r$ and $y = b$, about the y-axis. We then have

$$I_y = 2\pi k \int_0^r x^3 y \, dx = 2\pi k \int_0^r x^3 (b) \, dx$$

$$= 2\pi kb \int_0^r x^3 \, dx = 2\pi kb \left(\frac{1}{4} x^4 \right) \Bigg|_0^r = \frac{\pi kbr^4}{2}$$

The mass of the disk is $k(\pi r^2)b$. Rewriting the expression for I_y, we have

$$I_y = \frac{(\pi kbr^2)r^2}{2} = \frac{mr^2}{2}$$

This is the usual way of expressing the moment of inertia of a particular solid. That is, the mass is usually included in the formula for the moment of inertia.

Due to the limited methods of integration available at this point, we cannot integrate the expressions for the moments of inertia of circular areas or of a sphere. These will be introduced in Section 27–8 in the exercises, by which point the proper method of integration will have been developed.

Exercises 25–5

In Exercises 1 through 4 find the moment of inertia and the radius of gyration with respect to the origin of the given masses at the given points.

1. 5 units at $(2, 0)$ and 3 units at $(6, 0)$
2. 3 units at $(-1, 0)$, 6 units at $(3, 0)$, 2 units at $(4, 0)$
3. 4 units at $(-4, 0)$, 10 units at $(0, 0)$, 6 units at $(5, 0)$
4. 6 units at $(-3, 0)$, 5 units at $(-2, 0)$, 9 units at $(1, 0)$, 2 units at $(8, 0)$

In Exercises 5 through 20 find the indicated moment of inertia or radius of gyration.

5. Find the moment of inertia of the area bounded by $y^2 = x$, $x = 4$, and the x-axis with respect to the x-axis.
6. Find the moment of inertia of the area bounded by $y = 2x$, $x = 1$, $x = 2$, and the x-axis with respect to the y-axis.
7. Find the radius of gyration of the area bounded by $y = x^3$, $x = 2$, and the x-axis with respect to the y-axis.
8. Find the radius of gyration of the first-quadrant area bounded by the curve $y = 1 - x^2$ with respect to the y-axis.
9. Express the moment of inertia of the triangular area in Example C in terms of the mass of the area.
10. Find the moment of inertia of a rectangular area of sides a and b with respect to side a. Express the result in terms of the mass.
11. Find the radius of gyration of the area bounded by $y = x^2$, $x = 2$, and the x-axis with respect to the x-axis.
12. Find the radius of gyration of the area bounded by $y^2 = x^3$, $y = 8$, and the y-axis with respect to the y-axis.
13. Find the radius of gyration of the area of Exercise 12 with respect to the x-axis.
14. Find the radius of gyration of the first-quadrant area bounded by $x = 1$, $y = 2 - x$, and the y-axis with respect to the y-axis.
15. Find the moment of inertia with respect to its axis of the solid generated by rotating the area bounded by $y^2 = x$, $y = 2$, and the y-axis about the x-axis.
16. Find the radius of gyration with respect to its axis of the volume generated by rotating the first-quadrant area under the curve $y = 4 - x^2$ about the y-axis.
17. Find the radius of gyration with respect to its axis of the volume generated by rotating the area bounded by $y = 2x - x^2$ and the x-axis about the y-axis.
18. Find the radius of gyration with respect to its axis of the volume generated by rotating the area bounded by $y = 2x$ and $y = x^2$ about the y-axis.
19. Find the moment of inertia in terms of its mass of a right circular cone of radius r and height h with respect to its axis.
20. Find the moment of inertia in terms of its mass of a circular disk with a hole in the center with respect to its axis. Let the inner radius be r_1 and the outer radius be r_2.

25–6 Other Applications

To illustrate the great variety of applications of integration, we shall demonstrate two more types of applied problems in this section. Certain other applications are shown in the exercises. The first application is to the physical problem of work done by a variable force.

Work, *in its physical sense, is the product of a constant force times the distance through which it acts.* When we consider the work done in stretching a spring, the first thing we recognize is that the more the spring is stretched, the greater is the force necessary to stretch it. Thus, the force varies. However, if we are stretching the spring a distance Δx, where we are considering the limit as $\Delta x \to 0$, the force can be considered as approaching a constant. Adding the products of force$_1$ times Δx_1, force$_2$ times Δx_2, and so forth, we see that the total work is the sum of these products. Thus, the work can be expressed as a definite integral in the form

$$W = \int_a^b f(x) \ dx \qquad\qquad (25\text{–}22)$$

where $f(x)$ is the force as a function of the distance the spring is stretched. The limits a and b refer to the initial and the final distances the spring is stretched *from its normal length.*

One problem remains: We must find the function $f(x)$. From physics we learn that the force required to stretch a spring is proportional to the amount it is stretched (Hooke's law). If a spring is stretched x units from its normal length, then $f(x) = kx$. From conditions stated for a particular spring, the value of k may be determined. Thus, $W = \int_a^b kx \ dx$ is the formula for finding the total work done in stretching a spring.

Example A

A spring of natural length 12 in. requires a force of 6 lb to stretch it 2 in. (see Fig. 25–36). Find the work done in stretching it 6 in.

From Hooke's law, we have $F = kx$, or $6 = k2$, $k = 3$ lb/in. Thus,

$$W = \int_0^6 3x \ dx = \frac{3}{2}x^2 \bigg|_0^6 = 54 \text{ lb-in.}$$

Problems involving work by a variable force arise in many fields of technology. An illustration from electricity is found in the motion of an electric charge through an electric field created by another electric charge.

Electric charges are of two types, designated as positive and negative. A basic law is that charges of the same sign repel each other and charges of opposite signs attract each other. *The force between charges is pro-*

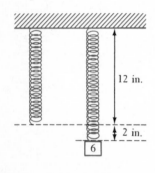

12 in.

2 in.

6

Figure 25–36

portional to the product of their charges, and inversely proportional to the square of the distance between them. Thus for this case

$$f(x) = \frac{kq_1 q_2}{x^2} \qquad \cdot \qquad (25\text{–}23)$$

where q_1 and q_2 are the charges (in coulombs), x is the distance (in meters), the force is in newtons, and $k = 9 \times 10^9$ N·m²/C². For other systems of units, the numerical value of k is different. We can find the work done when electric charges move toward each other or when they separate by use of Eq. (25–23) in Eq. (25–22).

Example B
Find the work done when two α-particles, $q = 0.32$ aC each, move until they are 10 nm apart, if they were originally separated by one meter.
   Applying Eqs. (25–22) and (25–23) we have, with $q = 0.32$ aC $= 0.32 \times 10^{-18}$ C $= 3.2 \times 10^{-19}$ C, and an upper limit of 10 nm $= 10 \times 10^{-9}$ m $= 10^{-8}$ m,

$$W = \int_1^{10^{-8}} \frac{9 \times 10^9 (3.2 \times 10^{-19})^2}{x^2}\, dx$$

$$= 9.2 \times 10^{-28} \int_1^{10^{-8}} \frac{dx}{x^2}$$

$$= 9.2 \times 10^{-28} \left(-\frac{1}{x}\right)\Big|_1^{10^{-8}} = -9.2 \times 10^{-28}(10^8 - 1)$$

$$= -9.2 \times 10^{-20} \text{ J}$$

Since $10^8 \gg 1$, where $\gg$ means "much greater than," the 1 may be neglected in the calculation. The meaning of the minus sign in the result is that work must be done *on* the system to move the particles together. If free to move, they would tend to separate.

   The second application of integration in this section is that of the **average value** of a function. In general *an average is found by summing up the quantities to be averaged, and then dividing by the total number of them.* Generalizing on this and using integration for the summation, the average value of a function y with respect to x from $x = a$ to $x = b$ is given by

$$y_{av} = \frac{\displaystyle\int_a^b y\, dx}{b - a} \qquad (25\text{–}24)$$

The following examples illustrate the applications of the average value of a function.

Example C

The velocity of an object falling under the influence of gravity as a function of time is given by $v = 32t$. What is the average velocity (in feet per second) with respect to time for the first three seconds?

In this case, the average velocity with respect to t is

$$v_{av} = \frac{\int_0^3 v \, dt}{3 - 0} = \frac{\int_0^3 32t \, dt}{3} = \frac{16t^2}{3}\Big|_0^3$$

$$= 48 \text{ ft/s}$$

This result can be interpreted as meaning that an average velocity of 48 ft/s for 3 s would result in the same distance, 144 ft, being traveled as that with the variable velocity. Since $s = \int v \, dt$, the numerator represents the distance traveled.

Example D

The power developed in a certain resistor as a function of current is $P = 6i^2$. What is the average power (in watts) with respect to the current as the current changes from 2 to 5 A?

The average value of P with respect to i is

$$P_{av} = \frac{\int_2^5 P \, di}{5 - 2} = \frac{6 \int_2^5 i^2 \, di}{3} = \frac{2}{3}i^3\Big|_2^5 = \frac{2}{3}(125 - 8) = 78 \text{ W}$$

In general, it might be noted that the average value of y with respect to x is that value of y, which when multiplied by the length of the interval for x, gives the same area as that under the curve of y as a function of x.

Exercises 25–6

1. A spring of natural length 10 in. measures 14 in. when a 12-lb weight is attached to it. How much work is required to stretch it 5 in. from its natural length?

2. How much work is required to stretch the spring in Exercise 1 from a length of 14 in. to a length of 19 in.?

3. It requires 500 lb to stretch a certain spring 1 ft. How much work is done on the spring if we attach a 300-lb weight to it?

4. If another 300-lb weight is attached to the spring in Exercise 3, how much work is done by the additional weight?

5. Find the work done in separating a 2 μC positive charge from a 4 μC negative charge to a distance of 6 cm from a distance of 2 cm.

6. Find the work done in separating an 0.8 nC negative charge from an 0.6 nC charge to a distance of 200 mm from a distance of 100 mm.

7. The gravitational force of attraction between two objects which are x units apart is given by $F = k/x^2$. If two objects are 1 cm apart, find the work done in separating them to a distance of 1 m apart. Express the answer in terms of k.

8. Given two objects which are 10 ft apart, find the work required to separate them further, until they are 100 ft apart. Express the result in terms of k. (See Exercise 7.)

9. Find the work done in winding up a 200-ft cable which weighs 100 lb.

10. In Exercise 9, if an object weighing 50 lb is attached at the end of the cable, find the work done in winding it up.

11. Find the work done in pumping the water out of the top of a cylindrical tank 3 ft in radius and 10 ft high, if water weighs 62.4 lb/ft³. (*Hint:* If horizontal slices dx ft thick are used, each element weighs 62.4(9π dx) lb, and each element must be raised $10 - x$ ft, if x is the distance from the base to the element. In this way the force, the weight of the slice, and distance are determined. Thus, the products of force and distance are summed by integration.)

12. A hemispherical tank of radius 10 ft is full of water. Find the work done in pumping the water to the top of the tank. (See Exercise 11. This problem is similar, except that the weight of each element is 62.4π(radius)²(thickness), where the radius of each element is different. If we let x be the radius of an element and y be the distance the element must be raised, we have $62.4\pi x^2 \, dy$, with $x^2 + y^2 = 100$.)

13. The electric current as a function of time for a certain circuit is given by $i = 4t - t^2$. Find the average value of the current (in amperes), with respect to time, for the first 4 s.

14. Find the average value of the function $y = \sqrt{x}$, with respect to x, from $x = 1$ to $x = 9$.

15. An object moves along the x-axis according to the relation $x = t^3 - 3t$. Find the average value of the velocity with respect to t, as t changes from 1 to 3.

16. Find the average value of the volume of a sphere with respect to the radius.

17. The length of arc of a curve is given by

$$s = \int_a^b \sqrt{1 + (dy/dx)^2} \, dx$$

Find the length of the curve $y = \frac{2}{3}x^{3/2}$ from $x = 0$ to $x = 3$.

18. Find the length of the curve $y = \frac{2}{3}(x^2 - 1)^{3/2}$ from $x = 1$ to $x = 3$.

19. The force on a vertical plane area below the surface of a liquid equals the product of the weight per unit volume of the liquid, the area of the plane, and the depth of the centroid of the area. Find the force on the area bounded by $y = x^2$ and $y = 4$ if the surface of the water is at $y = 20$. The weight per unit volume of water is 62.4 lb/ft³. Assume all distances are in feet.

20. Find the force on the area bounded by $x = 2y - y^2$ and the y-axis if the upper point of the area is at the surface of the water. Assume all distances are in feet.

21. The area of a surface of revolution generated by rotating an arc about the x-axis is given by $S = 2\pi \int_a^b y\sqrt{1 + (dy/dx)^2} \, dx$. Find the area of the surface generated by rotating $y = \frac{3}{4}x$ from $x = 0$ to $x = 2$ about the x-axis.

22. Find the area of the surface generated by rotating $y = x^3$ from $x = 1$ to $x = 3$ about the x-axis.

25–7 Exercises for Chapter 25

1. What is the velocity after 5 s of an object which is dropped and falls due to gravity?

2. Find an expression for the distance traveled by an object for which the velocity is given by $v = 3t/\sqrt{t^2 + 1}$, if $s = 0$ when $t = 0$.

3. A weather balloon is rising at the rate of 20 ft/s when a small metal part drops off. If the balloon is 200 ft high at this instant, when will the part hit the ground?

4. The acceleration of an object rolling down an inclined plane is 12 ft/s². If it starts from rest at the top of the plane, find an expression for the distance the object travels from the top as a function of the time.

5. The current in a certain electric resistor is given by $i = 5t^{2/3}$. Find the charge which passes a given point in the circuit during the third second.

6. The current in a certain electric circuit is given by $i = \sqrt{1 + 4t}$, where i is in amperes and t in seconds. Find the charge which passes a given point during the first two seconds.

7. The voltage across a 55.0 nF capacitor is zero. What is the voltage after 20 ms if a current of 10 mA charges the capacitor?

8. What is the capacitance of a capacitor if a current $i = \sqrt[6]{3t + 1}$ charges it such that the voltage across it is 200 V at $t = 0$ and 280 V when $t = 0.7$ s?

9. The distribution of weight on a cable is not uniform. If the slope of the cable at any point is given by $dy/dx = 20 + \frac{1}{40}x^2$, and if the origin of the coordinate system is at the lowest point, find the equation which gives the curve described by the cable.

10. The slope of a certain cable is given by $dy/dx = -0.1/\sqrt{x}$. If the point $(0, 0)$ is on the cable, find the equation which gives the curve described by the cable.

11. Find the area between $y = \sqrt{1 - x}$ and the coordinate axes.

12. Find the area bounded by $y = 3x^2 - x^3$ and the x-axis.

13. Find the area bounded by $y^2 = 2x$ and $y = x - 4$.

14. Find the area bounded by $y = 1/(2x + 1)^2$, $y = 0$, $x = 1$, and $x = 2$.

15. Find the volume generated by rotating the area bounded by $y = 3 + x^2$ and the line $y = 4$ about the x-axis.

16. Find the volume generated by rotating the area of Exercise 12 about the x-axis.

17. Find the volume generated by rotating the area of Exercise 12 about the y-axis.

18. Find the volume generated by rotating the area bounded by $y = x$ and $y = 3x - x^2$ about the y-axis.

19. Find the centroid of the area bounded by $y^2 = x^3$ and $y = 2x$.

20. Find the centroid of the area bounded by $y = 2x - 4$, $x = 1$, and $y = 0$.

21. Find the centroid of the volume generated by rotating the area bounded by $y = \sqrt{x}$, $x = 1$, $x = 4$, and $y = 0$ about the x-axis.

22. Find the centroid of the volume generated by rotating the area bounded by $yx^4 = 1$, $y = 1$, and $y = 4$ about the y-axis.

23. Find the moment of inertia of the area bounded by $y = 3x - x^2$ and $y = x$ with respect to the y-axis.

24. Find the radius of gyration of the first-quadrant area which is bounded by $y = 8 - x^3$ with respect to the y-axis.

25. Find the moment of inertia with respect to its axis of the solid generated by rotating the area bounded by $y = x^{1/2}$, $y = 0$, and $x = 8$ about the x-axis.

26. Find the radius of gyration with respect to its axis of the solid formed by rotating the area bounded by $xy = 1$, $x = 1$, $x = 3$, and $y = \frac{1}{3}$ about the x-axis.

27. A certain spring is 16 in. long when 12 lb are hung from it. It measures 12 in. when 4 lb are hung from it. How much work is done when the spring is stretched from its natural length by hanging 6 lb on it?

28. If the pressure of a gas is 400 kPa when the volume is 243 cm³, how much work is done in compressing the gas to 32 cm³, if it undergoes an adiabatic change such that $pV^{1.4} = c$?

29. The electric resistance of a wire is inversely proportional to the square of its radius. If a certain wire has a resistance of 0.3 Ω when its radius is 2.0 mm, find the average value of the resistance with respect to the radius if the radius changes from 2.0 mm to 2.1 mm.

30. The velocity of an object as a function of the distance fallen after being dropped is $v = 8\sqrt{s}$. Find the average value of v with respect to s during the third second, if $s = 16t^2$, and s is measured in feet.

31. The production rate (in pounds per hour) of a certain chemical plant is given by the following table. Using the trapezoidal rule, determine the approximate daily production.

Time	12 M	4 AM	8 AM	12 N	4 PM	8 PM	12 M
Production rate	2000	2000	7000	12,000	12,000	8000	2000

32. The current as a function of time was measured in a particular circuit, with the results tabulated below. What charge (in coulombs) was transmitted across a given point in the first 6 s? Use the trapezoidal rule.

i (amperes)	0.0	1.5	4.0	7.2	7.5	4.3	1.6
t (seconds)	0.0	1.0	2.0	3.0	4.0	5.0	6.0

26

Differentiation of Transcendental Functions

26–1 Derivatives of the Sine and Cosine Functions

In Chapters 22 and 24 we introduced the operations of differentiation and integration of functions. The functions we used, however, were in all cases algebraic functions. We did not use the trigonometric, inverse trigonometric, exponential, or logarithmic functions, for the derivative of each of these is a special form. In this chapter we shall find the formulas for the derivatives of these functions, which are the most important of the **transcendental** (non-algebraic) **functions.** In the next chapter we shall take up integration involving these functions.

If we find the derivative of the sine function, it is possible to use this derivative as a basis for finding the derivatives of the other trigonometric and inverse trigonometric functions. Therefore, we shall now find the derivative of the sine function, by using the Δ-process.

Let $y = \sin u$, where u is a function of x and is expressed in radians. If u changes by Δu, y then changes by Δy. Thus,

$$y + \Delta y = \sin(u + \Delta u)$$
$$\Delta y = \sin(u + \Delta u) - \sin u$$
$$\frac{\Delta y}{\Delta u} = \frac{\sin(u + \Delta u) - \sin u}{\Delta u}$$

Referring now to Eq. (19–17) we have

$$\frac{\Delta y}{\Delta u} = \frac{2 \sin \frac{1}{2}(u + \Delta u - u)\cos \frac{1}{2}(u + \Delta u + u)}{\Delta u}$$

$$= \frac{\sin(\Delta u/2)\cos[u + (\Delta u/2)]}{\Delta u/2}$$

Looking ahead to the next step of letting $\Delta u \to 0$, we see that the numerator and denominator both approach zero. This situation is precisely the same as that in which we were finding the derivatives of the algebraic functions. To find the limit, we must find

$$\lim_{\Delta u \to 0} \frac{\sin(\Delta u/2)}{\Delta u/2}$$

since these are the factors which cause the numerator and the denominator to approach zero.

In finding this limit, we shall let $\theta = \Delta u/2$ for convenience of notation. This means that we are to determine $\lim_{\theta \to 0} \frac{\sin \theta}{\theta}$. Of course, it would be convenient to know before proceeding if this limit does actually exist. Therefore, we now show a table of values of $\frac{\sin \theta}{\theta}$ as θ becomes very small.

θ (radians)	0.5	0.1	0.05	0.01
$\dfrac{\sin \theta}{\theta}$	0.958851	0.9983342	0.9995834	0.9999833

We see from this table that the limit of $\frac{\sin \theta}{\theta}$, as θ approaches zero, appears to be 1.

In order to prove that $\lim_{\theta \to 0} \frac{\sin \theta}{\theta} = 1$, we use a geometric approach. Considering Fig. 26–1, we can see that the following inequality is true for the indicated areas:

Area triangle OBD < area sector OBD < area triangle OBC

$$\frac{1}{2}r(r \sin \theta) < \frac{1}{2}r^2\theta < \frac{1}{2}r(r \tan \theta) \qquad \text{or} \qquad \sin \theta < \theta < \tan \theta$$

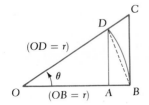

(OD = r)

θ

O   (OB = r)   A   B

Figure 26–1

Remembering that we want to find the limit of $(\sin \theta)/\theta$, we next divide through by $\sin \theta$, and then take reciprocals:

$$1 < \frac{\theta}{\sin \theta} < \frac{1}{\cos \theta} \qquad \text{or} \qquad 1 > \frac{\sin \theta}{\theta} > \cos \theta$$

When we consider the limit as $\theta \to 0$, we see that the left member remains 1 and the right member approaches 1. Thus, $(\sin \theta)/\theta$ must approach 1. This means

$$\lim_{\theta \to 0} \frac{\sin \theta}{\theta} = \lim_{\Delta u \to 0} \frac{\sin(\Delta u/2)}{\Delta u/2} = 1 \qquad\qquad (26\text{–}1)$$

Using the result expressed in Eq. (26–1) in the expression for $\Delta y/\Delta u$, we have

$$\lim_{\Delta u \to 0} \frac{\Delta y}{\Delta u} = \lim_{\Delta u \to 0} \left[\cos\left(u + \frac{\Delta u}{2}\right) \frac{\sin(\Delta u/2)}{\Delta u/2} \right] = \cos u$$

or

$$\frac{dy}{du} = \cos u \qquad\qquad (26\text{–}2)$$

However, we want the derivative of y with respect to x. This requires the use of the chain rule, Eq. 22–14, which we repeat here for reference.

$$\frac{dy}{dx} = \frac{dy}{du}\frac{du}{dx} \qquad\qquad (26\text{–}3)$$

Combining Eqs. (26–2) and (26–3), we have ($y = \sin u$)

$$\frac{d(\sin u)}{dx} = \cos u \frac{du}{dx} \qquad\qquad (26\text{–}4)$$

Example A
Find the derivative of $y = \sin 2x$.
   In this example $u = 2x$. Thus,

$$\frac{dy}{dx} = \cos 2x \left(\frac{d(2x)}{dx} \right)$$

$$= 2 \cos 2x$$

Example B
Find the derivative of $y = 2\sin(x^2)$.
   In this example $u = x^2$, which means that $du/dx = 2x$. Hence,

$$\frac{dy}{dx} = 2[\cos(x^2)](2x)$$

$$= 4x \cos(x^2)$$

It is important here, just as it is in finding the derivatives of powers of functions, to remember to include the factor du/dx.

Example C
Find the derivative of $y = \sin^2 x$.
   This example is a combination of the use of the power rule, Eq.

(22–15), and the derivative of the sine function, Eq. (26–4). Since $\sin^2 x$ means $(\sin x)^2$, in using the power rule we have $u = \sin x$. Thus,

$$\frac{dy}{dx} = 2(\sin x)\frac{d \sin x}{dx}$$

$$= 2 \sin x \cos x = \sin 2x$$

Example D
Find the derivative of $y = 2 \sin^3(2x^4)$.

In the general power rule, $u = \sin(2x^4)$. In the derivative of the sine function, $u = 2x^4$. Thus, we have

$$\frac{dy}{dx} = 2(3)\sin^2(2x^4)\frac{d(\sin 2x^4)}{dx}$$

$$= 6 \sin^2(2x^4)\cos(2x^4)\frac{d(2x^4)}{dx} = 48x^3\sin^2(2x^4)\cos(2x^4)$$

To find the derivative of the cosine function, we write it in the form $\cos u = \sin[(\pi/2) - u]$. Thus, if $y = \sin[(\pi/2) - u]$, we have

$$\frac{dy}{dx} = \cos\left(\frac{\pi}{2} - u\right)\frac{d[(\pi/2) - u]}{dx} = \cos\left(\frac{\pi}{2} - u\right)\left(-\frac{du}{dx}\right)$$

$$= -\cos\left(\frac{\pi}{2} - u\right)\frac{du}{dx}$$

Since $\cos[(\pi/2) - u] = \sin u$, we have

$$\frac{d(\cos u)}{dx} = -\sin u \frac{du}{dx} \qquad\qquad (26\text{–}5)$$

Example E
Find the derivative of $y = \cos^2(2x)$.

Using the power rule and Eq. (26–5), we have

$$\frac{dy}{dx} = 2 \cos(2x)(-\sin 2x)(2) = -4 \sin 2x \cos 2x$$

$$= -2 \sin 4x$$

Example F
Find the derivative of $y = \sin 2x \cos x^2$.

Using the product rule and the derivatives of the sine and cosine functions, we arrive at the following result:

$$\frac{dy}{dx} = \sin 2x(-\sin x^2)(2x) + \cos x^2(\cos 2x)(2)$$

$$= -2x \sin 2x \sin x^2 + 2 \cos 2x \cos x^2$$

Example G
Find the slope of a tangent line to the curve of $y = 5 \sin 3x$ where $x = 0.2$.

Here we are to find the derivative of $y = 5 \sin 3x$, and then evaluate the derivative for $x = 0.2$. Therefore, we have the following:

$$y = 5 \sin 3x$$

$$\frac{dy}{dx} = 5(\cos 3x)(3) = 15 \cos 3x$$

$$\left.\frac{dy}{dx}\right|_{x=0.2} = 15 \cos 3(0.2) = 15 \cos 0.6$$

$$= 15(0.8253) = 12.38$$

In evaluating the slope we must remember that $x = 0.2$ means the values are in radians. Therefore, the slope is 12.38.

Exercises 26–1

In Exercises 1 through 28 find the derivatives of the given functions.

1. $y = \sin(x + 2)$ 2. $y = 3 \sin 4x$ 3. $y = 2 \sin(2x - 1)$
4. $y = 5 \sin(3 - x)$ 5. $y = 5 \cos 2x$ 6. $y = \cos(1 - x)$
7. $y = 2 \cos(3x - 1)$ 8. $y = 4 \cos(6x^2 + 5)$ 9. $y = \sin^2 4x$
10. $y = 3 \sin^3(2x + 1)$ 11. $y = 3 \cos^3(5x + 2)$ 12. $y = 4 \cos^2 \sqrt{x}$
13. $y = x \sin 3x$ 14. $y = x^2 \sin 2x$ 15. $y = 3x^3 \cos 5x$
16. $y = x \cos 2x^3$ 17. $y = \sin x^2 \cos 2x$ 18. $y = \sin x \cos 4x$
19. $y = 2 \sin 2x \cos x$ 20. $y = 6 \sin 4x \cos 2x^2$

21. $y = \dfrac{\sin 3x}{x}$ 22. $y = \dfrac{2x}{\sin 4x}$ 23. $y = \dfrac{2 \cos x^2}{3x}$

24. $y = \dfrac{5x}{\cos x}$ 25. $y = \sin^3 x - \cos 2x$ 26. $y = x \sin x + \cos x$

27. $y = x - \frac{1}{3}\sin^3 4x$ 28. $y = 2x \sin x + 2 \cos x - x^2 \cos x$

In Exercises 29 through 36 solve the given problems.

29. Verify the values of $(\sin \theta)/\theta$ in the table where $(\sin \theta)/\theta$ is calculated for $\theta = 0.5, 0.1, 0.05,$ and 0.01.
30. Evaluate $\lim_{\theta \to 0}(\tan \theta)/\theta$. (Use the fact that $\lim_{\theta \to 0}(\sin \theta)/\theta = 1$.)
31. Show that $\dfrac{d^4 \sin x}{dx^4} = \sin x$
32. If $y = \cos 2x$, show that $\dfrac{d^2 y}{dx^2} = -4y$
33. Find the slope of a line tangent to the curve of $y = 3 \sin 2x$ where $x = \pi/8$.
34. Find the slope of a line tangent to the curve of $y = 6 \cos x^2$ where $x = 0.5$.
35. Find the velocity after 2 s, if the displacement from a reference point of a particle is given by $s = 2 \sin 3t + t^2$, where s is in feet.
36. The current (in amperes) in a certain electric circuit, as a function of time (in seconds), is given by $i = 10 \cos(120\pi t + \pi/6)$. Find the expression for the voltage across a 2-H inductor. (See Exercise 23 of Section 25–1.)

26–2 Derivatives of the Other Trigonometric Functions

We can find the derivatives of the other trigonometric functions by expressing the functions in terms of the sine and cosine. After we perform the differentiation, we use trigonometric relations to put the derivative in a convenient form.

We obtain the derivative of tan u by expressing tan u as sin $u/\cos u$. Therefore, letting $y = \sin u/\cos u$, by employing the quotient rule we have

$$\frac{dy}{dx} = \frac{\cos u[\cos u\,(du/dx)] - \sin u[-\sin u\,(du/dx)]}{\cos^2 u}$$

$$= \frac{\cos^2 u + \sin^2 u}{\cos^2 u}\frac{du}{dx} = \frac{1}{\cos^2 u}\frac{du}{dx} = \sec^2 u\,\frac{du}{dx}$$

$$\frac{d(\tan u)}{dx} = \sec^2 u\,\frac{du}{dx} \qquad (26\text{–}6)$$

We find the derivative of cot u by letting $y = \cos u/\sin u$, and again using the quotient rule:

$$\frac{dy}{dx} = \frac{\sin u[-\sin u\,(du/dx)] - \cos u[\cos u\,(du/dx)]}{\sin^2 u}$$

$$= \frac{-\sin^2 u - \cos^2 u}{\sin^2 u}\frac{du}{dx}$$

$$\frac{d(\cot u)}{dx} = -\csc^2 u\,\frac{du}{dx} \qquad (26\text{–}7)$$

To obtain the derivative of sec u, we let $y = 1/\cos u$. Then

$$\frac{dy}{dx} = -(\cos u)^{-2}\left[(-\sin u)\left(\frac{du}{dx}\right)\right] = \frac{1}{\cos u}\frac{\sin u}{\cos u}\frac{du}{dx}$$

$$\frac{d(\sec u)}{dx} = \sec u\,\tan u\,\frac{du}{dx} \qquad (26\text{–}8)$$

We obtain the derivative of csc u by letting $y = 1/\sin u$. And so

$$\frac{dy}{dx} = -(\sin u)^{-2}\left(\cos u\,\frac{du}{dx}\right) = -\frac{1}{\sin u}\frac{\cos u}{\sin u}\frac{du}{dx}$$

$$\frac{d(\csc u)}{dx} = -\csc u\,\cot u\,\frac{du}{dx} \qquad (26\text{–}9)$$

Example A

Find the derivative of $y = 2 \tan 8x$.

Using Eq. (26–6), we have

$$\frac{dy}{dx} = 2(\sec^2 8x)(8)$$

$$= 16 \sec^2 8x$$

Example B

Find the derivative of $y = \cot^4 2x$.

Using the power rule and Eq. (26–7), we have

$$\frac{dy}{dx} = 4(\cot^3 2x)(-\csc^2 2x)(2)$$

$$= -8 \cot^3 2x \csc^2 2x$$

Example C

Find the derivative of $y = 3 \sec^2 4x$.

Using the power rule and Eq. (26–8), we have

$$\frac{dy}{dx} = 3(2)(\sec 4x)(\sec 4x \tan 4x)(4)$$

$$= 24 \sec^2 4x \tan 4x$$

Example D

Find the derivative of $y = x \csc^3 2x$.

Using the power rule, the product rule, and Eq. (26–9), we have

$$\frac{dy}{dx} = x(3 \csc^2 2x)(-\csc 2x \cot 2x)(2) + \csc^3 2x$$

$$= \csc^3 2x(-6x \cot 2x + 1)$$

Example E

Find the derivative of $y = (\tan 2x + \sec 2x)^3$.

Using the power rule, Eqs. (26–6) and (26–8), we have

$$\frac{dy}{dx} = 3(\tan 2x + \sec 2x)^2[\sec^2 2x(2) + \sec 2x \tan 2x(2)]$$

$$= 6 \sec 2x(\tan 2x + \sec 2x)^3$$

Example F

Find the differential of $y = \sin 2x \tan x^2$.

Here we are to find the derivative of the given function and multiply by dx. Therefore, using the product rule along with Eqs. (26–4) and (26–6), we have the following:

$$dy = [(\sin 2x)(\sec^2 x^2)(2x) + (\tan x^2)(\cos 2x)(2)]\ dx$$

$$= (2x \sin 2x \sec^2 x^2 + 2 \cos 2x \tan x^2)\ dx$$

Exercises 26—2

In Exercises 1 through 28 find the derivatives of the given functions.

1. $y = \tan 5x$

2. $y = 3 \tan(3x + 2)$

3. $y = \cot(1 - x)^2$

4. $y = 3 \cot 6x$

5. $y = 3 \sec 2x$

6. $y = \sec\sqrt{1 - x}$

7. $y = -3 \csc\sqrt{x}$

8. $y = \csc(1 - 2x)$

9. $y = 5 \tan^2 3x$

10. $y = 2 \tan^2(x^2)$

11. $y = 2 \cot^4 3x$

12. $y = \cot^2(1 - x)$

13. $y = \sqrt{\sec 4x}$

14. $y = \sec^3 x$

15. $y = 3 \csc^4 7x$

16. $y = \csc^2(2x^2)$

17. $y = x^2 \tan x$

18. $y = 3x \sec 4x$

19. $y = 4 \cos x \csc x^2$

20. $y = \frac{1}{2}\sin 2x \sec x$

21. $y = \dfrac{\csc x}{x}$

22. $y = \dfrac{\cot 4x}{2x}$

23. $y = \dfrac{\cos 4x}{1 + \cot 3x}$

24. $y = \dfrac{\tan^2 3x}{2 + \sin x^2}$

25. $y = \frac{1}{3}\tan^3 x - \tan x$

26. $y = \csc 2x - 2 \cot 2x$

27. $y = \tan 2x - \sec 2x$

28. $y = x - \tan x + \sec^2 2x$

In Exercises 29 through 32 find the differentials of the given functions.

29. $y = 4 \tan^2 3x$

30. $y = 5 \sec^3 2x$

31. $y = \tan 4x \sec 4x$

32. $y = 2x \cot 3x$

In Exercises 33 through 36 solve the given problems.

33. Find the slope of a line tangent to the curve of $y = 2 \cot 3x$ where $x = \frac{\pi}{12}$.

34. If the displacement of an object is given by $s = 2t \sec \sqrt{t}$, find the velocity v when $t = 0.04$.

35. An observer to a rocket launch was 1000 ft from the take-off position. He found the angle of elevation of the rocket as a function of time to be $\theta = 3t/(2t + 10)$. Therefore, the height h of the rocket was

$$h = 1000 \tan \frac{3t}{2t + 10}$$

Find the time-rate of change of height after 5 s.

36. Show that $y = \cos^3 x \tan x$ satisfies the equation

$$\cos x \frac{dy}{dx} + 3y \sin x - \cos^2 x = 0$$

26–3 Derivatives of the Inverse Trigonometric Functions

To obtain the derivatives of $y = $ Arcsin u, we first solve for u in the form $u = \sin y$ and then take derivatives with respect to x:

$$\frac{du}{dx} = \cos y \frac{dy}{dx}$$

Solving this equation for dy/dx, we obtain

$$\frac{dy}{dx} = \frac{1}{\cos y}\frac{du}{dx} = \frac{1}{\sqrt{1 - \sin^2 y}}\frac{du}{dx} = \frac{1}{\sqrt{1 - u^2}}\frac{du}{dx}$$

We choose the positive square root since $\cos y > 0$ for $-\pi/2 < y < \pi/2$, which is the range of the principal values of Arcsin u. Therefore, we obtain the following result:

$$\frac{d(\text{Arcsin } u)}{dx} = \frac{1}{\sqrt{1 - u^2}}\frac{du}{dx} \qquad (26\text{--}10)$$

We note here that the derivative of the inverse sine function is an algebraic function.

We find the derivative of the inverse cosine function by letting $y = $ Arccos u, and by following the same procedure as that used in finding the derivative of Arcsin u:

$$u = \cos y, \qquad \frac{du}{dx} = -\sin y \frac{dy}{dx}$$

$$\frac{dy}{dx} = -\frac{1}{\sin y}\frac{du}{dx} = -\frac{1}{\sqrt{1 - \cos^2 y}}\frac{du}{dx}$$

Therefore,

$$\frac{d(\text{Arccos } u)}{dx} = -\frac{1}{\sqrt{1 - u^2}}\frac{du}{dx} \qquad (26\text{--}11)$$

The positive square root is chosen here since $\sin y > 0$ for $0 < y < \pi$, which is the range of the principal values of Arccos u. We note that the derivative of the inverse cosine is the negative of the derivative of the inverse sine.

By letting $y = $ Arctan u, solving for u, and taking derivatives, we find the derivative of the inverse tangent function:

$$u = \tan y, \qquad \frac{du}{dx} = \sec^2 y \frac{dy}{dx}, \qquad \frac{dy}{dx} = \frac{1}{\sec^2 y}\frac{du}{dx} = \frac{1}{1 + \tan^2 y}\frac{du}{dx}$$

Therefore,

$$\frac{d(\text{Arctan } u)}{dx} = \frac{1}{1 + u^2}\frac{du}{dx} \tag{26–12}$$

We can see that the derivative of the inverse tangent is an algebraic function also.

The inverse sine, inverse cosine, and inverse tangent prove to be of the greatest importance in applications and in further development of mathematics. Therefore, the formulas for the derivatives of the other inverse functions are not presented here, although they are included in the exercises.

Example A
Find the derivative of $y = \text{Arcsin } 4x$.
   Using Eq. (26–10), we have

$$\frac{dy}{dx} = \frac{1}{\sqrt{1 - (4x)^2}}(4)$$

$$= \frac{4}{\sqrt{1 - 16x^2}}$$

Example B
Find the derivative of $y = \text{Arccos}(1 - 2x)$.
   Using Eq. (26–11), we have

$$\frac{dy}{dx} = \frac{-1}{\sqrt{1 - (1 - 2x)^2}}(-2) = \frac{2}{\sqrt{1 - 1 + 4x - 4x^2}}$$

$$= \frac{2}{\sqrt{4x - 4x^2}} = \frac{1}{\sqrt{x - x^2}}$$

Example C
Find the derivative of $y = (x^2 + 1)\text{Arctan } x - x$.
   Using the product rule along with Eq. (26–12) on the first term, we have

$$\frac{dy}{dx} = (x^2 + 1)\frac{1}{1 + x^2} + \text{Arctan } x(2x) - 1$$

$$= 2x \text{ Arctan } x$$

Example D
Find the derivative of $y = x \text{ Arctan}^2 2x$.
   We have

$$\frac{dy}{dx} = x(2)(\text{Arctan } 2x)\left(\frac{1}{1 + 4x^2}\right)(2) + \text{Arctan}^2 2x$$

$$= \text{Arctan } 2x\left(\frac{4x}{1 + 4x^2} + \text{Arctan } 2x\right)$$

Example E

Find the derivative of $y = x \operatorname{Arcsin} 2x + \frac{1}{2}\sqrt{1 - 4x^2}$.

We write

$$\frac{dy}{dx} = x\left(\frac{2}{\sqrt{1 - 4x^2}}\right) + \operatorname{Arcsin} 2x + \frac{1}{2} \cdot \frac{1}{2}(1 - 4x^2)^{-1/2}(-8x)$$

$$= \frac{2x}{\sqrt{1 - 4x^2}} + \operatorname{Arcsin} 2x - \frac{2x}{\sqrt{1 - 4x^2}}$$

$$= \operatorname{Arcsin} 2x$$

Exercises 26–3

In Exercises 1 through 24 find the derivatives of the given functions.

1. $y = \operatorname{Arcsin}(x^2)$

2. $y = \operatorname{Arcsin}(1 - x^2)$

3. $y = 2 \operatorname{Arcsin} 3x^3$

4. $y = \operatorname{Arcsin} \sqrt{1 - x}$

5. $y = \operatorname{Arccos} \frac{1}{2}x$

6. $y = 3 \operatorname{Arccos} 5x$

7. $y = 2 \operatorname{Arccos} \sqrt{2 - x}$

8. $y = 3 \operatorname{Arccos}(x^2 + 1)$

9. $y = \operatorname{Arctan} \sqrt{x}$

10. $y = \operatorname{Arctan}(1 + x)$

11. $y = \operatorname{Arctan}\left(\frac{1}{x}\right)$

12. $y = 4 \operatorname{Arctan} 3x^4$

13. $y = x \operatorname{Arcsin} x$

14. $y = x^2 \operatorname{Arccos} x$

15. $y = 2x \operatorname{Arctan} 2x$

16. $y = (x^2 + 1)\operatorname{Arcsin} 4x$

17. $y = \dfrac{3x}{\operatorname{Arcsin} 2x}$

18. $y = \dfrac{\operatorname{Arctan} 2x}{x}$

19. $y = \operatorname{Arcsin}^2 4x$

20. $y = \sqrt{\operatorname{Arcsin} (x - 1)}$

21. $y = 3 \operatorname{Arctan}^3 x$

22. $y = \dfrac{1}{\operatorname{Arccos} 2x}$

23. $y = \dfrac{1}{1 + 4x^2} - \operatorname{Arctan} 2x$

24. $y = \operatorname{Arcsin} x - \sqrt{1 - x^2}$

In Exercises 25 through 32 solve the given problems.

25. Find the differential of the function $y = \operatorname{Arcsin}^3 x$.

26. Find the differential of the function $y = x^3 \operatorname{Arccos} x^2$.

27. Find the second derivative of the function $y = \operatorname{Arctan} 2x$.

28. Show that $\dfrac{d(\operatorname{Arccot} u)}{dx} = -\dfrac{1}{1 + u^2}\dfrac{du}{dx}$

29. Show that $\dfrac{d(\operatorname{Arcsec} u)}{dx} = \dfrac{1}{\sqrt{u^2(u^2 - 1)}}\dfrac{du}{dx}$

30. Show that $\dfrac{d(\operatorname{Arccsc} u)}{dx} = -\dfrac{1}{\sqrt{u^2(u^2 - 1)}}\dfrac{du}{dx}$

31. The angle with the x-axis of the x-component of a vector **R** is given by $\theta = \operatorname{Arccos} R_x/R$. If R is constant and R_x changes with time, find an expression for $d\theta/dt$.

32. As a person approaches a building of height h, the angle of elevation of the top of the building is a function of his distance from the building. Express the angle of elevation θ in terms of h and the distance x from the building and then find $d\theta/dx$. Assume the person's height is negligible to that of the building.

26–4 Applications

With our development of the formulas for the derivatives of the trigono-metric and inverse trigonometric functions, it is now possible for us to apply these derivatives in the same manner as we applied the derivatives of algebraic functions. We can now use trigonometric and inverse trigono-metric functions to solve time-rate of change, tangent and normal, curve tracing, maximum and minimum, and differential application problems. The following examples illustrate the use of these functions in these types of problems.

Example A

Sketch the curve $y = \sin^2 x - \dfrac{x}{2}$ $(0 \le x \le 2\pi)$.

First we write

$$\frac{dy}{dx} = 2 \sin x \cos x - \frac{1}{2} = \sin 2x - \frac{1}{2}, \qquad \frac{d^2y}{dx^2} = 2 \cos 2x$$

The only easily obtained intercept is $(0, 0)$. (See Fig. 26–2.) Maximum and minimum points will occur for $\sin 2x = \frac{1}{2}$. Thus, maximum and minimum points occur if

$$2x = \frac{\pi}{6}, \frac{5\pi}{6}, \frac{13\pi}{6}, \frac{17\pi}{6}$$

Thus, the x-values are

$$x = \frac{\pi}{12}, \frac{5\pi}{12}, \frac{13\pi}{12}, \frac{17\pi}{12}$$

Using the second derivative, we find that d^2y/dx^2 is positive for $x = \pi/12$ and $x = 13\pi/12$, and negative for $x = 5\pi/12$ and $x = 17\pi/12$. Thus, the maximum points are $(5\pi/12, 0.279)$ and $(17\pi/12, -1.29)$. The minimum points are $(\pi/12, -0.064)$ and $(13\pi/12, -1.64)$. Inflection points occur for $\cos 2x = 0$, or

$$2x = \frac{\pi}{2}, \frac{3\pi}{2}, \frac{5\pi}{2}, \frac{7\pi}{2}, \qquad \text{or} \qquad x = \frac{\pi}{4}, \frac{3\pi}{4}, \frac{5\pi}{4}, \frac{7\pi}{4}$$

Therefore, the points of inflection are $(\pi/4, 0.11)$, $(3\pi/4, -0.68)$, $(5\pi/4, -1.47)$, and $(7\pi/4, -2.25)$. Using this information, we sketch the curve.

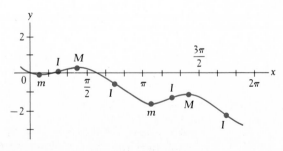

Figure 26–2

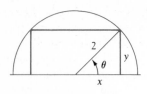

Figure 26–3

Example B

Find the area of the largest rectangle which can be inscribed in a semi-circle of radius 2.

From Fig. 26–3, we see that $A = 2xy$, where $x = 2 \cos \theta$ and $y = 2 \sin \theta$. Thus, $A = 8 \sin \theta \cos \theta = 4 \sin 2\theta$. We shall find the value of θ corresponding to the maximum area. Therefore,

$$\frac{dA}{d\theta} = 8 \cos 2\theta, \qquad 8 \cos 2\theta = 0, \qquad 2\theta = \frac{\pi}{2}, \qquad \theta = \frac{\pi}{4}$$

Thus, the maximum area is $A = 4 \sin 2(\pi/4) = 4$.

Example C

A rocket is taking off vertically at a distance of 600 ft from an observer. If, when the angle of elevation is $\pi/4$ it is changing at the rate of 0.5 rad/s, how fast is the rocket ascending?

From Fig. 26–4, we see that

$$\theta = \operatorname{Arctan}(x/600)$$

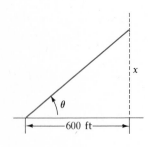

Figure 26–4

Taking derivatives with respect to time, we have

$$\frac{d\theta}{dt} = \frac{1}{1 + x^2/(3.6 \times 10^5)} \frac{dx/dt}{600} = \frac{600 \, dx/dt}{3.6 \times 10^5 + x^2}$$

From the given information, we find that $d\theta/dt = 0.5$ rad/s, $x = 600$ ft when $\theta = \pi/4$. Thus,

$$0.5 = \frac{600 \, dx/dt}{7.2 \times 10^5} \qquad \text{or} \qquad \frac{dx}{dt} = 600 \text{ ft/s}$$

Example D

A particle moves so that its x- and y-coordinates are given by $x = \cos 2t$ and $y = \sin 2t$. Find the magnitude and direction of its velocity when $t = \pi/8$.

We write

$$v_x = \frac{dx}{dt} = -2 \sin 2t \qquad v_y = \frac{dy}{dt} = 2 \cos 2t$$

$$v_x \big|_{t=\pi/8} = -2 \sin 2\left(\frac{\pi}{8}\right) = -2\left(\frac{\sqrt{2}}{2}\right) = -\sqrt{2}$$

$$v_y \big|_{t=\pi/8} = 2 \cos 2\left(\frac{\pi}{8}\right) = 2\left(\frac{\sqrt{2}}{2}\right) = \sqrt{2}$$

$$v = \sqrt{v_x^2 + v_y^2} = \sqrt{2 + 2} = 2 \qquad \tan \theta = \frac{v_y}{v_x} = \frac{\sqrt{2}}{-\sqrt{2}} = -1$$

Since v_x is negative and v_y is positive, $\theta = 135°$. Note that in this example θ is the angle, in standard position, between the horizontal and the resultant velocity.

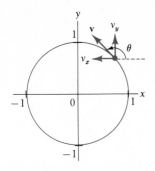

Figure 26–5

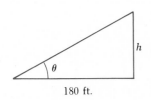

180 ft.

Figure 26–6

By plotting the curve we note that it is a circle (see Fig. 26–5). Thus, the object is moving about a circle in a counterclockwise direction. It can be determined in another way that the curve is a circle. If we square each of the expressions defining x and y, and then add these, we have the equation $x^2 + y^2 = \cos^2 2t + \sin^2 2t = 1$. This is a circle of radius 1.

Example E
At a point 180 ft from the base of a building on level ground, the angle of elevation of the top of the building is 30°. What would be the error in the height of the building due to an error of 15′ in this angle?

From Fig. 26–6 we see that $h = 180 \tan \theta$. Finding the differential of h, we have

$$dh = 180 \sec^2 \theta \; d\theta$$

Since $\theta = 30°$, $\sec \theta = 2/\sqrt{3}$, or $\sec^2 \theta = 4/3$. The possible error in θ is 15′ which is 0.25°, which in turn is 0.00436 rad, which is the value we must use in the calculation. Thus,

$$dh = (180)\left(\frac{4}{3}\right)(0.00436) = 1.05 \text{ ft}$$

We see that an error of 15′ in the angle can result in an error of over 1 ft in the calculation of the height.

Exercises 26–4

1. Show that the slopes of the sine and cosine curves are negatives of each other at the points of intersection.
2. Show that the tangent curve, when defined, is always increasing.
3. Show that $y = \text{Arctan } x$ is always increasing.
4. Sketch the graph of $y = \sin x + \cos x$ $(0 \leq x \leq 2\pi)$.
5. Sketch the graph of $y = x - \tan x$ $(-\pi/2 < x < \pi/2)$.
6. Sketch the graph of $y = \sin x + \sin 2x$ $(0 < x < 2\pi)$.
7. Find the equation of the line tangent to the curve of $y = \sin 2x$ at $x = 5\pi/8$.
8. Find the equation of a line normal to the curve of $y = \text{Arctan}(x/2)$ at $x = 3$.
9. The displacement (in feet) a certain object travels is given as a function of time (in seconds) by $s = \sin t + \cos 2t$. Find its velocity and acceleration when $t = \pi/6$ s.
10. The displacement s (in centimeters) of an object as a function of time is given by $s = t \sin t$. Find the velocity when $t = 0.2$ s.
11. The y-component of a vector of magnitude 20 mm is given by $A_y = 20 \sin \theta$. If $d\theta/dt = 0.05$ rad/s, determine the time rate of change of A_y in millimeters per second when $\theta = 0.3$.
12. The *apparent power* P_a of an electric circuit whose power is P and impedance phase angle is θ is given by $P_a = P \sec \theta$. Given P is constant at 12 W, find the time-rate of change of P_a if θ is changing at the rate of 0.05 rad/min, when $\theta = 2\pi/9$.

13. Find the magnitude and direction of the velocity of an object which moves so that its x- and y-coordinates are given by $x = \sin 2\pi t$ and $y = \cos 2\pi t$ when $t = \frac{1}{12}$.

14. Find the magnitude and direction of the velocity of an object which moves so that the x- and y-coordinates are given by $x = \cos 2t$ and $y = 2 \sin t$ when $t = \pi/6$.

15. Find the magnitude and direction of the acceleration of the particle in Example D when $t = \pi/8$.

16. Find the magnitude and direction of the acceleration of the object in Exercise 14.

17. A person observes an object dropped from the top of a building 100 ft away. If the building is 200 ft high, how fast is the angle of elevation of the object changing after 1 s? (The distance the object drops is given by $s = 16t^2$.)

18. An airplane flying horizontally at 500 ft/s passes directly over an observer 4000 ft below. How fast is the angle of elevation changing at this instant?

19. One leg of a right triangle is 6 units long. If the acute angle of which this side is the adjacent side is increasing at the rate of 0.1 rad/s, how fast is the hypotenuse changing when it is 10 units in length?

20. In a modern hotel, where the elevators are directly observable from the lobby area (and a person can see from the elevators), a person in the lobby observes one of the elevators rising at the rate of 12 ft/s. If the person was 50 ft from the elevator when it left the lobby, how fast is the angle of elevation of the line of sight to the elevator increasing 10 s later?

21. If a block is placed on a plane inclined with the horizontal at an angle θ such that the block just moves down the plane, the coefficient of friction μ is given by $\mu = \tan \theta$. Use differentials to find the change in μ if θ changes from 20° to 21°.

22. The current (in amperes) in a certain electric circuit is given by $i = \sin^2 2t$. Find the approximate change in i between $t = 1.00$ s and $t = 1.05$ s.

23. Two sides and the included angle of a triangle are measured to be 8.00 cm, 7.50 cm, and 40.0°. What is the error caused in the calculation of the third side by an error of 0.5° in the angle?

24. A firm determined that its total weekly profit from the production of x units is $P = 10,000 \sin 0.1x$, for $0 < x < 30$ and where P is measured in dollars. What is the approximate change in profit when production changes from 20 to 22 units?

25. Use trigonometric functions to find the area of the largest rectangle with a perimeter of 12 in.

26. Find the angle between the equal sides of an isosceles triangle of greatest area for a given perimeter.

27. A wall is 6 ft high and 4 ft from a building. What is the length of the shortest pole that can touch the building and the ground beyond the wall? [Hint: $y = 6 \csc \theta + 4 \sec \theta$ (see Fig. 26–7).]

28. A painting 8 ft high is hung such that the lower edge is 2 ft above an observer's eye level. Assuming that the best view is obtained when the angle subtended by the painting at the eye level is a maximum, how far from the wall should the observer stand?

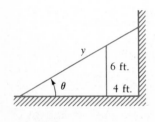

Figure 26–7

26–5 Derivative of the Logarithmic Function

The next function for which we shall find the derivative is the logarithmic function. Again, we shall use the Δ-process.

If we let $y = \log_b u$, where u is a function of x, we have

$$y + \Delta y = \log_b (u + \Delta u)$$

$$\Delta y = \log_b (u + \Delta u) - \log_b u = \log_b \left(\frac{u + \Delta u}{u} \right) = \log_b \left(1 + \frac{\Delta u}{u} \right)$$

$$\frac{\Delta y}{\Delta u} = \frac{\log_b (1 + \Delta u / u)}{\Delta u} = \frac{1}{u} \cdot \frac{u}{\Delta u} \log_b \left(1 + \frac{\Delta u}{u} \right)$$

$$= \frac{1}{u} \log_b \left(1 + \frac{\Delta u}{u} \right)^{u/\Delta u}$$

(We multiply and divide by u for purposes of evaluating the limit, as we shall now show.)

Before we can evaluate the $\lim_{\Delta u \to 0} \Delta y / \Delta u$ it is necessary to determine

$$\lim_{\Delta u \to 0} \left(1 + \frac{\Delta u}{u} \right)^{u/\Delta u}$$

We can see that the exponent becomes unbounded, but the number being raised to this exponent approaches 1. Therefore, we shall investigate this limiting value.

To find an approximate value, let us graph the function $y = (1 + x)^{1/x}$ (for purposes of graphing we let $\Delta u / u = x$). We construct a table of values and then graph the function (see Fig. 26–8).

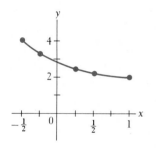

Figure 26–8

x	-0.5	-0.25	$+0.25$	$+0.50$	$+1.00$
y	4.00	3.16	2.44	2.25	2.00

Only these values are shown, since we are interested in the y-value corresponding to $x = 0$. We see from the graph that this value is approximately 2.7. Choosing very small values of x, we may obtain the following values:

x	0.1	0.01	0.001	0.0001
y	2.5937	2.7048	2.7169	2.71815

By methods developed in Chapter 28, it can be shown that this value is about 2.7182818. *The limiting value is the irrational number e.* This is the same number used in the exponential form of a complex number in Chapter 11 and as the base of natural logarithms in Chapter 12.

Returning to the derivative of the logarithmic function, we have

$$\lim_{\Delta u \to 0} \frac{\Delta y}{\Delta u} = \lim_{\Delta u \to 0} \left[\frac{1}{u} \log_b \left(1 + \frac{\Delta u}{u} \right)^{u/\Delta u} \right] = \frac{1}{u} \log_b e$$

Therefore,

$$\frac{dy}{du} = \frac{1}{u}\log_b e$$

Combining this equation with Eq. (26–3), we have

$$\frac{d(\log_b u)}{dx} = \frac{1}{u}\log_b e \frac{du}{dx} \qquad (26\text{--}13)$$

At this point we see that if we choose e as the basis of a system of logarithms, the above formula becomes

$$\frac{d(\ln u)}{dx} = \frac{1}{u}\frac{du}{dx} \qquad (26\text{--}14)$$

The choice of e as the base b makes $\log_e e = 1$; thus, this factor does not appear in Eq. (26–14). We can now see why the number e is chosen as the base for a system of logarithms, the natural logarithms. The notation $\ln u$ is the same as that used in Chapter 12 for natural logarithms.

Example A
Find the derivative of $y = \log 4x$.
    Using Eq. (26–13), we have

$$\frac{dy}{dx} = \frac{1}{4x}\log e\,(4) \qquad (\log e = 0.4343)$$

$$= \frac{1}{x}\log e$$

Example B
Find the derivative of $y = \ln 3x^4$.
    Using Eq. (26–14), we have

$$\frac{dy}{dx} = \frac{1}{3x^4}(12x^3)$$

$$= \frac{4}{x}$$

Example C
Find the derivative of $y = \ln \tan x$.
    Using Eq. (26–14), along with the derivative of the tangent, we have

$$\frac{dy}{dx} = \frac{1}{\tan x}\sec^2 x = \frac{\cos x}{\sin x}\frac{1}{\cos^2 x}$$

$$= \csc x \sec x$$

Example D

Find the derivative of $y = \ln \dfrac{x-1}{x+1}$.

In this example it is easier to write y in the form

$$y = \ln(x-1) - \ln(x+1)$$

(Often we can find a derivative more simply if we use the properties of logarithms to simplify the expression.) Hence,

$$\frac{dy}{dx} = \frac{1}{x-1} - \frac{1}{x+1} = \frac{x+1-x+1}{(x-1)(x+1)}$$

$$= \frac{2}{x^2-1}$$

Example E

Find the derivative of $y = \ln(1-2x)^3$.

First we may rewrite the equation as $y = 3\ln(1-2x)$. Then

$$\frac{dy}{dx} = 3\left(\frac{1}{1-2x}\right)(-2) = \frac{-6}{1-2x}$$

Be careful to distinguish this function from the function $y = [\ln(1-2x)]^3$ [which can also be written as $y = \ln^3(1-2x)$]. In this latter function it is the logarithm which is being cubed, whereas in the first function the quantity $(1-2x)$ is being cubed. The derivative in the latter case would be

$$\frac{dy}{dx} = 3[\ln(1-2x)]^2\left(\frac{1}{1-2x}\right)(-2) = -\frac{6[\ln(1-2x)]^2}{1-2x}$$

Exercises 26–5

In Exercises 1 through 24 find the derivatives of the given functions.

1. $y = \log x^2$

2. $y = \log_2 6x$

3. $y = 2\log_5(3x+1)$

4. $y = 3\log_7(x^2+1)$

5. $y = \ln(1-3x)$

6. $y = 2\ln(3x^2-1)$

7. $y = 2\ln\tan 2x$

8. $y = \ln\sin x$

9. $y = \ln\sqrt{x}$

10. $y = \ln\sqrt{4x-3}$

11. $y = x\ln x$

12. $y = x^2\ln 2x$

13. $y = \dfrac{3x}{\ln(2x+1)}$

14. $y = \dfrac{\ln x}{x}$

15. $y = \ln(\ln x)$

16. $y = \ln\cos x^2$

17. $y = \ln\dfrac{x}{1+x}$

18. $y = \ln(x\sqrt{x+1})$

19. $y = \sin\ln x$

20. $y = (\ln x)^3$

21. $y = \ln(x\tan x)$

22. $y = \ln(x+\sqrt{x^2-1})$

23. $y = \ln\dfrac{x^2}{x+2}$

24. $y = \sqrt{x^2+1} - \ln\dfrac{1+\sqrt{x^2+1}}{x}$

In Exercises 25 through 32 solve the given problems.

25. Verify the values of $(1 + x)^{1/x}$ in the table where $(1 + x)^{1/x}$ is calculated for $x = 0.1, 0.01, 0.001,$ and 0.0001.

26. Find the second derivative of the function $y = x^2 \ln x$.

27. Find the slope of a line tangent to the curve of $y = \ln \cos x$ at $x = \pi/4$.

28. Find the slope of a line tangent to the curve of $y = x \ln 2x$ at $x = 2$.

29. Find the derivative of $y = x^x$ by first taking logarithms of each side of the equation.

30. Find the derivative of $y = (\sin x)^x$ by first taking logarithms of each side of the equation.

31. If the loudness b (in decibels) of a sound of intensity I is given by $b = 10 \log(I/I_0)$ where I_0 is a constant, find the expression for db/dt in terms of dI/dt.

32. The temperature T from the inner surface of a certain steam pipe to the outer surface is given by

$$T = 100 - 40 \frac{\ln(r/5)}{\ln(6/5)}$$

where r is the distance from the center of the pipe. Determine the rate of change of temperature with respect to r. (This solution is valid only for $5 < r < 6$, since these are the inner and outer radii, respectively.)

26–6 Derivative of the Exponential Function

To obtain the derivative of the exponential function, we let $y = b^u$ and then write the equation in logarithmic form:

$$u = \log_b y, \qquad \frac{du}{dx} = \frac{1}{y} \log_b e \frac{dy}{dx}$$

$$\frac{dy}{dx} = \frac{1}{\log_b e} y \frac{du}{dx} = \frac{1}{\log_b e} b^u \left(\frac{du}{dx} \right)$$

Thus,

$$\frac{d(b^u)}{dx} = \frac{1}{\log_b e} b^u \left(\frac{du}{dx} \right) \qquad\qquad (26\text{--}15)$$

If we let $b = e$, Eq. (26–15) becomes

$$\frac{d(e^u)}{dx} = e^u \left(\frac{du}{dx} \right) \qquad\qquad (26\text{--}16)$$

The simplicity of Eq. (26–16) compared with Eq. (26–15) again shows the advantage of choosing e as the basis of natural logarithms. It is for this reason that e appears so often in applications of calculus.

Example A

Find the derivative of $y = e^x$.

Using Eq. (26–16), we have

$$\frac{dy}{dx} = e^x(1) = e^x$$

We see that the derivative of the function e^x equals itself. This exponential function is widely used in applications of calculus.

Example B

Find the derivative of $y = 2^{4x}$.

Using Eq. (26–15), we have

$$\frac{dy}{dx} = \frac{1}{\log_2 e} 2^{4x}(4)$$

$$= \frac{4}{\log_2 e} 2^{4x}$$

Example C

Find the derivative of $y = \ln \cos e^{2x}$.

Using Eq. (26–16) along with the derivatives of the logarithmic and cosine functions, we have

$$\frac{dy}{dx} = \frac{1}{\cos e^{2x}}(-\sin e^{2x})(e^{2x})(2)$$

$$= -2e^{2x}\tan e^{2x}$$

Example D

Find the derivative of $y = x\, e^{\tan x}$.

Using Eq. (26–16) along with the derivatives of a product and the tangent, we have

$$\frac{dy}{dx} = x(e^{\tan x})\sec^2 x + e^{\tan x}$$

$$= e^{\tan x}(x\sec^2 x + 1)$$

Example E

Find the derivative of $y = (e^{1/x})^2$.

In this example,

$$\frac{dy}{dx} = 2(e^{1/x})(e^{1/x})\left(-\frac{1}{x^2}\right) = \frac{-2(e^{1/x})^2}{x^2}$$

$$= \frac{-2e^{2/x}}{x^2}$$

This problem could have also been solved by first writing the function as $y = e^{2/x}$, which is an equivalent form determined by the laws of exponents. When we use this form, the derivative becomes

$$\frac{dy}{dx} = e^{2/x}\left(-\frac{2}{x^2}\right)$$

$$= \frac{-2e^{2/x}}{x^2}$$

We can see that this change in form of the function simplifies the steps necessary for finding the derivative.

Exercises 26-6

In Exercises 1 through 24 find the derivatives of the given functions.

1. $y = 3^{2x}$
2. $y = 3^{1-x}$
3. $y = 4^{6x}$
4. $y = 10^{x^2}$
5. $y = e^{6x}$
6. $y = 3e^{x^2}$
7. $y = e^{\sqrt{x}}$
8. $y = e^{2x^4}$
9. $y = xe^x$
10. $y = x^2e^{2x}$
11. $y = xe^{\sin x}$
12. $y = e^x\sin x$
13. $y = \dfrac{3e^{2x}}{x+1}$
14. $y = \dfrac{e^x}{x}$
15. $y = e^{-3x}\sin 4x$
16. $y = (\cos 2x)(e^{x^2-1})$
17. $y = e^{2x} - e^{-2x}$
18. $y = 4e^{-(2/x)}$
19. $y = e^{2x}\ln x$
20. $y = e^{x^2}\ln \cos x$
21. $y = \ln \sin 2e^{6x}$
22. $y = \tan e^{x+1}$
23. $y = 2 \operatorname{Arcsin} e^{2x}$
24. $y = \operatorname{Arctan} e^{3x}$

In Exercises 25 through 30 solve the given problems.

25. Show that $y = xe^{-x}$ satisfies the equation $(dy/dx) + y = e^{-x}$.

26. Show that $y = e^{-x}\sin x$ satisfies the equation

$$\frac{d^2y}{dx^2} + 2\frac{dy}{dx} + 2y = 0$$

27. Show that the slope of the curve $y = e^x$ at $x = a$ equals the ordinate at $x = a$.

28. The Beer–Lambert law of light absorption may be expressed as $I = I_0e^{-\alpha x}$ where I/I_0 is that fraction of the incident light beam which is transmitted, α is a constant, and x is the distance the light travels through the medium. Find the rate of change of I with respect to x.

29. In a particular electric circuit, the voltage is given by $V = 100(1 - e^{-0.1t})$. Find the expression for dV/dt.

30. The displacement s of an object is given by $s = e^{-0.02t}\cos 2t$. Find the expression for the velocity of the object.

In Exercises 31 and 32 use the following information:

The **hyperbolic sine** of u is defined as

$$\sinh u = \frac{1}{2}(e^u - e^{-u})$$

and the **hyperbolic cosine** of u is defined as

$$\cosh u = \frac{1}{2}(e^u + e^{-u})$$

These functions are called *hyperbolic* functions since, if $x = \cosh u$ and $y = \sinh u$, x and y satisfy the equation of the hyperbola $x^2 - y^2 = 1$.

31. Verify the fact that the expressions for the hyperbolic sine and hyperbolic cosine satisfy the equation of the hyperbola.

32. By using the above definitions, show that

$$\frac{d}{dx}\sinh u = \cosh u\,\frac{du}{dx} \quad \text{and} \quad \frac{d}{dx}\cosh u = \sinh u\,\frac{du}{dx}$$

where u is a function of x.

26–7 Applications

The following examples show applications of the logarithmic and exponential functions to curve-tracing and time-rate-of-change problems. Certain other applications are indicated in the exercises.

Example A

Sketch the graph of the function $y = x \ln x$.

The only intercept is $(1, 0)$ (there is no intercept for $x = 0$ since $\log_b 0$ is not defined). Thus,

$$\frac{dy}{dx} = x\left(\frac{1}{x}\right) + \ln x = 1 + \ln x, \qquad \frac{d^2y}{dx^2} = \frac{1}{x}$$

The first derivative is zero if $\ln x = -1$, or $x = e^{-1}$. The second derivative is positive for this value of x. Thus, there is a minimum point at $(1/e, -1/e)$. The only values of x which can be considered are positive values, since $\ln x$ is undefined for negative values. The second derivative indicates that the curve is always concave up. The graph is shown in Fig. 26–9.

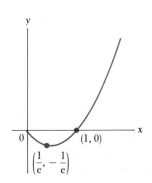

Figure 26–9

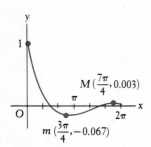

$M\left(\dfrac{7\pi}{4}, 0.003\right)$

$m\left(\dfrac{3\pi}{4}, -0.067\right)$

Figure 26–10

Example B

Sketch the graph of the function $y = e^{-x}\cos x$ $(0 \le x \le 2\pi)$.

This curve has intercepts for all values for which $\cos x$ is zero. Those values between 0 and 2π for which $\cos x = 0$ are $x = \pi/2$ and $x = 3\pi/2$. Also, $(0, 1)$ is an intercept. We next write

$$\frac{dy}{dx} = -e^{-x}\sin x - e^{-x}\cos x = -e^{-x}(\sin x + \cos x)$$

Setting the first derivative equal to zero, we obtain $\sin x + \cos x = 0$, or $\tan x = -1$. There are maximum or minimum points for $x = 3\pi/4$ and $x = 7\pi/4$ (see Fig. 26–10). We then write

$$\frac{d^2y}{dx^2} = 2e^{-x}\sin x$$

Thus, d^2y/dx^2 is positive when $x = 3\pi/4$ and negative when $x = 7\pi/4$. Hence, $(3\pi/4, -0.067)$ is a minimum and $(7\pi/4, 0.003)$ is a maximum. From the second derivative we find that points of inflection occur for $x = 0$, π, and 2π.

Example C

The population P, in terms of the population P_0 (at $t = 0$) and t, is given by $P = P_0 e^{kt}$. Show that the time-rate of change of population is proportional to the population present at time t.

By taking a derivative of P with respect to time, we have

$$\frac{dP}{dt} = kP_0 e^{kt} = kP$$

Example D

A rocket moving vertically, if the only force acting on it is due to gravity and its mass is decreasing at a constant rate r, has its velocity as a function of time given by

$$v = v_0 - gt - k \ln\left(1 - \frac{rt}{m_0}\right)$$

where v_0 is the initial velocity, m_0 is the initial mass, and k is a constant. Determine the expression for the acceleration.

Since the acceleration is the time-rate of change of the velocity, we must find dv/dt. Therefore,

$$\frac{dv}{dt} = -g - k\frac{1}{1 - \dfrac{rt}{m_0}}\left(\frac{-r}{m_0}\right) = -g + \frac{km_0}{m_0 - rt}\left(\frac{r}{m_0}\right)$$

$$= \frac{kr}{m_0 - rt} - g$$

Exercises 26–7

In Exercises 1 through 12 sketch the graphs of the given functions.

1. $y = \ln \cos x$ 2. $y = \dfrac{\ln x}{x}$ 3. $y = xe^{-x}$ 4. $y = xe^{x}$

5. $y = \ln \dfrac{1}{x^2 + 1}$ 6. $y = \ln \dfrac{1}{x}$ 7. $y = e^{-x^2}$ 8. $y = x - e^{x}$

9. $y = \ln x - x$ 10. $y = e^{-x}\sin x$

11. $y = \frac{1}{2}(e^{x} - e^{-x})$ (See Exercise 31 of Section 26–6.)

12. $y = \frac{1}{2}(e^{x} + e^{-x})$ (See Exercise 31 of Section 26–6.)

In Exercises 13 through 24 solve the given problems by finding the appropriate derivatives.

13. Find the equation of the line tangent to the curve $y = x^2\ln x$ at the point $(1, 0)$.

14. Find the equation of the line normal to the curve of $y = \dfrac{e^{2x}}{x}$ at $x = 1$.

15. The power supply P (in watts) in a satellite is given by $P = 100e^{-0.005t}$, where t is measured in days. Find the time-rate of change of power after 100 days.

16. The number N of atoms of radium at any time t is given in terms of the number at $t = 0$, N_0, by $N = N_0 e^{-kt}$. Show that the time-rate of change of N is proportional to N.

17. The velocity of an object moving through a resisting medium is found to obey the relation $v = 40(1 - e^{-0.05t})$. Find the acceleration when $t = 10$.

18. The charge on a capacitor in a circuit containing the capacitor of capacitance C, a resistance R, and voltage source of voltage E is given by $q = CE(1 - e^{-t/RC})$. Show that this equation satisfies the equation

$$R\frac{dq}{dt} + \frac{q}{C} = E$$

19. Assuming that force is proportional to acceleration, show that a particle moving along the x-axis, so that its displacement $x = ae^{kt} + be^{-kt}$, has a force acting on it which is proportional to its displacement.

20. The radius of curvature at any point on a curve is given by

$$R = \frac{[1 + (dy/dx)^2]^{3/2}}{d^2y/dx^2}$$

Determine the radius of curvature of $y = \ln \sec x$ when $x = \pi/4$.

21. An object on the end of a spring is moving so that its displacement in centimeters from the equilibrium position is given by

$$y = e^{-0.5t}(0.4 \cos 6t - 0.2 \sin 6t)$$

Find the expression for the velocity of the object. What is the velocity when $t = \pi/12$ s? The motion described by this equation is called *damped harmonic motion*. The general shape of the curve of displacement as a function of time is that of the curve in Example B.

22. In an electronic device, the maximum current density i_m as a function of the temperature T is given by $i_m = AT^2 e^{k/T}$, where A and k are constants. If the temperature is changing with time, find the expression for the time rate of change of i_m.

23. In developing the theory dealing with the friction between a pulley wheel and the pulley belt, the ratio of the tensions in the belt on either side of the wheel is given by

$$\frac{T_2}{T_1} = e^{k/\sin(\theta/2)}$$

where k is a constant and θ is the angle of the opening of the pulley wheel. Find the expression for a small change in the ratio of tensions for a small change in the angle θ.

24. Find the area of the largest rectangle which can be inscribed in the first quadrant under the curve of $y = e^{-x}$.

26-8 Exercises for Chapter 26

In Exercises 1 through 32 find the derivative of each of the given functions.

1. $y = 3 \cos(4x - 1)$
2. $y = 4 \sec(1 - x^3)$
3. $y = \tan \sqrt{3 - x}$
4. $y = 5 \sin(1 - 6x)$
5. $y = \csc^2(3x + 2)$
6. $y = \cot^2 5x$
7. $y = 3 \cos^4 x^2$
8. $y = 2 \sin^3 \sqrt{x}$
9. $y = (e^{x-3})^2$
10. $y = e^{\sin 2x}$
11. $y = 3 \ln(x^2 + 1)$
12. $y = \ln(3 + \sin x^2)$

13. $y = 3 \operatorname{Arctan}\left(\frac{x}{3}\right)$
14. $y = 4 \operatorname{Arccos}(2x + 3)$

15. $y = \ln \operatorname{Arcsin} 4x$
16. $y = \sin(\operatorname{Arctan} x)$
17. $y = \sqrt{\csc 4x + \cot 4x}$
18. $y = (1 + \sin 2x)^4$

19. $y = \dfrac{x^2}{\operatorname{Arctan} 2x}$
20. $y = \dfrac{\operatorname{Arcsin} x}{x}$

21. $y = \ln(\csc x^2)$
22. $y = e^{\sqrt{1-x}}$
23. $y = [\ln(3 + \sin x)]^2$
24. $y = \ln(3 + \sin x)^2$
25. $y = e^{-x} \sec x$
26. $y = e^{3x} \ln x$
27. $y = x \operatorname{Arccos} x - \sqrt{1 - x^2}$
28. $x + \sec^2(xy) = 1$
29. $y = x^2 (e^{\cos^2 x})^2$
30. $y = x \operatorname{Arcsin}^2 x + 2\sqrt{1 - x^2} \operatorname{Arcsin} x - 2x$
31. $x \sin 2y = y \cos 2x$
32. $x^2 \ln y = y + x$

In Exercises 33 through 36 sketch the graphs of the given functions.

33. $y = x - \cos x$
34. $y = 4 \sin x + \cos 2x$
35. $y = x (\ln x)^2$
36. $y = \ln(1 + x)$

In Exercises 37 through 40 find the equations of the indicated tangent or normal line.

37. Find the equation of the line tangent to the curve $y = 4 \cos^2(x^2)$ at $x = 1$.

38. Find the equation of the line tangent to the curve of $y = \ln \cos x$ at $x = \pi/6$.

39. Find the equation of the line normal to the curve of $y = e^{x^2}$ at $x = 1/2$.

40. Find the equation of the line normal to the curve $y = \text{Arctan } x$ at $x = 1$.

In Exercises 41 through 60 solve the given problems.

41. A particle moves on the circumference of a circle with constant angular velocity. The projection of its motion on the x-axis is given by

$$x = 4 \cos\left(\frac{\pi}{3}t + \frac{\pi}{6}\right)$$

Find the velocity of the projection when $t = 0$.

42. Find the acceleration of an object whose x- and y-components of displacement are given by $x = e^{-t}\sin 2t$ and $y = e^{-t}\cos 2t$ when $t = \pi/4$.

43. Power can be defined as the time rate of doing work. If work is being done in an electric circuit according to $W = 10 \cos 2t$, find P as a function of t.

44. A business firm estimates the value of a machine is given by $V = 25000e^{-0.5t}$, where V is measured in dollars and t in years. What is the time-rate of change of the value after 4 years?

45. The voltage in a certain circuit is given by $E = 170e^{-0.015t}$. Find the approximate change in E if t changes from 0.100 s to 0.105 s.

46. The force which acts on a pendulum bob to restore it to its equilibrium position is given by $F = -mg \sin \theta$ where m is the mass of the bob, g is the acceleration of gravity, and θ is the angle between the string and the vertical. For a bob of 10 g, find the approximate difference in the force for angles of $3.0°$ and $3.1°$ if $g = 980$ cm/s^2.

47. In the study of polarized light, under certain conditions, the intensity of transmitted light is given by $I = I_m\cos^2\theta$, where I_m is the maximum intensity and θ is the angle of transmission. If $I = 8$ units when $\theta = 60°$, and θ is changing at the rate of 0.03 rad/s, find the rate of change of I when $\theta = 60°$.

48. In optics, the index of refraction of a liquid is defined as $n = \sin i/\sin r$, where i is the angle at which the light is incident on the liquid from the air, and r is the angle of the refracted light in the liquid. If the angle of incidence is increasing at the rate of 0.05 rad/min, what is the rate of change of the angle r when $i = 10°$ for benzene, for which $n = 1.50$?

49. For a circuit containing a capacitance C and a resistance R, the current as a function of time is given by $i = i_0 e^{-t/RC}$. If $i_0 = 10$ A, $R = 2$ Ω, and $C = 50$ μF, find the approximate change in the current between 200 μs and 210 μs.

50. Under certain circumstances, 2% of a substance changes chemically each minute. If there are originally 100 g present, the amount at any time t which is present is $A = 100(0.98)^t$. What is the approximate change in the amount present during the fourth second?

51. Under certain conditions the work done by a gas in an expansion from a pressure P_1 to a pressure P_2 is given by $W = k \ln(P_1/P_2)$, where k is a constant depending on the gas. If P_2 varies with time at the rate of 2 kPa/min, find the rate (in joules per minute) at which the work changes when $P_2 = 300$ kPa, if $k = 2400$ J/mol and $P_1 = 200$ kPa.

52. The charge q on a certain capacitor as a function of time is given by

$$q = e^{-0.1t}(0.2 \sin 100t + \cos 100t).$$

The current i in the circuit is the time-rate of change of charge. Find the expression for i as a function of t.

53. An object is dropped from a weather balloon. The distance (in feet) it falls, assuming a resisting force of the air on the object, is given by

$$y = 320(t + 10e^{-0.1t} - 10).$$

Find the velocity after 10 s.

54. A revolving light 300 ft from a straight wall makes 6 r/min. Find the velocity of the beam along the wall at the instant it makes an angle of 45° with the wall.

55. Find the area of the largest rectangle which can be inscribed under the curve of $y = e^{-x^2}$ in the first quadrant.

56. A force P at an angle θ above the horizontal drags a 50-lb box across a level floor. The coefficient of friction between the floor and box is constant and equals 0.2. The magnitude of the force P is given by

$$P = \frac{(0.2)(50)}{(0.2)\sin\theta + \cos\theta}$$

Find θ such that P is a minimum.

57. The illuminance from a point source of light varies directly as the cosine of the angle of incidence (measured from the perpendicular), and inversely as the square of the distance r from the source. How high above the center of a circle of radius 10 in. should a light be placed so that the illuminance at the circumference will be a maximum? (See Fig. 26–11.)

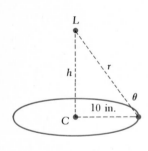

Figure 26–11

58. A metal bracket in the shape of a Y is to be made such that its height is 10 cm and width across the top is 6 cm. What shape will require the least amount of material?

59. A gutter is to be made from a sheet of metal 12 in. wide by turning up strips of width 4 in. along each side to make equal angles θ with the vertical. Sketch a graph of the cross-sectional area A as a function of θ (see Fig. 26–12).

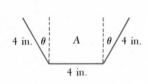

Figure 26–12

60. The curve formed by a uniform cable hanging from two points under its own weight is called the *catenary*. The equation of this curve is

$$y = \frac{H}{w} \cosh \frac{wx}{H}$$

where w and H are constants. Show that the equation of the catenary satisfies the equation

$$\frac{d^2y}{dx^2} = \frac{w}{H}\sqrt{1 + \left(\frac{dy}{dx}\right)^2}$$

(See Exercises 31 and 32 of Section 26–6.)

27

Methods of Integration

_____

27—1 The General Power Formula

Having developed the derivatives of the basic transcendental functions, we can now expand considerably the functions which we can integrate. Many algebraic functions which we were unable to integrate previously will now lend themselves to integration. The first integration formula to be discussed is the general power formula.

For reference, we repeat here the general power formula for integration, Eq. (24–5):

$$\int u^n du = \frac{u^{n+1}}{n+1} + C \qquad n \neq -1 \tag{27–1}$$

This formula will now be applied to transcendental functions as well as algebraic functions. When we are applying the general power formula, we must properly recognize the quantities u, n, and du. This requires familiarity with the differential forms presented in Chapters 22 and 26. The following examples illustrate the use of the general power formula.

Example A

Integrate: $\int \sin^3 x \cos x \, dx$.

Since $d(\sin x) = \cos x \, dx$, we note that this integral fits the form of Eq. (27–1) for $u = \sin x$. Thus, with $u = \sin x$, we have $du = \cos x \, dx$, which means that this integral is of the form $\int u^3 du$. Therefore, the integration can now be completed.

$$\int \sin^3 x \cos x \, dx = \int \sin^3 x (\cos x \, dx)$$

$$= \frac{1}{4}\sin^4 x + C$$

We note here that the factor $\cos x$ is a necessary part of the du in order to have the proper form of integration, and therefore does not appear in the final result.

Example B

Integrate: $\int \sqrt{1 + \tan x} \, \sec^2 x \, dx$.

Here we note that $d(\tan x) = \sec^2 x \, dx$, which means that the integral fits the form of Eq. (27–1) with $u = 1 + \tan x$, $du = \sec^2 x \, dx$, and $n = \frac{1}{2}$. The integral is of the form $\int u^{1/2} du$. Thus,

$$\int \sqrt{1 + \tan x} \, (\sec^2 x \, dx) = \frac{2}{3}(1 + \tan x)^{3/2} + C$$

Example C

Integrate: $\int \ln x \left(\frac{dx}{x}\right)$.

By noting that $d(\ln x) = \frac{dx}{x}$, we have $u = \ln x$, $du = \frac{dx}{x}$, and $n = 1$. This means that the integral is of the form $\int u \, du$. Thus,

$$\int \ln x \left(\frac{dx}{x}\right) = \frac{1}{2}(\ln x)^2 + C$$

Example D

Integrate: $\int (2 + 3e^{2t})^3 e^{2t} dt$.

In this case $u = 2 + 3e^{2t}$, $du = 6e^{2t}dt$, and $n = 3$. This means that we must introduce a 6 to complete the differential in order that the integral fits the proper form. Also, a $\frac{1}{6}$ must then be placed before the integral. This means that the integration is as follows:

$$\int (2 + 3e^{2t})^3 e^{2t} dt = \frac{1}{6}\int (2 + 3e^{2t})^3 (6e^{2t}dt)$$

$$= \frac{1}{6}\cdot\frac{1}{4}(2 + 3e^{2t})^4 + C$$

$$= \frac{1}{24}(2 + 3e^{2t})^4 + C$$

Example E

Find the value of

$$\int_0^{0.5} \frac{\text{Arcsin } x}{\sqrt{1-x^2}} \, dx; \quad n = 1, \quad u = \text{Arcsin } x, \quad \text{and} \quad du = \frac{dx}{\sqrt{1-x^2}}$$

Therefore,

$$\int_0^{0.5} \frac{\text{Arcsin } x}{\sqrt{1-x^2}} \, dx = \frac{(\text{Arcsin } x)^2}{2} \Big|_0^{0.5} = \frac{(\pi/6)^2}{2} - 0 = \frac{\pi^2}{72}$$

Exercises 27–1

In Exercises 1 through 24 integrate each of the given functions.

1. $\displaystyle\int \sin^4 x \, \cos x \, dx$

2. $\displaystyle\int \cos^5 x \, (-\sin x \, dx)$

3. $\displaystyle\int \sqrt{\cos x} \, \sin x \, dx$

4. $\displaystyle\int \sin^{1/3} x \, \cos x \, dx$

5. $\displaystyle\int \tan^2 x \, \sec^2 x \, dx$

6. $\displaystyle\int \sec^3 x \, (\sec x \, \tan x) \, dx$

7. $\displaystyle\int_0^{\pi/8} \cos 2x \, \sin 2x \, dx$

8. $\displaystyle\int_{\pi/6}^{\pi/4} \sqrt{\cot x} \, \csc^2 x \, dx$

9. $\displaystyle\int (\text{Arcsin } x)^3 \left(\frac{dx}{\sqrt{1-x^2}} \right)$

10. $\displaystyle\int \frac{(\text{Arccos } 2x)^4 \, dx}{\sqrt{1-4x^2}}$

11. $\displaystyle\int \frac{\text{Arctan } 5x}{1+25x^2} \, dx$

12. $\displaystyle\int \frac{\text{Arcsin } 4x \, dx}{\sqrt{1-16x^2}}$

13. $\displaystyle\int [\ln(x+1)]^2 \, \frac{dx}{x+1}$

14. $\displaystyle\int (3 + \ln 2x)^3 \, \frac{dx}{x}$

15. $\displaystyle\int_0^{1/2} \frac{\ln(2x+3)}{2x+3} \, dx$

16. $\displaystyle\int_1^e \frac{(1 - 2 \ln x) \, dx}{x}$

17. $\displaystyle\int (4 + e^x)^3 \, e^x \, dx$

18. $\displaystyle\int \sqrt{1 - e^{-x}} \, (e^{-x} dx)$

19. $\displaystyle\int (2e^{2x} - 1)^{1/3} \, e^{2x} dx$

20. $\displaystyle\int \frac{(1 + 3e^{-2x})^4 \, dx}{e^{2x}}$

21. $\displaystyle\int (1 + \sec^2 x)^4 \, (\sec^2 x \, \tan x \, dx)$

22. $\displaystyle\int (e^x + e^{-x})^{1/4} \, (e^x - e^{-x}) \, dx$

23. $\displaystyle\int_0^{\pi/6} \frac{\tan x}{\cos^2 x} \, dx$

24. $\displaystyle\int_{\pi/3}^{\pi/2} \frac{\sin \theta \, d\theta}{\sqrt{1 + \cos \theta}}$

In Exercises 25 through 28 solve the given problems by integration.

25. Find the area under the curve $y = \sin x \cos x$ from $x = 0$ to $x = \pi/2$.

26. In developing the expression for the total pressure P on a wall due to molecules with mass m and velocity v striking the wall, the following equation is found: $P = mnv^2 \int_0^{\pi/2} \sin \theta \cos^2\theta \, d\theta$. The symbol n represents the number of molecules per unit volume and θ represents the angle between a perpendicular to the wall and the direction of the molecule. Find the final expression for P.

27. The general expression for the slope of a given curve is $(\ln x)^2/x$. If the curve passes through $(1, 2)$, find its equation.

28. The voltage across a certain capacitor is given by $V = 10^6 \int e^t dt/\sqrt{3 + e^t}$. Determine the function which gives the voltage in terms of time.

27–2 The Basic Logarithmic Form

The general power formula for integration, Eq. (27–1), is valid for all values of n except $n = -1$. If n were set equal to -1, this would cause the result to be undefined. When we obtained the derivative of the logarithmic function, we found that

$$\frac{d(\ln u)}{dx} = \frac{1}{u} \frac{du}{dx}$$

which means the differential of the logarithmic form is $d(\ln u) = du/u$. Reversing the process, we then determine that $\int du/u = \ln u + C$. In other words, when the exponent of the expression being integrated is -1, it is a logarithmic form.

Logarithms are defined only for positive numbers. Thus, $\int du/u = \ln u + C$ is valid if $u > 0$. If $u < 0$, then $-u > 0$. In this case $d(-u) = -du$, or $\int (-du)/(-u) = \ln(-u) + C$. However, $\int du/u = \int (-du)/(-u)$. These results can be combined into the single form

$$\int \frac{du}{u} = \ln|u| + C \tag{27–2}$$

Example A

Integrate: $\displaystyle\int \frac{dx}{x + 1}$.

Since $d(x + 1) = dx$, this integral fits the form of Eq. (27–2) with $u = x + 1$ and $du = dx$. Therefore, we have

$$\int \frac{dx}{x + 1} = \ln|x + 1| + C$$

Example B

Integrate: $\displaystyle\int \frac{\cos x}{\sin x}\,dx.$

We note that $d(\sin x) = \cos x\,dx$. This means that this integral fits the form of Eq. (27–2) with $u = \sin x$ and $du = \cos x\,dx$. Thus,

$$\int \frac{\cos x}{\sin x}\,dx = \int \frac{\cos x\,dx}{\sin x}$$

$$= \ln|\sin x| + C$$

Example C

Integrate: $\displaystyle\int \frac{x\,dx}{4 - x^2}.$

This integral fits the form of Eq. (27–2) with $u = 4 - x^2$ and $du = -2x\,dx$. This means that we must introduce a factor of -2 into the numerator and a factor of $-\frac{1}{2}$ before the integral. Therefore,

$$\int \frac{x\,dx}{4 - x^2} = -\frac{1}{2}\int \frac{-2x\,dx}{4 - x^2}$$

$$= -\frac{1}{2}\ln|4 - x^2| + C$$

We should note that if the quantity $4 - x^2$ were raised to any power other than that in the example, we would have to employ the general power formula for integration. For example,

$$\int \frac{x\,dx}{(4 - x^2)^2} = -\frac{1}{2}\int \frac{-2x\,dx}{(4 - x^2)^2} = -\frac{1}{2}\frac{(4 - x^2)^{-1}}{-1} + C$$

$$= \frac{1}{2(4 - x^2)} + C$$

Example D

Integrate: $\displaystyle\int \frac{e^{4x}dx}{1 + 3e^{4x}};\ u = 1 + 3e^{4x},\ du = 12e^{4x}dx.$

We write
$$\int \frac{e^{4x}dx}{1 + 3e^{4x}} = \frac{1}{12}\int \frac{12e^{4x}dx}{1 + 3e^{4x}} = \frac{1}{12}\ln|1 + 3e^{4x}| + C$$

$$= \frac{1}{12}\ln(1 + 3e^{4x}) + C \qquad (\text{since } 1 + 3e^{4x} > 0)$$

Example E

Evaluate: $\displaystyle\int_0^{\pi/8} \frac{\sec^2 2x}{1 + \tan 2x}\, dx$; $u = 1 + \tan 2x$, $du = 2 \sec^2 2x\, dx$.

We write

$$\int_0^{\pi/8} \frac{\sec^2 2x}{1 + \tan 2x}\, dx = \frac{1}{2}\int_0^{\pi/8} \frac{2 \sec^2 2x\, dx}{1 + \tan 2x}$$

$$= \frac{1}{2}\ln|1 + \tan 2x|\Big|_0^{\pi/8} = \frac{1}{2}\ln|2| - \frac{1}{2}\ln|1|$$

$$= \frac{1}{2}\ln 2 - \frac{1}{2}\ln 1 = \frac{1}{2}\ln 2 - 0 = \frac{1}{2}\ln 2$$

Exercises 27–2

In Exercises 1 through 24 integrate each of the given functions.

1. $\displaystyle\int \frac{dx}{1 + 4x}$

2. $\displaystyle\int \frac{dx}{1 - 4x}$

3. $\displaystyle\int \frac{x\, dx}{4 - 3x^2}$

4. $\displaystyle\int \frac{x^2 dx}{1 - x^3}$

5. $\displaystyle\int \frac{\csc^2 2x}{\cot 2x}\, dx$

6. $\displaystyle\int \frac{\sin x}{\cos x}\, dx$

7. $\displaystyle\int_0^{\pi/2} \frac{\cos x\, dx}{1 + \sin x}$

8. $\displaystyle\int_0^{\pi/4} \frac{\sec^2 x\, dx}{4 + \tan x}$

9. $\displaystyle\int \frac{e^{-x}}{1 - e^{-x}}\, dx$

10. $\displaystyle\int \frac{e^{3x}}{1 - e^{3x}}\, dx$

11. $\displaystyle\int \frac{1 + e^x}{x + e^x}\, dx$

12. $\displaystyle\int \frac{e^x - e^{-x}}{e^x + e^{-x}}\, dx$

13. $\displaystyle\int \frac{\sec x \tan x\, dx}{1 + 4 \sec x}$

14. $\displaystyle\int \frac{\sin 2x}{1 - \cos^2 x}\, dx$

15. $\displaystyle\int_1^3 \frac{1 + x}{4x + 2x^2}\, dx$

16. $\displaystyle\int_1^2 \frac{4x + 6x^2}{x^2 + x^3}\, dx$

17. $\displaystyle\int \frac{dx}{x \ln x}$

18. $\displaystyle\int \frac{dx}{x(1 + 2 \ln x)}$

19. $\displaystyle\int \frac{2 + \sec^2 x}{2x + \tan x}\, dx$

20. $\displaystyle\int \frac{x + \cos 2x}{x^2 + \sin 2x}\, dx$

21. $\displaystyle\int \frac{dx}{\sqrt{1 - 2x}}$

22. $\displaystyle\int \frac{x\, dx}{(1 + x^2)^2}$

23. $\displaystyle\int_0^{\pi/12} \frac{\sec^2 3x}{4 + \tan 3x}\, dx$

24. $\displaystyle\int_1^2 \frac{x^2 + 1}{x^3 + 3x}\, dx$

In Exercises 25 through 32 solve the given problems by integration.

25. Find the area bounded by $y(x + 1) = 1$, $x = 0$, $y = 0$, and $x = 2$.

26. Find the area bounded by $xy = 1$, $x = 1$, $x = 2$, and $y = 0$.

27. Find the volume generated by rotating the area bounded by $y = 1/(x^2 + 1)$, $x = 0$, $x = 1$, $y = 0$ about the y-axis. (Use shells.)

28. Find the average value of the function $xy = 4$ from $x = 1$ to $x = 2$.

29. If the current (in amperes), as a function of time, in a certain electric circuit is given by $i = t/(1 + 2t^2)$, find the total charge to pass a given point in the circuit in the first two seconds.

30. The electric current in a certain inductor is given by

$$i = 5 \int \frac{e^t \cos e^t \, dt}{\sin e^t}$$

Find the function which expresses the current in terms of the time.

31. Under certain conditions an object is subject to a force (in pounds) given by $F = 1/(1 + 3x)$. Find the work done on the object in moving it 5 ft from the origin.

32. Conditions are often such that a force proportional to the velocity tends to retard the motion of an object. Under such conditions, the acceleration of a certain object moving down an inclined plane is given by $20 - v$. This leads to the equation $dv/(20 - v) = dt$. If the object starts from rest, find the expression for the velocity as a function of time.

27–3 The Exponential Form

In deriving the derivative for the exponential function we obtained the result $de^u/dx = e^u(du/dx)$. This means that the differential of the exponential form is $d(e^u) = e^u du$. Reversing this form to find the proper form of the integral for the exponential function, we have

$$\int e^u du = e^u + C \tag{27–3}$$

Example A

Integrate: $\int e^{5x} dx$.

Since $d(5x) = 5 \, dx$, this integral can be put in the form of Eq. (27–3) with $u = 5x$ and $du = 5 \, dx$. This means that we place a 5 with the dx and a $\frac{1}{5}$ before the integral. Therefore,

$$\int e^{5x} dx = \frac{1}{5} \int e^{5x}(5 \, dx)$$

$$= \frac{1}{5} e^{5x} + C$$

Example B

Integrate: $\int xe^{x^2} dx$.

Since $d(x^2) = 2x \, dx$, we can write this integral in the form of Eq. (27–3) with $u = x^2$ and $du = 2x \, dx$. Thus,

$$\int xe^{x^2} dx = \frac{1}{2} \int e^{x^2}(2x \, dx)$$

$$= \frac{1}{2} e^{x^2} + C$$

Example C

Integrate: $\int \dfrac{dx}{e^{3x}}$.

This integral can be put in proper form by writing it as $\int e^{-3x}dx$. In this form $u = -3x$, $du = -3\ dx$. Thus,

$$\int \frac{dx}{e^{3x}} = \int e^{-3x}dx = -\frac{1}{3}\int e^{-3x}(-3\ dx)$$

$$= -\frac{1}{3}e^{-3x} + C$$

Example D

Evaluate: $\int_0^{\pi/2} (\sin 2x)\,(e^{\cos 2x})\,dx$; $u = \cos 2x$, $du = -2 \sin 2x\ dx$.

We write

$$\int_0^{\pi/2} (\sin 2x)\,(e^{\cos 2x})\,dx = -\frac{1}{2}\int_0^{\pi/2} (e^{\cos 2x})\,(-2\ \sin 2x\ dx)$$

$$= -\frac{1}{2}e^{\cos 2x}\Big|_0^{\pi/2} = -\frac{1}{2}\Big(\frac{1}{e} - e\Big) = 1.175$$

Example E

Find the equation of the curve for which $\dfrac{dy}{dx} = \dfrac{e^{\sqrt{x+1}}}{\sqrt{x+1}}$ if the curve passes through $(0, 1)$.

The solution of this problem requires that we integrate the given function and then evaluate the constant of integration. Hence,

$$dy = \frac{e^{\sqrt{x+1}}}{\sqrt{x+1}}\,dx, \qquad \int dy = \int \frac{e^{\sqrt{x+1}}}{\sqrt{x+1}}\,dx$$

For purposes of integrating the righthand side,

$$u = \sqrt{x+1} \quad \text{and} \quad du = \frac{1}{2\sqrt{x+1}}\,dx$$

$$y = 2\int e^{\sqrt{x+1}}\Big(\frac{1}{2\sqrt{x+1}}\,dx\Big) = 2e^{\sqrt{x+1}} + C$$

Letting $x = 0$ and $y = 1$, we have $1 = 2e + C$, or $C = 1 - 2e$. This means that the equation is

$$y = 2e^{\sqrt{x+1}} + 1 - 2e$$

Exercises 27–3   In Exercises 1 through 20 integrate each of the given functions.

1. $\int e^{7x}(7\ dx)$

2. $\int e^{x^4}(4x^3\,dx)$

3. $\int e^{2x+5}\,dx$

4. $\int e^{-4x}\,dx$

5. $\int_0^2 e^{x/2}\,dx$

6. $\int_1^2 3e^{4x}\,dx$

7. $\int x^2 e^{x^3}\,dx$

8. $\int xe^{-x^2}\,dx$

9. $\int \dfrac{e^{\sqrt{x}}}{\sqrt{x}}dx$

10. $\int 4x^3 e^{2x^4}\,dx$

11. $\int (\sec x \tan x)\,e^{2\sec x}\,dx$

12. $\int (\sec^2 x)\,e^{\tan x}\,dx$

13. $\int \dfrac{(3-e^x)\,dx}{e^{2x}}$

14. $\int (e^x - e^{-x})^2\,dx$

15. $\int \dfrac{e^{\text{Arctan } x}}{x^2+1}dx$

16. $\int \dfrac{e^{\text{Arcsin } 2x}dx}{\sqrt{1-4x^2}}$

17. $\int \dfrac{e^{\cos 3x}\,dx}{\csc 3x}$

18. $\int \dfrac{e^{2/x}dx}{x^2}$

19. $\int_0^\pi (\sin 2x)\,e^{\cos^2 x}\,dx$

20. $\int_{-1}^1 \dfrac{dx}{e^{2-3x}}$

In Exercises 21 through 28 solve the given problems by integration.

21. Find the area bounded by $y = e^x$, $x = 0$, $y = 0$, and $x = 2$.

22. Prove that the area bounded by $x = a$, $x = b$, $y = 0$, and $y = e^x$ equals the difference of the ordinates at $x = b$ and $x = a$.

23. Find the volume generated by rotating the area bounded by $y = e^{x^2}$, $x = 1$, $y = 0$, and $x = 2$ about the y-axis.

24. Find the equation of the curve for which $dy/dx = \sqrt{e^{x+3}}$ if the curve passes through $(1, 0)$.

25. Find the average value of the function $y = e^{2x}$ from $x = 0$ to $x = 4$.

26. Find the moment of inertia with respect to the y-axis of the first quadrant area bounded by $y = e^{x^3}$, $x = 1$, and the axes.

27. If the current in a certain electric circuit is given by $i = 2e^{-30t}$, find the expression for the voltage across a 250 μF capacitor as a function of time. The initial voltage across the capacitor is 60 V.

28. A particle moves such that its velocity is given by $v = e^{\sin 2x}\cos 2x\ dx$. Find the equation for the displacement as a function of the time if $s = 2$ when $t = 0$.

27–4 Basic Trigonometric Forms

In this section we shall discuss the integrals of the six trigonometric functions and the trigonometric integrals which arise directly from reversing the formulas for differentiation. Other trigonometric forms will be discussed later.

By directly reversing the differentiation formulas, the following integral formulas are obtained:

$$\int \sin u \; du = -\cos u + C \qquad\qquad (27\text{--}4)$$

$$\int \cos u \; du = \sin u + C \qquad\qquad (27\text{--}5)$$

$$\int \sec^2 u \; du = \tan u + C \qquad\qquad (27\text{--}6)$$

$$\int \csc^2 u \; du = -\cot u + C \qquad\qquad (27\text{--}7)$$

$$\int \sec u \tan u \; du = \sec u + C \qquad\qquad (27\text{--}8)$$

$$\int \csc u \cot u \; du = -\csc u + C \qquad\qquad (27\text{--}9)$$

Example A

Integrate: $\int \sin 3x \; dx$; $u = 3x$, $du = 3 \; dx$.
We have

$$\int \sin 3x \; dx = \frac{1}{3} \int \sin 3x (3 \; dx)$$

$$= -\frac{1}{3} \cos 3x + C$$

Example B

Integrate: $\int x \sec^2 (x^2) \, dx$; $u = x^2$, $du = 2x \; dx$.
We write

$$\int x \sec^2 (x^2) \, dx = \frac{1}{2} \int \sec^2 (x^2) \cdot (2x \; dx)$$

$$= \frac{1}{2} \tan(x^2) + C$$

Example C

Integrate: $\displaystyle\int \frac{\tan 2x}{\cos 2x} \, dx$.

By using the basic identity $\sec \theta = 1/\cos \theta$, we can transform this integral to form $\int \sec 2x \tan 2x \; dx$. In this form $u = 2x$, $du = 2 \; dx$. Therefore,

$$\int \frac{\tan 2x}{\cos 2x} \, dx = \int \sec 2x \tan 2x \; dx = \frac{1}{2} \int \sec 2x \tan 2x (2 \; dx)$$

$$= \frac{1}{2} \sec 2x + C$$

To find the integrals for the other trigonometric functions, we must change them to a form for which the integral can be determined by methods previously discussed. We can accomplish this by use of the basic trigonometric relations.

The formula for $\int \tan u \, du$ is found by expressing the integral in the form $\int [(\sin u)/(\cos u)] \, du$. We recognize this as being a logarithmic form, where the u of the logarithmic form is $\cos u$ in this integral. The differential of $\cos u$ is $-\sin u \, du$. Therefore, we have

$$\int \tan u \, du = \int \frac{\sin u}{\cos u} \, du = -\int \frac{-\sin u \, du}{\cos u} = -\ln |\cos u| + C$$

The formula for $\int \cot u \, du$ is found by writing it in the form

$$\int \frac{\cos u}{\sin u} \, du$$

In this manner we obtain the following result:

$$\int \cot u \, du = \int \frac{\cos u}{\sin u} \, du = \int \frac{\cos u \, du}{\sin u} = \ln |\sin u| + C$$

The formula for $\int \sec u \, du$ is found by writing it in the form

$$\int \frac{\sec u (\sec u + \tan u)}{\sec u + \tan u} \, du$$

In this form we recognize this as being also a logarithmic form, since

$$d(\sec u + \tan u) = (\sec u \tan u + \sec^2 u) \, du$$

The right side of this equation is the expression appearing in the numerator of the integral. Thus,

$$\int \sec u \, du = \int \frac{\sec u (\sec u + \tan u) \, du}{\sec u + \tan u} = \int \frac{\sec u \tan u + \sec^2 u}{\sec u + \tan u} \, du$$

$$= \ln |\sec u + \tan u| + C$$

To obtain the formula for $\int \csc u \, du$, we write it in the form

$$\int \frac{\csc u (\csc u - \cot u) \, du}{\csc u - \cot u}$$

Thus, we have

$$\int \csc u \, du = \int \frac{\csc u (\csc u - \cot u)}{\csc u - \cot u} \, du$$

$$= \int \frac{(-\csc u \cot u + \csc^2 u) \, du}{\csc u - \cot u}$$

$$= \ln |\csc u - \cot u| + C$$

Summarizing these results, we have the following integrals:

$$\int \tan u \ du = -\ln|\cos u| + C \tag{27-10}$$

$$\int \cot u \ du = \ln|\sin u| + C \tag{27-11}$$

$$\int \sec u \ du = \ln|\sec u + \tan u| + C \tag{27-12}$$

$$\int \csc u \ du = \ln|\csc u - \cot u| + C \tag{27-13}$$

Example D
Integrate: $\int \tan 4x \ dx$; $u = 4x$, $du = 4 \ dx$.
   We write

$$\int \tan 4x \ dx = \frac{1}{4}\int \tan 4x(4 \ dx) = -\frac{1}{4}\ln|\cos 4x| + C$$

Example E
Integrate: $\displaystyle\int \frac{\sec e^{-x}dx}{e^x}$; $u = e^{-x}$, $du = -e^{-x}dx$.

   We have

$$\int \frac{\sec e^{-x}dx}{e^x} = -\int (\sec e^{-x})(-e^{-x}dx)$$

$$= -\ln|\sec e^{-x} + \tan e^{-x}| + C$$

Example F
Evaluate: $\displaystyle\int_{\pi/6}^{\pi/4} \frac{1 + \cos x}{\sin x} \ dx$.

   We write

$$\int_{\pi/6}^{\pi/4} \frac{1 + \cos x}{\sin x} \ dx = \int_{\pi/6}^{\pi/4} \csc x \ dx + \int_{\pi/6}^{\pi/4} \cot x \ dx$$

$$= \ln|\csc x - \cot x| \Big|_{\pi/6}^{\pi/4} + \ln|\sin x| \Big|_{\pi/6}^{\pi/4}$$

$$= \ln|\sqrt{2} - 1| - \ln|2 - \sqrt{3}| + \ln\left|\frac{1}{2}\sqrt{2}\right| - \ln\left|\frac{1}{2}\right|$$

$$= \ln \frac{\left(\frac{1}{2}\sqrt{2}\right)(\sqrt{2} - 1)}{\left(\frac{1}{2}\right)(2 - \sqrt{3})} = \ln\frac{2 - \sqrt{2}}{2 - \sqrt{3}} = \ln\frac{0.586}{0.268}$$

$$= \ln 2.19 = 0.784$$

Exercises 27-4

In Exercises 1 through 24 integrate each of the given functions.

1. $\displaystyle\int \cos 2x\ dx$

2. $\displaystyle\int \sin(2-x)\ dx$

3. $\displaystyle\int \sec^2 3x\ dx$

4. $\displaystyle\int \csc 2x \cot 2x\ dx$

5. $\displaystyle\int \sec \tfrac{1}{2}x \tan \tfrac{1}{2}x\ dx$

6. $\displaystyle\int e^x \csc^2(e^x)\ dx$

7. $\displaystyle\int x^2 \cot x^3\ dx$

8. $\displaystyle\int \tan \tfrac{1}{2}x\ dx$

9. $\displaystyle\int x \sec x^2\ dx$

10. $\displaystyle\int 2 \csc 3x\ dx$

11. $\displaystyle\int \frac{\sin(1/x)}{x^2}\ dx$

12. $\displaystyle\int \frac{dx}{\sin 4x}$

13. $\displaystyle\int_0^{\pi/6} \frac{dx}{\cos^2 2x}$

14. $\displaystyle\int_0^1 e^x \cos\ e^x\ dx$

15. $\displaystyle\int \frac{\sec 5x}{\cot 5x}\ dx$

16. $\displaystyle\int \frac{\sin 2x}{\cos^2 x}\ dx$

17. $\displaystyle\int \sqrt{\tan^2 2x + 1}\ dx$

18. $\displaystyle\int (1 + \cot x)^2\ dx$

19. $\displaystyle\int_0^{\pi/9} \sin 3x(\csc 3x + \sec 3x)\ dx$

20. $\displaystyle\int_{\pi/4}^{\pi/3} (1 + \sec x)^2\ dx$

21. $\displaystyle\int \frac{1 - \sin x}{1 + \cos x}\ dx$

22. $\displaystyle\int \frac{1 + \sec^2 x}{x + \tan x}\ dx$

23. $\displaystyle\int \frac{1 + \sin 2x}{\tan 2x}\ dx$

24. $\displaystyle\int \frac{1 - \cot^2 x}{\cos^2 x}\ dx$

In Exercises 25 through 32 solve the given problems by integration.

25. Find the area bounded by $y = \tan x$, $x = \pi/4$, and $y = 0$.

26. Find the area under the curve $y = \sin x$ from $x = 0$ to $x = \pi$.

27. Find the volume generated by rotating the area bounded by $y = \sec x$, $x = 0$, $x = \pi/3$, and $y = 0$ about the x-axis.

28. Find the volume generated by rotating the area bounded by $y = \cos x^2$, $x = 0$, $y = 0$, and $x = 1$ about the y-axis.

29. Under certain conditions the expression for the velocity as a function of time for a particle is $v = 4 \cos 3t$. Determine the distance as a function of the time, if $s = 0$ when $t = 0$.

30. If the current in a certain electric circuit is $i = 110 \cos 377t$, find the expression for the voltage across a 500 μF capacitor as a function of time. The initial voltage is zero. Show that the voltage across the capacitor is 90° out of phase with the current.

31. If the area bounded by $y = \sin(x^2)$, $y = 0$, and $x = 1$ is 0.3103, find the x-coordinate of the centroid of this area.

32. A force is given as a function of the distance from the origin by

$$F = \frac{\sin x}{2 + \cos x}$$

Express the work done by this force as a function of x, if $W = 0$ when $x = 0$.

27–5 Other Trigonometric Forms

The basic trigonometric relations developed in Chapter 19 provide the means by which many other integrals involving trigonometric functions may be integrated. By use of the square relations, Eqs. (19–6), (19–7), and (19–8), and the equations for the cosine of the double angle, Eqs. (19–21), (19–22), and (19–23), it is possible to transform integrals involving powers of the trigonometric functions into integrable form. We repeat these equations here for reference:

$$\cos^2 x + \sin^2 x = 1 \qquad (27\text{–}14)$$
$$1 + \tan^2 x = \sec^2 x \qquad (27\text{–}15)$$
$$1 + \cot^2 x = \csc^2 x \qquad (27\text{–}16)$$
$$2\cos^2 x = 1 + \cos 2x \qquad (27\text{–}17)$$
$$2\sin^2 x = 1 - \cos 2x \qquad (27\text{–}18)$$

To integrate a product of powers of the sine and cosine, we use Eq. (27–14) if at least one of the powers is odd. The method is based on transforming the integral so that it is made up of powers of either the sine or cosine and the first power of the other. In this way this first power becomes a factor of du.

Example A

Integrate: $\int \sin^3 x \cos^2 x \, dx$.

Since $\sin^3 x = \sin^2 x \sin x = (1 - \cos^2 x)\sin x$, it is possible to write this integral with powers of $\cos x$ along with $\sin x \, dx$. Thus, $\sin x \, dx$ becomes the necessary du of the integral. Therefore,

$$\int \sin^3 x \cos^2 x \, dx = \int (1 - \cos^2 x)(\sin x)(\cos^2 x) \, dx$$

$$= \int (\cos^2 x - \cos^4 x)(\sin x \, dx)$$

$$= \int \cos^2 x (\sin x \, dx) - \int \cos^4 x (\sin x \, dx)$$

$$= -\int \cos^2 x (-\sin x \, dx) + \int \cos^4 x (-\sin x \, dx)$$

$$= -\frac{1}{3}\cos^3 x + \frac{1}{5}\cos^5 x + C$$

Example B

Integrate: $\int \cos^5 2x \, dx$.

Since $\cos^5 2x = \cos^4 2x \cos 2x = (1 - \sin^2 2x)^2 \cos 2x$, it is possible to write this integral with powers of $\sin 2x$ along with $\cos 2x \, dx$. Thus, with the introduction of a factor of 2, $(\cos 2x)(2\ dx)$ is the necessary du of the integral. Thus,

$$\int \cos^5 2x\ dx = \int (1 - \sin^2 2x)^2 \cos 2x\ dx$$

$$= \int (1 - 2\ \sin^2 2x + \sin^4 2x)\cos 2x\ dx$$

$$= \int \cos 2x\ dx - \int 2\ \sin^2 2x\ \cos 2x\ dx + \int \sin^4 2x\ \cos 2x\ dx$$

$$= \frac{1}{2}\int \cos 2x(2\ dx) - \int \sin^2 2x(2\ \cos 2x\ dx)$$

$$+ \frac{1}{2}\int \sin^4 2x(2\ \cos 2x\ dx)$$

$$= \frac{1}{2}\sin 2x - \frac{1}{3}\sin^3 2x + \frac{1}{10}\sin^5 2x + C$$

In products of powers of the sine and cosine, if the powers to be integrated are even, we use Eqs. (27–17) and (27–18) to transform the integral. Those most commonly met are $\int \cos^2 u\ du$ and $\int \sin^2 u\ du$.

Example C
Integrate: $\int \sin^2 2x\ dx$.

Using Eq. (27–18) in the form $\sin^2 2x = \frac{1}{2}(1 - \cos 4x)$, this integral can be transformed into a form which can be integrated. (Here we note the x of Eq. (27–18) is treated as $2x$ for this integral.) Therefore, we write

$$\int \sin^2 2x\ dx = \int \left[\frac{1}{2}(1 - \cos 4x) \right] dx = \frac{1}{2}\int dx - \frac{1}{8}\int \cos 4x(4\ dx)$$

$$= \frac{x}{2} - \frac{1}{8}\sin 4x + C$$

To integrate even powers of the secant, powers of the tangent, or products of the secant and tangent, we use Eq. (27–15) to transform the integral. In transforming, the forms we look for are powers of the tangent with $\sec^2 x$, which becomes part of du, or powers of the secant along with $\sec x\ \tan x$, which becomes part of du in this case. Similar transformations are made when we integrate powers of the cotangent and cosecant, with the use of Eq. (27–16).

Example D

Integrate: $\int \tan^5 x \, dx$.

Since $\tan^5 x = \tan^3 x \tan^2 x = \tan^3 x (\sec^2 x - 1)$, we can write this integral with powers of tan x along with $\sec^2 x \, dx$. Thus, $\sec^2 x \, dx$ becomes the necessary du of the integral. It is necessary to replace $\tan^2 x$ with $\sec^2 x - 1$ twice during the integration. Therefore,

$$\int \tan^5 x \, dx = \int \tan^3 x (\sec^2 x - 1) \, dx$$

$$= \int \tan^3 x (\sec^2 x \, dx) - \int \tan^3 x \, dx$$

$$= \frac{1}{4} \tan^4 x - \int \tan x (\sec^2 x - 1) \, dx$$

$$= \frac{1}{4} \tan^4 x - \int \tan x (\sec^2 x \, dx) + \int \tan x \, dx$$

$$= \frac{1}{4} \tan^4 x - \frac{1}{2} \tan^2 x - \ln|\cos x| + C$$

Example E

Integrate: $\int \sec^3 x \tan x \, dx$.

By writing $\sec^3 x \tan x$ as $\sec^2(\sec x \tan x)$, we can use the $\sec x \tan x \, dx$ as the du of the integral. Thus,

$$\int \sec^3 x \tan x \, dx = \int (\sec^2 x)(\sec x \tan x \, dx)$$

$$= \frac{1}{3} \sec^3 x + C$$

Example F

Integrate: $\int \csc^4 2x \, dx$.

By writing $\csc^4 2x = \csc^2 2x \csc^2 2x = \csc^2 2x (1 + \cot^2 2x)$, we can write this integral with powers of cot $2x$ along with $\csc^2 2x \, dx$, which becomes part of the necessary du of the integral. Thus,

$$\int \csc^4 2x \, dx = \int \csc^2 2x (1 + \cot^2 2x) \, dx$$

$$= \frac{1}{2} \int \csc^2 2x (2 \, dx) - \frac{1}{2} \int \cot^2 2x (-2 \csc^2 2x \, dx)$$

$$= -\frac{1}{2} \cot 2x - \frac{1}{6} \cot^3 2x + C$$

Example G

Integrate: $\displaystyle\int_0^{\pi/4} \frac{\tan^3 x}{\sec^3 x} dx$.

This integral requires the use of several trigonometric relationships to obtain integrable forms.

$$\int_0^{\pi/4} \frac{\tan^3 x}{\sec^3 x} dx = \int_0^{\pi/4} \frac{(\sec^2 x - 1)\tan x}{\sec^3 x} dx$$

$$= \int_0^{\pi/4} \frac{\tan x}{\sec x} dx - \int_0^{\pi/4} \frac{\tan x \ dx}{\sec^3 x}$$

$$= \int_0^{\pi/4} \sin x \ dx - \int_0^{\pi/4} \frac{\sec x \tan x \ dx}{\sec^4 x}$$

$$= -\cos x + \frac{1}{3}\cos^3 x \Big|_0^{\pi/4}$$

$$= -\frac{\sqrt{2}}{2} + \frac{1}{3}\left(\frac{\sqrt{2}}{2}\right)^3 - \left(-1 + \frac{1}{3}\right) = \frac{8 - 5\sqrt{2}}{12}$$

Example H

The *root-mean-square value of a function* with respect to x is defined by

$$y_{\text{rms}} = \sqrt{\frac{1}{T} \int_0^T y^2 dx} \tag{27–19}$$

Usually the value of T which is of importance is the period of the function. Find the root-mean-square value of the current in an electric circuit for one period, if $i = 3 \cos \pi t$.

The period of the current is $2\pi/\pi = 2$ s. Therefore, we must find the square root of the integral

$$\frac{1}{2} \int_0^2 (3 \cos \pi t)^2 dt = \frac{9}{2} \int_0^2 \cos^2 \pi t \ dt$$

Hence, we have

$$\frac{9}{2} \int_0^2 \cos^2 \pi t \ dt = \frac{9}{4} \int_0^2 (1 + \cos 2\pi t) \ dt$$

after substituting $\frac{1}{2}(1 + \cos 2\pi t)$ for $\cos^2 \pi t$. Evaluating, we have

$$\frac{9}{4} \int_0^2 (1 + \cos 2\pi t) \ dt = \frac{9}{4} t \Big|_0^2 + \frac{9}{8\pi} \int_0^2 \cos 2\pi t (2\pi \ dt)$$

$$= \frac{9}{2} + \frac{9}{8\pi} \sin 2\pi t \Big|_0^2 = \frac{9}{2}$$

Thus, the root-mean-square current is

$$i_{\text{rms}} = \sqrt{\frac{9}{2}} = \frac{3\sqrt{2}}{2} = 2.12 \text{ A}$$

This value of the current, often referred to as the *effective current*, is the value of the direct current which would develop the same quantity of heat in the same time.

Exercises 27–5

In Exercises 1 through 28 integrate each of the given functions.

1. $\int \sin^2 x \cos x\, dx$

2. $\int \sin x \cos^5 x\, dx$

3. $\int \sin^3 2x\, dx$

4. $\int \cos^3 x\, dx$

5. $\int \sin^2 x \cos^3 x\, dx$

6. $\int \sin^3 x \cos^6 x\, dx$

7. $\int_0^{\pi/4} \sin^5 x\, dx$

8. $\int_{\pi/3}^{\pi/2} \sqrt{\cos x}\, \sin^3 x\, dx$

9. $\int \sin^2 x\, dx$

10. $\int \cos^2 2x\, dx$

11. $\int \cos^2 3x\, dx$

12. $\int \sin^2 4x\, dx$

13. $\int \tan^3 x\, dx$

14. $\int \cot^3 x\, dx$

15. $\int \tan x \sec^4 x\, dx$

16. $\int \cot 4x \csc^4 4x\, dx$

17. $\int \tan^4 2x\, dx$

18. $\int \cot^4 x\, dx$

19. $\int \tan^3 3x \sec^3 3x\, dx$

20. $\int \sqrt{\tan x}\, \sec^4 x\, dx$

21. $\int (\sin x + \cos x)^2\, dx$

22. $\int (\tan 2x + \cot 2x)^2\, dx$

23. $\int \dfrac{1 - \cot x}{\sin^4 x}\, dx$

24. $\int \dfrac{1 + \sin x}{\cos^4 x}\, dx$

25. $\int_{\pi/6}^{\pi/4} \cot^5 x\, dx$

26. $\int_{\pi/6}^{\pi/3} \dfrac{dx}{1 + \sin x}$

27. $\int \sec^6 x\, dx$

28. $\int \tan^7 x\, dx$

In Exercises 29 through 36 solve the given problems by integration.

29. Find the volume generated by rotating the area bounded by $y = \sin x$ and $y = 0$, from $x = 0$ to $x = \pi$, about the x-axis.

30. Find the volume generated by rotating the area bounded by $y = \tan^3(x^2)$, $y = 0$, and $x = \pi/4$ about the y-axis.

31. In determining the rate of radiation by an accelerated charge, the following integral must be evaluated: $\int_0^\pi \sin^3\theta\, d\theta$. Find the value of the integral.

32. Find the length of the curve $y = \ln \cos x$ from $x = 0$ to $x = \pi/3$. (See Exercise 17 of Section 25–6.)

33. Find the root-mean-square value of the current for one period if $i = 2 \sin t$.

34. Find the root-mean-square value of the voltage for one period if the voltage is given by $V = 100 \sin 120\pi t$.

35. For a current $i = i_0 \sin \omega t$, show that the root-mean-square value of the current for one period is $i_0/\sqrt{2}$.

36. Show that $\int \sec^2 x \tan x\, dx$ can be integrated in two ways. Explain the difference in the answers.

27–6 Inverse Trigonometric Forms

Referring to Eq. (26–10), we can find the differential of $\mathrm{Arcsin}(u/a)$, where a is a constant:

$$d\left(\mathrm{Arcsin}\,\frac{u}{a}\right) = \frac{1}{\sqrt{1 - (u/a)^2}}\frac{du}{a} = \frac{a}{\sqrt{a^2 - u^2}}\frac{du}{a} = \frac{du}{\sqrt{a^2 - u^2}}$$

Reversing this differentiation formula, we have the important integration formula:

$$\int \frac{du}{\sqrt{a^2 - u^2}} = \text{Arcsin} \frac{u}{a} + C \qquad\qquad (27\text{-}20)$$

By finding the differential of Arctan(u/a), and then reversing the equation, we derive another important integration formula:

$$d\left(\text{Arctan} \frac{u}{a}\right) = \frac{1}{1 + (u/a)^2} \frac{du}{a} = \frac{a^2}{a^2 + u^2} \frac{du}{a} = \frac{a\, du}{a^2 + u^2}$$

Thus,

$$\int \frac{du}{a^2 + u^2} = \frac{1}{a} \text{Arctan} \frac{u}{a} + C \qquad\qquad (27\text{-}21)$$

This shows one of the principal uses of the inverse trigonometric functions: they provide a solution to the integration of important algebraic functions.

Example A

Integrate: $\displaystyle\int \frac{dx}{\sqrt{9 - x^2}}$.

This integral fits the form of Eq. (27–20) with $u = x$, $du = dx$, and $a = 3$. Thus,

$$\int \frac{dx}{\sqrt{9 - x^2}} = \text{Arcsin} \frac{x}{3} + C$$

Example B

Integrate: $\displaystyle\int \frac{dx}{4x^2 + 25}$.

This integral fits the form of Eq. (27–21) with $u = 2x$, $du = 2\, dx$, and $a = 5$. Thus, we must insert a 2 in the numerator and a $\frac{1}{2}$ before the integral. Therefore,

$$\int \frac{dx}{4x^2 + 25} = \frac{1}{2} \int \frac{2\, dx}{(2x)^2 + (5)^2}$$

$$= \frac{1}{2} \cdot \frac{1}{5} \text{Arctan} \frac{2x}{5} + C$$

$$= \frac{1}{10} \text{Arctan} \frac{2x}{5} + C$$

Example C

Integrate: $\displaystyle\int_{-1}^{3} \frac{dx}{x^2 + 6x + 13}$.

At first glance it does not appear that this integral fits any of the forms presented up to this point. However, by writing the denominator in the form $(x^2 + 6x + 9) + 4 = (x + 3)^2 + 2^2$, we recognize that $u = x + 3$, $du = dx$, and $a = 2$. Thus,

$$\int_{-1}^{3} \frac{dx}{x^2 + 6x + 13} = \int_{-1}^{3} \frac{dx}{(x+3)^2 + 2^2} = \frac{1}{2} \text{Arctan} \frac{x+3}{2}\Bigg|_{-1}^{3}$$

$$= \frac{1}{2}(\text{Arctan } 3 - \text{Arctan } 1) = \frac{1}{2}(1.249 - 0.785)$$

$$= 0.232$$

Now we can see the use of completing the square when we are transforming integrals into proper form. We also see the use of inverse trigonometric functions in evaluating integrals.

Example D

Integrate: $\displaystyle\int \frac{2x + 5}{x^2 + 9} dx$.

By writing this integral as the sum of two integrals, we may integrate each of these separately:

$$\int \frac{2x + 5}{x^2 + 9} dx = \int \frac{2x \, dx}{x^2 + 9} + \int \frac{5 \, dx}{x^2 + 9}$$

The first of these integrals is seen to be a logarithmic form, and the second the inverse tangent form. For the first, $u = x^2 + 9$, $du = 2x \, dx$. For the second, $u = x$, $du = dx$, $a = 3$. Thus,

$$\int \frac{2x \, dx}{x^2 + 9} + 5 \int \frac{dx}{x^2 + 9} = \ln|x^2 + 9| + \frac{5}{3}\text{Arctan}\frac{x}{3} + C$$

The inverse trigonometric integral forms of this section show very well the importance of proper recognition of the form of the integral. It is important that these forms are not confused with those of the general power rule or the logarithmic form. Consider the following example.

Example E

The integral $\displaystyle\int \frac{dx}{\sqrt{1 - x^2}}$ is of the inverse sine form, with $u = x$, $du = dx$, and $a = 1$. Thus,

$$\int \frac{dx}{\sqrt{1 - x^2}} = \text{Arcsin } x + C$$

The integral $\int \dfrac{x\,dx}{\sqrt{1-x^2}}$ is not of the inverse sine form due to the factor of x in the numerator. It is integrated by use of the general power rule, with $u = 1 - x^2$, $du = -2x\,dx$, and $n = -\frac{1}{2}$. Thus,

$$\int \frac{x\,dx}{\sqrt{1-x^2}} = -\sqrt{1-x^2} + C$$

In the same way, the integral $\int \dfrac{dx}{1+x^2}$ is of the inverse tangent form, whereas the integral $\int \dfrac{x\,dx}{1+x^2}$ is of logarithmic form due to the factor of x.

Exercises 27–6

In Exercises 1 through 24 integrate each of the given functions.

1. $\int \dfrac{dx}{\sqrt{4-x^2}}$

2. $\int \dfrac{dx}{\sqrt{49-x^2}}$

3. $\int \dfrac{dx}{64+x^2}$

4. $\int \dfrac{dx}{4+x^2}$

5. $\int \dfrac{dx}{\sqrt{1-16x^2}}$

6. $\int \dfrac{dx}{\sqrt{9-4x^2}}$

7. $\int_0^2 \dfrac{dx}{1+9x^2}$

8. $\int_1^3 \dfrac{dx}{49+4x^2}$

9. $\int \dfrac{2\,dx}{\sqrt{4-5x^2}}$

10. $\int \dfrac{4x\,dx}{\sqrt{3-2x^2}}$

11. $\int \dfrac{8x\,dx}{9x^2+16}$

12. $\int \dfrac{3\,dx}{25+16x^2}$

13. $\int_1^2 \dfrac{dx}{5x^2+7}$

14. $\int_0^1 \dfrac{x\,dx}{1+x^4}$

15. $\int \dfrac{e^x\,dx}{\sqrt{1-e^{2x}}}$

16. $\int \dfrac{\sec^2 x\,dx}{\sqrt{1-\tan^2 x}}$

17. $\int \dfrac{dx}{x^2+2x+2}$

18. $\int \dfrac{2\,dx}{x^2+8x+17}$

19. $\int \dfrac{4\,dx}{\sqrt{-4x-x^2}}$

20. $\int \dfrac{dx}{\sqrt{2x-x^2}}$

21. $\int_{\pi/6}^{\pi/2} \dfrac{\cos 2x}{1+\sin^2 2x}\,dx$

22. $\int_{-4}^0 \dfrac{dx}{x^2+4x+5}$

23. $\int \dfrac{2-x}{\sqrt{4-x^2}}\,dx$

24. $\int \dfrac{3-2x}{1+4x^2}\,dx$

In Exercises 25 through 32 solve the given problems by integration.

25. Find the area bounded by the curve $y(1+x^2) = 1$, $x = 0$, $y = 0$, and $x = 2$.

26. Find the area bounded by $y^2(4-x^2) = 1$, $x = 0$, $y = 0$, and $x = 1$.

27. Find the volume generated by rotating the area bounded by $y = 1/\sqrt{16+x^2}$, $x = 0$, $y = 0$, and $x = 3$ about the x-axis.

28. The electric current in a certain inductor is given by $i = 8\int \dfrac{dt}{100+t^2}$. Determine the function for i in terms of t.

29. In dealing with the theory for simple harmonic motion, it is necessary to solve the following equation:

$$\frac{dx}{\sqrt{A^2 - x^2}} = \sqrt{\frac{k}{m}}\, dt \quad \text{with } k, m, \text{ and } A \text{ as constants}$$

Determine the solution to this equation if $x = x_0$ when $t = 0$.

30. Determine the function of distance in terms of time if the velocity of an object is given by $v = 1/(4 + t^2)$ and $s = 0$ when $t = 0$.

31. Find the moment of inertia with respect to the y-axis for the area bounded by $y = 1/(1 + x^6)$, the x-axis, $x = 1$, and $x = 2$.

32. Find the length of arc along the curve $y = \sqrt{1 - x^2}$ between $x = 0$ and $x = 1$. (See Exercise 17 of Section 25–6).

27–7 Integration by Parts

There are many methods of transforming integrals into a form which can be integrated by one of the basic formulas. In the preceding section we saw that completing the square and trigonometric identities can be used for this purpose. In this section and the following one, we shall develop two general methods. The method of integration by parts is discussed in this section.

Since the derivative of a product of functions is found by use of the formula

$$\frac{d(uv)}{dx} = u\frac{dv}{dx} + v\frac{du}{dx}$$

the differential of a product of functions is given by $d(uv) = u\, dv + v\, du$. Integrating both sides of this equation we have $uv = \int u\, dv + \int v\, du$. Solving for $\int u\, dv$, we obtain

$$\int u\, dv = uv - \int v\, du \tag{27–22}$$

Integration by use of Eq. (27–22) is called **integration by parts.**

Example A

Integrate: $\int x \sin x\, dx$.

This integral does not fit any of the previous forms which have been discussed, since neither x nor $\sin x$ can be made a factor of a proper du. However, by choosing $u = x$, and $dv = \sin x\, dx$, the above formula may be used. Thus,

$$u = x, \qquad dv = \sin x\, dx$$
$$du = dx, \qquad v = -\cos x$$

(The constant of integration will be included in the final result.) Hence,

$$\int (x)(\sin x\, dx) = -x \cos x - \int (-\cos x)\, dx$$
$$= -x \cos x + \sin x + C$$

Other choices of u and dv may be made in the preceding example, but they do not prove useful. If $u = \sin x$ and $dv = x\ dx$, v becomes $x^2/2$, which makes the integral $\int v\ du$ more complex than the original problem. The choice of $u = x \sin x$ makes the expression for du/dx more complex than u itself. There are no set rules which may be stated for the best choice of u and dv, but there are two guidelines which may be stated: (1) *The quantity u is normally chosen such that du/dx is of simpler form than u.* (2) *The differential dv is normally chosen such that $\int dv$ is easily obtained.* Working examples, and thereby gaining experience in methods of integrating, is the best way to determine when this method should be used and how to employ it.

Example B
Integrate: $\int_0^1 xe^{-x}dx$. Let $u = x$, $dv = e^{-x}dx$, $du = dx$, $v = -e^{-x}$.
We write

$$\int_0^1 xe^{-x}dx = -xe^{-x}\Big|_0^1 + \int_0^1 e^{-x}dx = -xe^{-x} - e^{-x}\Big|_0^1$$

$$= -e^{-1} - e^{-1} + 1$$

$$= 1 - \frac{2}{e}$$

Example C
Integrate: $\int \text{Arcsin } x\ dx$. Let $u = \text{Arcsin } x$, $dv = dx$, $du = \dfrac{dx}{\sqrt{1 - x^2}}$, $v = x$.
We obtain

$$\int \text{Arcsin } x\ dx = x \text{ Arcsin } x - \int \frac{x\ dx}{\sqrt{1 - x^2}}$$

$$= x \text{ Arcsin } x + \frac{1}{2}\int \frac{-2x\ dx}{\sqrt{1 - x^2}}$$

$$= x \text{ Arcsin } x + \sqrt{1 - x^2} + C$$

Example D
Integrate: $\int \sqrt{x} \ln x\ dx$. Let $u = \ln x$, $dv = x^{1/2}dx$, $du = \dfrac{1}{x}dx$, $v = \frac{2}{3}x^{3/2}$.
We write

$$\int \sqrt{x} \ln x\ dx = \frac{2}{3}x^{3/2}\ln x - \frac{2}{3}\int x^{1/2}dx$$

$$= \frac{2}{3}x^{3/2}\ln x - \frac{4}{9}x^{3/2} + C$$

Example E

Integrate: $\int e^x \sin x \, dx$. Let $u = \sin x$, $dv = e^x dx$, $du = \cos x \, dx$, $v = e^x$. We write

$$\int e^x \sin x \, dx = e^x \sin x - \int e^x \cos x \, dx$$

At first glance it appears that we have made no progress in applying the method of integration by parts. We note, however, that when we integrated $\int e^x \sin x \, dx$, part of the result was a term of $\int e^x \cos x \, dx$. This implies that if $\int e^x \cos x \, dx$ were integrated, a term of $\int e^x \sin x \, dx$ might result. Thus, the method of integration by parts is now applied to the integral $\int e^x \cos x \, dx$:

$$u = \cos x, \qquad dv = e^x dx, \qquad du = -\sin x \, dx, \qquad v = e^x$$

And so $\int e^x \cos x \, dx = e^x \cos x + \int e^x \sin x \, dx$. Substituting this expression into the expression for $\int e^x \sin x \, dx$, we obtain

$$\int e^x \sin x \, dx = e^x \sin x - \left(e^x \cos x + \int e^x \sin x \, dx \right)$$

$$= e^x \sin x - e^x \cos x - \int e^x \sin x \, dx$$

$$2 \int e^x \sin x \, dx = e^x (\sin x - \cos x) + 2C$$

$$\int e^x \sin x \, dx = \frac{e^x}{2} (\sin x - \cos x) + C$$

Thus, by combining integrals of like form, we obtain the desired result.

Exercises 27–7

In Exercises 1 through 16 integrate each of the given functions.

1. $\int x \cos x \, dx$

2. $\int x \sin 2x \, dx$

3. $\int x e^{2x} dx$

4. $\int x e^x dx$

5. $\int x \sec^2 x \, dx$

6. $\int_0^{\pi/4} x \sec x \tan x \, dx$

7. $\int \text{Arctan } x \, dx$

8. $\int \ln x \, dx$

9. $\int \frac{x \, dx}{\sqrt{1-x}}$

10. $\int x \sqrt{x+1} \, dx$

11. $\int x \ln x \, dx$

12. $\int x^2 \ln 4x \, dx$

13. $\int x^2 \sin 2x \, dx$

14. $\int x^2 e^{2x} dx$

15. $\int_0^{\pi/2} e^x \cos x \, dx$

16. $\int e^{-x} \sin 2x \, dx$

In Exercises 17 through 24 solve the given problems by integration.

17. Find the area bounded by $y = xe^{-x}$, $y = 0$, and $x = 2$.

18. Find the volume generated by rotating the area bounded by $y = \sin x$ and $y = 0$ (from $x = 0$ to $x = \pi$) about the y-axis.

19. Find the x-coordinate of the centroid of the area bounded by $y = \cos x$ and $y = 0$ for $0 \leq x \leq \pi/2$.

20. Find the moment of inertia with respect to its axis of the volume generated by rotating the area bounded by $y = e^x$, $x = 1$, and the coordinate axes about the y-axis.

21. Find the root-mean-square value of the function $y = \sqrt{\text{Arcsin } x}$ between $x = 0$ and $x = 1$. (See Example H of Section 27–5.)

22. The general expression for the slope of a curve is $dy/dx = x^3\sqrt{1 + x^2}$. Find the equation of the curve if it passes through the origin.

23. The current in a given circuit is given by $i = e^{-2t}\cos t$. Find an expression for the amount of charge which passes a given point in the circuit as a function of the time, if $q_0 = 0$.

24. A particle moves such that its velocity v is given by $v = t\sqrt{t + 1}$. Find the expression for the displacement s as a function of time if $s = 0$ when $t = 0$.

27–8 Integration by Trigonometric Substitution

Trigonometric relations not only provide a means of transforming trigonometric integrals, but they can also be used to transform algebraic integrals. Substitutions based on Eqs. (27–14), (27–15), and (27–16) prove to be particularly useful for integrals involving radicals. The method is illustrated in the following examples.

Example A

Integrate: $\displaystyle\int \frac{dx}{x^2\sqrt{1 - x^2}}$.

If we let $x = \sin \theta$, the radical becomes $\sqrt{1 - \sin^2\theta} = \cos \theta$. Therefore, by making this substitution, the integral can be transformed into a trigonometric integral. We must be careful to replace all factors of the integral by proper expressions in terms of θ; $x = \sin \theta$, $dx = \cos \theta \, d\theta$. And so we obtain

$$\int \frac{dx}{x^2\sqrt{1 - x^2}} = \int \frac{\cos \theta \, d\theta}{\sin^2\theta\sqrt{1 - \sin^2\theta}} = \int \frac{\cos \theta \, d\theta}{\sin^2\theta \, \cos \theta} = \int \csc^2\theta \, d\theta$$

$$= -\cot \theta + C$$

We have now performed the integration, but the answer is expressed in terms of a variable different from the original integral. We must express the result in terms of x. Making a right triangle with an angle θ such that

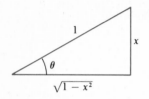

Figure 27–1

$\sin \theta = x/1$ (see Fig. 27–1), we may express any of the trigonometric functions in terms of x. (This is the method used with inverse trigonometric functions.) Thus,

$$\cot \theta = \frac{\sqrt{1 - x^2}}{x}$$

Therefore, the result of the integration becomes

$$\int \frac{dx}{x^2 \sqrt{1 - x^2}} = -\cot \theta + C = -\frac{\sqrt{1 - x^2}}{x} + C$$

Example B

Integrate: $\displaystyle\int \frac{dx}{\sqrt{x^2 + 4}}$. Let $x = 2 \tan \theta$, $dx = 2 \sec^2\theta \; d\theta$.

We write

$$\int \frac{dx}{\sqrt{x^2 + 4}} = \int \frac{2 \sec^2\theta \; d\theta}{\sqrt{4 \tan^2\theta + 4}}$$

$$= \int \frac{2 \sec^2\theta \; d\theta}{2\sqrt{\tan^2\theta + 1}} = \int \frac{2 \sec^2\theta \; d\theta}{2 \sec \theta}$$

$$= \int \sec \theta \; d\theta = \ln |\sec \theta + \tan \theta| + C$$

$$= \ln \left| \frac{\sqrt{x^2 + 4}}{2} + \frac{x}{2} \right| + C$$

$$= \ln \left| \frac{\sqrt{x^2 + 4} + x}{2} \right| + C$$

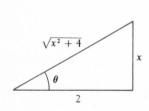

Figure 27–2

See Fig. 27–2. This answer is acceptable, but there is another form. By using the properties of logarithms, we have

$$\ln \left| \frac{\sqrt{x^2 + 4} + x}{2} \right| + C = \ln |\sqrt{x^2 + 4} + x| + (C - \ln 2)$$
$$= \ln | \sqrt{x^2 + 4} + x| + C'$$

Here C' is another arbitrary constant, which combines all constants of the previous expression. Combining constants in this manner is a common practice in integration problems.

Example C

Integrate: $\displaystyle\int \frac{dx}{x\sqrt{x^2 - 9}}$. Let $x = 3 \sec \theta$, $dx = 3 \sec \theta \tan \theta \; d\theta$.

$$\int \frac{dx}{x\sqrt{x^2 - 9}} = \int \frac{3 \sec \theta \tan \theta \; d\theta}{3 \sec \theta \sqrt{9 \sec^2\theta - 9}}$$

$$= \int \frac{\tan \theta \, d\theta}{3\sqrt{\sec^2\theta - 1}} = \int \frac{\tan \theta \, d\theta}{3 \tan \theta}$$

$$= \frac{1}{3} \int d\theta = \frac{1}{3}\theta + C = \frac{1}{3}\text{Arcsec}\frac{x}{3} + C$$

In this solution it is not necessary to refer to a triangle to express the integral in terms of x. This solution is found by solving $x = 3 \sec \theta$ for θ, as indicated.

From the preceding examples we can see that by making the proper substitution in terms of a trigonometric function, we can solve integrals of algebraic functions by means of simpler equivalent trigonometric forms, due to the properties of the square relations of the trigonometric functions. In summary, for the indicated radical form the following trigonometric substitutions are used.

For	$\sqrt{a^2 - x^2}$	use	$x = a \sin \theta$
For	$\sqrt{a^2 + x^2}$	use	$x = a \tan \theta$
For	$\sqrt{x^2 - a^2}$	use	$x = a \sec \theta$

Example D

Evaluate: $\displaystyle\int_{1}^{4} \frac{\sqrt{9 + x^2}}{x} dx.$

Since we have the form $\sqrt{a^2 + x^2}$, where $a = 3$, we make the substitution $x = 3 \tan \theta$. This means that $dx = 3 \sec^2\theta \, d\theta$. Thus,

$$\int \frac{\sqrt{9 + x^2}}{x} dx = \int \frac{\sqrt{9 + 9 \tan^2\theta}}{3 \tan \theta} (3 \sec^2\theta \, d\theta)$$

$$= 3 \int \frac{\sqrt{1 + \tan^2\theta}}{\tan \theta} \sec^2\theta \, d\theta = 3 \int \frac{\sec^3\theta}{\tan \theta} d\theta$$

$$= 3 \int \frac{1 + \tan^2\theta}{\tan \theta} \sec \theta \, d\theta$$

$$= 3\left(\int \frac{\sec \theta}{\tan \theta} d\theta + \int \tan \theta \sec \theta \, d\theta \right)$$

$$= 3\left(\int \csc \theta \, d\theta + \int \tan \theta \sec \theta \, d\theta \right)$$

$$= 3[\ln |\csc \theta - \cot \theta| + \sec \theta]$$

$$= 3\left[\ln \left| \frac{\sqrt{x^2 + 9}}{x} - \frac{3}{x} \right| + \frac{\sqrt{x^2 + 9}}{3} \right]$$

$$= 3\left[\ln \left| \frac{\sqrt{x^2 + 9} - 3}{x} \right| + \frac{\sqrt{x^2 + 9}}{3} \right]$$

Limits have not been included, due to the changes in variables. The actual evaluation may now be completed:

$$\int_1^4 \frac{\sqrt{9+x^2}}{x}\,dx = 3\left[\ln\left|\frac{\sqrt{x^2+9}-3}{x}\right| + \frac{\sqrt{x^2+9}}{3}\right]_1^4$$

$$= 3\left[\left(\ln\frac{1}{2} - \ln\frac{\sqrt{10}-3}{1}\right) + \left(\frac{5}{3} - \frac{\sqrt{10}}{3}\right)\right]$$

$$= 3\left[\ln\frac{1}{2(\sqrt{10}-3)} + \frac{5-\sqrt{10}}{3}\right]$$

$$= 3\ln\frac{\sqrt{10}+3}{2} + \frac{5-\sqrt{10}}{1} = 5.21$$

Exercises 27–8

In Exercises 1 through 16 integrate each of the given functions.

1. $\displaystyle\int \frac{\sqrt{1-x^2}}{x^2}\,dx$

2. $\displaystyle\int \frac{dx}{(x^2+9)^{3/2}}$

3. $\displaystyle\int \frac{dx}{\sqrt{x^2-4}}$

4. $\displaystyle\int \frac{\sqrt{x^2-25}}{x}\,dx$

5. $\displaystyle\int \frac{dx}{x^2\sqrt{x^2+9}}$

6. $\displaystyle\int \frac{dx}{x\sqrt{4-x^2}}$

7. $\displaystyle\int \frac{dx}{(4-x^2)^{3/2}}$

8. $\displaystyle\int \frac{x^3\,dx}{\sqrt{9+x^2}}$

9. $\displaystyle\int_0^{0.5} \frac{x^3\,dx}{\sqrt{1-x^2}}$

10. $\displaystyle\int_4^5 \frac{\sqrt{x^2-16}}{x^2}\,dx$

11. $\displaystyle\int \frac{dx}{\sqrt{x^2+2x+2}}$

12. $\displaystyle\int \frac{dx}{\sqrt{x^2+2x}}$

13. $\displaystyle\int \frac{dx}{x\sqrt{4x^2-9}}$

14. $\displaystyle\int \sqrt{16-x^2}\,dx$

15. $\displaystyle\int \frac{dx}{\sqrt{e^{2x}-1}}$

16. $\displaystyle\int \frac{\sec^2 x\,dx}{(4-\tan^2 x)^{3/2}}$

In Exercises 17 through 22 solve the given problems by integration.

17. Find the area of a circle of radius 1 by integration.

18. Find the area bounded by $y = \dfrac{1}{x^2\sqrt{x^2-1}}$, $x = \sqrt{2}$, $x = \sqrt{5}$, and $y = 0$.

19. Find the moment of inertia with respect to the y-axis of the first-quadrant area under the circle $x^2 + y^2 = a^2$ in terms of its mass.

20. Find the moment of inertia of a sphere of radius a with respect to its axis in terms of its mass.

21. Find the volume generated by rotating the area bounded by $y = \dfrac{\sqrt{x^2-16}}{x^2}$, $y = 0$, and $x = 5$ about the y-axis.

22. Find the length of arc along the curve of $y = \ln x$ from $x = 1$ to $x = 3$. (See Exercise 17 of Section 25–6.)

27–9 Integration by Use of Tables

In this chapter we have introduced certain basic integrals, and have also brought in some methods of reducing other integrals to these basic forms. Often this transformation and integration require a number of steps to be performed, and therefore integrals are tabulated for reference. The integrals found in tables have been derived by using the methods introduced thus far, as well as many other methods which can be used. Therefore, an understanding of the basic forms and some of the basic methods is very useful in finding integrals from tables. Such an understanding forms a basis for proper recognition of the forms which are used in the tables, as well as the types of results which may be expected. Therefore, *the use of the tables depends on the proper recognition of the form, and the variables and constants of the integral.* The following examples illustrate the use of the table of integrals found in Table 8 in Appendix E.

Example A

Integrate: $\displaystyle \int \frac{x \, dx}{\sqrt{2 + 3x}}$.

We first note that this integral fits the form of formula 6 of Table 8, with $u = x$, $a = 2$, and $b = 3$. Therefore,

$$\int \frac{x \, dx}{\sqrt{2 + 3x}} = -\frac{2(4 - 3x)\sqrt{2 + 3x}}{27} + C$$

Example B

Integrate: $\displaystyle \int \frac{\sqrt{4 - 9x^2}}{x} \, dx$.

This fits the form of formula 18, with proper identification of constants; $u = 3x$, $du = 3 \, dx$, $a = 2$. Hence,

$$\int \frac{\sqrt{4 - 9x^2}}{x} \, dx = \int \frac{\sqrt{4 - 9x^2}}{3x} 3 \, dx$$

$$= \sqrt{4 - 9x^2} - 2 \ln\left(\frac{2 + \sqrt{4 - 9x^2}}{3x}\right) + C$$

Example C

Integrate: $\int \sec^3 2x \, dx$.

This fits the form of formula 37; $n = 3$, $u = 2x$, $du = 2 \, dx$. And so

$$\int \sec^3 2x \, dx = \frac{1}{2} \int \sec^3 2x (2 \, dx)$$

$$= \frac{1}{2} \frac{\sec 2x \tan 2x}{2} + \frac{1}{2} \cdot \frac{1}{2} \int \sec 2x (2 \, dx)$$

[To complete this integral, we must use the basic form of Eq. (27–12).] Thus, we complete it by

$$\int \sec^3 2x \, dx = \frac{\sec 2x \tan 2x}{4} + \frac{1}{4} \ln |\sec 2x + \tan 2x| + C$$

Example D
Evaluate: $\int_1^e x^2 \ln 2x \, dx$.

This integral fits the form of formula 46 if $u = 2x$. Thus, we have

$$\frac{1}{8} \int_1^e (2x)^2 \ln 2x(2 \, dx) = \frac{1}{8} (2x)^3 \left[\frac{\ln 2x}{3} - \frac{1}{9} \right]_1^e$$

$$= e^3 \left(\frac{\ln 2e}{3} - \frac{1}{9} \right) - \left(\frac{\ln 2}{3} - \frac{1}{9} \right)$$

$$= e^3 \left(\frac{\ln 2}{3} + \frac{\ln e}{3} - \frac{1}{9} \right) - \left(\frac{\ln 2}{3} - \frac{1}{9} \right)$$

$$= \frac{\ln 2}{3} (e^3 - 1) + \frac{1}{9} (2e^3 + 1) = 8.99$$

We summarize briefly here the approach to integrating a function. There are two methods of approaching a problem in order to obtain the exact result for the integral.

(1) Write the integral such that it fits an integral form. Either a basic form as developed in this chapter, or a form from a table of integrals may be used.

(2) Use a technique of transforming the integral such that an integral form may be used. Techniques include integration by parts, trigonometric substitution, and others which may be found in other sources.

Also, definite integrals may be approximated by methods such as the trapezoidal rule. There are other approximation methods, one of which is developed in the next chapter.

Exercises 27–9

In Exercises 1 through 32 integrate each function by use of Table 8 in Appendix E.

1. $\int \dfrac{3x \, dx}{2 + 5x}$

2. $\int \dfrac{4x \, dx}{(1 + x)^2}$

3. $\int 5x \sqrt{2 + 3x} \, dx$

4. $\int \dfrac{dx}{x^2 - 4}$

5. $\int \sqrt{4 - x^2} \, dx$

6. $\int_0^{\pi/3} \sin^3 x \, dx$

7. $\int \sin 2x \sin 3x \, dx$

8. $\int \text{Arcsin } 3x \, dx$

9. $\int \dfrac{\sqrt{4x^2 - 9}}{x} dx$

10. $\int \dfrac{(9x^2 + 16)^{3/2}}{x} dx$

11. $\int \cos^5 4x \, dx$

12. $\int \tan^2 x \, dx$

13. $\int \text{Arctan } x^2 (x \, dx)$

14. $\int xe^{4x} dx$

15. $\int_1^2 (4 - x^2)^{3/2} dx$

16. $\int \dfrac{dx}{9 - 16x^2}$

17. $\int \dfrac{dx}{x\sqrt{4x^2 + 1}}$

18. $\int \dfrac{\sqrt{4 + x^2}}{x} dx$

19. $\displaystyle \int \frac{dx}{x\sqrt{1-4x^2}}$ **20.** $\displaystyle \int \frac{dx}{x(1+4x)^2}$ **21.** $\displaystyle \int \sin x \cos 5x \, dx$

22. $\displaystyle \int_0^2 x^2 e^{3x} dx$ **23.** $\displaystyle \int x^5 \cos x^3 \, dx$ **24.** $\displaystyle \int \sin^3 x \cos^2 x \, dx$

25. $\displaystyle \int \frac{x \, dx}{(1-x^4)^{3/2}}$ **26.** $\displaystyle \int \frac{dx}{x(1-4x)}$ **27.** $\displaystyle \int_1^3 \frac{\sqrt{3+5x^2}}{x} dx$

28. $\displaystyle \int_0^1 \frac{\sqrt{9-4x^2}}{x} dx$ **29.** $\displaystyle \int x^3 \ln x^2 \, dx$ **30.** $\displaystyle \int \frac{x \, dx}{x^2 \sqrt{x^4-9}}$

31. $\displaystyle \int \frac{x^2 \, dx}{(x^6-1)^{3/2}}$ **32.** $\displaystyle \int x^7 \sqrt{x^4+4} \, dx$

In Exercises 33 through 38 evaluate the integrals by use of Table 8.

33. Find the length of arc of the curve $y = x^2$ from $x = 0$ to $x = 1$. (See Exercise 17 of Section 25–6.)

34. Find the moment of inertia with respect to its axis of the volume generated by rotating the area bounded by $y = \ln x$, $x = e$, and the x-axis about the y-axis.

35. Find the force on the area bounded by $x = 1/\sqrt{1+y}$, $y = 0$, $y = 3$, and the y-axis, if the surface of the water is at the upper edge of the area. (See Exercise 19 of Section 25–6.)

36. Under certain conditions the velocity as a function of time for a particle is given by $v = t/\sqrt{1+2t}$. Determine the displacement as a function of time if $s = 6$ when $t = 0$.

37. Find the volume of the solid generated by rotating the area bounded by $y = e^x \sin x$ and the x-axis between $x = 0$ and $x = \pi$ about the x-axis.

38. The electric current in a certain circuit as a function of time is given by $i = 2/\sqrt{t^2+100}$. Find the expression for the charge which passes a given point if $q = 0$ when $t = 0$.

27–10 Exercises for Chapter 27

In Exercises 1 through 32 integrate the given functions without the use of Table 8.

1. $\displaystyle \int e^{-2x} dx$ **2.** $\displaystyle \int e^{\cos 2x} \sin x \cos x \, dx$ **3.** $\displaystyle \int \frac{dx}{x(\ln 2x)^2}$

4. $\displaystyle \int x^{1/3} \sqrt{x^{4/3}+1} \, dx$ **5.** $\displaystyle \int \frac{\cos x \, dx}{1+\sin x}$ **6.** $\displaystyle \int \frac{\sec^2 x \, dx}{2+\tan x}$

7. $\displaystyle \int \frac{2 \, dx}{25+49x^2}$ **8.** $\displaystyle \int \frac{dx}{\sqrt{1-4x^2}}$ **9.** $\displaystyle \int_0^{\pi/2} \cos^3 2x \, dx$

10. $\displaystyle \int_0^{\pi/8} \sec^3 2x \tan 2x \, dx$ **11.** $\displaystyle \int_0^2 \frac{x \, dx}{4+x^2}$ **12.** $\displaystyle \int_1^e \frac{\ln x^2 \, dx}{x}$

13. $\displaystyle\int \sec^4 3x \, \tan 3x \, dx$

14. $\displaystyle\int \frac{\sin^3 x \, dx}{\sqrt{\cos x}}$

15. $\displaystyle\int \tan 3x \, dx$

16. $\displaystyle\int \sec 4x \, dx$

17. $\displaystyle\int \sec^4 3x \, dx$

18. $\displaystyle\int \frac{\sin^2 2x \, dx}{1 + \cos 2x}$

19. $\displaystyle\int \frac{dx}{\sqrt{4x^2 - 9}}$

20. $\displaystyle\int \frac{x^2 \, dx}{\sqrt{9 - x^2}}$

21. $\displaystyle\int \frac{e^{2x} dx}{\sqrt{e^{2x} + 1}}$

22. $\displaystyle\int \frac{(4 + \ln 2x)^3 dx}{x}$

23. $\displaystyle\int \sin^2 3x \, dx$

24. $\displaystyle\int \sin^4 x \, dx$

25. $\displaystyle\int x \csc^2 2x \, dx$

26. $\displaystyle\int x \, \text{Arctan} \, x \, dx$

27. $\displaystyle\int e^{2x} \cos e^{2x} dx$

28. $\displaystyle\int \frac{dx}{x^2 + 6x + 10}$

29. $\displaystyle\int_1^e \frac{(\ln x)^2 dx}{x}$

30. $\displaystyle\int_1^3 \frac{dx}{x^2 - 2x + 5}$

31. $\displaystyle\int \frac{x^2 - 1}{x + 2} dx$

32. $\displaystyle\int [\cos(\ln x)] \frac{dx}{x}$

In Exercises 33 through 44 solve the given problems by integration.

33. The change in the thermodynamic entity of entropy ΔS may be expressed as $\Delta S = \int (c_v / T) \, dT$, where c_v is the heat capacity at constant volume and T is the temperature. For increased accuracy, c_v is often given by the equation $c_v = a + bT + cT^2$, where a, b, and c are constants. Express ΔS as a function of temperature.

34. A second-order chemical reaction leads to the equation

$$dt = \frac{k_1 \, dx}{a - x} + \frac{k_2 \, dx}{b - x}$$

where k_1 and k_2 are constants, a and b are initial concentrations, t is the time, and x is the decrease in concentration. Solve for t as a function of x.

35. Assuming a resisting force, the velocity (in feet per second) of a falling object in terms of time (in seconds) is given by

$$\frac{dv}{32 - 0.1v} = dt$$

If $v = 0$ when $t = 0$, find v as a function of t.

36. Find the area bounded by $y = x/(1 + x)^2$, the x-axis, and the line $x = 4$.

37. Find the equation of the curve for which $dy/dx = \sec^4 x$, if the curve passes through the origin.

38. Find the distance from the origin after $\frac{1}{2}$ second of the particle for which the velocity as a function of time is given by $v = 4 \cos 3t$. (That is, integrate the function with limits of 0 and $\frac{1}{2}$, or substitute into the function derived in Exercise 29 of Section 27–4). Compare the results of these two methods.)

39. Find the area inside the circle $x^2 + y^2 = 25$ and to the right of the line $x = 3$.

40. Find the area bounded by $y = x\sqrt{x + 4}$, $y = 0$, and $x = 5$.

41. Find the volume generated by rotating the area bounded by $y = xe^x$, $y = 0$, and $x = 2$ about the y-axis.

42. Find the volume generated by rotating about the y-axis the area bounded by $y = x + \sqrt{x + 1}$, $x = 3$, and the axes.

43. Find the centroid of the area bounded by $y = \ln x$, $x = 2$, and the x-axis.

44. The power delivered to an electric circuit is given by $P = ei$, where e and i are, respectively, the instantaneous voltage and the instantaneous current in the circuit. The mean power, averaged over a period $2\pi/\omega$, is given by

$$P_{av} = \frac{\omega}{2\pi} \int_0^{2\pi/\omega} ei \, dt$$

If $e = 20 \cos 2t$ and $i = 3 \sin 2t$, find the average power over a period of $\pi/4$.

28

Expansion of Functions in Series

28—1 Maclaurin Series

The values of the trigonometric functions can be determined exactly only for a few certain angles. The value of e can only be approximated in decimal form. The question arises as to how these values may be determined, particularly if a specified degree of accuracy is necessary. In this chapter we shall show how a given function may be expressed in terms of a polynomial. Once the polynomial which approximates a given function has been found, we will be able to evaluate the function to any desired accuracy. Also, a number of other uses of this polynomial form will be shown.

Example A

By using long division, we have

$$\frac{2}{2-x} = 1 + \frac{1}{2}x + \frac{1}{4}x^2 + \cdots + \left(\frac{1}{2}x\right)^{n-1} + \cdots \tag{1}$$

where n is the number of the term of the expression on the right. Since x represents a number, the right-hand side of Eq. (1) becomes a geometric progression of infinitely many terms.

From Chapter 18 we know that the sum of a geometric progression of infinitely many terms is

$$S = \frac{a}{1-r} \tag{2}$$

where a is the first term and r is the common ratio of successive terms.

If $x = 1$, the righthand side of Eq. (1) is

$$1 + \frac{1}{2} + \frac{1}{4} + \cdots + \left(\frac{1}{2}\right)^{n-1}$$

For this progression, $r = \frac{1}{2}$ and $a = 1$, which means $S = 2$. If $x = 3$, the righthand side of Eq. (1) is

$$1 + \frac{3}{2} + \frac{9}{4} + \cdots + \left(\frac{3}{2}\right)^{n-1} + \cdots$$

and this we know does not have a defined value, since $r > 1$. Referring back to the left side of Eq. (1), we see that it also equals 2 when $x = 1$. If $x = 3$, the left side is -2. Thus, we see that the two sides agree for $x = 1$, and do not give consistent results for $x = 3$. In fact, as long as $|x| < 2$, the values will agree. From this we can conclude that the expression on the righthand side properly represents that on the left side, as long as $|x| < 2$.

From Example A, we see that an algebraic function may be properly represented by a function of the form

$$f(x) = a_0 + a_1 x + a_2 x^2 + \cdots + a_n x^n + \cdots \tag{28–1}$$

Equation (28–1) is known as a **power-series expansion** *of the function* $f(x)$. The problem now arises as to whether or not functions in general may be represented in this form. If such a representation were possible, it would provide a means of evaluating the transcendental functions for the purpose of making tables of values. Also, since a power-series expansion is in the form of a polynomial, it makes algebraic operations much simpler due to the properties of polynomials. A further study of calculus shows many other uses of power series.

At this point we shall assume that unless otherwise noted the functions with which we shall be dealing may be properly represented by a power-series expansion (it takes more advanced methods to prove that this is generally possible), for specified ranges of values of x. We shall find that the methods of calculus are very useful in developing the method of general representation. Thus, writing a general power series, along with the first few derivatives, we have

$$
\begin{aligned}
f(x) &= a_0 + a_1 x + a_2 x^2 + a_3 x^3 + a_4 x^4 + a_5 x^5 + \cdots + a_n x^n + \cdots \\
f'(x) &= a_1 + 2a_2 x + 3a_3 x^2 + 4a_4 x^3 + 5a_5 x^4 + \cdots + na_n x^{n-1} + \cdots \\
f''(x) &= 2a_2 + 2\cdot 3 a_3 x + 3\cdot 4 a_4 x^2 + 4\cdot 5 a_5 x^3 + \cdots \\
&\qquad\qquad\qquad\qquad\qquad\qquad + (n-1)na_n x^{n-2} + \cdots
\end{aligned}
$$

$$f'''(x) = 2 \cdot 3a_3 + 2 \cdot 3 \cdot 4a_4 x + 3 \cdot 4 \cdot 5a_5 x^2 + \cdots$$
$$+ (n - 2)(n - 1)na_n x^{n-3} + \cdots$$
$$f^{iv}(x) = 2 \cdot 3 \cdot 4a_4 + 2 \cdot 3 \cdot 4 \cdot 5a_5 x + \cdots$$
$$+ (n - 3)(n - 2)(n - 1)na_n x^{n-4} + \cdots$$

Regardless of the values of the constants a_n for any power series, if $x = 0$, the left and right sides must be equal, and all the terms on the right are zero except the first. Thus, setting $x = 0$ in each of the above equations, we have

$$f(0) = a_0 \qquad\qquad f'(0) = a_1$$
$$f''(0) = 2a_2 \qquad\qquad f'''(0) = 2 \cdot 3a_3$$
$$f^{iv}(0) = 2 \cdot 3 \cdot 4a_4$$

Solving each of these for the constants a_n, we have

$$a_0 = f(0) \qquad a_1 = f'(0) \qquad a_2 = \frac{f''(0)}{2!} \qquad a_3 = \frac{f'''(0)}{3!} \qquad a_4 = \frac{f^{iv}(0)}{4!}$$

Substituting these back into the expression for $f(x)$, we have

$$f(x) = f(0) + f'(0)x + \frac{f''(0)x^2}{2!} + \frac{f'''(0)x^3}{3!} + \cdots + \frac{f^{(n)}(0)x^n}{n!} + \cdots \qquad (28\text{--}2)$$

Equation (28–2) is known as the **Maclaurin series expansion** *of a function.* For a function to be represented by a Maclaurin expansion, the function and all its derivatives must exist at $x = 0$. Also, we note that the factorial notation introduced in Section 18–4 is used in writing the Maclaurin series expansion.

As we mentioned at the beginning of this section, one of the uses we will make of series expansions is that of determining the values of functions for particular values of x. If x is sufficiently small, successive terms become smaller and smaller and the series will converge rapidly. This is considered in the sections which follow.

Example B
Find the first four terms of the Maclaurin series expansion of $f(x) = \dfrac{2}{2 - x}$.

This is written as

$$f(x) = \frac{2}{2 - x} \qquad\qquad f(0) = 1$$

$$f'(x) = \frac{2}{(2 - x)^2} \qquad\qquad f'(0) = \frac{1}{2}$$

$$f''(x) = \frac{4}{(2-x)^3} \qquad f''(0) = \frac{1}{2}$$

$$f'''(x) = \frac{12}{(2-x)^4} \qquad f'''(0) = \frac{3}{4}$$

$$f(x) = 1 + \frac{1}{2}x + \frac{1}{2}\left(\frac{x^2}{2!}\right) + \frac{3}{4}\left(\frac{x^3}{3!}\right) + \cdots$$

or

$$\frac{2}{2-x} = 1 + \frac{1}{2}x + \frac{1}{4}x^2 + \frac{1}{8}x^3 + \cdots$$

We see that this result agrees with that obtained by direct division.

Example C

Find the first four terms of the Maclaurin series expansion of $f(x) = e^{-x}$.
  We write

$$
\begin{array}{ll}
f(x) = e^{-x} & f(0) = 1 \\
f'(x) = -e^{-x} & f'(0) = -1 \\
f''(x) = e^{-x} & f''(0) = 1 \\
f'''(x) = -e^{-x} & f'''(0) = -1
\end{array}
$$

$$f(x) = 1 + (-1)x + 1\left(\frac{x^2}{2!}\right) + (-1)\left(\frac{x^3}{3!}\right) + \cdots$$

or

$$e^{-x} = 1 - x + \frac{x^2}{2!} - \frac{x^3}{3!} + \cdots$$

Example D

Find the first three nonzero terms of the Maclaurin series expansion of $f(x) = \sin 2x$.
  We have

$$
\begin{array}{llll}
f(x) = \sin 2x, & f(0) = 0, & f'(x) = 2\cos 2x, & f'(0) = 2 \\
f''(x) = -4\sin 2x, & f''(0) = 0, & f'''(x) = -8\cos 2x, & f'''(0) = -8 \\
f^{iv}(x) = 16\sin 2x, & f^{iv}(0) = 0, & f^{v}(x) = 32\cos 2x, & f^{v}(0) = 32
\end{array}
$$

$$f(x) = 0 + 2x + 0 + (-8)\frac{x^3}{3!} + 0 + 32\frac{x^5}{5!} + \cdots$$

or

$$\sin 2x = 2x - \frac{4}{3}x^3 + \frac{4}{15}x^5 - \cdots$$

This series is called an **alternating series,** since precisely every other term is negative.

Example E

Find the first three nonzero terms of the Maclaurin expansion of the function $f(x) = xe^{2x}$.

$$
\begin{aligned}
f(x) &= xe^{2x} & f(0) &= 0 \\
f'(x) &= e^{2x}(1 + 2x) & f'(0) &= 1 \\
f''(x) &= e^{2x}(4 + 4x) & f''(0) &= 4 \\
f'''(x) &= e^{2x}(12 + 8x) & f'''(0) &= 12
\end{aligned}
$$

$$
f(x) = 0 + x + 4\left(\frac{x^2}{2!}\right) + 12\left(\frac{x^3}{3!}\right) + \cdots
$$

or

$$
xe^{2x} = x + 2x^2 + 2x^3 + \cdots
$$

Exercises 28–1

In Exercises 1 through 12 find in each case the first three nonzero terms of the Maclaurin expansion of the given functions.

1. $f(x) = e^x$

2. $f(x) = \sin x$

3. $f(x) = \cos x$

4. $f(x) = \ln(1 + x)$

5. $f(x) = \sqrt{1 + x}$

6. $f(x) = \dfrac{1}{(1 - x)^{1/3}}$

7. $f(x) = e^{-2x}$

8. $f(x) = \frac{1}{2}(e^x + e^{-x})$

9. $f(x) = \cos 4x$

10. $f(x) = e^x \sin x$

11. $f(x) = \dfrac{1}{1 - x}$

12. $f(x) = \dfrac{1}{(1 + x)^2}$

In Exercises 13 through 20 find the first two nonzero terms of the Maclaurin expansion of the given functions.

13. $f(x) = \text{Arctan } x$

14. $f(x) = \cos x^2$

15. $f(x) = \tan x$

16. $f(x) = \sec x$

17. $f(x) = \ln \cos x$

18. $f(x) = xe^{\sin x}$

19. $f(x) = \sin^2 x$

20. $f(x) = e^{-x^2}$

In Exercises 21 through 24 solve the given problems.

21. Why is it not possible to find a Maclaurin expansion of the following functions?

   (a) $f(x) = \csc x$,   (b) $f(x) = \sqrt{x}$,   (c) $f(x) = \ln x$.

22. By finding the Maclaurin expansion, show that

$$
(1 + x)^n = 1 + nx + \frac{n(n - 1)}{2!}x^2 + \cdots \tag{28–3}
$$

   This series is the **binomial series**, which was first introduced in Section 18–4. It can be shown to be valid for $|x| < 1$ for all values of n.

23. If $f(x) = x^3$, show that this function is obtained when a Maclaurin expansion is found.

24. If $f(x) = x^4 + 2x^2$, show that this function is obtained when a Maclaurin expansion is found.

28–2 Certain Operations with Series

The series found in the first four exercises of Section 28–1 are of particular importance. They are used to find values of exponential functions, trigonometric functions, and logarithms. Also they may be used to develop other series. We shall give them here for reference:

$$e^x = 1 + x + \frac{x^2}{2!} + \frac{x^3}{3!} + \cdots \qquad (28\text{--}4)$$

$$\sin x = x - \frac{x^3}{3!} + \frac{x^5}{5!} - \cdots \qquad (28\text{--}5)$$

$$\cos x = 1 - \frac{x^2}{2!} + \frac{x^4}{4!} - \cdots \qquad (28\text{--}6)$$

$$\ln(1 + x) = x - \frac{x^2}{2} + \frac{x^3}{3} - \frac{x^4}{4} + \cdots \qquad (28\text{--}7)$$

In the next section we shall see how we make use of these series in the development of tables. In this section we shall not only see how new series may be developed by using the above basic series, but also we shall see certain other uses of series.

When we discussed functions in Chapter 2, we made brief mention of functions such as $f(2x)$, $f(x^2)$, and $f(-x)$. *By using this functional notation and the preceding series, we can find the series expansions of a great many other series without the use of direct expansion.* This can often save a great deal of time in finding a desired series. The examples given here illustrate the use of series for this purpose.

Example A
Find the Maclaurin expansion of e^{2x}.
From Eq. (28–4), we know the expansion of e^x. Hence,

$$f(x) = 1 + x + \frac{x^2}{2!} + \frac{x^3}{3!} + \cdots$$

Since $e^{2x} = f(2x)$, we have

$$f(2x) = 1 + (2x) + \frac{(2x)^2}{2!} + \frac{(2x)^3}{3!} + \cdots$$

or

$$e^{2x} = 1 + 2x + 2x^2 + \frac{4x^3}{3} + \cdots$$

Example B

Find the Maclaurin expansion of sin x^2.

From Eq. (28–5), we know the expansion of sin x. Therefore,

$$f(x) = x - \frac{x^3}{3!} + \frac{x^5}{5!} - \cdots$$

Since sin $x^2 = f(x^2)$, we have

$$f(x^2) = (x^2) - \frac{(x^2)^3}{3!} + \frac{(x^2)^5}{5!} - \cdots$$

or

$$\sin x^2 = x^2 - \frac{x^6}{3!} + \frac{x^{10}}{5!} - \cdots$$

Direct expansion of this series is quite lengthy.

The basic algebraic operations may be applied to series in the same manner they are applied to polynomials. That is, we may add, subtract, multiply, or divide series in order to obtain other series. This is illustrated in the following example.

Example C

Multiply the series for e^x and cos x in order to obtain the series expansion for e^xcos x.

Using the series expansions for e^x and cos x as shown in Eqs. (28–4) and (28–6), we have the following indicated multiplication.

$$e^x\cos x = \left(1 + x + \frac{x^2}{2!} + \frac{x^3}{3!} + \frac{x^4}{4!} + \cdots\right)\left(1 - \frac{x^2}{2!} + \frac{x^4}{4!} - \cdots\right)$$

By multiplying the series on the right, we have the following result, considering through the x^4 terms in the product.

$$e^x\cos x = 1 - \frac{x^2}{2} + \frac{x^4}{24} + x - \frac{x^3}{2} + \frac{x^2}{2} - \frac{x^4}{4} + \frac{x^3}{6} + \frac{x^4}{24} + \cdots$$

$$= 1 + x - \frac{1}{3}x^3 - \frac{1}{6}x^4 + \cdots$$

We can use algebraic operations on series to verify that the definition of the exponential form of a complex number, as shown in Eq. (11–9), is consistent with other definitions. The only assumption required here is that the Maclaurin expansions for e^x, sin x, and cos x are also valid for complex numbers. This is shown in advanced calculus. Thus,

$$e^{j\theta} = 1 + j\theta + \frac{(j\theta)^2}{2!} + \frac{(j\theta)^3}{3!} + \cdots = 1 + j\theta - \frac{\theta^2}{2!} - j\frac{\theta^3}{3!} + \cdots \qquad (28\text{–}8)$$

$$j \sin \theta = j\theta - j\frac{\theta^3}{3!} + \cdots \qquad (28\text{–}9)$$

$$\cos \theta = 1 - \frac{\theta^2}{2!} + \cdots \qquad (28\text{–}10)$$

When we add the terms of Eq. (28–9) to those of Eq. (28–10), the result is the series given in Eq. (28–8). Thus,

$$e^{j\theta} = \cos\theta + j\sin\theta \tag{28–11}$$

A comparison of Eqs. (11–9) and (28–11) indicates the reason for the choice of the definition of the exponential form of a complex number.

An additional use of power series is now shown. Many integrals which occur in practice cannot be integrated by methods given in the preceding chapters. However, power series can be very useful in giving excellent approximations to some definite integrals.

Example D

Evaluate: $\int_0^1 \dfrac{\sin x}{x}\,dx.$

By using the series expansion of $\sin x$, and dividing each term by x, we have

$$\int_0^1 \frac{\sin x}{x}\,dx = \int_0^1 \left(1 - \frac{x^2}{3!} + \frac{x^4}{5!} - \cdots\right) dx$$

$$= \left(x - \frac{x^3}{3(3!)} + \frac{x^5}{5(5!)} - \cdots\right)\Bigg|_0^1$$

$$= 1 - \frac{1}{3(6)} + \frac{1}{5(120)} - \cdots$$

Combining the values indicated, we have $1 - 0.05556 + 0.00166 = 0.9461$. We can see that each of the terms omitted was very small. The correct answer, to 4 decimal places, is 0.9461, which is precisely the value found by using only three terms of the expansion for $\sin x$.

Example E

Evaluate: $\int_0^{0.1} e^{-x^2}\,dx.$

We write

$$e^{-x^2} = 1 + (-x^2) + \frac{(-x^2)^2}{2!} + \cdots$$

Thus,

$$\int_0^{0.1} e^{-x^2}\,dx = \int_0^{0.1} \left(1 - x^2 + \frac{x^4}{2} - \cdots\right) dx$$

$$= \left(x - \frac{x^3}{3} + \frac{x^5}{10} - \cdots\right)\Bigg|_0^{0.1}$$

$$= 0.1 - \frac{0.001}{3} + \frac{0.00001}{10} = 0.09967$$

This answer is correct to the indicated accuracy.

The question of accuracy now arises. The integrals just evaluated indicate that the more terms used, the greater the accuracy of the result. As an indication of the accuracy involved, Fig. 28–1 shows the graphs of $y = \sin x$ and the graphs of

$$y = x, \qquad y = x - \frac{x^3}{3!} \quad \text{and} \quad y = x - \frac{x^3}{3!} + \frac{x^5}{5!}$$

which are the first three approximations. We can see that each term added gives a better fit to the sin x. Also, a graphical representation of the meaning of series expansions is indicated.

We have just shown that the more terms included, the more accurate the result. For small values of x, a Maclaurin series can give good accuracy with a very few terms. When this happens, we see that the series *converges* rapidly, as we mentioned earlier. For this reason a Maclaurin series is of particular use if small values of x are being considered. If larger values of x are being considered, usually a function is expanded in a Taylor series (see Section 28–4). Of course, when we omit the later terms in any series, a certain error is inherent in the calculation.

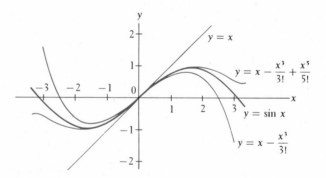

Figure 28–1

Exercises 28–2

In Exercises 1 through 8 find in each case the first four nonzero terms of the Maclaurin expansions of the given functions by using Eqs. (28–4) to (28–7).

1. $f(x) = e^{3x}$
2. $f(x) = e^{-2x}$
3. $f(x) = \sin \tfrac{1}{2}x$
4. $f(x) = \sin x^4$
5. $f(x) = \cos 4x$
6. $f(x) = \cos 2x^3$
7. $f(x) = \ln(1 + x^2)$
8. $f(x) = \ln(1 - x)$

In Exercises 9 through 12 evaluate the given integrals by use of three terms of the appropriate series.

9. $\displaystyle\int_0^1 \sin x^2 \, dx$

10. $\displaystyle\int_0^{0.1} \frac{e^x - 1}{x} \, dx$

11. $\displaystyle\int_0^{0.2} \cos \sqrt{x} \, dx$

12. $\displaystyle\int_0^1 \frac{\cos x - 1}{x} \, dx$

In Exercises 13 through 20 determine the indicated series by the given operation.

13. Find the first four nonzero terms of the expansion of the function $f(x) = \frac{1}{2}(e^x + e^{-x})$ by adding the terms of the appropriate series. The result is the series for cosh x. (See Exercise 31 of Section 26–6.)

14. Find the first four nonzero terms of the expansion of the function $f(x) = \frac{1}{2}(e^x - e^{-x})$ by subtracting the terms of the appropriate series. The result is the series for sinh x. (See Exercise 31 of Section 26–6.)

15. Find the first three terms of the expansion for $e^x \sin x$ by multiplying the proper expansions together, term by term.

16. Find the first three nonzero terms of the expansion for $f(x) = \tan x$, by dividing the series for sin x by that for cos x.

17. Show that by differentiating term by term the expansion for sin x, the result is the expansion for cos x.

18. Show that by differentiating term by term the expansion for e^x, the result is also the expansion for e^x.

19. Show that by integrating term by term the expansion for cos x, the result is the expansion for sin x.

20. Show that by integrating term by term the expansion for $-1/(1 - x)$ (see Exercise 11 of Section 28–1), the result is the expansion for $\ln(1 - x)$.

In Exercises 21 through 24 solve the given problems.

21. Find the approximate value of the area bounded by $y = x^2 e^x$, $x = 0.2$, and the x-axis, by use of three terms of the appropriate Maclaurin series.

22. Find the approximate area bounded by $y = e^{-x^2}$, $x = 0$, $x = 1$, and $y = 0$, by use of three terms of the appropriate series.

23. Find the volume generated by rotating the area bounded by $y = \sin x$ from $x = 0$ to $x = \frac{1}{8}\pi$ and the x-axis, about the y-axis by use of two terms of the appropriate series.

24. Find the volume generated by rotating the area bounded by $y = e^{-x}$, $y = 0$, $x = 0$, and $x = 0.1$ about the y-axis, by use of three terms of the appropriate series.

28–3 Computations by Use of Series Expansions

As mentioned previously, power-series expansions can be used to compute numerical values of transcendental functions. By including a sufficient number of terms, we can calculate the values to any desired degree of accuracy. It is through such calculations that tables of values are made, and values of such numbers as e and π can be found. Also, some of the values determined by a calculator are found by use of series expansions.

Example A
Calculate the value of $e^{0.1}$.

$$e^x = 1 + x + \frac{x^2}{2!} + \cdots$$

Let $x = 0.1$. Thus,

$$e^{0.1} = 1 + 0.1 + \frac{(0.1)^2}{2} + \cdots = 1.105 \quad \text{(using 3 terms)}$$

To 5 significant digits, $e^{0.1} = 1.1052$, which indicates that our answer is valid to four significant digits.

Example B
Calculate the value of $\sin 2°$.

$$\sin x = x - \frac{x^3}{3!} + \cdots$$

Let $x = \pi/90$ (converting to radians). Hence,

$$\sin 2° = \left(\frac{\pi}{90}\right) - \frac{(\pi/90)^3}{6} + \cdots = 0.034907 - 0.000007 = 0.034900$$

This result is accurate to the number of digits shown. Here we note that the second term is much smaller than the first. In fact, a good approximation can be found by use of one term. Thus, we see the reason that $\sin \theta = \theta$ for small values of θ, as we noted in Section 7–4.

Example C
Calculate the value of $\ln(1.2)$.

$$\ln(1 + x) = x - \frac{x^2}{2} + \frac{x^3}{3} - \cdots$$

Let $x = 0.2$. Therefore,

$$\ln(1 + 0.2) = 0.2 - \frac{(0.2)^2}{2} + \frac{(0.2)^3}{3} - \cdots = 0.183$$

To 4 significant digits, $\ln(1.2) = 0.1823$. One more term is required to obtain this accuracy.

Series approximations can also be used to determine errors in calculated values due to errors in measured values. We have already discussed this topic as an application of differentials. A series solution of this kind of problem allows as close a calculation of the error as needed. With differentials, only one term can be found. Of course, for many purposes this is sufficient. The following example illustrates the use of series in error calculations.

Example D

The velocity v attained by an object that has fallen h ft is given by

$$v = 8\sqrt{h}$$

Find the approximate error in the velocity of an object which has been found to drop 100 ft, with a possible error of 2 ft.

If we let

$$v = 8\sqrt{100 + x}$$

where x is the error in h, we may express v as a Maclaurin series in x.

$$\begin{aligned}
f(x) &= 8(100 + x)^{1/2} & f(0) &= 80 \\
f'(x) &= 4(100 + x)^{-1/2} & f'(0) &= 0.4 \\
f''(x) &= -2(100 + x)^{-3/2} & f''(0) &= -0.002
\end{aligned}$$

Therefore,

$$v = 8\sqrt{100 + x} = 80 + 0.4x - 0.001x^2 + \cdots$$

Since the actual calculated value of v, for $x = 0$, is 80, the error e in the value of v is

$$e = 0.4x - 0.001x^2 + \cdots$$

Calculating the error for $x = 2$ is

$$e = 0.4(2) - 0.001(4) = 0.800 - 0.004 = 0.796 \text{ ft/s}$$

The value 0.8 is the same as that which would be found by use of differentials, since it is a result of evaluating the first derivative term of the expansion. The additional terms are corrections to this term. The one additional term in this case indicates that the first term is a good approximation to the error. Although this problem could be done numerically, a series solution, once set up, allows us to easily calculate the error for a given value of x.

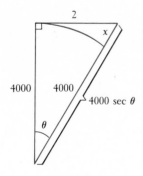

Figure 28–2

Example E

Assuming that the earth is a perfect sphere 4000 mi in radius, find how far the end of a tangent line 2 mi long is from the surface.

From Fig. 28–2 we see that

$$x = 4000 \sec \theta - 4000$$

Finding the series for $\sec \theta$, we have

$$\begin{aligned}
f(\theta) &= \sec \theta & f(0) &= 1 \\
f'(\theta) &= \sec \theta \tan \theta & f'(0) &= 0 \\
f''(\theta) &= \sec^3\theta + \sec \theta \tan^2\theta & f''(0) &= 1
\end{aligned}$$

Thus, the first two nonzero terms are $\sec \theta = 1 + (\theta^2/2)$. Thus,

$$x = 4000(\sec \theta - 1)$$

$$= 4000\left(1 + \frac{\theta^2}{2} - 1\right) = 2000\,\theta^2$$

The first two terms of the expansion for $\tan \theta$ are $\theta + \theta^3/3$, which means that if θ is small, which it is in this case, then $\tan \theta = \theta$. Again, this was noted in Section 7–4. From Fig. 28–2, $\tan \theta = 2/4000$. Therefore, $\theta = 1/2000$. Substituting this value into the expression for x, we have

$$x = 2000\left(\frac{1}{2000}\right)^2 = \frac{1}{2000} = 0.0005 \text{ mi}$$

This means that the end of a two-mile line drawn tangent to the earth would only be about 2.6 ft from the surface!

Exercises 28–3

In Exercises 1 through 16 calculate the value of each of the given functions. Use the indicated number of terms of the appropriate series.

1. $e^{0.2}$ (3)   2. $e^{-0.5}$ (3)   3. $\sin(0.1)$ (2)

4. $\sin 4°$ (2)   5. e (7)   6. $\sqrt{e}$ (5)

7. $\cos 3°$ (2)   8. $\cos(0.5°)$ (2)   9. $\ln(1.4)$ (4)

10. $\ln(0.95)$ (4)   11. $\sin 1$ (4)   12. $\ln(0.5)$ (5)

13. $\sqrt{1.1}$ (3) (use the expansion for $\sqrt{1+x}$)

14. $\sqrt{0.8}$ (3)   15. $\sqrt[3]{1.1}$ (3)   16. $\sqrt[3]{0.95}$ (3)

In Exercises 17 through 20 calculate the maximum error of the values indicated. If a series is alternating (every other term is negative) the maximum possible error in the calculated value is the value of the first term omitted.

17. The value found in Exercise 2   18. The value found in Exercise 3

19. The value found in Exercise 8   20. The value found in Exercise 9

In Exercises 21 through 26 solve the given problems by the use of series expansions.

21. The electric current in a circuit containing a resistance R, an inductance L, and a battery whose voltage is E is given by

$$i = \frac{E}{R}(1 - e^{-Rt/L})$$

where t is the time. Approximate this expression by using the first three terms of the appropriate exponential series. Under what conditions will this approximation be valid?

22. The image distance q from a certain lens as a function of the object distance p is given by $q = 20p/(p - 20)$. Find the first three nonzero terms of the expansion of the right side. From this expression calculate q for $p = 2$ cm and compare with the value found by substituting 2 in the original expression.

23. From what height can a person see a point 10 mi distant along the earth's surface?

24. The efficiency (in percent) of an internal combustion engine in terms of its compression ratio c is given approximately by $E = 100(1 - c^{-0.4})$. Determine the possible approximate error in the efficiency for a compression ratio measured to be 6.00 with a possible error of 0.50. [*Hint:* Set up a series for $(6 + x)^{-0.4}$.]

25. We can evaluate π by use of the fact that $\frac{1}{4}\pi = \text{Arctan}\frac{1}{2} + \text{Arctan}\frac{1}{3}$ (see

Exercise 34 of Section 19–7), along with the use of the series expansion for Arctan x. The first three terms are Arctan $x = x - \frac{1}{3}x^3 + \frac{1}{5}x^5$. Using these terms, expand Arctan$\frac{1}{2}$ and Arctan$\frac{1}{3}$, and thereby approximate the value of π.

26. Use the fact that $\frac{1}{4}\pi = $ Arctan$\frac{1}{7} + 2$ Arctan$\frac{1}{3}$ to approximate the value of π. (See Exercise 25.)

28–4 Taylor's Series

If values of certain functions are determined for values of x which are not close to zero, it is usually necessary to use many terms of a Maclaurin expansion to obtain accurate values. Therefore, another type of series, called a **Taylor's series**, is used for this purpose. Also, functions for which a Maclaurin series may not be found may have a Taylor's series. Actually, a Taylor's series is a more general expansion than a Maclaurin expansion.

The basic assumption in formulating a Taylor's expansion is that a function may be expanded in a polynomial of the form

$$f(x) = c_0 + c_1(x - a) + c_2(x - a)^2 + \cdots \qquad (28\text{–}12)$$

By following precisely the same line of reasoning that we did in deriving the general Maclaurin expansion, we may find the constants c_0, c_1, c_2, and so forth. That is, derivatives of Eq. (28–12) are taken, and the function and its derivatives are evaluated for $x = a$. This leads to

$$f(x) = f(a) + f'(a)(x - a) + \frac{f''(a)(x - a)^2}{2!} + \cdots \qquad (28\text{–}13)$$

This series converges rapidly for values of x which are close to a, and this is illustrated in Examples C and D. The following examples illustrate the derivation of Taylor's series.

Example A
Expand $f(x) = e^x$ in a Taylor's series with $a = 1$.

$$
\begin{aligned}
f(x) &= e^x & f(1) &= e \\
f'(x) &= e^x & f'(1) &= e \\
f''(x) &= e^x & f''(1) &= e \\
f'''(x) &= e^x & f'''(1) &= e, \text{ etc.}
\end{aligned}
$$

$$f(x) = e + e(x - 1) + e\frac{(x - 1)^2}{2!} + e\frac{(x - 1)^3}{3!} + \cdots$$

$$e^x = e\left[1 + (x - 1) + \frac{(x - 1)^2}{2} + \frac{(x - 1)^3}{6} + \cdots\right]$$

This series is of particular use in evaluating e^x for values of x near 1.

Example B

Expand $f(x) = \sqrt{x}$ in powers of $(x - 4)$.

Another way of stating this is to find the Taylor's series for $f(x) = \sqrt{x}$, with $a = 4$. Thus,

$$f(x) = x^{1/2} \qquad\qquad f(4) = 2$$

$$f'(x) = \frac{1}{2x^{1/2}} \qquad\qquad f'(4) = \frac{1}{4}$$

$$f''(x) = -\frac{1}{4x^{3/2}} \qquad\qquad f''(4) = -\frac{1}{32}$$

$$f'''(x) = \frac{3}{8x^{5/2}} \qquad\qquad f'''(4) = \frac{3}{256}, \text{ etc.}$$

$$f(x) = 2 + \frac{1}{4}(x - 4) - \frac{1}{32}\frac{(x - 4)^2}{2!} + \frac{3}{256}\frac{(x - 4)^3}{3!} - \cdots$$

$$\sqrt{x} = 2 + \frac{(x - 4)}{4} - \frac{(x - 4)^2}{64} + \frac{(x - 4)^3}{512} - \cdots$$

This series would be used to evaluate square roots of numbers near 4.

In the last section we evaluated functions by the use of Maclaurin series. In the following examples we shall use Taylor's series to determine certain values of functions.

Example C

By use of a Taylor series, evaluate $\sqrt{4.5}$.

Using the four terms of the series found in Example B, we have

$$\sqrt{4.5} = 2 + \frac{(4.5 - 4)}{4} - \frac{(4.5 - 4)^2}{64} + \frac{(4.5 - 4)^3}{512}$$

$$= 2 + \frac{(0.5)}{4} - \frac{(0.5)^2}{64} + \frac{(0.5)^3}{512}$$

$$= 2.0000 + 0.1250 - 0.0039 + 0.0002$$

$$= 2.1213$$

The result is accurate to the number of places shown.

In Example C we see that successive terms become small rapidly. If a value of x is chosen such that $x - a$ is larger, the successive terms may not become small rapidly and many terms may be required. Therefore, we should choose a as conveniently close to the x-values which will be used. Also, we should note that a Maclaurin expansion for $\sqrt{x}$ cannot be used since the derivatives of $\sqrt{x}$ are not defined for $x = 0$.

Example D

Calculate the approximate value of sin 29° by means of the appropriate Taylor expansion.

Since the value of sin 30° is known to be $\frac{1}{2}$, if we let $a = \pi/6$ (remember, we must use values expressed in radians), when we evaluate the expansion for $x = 29°$ (when expressed in radians) the quantity $(x - a)$ is $-\pi/180$ (equivalent to $-1°$). This means that its numerical values are small and become smaller when it is raised to higher powers.

Therefore,

$$f(x) = \sin x, \qquad f\left(\frac{\pi}{6}\right) = \frac{1}{2}$$

$$f'(x) = \cos x, \qquad f'\left(\frac{\pi}{6}\right) = \frac{\sqrt{3}}{2}$$

$$f''(x) = -\sin x, \qquad f''\left(\frac{\pi}{6}\right) = -\frac{1}{2}$$

$$f(x) = \frac{1}{2} + \frac{\sqrt{3}}{2}\left(x - \frac{\pi}{6}\right) - \frac{1}{4}\left(x - \frac{\pi}{6}\right)^2 - \cdots$$

or

$$\sin x = \frac{1}{2} + \frac{\sqrt{3}}{2}\left(x - \frac{\pi}{6}\right) - \frac{1}{4}\left(x - \frac{\pi}{6}\right)^2 - \cdots$$

$$\sin 29° = \sin\left(\frac{\pi}{6} - \frac{\pi}{180}\right)$$

$$= \frac{1}{2} + \frac{\sqrt{3}}{2}\left(\frac{\pi}{6} - \frac{\pi}{180} - \frac{\pi}{6}\right) - \frac{1}{4}\left(\frac{\pi}{6} - \frac{\pi}{180} - \frac{\pi}{6}\right)^2 - \cdots$$

$$= \frac{1}{2} + \frac{\sqrt{3}}{2}\left(-\frac{\pi}{180}\right) - \frac{1}{4}\left(-\frac{\pi}{180}\right)^2 - \cdots$$

$$= 0.500000 - (0.86603)(0.017453) - (0.25000)(0.017453)^2$$

$$= 0.500000 - 0.015115 - 0.000076 = 0.48481$$

This result is accurate to the number of places shown.

Exercises 28–4

In Exercises 1 through 8 evaluate the given functions by use of the series developed in the examples of this section.

1. $e^{1.2}$ (use $e = 2.718$)   2. $e^{0.7}$ 3. $\sqrt{4.2}$ 4. $\sqrt{3.5}$
5. $\sin 31°$ 6. $\sin 28°$ 7. $\sin 29.5°$ 8. $\sqrt{3.85}$

In Exercises 9 through 16 find the first three nonzero terms of the Taylor expansion for the given function and given value of a.

9. e^{-x} $(a = 2)$ 10. $\cos x$ $(a = \pi/4)$ 11. $\sin x$ $(a = \pi/3)$

12. $\ln x$ $(a = 3)$ 13. $\sqrt[3]{x}$ $(a = 8)$ 14. $\dfrac{1}{x}$ $(a = 2)$

15. $\tan x$ $(a = \pi/4)$ 16. $\ln \sin x$ $(a = \pi/2)$

In Exercises 17 through 24 evaluate the given functions by use of three terms of the appropriate Taylor series.

17. $e^{-2.2}$ (use $e^{-2} = 0.1353$)   **18.** $\ln(3.1)$ (use $\ln 3 = 1.0986$)

19. $\sqrt{9.3}$ **20.** $\sqrt{17}$ **21.** $\sqrt[3]{8.3}$

22. $\tan 46°$ **23.** $\sin 61°$ **24.** $\cos 42°$

In Exercises 25 and 26 solve the given problems.

25. By completing the steps indicated before Eq. (28–13) in the text, complete the derivation of Eq. (28–13).

26. Calculate $e^{0.9}$ by use of four terms of the Maclaurin expansion for e^x. Also calculate $e^{0.9}$ by using the first three terms of the Taylor's expansion in Example A, using $e = 2.718$. Compare the accuracy of the values obtained with that in Table 7 in the Appendix.

28–5 Fourier Series

Many problems which are encountered in the various fields of science and technology involve functions which are periodic. *A periodic function is one for which $F(x + P) = F(x)$, where P is the period.* We noted that the trigonometric functions are periodic when we discussed their graphs in Chapter 9. Illustrations of applied problems which involve periodic functions are alternating-current voltages and mechanical oscillations.

Therefore, in this section we use a series made of terms of sines and cosines. This allows us to represent complicated periodic functions in terms of the simpler sines and cosines. It also provides a good approximation over a greater interval than Maclaurin and Taylor series, which give good approximations with a few terms only near a specific value. Illustrations of applications of this type of series are given in Example C and Exercise 11.

We shall assume that a function $f(x)$ may be represented by the series of sines and cosines as indicated:

$$f(x) = a_0 + a_1 \cos x + a_2 \cos 2x + \cdots + a_n \cos nx + \cdots$$
$$+ b_1 \sin x + b_2 \sin 2x + \cdots + b_n \sin nx + \cdots \quad (28\text{–}14)$$

Since all of the sines and cosines indicated in this expansion have a period of 2π (the period of any given term may be less than 2π, but all do repeat every 2π units—e.g., $\sin 2x$ has a period of π, but it also repeats every 2π), the series expansion indicated in Eq. (28–14) will also have a period of 2π. This series is called a **Fourier series.**

The principal problem to be solved is that of finding the coefficients a_n and b_n. Derivatives proved to be useful in finding the coefficients for a Maclaurin expansion. We use the properties of certain integrals to find the coefficients of a Fourier series. To utilize these properties, we multiply all terms of Eq. (28–14) by $\cos mx$, and then evaluate from $-\pi$ to π (in this way we take advantage of the period 2π). Thus, we have

$$\int_{-\pi}^{\pi} f(x)\cos mx\ dx = \int_{-\pi}^{\pi} (a_0 + a_1\cos x + a_2\cos 2x + \cdots)(\cos mx)\ dx$$

$$+ \int_{-\pi}^{\pi} (b_1\sin x + b_2\sin 2x + \cdots)(\cos mx)\ dx \qquad (28\text{–}15)$$

Using the methods of integration of Chapter 27, we have the following:

$$\int_{-\pi}^{\pi} a_0\cos mx\ dx = \frac{a_0}{m}\sin mx\Big|_{-\pi}^{\pi} = \frac{a_0}{m}(0 - 0) = 0 \qquad (28\text{–}16)$$

$$\int_{-\pi}^{\pi} a_n\cos nx \cos mx\ dx =$$

$$a_n\left(\frac{\sin(n - m)x}{2(n - m)} + \frac{\sin(n + m)x}{2(n + m)}\right)\Big|_{-\pi}^{\pi} = 0 \qquad n \neq m; \qquad (28\text{–}17)$$

(since the sine of any multiple of π is zero),

$$\int_{-\pi}^{\pi} a_n\cos nx \cos nx\ dx = \int_{-\pi}^{\pi} a_n\cos^2 nx\ dx$$

$$= \left(\frac{a_n x}{2} + \frac{1}{2n}\sin nx \cos nx\right)\Big|_{-\pi}^{\pi} = \frac{a_n x}{2}\Big|_{-\pi}^{\pi} = \pi a_n \qquad (28\text{–}18)$$

as well as

$$\int_{-\pi}^{\pi} b_n\sin nx \cos mx\ dx = b_n\left(-\frac{\cos(n - m)x}{2(n - m)} - \frac{\cos(n + m)x}{2(n + m)}\right)\Big|_{-\pi}^{\pi}$$

$$= b_n\left(-\frac{\cos(n - m)\pi}{2(n - m)} - \frac{\cos(n + m)\pi}{2(n + m)}\right.$$

$$\left.+ \frac{\cos(n - m)(-\pi)}{2(n - m)} + \frac{\cos(n + m)(-\pi)}{2(n + m)}\right)$$

$$= 0 \text{ [since } \cos \theta = \cos(-\theta)\text{]}, \qquad n \neq m \qquad (28\text{–}19)$$

$$\int_{-\pi}^{\pi} b_n\sin nx \cos nx\ dx = \frac{b_n}{2n}\sin^2 nx\Big|_{-\pi}^{\pi} = 0 \qquad (28\text{–}20)$$

These integrals are seen to be zero, except for the one specific case of $\int_{-\pi}^{\pi} a_n \cos nx \cos mx\ dx$ when $n = m$, for which the result is indicated in Eq. (28–18). Using these results in Eq. (28–15), we have

$$\int_{-\pi}^{\pi} f(x)\cos nx\ dx = a_n\int_{-\pi}^{\pi} \cos^2 nx\ dx = \pi a_n$$

or

$$a_n = \frac{1}{\pi}\int_{-\pi}^{\pi} f(x)\cos nx\ dx \qquad (28\text{–}21)$$

This equation allows us to find the coefficients a_n, except a_0. We find the term a_0 by direct integration of Eq. (28–14) from $-\pi$ to π. When we perform this integration, all the sine and cosine terms integrate to zero, thereby giving the result

$$\int_{-\pi}^{\pi} f(x)\,dx = \int_{-\pi}^{\pi} a_0\,dx = a_0 x \Big|_{-\pi}^{\pi} = 2\pi a_0$$

or

$$a_0 = \frac{1}{2\pi}\int_{-\pi}^{\pi} f(x)\,dx \tag{28–22}$$

By multiplying all terms of Eq. (28–14) by $\sin mx$ and then integrating from $-\pi$ to π, we find the coefficients b_n. We obtain the result

$$b_n = \frac{1}{\pi}\int_{-\pi}^{\pi} f(x)\sin nx\,dx \tag{28–23}$$

We can restate our equations for the Fourier series of a function $f(x)$:

$$f(x) = a_0 + a_1\cos x + a_2\cos 2x + \cdots + a_n\cos nx + \cdots$$
$$+ b_1\sin x + b_2\sin 2x + \cdots + b_n\sin nx + \cdots \tag{28–14}$$

where the coefficients are found by

$$a_0 = \frac{1}{2\pi}\int_{-\pi}^{\pi} f(x)\,dx \tag{28–22}$$

$$a_n = \frac{1}{\pi}\int_{-\pi}^{\pi} f(x)\cos nx\,dx \tag{28–21}$$

and

$$b_n = \frac{1}{\pi}\int_{-\pi}^{\pi} f(x)\sin nx\,dx \tag{28–23}$$

Example A
Find the Fourier series for the square wave function

$$f(x) = \begin{cases} 0 & \text{for } -\pi \le x < 0 \\ 1 & \text{for } 0 \le x < \pi \end{cases}$$

(Many of the functions which we shall expand in Fourier series are discontinuous (not continuous) like this one. See Section 22–1 for a discussion of continuity.)

Since $f(x)$ is defined differently for the ranges of x indicated, it requires two integrals for each coefficient:

$$a_0 = \frac{1}{2\pi} \int_{-\pi}^{0} 0 \; dx + \frac{1}{2\pi} \int_{0}^{\pi} (1) \; dx = \frac{1}{2}$$

$$a_n = \frac{1}{\pi} \int_{-\pi}^{0} 0 \; dx + \frac{1}{\pi} \int_{0}^{\pi} (1) \cos nx \; dx = \frac{1}{n\pi} (\sin nx) \Big|_{0}^{\pi} = 0$$

for all values of n, since $\sin n\pi = 0$;

$$b_1 = \frac{1}{\pi} \int_{-\pi}^{0} 0 \; dx + \frac{1}{\pi} \int_{0}^{\pi} \sin x \; dx = -\frac{1}{\pi} (\cos x) \Big|_{0}^{\pi}$$

$$= -\frac{1}{\pi}(-1 - 1) = \frac{2}{\pi}$$

$$b_2 = \frac{1}{\pi} \int_{-\pi}^{0} 0 \; dx + \frac{1}{\pi} \int_{0}^{\pi} \sin 2x \; dx = -\frac{1}{2\pi} (\cos 2x) \Big|_{0}^{\pi}$$

$$= -\frac{1}{2\pi}(1 - 1) = 0$$

$$b_3 = \frac{1}{\pi} \int_{-\pi}^{0} 0 \; dx + \frac{1}{\pi} \int_{0}^{\pi} \sin 3x \; dx = -\frac{1}{3\pi} (\cos 3x) \Big|_{0}^{\pi}$$

$$= -\frac{1}{3\pi}(-1 - 1) = \frac{2}{3\pi}$$

In general we can see that if n is even, $b_n = 0$, and if n is odd, then $b_n = 2/n\pi$. Thus, we have

$$f(x) = \frac{1}{2} + \frac{2}{\pi} \sin x + \frac{2}{3\pi} \sin 3x + \cdots$$

A graph of the function as defined, and the curve found by using the first three terms of the Fourier series, is shown in Fig. 28–3.

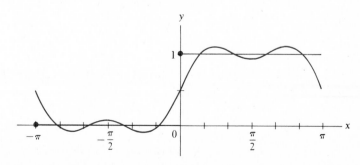

Figure 28–3

Since functions found by Fourier series have a period of 2π, they can represent functions with this period. If the function $f(x)$ were defined to be periodic with period 2π, with the same definitions as originally indicated, we would graph the function as shown in Fig. 28–4. The Fourier series representation would follow it as in Fig. 28–3. If more terms were used, the fit would be closer.

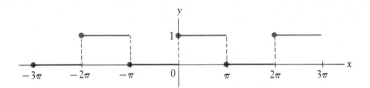

Figure 28–4

Example B

Find the Fourier series for the function

$$f(x) = \begin{cases} 1 & \text{for } -\pi \leq x < 0 \\ x & \text{for } 0 \leq x < \pi \end{cases}$$

For the periodic function, let $f(x + 2\pi) = f(x)$ for all x.

A graph of three periods of this function is shown in Fig. 28–5. The coefficients are

$$a_0 = \frac{1}{2\pi}\int_{-\pi}^{0} dx + \frac{1}{2\pi}\int_{0}^{\pi} x\, dx = \frac{x}{2\pi}\Big|_{-\pi}^{0} + \frac{x^2}{4\pi}\Big|_{0}^{\pi} = \frac{1}{2} + \frac{\pi}{4} = \frac{2 + \pi}{4}$$

$$a_1 = \frac{1}{\pi}\int_{-\pi}^{0} \cos x\, dx + \frac{1}{\pi}\int_{0}^{\pi} x \cos x\, dx$$

$$= \frac{1}{\pi}\sin x\Big|_{-\pi}^{0} + \frac{1}{\pi}(\cos x + x \sin x)\Big|_{0}^{\pi} = -\frac{2}{\pi}$$

$$a_2 = \frac{1}{\pi}\int_{-\pi}^{0} \cos 2x\, dx + \frac{1}{\pi}\int_{0}^{\pi} x \cos 2x\, dx$$

$$= \frac{1}{2\pi}\sin 2x\Big|_{-\pi}^{0} + \frac{1}{4\pi}(\cos 2x + 2x \sin 2x)\Big|_{0}^{\pi} = 0$$

$$a_3 = \frac{1}{\pi}\int_{-\pi}^{0} \cos 3x\, dx + \frac{1}{\pi}\int_{0}^{\pi} x \cos 3x\, dx$$

$$= \frac{1}{3\pi}\sin 3x\Big|_{-\pi}^{0} + \frac{1}{9\pi}(\cos 3x + 3x \sin 3x)\Big|_{0}^{\pi} = -\frac{2}{9\pi}$$

$$b_1 = \frac{1}{\pi}\int_{-\pi}^{0} \sin x\, dx + \frac{1}{\pi}\int_{0}^{\pi} x \sin x\, dx$$

$$= -\frac{1}{\pi}\cos x \Big|_{-\pi}^{0} + \frac{1}{\pi}(\sin x - x \cos x)\Big|_{0}^{\pi} = \frac{\pi - 2}{\pi}$$

$$b_2 = \frac{1}{\pi}\int_{-\pi}^{0} \sin 2x \; dx + \frac{1}{\pi}\int_{0}^{\pi} x \sin 2x \; dx$$

$$= -\frac{\cos 2x}{2\pi}\Big|_{-\pi}^{0} + \frac{\sin 2x - 2x \cos 2x}{4\pi}\Big|_{0}^{\pi} = -\frac{1}{2}$$

Thus,

$$f(x) = \frac{2 + \pi}{4} - \frac{2}{\pi}\cos x - \frac{2}{9\pi}\cos 3x - \cdots$$

$$+ \left(\frac{\pi - 2}{\pi}\right)\sin x - \frac{1}{2}\sin 2x + \cdots$$

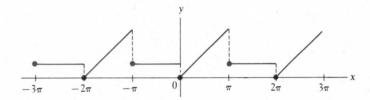

Figure 28–5

Example C

Certain electronic devices allow an electric current to pass through in only one direction. The result is that when an alternating current is applied, the current exists for only half the cycle. Figure 28–6 is a representation for such a current as a function of time. A device which causes this effect is called a half-wave rectifier. Derive the Fourier series for the half of the rectified wave for which $f(t) = \sin t$ $(0 \le t \le \pi)$, and $f(t) = 0$ for the other half of the cycle.

We write

$$a_0 = \frac{1}{2\pi}\int_{0}^{\pi} \sin t \; dt = \frac{1}{2\pi}(-\cos t)\Big|_{0}^{\pi} = \frac{1}{2\pi}(1 + 1) = \frac{1}{\pi}$$

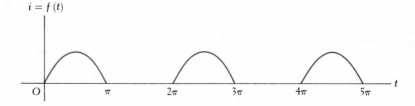

$i = f(t)$

Figure 28–6

In the previous example we evaluated each of the coefficients individually. Here we shall show how to set up a general expression for a_n and another for b_n. Once we have determined these, we can substitute values of n in the formula to obtain the individual coefficients:

$$a_n = \frac{1}{\pi} \int_0^\pi \sin t \cos nt \, dt = -\frac{1}{2\pi} \left[\frac{\cos(1-n)t}{1-n} + \frac{\cos(1+n)t}{1+n} \right]_0^\pi$$

$$= -\frac{1}{2\pi} \left[\frac{\cos(1-n)\pi}{1-n} + \frac{\cos(1+n)\pi}{1+n} - \frac{1}{1-n} - \frac{1}{1+n} \right]$$

See formula 40 of Table 8 in Appendix E. This formula is valid for all values of n except $n = 1$. Now we write

$$a_1 = \frac{1}{\pi} \int_0^\pi \sin t \cos t \, dt = \frac{1}{2\pi} \sin^2 t \bigg|_0^\pi = 0$$

$$a_2 = -\frac{1}{2\pi} \left(\frac{-1}{-1} + \frac{-1}{3} - \frac{1}{-1} - \frac{1}{3} \right) = -\frac{2}{3\pi}$$

$$a_3 = -\frac{1}{2\pi} \left(\frac{1}{-2} + \frac{1}{4} - \frac{1}{-2} - \frac{1}{4} \right) = 0$$

$$a_4 = -\frac{1}{2\pi} \left(\frac{-1}{-3} + \frac{-1}{5} - \frac{1}{-3} - \frac{1}{5} \right) = -\frac{2}{15\pi}$$

$$b_n = \frac{1}{\pi} \int_0^\pi \sin t \sin nt \, dt = \frac{1}{2\pi} \left[\frac{\sin(1-n)t}{1-n} - \frac{\sin(1+n)t}{1+n} \right]_0^\pi$$

$$= \frac{1}{2\pi} \left[\frac{\sin(1-n)\pi}{1-n} - \frac{\sin(1+n)\pi}{1+n} \right]$$

See formula 39 of Table 8. This formula is valid for all values of n except $n = 1$. Thus,

$$b_1 = \frac{1}{\pi} \int_0^\pi \sin t \sin t \, dt = \frac{1}{\pi} \int_0^\pi \sin^2 t \, dt$$

$$= \frac{1}{2\pi} (t - \sin t \cos t) \bigg|_0^\pi = \frac{1}{2}$$

We see that $b_n = 0$ if $n > 1$, since each is evaluated in terms of the sine of a multiple of π.

Therefore, the Fourier series for the rectified wave is

$$f(t) = \frac{1}{\pi} + \frac{1}{2} \sin t - \frac{2}{\pi} \left(\frac{1}{3} \cos 2t + \frac{1}{15} \cos 4t + \cdots \right)$$

All the types of periodic functions included in this section (as well as many others) may actually be seen on an oscilloscope when the proper signal is sent into it. In this way the oscilloscope may be used to analyze the periodic nature of such phenomena as sound waves and electric currents.

Exercises 28–5

In the following exercises find a few terms of the Fourier series for the given functions, and sketch at least three periods of the function $f(x + 2\pi) = f(x)$.

1. $f(x) = \begin{cases} 1, & -\pi \le x < 0 \\ 0, & 0 \le x < \pi \end{cases}$ 2. $f(x) = \begin{cases} -1, & -\pi \le x < 0 \\ 1, & 0 \le x < \pi \end{cases}$

3. $f(x) = \begin{cases} 1, & -\pi \le x < 0 \\ 2, & 0 \le x < \pi \end{cases}$ 4. $f(x) = \begin{cases} 0, & -\pi \le x < 0, \; \pi/2 < x < \pi \\ 1, & 0 \le x \le \pi/2 \end{cases}$

5. $f(x) = \begin{cases} 0, & -\pi \le x < 0 \\ x, & 0 \le x < \pi \end{cases}$ 6. $f(x) = x, \; -\pi \le x < \pi$

7. $f(x) = \begin{cases} -1, & -\pi \le x < 0 \\ 0, & 0 \le x < \pi/2 \\ 1, & \pi/2 \le x < \pi \end{cases}$ 8. $f(x) = x^2, \; -\pi \le x < \pi$

9. $f(x) = \begin{cases} -x, & -\pi \le x < 0 \\ x, & 0 \le x < \pi \end{cases}$ 10. $f(x) = \begin{cases} 0, & -\pi \le x < 0 \\ x^2, & 0 \le x < \pi \end{cases}$

11. Find the Fourier expansion of the electronic device known as a *full-wave rectifier*. This is found by using as the function for the current $f(t) = -\sin t$ for $-\pi \le t \le 0$ and $f(t) = \sin t$ for $0 < t \le \pi$. See Fig. 28–7. The portion of the curve to the left of the $f(t)$-axis is dashed, since from a physical point of view we can give no significance to this part of the wave; although mathematically we can derive the proper form of the expansion by using it.

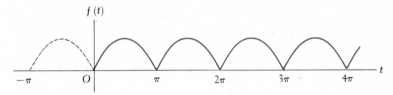

Figure 28–7

12. A function for which $f(x) = f(-x)$ is called an **even function,** and a function for which $f(x) = -f(-x)$ is called an **odd function.** From our discussion of symmetry in Section 2–3, we see that an even function is symmetrical to the y-axis, and an odd function is symmetrical to the origin. It can be proven that the Fourier expansion of an even function contains no sine terms, and that of an odd function contains no cosine terms. Show that the function in Exercise 9 is an even function, and the function in Exercise 6 is an odd function. Note and compare the Fourier expansions of these functions.

28–6 Exercises for Chapter 28

In Exercises 1 through 8 find the first three nonzero terms of the Maclaurin expansion of the given functions.

1. $f(x) = \dfrac{1}{1 + e^x}$ 2. $f(x) = \frac{1}{2}(e^x - e^{-x})$ 3. $f(x) = \sin 2x^2$ 4. $f(x) = \dfrac{1}{(1 - x)^2}$

5. $f(x) = (x + 1)^{1/3}$ **6.** $f(x) = \dfrac{2}{2 - x^2}$ **7.** $f(x) = \text{Arcsin } x$ **8.** $f(x) = \dfrac{1}{1 - \sin x}$

In Exercises 9 through 16 calculate the value of each of the given functions. Use three terms of the appropriate series.

9. $e^{-0.2}$ **10.** $\ln(1.10)$ **11.** $\sqrt[3]{1.3}$ **12.** $\sin 58°$

13. $\tan 43°$ **14.** $\sqrt[4]{260}$ **15.** $\sqrt{148}$ **16.** $\cos 47°$

In Exercises 17 and 18 evaluate the given integrals by using three terms of the appropriate series.

17. $\displaystyle\int_{0.1}^{0.2} \dfrac{\cos x}{\sqrt{x}}\,dx$ **18.** $\displaystyle\int_{0}^{0.1} \sqrt[3]{1 + x^2}\,dx$

In Exercises 19 and 20 find the first three nonzero terms of the Taylor expansion for the given function and the given value of a.

19. $\cos x$ $(a = \pi/3)$ **20.** $\ln \cos x$ $(a = \pi/4)$

In Exercises 21 and 22 find a few terms of the Fourier series for the given functions. Sketch three periods of the function $f(x + 2\pi) = f(x)$.

21. $f(x) = \begin{cases} 0, & -\pi \le x < -\pi/2 \quad \text{and} \quad \pi/2 < x < \pi \\ 1, & -\pi/2 \le x \le \pi/2 \end{cases}$

22. $f(x) = \begin{cases} -x, & -\pi \le x < 0 \\ 0, & 0 \le x < \pi \end{cases}$

In Exercises 23 and 24 determine the indicated series by the given operation.

23. If h is small, show that $\sin(x + h) - \sin(x - h) = 2h \cos x$.

24. Find the first three nonzero terms of the Maclaurin expansion of the function $\sin x + x \cos x$ by differentiating the expansion term by term for $x \sin x$.

In Exercises 25 through 30 solve the given problems.

25. On the same diagram plot the graphs of $y = e^x$, $y = 1$, $y = 1 + x$, and $y = 1 + x + \frac{1}{2}x^2$.

26. Show that the Maclaurin expansion for the polynomial $f(x) = ax^2 + bx + c$ is the polynomial itself.

27. Find the approximate area between the curve $y = \dfrac{x - \sin x}{x^2}$ and the x-axis between $x = 0$ and $x = 0.1$.

28. Find the approximate value of the moment of inertia with respect to its axis of the volume generated by rotating the area bounded by $y = \sin x$, $x = 0.3$, and the x-axis about the y-axis. Use two terms of the appropriate series.

29. A certain electric current is pulsating so that the current as a function of time is given by $f(t) = 0$ if $-\pi \le t < 0$ and $\pi/2 < t < \pi$. If $0 < t < \pi/2$, $f(t) = \sin t$. Find the Fourier expansion for this pulsating current, and plot three periods.

30. By use of series, find the greatest distance that a great circle arc 200 miles long on the earth is from its chord.

29

Differential Equations

29–1 Solutions of Differential Equations

Various relationships which occur in science and technology give rise to equations involving the rate of change of one variable with respect to another. When this type of relationship exists, we express the rate of change as a derivative. This results in an equation which contains a derivative. *Any such equation containing derivatives, or differentials, is called a* **differential equation.** It is our purpose in this section to introduce the basic meaning of the solution of a differential equation. In the sections which follow we consider certain methods of finding such solutions. We shall see that finding the solutions of differential equations is a powerful method of solving many types of applied problems.

The basic types of differential equations we shall consider are those which contain first and second derivatives. *If an equation contains only first derivatives, it is called a* **first-order** *differential equation. If the equation contains second derivatives, and possibly first derivatives, it is called a* **second-order** *differential equation. In general, the* **order** *of a differential equation is that of the highest derivative in the equation. The* **degree** *of a differential equation is the highest power of that derivative.*

Example A

The equation $\dfrac{dy}{dx} + x = y$ is a first-order differential equation. The equations

$$\frac{d^2y}{dx^2} + y = 3x^2 \quad \text{and} \quad \frac{d^2y}{dx^2} + 2\frac{dy}{dx} = x$$

are both second-order equations. The presence of dy/dx in the second equation does not affect the order.

Example B

The equation

$$\frac{d^2y}{dx^2} + \left(\frac{dy}{dx}\right)^4 - y = 6$$

is a differential equation of the second order and the first degree. That is, the highest derivative which appears is the second, and it is raised to the first power. Since the second derivative appears, the fourth power of the first derivative does not affect the degree.

In our discussion of differential equations, we shall restrict our attention to equations of the first degree.

A **solution** *of a differential equation is a relation between the variables which satisfies the differential equation. That is, when this relation is substituted into the differential equation, an algebraic identity results. A solution containing a number of independent arbitrary constants equal to the order of the differential equation is called the* **general solution** *of the equation. When specific values are given to at least one of these constants, the solution is called a* **particular solution.**

Example C

Any coefficients which are not specified numerically after like terms have been combined are independent arbitrary constants. The expression $c_1x + c_2 + c_3x$ has only two independent constants, in that the two x terms may be combined with one arbitrary constant; $c_2 + c_4x$ is an equivalent expression, with $c_4 = c_1 + c_3$.

Example D

The equation $y = c_1e^{-x} + c_2e^{2x}$ is the general solution of the differential equation

$$\frac{d^2y}{dx^2} - \frac{dy}{dx} = 2y$$

The order of this differential equation is two, and there are two independent arbitrary constants in the equation representing the solution. The equation $y = 4e^{-x}$ is a particular solution. This particular solution can be derived from the general solution by letting $c_1 = 4$ and $c_2 = 0$. Each of these solutions can be shown to satisfy the differential equation by taking two derivatives and substituting.

To solve a differential equation, we have to find some method of transforming the equation so that the terms may be integrated. This will be the topic of principal concern after this section. The purpose here is to show that a given equation is a solution of a differential equation by taking the required derivatives, and to show that an algebraic identity results after substitution.

Example E

Show that $y = c_1 \sin x + c_2 \cos x$ is the general solution of the differential equation $y'' + y = 0$.

The function and its first two derivatives are

$$y = c_1 \sin x + c_2 \cos x, \qquad y' = c_1 \cos x - c_2 \sin x$$

and

$$y'' = -c_1 \sin x - c_2 \cos x$$

Substituting these into the differential equation, we have

$$(-c_1 \sin x - c_2 \cos x) + (c_1 \sin x + c_2 \cos x) = 0 \quad \text{or} \quad 0 = 0$$

We know that this must be the general solution, since there are two independent arbitrary constants, and the order of the differential equation is two.

Example F

Show that $y = cx + x^2$ is a solution of the differential equation $xy' - y = x^2$.

Taking one derivative, $y' = c + 2x$, and substituting, we have

$$x(c + 2x) - (cx + x^2) = x^2 \quad \text{or} \quad x^2 = x^2$$

Exercises 29–1

In the following exercises show that each equation is a solution of the indicated differential equation.

1. $\dfrac{dy}{dx} = 1, \quad y = x + 3$

2. $\dfrac{dy}{dx} = 2x, \quad y = x^2 + 1$

3. $\dfrac{dy}{dx} - y = 1, \quad y = e^x - 1$

4. $\dfrac{dy}{dx} - 3 = 2x, \quad y = x^2 + 3x$

5. $xy' = 2y, \quad y = cx^2$

6. $y' = 2xy^2, \quad y = -\dfrac{1}{x^2 + c}$

7. $y' + 2y = 2x, \quad y = ce^{-2x} + x - \frac{1}{2}$

8. $y' - 3x^2 = 1, \quad y = x^3 + x + c$

9. $\dfrac{d^2y}{dx^2} + 4y = 0, \quad y = 3 \cos 2x$

10. $y'' + 9y = 4 \cos x, \quad 2y = \cos x$

11. $y'' - 4y' + 4y = e^{2x}, \quad y = e^{2x}\left(c_1 + c_2 x + \dfrac{x^2}{2}\right)$

12. $\dfrac{d^3y}{dx^3} = \dfrac{d^2y}{dx^2}, \quad y = c_1 + c_2 x + c_3 e^x$

13. $x^2 y' + y^2 = 0, \quad xy = cx + cy$

14. $xy' - 3y = x^2, \quad y = cx^3 - x^2$
15. $x\dfrac{d^2y}{dx^2} + \dfrac{dy}{dx} = 0, \quad y = c_1 \ln x + c_2$

16. $\dfrac{d^3y}{dx^3} + 4\dfrac{d^2y}{dx^2} + 4\dfrac{dy}{dx} = 0, \quad y = c_1 + c_2 e^{-2x} + xe^{-2x}$

17. $y' + y = 2\cos x, \quad y = \sin x + \cos x - e^{-x}$
18. $(x + y) - xy' = 0, \quad y = x \ln x - cx$
19. $y'' + y' = 6 \sin 2x, \quad y = e^{-x} - \frac{3}{5}\cos 2x - \frac{6}{5}\sin 2x$
20. $xy''' + 2y'' = 0, \quad y = c_1 x + c_2 \ln x$

21. $\cos x \dfrac{dy}{dx} + \sin x = 1 - y, \quad y = \dfrac{x + c}{\sec x + \tan x}$

22. $2xyy' + x^2 = y^2, \quad x^2 + y^2 = cx$
23. $(y')^2 + xy' = y, \quad y = cx + c^2$

24. $x^4(y')^2 - xy' = y, \quad y = c^2 + \dfrac{c}{x}$

29–2 Separation of Variables

The first type of differential equation we shall solve is one of the first order and first degree. There are many methods for solving such equations, a few of which are presented in this and the following two sections. The first of these is the method of **separation of variables.**

By the definitions of order and degree, a differential equation of the first order and first degree contains the first derivative to the first power. That is, it may be written as $dy/dx = f(x, y)$. This type of equation is more commonly expressed in its differential form

$$M(x, y)\,dx + N(x, y)\,dy = 0 \qquad (29\text{–}1)$$

where $M(x, y)$ and $N(x, y)$ may represent constants, functions of either x or y, or functions of x and y.

To find the solution of an equation of the form (29–1), it is necessary to integrate. However, if $M(x, y)$ is a function of x and y, we cannot integrate this term. It is integrable only if $M(x, y)$ is a function of x only. For the same reason, the second term is integrable only if $N(x, y)$ is a function of y only. If it is possible, by some algebraic means, to rewrite Eq. (29–1) in the form

$$A(x)\,dx + B(y)\,dy = 0 \qquad (29\text{–}2)$$

where $A(x)$ is a function of x alone and $B(y)$ is a function of y alone, then we may find the solution by integrating each term and adding the constant of integration. (If division is involved, we must remember that the solution is not valid for values which make the divisor zero.) Many elementary differential equations lend themsleves to this type of solution.

Example A

Solve the differential equation $dx - 4xy^3 dy = 0$; $M(x, y) = 1$ (a function of neither x nor y), $N(x, y) = -4xy^3$ (a function of both x and y).

We have to remove the x from the coefficient of dy, without introducing any factor of y in the coefficient of dx. This can be done by dividing each term of the equation by x, which results in the equation $(dx/x) - 4y^3 dy = 0$. It is now possible to integrate each term separately. Performing this integration, we have $\ln x - y^4 = c$, which is the desired solution. The c is the constant of integration, which becomes the arbitrary constant of the solution.

Example B

Solve the differential equation $xy\,dx + (x^2 + 1)\,dy = 0$.

In order to integrate each term, it is necessary to divide each term by $y(x^2 + 1)$. When this is done, we have

$$\frac{x\,dx}{x^2 + 1} + \frac{dy}{y} = 0$$

Integrating, we have $\frac{1}{2}\ln(x^2 + 1) + \ln y = c$, which is the desired solution. It is possible to make use of the properties of logarithms to make the form of this solution much neater. If we write the constant of integration as $\ln c_1$, rather than c, we have $\frac{1}{2}\ln(x^2 + 1) + \ln y = \ln c_1$. Multiplying through by 2, and using the property of logarithms given by Eq. (12–9), we have $\ln(x^2 + 1) + \ln y^2 = \ln c_1^2$. Next, using the property of logarithms given by Eq. (12–7), we have $\ln(x^2 + 1)y^2 = \ln c_1^2$, which means $(x^2 + 1)y^2 = c_1^2$. This form of the solution is much more compact, and would be generally preferred. However, it must be pointed out that any expression representing a constant may be chosen as the constant of integration and leads to a proper result. In checking answers, we must remember that a different choice of constant will lead to a different form of the answer. Thus, two different-appearing answers may both be correct. *It often happens that there is more than one reasonable choice of a constant, and different forms of the answer may be expected.*

Example C

Solve the differential equation

$$\frac{dy}{dx} = \frac{y}{x^2 + 4}$$

The variables are separated by writing the equation in the form

$$\frac{dy}{y} = \frac{dx}{x^2 + 4}$$

Integrating, we have

$$\ln y = \frac{1}{2}\text{Arctan}\frac{x}{2} + \frac{c}{2} \quad \text{or} \quad 2 \ln y = \text{Arctan}\frac{x}{2} + c$$

The choice of $\ln c$ as the constant of integration (on the left) is also reasonable. It would lead to the result $2 \ln cy = \text{Arctan}(x/2)$.

Example D

Solve the differential equation $3x^2y^2\,dx + y^2\,dx + dy = 0$, subject to the condition that $x = 2$ when $y = 1$.

Dividing through by y^2 and combining terms, we have

$$(3x^2 + 1)\,dx + \frac{dy}{y^2} = 0$$

Integrating, the result is $x^3 + x - (1/y) = c$. Since a specified set of values is given, we can evaluate the constant of integration. Using these values, we have $8 + 2 - 1 = c$, or $c = 9$. Thus, the solution may be written as

$$x^3 + x - \frac{1}{y} = 9 \quad \text{or} \quad yx^3 + yx = 1 + 9y$$

Example E

If the function in Example B is subject to the condition that $x = 0$ when $y = e$, we have

$$\frac{1}{2}\ln(0 + 1) + \ln e = c, \quad \frac{1}{2}\ln 1 + 1 = c \quad \text{or} \quad c = 1$$

The solution is then

$$\frac{1}{2}\ln(x^2 + 1) + \ln y = 1, \quad \ln(x^2 + 1) + 2\ln y = 2$$

$$\ln y^2(x^2 + 1) = 2 \quad \text{or} \quad y^2(x^2 + 1) = e^2$$

If we use the solution with c_1, we have $(0 + 1)e^2 = c_1^2$, or $c_1^2 = e^2$. The solution is then $y^2(x^2 + 1) = e^2$. We see, therefore, that the choice of the constant does not affect the final result, and therefore is truly arbitrary.

Exercises 29–2

In Exercises 1 through 20 solve the given differential equations.

1. $2x\,dx + dy = 0$
2. $y^2\,dy + x^3\,dx = 0$
3. $y^2\,dx + dy = 0$
4. $y\,dx + x\,dy = 0$
5. $x^2 + (x^3 + 5)y' = 0$
6. $xyy' + \sqrt{1 + y^2} = 0$
7. $e^{x^2}\,dy = x\sqrt{1 - y}\,dx$
8. $(x^3 + x^2)\,dx + (x + 1)y\,dy = 0$
9. $e^{x+y}\,dx + dy = 0$
10. $e^{2x}\,dy + e^x\,dx = 0$
11. $y\tan x\,dx + \cos^2 x\,dy = 0$
12. $\sin x\sec y\,dx = dy$
13. $yx^2\,dx = y\,dx - x^2\,dy$
14. $2y(x^3 + 1)\,dy + 3x^2(y^2 - 1)\,dx = 0$
15. $y\sqrt{1 - x^2}\,dy + 2\,dx = 0$
16. $\sqrt{1 + 4x^2}\,dy = y^3x\,dx$
17. $2\ln x\,dx + x\,dy = 0$
18. $e^{\sin x}\,dx + \sec x\,dy = 0$
19. $y^2e^x + (e^x + 1)\dfrac{dy}{dx} = 0$
20. $y + 1 + \sec x(\sin x + 1)\dfrac{dy}{dx} = 0$

In Exercises 21 through 24 in each case find the particular solution of the given differential equations for the indicated conditions.

21. $\dfrac{dy}{dx} + yx^2 = 0,\quad x = 0$ when $y = 1$

22. $y' = \sec y,\quad x = 0$ when $y = 0$

23. $y' = (1 - y)\cos x,\quad x = \pi/6$ when $y = 0$

24. $2y \cos y\, dy - \sin y\, dy = y \sin y\, dx,\quad x = 0$ when $y = \pi/2$

29–3 Integrable Combinations

Many differential equations cannot be solved by the method of separation of variables. Many other methods have been developed for solving such equations. One of these methods is based on the fact that certain combinations of basic differentials can be integrated together as a unit. The following differentials suggest some of these combinations which may occur:

$$d(xy) = x\, dy + y\, dx \tag{29–3}$$
$$d(x^2 + y^2) = 2(x\, dx + y\, dy) \tag{29–4}$$
$$d\left(\frac{y}{x}\right) = \frac{x\, dy - y\, dx}{x^2} \tag{29–5}$$
$$d\left(\frac{x}{y}\right) = \frac{y\, dx - x\, dy}{y^2} \tag{29–6}$$

Equation (29–3) suggests that, if the combination of $x\, dy + y\, dx$ occurs in a differential equation, we look for functions of xy. Equation (29–4) suggests that, if the combination $x\, dx + y\, dy$ occurs, we look for functions of $x^2 + y^2$. Equations (29–5) and (29–6) suggest that, if the combination $x\, dy - y\, dx$ occurs, we look for functions of y/x or x/y. The following examples illustrate the use of these combinations.

Example A
Solve the differential equation $x\, dy + y\, dx + xy\, dy = 0$.
  By dividing through by xy, we have

$$\frac{x\, dy + y\, dx}{xy} + dy = 0$$

The left term is the differential of xy divided by xy. This means it integrates to $\ln xy$. Thus, we have

$$\ln xy + y = c$$

Example B
Solve the differential equation $y\,dx - x\,dy + x\,dx = 0$.

The combination of $y\,dx - x\,dy$ suggests that this equation might make use of either Eq. (29–5) or (29–6). This would require dividing through by x^2 or y^2. If we divide by y^2, the last term cannot be integrated, but division by x^2 still allows integration of the last term. Performing this division, we obtain

$$\frac{y\,dx - x\,dy}{x^2} + \frac{dx}{x} = 0$$

This left combination is the negative of Eq. (29–5). The result then becomes

$$-\frac{y}{x} + \ln x = c$$

Example C
Solve the differential equation $(x^2 + y^2 + x)\,dx + y\,dy = 0$.

Regrouping the terms of this equation, we have

$$(x^2 + y^2)\,dx + (x\,dx + y\,dy) = 0$$

By dividing through by $x^2 + y^2$, we have

$$dx + \frac{x\,dx + y\,dy}{x^2 + y^2} = 0$$

The right term now can be put in the form of du/u (with $u = x^2 + y^2$) by multiplying each of the terms of the numerator by 2. The result then becomes

$$x + \frac{1}{2}\ln(x^2 + y^2) = \frac{c}{2} \quad \text{or} \quad 2x + \ln(x^2 + y^2) = c$$

Example D
Find the particular solution of the differential equation

$$(x^3 + xy^2 + 2y)\,dx + (y^3 + x^2y + 2x)\,dy = 0$$

which satisfies the condition that $x = 1$ when $y = 0$.

Regrouping the terms of the equation, we have

$$x(x^2 + y^2)\,dx + y(x^2 + y^2)\,dy + 2(y\,dx + x\,dy) = 0$$

Factoring $x^2 + y^2$ from each of the first two terms gives

$$(x^2 + y^2)(x\,dx + y\,dy) + 2(y\,dx + x\,dy) = 0$$

The first term can be integrated by multiplying the $x\,dx + y\,dy$ factor by 2. The result is

$$\frac{1}{2}\cdot\frac{1}{2}(x^2 + y^2)^2 + 2xy + \frac{c}{4} = 0 \quad \text{or} \quad (x^2 + y^2)^2 + 8xy + c = 0$$

Using the given condition gives $(1 + 0)^2 + 0 + c = 0$, or $c = -1$. The particular solution is then

$$(x^2 + y^2)^2 + 8xy = 1$$

The use of these integrable combinations depends on proper recognition of the forms. At times it may take two or three arrangements to arrive at the combination which leads to the result. Of course, many problems cannot be so arranged as to give integrable combinations in all terms.

Exercises 29–3

In Exercises 1 through 16 solve the given differential equations.

1. $x\,dy + y\,dx + x\,dx = 0$
2. $(2y + x)\,dy + y\,dx = 0$
3. $y\,dx - x\,dy + x^3\,dx = 2\,dx$
4. $x\,dy - y\,dx + y^2\,dx = 0$
5. $x^3\,dy + x^2y\,dx + y\,dx - x\,dy = 0$
6. $\sec(xy)\,dx + (x\,dy + y\,dx) = 0$
7. $x^3y^4(x\,dy + y\,dx) = dy$
8. $x\,dy + y\,dx + 4xy^3\,dy = 0$
9. $\sqrt{x^2 + y^2}\,dx - 2y\,dy = 2x\,dx$
10. $x\,dx + (x^2 + y^2 + y)\,dy = 0$
11. $\tan(x^2 + y^2)\,dy + x\,dx + y\,dy = 0$
12. $(x^2 + y^3)^2\,dy + 2x\,dx + 3y^2\,dy = 0$
13. $y\,dy - x\,dx + (y^2 - x^2)\,dx = 0$
14. $e^{x+y}(dx + dy) + 4x\,dx = 0$
15. $10x\,dy + 5y\,dx + 3y\,dy = 0$
16. $x^2\,dy + 3xy\,dx + 2\,dx = 0$

In Exercises 17 through 20 find the particular solutions to the given differential equations which satisfy the given conditions.

17. $2(x\,dy + y\,dx) + 3x^2\,dx = 0$, $x = 1$ when $y = 2$
18. $x\,dx + y\,dy = 2(x^2 + y^2)\,dx$, $x = 1$ when $y = 0$
19. $y\,dx - x\,dy = y^3\,dx + y^2x\,dy$, $x = 2$ when $y = 4$
20. $e^{x/y}(x\,dy - y\,dx) = y^4\,dy$, $x = 0$ when $y = 2$

29–4 The Linear Differential Equation of the First Order

There is one type of differential equation of the first order and first degree for which an integrable combination can always be determined. It is the linear differential equation of the first order and is of the form

$$dy + Py\,dx = Q\,dx \qquad (29\text{–}7)$$

where P and Q are functions of x only. We shall find that this type of equation occurs widely in applications.

If each side of Eq. (29–7) is multiplied by $e^{\int P\,dx}$ it becomes integrable, since the left side becomes of the form du with $u = ye^{\int P\,dx}$ and the right side is a function of x only. This is shown by finding the differential of $ye^{\int P\,dx}$. Thus,

$$d(ye^{\int P\,dx}) = e^{\int P\,dx}(dy + Py\,dx)$$

(In finding the differential of $\int P \, dx$ we use the fact that, by definition, these are reverse processes. Thus, $d(\int P \, dx) = P \, dx$.) Therefore, if each side is multiplied by $e^{\int P dx}$, the left side may be immediately integrated to $ye^{\int P dx}$ and the right-side integration may be indicated. The solution becomes

$$ye^{\int P dx} = \int Q e^{\int P dx} dx + c \qquad (29\text{–}8)$$

Example A

Solve the differential equation $dy + \left(\dfrac{2}{x}\right) y \, dx = 4x \, dx$. Here we note that $P = 2/x$ and $Q = 4x$.

The first expression to find is $e^{\int P dx}$. In this case this is $e^{\int (2/x) dx} = e^{2 \ln x} = e^{\ln x^2} = x^2$ (see below). This means that the left side integrates to yx^2, while the right side becomes $\int 4x(x^2) \, dx$. Thus,

$$yx^2 = \int 4x^3 dx + c \quad \text{or} \quad yx^2 = x^4 + c$$

In finding solutions to this type of differential equation, we often have need of certain basic properties of logarithms and exponential functions. In finding the factor $e^{\int P dx}$, we often obtain an expression of the form $e^{\ln u}$. It is easily shown that this equals u.

Let $y = e^{\ln u}$. Writing this in logarithmic form, we have $\ln y = \ln u$. Thus, $y = u$ or $u = e^{\ln u}$.

Example B

Solve the differential equation $x \, dy - 3y \, dx = x^3 dx$.

This equation can be put in the form of Eq. (29–7) by dividing through by x. This gives $dy - (3/x)y \, dx = x^2 dx$. Here, $P = -3/x$, $Q = x^2$, and the factor $e^{\int P dx}$ becomes

$$e^{\int (-3/x) dx} = e^{-3 \ln x} = e^{\ln x^{-3}} = x^{-3}$$

The original equation can now be written as

$$yx^{-3} = \int x^2 (x^{-3}) \, dx = \int x^{-1} dx = \ln x + c$$

A better form of the answer is $y = x^3 (\ln x + c)$.

Example C

Solve the differential equation $dy + y \, dx = x \, dx$. Here, $P = 1$, $Q = x$, and

$$e^{\int P dx} = e^{\int (1) dx} = e^x$$

Therefore,

$$ye^x = \int xe^x dx + c = e^x(x - 1) + c$$

or

$$y = x - 1 + ce^{-x}$$

Example D
Solve the differential equation $x^2 dy + 2xy\, dx = \sin x\, dx$.
   This equation is first written in the form of Eq. (29–7). This gives us

$$dy + \left(\frac{2}{x}\right) y\, dx = \frac{1}{x^2} \sin x\, dx \qquad \text{thus} \quad e^{\int P\, dx} = e^{\int (2/x)dx} = x^2$$

The solution to the equation then becomes

$$yx^2 = \int \sin x\, dx + c = -\cos x + c$$

or

$$yx^2 + \cos x = c$$

Example E
Find the particular solution of the differential equation

$$dy = (1 - 2y)x\, dx$$

such that $x = 0$ when $y = 2$.
   First writing this equation in the form of Eq. (29–7), we have

$$dy + 2xy\, dx = x\, dx$$

Therefore, in this case

$$e^{\int P\, dx} = e^{\int 2x\, dx} = e^{x^2}$$

This leads to

$$ye^{x^2} = \int xe^{x^2}\, dx$$

Integrating the right side, we have

$$ye^{x^2} = \frac{1}{2}e^{x^2} + c$$

Using the condition that $x = 0$ when $y = 2$ gives

$$(2)(e^0) = \frac{1}{2}(e^0) + c$$

$$2 = \frac{1}{2} + c$$

$$c = \frac{3}{2}$$

Therefore, the required solution is

$$ye^{x^2} = \frac{1}{2}e^{x^2} + \frac{3}{2}$$

$$y = \frac{1}{2}(1 + 3e^{-x^2})$$

Exercises 29—4

In Exercises 1 through 20 solve the given differential equations.

1. $dy + y\,dx = e^{-x}dx$

2. $dy + 3y\,dx = e^{-3x}dx$

3. $dy + 2y\,dx = e^{-4x}dx$

4. $dy + y\,dx = e^{-x}\cos x\,dx$

5. $x\,dy - y\,dx = 3x\,dx$

6. $x\,dy + 3y\,dx = dx$

7. $2x\,dy + y\,dx = 8x^3dx$

8. $3x\,dy - y\,dx = 9x\,dx$

9. $dy + y\cot x\,dx = dx$

10. $dy + y\tan x\,dx + \sin x\,dx = 0$

11. $y' + y = 3$

12. $y' + 2y = \sin x$

13. $\dfrac{dy}{dx} = xe^{4x} + 4y$

14. $y' - 2y = 2e^{2x}$

15. $y' = x^3(1 - 4y)$

16. $y' = x^2y + 3x^2$

17. $x\dfrac{dy}{dx} = y + (x^2 - 1)^2$

18. $2x(dy - dx) + y\,dx = 0$

19. $x\,dy + (1 - 3x)y\,dx = 3x^2e^{3x}dx$

20. $(1 + x^2)\,dy + xy\,dx = x\,dx$

In Exercises 21 through 24 find the indicated particular solutions of the given differential equations.

21. $\dfrac{dy}{dx} + 2y = e^{-x}, \quad x = 0$ when $y = 1$

22. $y' - 3y = e^{4x}, \quad x = 0$ when $y = 2$

23. $y' + \dfrac{1}{2\sqrt{x}}y = e^{\sqrt{x}}, \quad x = 1$ when $y = 3$

24. $(\sin x)\,y' + y = \tan x, \quad x = \pi/4$ when $y = 0$

29—5 Elementary Applications

The differential equations of the first order and first degree which we have discussed thus far have a number of applications in geometry and the various fields of technology. In this section we shall illustrate some of these applications, both through the examples and the exercises.

Example A

The slope of a given curve is given by the expression $6xy$. Find the equation of the curve if it passes through the point $(2, 1)$.

Since the slope is $6xy$, we know that $dy/dx = 6xy$. Writing this differential equation as

$$dy = 6xy\,dx$$

we divide through by y to separate the variables. Then integrating we find the general form of the solution.

$$\frac{dy}{y} = 6x\,dx$$

$$\ln y = 3x^2 + c$$

Using the fact that the curve passes through (2, 1), we evaluate the constant.

$$\ln 1 = 3(2^2) + c$$
$$0 = 12 + c$$
$$c = -12$$

Thus, the required particular solution is

$$\ln y = 3x^2 - 12$$

Example B

A curve which intersects all the members of a family of curves at right angles is called an **orthogonal trajectory** *of the family. By a family of curves is meant a specified set of curves which satisfy given conditions. Find the equations of the orthogonal trajectories of the parabolas* $x^2 = cy$.

Finding the derivative of the given equation, we have $dy/dx = 2x/c$. This equation contains the constant c, which depends on the point (x, y) on the parabola. Eliminating this constant between the equations of the parabola and derivative, we have $dy/dx = 2y/x$. This expression gives us a general expression for the slope of any of the members of the family. For a curve to be perpendicular, its slope must equal the negative reciprocal of this expression, or the slope of the orthogonal trajectories must be given by

$$\frac{dy}{dx}\bigg|_{OT} = -\frac{x}{2y}$$

Solving this differential equation gives the family of orthogonal trajectories:

$$2y \, dy = -x \, dx, \quad y^2 = -\frac{x^2}{2} + \frac{c}{2} \quad \text{or} \quad 2y^2 + x^2 = c$$

Thus, the orthogonal trajectories are a family of ellipses. Note in Fig. 29–1 that each parabola intersects each ellipse at right angles.

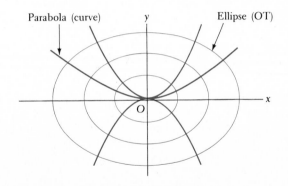

Parabola (curve)   *y*   Ellipse (OT)

O

x

Figure 29–1

Example C

Radioactive elements decay at rates proportional to the amount present. A certain isotope of uranium decays so that half of the original amount disappears in 20 d. Determine the equation relating the amount present with the time and the fraction remaining after 50 d.

Let N_0 be the original amount and N the amount present at any time t (in days). Since we know information related to the rate of decay, we express this rate as a derivative. This means that we have the equation

$$\frac{dN}{dt} = kN$$

Separating variables, and integrating, we have

$$\frac{dN}{N} = k\, dt$$

$$\ln N = kt + \ln c$$

Evaluating the constant, we know that $N = N_0$ for $t = 0$. Thus,

$$\ln N_0 = k(0) + \ln c$$
$$c = N_0$$

Therefore, we have the solution

$$\ln N = kt + \ln N_0$$
$$\ln N - \ln N_0 = kt$$

$$\ln \frac{N}{N_0} = kt$$

$$N = N_0 e^{kt}$$

Now using the condition that half this isotope decays in 20 d, we have $N = N_0/2$ when $t = 20$ d. This gives

$$\frac{N_0}{2} = N_0 e^{20k} \quad \text{or} \quad \frac{1}{2} = (e^k)^{20} \quad \text{or} \quad e^k = \left(\frac{1}{2}\right)^{1/20}$$

Therefore, the equation relating N and t is

$$N = N_0 \left(\frac{1}{2}\right)^{t/20}$$

In order to determine the fraction remaining after 50 d, we evaluate this equation for $t = 50$. This gives

$$N = N_0 \left(\frac{1}{2}\right)^{5/2} = 0.177\, N_0$$

This means that 17.7% remains after 50 d.

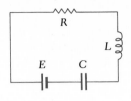

Figure 29–2

Example D

The general equation relating the current i, voltage E, inductance L, capacitance C, and resistance R of a simple electric circuit (see Fig. 29–2) is given by

$$L\frac{di}{dt} + Ri + \frac{q}{C} = E$$

where q is the charge on the capacitor. Find the general expression for the current in a circuit containing an inductance, resistance, and voltage source if $i = 0$ when $t = 0$.

The differential equation for this circuit is

$$L\frac{di}{dt} + Ri = E \qquad (29\text{–}9)$$

Using the method of the linear differential equation of the first order, we have the equation

$$di + \frac{R}{L}i\,dt = \frac{E}{L}dt$$

The factor $e^{\int P\,dt}$ is $e^{\int (R/L)dt} = e^{(R/L)t}$. This gives

$$ie^{(R/L)t} = \frac{E}{L}\int e^{(R/L)t}dt = \frac{E}{R}e^{(R/L)t} + c$$

Letting the current be zero for $t = 0$, we have $c = -E/R$. The result is

$$ie^{(R/L)t} = \frac{E}{R}e^{(R/L)t} - \frac{E}{R}$$

or

$$i = \frac{E}{R}(1 - e^{-(R/L)t})$$

(We can see that $i \to E/R$ as $t \to \infty$. In practice the exponential term becomes negligible very quickly.)

Example E

Fifty gallons of brine originally containing 20 lb of salt are in a tank into which 2 gal of water run each minute with the same amount of mixture running out each minute. How much salt remains in the tank after 10 min?

Let $x =$ the number of pounds of salt in the tank at time t. Each gallon of brine contains $x/50$ lb of salt, and in time dt, $2\,dt$ gal of mixture leave the tank with $(x/50)(2\,dt)$ lb of salt. The amount of salt which is leaving may also be written as $-dx$ (the minus sign is included to show that x is decreasing). This means

$$-dx = \frac{2x\,dt}{50} \quad \text{or} \quad \frac{dx}{x} = -\frac{dt}{25}$$

This leads to $\ln x = -(t/25) + \ln c$. Using the fact that $x = 20$ when $t = 0$, we have $\ln 20 = \ln c$, or $c = 20$. Therefore,

$$x = 20e^{-t/25}$$

is the general expression for the amount of salt in the tank at time t. When $t = 10$ we have

$$x = 20e^{-10/25} = 20e^{-0.4} = 20(0.670) = 13.4$$

There are 13.4 lb of salt in the tank after 10 min.

Example F

An object moving through (or across) a resisting medium often experiences a retarding force approximately proportional to the velocity. Applying Newton's laws of motion to this leads to the equation

$$m\frac{dv}{dt} = F - kv \qquad\qquad (29\text{--}10)$$

where m is the mass of the object, v is the velocity, t is the time, F is the force causing the motion, and k is a constant $(k > 0)$. The quantity kv is the retarding force.

We assume these conditions hold for a certain falling object whose mass is 1 slug (this is the unit of mass when force is expressed in pounds) and which experiences a force (its weight) of 32 lb, if the object starts from rest and the retarding force is numerically equal to 0.2 times the velocity.

Substituting in Eq. (29–10), we have

$$\frac{dv}{dt} = 32 - 0.2v$$

Separating variables, we have

$$\frac{dv}{32 - 0.2v} = dt$$

Integrating, we find

$$-5 \ln(32 - 0.2v) = t - 5 \ln c$$

Solving for v, we obtain

$$\ln(32 - 0.2v) = -\frac{t}{5} + \ln c$$

$$32 - 0.2v = ce^{-t/5}$$

$$v = 5(32 - ce^{-t/5})$$

Since the object started from rest, $v = 0$ when $t = 0$. Thus,

$$0 = 5(32 - c) \quad \text{or} \quad c = 32$$

Therefore, the expression giving the velocity as a function of time is

$$v = 160(1 - e^{-t/5})$$

Evaluating v for $t = 5$ s, we have

$$v = 160(1 - e^{-1}) = 160(1 - 0.368) = 101 \text{ ft/s}$$

Exercises 29–5

1. Find the equation of the curve for which the slope is given by $2x/y$ and which passes through $(2, 3)$.

2. Find the equation of the curve for which the slope is given by $-y/(x + y)$ and which passes through $(-1, 3)$.

3. Find the equation of the curve for which the slope is given by $y + x$ and which passes through $(0, 1)$.

4. Find the equation of the curve for which the slope is given by $-2y + e^{-x}$ and which passes through $(0, 2)$.

5. Find the equation of the orthogonal trajectories of the exponential curves $y = ce^x$.

6. Find the equation of the orthogonal trajectories of the cubic curves $y = cx^3$.

7. Find the equation of the orthogonal trajectories of the curves $y = c(\sec x + \tan x)$.

8. Find the equation of the orthogonal trajectories of the family of circles, all with centers at the origin.

9. A certain isotope decays in such a way that half the original amount disappears in 2 h. Find the proportion remaining after 3 h, by first finding the relation between the amount present and the time.

10. Radium decays in such a way that half of the original amount disappears in 1800 yr. What part of the original amount disappears in 100 yr?

11. If interest in a bank account is compounded continuously, the amount grows at a rate which is proportional to the amount which is present. A "day of deposit to day of withdrawal" account very closely approximates this situation. Determine the amount in an account after one year if $1000 is placed in an account which pays 4% interest per year, compounded continuously.

12. If interest is compounded continuously (see Exercise 11), how long will it take a bank account to double in value if the rate of interest is 5% per year?

13. If the current in an RL circuit with a voltage source E is zero when $t = 0$ (see Example D), show that $\lim_{t \to \infty} i = E/R$.

14. If a circuit contains only an inductance and a resistance, find the expression for the current in terms of the time, if $i = i_0$ when $t = 0$.

15. For the circuit in Example D, find the current after 1 ms, if $R = 10 \ \Omega$, $L = 0.1$ H, and $E = 100$ V.

16. For the circuit in Exercise 14, what is the value of the current after 1 ms, if $R = 10 \ \Omega$, $L = 0.1$ H, and $i_0 = 2$ A?

17. If a circuit contains only a resistance and a capacitance, find the expression relating the charge on the capacitor in terms of the time, if $i = dq/dt$ and $q = q_0$ when $t = 0$.

18. If the charge on a capacitor in an RC circuit (see Exercise 17) is 2 mC when $t = 0$, find the charge on the capacitor after 0.01 s if the resistance is 450 Ω and the capacitance is 4 μF.

19. One hundred gallons of brine originally containing 30 lb of salt are in a tank into which 5 gal of water run each minute. The same amount of mixture from the tank leaves each minute. How much salt is in the tank after 20 min?

20. Repeat Exercise 19 with the change that the water entering the tank contains 1 lb of salt per gallon.

21. An object falling under the influence of gravity has a variable acceleration given by $32 - v$, where v represents the velocity. If the object starts from rest, find an expression for the velocity in terms of the time. Also find the limiting value of the velocity (find $\lim_{t \to \infty} v$).

22. A mass of 2 slugs (weight is 64 lb) is falling under the influence of gravity. Find its velocity after 5 s if it starts from rest and experiences a retarding force numerically equal to 0.1 times the velocity.

23. A boat with a mass of 10 slugs is being towed at 8 mi/h. The tow rope is then cut and a motor which exerts a force of 20 lb on the boat is started. If the water exerts a retarding force which numerically equals twice the velocity, what is the velocity of the boat 3 min later?

24. A parachutist is falling at the rate of 200 ft/s when her parachute opens. If the air resists the fall with a force equal to $0.5v^2$, find the velocity as a function of time. The woman and her equipment have a mass of 5 slugs (weight is 160 lb).

25. A particle moves on the hyperbola $xy = 1$ such that $dx/dt = 2t$. Express x and y in terms of t if $x = 1$, $y = 1$ when $t = 0$.

26. A particle moves on the parabola $y = x^2 + x$, such that $dx/dt = 4t + 1$. Express x and y in terms of t, if both x and y are zero when $t = 0$.

27. The rate of change of air pressure with respect to height is approximately proportional to the pressure. If the pressure is 15 lb/in.^{2} when $h = 0$, and $p = 10$ lb/in.^{2} when $h = 10,000$ ft, find the expression relating pressure and height.

28. When a gas undergoes an adiabatic change (no gain or loss of heat), the rate of change of pressure with respect to volume is directly proportional to the pressure and inversely proportional to the volume. Express the pressure in terms of the volume.

29. Assume that the rate of depreciation of an object is proportional to its value at any time t. If a car costs $4000 new, and its value is $2000 three years later, what is its value five years after it was purchased?

30. Assume that sugar dissolves at a rate proportional to the undissolved amount. If there are initially 500 g of sugar, and 200 g remain after 4 min, how long does it take to dissolve 400 g?

31. Under certain conditions, the rate of change of temperature of an object is proportional to the difference in temperature between it and the surrounding medium. (Remember, the temperature decreases with time.) Find the temperature after an hour of an object originally 100°C, if it cools to 90°C in 5 min in air which is at 20°C.

32. The lines of equal potential in a field of force are all at right angles to the lines of force. In an electric field of force caused by charged particles, the lines of force are given by

$$x^2 + y^2 = cx$$

Find the equation of the lines of equal potential. Sketch a few members of the lines of force and those of equal potential.

29–6 Linear Differential Equations of Higher Order

Until now we have restricted our attention to differential equations of the first order and first degree. Another important type of differential equation is the linear differential equation of the second order with constant coefficients. Before restricting our attention to this specific type, we shall briefly describe the general higher-order equation and the notation we shall use with this type of equation.

The general linear differential equation of the nth order is of the form

$$a_0 \frac{d^n y}{dx^n} + a_1 \frac{d^{n-1} y}{dx^{n-1}} + \cdots + a_{n-1} \frac{dy}{dx} + a_n y = b \qquad (29\text{--}11)$$

where the a's and b are either functions of x or are constants.

For convenience of notation, the nth derivative with respect to x will be denoted by D^n. The other derivatives are denoted in a similar manner. In this notation, the general linear differential equation becomes

$$a_0 D^n y + a_1 D^{n-1} y + \cdots + a_{n-1} Dy + a_n y = b \qquad (29\text{--}12)$$

If $b = 0$ the general linear equation is called **homogeneous,** and if $b \neq 0$ it is called **nonhomogeneous.** Both these types have important applications.

Although the a's may be functions of x, we shall restrict our attention to the case where they are all constants. We shall, however, consider both homogeneous and nonhomogeneous equations. Since second-order equations occur widely in applied problems, the remainder of the chapter will be devoted to this type of equation. The methods developed, however, may be applied to equations of higher order.

29–7 Second-Order Homogeneous Equations with Constant Coefficients

In accordance with the notation and terminology introduced in Section (29–6), a second-order, linear, homogeneous differential equation with constant coefficients is one of the form

$$a_0 D^2 y + a_1 Dy + a_2 y = 0 \qquad\qquad (29\text{–}13)$$

where the a's are now constants. The following example will indicate the kind of solution we should expect for this type of equation.

Example A

Solve the differential equation $D^2 y - Dy - 2y = 0$.

First we put this equation in the form $(D^2 - D - 2)y = 0$. This is another way of saying that we are to take the second derivative of y, subtract the first derivative, and finally subtract twice the function. This may now be factored as $(D - 2)(D + 1)y = 0$. (We shall not develop the algebra of the operator D. However, most such algebraic operations can be shown to be valid.) This formula tells us to find the first derivative of the function and add this to the function. Then twice this result is to be subtracted from the derivative of this result. If we let $z = (D + 1)y$, which is valid since $(D + 1)y$ is a function of x, we have $(D - 2)z = 0$. This equation is easily solved by separation of variables. Thus,

$$\frac{dz}{dx} - 2z = 0, \qquad \frac{dz}{z} - 2\,dx = 0, \qquad \ln z - 2x = \ln c_1$$

$$\ln \frac{z}{c_1} = 2x \quad \text{or} \quad z = c_1 e^{2x}$$

Replacing z by $(D + 1)y$, we have

$$(D + 1)y = c_1 e^{2x}$$

This is a linear equation of the first order. Then

$$dy + y\,dx = c_1 e^{2x} dx$$

The factor $e^{\int P\,dx}$ is $e^{\int dx} = e^x$. And so

$$ye^x = \int c_1 e^{3x} dx = \frac{c_1}{3} e^{3x} + c_2$$

or

$$y = c_1' e^{2x} + c_2 e^{-x}$$

where $c_1' = \frac{1}{3}c_1$. This example indicates that solutions of the form e^{mx} result for this equation.

Basing our reasoning on the result of this example, let us assume that an equation of the form (29–13) has a particular solution e^{mx}. Substituting this into Eq. (29–13) gives

$$a_0 m^2 e^{mx} + a_1 m e^{mx} + a_2 e^{mx} = 0$$

The exponential function e^{mx} is never zero, which means that the polynomial in m is zero, or

$$a_0 m^2 + a_1 m + a_2 = 0 \qquad (29\text{–}14)$$

Equation (29–14) is called the **auxiliary equation** *of Eq. (29–13). Note that it may be formed directly by inspection from Eq. (29–13).*

There are two roots of the auxiliary Eq. (29–14) and there are two arbitrary constants in the solution of Eq. (29–13). These factors lead us to the general solution of Eq. (29–13), which is

$$y = c_1 e^{m_1 x} + c_2 e^{m_2 x} \qquad (29\text{–}15)$$

where m_1 and m_2 are the solutions of Eq. (29–14). This is seen to be in agreement with the results of Example A.

Example B
Solve the differential equation $D^2 y - 5Dy + 6y = 0$. The auxiliary equation is $m^2 - 5m + 6 = 0$.

Factoring this equation, we have $(m - 3)(m - 2) = 0$, which gives the roots $m_1 = 3$ and $m_2 = 2$. Thus, the solution is

$$y = c_1 e^{3x} + c_2 e^{2x}$$

Obviously it makes no difference which constant is written with each exponential function.

Example C
Solve the differential equation $D^2 y - 6Dy = 0$. Here

$$m^2 - 6m = 0$$
$$m(m - 6) = 0$$

or

$$m_1 = 0 \quad \text{and} \quad m_2 = 6$$

The solution is $y = c_1 e^{0x} + c_2 e^{6x}$. Since $e^{0x} = 1$, we have

$$y = c_1 + c_2 e^{6x}$$

Example D

Solve the differential equation $2y'' + 7y' = 4y$.

We first rewrite this equation using the D notation for derivatives. Also, we want to write in the proper form of a homogeneous equation. This gives us

$$2D^2y + 7Dy - 4y = 0$$

Now writing the auxiliary equation and solving it, we have

$$2m^2 + 7m - 4 = 0, \qquad (2m - 1)(m + 4) = 0$$

or

$$m_1 = \frac{1}{2} \quad \text{and} \quad m_2 = -4$$

The solution is

$$y = c_1 e^{x/2} + c_2 e^{-4x}$$

Example E

Solve the differential equation $2\dfrac{d^2y}{dx^2} + \dfrac{dy}{dx} - 7y = 0$.

First writing this equation with the D notation for derivatives, we have

$$2D^2y + Dy - 7y = 0$$

This means that the auxiliary equation is

$$2m^2 + m - 7 = 0$$

Since this equation is not factorable, we solve it by use of the quadratic formula. This gives us

$$m = \frac{-1 \pm \sqrt{1 + 56}}{4} = \frac{-1 \pm \sqrt{57}}{4}$$

Therefore, the solution is

$$y = c_1 e^{\frac{-1+\sqrt{57}}{4}x} + c_2 e^{\frac{-1-\sqrt{57}}{4}x}$$

A somewhat better form can be obtained by observing that $e^{-x/4}$ is a factor of each term. Thus,

$$y = e^{-x/4}(c_1 e^{\sqrt{57}x/4} + c_2 e^{-\sqrt{57}x/4})$$

Example F

Solve the differential equation $D^2y - 2Dy - 15y = 0$, and find the particular solution which satisfies the conditions that $Dy = 2$, $y = -1$ when $x = 0$. (It is necessary to give two conditions since there are two constants to evaluate.)

We have

$$m^2 - 2m - 15 = 0, \qquad (m - 5)(m + 3) = 0$$

or

$$m_1 = 5 \quad \text{and} \quad m_2 = -3$$

$$y = c_1 e^{5x} + c_2 e^{-3x}$$

This equation is the general solution. In order to evaluate the constants c_1 and c_2, we use the given conditions to find two simultaneous equations in c_1 and c_2. These are then solved to determine the particular solution. Thus,

$$y' = 5c_1 e^{5x} - 3c_2 e^{-3x}$$

Using the given conditions in the general solution and its derivative, we have

$$c_1 + c_2 = -1, \qquad 5c_1 - 3c_2 = 2$$

The solution to this system of equations is $c_1 = -\frac{1}{8}$ and $c_2 = -\frac{7}{8}$. The particular solution becomes

$$y = -\frac{1}{8}e^{5x} - \frac{7}{8}e^{-3x}$$

or

$$8y + e^{5x} + 7e^{-3x} = 0$$

To solve the auxiliary equation for differential equations of higher order, we use methods developed in the theory of equations. With these roots, we form the solutions in the same way as we do those for second-order equations.

Exercises 29–6, 29–7   In Exercises 1 through 20 solve the given differential equations.

1. $\dfrac{d^2y}{dx^2} - \dfrac{dy}{dx} - 6y = 0$

2. $\dfrac{d^2y}{dx^2} + \dfrac{dy}{dx} = 0$

3. $\dfrac{d^2y}{dx^2} + 4\dfrac{dy}{dx} + 3y = 0$

4. $\dfrac{d^2y}{dx^2} - 2\dfrac{dy}{dx} - 8y = 0$

5. $D^2y - Dy = 0$

6. $D^2y + 7Dy + 6y = 0$

7. $3D^2y + 12y = 20Dy$

8. $4D^2y + 12Dy = 7y$

9. $3y'' + 8y' - 3y = 0$

10. $8y'' + 6y' - 9y = 0$

11. $3y'' + 2y' - y = 0$

12. $2y'' - 7y' + 6y = 0$

13. $2\dfrac{d^2y}{dx^2} - 4\dfrac{dy}{dx} + y = 0$

14. $\dfrac{d^2y}{dx^2} + \dfrac{dy}{dx} - 5y = 0$

15. $4D^2y - 3Dy - 2y = 0$

16. $2D^2y - 3Dy - y = 0$

17. $y'' = 3y' + y$

18. $5y'' - y' = 3y$

19. $y'' + y' = 8y$

20. $8y'' = y' + y$

In Exercises 21 through 24 find the particular solutions of the differential equations which satisfy the given conditions.

21. $D^2y - 4Dy - 21y = 0$, $Dy = 0$ and $y = 2$ when $x = 0$

22. $D^2y - 4Dy = 0$, $Dy = 2$ and $y = 4$ when $x = 0$

23. $D^2y - Dy - 12y = 0$, $y = 0$ when $x = 0$ and $y = 1$ when $x = 1$

24. $2D^2y + 5Dy = 0$, $y = 0$ when $x = 0$ and $y = 2$ when $x = 1$

In Exercises 25 and 26 solve the given differential equations. The auxiliary equations will be third degree, and the solutions will have three arbitrary constants.

25. $y''' - 2y'' - 3y' = 0$ **26.** $y''' - 2y'' - y' + 2y = 0$

29–8 Auxiliary Equations with Repeated or Complex Roots

The way to solve a second-order homogeneous linear differential equation was shown in the last section. However, we purposely avoided repeated and complex roots of the auxiliary equation. In this section we shall develop the solutions for such equations. The following example indicates the type of solution which results from the case of repeated roots.

Example A

Solve the differential equation $D^2y - 4Dy + 4y = 0$.

Using the method of Example A of the previous section, we have the following steps:

$$(D^2 - 4D + 4)y = 0, \qquad (D - 2)(D - 2)y = 0, \qquad (D - 2)z = 0$$

where $z = (D - 2)y$. The solution to $(D - 2)z = 0$ is found by separation of variables. And so

$$\frac{dz}{dx} - 2z = 0, \qquad \frac{dz}{z} - 2\,dx = 0$$

$$\ln z - 2x = \ln c_1 \quad \text{or} \quad z = c_1 e^{2x}$$

Substituting back, we have $(D - 2)y = c_1 e^{2x}$, which is a linear equation of the first order. Then

$$dy - 2y\,dx = c_1 e^{2x}dx, \qquad e^{\int -2\,dx} = e^{-2x}$$

This leads to

$$ye^{-2x} = c_1 \int dx = c_1 x + c_2 \quad \text{or} \quad y = c_1 xe^{2x} + c_2 e^{2x}$$

This example indicates the type of solution which results when the auxiliary equation has repeated roots. If the method of the previous section were to be used, the solution of the above example would be $y = c_1 e^{2x} + c_2 e^{2x}$. This would not be the general solution, since both terms are similar, which means that there is only one independent constant. The constants can be combined to give a solution of the form $y = ce^{2x}$, where $c = c_1 + c_2$. This solution would contain only one constant for a second-order equation.

Based on the above example, the solution to Eq. (29–13) when the auxiliary Eq. (29–14) has repeated roots is

$$y = e^{mx}(c_1 + c_2 x) \tag{29–16}$$

where m is the double root. (In Example A, this double root is 2.)

Example B
Solve the differential equation $(D + 2)^2y = 0$.

The auxiliary equation is $(m + 2)^2 = 0$, for which the solutions are $m = -2, -2$. Since we have repeated roots, the solution of the differential equation is

$$y = e^{-2x}(c_1 + c_2x)$$

Example C

Solve the differential equation $\dfrac{d^2y}{dx^2} - 10\dfrac{dy}{dx} + 25y = 0$.

Using the D notation, this equation becomes

$$D^2y - 10Dy + 25y = 0$$

Therefore, the auxiliary equation is

$$m^2 - 10m + 25 = 0$$

Factoring, we have

$$(m - 5)^2 = 0$$

This means the roots are $m = 5, 5$. We therefore have the solution

$$y = e^{5x}(c_1 + c_2x)$$

When the auxiliary equation has complex roots, it can be solved by the method developed in the preceding section. However, the solution can be put in a special, more useful form. Let us assume that the quadratic formula is used to find the roots of the auxiliary equation, and that these roots are complex of the form $m = \alpha \pm j\beta$. This means that the solution is of the form

$$y = c_1e^{(\alpha+j\beta)x} + c_2e^{(\alpha-j\beta)x} = e^{\alpha x}(c_1e^{j\beta x} + c_2e^{-j\beta x})$$

Using the exponential form of a complex number, Eq. (11–9), we have

$$\begin{aligned} y &= e^{\alpha x}[c_1\cos \beta x + jc_1\sin \beta x + c_2\cos(-\beta x) + jc_2\sin(-\beta x)] \\ &= e^{\alpha x}(c_1\cos \beta x + c_2\cos \beta x + jc_1\sin \beta x - jc_2\sin \beta x) \\ &= e^{\alpha x}(c_3\cos \beta x + c_4\sin \beta x) \end{aligned}$$

where $c_3 = c_1 + c_2$ and $c_4 = jc_1 - jc_2$.

Summarizing these results, we see that if the auxiliary equation has complex roots of the form $\alpha \pm j\beta$, the solution to Eq. (29–13) is

$$y = e^{\alpha x}(c_1\sin \beta x + c_2\cos \beta x) \qquad\qquad (29\text{--}17)$$

The c_1 and c_2 here are not the same as those above. They are simply the two arbitrary constants of the solution.

Example D

Solve the differential equation $D^2y - Dy + y = 0$. Thus,

$$m^2 - m + 1 = 0 \quad \text{or} \quad m = \frac{1 \pm \sqrt{3}j}{2}$$

$$\alpha = \frac{1}{2} \quad \text{and} \quad \beta = \frac{\sqrt{3}}{2}$$

We have

$$y = e^{x/2}\left(c_1 \sin \frac{\sqrt{3}}{2}x + c_2 \cos \frac{\sqrt{3}}{2}x\right)$$

Example E

Solve the differential equation $D^2y + 4y = 0$. In this case we have $m^2 + 4 = 0$, $m_1 = 2j$, and $m_2 = -2j$; $\alpha = 0$ and $\beta = 2$.
The solution is

$$y = e^{0x}(c_1 \sin 2x + c_2 \cos 2x)$$

or

$$y = c_1 \sin 2x + c_2 \cos 2x$$

Example F

Solve the differential equation $y'' - 2y' + 12y = 0$, if $y' = 2$ and $y = 1$ when $x = 0$.

Rewriting this equation as

$$D^2y - 2Dy + 12y = 0$$

we have the auxiliary equation

$$m^2 - 2m + 12 = 0$$

Solving for m, we have

$$m = \frac{2 \pm \sqrt{4 - 48}}{2} = \frac{2 \pm 2\sqrt{11}j}{2} = 1 \pm \sqrt{11}j$$

This means that the general solution is

$$y = e^x(c_1 \cos \sqrt{11}x + c_2 \sin \sqrt{11}x)$$

Using the condition that $y = 1$ when $x = 0$, we have

$$1 = e^0(c_1 \cos 0 + c_2 \sin 0)$$

or

$$c_1 = 1$$

Since the other condition involves the value of y', we find the derivative.

$$y' = e^x(c_1 \cos \sqrt{11}x + c_2 \sin \sqrt{11}x - \sqrt{11}c_1 \sin \sqrt{11}x + \sqrt{11}c_2 \cos \sqrt{11}x)$$
$$2 = e^0(\cos 0 + c_2 \sin 0 - \sqrt{11} \sin 0 + \sqrt{11}c_2 \cos 0)$$

$$2 = 1 + \sqrt{11}c_2$$

$$c_2 = \frac{1}{11}\sqrt{11}$$

Therefore, the solution is

$$y = e^x\left(\cos \sqrt{11}x + \frac{1}{11}\sqrt{11} \sin \sqrt{11}x\right)$$

Exercises 29-8

In Exercises 1 through 24 solve the given differential equations.

1. $\dfrac{d^2y}{dx^2} - 2\dfrac{dy}{dx} + y = 0$ 2. $\dfrac{d^2y}{dx^2} - 6\dfrac{dy}{dx} + 9y = 0$

3. $D^2y + 12Dy + 36y = 0$ 4. $D^2y + 8Dy + 16y = 0$

5. $\dfrac{d^2y}{dx^2} + 9y = 0$ 6. $\dfrac{d^2y}{dx^2} + y = 0$

7. $D^2y + Dy + 2y = 0$ 8. $D^2y - 2Dy + 4y = 0$
9. $D^2y = 0$ 10. $4D^2y - 12Dy + 9y = 0$
11. $4D^2y + y = 0$ 12. $9D^2y + 4y = 0$
13. $16y'' - 24y' + 9y = 0$ 14. $9y'' - 24y' + 16y = 0$
15. $25y'' + 2y = 0$ 16. $y'' - 4y' + 5y = 0$
17. $2D^2y + 5y = 4Dy$ 18. $D^2y + 4Dy + 6y = 0$
19. $25y'' + 16y = 40y'$ 20. $9y'' + 0.6y' + 0.01y = 0$
21. $2D^2y - 3Dy - y = 0$ 22. $D^2y - 5Dy - 4y = 0$
23. $3D^2y + 12Dy = 2y$ 24. $36D^2y = 25y$

In Exercises 25 through 28 find the particular solutions of the given differential equations which satisfy the stated conditions.

25. $y'' + 2y' + 10y = 0$, $y = 0$ when $x = 0$ and $y = e^{-1}$ when $x = \pi/6$
26. $D^2y + 16y = 0$, $Dy = 0$ and $y = 2$ when $x = \pi/2$
27. $D^2y - 8Dy + 16y = 0$, $Dy = 2$ and $y = 4$ when $x = 0$
28. $4y'' + 20y' + 25y = 0$, $y = 0$ when $x = 0$ and $y = e$ when $x = -2/5$

29-9 Solutions of Nonhomogeneous Equations

To solve a nonhomogeneous linear equation of the form

$$a_0D^2y + a_1Dy + a_2y = b \tag{29-18}$$

where b is a function of x or is a constant, the solution must be such that if we substitute it into the left side, we obtain the right side. Solutions obtained from the methods of Sections 29-7 and 29-8 will give zero when substituted into the left side, but they do contain the arbitrary constants necessary in the solution. If we could find some particular solution, which when substituted into the left side produced the expression of the right, it could be added to the solution containing the arbitrary constants. Thus, the solution is of the form

$$y = y_c + y_p \tag{29–19}$$

where y_c, called the **complementary solution**, is obtained by solving the corresponding homogeneous equation and where y_p is the particular solution necessary to produce the expression b of Eq. (29–18).

Example A
The differential equation $D^2y - Dy - 6y = e^x$ has the solution $y = c_1e^{3x} + c_2e^{-2x} - \frac{1}{6}e^x$, with $y_c = c_1e^{3x} + c_2e^{-2x}$ and $y_p = -\frac{1}{6}e^x$. The complementary solution y_c is obtained by solving the corresponding homogeneous equation $D^2y - Dy - 6y = 0$, and we shall discuss below the method of finding y_p. We can see that y_p alone will satisfy the differential equation, but since it has no arbitrary constants, it cannot be the general solution.

By inspection of the form of the expression b on the right side of the equation, we can determine the form which the particular solution must have. Since a combination of the derivatives and the function itself must form the function b, **we assume an expression which contains all possible forms of b and its derivatives will form y_p.** The method used to find the exact form of this expression is called the **method of undetermined coefficients.**

Example B
If the function b on the right of Eq. (29–18) is $4x + e^{2x}$, we would assume the particular solution is of the form $y_p = A + Bx + Ce^{2x}$. The Bx term is included to account for the presence of the $4x$. The A term is included to account for any derivatives of the Bx term which may occur. The Ce^{2x} is included to account for the presence of the e^{2x} term. Since all the derivatives of Ce^{2x} are of this same form, it is not necessary to include any other terms for the e^{2x}.

Example C
If the expression b is of the form $x^2 + e^{-x}$, we would assume y_p to be of the form $y_p = A + Bx + Cx^2 + Ee^{-x}$.
    If b is of the form $xe^x - 5$, we would assume y_p to be of the form $y_p = Ae^x + Bxe^x + C$.
    If b is of the form $x \sin x$, we would assume y_p to be of the form $A \sin x + B \cos x + Cx \sin x + Ex \cos x$. All of these terms occur in the derivatives of $x \sin x$.

Once we have determined the form of the particular solution, we have to find the numerical values of the coefficients A, B, ... *This is done by substituting the form into the differential equation, and equating coefficients of like terms.*

Example D

Solve the differential equation $D^2y - Dy - 6y = e^x$.

In this case the solution of the auxiliary equation $m^2 - m - 6 = 0$, gives us the roots $m_1 = 3$ and $m_2 = -2$. Thus,

$$y_c = c_1e^{3x} + c_2e^{-2x}$$

The proper form of y_p is $y_p = Ae^x$. Substituting y_p and its derivatives into the differential equation, we have

$$Ae^x - Ae^x - 6Ae^x = e^x$$

To produce equality, the coefficients of e^x must be the same on each side of the equation. Thus,

$$-6A = 1 \quad \text{or} \quad A = -\frac{1}{6}$$

This gives the complete solution $y = y_c + y_p$, where $y_p = -\frac{1}{6}e^x$. Thus,

$$y = c_1e^{3x} + c_2e^{-2x} - \frac{1}{6}e^x$$

See Example A.

Example E

Solve the differential equation $D^2y + 4y = x - 4e^{-x}$.

In this case we have $m^2 + 4 = 0$, which gives us $m_1 = 2j$ and $m_2 = -2j$. Therefore,

$$y_c = c_1\sin 2x + c_2\cos 2x$$

The proper form of the particular solution is $y_p = A + Bx + Ce^{-x}$. Finding two derivatives and then substituting gives the following result:

$$y_p' = B - Ce^{-x}, \quad y_p'' = Ce^{-x}$$
$$(Ce^{-x}) + 4(A + Bx + Ce^{-x}) = x - 4e^{-x}$$
$$Ce^{-x} + 4A + 4Bx + 4Ce^{-x} = x - 4e^{-x}$$
$$(4A) + (4Bx) + (5Ce^{-x}) = 0 + (1)x + (-4e^{-x})$$

Equating the coefficients of the constants, x and e^{-x}, gives $4A = 0$, $4B = 1$, and $5C = -4$. Thus, $A = 0$, $B = \frac{1}{4}$, and $C = -\frac{4}{5}$. This means that the particular solution is $y_p = \frac{1}{4}x - \frac{4}{5}e^{-x}$. In turn this tells us that the complete solution is

$$y = c_1\sin 2x + c_2\cos 2x + \frac{1}{4}x - \frac{4}{5}e^{-x}$$

Example F

Solve the differential equation $D^2y - 3Dy + 2y = 2 \sin x$. Thus,

$$m^2 - 3m + 2 = 0, \qquad (m - 1)(m - 2) = 0, \qquad m_1 = 1 \quad \text{and} \quad m_2 = 2$$

We write $y_c = c_1 e^x + c_2 e^{2x}$. Then

$$y_p = A \sin x + B \cos x$$
$$Dy_p = A \cos x - B \sin x$$
$$D^2y_p = -A \sin x - B \cos x$$
$$(-A \sin x - B \cos x) - 3(A \cos x - B \sin x) + 2(A \sin x + B \cos x)$$
$$= 2 \sin x$$

$$(A + 3B)\sin x + (B - 3A)\cos x = 2 \sin x$$
$$A + 3B = 2, \qquad -3A + B = 0$$

The solution of this system is $A = \frac{1}{5}$ and $B = \frac{3}{5}$. Thus,

$$y_p = \frac{1}{5}\sin x + \frac{3}{5}\cos x$$

Therefore,

$$y = c_1 e^x + c_2 e^{2x} + \frac{1}{5}\sin x + \frac{3}{5}\cos x$$

Exercises 29—9

In Exercises 1 through 8 solve the given differential equations. The form of y_p is indicated.

1. $D^2y - Dy - 2y = 4$ (Let $y_p = A$.)
2. $D^2y - Dy - 6y = 4x$ (Let $y_p = A + Bx$.)
3. $D^2y + y = x^2$ (Let $y_p = A + Bx + Cx^2$.)
4. $D^2y + Dy + y = e^{2x}$ (Let $y_p = Ae^{2x}$.)
5. $D^2y + 4Dy + 3y = 2 + e^x$ (Let $y_p = A + Be^x$.)
6. $D^2y - y = \sin x$ (Let $y_p = A \sin x + B \cos x$.)
7. $D^2y - Dy - 20y = 8 + e^{2x}$ (Let $y_p = A + Be^{2x}$.)
8. $D^2y + 4y = \sin x + 4$ (Let $y_p = A + B \sin x + C \cos x$.)

In Exercises 9 through 20 solve the given differential equations.

9. $\dfrac{d^2y}{dx^2} - \dfrac{dy}{dx} - 30y = 10$

10. $2\dfrac{d^2y}{dx^2} + 11\dfrac{dy}{dx} - 6y = 8x$

11. $3\dfrac{d^2y}{dx^2} + 13\dfrac{dy}{dx} - 10y = 14e^{3x}$

12. $\dfrac{d^2y}{dx^2} + 4y = 2 \sin 3x$

13. $D^2y - 4y = \sin x + 2 \cos x$

14. $2D^2y + 5Dy - 3y = e^x + 4e^{2x}$

15. $D^2y + y = 4 + \sin 2x$

16. $D^2y - Dy + y = x + \sin x$

17. $D^2y + 5Dy + 4y = xe^x + 4$

18. $3D^2y + Dy - 2y = 4 + 2x + e^x$

19. $y'' + 6y' + 9y = e^{2x} - e^{-2x}$

20. $y'' + 8y' - y = x^2 + 4e^{-2x}$

In Exercises 21 through 24 find the particular solution of each differential equation for the given conditions.

21. $D^2y - Dy - 6y = 5 - e^x$, $Dy = 4$ and $y = 2$ when $x = 0$.

22. $3y'' - 10y' + 3y = xe^{-2x}$, $Dy = 0$ and $y = -1$ when $x = 0$.

23. $y'' + y = x + \sin 2x$, $Dy = 1$ and $y = 0$ when $x = \pi$.

24. $D^2y - 2Dy + y = xe^{2x} - e^{2x}$, $Dy = 4$ and $y = -2$ when $x = 0$.

29–10 Applications of Second-Order Equations

Linear differential equations of the second order have many important applications. We shall restrict our attention to two of these. In this section, we shall apply second-order differential equations to solving problems in simple harmonic motion and simple electric circuits.

Example A

Simple harmonic motion may be defined as motion in a straight line for which the acceleration is proportional to the displacement and in the opposite direction. Examples of this type of motion are a weight on a spring, a simple pendulum, and an object bobbing in water. If x represents the displacement, d^2x/dt^2 is the acceleration.

Using the definition of simple harmonic motion, we have

$$\frac{d^2x}{dt^2} = -k^2x$$

(We chose k^2 for convenience of notation in the solution.) Writing this equation in the form

$$D^2x + k^2x = 0 \quad \left(\text{here, } D = \frac{d}{dt}\right)$$

we find that its solution is $x = c_1\sin kt + c_2\cos kt$. This solution indicates an oscillating motion, which is known to be the case. If, for example, $k = 4$ and we know that $x = 2$ and $Dx = 0$ (which means the velocity is zero) for $t = 0$, we have

$$Dx = 4c_1\cos 4t - 4c_2\sin 4t$$

or (substituting into the expression for x and that for Dx)

$$2 = c_1(0) + c_2(1) \quad \text{and} \quad 0 = 4c_1(1) - 4c_2(0)$$

which gives $c_1 = 0$ and $c_2 = 2$. Therefore,

$$x = 2 \cos 4t$$

is the equation relating the displacement and time; Dx is the velocity and D^2x is the acceleration.

Example B

In practice an object will in time cease to move due to unavoidable frictional forces. It is found that a "freely" oscillating object has a force retarding it which is approximately proportional to the velocity. The differential equation for this case is $D^2x = -k^2x - bDx$, which results from applying (from physics) Newton's second law of motion, which states that the net force acting on an object equals its mass times its acceleration. The term D^2x represents the acceleration, the term $-k^2x$ is a measure of the restoring force (of the spring, for example), and the term $-bDx$ is the term which represents the retarding (damping) force. This equation can be written as

$$D^2x + bDx + k^2x = 0$$

The auxiliary equation is $m^2 + bm + k^2 = 0$, for which the roots are

$$m = \frac{-b \pm \sqrt{b^2 - 4k^2}}{2}$$

If $k = 3$ and $b = 4$, $m = -2 \pm \sqrt{5}j$, which means the solution is

$$x = e^{-2t}(c_1 \sin \sqrt{5}t + c_2 \cos \sqrt{5}t) \tag{1}$$

Here, $4k^2 > b^2$, and this case is called **underdamped**.

If $k = 2$ and $b = 5$, $m = -1, -4$, which means the solution is

$$x = c_1 e^{-t} + c_2 e^{-4t} \tag{2}$$

Here $4k^2 < b^2$, and the case is called **overdamped**. It will be noted that the motion is not oscillatory, since no sine or cosine terms appear.

If $k = 2$ and $b = 4$, $m = -2, -2$, which means the solution is

$$x = e^{-2t}(c_1 + c_2 t) \tag{3}$$

Here $4k^2 = b^2$, and the case is called **critically damped**. Again the motion is not oscillatory. See Fig. 29–3 in which Eqs. (1), (2), and (3) are represented in general. Of course, the actual values depend upon c_1 and c_2, which in turn depend upon the conditions imposed on the motion.

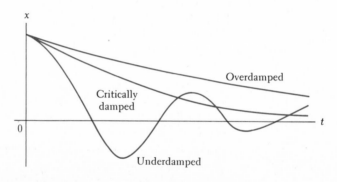

Figure 29–3

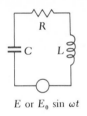

E or $E_0 \sin \omega t$

Figure 29–4

Example C

A basic relation for electric circuits is that the impressed voltage of a circuit equals the sums of the voltages across the components of the circuit. When this principle is applied to a circuit containing a resistance, inductance, capacitance, and a voltage source (see Fig. 29–4), the differential equation is

$$L\frac{d^2q}{dt^2} + R\frac{dq}{dt} + \frac{q}{C} = E \tag{29-20}$$

By definition q represents the electric charge, $dq/dt = i$ is the current and d^2q/dt^2 is the time-rate of change of current. This equation may be written as $LD^2q + RDq + q/C = E$. The auxiliary equation is $Lm^2 + Rm + 1/C = 0$. The roots are

$$m = \frac{-R \pm \sqrt{R^2 - 4L/C}}{2L} = -\frac{R}{2L} \pm \sqrt{\frac{R^2}{4L^2} - \frac{1}{LC}}$$

If we let $a = R/2L$ and $\omega = \sqrt{1/LC - R^2/4L^2}$, we have (assuming complex roots, which corresponds to realistic values of R, L, and C)

$$q_c = e^{-at}(c_1 \sin \omega t + c_2 \cos \omega t)$$

This indicates an oscillating charge, or an alternating current. However, the exponential term usually is such that the current dies out rapidly unless there is a source of voltage in the circuit.

If there is no source of voltage in the circuit of Example C, we have a homogeneous differential equation to solve. If we have a constant voltage source, the particular solution is of the form $q_p = A$. If there is an alternating voltage source, the particular solution is of the form $q_p = A \sin \omega_1 t + B \cos \omega_1 t$, where ω_1 is the angular velocity of the source. After a very short time, the exponential factor in the complementary solution makes it negligible. For this reason it is referred to as the **transient** term, and the particular solution is the **steady-state** solution. Therefore, to find the steady-state solution, we need find only the particular solution.

Example D

Find the steady-state solution for the current in a circuit containing the following elements: $C = 400\ \mu\text{F}$, $L = 1$ H, $R = 10\ \Omega$, and a voltage source of $500 \sin 100t$.

This means the differential equation to be solved is

$$\frac{d^2q}{dt^2} + 10\frac{dq}{dt} + \frac{10^4}{4}q = 500 \sin 100t$$

Since we wish to find the steady-state solution, we must find q_p from which we may find i_p by finding a derivative. Thus,

$$q_p = A \sin 100t + B \cos 100t$$

$$\frac{dq_p}{dt} = 100A \cos 100t - 100B \sin 100t$$

$$\frac{d^2q_p}{dt^2} = -10^4 A \sin 100t - 10^4 B \cos 100t$$

Substituting into the differential equation, we have

$$-10^4 A \sin 100t - 10^4 B \cos 100t + 10^3 A \cos 100t - 10^3 B \sin 100t$$

$$+ \frac{10^4}{4} A \sin 100t + \frac{10^4}{4} B \cos 100t = 500 \sin 100t$$

This means

$$(-0.75 \times 10^4 A - 10^3 B)\sin 100t + (-0.75 \times 10^4 B + 10^3 A)\cos 100t$$
$$= 500 \sin 100t$$

or

$$-7.5 \times 10^3 A - 10^3 B = 500, \qquad 10^3 A - 7.5 \times 10^3 B = 0$$

Solving these equations, we obtain

$$B = -8.73 \times 10^{-3} \quad \text{and} \quad A = -65.5 \times 10^{-3}$$

Therefore,

$$q_p = -65.5 \times 10^{-3}\sin 100t - 8.73 \times 10^{-3}\cos 100t$$

and

$$i_p = \frac{dq_p}{dt} = -6.55 \cos 100t + 0.87 \sin 100t$$

which is the desired solution.

It should be noted that the complimentary solutions of the mechanical and electrical cases are of the same identical form. There is also an equivalent mechanical case to that of an impressed sinusoidal voltage source in the electrical case. This arises when an external force affecting the vibrations is applied to the system. Such cases are called **forced vibrations**. Thus, we may have transient and steady-state solutions to mechanical and other nonelectrical situations.

Exercises 29–10

1. An object moves with simple harmonic motion according to the equation $D^2x + 100x = 0$; $D = d/dt$. Find the displacement x as a function of the time, subject to the conditions that $x = 4$ and $Dx = 0$ when $t = 0$. Sketch the resulting curve.

2. Solve the problem given in Exercise 1 if the motion is in accordance with the equation $D^2x + 0.2Dx + 100x = 0$. Sketch the curve.

3. What must be the value of b so that the motion given by the equation $D^2x + bDx + 100x = 0$ is critically damped?

4. What is the displacement of an object which moves according to the equation $D^2x + 4Dx + 6x = 0$ (if $x = 6$ and $Dx = 0$ when $t = 0$) after 2 s?

5. For a certain spring a 4-lb weight stretches it 1/8 ft. With this weight attached, the spring is pulled three inches longer than its equilibrium length and released. Find the equation of the resulting motion, assuming no damping.

6. Find the solution for the spring of Exercise 5 if a damping force numerically equal to the velocity is present.

7. Find the solution for the spring of Exercise 5 if no damping is present, but an external force of $4 \sin 2t$ is acting on the spring.

8. Find the solution for the spring of Exercise 5 if the damping force of Exercise 6 and the impressed force of Exercise 7 are both acting.

9. Find the equation relating the charge and time in an electric circuit with the following elements: $L = 0.2$ H, $R = 8$ Ω, $C = 1$ μF, $E = 0$. In this circuit $q = 0$ and $i = 0.5$ A when $t = 0$.

10. For a given electric circuit, $L = 0.5$ H, $R = 0$, $C = 200$ μF, and $E = 0$. Find the equation relating charge and time if $q = 100$ μC and $i = 0$ when $t = 0$.

11. For a given circuit, $L = 0.1$ H, $R = 0$, $C = 100$ μF, and $E = 100$ V. Find the equation relating the charge and time if $q = 0$ and $i = 0$ when $t = 0$.

12. Find the relation between the current and time for the circuit of Exercise 11.

13. For a given circuit, $L = 0.5$ H, $R = 10$ Ω, $C = 200$ μF, and $E = 100 \sin 200t$. Find the expression relating charge and time.

14. Find the relation between the current and time for the circuit of Exercise 13.

15. Find the steady-state current for the circuit of Exercise 13.

16. Find the steady-state current for a circuit with $L = 1$ H, $R = 5$ Ω, $C = 150$ μF, and $E = 120 \sin 100t$ V.

17. Find the steady-state solution for the charge of the circuit of Exercise 16 if the voltage source is changed to $100 \sin 400t$.

18. Find the steady-state solution for the current in an electric circuit containing the following elements: $C = 20$ μF, $L = 2$ H, $R = 20$ Ω, and $E = 200 \sin 10t$.

29—11 Laplace Transforms

The final method of solving differential equations which we shall discuss is by means of **Laplace transforms**. As we shall see, *Laplace transforms provide an algebraic method of obtaining a **particular** solution of a differential equation from stated initial conditions.* Since this is often what is desired in practice, Laplace transforms are often preferred for the solution of differential equations in engineering and electronics. In this section we shall discuss the meaning of the transform and its operations.

These methods will be applied to solving differential equations in the following section. The treatment in this text is intended only as an introduction to the topic of Laplace transforms.

The Laplace transform of a function $f(t)$ is defined as the function $F(s)$ by the equation

$$F(s) = \int_0^\infty e^{-st} f(t)\, dt \tag{29-21}$$

By writing the transform as $F(s)$ we show that the result of integrating and evaluating is a function of s. To denote that we are dealing with "the Laplace transform of the function $f(t)$" the notation $L(f)$ is used. Thus,

$$F(s) = L(f) = \int_0^\infty e^{-st} f(t)\, dt \tag{29-22}$$

We will see that both notations are quite useful.

A note regarding the form of the integral in Eqs. (29–21) and (29–22) is in order at this point. Since the upper limit is ∞, which means that it is unbounded, this integral is one type of what is known as an **improper integral**. In evaluating this integral at the upper limit, it is necessary to find the limit of the resulting function as the upper limit approaches infinity. This may be denoted by

$$\lim_{c \to \infty} \int_0^c e^{-st} f(t)\, dt$$

where we substitute c for t in the resulting function and determine the limit as $c \to \infty$ to determine the result for the upper limit.

Example A

Find the Laplace transform of the function $f(t) = t, \; t > 0$.

By the definition of the Laplace transform

$$L(f) = L(t) = \int_0^\infty e^{-st} t \, dt$$

This may be integrated by parts or by formula (44) of Table 8 in Appendix E. Using the formula we have

$$L(t) = \int_0^\infty t e^{-st} dt = \lim_{c \to \infty} \int_0^c t e^{-st} dt = \lim_{c \to \infty} \left. \frac{e^{-st}(-st-1)}{s^2} \right|_0^c$$

$$= \lim_{c \to \infty} \left[\frac{e^{-sc}(-sc-1)}{s^2} \right] + \frac{1}{s^2}$$

Now, as $c \to \infty$, $e^{-sc} \to 0$ and $sc \to \infty$. However, although we cannot prove it here, $e^{-sc} \to 0$ much faster than $sc \to \infty$. We can see that this is reasonable, for $ce^{-c} = 4.5 \times 10^{-4}$ for $c = 10$ and $ce^{-c} = 3.7 \times 10^{-42}$ for $c = 100$. Thus, the value at the upper limit approaches zero, which means the limit is zero. This means that

$$L(t) = \frac{1}{s^2}$$

Example B
Find the Laplace transform of the function $f(t) = \cos at$.
  By definition

$$L(f) = L(\cos at) = \int_0^\infty e^{-st} \cos at \, dt$$

Using formula (50) of Table 8 we have

$$L(\cos at) = \int_0^\infty e^{-st} \cos at \, dt = \lim_{c \to \infty} \int_0^c e^{-st} \cos at \, dt$$

$$= \lim_{c \to \infty} \frac{e^{-st}(-s \cos at + a \sin at)}{s^2 + a^2} \bigg|_0^c$$

$$= \lim_{c \to \infty} \frac{e^{-sc}(-s \cos ac + a \sin ac)}{s^2 + a^2} - \left(-\frac{s}{s^2 + a^2} \right)$$

$$= 0 + \frac{s}{s^2 + a^2}$$

$$= \frac{s}{s^2 + a^2}$$

Therefore, the Laplace transform of the function $\cos at$ is

$$L(\cos at) = \frac{s}{s^2 + a^2}$$

In both examples the resulting transform was an algebraic function of s.

  An important property of Laplace transforms is the **linearity property**

$$L[af(t) + bg(t)] = aL(f) + bL(g) \tag{29–23}$$

We state this property here since it determines that the transform of a sum of functions is the sum of the transforms. This is of definite importance when dealing with a sum of functions. This property is a direct result of the definition of the Laplace transform.
  We now present a short table of Laplace transforms. The transforms given here are sufficient for our remaining work in this chapter. Much more complete tables are available in many standard reference sources.

Table of Laplace Transforms

	$f(t) = L^{-1}(F)$	$L(f) = F(s)$
1.	1	$\dfrac{1}{s}$
2.	$\dfrac{t^{n-1}}{(n-1)!}$	$\dfrac{1}{s^n} (n = 1, 2, 3, \ldots)$
3.	e^{-at}	$\dfrac{1}{s + a}$
4.	$1 - e^{-at}$	$\dfrac{a}{s(s + a)}$
5.	$\cos at$	$\dfrac{s}{s^2 + a^2}$
6.	$\sin at$	$\dfrac{a}{s^2 + a^2}$
7.	$1 - \cos at$	$\dfrac{a^2}{s(s^2 + a^2)}$
8.	$at - \sin at$	$\dfrac{a^3}{s^2(s^2 + a^2)}$
9.	$e^{-at} - e^{-bt}$	$\dfrac{b - a}{(s + a)(s + b)}$
10.	$ae^{-at} - be^{-bt}$	$\dfrac{s(a - b)}{(s + a)(s + b)}$
11.	te^{-at}	$\dfrac{1}{(s + a)^2}$
12.	$t^{n-1}e^{-at}$	$\dfrac{(n - 1)!}{(s + a)^n}$
13.	$e^{-at}(1 - at)$	$\dfrac{s}{(s + a)^2}$
14.	$[(b - a)t + 1]e^{-at}$	$\dfrac{s + b}{(s + a)^2}$
15.	$\sin at - at \cos at$	$\dfrac{2a^3}{(s^2 + a^2)^2}$
16.	$t \sin at$	$\dfrac{2as}{(s^2 + a^2)^2}$

	$f(t) = L^{-1}(F)$	$L(f) = F(s)$
17.	$\sin at + at \cos at$	$\dfrac{2as^2}{(s^2 + a^2)^2}$
18.	$t \cos at$	$\dfrac{s^2 - a^2}{(s^2 + a^2)^2}$
19.	$e^{-at}\sin bt$	$\dfrac{b}{(s + a)^2 + b^2}$
20.	$e^{-at}\cos bt$	$\dfrac{s + a}{(s + a)^2 + b^2}$

Another Laplace transform important to the solution of a differential equation is the transform of the derivative of a function. Let us first find the Laplace transform of the first derivative of a function.

By definition

$$L(f') = \int_0^\infty e^{-st} f'(t)\,dt$$

To integrate by parts, let $u = e^{-st}$ and $dv = f'(t)\,dt$, so $du = -se^{-st}\,dt$ and $v = f(t)$ [the integral of the derivative of a function is the function]. Therefore,

$$L(f') = e^{-st} f(t)\Big|_0^\infty + s \int_0^\infty e^{-st} f(t)\,dt$$

$$= 0 - f(0) + sL(f)$$

It is noted that the integral in the second term on the right is the Laplace transform of $f(t)$ by definition. Therefore, the Laplace transform of the first derivative of a function is

$$L(f') = sL(f) - f(0) \qquad\qquad (29\text{–}24)$$

Applying the same analysis, we may find the Laplace transform of the second derivative of a function. It is

$$L(f'') = s^2 L(f) - sf(0) - f'(0) \qquad\qquad (29\text{–}25)$$

Here it is necessary to integrate by parts twice to derive the result. The transforms of higher derivatives are found in a similar manner.

Equations (29–24) and (29–25) allow us to express the transform of each derivative in terms of s and the transform of the function itself.

Example C

If $f(0) = 0$ and $f'(0) = 1$, express the transform of $f''(t) - 2f'(t)$ in terms of s and the transform of $f(t)$.

By the linearity property, Eq. (29–23), we have

$$
\begin{aligned}
L[f''(t) - 2f'(t)] &= L(f'') - 2L(f') \\
&= [s^2 L(f) - s(0) - 1] - 2[sL(f) - 0] \\
&= (s^2 - 2s)L(f) - 1
\end{aligned}
$$

If the Laplace transform of a function is known, it is then possible to find the function by finding the **inverse transform**

$$
\overline{L^{-1}(F) = f(t)} \tag{29–26}
$$

where L^{-1} denotes the inverse transform.

Example D

If $F(s) = \dfrac{s}{(s^2 + a^2)}$, from transform (5) of the table we see that

$$
L^{-1}(F) = L^{-1}\left(\frac{s}{s^2 + a^2}\right) = \cos at
$$

or

$$
f(t) = \cos at
$$

Example E

If $(s^2 - 2s)L(f) - 1 = 0$, then

$$
L(f) = \frac{1}{s^2 - 2s} \quad \text{or} \quad F(s) = \frac{1}{s(s - 2)}
$$

Thus,

$$
L^{-1}(F) = f(t) = -\frac{1}{2}(1 - e^{2t})
$$

from transform (4) of the table.

Example F

If $F(s) = \dfrac{(s + 5)}{(s^2 + 6s + 10)}$, then

$$
L^{-1}(F) = L^{-1}\left[\frac{s + 5}{s^2 + 6s + 10}\right]
$$

It appears that this function does not fit any of the forms given. However,

$$
s^2 + 6s + 10 = (s^2 + 6s + 9) + 1 = (s + 3)^2 + 1
$$

By writing $F(s)$ as

$$F(s) = \frac{(s+3)+2}{(s+3)^2+1} = \frac{s+3}{(s+3)^2+1} + \frac{2}{(s+3)^2+1}$$

we can find the inverse of each term. Therefore,

$$L^{-1}(F) = e^{-3t}\cos t + 2e^{-3t}\sin t$$

or

$$f(t) = e^{-3t}(\cos t + 2 \sin t)$$

Exercises 29–11

In Exercises 1 through 4 verify the indicated transforms given in the table.

1. Transform 1   **2.** Transform 3   **3.** Transform 6   **4.** Transform 11

In Exercises 5 through 12 find the transforms of the given functions by use of the table.

5. $f(t) = e^{3t}$

6. $f(t) = 1 - \cos 2t$

7. $f(t) = t^3 e^{-2t}$

8. $f(t) = 2e^{-3t}\sin 4t$

9. $f(t) = \cos 2t - \sin 2t$

10. $f(t) = 2t \sin 3t + e^{-3t}\cos t$

11. $f(t) = 3 + 2t \cos 3t$

12. $f(t) = t^3 - 3te^{-t}$

In Exercises 13 through 16 express the transforms of the given expressions in terms of s and $L(f)$.

13. $y'' + y'$, $f(0) = 0$, $f'(0) = 0$

14. $y'' - 3y'$, $f(0) = 2$, $f'(0) = -1$

15. $2y'' - y' + y$, $f(0) = 1$, $f'(0) = 0$

16. $y'' - 3y' + 2y$, $f(0) = -1$, $f'(0) = 2$

In Exercises 17 through 24 find the inverse transform of the given functions of s.

17. $F(s) = \dfrac{2}{s^3}$

18. $F(s) = \dfrac{3}{s^2+4}$

19. $F(s) = \dfrac{1}{s+5}$

20. $F(s) = \dfrac{3}{s^4+4s^2}$

21. $F(s) = \dfrac{1}{s^3+3s^2+3s+1}$

22. $F(s) = \dfrac{s^2-1}{s^4+2s^2+1}$

23. $F(s) = \dfrac{s+2}{(s^2+9)^2}$

24. $F(s) = \dfrac{s+3}{s^2+4s+13}$

29-12 Solving Differential Equations by Laplace Transforms

We will now show how certain differential equations can be solved by the use of Laplace transforms. It must be remembered that these solutions are the *particular* solutions of the equations subject to the given conditions. The necessary operations were developed in the preceding section. The following examples illustrate the method.

Example A

Solve the differential equation $2y' - y = 0$ if $y(0) = 1$. (Note that we are using y to denote the function.)

Taking transforms of each term in the equation, we have

$$L(2y') - L(y) = L(0), \qquad 2L(y') - L(y) = 0$$

$L(0) = 0$ by direct use of the definition of the transform. Now using Eq. (29-24), $L(y') = sL(y) - y(0)$, we have

$$2[sL(y) - 1] - L(y) = 0$$

Solving for $L(y)$ we obtain

$$2sL(y) - L(y) = 2, \qquad L(y) = \frac{2}{2s - 1} = \frac{1}{s - 1/2}$$

Finding the inverse transform of $F(s) = 1/(s - \frac{1}{2})$, we have

$$y = e^{t/2}$$

The reader should check this solution with that obtained by methods developed earlier. Also it should be noted that the solution was essentially an algebraic one. This points out the power and usefulness of Laplace transforms. We are able to translate a differential equation into an algebraic form, which can in turn be translated into the solution of the differential equation. Thus, we are able to solve a differential equation by use of algebra and specific algebraic forms.

Example B

Solve the differential equation $y'' + 2y' + 2y = 0$, if $y(0) = 0$ and $y'(0) = 1$.

Using the same steps as outlined in Example A, we have the following solution:

$$L(y'') + 2L(y') + 2L(y) = 0$$
$$[s^2L(y) - sy(0) - y'(0)] + 2[sL(y) - y(0)] + 2L(y) = 0$$
$$s^2L(y) - 1 + 2sL(y) + 2L(y) = 0$$
$$(s^2 + 2s + 2)L(y) = 1$$

$$L(y) = \frac{1}{s^2 + 2s + 2} = \frac{1}{(s + 1)^2 + 1}$$

$$y = e^{-t}\sin t$$

Example C
Solve the differential equation $y'' + y = \cos t$ if $y(0) = 1$ and $y'(0) = 2$.
  Taking transforms of each term in the equation, we have

$$L(y'') + L(y) = L(\cos t)$$

The transform on the left is found by use of Eq. (29–25) and the transform on the right is found by direct use of transform (5) in the table. Therefore, we have

$$[s^2 L(y) - s(1) - 2] + L(y) = \frac{s}{s^2 + 1}$$

Solving for $L(y)$, we obtain

$$(s^2 + 1)L(y) = \frac{s}{s^2 + 1} + s + 2$$

$$L(y) = \frac{s}{(s^2 + 1)^2} + \frac{s}{s^2 + 1} + \frac{2}{s^2 + 1}$$

Finding the inverse-transform of each expression completes the solution:

$$y = \frac{t}{2}\sin t + \cos t + 2 \sin t$$

Example D
An electric circuit contains a 1-H inductor, a 10-Ω resistor, and a battery whose voltage is 6 V. Find the current as a function of the time, if the initial current is zero.
  The differential equation for this circuit is

$$\frac{di}{dt} + 10i = 6$$

Following the procedures outlined in the previous examples, the solution is found:

$$L\left(\frac{di}{dt}\right) + 10L(i) = L(6), \qquad [sL(i) - 0] + 10L(i) = \frac{6}{s}$$

$$L(i) = \frac{6}{s(s + 10)}$$

$$i = 0.6(1 - e^{-10t})$$

Example E
A spring is stretched 1 ft by a weight of 16 lb (mass of $\frac{1}{2}$ slug). The medium resists the motion of the object with a force of $4v$, where v is the velocity. The differential equation describing the motion is

$$\frac{1}{2}\frac{d^2y}{dt^2} + 4\frac{dy}{dt} + 16y = 0$$

Find y, the displacement of the object, as a function of the time if $y(0) = 1$ and $dy/dt = 0$ when $t = 0$.

Clearing fractions and denoting the derivatives by y'' and y', the differential equation becomes

$$y'' + 8y' + 32y = 0$$

Following the procedure of solving this type of an equation by use of Laplace transforms, we have the following steps:

$$L(y'') + 8L(y') + 32L(y) = 0$$
$$[s^2 L(y) - s(1) - 0] + 8[sL(y) - 1] + 32L(y) = 0$$
$$(s^2 + 8s + 32)L(y) = s + 8$$

$$L(y) = \frac{s + 8}{(s + 4)^2 + 4^2} = \frac{s + 4}{(s + 4)^2 + 4^2} + \frac{4}{(s + 4)^2 + 4^2}$$

$$y = e^{-4t}\cos 4t + e^{-4t}\sin 4t$$
$$= e^{-4t}(\cos 4t + \sin 4t)$$

Exercises 29 — 12

In the following exercises solve the given differential equations by use of Laplace transforms, where the function is subject to the given conditions.

1. $y' + y = 0$, $y(0) = 1$
2. $y' - 2y = 0$, $y(0) = 2$
3. $2y' - 3y = 0$, $y(0) = -1$
4. $y' + 2y = 1$, $y(0) = 0$
5. $y' + 3y = e^{-3t}$, $y(0) = 1$
6. $y' + 2y = te^{-2t}$, $y(0) = 0$
7. $y'' + 4y = 0$, $y(0) = 0$, $y'(0) = 1$
8. $y'' - 4y = 0$, $y(0) = 2$, $y'(0) = 0$
9. $y'' + 2y' = 0$, $y(0) = 0$, $y'(0) = 2$
10. $y'' + 2y' + y = 0$, $y(0) = 0$, $y'(0) = -2$
11. $y'' - 4y' + 5y = 0$, $y(0) = 1$, $y'(0) = 2$
12. $4y'' + 4y' + y = 0$, $y(0) = 1$, $y'(0) = 0$
13. $y'' + y = 1$, $y(0) = 1$, $y'(0) = 1$
14. $y'' + 4y = 2t$, $y(0) = 0$, $y'(0) = 0$
15. $y'' + 2y' + y = e^{-t}$, $y(0) = 1$, $y'(0) = 2$
16. $y'' + 4y = \sin 2t$, $y(0) = 0$, $y'(0) = 0$

17. A constant force of 6 lb moves a 2-slug mass through a medium which resists the motion with a force equal to v, where v is the velocity. The differential equation relating the velocity and time is

$$2\frac{dv}{dt} = 6 - v$$

Find v as a function of time if the object starts from rest.

18. A 100-Ω resistor and a 400-μF capacitor make up an electric circuit. If the initial charge on the capacitor is 10 mC, find the charge on the capacitor as a function of time after the switch is closed.

19. A 50-Ω resistor, a 4-μF capacitor, and a 40-V battery are connected in series. Find the charge as a function of the time if the initial charge on the capacitor is zero.

20. A 2-H inductor, an 80-Ω resistor, and an 8-V battery are connected in series. Find the current in the circuit as a function of time if the initial current is zero.

21. A 10-H inductor, a 40-μF capacitor, and a voltage supply whose voltage is given by $100 \sin 50t$ are connected in series in an electric circuit. Find the current as a function of time if the initial charge on the capacitor and the initial current are zero.

22. A spring is such that it is stretched 6 in. by an 8-lb (1/4-slug) weight. The spring is stretched 6 in. below the equilibrium position with the weight on it and is given a velocity of 4 ft/s. If the spring is in a medium which resists the motion with a force equal to $4v$, find the displacement y as a function of time.

29–13 Exercises for Chapter 29

In Exercises 1 through 24 find the general solution of each of the given differential equations.

1. $4xy^3\,dx + (x^2 + 1)\,dy = 0$

2. $\dfrac{dy}{dx} = e^{x-y}$

3. $\dfrac{dy}{dx} + 2y = e^{-2x}$

4. $x\,dy + y\,dx = y\,dy$

5. $2D^2y + Dy = 0$

6. $2D^2y - 5Dy + 2y = 0$

7. $y'' + 2y' + y = 0$

8. $y'' + 2y' + 2y = 0$

9. $(x + y)\,dx + (x + y^3)\,dy = 0$

10. $y \ln x\,dx = x\,dy$

11. $x\dfrac{dy}{dx} - 3y = x^2$

12. $dy - 2y\,dx = (x - 2)e^x dx$

13. $dy = 2y\,dx + y^2 dx$

14. $x^2 y\,dy = (1 + x)\csc y\,dx$

15. $\sin x\dfrac{dy}{dx} + y \cos x + x = 0$

16. $y\,dy = (x^2 + y^2 - x)\,dx$

17. $2\dfrac{d^2y}{dx^2} + \dfrac{dy}{dx} - 3y = 6$

18. $\dfrac{d^2y}{dx^2} + 6\dfrac{dy}{dx} + 9y = 3x$

19. $9D^2y - 18Dy + 8y = 16 + 4x$

20. $D^2y + 9y = xe^x$

21. $y'' + y' - y = 2e^x$

22. $y'' + y = 4 \cos 2x$

23. $D^2y + 9y = \sin x$

24. $y'' + y' = e^x + \cos 2x$

In Exercises 25 through 32 find the particular solution of each of the given differential equations.

25. $y' = 2y \cot x$, $x = \pi/2$ when $y = 2$

26. $x\,dy - y\,dx = y^3 dy$, $x = 1$ when $y = 3$

27. $y' = 4x - 2y$, $x = 0$ when $y = -2$

28. $xy^2\,dx + e^x dy = 0$, $x = 0$ when $y = 2$

29. $\dfrac{d^2y}{dx^2} + \dfrac{dy}{dx} + 4y = 0$, $Dy = \sqrt{15}$, $y = 0$ when $x = 0$

30. $5y'' + 7y' - 6y = 0$, $y' = 10$, $y = 2$ when $x = 0$

31. $(D^2 + 4D + 4)y = 4 \cos x$, $Dy = 1$, $y = 0$ when $x = 0$

32. $y'' - 2y' + y = e^{2x} + x$, $Dy = 0$, $y = 2$ when $x = 0$

In Exercises 33 through 40 solve the given differential equations by use of Laplace transforms where the function is subject to the given conditions.

33. $4y' - y = 0$, $y(0) = 1$ 34. $2y' - y = 4$, $y(0) = 1$

35. $y' - 3y = e^t$, $y(0) = 0$ 36. $y' + 2y = e^{-2t}$, $y(0) = 2$

37. $y'' + y = 0$, $y(0) = 0$, $y'(0) = -4$

38. $y'' + 4y' + 5y = 0$, $y(0) = 1$, $y'(0) = 1$

39. $y'' + 9y = 3t$, $y(0) = 0$, $y'(0) = -1$

40. $y'' + 4y' + 4y = 2e^{-2t}$, $y(0) = 0$, $y'(0) = 1$

In Exercises 41 through 60 solve the given problems.

41. The time rate of change of volume of an evaporating substance is proportional to the surface area. Express the radius of a sphere as a function of time. Let $r = r_0$ when $t = 0$. [Hint: Express both V and A in terms of the radius r.]

42. An insulated tank is filled with a solution containing radioactive cobalt. Due to the radioactivity, energy is released and the temperature T (in degrees Fahrenheit) of the solution rises with the time t (in hours). The following differential equation expresses the relation between temperature and time for a specific case:

$$56{,}600 = 262(T - 70) + 20{,}200\frac{dT}{dt}$$

If the initial temperature is 70°F, what is the temperature 24 h later?

43. An object with a mass of 1 kg slides down a long inclined plane. The effective force of gravity is 4 N, and the motion is retarded by a force numerically equal to the velocity. If the object starts from rest, what is its velocity (in meters per second) 4 s later?

44. Under proper conditions, bacteria grow at a rate proportional to the number present. In a certain culture there were 10^5 present at a given time, and there were 3.0×10^5 present after 10 h. How many were present after 5 h?

45. Find the orthogonal trajectories of the family of curves $y = cx^5$.

46. Find the equation of the curves such that their normals at all points are in the direction with the lines connecting the points and the origin.

47. If a circuit contains a resistance R, capacitance C, and a source of voltage E, express the charge q on the capacitor as a function of time.

48. A 2-H inductor, a 40-Ω resistor, and a 20-V battery are connected in series. Find the current in the circuit as a function of time if the initial current is zero.

49. A certain spring stretches 2 ft by a 64-lb weight. With this weight suspended on it, the spring is stretched 2 ft beyond the equilibrium position and released. Find the equation of the resulting motion if the medium in which the weight is suspended retards the motion with a force equal to 16 times the velocity. Classify the motion as underdamped, critically damped, or overdamped.

50. What is the expression for the steady-state displacement x of an object which moves according to the relation $D^2x - 2Dx - 8x = 5 \sin 2t$?

51. An 0.5-H inductor, a 6-Ω resistor, and a 20-mF capacitor are connected in series with a generator for which $E = 24 \sin 10t$. Find the charge on the capacitor as a function of time if the initial charge and current are zero.

52. Find the general expression for the charge on the capacitor in a circuit having a resistance of 20 Ω, an inductor of 4 H, and a capacitor of 100 μF if the initial charge on the capacitor is 10 mC and the current is zero when $t = 0$.

53. Find the solution for the current for the circuit of Exercise 52 if a battery of 100 V is placed in the circuit.

54. If a circuit contains an inductor of inductance L, capacitor of capacitance C, and a sinusoidal source of voltage $E_0 \sin \omega t$, express the charge q on the capacitor as a function of the time. Assume $q = 0$, $i = 0$ when $t = 0$.

55. The differential equation relating current and time for a certain electric circuit is

$$2\frac{di}{dt} + i = 12$$

Solve this equation by use of Laplace transforms given that the initial current is zero. Evaluate the current for $t = 0.3$ s.

56. A 6-H inductor and a 30-Ω resistor are connected in series. Find the current as a function of time if the initial current is 15 A. Use Laplace transforms.

57. A 0.25-H inductor, a 4-Ω resistor, and a 100-μF capacitor are connected in series. If the initial charge on the capacitor is 400 μC and the initial current is zero, find the charge on the capacitor as a function of time. Use Laplace transforms.

58. An inductor of 0.5 H, a resistor of 6 Ω, and a capacitor of 200 μF are connected in series. If the initial charge on the capacitor is 10 mC and the initial current is zero, find the charge on the capacitor as a function of time after the switch is closed. Use Laplace transforms.

59. The approximate differential equation relating the displacement y of a beam at a horizontal distance x from one end is

$$EI\frac{d^2y}{dx^2} = M$$

where E is the modulus of elasticity, I is the moment of inertia of the cross section of the beam perpendicular to its axis, and M is the bending moment at the cross section. If $M = 2000x - 40x^2$ for a particular beam of length L for which $y = 0$ when $x = 0$ and when $x = L$, express y in terms of x. Consider E and I as constants.

60. When a circular disk of mass m and radius r is suspended by a wire at the center on one of its flat faces and the disk is twisted through an angle θ, torsion in the wire tends to turn the disk back in the opposite direction. The differential equation for this case is

$$\frac{1}{2}mr^2 \cdot \frac{d^2\theta}{dt^2} = -k\theta$$

where k is a constant. Determine the equation of motion if $\theta = \theta_0$ and $d\theta/dt = \omega_0$ when $t = 0$.

Appendix A

Study Aids

A—1 Introduction

The primary objective of this text is to give you an understanding of mathematics so that you can use it effectively as a tool in your technology. Without understanding of the basic methods, knowledge is usually short-lived. However, if you do understand, you will find your work much more enjoyable and rewarding. This is true in any course you may take, be it in mathematics or in any other field.

Mathematics is an indispensable tool in almost all scientific fields of study. You will find it used to a greater and greater degree as you work in your chosen field. Generally, in the introductory portions of allied courses, it is enough to have a grasp of elementary concepts in algebra and geometry. However, as you develop in your field, the need for more mathematics will be apparent. This text is designed to develop these necessary tools so they will be available in your allied courses. You cannot derive the full benefit from your mathematics course unless you devote the necessary amount of time to developing a sound understanding of the subject.

It is assumed in this text that you have a background which includes geometry and some algebra. Therefore, many of the topics covered in this book may seem familiar to you, especially in the earlier chapters. However, it is likely that your background in some of these areas is not complete, either because you have not studied mathematics for a year or two or because you did not understand the topics when you first encountered them. If a topic is familiar, do not reason that there is no sense in studying it again, but take the opportunity to clarify any points on which you are not certain. In almost every topic you will probably find certain points which can use further study. If the topic is new, use your time effectively to develop an understanding of the methods involved, and do not simply memorize problems of a certain type.

There is only one good way to develop the understanding and working knowledge necessary in any course, and that is to *work with it.* Many students consider mathematics difficult. They will tell you that it is their lack of mathematical ability and the complexity of the material itself which make it difficult. Some topics in mathematics, especially in the more advanced areas, do require a certain aptitude for full comprehension. However, a large proportion of poor grades in elementary mathematics courses result from the fact that the student is not willing to put in the necessary time to develop the understanding. The student takes a quick glance through the material, tries a few exercises, is largely unsuccessful, and then decides that the material is "impossible." A detailed reading of the text, a careful following of the illustrative examples, and then solving the exercises would lead to more success and therefore would make the work much more enjoyable and rewarding. No matter what text is used or what methods are used in the course or what other variables may be introduced, if you do not put in an adequate amount of time for studying, you will not derive the proper results. More detailed suggestions for study are included in the following section.

If you consider these suggestions carefully, and follow good study habits, you should enjoy a successful learning experience in this as well as in other courses.

A—2 Suggestions for Study

When you are studying the material presented in this text, the following suggestions may help you to derive full benefit from the time you devote to it.

(1) Before attempting to do the exercises, read through the material preceding them.

(2) Follow the illustrative examples carefully, being certain that you know how to proceed from step to step. You should then have a good idea of the methods involved.

(3) Work through the exercises, spending a reasonable amount of time on each problem. If you cannot solve a certain problem in a reasonable amount of time, leave it and go to the next. Return to this problem later. If you find many problems difficult, you should reread the explanatory material and the examples to determine what point or points you have not understood.

(4) When you have completed the exercises, or at least most of them, glance back through the explanatory material to be sure you understand the methods and principles.

(5) If you have gone through the first four steps and certain points still elude you, ask to have these points clarified in class. Do not be afraid to ask questions; only be sure that you have made a sincere effort on your own before you ask them.

Some study habits which are useful not only here but in all of your other subjects are the following:

(1) Put in the time required to develop the material fully, being certain that you are making effective use of your time. A good place to study helps immeasurably.

(2) Learn the *methods and principles* being presented. Memorize as little as possible, for although there are certain basic facts which are more expediently learned by memorization, these should be kept to a minimum.

(3) Keep up with the material in all of your courses. Do not let yourself get so behind in your studies that it becomes difficult to make up the time. Usually the time is never really made up. Studying only before tests is a poor way of learning, and is usually rather ineffective.

(4) When you are taking examinations, always read each question carefully before attempting the solution. Solve those you find easiest first, and do not spend too much time on any one problem. Also, use all the time available for the examination. If you finish early, use the remainder of the time to check your work.

A—3 Problem Analysis

Drill-type problems require a working knowledge of the methods presented. However, they do not require, in general, much analysis before being put in proper form for solution. Stated problems, on the other hand, do require proper interpretation before they can be put in a form for solution. The remainder of this section is devoted to some suggestions for solving stated problems.

We have to put stated problems in symbolic form before we attempt to solve them. It is this step which most students find difficult. Because such problems require the student to do more than merely go through a certain routine, they demand more analysis and thus appear more "difficult." There are several reasons for the student's difficulty, some of them being: (1) unsuccessful previous attempts at solving such problems, leading the student to believe that all stated problems are "impossible"; (2) failure to read the problem carefully; (3) a poorly organized approach to the solution; and (4) improper and incomplete interpretation of the statements given. The first two of these can be overcome only with the proper attitude and care.

There are over 70 completely worked examples of stated problems (as well as many other problems which indicate a similar analysis) throughout this text, illustrating proper interpretations and approaches to these problems. Therefore, we shall not include specific examples here. However, we shall set forth the method of analysis of any stated problem. Such an analysis generally follows these steps:

(1) Read the problem carefully.
(2) Carefully identify known and unknown quantities.
(3) Draw a figure when appropriate (which is quite often the case).
(4) Write, in symbols, the relations given in the statements.
(5) Solve for the desired quantities.

If you follow this step-by-step method, and write out the solution neatly, you should find that stated problems lend themselves to solution more readily than you had previously found.

Appendix B

Units of Measurement
and Approximate Numbers

B—1 Units of Measurement; the Metric System

The solution of most technical problems involves the use of the basic operations on numbers, where many of these numbers represent some sort of measurement or calculation. Therefore, associated with these numbers are *units of measurement,* and for the calculations to be meaningful, we must know these units. For example, if we measure the length of an object to be 12, we must know whether it is being measured in feet, yards, or some other specified unit of length.

Certain universally accepted **base units** *are used to measure fundamental quantities.* The units for numerous other quantities are expressed in terms of the base units. Fundamental quantities for which base units are defined are (1) length, (2) mass or force, depending on the system of units being used, (3) time, (4) electric current, (5) temperature, (6) amount of substance, and (7) luminous intensity. *Other units, referred to as* **derived units,** *are expressible in terms of the units for these quantities.*

Even though all other quantities can be expressed in terms of the fundamental ones, many have units which are given a specified name. This is done primarily for those quantities which are used very commonly, although it is not done for all such quantities. For example, the volt is defined as a meter^2-kilogram/second^3-ampere, which is in terms of (a unit of length)^2(a unit of mass)/(a unit of time)^3(a unit of electric current). The unit for acceleration has no special name, and is left in terms of the base units; for example, feet/second². For convenience, special symbols are usually used to designate units. The units for acceleration would be written as ft/s².

Although two basic systems of units, the **metric system** and the **British system,** are in use today, nearly every country in the world is either using or in the process of converting to the metric system. This includes the United States and all of the other countries which now use or previously used the British system. In the United States the conversion is presently coordinated by a national committee, but it is voluntary. However, most major United States' industrial firms will have completed the conversion for their products by the mid 1980s. In most of the other countries in the process of converting, specific timetables of completing various phases of the conversion have been set up.

Therefore, for the present, both systems are of importance although the metric system will eventually be used almost universally. For that reason, where units are used, some of the exercises and examples have

Table B—1. Quantities and Their Associated Units

Quantity	Quantity Symbol	Unit				
		British		Metric (SI)		
		Name	Symbol	Name	Symbol	In terms of other SI units
Length	s	foot	ft	**meter**	m	
Mass	m	slug		**kilogram**	kg	
Force	F	pound	lb	newton	N	$m \cdot kg/s^2$
Time	t	second	s	**second**	s	
Area	A		ft²		m²	
Volume	V		ft³		m³	
Capacity	V	gallon	gal	liter	L	(1 L = 1 dm³)
Velocity	v		ft/s		m/s	
Acceleration	a		ft/s²		m/s²	
Density	d, ρ		lb/ft³		kg/m³	
Pressure	p		lb/ft²	pascal	Pa	N/m²
Energy, work	E, W		ft · lb	joule	J	N · m
Power	P	horsepower	hp	watt	W	J/s
Period	T		s		s	
Frequency	f		1/s	hertz	Hz	1/s
Angle	θ	radian	rad	radian	rad	

Special Notes:
1. The SI base units are shown in boldface type.
2. The units symbols shown above are those which are used in the text. Many of them were adopted with the adoption of the SI system. This means, for example, that we use s rather than sec for seconds, and A rather than amp for amperes. Also, other units such as volt are not spelled out, a common practice in the past. When a given unit is used with both systems, we use the SI symbol for the unit.
3. The liter and degree Celsius are not actually SI units. However, they are recognized for use with the SI system due to their practical importance. Also, the symbol for liter has several variations. Presently L is recognized for use in the United States and Canada, l is recognized by the International Committee of Weights and Measures, and ℓ is also recognized for use in several countries.

metric units and others have British units. Technicians and engineers need to have some knowledge of both systems.

Although more than one system has been developed in which metric units are used, the system which is now becoming accepted as the metric system is the **International System of Units (SI)**. This was established in 1960, and uses some different definitions for base units than the previously developed metric units. However, the measurement of the base units in the SI system is more accessible, and the differences are very slight. *Therefore, when we refer to the metric system, we are using SI units.*

As we have stated, in each system the base units are specified, and all other units are then expressible in terms of these. Table B-1 lists the fundamental quantities, as well as many other commonly used quantities, along with the symbols and names of units used to represent each quantity shown.

Table B—1. Continued

| Quantity | Quantity Symbol | Unit | | | | |
| | | British | | Metric (SI) | | |
		Name	Symbol	Name	Symbol	In terms of other SI units
Electric current	I, i	ampere	A	ampere	A	
Electric charge	q	coulomb	C	coulomb	C	$A \cdot s$
Electric potential	V, E	volt	V	volt	V	$J/A \cdot s$
Capacitance	C	farad	F	farad	F	s/Ω
Inductance	L	henry	H	henry	H	$\Omega \cdot s$
Resistance	R	ohm	Ω	ohm	Ω	V/A
Thermodynamic Temperature	T			kelvin	K	
Temperature	T	Fahrenheit degree	°F	degree Celsius	°C	$(1°C = 1\ K)$
Quantity of heat	Q	British thermal unit	Btu	joule	J	
Amount of substance	n			mole	mol	
Luminous intensity	I	candlepower	cp	candela	cd	

4. Other units of time, along with their symbols, which are recognized for use with the SI system and are used in this text are: minute, min; hour, h; day, d.

5. There are many additional specialized units which are used with the SI system. However, most of those which appear in this text are shown in the table. A few of the specialized units are noted when used in the text. One which is frequently used is that for revolution, r.

6. Other common British units which are used in the text are as follows: inch, in.; yard, yd; mile, mi; ounce, oz.; ton; quart, qt; acre.

7. There are a number of units which were used with the metric system prior to the development of the SI system. However, many of these are not to be used with the SI system. Among these which were commonly used are the dyne, erg, and calorie.

In the British system, the base unit of length is the foot, and that of force is the pound. In the metric system, the base unit of length is the meter, and that of mass is the kilogram. Here, we see a difference in the definition of the systems which causes some difficulty when units are converted from one system to the other. That is, a base unit in the British system is a unit of force, and a base unit in the metric system is a unit of mass.

The distinction between mass and force is very significant in physics, and the weight of an object is the force with which it is attracted to the earth. Weight, which is therefore a force, is different from mass, which is a measure of the inertia an object exhibits. Although they are different quantities, mass and weight are however very closely related. In fact, the weight of an object equals its mass multiplied by the acceleration due to gravity. Near the surface of the earth the acceleration due to gravity is nearly constant, although it decreases as the distance from the earth increases. Therefore, near the surface of the earth the weight of an object is directly proportional to its mass. However, at great distances from the earth the weight of an object will be zero, whereas its mass does not change.

Since force and mass are different, it is not strictly correct to convert pounds to kilograms. However, since the pound is the base unit in the British system, and the kilogram is the base unit in the metric system, at the earth's surface 1 kg corresponds to 2.21 lb, in the sense that the force of gravity on a 1 kg mass is 2.21 lb.

When designating units for weight, we use pounds in the British system. In the metric system, although kilograms are used for weight, it is preferable to specify the mass of an object in kilograms. The force of gravity on an object is designated in newtons, and the use of the term weight is avoided unless its meaning is completely clear.

As for the other fundamental quantities, both systems use the second as the base unit of time. In the SI system the ampere is defined as the base unit of electric current, and this can also be used in the British system. As for temperature, degrees Fahrenheit are used with the British system, and degrees Celsius (formerly Centigrade) are used with the metric system (actually, the kelvin is defined as the base unit, where the temperature in kelvins is the temperature in degrees Celsius plus 273.16). In the SI system, the base unit for the amount of a substance is the mole, and the base unit of luminous intensity is the candela. These last two are of limited importance to our use in this text.

Due to greatly varying sizes of certain quantities, the metric system employs certain prefixes to units to denote different orders of magnitude. These prefixes, with their meanings and symbols, are shown in Table B–2 on the following page.

Table B–2. Metric Prefixes

Prefix	Factor	Symbol	Prefix	Factor	Symbol
exa	10^{18}	E	deci	10^{-1}	d
peta	10^{15}	P	centi	10^{-2}	c
tera	10^{12}	T	milli	10^{-3}	m
giga	10^{9}	G	micro	10^{-6}	μ
mega	10^{6}	M	nano	10^{-9}	n
kilo	10^{3}	k	pico	10^{-12}	p
hecto	10^{2}	h	femto	10^{-15}	f
deca	10^{1}	da	atto	10^{-18}	a

Example A

Some of the more commonly used units which use the prefixes in Table B–2, along with their meanings, are shown below.

Unit	Symbol	Meaning	Unit	Symbol	Meaning
megohm	MΩ	10^{6} ohms	milligram	mg	10^{-3} gram
kilometer	km	10^{3} meters	microfarad	μF	10^{-6} farad
centimeter	cm	10^{-2} meter	nanosecond	ns	10^{-9} second

(Mega is shortened to meg when used with "ohm.")

When designating units of area or volume, where square or cubic units are used, we use exponents in the designation. For example, we use m² rather than sq m, or in.³ rather than cu in.

When we are working with numbers that represent units of measurement (referred to as **denominate numbers**), it is sometimes necessary to change from one set of units to another. A change within a given system is called a **reduction,** and a change from one system to another is called a **conversion.** Table B–3 gives some basic reduction and conversion factors. It should be noted that some calculators are programmed to do conversions.

Table B–3. Reduction and Conversion Factors

1 in. = 2.54 cm (exact)	1 ft³ = 28.32 L
1 km = 0.6214 mi	1 L = 1.057 qt
1 lb = 453.6 g	1 Btu = 778.0 ft · lb
1 kg = 2.205 lb	1 hp = 550 ft · lb/s (exact)
1 lb = 4.448 N	1 hp = 746.0 W

The advantages of the metric system should be evident. Reductions within the system are made by use of powers of ten, which amounts to moving the decimal point. Reductions in the British system have numerous different multiples that must be used. Thus, comparisons and changes within the metric system are much simpler than in the British system.

To change a given number of one set of units into another set of units, *we perform algebraic operations with units in the same manner as we do with any algebraic symbol.* Consider the following example.

Example B

If we had a number representing feet per second to be multiplied by another number representing seconds per minute, as far as the units are concerned, we have

$$\frac{\text{ft}}{\text{s}} \times \frac{\text{s}}{\text{min}} = \frac{\text{ft} \times \text{s}}{\text{s} \times \text{min}} = \frac{\text{ft}}{\text{min}}$$

This means that the final result would be in feet per minute.

In changing a number of one set of units to another set of units, we use reduction and conversion factors and the principle illustrated in Example B. The convenient way to use the values in the tables is in the form of fractions. Since the given values are equal to each other, their quotient is 1. For example, since 1 in. = 2.54 cm,

$$\frac{1 \text{ in.}}{2.54 \text{ cm}} = 1 \qquad \text{or} \qquad \frac{2.54 \text{ cm}}{1 \text{ in.}} = 1$$

since each represents the division of a certain length by itself. Multiplying a quantity by 1 does not change its value. The following examples illustrate reduction and conversion of units.

Example C

Reduce 20 kg to milligrams.

$$20 \text{ kg} = 20 \text{ kg}\left(\frac{10^3 \text{ g}}{1 \text{ kg}}\right)\left(\frac{10^3 \text{ mg}}{1 \text{ g}}\right)$$

$$= 20 \times 10^6 \text{ mg}$$

$$= 2.0 \times 10^7 \text{ mg}$$

We note that this result is found essentially by moving the decimal point three places when changing from kilograms to grams, and another three places when changing from grams to milligrams.

Example D

Change 30 mi/h to feet per second.

$$30\frac{\text{mi}}{\text{h}} = \left(30\frac{\text{mi}}{\text{h}}\right)\left(\frac{5280 \text{ ft}}{1 \text{ mi}}\right)\left(\frac{1 \text{ h}}{60 \text{ min}}\right)\left(\frac{1 \text{ min}}{60 \text{ s}}\right) = \frac{(30)(5280) \text{ ft}}{(60)(60) \text{ s}} = 44\frac{\text{ft}}{\text{s}}$$

Note that the only units remaining after the division are those required.

Example E

Change 575 g/cm³ to kilograms per cubic meter.

$$575\frac{g}{cm^3} = \left(575\frac{g}{cm^3}\right)\left(\frac{100\ cm}{1\ m}\right)^3\left(\frac{1\ kg}{1000\ g}\right)$$

$$= \left(575\frac{g}{cm^3}\right)\left(\frac{10^6\ cm^3}{1\ m^3}\right)\left(\frac{1\ kg}{10^3\ g}\right)$$

$$= 575 \times 10^3\frac{kg}{m^3} = 5.75 \times 10^5\frac{kg}{m^3}$$

Example F

Change 62.8 lb/in.² to newtons per square meter.

$$62.8\frac{lb}{in.^2} = \left(62.8\frac{lb}{in.^2}\right)\left(\frac{4.448\ N}{1\ lb}\right)\left(\frac{1\ in.}{2.54\ cm}\right)^2\left(\frac{100\ cm}{1\ m}\right)^2$$

$$= \left(62.8\frac{lb}{in.^2}\right)\left(\frac{4.448\ N}{1\ lb}\right)\left(\frac{1\ in.^2}{2.54^2\ cm^2}\right)\left(\frac{10^4\ cm^2}{1\ m^2}\right)$$

$$= \frac{(62.8)(4.448)(10^4)}{(2.54)^2}\frac{N}{m^2} = 4.33 \times 10^5\frac{N}{m^2}$$

Exercises B—1

In Exercises 1 through 4 give the symbol and meaning for the given unit.

1. megahertz   2. kilowatt   3. millimeter   4. picosecond

In Exercises 5 through 8 give the name and meaning for the units whose symbols are given.

5. kV   6. GΩ   7. mA   8. pF

In Exercises 9 through 38 make the indicated reductions or conversions.

9. Reduce 1 km to centimeters   10. Reduce 1 kg to milligrams
11. Reduce 1 mi to inches   12. Reduce 1 gal to pints (2 pt = 1 qt)
13. Convert 5.25 in. to centimeters   14. Convert 6.50 kg to pounds
15. Convert 15.7 qt to liters   16. Convert 100 km to miles
17. Reduce 1 ft² to square inches   18. Reduce 1 yd³ to cubic feet
19. Reduce 250 mm² to square meters   20. Reduce 30.8 kL to milliliters
21. Convert 4.50 lb to grams   22. Convert 0.360 in. to meters
23. Convert 829 in.³ to liters
24. Convert 0.0680 kL to cubic feet
25. Convert 1 hp to kilogram centimeters per second
26. Convert 8.75 Btu to joules
27. Convert 75.0 W to horsepower
28. Convert 300 mL to quarts
29. An Atlas rocket weighed 260,000 lb at takeoff. How many tons is this?
30. An airplane is flying at 37,000 ft. What is its altitude in miles?
31. The speedometer of a car is calibrated in kilometers per hour. If the speedometer of such a car reads 60, how fast in miles per hour is the car traveling?

32. The acceleration due to gravity is about 980 cm/s². Convert this to feet per second squared.

33. The speed of sound is about 1130 ft/s. Change this speed to miles per hour.

34. The density of water is about 62.4 lb/ft³. Convert this to kilograms per cubic meter.

35. The average density of the earth is about 5.52 g/cm³. Convert this to pounds per cubic foot.

36. The moon travels about 1,500,000 miles in about 28 d in one rotation about the earth. Express its velocity in feet per second.

37. At sea level, atmospheric pressure is about 14.7 lb/in.². Express this pressure in pascals.

38. The earth's surface receives energy from the sun at the rate of 1.35 kW/m². Reduce this to joules per second square centimeter.

B—2 Approximate Numbers and Significant Digits

When we perform calculations on numbers, we must consider the accuracy of these numbers, since this affects the accuracy of the results obtained. Most of the numbers involved in technical and scientific work are *approximate*, having been arrived at through some process of measurement. However, certain other numbers are *exact*, having been arrived at through some definition or counting process. We can determine whether or not a number is approximate or exact if we know how the number was determined.

Example A
If we measure the length of a rope to be 15.3 ft, we know that the 15.3 is approximate. A more precise measuring device may cause us to determine the length as 15.28 ft. However, regardless of the method of measurement used, we shall not be able to determine this length exactly.

If a voltage shown on a voltmeter is read as 116 V, the 116 is approximate. A more precise voltmeter may show the voltage as 115.7 V. However, this voltage cannot be determined exactly.

Example B
If a computer counts the cards it has processed and prints this number as 837, this 837 is exact. We know the number of cards was not 836 or 838. Since 837 was determined through a counting process, it is exact.

When we say that 60 s = 1 min, the 60 is exact, since this is a definition. By this definition there are exactly sixty seconds in one minute.

When we are writing approximate numbers we often have to include some zeros so that the decimal point will be properly located. However, except for these zeros, all other digits are considered to be **significant digits.** When we make computations with approximate numbers, we must know the number of significant digits. The following example illustrates how we determine this.

Example C

All numbers in this example are assumed to be approximate.

34.7 has three significant digits.

8900 has two significant digits. We assume that the two zeros are place holders (unless we have specific knowledge to the contrary.)

0.039 has two significant digits. The zeros are for proper location of the decimal point.

706.1 has four significant digits. The zero is not used for the location of the decimal point. It shows specifically the number of tens in the number.

5.90 has three significant digits. The zero is not necessary as a place holder, and should not be written unless it is significant.

Other approximate numbers with the proper number of significant digits are listed below.

96000	two	0.0709	three	1.070	four
30900	three	6.000	four	700.00	five
4.006	four	0.0005	one	20008	five

Note from the example above that *all nonzero digits are significant. Zeros, other than those used as place holders for proper positioning of the decimal point, are also significant.*

In computations involving approximate numbers, the position of the decimal point as well as the number of significant digits is important. *The* **precision** *of a number refers directly to the decimal position of the last significant digit, whereas the* **accuracy** *of a number refers to the number of significant digits in the number.* Consider the illustrations in the following example.

Example D

Suppose that you are measuring an electric current with two ammeters. One ammeter reads 0.031 A and the second ammeter reads 0.0312 A. The second reading is more precise, in that the last significant digit is the number of ten-thousandths, and the first reading is expressed only to thousandths. The second reading is also more accurate, since it has three significant digits rather than two.

A machine part is measured to be 2.5 cm long. It is coated with a film 0.025 cm thick. The thickness of the film has been measured to a greater precision, although the two measurements have the same accuracy: two significant digits.

A segment of a newly completed highway is 9270 ft long. The concrete surface is 0.8 ft thick. Of these two numbers, 9270 is more accurate, since it contains three significant digits, and 0.8 is more precise, since it is expressed to tenths.

The last significant digit of an approximate number is known not to be completely accurate. It has usually been determined by estimation or **rounding off.** However, we do know that it is at most in error by one-half of a unit in its place value.

Example E
When we measure the length of the rope referred to in Example A to be 15.3 ft, we are saying that the length is at least 15.25 ft and no longer than 15.35 ft. Any value between these two, rounded off to tenths, would be expressed as 15.3 ft.

In converting the fraction $\frac{2}{3}$ to the decimal form 0.667, we are saying that the value is between 0.6665 and 0.6675.

The principle of rounding off a number is to write the closest approximation, with the last significant digit in a specified position, or with a specified number of significant digits. We shall now formalize the process of rounding off as follows: If we want a certain number of significant digits, we examine the digit in the next place to the right. If this digit is less than 5, we accept the digit in the last place. If the next digit is 5 or greater, we increase the digit in the last place by 1, and this resulting digit becomes the final significant digit of the approximation. If necessary, we use zeros to replace other digits in order to locate the decimal point properly. Except when the next digit is a 5, and no other nonzero digits are discarded, we have the closest possible approximation with the desired number of significant digits.

Example F
70360 rounded off to three significant digits is 70400.
70430 rounded off to three significant digits is 70400.
187.35 rounded off to four significant digits is 187.4.
71500 rounded off to two significant digits is 72000.

With the advent of computers and calculators, another method of reducing numbers to a specified number of significant digits is used. This is the process of **truncation,** in which the digits beyond a certain place are discarded. For example, 3.17482 truncated to thousandths is 3.174. For our purposes in this text, when working with approximate numbers, we have used only rounding off.

Exercises B—2

In Exercises 1 through 8 determine whether the numbers given are exact or approximate.

1. There are 24 hours in one day.
2. The velocity of light is 186,000 mi/s.
3. The 3-stage rocket took 74.6 hours to reach the moon.
4. A man bought 5 lb of nails for $1.56.
5. The melting point of gold is 1063°C.
6. The 21 students had an average test grade of 81.6.
7. A building lot 100 ft by 200 ft cost $3200.
8. In a certain city 5% of the people have their money in a bank that pays 5% interest.

In Exercises 9 through 16 determine the number of significant digits in the given approximate numbers.

9. 37.2; 6844   10. 3600; 730   11. 107; 3004   12. 0.8735; 0.0075

13. 6.80; 6.08   14. 90050; 105040   15. 30000; 30000.0   16. 1.00; 0.01

In Exercises 17 through 24 determine which of each pair of approximate numbers is (a) more precise and (b) more accurate.

17. 3.764, 2.81   18. 0.041, 7.673   19. 30.8, 0.01   20. 70,370, 50,400

21. 0.1, 78.0   22. 7040, 37.1   23. 7000, 0.004   24. 50.060, 8.914

In Exercises 25 through 32 round off each of the given approximate numbers (a) to three significant digits, and (b) to two significant digits.

25. 4.933   26. 80.53   27. 57893   28. 30490

29. 861.29   30. 9555   31. 0.30505   32. 0.7350

B—3 Arithmetic Operations with Approximate Numbers

When performing arithmetic operations on approximate numbers we must be careful not to express the result to a precision or accuracy which is not warranted. The following two examples illustrate how a false indication of the accuracy of a result could be obtained when using approximate numbers.

Example A

A pipe is made in two sections. The first is measured to be 16.3 ft long and the second is measured to be 0.927 ft long. A plumber wants to know what the total length will be when the two sections are put together.

At first, it appears we might simply add the numbers as follows to obtain the necessary result.

$$
\begin{array}{r}
16.3 \ \text{ft} \\
\underline{0.927 \ \text{ft}} \\
17.227 \ \text{ft}
\end{array}
$$

However, the first length is precise only to tenths, and the digit in this position was obtained by rounding off. It might have been as small as 16.25 ft or as large as 16.35 ft. If we consider only the precision of this first number, the total length might be as small as 17.177 ft or as large as 17.277 ft. These two values agree when rounded off to two significant digits (17). They vary by 0.1 when rounded off to tenths (17.2 and 17.3). When rounded to hundredths, they do not agree at all, since the third significant digit is different (17.18 and 17.28). Therefore there is no agreement at all in the digits after the third when these two numbers are rounded off to a precision beyond tenths. This may also be deemed reasonable, since the first length is not expressed beyond tenths. The second number does not further change the precision of the result, since it is expressed to thousandths. Therefore we may conclude that the result must be rounded off at least to tenths, the precision of the first number.

Example B
We can find the area of a rectangular piece of land by multiplying the length, 207.54 ft, by the width, 81.4 ft. Performing the multiplication, we find the area to be (207.54 ft)(81.4 ft) = 16893.756 ft².

However, we know this length and width were found by measurement and that the least each could be is 207.535 ft and 81.35 ft. Multiplying these values, we find the least value for the area to be

$$(207.535 \text{ ft})(81.35 \text{ ft}) = 16882.97225 \text{ ft}^2$$

The greatest possible value for the area is

$$(207.545 \text{ ft})(81.45 \text{ ft}) = 16904.54025 \text{ ft}^2$$

We now note that the least possible and greatest possible values of the area agree when rounded off to three significant digits (16900 ft²) and there is no agreement in digits beyond this if the two values are rounded off to a greater accuracy. Therefore we can conclude that the accuracy of the result is good to three significant digits, or certainly no more than four. We also note that the width was accurate to three significant digits, and the length to five significant digits.

The following rules are based on reasoning similar to that in Examples A and B; we go by these rules when we perform the basic arithmetic operations on approximate numbers.

(1) When approximate numbers are added or subtracted, the result is expressed with the precision of the least precise number.

(2) When approximate numbers are multiplied or divided, the result is expressed with the accuracy of the least accurate number.

(3) When the root of an approximate number is found, the result is accurate to the accuracy of the number.

(4) Before and during the calculation all numbers except the least precise or least accurate may be rounded off to one place beyond that of the least precise or least accurate. This procedure is helpful when a calculator is not used. It need not be followed when a calculator is used.

Example C
Add the approximate numbers 73.2, 8.0627, 93.57, 66.296.

The least precise of these numbers is 73.2. Therefore, before performing the addition we may round off the other number to hundredths. If we are using a calculator, this need not be done. In either case, after the addition we round off the result to tenths. This leads to

73.2		73.2
8.06	or	8.0627
93.57		93.57
66.30		66.296
241.13		241.1287

Therefore, the final result is 241.1.

Example D
If we multiply 2.4832 by 30.5, we obtain 75.7376. However, since 30.5 has only three significant digits, we express the product and result as $(2.4832)(30.5) = 75.7$.

To find the square root of 3.7, we may express the result only to two significant digits. Thus, $\sqrt{3.7} = 1.9$.

The rules stated in this section are usually sufficiently valid for the computations encountered in technical work. They are intended only as good practical rules for working with approximate numbers. It was recognized in Examples A and B that the last significant digit obtained by these rules is subject to some possible error. Therefore it is possible that the most accurate result is not obtained by their use, although this is not often the case.

If an exact number is included in a calculation, there is no limitation to the number of decimal positions it may take on. The accuracy of the result is limited only by the approximate numbers involved.

Exercises B—3

In Exercises 1 through 4 add the given approximate numbers.

1.	3.8	2.	26	3.	0.36294	4.	56.1
	0.154		5.806		0.086		3.0645
	47.26		147.29		0.5056		127.38
					0.74		0.055

In Exercises 5 through 8 subtract the given approximate numbers.

5.	468.14	6.	1.03964	7.	57.348	8.	8.93
	36.7		0.69		26.5		6.8947

In Exercises 9 through 12 multiply the given approximate numbers.

9. $(3.64)(17.06)$ 10. $(0.025)(70.1)$

11. $(704.6)(0.38)$ 12. $(0.003040)(6079.52)$

In Exercises 13 through 16 divide the given approximate numbers.

13. $608 \div 3.9$ 14. $0.4962 \div 827$

15. $\dfrac{596000}{22}$ 16. $\dfrac{53.267}{0.3002}$

In Exercises 17 through 20 find the indicated square roots of the given approximate numbers.

17. $\sqrt{32}$ 18. $\sqrt{6.5}$ 19. $\sqrt{19.3}$ 20. $\sqrt{0.0694}$

In Exercises 21 through 24 evaluate the given expression. All numbers are approximate.

21. $3.862 + 14.7 - 8.3276$ 22. $(3.2)(0.386) + 6.842$

23. $\dfrac{8.60}{0.46} + (0.9623)(3.86)$ 24. $9.6 - 0.1962(7.30)$

In Exercises 25 through 28 perform the indicated operations. The first number given is approximate and the second number is exact.

25. $3.62 + 14$

26. $17.382 - 2.5$

27. $(0.3142)(60)$

28. $8.62 \div 1728$

In Exercises 29 through 38 the solution to some of the problems will require the use of reduction and conversion factors.

29. Two forces, 18.6 lb and 2.382 lb, are acting on an object. What is the sum of these forces?

30. Three sections of a bridge are measured to be 52.3 ft, 36.38 ft, and 38 ft, respectively. What is the total length of these three sections?

31. Two planes are reported to have flown at speeds of 938 mi/h and 1400 km/h, respectively. Which plane is faster, and by how many miles per hour?

32. The density of a certain type of iron is 7.10 g/cm³. The density of a type of tin is 448 lb/ft³. Which is greater?

33. If the temperature of water is raised from 4°C to 30°C, its density reduces by 0.420%. If the density of water at 4°C is 62.4 lb/ft³, what is its density at 30°C?

34. The power (in watts) developed in an electric circuit is found by multiplying the current (in amperes) by the voltage. In a certain circuit the current is 0.0125 A and the voltage is 12.68 V. What is the power that is developed?

35. A certain ore is 5.3% iron. How many tons of ore must be refined to obtain 45,000 lb of iron?

36. An electric data-processing card sorter sorts 32,000 cards, by count, in 10.25 min. At what rate does the sorter operate?

37. In order to find the velocity (in feet per second) of an object which has fallen a certain height, we calculate the square root of the product of 64.4 (an approximate number) and the height in feet. What is the velocity of an object which has fallen 63 m?

38. A student reports the current in a certain experiment to be 0.02 A at one time and later notes that it is 0.023 A. He then states that the change in current is 0.003 A. What is wrong with his conclusion?

Appendix C

The Scientific Calculator

C—1 Introduction

Until the early 1970s the personal calculational device of many technicians and scientists was the slide rule. However, with the development of the microprocessor chip, the pocket scientific electronic calculator has become readily available. Since the calculator is easier to use, and has much greater accuracy and calculational ability than the slide rule, it is now the personal calculational device of technicians and scientists, as well as many others.

The scientific calculator can be used for any of the calculations which may be found in this text. Many of the more sophisticated models can also perform many types of calculations beyond the needs of this text. However, the calculator which performs only the basic arithmetic operations is in itself not sufficient for many of the calculations which are encountered.

There are a great many types and models of calculators. However, the discussions in this appendix are based on the operations which can be performed by use of the basic keys of a scientific, or slide-rule, calculator, and are general enough to apply to most such calculators. Some of the variations and special features which may be found are noted. Nearly all calculators come with a manual that can be used to learn the operations of the calculator, or to supplement this material on the variations and special features which a given calculator may have.

In this appendix we discuss the calculator keys and operations which are used for data entry, arithmetic operations, special functions (squares, square roots, reciprocals, powers and roots), trigonometric and inverse trigonometric functions, exponential and logarithmic functions, and calculator memory.

Many models of calculators use individual keys for each number and function. Other models use many of the keys for more than one purpose. However, the difference in operation is generally minor. Also, the labeling of many keys varies from one model to another. Some of these variations which are used are noted.

When a number is entered, or a result calculated, it shows on the display at the top of the calculator. We will use an eight-digit display, as well as a possible display of the exponent of ten for scientific notation, in our discussion of the calculator. Many calculators do display ten or more digits, although the operation is essentially the same. If the result contains more significant digits than the display can show, the result shown will be truncated or rounded off. We will use a rounded off display.

In making entries or calculations, it is possible that a number with too many digits or one which is too large or too small cannot be entered. Also, certain operations do not have defined results. If such an entry or calculation is attempted, the calculator will make an error indication such as E, ⊓, or a flashing display. Operations which can result in an error indication include division by zero, square root of a negative number, logarithm of a negative number, and the inverse trigonometric function of a value outside of the appropriate interval.

Also, some calculators will give a special display when batteries need charging or replacement. The user should become acquainted with any special displays the calculator may use.

The way in which a calculator must be used for various types of calculations depends on the type of calculation and the logic used by the particular calculator. The logic refers to the order in which entries and operations must be made in order to obtain the required result. Some discussion of this is found in the final section of this appendix, but *the user must become acquainted with the logic of the particular calculator being used.*

C—2 Calculator Data Entry and Function Keys

Following is a listing of certain basic keys which may be found on a scientific calculator, and which can be of use in performing calculations for the exercises in this text. There are other keys and other calculational capabilities which are also found on many calculators. Along with each key is a description and an illustration of its basic use. In the examples shown at the right, to perform the indicated entry or calculation, use the given sequence of entries and operations. The final display is also shown.

It should be emphasized that not all the keys which will probably be found on a given calculator are discussed, but that a scientific calculator will have most of the keys which are described. It should be noted that the use of many of the keys may be beyond the scope of the reader until the appropriate text material has been covered.

Keys	Examples
$\boxed{0}$, $\boxed{1}$, . . . , $\boxed{9}$ **Digit Keys** These keys are used to enter the digits 0 through 9 in the display, or to enter an exponent of 10 when scientific notation is used.	To enter: 37514 Sequence: $\boxed{3}$, $\boxed{7}$, $\boxed{5}$, $\boxed{1}$, $\boxed{4}$ Display: 37514.
$\boxed{\cdot}$ **Decimal Point Key** This key is used to enter a decimal point.	To enter: 375.14 Sequence: $\boxed{3}$, $\boxed{7}$, $\boxed{5}$, $\boxed{\cdot}$, $\boxed{1}$, $\boxed{4}$ Display: 375.14
$\boxed{+/-}$ **Change Sign Key** This key is used to change the sign of the number on the display, or to change the sign of the exponent when scientific notation is used. (May be designated as $\boxed{\text{CHS}}$.)	To enter: -375.14 Sequence: $\boxed{3}$, $\boxed{7}$, $\boxed{5}$, $\boxed{\cdot}$, $\boxed{1}$, $\boxed{4}$, $\boxed{+/-}$ Display: -375.14 (From here on the entry of a number will be shown as one operation.)
$\boxed{\pi}$ **Pi Key** This key is used to enter π to the number of digits of the display.	To enter: π Sequence: $\boxed{\pi}$ Display: 3.1415927 (For calculators with dual purpose keys, see $\boxed{F}$ key.)
$\boxed{\text{EE}}$ **Enter Exponent Key** This key is used to enter an exponent when scientific notation is used. After the key is pressed, the exponent is entered. For a negative exponent, the $\boxed{+/-}$ key is pressed after the exponent is entered. (May be designated as $\boxed{\text{E EX}}$.)	To enter: 2.936×10^8 Sequence: 2.936, $\boxed{\text{EE}}$, 8 Display: 2.936 08 To enter: -2.936×10^{-8} Sequence: 2.936, $\boxed{+/-}$, $\boxed{\text{EE}}$, 8, $\boxed{+/-}$ Display: -2.936 -08
$\boxed{=}$ **Equals Key** This key is used to complete a calculation to give the required result.	See the following examples for illustrations of the use of this key.

Keys	Examples
$+$ **Add Key** This key is used to add the next entry to the previous entry or result.	Evaluate: $37.56 + 241.9$ Sequence: 37.56, $+$, 241.9, $=$ Display: 279.46
$-$ **Subtract Key** This key is used to subtract the next entry from the previous entry or result.	Evaluate: $37.56 - 241.9$ Sequence: 37.56, $-$, 241.9, $=$ Display: -204.34
$\times$ **Multiply Key** This key is used to multiply the previous entry or result by the next entry.	Evaluate: 8.75×30.92 Sequence: 8.75, $\times$, 30.92, $=$ Display: 270.55
$\div$ **Divide Key** This key is used to divide the previous entry or result by the next entry.	Evaluate: $8.75 \div 30.92$ Sequence: 8.75, $\div$, 30.92, $=$ Display: 0.2829884 (truncated or rounded off)
CE **Clear Entry Key** This key is used to clear the last entry. Its use will not affect any other part of a calculation. On some calculators one press of the C/CE or CL key is used for this purpose.	Evaluate: $37.56 + 241.9$, with an improper entry of 242.9 Sequence: 37.56, $+$, 242.9, CE, 241.9, $=$ Display: 279.46
C **Clear Key** This key is used to clear the display and information being calculated (not including memory), so that a new calculation may be started. For calculators with a C/CE or C key, and no CE key, a second press on these keys is used for this purpose.	To clear previous calculation Sequence: C Display: $0.$
F **Function Key** This key is used on calculators on which many of the keys serve dual purposes. It is pressed before the second key functions are activated.	Evaluate: 37.4^2 on a calculator where x^2 is a second use of a key Sequence: 37.4, F, x^2 Display: 1398.76

Keys	Examples
x^2 **Square Key** This key is used to square the number on the display.	Evaluate: 37.4^2 Sequence: 37.4, x^2 Display: 1398.76
$\sqrt{x}$ **Square Root Key** This key is used to find the square root of the number on the display.	Evaluate: $\sqrt{37.4}$ Sequence: 37.4, $\sqrt{x}$ Display: 6.1155539
$1/x$ **Reciprocal Key** This key is used to find the reciprocal of the number on the display.	Evaluate: $\dfrac{1}{37.4}$ Sequence: 37.4, $1/x$ Display: 0.0267380
x^y **x to the y Power Key** This key is used to raise x, the first entry, to the y power, the second entry.	Evaluate: $3.73^{1.5}$ Sequence: 3.73, x^y, 1.5, $=$ Display: 7.2038266
RD DG **Degree-Radian Switch** This switch is used to designate a displayed angle as being measured in degrees or in radians.	This switch should be on the appropriate angle measurement before the calculation is started.
SIN **Sine Key** This key is used to find the sine of the angle on the display.	Evaluate: $\sin 37.4°$ Sequence: RD (DG), 37.4, SIN Display: 0.6073758
COS **Cosine Key** This key is used to find the cosine of the angle on the display.	Evaluate: $\cos 2.475$ (rad) Sequence: (RD) DG, 2.475, COS Display: -0.7859330
TAN **Tangent Key** This key is used to find the tangent of the angle on the display.	Evaluate: $\tan(-24.9°)$ Sequence: RD (DG), 24.9, $+/-$, TAN Display: -0.4641845
Cotangent, Secant, Cosecant These functions are found through their reciprocal relation with the tangent, cosine, and sine functions, respectively. See Section 19–1.	Evaluate: $\cot 2.841$ (rad) Sequence: (RD) DG, 2.841, TAN, $1/x$ Display: -3.2259549

Keys	Examples
ARC **Inverse Trigonometric** or INV **Function Key** This key is used (prior to the appropriate trigonometric function key) to find the angle whose trigonometric function is on the display. Some calculators have separate SIN⁻¹ , COS⁻¹ , TAN⁻¹ keys for this purpose.	Evaluate: Arcsin 0.1758 (the angle whose sine is 0.1758) in degrees Sequence: RD DG , .1758, ARC , SIN Display: 10.125217
LOG **Common Logarithm Key** This key is used to find the logarithm to the base 10 of the number on the display.	Evaluate: log 37.45 Sequence: 37.45, LOG Display: 1.5734518
LN **Natural Logarithm Key** This key is used to find the logarithm to the base e of the number on the display.	Evaluate: ln 0.8421 Sequence: 0.8421, LN Display: -0.1718565
10^x **10 to the x Power Key** This key is used to find anti-logarithms (base 10) of the number on the display. (The x^y key can be used for this purpose, if the calculator does not have this key.)	Evaluate: Antilog 0.7265 (or $10^{0.7265}$) Sequence: .7265, 10^x Display: 5.3272122
e^x **e to the x Power Key** This key is used to raise e to the power on the display.	Evaluate: $e^{-4.05}$ Sequence: 4.05, $+/-$, e^x Display: 0.0174224
STO **Store in Memory Key** This key is used to store the number on the display in the memory.	Store in memory: 56.02 Sequence: 56.02, STO
RCL **Recall from Memory Key** This key is used to recall the number in the memory to the display. (May be designated as MR .)	Recall from memory: 56.02 Sequence: RCL Display: 56.02

Keys

M   **Other Memory Keys**

Some calculators use an M key
to store a number in the memory.
It also may add the entry to the
number in the memory. On such
calculators CM (Clear Memory)
is used to clear the memory.
There are also keys for other
operations on the number in the
memory.

C—3 Combined Operations

Throughout the text there are numerous exercises in which numerical calculations are required. The use of a calculator can save a great deal of time in making these calculations. Many of them require only a basic two- or three-step use of the calculator, and a knowledge of the basic use of the keys is generally sufficient to make such calculations.

There are also many types of problems which require more extensive calculations for their solutions, and for many of these problems a calculator is the only practical way of making the necessary calculations. In this section we discuss some of these calculations which require a combination of operations, and how they may be solved on a calculator. However, it is not possible to cover all such types of calculations, since there are too many types, and there are many variations in calculator logic.

In the illustrations which are presented a sequence of calculator entries and operations is given. Due to the variations in calculator logic which are used in different models, it is not possible to give a sequence which is usable on all models. Therefore, a simple algebraic logic is assumed, and the sequences shown can be used, at least to a great extent, on most models. It is possible that a knowledge of the logic of a given calculator would allow for a reduction in the number of steps, and a change in the order of some of the steps, needed to complete a given calculation.

It should be noted that some of the following examples may be beyond the scope of the reader until the appropriate text material has been covered.

Example A

Evaluate $(30.45 + 75.76) \div 8.27 \times 10^7$

On calculations involving arithmetic operations, *we will assume that calculator logic is such that multiplications and divisions are performed before additions and subtractions, until the* = *key is pressed.* This is generally referred to as *algebraic logic.* Therefore, in this illustration, it is necessary to use the $\boxed{=}$ key after summing 30.45 and 75.76. Thus, the sequence is

$$30.45, \boxed{+}, 75.76, \boxed{=}, \boxed{\div}, 8.27, \boxed{EE}, 7, \boxed{=}$$
Display: 1.2842805 -06

If the $\boxed{=}$ key had not been pressed after 75.76, the evaluation being made would have been 30.45 + (75.76 ÷ 8.27 × 10⁷). It should be carefully noted, however, that on many calculators it is not necessary to press the $\boxed{=}$ key after 75.76 in order to perform the above evaluation. On such calculators one arithmetic operation is completed when another is entered. Therefore *check the manual of the calculator being used to determine the type of logic the calculator uses, and how such calculations should be entered.*

Example B
Given $f(x) = \sqrt{x^4 - 3x}$, evaluate $f(2.37)$.

$$f(2.37) = \sqrt{2.37^4 - 3(2.37)}$$

Sequence: $2.37, \boxed{x^y}, 4, \boxed{-}, 3, \boxed{\times}, 2.37, \boxed{=}, \boxed{\sqrt{x}}$
Display: 4.9436389

Therefore, $f(2.37) = 4.94$. Normally, the result is expressed to the number of significant digits of the given data. See Appendix B.

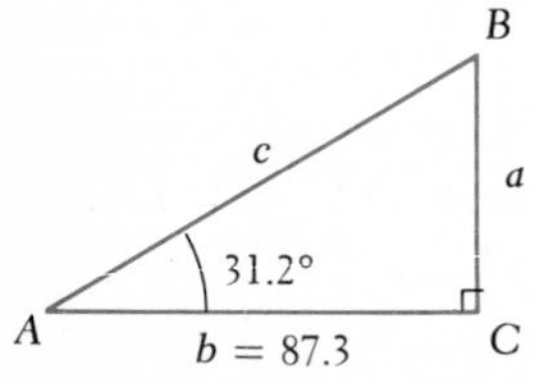

Example C
Solve the right triangle with $b = 87.3$, $A = 31.2°$. Use Figure C–1.

$$B = 90.0° - 31.2°$$

$$\frac{a}{87.3} = \tan 31.2°, \qquad a = 87.3 \tan 31.2°$$

$$\frac{87.3}{c} = \cos 31.2°, \qquad c = \frac{87.3}{\cos 31.2°}$$

Sequences: for B: $90, \boxed{-}, 31.2, \boxed{=}$
Display: 58.8

for a: $\boxed{RD} \ (DG), 31.2, \boxed{TAN}, \boxed{\times}, 87.3, \boxed{=}$
Display: 52.870759

for c: $\boxed{RD} \ (DG), 31.2, \boxed{COS}, \boxed{\div}, 87.3, \boxed{=}, \boxed{1/x}$
Display: 102.06178

Therefore, $B = 58.8°$, $a = 52.9$, $c = 102$. Again, we should carefully note the number of significant digits in the given data. In this case three significant digits are used. Therefore, we should write $a = 52.9$ (not 52.870759) and $c = 102$ (not 102.06178) in giving the results.

B

c

a

31.2°

A $b = 87.3$ C

Figure C–1

Example D

Solve $2x^2 - 3x - 7 = 0$ by the quadratic formula.

$$x = \frac{3 \pm \sqrt{3^2 - 4(2)(-7)}}{4}$$

Sequence: 3, $\boxed{x^2}$, $\boxed{-}$, 4, $\boxed{\times}$, 2, $\boxed{\times}$, 7, $\boxed{+/-}$,
$\boxed{=}$, $\boxed{\sqrt{x}}$, $\boxed{\text{STO}}$ (the value of $\sqrt{3^2 - 4(2)(-7)}$
is now in memory).

for first root: $\boxed{+}$, 3, $\boxed{=}$, $\boxed{\div}$, 4, $\boxed{=}$

Display: 2.7655644

for second root: 3, $\boxed{-}$, $\boxed{\text{RCL}}$, $\boxed{=}$, $\boxed{\div}$, 4, $\boxed{=}$

Display: -1.2655644

Therefore, the roots are (to four significant digits) 2.766 and -1.266.

Example E

Convert $123.7°$ to radians.

$$123.7° = \frac{(123.7)(\pi)}{180} \text{ rad}$$

Sequence: 123.7, $\boxed{\times}$, $\boxed{\pi}$, $\boxed{\div}$, 180, $\boxed{=}$

Display: 2.1589723

Therefore, $123.7° = 2.159$ rad.

Example F

Add the vectors shown in Figure C–2.

We are to find the magnitude and direction of vector **R**, with magnitude R, and components of R_x and R_y.

$$R_x = 36.9 \cos 10.4° + 48.5 \cos 109.8°$$
$$R_y = 36.9 \sin 10.4° + 48.5 \sin 109.8°$$
$$R = \sqrt{R_x^2 + R_y^2}$$
$$\tan \theta = \frac{R_y}{R_x} \quad \text{or} \quad \theta = \text{Arctan} \frac{R_y}{R_x}$$

Sequences: $\boxed{\text{RD} \ \ \textcircled{DG}}$ for all calculations

for R_x: 10.4, $\boxed{\text{COS}}$, $\boxed{\times}$, 36.9, $\boxed{+}$, 109.8, $\boxed{\text{COS}}$, $\boxed{\times}$,
48.5, $\boxed{=}$

Display: 19.864998

for R_y: 10.4, $\boxed{\text{SIN}}$, $\boxed{\times}$, 36.9, $\boxed{+}$, 109.8, $\boxed{\text{SIN}}$, $\boxed{\times}$,
48.5, $\boxed{=}$

Display: 52.293874

for R: 19.87, $\boxed{x^2}$, $\boxed{+}$, 52.29, $\boxed{x^2}$, $\boxed{=}$, $\boxed{\sqrt{x}}$

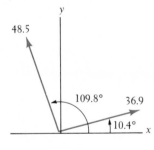

Figure C–2

Display: 55.938010

for θ: 52.29, $\boxed{\div}$, 19.86, $\boxed{=}$, $\boxed{\text{ARC}}$, $\boxed{\text{TAN}}$

Display: 69.202976

Therefore, $R = 55.9$ and $\theta = 69.2°$

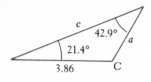

Figure C–3

Example G

Solve the triangle in Fig. C–3 by the law of sines.

$$C = 180° - (21.4° + 42.9°)$$

$$\frac{a}{\sin 21.4°} = \frac{3.86}{\sin 42.9°} = \frac{c}{\sin 115.7°}$$

$$a = \frac{3.86 \sin 21.4°}{\sin 42.9°}, \qquad c = \frac{3.86 \sin 115.7°}{\sin 42.9°}$$

Sequences: for C: 180, $\boxed{-}$, 21.4, $\boxed{-}$, 42.9, $\boxed{=}$

Display: 115.7

for a: $\boxed{\text{RD}\ \text{DG}}$, 3.86, $\boxed{\div}$, 42.9, $\boxed{\text{SIN}}$, $\boxed{=}$

$\boxed{\text{STO}}$ (the quotient 3.86/sin 42.9° is now in memory), $\boxed{\times}$, 21.4, $\boxed{\text{SIN}}$, $\boxed{=}$

Display: 2.0690190

for c: 115.7, $\boxed{\text{SIN}}$, $\boxed{\times}$, $\boxed{\text{RCL}}$, $\boxed{=}$

Display: 5.1095206

Therefore, $a = 2.07$, $c = 5.11$, $C = 115.7°$

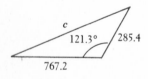

Figure C–4

Example H

Solve for side c in the triangle shown in Fig. C–4 by the law of cosines.

$$c = \sqrt{285.4^2 + 767.2^2 - 2(285.4)(767.2)\cos 121.3°}$$

Sequence: 285.4, $\boxed{x^2}$, $\boxed{+}$, 767.2, $\boxed{x^2}$, $\boxed{-}$, $\boxed{\text{RD}\ \text{DG}}$, 121.3, $\boxed{\text{COS}}$, $\boxed{\times}$, 2, $\boxed{\times}$, 285.4, $\boxed{\times}$, 767.2, $\boxed{=}$, $\boxed{\sqrt{x}}$

Display: 947.39413

Therefore, $c = 947.4$.

Example I

Evaluate $(709.6)^{2/7}$.

Sequence: 2, $\boxed{\div}$, 7, $\boxed{=}$, $\boxed{\text{STO}}$, 709.6, $\boxed{x^y}$, RCL , $\boxed{=}$

Display: 6.5249219

Therefore, $(709.6)^{2/7} = 6.525$.

Example J

Change the complex number $8.62 + 5.67j$ to exponential form.

$$r = \sqrt{8.62^2 + 5.67^2}; \qquad \tan\theta = \frac{5.67}{8.62}, \qquad \theta = \text{Arctan}\frac{5.67}{8.62}$$

Sequences: for r: 8.62, $\boxed{x^2}$, $\boxed{+}$, 5.67, $\boxed{x^2}$, $\boxed{=}$, $\boxed{\sqrt{x}}$

Display: 10.317621

for θ: 5.67, $\boxed{\div}$, 8.62, $\boxed{=}$, $\boxed{\text{RD}}$ DG , $\boxed{\text{ARC}}$, $\boxed{\text{TAN}}$

Display: 0.5818199

Therefore, the exponential form of $8.62 + 5.67j$ is $10.3e^{0.582j}$.

Example K

Solve for x: $2^{x+2} = 7$.

$$(x + 2)\log 2 = \log 7,$$

$$x = \frac{\log 7}{\log 2} - 2.$$

Sequence: 7, $\boxed{\text{LOG}}$, $\boxed{\div}$, 2, $\boxed{\text{LOG}}$, $\boxed{=}$, $\boxed{-}$, 2, $\boxed{=}$

Display: 0.8073549

Therefore, to four significant digits, $x = 0.8074$.

Example L

Find the sum of the geometric progression for which $a = 175$, $r = 1.05$, $n = 20$.

$$s = \frac{175[1 - (1.05)^{20}]}{1 - 1.05} = \frac{175[1 - (1.05)^{20}]}{-0.05}$$

Sequence: 1.05, $\boxed{x^y}$, 20, $\boxed{=}$, $\boxed{-}$, 1, $\boxed{=}$, $\boxed{+/-}$, $\boxed{\times}$, 175, $\boxed{\div}$, .05, $\boxed{+/-}$, $\boxed{=}$

Display: 5786.5420

The number of significant digits which are kept in the result depends on the meaning of a, r, and n.

We can see from the above illustrations that a calculator is especially valuable in trigonometry problems where we use the Pythagorean theorem and the trigonometric and inverse trigonometric functions. Also, it is of great value in other problems where special functions, such as square roots, exponentials, and logarithms are used.

The calculator is also valuable in many calculations which are primarily arithmetic. Some are illustrated in the previous examples, but there are many others. In fact, there are some exercises in which the calculator plays an important role in developing a concept. Without the calculator it is difficult to make such calculations quickly and easily.

With a little practice, which with a calculator usually is interesting and enjoyable, you will find that a scientific calculator will save you a great deal of time in making necessary calculations. However, it is still necessary to learn the mathematics which is required to set up and understand a given problem.

C–4 Exercises for Appendix C

In Exercises 1 through 36, perform all calculations on a scientific calculator.

1. $47.08 + 8.94$
2. $654.1 + 407.7$
3. $4724 - 561.9$
4. $0.9365 - 8.077$
5. 0.0396×471
6. 26.31×0.9393
7. $76.7 \div 194$
8. $52060 \div 75.09$
9. 3.76^2
10. 0.986^2
11. $\sqrt{0.2757}$
12. $\sqrt{60.36}$
13. $\dfrac{1}{0.0749}$
14. $\dfrac{1}{607.9}$
15. $(19.66)^{2.3}$
16. $(8.455)^{1.75}$
17. $\sin 47.3°$
18. $\sin 1.15$
19. $\cos 3.85$
20. $\cos 119.1°$
21. $\tan 306.8°$
22. $\tan 0.537$
23. $\sec 6.11$
24. $\csc 242.0°$
25. Arcsin 0.6607 (in degrees)
26. Arccos(-0.8311) (in radians)
27. Arctan(-2.441) (in radians)
28. Arcsin 0.0737 (in degrees)
29. $\log 3.857$
30. $\log 0.9012$
31. $\ln 808$
32. $\ln 70.5$
33. $10^{0.545}$
34. $10^{-0.0915}$
35. $e^{-5.17}$
36. $e^{1.672}$

In Exercises 37 through 90, perform all calculations on a scientific calculator. In these exercises some of the combined operations which are encountered in certain types of problems are given. For specific types of applications, such as those demonstrated in Section C–3, solve problems from the appropriate sections of the text.

37. $(4.38 + 9.07) \div 6.55$
38. $(382 + 964) \div 844$
39. $4.38 + (9.07 \div 6.55)$
40. $382 + (964 \div 844)$
41. $\dfrac{5.73 \times 10^{11}}{20.61 - 7.88}$
42. $\dfrac{7.09 \times 10^{23}}{284 + 839}$
43. $50.38\pi^2$
44. $\dfrac{5\pi}{14.6}$
45. $\sqrt{1.65^2 + 6.44^2}$
46. $\sqrt{0.735^2 + 0.409^2}$
47. $3(3.5)^4 - 4(3.5)^2$
48. $\dfrac{3(-1.86)}{(-1.86)^2 + 1}$
49. $29.4 \cos 72.5°$
50. $\dfrac{477}{\sin 58.7°}$
51. $\dfrac{4 + \sqrt{(-4)^2 - 4(3)(-9)}}{2(3)}$
52. $\dfrac{-5 - \sqrt{5^2 - 4(4)(-7)}}{2(4)}$
53. $\dfrac{0.176(180)}{\pi}$
54. $\dfrac{209.6\pi}{180}$
55. $\dfrac{1}{2}\left(\dfrac{51.4\pi}{180}\right)(7.06)^2$

56. $\dfrac{1}{2}\left(\dfrac{148.2\pi}{180}\right)(49.13)^2$ 57. $\text{Arcsin}\dfrac{27.3 \sin 36.5°}{46.8}$ 58. $\dfrac{0.684 \sin 76.1°}{\sin 39.5°}$

59. $\sqrt{3924^2 + 1762^2 - 2(3924)(1762)\cos 106.2°}$

60. $\text{Arccos}\dfrac{8.09^2 + 4.91^2 - 9.81^2}{2(8.09)(4.91)}$ 61. $\sqrt{5.81 \times 10^8} + \sqrt[3]{7.06 \times 10^{11}}$

62. $(6.074 \times 10^{-7})^{2/5} - (1.447 \times 10^{-5})^{4/9}$

63. $\dfrac{3}{2\sqrt{7} - \sqrt{6}}$ 64. $\dfrac{7\sqrt{5}}{4\sqrt{5} - \sqrt{11}}$ 65. $\text{Arctan}\dfrac{7.37}{5.06}$ 66. $\text{Arctan}\dfrac{46.3}{-25.5}$

67. $2 + \dfrac{\log 12}{\log 7}$ 68. $\dfrac{10^{0.4115}}{\pi}$

69. $\dfrac{26}{2}(-1.450 + 2.075)$ 70. $\dfrac{4.55(1 - 1.08^{15})}{1 - 1.08}$

71. $\sin^2\left(\dfrac{\pi}{7}\right) + \cos^2\left(\dfrac{\pi}{7}\right)$ 72. $\sec^2\left(\dfrac{2}{9}\pi\right) - \tan^2\left(\dfrac{2}{9}\pi\right)$

73. $\sin 31.6° \cos 58.4° + \sin 58.4° \cos 31.6°$

74. $\cos^2 296.7° - \sin^2 296.7°$ 75. $\sqrt{(1.54 - 5.06)^2 + (-4.36 - 8.05)^2}$

76. $\sqrt{(7.03 - 2.94)^2 + (3.51 - 6.44)^2}$ 77. $\dfrac{(4.001)^2 - 16}{4.001 - 4}$

78. $\dfrac{(2.001)^2 + 3(2.001) - 10}{2.001 - 2}$ 79. $\dfrac{4\pi}{3}(8.01^3 - 8.00^3)$

80. $4\pi(76.3^2 - 76.0^2)$

81. $0.01\left(\dfrac{1}{2}\sqrt{2} + \sqrt{2.01} + \sqrt{2.02} + \dfrac{1}{2}\sqrt{2.03}\right)$

82. $0.2\left[\dfrac{1}{2}(3.5)^2 + 3.7^2 + 3.9^2 + \dfrac{1}{2}(4.1)^2\right]$

83. $\dfrac{e^{0.45} - e^{-0.45}}{e^{0.45} + e^{-0.45}}$ 84. $\ln \sin 2e^{-0.055}$

85. $\ln\dfrac{2 - \sqrt{2}}{2 - \sqrt{3}}$ 86. $\sqrt{\dfrac{9}{2} + \dfrac{9 \sin 0.2\pi}{8\pi}}$

87. $2 + \dfrac{0.3}{4} - \dfrac{(0.3)^2}{64} + \dfrac{(0.3)^3}{512}$ 88. $\dfrac{1}{2} + \dfrac{\pi\sqrt{3}}{360} - \dfrac{1}{4}\left(\dfrac{\pi}{180}\right)^2$

89. $160(1 - e^{-1.50})$ 90. $e^{-3.60}(\cos 1.20 + 2 \sin 1.20)$

Appendix D

Review of Geometry

D–1 Basic Geometric Figures and Definitions

In this appendix we shall present geometric terminology and formulas that are related to the basic geometric figures. It is intended only as a brief summary of basic geometry.

Geometry deals with the properties and measurement of angles, lines, surfaces, and volumes, and the basic figures that are formed. We shall restrict our attention in this appendix to the basic figures and concepts which are related to these figures.

Repeating the definition in Section 3–1, an **angle** is generated by rotating a half-line about its endpoint from an initial position to a terminal position. One complete rotation of a line about a point is defined to be an angle of 360 **degrees**, written as 360°. A **straight angle** contains 180°, and a **right angle** contains 90°. If two lines meet so that the angle between them is 90°, the lines are said to be **perpendicular**.

An angle less than 90° is an **acute angle**. An angle greater than 90°, but less than 180°, is an **obtuse angle**. **Supplementary** angles are two angles whose sum is 180°, and **complementary angles** are two angles whose sum equals 90°.

In a plane, if a line crosses two **parallel** or nonparallel lines, it is called a **transversal**. If a transversal crosses a pair of parallel lines, certain pairs of equal angles result. In Fig. D–1, the **corresponding angles** are equal. (That is, $\angle 1 = \angle 5$, $\angle 2 = \angle 6$, $\angle 3 = \angle 7$, $\angle 4 = \angle 8$.) Also, the **alternate interior angles** are equal ($\angle 3 = \angle 6$ and $\angle 4 = \angle 5$).

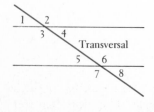

Figure D–1

Adjacent angles have a common vertex and a side common to them. For example, in Fig. D–1, ∡ 5 and 6 are adjacent angles. Vertical angles are equal angles formed "across" the point of intersection of two intersecting lines. In Fig. D–1, ∡ 5 and 8 are vertical angles.

When a part of the plane is bounded and closed by straight line segments, it is called a polygon. In general, polygons are named according to the number of sides they contain. A triangle has three sides, a quadrilateral has four sides, a pentagon has five sides, and so on. In a regular polygon, all of the sides are equal in length, and all of the interior angles are equal.

There are several important types of triangles. In an equilateral triangle the three sides are equal, and the three angles are also equal, each being 60°. In an isosceles triangle two of the sides are equal, as are the two base angles (the angles opposite the equal sides). In a scalene triangle no two sides are equal, and none of the angles is a right angle. In a right triangle one of the angles is a right angle. The side opposite the right angle is called the hypotenuse.

There are certain basic properties of triangles which we shall mention here. One very important property is that the sum of the three angles of any triangle is 180°. Also, the three medians (line segments drawn from a vertex to the midpoint of the opposite side) meet at a single point. This point of intersection of the medians is called the centroid of the triangle. It is also true that the three angle bisectors meet at a common point, as do the three altitudes (heights) which are drawn from a vertex perpendicular to the opposite side (or the extension of the opposite side).

Of particular importance is the Pythagorean theorem, which states that in a right triangle, the square of the length of the hypotenuse equals the sum of the squares of the lengths of the other two sides.

Two triangles are said to be congruent if corresponding angles are equal and if corresponding sides are equal. Two triangles are said to be similar if corresponding angles are equal. In similar triangles corresponding sides are proportional.

A quadrilateral is a plane figure having four sides and therefore four interior angles. A parallelogram is a quadrilateral with opposite sides parallel (extension of the sides will not intersect). Also opposite sides and opposite angles of a parallelogram are equal. A rectangle is a parallelogram with intersecting sides perpendicular, which means that all four angles are right angles. It also means that opposite sides of a rectangle are equal and parallel. A square is a rectangle all sides of which are equal. A trapezoid is a quadrilateral with two of the sides parallel. These parallel sides are called the bases of the trapezoid. A rhombus is a parallelogram all four sides of which are equal.

All of the points on a circle are the same distance from a fixed point in the plane. This point is the center of the circle. The distance from the center to a point on the circle is the radius of the circle. The distance between two points on the circle and on a line passing through the center of the circle is the diameter of the circle. Thus the diameter is twice the radius.

Also associated with the circle is the **chord,** which is a line segment having its endpoints on the circle. A **tangent** is a line that touches a circle (does not pass through) at one point. A **secant** is a line that passes through two points of a circle. An **arc** is a part of the circle. When two radii form an angle at the center, the angle is called a **central angle.** An **inscribed angle** of an arc is one for which the endpoints of the arc are points on the sides of the angle, and for which the vertex is a point of the arc, although not an endpoint.

There arc two important properties of a circle which we shall mention here.

(1) *A tangent to a circle is perpendicular to the radius drawn to the point of contact.*

(2) *An angle inscribed in a semicircle is a right angle.*

D–2 Basic Geometric Formulas

For the indicated figures, the following symbols are used: A = area, B = area of base, c = circumference, S = lateral area, V = volume.

1. **Triangle.** $A = \frac{1}{2}bh$ (Fig. D–2)
2. **Pythagorean theorem.** $c^2 = a^2 + b^2$ (Fig. D–3)
3. **Parallelogram.** $A = bh$ (Fig. D–4)
4. **Trapezoid.** $A = \frac{1}{2}(a + b)h$ (Fig. D–5)
5. **Circle.** $A = \pi r^2$, $c = 2\pi r$ (Fig. D–6)
6. **Rectangular solid.** $A = 2(lw + lh + wh)$, $V = lwh$ (Fig. D–7)
7. **Cube.** $A = 6e^2$, $V = e^3$ (Fig. D–8)
8. **Any cylinder or prism with parallel bases.** $V = Bh$ (Fig. D–9)
9. **Right circular cylinder.** $S = 2\pi rh$, $V = \pi r^2 h$ (Fig. D–10)
10. **Any cone or pyramid.** $V = \frac{1}{3}Bh$ (Fig. D–11)
11. **Right circular cone.** $S = \pi rs$, $V = \frac{1}{3}\pi r^2 h$ (Fig. D–12)
12. **Sphere.** $A = 4\pi r^2$, $V = \frac{4}{3}\pi r^3$ (Fig. D–13)

Also, the **perimeter** of a plane figure is the distance around it. For example, the perimeter p of the triangle in Fig. D–3 is $p = a + b + c$. (∟ denotes a right angle.)

Figure D–2

Figure D–3

Figure D–4

Figure D–5

Figure D—6

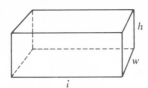

Figure D—7

Figure D—8

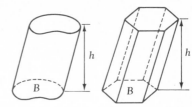

Figure D—9

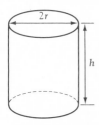

Figure D—10

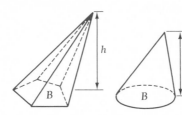

Figure D—11

Figure D—12

Figure D—13

D—3 Exercises for Appendix D

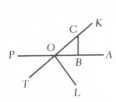

Figure D—14

In Exercises 1 through 4 refer to Fig. D–14, in which AOP and KOT are straight lines. $\angle TOP = 40°$, $\angle OBC = 90°$, and $\angle AOL = 55°$. Determine the indicated angles.

1. $\angle LOT$ 2. $\angle POL$ 3. $\angle KOP$ 4. $\angle KCB$

In Exercises 5 through 8 refer to Fig. D–15, where AB is a diameter, line TB is tangent to the circle at B, and $\angle ABC = 65°$. Determine the indicated angles.

5. $\angle CBT$ 6. $\angle BCT$ 7. $\angle CAB$ 8. $\angle BTC$

In Exercises 9 through 12, refer to Figs. D–16 and D–17. In each $ABCD$ is a parallelogram. In Fig. D–16, $\triangle BEC$ is isosceles, and $\angle BCE = 40°$. In Fig. D–17, $\angle FED = 34°$. Determine the indicated angles.

9. In Fig. D–16, $\angle CEB$ 10. In Fig. D–16, $\angle ADC$
11. In Fig. D–17, $\angle EDF$ 12. In Fig. D–17, $\angle DCB$

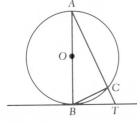

Figure D—15

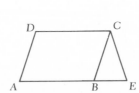

Figure D—16

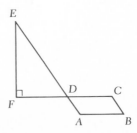

Figure D—17

In Exercises 13 through 24 use the Pythagorean theorem to solve for the unknown side of the right triangle. In each case c is the hypotenuse. If the answer is not exact, round off to three digits.

	a	b	c		a	b	c
13.	3	4	?	14.	9	12	?
15.	8	15	?	16.	24	10	?
17.	6	?	10	18.	2	?	4
19.	5	?	7	20.	3	?	9
21.	?	12	16	22.	?	10	18
23.	?	15	32	24.	?	5	36

In Exercises 25 through 28 determine the required values. In Fig. D–18, $AD = 9$ and $AC = 12$.

25. Two triangles are similar, and the sides of the larger triangle are 3.0 in., 5.0 in., and 6.0 in., and the shortest side of the other triangle is 2.0 in. Find the remaining sides of the smaller triangle.

26. Two triangles are similar. The angles of the smaller triangle are 50°, 100°, and 30°, and the sides of the smaller triangle are 7.00 in., 9.00 in., and 4.57 in. The longest side of the larger triangle is 15.00 in. Find the other two sides and the three angles of the larger triangle.

27. In Fig. D–18, find AB
28. In Fig. D–18, find BC

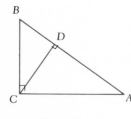

Figure D–18

In Exercises 29 through 40 find the perimeters (or circumferences) of the indicated figures.

29. A triangle with sides 6 ft, 8 ft, and 11 ft
30. A quadrilateral with sides 3 in., 4 in., 6 in., and 9 in.
31. An isosceles triangle whose equal sides are 3 yd long and whose third side is 4 yd long
32. An equilateral triangle whose sides are 7 ft long
33. A rectangle 5.14 in. long and 4.09 in. wide
34. A parallelogram whose longer side is 13.7 mm and whose shorter side is 11.9 mm
35. A square of side 8.18 cm
36. A rhombus of side 15.6 m
37. A trapezoid whose longer base is 74.2 cm, whose shorter base is 46.8 cm, and whose other sides are equal to the shorter base
38. A trapezoid whose bases are 17.8 in. and 7.4 in., and whose other sides are each 8.1 in.
39. A circle of radius 10.0 ft
40. A circle of diameter 7.06 cm

In Exercises 41 through 48 find the areas of the indicated figures.

41. A parallelogram of base 7.0 in. and height 4.0 in.
42. A rectangle of base 8.2 ft and height 2.5 ft
43. A triangle of base 6.3 yd and height 4.1 yd
44. A triangle of base 14.2 cm and height 6.83 cm

45. A trapezoid of bases 3.0 m and 9.0 m and height 4.0 m

46. A trapezoid of bases 18.5 in. and 26.3 in. and height 10.5 in.

47. A circle of radius 7.00 in.

48. A semicircle of diameter 8.24 cm

In Exercises 49 through 60 evaluate the volume of the indicated figures.

49. A rectangular solid of length 9.00 ft, width 6.00 ft, and height 4.00 ft

50. A cube of edge 7.15 ft

51. Prism, square base of side 2.0 mm, altitude 5.0 mm

52. Prism, trapezoidal base (bases 4.0 in. and 6.0 in. and height 3.0 in.), altitude 6.0 in.

53. Cylinder, radius of base 7.00 in., altitude 6.00 in.

54. Cylinder, diameter of base 6.36 cm, altitude 18.0 cm

55. Pyramid, rectangular base 12.5 in. by 8.75 in., altitude 4.20 in.

56. Pyramid, equilateral triangular base of side 3.00 in., altitude 4.25 in.

57. Cone, radius of base 2.66 ft, altitude 1.22 yd

58. Cone, diameter of base 16.3 cm, altitude 18.4 cm

59. Sphere, radius 5.48 m

60. Sphere, diameter 15.7 yd

In Exercises 61 through 68 find the total surface area of the indicated figure for the given values.

61. Prism, rectangular base 4.00 in. by 8.00 in., altitude 6.00 in.

62. Prism, parallelogram base (base 16.0 in., height 4.25 in., perimeter 42.0 in.), altitude 6.45 in.

63. Cylinder, radius of base 8.58 m, altitude 1.38 m

64. Cylinder, diameter of base 12.5 ft, altitude 4.60 ft

65. Sphere, radius 16.0 ft

66. Sphere, diameter 15.3 in.

67. Cone, radius of base 7.05 cm, slant height 8.45 cm

68. Cone, diameter of base 76.2 mm, altitude 22.1 mm

Appendix E

Tables

Table 1   A—39

Table 1. Powers and Roots

No.	Sq.	Sq. Root	Cube	Cube Root	No.	Sq.	Sq. Root	Cube	Cube Root
1	1	1.000	1	1.000	51	2,601	7.141	132,651	3.708
2	4	1.414	8	1.260	52	2,704	7.211	140,608	3.733
3	9	1.732	27	1.442	53	2,809	7.280	148,877	3.756
4	16	2.000	64	1.587	54	2,916	7.348	157,464	3.780
5	25	2.236	125	1.710	55	3,025	7.416	166,375	3.803
6	36	2.449	216	1.817	56	3,136	7.483	175,616	3.826
7	49	2.646	343	1.913	57	3,249	7.550	185,193	3.849
8	64	2.828	512	2.000	58	3,364	7.616	195,112	3.871
9	81	3.000	729	2.080	59	3,481	7.681	205,379	3.893
10	100	3.162	1,000	2.154	60	3,600	7.746	216,000	3.915
11	121	3.317	1,331	2.224	61	3,721	7.810	226,981	3.936
12	144	3.464	1,728	2.289	62	3,844	7.874	238,328	3.958
13	169	3.606	2,197	2.351	63	3,969	7.937	250,047	3.979
14	196	3.742	2,744	2.410	64	4,096	8.000	262,144	4.000
15	225	3.873	3,375	2.466	65	4,225	8.062	274,625	4.021
16	256	4.000	4,096	2.520	66	4,356	8.124	287,496	4.041
17	289	4.123	4,913	2.571	67	4,489	8.185	300,763	4.062
18	324	4.243	5,832	2.621	68	4,624	8.246	314,432	4.082
19	361	4.359	6,859	2.668	69	4,761	8.307	328,509	4.102
20	400	4.472	8,000	2.714	70	4,900	8.367	343,000	4.121
21	441	4.583	9,261	2.759	71	5,041	8.426	357,911	4.141
22	484	4.690	10,648	2.802	72	5,184	8.485	373,248	4.160
23	529	4.796	12,167	2.844	73	5,329	8.544	389,017	4.179
24	576	4.899	13,824	2.884	74	5,476	8.602	405,224	4.198
25	625	5.000	15,625	2.924	75	5,625	8.660	421,875	4.217
26	676	5.099	17,576	2.962	76	5,776	8.718	438,976	4.236
27	729	5.196	19,683	3.000	77	5,929	8.775	456,533	4.254
28	784	5.292	21,952	3.037	78	6,084	8.832	474,552	4.273
29	841	5.385	24,389	3.072	79	6,241	8.888	493,039	4.291
30	900	5.477	27,000	3.107	80	6,400	8.944	512,000	4.309
31	961	5.568	29,791	3.141	81	6,561	9.000	531,441	4.327
32	1,024	5.657	32,768	3.175	82	6,724	9.055	551,368	4.344
33	1,089	5.745	35,937	3.208	83	6,889	9.110	571,787	4.362
34	1,156	5.831	39,304	3.240	84	7,056	9.165	592,704	4.380
35	1,225	5.916	42,875	3.271	85	7,225	9.220	614,125	4.397
36	1,296	6.000	46,656	3.302	86	7,396	9.274	636,056	4.414
37	1,369	6.083	50,653	3.332	87	7,569	9.327	658,503	4.431
38	1,444	6.164	54,872	3.362	88	7,744	9.381	681,472	4.448
39	1,521	6.245	59,319	3.391	89	7,921	9.434	704,969	4.465
40	1,600	6.325	64,000	3.420	90	8,100	9.487	729,000	4.481
41	1,681	6.403	68,921	3.448	91	8,281	9.539	753,571	4.498
42	1,764	6.481	74,088	3.476	92	8,464	9.592	778,688	4.514
43	1,849	6.557	79,507	3.503	93	8,649	9.644	804,357	4.531
44	1,936	6.633	85,184	3.530	94	8,836	9.695	830,584	4.547
45	2,025	6.708	91,125	3.557	95	9,025	9.747	857,375	4.563
46	2,116	6.782	97,336	3.583	96	9,216	9.798	884,736	4.579
47	2,209	6.856	103,823	3.609	97	9,409	9.849	912,673	4.595
48	2,304	6.928	110,592	3.634	98	9,604	9.899	941,192	4.610
49	2,401	7.000	117,649	3.659	99	9,801	9.950	970,299	4.626
50	2,500	7.071	125,000	3.684	100	10,000	10.000	1,000,000	4.642

Table 2. Four-Place Logarithms of Numbers

N	0	1	2	3	4	5	6	7	8	9
10	0000	0043	0086	0128	0170	0212	0253	0294	0334	0374
11	0414	0453	0492	0531	0569	0607	0645	0682	0719	0755
12	0792	0828	0864	0899	0934	0969	1004	1038	1072	1106
13	1139	1173	1206	1239	1271	1303	1335	1367	1399	1430
14	1461	1492	1523	1553	1584	1614	1644	1673	1703	1732
15	1761	1790	1818	1847	1875	1903	1931	1959	1987	2014
16	2041	2068	2095	2122	2148	2175	2201	2227	2253	2279
17	2304	2330	2355	2380	2405	2430	2455	2480	2504	2529
18	2553	2577	2601	2625	2648	2672	2695	2718	2742	2765
19	2788	2810	2833	2856	2878	2900	2923	2945	2967	2989
20	3010	3032	3054	3075	3096	3118	3139	3160	3181	3201
21	3222	3243	3263	3284	3304	3324	3345	3365	3385	3404
22	3424	3444	3464	3483	3502	3522	3541	3560	3579	3598
23	3617	3636	3655	3674	3692	3711	3729	3747	3766	3784
24	3802	3820	3838	3856	3874	3892	3909	3927	3945	3962
25	3979	3997	4014	4031	4048	4065	4082	4099	4116	4133
26	4150	4166	4183	4200	4216	4232	4249	4265	4281	4298
27	4314	4330	4346	4362	4378	4393	4409	4425	4440	4456
28	4472	4487	4502	4518	4533	4548	4564	4579	4594	4609
29	4624	4639	4654	4669	4683	4698	4713	4728	4742	4757
30	4771	4786	4800	4814	4829	4843	4857	4871	4886	4900
31	4914	4928	4942	4955	4969	4983	4997	5011	5024	5038
32	5051	5065	5079	5092	5105	5119	5132	5145	5159	5172
33	5185	5198	5211	5224	5237	5250	5263	5276	5289	5302
34	5315	5328	5340	5353	5366	5378	5391	5403	5416	5428
35	5441	5453	5465	5478	5490	5502	5514	5527	5539	5551
36	5563	5575	5587	5599	5611	5623	5635	5647	5658	5670
37	5682	5694	5705	5717	5729	5740	5752	5763	5775	5786
38	5798	5809	5821	5832	5843	5855	5866	5877	5888	5899
39	5911	5922	5933	5944	5955	5966	5977	5988	5999	6010
40	6021	6031	6042	6053	6064	6075	6085	6096	6107	6117
41	6128	6138	6149	6160	6170	6180	6191	6201	6212	6222
42	6232	6243	6253	6263	6274	6284	6294	6304	6314	6325
43	6335	6345	6355	6365	6375	6385	6395	6405	6415	6425
44	6435	6444	6454	6464	6474	6484	6493	6503	6513	6522
45	6532	6542	6551	6561	6571	6580	6590	6599	6609	6618
46	6628	6637	6646	6656	6665	6675	6684	6693	6702	6712
47	6721	6730	6739	6749	6758	6767	6776	6785	6794	6803
48	6812	6821	6830	6839	6848	6857	6866	6875	6884	6893
49	6902	6911	6920	6928	6937	6946	6955	6964	6972	6981
50	6990	6998	7007	7016	7024	7033	7042	7050	7059	7067
51	7076	7084	7093	7101	7110	7118	7126	7135	7143	7152
52	7160	7168	7177	7185	7193	7202	7210	7218	7226	7235
53	7243	7251	7259	7267	7275	7284	7292	7300	7308	7316
54	7324	7332	7340	7348	7356	7364	7372	7380	7388	7396

Table 2   A—41

Table 2. Continued

N	0	1	2	3	4	5	6	7	8	9
55	7404	7412	7419	7427	7435	7443	7451	7459	7466	7474
56	7482	7490	7497	7505	7513	7520	7528	7536	7543	7551
57	7559	7566	7574	7582	7589	7597	7604	7612	7619	7627
58	7634	7642	7649	7657	7664	7672	7679	7686	7694	7701
59	7709	7716	7723	7731	7738	7745	7752	7760	7767	7774
60	7782	7789	7796	7803	7810	7818	7825	7832	7839	7846
61	7853	7860	7868	7875	7882	7889	7896	7903	7910	7917
62	7924	7931	7938	7945	7952	7959	7966	7973	7980	7987
63	7993	8000	8007	8014	8021	8028	8035	8041	8048	8055
64	8062	8069	8075	8082	8089	8096	8102	8109	8116	8122
65	8129	8136	8142	8149	8156	8162	8169	8176	8182	8189
66	8195	8202	8209	8215	8222	8228	8235	8241	8248	8254
67	8261	8267	8274	8280	8287	8293	8299	8306	8312	8319
68	8325	8331	8338	8344	8351	8357	8363	8370	8376	8382
69	8388	8395	8401	8407	8414	8420	8426	8432	8439	8445
70	8451	8457	8463	8470	8476	8482	8488	8494	8500	8506
71	8513	8519	8525	8531	8537	8543	8549	8555	8561	8567
72	8573	8579	8585	8591	8597	8603	8609	8615	8621	8627
73	8633	8639	8645	8651	8657	8663	8669	8675	8681	8686
74	8692	8698	8704	8710	8716	8722	8727	8733	8739	8745
75	8751	8756	8762	8768	8774	8779	8785	8791	8797	8802
76	8808	8814	8820	8825	8831	8837	8842	8848	8854	8859
77	8865	8871	8876	8882	8887	8893	8899	8904	8910	8915
78	8921	8927	8932	8938	8943	8949	8954	8960	8965	8971
79	8976	8982	8987	8993	8998	9004	9009	9015	9020	9025
80	9031	9036	9042	9047	9053	9058	9063	9069	9074	9079
81	9085	9090	9096	9101	9106	9112	9117	9122	9128	9133
82	9138	9143	9149	9154	9159	9165	9170	9175	9180	9186
83	9191	9196	9201	9206	9212	9217	9222	9227	9232	9238
84	9243	9248	9253	9258	9263	9269	9274	9279	9284	9289
85	9294	9299	9304	9309	9315	9320	9325	9330	9335	9340
86	9345	9350	9355	9360	9365	9370	9375	9380	9385	9390
87	9395	9400	9405	9410	9415	9420	9425	9430	9435	9440
88	9445	9450	9455	9460	9465	9469	9474	9479	9484	9489
89	9494	9499	9504	9509	9513	9518	9523	9528	9533	9538
90	9542	9547	9552	9557	9562	9566	9571	9576	9581	9586
91	9590	9595	9600	9605	9609	9614	9619	9624	9628	9633
92	9638	9643	9647	9652	9657	9661	9666	9671	9675	9680
93	9685	9689	9694	9699	9703	9708	9713	9717	9722	9727
94	9731	9736	9741	9745	9750	9754	9759	9763	9768	9773
95	9777	9782	9786	9791	9795	9800	9805	9809	9814	9818
96	9823	9827	9832	9836	9841	9845	9850	9854	9859	9863
97	9868	9872	9877	9881	9886	9890	9894	9899	9903	9908
98	9912	9917	9921	9926	9930	9934	9939	9943	9948	9952
99	9956	9961	9965	9969	9974	9978	9983	9987	9991	9996

Table 3. Four-Place Values of Functions and Radians
Angle θ in Degrees and Minutes

Degrees	Radians	Sin θ	Cos θ	Tan θ	Cot θ	Sec θ	Csc θ		
0°00′	0.0000	0.0000	1.0000	0.0000	—	1.000	—	1.5708	90°00′
10	0.0029	0.0029	1.0000	0.0029	343.8	1.000	343.8	1.5679	50
20	0.0058	0.0058	1.0000	0.0058	171.9	1.000	171.9	1.5650	40
30	0.0087	0.0087	1.0000	0.0087	114.6	1.000	114.6	1.5621	30
40	0.0116	0.0116	0.9999	0.0116	85.94	1.000	85.95	1.5592	20
50	0.0145	0.0145	0.9999	0.0145	68.75	1.000	68.76	1.5563	10
1°00′	0.0175	0.0175	0.9998	0.0175	57.29	1.000	57.30	1.5533	89°00′
10	0.0204	0.0204	0.9998	0.0204	49.10	1.000	49.11	1.5504	50
20	0.0233	0.0233	0.9997	0.0233	42.96	1.000	42.98	1.5475	40
30	0.0262	0.0262	0.9997	0.0262	38.19	1.000	38.20	1.5446	30
40	0.0291	0.0291	0.9996	0.0291	34.37	1.000	34.38	1.5417	20
50	0.0320	0.0320	0.9995	0.0320	31.24	1.001	31.26	1.5388	10
2°00′	0.0349	0.0349	0.9994	0.0349	28.64	1.001	28.65	1.5359	88°00′
10	0.0378	0.0378	0.9993	0.0378	26.43	1.001	26.45	1.5330	50
20	0.0407	0.0407	0.9992	0.0407	24.54	1.001	24.56	1.5301	40
30	0.0436	0.0436	0.9990	0.0437	22.90	1.001	22.93	1.5272	30
40	0.0465	0.0465	0.9989	0.0466	21.47	1.001	21.49	1.5243	20
50	0.0495	0.0494	0.9988	0.0495	20.21	1.001	20.23	1.5213	10
3°00′	0.0524	0.0523	0.9986	0.0524	19.08	1.001	19.11	1.5184	87°00′
10	0.0553	0.0552	0.9985	0.0553	18.07	1.002	18.10	1.5155	50
20	0.0582	0.0581	0.9983	0.0582	17.17	1.002	17.20	1.5126	40
30	0.0611	0.0610	0.9981	0.0612	16.35	1.002	16.38	1.5097	30
40	0.0640	0.0640	0.9980	0.0641	15.60	1.002	15.64	1.5068	20
50	0.0669	0.0669	0.9978	0.0670	14.92	1.002	14.96	1.5039	10
4°00′	0.0698	0.0698	0.9976	0.0699	14.30	1.002	14.34	1.5010	86°00′
10	0.0727	0.0727	0.9974	0.0729	13.73	1.003	13.76	1.4981	50
20	0.0756	0.0756	0.9971	0.0758	13.20	1.003	13.23	1.4952	40
30	0.0785	0.0785	0.9969	0.0787	12.71	1.003	12.75	1.4923	30
40	0.0814	0.0814	0.9967	0.0816	12.25	1.003	12.29	1.4893	20
50	0.0844	0.0843	0.9964	0.0846	11.83	1.004	11.87	1.4864	10
5°00′	0.0873	0.0872	0.9962	0.0875	11.43	1.004	11.47	1.4835	85°00′
10	0.0902	0.0901	0.9959	0.0904	11.06	1.004	11.10	1.4806	50
20	0.0931	0.0929	0.9957	0.0934	10.71	1.004	10.76	1.4777	40
30	0.0960	0.0958	0.9954	0.0963	10.39	1.005	10.43	1.4748	30
40	0.0989	0.0987	0.9951	0.0992	10.08	1.005	10.13	1.4719	20
50	0.1018	0.1016	0.9948	0.1022	9.788	1.005	9.839	1.4690	10
6°00′	0.1047	0.1045	0.9945	0.1051	9.514	1.006	9.567	1.4661	84°00′
10	0.1076	0.1074	0.9942	0.1080	9.255	1.006	9.309	1.4632	50
20	0.1105	0.1103	0.9939	0.1110	9.010	1.006	9.065	1.4603	40
30	0.1134	0.1132	0.9936	0.1139	8.777	1.006	8.834	1.4573	30
40	0.1164	0.1161	0.9932	0.1169	8.556	1.007	8.614	1.4544	20
50	0.1193	0.1190	0.9929	0.1198	8.345	1.007	8.405	1.4515	10
7°00′	0.1222	0.1219	0.9925	0.1228	8.144	1.008	8.206	1.4486	83°00′
10	0.1251	0.1248	0.9922	0.1257	7.953	1.008	8.016	1.4457	50
20	0.1280	0.1276	0.9918	0.1287	7.770	1.008	7.834	1.4428	40
30	0.1309	0.1305	0.9914	0.1317	7.596	1.009	7.661	1.4399	30
40	0.1338	0.1334	0.9911	0.1346	7.429	1.009	7.496	1.4370	20
50	0.1367	0.1363	0.9907	0.1376	7.269	1.009	7.337	1.4341	10
8°00′	0.1396	0.1392	0.9903	0.1405	7.115	1.010	7.185	1.4312	82°00′
		Cos θ	Sin θ	Cot θ	Tan θ	Csc θ	Sec θ	Radians	Degrees

Table 3   A—43

Table 3. Continued

Degrees	Radians	Sin θ	Cos θ	Tan θ	Cot θ	Sec θ	Csc θ		
8°00′	0.1396	0.1392	0.9903	0.1405	7.115	1.010	7.185	1.4312	82°00′
10	0.1425	0.1421	0.9899	0.1435	6.968	1.010	7.040	1.4283	50
20	0.1454	0.1449	0.9894	0.1465	6.827	1.011	6.900	1.4254	40
30	0.1484	0.1478	0.9890	0.1495	6.691	1.011	6.765	1.4224	30
40	0.1513	0.1507	0.9886	0.1524	6.561	1.012	6.636	1.4195	20
50	0.1542	0.1536	0.9881	0.1554	6.435	1.012	6.512	1.4166	10
9°00′	0.1571	0.1564	0.9877	0.1584	6.314	1.012	6.392	1.4137	81°00′
10	0.1600	0.1593	0.9872	0.1614	6.197	1.013	6.277	1.4108	50
20	0.1629	0.1622	0.9868	0.1644	6.084	1.013	6.166	1.4079	40
30	0.1658	0.1650	0.9863	0.1673	5.976	1.014	6.059	1.4050	30
40	0.1687	0.1679	0.9858	0.1703	5.871	1.014	5.955	1.4021	20
50	0.1716	0.1708	0.9853	0.1733	5.769	1.015	5.855	1.3992	10
10°00′	0.1745	0.1736	0.9848	0.1763	5.671	1.015	5.759	1.3963	80°00′
10	0.1774	0.1765	0.9843	0.1793	5.576	1.016	5.665	1.3934	50
20	0.1804	0.1794	0.9838	0.1823	5.485	1.016	5.575	1.3904	40
30	0.1833	0.1822	0.9833	0.1853	5.396	1.017	5.487	1.3875	30
40	0.1862	0.1851	0.9827	0.1883	5.309	1.018	5.403	1.3846	20
50	0.1891	0.1880	0.9822	0.1914	5.226	1.018	5.320	1.3817	10
11°00′	0.1920	0.1908	0.9816	0.1944	5.145	1.019	5.241	1.3788	79°00′
10	0.1949	0.1937	0.9811	0.1974	5.066	1.019	5.164	1.3759	50
20	0.1978	0.1965	0.9805	0.2004	4.989	1.020	5.089	1.3730	40
30	0.2007	0.1994	0.9799	0.2035	4.915	1.020	5.016	1.3701	30
40	0.2036	0.2022	0.9793	0.2065	4.843	1.021	4.945	1.3672	20
50	0.2065	0.2051	0.9787	0.2095	4.773	1.022	4.876	1.3643	10
12°00′	0.2094	0.2079	0.9781	0.2126	4.705	1.022	4.810	1.3614	78°00′
10	0.2123	0.2108	0.9775	0.2156	4.638	1.023	4.745	1.3584	50
20	0.2153	0.2136	0.9769	0.2186	4.574	1.024	4.682	1.3555	40
30	0.2182	0.2164	0.9763	0.2217	4.511	1.024	4.620	1.3526	30
40	0.2211	0.2193	0.9757	0.2247	4.449	1.025	4.560	1.3497	20
50	0.2240	0.2221	0.9750	0.2278	4.390	1.026	4.502	1.3468	10
13°00′	0.2269	0.2250	0.9744	0.2309	4.331	1.026	4.445	1.3439	77°00′
10	0.2298	0.2278	0.9737	0.2339	4.275	1.027	4.390	1.3410	50
20	0.2327	0.2306	0.9730	0.2370	4.219	1.028	4.336	1.3381	40
30	0.2356	0.2334	0.9724	0.2401	4.165	1.028	4.284	1.3352	30
40	0.2385	0.2363	0.9717	0.2432	4.113	1.029	4.232	1.3323	20
50	0.2414	0.2391	0.9710	0.2462	4.061	1.030	4.182	1.3294	10
14°00′	0.2443	0.2419	0.9703	0.2493	4.011	1.031	4.134	1.3265	76°00′
10	0.2473	0.2447	0.9696	0.2524	3.962	1.031	4.086	1.3235	50
20	0.2502	0.2476	0.9689	0.2555	3.914	1.032	4.039	1.3206	40
30	0.2531	0.2504	0.9681	0.2586	3.867	1.033	3.994	1.3177	30
40	0.2560	0.2532	0.9674	0.2617	3.821	1.034	3.950	1.3148	20
50	0.2589	0.2560	0.9667	0.2648	3.776	1.034	3.906	1.3119	10
15°00′	0.2618	0.2588	0.9659	0.2679	3.732	1.035	3.864	1.3090	75°00′
10	0.2647	0.2616	0.9652	0.2711	3.689	1.036	3.822	1.3061	50
20	0.2676	0.2644	0.9644	0.2742	3.647	1.037	3.782	1.3032	40
30	0.2705	0.2672	0.9636	0.2773	3.606	1.038	3.742	1.3003	30
40	0.2734	0.2700	0.9628	0.2805	3.566	1.039	3.703	1.2974	20
50	0.2763	0.2728	0.9621	0.2836	3.526	1.039	3.665	1.2945	10
16°00′	0.2793	0.2756	0.9613	0.2867	3.487	1.040	3.628	1.2915	74°00′
		Cos θ	Sin θ	Cot θ	Tan θ	Csc θ	Sec θ	Radians	Degrees

Table 3. Continued

Degrees	Radians	Sin θ	Cos θ	Tan θ	Cot θ	Sec θ	Csc θ		
16°00′	0.2793	0.2756	0.9613	0.2867	3.487	1.040	3.628	1.2915	74°00′
10	0.2822	0.2784	0.9605	0.2899	3.450	1.041	3.592	1.2886	50
20	0.2851	0.2812	0.9596	0.2931	3.412	1.042	3.556	1.2857	40
30	0.2880	0.2840	0.9588	0.2962	3.376	1.043	3.521	1.2828	30
40	0.2909	0.2868	0.9580	0.2994	3.340	1.044	3.487	1.2799	20
50	0.2938	0.2896	0.9572	0.3026	3.305	1.045	3.453	1.2770	10
17°00′	0.2967	0.2924	0.9563	0.3057	3.271	1.046	3.420	1.2741	73°00′
10	0.2996	0.2952	0.9555	0.3089	3.237	1.047	3.388	1.2712	50
20	0.3025	0.2979	0.9546	0.3121	3.204	1.048	3.356	1.2683	40
30	0.3054	0.3007	0.9537	0.3153	3.172	1.049	3.326	1.2654	30
40	0.3083	0.3035	0.9528	0.3185	3.140	1.049	3.295	1.2625	20
50	0.3113	0.3062	0.9520	0.3217	3.108	1.050	3.265	1.2595	10
18°00′	0.3142	0.3090	0.9511	0.3249	3.078	1.051	3.236	1.2566	72°00′
10	0.3171	0.3118	0.9502	0.3281	3.047	1.052	3.207	1.2537	50
20	0.3200	0.3145	0.9492	0.3314	3.018	1.053	3.179	1.2508	40
30	0.3229	0.3173	0.9483	0.3346	2.989	1.054	3.152	1.2479	30
40	0.3258	0.3201	0.9474	0.3378	2.960	1.056	3.124	1.2450	20
50	0.3287	0.3228	0.9465	0.3411	2.932	1.057	3.098	1.2421	10
19°00′	0.3316	0.3256	0.9455	0.3443	2.904	1.058	3.072	1.2392	71°00′
10	0.3345	0.3283	0.9446	0.3476	2.877	1.059	3.046	1.2363	50
20	0.3374	0.3311	0.9436	0.3508	2.850	1.060	3.021	1.2334	40
30	0.3403	0.3338	0.9426	0.3541	2.824	1.061	2.996	1.2305	30
40	0.3432	0.3365	0.9417	0.3574	2.798	1.062	2.971	1.2275	20
50	0.3462	0.3393	0.9407	0.3607	2.773	1.063	2.947	1.2246	10
20°00′	0.3491	0.3420	0.9397	0.3640	2.747	1.064	2.924	1.2217	70°00′
10	0.3520	0.3448	0.9387	0.3673	2.723	1.065	2.901	1.2188	50
20	0.3549	0.3475	0.9377	0.3706	2.699	1.066	2.878	1.2159	40
30	0.3578	0.3502	0.9367	0.3739	2.675	1.068	2.855	1.2130	30
40	0.3607	0.3529	0.9356	0.3772	2.651	1.069	2.833	1.2101	20
50	0.3636	0.3557	0.9346	0.3805	2.628	1.070	2.812	1.2072	10
21°00′	0.3665	0.3584	0.9336	0.3839	2.605	1.071	2.790	1.2043	69°00′
10	0.3694	0.3611	0.9325	0.3872	2.583	1.072	2.769	1.2014	50
20	0.3723	0.3638	0.9315	0.3906	2.560	1.074	2.749	1.1985	40
30	0.3752	0.3665	0.9304	0.3939	2.539	1.075	2.729	1.1956	30
40	0.3782	0.3692	0.9293	0.3973	2.517	1.076	2.709	1.1926	20
50	0.3811	0.3719	0.9283	0.4006	2.496	1.077	2.689	1.1897	10
22°00′	0.3840	0.3746	0.9272	0.4040	2.475	1.079	2.669	1.1868	68°00′
10	0.3869	0.3773	0.9261	0.4074	2.455	1.080	2.650	1.1839	50
20	0.3898	0.3800	0.9250	0.4108	2.434	1.081	2.632	1.1810	40
30	0.3927	0.3827	0.9239	0.4142	2.414	1.082	2.613	1.1781	30
40	0.3956	0.3854	0.9228	0.4176	2.394	1.084	2.595	1.1752	20
50	0.3985	0.3881	0.9216	0.4210	2.375	1.085	2.577	1.1723	10
23°00′	0.4014	0.3907	0.9205	0.4245	2.356	1.086	2.559	1.1694	67°00′
10	0.4043	0.3934	0.9194	0.4279	2.337	1.088	2.542	1.1665	50
20	0.4072	0.3961	0.9182	0.4314	2.318	1.089	2.525	1.1636	40
30	0.4102	0.3987	0.9171	0.4348	2.300	1.090	2.508	1.1606	30
40	0.4131	0.4014	0.9159	0.4383	2.282	1.092	2.491	1.1577	20
50	0.4160	0.4041	0.9147	0.4417	2.264	1.093	2.475	1.1548	10
24°00′	0.4189	0.4067	0.9135	0.4452	2.246	1.095	2.459	1.1519	66°00′
		Cos θ	Sin θ	Cot θ	Tan θ	Csc θ	Sec θ	Radians	Degrees

Table 3   A—45

Table 3. Continued

Degrees	Radians	Sin θ	Cos θ	Tan θ	Cot θ	Sec θ	Csc θ		
24°00′	0.4189	0.4067	0.9135	0.4452	2.246	1.095	2.459	1.1519	66°00′
10	0.4218	0.4094	0.9124	0.4487	2.229	1.096	2.443	1.1490	50
20	0.4247	0.4120	0.9112	0.4522	2.211	1.097	2.427	1.1461	40
30	0.4276	0.4147	0.9100	0.4557	2.194	1.099	2.411	1.1432	30
40	0.4305	0.4173	0.9088	0.4592	2.177	1.100	2.396	1.1403	20
50	0.4334	0.4200	0.9075	0.4628	2.161	1.102	2.381	1.1374	10
25°00′	0.4363	0.4226	0.9063	0.4663	2.145	1.103	2.366	1.1345	65°00′
10	0.4392	0.4253	0.9051	0.4699	2.128	1.105	2.352	1.1316	50
20	0.4422	0.4279	0.9038	0.4734	2.112	1.106	2.337	1.1286	40
30	0.4451	0.4305	0.9026	0.4770	2.097	1.108	2.323	1.1257	30
40	0.4480	0.4331	0.9013	0.4806	2.081	1.109	2.309	1.1228	20
50	0.4509	0.4358	0.9001	0.4841	2.066	1.111	2.295	1.1199	10
26°00′	0.4538	0.4384	0.8988	0.4877	2.050	1.113	2.281	1.1170	64°00′
10	0.4567	0.4410	0.8975	0.4913	2.035	1.114	2.268	1.1141	50
20	0.4596	0.4436	0.8962	0.4950	2.020	1.116	2.254	1.1112	40
30	0.4625	0.4462	0.8949	0.4986	2.006	1.117	2.241	1.1083	30
40	0.4654	0.4488	0.8936	0.5022	1.991	1.119	2.228	1.1054	20
50	0.4683	0.4514	0.8923	0.5059	1.977	1.121	2.215	1.1025	10
27°00′	0.4712	0.4540	0.8910	0.5095	1.963	1.122	2.203	1.0996	63°00′
10	0.4741	0.4566	0.8897	0.5132	1.949	1.124	2.190	1.0966	50
20	0.4771	0.4592	0.8884	0.5169	1.935	1.126	2.178	1.0937	40
30	0.4800	0.4617	0.8870	0.5206	1.921	1.127	2.166	1.0908	30
40	0.4829	0.4643	0.8857	0.5243	1.907	1.129	2.154	1.0879	20
50	0.4858	0.4669	0.8843	0.5280	1.894	1.131	2.142	1.0850	10
28°00′	0.4887	0.4695	0.8829	0.5317	1.881	1.133	2.130	1.0821	62°00′
10	0.4916	0.4720	0.8816	0.5354	1.868	1.134	2.118	1.0792	50
20	0.4945	0.4746	0.8802	0.5392	1.855	1.136	2.107	1.0763	40
30	0.4974	0.4772	0.8788	0.5430	1.842	1.138	2.096	1.0734	30
40	0.5003	0.4797	0.8774	0.5467	1.829	1.140	2.085	1.0705	20
50	0.5032	0.4823	0.8760	0.5505	1.816	1.142	2.074	1.0676	10
29°00′	0.5061	0.4848	0.8746	0.5543	1.804	1.143	2.063	1.0647	61°00′
10	0.5091	0.4874	0.8732	0.5581	1.792	1.145	2.052	1.0617	50
20	0.5120	0.4899	0.8718	0.5619	1.780	1.147	2.041	1.0588	40
30	0.5149	0.4924	0.8704	0.5658	1.767	1.149	2.031	1.0559	30
40	0.5178	0.4950	0.8689	0.5696	1.756	1.151	2.020	1.0530	20
50	0.5207	0.4975	0.8675	0.5735	1.744	1.153	2.010	1.0501	10
30°00′	0.5236	0.5000	0.8660	0.5774	1.732	1.155	2.000	1.0472	60°00′
10	0.5265	0.5025	0.8646	0.5812	1.720	1.157	1.990	1.0443	50
20	0.5294	0.5050	0.8631	0.5851	1.709	1.159	1.980	1.0414	40
30	0.5323	0.5075	0.8616	0.5890	1.698	1.161	1.970	1.0385	30
40	0.5352	0.5100	0.8601	0.5930	1.686	1.163	1.961	1.0356	20
50	0.5381	0.5125	0.8587	0.5969	1.675	1.165	1.951	1.0327	10
31°00′	0.5411	0.5150	0.8572	0.6009	1.664	1.167	1.942	1.0297	59°00′
10	0.5440	0.5175	0.8557	0.6048	1.653	1.169	1.932	1.0268	50
20	0.5469	0.5200	0.8542	0.6088	1.643	1.171	1.923	1.0239	40
30	0.5498	0.5225	0.8526	0.6128	1.632	1.173	1.914	1.0210	30
40	0.5527	0.5250	0.8511	0.6168	1.621	1.175	1.905	1.0181	20
50	0.5556	0.5275	0.8496	0.6208	1.611	1.177	1.896	1.0152	10
32°00′	0.5585	0.5299	0.8480	0.6249	1.600	1.179	1.887	1.0123	58°00′
		Cos θ	Sin θ	Cot θ	Tan θ	Csc θ	Sec θ	Radians	Degrees

Table 3. Continued

Degrees	Radians	Sin θ	Cos θ	Tan θ	Cot θ	Sec θ	Csc θ		
32°00′	0.5585	0.5299	0.8480	0.6249	1.600	1.179	1.887	1.0123	58°00′
10	0.5614	0.5324	0.8465	0.6289	1.590	1.181	1.878	1.0094	50
20	0.5643	0.5348	0.8450	0.6330	1.580	1.184	1.870	1.0065	40
30	0.5672	0.5373	0.8434	0.6371	1.570	1.186	1.861	1.0036	30
40	0.5701	0.5398	0.8418	0.6412	1.560	1.188	1.853	1.0007	20
50	0.5730	0.5422	0.8403	0.6453	1.550	1.190	1.844	0.9977	10
33°00′	0.5760	0.5446	0.8387	0.6494	1.540	1.192	1.836	0.9948	57°00′
10	0.5789	0.5471	0.8371	0.6536	1.530	1.195	1.828	0.9919	50
20	0.5818	0.5495	0.8355	0.6577	1.520	1.197	1.820	0.9890	40
30	0.5847	0.5519	0.8339	0.6619	1.511	1.199	1.812	0.9861	30
40	0.5876	0.5544	0.8323	0.6661	1.501	1.202	1.804	0.9832	20
50	0.5905	0.5568	0.8307	0.6703	1.492	1.204	1.796	0.9803	10
34°00′	0.5934	0.5592	0.8290	0.6745	1.483	1.206	1.788	0.9774	56°00′
10	0.5963	0.5616	0.8274	0.6787	1.473	1.209	1.781	0.9745	50
20	0.5992	0.5640	0.8258	0.6830	1.464	1.211	1.773	0.9716	40
30	0.6021	0.5664	0.8241	0.6873	1.455	1.213	1.766	0.9687	30
40	0.6050	0.5688	0.8225	0.6916	1.446	1.216	1.758	0.9657	20
50	0.6080	0.5712	0.8208	0.6959	1.437	1.218	1.751	0.9628	10
35°00′	0.6109	0.5736	0.8192	0.7002	1.428	1.221	1.743	0.9599	55°00′
10	0.6138	0.5760	0.8175	0.7046	1.419	1.223	1.736	0.9570	50
20	0.6167	0.5783	0.8158	0.7089	1.411	1.226	1.729	0.9541	40
30	0.6196	0.5807	0.8141	0.7133	1.402	1.228	1.722	0.9512	30
40	0.6225	0.5831	0.8124	0.7177	1.393	1.231	1.715	0.9483	20
50	0.6254	0.5854	0.8107	0.7221	1.385	1.233	1.708	0.9454	10
36°00′	0.6283	0.5878	0.8090	0.7265	1.376	1.236	1.701	0.9425	54°00′
10	0.6312	0.5901	0.8073	0.7310	1.368	1.239	1.695	0.9396	50
20	0.6341	0.5925	0.8056	0.7355	1.360	1.241	1.688	0.9367	40
30	0.6370	0.5948	0.8039	0.7400	1.351	1.244	1.681	0.9338	30
40	0.6400	0.5972	0.8021	0.7445	1.343	1.247	1.675	0.9308	20
50	0.6429	0.5995	0.8004	0.7490	1.335	1.249	1.668	0.9279	10
37°00′	0.6458	0.6018	0.7986	0.7536	1.327	1.252	1.662	0.9250	53°00′
10	0.6487	0.6041	0.7969	0.7581	1.319	1.255	1.655	0.9221	50
20	0.6516	0.6065	0.7951	0.7627	1.311	1.258	1.649	0.9192	40
30	0.6545	0.6088	0.7934	0.7673	1.303	1.260	1.643	0.9163	30
40	0.6574	0.6111	0.7916	0.7720	1.295	1.263	1.636	0.9134	20
50	0.6603	0.6134	0.7898	0.7766	1.288	1.266	1.630	0.9105	10
38°00′	0.6632	0.6157	0.7880	0.7813	1.280	1.269	1.624	0.9076	52°00′
10	0.6661	0.6180	0.7862	0.7860	1.272	1.272	1.618	0.9047	50
20	0.6690	0.6202	0.7844	0.7907	1.265	1.275	1.612	0.9018	40
30	0.6720	0.6225	0.7826	0.7954	1.257	1.278	1.606	0.8988	30
40	0.6749	0.6248	0.7808	0.8002	1.250	1.281	1.601	0.8959	20
50	0.6778	0.6271	0.7790	0.8050	1.242	1.284	1.595	0.8930	10
39°00′	0.6807	0.6293	0.7771	0.8098	1.235	1.287	1.589	0.8901	51°00′
		Cos θ	Sin θ	Cot θ	Tan θ	Csc θ	Sec θ	Radians	Degrees

Table 3   A—47

Table 3. Continued

Degrees	Radians	Sin θ	Cos θ	Tan θ	Cot θ	Sec θ	Csc θ		
39°00′	0.6807	0.6293	0.7771	0.8098	1.235	1.287	1.589	0.8901	51°00′
10	0.6836	0.6316	0.7753	0.8146	1.228	1.290	1.583	0.8872	50
20	0.6865	0.6338	0.7735	0.8195	1.220	1.293	1.578	0.8843	40
30	0.6894	0.6361	0.7716	0.8243	1.213	1.296	1.572	0.8814	30
40	0.6923	0.6383	0.7698	0.8292	1.206	1.299	1.567	0.8785	20
50	0.6952	0.6406	0.7679	0.8342	1.199	1.302	1.561	0.8756	10
40°00′	0.6981	0.6428	0.7660	0.8391	1.192	1.305	1.556	0.8727	50°00′
10	0.7010	0.6450	0.7642	0.8441	1.185	1.309	1.550	0.8698	50
20	0.7039	0.6472	0.7623	0.8491	1.178	1.312	1.545	0.8668	40
30	0.7069	0.6494	0.7604	0.8541	1.171	1.315	1.540	0.8639	30
40	0.7098	0.6517	0.7585	0.8591	1.164	1.318	1.535	0.8610	20
50	0.7127	0.6539	0.7566	0.8642	1.157	1.322	1.529	0.8581	10
41°00′	0.7156	0.6561	0.7547	0.8693	1.150	1.325	1.524	0.8552	49°00′
10	0.7185	0.6583	0.7528	0.8744	1.144	1.328	1.519	0.8523	50
20	0.7214	0.6604	0.7509	0.8796	1.137	1.332	1.514	0.8494	40
30	0.7243	0.6626	0.7490	0.8847	1.130	1.335	1.509	0.8465	30
40	0.7272	0.6648	0.7470	0.8899	1.124	1.339	1.504	0.8436	20
50	0.7301	0.6670	0.7451	0.8952	1.117	1.342	1.499	0.8407	10
42°00′	0.7330	0.6691	0.7431	0.9004	1.111	1.346	1.494	0.8378	48°00′
10	0.7359	0.6713	0.7412	0.9057	1.104	1.349	1.490	0.8348	50
20	0.7389	0.6734	0.7392	0.9110	1.098	1.353	1.485	0.8319	40
30	0.7418	0.6756	0.7373	0.9163	1.091	1.356	1.480	0.8290	30
40	0.7447	0.6777	0.7353	0.9217	1.085	1.360	1.476	0.8261	20
50	0.7476	0.6799	0.7333	0.9271	1.079	1.364	1.471	0.8232	10
43°00′	0.7505	0.6820	0.7314	0.9325	1.072	1.367	1.466	0.8203	47°00′
10	0.7534	0.6841	0.7294	0.9380	1.066	1.371	1.462	0.8174	50
20	0.7563	0.6862	0.7274	0.9435	1.060	1.375	1.457	0.8145	40
30	0.7592	0.6884	0.7254	0.9490	1.054	1.379	1.453	0.8116	30
40	0.7621	0.6905	0.7234	0.9545	1.048	1.382	1.448	0.8087	20
50	0.7650	0.6926	0.7214	0.9601	1.042	1.386	1.444	0.8058	10
44°00′	0.7679	0.6947	0.7193	0.9657	1.036	1.390	1.440	0.8029	46°00′
10	0.7709	0.6967	0.7173	0.9713	1.030	1.394	1.435	0.7999	50
20	0.7738	0.6988	0.7153	0.9770	1.024	1.398	1.431	0.7970	40
30	0.7767	0.7009	0.7133	0.9827	1.018	1.402	1.427	0.7941	30
40	0.7796	0.7030	0.7112	0.9884	1.012	1.406	1.423	0.7912	20
50	0.7825	0.7050	0.7092	0.9942	1.006	1.410	1.418	0.7883	10
45°00′	0.7854	0.7071	0.7071	1.000	1.000	1.414	1.414	0.7854	45°00′
		Cos θ	Sin θ	Cot θ	Tan θ	Csc θ	Sec θ	Radians	Degrees

Table 4. Four-Place Values of Trigonometric Functions
Angle θ in Degrees and Tenths

Degrees	Sin θ	Cos θ	Tan θ	Cot θ		Degrees	Sin θ	Cos θ	Tan θ	Cot θ	
0.0	0.0000	1.0000	0.0000	—	90.0	5.0	0.0872	0.9962	0.0875	11.43	85.0
.1	0.0017	1.0000	0.0017	573.0	.9	.1	0.0889	0.9960	0.0892	11.20	.9
.2	0.0035	1.0000	0.0035	286.4	.8	.2	0.0906	0.9959	0.0910	10.99	.8
.3	0.0052	1.0000	0.0052	191.0	.7	.3	0.0924	0.9957	0.0928	10.78	.7
.4	0.0070	1.0000	0.0070	143.2	.6	.4	0.0941	0.9956	0.0945	10.58	.6
.5	0.0087	1.0000	0.0087	114.6	.5	.5	0.0958	0.9954	0.0963	10.39	.5
.6	0.0105	0.9999	0.0105	95.49	.4	.6	0.0976	0.9952	0.0981	10.20	.4
.7	0.0122	0.9999	0.0122	81.85	.3	.7	0.0993	0.9951	0.0998	10.02	.3
.8	0.0140	0.9999	0.0140	71.62	.2	.8	0.1011	0.9949	0.1016	9.845	.2
.9	0.0157	0.9999	0.0157	63.66	.1	.9	0.1028	0.9947	0.1033	9.677	.1
1.0	0.0175	0.9998	0.0175	57.29	89.0	6.0	0.1045	0.9945	0.1051	9.514	84.0
.1	0.0192	0.9998	0.0192	52.08	.9	.1	0.1063	0.9943	0.1069	9.357	.9
.2	0.0209	0.9998	0.0209	47.74	.8	.2	0.1080	0.9942	0.1086	9.205	.8
.3	0.0227	0.9997	0.0227	44.07	.7	.3	0.1097	0.9940	0.1104	9.058	.7
.4	0.0244	0.9997	0.0244	40.92	.6	.4	0.1115	0.9938	0.1122	8.915	.6
.5	0.0262	0.9997	0.0262	38.19	.5	.5	0.1132	0.9936	0.1139	8.777	.5
.6	0.0279	0.9996	0.0279	35.80	.4	.6	0.1149	0.9934	0.1157	8.643	.4
.7	0.0297	0.9996	0.0297	33.69	.3	.7	0.1167	0.9932	0.1175	8.513	.3
.8	0.0314	0.9995	0.0314	31.82	.2	.8	0.1184	0.9930	0.1192	8.386	.2
.9	0.0332	0.9995	0.0332	30.14	.1	.9	0.1201	0.9928	0.1210	8.264	.1
2.0	0.0349	0.9994	0.0349	28.64	88.0	7.0	0.1219	0.9925	0.1228	8.144	83.0
.1	0.0366	0.9993	0.0367	27.27	.9	.1	0.1236	0.9923	0.1246	8.028	.9
.2	0.0384	0.9993	0.0384	26.03	.8	.2	0.1253	0.9921	0.1263	7.916	.8
.3	0.0401	0.9992	0.0402	24.90	.7	.3	0.1271	0.9919	0.1281	7.806	.7
.4	0.0419	0.9991	0.0419	23.86	.6	.4	0.1288	0.9917	0.1299	7.700	.6
.5	0.0436	0.9990	0.0437	22.90	.5	.5	0.1305	0.9914	0.1317	7.596	.5
.6	0.0454	0.9990	0.0454	22.02	.4	.6	0.1323	0.9912	0.1334	7.495	.4
.7	0.0471	0.9989	0.0472	21.20	.3	.7	0.1340	0.9910	0.1352	7.396	.3
.8	0.0488	0.9988	0.0489	20.45	.2	.8	0.1357	0.9907	0.1370	7.300	.2
.9	0.0506	0.9987	0.0507	19.74	.1	.9	0.1374	0.9905	0.1388	7.207	.1
3.0	0.0523	0.9986	0.0524	19.08	87.0	8.0	0.1392	0.9903	0.1405	7.115	82.0
.1	0.0541	0.9985	0.0542	18.46	.9	.1	0.1409	0.9900	0.1423	7.026	.9
.2	0.0558	0.9984	0.0559	17.89	.8	.2	0.1426	0.9898	0.1441	6.940	.8
.3	0.0576	0.9983	0.0577	17.34	.7	.3	0.1444	0.9895	0.1459	6.855	.7
.4	0.0593	0.9982	0.0594	16.83	.6	.4	0.1461	0.9893	0.1477	6.772	.6
.5	0.0610	0.9981	0.0612	16.35	.5	.5	0.1478	0.9890	0.1495	6.691	.5
.6	0.0628	0.9980	0.0629	15.89	.4	.6	0.1495	0.9888	0.1512	6.612	.4
.7	0.0645	0.9979	0.0647	15.46	.3	.7	0.1513	0.9885	0.1530	6.535	.3
.8	0.0663	0.9978	0.0664	15.06	.2	.8	0.1530	0.9882	0.1548	6.460	.2
.9	0.0680	0.9977	0.0682	14.67	.1	.9	0.1547	0.9880	0.1566	6.386	.1
4.0	0.0698	0.9976	0.0699	14.30	86.0	9.0	0.1564	0.9877	0.1584	6.314	81.0
.1	0.0715	0.9974	0.0717	13.95	.9	.1	0.1582	0.9874	0.1602	6.243	.9
.2	0.0732	0.9973	0.0734	13.62	.8	.2	0.1599	0.9871	0.1620	6.174	.8
.3	0.0750	0.9972	0.0752	13.30	.7	.3	0.1616	0.9869	0.1638	6.107	.7
.4	0.0767	0.9971	0.0769	13.00	.6	.4	0.1633	0.9866	0.1655	6.041	.6
.5	0.0785	0.9969	0.0787	12.71	.5	.5	0.1650	0.9863	0.1673	5.976	.5
.6	0.0802	0.9968	0.0805	12.43	.4	.6	0.1668	0.9860	0.1691	5.912	.4
.7	0.0819	0.9966	0.0822	12.16	.3	.7	0.1685	0.9857	0.1709	5.850	.3
.8	0.0837	0.9965	0.0840	11.91	.2	.8	0.1702	0.9854	0.1727	5.789	.2
.9	0.0854	0.9963	0.0857	11.66	.1	.9	0.1719	0.9851	0.1745	5.730	.1
5.0	0.0872	0.9962	0.0875	11.43	85.0	10.0	0.1736	0.9848	0.1763	5.671	80.0
	Cos θ	Sin θ	Cot θ	Tan θ	Degrees		Cos θ	Sin θ	Cot θ	Tan θ	Degrees

Table 4   A—49

Table 4. Continued

Degrees	Sin θ	Cos θ	Tan θ	Cot θ		Degrees	Sin θ	Cos θ	Tan θ	Cot θ	
10.0	0.1736	0.9848	0.1763	5.671	80.0	15.0	0.2588	0.9659	0.2679	3.732	75.0
.1	0.1754	0.9845	0.1781	5.614	.9	.1	0.2605	0.9655	0.2698	3.706	.9
.2	0.1771	0.9842	0.1799	5.558	.8	.2	0.2622	0.9650	0.2717	3.681	.8
.3	0.1788	0.9839	0.1817	5.503	.7	.3	0.2639	0.9646	0.2736	3.655	.7
.4	0.1805	0.9836	0.1835	5.449	.6	.4	0.2656	0.9641	0.2754	3.630	.6
.5	0.1822	0.9833	0.1853	5.396	.5	.5	0.2672	0.9636	0.2773	3.606	.5
.6	0.1840	0.9829	0.1871	5.343	.4	.6	0.2689	0.9632	0.2792	3.582	.4
.7	0.1857	0.9826	0.1890	5.292	.3	.7	0.2706	0.9627	0.2811	3.558	.3
.8	0.1874	0.9823	0.1908	5.242	.2	.8	0.2723	0.9622	0.2830	3.534	.2
.9	0.1891	0.9820	0.1926	5.193	.1	.9	0.2740	0.9617	0.2849	3.511	.1
11.0	0.1908	0.9816	0.1944	5.145	79.0	16.0	0.2756	0.9613	0.2867	3.487	74.0
.1	0.1925	0.9813	0.1962	5.097	.9	.1	0.2773	0.9608	0.2886	3.465	.9
.2	0.1942	0.9810	0.1980	5.050	.8	.2	0.2790	0.9603	0.2905	3.442	.8
.3	0.1959	0.9806	0.1998	5.005	.7	.3	0.2807	0.9598	0.2924	3.420	.7
.4	0.1977	0.9803	0.2016	4.959	.6	.4	0.2823	0.9593	0.2943	3.398	.6
.5	0.1994	0.9799	0.2035	4.915	.5	.5	0.2840	0.9588	0.2962	3.376	.5
.6	0.2011	0.9796	0.2053	4.872	.4	.6	0.2857	0.9583	0.2981	3.354	.4
.7	0.2028	0.9792	0.2071	4.829	.3	.7	0.2874	0.9578	0.3000	3.333	.3
.8	0.2045	0.9789	0.2089	4.787	.2	.8	0.2890	0.9573	0.3019	3.312	.2
.9	0.2062	0.9785	0.2107	4.745	.1	.9	0.2907	0.9568	0.3038	3.291	.1
12.0	0.2079	0.9781	0.2126	4.705	78.0	17.0	0.2924	0.9563	0.3057	3.271	73.0
.1	0.2096	0.9778	0.2144	4.665	.9	.1	0.2940	0.9558	0.3076	3.251	.9
.2	0.2113	0.9774	0.2162	4.625	.8	.2	0.2957	0.9553	0.3096	3.230	.8
.3	0.2130	0.9770	0.2180	4.586	.7	.3	0.2974	0.9548	0.3115	3.211	.7
.4	0.2147	0.9767	0.2199	4.548	.6	.4	0.2990	0.9542	0.3134	3.191	.6
.5	0.2164	0.9763	0.2217	4.511	.5	.5	0.3007	0.9537	0.3153	3.172	.5
.6	0.2181	0.9759	0.2235	4.474	.4	.6	0.3024	0.9532	0.3172	3.152	.4
.7	0.2198	0.9755	0.2254	4.437	.3	.7	0.3040	0.9527	0.3191	3.133	.3
.8	0.2215	0.9751	0.2272	4.402	.2	.8	0.3057	0.9521	0.3211	3.115	.2
.9	0.2233	0.9748	0.2290	4.366	.1	.9	0.3074	0.9516	0.3230	3.096	.1
13.0	0.2250	0.9744	0.2309	4.331	77.0	18.0	0.3090	0.9511	0.3249	3.078	72.0
.1	0.2267	0.9740	0.2327	4.297	.9	.1	0.3107	0.9505	0.3269	3.060	.9
.2	0.2284	0.9736	0.2345	4.264	.8	.2	0.3123	0.9500	0.3288	3.042	.8
.3	0.2300	0.9732	0.2364	4.230	.7	.3	0.3140	0.9494	0.3307	3.024	.7
.4	0.2317	0.9728	0.2382	4.198	.6	.4	0.3156	0.9489	0.3327	3.006	.6
.5	0.2334	0.9724	0.2401	4.165	.5	.5	0.3173	0.9483	0.3346	2.989	.5
.6	0.2351	0.9720	0.2419	4.134	.4	.6	0.3190	0.9478	0.3365	2.971	.4
.7	0.2368	0.9715	0.2438	4.102	.3	.7	0.3206	0.9472	0.3385	2.954	.3
.8	0.2385	0.9711	0.2456	4.071	.2	.8	0.3223	0.9466	0.3404	2.937	.2
.9	0.2402	0.9707	0.2475	4.041	.1	.9	0.3239	0.9461	0.3424	2.921	.1
14.0	0.2419	0.9703	0.2493	4.011	76.0	19.0	0.3256	0.9455	0.3443	2.904	71.0
.1	0.2436	0.9699	0.2512	3.981	.9	.1	0.3272	0.9449	0.3463	2.888	.9
.2	0.2453	0.9694	0.2530	3.952	.8	.2	0.3289	0.9444	0.3482	2.872	.8
.3	0.2470	0.9690	0.2549	3.923	.7	.3	0.3305	0.9438	0.3502	2.856	.7
.4	0.2487	0.9686	0.2568	3.895	.6	.4	0.3322	0.9432	0.3522	2.840	.6
.5	0.2504	0.9681	0.2586	3.867	.5	.5	0.3338	0.9426	0.3541	2.824	.5
.6	0.2521	0.9677	0.2605	3.839	.4	.6	0.3355	0.9421	0.3561	2.808	.4
.7	0.2538	0.9673	0.2623	3.812	.3	.7	0.3371	0.9415	0.3581	2.793	.3
.8	0.2554	0.9668	0.2642	3.785	.2	.8	0.3387	0.9409	0.3600	2.778	.2
.9	0.2571	0.9664	0.2661	3.758	.1	.9	0.3404	0.9403	0.3620	2.762	.1
15.0	0.2588	0.9659	0.2679	3.732	75.0	20.0	0.3420	0.9397	0.3640	2.747	70.0
	Cos θ	Sin θ	Cot θ	Tan θ	Degrees		Cos θ	Sin θ	Cot θ	Tan θ	Degrees

Table 4. Continued

Degrees	Sin θ	Cos θ	Tan θ	Cot θ		Degrees	Sin θ	Cos θ	Tan θ	Cot θ	
20.0	0.3420	0.9397	0.3640	2.747	70.0	25.0	0.4226	0.9063	0.4663	2.145	65.0
.1	0.3437	0.9391	0.3659	2.733	.9	.1	0.4242	0.9056	0.4684	2.135	.9
.2	0.3453	0.9385	0.3679	2.718	.8	.2	0.4258	0.9048	0.4706	2.125	.8
.3	0.3469	0.9379	0.3699	2.703	.7	.3	0.4274	0.9041	0.4727	2.116	.7
.4	0.3486	0.9373	0.3719	2.689	.6	.4	0.4289	0.9033	0.4748	2.106	.6
.5	0.3502	0.9367	0.3739	2.675	.5	.5	0.4305	0.9026	0.4770	2.097	.5
.6	0.3518	0.9361	0.3759	2.660	.4	.6	0.4321	0.9018	0.4791	2.087	.4
.7	0.3535	0.9354	0.3779	2.646	.3	.7	0.4337	0.9011	0.4813	2.078	.3
.8	0.3551	0.9348	0.3799	2.633	.2	.8	0.4352	0.9003	0.4834	2.069	.2
.9	0.3567	0.9342	0.3819	2.619	.1	.9	0.4368	0.8996	0.4856	2.059	.1
21.0	0.3584	0.9336	0.3839	2.605	69.0	26.0	0.4384	0.8988	0.4877	2.050	64.0
.1	0.3600	0.9330	0.3859	2.592	.9	.1	0.4399	0.8980	0.4899	2.041	.9
.2	0.3616	0.9323	0.3879	2.578	.8	.2	0.4415	0.8973	0.4921	2.032	.8
.3	0.3633	0.9317	0.3899	2.565	.7	.3	0.4431	0.8965	0.4942	2.023	.7
.4	0.3649	0.9311	0.3919	2.552	.6	.4	0.4446	0.8957	0.4964	2.014	.6
.5	0.3665	0.9304	0.3939	2.539	.5	.5	0.4462	0.8949	0.4986	2.006	.5
.6	0.3681	0.9298	0.3959	2.526	.4	.6	0.4478	0.8942	0.5008	1.997	.4
.7	0.3697	0.9291	0.3979	2.513	.3	.7	0.4493	0.8934	0.5029	1.988	.3
.8	0.3714	0.9285	0.4000	2.500	.2	.8	0.4509	0.8926	0.5051	1.980	.2
.9	0.3730	0.9278	0.4020	2.488	.1	.9	0.4524	0.8918	0.5073	1.971	.1
22.0	0.3746	0.9272	0.4040	2.475	68.0	27.0	0.4540	0.8910	0.5095	1.963	63.0
.1	0.3762	0.9265	0.4061	2.463	.9	.1	0.4555	0.8902	0.5117	1.954	.9
.2	0.3778	0.9259	0.4081	2.450	.8	.2	0.4571	0.8894	0.5139	1.946	.8
.3	0.3795	0.9252	0.4101	2.438	.7	.3	0.4586	0.8886	0.5161	1.937	.7
.4	0.3811	0.9245	0.4122	2.426	.6	.4	0.4602	0.8878	0.5184	1.929	.6
.5	0.3827	0.9239	0.4142	2.414	.5	.5	0.4617	0.8870	0.5206	1.921	.5
.6	0.3843	0.9232	0.4163	2.402	.4	.6	0.4633	0.8862	0.5228	1.913	.4
.7	0.3859	0.9225	0.4183	2.391	.3	.7	0.4648	0.8854	0.5250	1.905	.3
.8	0.3875	0.9219	0.4204	2.379	.2	.8	0.4664	0.8846	0.5272	1.897	.2
.9	0.3891	0.9212	0.4224	2.367	.1	.9	0.4679	0.8838	0.5295	1.889	.1
23.0	0.3907	0.9205	0.4245	2.356	67.0	28.0	0.4695	0.8829	0.5317	1.881	62.0
.1	0.3923	0.9198	0.4265	2.344	.9	.1	0.4710	0.8821	0.5340	1.873	.9
.2	0.3939	0.9191	0.4286	2.333	.8	.2	0.4726	0.8813	0.5362	1.865	.8
.3	0.3955	0.9184	0.4307	2.322	.7	.3	0.4741	0.8805	0.5384	1.857	.7
.4	0.3971	0.9178	0.4327	2.311	.6	.4	0.4756	0.8796	0.5407	1.849	.6
.5	0.3987	0.9171	0.4348	2.300	.5	.5	0.4772	0.8788	0.5430	1.842	.5
.6	0.4003	0.9164	0.4369	2.289	.4	.6	0.4787	0.8780	0.5452	1.834	.4
.7	0.4019	0.9157	0.4390	2.278	.3	.7	0.4802	0.8771	0.5475	1.827	.3
.8	0.4035	0.9150	0.4411	2.267	.2	.8	0.4818	0.8763	0.5498	1.819	.2
.9	0.4051	0.9143	0.4431	2.257	.1	.9	0.4833	0.8755	0.5520	1.811	.1
24.0	0.4067	0.9135	0.4452	2.246	66.0	29.0	0.4848	0.8746	0.5543	1.804	61.0
.1	0.4083	0.9128	0.4473	2.236	.9	.1	0.4863	0.8738	0.5566	1.797	.9
.2	0.4099	0.9121	0.4494	2.225	.8	.2	0.4879	0.8729	0.5589	1.789	.8
.3	0.4115	0.9114	0.4515	2.215	.7	.3	0.4894	0.8721	0.5612	1.782	.7
.4	0.4131	0.9107	0.4536	2.204	.6	.4	0.4909	0.8712	0.5635	1.775	.6
.5	0.4147	0.9100	0.4557	2.194	.5	.5	0.4924	0.8704	0.5658	1.767	.5
.6	0.4163	0.9092	0.4578	2.184	.4	.6	0.4939	0.8695	0.5681	1.760	.4
.7	0.4179	0.9085	0.4599	2.174	.3	.7	0.4955	0.8686	0.5704	1.753	.3
.8	0.4195	0.9078	0.4621	2.164	.2	.8	0.4970	0.8678	0.5727	1.746	.2
.9	0.4210	0.9070	0.4642	2.154	.1	.9	0.4985	0.8669	0.5750	1.739	.1
25.0	0.4226	0.9063	0.4663	2.145	65.0	30.0	0.5000	0.8660	0.5774	1.732	60.0
	Cos θ	Sin θ	Cot θ	Tan θ	Degrees		Cos θ	Sin θ	Cot θ	Tan θ	Degrees

Table 4 A—51

Table 4. Continued

Degrees	Sin θ	Cos θ	Tan θ	Cot θ		Degrees	Sin θ	Cos θ	Tan θ	Cot θ	
30.0	0.5000	0.8660	0.5774	1.732	60.0	35.0	0.5736	0.8192	0.7002	1.428	55.0
.1	0.5015	0.8652	0.5797	1.725	.9	.1	0.5750	0.8181	0.7028	1.423	.9
.2	0.5030	0.8643	0.5820	1.718	.8	.2	0.5764	0.8171	0.7054	1.418	.8
.3	0.5045	0.8634	0.5844	1.711	.7	.3	0.5779	0.8161	0.7080	1.412	.7
.4	0.5060	0.8625	0.5867	1.704	.6	.4	0.5793	0.8151	0.7107	1.407	.6
.5	0.5075	0.8616	0.5890	1.698	.5	.5	0.5807	0.8141	0.7133	1.402	.5
.6	0.5090	0.8607	0.5914	1.691	.4	.6	0.5821	0.8131	0.7159	1.397	.4
.7	0.5105	0.8599	0.5938	1.684	.3	.7	0.5835	0.8121	0.7186	1.392	.3
.8	0.5120	0.8590	0.5961	1.678	.2	.8	0.5850	0.8111	0.7212	1.387	.2
.9	0.5135	0.8581	0.5985	1.671	.1	.9	0.5864	0.8100	0.7239	1.381	.1
31.0	0.5150	0.8572	0.6009	1.664	59.0	36.0	0.5878	0.8090	0.7265	1.376	54.0
.1	0.5165	0.8563	0.6032	1.658	.9	.1	0.5892	0.8080	0.7292	1.371	.9
.2	0.5180	0.8554	0.6056	1.651	.8	.2	0.5906	0.8070	0.7319	1.366	.8
.3	0.5195	0.8545	0.6080	1.645	.7	.3	0.5920	0.8059	0.7346	1.361	.7
.4	0.5210	0.8536	0.6104	1.638	.6	.4	0.5934	0.8049	0.7373	1.356	.6
.5	0.5225	0.8526	0.6128	1.632	.5	.5	0.5948	0.8039	0.7400	1.351	.5
.6	0.5240	0.8517	0.6152	1.625	.4	.6	0.5962	0.8028	0.7427	1.347	.4
.7	0.5255	0.8508	0.6176	1.619	.3	.7	0.5976	0.8018	0.7454	1.342	.3
.8	0.5270	0.8499	0.6200	1.613	.2	.8	0.5990	0.8007	0.7481	1.337	.2
.9	0.5284	0.8490	0.6224	1.607	.1	.9	0.6004	0.7997	0.7508	1.332	.1
32.0	0.5299	0.8480	0.6249	1.600	58.0	37.0	0.6018	0.7986	0.7536	1.327	53.0
.1	0.5314	0.8471	0.6273	1.594	.9	.1	0.6032	0.7976	0.7563	1.322	.9
.2	0.5329	0.8462	0.6297	1.588	.8	.2	0.6046	0.7965	0.7590	1.317	.8
.3	0.5344	0.8453	0.6322	1.582	.7	.3	0.6060	0.7955	0.7618	1.313	.7
.4	0.5358	0.8443	0.6346	1.576	.6	.4	0.6074	0.7944	0.7646	1.308	.6
.5	0.5373	0.8434	0.6371	1.570	.5	.5	0.6088	0.7934	0.7673	1.303	.5
.6	0.5388	0.8425	0.6395	1.564	.4	.6	0.6101	0.7923	0.7701	1.299	.4
.7	0.5402	0.8415	0.6420	1.558	.3	.7	0.6115	0.7912	0.7729	1.294	.3
.8	0.5417	0.8406	0.6445	1.552	.2	.8	0.6129	0.7902	0.7757	1.289	.2
.9	0.5432	0.8396	0.6469	1.546	.1	.9	0.6143	0.7891	0.7785	1.285	.1
33.0	0.5446	0.8387	0.6494	1.540	57.0	38.0	0.6157	0.7880	0.7813	1.280	52.0
.1	0.5461	0.8377	0.6519	1.534	.9	.1	0.6170	0.7869	0.7841	1.275	.9
.2	0.5476	0.8368	0.6544	1.528	.8	.2	0.6184	0.7859	0.7869	1.271	.8
.3	0.5490	0.8358	0.6569	1.522	.7	.3	0.6198	0.7848	0.7898	1.266	.7
.4	0.5505	0.8348	0.6594	1.517	.6	.4	0.6211	0.7837	0.7926	1.262	.6
.5	0.5519	0.8339	0.6619	1.511	.5	.5	0.6225	0.7826	0.7954	1.257	.5
.6	0.5534	0.8329	0.6644	1.505	.4	.6	0.6239	0.7815	0.7983	1.253	.4
.7	0.5548	0.8320	0.6669	1.499	.3	.7	0.6252	0.7804	0.8012	1.248	.3
.8	0.5563	0.8310	0.6694	1.494	.2	.8	0.6266	0.7793	0.8040	1.244	.2
.9	0.5577	0.8300	0.6720	1.488	.1	.9	0.6280	0.7782	0.8069	1.239	.1
34.0	0.5592	0.8290	0.6745	1.483	56.0	39.0	0.6293	0.7771	0.8098	1.235	51.0
.1	0.5606	0.8281	0.6771	1.477	.9	.1	0.6307	0.7760	0.8127	1.230	.9
.2	0.5621	0.8271	0.6796	1.471	.8	.2	0.6320	0.7749	0.8156	1.226	.8
.3	0.5635	0.8261	0.6822	1.466	.7	.3	0.6334	0.7738	0.8185	1.222	.7
.4	0.5650	0.8251	0.6847	1.460	.6	.4	0.6347	0.7727	0.8214	1.217	.6
.5	0.5664	0.8241	0.6873	1.455	.5	.5	0.6361	0.7716	0.8243	1.213	.5
.6	0.5678	0.8231	0.6899	1.450	.4	.6	0.6374	0.7705	0.8273	1.209	.4
.7	0.5693	0.8221	0.6924	1.444	.3	.7	0.6388	0.7694	0.8302	1.205	.3
.8	0.5707	0.8211	0.6950	1.439	.2	.8	0.6401	0.7683	0.8332	1.200	.2
.9	0.5721	0.8202	0.6976	1.433	.1	.9	0.6414	0.7672	0.8361	1.196	.1
35.0	0.5736	0.8192	0.7002	1.428	55.0	40.0	0.6428	0.7660	0.8391	1.192	50.0
	Cos θ	Sin θ	Cot θ	Tan θ	Degrees		Cos θ	Sin θ	Cot θ	Tan θ	Degrees

Table 4. Continued

Degrees	Sin θ	Cos θ	Tan θ	Cot θ	
40.0	0.6428	0.7660	0.8391	1.192	50.0
.1	0.6441	0.7649	0.8421	1.188	.9
.2	0.6455	0.7638	0.8451	1.183	.8
.3	0.6468	0.7627	0.8481	1.179	.7
.4	0.6481	0.7615	0.8511	1.175	.6
.5	0.6494	0.7604	0.8541	1.171	.5
.6	0.6508	0.7593	0.8571	1.167	.4
.7	0.6521	0.7581	0.8601	1.163	.3
.8	0.6534	0.7570	0.8632	1.159	.2
.9	0.6547	0.7559	0.8662	1.154	.1
41.0	0.6561	0.7547	0.8693	1.150	49.0
.1	0.6574	0.7536	0.8724	1.146	.9
.2	0.6587	0.7524	0.8754	1.142	.8
.3	0.6600	0.7513	0.8785	1.138	.7
.4	0.6613	0.7501	0.8816	1.134	.6
.5	0.6626	0.7490	0.8847	1.130	.5
.6	0.6639	0.7478	0.8878	1.126	.4
.7	0.6652	0.7466	0.8910	1.122	.3
.8	0.6665	0.7455	0.8941	1.118	.2
.9	0.6678	0.7443	0.8972	1.115	.1
42.0	0.6691	0.7431	0.9004	1.111	48.0
.1	0.6704	0.7420	0.9036	1.107	.9
.2	0.6717	0.7408	0.9067	1.103	.8
.3	0.6730	0.7396	0.9099	1.099	.7
.4	0.6743	0.7385	0.9131	1.095	.6
.5	0.6756	0.7373	0.9163	1.091	.5
.6	0.6769	0.7361	0.9195	1.087	.4
.7	0.6782	0.7349	0.9228	1.084	.3
.8	0.6794	0.7337	0.9260	1.080	.2
.9	0.6807	0.7325	0.9293	1.076	.1
43.0	0.6820	0.7314	0.9325	1.072	47.0
.1	0.6833	0.7302	0.9358	1.069	.9
.2	0.6845	0.7290	0.9391	1.065	.8
.3	0.6858	0.7278	0.9424	1.061	.7
.4	0.6871	0.7266	0.9457	1.057	.6
.5	0.6884	0.7254	0.9490	1.054	.5
.6	0.6896	0.7242	0.9523	1.050	.4
.7	0.6909	0.7230	0.9556	1.046	.3
.8	0.6921	0.7218	0.9590	1.043	.2
.9	0.6934	0.7206	0.9623	1.039	.1
44.0	0.6947	0.7193	0.9657	1.036	46.0
.1	0.6959	0.7181	0.9691	1.032	.9
.2	0.6972	0.7169	0.9725	1.028	.8
.3	0.6984	0.7157	0.9759	1.025	.7
.4	0.6997	0.7145	0.9793	1.021	.6
.5	0.7009	0.7133	0.9827	1.018	.5
.6	0.7022	0.7120	0.9861	1.014	.4
.7	0.7034	0.7108	0.9896	1.011	.3
.8	0.7046	0.7096	0.9930	1.007	.2
.9	0.7059	0.7083	0.9965	1.003	.1
45.0	0.7071	0.7071	1.0000	1.000	45.0
	Cos θ	Sin θ	Cot θ	Tan θ	Degrees

Table 5   A—53

Table 5. Four-Place Logarithms of Trigonometric Functions
Angle θ in Degrees. Attach -10 to Logarithms Obtained from This Table

Angle θ	log sin θ	log cos θ	log tan θ	log cot θ		Angle θ	log sin θ	log cos θ	log tan θ	log cot θ	
0°00′	No value	10.0000	No value	No value	90°00′	8°00′	9.1436	9.9958	9.1478	10.8522	82°00′
10	7.4637	10.0000	7.4637	12.5363	50	10	9.1525	9.9956	9.1569	10.8431	50
20	7.7648	10.0000	7.7648	12.2352	40	20	9.1612	9.9954	9.1658	10.8342	40
30	7.9408	10.0000	7.9409	12.0591	30	30	9.1697	9.9952	9.1745	10.8255	30
40	8.0658	10.0000	8.0658	11.9342	20	40	9.1781	9.9950	9.1831	10.8169	20
50	8.1627	10.0000	8.1627	11.8373	10	50	9.1863	9.9948	9.1915	10.8085	10
1°00′	8.2419	9.9999	8.2419	11.7581	89°00′	9°00′	9.1943	9.9946	9.1997	10.8003	81°00′
10	8.3088	9.9999	8.3089	11.6911	50	10	9.2022	9.9944	9.2078	10.7922	50
20	8.3668	9.9999	8.3669	11.6331	40	20	9.2100	9.9942	9.2158	10.7842	40
30	8.4179	9.9999	8.4181	11.5819	30	30	9.2176	9.9940	9.2236	10.7764	30
40	8.4637	9.9998	8.4638	11.5362	20	40	9.2251	9.9938	9.2313	10.7687	20
50	8.5050	9.9998	8.5053	11.4947	10	50	9.2324	9.9936	9.2389	10.7611	10
2°00′	8.5428	9.9997	8.5431	11.4569	88°00′	10°00′	9.2397	9.9934	9.2463	10.7537	80°00′
10	8.5776	9.9997	8.5779	11.4221	50	10	9.2468	9.9931	9.2536	10.7464	50
20	8.6097	9.9996	8.6101	11.3899	40	20	9.2538	9.9929	9.2609	10.7391	40
30	8.6397	9.9996	8.6401	11.3599	30	30	9.2606	9.9927	9.2680	10.7320	30
40	8.6677	9.9995	8.6682	11.3318	20	40	9.2674	9.9924	9.2750	10.7250	20
50	8.6940	9.9995	8.6945	11.3055	10	50	9.2740	9.9922	9.2819	10.7181	10
3°00′	8.7188	9.9994	8.7194	11.2806	87°00′	11°00′	9.2806	9.9919	9.2887	10.7113	79°00′
10	8.7423	9.9993	8.7429	11.2571	50	10	9.2870	9.9917	9.2953	10.7047	50
20	8.7645	9.9993	8.7652	11.2348	40	20	9.2934	9.9914	9.3020	10.6980	40
30	8.7857	9.9992	8.7865	11.2135	30	30	9.2997	9.9912	9.3085	10.6915	30
40	8.8059	9.9991	8.8067	11.1933	20	40	9.3058	9.9909	9.3149	10.6851	20
50	8.8251	9.9990	8.8261	11.1739	10	50	9.3119	9.9907	9.3212	10.6788	10
4°00′	8.8436	9.9989	8.8446	11.1554	86°00′	12°00′	9.3179	9.9904	9.3275	10.6725	78°00′
10	8.8613	9.9989	8.8624	11.1376	50	10	9.3238	9.9901	9.3336	10.6664	50
20	8.8783	9.9988	8.8795	11.1205	40	20	9.3296	9.9899	9.3397	10.6603	40
30	8.8946	9.9987	8.8960	11.1040	30	30	9.3353	9.9896	9.3458	10.6542	30
40	8.9104	9.9986	8.9118	11.0882	20	40	9.3410	9.9893	9.3517	10.6483	20
50	8.9256	9.9985	8.9272	11.0728	10	50	9.3466	9.9890	9.3576	10.6424	10
5°00′	8.9403	9.9983	8.9420	11.0580	85°00′	13°00′	9.3521	9.9887	9.3634	10.6366	77°00′
10	8.9545	9.9982	8.9563	11.0437	50	10	9.3575	9.9884	9.3691	10.6309	50
20	8.9682	9.9981	8.9701	11.0299	40	20	9.3629	9.9881	9.3748	10.6252	40
30	8.9816	9.9980	8.9836	11.0164	30	30	9.3682	9.9878	9.3804	10.6196	30
40	8.9945	9.9979	8.9966	11.0034	20	40	9.3734	9.9875	9.3859	10.6141	20
50	9.0070	9.9977	9.0093	10.9907	10	50	9.3786	9.9872	9.3914	10.6086	10
6°00′	9.0192	9.9976	9.0216	10.9784	84°00′	14°00′	9.3837	9.9869	9.3968	10.6032	76°00′
10	9.0311	9.9975	9.0336	10.9664	50	10	9.3887	9.9866	9.4021	10.5979	50
20	9.0426	9.9973	9.0453	10.9547	40	20	9.3937	9.9863	9.4074	10.5926	40
30	9.0539	9.9972	9.0567	10.9433	30	30	9.3986	9.9859	9.4127	10.5873	30
40	9.0648	9.9971	9.0678	10.9322	20	40	9.4035	9.9856	9.4178	10.5822	20
50	9.0755	9.9969	9.0786	10.9214	10	50	9.4083	9.9853	9.4230	10.5770	10
7°00′	9.0859	9.9968	9.0891	10.9109	83°00′	15°00′	9.4130	9.9849	9.4281	10.5719	75°00′
10	9.0961	9.9966	9.0995	10.9005	50	10	9.4177	9.9846	9.4331	10.5669	50
20	9.1060	9.9964	9.1096	10.8904	40	20	9.4223	9.9843	9.4381	10.5619	40
30	9.1157	9.9963	9.1194	10.8806	30	30	9.4269	9.9839	9.4430	10.5570	30
40	9.1252	9.9961	9.1291	10.8709	20	40	9.4314	9.9836	9.4479	10.5521	20
50	9.1345	9.9959	9.1385	10.8615	10	50	9.4359	9.9832	9.4527	10.5473	10
8°00′	9.1436	9.9958	9.1478	10.8522	82°00′	16°00′	9.4403	9.9828	9.4575	10.5425	74°00′
	log cos θ	log sin θ	log cot θ	log tan θ	Angle θ		log cos θ	log sin θ	log cot θ	log tan θ	Angle θ

Table 5. Continued

Angle θ	log sin θ	log cos θ	log tan θ	log cot θ		Angle θ	log sin θ	log cos θ	log tan θ	log cot θ	
16°00′	9.4403	9.9828	9.4575	10.5425	74°00′	24°00′	9.6093	9.9607	9.6486	10.3514	66°00′
10	9.4447	9.9825	9.4622	10.5378	50	10	9.6121	9.9602	9.6520	10.3480	50
20	9.4491	9.9821	9.4669	10.5331	40	20	9.6149	9.9596	9.6553	10.3447	40
30	9.4533	9.9817	9.4716	10.5284	30	30	9.6177	9.9590	9.6587	10.3413	30
40	9.4576	9.9814	9.4762	10.5238	20	40	9.6205	9.9584	9.6620	10.3380	20
50	9.4618	9.9810	9.4808	10.5192	10	50	9.6232	9.9579	9.6654	10.3346	10
17°00′	9.4659	9.9806	9.4853	10.5147	73°00′	25°00′	9.6259	9.9573	9.6687	10.3313	65°00′
10	9.4700	9.9802	9.4898	10.5102	50	10	9.6286	9.9567	9.6720	10.3280	50
20	9.4741	9.9798	9.4943	10.5057	40	20	9.6313	9.9561	9.6752	10.3248	40
30	9.4781	9.9794	9.4987	10.5013	30	30	9.6340	9.9555	9.6785	10.3215	30
40	9.4821	9.9790	9.5031	10.4969	20	40	9.6366	9.9549	9.6817	10.3183	20
50	9.4861	9.9786	9.5075	10.4925	10	50	9.6392	9.9543	9.6850	10.3150	10
18°00′	9.4900	9.9782	9.5118	10.4882	72°00′	26°00′	9.6418	9.9537	9.6882	10.3118	64°00′
10	9.4939	9.9778	9.5161	10.4839	50	10	9.6444	9.9530	9.6914	10.3086	50
20	9.4977	9.9774	9.5203	10.4797	40	20	9.6470	9.9524	9.6946	10.3054	40
30	9.5015	9.9770	9.5245	10.4755	30	30	9.6495	9.9518	9.6977	10.3023	30
40	9.5052	9.9765	9.5287	10.4713	20	40	9.6521	9.9512	9.7009	10.2991	20
50	9.5090	9.9761	9.5329	10.4671	10	50	9.6546	9.9505	9.7040	10.2960	10
19°00′	9.5126	9.9757	9.5370	10.4630	71°00′	27°00′	9.6570	9.9499	9.7072	10.2928	63°00′
10	9.5163	9.9752	9.5411	10.4589	50	10	9.6595	9.9492	9.7103	10.2897	50
20	9.5199	9.9748	9.5451	10.4549	40	20	9.6620	9.9486	9.7134	10.2866	40
30	9.5235	9.9743	9.5491	10.4509	30	30	9.6644	9.9479	9.7165	10.2835	30
40	9.5270	9.9739	9.5531	10.4469	20	40	9.6668	9.9473	9.7196	10.2804	20
50	9.5306	9.9734	9.5571	10.4429	10	50	9.6692	9.9466	9.7226	10.2774	10
20°00′	9.5341	9.9730	9.5611	10.4389	70°00′	28°00′	9.6716	9.9459	9.7257	10.2743	62°00′
10	9.5375	9.9725	9.5650	10.4350	50	10	9.6740	9.9453	9.7287	10.2713	50
20	9.5409	9.9721	9.5689	10.4311	40	20	9.6763	9.9446	9.7317	10.2683	40
30	9.5443	9.9716	9.5727	10.4273	30	30	9.6787	9.9439	9.7348	10.2652	30
40	9.5477	9.9711	9.5766	10.4234	20	40	9.6810	9.9432	9.7378	10.2622	20
50	9.5510	9.9706	9.5804	10.4196	10	50	9.6833	9.9425	9.7408	10.2592	10
21°00′	9.5543	9.9702	9.5842	10.4158	69°00′	29°00′	9.6856	9.9418	9.7438	10.2562	61°00′
10	9.5576	9.9697	9.5879	10.4121	50	10	9.6878	9.9411	9.7467	10.2533	50
20	9.5609	9.9692	9.5917	10.4083	40	20	9.6901	9.9404	9.7497	10.2503	40
30	9.5641	9.9687	9.5954	10.4046	30	30	9.6923	9.9397	9.7526	10.2474	30
40	9.5673	9.9682	9.5991	10.4009	20	40	9.6946	9.9390	9.7556	10.2444	20
50	9.5704	9.9677	9.6028	10.3972	10	50	9.6968	9.9383	9.7585	10.2415	10
22°00′	9.5736	9.9672	9.6064	10.3936	68°00′	30°00′	9.6990	9.9375	9.7614	10.2386	60°00′
10	9.5767	9.9667	9.6100	10.3900	50	10	9.7012	9.9368	9.7644	10.2356	50
20	9.5798	9.9661	9.6136	10.3864	40	20	9.7033	9.9361	9.7673	10.2327	40
30	9.5828	9.9656	9.6172	10.3828	30	30	9.7055	9.9353	9.7701	10.2299	30
40	9.5859	9.9651	9.6208	10.3792	20	40	9.7076	9.9346	9.7730	10.2270	20
50	9.5889	9.9646	9.6243	10.3757	10	50	9.7097	9.9338	9.7759	10.2241	10
23°00′	9.5919	9.9640	9.6279	10.3721	67°00′	31°00′	9.7118	9.9331	9.7788	10.2212	59°00′
10	9.5948	9.9635	9.6314	10.3686	50	10	9.7139	9.9323	9.7816	10.2184	50
20	9.5978	9.9629	9.6348	10.3652	40	20	9.7160	9.9315	9.7845	10.2155	40
30	9.6007	9.9624	9.6383	10.3617	30	30	9.7181	9.9308	9.7873	10.2127	30
40	9.6036	9.9618	9.6417	10.3583	20	40	9.7201	9.9300	9.7902	10.2098	20
50	9.6065	9.9613	9.6452	10.3548	10	50	9.7222	9.9292	9.7930	10.2070	10
24°00′	9.6093	9.9607	9.6486	10.3514	66°00′	32°00′	9.7242	9.9284	9.7958	10.2042	58°00′
	log cos θ	log sin θ	log cot θ	log tan θ	Angle θ		log cos θ	log sin θ	log cot θ	log tan θ	Angle θ

Table 5   A—55

Table 5. Continued

Angle θ	log sin θ	log cos θ	log tan θ	log cot θ		Angle θ	log sin θ	log cos θ	log tan θ	log cot θ	
32°00′	9.7242	9.9284	9.7958	10.2042	58°00′	39°00′	9.7989	9.8905	9.9084	10.0916	51°00′
10	9.7262	9.9276	9.7986	10.2014	50	10	9.8004	9.8895	9.9110	10.0890	50
20	9.7282	9.9268	9.8014	10.1986	40	20	9.8020	9.8884	9.9135	10.0865	40
30	9.7302	9.9260	9.8042	10.1958	30	30	9.8035	9.8874	9.9161	10.0839	30
40	9.7322	9.9252	9.8070	10.1930	20	40	9.8050	9.8864	9.9187	10.0813	20
50	9.7342	9.9244	9.8097	10.1903	10	50	9.8066	9.8853	9.9212	10.0788	10
33°00′	9.7361	9.9236	9.8125	10.1875	57°00′	40°00′	9.8081	9.8843	9.9238	10.0762	50°00′
10	9.7380	9.9228	9.8153	10.1847	50	10	9.8096	9.8832	9.9264	10.0736	50
20	9.7400	9.9219	9.8180	10.1820	40	20	9.8111	9.8821	9.9289	10.0711	40
30	9.7419	9.9211	9.8208	10.1792	30	30	9.8125	9.8810	9.9315	10.0685	30
40	9.7438	9.9203	9.8235	10.1765	20	40	9.8140	9.8800	9.9341	10.0659	20
50	9.7457	9.9194	9.8263	10.1737	10	50	9.8155	9.8789	9.9366	10.0634	10
34°00′	9.7476	9.9186	9.8290	10.1710	56°00′	41°00′	9.8169	9.8778	9.9392	10.0608	49°00′
10	9.7494	9.9177	9.8317	10.1683	50	10	9.8184	9.8767	9.9417	10.0583	50
20	9.7513	9.9169	9.8344	10.1656	40	20	9.8198	9.8756	9.9443	10.0557	40
30	9.7531	9.9160	9.8371	10.1629	30	30	9.8213	9.8745	9.9468	10.0532	30
40	9.7550	9.9151	9.8398	10.1602	20	40	9.8227	9.8733	9.9494	10.0506	20
50	9.7568	9.9142	9.8425	10.1575	10	50	9.8241	9.8722	9.9519	10.0481	10
35°00′	9.7586	9.9134	9.8452	10.1548	55°00′	42°00′	9.8255	9.8711	9.9544	10.0456	48°00′
10	9.7604	9.9125	9.8479	10.1521	50	10	9.8269	9.8699	9.9570	10.0430	50
20	9.7622	9.9116	9.8506	10.1494	40	20	9.8283	9.8688	9.9595	10.0405	40
30	9.7640	9.9107	9.8533	10.1467	30	30	9.8297	9.8676	9.9621	10.0379	30
40	9.7657	9.9098	9.8559	10.1441	20	40	9.8311	9.8665	9.9646	10.0354	20
50	9.7675	9.9089	9.8586	10.1414	10	50	9.8324	9.8653	9.9671	10.0329	10
36°00′	9.7692	9.9080	9.8613	10.1387	54°00′	43°00′	9.8338	9.8641	9.9697	10.0303	47°00′
10	9.7710	9.9070	9.8639	10.1361	50	10	9.8351	9.8629	9.9722	10.0278	50
20	9.7727	9.9061	9.8666	10.1334	40	20	9.8365	9.8618	9.9747	10.0253	40
30	9.7744	9.9052	9.8692	10.1308	30	30	9.8378	9.8606	9.9772	10.0228	30
40	9.7761	9.9042	9.8718	10.1282	20	40	9.8391	9.8594	9.9798	10.0202	20
50	9.7778	9.9033	9.8745	10.1255	10	50	9.8405	9.8582	9.9823	10.0177	10
37°00′	9.7795	9.9023	9.8771	10.1229	53°00′	44°00′	9.8418	9.8569	9.9848	10.0152	46°00′
10	9.7811	9.9014	9.8797	10.1203	50	10	9.8431	9.8557	9.9874	10.0126	50
20	9.7828	9.9004	9.8824	10.1176	40	20	9.8444	9.8545	9.9899	10.0101	40
30	9.7844	9.8995	9.8850	10.1150	30	30	9.8457	9.8532	9.9924	10.0076	30
40	9.7861	9.8985	9.8876	10.1124	20	40	9.8469	9.8520	9.9949	10.0051	20
50	9.7877	9.8975	9.8902	10.1098	10	50	9.8482	9.8507	9.9975	10.0025	10
38°00′	9.7893	9.8965	9.8928	10.1072	52°00′	45°00′	9.8495	9.8495	10.0000	10.0000	45°00′
10	9.7910	9.8955	9.8954	10.1046	50						
20	9.7926	9.8945	9.8980	10.1020	40						
30	9.7941	9.8935	9.9006	10.0994	30						
40	9.7957	9.8925	9.9032	10.0968	20						
50	9.7973	9.8915	9.9058	10.0942	10						
39°00′	9.7989	9.8905	9.9084	10.0916	51°00′						
	log cos θ	log sin θ	log cot θ	log tan θ	Angle θ		log cos θ	log sin θ	log cot θ	log tan θ	Angle θ

Table 6. Natural Logarithms of Numbers

n	$\ln n$	n	$\ln n$	n	$\ln n$
0.0	——— *	4.5	1.5041	9.0	2.1972
0.1	7.6974	4.6	1.5261	9.1	2.2083
0.2	8.3906	4.7	1.5476	9.2	2.2192
0.3	8.7960	4.8	1.5686	9.3	2.2300
0.4	9.0837	4.9	1.5892	9.4	2.2407
0.5	9.3069	5.0	1.6094	9.5	2.2513
0.6	9.4892	5.1	1.6292	9.6	2.2618
0.7	9.6433	5.2	1.6487	9.7	2.2721
0.8	9.7769	5.3	1.6677	9.8	2.2824
0.9	9.8946	5.4	1.6864	9.9	2.2925
1.0	0.0000	5.5	1.7047	10	2.3026
1.1	0.0953	5.6	1.7228	11	2.3979
1.2	0.1823	5.7	1.7405	12	2.4849
1.3	0.2624	5.8	1.7579	13	2.5649
1.4	0.3365	5.9	1.7750	14	2.6391
1.5	0.4055	6.0	1.7918	15	2.7081
1.6	0.4700	6.1	1.8083	16	2.7726
1.7	0.5306	6.2	1.8245	17	2.8332
1.8	0.5878	6.3	1.8405	18	2.8904
1.9	0.6419	6.4	1.8563	19	2.9444
2.0	0.6931	6.5	1.8718	20	2.9957
2.1	0.7419	6.6	1.8871	25	3.2189
2.2	0.7885	6.7	1.9021	30	3.4012
2.3	0.8329	6.8	1.9169	35	3.5553
2.4	0.8755	6.9	1.9315	40	3.6889
2.5	0.9163	7.0	1.9459	45	3.8067
2.6	0.9555	7.1	1.9601	50	3.9120
2.7	0.9933	7.2	1.9741	55	4.0073
2.8	1.0296	7.3	1.9879	60	4.0943
2.9	1.0647	7.4	2.0015	65	4.1744
3.0	1.0986	7.5	2.0149	70	4.2485
3.1	1.1314	7.6	2.0281	75	4.3175
3.2	1.1632	7.7	2.0412	80	4.3820
3.3	1.1939	7.8	2.0541	85	4.4427
3.4	1.2238	7.9	2.0669	90	4.4998
3.5	1.2528	8.0	2.0794	95	4.5539
3.6	1.2809	8.1	2.0919	100	4.6052
3.7	1.3083	8.2	2.1041		
3.8	1.3350	8.3	2.1163		
3.9	1.3610	8.4	2.1282		
4.0	1.3863	8.5	2.1401	$\ln 10$	2.3026
4.1	1.4110	8.6	2.1518	$2 \ln 10$	4.6052
4.2	1.4351	8.7	2.1633	$3 \ln 10$	6.9078
4.3	1.4586	8.8	2.1748	$4 \ln 10$	9.2103
4.4	1.4816	8.9	2.1861	$5 \ln 10$	11.5129

*Attach -10 to these logarithms.

Table 7   A—57

Table 7. Exponential Functions

x	e^x	e^{-x}	x	e^x	e^{-x}
0.00	1.0000	1.0000	2.5	12.182	0.0821
0.05	1.0513	0.9512	2.6	13.464	0.0743
0.10	1.1052	0.9048	2.7	14.880	0.0672
0.15	1.1618	0.8607	2.8	16.445	0.0608
0.20	1.2214	0.8187	2.9	18.174	0.0550
0.25	1.2840	0.7788	3.0	20.086	0.0498
0.30	1.3499	0.7408	3.1	22.198	0.0450
0.35	1.4191	0.7047	3.2	24.533	0.0408
0.40	1.4918	0.6703	3.3	27.113	0.0369
0.45	1.5683	0.6376	3.4	29.964	0.0334
0.50	1.6487	0.6065	3.5	33.115	0.0302
0.55	1.7333	0.5769	3.6	36.598	0.0273
0.60	1.8221	0.5488	3.7	40.447	0.0247
0.65	1.9155	0.5220	3.8	44.701	0.0224
0.70	2.0138	0.4966	3.9	49.402	0.0202
0.75	2.1170	0.4724	4.0	54.598	0.0183
0.80	2.2255	0.4493	4.1	60.340	0.0166
0.85	2.3369	0.4274	4.2	66.686	0.0150
0.90	2.4596	0.4066	4.3	73.700	0.0136
0.95	2.5857	0.3867	4.4	81.451	0.0123
1.0	2.7183	0.3679	4.5	90.017	0.0111
1.1	3.0042	0.3329	4.6	99.484	0.0101
1.2	3.3201	0.3012	4.7	109.95	0.0091
1.3	3.6693	0.2725	4.8	121.51	0.0082
1.4	4.0552	0.2466	4.9	134.29	0.0074
1.5	4.4817	0.2231	5	148.41	0.0067
1.6	4.9530	0.2019	6	403.43	0.0025
1.7	5.4739	0.1827	7	1096.6	0.0009
1.8	6.0496	0.1653	8	2981.0	0.0003
1.9	6.6859	0.1496	9	8103.1	0.0001
2.0	7.3891	0.1353	10	22026	0.00005
2.1	8.1662	0.1225			
2.2	9.0250	0.1108			
2.3	9.9742	0.1003			
2.4	11.023	0.0907			

Table 8. A Short Table of Integrals

The basic forms of Chapter 25 are not included. The constant of integration is omitted.

Forms containing $a + bu$ and $\sqrt{a + bu}$

1. $\displaystyle\int \frac{u\,du}{a + bu} = \frac{1}{b^2}[(a + bu) - a\ln(a + bu)]$

2. $\displaystyle\int \frac{du}{u(a + bu)} = -\frac{1}{a}\ln\frac{a + bu}{u}$

3. $\displaystyle\int \frac{u\,du}{(a + bu)^2} = \frac{1}{b^2}\left(\frac{a}{a + bu} + \ln(a + bu)\right)$

4. $\displaystyle\int \frac{du}{u(a + bu)^2} = \frac{1}{a(a + bu)} - \frac{1}{a^2}\ln\frac{a + bu}{u}$

5. $\displaystyle\int u\sqrt{a + bu}\,du = -\frac{2(2a - 3bu)(a + bu)^{3/2}}{15b^2}$

6. $\displaystyle\int \frac{u\,du}{\sqrt{a + bu}} = -\frac{2(2a - bu)\sqrt{a + bu}}{3b^2}$

7. $\displaystyle\int \frac{du}{u\sqrt{a + bu}} = \frac{1}{\sqrt{a}}\ln\left(\frac{\sqrt{a + bu} - \sqrt{a}}{\sqrt{a + bu} + \sqrt{a}}\right), \qquad a > 0$

8. $\displaystyle\int \frac{\sqrt{a + bu}}{u}\,du = 2\sqrt{a + bu} + a\int \frac{du}{u\sqrt{a + bu}}$

Forms containing $\sqrt{u^2 \pm a^2}$ and $\sqrt{a^2 - u^2}$

9. $\displaystyle\int \frac{du}{u^2 - a^2} = \frac{1}{2a}\ln\frac{u - a}{u + a}$

10. $\displaystyle\int \frac{du}{\sqrt{u^2 \pm a^2}} = \ln(u + \sqrt{u^2 \pm a^2})$

11. $\displaystyle\int \frac{du}{u\sqrt{u^2 + a^2}} = -\frac{1}{a}\ln\left(\frac{a + \sqrt{u^2 + a^2}}{u}\right)$

12. $\displaystyle\int \frac{du}{u\sqrt{u^2 - a^2}} = \frac{1}{a}\text{Arcsec}\frac{u}{a}$

13. $\displaystyle\int \frac{du}{u\sqrt{a^2 - u^2}} = -\frac{1}{a}\ln\left(\frac{a + \sqrt{a^2 - u^2}}{u}\right)$

14. $\displaystyle\int \sqrt{u^2 \pm a^2}\,du = \frac{u}{2}\sqrt{u^2 \pm a^2} \pm \frac{a^2}{2}\ln(u + \sqrt{u^2 \pm a^2})$

15. $\displaystyle\int \sqrt{a^2 - u^2}\,du = \frac{u}{2}\sqrt{a^2 - u^2} + \frac{a^2}{2}\text{Arcsin}\frac{u}{a}$

16. $\displaystyle\int \frac{\sqrt{u^2 + a^2}}{u}\,du = \sqrt{u^2 + a^2} - a\ln\left(\frac{a + \sqrt{u^2 + a^2}}{u}\right)$

17. $\displaystyle\int \frac{\sqrt{u^2 - a^2}}{u}\,du = \sqrt{u^2 - a^2} - a\,\text{Arcsec}\frac{u}{a}$

Table 8   A—59

Table 8. Continued

18. $\displaystyle\int \frac{\sqrt{a^2 - u^2}}{u}du = \sqrt{a^2 - u^2} - a\,\ln\!\left(\frac{a + \sqrt{a^2 - u^2}}{u}\right)$

19. $\displaystyle\int (u^2 \pm a^2)^{3/2}\,du = \frac{u}{4}(u^2 \pm a^2)^{3/2} \pm \frac{3a^2 u}{8}\sqrt{u^2 \pm a^2} + \frac{3a^4}{8}\ln(u + \sqrt{u^2 \pm a^2})$

20. $\displaystyle\int (a^2 - u^2)^{3/2}\,du = \frac{u}{4}(a^2 - u^2)^{3/2} + \frac{3a^2 u}{8}\sqrt{a^2 - u^2} + \frac{3a^4}{8}\mathrm{Arcsin}\frac{u}{a}$

21. $\displaystyle\int \frac{(u^2 + a^2)^{3/2}}{u}du = \frac{1}{3}(u^2 + a^2)^{3/2} + a^2\sqrt{u^2 + a^2} - a^3\ln\!\left(\frac{a + \sqrt{u^2 + a^2}}{u}\right)$

22. $\displaystyle\int \frac{(u^2 - a^2)^{3/2}}{u}du = \frac{1}{3}(u^2 - a^2)^{3/2} - a^2\sqrt{u^2 - a^2} + a^3\mathrm{Arcsec}\frac{u}{a}$

23. $\displaystyle\int \frac{(a^2 - u^2)^{3/2}}{u}du = \frac{1}{3}(a^2 - u^2)^{3/2} - a^2\sqrt{a^2 - u^2} + a^3\ln\!\left(\frac{a + \sqrt{a^2 - u^2}}{u}\right)$

24. $\displaystyle\int \frac{du}{(u^2 \pm a^2)^{3/2}} = \pm\frac{u}{a^2\sqrt{u^2 \pm a^2}}$

25. $\displaystyle\int \frac{du}{(a^2 - u^2)^{3/2}} = \frac{u}{a^2\sqrt{a^2 - u^2}}$

26. $\displaystyle\int \frac{du}{u(u^2 + a^2)^{3/2}} = \frac{1}{a^2\sqrt{u^2 + a^2}} - \frac{1}{a^3}\ln\!\left(\frac{a + \sqrt{u^2 + a^2}}{u}\right)$

27. $\displaystyle\int \frac{du}{u(u^2 - a^2)^{3/2}} = -\frac{1}{a^2\sqrt{u^2 - a^2}} - \frac{1}{a^3}\mathrm{Arcsec}\frac{u}{a}$

28. $\displaystyle\int \frac{du}{u(a^2 - u^2)^{3/2}} = \frac{1}{a^2\sqrt{a^2 - u^2}} - \frac{1}{a^3}\ln\!\left(\frac{a + \sqrt{a^2 - u^2}}{u}\right)$

Trigonometric forms

29. $\displaystyle\int \sin^2 u\,du = \frac{u}{2} - \frac{1}{2}\sin u\,\cos u$

30. $\displaystyle\int \sin^3 u\,du = -\cos u + \frac{1}{3}\cos^3 u$

31. $\displaystyle\int \sin^n u\,du = -\frac{1}{n}\sin^{n-1} u\,\cos u + \frac{n-1}{n}\int \sin^{n-2} u\,du$

32. $\displaystyle\int \cos^2 u\,du = \frac{u}{2} + \frac{1}{2}\sin u\,\cos u$

33. $\displaystyle\int \cos^3 u\,du = \sin u - \frac{1}{3}\sin^3 u$

34. $\displaystyle\int \cos^n u\,du = \frac{1}{n}\cos^{n-1} u\,\sin u + \frac{n-1}{n}\int \cos^{n-2} u\,du$

35. $\displaystyle\int \tan^n u\,du = \frac{\tan^{n-1} u}{n-1} - \int \tan^{n-2} u\,du$

Table 8. Continued

36. $\int \cot^n u \ du = -\dfrac{\cot^{n-1}u}{n-1} - \int \cot^{n-2}u \ du$

37. $\int \sec^n u \ du = \dfrac{\sec^{n-2}u \tan u}{n-1} + \dfrac{n-2}{n-1}\int \sec^{n-2}u \ du$

38. $\int \csc^n u \ du = -\dfrac{\csc^{n-2}u \cot u}{n-1} + \dfrac{n-2}{n-1}\int \csc^{n-2}u \ du$

39. $\int \sin au \sin bu \ du = \dfrac{\sin(a-b)u}{2(a-b)} - \dfrac{\sin(a+b)u}{2(a+b)}$

40. $\int \sin au \cos bu \ du = -\dfrac{\cos(a-b)u}{2(a-b)} - \dfrac{\cos(a+b)u}{2(a+b)}$

41. $\int \cos au \cos bu \ du = \dfrac{\sin(a-b)u}{2(a-b)} + \dfrac{\sin(a+b)u}{2(a+b)}$

42. $\int \sin^m u \cos^n u \ du = \dfrac{\sin^{m+1}u \cos^{n-1}u}{m+n} + \dfrac{n-1}{m+n}\int \sin^m u \cos^{n-2}u \ du$

43. $\int \sin^m u \cos^n u \ du = -\dfrac{\sin^{m-1}u \cos^{n+1}u}{m+n} + \dfrac{m-1}{m+n}\int \sin^{m-2}u \cos^n u \ du$

Other forms

44. $\int ue^{au}du = \dfrac{e^{au}(au-1)}{a^2}$

45. $\int u^2 e^{au}du = \dfrac{e^{au}}{a^3}(a^2u^2 - 2au + 2)$

46. $\int u^n \ln u \ du = u^{n+1}\left(\dfrac{\ln u}{n+1} - \dfrac{1}{(n+1)^2}\right)$

47. $\int u \sin u \ du = \sin u - u \cos u$

48. $\int u \cos u \ du = \cos u + u \sin u$

49. $\int e^{au}\sin bu \ du = \dfrac{e^{au}(a \sin bu - b \cos bu)}{a^2 + b^2}$

50. $\int e^{au}\cos bu \ du = \dfrac{e^{au}(a \cos bu + b \sin bu)}{a^2 + b^2}$

51. $\int \mathrm{Arcsin}\ u \ du = u \ \mathrm{Arcsin}\ u + \sqrt{1 - u^2}$

52. $\int \mathrm{Arctan}\ u \ du = u \ \mathrm{Arctan}\ u - \dfrac{1}{2}\ln(1 + u^2)$

Answers to Odd-numbered Exercises

Exercises 1–1, p. 5

1. integer, rational, real; irrational, real   3. imaginary; irrational, real   5. $3, \frac{7}{2}$ 7. $\frac{6}{7}, \sqrt{3}$

9. $6 < 8$ 11. $\pi > 1$ 13. $-4 < -3$ 15. $-\frac{1}{3} > -\frac{1}{2}$ 17. $\frac{1}{3}, -\frac{1}{2}$ 19. $-\frac{\pi}{5}, \frac{1}{x}$

21.   23.

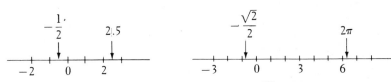

25. $-18, -|-3|, -1, \sqrt{5}, \pi, |-8|, 9$ 27. $\frac{22}{7} = 3.1429$ 29. (a) to right of origin, (b) to left of -4

31. P, V variables; c constant   33. Yes

Exercises 1–2, 1–3, p. 10

1. 11   3. -11 5. 9   7. -3 9. -24 11. 35   13. -3 15. 20   17. 40   19. -1
21. 9   23. undefined   25. 20   27. -5 29. -9 31. 24   33. -6 35. 3
37. commutative law of multiplication   39. distributive law   41. associative law of addition
43. associative law of multiplication   45. positive   47. No; $6 - 2 \neq 2 - 6$

Exercises 1–4, p. 14

1. x^7 3. $2b^6$ 5. m^2 7. $\frac{1}{n^4}$ 9. a^8 11. t^{20} 13. $8n^3$ 15. a^2x^8 17. $\frac{8}{b^3}$ 19. $\frac{x^8}{16}$

21. 1   23. 3   25. $\frac{1}{6}$ 27. s^2 29. $-t^{14}$ 31. $64x^{12}$ 33. 1   35. b^2 37. $\frac{a}{x^2}$ 39. $\frac{x^3}{64a^3}$

41. $64g^2s^6$ 43. $\frac{5a}{n}$ 45. -53 47. -10 49. $\frac{-wLx^3}{12EI}$

Exercises 1–5, p. 17

1. 45,000   3. 0.00201   5. 3.23   7. 18.6   9. 4×10^4 11. 8.7×10^{-3} 13. 6.89×10^0
15. 6.3×10^{-2} 17. 2.04×10^{14} 19. 4.85×10^9 21. 9.77×10^6 23. 3.66×10^5
25. 2.25×10^4 lb/in.^{2} 27. 0.0000000000016 W   29. 4500 Ω 31. 1.1×10^{-5} ft
33. 0.00000000000000000000000167 g   35. 8.64×10^5 mi   37. 2.59×10^{10} cm^2
39. 6.03×10^{-3} Ω = 6.03 mΩ

Exercises 1–6, p. 20

1. 5   3. -11 5. 5   7. -6 9. $3\sqrt{2}$ 11. $2\sqrt{3}$ 13. $3j$ 15. $2\sqrt{3}j$ 17. 5   19. 31
21. $8\sqrt{3}$ 23. 7   25. 20 m   27. No; true for $a \geq 0$

Exercises 1–7, p. 23

1. $8x$ 3. $4x + y$ 5. $a + c - e$ 7. $-a^2b - a^2b^2$ 9. $-v + 5x - 4$ 11. $5a - 5$
13. $-5a + 2$ 15. $7r - 2s$ 17. $19c - 50$ 19. $3n - 9$ 21. $-2t^2 + 18$ 23. $6a$
25. $2a\sqrt{xy} + 1$ 27. $4c - 6$ 29. $8p - 5q$ 31. $-4x^2 + 22$ 33. $R + 1$ 35. $3x - 395$

Exercises 1–8, p. 25

1. a^3x 3. $-a^2c^3x^3$ 5. $-8a^3x^5$ 7. $2a^4x$ 9. $a^2x + a^2y$ 11. $-3s^3 + 15st$
13. $5m^3n + 15m^2n$ 15. $3x^2 - 3xy + 6x$ 17. $a^2b^2c^5 - ab^3c^5 - a^2b^3c^4$ 19. $acx^4 + acx^3y^3$
21. $x^2 + 2x - 15$ 23. $2x^2 + 9x - 5$ 25. $6a^2 - 7ab + 2b^2$ 27. $x^3 + 2x^2 - 8x$
29. $x^3 - 2x^2 - x + 2$ 31. $x^5 - x^4 - 6x^3 + 4x^2 + 8x$ 33. $2x^3 + 6x^2 - 8x$ 35. $4x^2 - 20x + 25$
37. $x^2 + 6ax + 9a^2$ 39. $x^2y^2z^2 - 4xyz + 4$ 41. $-x^3 + 2x^2 + 5x - 6$ 43. $6x^4 + 21x^3 + 12x^2 - 12x$
45. $8000 - 5x^2$ 47. $wl^4 - 2wl^2x^2 + wx^4$

Exercises 1–9, p. 27

1. $-4x^2y$ 3. $4t^4/r^2$ 5. $4x^2$ 7. $-6a$ 9. $a^2 + 4y$ 11. $t - 2rt^2$ 13. $q + 2p - 4q^3$

15. $\dfrac{2L}{R} - R$ 17. $\dfrac{1}{3a} - \dfrac{2b}{3a} + \dfrac{a}{b}$ 19. $x^2 + a$ 21. $x - 1$ 23. $4x^2 - x - 1, R = -3$

25. $x^2 + x - 6$ 27. $2x^2 + 4x + 2, R = 4x + 4$ 29. $x^2 - 2x + 4$ 31. $x - y$ 33. $V^2 - \dfrac{aV}{RT} + \dfrac{ab}{RT}$

35. $\dfrac{1}{R_1} + \dfrac{1}{R_2} + \dfrac{1}{R_3}$

Exercises 1–10, p. 31

1. 9   3. -1 5. 10   7. 5   9. -3 11. 1   13. $-\dfrac{7}{2}$ 15. 8   17. $\dfrac{10}{3}$ 19. $-\dfrac{13}{3}$

21. 2   23. 0   25. True for all values of x, identity   27. Left side 2 greater than right side for all values of x

Exercises 1–11, p. 35

1. $\dfrac{b}{a}$ 3. $\dfrac{m-1}{4}$ 5. $\dfrac{c+6}{a}$ 7. $2a + 8$ 9. $\dfrac{E}{I}$ 11. $\dfrac{M}{Afj}$ 13. $\dfrac{v_0 - v}{t}$ 15. $\dfrac{a + PV^2}{V}$

17. $\dfrac{l - a + d}{d}$ 19. $\dfrac{L - 2d - \pi r_2}{\pi}$ 21. $\dfrac{5F - 160}{9}$ 23. $\dfrac{wL}{R(w + L)}$ 25. \$5200, \$2600

27. 2 A, 4 A, 6 A   29. 6.5 ft   31. 36 in., 54 in.   33. 6 h after second car leaves   35. 6 L

Exercises for Chapter 1, p. 36

1. -10 3. -20 5. 2   7. -25 9. -4 11. 5   13. $4r^2t^4$ 15. $\dfrac{6m^2}{nt^2}$ 17. $\dfrac{z^6}{y^2}$ 19. $\dfrac{8t^3}{s^2}$

21. $3\sqrt{5}$ 23. $2\sqrt{5}j$ 25. $-a - 2ab$ 27. $5xy + 3$ 29. $2x^2 + 9x - 5$ 31. $hk - 3h^2k^4$
33. $7a - 6b$ 35. $13xy - 10z$ 37. $2x^3 - x^2 - 7x - 3$ 39. $-3x^2y + 24xy^2 - 48y^3$

41. $-9p^2 + 3pq + 18p^2q$ 43. $\dfrac{6q}{p} - 2 + \dfrac{3q^4}{p^3}$ 45. $x^2 - 2x + 3$ 47. $4x^3 - 2x^2 + 6x, R = -1$

49. $15r - 3s - 3t$ 51. $y^2 + 5y - 1, R = 4$ 53. 4   55. -5 57. $-\dfrac{7}{3}$ 59. 3

61. 2.5×10^4 mi/h   63. 176,000,000,000 C/kg   65. 2×10^{-7} in.   67. 0.000018 Pa·s   69. $\dfrac{5a - 2}{3}$

71. $\dfrac{4 - 2n}{3}$ 73. ϕ/B 75. $\dfrac{I - P}{Pr}$ 77. $\dfrac{nE - Ir}{In}$ 79. $\dfrac{D_p(N + 2)}{N}$ 81. $\dfrac{L - L_0}{L_0(t_2 - t_1)}$

83. $\dfrac{2S - n^2d + nd}{2n}$ 85. $12x - 2x^2$ 87. $252T_f - 6250$

89. 450 kg to carbon dioxide, 50 kg to carbon monoxide   91. \$15, \$5   93. after 7.5 h

Exercises 2—1, p. 44

1. $A = \pi r^2$ 3. $c = 2\pi r$ 5. $A = 5l$ 7. $A = s^2; s = \sqrt{A}$ 9. $w = 3000 - 10t$ 11. $I = 8t$

13. $3, -1$ 15. $11, -7$ 17. $-18, 70$ 19. $\dfrac{5}{2}, -\dfrac{1}{2}$ 21. $\dfrac{1}{4}a + \dfrac{1}{2}a^2, 0$

23. $3s^2 + s + 6, 12s^2 - 2s + 6$ 25. Square the value of the independent variable and add 2.

27. Cube the value of the independent variable and subtract this value from 6 times the value of the independent variable.

29. $y \neq 1$ 31. $y \geq 1$ 33. 54π cm³   35. $\dfrac{39}{5}$ 37. $\dfrac{1}{16}p^2 + \dfrac{(60 - p)^2}{4\pi}$

39. $0.006T^2 + 2.8T + 0.012hT + 2.8h + 0.006h^2$

Exercises 2—2, p. 47

1. $(2, 1), (-1, 2), (-2, -3)$ 3.   5. Isosceles triangle   7. $(5,4)$

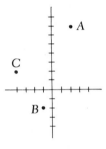

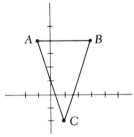

9. On a line parallel to the y-axis, one unit to the right
11. On a line bisecting the first and third quadrants
13. 0   15. To the right of the y-axis   17. Fourth quadrant   19. First, third

Exercises 2—3, p. 52

1.   3.    5. 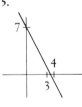   7.   9.

11.   13.   15.   17.   19.

21.   **23.** 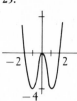 **25.**   **27.**   **29.**

31.   **33.** 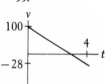 **35.**   **37.** 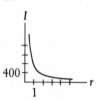 **39.**

41.   **43.** 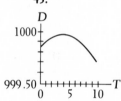 **45.**

Exercises 2—4, p. 56

1. 2   **3.** $-\dfrac{9}{4}$ **5.** 3.5   **7.** 0.5   **9.** 0.0, -1.0 **11.** No (real) zeros   **13.** 0.7   **15.** 2.5

17. -2.8, 1.8   **19.** -1.2, 0.5   **21.** -1.6, 5.6   **23.** -2.0, 0.0, 2.0   **25.** 0.0, 1.3   **27.** No zeros
29. After 1.9 s   **31.** 13, 77   **33.** 1.8 ft from end

Exercises for Chapter 2, p. 57

1. $V = 8\pi r^2$ **3.** $F = \dfrac{9}{5}C + 32$ **5.** 16, -47 **7.** -5, -27 **9.** 3, $\sqrt{1 - 4h}$ **11.** $6hx + 3h^2 - 2h$

13.   **15.** **17.** **19.** **21.** **23.**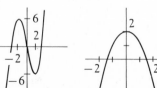

25. -0.5 **27.** 0, 4   **29.** -1.5, 1.0   **31.** -2.4, 0.0, 2.4   **33.** -1.2, 1.2   **35.** 0   **37.** 0.4
39. 0.2, 5.8   **41.** 1.3   **43.** -0.7, 0.7   **45.** On a line in the first quadrant, one unit to the right of y-axis
47. 19

49.

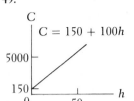

$C = 150 + 100h$

51.

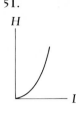

53.

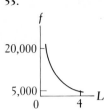

55.

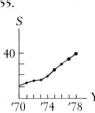

57.

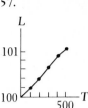

Exercises 3–1, p. 63

1.

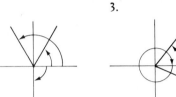

3.

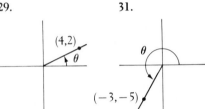

5. 405°, −315°   **7.** 510°, −210°   **9.** 430°30′, −289°30′
11. 638.1°, −81.9°   **13.** 15.2°   **15.** 86.05°   **17.** 301.27°
19. 96.13°   **21.** 47°30′   **23.** 19°45′   **25.** 5°37′   **27.** 24°55′

29.

31.

Exercises 3–2, p. 65

1. $\sin \theta = \frac{3}{5}$, $\cos \theta = \frac{4}{5}$, $\tan \theta = \frac{3}{4}$, $\cot \theta = \frac{4}{3}$, $\sec \theta = \frac{5}{4}$, $\csc \theta = \frac{5}{3}$

3. $\sin \theta = \frac{8}{17}$, $\cos \theta = \frac{15}{17}$, $\tan \theta = \frac{8}{15}$, $\cot \theta = \frac{15}{8}$, $\sec \theta = \frac{17}{15}$, $\csc \theta = \frac{17}{8}$

5. $\sin \theta = \frac{\sqrt{15}}{4}$, $\cos \theta = \frac{1}{4}$, $\tan \theta = \sqrt{15}$, $\cot \theta = \frac{1}{\sqrt{15}}$, $\sec \theta = 4$, $\csc \theta = \frac{4}{\sqrt{15}}$

7. $\sin \theta = \frac{5}{\sqrt{29}}$, $\cos \theta = \frac{2}{\sqrt{29}}$, $\tan \theta = \frac{5}{2}$, $\cot \theta = \frac{2}{5}$, $\sec \theta = \frac{\sqrt{29}}{2}$, $\csc \theta = \frac{\sqrt{29}}{5}$

9. $\frac{1}{\sqrt{2}}$, $\sqrt{2}$ **11.** $\frac{\sqrt{5}}{2}$, $\frac{2}{\sqrt{5}}$ **13.** $\frac{\sqrt{51}}{7}$, $\frac{10}{7}$ **15.** $\frac{1}{\sqrt{17}}$, $\frac{\sqrt{17}}{4}$ **17.** $\sin \theta = \frac{4}{5}$, $\tan \theta = \frac{4}{3}$

19. $\tan \theta = \frac{1}{2}$, $\sec \theta = \frac{\sqrt{5}}{2}$ **21.** $\sec \theta$, $\tan \theta$

Exercises 3–3, p. 70

1. $\sin 40° = 0.64$, $\cos 40° = 0.77$, $\tan 40° = 0.84$, $\cot 40° = 1.19$, $\sec 40° = 1.30$, $\csc 40° = 1.56$
3. $\sin 15° = 0.26$, $\cos 15° = 0.97$, $\tan 15° = 0.27$, $\cot 15° = 3.73$, $\sec 15° = 1.04$, $\csc 15° = 3.86$
5. 0.3256   **7.** 2.356   **9.** 0.9250   **11.** 1.812   **13.** 0.8988   **15.** 0.9686   **17.** 0.1299   **19.** 0.8601
21. 40°10′   **23.** 61°50′   **25.** 45°10′   **27.** 66°30′   **29.** 13.8°   **31.** 78.4°   **33.** 47.2°   **35.** 49.9°
37. 0.5528   **39.** 0.8930   **41.** 7.991   **43.** 0.4097   **45.** 72°47′   **47.** 35°8′   **49.** 39°7′   **51.** 51°6′
53. 0.7582   **55.** 144.4°   **57.** 162 ft/s   **59.** 67.8 V

Exercises 3–4, p. 75

1.

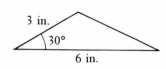

3.

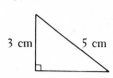

5. $a = 30.1$, $B = 58°0'$, $b = 48.2$
7. $A = 52°10'$, $B = 37°50'$, $c = 71.8$
9. $A = 52°20'$, $b = 0.684$, $c = 1.12$
11. $A = 58°40'$, $a = 143$, $B = 31°20'$
13. $B = 12.2°$, $b = 1450$, $c = 6850$
15. $A = 25.8°$, $B = 64.2°$, $b = 311$

17. $A = 57.9°$, $a = 20.2$, $h = 12.6$ 19. $A = 22.1°$, $a = 33.3$, $B = 67.9°$
21. $a = c \sin A$, $b = c \cos A$, $B = 90° - A$ 23. $A = 90° - B$, $b = a \tan B$, $c = a/\cos B$

Exercises 3–5, p. 78

1. 384 ft   3. 503 ft   5. 255 m   7. 34.3°   9. 5.26 mi   11. 9.50 ft   13. 8.8°   15. 1080 mi
17. 6.68 cm   19. 103°30'   21. 29.6 m

Exercises for Chapter 3, p. 79

1. $377°0'$, $-343°0'$ 3. $142.5°$, $-577.5°$ 5. $31.9°$ 7. $38.1°$ 9. $17°30'$ 11. $49°42'$

13. $\sin \theta = \dfrac{7}{25}$, $\cos \theta = \dfrac{24}{25}$, $\tan \theta = \dfrac{7}{24}$, $\cot \theta = \dfrac{24}{7}$, $\sec \theta = \dfrac{25}{24}$, $\csc \theta = \dfrac{25}{7}$

15. $\sin \theta = \dfrac{1}{\sqrt{2}}$, $\cos \theta = \dfrac{1}{\sqrt{2}}$, $\tan \theta = 1$, $\cot \theta = 1$, $\sec \theta = \sqrt{2}$, $\csc \theta = \sqrt{2}$ 17. $\cos \theta = \dfrac{12}{13}$, $\cot \theta = \dfrac{12}{5}$

19. $\cos \theta = \dfrac{1}{\sqrt{5}}$, $\csc \theta = \dfrac{\sqrt{5}}{2}$ 21. 0.9511   23. 1.829   25. 0.7590   27. 0.9751   29. 6.498

31. 0.5609   33. $32°10'$ 35. $62°10'$ 37. $18.2°$ 39. $57.6°$ 41. $41°27'$ 43. $60°4'$
45. $a = 1.83$, $B = 73°0'$, $c = 6.27$ 47. $A = 51°30'$, $B = 38°30'$, $c = 104$
49. $B = 52.5°$, $b = 15.6$, $c = 19.7$ 51. $A = 31.2°$, $a = 3.94$, $B = 58.8°$
53. $a = 0.626$, $B = 40°17'$, $b = 0.530$ 55. $A = 41°49'$, $B = 48°11'$, $b = 11.2$ 57. $0.611\ \mu m$
59. $30.6°$ 61. 1.58 mi   63. 4820 ft   65. 47.2 ft   67. $25°10'$ 69. 154 m   71. 67.1 ft
73. 1.75 km

Exercises 4–1, p. 85

1. Yes, No   3. Yes, Yes   5. -1, -16 7. $-\dfrac{9}{5}$, $-\dfrac{16}{5}$ 9. Yes   11. No   13. No   15. Yes
17. Yes   19. Yes

Exercises 4–2, p. 89

1. $x = 3.0$, $y = 1.0$ 3. $x = 3.0$, $y = 0.0$ 5. $x = 2.2$, $y = -0.3$ 7. $x = -0.9$, $y = -2.3$
9. $x = 6.2$, $y = -0.5$ 11. $x = 0.0$, $y = 3.0$ 13. $x = -14.0$, $y = -5.0$ 15. $x = 4.0$, $y = 7.5$
17. $x = -3.6$, $y = -1.4$ 19. $x = 1.1$, $y = 0.8$ 21. Dependent   23. $x = -1.2$, $y = -3.6$
25. $x = 1.5$, $y = 4.5$ 27. Inconsistent   29. $l = 9.0$ km, $w = 3.0$ km   31. $I = 0.9$ A, $E = 3.6$ V

Exercises 4–3, p. 95

1. $x = 1$, $y = -2$ 3. $x = 7$, $y = 3$ 5. $x = -1$, $y = -4$ 7. $x = \dfrac{1}{2}$, $y = 2$ 9. $x = -\dfrac{1}{3}$, $y = 4$

11. $x = \dfrac{9}{22}$, $y = -\dfrac{16}{11}$ 13. $x = 3$, $y = 1$ 15. $x = -1$, $y = -2$ 17. $x = 1$, $y = 2$ 19. Inconsistent

21. $x = -\dfrac{14}{5}$, $y = -\dfrac{16}{5}$ 23. $x = \dfrac{69}{29}$, $y = \dfrac{13}{29}$ 25. $x = \dfrac{1}{2}$, $y = -4$ 27. $x = -\dfrac{2}{3}$, $y = 0$

29. Dependent   31. $x = -1$, $y = -2$ 33. $i_1 = 1.5$ A, $i_2 = 0.5$ A   35. 7.5 m from one end
37. 2800 cards/min, 1900 cards/min   39. 105 mL and 75 mL   41. $300 fixed cost, $0.25 per booklet

Exercises 4–4, p. 101

1. -10 **3.** 29   **5.** 32   **7.** 93   **9.** 83   **11.** 96   **13.** $x = 3, y = 1$ **15.** $x = -1, y = -2$

17. $x = 1, y = 2$ **19.** Inconsistent   **21.** $x = -\dfrac{14}{5}, y = -\dfrac{16}{5}$ **23.** $x = \dfrac{69}{29}, y = \dfrac{13}{29}$

25. $x = \dfrac{1}{2}, y = -4$ **27.** $x = -\dfrac{2}{3}, y = 0$ **29.** Dependent   **31.** $x = -1, y = -2$

33. $s_0 = 15$ ft, $v = 10$ ft/s   **35.** 58, 31   **37.** $600, $7400   **39.** $R = \dfrac{1}{600} T + \dfrac{11}{30}$

Exercises 4–5, p. 106

1. $x = 2, y = -1, z = 1$ **3.** $x = 4, y = -3, z = 3$ **5.** $x = \dfrac{1}{2}, y = \dfrac{2}{3}, z = \dfrac{1}{6}$

7. $x = \dfrac{2}{3}, y = -\dfrac{1}{3}, z = 1$ **9.** $x = \dfrac{4}{15}, y = -\dfrac{3}{5}, z = \dfrac{1}{3}$ **11.** $x = -2, y = \dfrac{2}{3}, z = \dfrac{1}{3}$

13. $x = \dfrac{3}{4}, y = 1, z = -\dfrac{1}{2}$ **15.** $r = 0, s = 0, t = 0, u = -1$

17. Eliminate x and the resulting system with y and z is dependent; choose a z value, then find y, then x; if $z = 0, y = -6, x = -10$; if $z = 1, y = -11, x = -17$

19. $-\dfrac{3}{22}$ A, $-\dfrac{39}{110}$ A, $\dfrac{27}{55}$ A   **21.** 16 parts/h, 20 parts/h, 28 parts/h   **23.** 50 cm³ alloy B, 50 cm³ alloy C

Exercises 4–6, p. 113

1. 122   **3.** 651   **5.** -439 **7.** 202   **9.** 128   **11.** 0.128   **13.** $x = -1, y = 2, z = 0$

15. $x = 2, y = -1, z = 1$ **17.** $x = 4, y = -3, z = 3$ **19.** $x = \dfrac{1}{2}, y = \dfrac{2}{3}, z = \dfrac{1}{6}$

21. $x = \dfrac{2}{3}, y = -\dfrac{1}{3}, z = 1$ **23.** $x = \dfrac{4}{15}, y = -\dfrac{3}{5}, z = \dfrac{1}{3}$ **25.** $x = -2, y = \dfrac{2}{3}, z = \dfrac{1}{3}$

27. $x = \dfrac{3}{4}, y = 1, z = -\dfrac{1}{2}$ **29.** $A = 125$ lb, $B = 60$ lb, $F = 75$ lb   **31.** $12,000, $7000, $1000

Exercises for Chapter 4, p. 115

1. -17 **3.** -35 **5.** $x = 2.0, y = 0.0$ **7.** $x = 2.2, y = 2.7$ **9.** $x = 1.5, y = -1.9$

11. $x = 1.5, y = 0.3$ **13.** $x = 1, y = 2$ **15.** $x = \dfrac{1}{2}, y = -2$ **17.** $x = -\dfrac{1}{3}, y = \dfrac{7}{4}$

19. $x = \dfrac{11}{19}, y = -\dfrac{26}{19}$ **21.** $x = -\dfrac{6}{19}, y = \dfrac{36}{19}$ **23.** $x = \dfrac{43}{39}, y = \dfrac{7}{13}$ **25.** $x = 1, y = 2$

27. $x = \dfrac{1}{2}, y = -2$ **29.** $x = -\dfrac{1}{3}, y = \dfrac{7}{4}$ **31.** $x = \dfrac{11}{19}, y = -\dfrac{26}{19}$ **33.** $x = -\dfrac{6}{19}, y = \dfrac{36}{19}$

35. $x = \dfrac{43}{39}, y = \dfrac{7}{13}$ **37.** -115 **39.** 220   **41.** $x = 2, y = -1, z = 1$ **43.** $x = \dfrac{2}{3}, y = -\dfrac{1}{2}, z = 0$

45. $r = 3, s = -1, t = \dfrac{3}{2}$ **47.** $x = -\dfrac{1}{2}, y = \dfrac{1}{2}, z = 3$ **49.** $x = 2, y = -1, z = 1$

51. $x = \dfrac{2}{3}, y = -\dfrac{1}{2}, z = 0$ **53.** $r = 3, s = -1, t = \dfrac{3}{2}$ **55.** $x = -\dfrac{1}{2}, y = \dfrac{1}{2}, z = 3$ **57.** $x = \dfrac{8}{3}, y = -8$

59. $x = 1, y = 3$ **61.** -6 **63.** $-\dfrac{4}{3}$ **65.** $R_1 = 5\Omega, R_2 = 2\Omega$ **67.** $a = 10, b = 12, c = 15$

69. 450 mi/h, 50 mi/h   **71.** 11 ft   **73.** 79% nickel, 16% iron, 5% molybdenum

Exercises 5—1, p. 121

1. $40x - 40y$ 3. $2x^3 - 8x^2$ 5. $y^2 - 36$ 7. $9v^2 - 4$ 9. $25f^2 + 40f + 16$ 11. $4x^2 + 28x + 49$
13. $x^2 - 4xy + 4y^2$ 15. $36s^2 - 12st + t^2$ 17. $x^2 + 6x + 5$ 19. $c^2 + 9c + 18$ 21. $6x^2 + 13x - 5$
23. $20x^2 - 21x - 5$ 25. $20v^2 + 13v - 15$ 27. $6x^2 - 13xy - 63y^2$ 29. $2x^2 - 8$ 31. $8a^3 - 2a$
33. $6ax^2 + 24abx + 24ab^2$ 35. $16a^3 - 48a^2 + 36a$ 37. $x^2 + y^2 + 2xy + 2x + 2y + 1$
39. $x^2 + y^2 + 2xy - 6x - 6y + 9$ 41. $125 - 75t + 15t^2 - t^3$ 43. $8x^3 + 60x^2t + 150xt^2 + 125t^3$
45. $x^3 + 8$ 47. $64 - 27x^3$ 49. $Ri_1^2 + 2Ri_1i_2 + Ri_2^2$ 51. $192 - 16t - 16t^2$
53. $P + 3Pr + 3Pr^2 + Pr^3$

Exercises 5—2, p. 125

1. $6(x + y)$ 3. $5(a - 1)$ 5. $3x(x - 3)$ 7. $7b(by - 4)$ 9. $2(x + 2y - 4z)$
11. $3ab(b - 2 + 4b^2)$ 13. $4pq(3q - 2 - 7q^2)$ 15. $2(a^2 - b^2 + 2c^2 - 3d^2)$ 17. $(x + 2)(x - 2)$
19. $(10 + y)(10 - y)$ 21. $(9s + 5t)(9s - 5t)$ 23. $(12n + 13p^2)(12n - 13p^2)$ 25. $2(x + 2)(x - 2)$
27. $3(x + 3z)(x - 3z)$ 29. $(x^2 + 4)(x + 2)(x - 2)$ 31. $(x^4 + 1)(x^2 + 1)(x + 1)(x - 1)$
33. $(3 + b)(x - y)$ 35. $(a - b)(a + x)$ 37. $i(R_1 + R_2 + r)$ 39. $m(v_1 + v_2)(v_1 - v_2)$
41. $e(e + 2)(e - 2)$

Exercises 5—3, p. 128

1. $(x + 1)(x + 4)$ 3. $(s - 7)(s + 6)$ 5. $(x + 1)^2$ 7. $(x - 2)^2$ 9. $(3x + 1)(x - 2)$
11. $(3y + 1)(y - 3)$ 13. $(3t - 4u)(t - u)$ 15. $(9x - 2y)(x + y)$ 17. $(2m + 5)^2$ 19. $(2x - 3)^2$
21. $(3t - 4)(3t - 1)$ 23. $(8b - 1)(b + 4)$ 25. $(4p - q)(p - 6q)$ 27. $(12x - y)(x + 4y)$
29. $2(x - 1)(x - 6)$ 31. $2(2x - 1)(x + 4)$ 33. $(x + 1)^3$ 35. $(2x + 1)(4x^2 - 2x + 1)$
37. $16(t - 4)(t + 2)$ 39. $(x - 2L)(x - L)$

Exercises 5—4, p. 131

1. $\dfrac{14}{21}$ 3. $\dfrac{2ax^2}{2xy}$ 5. $\dfrac{2x - 4}{x^2 + x - 6}$ 7. $\dfrac{ax^2 - ay^2}{x^2 - xy - 2y^2}$ 9. $\dfrac{7}{11}$ 11. $\dfrac{2xy}{4y^2}$ 13. $\dfrac{2}{x + 1}$ 15. $\dfrac{x - 5}{2x - 1}$

17. $\dfrac{1}{4}$ 19. $\dfrac{3}{4}x$ 21. $\dfrac{1}{5a}$ 23. $\dfrac{3a - 2b}{2a - b}$ 25. $\dfrac{4x^2 + 1}{(2x + 1)(2x - 1)}$ (cannot be reduced)   27. $\dfrac{x - 4}{x + 4}$

29. $\dfrac{2x - 1}{x + 8}$ 31. $(x^2 + 4)(x - 2)$ 33. $\dfrac{x^2y^2(y + x)}{y - x}$ 35. $\dfrac{x + 3}{x - 3}$ 37. $-\dfrac{1}{2}$ 39. $\dfrac{(x + 5)(x - 3)}{(5 - x)(x + 3)}$

41. $\dfrac{x^2 + xy + y^2}{x + y}$ 43. $\dfrac{(x + 1)^2}{x^2 - x + 1}$ 45. (a)   47. (a)

Exercises 5—5, p. 135

1. $\dfrac{3}{28}$ 3. $6xy$ 5. $\dfrac{7}{18}$ 7. $\dfrac{xy^2}{bz^2}$ 9. $4t$ 11. $3(u + v)$ 13. $\dfrac{10}{3(a + 4)}$ 15. $\dfrac{x - 3}{x(x + 3)}$ 17. $\dfrac{3x}{5a}$

19. $\dfrac{(x + 1)(x - 1)(x - 4)}{4(x + 2)}$ 21. $\dfrac{x^2}{a + x}$ 23. $\dfrac{15}{4}$ 25. $\dfrac{7x^4}{3a^4}$ 27. $\dfrac{4t(2t - 1)(t + 5)}{(2t + 1)^2}$ 29. $\dfrac{1}{2}(x + y)$

31. $(x + y)(3p + 7q)$ 33. $\dfrac{(T + 100)(T - 400)}{2(T - 40)}$

Exercises 5—6, p. 140

1. $\dfrac{9}{5}$ 3. $\dfrac{8}{x}$ 5. $\dfrac{5}{4}$ 7. $\dfrac{3 + 7ax}{4x}$ 9. $\dfrac{ax - b}{x^2}$ 11. $\dfrac{30 + ax^2}{25x^3}$ 13. $\dfrac{14 - a^2}{10a}$

15. $\dfrac{-x^2 + 4x + xy + y - 2}{xy}$ 17. $\dfrac{7}{2(2x - 1)}$ 19. $\dfrac{5 - 3x}{2x(x + 1)}$ 21. $\dfrac{-3}{4(s - 3)}$ 23. $\dfrac{x + 6}{x^2 - 9}$

25. $\dfrac{2x - 5}{(x - 4)^2}$ 27. $\dfrac{x + 27}{(x - 5)(x + 5)(x - 6)}$ 29. $\dfrac{9x^2 + x - 2}{(3x - 1)(x - 4)}$ 31. $\dfrac{13t^2 + 27t}{(t - 3)(t + 2)(t + 3)^2}$

33. $\dfrac{x+1}{x-1}$ **35.** $-\dfrac{(x+1)(x^3+x^2-3x-1)}{x^2(x+2)}$ **37.** $\dfrac{h}{(x+1)(x+h+1)}$ **39.** $\dfrac{-2hx-h^2}{x^2(x+h)^2}$

41. $\dfrac{y^2-rx+r^2}{r^2}$ **43.** $\dfrac{2a-1}{a^2}$ **45.** $\dfrac{3\pi l^3-12cl^2+\pi c^3}{3\pi l^3}$ **47.** $\dfrac{b(y^2-x^2)}{(x^2+y^2)^2}$

Exercises 5–7, p. 145

1. 4   **3.** −3   **5.** $\dfrac{7}{2}$ **7.** $\dfrac{16}{21}$ **9.** −9   **11.** $-\dfrac{6}{5}$ **13.** $\dfrac{5}{3}$ **15.** −2   **17.** $\dfrac{3}{4}$ **19.** 6   **21.** −5

23. $-\dfrac{7}{8}$ **25.** No solution   **27.** $\dfrac{2}{3}$ **29.** $\dfrac{3b}{1-2b}$ **31.** $\dfrac{(2b-1)(b+6)}{2(b-1)}$

33. $\dfrac{2EI-2IV_0-p^2-ImV^2}{IV^2}$ **35.** $\dfrac{rR-rR_2+RR_2}{R_2-R}$ **37.** 3.6 min   **39.** 60 m, 36 m

Exercises for Chapter 5, p. 147

1. $12ax+15a^2$ **3.** $4a^2-49b^2$ **5.** $4a^2+4a+1$ **7.** $b^2+3b-28$ **9.** $2x^2-13x-45$

11. $16c^2+6cd-d^2$ **13.** $3(s+3t)$ **15.** $a^2(x^2+1)$ **17.** $(x+12)(x-12)$

19. $(20r+t^2)(20r-t^2)$ **21.** $(3t-1)^2$ **23.** $(5t+1)^2$ **25.** $(x+8)(x-7)$ **27.** $(t-9)(t+4)$

29. $(2x-9)(x+4)$ **31.** $(2x+5)(2x-7)$ **33.** $(5b-1)(2b+5)$ **35.** $4(x+4)(x-4)$

37. $(x+3)^3$ **39.** $(2x+3)(4x^2-6x+9)$ **41.** $(a-3)(b^2+1)$ **43.** $(x+5)(n-x+5)$ **45.** $\dfrac{16x^2}{3a^2}$

47. $\dfrac{3x+1}{2x-1}$ **49.** $\dfrac{16}{5x(x-y)}$ **51.** $\dfrac{6}{5-x}$ **53.** $\dfrac{x+2}{2x(7x-1)}$ **55.** $\dfrac{1}{x-1}$ **57.** $\dfrac{16x-15}{36x^2}$

59. $\dfrac{5y+6}{2xy}$ **61.** $\dfrac{-2(2a+3)}{a(a+2)}$ **63.** $\dfrac{x^3+6x^2-2x+2}{x(x-1)(x+3)}$ **65.** 2   **67.** $\dfrac{7}{2c+4}$ **69.** $-\dfrac{(a-1)^2}{2a}$

71. 6   **73.** $\dfrac{1}{4}[(x+y)^2-(x-y)^2]=\dfrac{1}{4}(x^2+2xy+y^2-x^2+2xy-y^2)=\dfrac{1}{4}(4xy)=xy$

75. $(x-10)(x+7)$ **77.** $4(3x^2+12x+16)$ **79.** $\dfrac{1-t}{(t+1)^3}$ **81.** $\dfrac{120T^4+20w^2x^2T^2-3w^4x^4}{120T^4}$

83. $\dfrac{\mu R}{r+R+\mu R}$ **85.** $\dfrac{fp}{p-f}$ **87.** $\dfrac{wL^3-24\theta EI}{4L}$ **89.** $\dfrac{48}{7}$ days   **91.** 3.1 days

Exercises 6–1, p. 155

1. $a=1, b=-8, c=5$ **3.** $a=1, b=-2, c=-4$ **5.** Not quadratic   **7.** $a=1. b=-1, c=0$

9. 2, −2   **11.** −1, 9   **13.** 3, 4   **15.** 0, −2   **17.** $\dfrac{3}{2}, -\dfrac{3}{2}$ **19.** $\dfrac{1}{3}, 4$ **21.** −4, −4   **23.** $\dfrac{2}{3}, \dfrac{3}{2}$

25. $\dfrac{1}{2}, -\dfrac{3}{2}$ **27.** 2, −1   **29.** $2b, -2b$ **31.** $0, \dfrac{5}{2}$ **33.** 0, −2   **35.** $b-a, -b-a$ **37.** 10 s

39. −3, −5   **41.** 4Ω, 12Ω   **43.** 12 mm, 8 mm

Exercises 6–2, p. 159

1. −5, 5   **3.** $-\sqrt{7}, \sqrt{7}$ **5.** −3, 7   **7.** $-3 \pm \sqrt{7}$ **9.** 2, −4   **11.** −2, −1   **13.** $2 \pm \sqrt{2}$

15. −5, 3   **17.** $-3, \dfrac{1}{2}$ **19.** $\dfrac{1}{6}(3 \pm \sqrt{33})$ **21.** $\dfrac{1}{4}(1 \pm \sqrt{17})$ **23.** $-b \pm \sqrt{b^2-c}$

Exercises 6—3, p. 162

1. $2, -4$ **3.** $-2, -1$ **5.** $2 \pm \sqrt{2}$ **7.** $-5, 3$ **9.** $-3, \dfrac{1}{2}$ **11.** $\dfrac{1}{6}(3 \pm \sqrt{33})$ **13.** $\dfrac{1}{4}(1 \pm \sqrt{17})$

15. $\dfrac{1}{4}(7 \pm \sqrt{17})$ **17.** $\dfrac{1}{2}(-5 \pm \sqrt{-5})$ **19.** $\dfrac{1}{6}(1 \pm \sqrt{109})$ **21.** $\dfrac{3}{2}, -\dfrac{3}{2}$ **23.** $\dfrac{3}{4}, -\dfrac{5}{8}$

25. $-c \pm \sqrt{c^2 + 1}$ **27.** $\dfrac{b + 1 \pm \sqrt{-3b^2 + 2b + 4b^2 a + 1}}{2b^2}$ **29.** 0.382 Pa   **31.** 0.915 in.

33. 3.06 mm   **35.** 8 in.

Exercises for Chapter 6, p. 163

1. $-4, 1$ **3.** $2, 8$ **5.** $\dfrac{1}{3}, -4$ **7.** $\dfrac{1}{2}, \dfrac{5}{3}$ **9.** $0, \dfrac{25}{6}$ **11.** $-\dfrac{3}{2}, \dfrac{7}{2}$ **13.** $-10, 11$ **15.** $-1 \pm \sqrt{6}$

17. $-4, \dfrac{9}{2}$ **19.** $\dfrac{1}{8}(3 \pm \sqrt{41})$ **21.** $\dfrac{1}{2}(-1 \pm 3\sqrt{-1})$ **23.** $\dfrac{1}{6}(-2 \pm \sqrt{58})$ **25.** $-2 \pm 2\sqrt{2}$

27. $\dfrac{1}{3}(-4 \pm \sqrt{10})$ **29.** $-1, \dfrac{5}{4}$ **31.** $\dfrac{1}{4}(-3 \pm \sqrt{-47})$ **33.** $\dfrac{-1 \pm \sqrt{-1}}{a}$ **35.** $\dfrac{-3 \pm \sqrt{9 + 4a^3}}{2a}$

37. $-5, 6$ **39.** $\dfrac{1}{4}(1 \pm \sqrt{33})$ **41.** $3 \pm \sqrt{7}$ **43.** 0 (3 is not a solution)   **45.** $-5 \pm 5\sqrt{-79}$

47. $\dfrac{E \pm \sqrt{E^2 - 4PR}}{2R}$ **49.** 10 units   **51.** $\dfrac{2.5 \pm \sqrt{6.25 - 50.4n}}{25.2}$ **53.** 8 V, 12 V   **55.** 0.106 in.

57. 25

Exercises 7—1, p. 168

1. $+, -, -$ **3.** $+, +, -$ **5.** $+, +, +$ **7.** $+, -, +$

9. $\sin \theta = \dfrac{1}{\sqrt{5}}$, $\cos \theta = \dfrac{2}{\sqrt{5}}$, $\tan \theta = \dfrac{1}{2}$, $\cot \theta = 2$, $\sec \theta = \dfrac{1}{2}\sqrt{5}$, $\csc \theta = \sqrt{5}$

11. $\sin \theta = -\dfrac{3}{\sqrt{13}}$, $\cos \theta = -\dfrac{2}{\sqrt{13}}$, $\tan \theta = \dfrac{3}{2}$, $\cot \theta = \dfrac{2}{3}$, $\sec \theta = -\dfrac{1}{2}\sqrt{13}$, $\csc \theta = -\dfrac{1}{3}\sqrt{13}$

13. $\sin \theta = \dfrac{12}{13}$, $\cos \theta = -\dfrac{5}{13}$, $\tan \theta = -\dfrac{12}{5}$, $\cot \theta = -\dfrac{5}{12}$, $\sec \theta = -\dfrac{13}{5}$, $\csc \theta = \dfrac{13}{12}$

15. $\sin \theta = -\dfrac{2}{\sqrt{29}}$, $\cos \theta = \dfrac{5}{\sqrt{29}}$, $\tan \theta = -\dfrac{2}{5}$, $\cot \theta = -\dfrac{5}{2}$, $\sec \theta = \dfrac{1}{5}\sqrt{29}$, $\csc \theta = -\dfrac{1}{2}\sqrt{29}$

17. II   **19.** II   **21.** IV   **23.** III

Exercises 7—2, p. 173

1. $\sin 20°$; $-\cos 40°$ **3.** $-\tan 75°$; $-\csc 58°$ **5.** $-\sin 57°$; $-\cot 6°$ **7.** $\cos 40°$; $-\tan 40°$
9. -0.2588 **11.** -0.2756 **13.** -1.036 **15.** -2.366 **17.** 0.2339 **19.** -0.8936 **21.** 0.9732
23. -1.767 **25.** -0.6036 **27.** -1.838 **29.** $28°, 208°$ **31.** $201°20', 338°40'$ **33.** $60°30', 119°30'$
35. $97°0', 263°0'$ **37.** $238.0°, 302.0°$ **39.** $66.4°, 293.6°$ **41.** $62.2°, 242.2°$ **43.** $15.8°, 195.8°$
45. $252°19', 287°41'$ **47.** $125°33', 305°33'$ **49.** -0.7002 **51.** -0.7771 **53.** $<$ **55.** $=$
57. 72.7 lb   **59.** 416
61. $\cos(-\theta) = x/r$; $\tan(-\theta) = -y/x$; $\cot(-\theta) = x/-y$; $\sec(-\theta) = r/x$; $\csc(-\theta) = r/-y$
63. (a) 5.671 (b) -1.428

Exercises 7—3, p. 178

1. $\dfrac{\pi}{12}, \dfrac{5\pi}{6}$ 3. $\dfrac{5\pi}{12}, \dfrac{11\pi}{6}$ 5. $\dfrac{7\pi}{6}, \dfrac{3\pi}{2}$ 7. $\dfrac{8\pi}{9}, \dfrac{13\pi}{9}$ 9. 72°, 270°   11. 10°, 315°   13. 170°, 300°

15. 15°, 27°   17. 0.401   19. 4.40   21. 5.82   23. 3.11   25. 43.0°   27. 172°   29. 140.4°
31. 939.7°   33. 0.7071   35. 3.732   37. −1.732   39. −0.1219   41. 0.9057   43. 0.5132
45. −0.4797   47. −0.9890   49. 0.3142, 2.8278   51. 2.9326, 6.0736   53. 0.8308, 5.4522
55. 2.4432, 3.8408   57. $1.43\pi = 4.49$ (closest 3 sig. digit result)   59. 15.7 cm/s

Exercises 7—4, p. 182

1. 10.5 in.   3. 52.4 in.²   5. 12 rad   7. 2.88 in.²   9. 0.262 ft   11. 129 km   13. 5650 cm
15. 21.5 ft   17. 75.4 rad/s   19. 3700 mi/h   21. 5200 ft/min   23. 2.60×10^{-6} rad/s
25. 25,100 ft/min   27. 66.0 cm²   29. 4.847×10^{-6} 31. 7.13×10^{7} mi

Exercises for Chapter 7, p. 183

1. $\sin \theta = \dfrac{4}{5}$, $\cos \theta = \dfrac{3}{5}$, $\tan \theta = \dfrac{4}{3}$, $\cot \theta = \dfrac{3}{4}$, $\sec \theta = \dfrac{5}{3}$, $\csc \theta = \dfrac{5}{4}$

3. $\sin \theta = -\dfrac{2}{\sqrt{53}}$, $\cos \theta = \dfrac{7}{\sqrt{53}}$, $\tan \theta = -\dfrac{2}{7}$, $\cot \theta = -\dfrac{7}{2}$, $\sec \theta = \dfrac{\sqrt{53}}{7}$, $\csc \theta = -\dfrac{\sqrt{53}}{2}$

5. $-\cos 48°$; $\tan 14°$ 7. $-\sin 71°$; $\sec 15°$ 9. $\dfrac{2\pi}{9}$; $\dfrac{17\pi}{20}$ 11. $\dfrac{4\pi}{15}$; $\dfrac{9\pi}{8}$ 13. 252°; 130°

15. 12°; 330°   17. 32.1°   19. 206.3°   21. 1.745   23. 0.358   25. 4.58   27. 2.38
29. −0.4226   31. −0.4663   33. −1.082   35. −0.4258   37. −1.638   39. 4.230   41. −0.5878
43. −0.8660   45. 0.5564   47. 1.195   49. 10.3°, 190.3°   51. 118.2°, 241.8°   53. 17°0′, 163°0′
55. 70°40′, 289°20′   57. 136°14′, 223°46′   59. 66°47′, 246°47′   61. 0.5760, 5.707   63. 4.186, 5.239
65. −120 V   67. 0.393 ft   69. 108 cm²   71. 188 in./s   73. 18.0 cm   75. 6.66 mm
77. 84,800 cm/min   79. 3.58×10^{5} km

Exercises 8—1, p. 191

1.   3.   5.   7.   9.   11.   13.

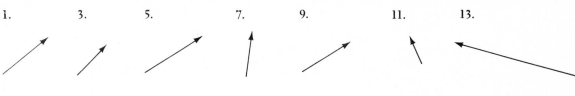

15.   17.   19.   21.   23.

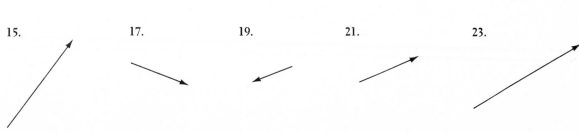

25. 3.22, 7.97   27. −62.9, 44.1   29. 2.11, −8.79   31. −2.53, 0.788

Exercises 8—2, p. 195

1. $R = 24.2$, $\theta = 52.6°$ with **A**   3. $R = 7.78$, $\theta = 66.7°$ with **A**   5. $R = 10.0$, $\theta = 58.8°$
7. $R = 2.74$, $\theta = 111.0°$ 9. $R = 2130$, $\theta = 107.7°$ 11. $R = 1.42$, $\theta = 299.2°$
13. $R = 29.2$, $\theta = 10.8°$ 15. $R = 47.0$, $\theta = 101.0°$ 17. $R = 27.2$, $\theta = 33.0°$
19. $R = 12.7$, $\theta = 25.2°$ 21. $R = 50.1$, $\theta = 50.3°$ 23. $R = 230$, $\theta = 125.1°$

Exercises 8—3, p. 199

1. 81.5 lb, 56.5° from 45-lb force   3. 850 N, 10.0° from 520-N force   5. 510 mi, 28.1° S of W
7. 37.7 mi, 77.4° S of W   9. $v_H = 80.3$ ft/s, $v_V = 89.2$ ft/s   11. 31.6 lb, 71.6°
13. 206 m/s, 14.0° from direction of plane   15. Yes   17. 94.0 lb   19. 128 m/s, 20.2° from horizontal
21. 6.96 rad/s²   23. 4.06 A/m, 11.6° with magnet

Exercises 8—4, p. 205

1. $b = 38.1$, $C = 66°0'$, $c = 46.1$ 3. $a = 2790$, $B = 2590$, $C = 109.0°$
5. $B = 12.0°$, $C = 150.0°$, $c = 7.44$ 7. $a = 111$, $A = 149.7°$, $C = 9.6°$
9. $A = 125°30'$, $a = 0.0777$, $c = 0.00583$ 11. $A = 99.4°$, $b = 55.1$, $c = 24.4$
13. $A = 68.1°$, $a = 552$, $c = 537$
15. $A_1 = 61.5°$, $C_1 = 70.4°$, $c_1 = 5.62$; $A_2 = 118.5°$, $C_2 = 13.4°$, $c_2 = 1.38$
17. $A_1 = 107.3°$, $a_1 = 5280$, $C_1 = 41.3°$; $A_2 = 9.9°$, $a_2 = 950$, $C_2 = 138.7°$ 19. No solution
21. 7.83 ft, 10.6 ft   23. 25,200 ft   25. 23.0 mi   27. 27,300 km

Exercises 8—5, p. 209

1. $A = 50.3°$, $B = 75.7°$, $c = 6.31$ 3. $A = 70.9°$, $B = 11.1°$, $c = 4750$
5. $A = 34.7°$, $B = 40.7°$, $C = 104.6°$ 7. $A = 18.2°$, $B = 22.2°$, $C = 139.6°$
9. $A = 6.0°$, $B = 16.0°$, $c = 1150$ 11. $A = 82.3°$, $b = 21.6$, $C = 11.4°$
13. $A = 38.8°$, $B = 36.6°$, $b = 97.9$ 15. $A = 46.9°$, $B = 61.8°$, $C = 71.3°$
17. $A = 137.9°$, $B = 33.7°$, $C = 8.4°$ 19. $b = 29.5$, $C = 14.4°$, $c = 14.9$ 21. 1140 ft   23. 96.9°
25. 8.88 km/h, 13.4° with bank   27. 42.4°

Exercises for Chapter 8, p. 209

1. $A_x = 57.4$, $A_y = 30.5$ 3. $A_x = -0.754$, $A_y = -0.528$ 5. $R = 602$, $\theta = 57.1°$ with **A**
7. $R = 5950$, $\theta = 33.6°$ with **A**   9. $R = 965$, $\theta = 8.6°$ 11. $R = 26.1$, $\theta = 146°0'$
13. $R = 71.9$, $\theta = 336.4°$ 15. $R = 99.2$, $\theta = 359.3°$ 17. $b = 18.1$, $C = 64°0'$, $c = 17.5$
19. $A = 21.2°$, $b = 34.8$, $c = 51.5$ 21. $A = 39.9°$, $a = 51.9$, $C = 30.1°$
23. $A_1 = 54.8°$, $a_1 = 12.7$, $B_1 = 68.6°$; $A_2 = 12.0°$, $a_2 = 3.24$, $B_2 = 111.4°$
25. $A = 32.3°$, $b = 267$, $C = 17.7°$ 27. $A = 148.7°$, $B = 9.3°$, $c = 5.66$
29. $A = 36.9°$, $B = 25.0°$, $C = 118.1°$ 31. $A = 20.6°$, $B = 35.6°$, $C = 123.8°$ 33. 0.0291 h
35. 27.0 ft/s, 33.7° with horizontal   37. 268 mi/h, 465 mi/h   39. 770 ft   41. 174 m   43. 57.7°
45. 186 mi   47. 110 lb   49. 310 lb, 330.4°

Exercises 9—1, p. 215

1. $0, -0.7, -1, -0.7, 0, 0.7, 1, 0.7, 0, -0.7, -1, -0.7, 0, 0.7, 1, 0.7, 0$

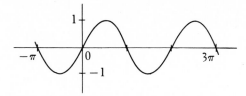

3. $-3, -2.1, 0, 2.1, 3, 2.1, 0, -2.1, -3, -2.1, 0, 2.1, 3, 2.1, 0, -2.1, -3$

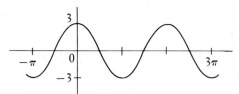

5.

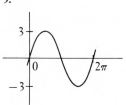

7.

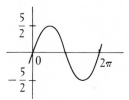

9.

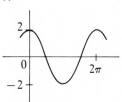

11.

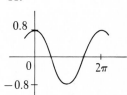

13.

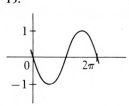

15.

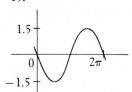

17.

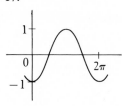

19.

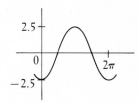

21. $0, 0.84, 0.91, 0.14, -0.76, -0.96, -0.28, 0.66$

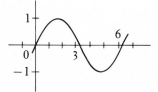

23. $1, 0.54, -0.42, -0.99, -0.65, 0.28, 0.96, 0.75$

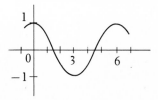

Exercises 9—2, p. 218

1. $\dfrac{\pi}{3}$ 3. $\dfrac{\pi}{4}$ 5. $\dfrac{\pi}{6}$ 7. $\dfrac{\pi}{8}$ 9. 1   11. $\dfrac{1}{2}$ 13. 6π 17. 3   19. $\dfrac{2}{\pi}$

21.
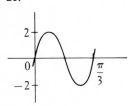

23.
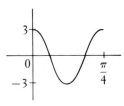

25.
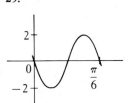

27.
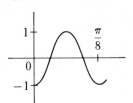

29.
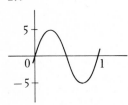

31.
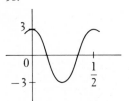

33.
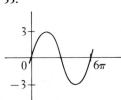

35.
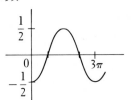

37.
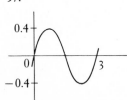

39.
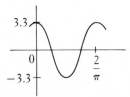

41.
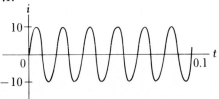

43.
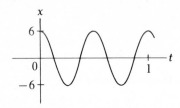

Exercises 9—3, p. 221

1. $1, 2\pi, \dfrac{\pi}{6}(R)$

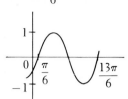

3. $1, 2\pi, \dfrac{\pi}{6}(L)$

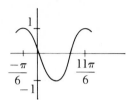

5. $2, \pi, \dfrac{\pi}{4}(L)$

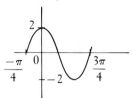

7. $1, \pi, \dfrac{\pi}{2}(R)$

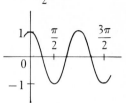

9. $\dfrac{1}{2}, 4\pi, \dfrac{\pi}{2}(R)$

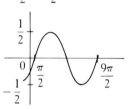

11. $3, 6\pi, \pi(L)$

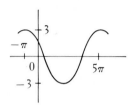

13. $1, 2, \dfrac{1}{8}(L)$

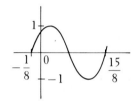

15. $\dfrac{3}{4}, \dfrac{1}{2}, \dfrac{1}{20}(R)$

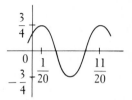

17. $0.6, 1, \dfrac{1}{2\pi}(R)$

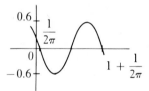

19. $4, \dfrac{2}{3}, \dfrac{2}{3\pi}(L)$

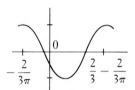

21. $1, \dfrac{2}{\pi}, \dfrac{1}{\pi}(R)$

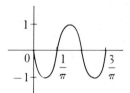

23. $\dfrac{3}{2}, 2, \dfrac{\pi}{6}(L)$

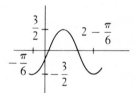

25.

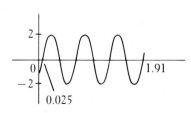

27.

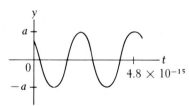

Exercises 9—4, p. 226

1. undef., −1.7, −1, −0.58, 0, 0.58,
   1, 1.7, undef., −1.7 −1, −0.58, 0

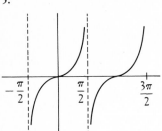

3. undef., 2, 1.4, 1.2, 1, 1.2, 1.4, 2,
   undef., −2, −1.4, −1.2, −1

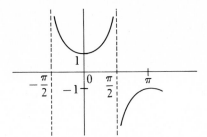

5.

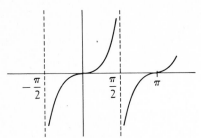

7.

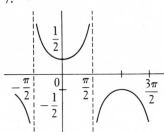

9.

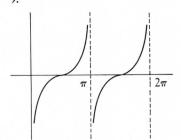

11.

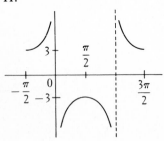

13.

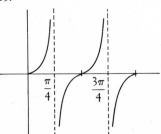

15.

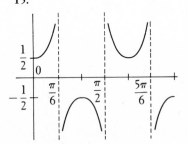

17.

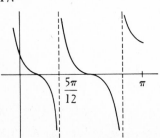

19.

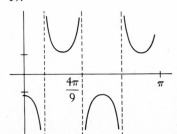

21.

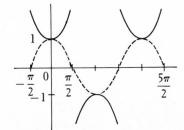

23.

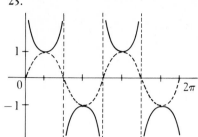

25.

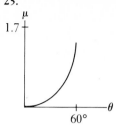

27.

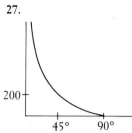

Exercises 9–5, p. 230

1.

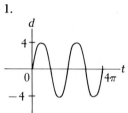

3.

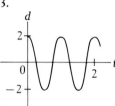

5.

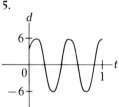

7.

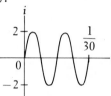

9.

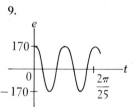

11.

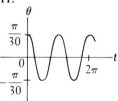

13.

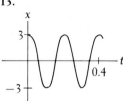

15.

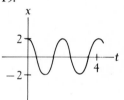

Exercises 9–6, p. 235

1.

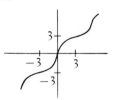

3.

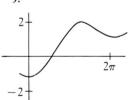

5.

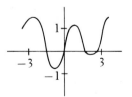

7.

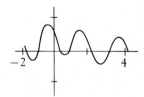

9.

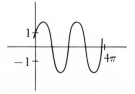

11.

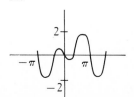

13.

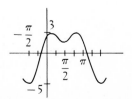

15.

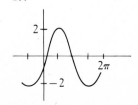

17.

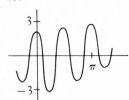

19.

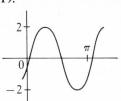

21.

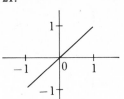

23.

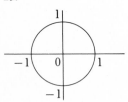

25.

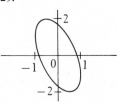

27.

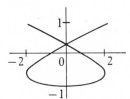

29.

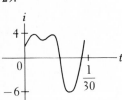

31.

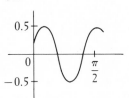

33.

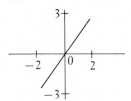

Exercises for Chapter 9, p. 236

1.

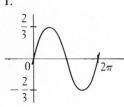

3.

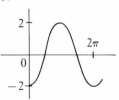

5.

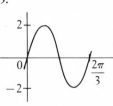

7.

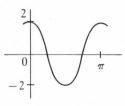

9.

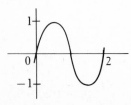

11.

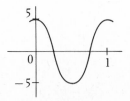

13.

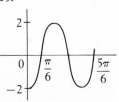

15.

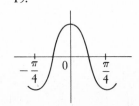

17.

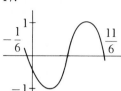

19.

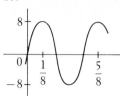

21.

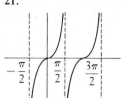

23.

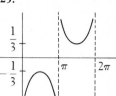

25.

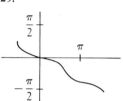

27.

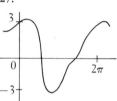

29.

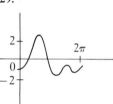

31.

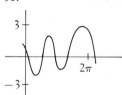

33.

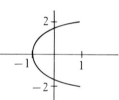

35.

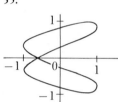

37.

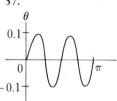

39.

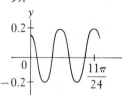

41.

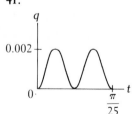

43.

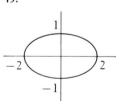

45.

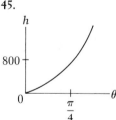

47.

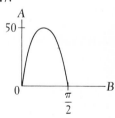

Exercises 10–1, p. 242

1. x^3 **3.** $\dfrac{1}{a^4}$ **5.** $\dfrac{4a^2}{x^2}$ **7.** $\dfrac{n^2}{5a}$ **9.** 1 **11.** -7 **13.** $\dfrac{n^3}{2}$ **15.** $\dfrac{1}{a^3 b^6}$ **17.** $\dfrac{1}{a+b}$

19. $\dfrac{2x^2 + 3y^2}{x^2 y^2}$ **21.** $\dfrac{b^3}{432a}$ **23.** $\dfrac{4}{t^4 v^4}$ **25.** $\dfrac{x^8 - y^2}{x^4 y^2}$ **27.** $\dfrac{36a^2 + 1}{9a^4}$ **29.** $\dfrac{ab}{a+b}$ **31.** $\dfrac{4n^2 - 4n + 1}{n^4}$

33. $-\dfrac{x}{y}$ **35.** $\dfrac{a^2 - ax + x^2}{ax}$ **37.** $\dfrac{t^2 + t + 2}{t^2}$ **39.** $\dfrac{2x}{(x+1)(x-1)}$ **41.** $\dfrac{rR}{r+R}$ **43.** $\dfrac{r_1 r_2}{(\mu - 1)(r_2 - r_1)}$

Exercises 10–2, p. 246

1. 5   3. 3   5. 16   7. 10^{25} 9. $\dfrac{1}{2}$ 11. $\dfrac{1}{16}$ 13. 25   15. 4096   17. $\dfrac{1}{110}$ 19. $\dfrac{6}{7}$

21. $-\dfrac{1}{2}$ 23. 24   25. $\dfrac{39}{1000}$ 27. $\dfrac{9}{100}$ 29. $a^{7/6}$ 31. $\dfrac{1}{y^{9/10}}$ 33. $s^{23/12}$ 35. $\dfrac{1}{y^{13/12}}$ 37. $2ab^2$

39. $\dfrac{1}{8a^3b^{9/4}}$ 41. $\dfrac{27}{64t^3}$ 43. $\dfrac{b^{11/10}}{2a^{1/12}}$ 45. $\dfrac{2}{3}x^{1/6}y^{11/12}$ 47. $\dfrac{x}{(x+2)^{1/2}}$ 49. $\dfrac{a^2+1}{a^4}$ 51. $\dfrac{a+1}{a^{1/2}}$

53. 4.72×10^{22} 55. 55.6%

Exercises 10–3, p. 250

1. $2\sqrt{6}$ 3. $3\sqrt{5}$ 5. $xy^2\sqrt{y}$ 7. $qr^3\sqrt{pr}$ 9. $x\sqrt{5}$ 11. $3ac^2\sqrt{2ab}$ 13. $2\sqrt[3]{2}$ 15. $2\sqrt[5]{3}$

17. $2\sqrt[3]{a^2}$ 19. $2st\sqrt[4]{4r^3t}$ 21. 2   23. $ab\sqrt[5]{b^2}$ 25. $\dfrac{1}{2}\sqrt{6}$ 27. $\dfrac{\sqrt{ab}}{b}$ 29. $\dfrac{1}{2}\sqrt[3]{6}$ 31. $\dfrac{1}{3}\sqrt[5]{27}$

33. $2\sqrt{5}$ 35. 2   37. 200   39. 2000   41. $\sqrt{2a}$ 43. $\dfrac{1}{2}\sqrt{2}$ 45. $\sqrt[3]{2}$ 47. $\sqrt[8]{2}$ 49. $\dfrac{1}{6}\sqrt{6}$

51. $\dfrac{\sqrt{b(a^2+b)}}{ab}$ 53. $a+b$ 55. $\dfrac{1}{2}\sqrt{4x^2+1}$ 57. $\dfrac{\pi\sqrt{6}}{4} = 1.92s$ 59. $\dfrac{\sqrt[3]{4MN^2\rho^2}}{2N\rho}$

Exercises 10–4, p. 252

$7\sqrt{3}$ 1. ~~$4\sqrt{5}$~~   3. $\sqrt{5}-\sqrt{7}$ 5. $3\sqrt{5}$ 7. $-4\sqrt{3}$ 9. $-2\sqrt{2}$ 11. $19\sqrt{7}$ 13. $-4\sqrt{5}$

15. $23\sqrt{3}-6\sqrt{2}$ 17. $\dfrac{7}{3}\sqrt{15}$ 19. 0   21. $13\sqrt[3]{3}$ 23. $\sqrt[4]{2}$ 25. $(a-2b^2)\sqrt{ab}$

27. $(3-2a)\sqrt{10}$ 29. $(2b-a)\sqrt[3]{3a^2b}$ 31. $\dfrac{(a^2-c^3)\sqrt{ac}}{a^2c^3}$ 33. $\dfrac{(a-2b)\sqrt[3]{ab^2}}{ab}$ 35. $\dfrac{2b\sqrt{a^2-b^2}}{b^2-a^2}$

37. $-\dfrac{b}{a}$

Exercises 10–5, p. 254

1. $\sqrt{30}$ 3. $2\sqrt{3}$ 5. 2   7. $2\sqrt[5]{2}$ 9. 50   11. 16   13. $\dfrac{1}{3}\sqrt{30}$ 15. $\dfrac{1}{33}\sqrt{165}$

17. $\sqrt{6}-\sqrt{15}$ 19. $8-12\sqrt{3}$ 21. -1 23. $39-12\sqrt{3}$ 25. $48+9\sqrt{15}$ 27. $36+13\sqrt{66}$
29. $a\sqrt{b}+\sqrt{ac}$ 31. $5n\sqrt{3}+10\sqrt{mn}$ 33. $2a-3b+2\sqrt{2ab}$ 35. $\sqrt{6}-\sqrt{10}-2$ 37. $\sqrt[8]{72}$

39. $\sqrt[12]{a^3b^7c^4}$ 41. 1   43. $2x\sqrt{x}+x\sqrt[6]{8y^4}-2\sqrt[6]{x^3y^2}-y$ 45. $\dfrac{2-a-a^2}{2a}$

47. $4x^2+x-2y-4x\sqrt{x-2y}$ (valid for $x \geq 2y$)   49. $\dfrac{c}{a}$

Exercises 10–6, p. 256

1. $\sqrt{7}$ 3. $\dfrac{1}{2}\sqrt{14}$ 5. $\dfrac{1}{6}\sqrt[3]{9x^2}$ 7. $\dfrac{1}{2}\sqrt[6]{4a^3}$ 9. $\dfrac{a\sqrt{2}-b\sqrt{a}}{a}$ 11. $\dfrac{3\sqrt{a}-\sqrt{3b}}{3}$

13. $\dfrac{1}{4}(\sqrt{7}-\sqrt{3})$ 15. $\dfrac{1}{3}(\sqrt{35}+\sqrt{14})$ 17. $-\dfrac{3}{8}(\sqrt{5}+3)$ 19. $\dfrac{1}{13}(9+\sqrt{3})$

21. $\dfrac{1}{11}(\sqrt{7}+3\sqrt{2}-6-\sqrt{14})$ 23. $\dfrac{1}{13}(4-\sqrt{3})$ 25. $\dfrac{1}{17}(-56+9\sqrt{15})$ 27. $\dfrac{1}{14}(4-\sqrt{2})$

29. $\dfrac{8(3\sqrt{a}+2\sqrt{b})}{9a-4b}$ 31. $-\dfrac{\sqrt{x^2-y^2}+\sqrt{x^2+xy}}{y}$ 33. $km\sqrt{m(E-E_1)}$

Exercises for Chapter 10, p. 257

1. $\dfrac{2}{a^2}$ 3. $\dfrac{2d^3}{c}$ 5. 375 7. $\dfrac{1}{8000}$ 9. $\dfrac{t^4}{9}$ 11. -28 13. $64a^2b^5$ 15. $-8m^9n^6$

17. $\dfrac{2y-x^2}{x^2y}$ 19. $\dfrac{2y}{x+y}$ 21. $\dfrac{b}{ab-3}$ 23. $\dfrac{(x^3y^3-1)^{1/3}}{y}$ 25. $4a(a^2+4)^{1/2}$ 27. $\dfrac{-2(x+1)}{(x-1)^3}$

29. $2\sqrt{17}$ 31. $b^2c\sqrt{ab}$ 33. $3ab^2\sqrt{a}$ 35. $2tu\sqrt{21st}$ 37. $\dfrac{5\sqrt{2s}}{2s}$ 39. $\dfrac{1}{9}\sqrt{33}$ 41. $mn^2\sqrt[4]{8m^2n}$

43. $\sqrt{2}$ 45. $14\sqrt{2}$ 47. $-7\sqrt{7}$ 49. $3ax\sqrt{2x}$ 51. $(2a+b)\sqrt[3]{a}$ 53. $10-\sqrt{55}$

55. $4\sqrt{3}-4\sqrt{5}$ 57. $-45-7\sqrt{17}$ 59. $33-7\sqrt{21}$ 61. $-\dfrac{8+\sqrt{6}}{29}$ 63. $\dfrac{13-2\sqrt{35}}{29}$

65. $\dfrac{\sqrt{a^2b^2+a}}{a}$ 67. $\dfrac{15-2\sqrt{15}}{4}$ 69. $\dfrac{e^{i\alpha t}}{e^{i\omega t}}$ 71. 0.011 73. $\dfrac{\sqrt{3RMT}}{M}$; 483 m/s

75. $\dfrac{\sqrt{LC_1C_2(C_1+C_2)}}{2\pi LC_1C_2}$

Exercises 11—1, p. 263

1. $9j$ 3. $-2j$ 5. $2\sqrt{2}j$ 7. $\dfrac{1}{2}\sqrt{7}j$ 9. $-7;7$ 11. $4;-4$ 13. $-j$ 15. 1 17. 0

19. $-2j$ 21. $2+3j$ 23. $-2+3j$ 25. $3\sqrt{2}-2\sqrt{2}j$ 27. -1 29. $6+7j$ 31. $-2j$
33. $x=2,\ y=-2$ 35. $x=10,\ y=-6$ 37. $x=0,\ y=-1$ 39. $x=-2,\ y=3$
41. It is a real number.

Exercises 11—2, p. 266

1. $5-8j$ 3. $-9+6j$ 5. $7-5j$ 7. $-5j$ 9. -1 11. $-8+21j$ 13. $7+49j$
15. $-8-20j$ 17. $22+3j$ 19. $-42-6j$ 21. $-18\sqrt{2}j$ 23. $25\sqrt{5}j$ 25. $3\sqrt{3}j$
27. $-4\sqrt{3}+6\sqrt{7}j$ 29. $-28j$ 31. $3\sqrt{7}+3j$ 33. $-40-42j$ 35. $-2-2j$

37. $\dfrac{1}{29}(-30+12j)$ 39. $\dfrac{2}{37}(6+j)$ 41. $-j$ 43. $\dfrac{1}{11}(-13+8\sqrt{2}j)$ 45. $\dfrac{1}{3}(2+5\sqrt{2}j)$

47. $\dfrac{1}{5}(-1+3j)$ 49. $(a+bj)+(a-bj)=2a$ 51. $(a+bj)-(a-bj)=2bj$

Exercises 11—3, p. 268

1.   3.   5. $8+j$ 7. $-3j$

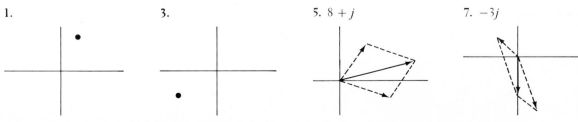

9. $-1 + 4j$

11. $-2 + 3j$

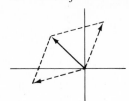

13. $7 + j$

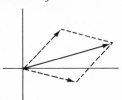

15. $4 - 11j$

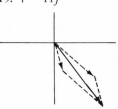

17. $-2j$

19. $-13 + j$

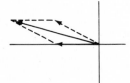

21.

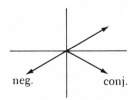

neg.   conj.

23.

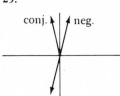

conj.   neg.

Exercises 11—4, p. 271

1. $10(\cos 36.9° + j \sin 36.9°)$

3. $5(\cos 306.9° + j \sin 306.9°)$

5. $3.61(\cos 123.7° + j \sin 123.7°)$

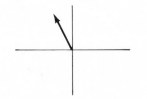

7. $6.00(\cos 203.6° + j \sin 203.6°)$

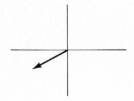

9. $2(\cos 60° + j \sin 60°)$

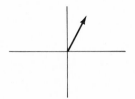

11. $8.01(\cos 295.9° + j \sin 295.9°)$

13. $3(\cos 180° + j \sin 180°)$

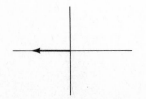

15. $9(\cos 90° + j \sin 90°)$

17. $2.94 + 4.05j$

19. $-1.39 + 0.80j$

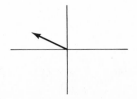

21. $9.68 - 2.50j$

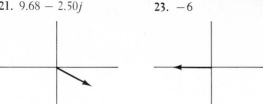

23. -6

25. 8

27. $-4.71 - 0.595j$

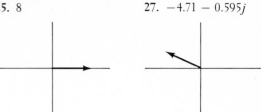

29. $-0.500 - 0.866j$

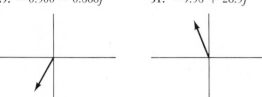

31. $-9.58 + 26.3j$

33. $R = 0.200,\ \theta = 53.1°$

Exercises 11—5, p. 274

1. $3.00e^{1.05j}$ **3.** $4.50e^{4.92j}$ **5.** $375e^{1.66j}$ **7.** $0.515e^{3.46j}$ **9.** $5.00e^{5.36j}$ **11.** $3.61e^{2.55j}$
13. $6.37e^{0.386j}$ **15.** $825e^{3.84j}$ **17.** $3.00(\cos 28.6° + j \sin 28.6°);\ 2.63 + 1.44j$
19. $4.64(\cos 106.0° + j \sin 106.0°);\ -1.28 + 4.46j$ **21.** $3.20(\cos 310.0° + j \sin 310.0°);\ 2.06 - 2.45j$
23. $0.172(\cos 136.9° + j \sin 136.9°);\ -0.126 + 0.117j$ **25.** $0.546e^{0.415j};\ 0.546$ A

Exercises 11—6, p. 279

1. $8(\cos 80° + j \sin 80°)$ **3.** $3(\cos 250° + j \sin 250°)$ **5.** $2(\cos 35° + j \sin 35°)$
7. $2.4(\cos 110° + j \sin 110°)$ **9.** $8(\cos 105° + j \sin 105°)$ **11.** $256(\cos 0° + j \sin 0°)$
13. $65.0(\cos 345.7° + j \sin 345.7°) = 63 - 16j$ **15.** $61.4(\cos 343.9° + j \sin 343.9°);\ 59 - 17j$

17. $2.21(\cos 71.6° + j \sin 71.6°);\ \dfrac{7}{10} + \dfrac{21}{10}j$ **19.** $0.385(\cos 120.5° + j \sin 120.5°) = \dfrac{1}{169}(-33 + 56j)$

21. $625(\cos 212.4° + j \sin 212.4°) = -527 - 336j$ **23.** $609(\cos 281.5° + j \sin 281.5°);\ 122 - 597j$
25. $2(\cos 30° + j \sin 30°),\ 2(\cos 210° + j \sin 210°)$ **27.** $-0.364 + 1.67j,\ -1.26 - 1.15j,\ 1.63 - 0.520j$

29. $1,\ -1,\ j,\ -j$ **31.** $j,\ \dfrac{1}{2}(\sqrt{3} + j),\ \dfrac{1}{2}(\sqrt{3} - j)$

33. $\left[\dfrac{1}{2}(1 - \sqrt{3}j)\right]^{3} = \dfrac{1}{8}[1 - 3(\sqrt{3}j) + 3(\sqrt{3}j)^{2} - (\sqrt{3}j)^{3}] = \dfrac{1}{8}[1 - 3\sqrt{3}j - 9 + 3\sqrt{3}j] = \dfrac{1}{8}(-8) = -1$

Exercises 11—7, p. 285

1. 30.0 V   **3.** (a) $18.0\ \Omega$ (b) $-56.3°$ (c) 54.0 V   **5.** (a) $3.00\ \Omega$ (b) $-90.0°$
7. (a) $11.7\ \Omega$ (b) $-59.0°$ **9.** 44.4 V   **11.** $2000\ \Omega,\ 3000\ \Omega$ **13.** $4120\ \Omega$, voltage leads by $14.0°$
15. 2.00×10^{-4} F   **17.** 9.41 W

Exercises for Chapter 11, p. 286

1. $10 - j$ **3.** $6 + 2j$ **5.** $9 + 2j$ **7.** $-12 + 66j$ **9.** $\dfrac{1}{85}(21 + 18j)$ **11.** $\dfrac{1}{53}(50 - 16j)$

13. $\dfrac{1}{10}(-12 + 9j)$ **15.** $\dfrac{1}{5}(13 + 11j)$ **17.** $x = -\dfrac{2}{3},\ y = -2$ **19.** $x = -\dfrac{1}{2},\ y = \dfrac{1}{2}$

21. $3 + 11j$

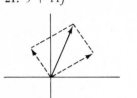

23. $4 + 8j$

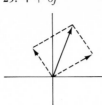

25. $1.41(\cos 315° + j \sin 315°) = 1.41e^{5.50j}$
27. $7.28(\cos 254.1° + j \sin 254.1°) = 7.28e^{4.43j}$
29. $4.67(\cos 76.7° + j \sin 76.7°); 4.67e^{1.34j}$
31. $10(\cos 0° + j \sin 0°); 10e^{0j}$ **33.** $-1.41 - 1.41j$
35. $-2.72 + 4.19j$ **37.** $1.94 + 0.495j$
39. $-1.74 + 5.08j$

41. $15.0(\cos 84.0° + j \sin 84.0°)$ **43.** $8.00(\cos 59.0° + j \sin 59.0°)$
45. $1020(\cos 160.0° + j \sin 160.0°)$ **47.** $27.0(\cos 331.5° + j \sin 331.5°)$

49. $32(\cos 270° + j \sin 270°) = -32j$ **51.** $\dfrac{625}{2}(\cos 270° + j \sin 270°) = -\dfrac{625}{2}j$

53. $1.00 + 1.73j, -2, 1.00 - 1.73j$
55. $\cos 67.5° + j \sin 67.5°, \cos 157.5° + j \sin 157.5°, \cos 247.5° + j \sin 247.5°, \cos 337.5° + j \sin 337.5°$
57. $12.5 \ \Omega, -53.1°$ **59.** $10.2 \ \Omega$, current leads by $8.7°$ **61.** 36.1 V   **63.** 814 lb, $317.5°$

6.5 $\dfrac{u - j\omega n}{u^2 - \omega^2 n^2}$ **67.** $e^{j\pi} = \cos \pi + j \sin \pi = -1$

Exercises 12–1, p. 290

1. $\log_3 27 = 3$ **3.** $\log_4 256 = 4$ **5.** $\log_4\left(\dfrac{1}{16}\right) = -2$ **7.** $\log_2\left(\dfrac{1}{64}\right) = -6$ **9.** $\log_8 2 = \dfrac{1}{3}$

11. $\log_{1/4}\left(\dfrac{1}{16}\right) = 2$ **13.** $81 = 3^4$ **15.** $9 = 9^1$ **17.** $5 = 25^{1/2}$ **19.** $3 = 243^{1/5}$ **21.** $0.1 = 10^{-1}$

23. $16 = (0.5)^{-4}$ **25.** 2 **27.** -2 **29.** 343 **31.** $\dfrac{1}{4}$ **33.** 9 **35.** $\dfrac{1}{64}$ **37.** 0.2 **39.** -3

41. $t = \log_2\left(\dfrac{N}{1000}\right)$ **43.** $N = N_0 e^{-kt}$

Exercises 12–2, p. 293

1.

3.

5.

7.

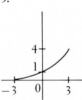

9.

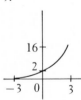

11.

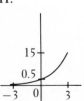

13.

15.

17.

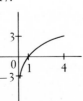

19.

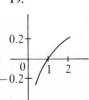

21.   **23.** 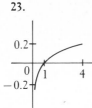   **25.** 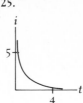   **27.**

Exercises 12–3, p. 297

1. $\log_5 x + \log_5 y$ **3.** $\log_7 5 - \log_7 a$ **5.** $3\log_2 a$ **7.** $\log_6 a + \log_6 b + \log_6 c$ **9.** $\frac{1}{4}\log_5 y$

11. $\frac{1}{2}\log_2 x - 2\log_2 a$ **13.** $\log_b ac$ **15.** $\log_5 3$ **17.** $\log_b x^{3/2}$ **19.** $\log_e 4n^3$ **21.** -5 **23.** 2.5

25. $\frac{1}{2}$ **27.** $\frac{3}{4}$ **29.** $2 + \log_3 2$ **31.** $-1 - \log_2 3$ **33.** $\frac{1}{2}(1 + \log_3 2)$ **35.** $4 + 3\log_2 3$

37. $3 + \log_{10} 3$ **39.** $-2 + 3\log_{10} 3$ **41.** $y = 2x$ **43.** $y = \dfrac{x}{5}$ **45.** $y = \dfrac{49}{x^3}$ **47.** $y = 2(2ax)^{1/5}$

49. 0.602 **51.** -0.301 **53.**   **55.** $S = \log_e\left(\dfrac{T^c}{P^{nR}}\right)$

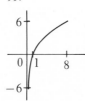

Exercises 12–4, p. 301

1. 2.7536   **3.** 8.8062 − 10   **5.** 6.9657   **7.** 6.0682 − 10   **9.** 0.0224   **11.** 9.3773 − 10
13. 8.8652   **15.** 8.6512 − 10   **17.** 27400   **19.** 0.0496   **21.** 2000   **23.** 0.724   **25.** 89.02
27. 4.065×10^{-4} **29.** 1.427   **31.** 0.005788   **33.** 9.0607   **35.** 8.8751 − 10   **37.** 5.1761
39. 1.0000 − 100

Exercises 12–5, p. 304

1. 85.50   **3.** 0.03742   **5.** 98.50   **7.** 1.757   **9.** 94,580   **11.** 3.844   **13.** 1.500   **15.** 25.33
17. 3.740   **19.** 0.003190   **21.** 1011   **23.** 1.46×10^9 **25.** 80,200 m²   **27.** 752.6 mg
29. 332.4 m/s   **31.** 0.2319 ft

Exercises 12–6, p. 308

1. 9.5767 − 10   **3.** 0.1072   **5.** 9.7916 − 10   **7.** 0.1834   **9.** 9.8982 − 10   **11.** 9.2766 − 10
13. 28°10′   **15.** 50°16′   **17.** 48°56′   **19.** 1°40′ or 1°50′   **21.** $a = 85.35$, $b = 11.87$, $B = 7°55′$
23. $b = 9506$, $C = 42°10′$, $c = 6703$ **25.** $a = 12.22$, $C = 68°8′$, $c = 12.31$
27. $A = 37°58′$, $B = 52°2′$, $c = 485.3$ **29.** 30.50 ft   **31.** 21.85 lb, 52.24 lb   **33.** 19.97 lb
35. −32°3′

Exercises 12–7, p. 312

1. 3.258   **3.** 0.4447   **5.** −0.6912   **7.** −4.917   **9.** 1.921   **11.** 3.418   **13.** 1.933   **15.** 1.795
17. 3.940   **19.** 0.3293   **21.** −0.00904   **23.** −4.351   **25.** 0.4314   **27.** 1.940   **29.** 12,000
31. 8.7   **33.** $y = 3x$ **35.** 8.155%   **37.** 0.384 s   **39.** 257 yr

Exercises 12—8, p. 316

1. 4   3. 0.861   5. 0.285   7. 0.203   9. 4.11   11. 14.2   13. 4   15. 4   17. −0.162
19. 2   21. 4   23. 1.42   25. 0.00203 s   27. 10^{11} 29. $10^{8.25} = 1.78 \times 10^8$ 31. 3.922×10^{-4}
33. 3.35   35. 3.50

Exercises 12—9, p. 320

1.

3.

5.

7.

9.

11.

13.

15.

17.

19.

21.

23.

25.
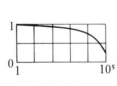

27.

29.

Exercises for Chapter 12, p. 321

1. 10,000   3. $\frac{1}{5}$ 5. 6   7. $\frac{5}{3}$ 9. 6   11. 100   13. $\log_3 2 + \log_3 x$ 15. $2 \log_3 t$

17. $2 + \log_2 7$ 19. $2 - \log_3 x$ 21. $1 + \frac{1}{2}\log_4 3$ 23. $3 + 4 \log_{10} x$ 25. $y = \frac{4}{x}$ 27. $y = \frac{8}{x}$

29.

31.

33. 9.423   35. 122.8   37. 1.180×10^{15}
39. 2.037   41. 4.771   43. 29.28
45. $B = 53°40'$, $b = 21.21$, $c = 26.33$
47. $b = 99.73$, $C = 67°50'$, $c = 100.2$ 49. 2.181
51. 0.7276   53. 4.30   55. 30

57.

59.

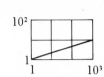

61. $P = 10^{(a+bT)/T}$ **63.** 2166 cm³   **65.** 52.84%
67. 477.1 ft   **69.** −2.506 V   **71.** 4.67 g-mol/L
73. 1985   **75.**

Exercises 13—1, p. 328

1. $x = 1.8$, $y = 3.6$; $x = -1.8$, $y = -3.6$ **3.** $x = 0.0$, $y = -2.0$; $x = 2.7$, $y = -0.7$
5. $x = 1.5$, $y = 0.2$ **7.** $x = 1.6$, $y = 2.5$
9. $x = 1.1$, $y = 2.8$; $x = -1.1$, $y = 2.8$; $x = 2.4$, $y = -1.8$; $x = -2.4$, $y = -1.8$ **11.** No solution
13. $x = -2.8$, $y = -1.0$; $x = 2.8$, $y = 1.0$; $x = 2.8$, $y = -1.0$; $x = -2.8$, $y = -1.0$
15. $x = 0.7$, $y = 0.7$; $x = -0.7$, $y = -0.7$ **17.** $x = 0.0$, $y = 0.0$; $x = 0.9$, $y = 0.8$ **19.** $x = 1.0$, $y = 0.0$
21. 360 m, 210 m   **23.** 1.3 A, 1.7 A; 2.3 A, 0.7 A

Exercises 13—2, p. 331

1. $x = 0$, $y = 1$; $x = 1$, $y = 2$ **3.** $x = -\dfrac{19}{5}$, $y = \dfrac{17}{5}$; $x = 5$, $y = -1$

5. $x = \dfrac{2}{7}(3 + \sqrt{2})$, $y = \dfrac{2}{7}(-1 + 2\sqrt{2})$; $x = \dfrac{2}{7}(3 - \sqrt{2})$, $y = \dfrac{2}{7}(-1 - 2\sqrt{2})$

7. $x = \dfrac{2}{3}$, $y = \dfrac{9}{2}$; $x = -3$, $y = -1$ **9.** $x = -2$, $y = 4$; $x = 2$, $y = 4$

11. $x = 1$, $y = 2$; $x = -1$, $y = 2$ **13.** $x = 1$, $y = 0$; $x = -1$, $y = 0$; $x = \dfrac{1}{2}\sqrt{6}$, $y = \dfrac{1}{2}$; $x = -\dfrac{1}{2}\sqrt{6}$, $y = \dfrac{1}{2}$

15. $x = \sqrt{19}$, $y = \sqrt{6}$; $x = \sqrt{19}$, $y = -\sqrt{6}$; $x = -\sqrt{19}$; $y = \sqrt{6}$; $x = -\sqrt{19}$, $y = -\sqrt{6}$

17. $x = \dfrac{1}{11}\sqrt{22}$, $y = \dfrac{1}{11}\sqrt{770}$; $x = \dfrac{1}{11}\sqrt{22}$, $y = -\dfrac{1}{11}\sqrt{770}$; $x = -\dfrac{1}{11}\sqrt{22}$, $y = \dfrac{1}{11}\sqrt{770}$; $x = -\dfrac{1}{11}\sqrt{22}$,

$y = -\dfrac{1}{11}\sqrt{770}$ **19.** $x = -5$, $y = -2$; $x = -5$, $y = 2$; $x = 5$, $y = -2$; $x = 5$, $y = 2$

21. 0 ft, 0 s; 50 ft, 1.25 s   **23.** 8, 13   **25.** 4.47 km, 8.94 km   **27.** 8 in., 12 in.

Exercises 13—3, p. 335

1. $-3, -2, 2, 3$ **3.** $1, -2$ **5.** $-\dfrac{1}{2}, \dfrac{1}{4}$ **7.** $-\dfrac{1}{2}, \dfrac{1}{2}$ **9.** $1, 9$ **11.** $\dfrac{64}{729}, 1$

13. $-27, 125$ **15.** 5 **17.** $-2, -1, 3, 4$ **19.** 18 **21.** 2000 **23.** 24 ft by 32 ft

Exercises 13—4, p. 338

1. 12 **3.** 2 **5.** $2, 3$ **7.** 32 **9.** 16 **11.** 9 **13.** 12 **15.** $7, -1$ **17.** 0 **19.** 5 **21.** 6

23. 258 **25.** $\dfrac{v_0^2 - v^2}{2g}$ **27.** $\dfrac{k^2}{2n(1-k)}$ **29.** 9 in. by 12 in.   **31.** 5.20 mi

Exercises for Chapter 13, p. 339

1. $x = -0.9$, $y = 3.5$; $x = 0.8$, $y = 2.6$ **3.** $x = 2.0$, $y = 0.0$; $x = 1.6$, $y = 0.6$
5. $x = 0.8$, $y = 1.6$; $x = -0.8$, $y = 1.6$ **7.** $x = -1.2$, $y = 2.7$; $x = 1.2$, $y = 2.7$
9. $x = 0$, $y = 0$; $x = 2$, $y = 16$

11. $x = \dfrac{1}{12}(1 + \sqrt{97})$, $y = \dfrac{1}{18}(5 - \sqrt{97})$; $x = \dfrac{1}{12}(1 - \sqrt{97})$, $y = \dfrac{1}{18}(5 + \sqrt{97})$

13. $x = 7$, $y = 5$; $x = 7$, $y = -5$; $x = -7$, $y = 5$; $x = -7$, $y = -5$

15. $x = -2$, $y = 2$; $x = \dfrac{2}{3}$, $y = \dfrac{10}{9}$ **17.** $-4, -2, 2, 4$ **19.** $1, 16$ **21.** $\dfrac{1}{3}, -\dfrac{1}{7}$ **23.** $\dfrac{25}{4}$ **25.** 11

27. 8 **29.** $\dfrac{9}{16}$ **31.** 7 **33.** $r = \sqrt{\dfrac{\sqrt{\pi^2 h^4 + 4S^2} - \pi h^2}{2\pi}}$ **35.** 0.75 s, 1.5 s   **37.** $x = 16$, $y = 15$

39. 493 ft

Exercises 14–1, p. 343

1. 0 **3.** 0 **5.** -40 **7.** 46 **9.** 8 **11.** 183 **13.** -28 **15.** 14 **17.** Yes   **19.** Yes
21. No   **23.** No   **25.** Yes   **27.** Yes   **29.** $(4x^3 + 8x^2 - x - 2) \div (2x - 1) = 2x^2 + 5x + 2$; No

Exercises 14–2, p. 347

1. $x^2 + 3x + 2$, $R = 0$ **3.** $x^2 - x + 3$, $R = 0$ **5.** $2x^4 - 4x^3 + 8x^2 - 17x + 42$, $R = -40$
7. $2x^3 + 3x^2 + 5x + 17$, $R = 46$ **9.** $x^2 + x - 4$, $R = 8$ **11.** $x^3 - 3x^2 + 10x - 45$, $R = 183$
13. $2x^3 - x^2 - 4x - 12$, $R = -28$ **15.** $x^4 + 2x^3 + x^2 + 7x + 4$, $R = 14$
17. $x^5 + 2x^4 + 4x^3 + 8x^2 + 18x + 36$, $R = 66$ **19.** $x^6 + 2x^5 + 4x^4 + 8x^3 + 16x^2 + 32x + 64$, $R = 0$
21. Yes   **23.** No   **25.** Yes   **27.** No   **29.** Yes   **31.** No

Exercises 14–3, p. 351
(Note: Unknown roots listed)

1. $-2, -1$ **3.** $-2, 3$ **5.** $-2, -2$ **7.** $-j, -\dfrac{2}{3}$ **9.** $2j, -2j$ **11.** $3, -1$ **13.** $-2, 1$

15. $\dfrac{1}{4}(1 + \sqrt{17})$, $\dfrac{1}{4}(1 - \sqrt{17})$ **17.** $2, -3$ **19.** $-2j, 3, -3$

Exercises 14–4, p. 356

1. $1, -1, -2$ **3.** $2, -1, -3$ **5.** $\dfrac{1}{2}, 5, -3$ **7.** $\dfrac{1}{3}, -3, -1$ **9.** $-2, -2, 2 \pm \sqrt{3}$

11. $2, 4, -1, -3$ **13.** $1, -\dfrac{1}{2}, 1 \pm \sqrt{3}$ **15.** $\dfrac{1}{2}, -\dfrac{2}{3}, -3, -\dfrac{1}{2}$ **17.** $2, 2, -1, -1, -3$

19. $\dfrac{1}{2}, 1, 1, j, -j$ **21.** $\dfrac{15}{2}$ **23.** a **25.** 2 in., or 1.17 in.

Exercises 14–5, p. 359

1. 0.59 **3.** 0.38 **5.** 1.71 **7.** -0.77 **9.** 2.56 **11.** 3.24 **13.** 0.68 **15.** $2.30, -1.30$
17. 7.01 ft   **19.** 0.0763 m

Exercises for Chapter 14, p. 359

1. 1 **3.** -107 **5.** Yes   **7.** No   **9.** $x^2 + 4x + 10$, $R = 11$ **11.** $2x^2 - 7x + 10$, $R = -17$
13. $x^3 - 3x^2 - 4$, $R = -4$ **15.** $2x^4 + 10x^3 + 4x^2 + 21x + 105$, $R = 516$ **17.** No   **19.** Yes
21. (Unlisted roots) $\dfrac{1}{2}(-5 \pm \sqrt{17})$ **23.** (Unlisted roots) $\dfrac{1}{3}(-1 \pm \sqrt{14}j)$ **25.** (Unlisted roots), $4, -3$

27. (Unlisted roots) $2, -2$ **29.** $1, 2, -4$ **31.** $-1, -1, \dfrac{5}{2}$ **33.** $\dfrac{5}{3}, -\dfrac{1}{2}, -1$

35. $\frac{1}{2}, -1, \sqrt{2}j, -\sqrt{2}j$ **37.** 0.75   **39.** 1.91   **41.** 5   **43.** 1, 0.4, 1.5, -0.6 (last three are irrational)

45. 7 ft, 7 ft, 11 ft   **47.** 1.64

Exercises 15—1, p. 368
1. 39   **3.** 30   **5.** 50   **7.** -40 **9.** -6 **11.** -86 **13.** 118   **15.** -2

17. $x = 2, y = -1, z = 3$ **19.** $x = -1, y = \frac{1}{3}, z = -\frac{1}{2}$ **21.** $x = -1, y = 0, z = 2, t = 1$

23. $x = 1, y = 2, z = -1, t = 3$ **25.** $\frac{2}{7}$ A, $\frac{18}{7}$ A, $-\frac{8}{7}$ A, $-\frac{12}{7}$ A   **27.** 100 mL, 300 mL, 200 mL

Exercises 15—2, p. 375
1. -60 **3.** -56 **5.** 0   **7.** 0   **9.** 57   **11.** -124 **13.** -13 **15.** -118 **17.** -72

19. 0   **21.** $x = 0, y = -1, z = 4$ **23.** $x = \frac{1}{3}, y = -\frac{1}{2}, z = 1$ **25.** $x = 2, y = -1, z = -1, t = 3$

27. $x = 1, y = 2, z = -1, t = -2$ **29.** $\frac{33}{16}$ A, $\frac{11}{8}$ A, $-\frac{5}{8}$ A, $-\frac{15}{8}$ A, $-\frac{15}{16}$ A

31. PPM of SO_2, NO, NO_2, CO: 0.5, 0.3, 0.2, 5.0

Exercises 15—3, p. 380

1. $a = 1, b = -3, c = 4, d = 7$ **3.** $x = 2, y = 3$ **5.** $\begin{pmatrix} 1 & 10 \\ 0 & 2 \end{pmatrix}$ **7.** $\begin{pmatrix} 0 & 0 \\ 1 & 6 \\ 1 & 0 \end{pmatrix}$

9. $\begin{pmatrix} 0 & 9 & -13 & 3 \\ 6 & -7 & 7 & 0 \end{pmatrix}$ **11.** Cannot be added   **13.** $\begin{pmatrix} -1 & 13 & -20 & 3 \\ 8 & -13 & 6 & 2 \end{pmatrix}$

15. $\begin{pmatrix} -3 & -6 & 5 & -6 \\ -6 & -4 & -17 & 6 \end{pmatrix}$ **17.** $A + B = B + A = \begin{pmatrix} 3 & 1 & 0 & 7 \\ 5 & -3 & -2 & 5 \\ 10 & 10 & 8 & 0 \end{pmatrix}$

19. $-(A - B) = B - A = \begin{pmatrix} 5 & -3 & -6 & -7 \\ 5 & 3 & 0 & -3 \\ -8 & 12 & 8 & 4 \end{pmatrix}$ **21.** $\begin{pmatrix} 24 & 18 & 0 & 0 \\ 15 & 12 & 9 & 0 \\ 0 & 9 & 15 & 18 \end{pmatrix}$

Exercises 15—4, p. 386

1. $(-8 \quad -12)$ **3.** $\begin{pmatrix} -15 & 15 & -26 \\ 8 & 5 & -13 \end{pmatrix}$ **5.** $\begin{pmatrix} 29 \\ -29 \end{pmatrix}$ **7.** $\begin{pmatrix} -7 & -5 \\ 8 & 0 \\ 14 & 10 \end{pmatrix}$

9. $\begin{pmatrix} 33 & -22 & 7 \\ 31 & -12 & 5 \\ 15 & 13 & -1 \\ 50 & -41 & 12 \end{pmatrix}$ **11.** $\begin{pmatrix} -62 & 68 \\ 73 & -27 \end{pmatrix}$ **13.** $AB = (40)$, $BA = \begin{pmatrix} -1 & 3 & -8 \\ 5 & -15 & 40 \\ 7 & -21 & 56 \end{pmatrix}$

15. $AB = \begin{pmatrix} -5 \\ 10 \end{pmatrix}$, BA not defined   17. $AI = IA = A$ 19. $AI = IA = A$ 21. $B = A^{-1}$

23. $B = A^{-1}$ 25. Yes   27. No

29. 800 ft of brass pipe, 1000 ft of steel pipe; 1620 ft of brass pipe, 2240 ft of steel pipe

Exercises 15—5, p. 393

1. $\begin{pmatrix} -2 & -\dfrac{5}{2} \\ -1 & -1 \end{pmatrix}$
3. $\begin{pmatrix} -\dfrac{1}{3} & \dfrac{1}{6} \\ \dfrac{2}{15} & \dfrac{1}{30} \end{pmatrix}$
5. $\begin{pmatrix} \dfrac{3}{4} & \dfrac{1}{2} \\ -\dfrac{1}{4} & 0 \end{pmatrix}$
7. $\begin{pmatrix} -\dfrac{1}{4} & -\dfrac{1}{8} \\ \dfrac{1}{16} & \dfrac{5}{32} \end{pmatrix}$
9. $\begin{pmatrix} -3 & 2 \\ 2 & -1 \end{pmatrix}$

11. $\begin{pmatrix} -\dfrac{1}{2} & -2 \\ \dfrac{1}{2} & 1 \end{pmatrix}$
13. $\begin{pmatrix} \dfrac{2}{9} & -\dfrac{5}{9} \\ \dfrac{1}{9} & \dfrac{2}{9} \end{pmatrix}$
15. $\begin{pmatrix} \dfrac{3}{8} & \dfrac{1}{16} \\ -\dfrac{1}{4} & \dfrac{1}{8} \end{pmatrix}$
17. $\begin{pmatrix} -18 & -7 & 5 \\ -3 & -1 & 1 \\ -5 & -2 & 1 \end{pmatrix}$
19. $\begin{pmatrix} 3 & -4 & -1 \\ -4 & 5 & 2 \\ 2 & -3 & -1 \end{pmatrix}$

21. $\begin{pmatrix} 2 & 4 & \dfrac{7}{2} \\ -1 & -2 & -\dfrac{3}{2} \\ 1 & 1 & \dfrac{1}{2} \end{pmatrix}$
23. $\begin{pmatrix} \dfrac{5}{2} & -2 & -2 \\ -1 & 1 & 1 \\ \dfrac{7}{4} & -\dfrac{3}{2} & -1 \end{pmatrix}$
25. $\begin{pmatrix} 2 & 4 & \dfrac{7}{2} \\ -1 & -2 & -\dfrac{3}{2} \\ 1 & 1 & \dfrac{1}{2} \end{pmatrix}$
27. $\begin{pmatrix} \dfrac{5}{2} & -2 & -2 \\ -1 & 1 & 1 \\ \dfrac{7}{4} & -\dfrac{3}{2} & -1 \end{pmatrix}$

Exercises 15—6, p. 398

1. $x = \dfrac{1}{2}$, $y = 3$ 3. $x = \dfrac{1}{2}$, $y = -\dfrac{5}{2}$ 5. $x = -4$, $y = 2$, $z = -1$ 7. $x = -1$, $y = 0$, $z = 3$

9. $x = 1$, $y = 2$ 11. $x = -\dfrac{3}{2}$, $y = -2$ 13. $x = -3$, $y = -\dfrac{1}{2}$ 15. $x = \dfrac{4}{5}$, $y = -\dfrac{22}{5}$

17. $x = 2$, $y = -4$, $x = 1$ 19. $x = 2$, $y = -\dfrac{1}{2}$, $x = 3$ 21. 10 lb, 8 lb, 12 lb

23. 8 of type A, 10 of type B

Exercises for Chapter 15, p. 399

1. 6   3. 186   5. −438   7. 44   9. 6   11. 186   13. −438   15. 44   17. −9

19. −44   21. $\begin{pmatrix} 1 & -3 \\ 8 & -5 \\ -8 & -2 \\ 3 & -10 \end{pmatrix}$
23. $\begin{pmatrix} 3 & 0 \\ -12 & 18 \\ 9 & 6 \\ -3 & 21 \end{pmatrix}$
25. Cannot be subtracted   27. $\begin{pmatrix} 7 & -6 \\ -4 & 20 \\ -1 & 6 \\ 1 & 15 \end{pmatrix}$

29. $\begin{pmatrix} 13 \\ -13 \end{pmatrix}$
31. $\begin{pmatrix} 34 & 11 & -5 \\ 2 & -8 & 10 \\ -1 & -17 & 20 \end{pmatrix}$
33. $\begin{pmatrix} -2 & \dfrac{5}{2} \\ -1 & 1 \end{pmatrix}$
35. $\begin{pmatrix} \dfrac{2}{15} & \dfrac{1}{60} \\ -\dfrac{1}{15} & \dfrac{7}{60} \end{pmatrix}$
37. $\begin{pmatrix} 11 & 10 & 3 \\ -4 & -4 & -1 \\ 3 & 3 & 1 \end{pmatrix}$

39. $\begin{pmatrix} \frac{1}{2} & -\frac{1}{2} & -1 \\ -3 & 2 & 1 \\ -4 & 3 & 2 \end{pmatrix}$ **41.** $x = -3, y = 1$ **43.** $x = -2, y = 7$ **45.** $x = -1, y = -3, z = 0$

47. $x = 1, y = \frac{1}{2}, x = -\frac{1}{3}$ **49.** $x = 3, y = 1, z = -1$ **51.** $x = 1, y = 2, z = -3, t = 1$

53. $F_1 = 12$ lb, $F_2 = 9$ lb   **55.** $p_1 = \$60, p_2 = \$50, p_3 = \$90$ **57.** 30 g, 50 g, 20 g

59. 70°, 80°, 100°, 110°   **61.** $\begin{pmatrix} 15{,}000 & 10{,}000 \\ 20{,}000 & 18{,}000 \\ 8{,}000 & 30{,}000 \end{pmatrix} + \begin{pmatrix} 18{,}000 & 12{,}000 \\ 30{,}000 & 22{,}000 \\ 12{,}000 & 40{,}000 \end{pmatrix} = \begin{pmatrix} 33{,}000 & 22{,}000 \\ 50{,}000 & 40{,}000 \\ 20{,}000 & 70{,}000 \end{pmatrix}$

Exercises 16—1, p. 407

1. $7 < 12$ **3.** $20 < 45$ **5.** $-4 > -9$ **7.** $16 < 81$ **9.** $x > -2$ **11.** $x \le 4$ **13.** $1 < x < 7$
15. $x < -9, x \ge -4$ **17.** $x > 0$ **19.** $x \le 0$ **21.** $x < -1, x > 1$ **23.** $2 \le x < 6$
25. $x > 0, y > 0$ **27.** $y > x$ **29.** Multiply both members by x.
31. Multiply by y and take square roots.   **33.** $3.5 < t < 15.3$ s, $h > 200$ m

35. $V = \frac{k}{a}$ if $r < a$, $V = \frac{k}{r}$ if $r > a$

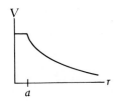

Exercises 16—2, p. 411
[*Note:* Solid portion of curves gives desired values of x.]

1.

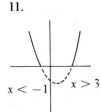

$x > 2$

3.

$x > \frac{1}{2}$

5.

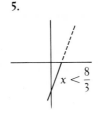

$x < \frac{8}{3}$

7.

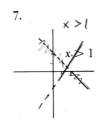

$x > l$ $x > 1$

9.

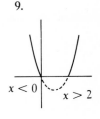

$x < 0$ $x > 2$

11.

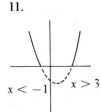

$x < -1$ $x > 3$

13.

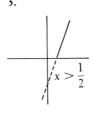

$1 < x < 4$

15.

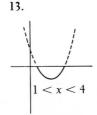

$x < -\frac{4}{3}$ $x > 2$

17.

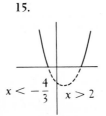

$x > 1$

19.

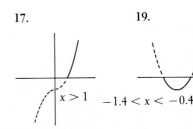

$-1.4 < x < -0.4$

21.

All x

23.

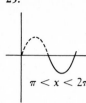

$\pi < x < 2\pi$

25.

$s \geq 2000$

27.

$T > 10$

Exercises 16–3, p. 415

1. $x > -3$ 3. $x < \dfrac{12}{7}$ 5. $x \leq 4$ 7. $x > \dfrac{9}{2}$ 9. $-1 < x < 1$ 11. $x \leq -2, x \geq \dfrac{1}{3}$

13. $\dfrac{1}{3} < x < \dfrac{1}{2}$ 15. All x 17. $-2 < x < 0, x > 1$ 19. $-2 < x < -1, x > 1$ 21. $x < 3, x > 8$

23. $-6 < x \leq \dfrac{3}{2}$ 25. $-1 < x < 2$ 27. $-5 < x < -1, x > 7$ 29. $-1 < x < \dfrac{3}{4}, x > 6$

31. $2 < x < 4, 5 < x < 9$ 33. $x \leq -2, x \geq 1$ 35. $-1 \leq x \leq 0$ 37. $t < \dfrac{15}{4}$ s
39. $-20°C < T < -16.2°C, 6.2°C < T < 10°C$

Exercises 16–4, p. 418

1. $3 < x < 5$ 3. $x < 1, x > \dfrac{7}{3}$ 5. $\dfrac{1}{6} \leq x \leq \dfrac{3}{2}$ 7. $x < -\dfrac{3}{2}, x > 0$ 9. $-10 < x < 2$

11. $x \leq -1, x \geq \dfrac{7}{3}$ 13. $-4 < x < -2, -1 < x < 1$ 15. $x < -4, -2 < x < -1, x > 1$

17. $y < \dfrac{1}{4}$ ft if $|x - 6| < 2$ ft   19. 5 in., 11 in.

Exercises 16–5, p. 424

1.

3.

5.

7.

9.

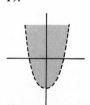

11.

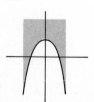

13.

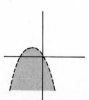

15.

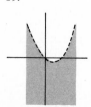

17.

19.

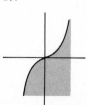

21.

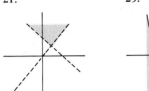

23.

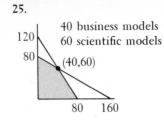

25.

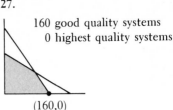
40 business models
60 scientific models
(40,60)
120
80
80   160

27.

160 good quality systems
0 highest quality systems
(160,0)

Exercises for Chapter 16, p. 425

1. $x > 6$ **3.** $x \leq -\dfrac{5}{3}$ **5.** $x < -9, x > 7$ **7.** $-4 < x < -1, x > 1$ **9.** $-\dfrac{1}{2} < x \leq 8$

11. $x < -4, \dfrac{1}{2} < x < 3$ **13.** No values   **15.** $x < 0, x > \dfrac{1}{2}$ **17.** $x < -1, x > 5$ **19.** $-2 \leq x \leq \dfrac{2}{3}$

21.

$x < \dfrac{1}{2}$

23.

$x > 1$

25.

All x

27.

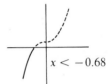
$x < -0.68$

29. $x \leq 3$

31. $x \leq -4, x \geq 0$ **33.**

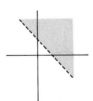

35.

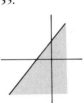

37.

39.

41. $\dfrac{1}{x} > 0, \dfrac{1}{y} < 0, \dfrac{1}{x} > \dfrac{1}{y}$ **43.** Divide both members by y; then use $x - 1 > y$.   **45.** $x \leq 80$ ft

47. $x > 16$ in.   **49.** $0 < R_1 < 0.73 \ \Omega, 7.27 \ \Omega < R_2 < 8 \ \Omega$ **51.**

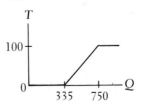
T
100
0   335   750   Q

Exercises 17–1, p. 429

1. 6   **3.** $\dfrac{4}{3}$ **5.** $\dfrac{1}{5}$ **7.** 40   **9.** 40,000 to 1   **11.** 19.2   **13.** 900 lb/in.^{2} **15.** 24 calculations/min

17. 1.44 Ω **19.** 32.5 r/min   **21.** 9080 g   **23.** 12.5 m/s   **25.** 27.0 in.   **27.** 72,000 ft-lb
29. 100 mg, 120 mg   **31.** 4.0 ft, 6.0 ft

Exercises 17–2, p. 434

1. $y = kz$ 3. $w = kxy^3$ 5. $r = \dfrac{16}{y}$ 7. $p = \dfrac{16q}{r^3}$ 9. 25   11. 50   13. 180   15. 2.56×10^5

17. 2.40 in.   19. $p = 30000t$ 21. 7680 BTU/h   23. $I = \dfrac{10^6}{r^2}$ 25. $R = \dfrac{5.00 \times 10^{-5}\, l}{A}$

27. $f = k\sqrt{T}$, 738 Hz   29. 135 ft   31. 23.3 cm/s   33. $P = \dfrac{k}{V^{3/2}}$, 2400 kPa   35. 0 ft/s²

Exercises for Chapter 17, p. 436

1. 10^6 3. $\dfrac{50}{7}$ 5. 3.25 in.   7. 1.72 L   9. 675,000   11. 492 g   13. $y = 3x^2$ 15. $v = \dfrac{128x}{y^3}$

17. 71.9 ft   19. 11.7 in.   21. 22.5 hp   23. $R = \dfrac{5}{18}$ A   25. $64.0\,\pi$ unit²   27. 551 ft/s
29. $\dfrac{1.28 \times 10^5 r^4}{\ell^2}$ 31. 50.0 Hz   33. 67.3×10^6 mi   35. $E = 0.139\,(\tan\theta + \cot\theta)$

Exercises 18–1, p. 444

1. 4, 6, 8, 10, 12   3. 13, 9, 5, 1, -3 5. 22   7. -62 9. 37   11. $49b$ 13. 440   15. $-\dfrac{85}{2}$

17. $n = 6$, $s = 150$ 19. $d = -\dfrac{2}{19}$, $l = -\dfrac{1}{3}$ 21. $a = 19$, $\ell = 106$ 23. $n = 62$, $s = -4867$

25. $n = 23$, $\ell = 6$ 27. $n = 8$, $d = \dfrac{1}{14}(b + 2c)$ 29. $a = 36$, $d = 4$, $s = 540$

31. $a = 3$, $d = -\dfrac{1}{3}$, $s = 15$ 33. 5050   35. 100, 500   37. 19th   39. 6400 ft

Exercises 18–2, p. 447

1. $45, 15, 5, \dfrac{5}{3}, \dfrac{5}{9}$ 3. 2, 6, 18, 54, 162   5. 16   7. $\dfrac{1}{125}$ 9. $\dfrac{1}{9}$ 11. 2×10^6 13. 248   15. 255

17. $a = 1$, $r = 3$ 19. $n = 5$, $s = \dfrac{2343}{25}$ 21. 32   23. \$219.11   25. \$597.53   27. 3.80 years

29. 22.0 ft   31. 0.0319 in.   33. 64.6%   35. $s = \dfrac{a - r\ell}{1 - r}$

Exercises 18–3, p. 451

1. 8   3. $\dfrac{25}{4}$ 5. $\dfrac{400}{21}$ 7. 8   9. $\dfrac{10000}{9999}$ 11. $\dfrac{1}{2}(5 + 3\sqrt{3})$ 13. $\dfrac{1}{3}$ 15. $\dfrac{40}{99}$ 17. $\dfrac{2}{11}$

19. $\dfrac{91}{333}$ 21. $\dfrac{11}{30}$ 23. $\dfrac{100741}{999000}$ 25. 24.0 ft

Exercises 18–4, p. 455

1. $t^3 + 3t^2 + 3t + 1$ 3. $16x^4 - 32x^3 + 24x^2 - 8x + 1$ 5. $x^5 + 10x^4 + 40x^3 + 80x^2 + 80x + 32$
7. $64a^6 - 192a^5b^2 + 240a^4b^4 - 160a^3b^6 + 60a^2b^8 - 12ab^{10} + b^{12}$
9. $625x^4 - 1500x^3 + 1350x^2 - 540x + 81$ 11. $64a^6 + 192a^5 + 240a^4 + 160a^3 + 60a^2 + 12a + 1$
13. $x^{10} + 20x^9 + 180x^8 + 960x^7 + \cdots$ 15. $1 + 2x + 3x^2 + 4x^3 + \cdots$ 17. 1.049   19. 9.980
21. 3.009   23. 0.924   25. $56a^3b^5$ 27. $10264320x^8b^4$ 29. $T^4 + 4T^3h + 6T^2h^2 + 4Th^3 + h^4$

31. $1 - \dfrac{x}{a} + \dfrac{x^3}{2a^3} - \cdots$

Exercises for Chapter 18, p. 456

1. 81   **3.** 1.28×10^{-6} **5.** $-\dfrac{119}{2}$ **7.** $\dfrac{16}{243}$ **9.** $\dfrac{195}{2}$ **11.** $\dfrac{1023}{96}$ **13.** 81   **15.** $\dfrac{9}{16}$ **17.** 186

19. $\dfrac{455}{2}$ (AP), 127 (GP) or 43 (GP)   **21.** $\dfrac{7}{9}$ **23.** $\dfrac{75}{99}$ **25.** $\dfrac{41}{333}$ **27.** $\dfrac{1}{6}$

29. $x^4 - 8x^3 + 24x^2 - 32x + 16$ **31.** $x^{10} + 5x^8 + 10x^6 + 10x^4 + 5x^2 + 1$

33. $1 - \dfrac{1}{2}a^2 - \dfrac{1}{8}a^4 - \dfrac{1}{16}a^6 - \cdots$ **35.** $1 + 6x + 24x^2 + 80x^3 + \cdots$ **37.** 30.133   **39.** 0.124

41. 1,001,000   **43.** 195   **45.** \$6633.24   **47.** \$8590   **49.** 20.0 s   **51.** 10^{-20} atm

Exercises 19–1, p. 465

[*Note:* "Answers" to trigonometric identities are intermediate steps of suggested reductions of the left member.]

1. $1.483 = \dfrac{1}{0.6745}$ **3.** $\left(\dfrac{1}{2}\sqrt{3}\right)^2 + \left(-\dfrac{1}{2}\right)^2 = \dfrac{3}{4} + \dfrac{1}{4} = 1$ **5.** $\dfrac{\cos\theta}{\sin\theta} \cdot \dfrac{1}{\cos\theta} = \dfrac{1}{\sin\theta}$

7. $\dfrac{\sin x}{\dfrac{\sin x}{\cos x}} = \dfrac{\sin x}{1} \cdot \dfrac{\cos x}{\sin x}$ **9.** $\sin y\left(\dfrac{\cos y}{\sin y}\right)$ **11.** $\sin x\left(\dfrac{1}{\cos x}\right)$ **13.** $\csc^2 x (\sin^2 x)$

15. $\sin x (\csc^2 x) = (\sin x)(\csc x)(\csc x) = \sin x\left(\dfrac{1}{\sin x}\right)\csc x$ **17.** $\sin x \csc x - \sin^2 x = 1 - \sin^2 x$

19. $\tan y \cot y + \tan^2 y = 1 + \tan^2 y$ **21.** $\sin x\left(\dfrac{\sin x}{\cos x}\right) + \cos x = \dfrac{\sin^2 x + \cos^2 x}{\cos x} = \dfrac{1}{\cos x}$

23. $\cos\theta\left(\dfrac{\cos\theta}{\sin\theta}\right) + \sin\theta = \dfrac{\cos^2\theta + \sin^2\theta}{\sin\theta} = \dfrac{1}{\sin\theta}$

25. $\sec\theta\left(\dfrac{\sin\theta}{\cos\theta}\right)\csc\theta = \sec\theta\left(\dfrac{1}{\cos\theta}\right)(\sin\theta\csc\theta) = \sec\theta(\sec\theta)(1)$

27. $\cot\theta(\sec^2\theta - 1) = \cot\theta\tan^2\theta = (\cot\theta\tan\theta)\tan\theta$ **29.** $\dfrac{\sin x}{\cos x} + \dfrac{\cos x}{\sin x} = \dfrac{\sin^2 x + \cos^2 x}{\cos x \sin x} = \dfrac{1}{\cos x \sin x}$

31. $(1 - \sin^2 x) - \sin^2 x$ **33.** $\dfrac{\sin x(1 + \cos x)}{1 - \cos^2 x} = \dfrac{1 + \cos x}{\sin x}$

35. $\dfrac{(1/\cos x) + (1/\sin x)}{1 + (\sin x/\cos x)} = \dfrac{(\sin x + \cos x)/\cos x \sin x}{(\cos x + \sin x)/\cos x} = \dfrac{\cos x}{\cos x \sin x}$

37. $\dfrac{\sin^2 x}{\cos^2 x}\cos^2 x + \dfrac{\cos^2 x}{\sin^2 x}\sin^2 x = \sin^2 x + \cos^2 x$ **39.** $4\sin x + \dfrac{\sin x}{\cos x} = \sin x\left(4 + \dfrac{1}{\cos x}\right)$

41. $\dfrac{1}{\cos x} + \dfrac{\sin x}{\cos x} + \dfrac{\cos x}{\sin x} = \dfrac{\sin x + \cos^2 x + \sin^2 x}{\sin x \cos x}$ **43.** $(2\sin^2 x - 1)(\sin^2 x - 1)$

45. $\dfrac{\cot 2y(\sec 2y + \tan 2y) - \cos 2y(\sec 2y - \tan 2y)}{\sec^2 2y - \tan^2 2y} = \dfrac{\cot 2y \sec 2y + 1 - 1 + \cos 2y \tan 2y}{1}$

47. Infinite GP: $\dfrac{1}{1 - \sin^2 x} = \dfrac{1}{\cos^2 x}$ **49.** $\mu = \dfrac{w\sin\theta}{w\cos\theta} = \dfrac{\sin\theta}{\cos\theta} = \tan\theta$ **51.** $\sin\theta(\sin^2\theta) = \sin\theta(1 - \cos^2\theta)$

Exercises 19–2, p. 470

1. $\sin 105° = \sin 60° \cos 45° + \cos 60° \sin 45° = \dfrac{\sqrt{3}}{2} \cdot \dfrac{\sqrt{2}}{2} + \dfrac{1}{2} \cdot \dfrac{\sqrt{2}}{2} = 0.9659$

3. $\cos 15° = \cos(60° - 45°) = \cos 60° \cos 45° + \sin 60° \sin 45°$

$$= \left(\dfrac{1}{2}\right)\left(\dfrac{1}{2}\sqrt{2}\right) + \left(\dfrac{1}{2}\sqrt{3}\right)\left(\dfrac{1}{2}\sqrt{2}\right) = \dfrac{1}{4}\sqrt{2} + \dfrac{1}{4}\sqrt{6} = \dfrac{1}{4}(\sqrt{2} + \sqrt{6}) = 0.9659$$

5. $-\dfrac{33}{65}$ 7. $-\dfrac{56}{65}$ 9. $\sin 3x$ 11. $\cos x$

13. $\sin(270° - x) = \sin 270° \cos x - \cos 270° \sin x = (-1)\cos x - 0(\sin x)$

15. $\cos\left(\dfrac{1}{2}\pi - x\right) = \cos \dfrac{1}{2}\pi \cos x + \sin \dfrac{1}{2}\pi \sin x = 0(\cos x) + 1(\sin x)$

17. $\cos(30° + x) = \cos 30° \cos x - \sin 30° \sin x = \dfrac{1}{2}\sqrt{3} \cos x - \dfrac{1}{2}\sin x = \dfrac{1}{2}(\sqrt{3} \cos x - \sin x)$

19. $\sin\left(\dfrac{\pi}{4} + x\right) = \sin \dfrac{\pi}{4} \cos x + \cos \dfrac{\pi}{4} \sin x = \dfrac{1}{2}\sqrt{2} \cos x + \dfrac{1}{2}\sqrt{2} \sin x$

21. $(\sin x \cos y + \cos x \sin y)(\sin x \cos y - \cos x \sin y) = \sin^2 x \cos^2 y - \cos^2 x \sin^2 y$
$$= \sin^2 x (1 - \sin^2 y) - (1 - \sin^2 x)\sin^2 y$$

23. $(\cos \alpha \cos \beta - \sin \alpha \sin \beta) + (\cos \alpha \cos \beta + \sin \alpha \sin \beta)$ 25. 27. 29. 31. Use indicated method.

33. $\sin(\omega t + \phi - \alpha)$ 35. $2A \sin \dfrac{2\pi t}{T} \cos \dfrac{2\pi x}{\lambda}$

Exercises 19–3, p. 474

1. $\sin 60° = \sin 2(30°) = 2 \sin 30° \cos 30° = 2\left(\dfrac{1}{2}\right)\left(\dfrac{1}{2}\sqrt{3}\right) = \dfrac{1}{2}\sqrt{3}$

3. $\cos 120° = \cos 2(60°) = \cos^2 60° - \sin^2 60° = \left(\dfrac{1}{2}\right)^2 - \left(\dfrac{1}{2}\sqrt{3}\right)^2 = -\dfrac{1}{2}$

5. $\dfrac{24}{25}$ 7. $\dfrac{3}{5}$ 9. $2 \sin 8x$ 11. $\cos 8x$ 13. $\cos^2 \alpha - (1 - \cos^2 \alpha)$

15. $(\cos^2 x - \sin^2 x)(\cos^2 x + \sin^2 x) = (\cos^2 x - \sin^2 x)(1)$ 17. $\dfrac{2 \tan x}{\sin 2x} = \dfrac{2(\sin x/\cos x)}{2 \sin x \cos x} = \dfrac{1}{\cos^2 x}$

19. $\dfrac{\sin 3x \cos x - \cos 3x \sin x}{\sin x \cos x} = \dfrac{\sin 2x}{\dfrac{1}{2}\sin 2x}$

21. $\sin(2x + x) = \sin 2x \cos x + \cos 2x \sin x = (2 \sin x \cos x)(\cos x) + (\cos^2 x - \sin^2 x)\sin x$
23. Use indicated method.

25. $\cos \alpha = \dfrac{A}{C}, \sin \alpha = \dfrac{B}{C}$:

$A \sin 2t + B \cos 2t = C\left(\dfrac{A}{C}\sin 2t + \dfrac{B}{C}\cos 2t\right)$

$= C(\cos \alpha \sin 2t + \sin \alpha \cos 2t)$

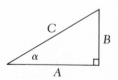

Exercises 19–4, p. 476

1. $\cos 15° = \cos\dfrac{1}{2}(30°) = \sqrt{\dfrac{1 + \cos 30°}{2}} = \sqrt{\dfrac{1.8660}{2}} = 0.9659$

3. $\sin 75° = \sin \frac{1}{2}(150°) = \sqrt{\dfrac{1 - \cos 150°}{2}} = \sqrt{\dfrac{1.8660}{2}} = 0.9659$

5. $\sin 3x$ 7. $4 \cos 2x$ 9. $\dfrac{1}{26}\sqrt{26}$ 11. $\dfrac{1}{10}\sqrt{2}$ 13. $\pm\sqrt{\dfrac{2}{1 - \cos \alpha}}$

15. $\tan\dfrac{1}{2}\alpha = \dfrac{1 - \cos \alpha}{\sin \alpha} = \dfrac{\sin \alpha}{1 + \cos \alpha}$ 17. $\dfrac{1 - \cos \alpha}{2 \sin \dfrac{1}{2}\alpha} = \dfrac{1 - \cos \alpha}{2\sqrt{\dfrac{1}{2}(1 - \cos \alpha)}} = \sqrt{\dfrac{1 - \cos \alpha}{2}}$

19. $2\left(\dfrac{1 - \cos \alpha}{2}\right) - \left(\dfrac{1 + \cos \alpha}{2}\right)$ 21. $\sqrt{1 - 2 \cos \alpha + \cos^2\alpha + \sin^2\alpha\,(\cos^2\beta + \sin^2\beta)} = \sqrt{2 - 2 \cos \alpha}$

Exercises 19–5, p. 480

1. $\dfrac{\pi}{2}$ 3. $\dfrac{3\pi}{4}, \dfrac{7\pi}{4}$ 5. $\dfrac{\pi}{3}, \dfrac{2\pi}{3}, \dfrac{4\pi}{3}, \dfrac{5\pi}{3}$ 7. $\dfrac{\pi}{3}, \dfrac{2\pi}{3}, \dfrac{4\pi}{3}, \dfrac{5\pi}{3}$ 9. $0, \dfrac{\pi}{6}, \dfrac{5\pi}{6}, \pi$

11. $\dfrac{\pi}{12}, \dfrac{\pi}{4}, \dfrac{5\pi}{12}, \dfrac{3\pi}{4}, \dfrac{13\pi}{12}, \dfrac{5\pi}{4}, \dfrac{17\pi}{12}, \dfrac{7\pi}{4}$ 13. $\dfrac{\pi}{2}, \dfrac{3\pi}{2}$ 15. $0, \dfrac{\pi}{3}, \pi, \dfrac{5\pi}{3}$ 17. $\dfrac{\pi}{4}, \dfrac{3\pi}{4}, \dfrac{5\pi}{4}, \dfrac{7\pi}{4}$ 19. 3.57, 5.85

21. 0.26, 1.31, 3.40, 4.45   23. $0, \pi$ 25. $\dfrac{3\pi}{8}, \dfrac{7\pi}{8}, \dfrac{11\pi}{8}, \dfrac{15\pi}{8}$ 27. 0.79, 1.25, 3.93, 4.39   29. 0.955

31. 1.41 units of time   33. $-0.95, 0.00, 0.95$ 35. $0, 4.49$

Exercises 19–6, p. 483

1. y is the angle whose tangent is x.   3. y is the angle whose cotangent is $3x$.

5. y is twice the angle whose sine is x.   7. y is 5 times the angle whose cosine is $2x$.

9. $\dfrac{\pi}{3}$ 11. $\dfrac{3\pi}{4}$ 13. $\dfrac{3\pi}{4}$ 15. $\dfrac{\pi}{3}$ 17. $\dfrac{5\pi}{6}$ 19. $\dfrac{\pi}{4}$ 21. $x = \dfrac{1}{3}\arcsin y$ 23. $x = 4 \tan y$

25. $x = \dfrac{1}{3}\operatorname{arcsec}(y - 1)$ 27. $x = 1 - \cos(1 - y)$ 29. I, II   31. II, III   33. $k = \dfrac{m\left(\arccos\dfrac{x}{A}\right)^2}{t^2}$

35. $\theta = \dfrac{1}{2}\arcsin\dfrac{2E}{E_0}$

Exercises 19–7, p. 487

1. $\dfrac{\pi}{3}$ 3. 0   5. $-\dfrac{\pi}{3}$ 7. $\dfrac{\pi}{3}$ 9. $\dfrac{\pi}{6}$ 11. $-\dfrac{\pi}{4}$ 13. $\dfrac{\pi}{4}$ 15. -1.309 17. $\dfrac{1}{2}\sqrt{3}$ 19. $\dfrac{1}{2}\sqrt{2}$

21. -1.150 23. -1 25. $\dfrac{x}{\sqrt{1 - x^2}}$ 27. $\dfrac{1}{x}$ 29. $\dfrac{3x}{\sqrt{9x^2 - 1}}$ 31. $2x\sqrt{1 - x^2}$

33. $\sin\left[\operatorname{Arcsin}\dfrac{3}{5} + \operatorname{Arcsin}\dfrac{5}{13}\right] = \dfrac{3}{5} \cdot \dfrac{12}{13} + \dfrac{4}{5} \cdot \dfrac{5}{13} = \dfrac{56}{65}$

35. let $b = $ side opp. β; $\tan \beta = \dfrac{b}{1} = b$; $\tan \alpha = a + b = a + \tan \beta$; $\alpha = \operatorname{Arctan}(a + \tan \beta)$

Exercises for Chapter 19, p. 488

1. $\sin(90° + 30°) = \sin 90° \cos 30° + \cos 90° \sin 30° = (1)\left(\dfrac{1}{2}\sqrt{3}\right) + (0)\left(\dfrac{1}{2}\right) = \dfrac{1}{2}\sqrt{3}$

3. $\sin(180° - 45°) = \sin 180° \cos 45° - \cos 180° \sin 45° = 0\left(\dfrac{1}{2}\sqrt{2}\right) - (-1)\left(\dfrac{1}{2}\sqrt{2}\right) = \dfrac{1}{2}\sqrt{2}$

5. $\cos 2(90°) = \cos^2 90° - \sin^2 90° = 0 - 1 = -1$

7. $\sin \dfrac{1}{2}(90°) = \sqrt{\dfrac{1}{2}(1 - \cos 90°)} = \sqrt{\dfrac{1}{2}(1 - 0)} = \dfrac{1}{2}\sqrt{2}$ 9. $\sin 5x$ 11. $4 \sin 12x$ 13. $2 \cos 12x$

15. $2 \cos x$ 17. $-\dfrac{\pi}{2}$ 19. 0.2618 21. $-\dfrac{1}{3}\sqrt{3}$ 23. 0 25. $\dfrac{\dfrac{1}{\cos y}}{\dfrac{1}{\sin y}} = \dfrac{1}{\cos y} \cdot \dfrac{\sin y}{1}$

27. $\sin x \csc x - \sin^2 x = 1 - \sin^2 x$ 29. $\dfrac{1 - \sin^2 \theta}{\sin \theta} = \dfrac{\cos^2 \theta}{\sin \theta}$ 31. $\cos \theta\left(\dfrac{\cos \theta}{\sin \theta}\right) + \sin \theta = \dfrac{\cos^2 \theta + \sin^2 \theta}{\sin \theta}$

33. $\dfrac{(\sec^2 x - 1)(\sec^2 x + 1)}{\tan^2 x} = \sec^2 x + 1$ 35. $2\left(\dfrac{1}{\sin 2x}\right)\left(\dfrac{\cos x}{\sin x}\right) = 2\left(\dfrac{1}{2 \sin x \cos x}\right)\left(\dfrac{\cos x}{\sin x}\right) = \dfrac{1}{\sin^2 x}$

37. $\dfrac{1}{2}\left(2 \sin \dfrac{\theta}{2} \cos \dfrac{\theta}{2}\right)$ 39. $\dfrac{1}{\cos x} + \dfrac{\sin x}{\cos x} = \dfrac{(1 + \sin x)(1 - \sin x)}{\cos x(1 - \sin x)} = \dfrac{1 - \sin^2 x}{\cos x(1 - \sin x)}$

41. $\cos[(x - y) + y]$ 43. $\sin 4x(\cos 4x)$ 45. $\dfrac{\sin x}{\dfrac{1}{\sin x} - \dfrac{\cos x}{\sin x}} = \dfrac{\sin^2 x}{1 - \cos x} = \dfrac{1 - \cos^2 x}{1 - \cos x}$

47. $\dfrac{\sin x \cos y + \cos x \sin y + \sin x \cos y - \cos x \sin y}{\cos x \cos y - \sin x \sin y + \cos x \cos y + \sin x \sin y} = \dfrac{2 \sin x \cos y}{2 \cos x \cos y}$ 49. $x = \dfrac{1}{2}\arccos\dfrac{1}{2}y$

51. $x = \dfrac{1}{5}\sin\dfrac{1}{3}\left(\dfrac{1}{4}\pi - y\right)$ 53. $0, \dfrac{\pi}{2}, \pi, \dfrac{3\pi}{2}$ 55. $\dfrac{\pi}{6}, \dfrac{5\pi}{6}, \dfrac{7\pi}{6}, \dfrac{11\pi}{6}$ 57. $0, \pi$ 59. 0 61. $\dfrac{1}{x}$

63. $2x\sqrt{1 - x^2}$ 65. $\cos x(\cos^2 x) = \cos x(1 - \sin^2 x)$ 67. $2 \sin^2 x = 1 - \cos 2x; \sin^2 x = \dfrac{1}{2}(1 - \cos 2x)$

69. $t = 3 \arcsin \dfrac{y}{20}$ 71. $\cos^2 \alpha - 3 \sin^2 \alpha$

73. $1 - 2r^2 + r^4 + 2r^2 - 2r^2 \cos \beta = (1 - r^2)^2 + 4r^2\left(\dfrac{1 - \cos \beta}{2}\right)$ 75. $54.7°$

Exercises 20–1, p. 496

1. $2\sqrt{29}$ 3. 3 5. $\sqrt{85}$ 7. 7 9. $\dfrac{5}{2}$ 11. undefined   13. $-\dfrac{7}{6}$ 15. 0 17. $\dfrac{1}{3}\sqrt{3}$

19. $-\dfrac{1}{3}\sqrt{3}$ 21. $20.0°$ 23. $98.5°$ 25. parallel   27. perpendicular   29. $8, -2$ 31. -3

33. two sides equal $2\sqrt{10}$ 35. $m_1 = \dfrac{1}{4}, m_2 = \dfrac{5}{4}$ 37. 10

Exercises 20–2, p. 501

1. $4x - y + 20 = 0$ 3. $7x - 6y - 16 = 0$ 5. $x - y + 2 = 0$ 7. $y = -3$ 9. $x = -3$

11. $3x - 2y - 12 = 0$ 13. $x + 3y + 5 = 0$ 15. $x + 7y - 18 = 0$

17.   19.   21. $y = \dfrac{3}{2}x - \dfrac{1}{2}; m = \dfrac{3}{2}, b = -\dfrac{1}{2}$ 23. $y = \dfrac{5}{2}x + \dfrac{5}{2}; m = \dfrac{5}{2}, b = \dfrac{5}{2}$

25. -2 27. 1 29. $m_1 = m_2 = \dfrac{3}{2}$ 31. $m_1 = 2, m_2 = -\dfrac{1}{2}$

33. $x + 2y - 4 = 0$ 35. $3x + y - 18 = 0$ 37. $s = 50t + 10$

39. $L = \dfrac{2}{3}F + 15$ 41. 3.35 kJ   43. $n = \dfrac{7}{6}t + 10$; at 6:30, $n = 10$;
at 8:30, $n = 150$

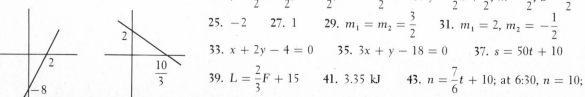

45.

47.

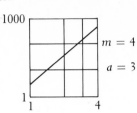

49.

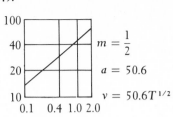

Exercises 20–3, p. 506

1. $(2, 1)$, $r = 5$ 3. $(-1, 0)$, $r = 2$ 5. $x^2 + y^2 = 9$ 7. $x^2 + y^2 - 4x - 4y - 8 = 0$
9. $x^2 + y^2 + 4x - 10y + 24 = 0$ 11. $x^2 + y^2 - 4x - 2y - 3 = 0$ 13. $x^2 + y^2 + 6x - 10y + 9 = 0$
15. $x^2 + y^2 - 4x - 10y + 4 = 0$; $x^2 + y^2 + 4x + 10y + 4 = 0$

17. $(0,0)$

$r = 5$

19. $(1,0)$

$r = 3$

21. $(-4,5)$

$r = 7$

23. $(1,2)$

$r = \dfrac{1}{2}\sqrt{22}$

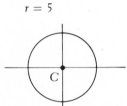

25. Symmetrical to both axes and origin   27. Symmetrical to y-axis   29. $(7, 0)$, $(-1, 0)$
31. $3x^2 + 3y^2 + 4x + 8y - 20 = 0$, circle
33. $x^2 + y^2 = 2.79$ (assume center of circle at center of coordinate system)
35. $x^2 + y^2 = 576$, $x^2 + y^2 + 36y + 315 = 0$ (in inches)

Exercises 20–4, p. 511

1. $F(1,0)$, $x = -1$ 3. $F(-1,0)$, $x = 1$ 5. $F(0,2)$, $y = -2$ 7. $F(0,-1)$, $y = 1$

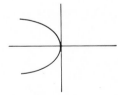

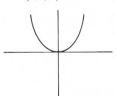

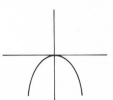

9. $F\left(\dfrac{1}{2},0\right)$, $x = -\dfrac{1}{2}$ 11. $F\left(0,\dfrac{1}{4}\right)$, $y = -\dfrac{1}{4}$

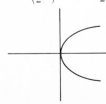

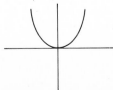

13. $y^2 = 12x$ 15. $x^2 = 16y$

17. $x^2 = 4y$ 19. $x^2 = \dfrac{1}{8}y$

21. $y^2 - 2y - 12x + 37 = 0$ **23.** $x^2 - 2x + 8y - 23 = 0$ **25.** $x = 2$, $y = 4$ **29.**

27. $x^2 = 100y$

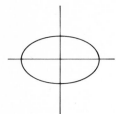

(3,1)

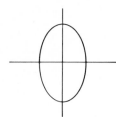

(1,3)

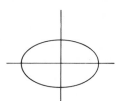

T

10^{-1}

C

250×10^{-6}

31. $y^2 = 8x$ or $x^2 = 8y$ with vertex midway between island and shore

Exercises 20—5, p. 517

1. $V(2,0)$, $V(-2,0)$
$F(\sqrt{3},0)$, $F(-\sqrt{3},0)$

3. $V(0,6)$, $V(0,-6)$
$F(0,\sqrt{11})$, $F(0,-\sqrt{11})$

5. $V(3,0)$, $V(-3,0)$
$F(\sqrt{5},0)$, $F(-\sqrt{5},0)$

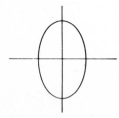

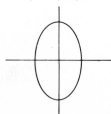

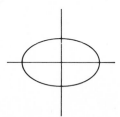

7. $V(0,7)$, $V(0,-7)$
$F(0)$, $\sqrt{45}$, $F(0, -\sqrt{45})$

9. $V(0,4)$, $V(0,-4)$
$F(0,\sqrt{14})$,
$F(0,-\sqrt{14})$

11. $V\left(\frac{5}{2},0\right)$, $V\left(-\frac{5}{2},0\right)$
$F(\sqrt{21}/2,0)$, $F(-\sqrt{21}/2,0)$

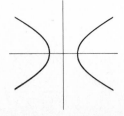

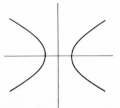

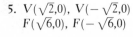

13. $144x^2 + 225y^2 = 32400$ **15.** $9x^2 + 5y^2 = 45$ **17.** $3x^2 + 20y^2 = 192$ **19.** $4x^2 + y^2 = 20$
21. $16x^2 + 25y^2 - 32x - 50y - 359 = 0$ **23.** $9x + 5y^2 - 18x - 20y - 16 = 0$
25. $2x^2 + 3y^2 - 8x - 4 = 2x^2 + 3(-y)^2 - 8x - 4$ **27.** $9x^2 + 25y^2 = 22500$
29. $2.70x^2 + 2.76y^2 = 7.45 \times 10^7$ **31.** major axis = 8.5 in., minor axis = 6.0 in.

Exercises 20—6, p. 523

1. $V(5,0)$, $V(-5,0)$
$F(13,0)$, $F(-13,0)$

3. $V(0,3)$, $V(0,-3)$
$F(0,\sqrt{10})$, $F(0,-\sqrt{10})$

5. $V(\sqrt{2},0)$, $V(-\sqrt{2},0)$
$F(\sqrt{6},0)$, $F(-\sqrt{6},0)$

7. $V(0,\sqrt{5})$, $V(0,-\sqrt{5})$
   $F(0,\sqrt{7})$, $F(0,-\sqrt{7})$

9. $V(0,2)$, $V(0,-2)$
   $F(0,\sqrt{5})$, $F(0,-\sqrt{5})$

11. $V(2,0)$, $V(-2,0)$
   $F\left(\frac{2}{3}\sqrt{13},0\right)$, $F\left(-\frac{2}{3}\sqrt{13},0\right)$

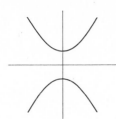
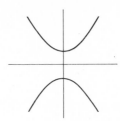
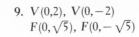

13. $16x^2 - 9y^2 = 144$ 15. $9y^2 - 25x^2 = 900$ 17. $3x^2 - y^2 = 3$ 19. $4x^2 - 5y^2 = 20$

21.

23.

25. $9x^2 - 16y^2 - 108x + 64y + 116 = 0$
27. $9x^2 - y^2 - 36x + 27 = 0$

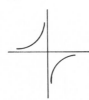

29.

R

Z

31.

f

λ

33. dist(rifle to P) − dist(target to P) = constant
    (related to distance from rifle to target)

Exercises 20–7, p. 528

1. Parabola, $(-1,2)$ 3. Hyperbola, $(1,2)$ 5. Ellipse, $(-1,0)$ 7. Parabola, $(-3,1)$

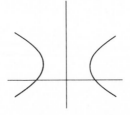
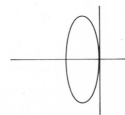
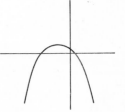

9. $y^2 - 6y - 16x - 7 = 0$ 11. $x^2 + 6x - 4y + 17 = 0$ 13. $16x^2 + 25y^2 + 64x - 100y - 236 = 0$
15. $16x^2 - 9y^2 + 32x + 18y - 137 = 0$

17. Parabola, $(-1,-1)$ **19.** Ellipse, $(-3,0)$ **21.** Hyperbola, $(0,4)$ **23.** Parabola, $(1,0)$

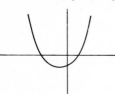   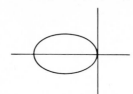   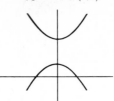   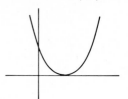

25. $x^2 - y^2 + 4x - 2y - 22 = 0$ **27.** $y^2 - 2y + 4x + 1 = 0$ **29.**   **31.** $\dfrac{(x - 0.9)^2}{3.7^2} + \dfrac{y^2}{3.6^2} = 1$

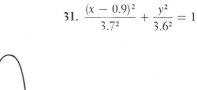

Exercises 20—8, p. 532

  1. Circle   **3.** Parabola   **5.** Hyperbola   **7.** Circle   **9.** Ellipse   **11.** Hyperbola   **13.** Ellipse
15. Parabola
17. Parabola; $V(-4,0)$; $F(-4,2)$ **19.** Hyperbola; $C(+1, -2)$;   **21.** Ellipse; $C(5,0)$; $V(5, \pm 2\sqrt{2})$
                                                    $V(+1, -2, \pm \sqrt{2})$

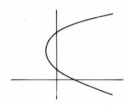   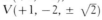   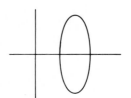

23. Parabola; $V\left(-\dfrac{1}{2}, \dfrac{5}{2}\right)$; $F\left(\dfrac{1}{2}, \dfrac{5}{2}\right)$ **25.** Hyperbola   **27.** Ellipse

Exercises 20—9, p. 535

1.   **3.**   **5.**   **7.**   **9.**   **11.**

13. $\left(2, \dfrac{\pi}{6}\right)$ 15. $\left(1, \dfrac{7\pi}{6}\right)$ 17. $(-4, -4\sqrt{3})$ 19. $(2.77, -1.15)$ 21. $r = 3 \sec \theta$ 23. $r = a$

25. $r = 4 \cot \theta \csc \theta$ 27. $r^2 = \dfrac{4}{1 + 3 \sin^2\theta}$ 29. $x^2 + y^2 - y = 0$ 31. $x = 4$

33. $x^4 + y^4 - 4x^3 + 2x^2y^2 - 4xy^2 - 4y^2 = 0$ 35. $(x^2 + y^2)^2 = 2xy$

37. $s = \sqrt{x^2 + y^2}\,\arctan\left(\dfrac{y}{x}\right)$, $A = \dfrac{1}{2}(x^2 + y^2)\arctan\left(\dfrac{y}{x}\right)$ 39. $r = 100 \tan \theta \sec \theta$

Exercises 20—10, p. 538

1.

3.

5.

7.

9.

11.
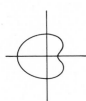

13.

15.

17.

19.
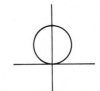

21.

23.

25.

27.

Exercises for Chapter 20, p. 539

1. $4x - y - 11 = 0$

3. $2x + 3y + 3 = 0$

5. $x^2 + y^2 - 2x + 4y - 5 = 0$

7. $y^2 = 12x$

9. $9x^2 + 25y^2 = 900$

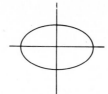

11. $144y^2 - 169x^2 = 24336$

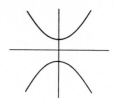

13. $(-3,0)$, $r = 4$

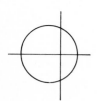

15. $(0,-5)$, $y = 5$

17. $V(0,4)$, $V(0,-4)$
$F(0,\sqrt{15})$, $F(0,-\sqrt{15})$

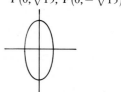

19. $V(2,0)$, $V(-2,0)$

$F\left(\dfrac{2}{5}\sqrt{35},0\right)$, $F\left(-\dfrac{2}{5}\sqrt{35},0\right)$

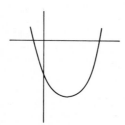

21. $V(4,-8)$, $F(4,-7)$

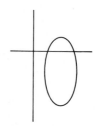

23. $(2,-1)$

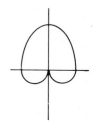

25.

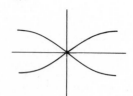...

27.

29.

31.

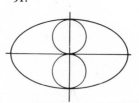

33. $\theta = \text{Arctan } 2 = 1.11$ **35.** $r^2\cos 2\theta = 16$ **37.** $(x^2 + y^2)^3 = 16x^2y^2$

39. $3x^2 + 4y^2 - 8x - 16 = 0$ **41.** $(1.90, 1.55)$, $(1.90, -1.55)$, $(-1.90, 1.55)$, $(-1.90, -1.55)$

43. $m_1 = -\dfrac{12}{5}$, $m_2 = \dfrac{5}{12}$; $d_1^2 = 169$, $d_2^2 = 169$, $d_3^2 = 338$ **45.** $x^2 - 6x - 8y + 1 = 0$

47.

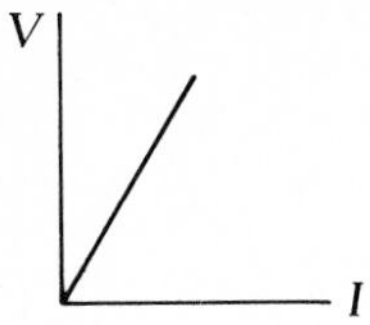

49.

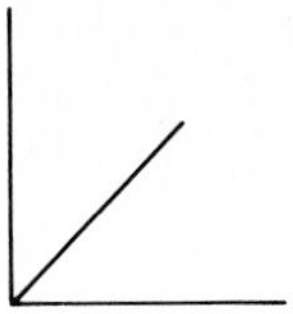

51. $x^2 = -80y$

53.

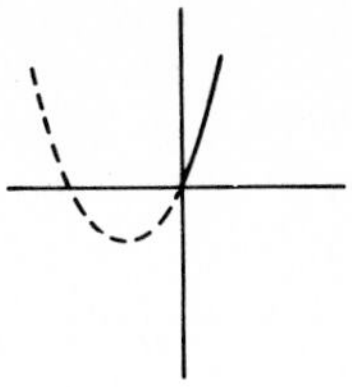

55. $225y = 450x - x^2$

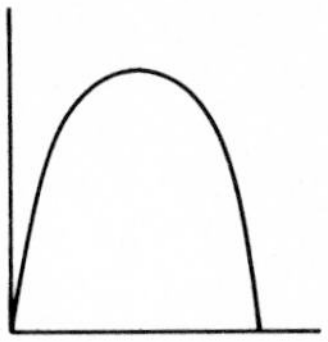

57. No

59. $\dfrac{x^2}{448900} + \dfrac{y^2}{445300} = 1$

61.

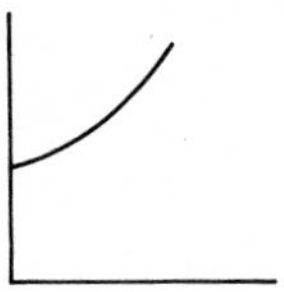

63. 13.5 ft

65. $r^2 = \dfrac{7.45 \times 10^7}{2.70 + 0.06 \sin^2\theta}$

67. $(a^2 - b^2)x^2 + a^2y^2 + 2bx - 1 = 0$; $a = b$, parabola; $a^2 > b^2$, ellipse; $a^2 < b^2$, hyperbola

Exercises 21–1, p. 548

1. $\frac{1}{4}$ 3. $\frac{3}{4}$ 5. $\frac{1}{16}$ 7. $\frac{1}{19}$ 9. $\frac{1}{6}$ 11. $\frac{5}{6}$ 13. $\frac{1}{36}$ 15. $\frac{11}{36}$ 17. $\frac{1}{36}$ 19. $\frac{1}{6}$ 21. $\frac{2}{9}$

23. $\frac{1}{36}$ 25. 0.470 27. 0.476 29. $\frac{7}{25}$ 31. $\frac{6}{25}$ 33. $\frac{1}{221}$ 35. 6 37. $\frac{4}{9}$ 39. 0.195

Exercises 21–2, p. 556

1.

No.	2	3	4	5	6	7
Freq.	1	3	4	2	3	2

3.

No.	45	46	47	48	49	50	51	52	53	54	55	56	57
Freq.	1	1	1	2	2	0	1	0	1	0	2	0	1

5.

Int.	2–3	4–5	6–7
Freq.	4	6	5

7.

Int.	43–45	46–48	49–51	52–54	55–57
Freq.	1	4	3	1	3

9.

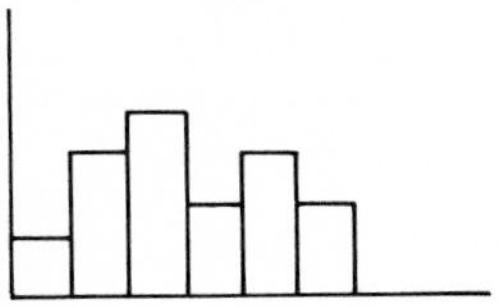

11.

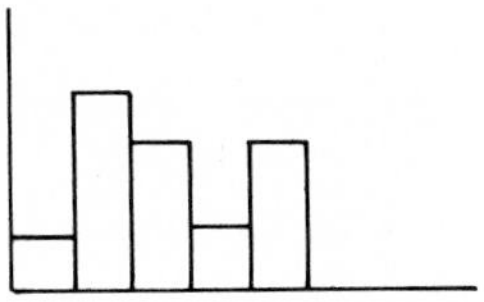

13.

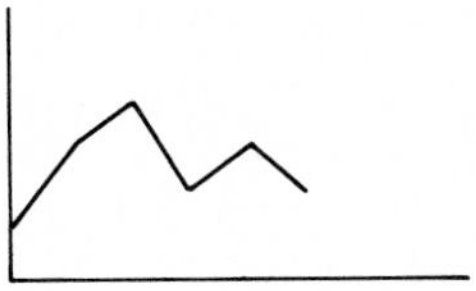

15.

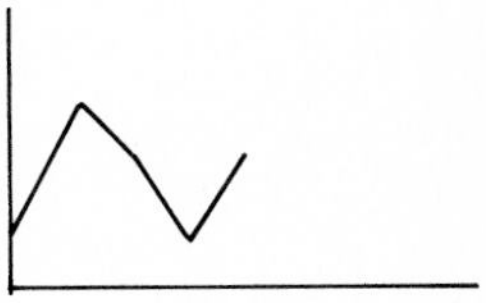

17. 4 19. 49 21. 4.6 23. 50.25

25.

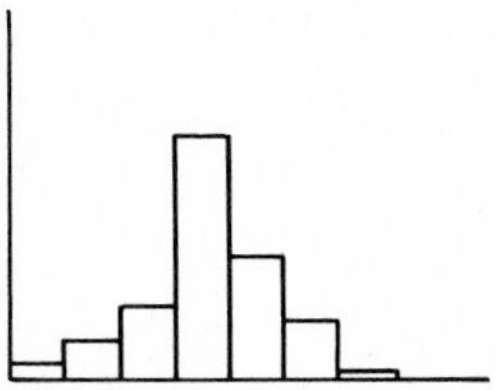

27.

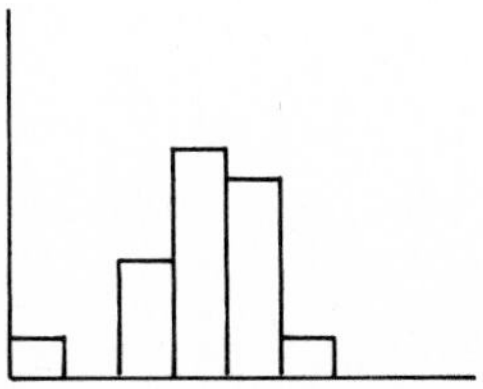

29. 3.435 A

31.

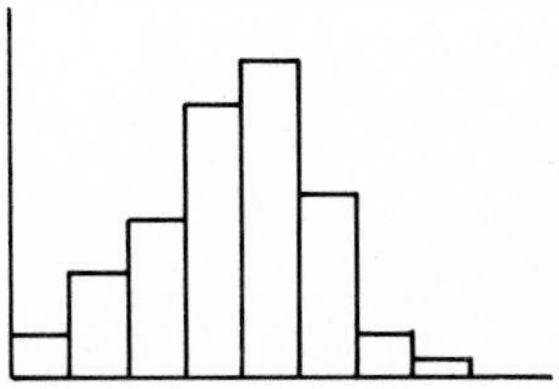

33. 20.9   **35.**   **37.** 4   **39.** 48, 49, 55

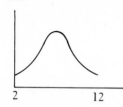

Exercises 21–3, p. 562

1. 1.50   **3.** 3.74   **5.** 1.50   **7.** 3.74   **9.** $60   **11.** 7   **13.** 60%   **15.** 58%   **17.** 64%
19. 75%

Exercises 21–4, p. 567

1. $y = (1,0)x - 2.6$    **3.** $y = -1.77x + 190$ **5.** $V = -0.6i + 11.3$

7. $V = 4.1 \times 10^{-15}f - 1.90$ **9.** 0.985   **11.** -0.90
  $f_0 = 0.460$ PHz

Exercises 21–5, p. 570

1. $y = 1.96x^2 + 5.2$ 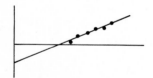   **3.** $y = 10.9/x$ **5.** $y = 5.95t^2 + 0.55$

        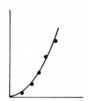

7. $y = 35.6(10^z) - 39.9$ **9.** $f = \dfrac{480}{\sqrt{L}} + 10$

Exercises for Chapter 21, p. 572

1. $\dfrac{5}{6}$ **3.** $\dfrac{1}{108}$ **5.** $\dfrac{5}{92}$ **7.** $\dfrac{10}{69}$ **9.** 3.6   **11.** 0.8   **13.** 0.264 Pa·s   **15.** 0.014 Pa·s

17. 0.927 in.   **19.** 0.012 in.   **21.** 4

23. 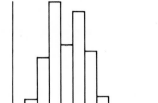   **25.** $\dfrac{4}{9}$ **27.** 3×10^{-5} **29.** $R = 0.0985T + 25.0$ **31.** $y = 1.04 \tan x - 0.02$

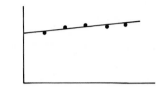

33. $y = 1.10\sqrt{x} + 0.01$

Exercises 22–1, p. 581

1. cont. all x **3.** not cont. $x = -3$, div. by zero   **5.** cont. all x
7. not cont. $x = 1$, small change
9. 10.000, 8.500, 7.300, 7.030, 7.003, 4.000, 5.500, 6.700, 6.970, 6.997; $\lim\limits_{x \to 3} f(x) = 7$
11. 1.71, 1.9701, 1.997001, 2.31, 2.0301, 2.003001; $\lim\limits_{x \to 1} f(x) = 2$ **13.** 7   **15.** 1   **17.** 1   **19.** -2
21. 2   **23.** -4 **25.** does not exist   **27.** 0   **29.** 3   **31.** 0
33. 1.26×10^{-6}, 1.827×10^{-6}, 1.8837×10^{-6}, 1.8963×10^{-6}, 1.953×10^{-6}, 2.52×10^{-6};
$\lim\limits_{i \to 0.3} B = 1.89 \times 10^{-6}$ W
35. 34.8°C, 0°C

Exercises 22–2, p. 587

1. (slopes) 3.5, 3.9, 3.99, 3.999; $m = 4$ **3.** (slopes) -3.5, -3.9, -3.99, -3.999; $m = -4$ **5.** 4   **7.** -4

   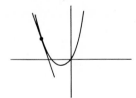

9. $m_{\tan} = 2x_1$; 4, -2 **11.** $m_{\tan} = 2x_1 + 2$; -4, 4   **13.** $m_{\tan} = 2x_1 + 4$; -2, 8

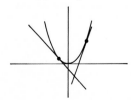   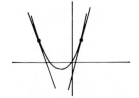   

15. $m_{\tan} = 6 - 2x_1$; 10, 0   **17.** $m_{\tan} = 3x_1^2 - 2$; 1, -2, 1   **19.** $m_{\tan} = 4x_1^3$; 0, 0.5, 4

Exercises 22–3, p. 592

1. 3   **3.** -2 **5.** $2x$ **7.** $10x$ **9.** $2x - 7$ **11.** $8 - 4x$ **13.** $3x^2 + 4$ **15.** $-\dfrac{1}{(x + 2)^2}$

17. $1 - \dfrac{1}{x^2}$ **19.** $-\dfrac{4}{x^3}$ **21.** $4x^3 + 3x^2 + 2x + 1$ **23.** $4x^3 + \dfrac{2}{x^2}$ **25.** $\dfrac{1}{2\sqrt{x + 1}}$ **27.** $-\dfrac{3}{2\sqrt{1 - 3x}}$

Exercises 22–4, p. 595

1. 4.00, 4.00, 4.00, 4.00, 4.00; $\lim\limits_{t \to 3} v = 4$ ft/s

3. -14, -15, -15.8, -15.98, -15.998; $\lim\limits_{t \to 4} v = -16$ ft/s

5. 4; 4   **7.** $-4t$; -16 **9.** $3 + \dfrac{2}{t^2}$ **11.** $3t^2 - 6$ **13.** $12t - 4$ **15.** -4 **17.** 2   **19.** $2\pi r$, 6π

21. 0.285 Ω/°C   **23.** $\dfrac{25}{(5 + R)^2}$, $\dfrac{25}{169}$ **25.** $4\pi r - \dfrac{200}{r^2}$ **27.** $-\dfrac{2}{r^3}$, -2×10^4 lx/m, -2×10^2 lx/m

Exercises 22–5, p. 601

1. $5x^4$ **3.** $36x^8$ **5.** $4x^3$ **7.** $2x + 2$ **9.** $15x^2 - 1$ **11.** $8x^7 - 28x^6 - 1$ **13.** $42x^6 - 15x^2$
15. $x^2 + x$ **17.** 360   **19.** 14   **21.** $30t^4 - 5$ **23.** $120 - 32t$ **25.** 1   **27.** $3\pi r^2$ **29.** 6.0 V/°C
31. 300 m/km

Exercises 22–6, p. 605

1. $9x^2 + 4x$ **3.** $54x^2 - 60x$ **5.** $4x - 1$ **7.** $-14x^6 + 30x^4 + 4x^3 - 18x^2 - 6x$ **9.** $\dfrac{3}{(2x + 3)^2}$

11. $\dfrac{-2x}{(x^2 + 1)^2}$ **13.** $\dfrac{6x - 2x^2}{(3 - 2x)^2}$ **15.** $\dfrac{-48x^5 - 12}{(4x^5 - 3x - 4)^2}$ **17.** $\dfrac{-12x^3 + 45x^2 - 14x}{(3x - 7)^2}$ **19.** 12   **21.** 1, -1

23. $-\dfrac{1}{16}$ ft/s   **25.** 0.925 V/s   **27.** $\dfrac{E^2(R - r)}{(R + r)^3}$

Exercises 22–7, p. 610

1. $\dfrac{1}{2x^{1/2}}$ **3.** $-\dfrac{2}{x^3}$ **5.** $-\dfrac{1}{x^{4/3}}$ **7.** $\dfrac{3}{2}x^{1/2} + \dfrac{1}{x^2}$ **9.** $10x(x^2 + 1)^4$ **11.** $-192x^2(7 - 4x^3)^7$

13. $\dfrac{2x^2}{(2x^3 - 3)^{2/3}}$ **15.** $\dfrac{8x}{(1 - x^2)^5}$ **17.** $\dfrac{24x^3}{(2x^4 - 5)^{1/4}}$ **19.** $\dfrac{-4x}{(1 - 8x^2)^{3/4}}$ **21.** $\dfrac{3x - 2}{2(x - 1)^{1/2}}$

23. $\dfrac{-16x - 22}{(4x + 3)^{1/2}(8x + 1)^2}$ **25.** $x = 0$ **27.** 1   **29.** $\dfrac{188}{3}\sqrt[3]{60}$ **31.** $\dfrac{-450000}{V^{5/2}}$, -4.50 kPa/cm³

33. $\dfrac{-kqx}{(x^2 + b^2)^{3/2}}$ **35.** $\dfrac{-3kr}{[r^2 + (l/2)^2]^{5/2}}$

Exercises 22–8, p. 614

1. $-\dfrac{3}{2}$ 3. $\dfrac{6x+1}{4}$ 5. $\dfrac{x}{y}$ 7. $\dfrac{2x}{5y^4}$ 9. $\dfrac{2x}{2y+1}$ 11. $\dfrac{-3y}{3x+1}$ 13. $\dfrac{-2x-y^3}{3xy^2+3}$

15. $\dfrac{-3x(x^2+1)^2}{y(y^2+1)}$ 17. 1   19. $-\dfrac{x}{y}$

Exercises for Chapter 22, p. 614

1. -4 3. $\dfrac{1}{4}$ 5. 1   7. $\dfrac{7}{3}$ 9. $\dfrac{2}{3}$ 11. -2 13. 5   15. $-4x$ 17. $-\dfrac{4}{x^3}$ 19. $\dfrac{1}{2\sqrt{x+5}}$

21. $14x^6-6x$ 23. $\dfrac{2}{x^{1/2}}+\dfrac{1}{x^2}$ 25. $\dfrac{1}{(1-x)^2}$ 27. $-12(2-3x)^3$ 29. $\dfrac{9x}{(5-2x^2)^{7/4}}$

31. $\dfrac{-15x^2+2x}{(1-6x)^{1/2}}$ 33. $\dfrac{-2x-3}{2x^2(4x+3)^{1/2}}$ 35. $\dfrac{2x-6(2x-3y)^2}{1-9(2x-3y)^2}$ 37. $\dfrac{1}{2}C_2,\ 0$ 39. -31 41. 0.01 m/s

43. 0.04 V   45. $-\dfrac{2k}{r^3}$ 47. $\dfrac{a}{\sqrt{v_0^2+2as}}$ 49. $\dfrac{40V_2^{0.4}}{V_1^{1.4}}$ 51. $4x-x^3,\ 4-3x^2$

Exercises 23–1, p. 620

1. $4x-y-2=0$ 3. $2y+x-2=0$ 5. $x+4y-21=0$ 7. $8x-4y-7=0$

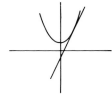   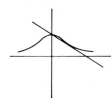      

9. $\sqrt{3}x+8y-7=0,\ 16x-2\sqrt{3}y-15\sqrt{3}=0$ 11. $2x-12y+37=0,\ 72x+12y+37=0$

13. $-\dfrac{1}{4}$ 15. $3x-5y-150=0$ 17. $x-\sqrt{3}y+4=0$

Exercises 23–2, p. 623

1. 3.16, 341.5°   3. 0.479, 211.5°   5. $a=0$ 7. 0.442, 58.1°   9. 301, 86.2°   11. 1.105
13. 136 ft/s, 28.1°; 124 ft/s, 345°   15. 32 ft/s²; 270.0° for both   17. 276 m/s, 43.5°; 2090 m/s, 16.7°
19. 21.2 mi/min, 296.6°

Exercises 23–3, p. 626

1. 0.0900 Ω/s   3. 1.22 \$/week   5. 0.00401   7. 0.375 V/s   9. 100 ft²/min   11. 10.2 ft/s
13. -3.75 kPa/min   15. 2.51×10^6 mm³/s   17. 128 km/h   19. 12.0 ft/s

Exercises 23–4, p. 633

1. Inc. $x > 1$, dec. $x < 1$ 3. Inc. $-2 < x < 2$, dec. $x < -2$, $x > 2$ 5. Min. $(1, -1)$
7. Min. $(-2, -16)$, max. $(2, 16)$ 9. Conc. up all x 11. Conc. up $x < 0$, conc. down $x > 0$, infl. $(0, 0)$
13.   15.   17. Max. $(3,18)$,   19. Max. $(-2,8)$, min. $(0,0)$,
                                                                    conc. down all x infl. $(-1,4)$

         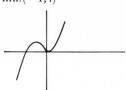

21. Max. $(-1,4)$, min. $(1,-4)$,   23. Max. $(1,1)$, infl. $(0,0)$,   25.   27. $V = 4x^3 - 40x^2 + 96x$
    infl. $(0, 0)$ $\left(\dfrac{2}{3}, \dfrac{16}{37}\right)$

   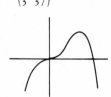   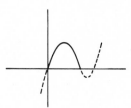

Exercises 23–5, p. 637

1. Inc. $x < 0$, dec. $x > 0$ 3. Int. $(-\sqrt[3]{2},0)$, min. $(1,3)$,   5. Int. $(0,0)$, max. $(-2,-4)$,
   Conc. up $x < 0$, $x > 0$ infl. $(-\sqrt[3]{2},0)$, asym. $x = 0$ min. $(0,0)$, asym. $x = -1$
   Asym. $x = 0$

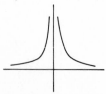   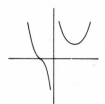   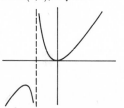

7. Int. $(0,-1)$, max. $(0,-1)$,   9. Asym. $R_T = 5$, inc. $R \geq 0$ 11. $A = 2\pi r^2 + \dfrac{40}{r}$, min. $(1.47,40.8)$
   asym. $x = 1$, $x = -1$ asym. $r = 0$

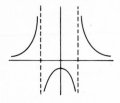   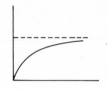

Exercises 23–6, p. 642

1. 196 ft   3. 0.250 ft/min   5. 8, 8   7. 5.00×10^5 ft²   9. $2\sqrt{2}$ by $2\sqrt{2}$ 11. $5\sqrt{2}$, $5\sqrt{2}$

13. 12   15. 3 km   17. $\dfrac{4}{3}$ in.   19. 2 units   21. 0.58 L   23. Width $= \dfrac{16\sqrt{3}}{3}$ in., depth $= \dfrac{16\sqrt{6}}{3}$ in.

Exercises for Chapter 23, p. 644

1. $5x - y + 1 = 0$ **3.** $27x - 3y - 26 = 0$ **5.** 31.0, 359.5°   **7.** 48.0, 0.3°

9. Min.$(-2, -16)$
    conc. up all x

11. Int: $(0, 0)$, $(\pm 3\sqrt{3}, 0)$
    Max:$(3, 54)$, Min:$(-3, -54)$
    Infl:$(0,0)$

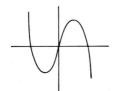

13. Min.$(2, -48)$
    Conc. up $x < 0$, $x > 0$

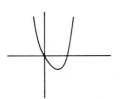

15. Int:$(0,0)$ Asym: $x = -1$, $y = 1$ **17.** $2x - y + 1 = 0$ **19.** 3180 cm/s, 292.2°   **21.**

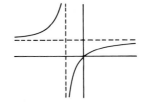

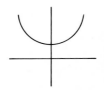

23. 3.27 in/min   **25.** $r = 1$, $\theta = 2$ **27.** Int:$(1,0)$ **29.** 13.0 km/h   **31.** 1.94 units

Max:$\left(2, \dfrac{1}{4}\right)$

Infl:$\left(3, \dfrac{2}{9}\right)$

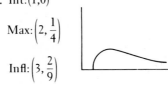

Asym: $P = 0$
$V \geq 1$ (only meaningful values)

Exercises 24—1, p. 649

1. $(5x^4 + 1)\,dx$ **3.** $8x(x^2 - 1)^3\,dx$ **5.** $x(1 - x)^2(-5x + 2)\,dx$ **7.** $\dfrac{dx}{(x + 1)^2}$ **9.** 12.28, 12

11. 1.71275, 1.675   **13.** -2.4, -2.473088 **15.** 0.6257, 0.62649   **17.** 16 in.²   **19.** 12.2 V

21. 2.64 in.³   **23.** $\dfrac{dA}{A} = \dfrac{2\,ds}{s}$

Exercises 24—2, p. 651

1. x^3 **3.** x^6 **5.** $\dfrac{3}{2}x^4 + x$ **7.** $\dfrac{2}{3}x^3 - \dfrac{1}{2}x^2$ **9.** $\dfrac{1}{x}$ **11.** $\dfrac{2}{x^3}$ **13.** $\dfrac{2}{5}x^5 + x$ **15.** $(2x + 1)^6$

17. $(x^2 - 1)^4$ **19.** $\dfrac{1}{40}(2x^4 + 1)^5$ **21.** $(6x + 1)^{3/2}$ **23.** $\dfrac{1}{4}(3x + 1)^{4/3}$

Exercises 24–3, p. 656

1. $x^2 + C$ 3. $\dfrac{1}{8}x^8 + C$ 5. $\dfrac{2}{5}x^{5/2} + C$ 7. $-\dfrac{1}{3x^3} + C$ 9. $\dfrac{1}{3}x^3 - \dfrac{1}{6}x^6 + C$ 11. $\dfrac{1}{6}(1 + 2x)^3 + C$

13. $\dfrac{2}{5}x^{5/2} - \dfrac{5}{3}x^3 + C$ 15. $\dfrac{2}{7}x^{7/2} - \dfrac{2}{5}x^{5/2} + C$ 17. $\dfrac{1}{6}(x^2 - 1)^6 + C$ 19. $\dfrac{1}{5}(x^4 + 3)^5 + C$

21. $\dfrac{1}{40}(x^5 + 4)^8 + C$ 23. $\dfrac{1}{12}(8x + 1)^{3/2} + C$ 25. $\dfrac{1}{6}\sqrt{6x^2 + 1} + C$ 27. $\sqrt{x^2 - 2x} + C$

29. $y = 2x^3 + 2$ 31. $y = 5 - \dfrac{1}{18}(1 - x^3)^6$ 33. $12y = 83 + (1 - 4x^2)^{3/2}$ 35. $y = 3x^2 + 2x - 3$

Exercises 24–4, p. 662

1. 9, 12.15   3. 1.92, 2.28   5. 7.625, 8.208   7. 0.464, 0.5995   9. 13.5   11. $\dfrac{8}{3}$ 13. 9   15. 0.8

Exercises 24–5, p. 665

1. 1   3. $\dfrac{254}{7}$ 5. $\dfrac{3}{2}\sqrt[3]{2}$ 7. $\dfrac{747}{20}$ 9. $\dfrac{4}{3}$ 11. $\dfrac{81}{4}$ 13. 2   15. 49   17. $\dfrac{364}{3}$ 19. $\dfrac{110}{3}$

Exercises 24–6, p. 668

1. $\dfrac{11}{2} = 5.50$, $\dfrac{16}{3} = 5.33$ 3. 7.661, $\dfrac{23}{3} = 7.667$ 5. 0.2042   7. 2.996   9. 0.5205   11. 21.74

13. 10.58   15. 31.70

Exercises for Chapter 24, p. 669

1. $x^4 - \dfrac{1}{2}x^2 + C$ 3. $x^2 - \dfrac{4}{5}x^{5/2} + C$ 5. $\dfrac{20}{3}$ 7. $\dfrac{16}{3}$ 9. $\dfrac{1}{5(2 - 5x)} + C$

11. $-\dfrac{2}{7}(7 - 2x)^{7/4} + C$ 13. $\dfrac{9}{8}(3\sqrt[3]{3} - 1)$ 15. $-\dfrac{1}{30}(1 - 2x^3)^5 + C$ 17. $-\dfrac{1}{(2x - x^3)} + C$

19. $\dfrac{3350}{3}$ 21. $\dfrac{-6x\,dx}{(x^2 - 1)^4}$ 23. $\dfrac{(1 - 4x)\,dx}{(1 - 3x)^{2/3}}$ 25. 0.061   27. $y = 3x - \dfrac{1}{3}x^3 + \dfrac{17}{3}$ 29. 22

31. 0.842   33. 1.01   35. 13.80   37. 4.02 mm³   39. $\dfrac{R\,dR}{R^2 + X^2}$

Exercises 25–1, p. 676

1. -96.0 ft/s   3. $v = 3t^2 + 5$ 5. 560 cm   7. 216 ft   9. 336 ft   11. 13.5 m   13. 1.0 C

15. 5.14 C   17. 667 V   19. 117 V   21. $\theta = 2t^2$ 23. 66.7 A   25. $\dfrac{k}{x_1}$

27. $m = 1002 - 2\sqrt{t + 1}$, 2.51×10^5 min

Exercises 25–2, p. 681

1. 2   3. $\dfrac{8}{3}$ 5. $\dfrac{3}{4}$ 7. $\dfrac{4}{3}\sqrt{2}$ 9. $\dfrac{1}{2}$ 11. $\dfrac{1}{6}$ 13. 12   15. $\dfrac{19}{3}$ 17. $\dfrac{15}{4}$ 19. $\dfrac{7}{6}$ 21. $\dfrac{256}{5}$

23. $\dfrac{2}{3}$ 25. $\dfrac{1}{3}$ 27. 1   29. 18.0 J   31. $\dfrac{128}{5}$ ft   33. 0.513 J

Exercises 25–3, p. 687

1. $\dfrac{1}{3}\pi$ 3. $\dfrac{1}{3}\pi$ 5. $\dfrac{348}{5}\pi$ 7. $\dfrac{768}{7}\pi$ 9. 72π 11. $\dfrac{16}{3}\pi$ 13. $\dfrac{1}{3}\pi$ 15. $\dfrac{1}{3}\pi$ 17. $\dfrac{2}{5}\pi$

19. $\dfrac{8}{3}\pi$ 21. $\dfrac{16}{15}\pi$ 23. $\dfrac{16}{3}\pi$ 25. $\dfrac{8}{3}\pi$ 27. $\dfrac{1}{3}\pi r^2 h$ 29. 3375π m³

Exercises 25–4, p. 694

1. $\left(\dfrac{10}{3}, 0\right)$ 3. $\left(\dfrac{14}{15}, 0\right)$ 5. $\left(-\dfrac{1}{2}, \dfrac{1}{2}\right)$ 7. $\left(\dfrac{7}{22}, \dfrac{5}{22}\right)$ 9. $\left(0, \dfrac{6}{5}\right)$ 11. $\left(\dfrac{4}{3}, \dfrac{4}{3}\right)$ 13. $\left(\dfrac{3}{5}, \dfrac{12}{35}\right)$

15. $\left(0, \dfrac{5}{6}\right)$ 17. $\left(\dfrac{2}{3}, 0\right)$ 19. $\left(\dfrac{9}{7}, 0\right)$ 21. $\left(\dfrac{2}{3}a, \dfrac{1}{3}b\right)$ 23. $\left(\dfrac{3}{8}a, 0\right)$

Exercises 25–5, p. 699

1. $128, 4$ 3. $214, 3.27$ 5. $\dfrac{64}{15}$ 7. $\dfrac{2}{3}\sqrt{6}$ 9. $\dfrac{1}{6}ma^2$ 11. $\dfrac{4}{7}\sqrt{7}$ 13. $\dfrac{8}{11}\sqrt{55}$ 15. $\dfrac{64}{3}\pi$

17. $\dfrac{2}{5}\sqrt{10}$ 19. $\dfrac{3}{10}mr^2$

Exercises 25–6, p. 702

1. 37.5 lb-in.   3. 90.0 ft-lb   5. 2.40 J   7. 0.99 kJ   9. 10,000 ft-lb   11. 8.82×10^4 ft-lb

13. 2.67 A   15. 10   17. $\dfrac{14}{3}$ 19. 11,700 lb   21. $\dfrac{15}{4}\pi$

Exercises for Chapter 25, p. 704

1. 160 ft/s   3. 4.22 s   5. 9.20 C   7. 3640 V   9. $y = 20x + \dfrac{1}{120}x^3$ 11. $\dfrac{2}{3}$ 13. 18

15. $\dfrac{48}{5}\pi$ 17. $\dfrac{243}{10}\pi$ 19. $\left(\dfrac{40}{21}, \dfrac{10}{3}\right)$ 21. $\left(\dfrac{14}{5}, 0\right)$ 23. $\dfrac{8}{5}$ 25. $\dfrac{256}{3}\pi$ 27. 9.00 ft-lb

29. 0.286 Ω   31. 172,000 lb

Exercises 26–1, p. 710

1. $\cos(x + 2)$ 3. $4\cos(2x - 1)$ 5. $-10\sin 2x$ 7. $-6\sin(3x - 1)$ 9. $8\sin 4x \cos 4x = 4\sin 8x$
11. $-45\cos^2(5x + 2)\sin(5x + 2)$ 13. $\sin 3x + 3x\cos 3x$ 15. $9x^2\cos 5x - 15x^3\sin 5x$

17. $2x\cos x^2\cos 2x - 2\sin x^2\sin 2x$ 19. $-2\sin 2x \sin x + 4\cos 2x \cos x$ 21. $\dfrac{3x\cos 3x - \sin 3x}{x^2}$

23. $\dfrac{-4x^2\sin x^2 - 2\cos x^2}{3x^2}$ 25. $3\sin^2 x \cos x + 2\sin 2x$ 27. $1 - 4\sin^2 4x \cos 4x$ 29. See table

31. $\dfrac{d \sin x}{dx} = \cos x, \dfrac{d^2 \sin x}{dx^2} = -\sin x, \dfrac{d^3 \sin x}{dx^3} = -\cos x, \dfrac{d^4 \sin x}{dx^4} = \sin x$ 33. $3\sqrt{2}$ 35. 9.77 ft/s

Exercises 26–2, p. 713

1. $5\sec^2 5x$ 3. $2(1 - x)\csc^2(1 - x)^2$ 5. $6\sec 2x \tan 2x$ 7. $\dfrac{3\csc\sqrt{x}\cot\sqrt{x}}{2\sqrt{x}}$ 9. $30\tan 3x \sec^2 3x$

11. $-24\cot^3 3x \csc^2 3x$ 13. $2\tan 4x\sqrt{\sec 4x}$ 15. $-84\csc^4 7x \cot 7x$ 17. $x^2\sec^2 x + 2x\tan x$

19. $-4 \csc x^2 (2x \cos x \cot x^2 + \sin x)$ **21.** $-\dfrac{\csc x (x \cot x + 1)}{x^2}$

23. $\dfrac{-4 \sin 4x - 4 \sin 4x \cot 3x + 3 \cos 4x \csc^2 3x}{(1 + \cot 3x)^2}$ **25.** $\sec^2 x (\tan^2 x - 1)$ **27.** $2 \sec 2x (\sec 2x - \tan 2x)$

29. $24 \tan 3x \sec^2 3x\ dx$ **31.** $4 \sec 4x (\tan^2 4x + \sec^2 4x)\ dx$ **33.** -12 **35.** 140 ft/s

Exercises 26–3, p. 716

1. $\dfrac{2x}{\sqrt{1 - x^4}}$ **3.** $\dfrac{18x^2}{\sqrt{1 - 9x^6}}$ **5.** $-\dfrac{1}{\sqrt{4 - x^2}}$ **7.** $\dfrac{1}{\sqrt{(x - 1)(2 - x)}}$ **9.** $\dfrac{1}{2\sqrt{x}(1 + x)}$ **11.** $-\dfrac{1}{x^2 + 1}$

13. $\dfrac{x}{\sqrt{1 - x^2}} + \text{Arcsin } x$ **15.** $\dfrac{4x}{1 + 4x^2} + 2 \text{ Arctan } 2x$ **17.** $\dfrac{3\sqrt{1 - 4x^2} \text{ Arcsin } 2x - 6x}{\sqrt{1 - 4x^2} \text{ Arcsin}^2 2x}$

19. $\dfrac{8 \text{ Arcsin } 4x}{\sqrt{1 - 16x^2}}$ **21.** $\dfrac{9 \text{ Arctan}^2 x}{1 + x^2}$ **23.** $\dfrac{-2(2x + 1)^2}{(1 + 4x^2)^2}$ **25.** $\dfrac{3 \text{ Arcsin}^2 x\ dx}{\sqrt{1 - x^2}}$ **27.** $\dfrac{-16x}{(1 + 4x^2)^2}$

29. Let $y = \text{Arcsec } u$; solve for u; take derivatives; substitute.   **31.** $-\dfrac{1}{\sqrt{R^2 - R_x^2}} \dfrac{dR_x}{dt}$

Exercises 26–4, p. 719

1. $d \sin x/dx = \cos x$ and $d \cos x/dx = -\sin x$, and $\sin x = \cos x$ at points of intersection.

3. $\dfrac{1}{x^2 + 1}$ is always positive   **5.** Dec., $x > 0, x < 0$;   Infl.$(0,0)$ Asym., $x = \pi/2, x = -\pi/2$ **7.** $8\sqrt{2}x + 8y + 4\sqrt{2} - 5\pi\sqrt{2} = 0$

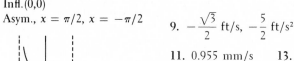

9. $-\dfrac{\sqrt{3}}{2}$ ft/s, $-\dfrac{5}{2}$ ft/s^2

11. 0.955 mm/s   **13.** $2\pi, 330°$
15. $4, 225°$ **17.** -0.0730 rad/s

19. $\dfrac{4}{3}$ unit/s   **21.** 0.02 **23.** 0.06 cm

25. 9.0 in.^{2} **27.** 14.05 ft

Exercises 26–5, p. 723

1. $\dfrac{2 \log e}{x}$ **3.** $\dfrac{6 \log_5 e}{3x + 1}$ **5.** $\dfrac{-3}{1 - 3x}$ **7.** $\dfrac{4 \sec^2 2x}{\tan 2x} = 4 \sec 2x \csc 2x$ **9.** $\dfrac{1}{2x}$ **11.** $1 + \ln x$

13. $\dfrac{3(2x + 1) \ln (2x + 1) - 6x}{(2x + 1)[\ln(2x + 1)]^2}$ **15.** $\dfrac{1}{x \ln x}$ **17.** $\dfrac{1}{x^2 + x}$ **19.** $\dfrac{\cos \ln x}{x}$ **21.** $\dfrac{x \sec^2 x + \tan x}{x \tan x}$

23. $\dfrac{x + 4}{x(x + 2)}$ **25.** See table.   **27.** -1 **29.** $x^x (\ln x + 1)$ **31.** $\dfrac{10 \log e}{I} \dfrac{dI}{dt}$

Exercises 26–6, p. 726

1. $\dfrac{2(3^{2x})}{\log_3 e}$ **3.** $\dfrac{6(4^{6x})}{\log_4 e}$ **5.** $6e^{6x}$ **7.** $\dfrac{e^{\sqrt{x}}}{2\sqrt{x}}$ **9.** $e^x(x + 1)$ **11.** $e^{\sin x}(x \cos x + 1)$ **13.** $\dfrac{3e^{2x}(2x + 1)}{(x + 1)^2}$

15. $e^{-3x}(4 \cos 4x - 3 \sin 4x)$ **17.** $2(e^{2x} + e^{-2x})$ **19.** $\dfrac{e^{2x}}{x} + 2e^{2x} \ln x$ **21.** $12e^{6x} \cot 2e^{6x}$

23. $\dfrac{4e^{2x}}{\sqrt{1 - e^{4x}}}$ **25.** $(-xe^{-x} + e^{-x}) + (xe^{-x}) = e^{-x}$ **27.** $\dfrac{dy}{dx} = e^a; y = e^a$ **29.** $10e^{-0.1t}$

31. Substitute and simplify.

Exercises 26—7, p. 729

1. Int.(0,0), max.(0,0),
not defined for cos $x < 0$,

asym. $x = -\frac{1}{2}\pi, \frac{1}{2}\pi, \ldots$

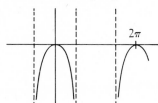

3. Int.(0,0), max. $\left(1, \frac{1}{e}\right)$,

infl. $\left(2, \frac{2}{e^2}\right)$, asym. $y = 0$

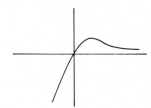

5. Int.(0,0), max.(0,0),
infl. $(-1, -\ln 2)$, $(1, -\ln 2)$

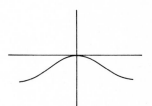

7. Int.(0,1), max.(0,1),

infl. $\left(\frac{1}{2}\sqrt{2}, \frac{1}{e}\sqrt{e}\right), \left(-\frac{1}{2}\sqrt{2}, \frac{1}{e}\sqrt{e}\right)$,

asym. $y = 0$

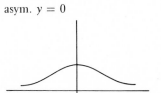

9. Max.$(1, -1)$, asym. $x = 0$

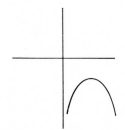

11. Int.(0,0), infl.(0,0),
inc. all x

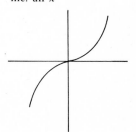

13. $y = x - 1$ **15.** -0.303 W/day   **17.** 1.21 **19.** $a = k^2 x$

21. $v = -e^{-0.5t}(1.4 \cos 6t + 2.3 \sin 6t)$, -2.02 **23.** $d\left(\frac{T_2}{T_1}\right) = -\frac{1}{2} e^{k/\sin(\theta/2)} \csc \frac{\theta}{2} \cos \frac{\theta}{2}\, d\theta$

Exercises for Chapter 26, p. 730

1. $-12 \sin(4x - 1)$ **3.** $-\dfrac{\sec^2 \sqrt{3 - x}}{2\sqrt{3 - x}}$ **5.** $-6 \csc^2(3x + 2)\cot(3x + 2)$ **7.** $-24x \cos^3 x^2 \sin x^2$

9. $2e^{2(x-3)}$ **11.** $\dfrac{6x}{x^2 + 1}$ **13.** $\dfrac{9}{9 + x^2}$ **15.** $\dfrac{4}{(\operatorname{Arcsin} 4x)(\sqrt{1 - 16x^2})}$

17. $(-2 \csc 4x)\sqrt{\csc 4x + \cot 4x}$ **19.** $\dfrac{2x(1 + 4x^2)\operatorname{Arctan} 2x - 2x^2}{(1 + 4x^2)(\operatorname{Arctan} 2x)^2}$ **21.** $-2x \cot x^2$

23. $\dfrac{2 \cos x \ln(3 + \sin x)}{3 + \sin x}$ **25.** $e^{-x}\sec x(\tan x - 1)$ **27.** $\operatorname{Arccos} x$ **29.** $2x(e^{2\cos^2 x})(1 - 2x \sin x \cos x)$

31. $\dfrac{2y \sin 2x + \sin 2y}{\cos 2x - 2x \cos 2y}$ **33.** Infl: $\left(\frac{1}{2}\pi, \frac{1}{2}\pi\right), \left(\frac{3}{2}\pi, \frac{3}{2}\pi\right)$ **35.** Max $(e^{-2}, 4e^{-2})$
Min $(1,0)$; Infl (e^{-1}, e^{-1})

37. $7.27x + y - 8.44 = 0$ **39.** $2x + 2.57y - 4.30 = 0$ **41.** $-\frac{2}{3}\pi$ **43.** $-20 \sin 2t$ **45.** -0.0127 V

47. -0.831 units/s   **49.** -0.135 A   **51.** -16.0 J/min   **53.** 202 ft/s   **55.** 0.429 **57.** 7.07 in.

59.

Exercises 27–1, p. 735

1. $\frac{1}{5}\sin^5 x + C$ **3.** $-\frac{2}{3}(\cos x)^{3/2} + C$ **5.** $\frac{1}{3}\tan^3 x + C$ **7.** $\frac{1}{8}$ **9.** $\frac{1}{4}(\text{Arcsin } x)^4 + C$

11. $\frac{1}{10}(\text{Arctan } 5x)^2 + C$ **13.** $\frac{1}{3}[\ln(x+1)]^3 + C$ **15.** 0.179 **17.** $\frac{1}{4}(4 + e^x)^4 + C$

19. $\frac{3}{16}(2e^{2x} - 1)^{4/3} + C$ **21.** $\frac{1}{10}(1 + \sec^2 x)^5 + C$ **23.** $\frac{1}{6}$ **25.** $\frac{1}{2}$ **27.** $y = \frac{1}{3}(\ln x)^3 + 2$

Exercises 27–2, p. 738

1. $\frac{1}{4}\ln|1 + 4x| + C$ **3.** $-\frac{1}{6}\ln|4 - 3x^2| + C$ **5.** $-\frac{1}{2}\ln|\cot 2x| + C$ **7.** $\ln 2 = 0.693$

9. $\ln|1 - e^{-x}| + C$ **11.** $\ln|x + e^x| + C$ **13.** $\frac{1}{4}\ln|1 + 4\sec x| + C$ **15.** $\frac{1}{4}\ln 5 = 0.402$

17. $\ln|\ln x| + C$ **19.** $\ln|2x + \tan x| + C$ **21.** $-\sqrt{1 - 2x} + C$ **23.** $\frac{1}{3}\ln\left(\frac{5}{4}\right) = 0.0744$ **25.** 1.10

27. $\pi \ln 2 = 2.18$ **29.** 0.549 C   **31.** 0.924 ft-lb

Exercises 27–3, p. 741

1. $e^{7x} + C$ **3.** $\frac{1}{2}e^{2x+5} + C$ **5.** $2(e - 1) = 3.44$ **7.** $\frac{1}{3}e^{x^3} + C$ **9.** $2e^{\sqrt{x}} + C$ **11.** $\frac{1}{2}e^{2\sec x} + C$

13. $\frac{2e^x - 3}{2e^{2x}} + C$ **15.** $e^{\text{Arctan } x} + C$ **17.** $-\frac{1}{3}e^{\cos 3x} + C$ **19.** 0 **21.** 6.389 **23.** $\pi(e^4 - e) = 163$

25. $\frac{1}{8}(e^8 - 1) = 372$ **27.** $V = \frac{1}{3}(980 - 800e^{-30t})$

Exercises 27–4, p. 745

1. $\frac{1}{2}\sin 2x + C$ **3.** $\frac{1}{3}\tan 3x + C$ **5.** $2\sec\frac{1}{2}x + C$ **7.** $\frac{1}{3}\ln|\sin x^3| + C$

9. $\frac{1}{2}\ln|\sec x^2 + \tan x^2| + C$ **11.** $\cos\left(\frac{1}{x}\right) + C$ **13.** $\frac{1}{2}\sqrt{3}$ **15.** $\frac{1}{5}\sec 5x + C$

17. $\frac{1}{2}\ln|\sec 2x + \tan 2x| + C$ **19.** $\frac{1}{9}\pi + \frac{1}{3}\ln 2 = 0.580$

21. $\csc x - \cot x - \ln|\csc x - \cot x| + \ln|\sin x| + C$ **23.** $\dfrac{1}{2}(\ln|\sin 2x| + \sin 2x) + C$ **25.** 0.347

27. $\pi\sqrt{3} = 5.44$ **29.** $s = \dfrac{4}{3}\sin 3t$ **31.** 0.742

Exercises 27–5, p. 750

1. $\dfrac{1}{3}\sin^3 x + C$ **3.** $-\dfrac{1}{2}\cos 2x + \dfrac{1}{6}\cos^3 2x + C$ **5.** $\dfrac{1}{3}\sin^3 x - \dfrac{1}{5}\sin^5 x + C$ **7.** $\dfrac{1}{120}(64 - 43\sqrt{2})$

9. $\dfrac{1}{2}x - \dfrac{1}{4}\sin 2x + C$ **11.** $\dfrac{1}{2}x + \dfrac{1}{12}\sin 6x + C$ **13.** $\dfrac{1}{2}\tan^2 x + \ln|\cos x| + C$ **15.** $\dfrac{1}{4}\sec^4 x + C$

17. $\dfrac{1}{6}\tan^3 2x - \dfrac{1}{2}\tan 2x + x + C$ **19.** $\dfrac{1}{15}\sec^5 3x - \dfrac{1}{9}\sec^3 3x + C$ **21.** $x - \dfrac{1}{2}\cos 2x + C$

23. $\dfrac{1}{4}\cot^4 x - \dfrac{1}{3}\cot^3 x + \dfrac{1}{2}\cot^2 x - \cot x + C$ **25.** $1 + \dfrac{1}{2}\ln 2 = 1.347$

27. $\dfrac{1}{5}\tan^5 x + \dfrac{2}{3}\tan^3 x + \tan x + C$ **29.** $\dfrac{1}{2}\pi^2 = 4.935$ **31.** $\dfrac{4}{3}$ **33.** $\sqrt{2}$

Exercises 27–6, p. 753

1. $\operatorname{Arcsin}\dfrac{1}{2}x + C$ **3.** $\dfrac{1}{8}\operatorname{Arctan}\dfrac{1}{8}x + C$ **5.** $\dfrac{1}{4}\operatorname{Arcsin} 4x + C$ **7.** $\dfrac{1}{3}\operatorname{Arctan} 6 = 0.469$

9. $\dfrac{2}{5}\sqrt{5}\operatorname{Arcsin}\dfrac{1}{2}\sqrt{5}x + C$ **11.** $\dfrac{4}{9}\ln|9x^2 + 16| + C$

13. $\dfrac{1}{35}\sqrt{35}\left(\operatorname{Arctan}\dfrac{2}{7}\sqrt{35} - \operatorname{Arctan}\dfrac{1}{7}\sqrt{35}\right) = 0.057$ **15.** $\operatorname{Arcsin} e^x + C$ **17.** $\operatorname{Arctan}(x + 1) + C$

19. $4\operatorname{Arcsin}\dfrac{1}{2}(x + 2) + C$ **21.** -0.357 **23.** $2\operatorname{Arcsin}\left(\dfrac{1}{2}x\right) + \sqrt{4 - x^2} + C$ **25.** $\operatorname{Arctan} 2 = 1.11$

27. $\dfrac{1}{4}\pi\operatorname{Arctan}\dfrac{3}{4} = 0.505$ **29.** $\operatorname{Arcsin}\dfrac{x}{A} = \sqrt{\dfrac{k}{m}}\,t + \operatorname{Arcsin}\dfrac{x_0}{A}$ **31.** 0.22

Exercises 27–7, p. 756

1. $\cos x + x\sin x + C$ **3.** $\dfrac{1}{2}xe^{2x} - \dfrac{1}{4}e^{2x} + C$ **5.** $x\tan x + \ln|\cos x| + C$

7. $x\operatorname{Arctan} x - \ln\sqrt{1 + x^2} + C$ **9.** $-2x\sqrt{1 - x} - \dfrac{4}{3}(1 - x)^{3/2} + C$ **11.** $\dfrac{1}{2}x^2\ln x - \dfrac{1}{4}x^2 + C$

13. $\dfrac{1}{2}x\sin 2x - \dfrac{1}{4}(2x^2 - 1)\cos 2x + C$ **15.** $\dfrac{1}{2}(e^{\pi/2} - 1) = 1.90$ **17.** $1 - \dfrac{3}{e^2} = 0.594$

19. $\dfrac{1}{2}\pi - 1 = 0.571$ **21.** $\dfrac{1}{2}\pi - 1 = 0.571$ **23.** $q = \dfrac{1}{5}[e^{-2t}(\sin t - 2\cos t) + 2]$

Exercises 27–8, p. 760

1. $-\dfrac{\sqrt{1 - x^2}}{x} - \operatorname{Arcsin} x + C$ **3.** $\ln|x + \sqrt{x^2 - 4}| + C$ **5.** $-\dfrac{\sqrt{x^2 + 9}}{9x} + C$ **7.** $\dfrac{x}{4\sqrt{4 - x^2}} + C$

9. $\dfrac{16 - 9\sqrt{3}}{24} = 0.017$ **11.** $\ln|\sqrt{x^2 + 2x + 2} + x + 1| + C$ **13.** $\dfrac{1}{3}\operatorname{Arcsec}\dfrac{2}{3}x + C$

15. $\operatorname{Arcsec} e^x + C$ **17.** π **19.** $\dfrac{1}{4}ma^2$ **21.** 2.68

Exercises 27–9, p. 762

1. $\dfrac{3}{25}[2 + 5x - 2 \ln |2 + 5x|] + C$ 3. $-\dfrac{2}{27}(4 - 9x)(2 + 3x)^{3/2} + C$

5. $\dfrac{1}{2}x\sqrt{4 - x^2} + 2 \, \text{Arcsin} \dfrac{1}{2}x + C$ 7. $\dfrac{1}{2}\sin x - \dfrac{1}{10}\sin 5x + C$ 9. $\sqrt{4x^2 - 9} - 3 \, \text{Arcsec}\left(\dfrac{2x}{3}\right) + C$

11. $\dfrac{1}{20}\cos^4 4x \sin 4x + \dfrac{1}{5}\sin 4x - \dfrac{1}{15}\sin^3 4x + C$ 13. $\dfrac{1}{2}x^2 \text{Arctan} \, x^2 - \dfrac{1}{4}\ln(1 + x^4) + C$

15. $\dfrac{1}{4}(8\pi - 9\sqrt{3}) = 2.386$ 17. $-\ln\left(\dfrac{1 + \sqrt{4x^2 + 1}}{2x}\right) + C$ 19. $-\ln\left(\dfrac{1 + \sqrt{1 - 4x^2}}{2x}\right) + C$

21. $\dfrac{1}{8}\cos 4x - \dfrac{1}{12}\cos 6x + C$ 23. $\dfrac{1}{3}(\cos x^3 + x^3 \sin x^3) + C$ 25. $\dfrac{x^2}{2\sqrt{1 - x^4}} + C$ 27. 2.42

29. $\dfrac{1}{4}x^4\left(\ln x^2 - \dfrac{1}{2}\right) + C$ 31. $-\dfrac{x^3}{3\sqrt{x^6 - 1}} + C$ 33. $\dfrac{1}{4}[2\sqrt{5} + \ln(2 + \sqrt{5})] = 1.479$ 35. 208 lb

37. 210

Exercises for Chapter 27, p. 763

1. $-\dfrac{1}{2}e^{-2x} + C$ 3. $-\dfrac{1}{\ln 2x} + C$ 5. $\ln(1 + \sin x) + C$ 7. $\dfrac{2}{35}\text{Arctan}\dfrac{7}{5}x + C$ 9. 0

11. $\dfrac{1}{2}\ln 2 = 0.3466$ 13. $\dfrac{1}{12}\sec^4 3x + C$ 15. $-\dfrac{1}{3}\ln|\cos 3x| + C$ 17. $\dfrac{1}{9}\tan^3 3x + \dfrac{1}{3}\tan 3x + C$

19. $\dfrac{1}{2}\ln|2x + \sqrt{4x^2 - 9}| + C$ 21. $\sqrt{e^{2x} + 1} + C$ 23. $\dfrac{1}{6}(3x - \sin 3x \cos 3x) + C$

25. $-\dfrac{1}{2}x \cot 2x + \dfrac{1}{4}\ln|\sin 2x| + C$ 27. $\dfrac{1}{2}\sin e^{2x} + C$ 29. $\dfrac{1}{3}$ 31. $\dfrac{1}{2}x^2 - 2x + 3 \ln|x + 2| + C$

33. $\Delta S = a \ln T + bT + \dfrac{1}{2}cT^2 + C$ 35. $v = 320(1 - e^{-t/10})$ 37. $y = \dfrac{1}{3}\tan^3 x + \tan x$ 39. 11.0

41. $4\pi(e^2 - 1) = 80.3$ 43. (1.65, 0.24)

Exercises 28–1, p. 770

1. $1 + x + \dfrac{1}{2}x^2 + \cdots$ 3. $1 - \dfrac{1}{2}x^2 + \dfrac{1}{24}x^4 - \cdots$ 5. $1 + \dfrac{1}{2}x - \dfrac{1}{8}x^2 + \cdots$

7. $1 - 2x + 2x^2 - \cdots$ 9. $1 - 8x^2 + \dfrac{32}{3}x^4 - \cdots$ 11. $1 + x + x^2 + \cdots$ 13. $x - \dfrac{1}{3}x^3 + \cdots$

15. $x + \dfrac{1}{3}x^3 + \cdots$ 17. $-\dfrac{1}{2}x^2 - \dfrac{1}{12}x^4 - \cdots$ 19. $x^2 - \dfrac{1}{3}x^4 + \cdots$

21. (a), (c) functions are not defined at $x = 0$; (b) derivatives are not defined at $x = 0$.
23. $f^n(0) = 0$ except $f'''(0) = 6$

Exercises 28—2, p. 774

1. $1 + 3x + \dfrac{9}{2}x^2 + \dfrac{9}{2}x^3 + \cdots$ 3. $\dfrac{x}{2} - \dfrac{x^3}{2^3 3!} + \dfrac{x^5}{2^5 5!} - \dfrac{x^7}{2^7 7!} + \cdots$ 5. $1 - 8x^2 + \dfrac{32}{3}x^4 - \dfrac{256}{45}x^6 + \cdots$

7. $x^2 - \dfrac{1}{2}x^4 + \dfrac{1}{3}x^6 - \dfrac{1}{4}x^8 + \cdots$ 9. 0.3103 11. 0.1901 13. $1 + \dfrac{1}{2}x^2 + \dfrac{1}{24}x^4 + \dfrac{1}{120}x^6 + \cdots$

15. $x + x^2 + \dfrac{1}{3}x^3 + \cdots$ 17. $\dfrac{d}{dx}\left(x - \dfrac{1}{6}x^3 + \dfrac{1}{120}x^5 - \cdots\right) = 1 - \dfrac{1}{2}x^2 + \dfrac{1}{24}x^4 - \cdots$

19. $\displaystyle\int \cos x \, dx = x - \dfrac{x^3}{3!} + \cdots$ 21. 0.003099 23. 0.1249

Exercises 28—3, p. 778

1. 1.22 3. 0.09983 5. 2.718 7. 0.9986 9. 0.335 11. 0.8415 13. 1.0488 15. 1.032

17. 0.021 19. 2.4×10^{-10} 21. $i = \dfrac{E}{L}\left(t - \dfrac{Rt^2}{2L}\right)$; small values of t 23. 66 ft   25. 3.146

Exercises 28—4, p. 781

1. 3.32 3. 2.049 5. 0.5150 7. 0.49242 9. $e^{-2}\left[1 - (x - 2) + \dfrac{(x - 2)^2}{2!} - \cdots\right]$

11. $\dfrac{1}{2}\left[\sqrt{3} + \left(x - \dfrac{1}{3}\pi\right) - \dfrac{\sqrt{3}}{2!}\left(x - \dfrac{1}{3}\pi\right)^2 - \cdots\right]$ 13. $2 + \dfrac{1}{12}(x - 8) - \dfrac{1}{288}(x - 8)^2 + \cdots$

15. $1 + 2\left(x - \dfrac{1}{4}\pi\right) + 2\left(x - \dfrac{1}{4}\pi\right)^2 + \cdots$ 17. 0.111 19. 3.049 21. 2.03 23. 0.8746

25. Use indicated method.

Exercises 28—5, p. 789

1. $f(x) = \dfrac{1}{2} - \dfrac{2}{\pi}\sin x - \dfrac{2}{3\pi}\sin 3x - \cdots$

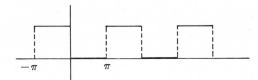

3. $f(x) = \dfrac{3}{2} + \dfrac{2}{\pi}\sin x + \dfrac{2}{3\pi}\sin 3x + \cdots$

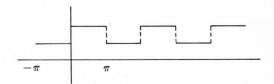

5. $f(x) = \dfrac{\pi}{4} - \dfrac{2}{\pi}\left(\cos x + \dfrac{1}{9}\cos 3x + \cdots\right) + \left(\sin x - \dfrac{1}{2}\sin 2x + \cdots\right)$

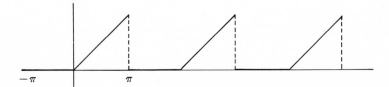

7. $f(x) = -\dfrac{1}{4} - \dfrac{1}{\pi}\cos x + \dfrac{1}{3\pi}\cos 3x - \cdots + \dfrac{3}{\pi}\sin x - \dfrac{1}{\pi}\sin 2x + \dfrac{1}{\pi}\sin 3x - \cdots$

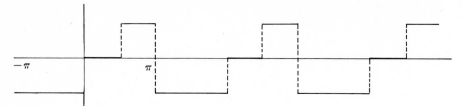

9. $f(x) = \dfrac{\pi}{2} - \dfrac{4}{\pi}\cos x - \dfrac{4}{9\pi}\cos 3x - \cdots$ 11. $f(t) = \dfrac{2}{\pi} - \dfrac{4}{3\pi}\cos 2t - \dfrac{4}{15\pi}\cos 4t - \cdots$

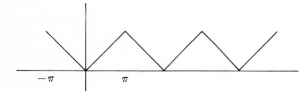

Exercises for Chapter 28, p. 789

1. $\dfrac{1}{2} - \dfrac{1}{4}x + \dfrac{1}{48}x^3 - \cdots$ 3. $2x^2 - \dfrac{4}{3}x^6 + \dfrac{4}{15}x^{10} - \cdots$ 5. $1 + \dfrac{1}{3}x - \dfrac{1}{9}x^2 + \cdots$

7. $x + \dfrac{1}{6}x^3 + \dfrac{3}{40}x^5 + \cdots$ 9. 0.82   11. 1.09   13. 0.9325   15. 12.1655   17. 0.259

19. $\dfrac{1}{2} - \dfrac{1}{2}\sqrt{3}\left(x - \dfrac{1}{3}\pi\right) - \dfrac{1}{4}\left(x - \dfrac{1}{3}\pi\right)^2 + \cdots$

21. $f(x) = \dfrac{1}{2} + \dfrac{2}{\pi}\left(\cos x - \dfrac{1}{3}\cos 3x + \cdots\right)$ 23. $(x + h) - \dfrac{(x + h)^3}{3!} + \cdots - (x - h) + \dfrac{(x - h)^3}{3!} - \cdots$

$$= 2h - \dfrac{2hx^2}{2!} + \cdots = 2h\left(1 - \dfrac{x^2}{2!} + \cdots\right)$$

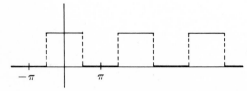

25.   27. 0.0008331   29. $f(t) = \dfrac{1}{2\pi} + \dfrac{1}{\pi}\left(\dfrac{1}{2}\cos t - \dfrac{1}{3}\cos 2t + \cdots\right)$

$+ \dfrac{1}{4}\sin t + \dfrac{2}{3\pi}\sin 2t + \cdots$

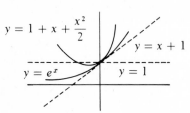

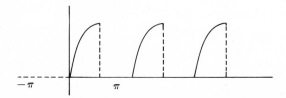

Exercises 29–1, p. 793
[*Note:* "Answers" in this section are the unsimplified expressions obtained by substituting functions and derivatives.]

1. $1 = 1$ 3. $e^x - (e^x - 1) = 1$ 5. $x(2cx) = 2(cx^2)$ 7. $(-2ce^{-2x} + 1) + 2\left(ce^{-2x} + x - \dfrac{1}{2}\right) = 2x$

9. $(-12 \cos 2x) + 4(3 \cos 2x) = 0$

11. $[e^{2x}(1 + 4c_1 + 4c_2 + 4x + 4c_2x + 2x^2)] - 4[e^{2x}(2c_1 + c_2 + x + 2c_2x + x^2)]$
$$+ 4[e^{2x}(c_1 + c_2x + x^2/2)] = e^{2x}$$

13. $x^2\left[-\dfrac{c^2}{(x-c)^2}\right] + \left[\dfrac{cx}{(x-c)}\right]^2 = 0$ 15. $x\left(-\dfrac{c_1}{x}\right) + \dfrac{c_1}{x} = 0$ 17. $(\cos x - \sin x + e^{-x})$
$$+ (\sin x + \cos x - e^{-x}) = 2 \cos x$$

19. $\left(e^{-x} + \dfrac{12}{5} \cos 2x + \dfrac{24}{5} \sin 2x\right) + \left(-e^{-x} + \dfrac{6}{5} \sin 2x - \dfrac{12}{5} \cos 2x\right) = 6 \sin 2x$

21. $\cos x\left[\dfrac{(\sec x + \tan x) - (x + c)(\sec x \tan x + \sec^2 x)}{(\sec x + \tan x)^2}\right] + \sin x = 1 - \dfrac{x + c}{\sec x + \tan x}$ 23. $c^2 + cx = cx + c^2$

Exercises 29–2, p. 796

1. $y = c - x^2$ 3. $x - \dfrac{1}{y} = c$ 5. $\ln(x^3 + 5) + 3y = c$ 7. $4\sqrt{1 - y} = e^{-x^2} + c$ 9. $e^x - e^{-y} = c$

11. $\tan^2 x + 2 \ln y = c$ 13. $x^2 + 1 + x \ln y + cx = 0$ 15. $y^2 + 4 \text{ Arcsin } x = c$ 17. $y = c - (\ln x)^2$

19. $\ln(e^x + 1) - \dfrac{1}{y} = c$ 21. $3 \ln y + x^3 = 0$ 23. $2 \ln(1 - y) = 1 - 2 \sin x$

Exercises 29–3, p. 799

1. $2xy + x^2 = c$ 3. $x^3 - 2y = cx - 4$ 5. $x^2y - y = cx$ 7. $(xy)^4 = 4 \ln y + c$

9. $2\sqrt{x^2 + y^2} = x + c$ 11. $y = c - \dfrac{1}{2}\ln \sin(x^2 + y^2)$ 13. $\ln(y^2 - x^2) + 2x = c$

15. $5xy^2 + y^3 = c$ 17. $2xy + x^3 = 5$ 19. $2x = 2xy^2 - 15y$

Exercises 29–4, p. 802

1. $y = e^{-x}(x + c)$ 3. $y = -\dfrac{1}{2}e^{-4x} + ce^{-2x}$ 5. $y = x(3 \ln x + c)$ 7. $y = \dfrac{8}{7}x^3 + \dfrac{c}{\sqrt{x}}$

9. $y = -\cot x + c \csc x$ 11. $y = 3 + ce^{-x}$ 13. $2y = e^{4x}(x^2 + c)$ 15. $y = \dfrac{1}{4} + ce^{-x^4}$

17. $3y = x^4 - 6x^2 - 3 + cx$ 19. $xy = (x^3 + c)e^{3x}$ 21. $y = e^{-x}$ 23. $ye^{\sqrt{x}} = e^x + 2e$

Exercises 29–5, p. 807

1. $y^2 = 2x^2 + 1$ 3. $y = 2e^x - x - 1$ 5. $y^2 = c - 2x$ 7. $y^2 = c - 2 \sin x$

9. $N = N_0 (0.5)^{t/2}, 0.354N_0$ 11. \$1040.81   13. $\lim\limits_{t \to \infty} \dfrac{E}{R}(1 - e^{-Rt/L}) = \dfrac{E}{R}$ 15. 0.952 A

17. $q = q_0 e^{-t/RC}$ 19. 11.04 lb   21. $v = 32(1 - e^{-t}), 32$ 23. 10.0 ft/s   25. $x = t^2 + 1, y = \dfrac{1}{t^2 + 1}$

27. $p = 15.0(0.667)^{10^{-4}h}$ 29. \$1260   31. 36.1°C

Exercises 29–6, 29–7, p. 813

1. $y = c_1 e^{3x} + c_2 e^{-2x}$ 3. $y = c_1 e^{-x} + c_2 e^{-3x}$ 5. $y = c_1 + c_2 e^x$ 7. $y = c_1 e^{6x} + c_2 e^{2x/3}$

9. $y = c_1 e^{x/3} + c_2 e^{-3x}$ 11. $y = c_1 e^{x/3} + c_2 e^{-x}$ 13. $y = e^x(c_1 e^{x\sqrt{2}/2} + c_2 e^{-x\sqrt{2}/2})$

15. $y = e^{3x/8}(c_1 e^{x\sqrt{41}/8} + c_2 e^{-x\sqrt{41}/8})$ 17. $y = e^{3x/2}(c_1 e^{x\sqrt{13}/2} + c_2 e^{-x\sqrt{13}/2})$

19. $y = e^{-x/2}(c_1 e^{x\sqrt{33}/2} + c_2 e^{-x\sqrt{33}/2})$ 21. $y = \frac{1}{5}(3e^{7x} + 7e^{-3x})$ 23. $y = \frac{e^3}{e^7 - 1}(e^{4x} - e^{-3x})$

25. $y = c_1 + c_2 e^{-x} + c_3 e^{3x}$

Exercises 29–8, p. 817

1. $y = (c_1 + c_2 x)e^x$ 3. $y = (c_1 + c_2 x)e^{-6x}$ 5. $y = c_1 \sin 3x + c_2 \cos 3x$

7. $y = e^{-x/2}\left(c_1 \sin \frac{1}{2}\sqrt{7}x + c_2 \cos \frac{1}{2}\sqrt{7}x\right)$ 9. $y = c_1 + c_2 x$ 11. $y = c_1 \sin \frac{1}{2}x + c_2 \cos \frac{1}{2}x$

13. $y = (c_1 + c_2 x)e^{3x/4}$ 15. $y = c_1 \sin \frac{1}{5}\sqrt{2}x + c_2 \cos \frac{1}{5}\sqrt{2}x$ 17. $y = e^x\left(c_1 \cos \frac{1}{2}\sqrt{6}x + c_2 \sin \frac{1}{2}\sqrt{6}x\right)$

19. $(c_1 + c_2 x)e^{4x/5}$ 21. $y = e^{3x/4}(c_1 e^{x\sqrt{17}/4} + c_2 e^{-x\sqrt{17}/4})$ 23. $y = c_1 e^{x(-6+\sqrt{42})/3} + c_2 e^{x(-6-\sqrt{42})/3}$

25. $y = e^{(\pi/6-1-x)}\sin 3x$ 27. $y = (4 - 14x)e^{4x}$

Exercises 29–9, p. 820

1. $y = c_1 e^{2x} + c_2 e^{-x} - 2$ 3. $y = c_1 \sin x + c_2 \cos x + x^2 - 2$ 5. $y = c_1 e^{-x} + c_2 e^{-3x} + \frac{1}{8}e^x + \frac{2}{3}$

7. $y = c_1 e^{5x} + c_2 e^{-4x} - \frac{1}{18}e^{2x} - \frac{2}{5}$ 9. $y = c_1 e^{-5x} + c_2 e^{6x} - \frac{1}{3}$ 11. $y = c_1 e^{2x/3} + c_2 e^{-5x} + \frac{1}{4}e^{3x}$

13. $y = c_1 e^{2x} + c_2 e^{-2x} - \frac{1}{5}\sin x - \frac{2}{5}\cos x$ 15. $y = c_1 \sin x + c_2 \cos x - \frac{1}{3}\sin 2x + 4$

17. $y = c_1 e^{-x} + c_2 e^{-4x} - \frac{7}{100}e^x + \frac{1}{10}xe^x + 1$ 19. $y = (c_1 + c_2 x)e^{-3x} + \frac{1}{25}e^{2x} - e^{-2x}$

21. $y = \frac{1}{6}(11e^{3x} + 5e^{-2x} + e^x - 5)$ 23. $y = -\frac{2}{3}\sin x + \pi \cos x + x - \frac{1}{3}\sin 2x$

Exercises 29–10, p. 824

1. $x = 4 \cos 10t$ 3. 20   5. $y = \frac{1}{4}\cos 16t$ 7. $y = \frac{1}{4}\cos 16t + \frac{8}{63}\sin 2t - \frac{1}{63}\sin 16t$

9. $q = 2.23 \times 10^{-4}e^{-20t}\sin 2240t$ 11. $q = \frac{1}{100}(1 - \cos 316t)$

13. $e^{-10t}(c_1 \sin 99.5t + c_2 \cos 99.5t) - \frac{1}{2290}(2 \cos 200t + 15 \sin 200t)$ 15. $\frac{20}{229}(2 \sin 200t - 15 \cos 200t)$

17. $-[(6.5 \times 10^{-4})\sin 400t + (8.5 \times 10^{-6})\cos 400t]$

Exercises 29–11, p. 831

1. $F(s) = \int_0^\infty e^{-st}\,dt = -\frac{1}{s}e^{-st}\Big|_0^\infty = \frac{1}{s}$

3. $F(s) = \int_0^\infty e^{-st}\sin at\,dt = \frac{e^{-st}(-s \sin at - a \cos at)}{s^2 + a^2}\Big|_0^\infty = \frac{a}{s^2 + a^2}$ 5. $\frac{1}{s - 3}$ 7. $\frac{6}{(s + 2)^4}$

9. $\dfrac{s-2}{s^2+4}$ 11. $\dfrac{3}{s} + \dfrac{2(s^2-9)}{(s^2+9)^2}$ 13. $s^2 L(f) + s L(f)$ 15. $(2s^2 - s + 1)L(f) - 2s + 1$ 17. t^2

19. e^{-5t} 21. $\dfrac{1}{2}t^2 e^{-t}$ 23. $\dfrac{1}{54}(9t \sin 3t + 2 \sin 3t - 6t \cos 3t)$

Exercises 29–12, p. 834

1. $y = e^{-t}$ 3. $y = -e^{3t/2}$ 5. $y = (1+t)e^{-3t}$ 7. $y = \dfrac{1}{2}\sin 2t$ 9. $y = 1 - e^{-2t}$

11. $y = e^{2t}\cos t$ 13. $y = 1 + \sin t$ 15. $y = e^{-t}\left(\dfrac{1}{2}t^2 + 3t + 1\right)$ 17. $v = 6(1 - e^{-t/2})$

19. $q = 1.6 \times 10^{-4}(1 - e^{-5 \times 10^{-3}t})$ 21. $i = 5t \sin 50t$

Exercises for Chapter 29, p. 835

1. $2 \ln(x^2 + 1) - \dfrac{1}{2y^2} = c$ 3. $ye^{2x} = x + c$ 5. $y = c_1 + c_2 e^{-x/2}$ 7. $y = (c_1 + c_2 x)e^{-x}$

9. $2x^2 + 4xy + y^4 = c$ 11. $y = cx^3 - x^2$ 13. $y = c(y + 2)e^{2x}$ 15. $y = \dfrac{1}{2}(c - x^2)\csc x$

17. $y = c_1 e^x + c_2 e^{-3x/2} - 2$ 19. $y = c_1 e^{2x/3} + c_2 e^{4x/3} + \dfrac{1}{2}x + \dfrac{25}{8}$

21. $y = e^{-x/2}(c_1 e^{x\sqrt{5}/2} + c_2 e^{-x\sqrt{5}/2}) + 2e^x$ 23. $y = c_1 \sin 3x + c_2 \cos 3x + \dfrac{1}{8}\sin x$ 25. $y = 2\sin^2 x$

27. $y = 2x - 1 - e^{-2x}$ 29. $y = 2e^{-x/2}\sin\left(\dfrac{1}{2}\sqrt{15}x\right)$ 31. $y = \dfrac{1}{25}[16 \sin x + 12 \cos x - 3e^{-2x}(4 + 5x)]$

33. $y = e^{t/4}$ 35. $y = \dfrac{1}{2}(e^{3t} - e^t)$ 37. $y = -4 \sin t$ 39. $y = \dfrac{1}{3}t - \dfrac{4}{9}\sin 3t$ 41. $r = r_0 + kt$

43. 3.93 m/s   45. $5y^2 + x^2 = c$ 47. $q = c_1 e^{-t/RC} + EC$ 49. $y = (2 + 8t)e^{-4t}$; critically damped
51. $q = e^{-6t}(0.4 \cos 8t + 0.3 \sin 8t) - 0.4 \cos 10t$ 53. $i = 0$ 55. $i = 12(1 - e^{-t/2})$; $i(0.3) = 1.67$ A

57. $q = 10^{-4}e^{-8t}(4.0 \cos 200t + 0.16 \sin 200t)$ 59. $y = \dfrac{10}{3EI}[100x^3 - x^4 + xL^2(L - 100)]$

Exercises B—1, p. A—11

1. MHz, 1 MHz $= 10^6$ Hz   3. mm, 1 mm $= 10^{-3}$ m   5. kilovolt, 1 kV $= 10^3$ V
7. milliampere, 1 mA $= 10^{-3}$ A   9. 10^5 cm   11. 63,360 in.   13. 13.3 cm   15. 14.9 L
17. 144 in.^{2} 19. 2.50×10^{-4} m^2 21. 2040 g   23. 13.6 L   25. 7600 kg · cm/s   27. 0.101 hp
29. 130 tons   31. 37 mi/h   33. 770 mi/h   35. 345 lb/ft^3 37. 101,000 Pa

Exercises B—2, p. A—14

1. 24 is exact.   3. 3 is exact, 74.6 is approx.   5. 1063 is approx.
7. 100 and 200 are approx., 3200 is exact.   9. 3, 4   11. 3, 4   13. 3, 3   15. 1, 6
17. (a) 3.764, (b) 3.764   19. (a) 0.01 (b) 30.8   21. (a) same (b) 78.0   23. (a) 0.004 (b) same
25. (a) 4.93 (b) 4.9   27. (a) 57900 (b) 58000   29. (a) 861 (b) 860   31. (a) 0.305 (b) 0.31

Exercises B—3, p. A—17

1. 51.2   3. 1.70   5. 431.4   7. 30.9   9. 62.1   11. 270   13. 160   15. 27,000   17. 5.7
19. 4.39   21. 10.2   23. 22   25. 17.62   27. 18.85   29. 21.0 lb   31. First plane, 70 mi/h
33. 62.1 lb/ft^3 35. 850,000 lb   37. 115 ft/s

Exercises C—4, p. A—30

1. 56.02   3. 4162.1   5. 18.65   7. 0.3954   9. 14.14   11. 0.5251   13. 13.35   15. 944.6
17. 0.7349   19. −0.7594   21. −1.337   23. 1.015   25. 41.35°   27. −1.182   29. 0.5862
31. 6.695   33. 3.508   35. 0.005685   37. 2.053   39. 5.765   41. 4.501×10^{10} 43. 497.2
45. 6.648   47. 401.2   49. 8.841   51. 2.523   53. 10.08   55. 22.36   57. 20.3°   59. 4729
61. 3.301×10^{4} 63. 1.056   65. 55.5°   67. 3.277   69. 8.125   71. 1.000   73. 1.000
75. 12.90   77. 8.001   79. 8.053   81. 0.04259   83. 0.4219   85. 0.7822   87. 2.0736465
89. 124.3

Exercises D—3, p. A—35

1. 85°   3. 140°   5. 25°   7. 25°   9. 70°   11. 56°   13. 5   15. 17   17. 8   19. 4.90
21. 10.6   23. 28.3   25. 3.3 in., 4.0 in.   27. 16   29. 25 ft   31. 10 yd   33. 18.46 in.
35. 32.7 cm   37. 214.6 cm   39. 62.8 ft   41. 28 in.²   43. 13 yd²   45. 24 m²   47. 154 in.²
49. 216 ft³   51. 20 mm³   53. 924 in.³   55. 153 in.³   57. 27.1 ft³   59. 689 m³   61. 208 in.²
63. 537 m²   65. 3220 ft²   67. 343 cm²

Index